GCSE
Mathematics
Higher Level

Q: What's the best way to get through GCSE Maths?
A: By taking the square route.

That was a joke, obviously. The real answer is: by using this CGP book.
It's packed with study notes, examples and indispensable tips for every topic,
plus enough practice questions to keep you happy throughout the course.

It even includes a free Online Edition to read on your PC, Mac or tablet!

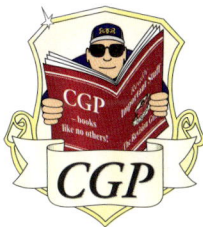

How to get your free Online Edition

Go to **cgpbooks.co.uk/extras** and enter this code...

2640 8781 9353 7139

Contents

Published by CGP

Editors:
Sammy El-Bahrawy, Shaun Harrogate, Samuel Mann, Tom Miles, Rosa Roberts, Caley Simpson, Dawn Wright

Contributors:
Katharine Brown, Eva Cowlishaw, Alastair Duncombe, Paul Garrett, Stephen Green, Philip Hale,
Phil Harvey, Judy Hornigold, Claire Jackson, Mark Moody, Charlotte O'Brien, Rosemary Rogers,
Manpreet Sambi, Neil Saunders, Jan Walker, Kieran Wardell, Jeanette Whiteman

Cover design by emc design ltd.

With thanks to Janet Dickinson, Allan Graham, Rosie Hanson, Samantha Krbacevic,
Simon Little, Glenn Rogers, David Ryan and Ruth Wilbourne for the proofreading.
With thanks to Ana Pungartnik for the copyright research.

Printed by Elanders Ltd, Newcastle upon Tyne.
Clipart from Corel®

About This Book

Sample Page (left)

30.1 Congruence and Similarity

Congruent shapes are exactly the same shape and size. Similar shapes are exactly the same shape but different sizes. Translated, rotated and reflected shapes are congruent to the original shapes, while enlarged shapes are similar to the original shapes.

Congruent Triangles

Learning Objectives — Spec Ref G5/G6:
- Know the congruence conditions for triangles.
- Be able to prove that two triangles are congruent.

Prior Knowledge Check:
Properties of parallel lines and polygons, circle theorems and Pythagoras' theorem (see Sections 20, 21 and 25).

In congruent shapes, all **side lengths** and **angles** on one shape are **identical** to the side lengths and angles on the other. However, you don't need to know **all** the side lengths and angles to show that two **triangles** are congruent — just show that they satisfy any **one** of the following 'congruence conditions'.

Tip: If any of these conditions are true then you can use trigonometry to show that the other sides and angles in the triangles must also be the same.

Side, Side, Side:
The three sides on one triangle are the same as the three sides on the other triangle.

Right angle, Hypotenuse, Side:
Both triangles have a right angle, both triangles have the same hypotenuse and one other side is the same.

Side, Angle, Side:
Two sides and the angle between them on one triangle are the same as two sides and the angle between them on the other triangle.

Angle, Angle, Side:
Two angles and an opposite side on one triangle are the same as two angles and the corresponding opposite side on the other triangle.

Be very careful when using 'Side, Angle, Side' and 'Angle, Angle, Side' — you have to have the **correct combination** of sides and angles.

Example 1

Are these two triangles congruent? Give a reason for your answer.
Look at the conditions listed above...
Two of the sides and the angle between them are the same on both triangles.
Condition SAS holds, so the triangles **are congruent**.

Tip: The conditions are often abbreviated to SAS, SSS, AAS and RHS.

Callouts (right)

Learning Objectives
Showing which bits of the specification are covered in each section.

Prior Knowledge Checks
Pointing you to the parts of the book that you should be familiar with before moving on to this topic.

Explanations
Clear explanations of every topic.

Tips
Lots of useful tips to help you get your head around the tricky bits.

Examples
Plenty of step-by-step worked examples.

Sample Page (bottom right)

Exercise 2

Q1   A recipe uses sugar and butter in the ratio 2 : 1.
　　a) How much butter is needed for 100 g of sugar?
　　b) Show the ratio as a graph on the axes on the right.
　　c) Use the graph to find the amount of sugar needed for 40 g of butter.

Q2   A wood has oak and beech trees in the ratio 2 : 9. 42 of the trees are oak. How many are beech?

Q3   The ages of a father and son are in the ratio 8 : 3. If the father is 48, how old is the son?

Q4   A cardboard cut-out of a footballer is 166 cm tall. The height of the cut-out and the footballer's actual height are in the ratio 5 : 6. How tall is the footballer?

Q5   Mai and Sid split their savings in the ratio 5 : 4.
　　a) Show this ratio on the axes on the right.
　　b) Use the graph to find how much Sid gets when Mai gets £400.
　　c) Sid gets £60. How much do they get in total?

Q6   A fruit punch is made by mixing pineapple juice, orange juice, and lemonade in the ratio 1 : 3 : 6. If 500 ml of pineapple juice is used, how much orange juice is needed?

Q7   Jem orders a pizza with olives, slices of courgette and slices of goat's cheese in the ratio 8 : 3 : 4. How many slices of courgette and how many slices of goat's cheese would she get with 24 olives?

Q8   Mo, Liz and Dee's heights are in the ratio 32 : 33 : 37. Mo is 144 cm tall. What is the combined height of the three people?

Q9   Max, Molly and Maisie are at a bus stop. The number of minutes they have waited can be represented by the ratio 3 : 7 : 2. Molly has been waiting for 1 hour and 10 minutes. How long have Max and Maisie been waiting?

Q10 The ratio of children to adults in a swimming pool must be 5 : 1 or less.
　　a) Draw a graph to show this ratio.
　　b) If there are 18 children, use your graph to work out how many adults there must be.

Q11 A TV-show producer is selecting a studio audience. He wants the ratio of under-30s to over-30s to be at least 8 : 1. If 100 under-30s are selected, find the maximum number of over-30s.

Q12 A recipe uses 1 aubergine for every 3 people. How many aubergines should you buy for 10 people?

Q13 Olga is allowed no more than 5 minutes of reality TV for every minute of news programmes she watches. She watches 45 minutes of reality TV. What's the least amount of news she should watch?

Exercises (with worked answers)
- Lots of practice for every topic, with fully worked answers at the back of the book.
- Throughout this book, the more challenging questions are marked like this: Q1

Problem Solving
Problem solving questions involve skills such as combining different areas of maths or interpreting given information to identify what's being asked for. Questions that involve problem solving are marked with stamps.

PROBLEM SOLVING

Review Exercise

Q1 A stone is dropped from the top of a 55 m-high tower. The distance in metres, h, between the stone and the ground after t seconds is given by the formula $h = 55 - 5t^2$.

    a) Draw a graph of the height of the stone for $h \geq 0$ and values of t between 0 and 5 seconds.

    b) Use your graph to estimate how long it takes the stone to fall to a height of 20 m above the ground.

    c) Find the exact time it takes for the stone to reach the ground.

Q2 By completing the square, calculate the position of the y-intercept and turning point of the graph of $y = x^2 - x - 1$.

Q3 Copy and complete the table below and draw a graph of the equation $y = x^3 - 3x^2 - x + 3$.

x	-3	-2	-1	0	1	2	3	4
y		-15				0		

Q4 Match each of these equations to one of the graphs below.

    (i) $y = x^3$ (ii) $y = \frac{1}{x}$ (iii) $y = 2^x - 1$ (iv) $x^2 + y^2 = 1$

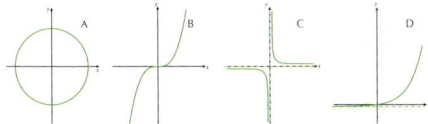

Q5 Susan uses a credit card to buy a computer that costs £500. The credit card company charges 3% interest per month. If she pays nothing back, the amount in pounds Susan will owe after t months is given by the formula $b = 500 \times (1.03)^t$.

    a) Explain what the numbers 500 and 1.03 represent in the formula.

    b) Draw a graph to show how this debt will increase during one year if Susan doesn't repay any money.

    c) Use your graph to estimate how long the interest on the credit card will take to reach £100.

Q6 Point P has coordinates (3, 4).

    a) Use Pythagoras' theorem to find the distance of P from the origin.

    b) Use your answer to a) to write the equation of the circle with centre (0, 0) that passes through point P.

Review Exercises
Mixed questions covering the whole section, with fully worked answers.

Non-Calculator Questions
There are some methods you'll have to be able to do without a calculator. Stamps on certain questions let you know that you can't use a calculator for them.

Exam-Style Questions
Questions in the same style as the ones you'll get in the exam, with worked solutions and mark schemes.

Mixed Exam-Style Questions
At the end of the book is a set of exam questions covering a mixture of different topics from across the GCSE 9-1 course.

Glossary
All the definitions you need to know for the exam, plus other useful words.

Formula Page
Contains all the formulas that you need to know for your GCSE exams. You'll find it inside the back cover.

Exam-Style Questions

Q1 A car is driven along a straight track to test its acceleration. The distance in metres, x, that the car has travelled after t seconds is given by $x = 2t^2 + t$.

    a) By calculating values of x at $t = 0, 1, 2, 3, 4$ and 5, draw a graph of x against t for $0 \leq t \leq 5$.
    [2 marks]

    b) Use your graph to find the distance the car will have travelled after 3.5 seconds.
    [1 mark]

    c) Describe how the speed of the car changes with time.
    [2 marks]

Q2 The graph below shows the relationship between the total amount in pounds (£), a, that a salesperson is paid in a year and the number of people, p, that they have signed up to a mobile phone contract in that year.

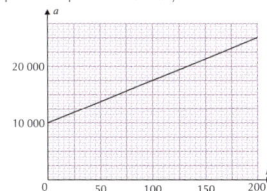

    a) How much would the salesperson earn in a year in which they signed up 100 people?
    [1 mark]

    b) Obtain a formula for a in terms of p.
    [3 marks]

Q3 The diagram on the right shows the cross-section of a container used to store chemicals. It is made from a cylinder joined to part of a sphere. The container has a total depth of d, as shown.

    A laboratory technician completely fills the container with water from a tap which is running at a constant rate. On a set of axes, sketch a graph to show how the depth of water in the container changes with time.
    [3 marks]

1.1 Calculations

Here's a nice gentle start to the book — non-calculator arithmetic (so put your calculator away).
All you have to do here is add, subtract, multiply and divide — but you need to know the order in which
to do them if you have to do multiple operations, and how to handle negative numbers and decimals.

Order of Operations

Learning Objective — Spec Ref N3:
Know the correct order in which to apply operations, including dealing with brackets.

BODMAS tells you the correct order to carry out mathematical operations:

Brackets, Other, Division, Multiplication, Addition, Subtraction

If there are two or more **consecutive divisions and/or multiplications** (e.g. $3 \times 6 \div 9 \times 5$),
do them in order **from left to right**. The same goes for **addition and subtraction**.

Example 1

Work out: a) $20 - 3 \times 4$ b) $30 \div (15 - 12)$

a) There are no brackets, 'other' operations
    or divisions in this calculation,
    so do the multiplication first $20 - 3 \times 4 = 20 - 12$
    then do the subtraction. $= \mathbf{8}$

Tip: Just ignore any bits of BODMAS that you don't need.

b) Do the bit in brackets first $30 \div (15 - 12) = 30 \div 3$
    then do the division. $= \mathbf{10}$

Exercise 1

Q1   Work out the following.
   a) $5 + 1 \times 3$ b) $11 - 2 \times 5$ c) $18 - 10 \div 5$
   d) $24 \div 4 + 2$ e) $35 \div 5 + 2$ f) $36 - 12 \div 4$

Q2   Work out the following.
   a) $2 \times (4 + 10)$ b) $(7 - 2) \times 3$ c) $4 + (48 \div 8)$
   d) $56 \div (2 \times 4)$ e) $(3 + 2) \times (9 - 4)$ f) $(8 - 7) \times (6 + 5)$

Q3   Work out the following.
   a) $2 \times (8 + 4) - 7$ b) $100 \div (8 + 3 \times 4)$ c) $7 + (10 - 9 \div 3)$
   d) $20 - (5 \times 3 + 2)$ e) $48 \div 3 - 7 \times 2$ f) $36 - (7 + 4 \times 4)$

Q4   Work out the following.   a) $4 - 5 + 2 - 1$ b) $5 \times 4 \div 10 \times 6$

Example 2

Work out: $\dfrac{18 - 2 \times 3}{8 \div 2}$

1. Think of the top and bottom of the fraction as having 'invisible brackets' around them.

2. To evaluate each set of brackets, apply BODMAS to the expression inside.

$$\frac{18 - 2 \times 3}{8 \div 2} = \frac{(18 - 2 \times 3)}{(8 \div 2)}$$

$$= \frac{(18 - 6)}{4}$$

$$= \frac{12}{4} = 3$$

Exercise 2

Q1 Work out the following.

a) $\dfrac{16}{4 \times (5 - 3)}$ b) $\dfrac{8 + 2}{15 \div 3}$ c) $\dfrac{4 \times (7 + 5)}{6 + 3 \times 2}$ d) $\dfrac{6 + (11 - 8)}{7 - 5}$

Q2 Work out the following.

a) $\dfrac{12 \div (9 - 5)}{25 \div 5}$ b) $\dfrac{8 \times 2 \div 4}{5 - 6 + 7}$ c) $\dfrac{3 \times 3}{21 \div (12 - 5)}$ d) $\dfrac{36 \div (11 - 2)}{8 - 8 \div 2}$

Negative Numbers

Learning Objective — Spec Ref N2:
Add, subtract, multiply and divide negative numbers.

Adding and Subtracting Negative Numbers

When adding and subtracting negative numbers, there are two rules you need to know:

1. Adding a negative number is the same as subtracting a positive number.
    So '+' next to '−' means **subtract**, i.e. $a + (-b) = a - b$.

2. Subtracting a negative number is the same as adding a positive number.
    So '−' next to '−' means **add**, i.e. $a - (-b) = a + b$.

Example 3

Work out: a) $1 - (-4)$ b) $-5 + (-2)$ c) $-3 - (-7)$

a) A '−' next to a '−' means add.

$$1 - (-4) = 1 + 4$$
$$= 5$$

b) A '+' next to a '−' means subtract.

$$-5 + (-2) = -5 - 2$$
$$= -7$$

c) A '−' next to a '−' means add.

$$-3 - (-7) = -3 + 7$$
$$= 4$$

Exercise 3

Q1 Work out the following.

a) $-4 + 3$ b) $-1 - 4$ c) $-12 + 15$

d) $6 - 17$ e) $4 - (-2)$ f) $-6 - (-2)$

Q2 Work out the following.

a) $-5 + (-5)$ b) $-5 - (-5)$ c) $-23 - (-35)$

d) $48 + (-22)$ e) $-27 + (-33)$ f) $61 - (-29)$

Multiplying and Dividing Negative Numbers

1. When you multiply or divide two numbers which have the **same** sign, the answer is **positive** — for example, $(-6) \times (-7) = 42$ and $(-40) \div (-8) = 5$.

2. When you multiply or divide two numbers which have **opposite** signs, the answer is **negative** — for example, $(-4) \times 9 = -36$ and $35 \div (-5) = -7$.

Example 4

Work out: a) $24 \div (-6)$ b) $(-5) \times (-8)$ c) $[(-27) \div 3] \times (-4)$

a) The signs are different so the answer will be negative. $24 \div (-6) = -4$

b) The signs are the same so the answer will be positive. $(-5) \times (-8) = 40$

c) You'll have to use BODMAS here:
1) Do the bit in brackets first (the signs are different so the result will be negative).
2) Then do the multiplication — because you ended up with a negative number inside the brackets, the signs are the same so the answer will be positive.

$$[(-27) \div 3] \times (-4) = (-9) \times (-4)$$
$$= 36$$

Exercise 4

Q1 Work out the following.

a) $(-15) \div (-3)$ b) $12 \div (-4)$ c) $(-72) \div (-6)$

d) $56 \div (-8)$ e) $(-16) \times (-3)$ f) $(-81) \div (-9)$

g) $(-13) \times (-3)$ h) $7 \times (-6)$ i) $(-34) \times 2$

Q2 Work out the following.

a) $[(-3) \times 7] \div (-21)$ b) $[(-24) \div 8] \div 3$ c) $[55 \div (-11)] \times (-9)$

d) $[(-63) \div (-9)] \times (-7)$ e) $[35 \div (-7)] \times (-8)$ f) $[(-12) \times 3] \times (-2)$

Q3 Copy the following calculations and fill in the blanks.

a) $(-3) \times \boxed{} = -6$ b) $(-14) \div \boxed{} = -2$ c) $\boxed{} \times 4 = -16$

d) $(-8) \times \boxed{} = -24$ e) $\boxed{} \times (-3) = 36$ f) $\boxed{} \div 11 = -7$

Decimals

Learning Objective — Spec Ref N2:
Add, subtract, multiply and divide decimals.

Adding and Subtracting Decimals

To add and subtract **decimals**, arrange them in columns like you would for normal numbers —
just make sure you line up the **decimal points**. You might have to **add in 0s** to fill in any gaps.

Example 5

Work out: a) **4.53 + 1.6**   b) **8.5 − 3.07**

1.   Set out the sum by lining up the decimal points.
2.   Fill in any gaps with 0s.
3.   Add or subtract the digits one column at a time,
     right to left. Carry or borrow digits as necessary.

$$\begin{array}{r} \textbf{a)} \quad 4\,.\,5\,3 \\ +\,1\,.\,6\,0 \\ \hline 6_1\,.\,1\,3 \end{array} \qquad \begin{array}{r} \textbf{b)} \quad 8\,.\,\overset{4}{\cancel{5}}{}^{1}0 \\ -\,3\,.\,0\,7 \\ \hline 5\,.\,4\,3 \end{array}$$

Exercise 5

Q1   Work out the following.
   a) 12.74 + 7   b) 0.8 − 0.03   c) 10.83 + 7.4   d) 91.7 + 0.492
   e) 6.474 + 0.92   f) 16.3 − 5.16   g) 9.241 − 2.8   h) 23 − 18.591

Q2   Copy the following calculations and fill in the blanks.

   a) $\quad \square\,.\,6\,\square$ b) $\quad 5\,.\,\square\,3$
      $+\ \ 0\,.\,\square\,0$ $-\ \ 2\,.\,1\,\square$
      $\overline{\ \ 8\,.\,2\,1\ \ }$ $\overline{\ \square\,.\,3\,1\ \ }$

Q3   Sunita buys a hat for £18.50 and a bag for £31.99. How much does she spend altogether?

Q4   Jay's meal costs £66.49. He uses a £15.25 off voucher. How much does he have left to pay?

Multiplying Decimals

To **multiply** decimals, multiply each decimal by a **power of 10** to get a whole number multiplication.
Do this multiplication in the normal way. Finally, **divide** your answer by the product of the powers of 10
to get back to the original multiplication.

Example 6

Work out 0.32 × 0.6

1.   Multiply each decimal by a power of 10
     to get a whole number multiplication.

2.   Multiply the whole numbers.

3.   Divide by the product of the powers
     of 10 you multiplied by in Step 1.

$$0.32 \times 0.6$$
$$\times 100 \overset{\curvearrowright}{} 32 \times 6 \overset{\curvearrowleft}{} \times 10$$

$$\begin{array}{r} 3\,2 \\ \times \quad 6 \\ \hline 1\,9_1\,2 \end{array}$$

So 0.32 × 0.6 = 192 ÷ 1000
                        = **0.192**

Exercise 6

Q1 132 × 238 = 31 416. Use this information to work out the following.
  a) 13.2 × 238
  b) 1.32 × 23.8
  c) 1.32 × 0.238
  d) 0.132 × 0.238

Q2 Work out the following.
  a) 16.7 × 8
  b) 31.2 × 6
  c) 3.1 × 40
  d) 0.7 × 600
  e) 0.05 × 0.04
  f) 0.08 × 0.5
  g) 2.1 × 0.6
  h) 1.6 × 0.04
  i) 5.2 × 0.09
  j) 3.9 × 8.3
  k) 0.16 × 3.3
  l) 0.64 × 0.42

Q3 1 litre is equal to 1.76 pints. What is 5 litres in pints?

Q4 Petrol costs £1.35 per litre. A car uses 9.2 litres during a journey. How much does this cost?

Dividing Decimals

1. To divide a decimal by a **whole number**, you can just treat it as a normal division, but set it out so that the decimal points in the question and answer are **lined up**.

2. To divide **by a decimal**, multiply both numbers by the **same power of 10** to turn the calculation into a division by a whole number. Then do the division, making sure you line up the decimal points in your calculation. You **don't** need to divide by the power of 10 at the end — you multiplied both numbers, so the answer wasn't affected.

Example 7

Work out 0.516 ÷ 0.8

1. Multiply both numbers by 10, so you're dividing by a whole number.

$$0.516 ÷ 0.8 = 5.16 ÷ 8$$

2. Line up the decimal points to set out the calculation, then divide.

$$\begin{array}{r} 0.\,6\;4\;5 \\ 8\overline{)5.^51^36^40} \end{array}$$

So 0.516 ÷ 0.8 = **0.645**

Tip: This example uses short division, but you can use long division here if you want.

Exercise 7

Q1 Work out the following.
  a) 8.52 ÷ 4
  b) 2.14 ÷ 4
  c) 8.62 ÷ 5
  d) 17.1 ÷ 6
  e) 0.081 ÷ 9
  f) 12.06 ÷ 8

Q2 Work out the following.
  a) 1.56 ÷ 0.2
  b) 0.624 ÷ 0.3
  c) 0.275 ÷ 0.5
  d) 16.42 ÷ 0.02
  e) 0.257 ÷ 0.05
  f) 7.665 ÷ 0.03
  g) 0.039 ÷ 0.06
  h) 50.4 ÷ 0.07
  i) 0.71 ÷ 0.002
  j) 108 ÷ 0.4
  k) 20.16 ÷ 0.007
  l) 1.44 ÷ 1.2

Q3 A 2.72 m ribbon is cut into equal pieces of length 0.08 m. How many pieces will there be?

Q4 It costs £6.93 to buy 3.5 kg of pears. How much do pears cost per kg?

1.2 Multiples and Factors

To tackle multiples and factors, you need to know your times tables — that's all there is to it.

Multiples

Learning Objective — Spec Ref N4:
Identify and find multiples and common multiples.

The **multiples** of a number are just the numbers in its **times table**. E.g. the multiples of 4 are 4, 8, 12, 16, ...

A **common multiple** of two (or more) numbers is a multiple of both (or all) of those numbers.

Example 1

a) **List the multiples of 5 between 23 and 43.**

These are the numbers between 23 and 43 that are in the 5 times table. **25, 30, 35, 40**

b) **Which of the numbers in the box on the right are common multiples of 2 and 7?**

| 24 | 7 | 28 | 42 | 35 |

The multiples of 2 in the box are 24, 28 and 42.
The multiples of 7 in the box are 7, 28, 35 and 42.
So 28 and 42 are multiples of both numbers. **28, 42**

Exercise 1

Q1 List the first five multiples of: a) 9   b) 13   c) 16

Q2 a) List the multiples of 12 between 20 and 100.

b) List the multiples of 14 between 25 and 90.

Q3 Write down the numbers from the box that are:

a) multiples of 10

b) multiples of 15

c) common multiples of 10 and 15

5	10	15	20	25	30	35
40	45	50	55	60	65	70
75	80	85	90	95	100	105

Q4 a) List the multiples of 3 between 19 and 35.   b) List the multiples of 4 between 19 and 35.

c) List the common multiples of 3 and 4 between 19 and 35.

Q5 List all the common multiples of 5 and 6 between 1 and 40.

Q6 List all the common multiples of 6, 8 and 10 between 1 and 100.

Q7 List all the common multiples of 9, 12 and 15 between 1 and 100.

Q8 List the first five common multiples of 3, 6 and 9.

Factors

Learning Objective — Spec Ref N4:
Identify and find factors and common factors.

The **factors** of a number are the numbers that divide into it exactly.
E.g. the factors of 8 are 1, 2, 4 and 8 — dividing 8 by any of these numbers gives a whole number.

A **common factor** of two numbers is a factor of both of those numbers.
E.g. the factors of 12 are 1, 2, 3, 4, 6 and 12, so the common factors of 8 and 12 are 1, 2 and 4.

Example 2

a) **Write down all the factors of: (i) 18   (ii) 30**

1. Check if 1, 2, 3, etc. divide into the number.

2. Stop when a factor is repeated — factors come in pairs, so once you get to one that's already been found, you'll have all the factors of the number.

(i) $1 \times 18 = 18$ — so 1 and 18 are factors
   $2 \times 9 = 18$ — so 2 and 9 are factors
   $3 \times 6 = 18$ — so 3 and 6 are factors
   $4 \times \text{—} = 18$ — so 4 is not a factor
   $5 \times \text{—} = 18$ — so 5 is not a factor
   $6 \times 3 = 18$ — 6 and 3 are repeated so stop
   So the factors of 18 are **1, 2, 3, 6, 9 and 18**.

(ii) $1 \times 30 = 30$ — so 1 and 30 are factors
   $2 \times 15 = 30$ — so 2 and 15 are factors
   $3 \times 10 = 30$ — so 3 and 10 are factors
   $4 \times \text{—} = 30$ — so 4 is not a factor
   $5 \times 6 = 30$ — so 5 and 6 are factors
   $6 \times 5 = 30$ — 6 and 5 are repeated so stop
   So the factors of 30 are **1, 2, 3, 5, 6, 10, 15 and 30**.

Tip: Any two factors that multiply to give the number are called a factor pair — e.g. 2 and 9 are a factor pair of 18.

b) **Write down the common factors of 18 and 30.**

These are the numbers which appear in both lists from part a).   **1, 2, 3 and 6**

Exercise 2

Q1 List all the factors of each of the following numbers.
a) 13   b) 25   c) 24   d) 35
e) 32   f) 40   g) 50   h) 49

Q2 A baker has 12 identical cakes. In how many different ways can he divide them up into equal packets? List all the possibilities.

Q3 In how many different ways can 100 people be arranged into groups of equal size? List all the possibilities.

Q4 a) List all the factors of: (i) 15   (ii) 21
b) Hence list the common factors of 15 and 21.

Q5 List the common factors of each of the following pairs of numbers.
a) 15, 20   b) 50, 90   c) 24, 32
d) 64, 80   e) 45, 81   f) 96, 108

Q6 List the common factors of each of the following sets of numbers.
a) 30, 45, 50   b) 8, 12, 20   c) 9, 27, 36   d) 24, 48, 96

1.3 Prime Numbers and Prime Factors

There's a key definition coming up — prime numbers. Once you know what they are (and how to find them), you can move on to prime factors — which will come in very handy later in this section.

Prime Numbers

Learning Objective — Spec Ref N4:
Identify and find prime numbers.

A **prime number** is a number that has no other factors except **itself** and **1**.

Here are a few things to note about prime numbers:

- 1 is **not** classed as a prime number — this is a common mistake.

- 2 is the only **even** prime number.

- Prime numbers end in **1**, **3**, **7** or **9** (2 and 5 are the only exceptions to this rule).
  But **not all** numbers ending in 1, 3, 7 or 9 are prime (e.g. 27 = 3 × 9, so it isn't prime).

Example 1

Which of the numbers in the box on the right are prime? | 15   16   17   18   19   20

1. Ignore any even numbers, and any ending in 5.

    16, 18 and 20 are even, so can't be prime.

    15 ends in 5, so can't be prime.

2. Look for factors of each of the remaining numbers. If there aren't any, it's prime.

    17 has no factors other than 1 and 17.

    19 has no factors other than 1 and 19.

    So the prime numbers are **17** and **19**.

Exercise 1

Q1 Consider the following list of numbers: 11, 13, 15, 17, 19
   a) Which number in the list is not prime?
   b) Find two factors greater than 1 that can be multiplied together to give this number.

Q2 a) Which three numbers in the box on the right are not prime? | 31   33   35   37   39
   b) Find a factor pair (where each factor is greater than 1) of each of your answers to (a).

Q3 Write down the prime numbers in this list: 5, 15, 22, 34, 47, 51, 59

Q4 a) Write down all the prime numbers less than 10.
   b) Find all the prime numbers between 20 and 50.

Q5 a) For each of the following, find a factor greater than 1 but less than the number itself.
      (i) 4   (ii) 14   (iii) 34   (iv) 74
   b) Explain why any number with last digit 4 cannot be prime.

Writing a Number as a Product of Prime Factors

Learning Objectives — Spec Ref N4:
- Understand the unique factorisation theorem.
- Find the prime factorisation of a number.
- Write a prime factorisation using product notation.

Prior Knowledge Check:
Be able to find factors, recognise prime numbers and use index notation. See p.8-9 and p.92.

Any integer greater than 1 can be broken down into a string of **prime numbers** all multiplied together — this is known as the **prime factorisation** of the number. For example, the prime factorisation of 112 is $2 \times 2 \times 2 \times 2 \times 7$. The prime factorisation of every number is **unique** — each number only has **one** prime factorisation, and no two numbers can have the **same one**.

If the prime factorisation has **repeated factors**, you can write it using **index notation** (i.e. as a product of powers). So the prime factorisation of 112 can be written as $2^4 \times 7$.

To find a prime factorisation, you can use a **factor tree**. A factor tree breaks a number into factors, then breaks these factors into smaller factors, and keeps going until all of the factors are prime.

Example 2

Write 20 as a product of prime factors. Give your answer in index form.

Make a factor tree.

1. Find any factor pair of 20.
   Circle any factors that are prime.

2. Repeat step 1 for any factors which aren't prime.

3. Stop when all the factor tree's branches end in a prime.

4. Give any repeated factors as a power, e.g. $2 \times 2 = 2^2$.

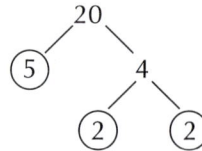

$20 = 2 \times 2 \times 5 = \mathbf{2^2 \times 5}$

Example 3

Write 450 as a product of prime factors. Give your answer in index form.

Make a factor tree.

1. Find any factor pair of 450.
   Neither of these are prime, so carry on to step 2.

2. Repeat step 1 for any factors which aren't prime. Circle any factors that are prime.

3. Stop when all the factor tree's branches end in a prime.

4. Give any repeated factors as a power.

Tip: You could have started with any two factors of 450 on the first branches — e.g. 9 and 50. The rest of the tree would look a bit different, but the final prime factorisation would be exactly the same.

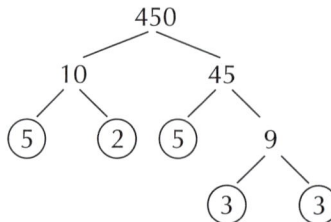

$450 = 2 \times 3 \times 3 \times 5 \times 5 = \mathbf{2 \times 3^2 \times 5^2}$

Exercise 2

Give your answers to these questions in index form where appropriate.

Q1 Write each of the following as the product of two prime factors.

a) 14   b) 55   c) 15   d) 21

e) 35   f) 39   g) 77   h) 121

Q2 a) Copy and complete the three factor trees below.

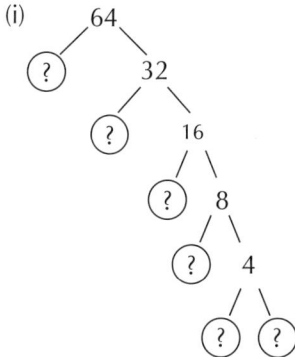

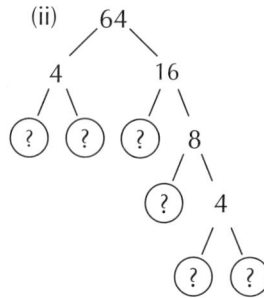

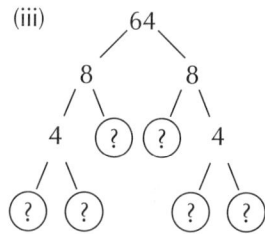

b) Use each of your factor trees to write down the prime factors of 64. What do you notice?

Q3 Copy and complete the factor tree on the right and use it to find the prime factorisation of 70.

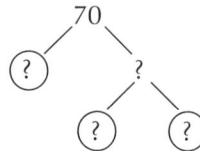

Q4 Write each of the following as the product of prime factors.

a) 30   b) 42   c) 66   d) 46

e) 78   f) 190   g) 210   h) 138

Q5 Write each of the following as the product of prime factors.

a) 44   b) 48   c) 72   d) 90

e) 50   f) 28   g) 98   h) 150

i) 132   j) 168   k) 325   l) 1000

Q6 Square numbers have all their prime factors raised to even powers.
For example, $36 = 2^2 \times 3^2$ and $64 = 2^6$.

a) Write 75 as a product of prime factors.

b) What is the smallest integer you could multiply 75 by to form a square number? Explain your answer.

Q7 By first writing each of the following as a product of prime factors, find the smallest integer that you could multiply each number by to give a square number.

a) 250   b) 416   c) 756   d) 1215

1.4. LCM and HCF

Now that you've got your head around multiples and factors, it's time to start looking at the 'lowest common multiple' and 'highest common factor' — or LCM and HCF for short.

LCM — 'Lowest Common Multiple'

Learning Objectives — Spec Ref N4:
- Understand the term 'lowest common multiple'.
- Be able to find the lowest common multiple of a set of numbers.

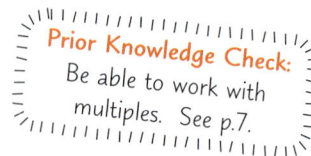

Prior Knowledge Check:
Be able to work with multiples. See p.7.

If you're given two (or more) numbers, then you can find their **lowest common multiple** (LCM). As the name suggests, the LCM is basically the smallest number that is a multiple of both (or all) of the numbers. In other words:

The LCM is the **smallest** number that will **divide** by **all** the numbers in a list.

Tip: The LCM is the smallest number in the times tables of all the numbers in the list.

If you're given a set of numbers, you can find their LCM by listing multiples of each number, then identifying the first one that appears in every list.

Example 1

Find the LCM of 4, 6 and 8.

1. List multiples of 4, 6 and 8.

   Multiples of 4:   4, 8, 12, 16, 20, (24) 28...

   Multiples of 6:   6, 12, 18, (24) 30...

   Multiples of 8:   8, 16, (24) 32...

2. The LCM is the smallest number that appears in all three lists.

   So the LCM of 4, 6 and 8 is **24**.

Exercise 1

Q1 Find the LCM of each of the following pairs of numbers.
   a) 3 and 5   b) 6 and 8   c) 2 and 10   d) 6 and 7

Q2 Find the LCM of each of the following sets of numbers.
   a) 3, 6, 8   b) 2, 5, 6   c) 4, 9, 12   d) 5, 7, 10

Q3 Mike visits Oscar every 4 days, while Narinda visits Oscar every 5 days.
   If they both visited today, how many days will it be before they visit on the same day again?

Q4 A garden centre has between 95 and 205 potted plants.
   They can be arranged exactly in rows of 25 and exactly in rows of 30.
   How many plants are there?

Q5 There are between 240 and 300 decorated plates hanging on a wall,
   and the number of plates divides exactly by both 40 and 70.
   How many plates are there?

HCF — 'Highest Common Factor'

Learning Objectives — Spec Ref N4:
- Understand the term 'highest common factor'.
- Be able to find the highest common factor of a set of numbers.

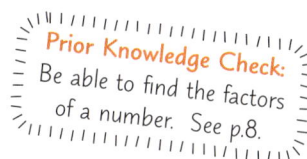

Prior Knowledge Check:
Be able to find the factors of a number. See p.8.

You may also be asked to find the **highest common factor** (HCF) of a list of numbers.
The HCF is just the largest value that is a factor of all the numbers in the list. In other words:

> HCF is the **largest** number that will **divide into all** the numbers in your list.

To find the HCF of a set of numbers, list the factors of each number, then pick the biggest one that's in every list.

Example 2

Find the HCF of 12 and 15.

1. Find the factors of 12 and the factors of 15.

2. Circle the common factors (the ones that appear in both lists).

3. The HCF is the biggest number that appears in both lists.

Factors of 12: (1), 2, (3), 4, 6, 12
Factors of 15: (1), (3), 5, 15

So the HCF of 12 and 15 is **3**.

Exercise 2

Q1 a) Find the common factors of 12 and 20.

   b) Hence find the highest common factor (HCF) of 12 and 20.

Q2 Find the HCF of each of the following pairs of numbers.

   a) 24 and 32   b) 36 and 60   c) 14 and 15   d) 12 and 36

Q3 Find the HCF of each of the following sets of numbers.

   a) 6, 8, 16   b) 12, 15, 18   c) 18, 36, 72   d) 36, 48, 60

Q4 An artist is using blue and pink ribbon in her artwork. *(PROBLEM SOLVING)*
She has 63 cm of blue ribbon and 91 cm of pink ribbon.
She wants to cut the ribbon into pieces so each piece
is of equal length and as long as possible.

   a) How long should she make each piece?

   b) How many pieces of each colour will she have in the end?

Q5 Kim is dividing counters into equal piles. *(PROBLEM SOLVING)*
She has 135 tangerine counters and 165 gold counters.
Each pile must contain only one of the colours.
What is the least number of piles she can make in total?

LCM and HCF using prime factors

Learning Objective — Spec Ref N4:
Be able to find the lowest common multiple and highest common factor
of a set of numbers using prime factorisation.

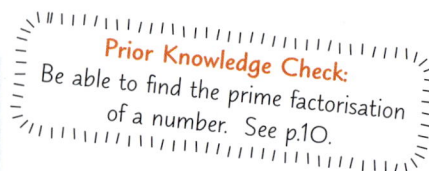

Prior Knowledge Check:
Be able to find the prime factorisation
of a number. See p.10.

You've just seen simple methods for finding the LCM and HCF by listing multiples and factors and looking
for the smallest or largest. Now it's time to see a more sophisticated method that uses **prime factorisation**.

LCM using prime factors

To find the LCM of a set of numbers using prime factors:

1. Write each number as a product of its prime factors.

2. For each prime factor, find the highest power of it that appears in **any** of the lists.

3. Multiply these numbers to find the lowest common multiple.

This method is especially useful when the numbers you're given are quite large
— then you're not stuck trying to write down huge lists of multiples.

Example 3

Use prime factors to find the LCM of 84 and 98.

1. Write 84 and 98 as products of prime factors.

2. Find the highest power of each prime factor
   that appears in either list. In this case, the
   highest power of 2 is 2^2, the highest power of
   3 is just 3, and the highest power of 7 is 7^2.

3. The LCM is the product of these numbers.

$84 = 2^2 \times 3 \times 7$

$98 = 2 \times 7^2$

So the LCM is:
$2^2 \times 3 \times 7^2 = \textbf{588}$

Tip: Remember to use
index form (powers) to
write any factors which
appear more than once.

The method works because a multiple of a number must contain **at least** its prime factorisation
(e.g. any multiple of 12 must be 'something' $\times 2^2 \times 3$). Therefore a common multiple of several numbers
must contain the prime factorisations of **all** the numbers — you have to take the highest power of each
factor to ensure this. You're not multiplying by anything else, so this is the LCM.

HCF using prime factors

A very similar method is used to find the HCF of a set of numbers using prime factors:

1. Write each number as a product of its prime factors.

2. For each prime factor, find the highest power of it that appears in **all** of the lists.

3. Multiply these numbers to find the highest common factor.

The only difference to the method for the LCM is that you only want factors that appear in **all** the
prime factorisations — if one prime factor only appears in one, it won't be part of the HCF.

Example 4

Use prime factors to find the HCF of 60 and 72.

1. Write 60 and 72 as products
    of prime factors.

    $60 = 2^2 \times 3 \times 5$

    $72 = 2^3 \times 3^2$

2. Find the highest power of each
    prime factor that appears in both lists.

    Both lists contain $2^2 \times 3$

3. The HCF is the product
    of these numbers.

    So the HCF is: $2^2 \times 3 = \mathbf{12}$

To understand how this method works, note that a factor of a number contains **some combination** of its prime factors. E.g. 2, 2^2 and 3×2 are factors of 12 because $12 = 2^2 \times 3$.

So a common factor of several numbers will be made up of some prime factors that the numbers have in common. And the **highest** common factor will contain **every** prime factor (including repeats) that the numbers have in common. Taking the highest power ensures that you catch all of them.

Exercise 3

Q1   a) Write 120 and 155 as products of their prime factors.
      b) Hence find the HCF of 120 and 155.

Q2   a) Write 76 and 88 as products of their prime factors.
      b) Hence find the LCM of 76 and 88.

Q3   Use prime factors to find (i) the HCF and (ii) the LCM of each of the following pairs of numbers.
      a) 60 and 75   b) 54 and 96   c) 108 and 144
      d) 200 and 240   e) 168 and 196   f) 150 and 180

Q4   Use prime factors to find the highest number that will divide into both 93 and 155.

Q5   Use prime factors to find the lowest number that divides exactly by both 316 and 408.

Q6   One day, Arran divides his action figures into equal groups of 26.
      The next day, he divides them up into equal groups of 12.
      Use prime factors to find the lowest possible number of action figures he owns.

Q7   Jess goes swimming every 21 days. Seamus goes swimming every 35 days.
      They both went swimming today. Use prime factors to find the number of days it will be
      until they both go swimming on the same day again.

Q8   a) Write 30, 140 and 210 as products of their prime factors.
      b) Hence find the HCF of 30, 140 and 210.

Q9   a) Write 121, 280 and 550 as products of their prime factors.
      b) Hence find the LCM of 121, 280 and 550.

Q10 Use prime factors to find (i) the HCF and (ii) the LCM of each of the following sets of numbers.
      a) 65, 143 and 231   b) 175, 245 and 1225   c) 104, 338 and 1078

Q11 a) Use prime factors to find a pair of numbers that have HCF = 12 and LCM = 120.
      b) Use prime factors to find a pair of numbers that have HCF = 20 and LCM = 300.

Review Exercise

Q1 Work out the following.   a) $5 \times 6 - 8 \div 2$ b) $18 \div (9 - 12 \div 4)$

Q2 At midday the temperature was 6 °C. By midnight, the temperature had decreased by 7 °C. What was the temperature at midnight?

Q3 Work out: $-4.2 - (1.5 \times -0.3)$

Q4 Milo spends £71.42 at the supermarket. His receipt says that he has saved £11.79 on special offers. How much would he have spent if there had been no special offers?

Q5 Asha bought 2 CDs each costing £11.95 and 3 CDs each costing £6.59. She paid with a £50 note. How much change did she receive?

Q6 It costs £31.85 to buy 7 identical DVDs. How much would it cost to buy 3 DVDs?

Q7 a) List all the common multiples of 12 and 16 between 1 and 100.
b) List all the common multiples of 6, 15 and 20 between 1 and 100.

Q8 A butcher has 54 identical sausages. In how many different ways can she divide them up into equal packets? List the possibilities.

Q9 List the common factors of each of the following pairs of numbers.
a) 12, 15   b) 25, 50   c) 36, 48   d) 64, 80

Q10 Without doing any calculations, explain how you can tell that none of the numbers in this box are prime.

20	30	40	50
70	90	110	130

Q11 Write each of the following as the product of prime factors.
a) 6   b) 40   c) 24   d) 110
e) 255   f) 60   g) 360   h) 225

Q12 Jill divides a pile of sweets into 5 equal piles. Kay then divides the same sweets into 7 equal piles. What is the smallest number of sweets there could be?

Q13 Find the HCF of each of the following pairs of numbers.
a) 8 and 12   b) 18 and 24   c) 35 and 42   d) 56 and 63

Q14 Use prime factors to find (i) the HCF and (ii) the LCM of each of the following sets of numbers.
a) 36 and 48   b) 210 and 308   c) 126, 150 and 1029

Exam-Style Questions

Q1 $539 \times 28 = 15\ 092$

Use the above result to work out the value of:

a) 539×14

[1 mark]

b) 5390×0.28

[1 mark]

c) $1\ 509\ 200 \div 53.9$

[2 marks]

Q2   Work out $3.774 \div 0.4$

[3 marks]

Q3   A single pack of salt and vinegar crisps costs 70p.
A single pack of cheese and onion crisps costs 65p.
A multipack of 3 salt and vinegar and 3 cheese and onion costs £3.19.
How much would you save buying a multipack instead of the equivalent amount
in individual packs?

[3 marks]

Q4   The lowest common multiple of two numbers is 60.
The highest common factor of the same two numbers is 4.
Neither of the numbers is 4 or 60. What are the numbers?

[3 marks]

Q5 p and q are prime numbers. $50 < p < 60$ and $60 < q < 70$.
If $pq = 3599$, find the values of p and q.

[3 marks]

Q6   When written as a product of prime factors in index form, $60 = 2^2 \times 3 \times 5$.

a) Write 135 as a product of prime factors in index form.

[2 marks]

b) Use the prime factor forms of 60 and 135 to show that $\sqrt{60 \times 135} = 90$.

[2 marks]

2.1 Rounding

Numbers can be approximated (or rounded) to the nearest whole number, 10, 100 etc. or to a given number of decimal places or significant figures. These approximations can be used to find an estimate of a tricky calculation.

Rounding

Learning Objectives — Spec Ref N15:
- Round numbers to a specified degree of accuracy.
- Round numbers to a given number of decimal places or significant figures.

Numbers can be **rounded** to make them easier to work with. This can be very handy when solving equations that don't have nice neat answers. For example, a number like 5468.9 could be rounded to:

- the nearest **whole number** (= 5469)
- the nearest **ten** (= 5470)
- the nearest **hundred** (= 5500)
- the nearest **thousand** (= 5000)

> **Tip:** Remember — a digit of 5 or more rounds up and a digit of 4 or less rounds down.

Be careful — if you round a number **too early** it can change the answer you get at the end. For example, if I had **£1.46** and rounded it to **the nearest 10p** I'd get **£1.50**. If I then rounded this to **the nearest pound** I'd get **£2**. But if I had rounded the original £1.46 to the nearest pound I'd get **£1**.

Exercise 1

Q1 Round the following to: (i) the nearest whole number   (ii) the nearest ten
                                         (iii) the nearest hundred   (iv) the nearest thousand

     a) 672.48   b) 2536.13   c) 8499.3   d) 3822.8

Q2 At its closest, Jupiter is about 390 682 810 miles from Earth.
Write this distance to the nearest million miles.

Decimal Places

You can also round to different numbers of **decimal places** (d.p.).
The method is as follows:

1. **Identify** the position of the '**last digit**' that you want to keep.
   E.g. when rounding to 2 d.p., it's the digit in the hundredths place.

2. Look at the next digit to the **right** — called **the decider**.

3. If the **decider** is **5 or more**, then **round up** the last digit.
   If the **decider** is **4 or less**, then **leave** the last digit as it is.

4. There must be **no more digits** after the last digit (not even zeros).

> **Tip:** When you're rounding up from 9 (to 10), replace the 9 with a 0 and **carry 1 to the left**. E.g. 4.98 rounded to 1 decimal place would be 5.0.

Round the number 8.9471 to: a) **1 d.p.**   b) **2 d.p.**   c) **3 d.p.**

a) The last digit is 9 and the decider is 4, so round down.   **8.9**

b) The decider is 7, so round up.   **8.95**

c) The decider is 1, so round down.   **8.947**

Exercise 2

Q1   Round the following numbers to: (i) 1 d.p.   (ii) 2 d.p.   (iii) 3 d.p.

a) 2.6893   b) 0.3249   c) 5.6023   d) 0.0525

e) 6.2571   f) 0.35273   g) 0.07953   h) 0.96734

Q2   The mass of a field vole is 0.0384 kilograms. Round this mass to two decimal places.

Significant Figures

You can also round a value to a number of **significant figures** (s.f.). The method for significant figures is **identical** to the one for decimal places — except it can be a little harder to locate the **last digit**.
There are a few key things to note:

- The **first significant figure** is the first digit that **isn't 0**, e.g. 2 is the first significant figure in 0.023.

- All the digits that follow are also **significant figures**, **regardless** of whether or not they're **0s**.

- Once you've **identified** your significant figures,
  round using the method on the previous page.

- After rounding the last digit, fill in all the places **up to the decimal point**
  with 0s. (You may need to **add a 0** after the decimal point to make up
  the correct number of significant figures — e.g. 32.0.)

> **Tip:** If the last
> significant figure is **after**
> the decimal point, you
> don't add any extra
> zeros at the end.

Round to three significant figures: a) **32 568**   b) **0.00097151**

a) 1.   The first significant figure is 3. The decider is 6, so round up.   32 568

   2.   As the significant figures are before the decimal point,
        fill the remaining places with zeros.   **32 600**

b) 1.   The first significant figure is the first digit that isn't
        zero — this is 9. The decider is 5, so round up.   0.000 971 51

   2.   As the significant figures are after the decimal point,
        no more zeros are needed.   **0.000972**

Exercise 3

Q1   Round the following numbers to: (i) 1 s.f.   (ii) 2 s.f.   (iii) 3 s.f.

a) 46.874   b) 5067   c) 35 722   d) 925 478

e) 0.08599   f) 0.10653   g) 0.00041769   h) 34.726

Q2   The speed of sound is 1236 km/h. Round this speed to two significant figures.

Estimates

- Find approximate answers using estimates.
- Estimate the value of square roots.

Prior Knowledge Check:
Be familiar with square numbers and their roots. See p.91.

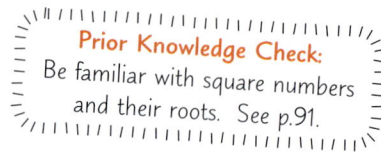

Using rounded numbers in a **calculation** gives an **estimate** of the actual answer — by **simplifying** in this way you get an **approximate value**. You can use this method to **check** calculations — i.e. to see if your actual answer looks 'about right'. The symbol used in estimating is '≈' and means 'is approximately equal to'.

You can usually figure out if your answer will be an **over-estimate** or an **under-estimate**.
- **Addition** or **multiplication** — if both numbers are **rounded up** you'll get an **over-estimate** and if both numbers are **rounded down** you'll get an **under-estimate**.
- **Subtraction** or **division** — you'll get an **over-estimate** if you rounded the **1st number up** and the **2nd number down**, and an **under-estimate** if you rounded the **1st number down** and the **2nd number up**.

Example 3

Estimate the value of $\dfrac{9.7 \times 326}{1.823 \times 5.325}$ by rounding each number to one significant figure.

1. Round each number to 1 s.f.

$$\frac{9.7 \times 326}{1.823 \times 5.325} \approx \frac{10 \times 300}{2 \times 5}$$

2. Work out the calculation in stages.

$$= \frac{3000}{10} = 300$$

Tip: The actual value is 325.748..., so this is a good approximation.

To estimate a **square root**, find the two square numbers it lies between and decide which it's closer to.

Example 4

Estimate the value of $\sqrt{61}$ to 1 d.p.

1. Find the closest square number on either side of the number in the question (61).

   61 lies between 49 (= 7^2) and 64 (= 8^2).

2. Decide which is closer to 61, then make a sensible estimate of the digit after the decimal point.

   61 is much closer to 64, so a sensible estimate is $\sqrt{61} \approx$ **7.8**.

Exercise 4

Q1 Estimate each of the following by rounding each number to one significant figure.
   a) 102.2×4.2 b) 288.7×7.8 c) $306.9 \div 6.4$ d) $3.9 \div 5.1$
   e) $494.27 \div 5.05$ f) $205.52 \div 8.44$ g) 142.75×9.56 h) $8.31 \div 1.86$

Q2 The following questions have three possible answers. Use estimation to decide which answer is correct.
   a) $101 \times 52 = (5252, 4606, 6304)$ b) $588 \div 12.4 = (75.78, 47.42, 69.86)$
   c) $0.79 \times 1594.3 = (1259.50, 864.80, 679.57)$ d) $0.94 \div 3.68 = (0.124, 0.901, 0.255)$

Q3 Estimate each of the following by rounding each number to one significant figure.
   a) $\dfrac{9.9 \times 285}{18.7 \times 3.2}$ b) $\dfrac{174.3 \times 3.45}{162.8 \times 10.63}$ c) $\dfrac{432.4 \times 2.75}{233.39 \times 0.81}$ d) $\dfrac{176.65 \div 8.84}{564.36 \div 2.78}$

Q4 Estimate the following roots to the nearest whole number.
   a) $\sqrt{18}$ b) $\sqrt{67}$ c) $\sqrt{51}$ d) $\sqrt{86}$

Q5 Estimate the following roots to 1 decimal place.
   a) $\sqrt{11}$ b) $\sqrt{30}$ c) $\sqrt{20}$ d) $\sqrt{37}$

2.2 Upper and Lower Bounds

Rounding leaves a level of uncertainty between the actual number and rounded number. This is where bounds come in. Upper and lower bounds show the maximum and minimum actual values a rounded number can be.

Upper and Lower Bounds

Learning Objective — Spec Ref N16:
Understand and use upper and lower bounds.

Upper and lower bounds show you where the actual value of a rounded number can lie.

- The **lower bound** is the **smallest actual value** the number can be and still be **rounded up** to the number in question. The actual value is **greater than or equal to** the lower bound.

- The **upper bound** is the **biggest actual value** the number can be and still be **rounded down** to the number in question. The actual value is **strictly less than** the upper bound (if it was exactly equal to the upper bound it would round up to the next unit).

For any given rounding unit, the actual value is anything up to **half a unit bigger or smaller**. For example, if you round a number to 1 d.p., the **rounding unit** is 0.1 so the actual value is anything up to **0.05 either side**. If you round a figure to the nearest whole number, the rounding unit is 1 so the actual value can be 0.5 either side — e.g. a weight of 72 kg to the nearest kilogram has a **lower bound** of 71.5 kg and an **upper bound** of 72.5 kg.

Example 1

A length is given as 12 m when rounded to the nearest metre. State its lower and upper bounds.

1. The rounding unit is metres, so the smallest number that would round to 12 m is half a metre less: 12 − 0.5 = 11.5 m.

   Lower bound = **11.5 m**

2. The biggest number that would round down to 12 is 12.49999999... so, by convention, we say that the upper bound is 12.5 m.

   Upper bound = **12.5 m**

Exercise 1

Q1   The following figures have been rounded to the nearest whole number.
State their lower and upper bounds.

    a) 10   b) 34   c) 76   d) 102

    e) 99   f) 999   g) 249   h) 2500

Q2   Give the lower and upper bounds of the following measurements.

    a) 645 kg (measured to the nearest kg)   b) 255 litres (measured to the nearest litre)

    c) 800 g (measured to the nearest 100 g)   d) 155 cm (measured to the nearest cm)

Q3   State the lower and upper bounds of the following prices.

    a) £15 (when rounded to the nearest £1)   b) £320 (when rounded to the nearest £10)

    c) £76.70 (when rounded to the nearest 10p)   d) £600 (when rounded to the nearest £50)

Calculating with rounded values will create a **discrepancy** between the **calculated value** and the **actual value**. This gives a **minimum** and **maximum value** of a calculation, found by using the lower and upper bounds.

If A and B are rounded numbers, then:

- **A + B**: Max = upper bound of A + upper bound of B, Min = lower bound of A + lower bound of B
- **A − B**: Max = upper bound of A − lower bound of B, Min = lower bound of A − upper bound of B

Example 2

Find the maximum and minimum perimeter of a rectangular garden measuring 38 m by 20 m to the nearest metre.

1. For the maximum perimeter, find the upper bounds of the garden's sides.

$38 + 0.5 = 38.5$ m
$20 + 0.5 = 20.5$ m

2. Use these to calculate the maximum possible perimeter.

$(38.5 \times 2) + (20.5 \times 2) = 77 + 41 = 118$ m
Maximum perimeter = **118 m**

3. For the minimum perimeter, find the lower bounds of the garden's sides.

$38 - 0.5 = 37.5$ m
$20 - 0.5 = 19.5$ m

4. Use these to calculate the minimum possible perimeter.

$(37.5 \times 2) + (19.5 \times 2) = 75 + 39 = 114$ m
Minimum perimeter = **114 m**

Exercise 2

Q1   Find the minimum and maximum possible perimeter of the following:

    a) A rectangle with sides 5 cm and 6 cm measured to the nearest cm.

    b) A square with sides of 4 m measured to the nearest m.

    c) An equilateral triangle with side length 7 cm measured to the nearest cm.

    d) A regular pentagon with sides 12.5 cm measured to the nearest mm.

Q2   Celia needs to sew a ribbon border onto a rectangular tablecloth. If the tablecloth measures 2.55 m by 3.45 m to the nearest cm, what is the longest length of ribbon that could be needed?

Q3   Last year Jack was 1.3 m tall, measured to the nearest 10 cm. This year he has grown 5 cm, to the nearest cm. Calculate the tallest height and the shortest height he could actually be this year.

Q4   Mr McGregor wants to build a fence around his garden, leaving a 2 m gap for the gate. The garden is a rectangle measuring 12 m by 15 m. All measurements are to the nearest metre.

    a) (i)   Give the minimum width of the gate.

        (ii) Give the maximum length and width of the garden.

        (iii) Hence find the maximum amount of fencing needed to fence the side of the garden with the gate on it.

        (iv) Calculate the maximum total length of fencing needed to fence the garden.

    b) By finding the maximum width of the gate and the minimum dimensions of the garden, calculate the minimum total length of fencing needed.

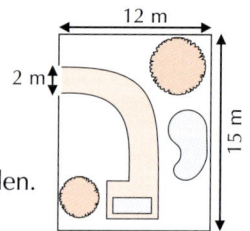

You can also work out the maximum and minimum values for multiplication and division:

- **A × B**: Max = upper bound of A × upper bound of B, Min = lower bound of A × lower bound of B
- **A ÷ B**: Max = upper bound of A ÷ lower bound of B, Min = lower bound of A ÷ upper bound of B

Example 3

Find the maximum and minimum possible area (A) of a rectangular room that measures 3.8 m by 4.6 m to the nearest 10 cm.

1. Find the upper bounds.	$3.8 + 0.05 = 3.85$ m $4.6 + 0.05 = 4.65$ m
2. Use these to work out the maximum possible area.	$A = 3.85 \times 4.65 = \mathbf{17.9025}$ **m²**
3. Find the lower bounds.	$3.8 - 0.05 = 3.75$ m $4.6 - 0.05 = 4.55$ m
4. Work out the minimum possible area.	$A = 3.75 \times 4.55 = \mathbf{17.0625}$ **m²**

Example 4

Find the maximum and minimum possible speeds of an ostrich if it runs 17.2 km in 0.3 hours (both to 1 d.p.).

1. Find the upper and lower bounds of the distance ($= d$) and time ($= t$).

Upper and lower bounds of 17.2 km are 17.25 km and 17.15 km, respectively.

Upper and lower bounds of 0.3 hours are 0.35 hours and 0.25 hours, respectively.

2. Use the formulas from the previous page.

$$\text{Maximum speed} = \frac{\text{upper bound of } d}{\text{lower bound of } t} = \frac{17.25}{0.25} = \mathbf{69 \text{ km/h}}$$

$$\text{Minimum speed} = \frac{\text{lower bound of } d}{\text{upper bound of } t} = \frac{17.15}{0.35} = \mathbf{49 \text{ km/h}}$$

Exercise 3

Q1 Calculate the maximum and minimum possible volumes for a storage unit that measures 4.00 m by 3.00 m by 1.90 m to the nearest cm.

Q2 Michael drove in his car for a measured time of 13 minutes at 34 km/h. If his time was measured to the nearest minute, calculate the maximum possible distance that he could have driven.

Q3 A snail travels a distance of 5 m in 45 minutes. If the distance is measured to the nearest cm and the time to the nearest minute, what is the maximum possible speed of the snail in cm/s to 3 s.f.?

Q4 Max wants to paint a wall that measures 2.1 m by 5.2 m. A tin of paint states that it will cover 3.5 m². If the tin contains exactly 0.5 litres of paint, and all other measurements are correct to 1 d.p., find the maximum volume of paint needed to paint the wall. *(PROBLEM SOLVING)*

Representing Bounds Using Intervals

Learning Objectives — Spec Ref N15:
- Be able to truncate numbers.
- Use inequality notation to describe error intervals.

Tip: Computers use truncation when converting data.

You **truncate** a number by **chopping off decimal places**, so if the mass of a cake was 2.468 kg, truncated to 1 d.p. it would be 2.4 kg. When a measurement is truncated to a given unit, the actual measurement can be up to a **whole unit bigger but no smaller**.

Truncate 17.65342 to 2 d.p.

Just delete any numbers after the second decimal place.   **17.65**

Exercise 4

Q1   Truncate the following values to 1 decimal place.
      a) 1.354   b) 55.73   c) 103.67183   d) 85.955

Q2   Truncate the following values to 2 decimal places.
      a) 2738.29109   b) 1.24692   c) 17.160   d) 100.0984

You can show the **upper and lower bounds** of a number using **inequalities**. These represent **the interval of possible values** the rounded or truncated number can be, from the smallest possible number (the lower bound) to the greatest possible number (the upper bound).

▪   The actual value is **greater than or equal to** the **lower bound**, so you use the inequality '≤'.
▪   The actual value is **strictly less than** the **upper bound**, so you use the inequality '<'.

For example, if a table was 1.2 m tall **truncated** to 1 d.p. the interval would be **1.2 m ≤ x < 1.3 m**.

Example 6

The approximate mass of a rock is 18.4 kg.
Find the interval within which the actual mass of the rock, m, lies if:
a) the mass has been rounded to 1 decimal place.
b) the mass has been truncated to 1 decimal place.

a)   The lower bound is 18.35 kg, and the upper bound is 18.45 kg.
         The actual value must be strictly less than 18.45 kg,   **18.35 kg ≤ m < 18.45 kg**
         or it would round to 18.5 kg.

b)   The lower bound is 18.4 kg, and the upper bound is 18.5 kg.
         The actual value must be strictly less than 18.5 kg,   **18.4 kg ≤ m < 18.5 kg**
          or it would truncate to 18.5 kg.

Exercise 5

Q1   Given that the following values have been rounded to 1 d.p., write down an inequality
      for each to show the range of possible actual values.
      a) $n = 15.2$ b) $p = 37.1$ c) $q = 109.9$ d) $r = 70.0$

Q2   Given that the following values have been truncated to 2 d.p., write down an inequality
      for each to show the range of possible actual values.
      a) $s = 6.57$ b) $t = 25.71$ c) $v = 99.99$ d) $w = 51.00$

Q3   Alasdair is canoeing down a river and says that he has travelled 10 km to the nearest 100 m.
      Write down the interval within which the actual distance in km, d, lies.
      Give your answer as an inequality.

Q4   Given that $x = 3.2$ to 1 d.p. and $y = 8.34$ to 2 d.p., find the interval that contains
      the actual value of $2x + y$. Give your answer as an inequality.

Review Exercise

Q1 The length of a snake is 1.245 metres. Round this length to one decimal place.

Q2 The table on the right shows the mass (in kg) of some mammals. Round each mass to two significant figures.

Mammal	Mass (kg)
Common vole	0.0279
Badger	9.1472
Meerkat	0.7751
Red squirrel	0.1998
Shrew	0.00612
Hare	3.6894

Q3 Jade buys four items costing £1.35, £8.52, £14.09 and £17.93 from a shop. Estimate how much she spent by rounding each price to the nearest pound.

Q4 Estimate each of the following by rounding each number to one significant figure.

a) $\dfrac{64.4 \times 5.6}{17 \times 9.5}$
b) $\dfrac{310.33 \times 2.68}{316.39 \times 0.82}$
c) $\dfrac{13.7 \times 5.2}{12.3 \div 3.9}$
d) $\dfrac{173.64 \times 10.6}{64.44 \div 5.58}$

Q5 Estimate to 1 d.p. the value of:

a) $\sqrt{14}$
b) $\sqrt{77}$
c) $\sqrt{130}$
d) $\sqrt{56}$

Q6 Josie took 24 minutes, to the nearest minute, to walk from the cinema to the museum. State the upper and lower bounds of this time.

Q7 Find the maximum and minimum possible values of the following.

a) The area of a rectangle with sides given as 5 cm and 6 cm, measured to the nearest cm.

b) The volume of a cube with sides given as 6 cm, measured to the nearest cm.

c) The volume of a cuboid with sides given as 3.5 cm, 4.4 cm and 5.6 cm, measured to the nearest mm. Give your answer in cm³, rounded to 2 d.p.

Q8 Kelly ran a 1500 m race in 260 s. The time was measured to the nearest second and the distance to the nearest 10 m. Calculate the maximum and minimum possible values of her average speed in m/s.

Q9 Truncate the following to 1 d.p.

a) 78.445
b) 32.510
c) 567.862
d) 999.999

Name	Height (m)
Lily	1.40
May	1.43
Isaac	1.60
Max	1.56
Daisy	1.28

Q10 A mum of 5 children has truncated her children's heights in metres to 2 decimal places, and written them down in the table on the left.

a) Write down the intervals within which each child's actual height lies.

b) Write down the minimum and maximum height difference between the tallest and shortest child.

PROBLEM SOLVING

Exam-Style Questions

Q1 A milliner measures the circumference of a hat he has made as 57 cm to the nearest cm.
What are the upper and lower bounds of the hat's circumference?

[1 mark]

Q2 One side of a square has a length (*l*) of 6.37 cm.

a) By rounding the length to 1 significant figure, estimate the area (*a*) of the square.

[1 mark]

The square is used as the base of a pyramid with a height, *h*, of 9.2 cm.

b) Using the formula $V = \frac{1}{3} \times$ base area $\times$ perpendicular height,
estimate the volume of the pyramid.

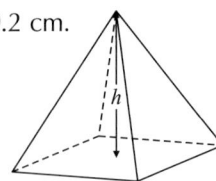

[2 marks]

Q3
$$x = \frac{628}{\sqrt{97} + 9.6}$$

a) By rounding each number to 1 significant figure, estimate the value of *x*.

[3 marks]

b) Explain why your answer to part a) is an under-estimate of the actual value of *x*.

[1 mark]

Q4 The audience at a rock concert, *r*, was estimated to be 7300, correct to the nearest hundred.
Write down the error interval for *r*.

[2 marks]

Q5 Delphine wants to know the perimeter of her bedroom. She measures the walls
to the nearest 10 cm. The lengths are shown on the floor plan below.

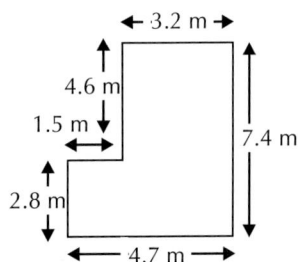

a) What is the maximum possible perimeter of Delphine's bedroom?

[2 marks]

b) Find the interval within which the actual perimeter lies.

[2 marks]

3.1 Equivalent Fractions

Fractions are a way of writing an amount as one number divided by another. Because different divisions can give the same answer (e.g. 1 ÷ 3 is the same as 2 ÷ 6), different fractions can have the same value.

Equivalent Fractions

Learning Objective — Spec Ref N3:
Be able to find equivalent fractions.

There are lots of **different ways** to write the **same amount** using fractions — these are known as **equivalent fractions**. For example, one half is the same as two quarters, three sixths, etc. You can show this visually:

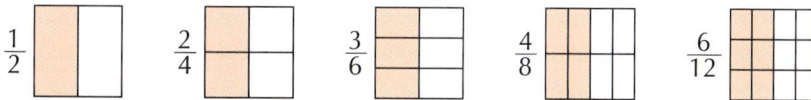

$\frac{1}{2}$ ☐ $\frac{2}{4}$ ☐ $\frac{3}{6}$ ☐ $\frac{4}{8}$ ☐ $\frac{6}{12}$ ☐

Tip: Notice that the same amount of the shape is shaded in each diagram.

To find an equivalent fraction, **multiply** or **divide** the **numerator** (the top number) and **denominator** (the bottom number) by the **same thing**.

Example 1

Find the value of *b* if $\frac{12}{30} = \frac{4}{b}$.

1. Find what you need to divide by to get from one numerator to the other.

   $$\overset{\div 3}{\frac{12}{30} = \frac{4}{b}}$$

2. Divide the denominator by the same number.

   $$\underset{\div 3}{\frac{12}{30} = \frac{4}{10}} \qquad \text{So } \textbf{\textit{b} = 10.}$$

Exercise 1

Q1 Find the values of the letters in the following fractions.

a) $\frac{1}{5} = \frac{a}{10}$ b) $\frac{1}{4} = \frac{b}{12}$ c) $\frac{3}{4} = \frac{c}{16}$ d) $\frac{1}{20} = \frac{d}{60}$

e) $\frac{1}{5} = \frac{5}{e}$ f) $\frac{1}{6} = \frac{3}{f}$ g) $\frac{7}{12} = \frac{35}{g}$ h) $\frac{9}{10} = \frac{81}{h}$

Q2 Find the values of the letters in the following fractions.

a) $\frac{1}{a} = \frac{5}{15}$ b) $\frac{3}{b} = \frac{12}{20}$ c) $\frac{c}{3} = \frac{10}{15}$ d) $\frac{d}{14} = \frac{9}{42}$

e) $\frac{e}{9} = \frac{15}{27}$ f) $\frac{f}{51} = \frac{9}{17}$ g) $\frac{11}{g} = \frac{55}{80}$ h) $\frac{1}{h} = \frac{11}{121}$

Q3 Sharon and Dev both take a test. Sharon gets $\frac{4}{6}$ of the questions right and Dev gets $\frac{37}{42}$ of the questions right. Who gets the most questions right?

Simplifying Fractions

Learning Objectives — Spec Ref N3/R3:
- Be able to write fractions in their simplest form.
- Be able to write one quantity as a fraction of another.

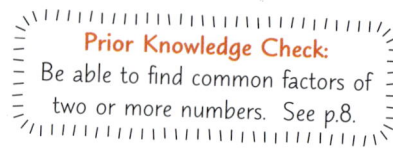

Prior Knowledge Check:
Be able to find common factors of two or more numbers. See p.8.

Simplifying a fraction means finding an equivalent fraction with the **smallest** possible whole numbers. This is also known as 'expressing a fraction in its **lowest terms**'.

To simplify a fraction, look for a **common factor** of the numerator and denominator. If they have one, **divide them both** by this factor, then **repeat** the process with the **new fraction**. Once the numerator and denominator have **no common factors** except **1**, the fraction is in its **simplest form**.

You can cut out some steps by finding the **highest common factor** (HCF) of the two numbers (see p.13) — dividing the numerator and denominator by the HCF will put the fraction in its simplest form **straight away**.

Example 2

Express $\frac{24}{30}$ as a fraction in its lowest terms.

1. Divide the numerator and denominator by any common factors. Repeat this until the numerator and denominator have no more common factors.

2. 4 and 5 have no common factors, so this fraction is in its lowest terms.

$$\overset{\div 3}{\frac{24}{30}} = \overset{\div 2}{\frac{8}{10}} = \frac{4}{5}$$
$$\underset{\div 3}{} \quad \underset{\div 2}{}$$

To express one number **as a fraction** of another, write the **first number** as the **numerator** and the **second** as the **denominator**, then **simplify**. E.g. to express 30 as a fraction of 100, write $\frac{30}{100}$, then simplify to $\frac{3}{10}$.

Example 3

A chef did a poll of customers at his restaurant. 96 of the 144 customers said they liked their food. Write this as a fraction in its simplest form.

1. 96 and 144 have a common factor of 12.

2. 8 and 12 have a common factor of 4.

3. 2 and 3 have no common factors.

$$\overset{\div 12}{\frac{96}{144}} = \overset{\div 4}{\frac{8}{12}} = \frac{2}{3}$$
$$\underset{\div 12}{} \quad \underset{\div 4}{}$$

Tip: You could skip straight to the answer by finding the HCF of 96 and 144, which is 48, and dividing both numbers by it.

Exercise 2

Q1 Write the following fractions in their lowest terms.

a) $\frac{9}{45}$ b) $\frac{15}{36}$ c) $\frac{24}{64}$ d) $\frac{72}{162}$

Q2 Write, in its lowest terms:

a) 21 as a fraction of 35   b) 36 as a fraction of 126   c) 70 as a fraction of 182

Q3 Simplify these fractions, then state which fraction is not equivalent to the other two.

a) $\frac{6}{18}, \frac{5}{20}, \frac{9}{27}$ b) $\frac{6}{8}, \frac{9}{15}, \frac{15}{25}$ c) $\frac{4}{18}, \frac{6}{33}, \frac{10}{45}$ d) $\frac{18}{24}, \frac{60}{80}, \frac{24}{40}$

Q4 There are 300 animals on a farm. 50 of them are cows, 70 are pigs and the rest are sheep. Find the fraction (in its simplest form) of the animals that are: a) cows,   b) sheep.

3.2 Mixed Numbers

So far, all of the fractions you've had to deal with have been smaller than 1. You can write numbers that are bigger than 1 as fractions in two different ways — either as mixed numbers or as improper fractions.

Learning Objective — Spec Ref N2:
Be able to convert between mixed numbers and improper fractions.

A **mixed number** has a whole number part and a fraction part — e.g. $2\frac{1}{2}$.

A fraction where the numerator is **bigger** than the denominator is called an **improper fraction** — e.g. $\frac{5}{2}$.

Example 1

Write: a) $4\frac{3}{5}$ as an improper fraction, b) $\frac{13}{5}$ as a mixed number.

a) 1. Find the fraction which is equivalent to 4 and has 5 as the denominator.

$$4 = \frac{4}{1} \overset{\times 5}{\underset{\times 5}{=}} \frac{20}{5}$$

2. Combine the two fractions into one improper fraction by adding the numerators.

So $4\frac{3}{5} = \frac{20}{5} + \frac{3}{5} = \frac{23}{5}$

b) 1. Split the numerator into a multiple of the denominator plus a remainder: $13 \div 5 = 2$ remainder 3, so $13 = (2 \times 5) + 3$.

$$\frac{13}{5} = \frac{(2 \times 5) + 3}{5} = \frac{10 + 3}{5}$$

2. Separate the fraction to write it as a mixed number.

$$= \frac{10}{5} + \frac{3}{5} = 2 + \frac{3}{5} = 2\frac{3}{5}$$

Exercise 1

Q1 Find the values of the letters by writing the following mixed numbers as improper fractions.

a) $1\frac{1}{3} = \frac{a}{3}$ b) $1\frac{2}{7} = \frac{b}{7}$ c) $4\frac{1}{2} = \frac{c}{2}$ d) $3\frac{4}{7} = \frac{d}{7}$

Q2 Write the following mixed numbers as improper fractions.

a) $1\frac{4}{5}$ b) $1\frac{5}{12}$ c) $2\frac{9}{10}$ d) $4\frac{3}{4}$

e) $12\frac{2}{5}$ f) $15\frac{5}{7}$ g) $3\frac{1}{9}$ h) $10\frac{3}{10}$

Q3 a) Simplify the improper fraction $\frac{26}{4}$.

b) Use your answer to write $\frac{26}{4}$ as a mixed number in its lowest terms.

Q4 Write, as a mixed number in its lowest terms:

a) $\frac{5}{3}$ b) $\frac{9}{5}$ c) $\frac{17}{10}$ d) $\frac{12}{7}$

e) $\frac{13}{6}$ f) $\frac{18}{12}$ g) $\frac{50}{15}$ h) $\frac{24}{18}$

i) 13 as a fraction of 11 j) 35 as a fraction of 25 k) 51 as a fraction of 12 l) 98 as a fraction of 8

Q5 Find the number in each list that is not equivalent to the other two.

a) $\frac{6}{4}, \frac{5}{2}, 1\frac{1}{2}$ b) $2\frac{1}{3}, \frac{7}{3}, 3\frac{1}{2},$ c) $\frac{15}{4}, \frac{19}{4}, 4\frac{3}{4}$ d) $2\frac{2}{3}, \frac{11}{3}, \frac{16}{6}$

3.3 Ordering Fractions

It can be difficult to figure out whether one fraction is bigger or smaller than another just by looking at them. The trick is to find equivalent fractions that have the same denominator — then comparing them is easy.

Learning Objective — Spec Ref N1:
Be able to order and compare fractions.

Prior Knowledge Check:
Be able to find the lowest common multiple of a set of numbers (p.12) and find equivalent fractions (p.27).

To **compare** or **order** two or more fractions, you need to put them over a **common denominator**.
To do this, find the **lowest common multiple** (LCM) of the **denominators** of the fractions.
Then, rewrite each fraction as an **equivalent fraction** with this number as its denominator.

When the fractions have a common denominator, you can use their **numerators** to put them in order.

Example 1

Rewrite $\frac{5}{6}$ and $\frac{7}{8}$ so they have a common denominator and say which is larger.

1. The lowest common multiple of 6 and 8 is 24, so find equivalent fractions that have a denominator of 24.

$$\frac{5}{6} \overset{\times 4}{=} \frac{20}{24} \qquad \frac{7}{8} \overset{\times 3}{=} \frac{21}{24}$$

2. Compare the numerators — $20 < 21$.

$$\frac{20}{24} < \frac{21}{24} \text{ so } \frac{7}{8} \text{ is } \textbf{larger}.$$

Example 2

Put the fractions $\frac{1}{2}$, $\frac{3}{8}$ and $\frac{3}{4}$ in order, from smallest to largest.

1. The LCM of 2, 8 and 4 is 8, so find equivalent fractions that have a denominator of 8.

$$\frac{1}{2} \overset{\times 4}{=} \frac{4}{8} \qquad \frac{3}{4} \overset{\times 2}{=} \frac{6}{8}$$

So the fractions are equivalent to $\frac{4}{8}$, $\frac{3}{8}$ and $\frac{6}{8}$.

2. Use the numerators to put the fractions in order.

From smallest to largest, these are: $\frac{3}{8}$, $\frac{4}{8}$, $\frac{6}{8}$.

3. Write the ordered fractions in their original form.

So in order, the original fractions are $\frac{3}{8}$, $\frac{1}{2}$, $\frac{3}{4}$.

Exercise 1

Q1 Rewrite each group of fractions so they have a common denominator and say which is largest.

a) $\frac{2}{9}$, $\frac{1}{3}$ b) $\frac{2}{3}$, $\frac{3}{4}$ c) $\frac{7}{8}$, $\frac{3}{10}$ d) $\frac{2}{5}$, $\frac{4}{9}$

e) $\frac{1}{5}$, $\frac{7}{10}$, $\frac{9}{20}$ f) $\frac{1}{7}$, $\frac{4}{21}$, $\frac{5}{14}$ g) $\frac{2}{5}$, $\frac{5}{12}$, $\frac{11}{30}$ h) $\frac{5}{18}$, $\frac{7}{24}$, $\frac{11}{30}$

Q2 Put each of the sets of fractions in order, from smallest to largest.

a) $\frac{1}{4}$, $\frac{5}{8}$ b) $\frac{5}{6}$, $\frac{3}{4}$ c) $\frac{2}{3}$, $\frac{3}{5}$ d) $\frac{7}{10}$, $\frac{3}{4}$

e) $\frac{1}{2}$, $\frac{5}{8}$, $\frac{7}{16}$ f) $\frac{5}{6}$, $\frac{11}{12}$, $\frac{19}{24}$ g) $\frac{4}{9}$, $\frac{5}{12}$, $\frac{2}{3}$ h) $\frac{9}{10}$, $\frac{11}{12}$, $\frac{4}{5}$

i) $\frac{7}{9}$, $\frac{4}{5}$, $\frac{13}{15}$ j) $\frac{3}{15}$, $\frac{7}{27}$, $\frac{12}{45}$ k) $\frac{5}{16}$, $\frac{7}{20}$, $\frac{9}{25}$ l) $\frac{11}{36}$, $\frac{4}{15}$, $\frac{9}{24}$

3.4 Adding and Subtracting Fractions

To be able to add and subtract fractions, you need to be able to do everything in this section so far. Finding common denominators is particularly important, so make sure you're happy with that.

Adding and Subtracting Fractions

Learning Objective — Spec Ref N2:
Be able to add and subtract fractions.

If you want to add or subtract fractions, you need to get them over a **common denominator** (see the previous page). Once they have the same denominator, you can just add or subtract the **numerators**. Finally, **simplify** the fraction if necessary.

Example 1

In a maths exam, $\frac{1}{8}$ of the questions are on number topics, $\frac{1}{3}$ of the questions are on algebra, and the rest are on geometry. What fraction of the questions are geometry questions?

1. The fractions of number questions, algebra questions and geometry questions must add up to 1.

   Fraction of geometry questions $= 1 - \frac{1}{8} - \frac{1}{3}$

2. Put the fractions over a common denominator.

   $1 = \frac{24}{24}$ $\frac{1}{8} = \frac{3}{24}$ ($\times 3$) $\frac{1}{3} = \frac{8}{24}$ ($\times 8$)

3. Subtract the numerators to find the fraction of geometry questions.

   $\frac{24}{24} - \frac{3}{24} - \frac{8}{24} = \frac{24 - 3 - 8}{24} = \frac{13}{24}$

Exercise 1

Q1 Work out the following. Give your answers as mixed numbers in their simplest form.

   a) $\frac{2}{3} - \frac{1}{4}$ b) $\frac{2}{3} + \frac{4}{5}$ c) $\frac{9}{10} - \frac{5}{6}$ d) $\frac{3}{7} + \frac{3}{4}$

   e) $\frac{6}{11} + \frac{7}{9}$ f) $\frac{8}{9} + \frac{12}{21}$ g) $\frac{6}{7} - \frac{3}{5}$ h) $\frac{15}{16} + \frac{3}{5}$

Q2 Work out the following. Give your answers as mixed numbers in their simplest form.

   a) $\frac{1}{9} + \frac{5}{9} + \frac{11}{18}$ b) $1 - \frac{2}{10} - \frac{2}{8}$ c) $\frac{3}{4} + \frac{1}{8} - \frac{7}{16}$ d) $\frac{6}{7} + \frac{1}{14} - \frac{1}{2}$

   e) $\frac{1}{4} + \frac{2}{3} + \frac{5}{6}$ f) $\frac{1}{5} + \frac{1}{3} + \frac{3}{15}$ g) $\frac{9}{10} - \frac{5}{6} + \frac{3}{12}$ h) $\frac{1}{6} + \frac{5}{7} - \frac{1}{3}$

Q3 In a school survey, $\frac{1}{2}$ of the pupils said they walk to school. $\frac{1}{5}$ said they catch the bus. The rest arrive by car. What fraction come to school by car?

Q4 A bag contains a mixture of sweets. $\frac{2}{9}$ of the sweets are white chocolates, $\frac{1}{12}$ of the sweets are milk chocolates, $\frac{1}{5}$ are toffees and the rest are mints. What fraction of the sweets are mints?

Adding and Subtracting Mixed Numbers

Learning Objective — Spec Ref N2:
Be able to add and subtract mixed numbers.

There are two methods for adding and subtracting mixed numbers:

- Change the mixed numbers into **improper fractions** (see p.29), then add or subtract like on the previous page by getting them over a **common denominator**.

- Add/subtract the number parts and the fraction parts **separately**, then **add** the results. (You might be adding a negative fraction if the question was a subtraction.)

The second method is useful when the **number part** is **large**, but it can be a bit **trickier** when the fractions go **past a whole number** — e.g. $2\frac{2}{5} - 1\frac{4}{5} \Rightarrow 2 - 1 = 1$, but $\frac{2}{5} - \frac{4}{5} = -\frac{2}{5}$, so the answer is $1 + \left(-\frac{2}{5}\right) = \frac{3}{5}$.

So, if you're **not sure** which method to use, it's generally **safer** to use the improper fractions method.

Example 2

Find $1\frac{1}{3} + 2\frac{5}{6}$ by: a) converting to improper fractions,
b) adding number parts and fraction parts separately.

a) 1. Write the mixed numbers as improper fractions.

$\frac{1}{1} = \frac{3}{3}$, so $1\frac{1}{3} = \frac{3}{3} + \frac{1}{3} = \frac{4}{3}$ $\frac{2}{1} = \frac{12}{6}$, so $2\frac{5}{6} = \frac{12}{6} + \frac{5}{6} = \frac{17}{6}$

2. Rewrite the improper fractions with a common denominator.

$\frac{4}{3} \overset{\times 2}{=} \frac{8}{6}$ $\frac{17}{6}$

3. Add the numerators. Give your answer as a mixed number in its simplest form.

$1\frac{1}{3} + 2\frac{5}{6} = \frac{8}{6} + \frac{17}{6} = \frac{25}{6} = \frac{24}{6} + \frac{1}{6} = \mathbf{4\frac{1}{6}}$

b) 1. Add the number parts.

$1 + 2 = 3$

2. Add the fraction parts in the usual way. The result is an improper fraction, so convert it to a mixed number.

$\frac{1}{3} + \frac{5}{6} = \frac{2}{6} + \frac{5}{6} = \frac{7}{6} = \frac{6}{6} + \frac{1}{6} = 1\frac{1}{6}$

3. Add the results.

$3 + 1\frac{1}{6} = \mathbf{4\frac{1}{6}}$

Exercise 2

For Q1-2, work out the calculations, giving your answers as mixed numbers in their simplest form.

Q1 a) $2\frac{3}{8} + \frac{3}{4}$ b) $4\frac{3}{14} + 1\frac{6}{7}$ c) $1\frac{3}{5} + \frac{3}{4}$ d) $2\frac{5}{8} + \frac{2}{3}$

e) $3\frac{1}{5} + 2\frac{3}{7}$ f) $1\frac{1}{6} + 4\frac{7}{15}$ g) $2\frac{5}{11} + 3\frac{2}{3}$ h) $5\frac{7}{12} + 3\frac{2}{5}$

Q2 a) $4\frac{5}{12} - 2\frac{5}{6}$ b) $7\frac{5}{9} - 1\frac{11}{18}$ c) $3\frac{3}{4} - \frac{5}{7}$ d) $5\frac{2}{5} - 3\frac{7}{9}$

e) $2\frac{1}{4} - 1\frac{6}{7}$ f) $4\frac{3}{8} - 1\frac{2}{9}$ g) $5\frac{3}{20} - 1\frac{7}{12}$ h) $3\frac{4}{7} - 3\frac{5}{9}$

Q3 The table shows the number of pies eaten by 3 contestants in a pie eating contest. Calculate the total number of pies eaten, as a mixed number in its simplest form.

Contestant	1	2	3
No. of pies eaten	$17\frac{7}{8}$	$9\frac{5}{12}$	$40\frac{5}{18}$

3.5 Multiplying and Dividing by Fractions

Being able to multiply by fractions is really useful — especially as it lets you find fractions of amounts. In order to divide by fractions, you need to be able to find the reciprocal of the fraction you're dividing by.

Multiplying Whole Numbers by Fractions

Learning Objectives — Spec Ref N2/N12:
- Be able to multiply whole numbers by fractions.
- Be able to find fractions of amounts.

Prior Knowledge Check:
Be able to simplify fractions (p.28) and convert between mixed numbers and improper fractions (p.29).

To multiply a whole number by a fraction, **multiply** it by the **numerator**, and **divide** it by the **denominator**. It doesn't matter what **order** you do the multiplication and division in — sometimes it's easier to do the **division first** as it keeps the numbers **smaller**, but either way will give the same result.

If the number **isn't a multiple** of the fraction's **denominator**, then your final answer will be a **fraction**. In this case, just multiply the number by the **numerator** of the fraction, write it **over the denominator**, then **simplify** as much as possible.

Finding **fractions of amounts** is the same as multiplying — e.g. $\frac{2}{3}$ of 60 is the same as $\frac{2}{3} \times 60$.

Example 1

Calculate $21 \times \frac{4}{7}$.

1. You need to multiply by 4 and divide by 7. Do the division first...
2. ...then multiply the answer by 4.

$$21 \times \frac{4}{7} = (21 \div 7) \times 4$$
$$= 3 \times 4 = \textbf{12}$$

Tip: If you multiplied first instead, you'd get:
$(21 \times 4) \div 7 = 84 \div 7 = 12$

Example 2

Find $\frac{3}{4}$ **of 18. Give your answer as a mixed number in its simplest form.**

1. You can replace 'of' with a multiplication sign.

$$\frac{3}{4} \text{ of } 18 = \frac{3}{4} \times 18$$

2. Since 18 isn't a multiple of 4, do 3×18 and write this on the top of the fraction.

$$\frac{3}{4} \times 18 = \frac{3 \times 18}{4} = \frac{54}{4}$$

3. Simplify the fraction and convert to a mixed number.

$$\frac{54}{4} = \frac{27}{2} = \frac{26 + 1}{2} = 13\frac{1}{2}$$

Exercise 1

Q1 Find the following.

a) $28 \times \frac{3}{4}$ b) $\frac{2}{9}$ of 36   c) $\frac{3}{8} \times 48$ d) $\frac{5}{12}$ of 60

e) $\frac{5}{6} \times 24$ f) $15 \times \frac{4}{5}$ g) $\frac{5}{6}$ of 54   h) $96 \times \frac{7}{12}$

Q2 Work out the following. Write your answers as mixed numbers in their simplest form.

a) $48 \times \frac{2}{7}$ b) $27 \times \frac{1}{6}$ c) $32 \times \frac{2}{3}$ d) $34 \times \frac{4}{5}$

e) $80 \times \frac{2}{9}$ f) $45 \times \frac{5}{12}$ g) $72 \times \frac{3}{11}$ h) $62 \times \frac{5}{8}$

Multiplying Fractions

Learning Objective — Spec Ref N2/N12:
Be able to multiply fractions together.

To multiply two or more fractions, multiply all the **numerators together** and all the **denominators together**. If they're **mixed numbers**, change them to improper fractions first (see p.29).

To make your calculations simpler when multiplying, you can **cancel factors** that appear in the numerator and denominator of **either** fraction — so if a number in the numerator of one fraction shares a **common factor** with the denominator of the other fraction, then you can **divide both** numbers by the factor first.

Example 3

Calculate $\frac{7}{25} \times \frac{15}{16}$.

1. 25 and 15 share a common factor (5), so divide both by 5.

    $\frac{7}{25\,5} \times \frac{15\,3}{16} = \frac{7}{5} \times \frac{3}{16}$

2. Multiply the numerators together and the denominators together.

    $= \frac{7 \times 3}{5 \times 16} = \frac{21}{80}$

> **Tip:** If you didn't cancel down, you'd get $\frac{7 \times 15}{25 \times 16} = \frac{105}{400}$, which then simplifies to $\frac{21}{80}$.

Example 4

Work out $4\frac{1}{2} \times 3\frac{3}{5}$. Give your answer as a mixed number in its simplest form.

1. Write the mixed numbers as improper fractions.

    $4\frac{1}{2} = \frac{8}{2} + \frac{1}{2} = \frac{9}{2}$ $\qquad$ $3\frac{3}{5} = \frac{15}{5} + \frac{3}{5} = \frac{18}{5}$

2. Multiply the two fractions, cancelling where possible.

    $4\frac{1}{2} \times 3\frac{3}{5} = \frac{9}{2\,1} \times \frac{18\,9}{5} = \frac{9}{1} \times \frac{9}{5} = \frac{9 \times 9}{1 \times 5} = \frac{81}{5}$

3. Finally, convert this into a mixed number.

    $\frac{81}{5} = \frac{80}{5} + \frac{1}{5} = 16\frac{1}{5}$

Exercise 2

Q1 Work out the following. Give your answers in their lowest terms.

a) $\frac{3}{5} \times \frac{1}{6}$ $\qquad$ b) $\frac{5}{6} \times \frac{2}{15}$ $\qquad$ c) $\frac{5}{12} \times \frac{3}{4}$

d) $\frac{6}{7} \times \frac{7}{8}$ $\qquad$ e) $\frac{7}{10} \times \frac{5}{14}$ $\qquad$ f) $\frac{9}{13} \times \frac{13}{9}$

Q2 Work out the following. Give your answers as mixed numbers in their lowest terms.

a) $1\frac{5}{6} \times \frac{2}{3}$ $\qquad$ b) $3\frac{3}{4} \times \frac{2}{5}$ $\qquad$ c) $2\frac{1}{7} \times \frac{2}{9}$

d) $1\frac{11}{12} \times \frac{1}{4}$ $\qquad$ e) $4\frac{3}{5} \times \frac{4}{5}$ $\qquad$ f) $2\frac{4}{9} \times \frac{3}{8}$

g) $2\frac{2}{5} \times 3\frac{1}{6}$ $\qquad$ h) $4\frac{4}{9} \times 1\frac{7}{10}$ $\qquad$ i) $5\frac{1}{13} \times 1\frac{32}{33}$

Dividing by Fractions

Learning Objectives — Spec Ref N2/N12:
- Be able to find the reciprocal of a number or fraction.
- Be able to divide by fractions and mixed numbers.

The **reciprocal** of a number is just **1 ÷ that number**. The reciprocal of a **whole number** is always a **unit fraction** (a fraction with a **numerator of 1**) — e.g. the reciprocal of 4 is $\frac{1}{4}$.
Also, if a is the reciprocal of b, then b is the reciprocal of a — so the reciprocal of $\frac{1}{4}$ is 4.

To find the reciprocal of a **fraction**, **swap the numerator and denominator** (i.e. turn it **upside down**) — for example, the reciprocal of $\frac{3}{5}$ is $\frac{5}{3}$. If it's a mixed number, convert it to an improper fraction first.

Dividing by a number is the same as **multiplying** by its **reciprocal** (e.g. $6 ÷ 4 = 6 \times \frac{1}{4}$).
So, to divide by a **fraction**, you can just turn the fraction **upside down** and change the ÷ into a ×. Once you've got a multiplication, the method is exactly the same as on the previous page.

Example 5

Find: a) $\frac{1}{4} ÷ \frac{6}{13}$,   b) $2\frac{2}{3} ÷ 1\frac{1}{5}$ as a mixed number in its lowest terms.

a) 1. Dividing by $\frac{6}{13}$ is the same as multiplying by its reciprocal, so flip the $\frac{6}{13}$ upside down and change the ÷ into a ×.

$$\frac{1}{4} ÷ \frac{6}{13} = \frac{1}{4} \times \frac{13}{6}$$

    2. Now you can multiply the fractions as on p.34.

$$= \frac{1 \times 13}{4 \times 6} = \frac{13}{24}$$

b) 1. Write both numbers as improper fractions.

$$2\frac{2}{3} = \frac{6+2}{3} = \frac{8}{3} \qquad 1\frac{1}{5} = \frac{5+1}{5} = \frac{6}{5}$$

    2. Flip the $\frac{6}{5}$ upside down and change the ÷ into a ×.

$$\frac{8}{3} ÷ \frac{6}{5} = \frac{\cancel{8}4}{3} \times \frac{5}{\cancel{6}3} = \frac{4}{3} \times \frac{5}{3} = \frac{4 \times 5}{3 \times 3} = \frac{20}{9}$$

    3. Convert your answer into a mixed number.

$$\frac{20}{9} = \frac{18+2}{9} = 2\frac{2}{9}$$

Exercise 3

Q1 Find the reciprocal of the following, giving your answer as an improper fraction where appropriate.

   a) 7   b) $\frac{1}{11}$ c) $\frac{7}{6}$ d) $\frac{3}{26}$

   e) $1\frac{11}{12}$ f) $2\frac{3}{4}$ g) $5\frac{2}{3}$ h) $4\frac{2}{7}$

Q2 Work out the following. Give your answers in their lowest terms.

   a) $\frac{4}{13} ÷ \frac{1}{3}$ b) $\frac{2}{25} ÷ \frac{1}{5}$ c) $\frac{2}{5} ÷ \frac{2}{3}$ d) $\frac{3}{4} ÷ \frac{9}{10}$

   e) $\frac{5}{7} ÷ \frac{11}{14}$ f) $\frac{2}{5} ÷ 3$ g) $\frac{3}{7} ÷ 6$ h) $7 ÷ \frac{15}{2}$

Q3 Work out the following. Give your answers as mixed numbers in their lowest terms.

   a) $2\frac{1}{2} ÷ \frac{1}{3}$ b) $1\frac{1}{6} ÷ \frac{1}{4}$ c) $2\frac{3}{7} ÷ 3$ d) $4\frac{4}{9} ÷ 6$

   e) $\frac{2}{3} ÷ 3\frac{2}{5}$ f) $4\frac{1}{6} ÷ \frac{15}{16}$ g) $1\frac{1}{4} ÷ 1\frac{1}{5}$ h) $3\frac{3}{10} ÷ 2\frac{1}{7}$

3.6 Fractions and Decimals

Fractions and decimals are different ways of writing a number, and there are a couple of methods you can use to switch between them. The tricky bit is when you get a recurring decimal — one that keeps repeating.

Converting Fractions to Decimals

Learning Objectives — Spec Ref N10:
- Convert fractions to decimals using a calculator.
- Convert fractions to decimals without using a calculator.

Prior Knowledge Check:
Be able to write a number as a product of prime factors. See p.10.

All fractions can be converted into **decimals**. Decimals can be either **terminating** or **recurring**. In a terminating decimal, the digits **stop** (e.g. 0.3, 0.625). A recurring decimal has a **repeating** pattern in its digits which goes on **forever** — these are shown using a dot above the first and last repeated digits.

For example: $0.111... = 0.\dot{1}$ $0.151515... = 0.\dot{1}\dot{5}$ $0.12341234... = 0.\dot{1}23\dot{4}$ $0.0232323... = 0.0\dot{2}\dot{3}$

When converting **mixed numbers** to decimals, just convert the **fraction** part, then add the whole number part on **afterwards**.

Converting Fractions to Decimals Using a Calculator

The easiest way to convert a fraction into a decimal is using a **calculator** — just type in '**numerator ÷ denominator**'. If your calculator gives answers as **fractions**, you might need to press a button to make it display it as a decimal (e.g. the $\boxed{\text{S} \Leftrightarrow \text{D}}$ button on some calculators).

Example 1

Use a calculator to convert the following fractions to decimals: a) $3\frac{2}{5}$ b) $\frac{41}{333}$

a) 1. Ignore the whole number for now and type 2 ÷ 5 into your calculator. $\frac{2}{5} = 2 \div 5 = 0.4$

   2. Now add the whole number back on. $3\frac{2}{5} = 3 + \frac{2}{5} = 3 + 0.4 = \mathbf{3.4}$

   Tip: If this was given as $\frac{17}{5}$, you could just divide straight away.

b) Your calculator gives 41 ÷ 333 = 0.123123123... so this is a recurring decimal. Show the repeating pattern by putting dots over the first and last digits of the repeated group.

   $\frac{41}{333} = 41 \div 333$
   $= 0.123123123...$
   $= \mathbf{0.\dot{1}2\dot{3}}$

Exercise 1

Q1 Use a calculator to convert the following fractions to decimals.

a) $\frac{329}{500}$ b) $2\frac{1}{8}$ c) $2\frac{37}{100}$ d) $4\frac{719}{1000}$

e) $\frac{4}{9}$ f) $\frac{4}{15}$ g) $\frac{1234}{9999}$ h) $\frac{88}{3}$

Q2 By first writing these fractions as decimals, put each of the following lists in order, from smallest to largest.

a) $\frac{167}{287}, \frac{87}{160}, \frac{196}{360}$ b) $\frac{96}{99}, \frac{16}{17}, \frac{5}{6}$ c) $\frac{963}{650}, \frac{13}{9}, \frac{77}{52}$

Converting Fractions to Decimals Without Using a Calculator

The method you use to convert a fraction to a decimal **without a calculator** depends on whether it's a **terminating** or a **recurring** decimal. You can tell which kind of decimal a fraction will give by looking at the **denominator** when the fraction is in its **lowest terms**:

- If the **only prime factors** (see p.10) of the denominator are **2 or 5** (or both), then the fraction will become a **terminating decimal**. For example, $20 = 2 \times 2 \times 5$, and $\frac{1}{20} = 0.05$.

- If the denominator has any prime factors **other than 2 or 5**, then the fraction will become a **recurring decimal**. For example, $6 = 2 \times 3$, and $\frac{1}{6} = 0.1\dot{6}$.

If you know that a fraction will give you a terminating decimal, then you can convert it to a decimal by finding an **equivalent fraction** with a **denominator** of **10, 100, 1000**, etc. Then the **numerator** of the equivalent fraction goes **after the decimal point**. Be careful with the **place value** — the decimal should have the same number of **decimal places** as the denominator has **zeros**. For example, a denominator of **1000** would give a decimal with **3 d.p.**: $\frac{1}{1000} = 0.001$, $\frac{12}{1000} = 0.012$, etc.

Example 2

Write $\frac{23}{250}$ as a decimal.

Multiply top and bottom to find an equivalent fraction with a denominator of 1000. Then put the numerator after the decimal point — it should have 3 d.p., so put a 0 in front of the 92.

$$\frac{23}{250} \overset{\times 4}{\underset{\times 4}{=}} \frac{92}{1000} = \mathbf{0.092}$$

You can also convert a fraction to a terminating decimal by **dividing** the numerator by the denominator using **short division**. You'll have to keep putting **zeros** after the decimal point until the answer terminates.

Example 3

Use short division to write $\frac{1}{8}$ as a decimal.

You need to work out $1 \div 8$.
1. 8 doesn't go into 1, so carry the 1.
2. 8 goes into 10 once, with remainder 2.
3. 8 goes into 20 twice, with remainder 4.
4. 8 goes into 40 exactly 5 times.

$$\begin{array}{r} 0.\ 1\ 2\ 5 \\ 8{\overline{)1.^10^20^40}} \end{array}$$

So $\frac{1}{8} = \mathbf{0.125}$

Tip: To do the division, write 1 as 1.000...

Exercise 2

Q1 Write the following fractions as decimals.

a) $\frac{46}{100}$
b) $\frac{492}{1000}$
c) $\frac{9}{30}$
d) $\frac{17}{50}$

e) $\frac{3}{5}$
f) $\frac{22}{25}$
g) $\frac{333}{500}$
h) $\frac{123}{200}$

Q2 Use short division to write the following fractions as decimals.

a) $\frac{1}{16}$
b) $\frac{7}{8}$
c) $\frac{1}{125}$
d) $\frac{3}{32}$

e) $\frac{21}{32}$
f) $\frac{7}{16}$
g) $\frac{9}{125}$
h) $\frac{53}{8}$

If you know that a fraction will give you a **recurring decimal**, you can still convert it to a decimal using **short division**, although eventually the digits of the answer will start to **repeat**. You can **stop** once you're sure that you've found the **entire repeating pattern**.

There's also a quicker method you can use. If you can find an **equivalent fraction** with a denominator of **9**, **99**, **999**, etc., then the **numerator** gives you the **repeating group**. The repeating group will always have the same **number of digits** as the **number of 9s** in the denominator, so you might need to put **zeros** at the start of the numerator. For example, $\frac{13}{99} = 0.\dot{1}\dot{3}$, $\frac{13}{999} = 0.\dot{0}1\dot{3}$, $\frac{13}{9999} = 0.\dot{0}01\dot{3}$, etc.

Example 4

Write $\frac{3}{11}$ as a decimal: a) using short division, b) by converting to a suitable equivalent fraction.

a) 1. Work out $3 \div 11$ using short division.

2. Eventually the digits of the answer and the remainders start to repeat.

3. Mark the start and end of the repeating group.

$$\begin{array}{r} 0.\ 2\ 7\ 2\ 7\ 2... \\ 11\overline{)3.^30^80^30^80^30} ... \end{array}$$

So $\frac{3}{11} = \mathbf{0.\dot{2}\dot{7}}$

Tip: You know you've found the whole pattern once the remainders start to repeat.

b) Convert to an equivalent fraction with a denominator of 99. There are two 9s in the denominator, so the repeating group has two digits. Since the numerator already has two digits, you don't need any extra zeros.

$$\frac{3}{11} \overset{\times 9}{\underset{\times 9}{=}} \frac{27}{99} = \mathbf{0.\dot{2}\dot{7}}$$

Exercise 3

Q1 Write, as a recurring decimal: a) $\frac{1}{9}$ b) $\frac{247}{999}$ c) $\frac{4}{9}$ d) $\frac{7}{99}$ e) $\frac{4}{999}$

Q2 Write, as a recurring decimal: a) $\frac{2}{3}$ b) $\frac{32}{33}$ c) $\frac{80}{111}$ d) $\frac{1}{11}$ e) $\frac{5}{333}$

Q3 Use short division to write the following fractions as recurring decimals.
   a) $\frac{1}{6}$ b) $\frac{7}{15}$ c) $\frac{1}{45}$ d) $\frac{27}{110}$

Converting Decimals to Fractions

Learning Objective — Spec Ref N10:
Be able to convert terminating and recurring decimals to fractions.

You can convert a **terminating decimal** to a fraction by doing the reverse of what you did on the previous page. Just write the bit after the decimal point as the **numerator**, and put a **power of 10** as the **denominator** with the same **number of zeros** as the number of **decimal places**. For example, 0.13 has 2 decimal places, so the denominator should have 2 zeros: $0.13 = \frac{13}{100}$.

Example 5

Write 0.025 as a fraction without using your calculator.

0.025 has 3 decimal places, so the denominator should have 3 zeros (1000). Don't forget to simplify the fraction.

$$0.025 = \overset{\div 5}{\underset{\div 5}{\frac{25}{1000}}} = \overset{\div 5}{\underset{\div 5}{\frac{5}{200}}} = \frac{1}{40}$$

Converting **recurring decimals** into fractions is a bit trickier, but there's a handy method you can use:

- Call the recurring decimal a **letter** — say, r.

- Multiply r by 10, 100, 1000, etc. to move **one whole chunk** of the repeated pattern **to the left** of the decimal point.

- **Subtract** r from this to get rid of the decimal part. You'll be left with 'multiple of r = whole number'.

- **Rearrange** so that you have 'r = fraction'. **Simplify** if necessary.

If the recurring pattern doesn't start **right after** the decimal point, **multiply** by a power of 10 so that it does. Then continue with the usual method, except you'll subtract **10r** (or 100r, 1000r, etc.) instead of just r.

Example 6

Convert the recurring decimal 0.171717... into a fraction.

1. Call the recurring decimal r. $\qquad r = 0.171717...$

2. Multiply r by 100 to get one whole chunk of the repeating pattern (a '17') on the left of the decimal point. $\qquad 100r = 17.171717...$

3. Subtract your original number, r. $\qquad 100r - r = 17.1717... - 0.1717...$

4. Rearrange the equation to find r as a fraction. $\qquad 99r = 17 \implies r = \dfrac{17}{99}$

Example 7

Convert the recurring decimal 0.1$\dot{2}\dot{3}$ into a fraction.

1. Call the recurring decimal r. $\qquad r = 0.1232323...$

2. Multiply by 10 so that the recurring part starts right after the decimal point. $\qquad 10r = 1.232323...$

3. Multiply again to get one whole chunk of the repeating pattern on the left of the decimal point. $\qquad 1000r = 123.232323...$

4. Subtract 10r from 1000r. $\qquad 1000r - 10r = 123.2323... - 1.2323...$
$\qquad 990r = 122$

5. Rearrange into a fraction and simplify. $\qquad r = \dfrac{122}{990} \overset{\div 2}{\underset{\div 2}{=}} \dfrac{61}{495}$

Exercise 4

Q1 Write each of the following decimals as a fraction in its simplest form.

    a) 0.12   b) 0.084   c) 0.375   d) 0.7654321

Q2 Write each of the following recurring decimals as a fraction in its simplest form.

    a) 0.$\dot{1}$ b) 0.3$\dot{4}$ c) 0.181818...

    d) 0.8$\dot{6}\dot{3}$ e) 0.207207207...   f) 0.72007200...

Q3 Write each of the following recurring decimals as a fraction in its simplest form.

    a) 0.5444444...   b) 0.8$\dot{7}\dot{2}$ c) 0.0121212...

    d) 0.07$\dot{5}$ e) 0.004$\dot{5}$ f) 0.3373737...

Q1 Simplify the following fractions: a) $\frac{21}{35}$ b) $\frac{36}{126}$ c) $\frac{70}{182}$

Q2 Convert each of the following to a mixed number:

  a) $\frac{37}{27}$ b) $\frac{89}{5}$ c) $\frac{230}{11}$ d) $\frac{135}{19}$

Q3 Which of the fractions on the right is closest to $\frac{3}{4}$? $\frac{11}{15}$ $\frac{7}{10}$ $\frac{4}{5}$ $\frac{5}{6}$

Q4 In Ancient Egypt, fractions were written using sums of unit fractions.
For example, instead of writing $\frac{3}{5}$, Ancient Egyptians would write $\frac{1}{2} + \frac{1}{10}$.

  a) Find the fractions that Ancient Egyptians could have written in the following ways.

  (i) $\frac{1}{3} + \frac{1}{12}$ (ii) $\frac{1}{2} + \frac{1}{3} + \frac{1}{7}$ (iii) $\frac{1}{2} + \frac{1}{5} + \frac{1}{20}$

  b) Find the values of the letters in the following equations.

  (i) $\frac{9}{20} = \frac{1}{a} + \frac{1}{5}$ (ii) $\frac{11}{18} = \frac{1}{3} + \frac{1}{b} + \frac{1}{9}$ (iii) $\frac{301}{600} = \frac{1}{2} + \frac{1}{c}$

Q5 A plank of wood is 20 inches long. Three pieces of length $7\frac{3}{4}$ inches, $5\frac{5}{16}$ inches and $2\frac{1}{8}$ inches are cut from the plank. What length of wood is left over?

Q6 It takes Ella $1\frac{1}{4}$ minutes to answer each question on her maths homework.
How many questions can she answer in 20 minutes?

Q7 What numbers need to go in the boxes to make the following true?

  a) $\frac{2}{5}$ of 100 = ☐ of 50 b) $\frac{3}{4}$ of ☐ = $\frac{2}{3}$ of 90 c) $\frac{1}{4}$ of 64 = $\frac{1}{7}$ of ☐

Q8 a) Convert the fractions $\frac{39}{100}$, $\frac{7}{20}$, $\frac{8}{25}$ and $\frac{3}{10}$ into decimals.
  b) Put the fractions in order, from smallest to largest.

Q9 Which of the fractions $\frac{5}{6}$, $\frac{4}{5}$, $\frac{2}{9}$, $\frac{9}{16}$ and $\frac{17}{40}$ are equivalent to recurring decimals?

Q10 Show that: a) $\frac{24}{112}$ is not equivalent to $\frac{3}{8}$. b) $\frac{1}{5}$ is not one third of $\frac{1}{15}$.

  c) $\frac{1}{4}$ is not halfway between $\frac{1}{2}$ and $\frac{1}{5}$. d) $\frac{8}{9}$ is greater than $0.8\dot{7}$.

Q11 Put each set of quantities in order, from smallest to largest.

  a) $\frac{8}{16}$, $\frac{6}{11}$, $0.5\dot{4}$, $\frac{5}{9}$ b) $\frac{51}{100}$, $\frac{102}{204}$, $0.4\dot{6}$, $\frac{8}{15}$

Exam-Style Questions

Q1 Show that $4\frac{1}{2} \div 1\frac{2}{5} = 3\frac{3}{14}$.

[3 marks]

Q2 $\frac{1}{6}$ of the flowers in a garden are roses, $\frac{3}{24}$ of the flowers are tulips, $\frac{3}{8}$ are daisies and the rest are daffodils. What fraction of the flowers are daffodils?

[2 marks]

Q3 A rectangle has a length of $3\frac{3}{5}$ cm and a width of $1\frac{5}{8}$ cm. Find, as a mixed number in its lowest terms:

a) the perimeter of the rectangle,

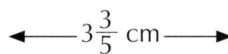

[2 marks]

b) the area of the rectangle.

[2 marks]

Q4 Use algebra to clearly show that $0.3181818... = \frac{7}{22}$.

[3 marks]

Q5 To go on a roller coaster ride, you must be at least $1\frac{1}{3}$ m tall.

- Jess is $1\frac{1}{2}$ m tall.
- Eric's height is $\frac{8}{9}$ of Jess's height.
- Xin is $\frac{2}{7}$ m shorter than Jess.
- Abbas is 1.3 m tall.

Which of the friends can go on the ride?

[4 marks]

Q6 Quinn buys a console game in a sale where he gets $\frac{1}{4}$ off the normal price. If he paid £21.99 in the sale, work out the normal price of the game.

[3 marks]

Q7 Work out the reciprocal of $2.\dot{6}$, giving your answer as a decimal.

[4 marks]

4.1 Ratios

Along with fractions and percentages, ratios are a way of showing proportion. They tell you how the size of one thing relates to the size of another thing, or sometimes several other things. They crop up in lots of practical problems, in real life as well as in maths exams — such as when using map scales and recipes.

Simplifying Ratios

Learning Objectives — Spec Ref R4:
- Write ratios in their simplest form.
- Write ratios in the form $1:n$.

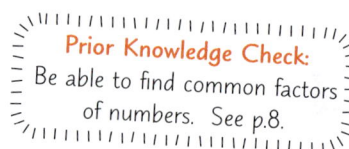

Prior Knowledge Check:
Be able to find common factors of numbers. See p.8.

Ratios are used to compare quantities. The ratio of quantity a compared to quantity b is written $a:b$. You can also compare more than two quantities in a ratio by writing them in the form $a:b:c$, or even $a:b:c:d$. You **simplify** ratios by dividing the numbers by a **common factor**, just like you do with fractions. E.g. $2:4$ simplifies to $1:2$ by dividing both sides by 2. A ratio is in its **simplest form** when all of the parts are **whole numbers** but they have **no common factors** greater than 1.

Example 1

There are 15 fiction books and 10 non-fiction books on a shelf.
Write down the ratio of fiction books to non-fiction books in its simplest form.

1. Write down the ratio with the quantities in the order that you're asked to give them.

   fiction : non-fiction
   $15 : 10$

2. Divide both sides by the same number — a common factor of the two sides.

   5 is a common factor:
   $\div 5 \left(\begin{array}{c} 15 : 10 \\ 3 : 2 \end{array} \right) \div 5$

3. Stop when you can't divide any further and leave whole numbers on each side.

   The simplest form is **3 : 2**

Tip: Dividing by the highest common factor (p.13) gets you to the simplest form in one go. Here, 5 is the HCF of 15 and 10.

Exercise 1

Q1 Write down each of the following ratios in its simplest form.

   a) $2:6$ b) $40:10$ c) $24:6$ d) $7:28$

   e) $16:12$ f) $48:36$ g) $80:32$ h) $121:33$

Q2 Write down each of the following ratios in its simplest form.

   a) $6:2:4$ b) $15:12:3$ c) $16:24:80$ d) $21:49:42$

Q3 There are 18 boys in a class of 33 pupils. Find the ratio of boys to girls in its simplest form.

Q4 Isabel and Sophia share a bag of 42 sweets. Isabel has 16 sweets, and Sophia has the rest. Find the ratio of Sophia's sweets to Isabel's sweets, giving your answer in its simplest form.

When you have a ratio involving **different units**, **convert** all quantities to the **same unit** before simplifying.

Example 2

Write the ratio 1 m : 40 cm in its simplest form.

1. Rewrite the ratio so that the units are the same.

2. Remove the units altogether.

3. Simplify as usual by dividing each side by the highest common factor, 20.

1 m = 100 cm, so 1 m : 40 cm is the same as 100 cm : 40 cm

$$\div 20 \left(\underset{5:2}{100:40} \right) \div 20$$

So the simplest form is **5 : 2**

Tip: You'll need to know your unit conversions — see Section 22.

Exercise 2

Q1 Write these ratios in their simplest form.

a) 10p : £1

b) 20 mm : 4 cm

c) 2 weeks : 7 days

d) 18 mins : 1 hour

e) 30 cm : 2 m

f) 1 m : 150 mm

g) 6 months : 5 years

h) 2.5 hours : 20 mins

i) 8 cm : 1.1 m

j) 9 g : 0.3 kg

k) 65 m : 1.56 km

l) 1.2 kg : 480 g

Q2 Emma's mass is 54 kg. Her award-winning pumpkin has a mass of 6000 g. Find the ratio of the pumpkin's mass to Emma's mass. Give your answer in its simplest form.

Q3 The icing for some cupcakes is made by mixing 1.6 kg of icing sugar with 640 g of butter. Find the ratio of butter to icing sugar. Give your answer in its simplest form.

Ratios in the form 1 : n

Practical problems can often involve finding 'how much of b is needed for **every one** of a'.

Instead of simplifying the ratio $a : b$ by dividing by a common factor, you need to divide both sides by a, to leave a ratio in the form $1 : \frac{b}{a} = \mathbf{1 : n}$, where n could be a **decimal** or a **fraction**.

You can also give the ratio $a : b$ in the form $\mathbf{n : 1}$ by dividing both sides by b.

Example 3

Nigel makes a smoothie by mixing half a litre of blueberry juice with 100 millilitres of plain yoghurt. How much yoghurt does he use for every millilitre of juice?

1. Write down the ratio of juice to yoghurt and remove the units.

1 litre = 1000 ml, so 0.5 litres : 100 ml is the same as 500 ml : 100 ml

2. You want the juice side (the left-hand side) to equal 1, so divide both sides by 500.

$$\div 500 \left(\underset{1:0.2}{500:100} \right) \div 500$$

3. Use this simplified ratio to answer the question.

So he uses **0.2 ml** of yoghurt for 1 ml of juice.

Exercise 3

Q1 Write down the following ratios in the form $1:n$.

   a) $2:6$ b) $30:120$ c) $8:26$ d) $6:21$

   e) $2:1$ f) $10:3$ g) $8:5$ h) $5:9$

   i) $10\,mm:5\,cm$ j) $30\,mins:2\,hours$ k) $90\,m:7.2\,km$ l) $50p:£6.25$

Q2 Two towns 4.8 km apart are shown on a map 12 cm apart.
   Find the ratio of the map distance to the true distance in the form $1:n$.

Q3 A recipe uses 125 ml of chocolate syrup and $2\frac{1}{2}$ litres of milk.
   How much milk does it use for every millilitre of chocolate syrup?

Ratios and Fractions

Learning Objectives — Spec Ref N11/R8:
- Use ratios to find fractions.
- Use fractions to find ratios.

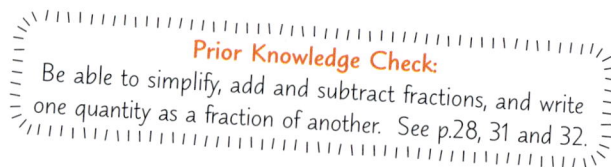

Prior Knowledge Check:
Be able to simplify, add and subtract fractions, and write one quantity as a fraction of another. See p.28, 31 and 32.

If you have a ratio of quantities $a:b$ (or $a:b:c$), you can find the **fraction** of one quantity out of the total:

- **Add up** all the quantities in the ratio to get the total number of parts, i.e. $a + b$, or $a + b + c$.

- Put the **number of parts** for the quantity you're interested in as the fraction's **numerator**, and the **total number of parts** as the fraction's **denominator**.

- **Simplify** the fraction if necessary.

Example 4

A box of doughnuts contains jam doughnuts and chocolate doughnuts in the ratio 3:5. What fraction of the doughnuts are chocolate flavoured?

1. Find the total number of 'parts'.

   There are $3 + 5 = 8$ parts altogether.

2. Write the number of parts that are chocolate as the numerator, and the total number of parts as the denominator. The fraction is in its simplest form.

   5 of the parts are chocolate, so $\frac{5}{8}$ of the doughnuts are chocolate flavoured.

Exercise 4

Q1 A recipe uses white flour and brown flour in the ratio $2:1$. What fraction of the flour is brown?

Q2 A tiled floor has blue and white tiles in the ratio $9:4$. What fraction of the tiles are blue?

Q3 In a tournament the numbers of home wins, away wins and draws were in the ratio $7:2:5$. What fraction of the games were home wins?

Q4   Amy's sock drawer has spotty, stripy and plain socks in the ratio $5:1:4$. What fraction are stripy?

Q5   At a birthday party there are purple, red and blue balloons in the ratio $3:8:11$.
   a) What fraction of the balloons are blue?
   b) What fraction of the balloons are purple?
   c) What fraction of the balloons aren't red?

You can also express quantities given as fractions in the form of a ratio.

- Work out the fractions of each quantity you're interested in and put them all over a **common denominator**.

- The **numerators** of each quantity are the numbers to go in the ratio.

> **Tip:** The denominator is the total number of parts for the ratio.

Example 5

Of Hannah's DVDs, $\frac{2}{7}$ are horror films and the rest are comedies. Find the ratio of comedies to horror films.

1. Find the fraction of films that are comedies by subtracting the fraction of horror films from 1.

$\frac{2}{7}$ are horror,
so $1 - \frac{2}{7} = \frac{5}{7}$ are comedies.

2. Write the numerators of the fractions in ratio form, making sure they're in the order you're asked for.

5 parts are comedies and 2 parts are horror.
So ratio of **comedies to horror** is **5:2**.

Exercise 5

Q1   In a bag of red sweets and green sweets, $\frac{1}{3}$ are red. Find the ratio of red to green sweets.

Q2   Al watched $\frac{3}{10}$ of the episodes in a series. Find the ratio of episodes he watched to ones he didn't.

Q3 $\frac{1}{8}$ of some counters are red, $\frac{5}{8}$ are blue and the rest are green. Find the ratio of red to blue to green.

Q4   Last season, a rugby team won half of their matches by more than 10 points and a quarter of their matches by less than 10 points. They lost the rest. Find the team's ratio of wins to losses.

Q5   A cycling challenge has three routes — A, B and C.
$\frac{7}{10}$ of the competitors choose route A, $\frac{1}{5}$ choose route B and the rest choose route C.
What is the ratio of competitors choosing route A to those choosing route C?

Q6   In a bowl of salad there are $\frac{5}{2}$ as many tomatoes as there are lettuce leaves.
   a) Write this as a ratio of tomatoes to lettuce leaves. Give your ratio in its simplest form. (PROBLEM SOLVING)
   b) There are 45 tomatoes in the salad bowl. How many lettuce leaves are there?

Q7   One third of the counters in a box are red, one fifth are white and the rest are beige. Find the ratio of red to beige counters.

4.2 Using Ratios

You should be feeling more confident with ratios by now, so it's time to see how they can be used to solve different types of problems. They're not too tricky as long as you understand what a given ratio tells you — so if you're not sure, have a look back at the earlier pages in this section.

Part : Whole Ratios

> **Learning Objective — Spec Ref R5:**
> Be able to change between part : part and part : whole ratios.

A **part : part** ratio $a:b$ contains a lot more information than it appears to at first. For example, you can use it to write **fractions** of each quantity out of the total $(a + b)$, as you saw on page 44. Similarly, you can form **part : whole** ratios of each quantity to the total — $a:a + b$ and $b:a + b$.

To turn a **part : whole** ratio back into a **part : part** ratio, you **subtract** the known 'part' from the 'whole' to find the **unknown part**. E.g. if the part : whole ratio is $3:9$ then the unknown part is $9 - 3 = 6$, and the part : part ratio is $3:6$.

Example 1

A shop only sells green and brown garden sheds.
The ratio of green sheds sold to the total number of sheds sold is **5 : 7**.
What is the ratio of green sheds sold to brown sheds sold?

1. Write down the ratio you know, and the one you need to know.

    green : total   green : brown
        5 : 7   5 : ?

2. The total (green + brown) is 7, so subtract to find the number of brown sheds for every 5 green.

    $7 - 5 = 2$ brown for every 5 green

3. Write this in ratio form, and simplify if necessary.

    green : brown = **5 : 2**

Exercise 1

Q1 A reptile house only contains snakes and lizards.
The ratio of snakes to the total number of reptiles is $3:8$.
   a) What fraction of the reptiles are snakes?   b) What is the ratio of snakes to lizards?

Q2 A biscuit tin only contains digestives and bourbons. The ratio of digestives to the total number of biscuits is $4:7$. What is the ratio of digestives to bourbons?

Q3 When Dan makes a cup of tea, the ratio of water to milk is $15:2$.
What is the ratio of water to the total amount of liquid in the cup?

Q4 A rock album is sold as a digital download and on a CD. The ratio of digital downloads to CDs sold is $53:46$. What is the ratio of CDs sold to the total number of albums sold?

Q5 Ashley and Cameron have a combined age of 20. Cameron is 10 years older than Ashley. (PROBLEM SOLVING)
   a) Find the ratio of Ashley's age to Cameron's age in its simplest form.
   b) Find the ratio of Cameron's age to their combined age in its simplest form.

Scaling Up Ratios

Learning Objectives — Spec Ref R5/R8:
- Be able to scale up ratios to find unknown amounts.
- Be able to connect ratios to linear graphs and use them for scaling.

Prior Knowledge Check:
Be able to draw the graphs of
straight lines. See Section 15.

You've seen how all the numbers in a ratio can be divided by the same number to get a **simplified** ratio. You can also **multiply** all the numbers in a ratio by the **same number** to get a scaled up **equivalent ratio**, which can then be used to solve problems. This is similar to finding equivalent fractions — see page 27.

Example 2

The ratio of men to women in an office is 3:4.
If there are 9 men in the office, how many women are there?

1. Write down the ratio as it is, and what you need it to be.
   Here you need the 'men' (left-hand) side to be 9.

   Ratio of men:women is 3:4
   But you need 9:?

2. Work out what you have to multiply
   the left-hand side by to get 9.

   $9 \div 3 = 3$
   So you need to multiply by 3.

3. Multiply the right-hand side by the same number.

   $\times 3 \binom{3:4}{9:12} \times 3$

4. Use your ratio to answer the question.

   So there are **12** women.

If two **variables** are in a **constant ratio**, they have a **linear** relationship. If you plot a graph of one variable against the other, you'll get a **straight line** that goes through the origin. To represent a ratio on a graph, follow these steps:

Tip: The two variables are in direct proportion — there's more about this on page 53.

- If the ratio between two variables is $a:b$, plot the **coordinate (a, b)**, where the first variable is along the horizontal axis, and the second variable is up the vertical axis.

- Draw a **straight line** through the **origin** and **(a, b)**.

For any value of one variable, you can now **read off the graph** to get the value of the other variable. Every **coordinate** on the graph can be written as a **scaled ratio** equivalent to $a:b$.

Example 3

A factory produces blue pens and black pens in the ratio 2:3.
a) Draw a graph to show this relationship.

1. Draw a set of axes with 'number of blue pens' on the horizontal axis, and 'number of black pens' on the vertical axis.

2. For every 2 blue pens, there are 3 black pens, so draw a straight line through the origin and the point (2, 3).

b) Use your graph to find the number of blue pens produced for every 12 black pens.

Read across from '12 black pens' on the vertical axis until you hit the graph, then read down to the horizontal axis to find the number of blue pens.

So there are **8 blue pens** for every 12 black pens (the ratio of blue pens:black pens = 8:12).

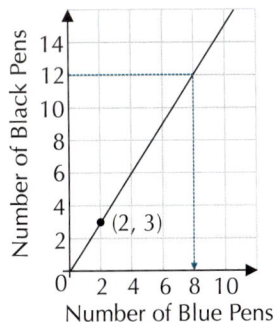

Exercise 2

Q1 A recipe uses sugar and butter in the ratio $2:1$.

a) How much butter is needed for 100 g of sugar?

b) Show the ratio as a graph on the axes on the right.

c) Use the graph to find the amount of sugar needed for 40 g of butter.

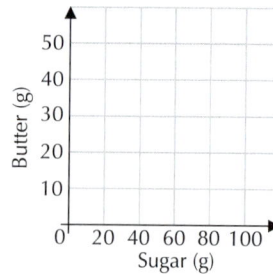

Q2 A wood has oak and beech trees in the ratio $2:9$. 42 of the trees are oak. How many are beech?

Q3 The ages of a father and son are in the ratio $8:3$. If the father is 48, how old is the son?

Q4 A cardboard cut-out of a footballer is 166 cm tall. The height of the cut-out and the footballer's actual height are in the ratio $5:6$. How tall is the footballer?

Q5 Mai and Sid split their savings in the ratio $5:4$.

a) Show this ratio on the axes on the right.

b) Use the graph to find how much Sid gets when Mai gets £400.

c) Sid gets £60. How much do they get in total?

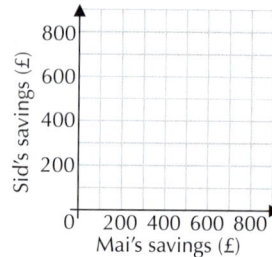

Q6 A fruit punch is made by mixing pineapple juice, orange juice, and lemonade in the ratio $1:3:6$. If 500 ml of pineapple juice is used, how much orange juice is needed?

Q7 Jem orders a pizza with olives, slices of courgette and slices of goat's cheese in the ratio $8:3:4$. How many slices of courgette and how many slices of goat's cheese would she get with 24 olives?

Q8 Mo, Liz and Dee's heights are in the ratio $32:33:37$. Mo is 144 cm tall. What is the combined height of the three people?

Q9 Max, Molly and Maisie are at a bus stop. The number of minutes they have waited can be represented by the ratio $3:7:2$. Molly has been waiting for 1 hour and 10 minutes. How long have Max and Maisie been waiting?

Q10 The ratio of children to adults in a swimming pool must be $5:1$ or less.

a) Draw a graph to show this ratio.

b) If there are 18 children, use your graph to work out how many adults there must be.

Q11 A TV-show producer is selecting a studio audience. He wants the ratio of under-30s to over-30s to be at least $8:1$. If 100 under-30s are selected, find the maximum number of over-30s.

Q12 A recipe uses 1 aubergine for every 3 people. How many aubergines should you buy for 10 people?

Q13 Olga is allowed no more than 5 minutes of reality TV for every minute of news programmes she watches. She watches 45 minutes of reality TV. What's the least amount of news she should watch?

Changing Between Ratios

Learning Objective — Spec Ref R5:
Be able to change ratios when one or more quantities change.

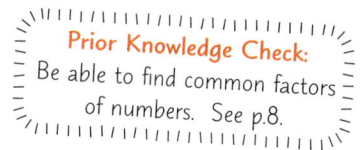

Prior Knowledge Check:
Be able to find common factors
of numbers. See p.8.

If two or more quantities are in a given ratio and some of the quantities **change**
then the **ratio** will change too.

Example 4

Ashraf collects coins. He has 45 gold coins, and the ratio of gold coins to silver coins is $9:13$.
If he sells 15 silver coins, what will be the new ratio of gold coins to silver coins?
Give your ratio in its simplest form.

1. First find the original number of silver coins.
   Write down the ratio as it is, and what it needs to be.
   Here you need the 'gold' (left-hand) side to be 45.

   Ratio of gold : silver is $9:13$
   But you need $45:?$

2. Work out what you have to multiply by to get 45.

   $45 \div 9 = 5$.
   So you need to multiply by 5.

3. Multiply the right-hand side by the same number
   to get the original number of silver coins.

   $\times 5 \left(\begin{array}{c} 9:13 \\ 45:65 \end{array} \right) \times 5$ So there were
   65 silver coins.

4. Work out the remaining number of silver coins
   by subtracting the number sold, 15.

   $65 - 15 = 50$ silver coins remaining

5. Write out the new amounts in ratio form.

   So new ratio gold : silver is $45:50$

6. Simplify your answer by dividing by the HCF, 5.

   This simplifies to: $\div 5 \left(\begin{array}{c} \\ \textbf{9:10} \end{array} \right) \div 5$

Exercise 3

For the following questions, give each ratio in its simplest form.

Q1 In a bakery, the ratio of fruit scones to cheese scones is $3:4$.

    a) Given that there are 24 cheese scones, how many fruit scones are there?

    b) If 6 fruit scones were sold, what would be the new ratio of fruit scones to cheese scones?

Q2 An aquarium contains fish and crabs. The ratio of fish to crabs is $4:1$. There are 16 fish in the
aquarium. If 2 more crabs were added to the aquarium, what would be the new ratio of fish to crabs?

Q3 A village has detached and terraced houses in the ratio $5:7$. There are 56 terraced houses. If 12 new
detached houses were built, what would be the new ratio of detached houses to terraced houses?

Q4 At the beginning of the week, a bookshop which sells school books has 48 English books
in stock and the ratio of English to Maths books is $2:5$. At the end of the week, the ratio of
English to Maths books is $3:4$. Given that the shop sold 12 English books during the week,
how many Maths books did it sell?

Q5 A bag contains black and white tokens in a ratio of $3:2$. 8 black tokens and
2 white tokens are added to the bag. The ratio of black to white tokens is now $2:1$.
How many white tokens were in the bag to begin with?

4.3 Dividing in a Given Ratio

Ratios can be used to share things out unequally. When dividing an amount in a given ratio, one side of the ratio gets a greater share than the other(s). You can use part:part and part:whole ratios for sharing, but you might have to come up with the ratio yourself from information about how things are related.

Learning Objectives — Spec Ref R5/R6:
- Divide quantities in a given part:part or part:whole ratio.
- Be able to write a ratio given a multiplicative relationship between two quantities.

Part:part ratios can be used to divide an amount into two or more **shares**. The numbers in the ratio show how many parts of the whole each share gets. To share an amount in a given ratio:

- **Add up** all the quantities in the ratio to get the total number of parts, i.e. $a + b$, or $a + b + c$.

- **Divide** the amount to be shared by the total number of parts to get the **amount for one part**.

- **Multiply** the amount for one part by the number of parts for each share.

Example 1

Divide £54 in the ratio 4:5.

1.	Find the total number of parts.	$4 + 5 = 9$ parts altogether
2.	Work out the amount for one part.	9 parts = £54. So 1 part = £54 ÷ 9 = £6
3.	Multiply by the number of parts for each share, i.e. the numbers in the ratio.	The ratio is 4:5. £6 × 4 = £24 £6 × 5 = £30
4.	Check that the shares add up to the initial amount.	The shares are **£24** and **£30**. (£24 + £30 = £54 ✔)

Tip: Using fractions, the shares are $\frac{4}{9}$ and $\frac{5}{9}$. So $\frac{4}{9} \times £54 = £24$, and $\frac{5}{9} \times £54 = £30$.

Exercise 1

Q1 Divide £48 in the following ratios.
    a) 2:1   b) 1:3   c) 5:1   d) 7:5

Q2 Divide 72 cm in the following ratios.
    a) 2:3:1   b) 2:2:5   c) 5:3:4   d) 7:6:5

Q3 Share £150 in these ratios.
    a) 1:4:5   b) 15:5:30   c) 6:7:2   d) 13:11:6

Q4 Find the smallest share when each amount below is divided in the given ratio.
    a) £22 in the ratio 5:6   b) 450 g in the ratio 22:28
    c) 45 kg in the ratio 2:3:4   d) 1800 ml in the ratio 5:6:7

Q5 Find the largest share when each amount below is divided in the given ratio.
    a) £30 in the ratio 1:3   b) 36 g in the ratio 3:2
    c) 150 kg in the ratio 10:7:3   d) 24 000 ml in the ratio 5:7:20

'Dividing into ratios' questions are often set in a **real-life context**, but you can still use the usual method if you're given a **part:part** ratio. If you're given a **part:whole** ratio, you can just **divide** the amount by the total number of parts (the whole) and **multiply** it by the number of parts in the share you're trying to find.

Example 2

A box of 45 chocolates contains dark and milk chocolates in the ratio 2:7.
How many of the chocolates are milk chocolate?

1. To find the number of chocolates that are milk chocolates, you need to divide 45 into the ratio 2:7.

2. First find the total number of parts... $2 + 7 = 9$ parts altogether

3. ... then divide to work out the amount for one part.   9 parts = 45 chocolates.
   1 part = $45 \div 9 = 5$ chocolates

4. Milk chocolates make up 7 parts, so multiply the amount for one part by 7.   5 chocolates × 7
   = **35 milk chocolates**

Tip: Using fractions, $\frac{7}{9}$ of the chocolates are milk chocolates, so find $\frac{7}{9}$ of 45 = 35.

Exercise 2

Q1   Share 32 sandwiches in the ratio 3:5.

Q2   Kat and Lindsay share 30 cupcakes in the ratio 3:2. How many do they each get?

Q3   There are 112 dogs at a show. For every 14 dogs at the show, 9 are male.
   How many male and how many female dogs are at the show?

Q4   Lauren is 16 and Cara is 14. Their grandad gives them £1200 to share in the ratio of their ages.
   How much money do they each get?

Q5   Tangerine paint is made from yellow and red paint. In a 42 litre batch of tangerine paint,
   the ratio of red paint to tangerine paint is 3:7. How many litres of yellow paint were used?

Q6   A quiz team share their £150 prize money in the ratio 2:4:6.
   What is the least amount that any team member receives?

Q7   a) The angles in a triangle are in the ratio 2:1:3. Find the sizes of the three angles.

   b) The four angles in a quadrilateral are in the ratio 1:5:2:4.
      Calculate the sizes of all four angles.

Q8   A rope is cut into three pieces in the ratio 2:1:4. The third piece is 22 cm longer   (PROBLEM SOLVING)
   than the first piece. What is the length of the second piece?

Q9   The length and width of a rectangle are in the ratio 5:1.
   If the perimeter of the rectangle is 72 cm,   (PROBLEM SOLVING)
   calculate the length and the width of the rectangle.

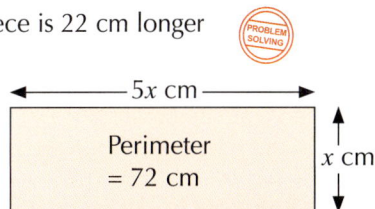

$\longleftarrow$ $5x$ cm $\longrightarrow$

| Perimeter = 72 cm |
x cm

Q10   Andrew, Pierre and Susanne are making tomato sauce.
   They share out tomatoes in the ratio 2:3:5. Susanne gets 16 more   (PROBLEM SOLVING)
   tomatoes than Pierre. How many tomatoes do they have in total?

Q11 Ali and Max earn £200 one weekend. They share it according to the hours they worked. Ali worked 10 hours and Max worked 6 hours. Ali had help from her brother Tim, so she splits her share with him in the ratio 4:1. How much do Ali, Max and Tim each get?

Q12 Dan, Stan and Jan are aged 8, 12 and 16. They get pocket money each week in the ratio of their ages. Dan normally gets £6, but has been naughty, so this week all the money is split between Stan and Jan, in the ratio of their ages, to the nearest penny. How much do they each receive this week?

You might be given the **multiplicative relationship** between two or more variables instead of a ratio. To divide an amount into shares using this relationship, you need to write it as a **ratio** first. E.g. if there are **five times** as many sparrows as pigeons then the ratio of sparrows to pigeons is 5:1.

Example 3

Mika is twice as old as Nate, and Nate's age is one third of Orla's age. They have a combined age of 72. Find each person's age.

1. You need to divide 72 in the ratio of their ages, so first find this ratio from what you're told.

2. For 1 year of Nate's age, Mika has 2.
    Mika:Nate = 2:1

    For 1 year of Orla's age, Nate has $\frac{1}{3}$.
    Nate:Orla = $\frac{1}{3}$:1

3. Multiply the second ratio by 3 so the parts for Nate are the same in both ratios (1).
    $\times 3$ $\qquad$ $\times 3$
    $= 1:3$

4. Now you can combine them into a single ratio.
    Mika:Nate:Orla = 2:1:3

5. As usual, find the total number of parts then divide to work out the amount for one part.
    2 + 1 + 3 = 6 parts altogether
    6 parts = 72. So 1 part = 72 ÷ 6 = 12

6. Finally, multiply by the number of parts for each share.
    Mika is 2 × 12 = **24**, Nate is 1 × 12 = **12**, and Orla is 3 × 12 = **36**.

7. Check the answer works with the given information.
    (24 is twice 12, 12 is one third of 36 ✔)

Exercise 3

Q1 In a school of 600 pupils, there are seven times as many right-handed pupils as left-handed pupils. How many pupils are right-handed?

Q2 Daniel puts one-and-a-half times as much money into a business as his partner Elsie. They share the profits in that same ratio. How much of a £5700 profit does Daniel get?

Q3 There are 28 passengers on a bus. There are two-and-a-half times as many passengers on the phone as not on the phone. How many passengers are on the phone?

Q4 A fruit salad has twice the mass of raspberries as redcurrants, and three times the mass of strawberries as redcurrants. How much of each fruit is needed to make 450 g of fruit salad?

Q5 Nicky, Jacinta and Charlie share a bag of 35 sweets so that Nicky gets half as much as Jacinta and Charlie gets half as much as Nicky. How many sweets do each of them get?

4.4 Proportion

Proportion tells you how one thing changes in relation to another — either in direct or inverse proportion.

Direct Proportion

> **Learning Objective — Spec Ref R7/R10:**
> Understand and use direct proportion.

If the ratio between two things (**variables**) is always the same, then they're in **direct proportion**. For example, the **diameter** and **circumference** of a circle are always in the ratio $1 : \pi$, so circumference is **directly proportional** to diameter. If two variables are in direct proportion then increasing one of the variables increases the other by the **same scale factor**, e.g. doubling one thing doubles the other. You can use the **unitary method** to solve simple problems where two things are in direct proportion:

- **Divide** to find the amount of one variable for every **one** of the other (e.g. the price per item).

- Then **multiply** to find the scaled amount.

> ### Example 1
>
> **If 8 chocolate bars cost £6, calculate the cost of 10 chocolate bars.**
>
> 1. Divide by 8 (the number of bars) to find the cost of one bar.   £6 ÷ 8 = £0.75
> 2. Multiply by 10 (the new number of bars) to find the cost of 10 bars.   £0.75 × 10 = **£7.50**

> ### Example 2
>
> **A bus travels 50 km in 40 minutes. Assuming the bus travels at the same average speed:**
> **a) Calculate the time it would take to travel 65 km.**
>
> 1. Divide by 50 to find the time taken for 1 km.   1 km takes 40 ÷ 50 = 0.8 minutes
> 2. Multiply by 65 to find the time taken for 65 km.   65 km takes 0.8 × 65 = **52 minutes**
>
> **b) Calculate the distance the bus would travel in 16 minutes.**
>
> 1. This time, divide by 40 to find the distance travelled in 1 minute.   In 1 minute it travels 50 ÷ 40 = 1.25 km
> 2. Multiply by 16 to find the distance travelled in 16 minutes.   In 16 minutes it travels 1.25 × 16 = **20 km**
>
> **Tip:** In a) you found the number of minutes per km, but in b) you need the km per minute, which is a measure of the average speed.

Exercise 1

Q1 The cost of 8 identical books is £36. What is the cost of 12 of these books?

Q2 If it takes 1.8 kg of flour to make 3 loaves of bread, how much flour is needed to make 5 loaves?

Q3 5 litres of water make 4 jugs of squash. How much water is needed to make 3 jugs of squash?

Q4 Ryan earns £192 for cleaning 12 cars. How much more will he earn if he cleans another 5 cars?

Q5 If 7 of the same DVDs cost £84, how many DVDs can be bought for £48?

Q6 A car uses 35 litres of petrol to travel 250 km. Assuming its fuel consumption stays constant:
   a) Calculate how far, to the nearest km, the car can travel on 50 litres of petrol.
   b) Calculate how many litres of petrol the car would use to travel 400 km.

Q7 A 58 g bar of chocolate contains 11 g of fat.
   a) To the nearest gram, how much fat is this per 100 g?
   b) If my diet only allows me to eat 5 g of fat, how many grams (to 1 d.p.) of the bar can I eat?

Exchange rates and **conversions** are also direct proportion problems, so you can use a similar method.

Example 3

Oliver has 30 euros (€) left over from a holiday in France.
Assuming the exchange rate is £1 = €1.14, how many pounds can he exchange his euros for?

1. Divide £1 by 1.14 to find the number of pounds per €1.   €1 = £1 ÷ 1.14 = £0.877...

2. Multiply by the number of euros, 30, and round to 2 d.p.   €30 = 30 × £0.877... = **£26.32** (2 d.p.)

Exercise 2

Q1 If £1 is worth €1.14, convert the following into pounds: a) €10   b) €100   c) €250

Q2 1 kg ≈ 2.2 pounds (lbs). Calculate the approximate mass of 8.25 lbs of sugar in kg.

Q3 Use the exchange rate £100 = 1055 Chinese yuan to convert 65 Chinese yuan into pounds.

Q4 Philip changed £50 into Swiss francs. The exchange rate was £1 = 1.47 Swiss francs.
   a) How many Swiss francs did he get?

   The next week, he changed 30 Swiss francs into pounds with exchange rate £1 = 1.50 Swiss francs.
   b) How many pounds did he get back?

When **two variables** are changing, split the calculation up into **stages** and change one variable at a time. Then use the **unitary method** on **each stage**.

Example 4

Two people can build four identical walls in three days.
At this rate, how many of these walls could ten people build in 3.75 days?

1. First change the number of people. Find how many walls 1 person, then 10 people, can build in 3 days, by dividing by 2 and multiplying by 10.

   In 3 days, 2 people can build 4 walls,
   1 person can build 4 ÷ 2 = 2 walls,
   and 10 people can build 2 × 10 = 20 walls.

2. Then change the number of days. Find how many walls 10 people can build in 1 day, then 3.75 days, by dividing by 3 then multiplying by 3.75.

   10 people can build 20 walls in 3 days.
   They can build 20 ÷ 3 = 6.66... walls in 1 day,
   and 6.66...× 3.75 = **25 walls** in 3.75 days.

Q1   It takes three people five minutes to eat six hotdogs.
      At this rate, how many hotdogs could five people eat in ten minutes?

Q2   Ten people can count 850 coins in two minutes.
      At this rate, how many coins could 22 people count in five minutes?

Q3   It takes two people 90 minutes to plant 66 flower bulbs.
      At this rate, how many bulbs could three people plant in three hours?

Q4   Two toy makers can make eight toy soldiers in two weeks.
      How many soldiers could seven toy makers make in three weeks if they work at the same rate?

Inverse Proportion

Learning Objective — Spec Ref R7/R10:
Understand and use inverse proportion.

If one variable decreases as another increases, then they're **inversely proportional**. E.g. **increasing** your
speed **decreases** the time it will take to finish your journey, so **speed** and **time** are inversely proportional.

With inverse proportion problems, think about whether the variable should be **increasing** (so you need to
multiply by something) or **decreasing** (so you need to **divide**). Then check that your answer **makes sense**.

Example 5

Four people take five and a quarter hours to dig a hole.
How long would it take seven people to dig the same sized hole at the same rate?

1.   Work out how long 1 person will take.   4 people take 5.25 hours
     The fewer people there are, the longer it will   1 person will take:
     take, so you need to multiply the time by 4. $5.25 \times 4 = 21$ hours

     Tip: Check it makes
     sense: 7 people
     should take less

2.   7 people will dig the hole in   7 people will take: time than 4, and
     a shorter time than 1 person, $21 \div 7 = \textbf{3 hours}$ 3 hours < 5.25 hours ✔
     so divide the time by 7.

Exercise 4

Q1   It takes three people two hours to paint a wall. How long
      would it take five people to paint the same wall at the same rate?

Q2   It takes four teachers two and a half hours to mark a set of test papers.
      How long would it take six teachers to mark the same set of papers at the same rate?

Q3   Four waiters can set all the tables in a restaurant in twenty minutes. They hire an extra waiter.
      How long will it take all five of them to set all of the tables, assuming they all work at the same rate?

Q4　A journey takes two and a quarter hours when travelling at an average speed of 30 mph.
How long would the same journey take when travelling at an average speed of 45 mph?

Q5　It will take five builders sixty-two days to complete a particular project.

a) At this rate, how long would the project take if there were only two builders?

b) If the project needed completing in under forty days,
what is the minimum number of builders that would be required?　(PROBLEM SOLVING)

As on page 54, when **two variables change**, split the calculation up into **stages** and change one variable at a time. Decide if the amount should go **up** or **down** at each stage of the calculation.

Example 6

It takes 5 examiners 4 hours to mark 125 exam papers.
How long would it take 8 examiners to mark 200 papers at the same rate?

1.　First change the number of examiners.
Find how long it will take 1 examiner,
then 8 examiners, to mark 125 papers,
by multiplying by 5 and dividing by 8.

5 examiners take 4 hours to mark 125 papers.
So 1 examiner will take $4 \times 5 = 20$ hours,
and 8 examiners will take $20 \div 8 = 2.5$ hours

2.　Then change the number of papers.
Find how long it will take 8 examiners
to mark 1 paper, then 200 papers, by
dividing by 125 and multiplying by 200.

8 examiners take 2.5 hours to mark 125 papers.
So they take $2.5 \div 125 = 0.02$ hours to mark 1 paper,
and $0.02 \times 200 = $ **4 hours** to mark 200 papers.

Exercise 5

Q1　It takes 45 minutes for two people to clean six identical rooms.
How long would it take five people to clean 20 of the same rooms at the same rate?

Q2　It takes two electricians nine days to rewire two identical houses.
How long would it take three electricians to rewire three of the same houses at the same rate?

Q3　It takes four bakers 144 minutes to ice 72 cakes.
How long would it six bakers to ice 96 of these cakes?

Q4　Three people take twenty minutes to shovel snow from two identical driveways.
At this rate, how long would it take two people to shovel snow from five of the same driveways?

Q5　Fourteen people can paint 35 identical plates in two hours. At this rate, how many people
would be needed to paint 60 of the same plates in two and a half hours?

Q6　It will take twelve workers three weeks to complete four stages of a ten-stage project.
Each stage of the project takes the same number of hours of work to complete.

a) At this rate, how long would it take fifteen workers to complete the whole project?

b) The project needs completing in under four weeks.
What is the minimum number of workers that would be required,　(PROBLEM SOLVING)
given that they all work at the same rate as the original workers?

Review Exercise

Q1 A builder mixes 10 kg of cement with 25 kg of sand. Give the ratio of cement to sand:
a) in its simplest form
b) in the form $1:n$
c) in the form $n:1$
d) Find the fraction of the mix that is cement, in its simplest form.

Q2 A picture has a width of 120 mm. An enlargement of the picture has a width of 35 cm.
Write the ratio of the original width to the enlarged width in its simplest form.

Q3 Alf has $\frac{2}{3}$ as many nails as screws.
a) Write the ratio of nails to screws in its simplest form.
b) Write the ratio of the number of screws to the total number of nails and screws.

Q4 Evie is making green cards and purple cards in the ratio $7:3$. She has made 12 green cards
so far and has 72 more green cards to make. How many purple cards will she need to make?

Q5 A farm has sheep, cows and pigs in the ratio $8:1:3$.
If there are 16 sheep on the farm, how many animals are there altogether?

Q6 On a normal day, 20 planes depart from an airport to go to Spain and Finland in the ratio $4:1$.
One day, 2 additional flights went to Spain, and the ratio of Spanish to Finnish
flights was $3:1$. How many additional flights departed for Finland on this day?

Q7 Luiz, Seth and Fran share £1440 in the ratio $3:5:7$. Fran splits her share with Ali in the ratio $4:1$.
a) How much does each person end up with?
b) Find the ratio of Fran's amount to Seth's amount in its simplest form.

Q8 The only ingredients of a particular cereal are raisins, nuts and oats. The mass of raisins
is three times the mass of nuts, and the mass of oats is half the mass of raisins.
How many grams of nuts, to the nearest gram, are needed to make 450 g of the cereal?

Q9 It costs £20 to put 12.5 litres of petrol in my car.
a) How much will it cost me for a full tank of petrol if the tank holds 60 litres?
b) How much petrol can I put in my car for £30?

Q10 Emma changed £500 into rand before going on holiday, at an exchange rate of £1 = 20.8 rand.
a) How many rand did she get for her £500?
Emma spent 8000 rand on holiday. When she got home, she changed her
leftover rand into pounds at the new exchange rate of £1 = 19.6 rand.
b) How much money did she get back in pounds?

Q11 Eight lumberjacks can chop down three identical trees in an hour and a half.
a) At this rate, how many trees could eight lumberjacks chop down in six hours?
b) How long would it take eight lumberjacks to chop down eight trees at the same rate?
c) At this same rate, how many trees could six lumberjacks chop down in an eight-hour day?

Exam-Style Questions

Q1 Write the ratio $0.75 : 1.2$ in the form $1 : n$ where n is a decimal.

[2 marks]

Q2 The ratio of female to male teachers at a school is $7 : 2$.
There are 40 more female teachers than male teachers.

Work out how many teachers there are in total at the school.

[2 marks]

Q3 At a wildlife park there are badgers, foxes and hares.
There are three times as many badgers as there are foxes.
For every two badgers there are five hares.

Work out the ratio of the number of badgers to the number of foxes to the number of hares.
Give your ratio in its simplest form, using integers.

[2 marks]

Q4 On the right is part of a recipe for cupcakes.

a) How much butter is needed to make 10 cupcakes?

[2 marks]

b) How much flour is needed to make 35 cupcakes?

[2 marks]

> **Makes 25 Cupcakes:**
> 200 g butter
> 250 g caster sugar
> 280 g flour
> 4 eggs

Q5 A team of eight people can deliver the morning newspapers in 20 minutes.
One morning, one of the team is off sick.

How long will it take the rest of them to deliver all the newspapers at their usual rate?
Assume each person delivers newspapers at the same rate.
Give your answer to the nearest second.

[3 marks]

Q6 A brand of apple juice has a sugar content of 12 grams per 100 millilitres.
A brand of lemonade has a sugar content of 2.5 grams per 100 millilitres.

Work out the ratio in which this apple juice and lemonade should be mixed
to produce a drink with a sugar content of 6 grams per 100 millilitres.

[4 marks]

5.1 Percentages

'Per cent' means 'out of 100' — so writing an amount as a percentage means writing it as a number out of 100. Percentages are written using the % symbol and are often used to show proportions.

Writing One Number as a Percentage of Another

Learning Objectives — Spec Ref R9:
- Be able to write one number as a percentage of another without a calculator.
- Be able to write one number as a percentage of another using a calculator.

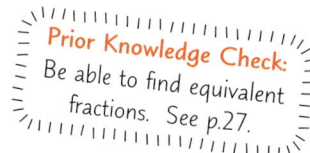

Prior Knowledge Check:
Be able to find equivalent fractions. See p.27.

Writing percentages is a good way to **compare** proportions of different amounts.
E.g. 2 out of 10 and 20 out of 100 are both represented by the **same percentage** — it's 20%.

A good method for writing one number as a percentage of another is to turn it into a fraction — the first number over the second. Then you need to find an **equivalent fraction** that has a **denominator of 100**. Once you've done this, the **numerator** of the equivalent fraction tells you the percentage.

Example 1

Express 15 as a percentage of 50.

1. Write '15 out of 50' as a fraction.

2. Write an equivalent fraction which is 'out of 100' by multiplying or dividing the top and bottom by the same number.

3. The numerator of the equivalent fraction tells you the percentage.

$$\overset{\times 2}{\frac{15}{50}} = \underset{\times 2}{\frac{30}{100}}$$

So 15 is **30%** of 50.

Tip: Sometimes the bigger number will be on top. If that's the case, you'll get a percentage greater than 100%.

Exercise 1

Q1 A chess club has 25 members. 12 of these members are female. Express the number of female members of the club as a percentage.

Q2 There are 300 counters in a bag, 45 of which are green. Express the amount of green counters as a percentage.

Q3 Write each of the following amounts as a percentage.
- a) 11 out of 25
- b) 33 out of 50
- c) 3 out of 20
- d) 100 out of 400
- e) 48 out of 32
- f) 200 out of 160

Q4 Out of 24 pupils in a class, 18 walk to school. What percentage of the class:
- a) walk to school?
- b) do not walk to school?

Q5 39 out of 65 people in a book club have blonde hair. What percentage do not have blonde hair?

If you've got a **calculator** then you don't need to find the equivalent fraction
— just **divide** the numerator by the denominator then **multiply by 100%**.

Example 2

Express 333 as a percentage of 360.

1. Write '333 out of 360' as a fraction.

    $$\frac{333}{360}$$

2. Divide the top number by the bottom number.

    $333 \div 360 = 0.925$

3. Multiply by 100% to write as a percentage.

    $0.925 \times 100\% = \mathbf{92.5\%}$

Tip: In this method we're converting from a fraction to a decimal, then to a percentage — see p.62.

Exercise 2

Q1 Write each of the following amounts as a percentage.

a) 15 out of 24

b) 221 out of 260

c) 661 out of 500

d) 258 out of 375

e) 323 out of 850

f) 301 out of 250

Q2 A school has 875 pupils. 525 are boys. What is this as a percentage?

Q3 171 out of 180 raffle tickets were sold for a summer fete. What percentage of the tickets were sold?

Q4 Write each of the following amounts as a percentage.

a) 116.6 out of 212

b) 53.5 out of 428

c) 226.8 out of 210

Q5 The jackpot for a lottery was £10 250. John won £1896.25.
What percentage of the total jackpot did he win?

Finding a Percentage of an Amount

Learning Objectives — Spec Ref N12:
- Be able to find a percentage of an amount without a calculator.
- Be able to find a percentage of an amount using a calculator.

You can find some percentages without a calculator using the following rules.

- **50%** $= \frac{1}{2}$, so find 50% of something by **dividing by 2** (which is the same as multiplying by $\frac{1}{2}$).

- **25%** $= \frac{1}{4}$, so find 25% of something by **dividing by 4** (which is the same as multiplying by $\frac{1}{4}$).

- **10%** $= \frac{1}{10}$, so find 10% of something by **dividing by 10** (which is the same as multiplying by $\frac{1}{10}$).

- **5%** $= \frac{1}{20}$, so find 5% of something by **dividing by 20** (or by dividing 10% of something by 2).

- **1%** $= \frac{1}{100}$, so find 1% of something by **dividing by 100** (which is the same as multiplying by $\frac{1}{100}$).

To find other percentages, add up combinations of the percentages above, e.g. 65% = 50% + 10% + 5%.

Example 3

Find 75% of 44.

1. First find 25% by dividing by 4.   25% of 44 = 44 ÷ 4 = 11

2. 75% = 3 × 25%, so multiply by 3.   So 75% of 44 = 3 × 11 = **33**

Tip: You could also find 75% by adding together 50%, 20%, and 5%.

Exercise 3

Q1   Find each of the following.

a) 50% of 24   b) 25% of 36   c) 10% of 270

d) 75% of 20   e) 5% of 140   f) 35% of 300

g) 65% of 120   h) 130% of 90   i) 180% of 70

Q2   Find each of the following.

a) 21% of 200   b) 3% of 260   c) 62% of 500

Q3   A wooden plank is 9 m long. 55% of the plank is cut off. What length of wood has been cut off?

To find a percentage of an amount using a calculator, change the percentage into a decimal (by dividing by 100) then **multiply** by the amount.

Example 4

Find 67% of 138.

1. Change 67% into a decimal by dividing by 100.   67% = 67 ÷ 100 = 0.67

2. Multiply the decimal by 138.   67% of 138 = 0.67 × 138 = **92.46**

Exercise 4

Q1   Find each of the following.

a) 17% of 200   b) 109% of 11   c) 68% of 320

d) 221% of 370   e) 79% of 615   f) 96% of 911

Q2   What is 12% of 68 kg?

Q3   Jeff is on a journey of 385 km. So far, he has completed 31% of his journey.
      How far has he travelled?

Q4   Which is larger, 22% of £57 or 161% of £8? By how much?

Q5   A jug can hold 2.4 litres of water. It is 34% full. How much more water will fit in the jug?

5.2 Percentages, Fractions and Decimals

Percentages, fractions and decimals are three different ways of showing a proportion of something. Being able to convert between them is crucial to understanding how closely related they are.

Converting between Percentages, Fractions and Decimals

> **Learning Objective — Spec Ref R9:**
> Be able to convert between percentages, fractions and decimals.

Prior Knowledge Check:
Be able to convert between fractions and decimals. See p.36-39.

You can switch between percentages, fractions and decimals in the following ways.

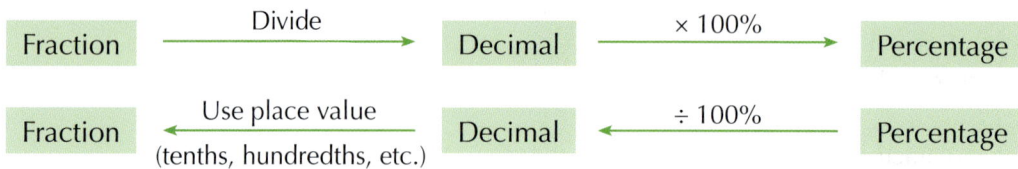

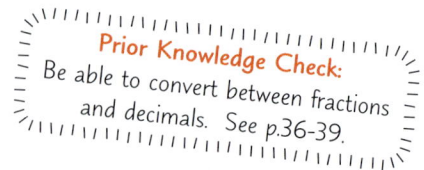

| Fraction | — Divide → | Decimal | — × 100% → | Percentage |

| Fraction | ← Use place value (tenths, hundredths, etc.) | Decimal | ← ÷ 100% | Percentage |

You can convert directly from a fraction to a percentage by finding an **equivalent fraction** with a **denominator of 100** — then the **numerator** gives the percentage.

You can also convert directly from a percentage to a fraction by writing the **percentage as the numerator** and **100 as the denominator** — then **simplify** the fraction.

Example 1

Write: a) 56% as a decimal   b) 0.24 as a fraction in its simplest terms.

1. Divide by 100% to write it as a decimal.   a) $56\% \div 100\% = \textbf{0.56}$

2. The final digit of 0.24 (the '4') is in the hundredths column, so write 0.24 as 24 hundredths, and then simplify.

   b) $0.24 = \dfrac{24}{100} = \dfrac{\textbf{6}}{\textbf{25}}$

Example 2

The probability of it raining tomorrow is 0.8. Write this probability as a percentage.

Multiply by 100% to turn the decimal into a percentage.

$0.8 \times 100\% = \textbf{80\%}$

Exercise 1

Q1 a) Find the fraction equivalent to $\dfrac{3}{20}$ which has 100 as the denominator.

   b) Write $\dfrac{3}{20}$ as: (i) a percentage   (ii) a decimal.

Q2 Write each of the following percentages as (i) a decimal, and (ii) a fraction in its simplest terms.

   a) 75%   b) 130%   c) 34%   d) 6%

Q3 Write each of the following fractions as (i) a decimal, and (ii) a percentage.

a) $\frac{3}{10}$ b) $\frac{7}{8}$ c) $\frac{1}{3}$ d) $\frac{7}{16}$

Q4 Write each of the following decimals as (i) a percentage, and (ii) a fraction in its simplest terms.

a) 0.35   b) 1.7   c) 0.6   d) 2.68

Q5 Raj answers 86% of the questions in a test correctly. Write this as a decimal.

Q6 $\frac{3}{5}$ of the pupils in a class are right-handed. What percentage of the class are right-handed?

Q7 Raphael eats 36% of a cake. Write this percentage as a fraction in its lowest terms.

Comparing Percentages, Fractions and Decimals

Learning Objective — Spec Ref R9:
Be able to compare and order percentages, fractions and decimals.

As percentages, fractions and decimals are all used to represent proportions it's useful to be able to **compare** them — you'll have to **convert** them all into the **same form** first.

Example 3

Put $\frac{7}{20}$, 33% and 0.3 in order, from smallest to largest.

Write the amounts in the same form. (Here, I've chosen to write them all as percentages.)

1. Write $\frac{7}{20}$ as a percentage by first writing it as a fraction out of 100. $\frac{7}{20} = \frac{35}{100} = 35\%$

2. Multiply 0.3 by 100% to turn it into a percentage. $0.3 \times 100\% = 30\%$

3. Put the percentages in order, from smallest to largest.   30%, 33%, 35%

4. Rewrite in their original forms.   **0.3, 33%, $\frac{7}{20}$**

Exercise 2

Q1 For each of the following pairs, write down which is larger.

a) 0.35, 32%   b) 0.58, 68%   c) 0.4, 4%   d) 0.09, 90%

e) 0.2, $\frac{21}{100}$ f) 0.6, $\frac{7}{10}$ g) 0.7, $\frac{3}{4}$ h) 0.55, $\frac{3}{5}$

Q2 Put the numbers in each of the following lists in order, from smallest to largest.

a) 0.42, 25%, $\frac{2}{5}$ b) 0.505, 45%, $\frac{1}{2}$ c) 0.37, 38%, $\frac{3}{8}$ d) 0.2, 22%, $\frac{2}{9}$

e) 0.13, 12.5%, $\frac{3}{20}$ f) 0.25, 23%, $\frac{9}{40}$ g) 0.4, 2.5%, $\frac{1}{25}$ h) 0.006, 0.06%, $\frac{3}{50}$

Q3 In a season, Team X won 14 out of the 20 matches they played. Team Y won 60% of their matches. Which team had the highest proportion of wins?

Q4 Oliver and Jen each try flicking a set of counters into a box. Oliver gets 65% of the counters into the box. Jen gets $\frac{11}{20}$ of the counters into the box. Who got more counters into the box?

Percentage, Fraction and Decimal Problems

You might need to convert between percentages, fractions and decimals before you're able to solve a problem. The first step will usually be **converting** them to the **same form**. Once you've converted them there's still work to do — don't forget to answer the question.

Example 4

$\frac{1}{4}$ of pupils in a school bring a packed lunch, 65% have school dinners, and the rest go home for lunch. **What percentage of pupils go home for lunch?**

1. Write $\frac{1}{4}$ as a percentage by writing it as a fraction out of 100.

    $\frac{1}{4} = \frac{25}{100} = 25\%$

2. Find the percentage that don't go home for lunch by adding the percentages for 'packed lunches' and 'school dinners'.

    $25\% + 65\% = 90\%$

3. Subtract this from 100% to find the percentage who do go home for lunch.

    $100\% - 90\% = \mathbf{10\%}$

Tip: Think about your final answer when deciding how to convert. If you need to find a percentage, it's usually best to convert everything to percentages.

Exercise 3

Q1 In a school survey of food preferences, 50% of pupils said they prefer pizza, $\frac{1}{5}$ said they prefer shepherd's pie and the rest said they prefer roast chicken.
Find the percentage that prefer roast chicken.

Q2 In a car park, $\frac{2}{5}$ of the cars are red, 0.12 of the total number are white, 15% are blue, and the rest are black.
Find the percentage of cars in the car park that are black.

Q3 $\frac{3}{4}$ of the people at a concert arrived by train, 5% walked, and the rest came by car.
What percentage came by car?

Q4 $\frac{1}{4}$ of Hattie's jackets are leather, $\frac{1}{5}$ are denim, 30% are suede, and the rest are corduroy.
Find the percentage of Hattie's jackets that are corduroy.

Q5 Ainslie is keeping a record of the birds in his garden. Of the birds he has seen this month, $\frac{3}{8}$ were sparrows, 41.5% were blackbirds, and the rest were robins.
What percentage were robins?

Q6 In a school, the ratio of girls to boys is $3:7$. 25% of the girls and $\frac{1}{5}$ of the boys are part of a sports club. In total what percentage of the school pupils are part of a sports club?

5.3 Percentage Increase and Decrease

Percentages are often used to describe the change in an amount. The change could be an increase or a decrease so make sure you know which one you're trying to find before tackling a question.

Calculating Amounts after a Percentage Increase or Decrease

Learning Objective — Spec Ref R9:
Be able to find amounts after a percentage increase or decrease.

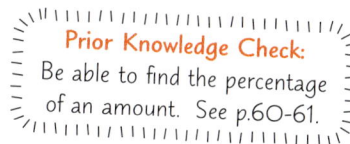

Prior Knowledge Check:
Be able to find the percentage of an amount. See p.60-61.

One method for increasing or decreasing by a percentage is to just work out the percentage of the amount and then:

- **add it** to the original amount if it's a **percentage increase**.
- **subtract it** from the original amount if it's a **percentage decrease**.

Example 1

Increase 450 by 15% without using your calculator.

1. Find 10% and 5% of 450.   10% of 450 = 450 ÷ 10 = 45
                                      5% of 450 = 45 ÷ 2 = 22.5

2. Add these to find 15% of 450.   So 15% of 450 = 45 + 22.5 = 67.5

3. Add this to the original amount.   450 + 67.5 = **517.5**

Exercise 1

Q1 Increase each of the following amounts by the percentage given.
   a) 90 by 10%   b) 60 by 25%   c) 80 by 75%   d) 270 by 20%

Q2 Increase each of the following amounts by the percentage given.
   a) 110 by 60%   b) 480 by 115%   c) 140 by 45%   d) 100 by 185%

Q3 Decrease each of the following amounts by the percentage given.
   a) 55 by 10%   b) 48 by 75%   c) 25 by 30%   d) 120 by 60%

Q4 Decrease each of the following amounts by the percentage given.
   a) 125 by 40%   b) 11 by 70%   c) 150 by 55%   d) 520 by 15%

Q5 A farmer has 380 acres of land. He sells 35% of his land. How much does he have left?

Q6 A population decreases by 15% from 2400. What is the new population?

Q7 A TV costs £485, plus 20% VAT. Find the total cost of the TV after the VAT is added.

Q8 Roberto's weekly wage of £400 is decreased by 5%. Mary's weekly wage of £350 is increased by 10%. Who now earns more? By how much?

You can also calculate a percentage increase or decrease in one go using a **multiplier**. To find the multiplier, turn the percentage into a **decimal** and **add it to 1** for an **increase** or **subtract it from 1** for a **decrease**. Then just **multiply** the original amount by your multiplier to get the answer.

Example 2

Fabian deposits £150 into an account which pays 7% interest per year.
How much will be in the account after one year?

1. Turn the percentage into a decimal. $7\% = 7 \div 100 = 0.07$

2. It's a percentage increase, so add it to 1. Multiplier $= 1 + 0.07 = 1.07$

3. Multiply the amount by the multiplier. £150 × 1.07 = **£160.50**

> **Tip:** A percentage increase has a multiplier greater than 1. A percentage decrease has a multiplier less than 1.

Example 3

A medium box of frosted flakes contains 500 g of cereal.
A small box of frosted flakes contains 45% less cereal than the medium box.
How many grams of cereal are in a small box?

1. Turn the percentage into a decimal. $45\% = 45 \div 100 = 0.45$

2. It's a percentage decrease, so subtract it from 1. Multiplier $= 1 - 0.45 = 0.55$

3. Multiply the amount by the multiplier. 500 g × 0.55 = **275 g**

Exercise 2

Use the multiplier method for the following questions.

Q1 Increase each of the following amounts by the percentage given.

a) 490 by 11% b) 101 by 16% c) 55 by 137% d) 89 by 61%

e) 139 by 28% f) 426 by 134% g) 854 by 89% h) 761 by 77%

Q2 Decrease each of the following amounts by the percentage given.

a) 77 by 8% b) 36 by 21% c) 82 by 13% d) 101 by 43%

e) 189 by 38% f) 313 by 62% g) 645 by 69% h) 843 by 91%

Q3 David's height increased by 20% between the ages of 6 and 10. He was 50 inches tall at age 6. How tall was he at age 10?

Q4 2 years ago, Alison earned £31 000 per year. Last year, she got a pay rise of 3%. This year, she got a pay cut of 2%. How much does she now earn per year?

Q5 A year ago, University A had 24 500 students and University B had 22 500 students. Over the last year, the number of students increased by 2% at University A, and by 9% at University B. Which University has more students now? By how much?

Q6 The population of Barton is 152 243, and increases by 12% each year. The population of Meristock is 210 059, and decreases by 8% each year. Which town has the larger population after one year? By how much? Give your answer to the nearest whole number.

Simple Interest

Learning Objective — Spec Ref R9:
Be able to solve problems involving simple interest.

Interest is a percentage of money that is added on to an initial figure over a period of time. Simple interest is when a certain percentage of the **original amount** is paid at regular intervals (usually once per year). Because the interest paid is always based on the original amount, and the original amount **never changes**, the amount of interest is the **same** every time it's paid.

Example 4

Kim invests £220 into an account which pays 1% simple interest per year. How much will be in the account after seven years?

1.	Start off by finding 1% of £220.	1% of 220 = 220 ÷ 100 = £2.20
2.	Multiply the interest each year by the number of years.	£2.20 × 7 = £15.40
3.	Add the total amount of interest to the original amount.	£220 + £15.40 = **£235.40**

Exercise 3

Q1 How much money would be in a savings account that pays simple interest if:

   a) £500 was invested for 2 years in an account which pays 10% interest each year.

   b) £900 was invested for 6 years in an account which pays 5% interest each year.

Q2 Gerry puts £5500 into an account that pays 6% simple interest each year.
How much will be in his account after: a) 3 years?   b) 10 years?

Q3 Raj invests £2300 into a bank account which pays simple interest. After 4 years there was £2852 in the account. What was the yearly interest rate on Raj's savings account?

Q4 Padma has forgotten how much money she invested into her savings account which pays simple interest. She knows that after 5 years there was £1440 and after 9 years there was £1632. How much did she originally invest into the account?

Finding the Original Value

Learning Objective — Spec Ref R9:
Be able to find the original value after a percentage increase or decrease.

You can find the **original value** after a percentage increase or decrease by following these steps:

1. Write the new amount as a percentage of the original value.

2. Divide the new amount by the percentage to find 1% of the original amount.

3. Multiply by 100 to give the original value (100%).

Penny donates 25% of her comic books to charity. She has 63 comic books left. How many comic books did she have before she donated any to charity?

1. Write 63 as a percentage of the original amount.	100% − 25% = 75% So 63 = 75% of the original amount
2. Divide 63 by 75 to work out 1% of the original amount.	1% of the original amount = 63 ÷ 75 = 0.84
3. Multiply by 100 to work out 100% of the original amount.	100% of the original amount = 0.84 × 100 = **84 comic books**

Exercise 4

Q1 A fridge costs £200 after a 50% reduction. Calculate the original price of the fridge.

Q2 Andy buys a top hat that has been reduced in a sale by 35%. If the sale price is £13.00, find the original price.

Q3 Find the original value of y if:

a) it is increased by 20% to give 24

b) it is decreased by 40% to give 30

c) it is decreased by 70% to give 99

d) it is increased by 180% to give 84

Q4 In the past year, the number of frogs living in a pond has increased by 10% to 528, and the number of newts living there has increased by 15% to 621. How many frogs and how many newts lived in the pond a year ago?

Q5 Monib is a stamp collector. The number of stamps in his collection increased by 5% in January and by 20% in February. How many stamps did Monib have at the beginning of January if he had 2268 at the beginning of March?

Finding a Change as a Percentage

Learning Objective — Spec Ref R9:
Be able to express the change in an amount as a percentage.

On p.65, you saw how to calculate a new amount after a percentage increase or decrease. Now we'll cover how to find the actual **percentage change** when you know the **original amount** and the **new amount**. To find the change in an amount as a percentage:

1. Calculate the **difference** between the new amount and the original amount.

2. Divide the difference by the **original amount** and multiply by 100%.

You can use this formula to help:

$$\text{Percentage Change} = \frac{\text{Difference in amounts}}{\text{Original amount}} \times 100\%$$

Tip: Percentage changes can come in different forms. You might see them given as percentage profits, discounts, errors, etc.

Example 6

A house price increases from £145 000 to £187 050. Find the percentage increase.

1. Calculate the difference.   187 050 − 145 000 = 42 050

2. Divide the difference
   by the original amount.   42 050 ÷ 145 000 = 0.29

3. Multiply by 100% to   0.29 × 100% = 29%
   write as a percentage.   So it is a **29%** increase.

> **Tip:** Always remember to divide by the original amount, not the new amount.

Example 7

In an experiment, the mass of a chemical drops from 75 g to 69 g. Find the percentage decrease.

1. Calculate the difference.   75 − 69 = 6 g

2. Divide the difference by the original amount.   6 ÷ 75 = 0.08

3. Multiply by 100% to write as a percentage.   0.08 × 100% = 8%
                                                  So it is an **8%** decrease.

Exercise 5

Q1   Find the percentage increase when:

   a) a price of £10 is increased to £12.   b) a price of £20 is increased to £52.

Q2   Find the percentage decrease when:

   a) a price of £10 is decreased to £8.   b) a price of £25 is decreased to £22.

Q3   The number of people working for a company increases from 45 to 72.
     Find the percentage increase in the number of people working for the company.

Q4   The price of a local newspaper increases from 80p to £1. Find the percentage increase.

Q5   Percy is on a healthy eating plan. His weight drops from 80 kg to 68 kg.
     Find the percentage decrease in Percy's weight.

Q6   In a sale, the price of a toaster is discounted from £50 to £30. Find the percentage discount.

Q7   What is the percentage error when 250 is rounded to the nearest 100?

Q8   Izzy buys an antique lamp for £450 and sells it for £315.
     She also buys an antique wardrobe for £980 and sells it for £1127.

   a) What was her percentage loss on the antique lamp?

   b) What was her percentage profit on the antique wardrobe?

   c) Overall Izzy made a profit on the two items.
      Calculate the percentage profit, giving your answer to 2 decimal places.

5.4 Compound Percentage Change

Compound percentage changes involve repeating a percentage increase or decrease. Each time you calculate the percentage change you apply it to the current amount, rather than the original amount.

Compound Growth

Learning Objective — Spec Ref R16:
Be able to solve compound growth problems.

Compound growth is when a quantity gets larger over time due to **successive percentage increases**. The percentage increases always use the **new value**, e.g. if £1000 is increased by 10% each year:

After 1 year: £1000 × 1.1 = £1100
After 2 years: £1100 × 1.1 = £1210 (which is the same as **£1000 × 1.1²**)
After 3 years: £1210 × 1.1 = £1331 (which is the same as **£1000 × 1.1³**)

So the formula for **compound growth** is: $P_n = P_0 \times \left(1 + \dfrac{r}{100}\right)^n$

Tip: The bit of the formula in brackets is just the percentage multiplier — see p.66.

P_n = amount after n periods, P_0 = initial amount,
n = number of periods, r = percentage rate of change

When applied to money, compound growth is know as **compound interest**.

Example 1

The number of bacteria in a sample increases at a rate of 60% each week. At the start of an experiment, a scientist placed 400 bacteria in a jar. How many bacteria will there be after 6 weeks?

1. Use the formula for compound growth: $P_n = P_0 \times \left(1 + \dfrac{r}{100}\right)^n$

2. Plug in the numbers...
$$P_0 = 400, \ r = 60, \ n = 6$$
$$P_n = 400 \times \left(1 + \dfrac{60}{100}\right)^6$$

Tip: Always check the context before giving your final answer. Here it makes sense to round it to a whole number.

3. Round the answer to an appropriate degree of accuracy.
$$= 400 \times 16.777\ldots\ldots$$
$$= 6710.88\ldots = \textbf{6711 bacteria}$$

Exercise 1

Q1   Calculate how much money you will have after 4 years if you invest £680 at 2.5% annual compound interest.

Q2   The population of the UK in 2017 was 66 000 000. The population is increasing by 0.6% every year. Assuming that this rate of growth continues, how many people will reside in the UK by 2025? Give your answer to two significant figures.

Q3   Find the value of a bar of gold, initially valued at £750, after seven months when gold prices are rising at a rate of 0.1% per month.

Q4   A colony of ants has set up home in Mr Murphy's shed. On Monday there were 250 ants. If the colony of ants grows at a rate of 5% per day, how many ants will there be on Saturday?

Q5 Mrs Honeybun is expecting twins. Every month her waistline increases by 5%. If her waist measurement started at 70 cm, what is her waistline, to the nearest cm, after nine months?

Q6 In 2018 a group called 'Smiley Faced People' decide to spread a little happiness around the world. The original group has 20 members. If their numbers increase by 70% every year as more and more people decide to join their happy band, how many people to the nearest 100 will be in the group in 2028?

Q7 Use the formula for compound growth to calculate the interest earned by the following investments:
a) £750 for 5 years at an annual rate of 3%.   b) £50 for 7 years at an annual rate of 5.5%.

Q8 During 2013 and 2014 a bank paid interest at a compound rate of 2% per annum. During 2015, 2016 and 2017 this rate rose to 3%. Calculate the total interest paid over five years if £650 was invested at the beginning of the year 2013.

Compound Decay

Learning Objective — Spec Ref R16:
Be able to solve compound decay problems.

Compound decay is the opposite of compound growth — it's when an amount gets smaller over time due to successive percentage decreases. When applied to money, compound decay is known as **depreciation**.

The formula for **compound decay** is the same as for growth except that there's a minus sign in the bracket so that the multiplier is less than 1:

$$P_n = P_0 \times \left(1 - \frac{r}{100}\right)^n$$

Example 2

A car was bought for £8000. Its value depreciates by 20% each year. What is the value of the car after 10 years to 2 significant figures?

1. Use the formula for compound decay: $P_n = P_0 \times \left(1 - \frac{r}{100}\right)^n$

2. Plug in the numbers... $P_0 = £8000, \ r = 20, \ n = 10$

3. Round the answer to 2 significant figures. $P_n = 8000 \times \left(1 - \frac{20}{100}\right)^{10}$
   $= 8000 \times 0.1073... = 858.9934... = $ **£860** (2 s.f.)

Exercise 2

Q1 Find, to the nearest £100, the value of a car that originally cost £10 000 after five years of depreciation at 15%.

Q2 Find, to the nearest £, the depreciation on £550 after 3 years at a depreciation rate of 3% per annum.

Q3 Mr Butterworth is on a diet and is losing weight at a rate of 2% of his total body weight every week. He weighed 110 kg when he started his diet. What is his weight, to the nearest kg, after 8 weeks?

Q4 Calculate the value of a car after seven years if it was bought for £6000 and depreciation is at a rate of 7.5% per annum. Give your answer to 2 significant figures.

Q5 The activity of a radioactive source decreases by 6% every hour. If the initial activity is 1200 Bq, calculate the activity after 10 hours to the nearest whole number.

For Questions 6-9, give your answers to 3 significant figures.

Q6 Calculate the depreciation of a house after five years if it was bought for £650 000 and house prices are falling at a rate of 2.5% per annum.

Q7 Claire the Evil Genius buys a laser for £68 000. It depreciates at 20% per annum for the first two years and at 15% per annum for the next three years. What is the value of the laser after 5 years?

Q8 Find the final value of an investment of £3500 that grows by 0.75% per annum for two years but then depreciates at a rate of 1.25% per annum for the next seven years

Q9 Find the final value of an investment of £1000 that grows by 5% per annum for the first two years but then depreciates at a rate of 2% per annum for the next three years.

More Compound Change Problems

Learning Objective — Spec Ref R16:
Find other unknowns in compound change scenarios such as the time period.

In some compound growth or decay questions, it isn't the amount after n periods that you're trying to find. You might have to **rearrange** the formulas or use **trial and improvement** to find a different value.

Example 3

A motorbike valued at £2500 depreciates at a rate of 5% per month.
After how many whole months will the value of the motorbike be less than £1150?

1. Using the compound decay formula, you know the values of P_n, P_0 and r, and you want to find the value of n:

$$P_n = P_0 \times \left(1 - \frac{r}{100}\right)^n \qquad P_n = £1150, \quad P_0 = £2500, \quad r = 5$$

$$1150 = 2500 \times \left(1 - \frac{5}{100}\right)^n = 2500 \times 0.95^n$$

2. Use trial and improvement to find the value of n.

Try $n = 10$: $2500 \times 0.95^{10} = 1496.84...$ n is too small
Try $n = 20$: $2500 \times 0.95^{20} = 896.21...$ n is too big
Try $n = 15$: $2500 \times 0.95^{15} = 1158.22...$ n is too small
Try $n = 16$: $2500 \times 0.95^{16} = 1100.31...$ correct

$n = $ **16 months**

Exercise 3

Q1 A beehive population is decreasing at a rate of 4.9% per year. If there are initially 500 bees in the hive, after how many whole years will the beehive population have halved?

Q2 Mr Quasar is visiting Las Vegas and has taken $1000 to gamble in the casino. He loses money at a rate of 15% a day. After how many whole days will Mr Quasar have less than $100 left?

Q3 The number of foxes in an area is falling by 12% per year. If there are 300 foxes in 2018, after how many whole years will the number have fallen below 200?

Q4 A biologist has a colony of 500 bacteria. When the population of bacteria grows over 1000, the biologist can start doing some tests on the colony. If the growth in the number of bacteria is at a rate of 15% per hour, after how many whole hours can the biologist start the tests?

Review Exercise

Q1 Write each of the following amounts as a percentage.

    a) 13 out of 50   b) 26 out of 40   c) 48 out of 120

Q2 Sandra gets paid £1385 per month. In December she gets a Christmas bonus of £83.10. What percentage of her monthly salary is this bonus?

Q3 Sarah has won 65% of her last 60 badminton games. How many games has she won?

Q4 Find each of the following:

    a) 3% of 210   b) 41% of 180   c) 4.5% of 900

Q5 For each of the following lists, write down which amount is not equal to the others.

    a) 0.5, 20%, $\frac{1}{2}$ b) 0.125, 1.25%, $\frac{1}{8}$ c) 0.22, 44%, $\frac{22}{50}$

Q6 Kelly and Nasir both had maths tests last week.
Kelly scored $\frac{47}{68}$ and Nasir scored $\frac{35}{52}$. Who got the higher percentage score?

Q7 The insurance for a car normally costs £356. With a no-claims discount, the cost is reduced by 27%. What is the reduced cost of the insurance?

Q8 House A is valued at £420 000 and House B is valued at £340 000.
After 5 years the value of House A has decreased by 15% and the value of House B has increased by 5%. What is the difference in value of the two houses after 5 years?

Q9 Ami deposits £650 into a savings account that pays 4% simple interest each year. How much will be in her savings account after 7 years?

Q10 Dave loves a bargain and buys a feather boa which has been reduced in price by 70%. If the sale price is £2.85, what was the original price of the boa?

Q11 A cafe sells 270 ice cream sundaes in April and 464 in May. Find the percentage increase in ice cream sales.

Q12 Murray invests £575 in a bank account that pays annual compound interest of 7.5%. How much money will he have in the bank after 3 years?

Q13 Find the final value of an investment of £2500 that earns 0.5% per month for the first six months but then depreciates at a rate of 0.25% per month for the next 9 months.

Q14 A balloon is deflating. Its volume decreases at a rate of 20% per hour. The balloon initially had a volume of 1500 cm³. How many whole hours will it be before the balloon's volume is less than 400 cm³?

Exam-Style Questions

Q1 A school has 1200 pupils. 50% of the pupils get the bus to school, 20% walk and 5% get to school by bicycle. The remaining pupils are driven to school by their parents. How many pupils are driven to school by their parents?

[2 marks]

Q2 In a sofa store 30% of the sofas are leather. 40% of the leather sofas are black. What percentage of the total number of sofas are made from black leather?

[2 marks]

Q3 Inflation is the percentage by which prices increase.
One year, the rate of inflation in England was 3.2%.

a) An item was worth £50 at the start of the year. If its value increased at the rate of inflation, how much was it worth at the end of the year?

[2 marks]

Venezuela is a South American country whose currency is the bolivar (Bs).
In that same year, the rate of inflation in Venezuela was 108%.

b) An item was worth 50Bs at the start of the year. If its value increased at the rate of inflation, how much was it worth at the end of the year?

[2 marks]

Q4 Gareth won the lottery and invested it all into a savings account at the start of 2014.
There was £4 862 025 in the account at the start of 2017. If compound interest is added to his account at a rate of 5% per annum, how much did he have in his account at the start of 2014?

[2 marks]

Q5 A type of new car loses 20% of its value in its first year.
It loses 15% per year in every year after the first. Graham buys one of these new cars.
Work out the percentage of its value that it has lost after five years.
Give your answer to the nearest whole number.

[4 marks]

6.1 Simplifying Expressions

Algebraic expressions involve variables (letters that represent numbers) and don't contain an equals sign — e.g. x, 3a + b, y² + z². You can simplify them to avoid writing the same thing over and over.

Learning Objective — Spec Ref A3/A4:
Simplify algebraic expressions by collecting like terms.

Terms are the individual parts of an expression separated by plus or minus signs. E.g. in the expression $2x + 6 - 3xy$, the terms are $2x$, 6 and $3xy$.

Expressions can sometimes be **simplified** by **collecting like terms** — these are terms that contain **exactly the same** combination of letters. For example, $4xy + xy - 3xy = 2xy$.

> **Tip:** Remember that in algebraic notation, $3xy$ means $3 \times x \times y$ — the × signs are left out to make it a bit clearer.

Example 1

Simplify the expression $x + x^2 + yx + 7 + 4x + 2xy - 3$

1. There are four sets of like terms:
   (i) terms involving just x
   (ii) terms involving x^2
   (iii) terms involving xy (or yx, which means the same)
   (iv) terms involving just numbers
2. Collect the different sets together separately.

$$x + x^2 + yx + 7 + 4x + 2xy - 3$$
$$= (x + 4x)$$
$$+ x^2$$
$$+ (xy + 2xy)$$
$$+ (7 - 3)$$
$$= 5x + x^2 + 3xy + 4$$

Exercise 1

Q1 Simplify these expressions by collecting like terms.

   a) $c + c + c + d + d$
   b) $x + y + x + y + x - y$
   c) $3m + m + 2n$

   d) $3a + 5b + 8a + 2b$
   e) $6p + q + p + 3q$
   f) $4b + 8c - b - 5c$

   g) $x + 7 + 4x + y + 5$
   h) $13m + 7 + 2n - 8m - 3$
   i) $13a - 5b + 8a + 12b + 7$

Q2 Simplify the following expressions by collecting like terms.

   a) $x^2 + 3x + 2 + 2x + 3$
   b) $x^2 + 4x + x^2 + 2x + 4$
   c) $x^2 + 2x^2 + 4x - 3x$

   d) $3p^2 + 6q + p^2 - 4q + 3p^2$
   e) $8 + 6p^2 - 5 + pq + p^2$
   f) $6b^2 + 7b + 9 - 4b^2 + 5b - 2$

Q3 Simplify the following expressions by collecting like terms.

   a) $ab + cd - xy + 3ab - 2cd + 3yx + 2x^2$
   b) $pq + 3pq + p^2 - 2qp + q^2$

   c) $3ab - 2b + ab + b^2 + 5b$
   d) $4abc - 3bc + 2ab + b^2 + 5b + 2abc$

Multiplying and Dividing Variables

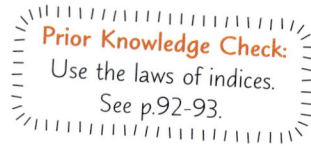

Learning Objectives — Spec Ref A1/A4:
- Use and interpret algebraic notation.
- Use the laws of indices when multiplying and dividing variables.

Prior Knowledge Check:
Use the laws of indices.
See p.92-93.

When you've got numbers and variables **multiplied** or **divided** by each other, you should deal with the numbers and each letter **separately**, using the **laws of indices** to write each term as **simply** as possible.

The most useful laws of indices are: $a^m \times a^n = a^{m+n}$ $(a^m)^n = a^{m \times n}$ $a^m \div a^n = a^{m-n}$

There are some more rules that might help you with these questions:

- $ab^2 = a \times b \times b$ — only the b is squared.
- $(ab)^2 = ab \times ab = a \times a \times b \times b = a^2b^2$ — the whole bracket is squared.
- $2ab \times 3a = 2 \times 3 \times a \times a \times b = 6a^2b$ — multiply the numbers and variables separately.
- $6a^5 \div 3a^3 = (6 \div 3)(a^5 \div a^3) = 2a^2$ — divide the numbers and variables separately.
- $(-a)^2 = (-a) \times (-a) = (-1) \times (-1) \times a \times a = a^2$ — squaring a negative makes it positive.

Example 2

Simplify: a) $b \times b \times b \times b$ b) $4a \times 5b$ c) $12x^3 \div 2x^2$ d) $(ab^2)^2$

a) If a variable is multiplied by itself, write it as a power. $b \times b \times b \times b = \textbf{\textit{b}}^\textbf{4}$

b) Multiply the numbers and variables separately. $4a \times 5b = 4 \times 5 \times a \times b = \textbf{20\textit{ab}}$

c) Divide the numbers and letters separately. $12x^3 \div 2x^2 = (12 \div 2)(x^3 \div x^2) = \textbf{6\textit{x}}$

d) Square the bracket and simplify the a's and the b's. $(ab^2)^2 = ab^2 \times ab^2 = a \times a \times b^2 \times b^2 = \textbf{\textit{a}}^\textbf{2}\textbf{\textit{b}}^\textbf{4}$

Exercise 2

Q1 Simplify the following expressions.

a) $x \times x \times x \times x$ b) $2y \times 3y$ c) $8p \times 2q$ d) $3a \times 7a$

e) $5x \times 3y$ f) $m \times m \times m \times m$ g) $12a \times 4b$ h) $6p \times 8p$

Q2 Simplify the following expressions.

a) $p \times pq$ b) $4a^2 \times 5a$ c) $4ab \times 2ab$ d) $3i^2 \times 8j^3$

e) $9n^2m \div 3n^2$ f) $12a^2 \div 4a$ g) $16p^3q \div 2p^2$ h) $6abc \times 5a^2b^3c^4$

Q3 Expand the brackets in these expressions.

a) $(2y)^2$ b) $(r^2)^2$ c) $(3c)^3$ d) $(2z^2)^3$

e) $a(2b)^2$ f) $5(q^2)^2$ g) $(xy)^2$ h) $i(jk)^2$

6.2 Expanding Brackets

Most of the time, when brackets show up, you want to expand them to get rid of them. Single brackets are pretty straightforward, but when you have two or three sets of brackets there's a lot to keep track of.

a(b + c)

Learning Objective — Spec Ref A4:
Be able to expand a single term multiplied by a bracket.

You can **expand** (or remove) brackets by multiplying everything **inside** the brackets by the letter or number **in front** — remember $a(b + c) = a \times (b + c) = (a \times b) + (a \times c)$.

$$a(b + c) = ab + ac$$
$$a(b - c) = ab - ac$$

Example 1

Expand the brackets in these expressions: a) $8(r - 3s)$ b) $m(n + 7)$ c) $-(q - 4)$

a)   Multiply each term in the brackets by 8. $8(r - 3s) = (8 \times r) - (8 \times 3s) = \mathbf{8r - 24s}$

b)   Multiply each term in the brackets by m. $m(n + 7) = (m \times n) + (m \times 7) = \mathbf{mn + 7m}$

c)   You can think of $-(q - 4)$ as $-1 \times (q - 4)$. $\begin{aligned} -(q - 4) &= (-1 \times q) - (-1 \times 4) \\ &= (-q) - (-4) \\ &= \mathbf{-q + 4} \end{aligned}$
A minus sign outside the brackets reverses
the sign of everything inside the brackets.

Example 2

Simplify the expression $3(x + 2) - 5(2x + 1)$.

1.   Multiply out both sets of brackets. $3(x + 2) - 5(2x + 1) = (3x + 6) - (10x + 5)$

$$= 3x + 6 - 10x - 5$$

2.   Collect like terms. $= (3x - 10x) + (6 - 5) = \mathbf{-7x + 1}$

Exercise 1

Q1   Expand the brackets in these expressions.
   a) $2(a + 5)$ b) $p(q + 2)$ c) $6(5 - r)$ d) $t(14 - t)$
   e) $3(u + 8v)$ f) $6(4x + 5y)$ g) $p(3q - 8)$ h) $r(5 - s)$

Q2   Expand the brackets in these expressions.
   a) $-(x + 2)$ b) $-(n - 11)$ c) $-y(4 + y)$ d) $-v(v - 5)$
   e) $-6(5g - 3)$ f) $-8(7 - w)$ g) $-4y(2y + 6)$ h) $-4p(u - 7)$

Q3   Simplify the following expressions.
   a) $2(z + 3) + 4(z + 2)$ b) $4(10b - 5) + (b - 2)$ c) $11(5x - 3) - (x + 2)$
   d) $5(2q + 5) - 2(q - 2)$ e) $4p(3p + 5) - 3(p + 1)$ f) $4(t + 1) - 7t(8t - 11)$

(a + b)(c + d)

When expanding **pairs** of brackets, multiply each term in the left bracket by each term in the right bracket. If each bracket contains two terms (brackets like this are called **binomials**), you can use **FOIL** to keep track of which terms you need to multiply:

F IRST — multiply the first term from each bracket
O UTSIDE — multiply the terms on the outside
I NSIDE — multiply the terms on the inside
L AST — multiply the last term from each bracket

$$(a + b)(c + d) = ac + ad + bc + bd$$

This method will give you **four terms**, but sometimes you'll be able to simplify by **collecting like terms**.

Example 3

Expand the brackets in the expression: a) $(q + 4)(p + 3)$ b) $3(x - 2)^2$

a) Multiply each term in the left bracket by each term in the right bracket using FOIL.

$$(q + 4)(p + 3) = (q \times p) + (q \times 3)$$
$$+ (4 \times p) + (4 \times 3)$$
$$= pq + 3q + 4p + 12$$

> **Tip:** Letters are typically written in alphabetical order — so put pq here, not qp.

b)
1. Write out $(x - 2)^2$ as two brackets. $3(x - 2)^2 = 3 \times (x - 2)(x - 2)$
2. Use FOIL to expand the brackets — leave the '3 ×' alone for now. $= 3 \times (x^2 - 2x - 2x + 4)$
   $= 3 \times (x^2 - 4x + 4)$
3. Collect like terms, then multiply each term in the brackets by 3. $= 3x^2 - 12x + 12$

> **Tip:** Be extra careful with the minus signs here — remember that $(-2) \times (-2) = 4$, not -4.

Exercise 2

Q1 Expand the brackets in the following expressions.

a) $(a + 2)(b + 3)$ b) $(j + 4)(k - 5)$ c) $3(j - 2)(k + 4)$ d) $(x + 6)(y + 2)$

e) $(x - 4)(y - 1)$ f) $8(g + 5)(h + 9)$ g) $2(w - 6)(z - 8)$ h) $7(a - 7)(b - 8)$

Q2 Expand the brackets in the following expressions. Simplify your answer.

a) $(x + 8)(x + 3)$ b) $(b + 2)(b - 4)$ c) $(a - 1)(a + 2)$ d) $(d + 7)(d + 6)$

e) $(c + 5)(3 - c)$ f) $(y - 8)(6 - y)$ g) $2(x + 2)(x + 1)$ h) $(z - 12)(z + 9)$

i) $3(y + 2)(3 - y)$ j) $4(b - 3)(b + 2)$ k) $6(x - 2)(x - 4)$ l) $12(a + 9)(a - 8)$

Q3 Expand and simplify the following expressions.

a) $(x + 1)^2$ b) $(x + 4)^2$ c) $(x - 2)^2$ d) $(x + 5)^2$

e) $(x - 3)^2$ f) $(x - 6)^2$ g) $4(x + 1)^2$ h) $2(x + 5)^2$

i) $3(x - 2)^2$ j) $2(x + 6)^2$ k) $5(x - 3)^2$ l) $2(x - 4)^2$

$(a + b)(c + d)(e + f)$

Learning Objective — Spec Ref A4:
Be able to expand three brackets multiplied together.

When expanding triple brackets, do it in **stages** — first multiply two brackets together, then multiply the result by the remaining bracket (it **doesn't matter** what **order** you multiply the brackets in). You can't use FOIL if you have **more than 2 terms** in a bracket, so in this case just multiply every term in one bracket by every term in the other bracket.

$$(a + b)(c + d)(e + f) = (a + b)(ce + cf + de + df)$$
$$= a(ce + cf + de + df) + b(ce + cf + de + df)$$
$$= ace + acf + ade + adf + bce + bcf + bde + bdf$$

Example 4

Expand and simplify the following expressions: a) $(r + 2)(s + 1)(t + 4)$ b) $5(y - 1)^3$

a)   1.   Multiply each term in the second bracket by each term in the third bracket.

      2.   Multiply the result by each term in the first bracket — simplify if possible.

$$(r + 2)(s + 1)(t + 4) = (r + 2)(st + 4s + t + 4)$$
$$= r(st + 4s + t + 4) + 2(st + 4s + t + 4)$$
$$= rst + 4rs + rt + 4r + 2st + 8s + 2t + 8$$

b)   1.   Write out $(y - 1)^3$ as $(y - 1)(y - 1)(y - 1)$.

      2.   Multiply each term in the second bracket by each term in the third bracket.

      3.   Collect like terms before multiplying the result by each term in the first bracket.

      4.   Collect like terms before multiplying everything in the bracket by 5.

$$5(y - 1)^3 = 5 \times (y - 1)(y - 1)(y - 1)$$
$$= 5 \times (y - 1)(y^2 - y - y + 1)$$
$$= 5 \times (y - 1)(y^2 - 2y + 1)$$
$$= 5 \times [y(y^2 - 2y + 1) - (y^2 - 2y + 1)]$$
$$= 5 \times (y^3 - 2y^2 + y - y^2 + 2y - 1)$$
$$= 5 \times (y^3 - 3y^2 + 3y - 1)$$
$$= 5y^3 - 15y^2 + 15y - 5$$

Exercise 3

Q1   Expand the brackets in the following expressions. Simplify where possible.

     a) $(a + 1)(b + 1)(c + 2)$ b) $(m - 5)(n - 1)(p - 3)$ c) $(-3 + z)(5 - 2a)(b + 3)$

     d) $(x + 3)(y + 1)(y + 2)$ e) $(y - 4)(y - 6)(y - 3)$ f) $(1 - 3z)(z + 5)(z - 3)$

     g) $2(z + 3)(z + 2)(z + 1)$ h) $4(w - 4)(w - 5)(w - 2)$ i) $3(3 - 2q)(3q + 3)(q - 2)$

Q2   Expand and simplify the following expressions.

     a) $(x + 3)^3$ b) $(x - 2)^3$ c) $(x + 4)^3$

     d) $(3 - x)^3$ e) $3(x - 1)^3$ f) $2(x + 5)^3$

     g) $-(5 - 2x)^3$ h) $-4(3 - 3x)^3$ i) $\frac{1}{4}(2x + 4)^3$

6.3 Factorising — Common Factors

Factorising an expression means adding brackets in where there weren't any before.
It's called factorising because you need to look for common factors of all the different terms.

Learning Objective — Spec Ref A4:
Factorise expressions by taking out common factors.

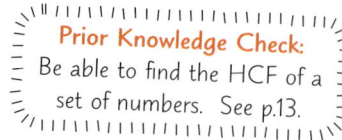

Prior Knowledge Check:
Be able to find the HCF of a set of numbers. See p.13.

Factorising is the opposite of expanding brackets. You look for a **common factor** of all the terms in an expression, and 'take it outside' the brackets. These common factors could be **numbers**, **variables**, or **both**. To factorise an expression **fully**, you need to find the **highest common factor** of all the terms.

When factorising **variables**, you'll need to remember how to **divide** two powers: $a^m \div a^n = \dfrac{a^m}{a^n} = a^{m-n}$

Example 1

Factorise the expression $12x - 18y$.

1. 6 is the highest common factor of $12x$ and $18y$. So 6 goes outside the brackets.

   $12x - 18y = 6(\quad - \quad)$

2. Divide each term by the common factor, and write the results inside the brackets.

   $12x \div 6 = 2x$
   and $18y \div 6 = 3y$
   So $12x - 18y = \mathbf{6(2x - 3y)}$

Tip: If you used 2 or 3 as the factor instead of the HCF of 6, you would get an expression that wasn't fully factorised, e.g. $2(6x - 9y)$.

Example 2

Factorise $3x^2 + 2x$.

1. x is the only common factor of $3x^2$ and $2x$. So x goes outside the brackets.

   $3x^2 + 2x = x(\quad + \quad)$

2. Divide each term by the common factor, and write the results inside the brackets.

   $3x^2 \div x = 3x$ and $2x \div x = 2$
   So $3x^2 + 2x = \mathbf{x(3x + 2)}$

Exercise 1

Q1 Factorise the following expressions.

a) $2a + 10$ b) $3b + 12$ c) $15 + 3y$ d) $28 + 7v$

e) $5a + 15b$ f) $9c - 12d$ g) $3x + 12y$ h) $21u - 7v$

i) $4a^2 - 12b$ j) $3c + 15d^2$ k) $5c^2 - 25f$ l) $6x - 12y^2$

Q2 Factorise the following expressions.

a) $3a^2 + 7a$ b) $4b^2 + 19b$ c) $2x^2 + 9x$ d) $4x^2 - 9x$

e) $21q^2 - 16q$ f) $15y - 7y^2$ g) $7y + 15y^2$ h) $27z^2 + 11z$

i) $10d^3 + 27d$ j) $4y^3 - 13y^2$ k) $11y^3 + 3y^4$ l) $22w - 5w^4$

Sometimes the highest common factor of all the terms might have both **numbers and variables** in it. It's usually best to work out the highest common factor for the numbers and for each letter **separately**.

E.g. if you need to factorise $2a^3b + 4a^2b^2$, find:

- the HCF of 2 and 4 (**2**)
- the HCF of a^3 and a^2 (**a^2**)
- the HCF of b and b^2 (**b**)

So the HCF of $2a^3b$ and $4a^2b^2$ is **$2a^2b$**, and $2a^3b + 4a^2b^2 = \mathbf{2a^2b(a + 2b)}$.

Example 3

Factorise the expression $15x^2 - 10xy$.

1. 5 and x are common factors of $15x^2$ and $10xy$. So $5x$ goes outside the brackets.

$15x^2 - 10xy = 5x(\quad - \quad)$

Tip: The first term doesn't have a y in it, so the HCF won't either.

2. Divide each term by the common factor.

$15x^2 \div 5x = 3x$
and $10xy \div 5x = 2y$

3. Write the results inside the brackets.

$15x^2 - 10xy = \mathbf{5x(3x - 2y)}$

Exercise 2

Q1 The expression $4x^3y^2 + 8xy^4$ contains two terms.

    a) What is the highest numerical common factor of both terms?

    b) What is the highest power of x that is common to both terms?

    c) What is the highest power of y that is common to both terms?

    d) Factorise the expression.

Q2 Factorise the following expressions.

    a) $15a + 10ab$ b) $12b + 9bc$ c) $16xy - 4y$ d) $21x + 3xy$

    e) $24uv + 6v$ f) $10p^2 + 15pq$ g) $12q^2 - 18pq$ h) $30ab^2 + 25ab$

    i) $14x^2 - 28xy^2$ j) $8ab^2 + 10a^2b$ k) $12pq - 8p^3$ l) $24x^3y - 16x^2$

Q3 Factorise the following expressions.

    a) $x^6 + x^4 - x^5$ b) $8a^2 + 17a^6$ c) $12y^6 + 6y^4$ d) $24b^2c^3 - 8c$

    e) $25z^2 + 13z^6$ f) $12p^2 + 15p^5q^3$ g) $9a^4b + 27ab^3$ h) $15b^4 - 21a^2 + 18ab$

    i) $22pq^2 - 11p^3q^3$ j) $16x^2y - 8xy^2 + 2x^3y^3$ k) $36x^7y^2 + 8x^2y^9$ l) $5x^4 + 3x^3y^4 - 25x^3y$

Q4 Factorise the following expressions.

    a) $13x^2y^2 + 22x^6y^3 + 20x^5y^3$ b) $16a^5b^5 - a^4b^5 - ab^3$ c) $21p^6q^2 - 14pq^2 + 7p^3q$

    d) $14xy^4 + 13x^5y^3 - 5x^6y^4$ e) $16c^6d^5 - 14c^3 + 8c^3d^6$ f) $18jk + 21j^3k^6 - 15j^2k^3$

    g) $36x^2y^4 - 72x^5y^7 + 18xy^3$ h) $20a^4b^5 + 4a^3b^4 - 5a^6b^{15}$ i) $11x^2y^3 + 11x^3y^2 + 66xy^5$

6.4 Factorising — Quadratics

So far, all the factorising you've seen has been the reverse of expanding brackets of the form a(b + c).
Doing the reverse of expanding double brackets is up next, and it's going to be pretty useful in later sections.

Quadratic Expressions

> **Learning Objective — Spec Ref A4:**
> Be able to factorise a quadratic expression.

A **quadratic expression** is an expression where the highest power of the variable (e.g. x) is **2** —
they have the form $ax^2 + bx + c$ where a, b and c are constants ($a \neq 0$).

You can **factorise** some quadratics into the form $(dx + e)(fx + g)$, but it's usually **not clear** what the values
of d, e, f, and g are. The method you use **depends** on whether or not there's a number in front of the x^2.

Factorising x² + bx + c

If there's **no number** in front of the x^2 (i.e. $a = 1$), then you can follow these steps to factorise the expression:

- Write out the brackets as $(x \quad)(x \quad)$ — don't put the signs or numbers in yet.
- Find pairs of numbers that **multiply to give** c — this is just like finding **factors** (see p.8).
  You can **ignore the sign** of c for now.
- Choose the pair of numbers that also **add** or **subtract** to give b (ignoring signs here too).
- Write one number in each bracket, then fill in the + or – signs so that b and c work out with the
  correct signs. Check they're right by expanding the brackets to get back to the original expression.

If c is **positive**, then the two brackets will have the same sign (both + or both –),
and if c is **negative** then the signs will be different (one + and one –).

> **Example 1**
>
> **Factorise: a)** $x^2 + 6x + 8$ **b)** $x^2 + 2x – 15$
>
> **a)** 1. Find all the pairs of numbers that multiply to give 8 1×8 or 2×4
>   (i.e. the factor pairs of 8).
>
>   2. Find the pair that add/subtract to give 6. $1 + 8 = 9$, $8 – 1 = 7$
> $2 + 4 = \boxed{6}$ $4 – 2 = 2$
>
>   3. You need +2 and +4 to give b (+6), so
>   both brackets should have + signs.   So $x^2 + 6x + 8 = (x + 4)(x + 2)$
>
>   4. Check your answer by expanding the brackets. $(x + 4)(x + 2) = x^2 + 2x + 4x + 8$
> $= x^2 + 6x + 8$
>
> **b)** 1. Find all the pairs of numbers that multiply to give 15. 1×15 or 3×5
>
>   2. Find the pair that add/subtract to give 2. $1 + 15 = 16$, $15 – 1 = 14$
> $3 + 5 = 8$, $5 – 3 = \boxed{2}$
>
>   3. You need +5 and –3 to give +2,
>   so put a + with the 5 and a – with the 3.   So $x^2 + 2x – 15 = (x + 5)(x – 3)$
>
>   4. Check your answer by expanding the brackets. $(x + 5)(x – 3) = x^2 – 3x + 5x – 15$
> $= x^2 + 2x – 15$

Q1 Factorise each of the following expressions.

a) $x^2 + 7x + 6$ b) $a^2 + 7a + 12$ c) $x^2 + 8x + 7$

d) $z^2 + 8z + 12$ e) $x^2 + 5x + 4$ f) $v^2 + 6v + 9$

Q2 Factorise each of the following expressions.

a) $x^2 + 4x + 3$ b) $x^2 - 6x + 8$ c) $x^2 - 7x + 10$

d) $x^2 - 5x + 4$ e) $y^2 + 3y - 10$ f) $x^2 + 2x - 8$

g) $s^2 + 3s - 18$ h) $x^2 - 2x - 15$ i) $t^2 - 4t - 12$

Factorising $ax^2 + bx + c$

When a is not 1, there are a few extra steps you need to do in order to factorise.

- Start by finding all the pairs of numbers that multiply to give a. Write out a separate set of brackets for each pair, writing the two numbers **in front of the x's**. For example, if you had $4x^2$, you would write out $(2x\quad)(2x\quad)$ and $(x\quad)(4x\quad)$.

- Now list all the pairs of numbers that **multiply to give c**, **ignoring the sign** of c for now.

- Here's the tricky bit — try putting **each pair** of numbers in the brackets until you find one that gives you the right value for b. Check them by working out the 'OI' bits from 'FOIL' — these are the bits that should **give you bx** when you add/subtract them. Make sure to try the pairs **both ways round**, e.g. if your factor pair was 2 and 3, try both $(x\quad3)(4x\quad2)$ and $(x\quad2)(4x\quad3)$.

- Once you've found the right combination, write the + or – signs so that it works out.

Example 2

Factorise: a) $2x^2 + 7x + 3$ b) $6x^2 - 11x - 10$

a) 1. The $2x^2$ has to come from $x \times 2x$. $(x\quad)(2x\quad)$

2. $c = 3$ has only got one pair of factors. 1×3

3. Find the potential values of b — try adding and subtracting.
     $(x\quad1)(2x\quad3)$: O $= 3x$, I $= 2x \Rightarrow 5x$ or x
     $(x\quad3)(2x\quad1)$: O $= x$, I $= 6x \Rightarrow$ ⑦x or $5x$

4. You need $+x$ and $+6x$ to make $+7x$, so both brackets should have + signs.
     So $2x^2 + 7x + 3 = (x + 3)(2x + 1)$

5. Check by expanding the brackets. $(x + 3)(2x + 1) = 2x^2 + x + 6x + 3 = 2x^2 + 7x + 3$

b) 1. The $6x^2$ could either come from $2x \times 3x$ or from $x \times 6x$.
     $(2x\quad)(3x\quad)$ or $(x\quad)(6x\quad)$

2. $c = 10$ has two pairs of factors. 1×10 or 2×5

3. Try pairs of numbers in the brackets until you get $11x$.
     $(2x\quad1)(3x\quad10)$: O $= 20x$, I $= 3x \Rightarrow 23x$ or $17x$
     $(2x\quad10)(3x\quad1)$: O $= 2x$, I $= 30x \Rightarrow 32x$ or $28x$
     $(2x\quad2)(3x\quad5)$: O $= 10x$, I $= 6x \Rightarrow 16x$ or $4x$
     $(2x\quad5)(3x\quad2)$: O $= 4x$, I $= 15x \Rightarrow 19x$ or ⑪x

4. You need $+4x$ and $-15x$ to make $-11x$, so put a + with the 2 $(2x \times +2 = +4x)$ and a – with the 5 $(3x \times -5 = -15x)$.
     So $6x^2 - 11x - 10 = (2x - 5)(3x + 2)$

5. Check by expanding the brackets. $(2x - 5)(3x + 2) = 6x^2 + 4x - 15x - 10 = 6x^2 - 11x - 10$

Exercise 2

Q1 Factorise the following expressions.

a) $2x^2 + 3x + 1$ b) $3x^2 - 16x + 5$ c) $5x^2 - 17x + 6$ d) $2t^2 - 5t - 12$

e) $2x^2 - 13x + 6$ f) $3b^2 - 7b - 6$ g) $5x^2 + 12x - 9$ h) $2x^2 - 3x + 1$

i) $7a^2 + 19a - 6$ j) $11x^2 - 62x - 24$ k) $7z^2 + 38z + 15$ l) $3y^2 - 26y + 16$

Q2 Factorise the following expressions.

a) $6x^2 + 7x + 1$ b) $6x^2 - 13x + 6$ c) $15x^2 - x - 2$

d) $10x^2 - 19x + 6$ e) $12u^2 - 5u - 3$ f) $14w^2 + 25w + 9$

Difference of Two Squares

> **Prior Knowledge Check:**
> Recognise square numbers and find square roots. See p.91.

Learning Objective — Spec Ref A4:
Be able to factorise a difference of two squares.

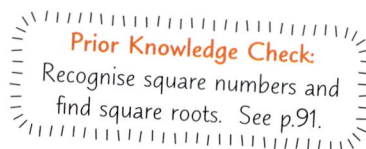

Some quadratic expressions have no middle term, e.g. $x^2 - 49$. When you factorise these, you get two brackets that are **the same**, except that one has a **+ sign** and one has a **– sign**. For example, $x^2 - 49$ factorises to $(x + 7)(x - 7)$.

More generally: $a^2 - b^2 = (a + b)(a - b)$ This is known as the **difference of two squares**.

Example 3

Factorise: a) $16x^2 - 9$ b) $5y^2 - 80z^2$ c) $x^2 - 7$

a) $16x^2 = 4x \times 4x$ — write the expression in the form $a^2 - b^2$, then use the formula.

$$16x^2 - 9 = (4x)^2 - 3^2$$
$$= (4x + 3)(4x - 3)$$

b)   You can take a factor of 5 out of the expression, so do this first. Then write the bit inside the bracket in the form $a^2 - b^2$ and use the formula.

$$5y^2 - 20z^2 = 5(y^2 - 4z^2)$$
$$= 5(y^2 - (2z)^2)$$
$$= 5(y + 2z)(y - 2z)$$

c)   7 isn't a square number, so write 7 as $(\sqrt{7})^2$ so that you can use the formula.

$$x^2 - 7 = x^2 - (\sqrt{7})^2$$
$$= (x + \sqrt{7})(x - \sqrt{7})$$

Tip: See p.99-103 for more about surds.

Exercise 3

Q1 Factorise each of the following expressions.

a) $x^2 - 25$ b) $x^2 - 9$ c) $x^2 - 36$ d) $x^2 - 81$

e) $x^2 - 64$ f) $b^2 - 121$ g) $x^2 - 1$ h) $c^2 - 4d^2$

Q2 Factorise each of the following expressions.

a) $4x^2 - 49$ b) $36x^2 - 4$ c) $9x^2 - 100$ d) $25x^2 - 16$

e) $16z^2 - 1$ f) $27t^2 - 12$ g) $98x^2 - 2$ h) $7p^2 - 175$

i) $x^2 - 11$ j) $n^2 - 51$ k) $4x^2 - 3$ l) $3x^2 - 15$

6.5 Algebraic Fractions

Algebraic fractions are just fractions with algebraic expressions in the numerator and/or denominator. You can do all the things you'd do with regular fractions — e.g. simplify, add, subtract, multiply or divide.

Simplifying Algebraic Fractions

Learning Objective — Spec Ref A4:
Be able to simplify algebraic fractions.

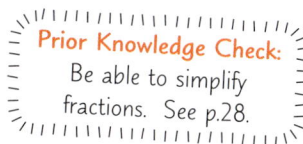

Prior Knowledge Check:
Be able to simplify
fractions. See p.28.

Algebraic fractions can be **simplified** just like regular fractions. Look for **common factors** between the numerator and denominator (you might have to do some **factorising**) and then **cancel** them. The only difference is that the factors may be **algebraic expressions** — e.g. $4x$, y^2z, $(5a + b)$, etc.

Example 1

Simplify the following algebraic fractions: a) $\dfrac{3x^2y^4}{21xy^5}$ b) $\dfrac{x^2 - 16}{x^2 + 8x + 16}$

a)   1.   3 is a common factor, so it can be cancelled.

     2. x is a common factor, so it can be cancelled.

     3. y^4 is a common factor, so it can be cancelled.

$$\frac{3x^2y^4}{21xy^5} = \frac{\cancel{3} \times x^2 \times y^4}{\cancel{3} \times 7 \times x \times y^5}$$

$$= \frac{x^2 \times y^4}{7 \times \cancel{x} \times y^5}$$

$$= \frac{x \times \cancel{y^4}}{7 \times y^{\cancel{5}}} = \frac{x}{7 \times y} = \frac{x}{7y}$$

b)   Factorise the numerator and the denominator, then cancel the common factor of $(x + 4)$.

$$\frac{x^2 - 16}{x^2 + 8x + 16} = \frac{\cancel{(x + 4)}(x - 4)}{\cancel{(x + 4)}(x + 4)}$$

$$= \frac{x - 4}{x + 4}$$

Tip: The numerator here is a difference of two squares.

Exercise 1

Q1   Simplify:

a) $\dfrac{2x}{x^2}$ b) $\dfrac{49x^3}{14x^4}$ c) $\dfrac{25s^3t}{5s}$ d) $\dfrac{26a^2b^3c^4}{52b^5c}$

Q2   Simplify:

a) $\dfrac{7x}{5x - x^2}$ b) $\dfrac{48t - 6t^2}{8s^2t}$ c) $\dfrac{3cd}{8c + 6c^2}$ d) $\dfrac{12}{4a^2b^3 + 8a^7b^9}$

Q3   Simplify:

a) $\dfrac{4st + 8s^2}{8t^2 + 16st}$ b) $\dfrac{15xz + 15z}{25xyz - 25yz}$ c) $\dfrac{6xy - 6x}{3y - 3}$ d) $\dfrac{3a^2b + 5ab^2}{7ab^3}$

Q4   Simplify each of the following fractions as far as possible.

a) $\dfrac{2x - 8}{x^2 - 5x + 4}$ b) $\dfrac{6a - 3}{5 - 10a}$ c) $\dfrac{x^2 + 7x + 10}{x^2 + 2x - 15}$

d) $\dfrac{x^2 - 7x + 12}{x^2 - 2x - 8}$ e) $\dfrac{x^2 + 4x}{x^2 + 7x + 12}$ f) $\dfrac{2t^2 + t - 45}{4t^2 - 81}$

Adding and Subtracting Algebraic Fractions

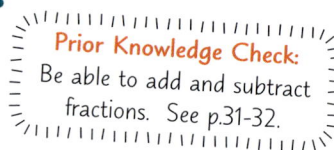

Learning Objective — Spec Ref A4:
Add and subtract algebraic fractions.

Prior Knowledge Check:
Be able to add and subtract fractions. See p.31-32.

Algebraic fractions should be treated in just the same way as numerical fractions. To add or subtract them, they need to have a **common denominator**. You can get them over a common denominator by multiplying the **top and bottom** of each fraction by the **denominators** of the others — this could be a **number**, or an **expression** like $3x + 2$.

Example 2

Express, as a single fraction: a) $\dfrac{x-2}{3} + \dfrac{2x+3}{4}$ **b)** $\dfrac{2x-1}{x+1} - \dfrac{3}{x-2}$

a)

1. First, find a common denominator — multiply the top and bottom of the first fraction by 4, and the second fraction by 3.

2. Expand the brackets in the numerator and simplify as much as possible.

$$\dfrac{x-2}{3} + \dfrac{2x+3}{4} = \dfrac{4 \times (x-2)}{4 \times 3} + \dfrac{3 \times (2x+3)}{3 \times 4}$$

$$= \dfrac{4(x-2)}{12} + \dfrac{3(2x+3)}{12}$$

$$= \dfrac{4(x-2) + 3(2x+3)}{12}$$

$$= \dfrac{4x - 8 + 6x + 9}{12} = \dfrac{10x+1}{12}$$

b)

1. Find a common denominator by multiplying the top and bottom of the first fraction by $(x - 2)$, and of the second fraction by $(x + 1)$.

2. Expand the brackets in the numerator and simplify.

3. $2x^2 - 8x - 1$ doesn't factorise nicely, so leave it as it is.

$$\dfrac{2x-1}{x+1} - \dfrac{3}{x-2} = \dfrac{(2x-1)(x-2)}{(x+1)(x-2)} - \dfrac{3(x+1)}{(x+1)(x-2)}$$

$$= \dfrac{(2x-1)(x-2) - 3(x+1)}{(x+1)(x-2)}$$

$$= \dfrac{2x^2 - 5x + 2 - 3x - 3}{(x+1)(x-2)}$$

$$= \dfrac{2x^2 - 8x - 1}{(x+1)(x-2)}$$

Exercise 2

Q1 Express each of these as a single fraction, simplified as far as possible.

a) $\dfrac{x}{4} + \dfrac{x}{5}$ b) $\dfrac{x}{2} - \dfrac{x}{3}$ c) $\dfrac{2b}{7} + \dfrac{b}{6}$ d) $\dfrac{5z}{6} - \dfrac{4z}{9}$

Q2 Express each of these as a single fraction, simplified as far as possible.

a) $\dfrac{x-2}{5} + \dfrac{x+1}{3}$ b) $\dfrac{2t+1}{4} + \dfrac{t-1}{3}$ c) $\dfrac{3x-1}{4} + \dfrac{2x+1}{6}$ d) $\dfrac{c+2}{c} + \dfrac{c+1}{2c}$

Q3 Express each of these as a single fraction, simplified as far as possible.

a) $\dfrac{2}{x+1} + \dfrac{1}{x-3}$ b) $\dfrac{x-2}{x-1} - \dfrac{x+1}{x+2}$ c) $\dfrac{2a-3}{a+2} + \dfrac{3a+2}{a+3}$

d) $\dfrac{x+2}{3x-2} + \dfrac{x-3}{2x+1}$ e) $\dfrac{s-2}{3s-1} + \dfrac{s+1}{3s+2}$ f) $\dfrac{x-2}{5} + \dfrac{x+1}{3x}$

g) $\dfrac{y+3}{y+1} - \dfrac{y+2}{y+3}$ h) $\dfrac{x-3}{x+2} - \dfrac{2x+1}{x+1}$ i) $\dfrac{2x+3}{x-2} - \dfrac{x-4}{x+3}$

If the denominators of the fractions you're adding or subtracting share a **common factor**, you **don't need to** multiply by that factor to get the fractions over a common denominator (similar to how you can do $\frac{1}{4} + \frac{1}{6} = \frac{3}{12} + \frac{2}{12}$ rather than using a common denominator of $4 \times 6 = 24$).

Example 3

Express $\dfrac{x}{x^2 + 3x + 2} + \dfrac{x-1}{x+2}$ **as a single fraction.**

1. Factorise the terms in each denominator.

$$\dfrac{x}{x^2 + 3x + 2} + \dfrac{x-1}{x+2}$$

2. Here the denominators have a common factor, $(x + 2)$. Multiply the top and bottom of the second fraction by $(x + 1)$ to make the denominators the same.

$$= \dfrac{x}{(x+1)(x+2)} + \dfrac{x-1}{x+2}$$

$$= \dfrac{x}{(x+1)(x+2)} + \dfrac{(x-1)(x+1)}{(x+2)(x+1)}$$

3. Simplify the numerator by expanding the brackets.

$$= \dfrac{x + (x^2 + x - x - 1)}{(x+1)(x+2)}$$

4. $x^2 + x - 1$ doesn't factorise, so leave it as it is.

$$= \dfrac{x^2 + x - 1}{(x+1)(x+2)}$$

Example 4

Show that $\dfrac{4}{(3x-5)(x+1)} - \dfrac{1}{(3x-5)(x-1)} = \dfrac{1}{x^2-1}.$

1. The denominators of the fractions are already factorised. They have a common factor of $(3x - 5)$, so you just need to multiply the left fraction by $(x - 1)$ and the right fraction by $(x + 1)$.

$$\dfrac{4}{(3x-5)(x+1)} - \dfrac{1}{(3x-5)(x-1)}$$

$$= \dfrac{4(x-1)}{(3x-5)(x+1)(x-1)} - \dfrac{x+1}{(3x-5)(x+1)(x-1)}$$

2. Simplify the numerator.

$$= \dfrac{4x - 4 - x - 1}{(3x-5)(x+1)(x-1)}$$

$$= \dfrac{3x - 5}{(3x-5)(x+1)(x-1)}$$

3. Cancel the factor of $(3x - 5)$ from the top and bottom. Expanding the brackets in the denominator gives you the final answer.

$$= \dfrac{1}{(x+1)(x-1)} = \dfrac{1}{x^2-1}$$

Exercise 3

Q1 Express each of the following as a single fraction, simplified as far as possible.

a) $\dfrac{x-2}{(x-3)(x+1)} + \dfrac{5}{(x-3)}$
b) $\dfrac{3x}{(x+1)(x+2)} + \dfrac{1}{x+2}$
c) $\dfrac{1}{(x+4)} - \dfrac{(x-2)}{(x+4)(x+3)}$

Q2 Express each of the following as a single fraction, simplified as far as possible.

a) $\dfrac{z}{z^2 + 3z + 2} + \dfrac{10}{z+1}$
b) $\dfrac{x-3}{x^2+x-6} + \dfrac{2x}{x-2}$
c) $\dfrac{x-3}{x^2+4x} + \dfrac{3}{x+4}$
d) $\dfrac{a+1}{a^2-2a-3} - \dfrac{a+1}{a^2+4a+3}$

Q3 Express each of the following as a single fraction, simplified as far as possible.

a) $\dfrac{2t+1}{t^2+3t} - \dfrac{t+2}{t^2+4t+3}$
b) $\dfrac{x}{x^2-9} + \dfrac{x+2}{x^2-5x+6}$
c) $\dfrac{3x}{x^2-4x+3} + \dfrac{x}{x^2-5x+4}$
d) $\dfrac{2y-1}{y^2-3y+2} - \dfrac{y-1}{y^2-5y+6}$

Multiplying and Dividing Algebraic Fractions

Learning Objective — Spec Ref A4:
Multiply and divide algebraic fractions.

Prior Knowledge Check:
Be able to multiply and divide fractions. See p.33-35.

The method for multiplying and dividing algebraic fractions is exactly the same as with numeric fractions — if you're multiplying, just times the **numerators together** and the **denominators together**.
If you're dividing, **flip the second fraction** and change it to a multiplication, e.g. $\frac{a}{b} \div \frac{c}{d} = \frac{a}{b} \times \frac{d}{c}$.

To make the calculations easier, it's usually best to **fully factorise** the numerator and denominator of each fraction, then **cancel any common factors**. Once you've got a multiplication, you can cancel factors that appear on the top and bottom of **either fraction** — so $\frac{(x+1)}{(x+2)} \times \frac{(x+3)}{2(x+1)}$ would cancel to $\frac{1}{(x+2)} \times \frac{(x+3)}{2}$.

Example 5

Express, as a single simplified fraction: a) $\dfrac{x-2}{x^2+6x+8} \times \dfrac{2x+4}{x^2+2x-8}$ b) $\dfrac{3x+6}{x^2-9} \div \dfrac{x+2}{x^2+4x+3}$

a) 1. Factorise each term as far as possible, then cancel any factor which appears on the top and bottom of either fraction.

$$\frac{x-2}{x^2+6x+8} \times \frac{2x+4}{x^2+2x-8} = \frac{\cancel{x-2}}{(x+2)(x+4)} \times \frac{2(x+2)}{\cancel{(x-2)}(x+4)}$$

2. Multiply the fractions like normal.

$$= \frac{1}{(x+4)} \times \frac{2}{(x+4)} = \frac{2}{(x+4)^2}$$

b) 1. This is a division, so turn the second fraction over and change the sign to a multiplication.

$$\frac{3x+6}{x^2-9} \div \frac{x+2}{x^2+4x+3} = \frac{3x+6}{x^2-9} \times \frac{x^2+4x+3}{x+2}$$

2. Factorise each term as far as possible. Cancel any factor which appears on both the top and the bottom of either fraction and multiply.

$$= \frac{3\cancel{(x+2)}}{\cancel{(x+3)}(x-3)} \times \frac{(x+1)\cancel{(x+3)}}{\cancel{(x+2)}}$$

$$= \frac{3}{(x-3)} \times \frac{(x+1)}{1} = \frac{3(x+1)}{x-3}$$

Exercise 4

Q1 Express each of the following as a single fraction, simplified as far as possible.

a) $\dfrac{x}{y} \times \dfrac{3}{x^2}$

b) $\dfrac{2a}{4b^2} \times \dfrac{5b}{a^3}$

c) $\dfrac{t}{2} \times \dfrac{24st}{6t}$

d) $\dfrac{64xy^2}{9y} \times \dfrac{3x^3}{16x^2y}$

e) $\dfrac{1}{x+4} \times \dfrac{3x}{x+2}$

f) $\dfrac{6a+b}{12} \times \dfrac{3a}{b+1}$

g) $\dfrac{1}{2z+5} \times \dfrac{3z}{z-1}$

h) $\dfrac{t^2+5}{12} \times \dfrac{4t^3}{1-t}$

Q2 Express each of the following as a single fraction, simplified as far as possible.

a) $\dfrac{4x}{3y^2} \div \dfrac{2x}{12y^4}$

b) $\dfrac{18a}{6b^2} \div \dfrac{a}{20b}$

c) $\dfrac{3x^3y^5}{4x^5y} \div \dfrac{xy}{28}$

d) $\dfrac{ab^2}{15} \div \dfrac{a^2}{5b}$

e) $\dfrac{y-2}{3y+2} \div \dfrac{y-2}{4y^4}$

f) $\dfrac{2c+1}{3d^2} \div \dfrac{cd}{18}$

g) $\dfrac{1-x^3y^5}{25x^2} \div \dfrac{y}{5}$

h) $\dfrac{12t^5}{6t+3t^3} \div \dfrac{18t^2}{9t}$

Q3 Express each of the following as a single fraction, simplified as far as possible.

a) $\dfrac{x^2-16}{x^2+5x+6} \times \dfrac{x+3}{x+4}$

b) $\dfrac{x^2+4x+3}{x^2+6x+8} \times \dfrac{x+4}{2x+6}$

c) $\dfrac{z^2+3z-10}{z^2+4z+3} \times \dfrac{z^2+6z+5}{z-2}$

d) $\dfrac{x^2-4x+3}{x^2+9x+20} \div \dfrac{x^2-x-6}{x^2+7x+12}$

e) $\dfrac{y^2-5y+6}{y^2+y-20} \div \dfrac{y-2}{3y-12}$

f) $\dfrac{t^2-9}{t^2+3t+2} \div \dfrac{t^2+6t+9}{t^2+8t+7}$

88   **Section 6** Expressions

Review Exercise

Q1 Simplify the following expressions.

a) $3x - 2x + 5x - x$

b) $3x + 2y - 4x + 5y$

c) $3x^2 + 5x - 7 + 2x^2 - 3x + 4$

d) $2a \times 3a$

e) $8p \times 2q$

f) $x^2 \times xy$

g) $22z^4 \div 2z$

h) $27g^3h^3 \div 9gh^2$

i) $x^2(4x)^3$

Q2 Simplify each of the following expressions by expanding the brackets.

a) $6(x + 3) + 3(x - 4)$

b) $4(a + 3) - 2(a + 2)$

c) $5(p + 2) - 3(p - 4)$

d) $2(2x + 3) - 3(2x + 1)$

e) $x(x + 3) + 2(x - 1)$

f) $2x(2x + 3) + 3(x - 4)$

Q3 Expand and simplify each of the following expressions.

a) $(x - 3)(x + 4)$

b) $(3x + 1)(2x + 5)$

c) $(x + 1)(x - 2)(x + 3)$

d) $-(t - 8)(t - 1)(t + 1)$

e) $(3x - 2)^2$

f) $(x + 2)^3$

Q4 Factorise each of the following expressions as far as possible.

a) $4x - 8$

b) $6a + 3$

c) $5t - 10$

d) $3x + 6xy$

e) $8xy - 12x^2$

f) $a^2b - 2ab + ab^2$

g) $16x^2 + 12x^2y - 8xy^2$

h) $14x^3 + 7x^2y - 7xy^4$

Q5 Factorise each quadratic.

a) $a^2 + 6a + 8$

b) $x^2 + 4x + 3$

c) $z^2 - 5z + 6$

d) $x^2 + 3x - 18$

e) $x^2 - 3x - 10$

f) $2x^2 + 5x + 2$

g) $3m^2 - 8m + 4$

h) $3x^2 - 5x - 2$

i) $4g^2 + 4g + 1$

j) $16a^2 - 25$

k) $4c^2 - 196$

l) $81t^2 - 121$

Q6 Express each of the following as a single, simplified, algebraic fraction.

a) $\dfrac{a}{4} + \dfrac{a}{8}$

b) $\dfrac{2x}{3} - \dfrac{2x}{7}$

c) $\dfrac{y - 1}{2} + \dfrac{y + 1}{3}$

d) $\dfrac{2}{x + 3} + \dfrac{4}{x - 1}$

e) $\dfrac{5}{z + 2} - \dfrac{3}{z + 3}$

f) $\dfrac{z + 2}{z + 3} + \dfrac{z + 1}{z - 2}$

g) $\dfrac{3}{x^2 + 4x + 3} + \dfrac{2}{x^2 + x - 6}$

h) $\dfrac{4}{t^2 + 6t + 9} - \dfrac{3}{t^2 + 3t}$

i) $\dfrac{x}{x^2 - 16} + \dfrac{x - 2}{x^2 - 5x + 4}$

Q7 Simplify the following expressions.

a) $\dfrac{x^2 - 4}{3x - 3} \times \dfrac{9}{2x - 4}$

b) $\dfrac{8}{x} \div \dfrac{6}{x^2}$

c) $\dfrac{y - 1}{2} \div \dfrac{x + 1}{3}$

d) $\dfrac{x^2 - 7x + 12}{x^2 + 3x + 2} \times \dfrac{x + 1}{x - 3}$

e) $\dfrac{s^2 + 4s + 3}{s^2 - 16} \div \dfrac{s + 1}{s + 4}$

f) $\dfrac{x^2 + x - 12}{x^2 - 4} \times \dfrac{x^2 + 2x}{3x + 12}$

g) $\dfrac{a^2 - 7a + 10}{a^2 + 5a + 6} \times \dfrac{a^2 + 2a - 3}{a^2 - 3a - 10}$

h) $\dfrac{b^2 + 5b + 6}{b^2 + 6b + 5} \div \dfrac{2b + 6}{3b + 3}$

i) $\dfrac{x^2 + 8x + 15}{x^2 + 4x - 12} \div \dfrac{x^2 + 4x + 3}{x^2 + 8x + 12}$

Exam-Style Questions

Q1 Expand and simplify:

    a) $x(x^2 - 4y) + 9xy$

[2 marks]

    b) $(2x - 7)^2$

[2 marks]

Q2 Use an algebraic equivalent to the expression $x^2 - y^2$ to work out the value of $145^2 - 55^2$.

[2 marks]

Q3 Two expressions A and B are defined such that:

$$A = 5x^2 + 9xy \qquad\qquad B = 3x(x - y)$$

Find an expression for $A - B$ in terms of x and y.
Give your answer in a fully factorised form.

[3 marks]

Q4 a) Factorise $5a^2 - 6a$

[1 mark]

    b) Use your answer to part a) to factorise $5(2x + 3)^2 - 6(2x + 3)$.
       Give your answer in its simplest form.

[2 marks]

Q5 $6(3x - y) - 4(x + 5y) = a(7x - by)$

Find the values of a and b.

[3 marks]

Q6 Simplify fully $\dfrac{6x^2 + 18x}{2x^2 - 4x - 30}$.

[3 marks]

Q7 Expand $(x + 3)(x - 2)^2$, simplifying your answer as much as possible.

[3 marks]

7.1 Squares, Cubes and Roots

The square of a number is the product of the number with itself — e.g. the square of 4 is 4 × 4 = 16.
Cubes are the product of a number with itself and then itself again — the cube of 4 is 4 × 4 × 4 = 64.
Roots are the inverse (or 'opposite') operation — i.e. the square root of x is the number which multiplies
with itself to give x and the cube root of y is the number which multiplies with itself and itself again to give y.

Learning Objective — Spec Ref N6:
Evaluate powers and roots.

Prior Knowledge Check:
Know how to multiply and divide by
negative numbers — see p.4.

You write squares and cubes using **powers** — '4 squared' would be written 4^2 and '4 cubed' would be 4^3.

If you square a number, you always get a **positive** answer because you're multiplying the **same signs** together. On the other hand, cubing a number can give a **positive or negative** answer since there are **three multiples** of the same sign — the answer will be the same sign as whatever the **original** number was.

Since squaring only gives positive results, **negative numbers** don't have square roots. However, positive numbers have **two** square roots — for instance, 4 × 4 = 16 and –4 × –4 = 16 so the square roots of 16 are 4 and –4. You can write the positive square root of x as $\sqrt{x}$ and the negative square root as $-\sqrt{x}$.
All numbers, either **positive or negative**, have **exactly one cube root**, which you can write as $\sqrt[3]{x}$.

Example 1

Find: a) 3^2 b) the square roots of 9   c) $\sqrt[3]{-27}$ d) $\sqrt{10^2 - 6^2}$

a) The square of 3 is 3 × 3.

$3^2 = 3 \times 3 = \mathbf{9}$

b) $3^2 = 9$ so a square root of 9 is 3.
But don't forget the second square root:
–3 × –3 = 9 so –3 is also a square root.

$\sqrt{9} = \mathbf{3}$

$-\sqrt{9} = \mathbf{-3}$

c) $(-3)^3 = -27$, so the cube root is –3.

$\sqrt[3]{-27} = \mathbf{-3}$

d) The square root and cube root symbols act like brackets. Evaluate the expression inside before you find the root.

$\sqrt{10^2 - 6^2} = \sqrt{100 - 36}$
$= \sqrt{64} = \mathbf{8}$

Tip: Always be careful with the sign of your answer when squaring and cubing. Anything squared is positive, and a positive number cubed is positive, but a negative number cubed is negative.

Exercise 1

Q1 Find (i) the square and (ii) the cube of the following numbers:
a) 5   b) 10   c) –2   d) 0.1

Q2 Find the square roots of:
a) 36   b) 10 000   c) –16   d) 81

Q3 Find the cube root of:
a) 8   b) –64   c) 1000   d) –1

Q4 Calculate: a) $\sqrt{4 \times 10^2}$ b) $\sqrt[3]{(3+5)^2}$ c) $\sqrt[3]{3^2 \times 2^3 + 12^2}$

7.2 Indices and Index Laws

Powers are a useful shorthand that allow you to write repeated multiplications with just two symbols — a base and an index (plural: indices). Squares and cubes are two simple examples of powers.

Indices

Learning Objective — Spec Ref A4:
Work with numbers in index notation.

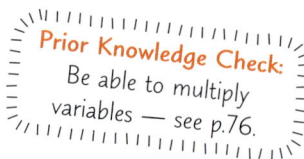

Prior Knowledge Check:
Be able to multiply variables — see p.76.

Index notation (i.e. powers) can be used to show **repeated multiplication** of a number or letter. For example, $2 \times 2 \times 2 \times 2 = 2^4$. This is read as "2 **to the power** 4".

In index notation, the **base** is the value that you're multiplying (here, 2) and the **index** is the number of instances of that value (here, 4).

base $\rightarrow 2^4 \leftarrow$ index

Example 1

Rewrite the following using index notation: a) $3 \times 3 \times 3 \times 3 \times 3$ b) $b \times b \times b \times b \times c \times c$

a) There are five lots of 3. $3 \times 3 \times 3 \times 3 \times 3 = \mathbf{3^5}$

b) There are four b's and two c's. $b \times b \times b \times b \times c \times c = b^4 \times c^2 = \boldsymbol{b^4 c^2}$

Example 2

Simplify $\left(\dfrac{2}{5}\right)^2$.

With powers of fractions, the power is applied to both the top and bottom of the fraction.

$$\left(\frac{2}{5}\right)^2 = \frac{2^2}{5^2} = \frac{\mathbf{4}}{\mathbf{25}}$$

Exercise 1

Q1 Using index notation, simplify the following.

a) $2 \times 2 \times 2 \times 2 \times 2$

b) $7 \times 7 \times 7 \times 7 \times 7 \times 7 \times 7$

c) $3 \times 3 \times x \times x \times x \times x \times y \times y$

Q2 Rewrite the following as powers of 10.

a) $10 \times 10 \times 10$

b) 10 million

c) 100 000 000

Q3 Using a calculator, evaluate the following.

a) 3^4

b) 2^8

c) 3^{10}

d) 9^5

e) 3×2^8

f) $8 + 2^5$

g) $8^7 \div 4^6$

h) $(9^3 + 4)^2$

Q4 Write the following as fractions without indices.

a) $\left(\dfrac{1}{2}\right)^2$

b) $\left(\dfrac{1}{2}\right)^3$

c) $\left(\dfrac{1}{4}\right)^2$

d) $\left(\dfrac{2}{3}\right)^2$

e) $\left(\dfrac{3}{10}\right)^2$

f) $\left(\dfrac{3}{2}\right)^3$

g) $\left(\dfrac{5}{3}\right)^4$

h) $\left(\dfrac{4}{3}\right)^3$

Laws of Indices

Learning Objective — Spec Ref N7/A4:
Use the laws of indices to simplify expressions.

The **laws of indices** let you **simplify** complicated-looking expressions that involve **powers**.

$a^m \times a^n = a^{m+n}$	When **multiplying** powers with the **same base**, you **add** the indices.
$a^m \div a^n = a^{m-n}$	When **dividing** powers with the **same base**, you **subtract** the indices.
$(a^m)^n = a^{m \times n}$	When **raising** one power to another, **multiply** the indices.
$a^1 = a$	Anything to the power **1** is just **itself**.
$a^0 = 1$	Anything to the power **0** is **1**.

Example 3

Simplify the following, leaving the answers in index form.

a) $3^8 \times 3^5$ b) $p^8 \div p^5$ c) $(17^7)^2$

a) You're multiplying two terms, so add the indices. $3^8 \times 3^5 = 3^{8+5} = \mathbf{3^{13}}$

b) You're dividing two terms, so subtract the indices. $p^8 \div p^5 = p^{8-5} = \mathbf{p^3}$

c) For one power raised to another power, multiply the indices.

$(17^7)^2 = 17^{7 \times 2} = \mathbf{17^{14}}$

Tip: The first two rules only work when the base numbers are the same. You can't simplify something like $3^2 \times 2^3$ using these rules.

Exercise 2

Q1 Simplify the following, leaving your answers in index form.

a) $3^2 \times 3^6$ b) $10^7 \div 10^3$ c) $a^6 \times a^4$ d) $(4^3)^3$

e) $8^6 \div 8^1$ f) 7×7^6 g) $(c^5)^4$ h) $\dfrac{b^8}{b^5}$

i) $f^{75} \div f^0$ j) $\dfrac{20^{228}}{20^{210}}$ k) $(g^{11})^8$ l) $(14^7)^d$

Q2 For each of the following, find the number that should replace the square.

a) $q^8 \div q^3 = q^\blacksquare$ b) $8^\blacksquare \times 8^{10} = 8^{12}$ c) $(6^{10})^4 = 6^\blacksquare$ d) $(15^6)^\blacksquare = 15^{24}$

e) $(9^\blacksquare)^{10} = 9^{30}$ f) $r^7 \times r^\blacksquare = r^{13}$ g) $5^\blacksquare \div 5^6 = 5^7$ h) $12^{14} \div 12^\blacksquare = 12^7$

Q3 Simplify each expression. Leave your answers in index form.

a) $3^2 \times 3^5 \times 3^7$ b) $5^4 \times 5 \times 5^8$ c) $(p^6)^2 \times p^5$ d) $(9^4 \times 9^3)^5$

e) $7^3 \times 7^5 \div 7^6$ f) $8^3 \div 8^9 \times 8^7$ g) $(12^8 \div 12^4)^3$ h) $(q^3)^6 \div q^4$

Q4 Simplify each expression. Leave your answers in index form.

a) $\dfrac{3^4 \times 3^5}{3^6}$ b) $\dfrac{s^8 \times s^4}{s^3 \times s^6}$ c) $\left(\dfrac{6^3 \times 6^9}{6^7}\right)^3$ d) $\dfrac{2^5 \times 2^5}{(2^3)^2}$

e) $\dfrac{5^5 \times 5^5}{5^8 \div 5^3}$ f) $\dfrac{10^8 \div 10^3}{10^4 \div 10^4}$ g) $\dfrac{(t^6 \div t^3)^4}{t^9 \div t^4}$ h) $\dfrac{(8^5)^7 \div 8^{12}}{8^6 \times 8^{10}}$

Q5 a) Write: (i) 4 as a power of 2   (ii) 4^5 as a power of 2   (iii) $2^3 \times 4^5$ as a power of 2

   b) Write: (i) 9×3^3 as a power of 3   (ii) $5 \times 25 \times 125$ as a power of 5   (iii) 16×2^6 as a power of 4

Negative Indices

Learning Objective — Spec Ref N7/A4:
Work with negative indices.

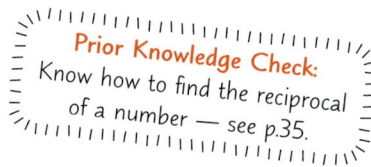

Prior Knowledge Check:
Know how to find the reciprocal of a number — see p.35.

You can evaluate powers that have a **negative index** by taking the **reciprocal** of the base (i.e. turning it upside down — the reciprocal of a is $\frac{1}{a}$ and the reciprocal of $\frac{a}{b}$ is $\frac{b}{a}$) and making the index **positive**.

$$a^{-m} = \frac{1}{a^m} \qquad \left(\frac{a}{b}\right)^{-m} = \left(\frac{b}{a}\right)^m = \frac{b^m}{a^m}$$

Example 4

Evaluate 5^{-3}

1. Take the reciprocal of the base and make the index positive.

2. Evaluate the index in the denominator.

$$5^{-3} = \frac{1}{5^3} = \frac{1}{5 \times 5 \times 5} = \frac{1}{125}$$

Exercise 3

Q1 Write the following as fractions.

    a) 4^{-1} b) 2^{-2} c) 3^{-3} d) 2×3^{-1}

Q2 Write the following in the form a^{-m}.

    a) $\frac{1}{5}$ b) $\frac{1}{11}$ c) $\frac{1}{3^2}$ d) $\frac{1}{2^7}$

Q3 Simplify the following.

    a) $\left(\frac{1}{2}\right)^{-1}$ b) $\left(\frac{1}{3}\right)^{-2}$ c) $\left(\frac{5}{2}\right)^{-3}$ d) $\left(\frac{7}{10}\right)^{-2}$

Example 5

Simplify the following: a) $y^4 \div \frac{1}{y^3}$ b) $z^8 \times (z^4)^{-2}$

a) 1. Rewrite $\frac{1}{y^3}$ as a negative index.

       2. Subtract the indices.

$$y^4 \div \frac{1}{y^3} = y^4 \div y^{-3}$$
$$= y^{4-(-3)} = y^7$$

Tip: You could also do part a) by dividing fractions: $y^4 \div \frac{1}{y^3} = y^4 \times y^3$

b) 1. Multiply the indices to simplify.

       2. Now add the indices.

       3. Anything to the power 0 is 1.

$$z^8 \times (z^4)^{-2} = z^8 \times z^{4 \times (-2)} = z^8 \times z^{-8}$$
$$= z^{8+(-8)} = z^0$$
$$= 1$$

Tip: Simplify $(z^4)^{-2}$ before combining the terms.

Example 6

Evaluate $2^4 \times 5^{-3}$. Give the answer as a fraction.

1. Turn the negative index into a fraction.

$$2^4 \times 5^{-3} = 2^4 \times \frac{1}{5^3}$$

2. Evaluate the powers.

$$= \frac{2^4}{5^3} = \frac{16}{125}$$

Q1 Simplify the following. Leave your answers in index form.

a) $5^4 \times 5^{-2}$ b) $g^6 \div g^{-6}$ c) $2^{16} \div \dfrac{1}{2^4}$ d) $k^{10} \times k^{-6} \div k^0$

e) $\left(\dfrac{1}{p^4}\right)^5$ f) $\left(\dfrac{l^{-5}}{l^6}\right)^{-3}$ g) $\dfrac{n^{-4} \times n}{(n^{-3})^6}$ h) $\left(\dfrac{10^7 \times 10^{-11}}{10^9 \div 10^4}\right)^{-5}$

Q2 a) Write the number 0.01 as:

(i) a fraction of the form $\dfrac{1}{a}$ (ii) a fraction of the form $\dfrac{1}{10^m}$ (iii) a power of 10.

b) Rewrite the following as powers of 10:

(i) 0.1   (ii) 0.00000001   (iii) 0.0001   (iv) 1

Q3 Evaluate the following. Write the answers as fractions.

a) $3^2 \times 5^{-2}$ b) $2^{-3} \times 7^1$ c) $\left(\dfrac{1}{2}\right)^{-2} \times \left(\dfrac{1}{3}\right)^2$ d) $6^{-4} \div 6^{-2}$

e) $(-9)^2 \times (-5)^{-3}$ f) $8^{-5} \times 8^3 \times 3^3$ g) $10^{-5} \div 10^6 \times 10^4$ h) $\left(\dfrac{3}{4}\right)^{-1} \div \left(\dfrac{1}{2}\right)^{-3}$

Fractional Indices

Learning Objective — Spec Ref N7/A4:
Work with fractional indices.

If the index is a **fraction**, you can rewrite the power as a **root**. The **denominator** tells you the root to use.

$a^{\frac{1}{2}} = \sqrt{a}$ If the index is $\dfrac{1}{2}$ then you replace the power with the **square** root $\sqrt{\ }$.

$a^{\frac{1}{3}} = \sqrt[3]{a}$ If the index is $\dfrac{1}{3}$ then you replace the power with the **cube** root $\sqrt[3]{\ }$.

$a^{\frac{1}{m}} = \sqrt[m]{a}$ More generally, if the index is $\dfrac{1}{m}$ then you replace the power with the **mth root** $\sqrt[m]{\ }$.

$a^{\frac{n}{m}} = \left(\sqrt[m]{a}\right)^n$ If the numerator **isn't 1**, i.e. the index is $\dfrac{n}{m}$, you still use the **mth root** $\sqrt[m]{\ }$ but you also need to **raise the root** to the power of n.

Example 7

Evaluate $27^{\frac{2}{3}}$

1. Split up the index using $(a^m)^n = a^{m \times n}$. $27^{\frac{2}{3}} = 27^{\frac{1}{3} \times 2} = \left(27^{\frac{1}{3}}\right)^2$
2. Write the fractional index as a root. $= \left(\sqrt[3]{27}\right)^2$
3. Evaluate the root — the cube root of 27 is 3. $= 3^2$
4. Evaluate the remaining power. $= \mathbf{9}$

Tip: You could write $\left(27^{\frac{1}{3}}\right)^2$ or $(27^2)^{\frac{1}{3}}$. Here, it's much easier to find the cube root of 27 first and then square it.

Exercise 5

Q1 Rewrite the following expressions in the form $\sqrt[m]{a}$ or $\left(\sqrt[m]{a}\right)^n$.

a) $a^{\frac{1}{5}}$ b) $a^{\frac{3}{5}}$ c) $a^{\frac{2}{5}}$ d) $a^{\frac{5}{2}}$

Evaluate the following expressions.

Q2 a) $64^{\frac{1}{2}}$ b) $64^{\frac{1}{3}}$ c) $16^{\frac{1}{4}}$ d) $1\,000\,000^{\frac{1}{2}}$

Q3 a) $125^{\frac{2}{3}}$ b) $9^{\frac{3}{2}}$ c) $1000^{\frac{5}{3}}$ d) $8000^{\frac{4}{3}}$

7.3 Standard Form

Standard form is useful for writing very big or very small numbers in a more convenient way —
e.g. 56 000 000 000 would be 5.6 × 10¹⁰ in standard form and 0.000 000 003 45 would be 3.45 × 10⁻⁹
in standard form. Any number can be written in standard form and they all follow the same rules.

Standard Form

> **Learning Objective — Spec Ref N9:**
> Write and interpret numbers in standard form.

Any number in **standard form** (or standard index form) must be written **exactly** like this:

$$ A \times 10^n $$

A can be **any number** between 1 and 10 (but not 10 itself) → A
n can be **any integer** ← n

There are a few vital things you need to know about numbers in standard form:

- The **front number**, A, must always be **between 1 and 10** (i.e. $1 \leq A < 10$).
- The power of 10, n, is how far the **decimal point moves**.
- n is **positive** for **BIG** numbers and n is **negative** for **SMALL** numbers.

Tip: It's handy to think of the decimal point moving, but it's the digits that shift around it.

Example 1

Write these numbers in standard form.

a) **360 000**   b) **0.000036**   c) **146.3 million**

a) 1. Move the decimal point until 360 000 becomes 3.6.
   The decimal point has moved 5 places.
   2. The number is big so n must be +5.

   $360000.0 = \mathbf{3.6 \times 10^5}$

b) 1. As in part a), the decimal point moves 5 places again to make the number 3.6.
   2. The number is small so n must be –5.

   $0.000036 = \mathbf{3.6 \times 10^{-5}}$

c) 1. Write the number out in full.
   2. Convert to standard form using the same method as in parts a) and b).

   146.3 million = 146.3 × 1 000 000
   = 146 300 000
   = $\mathbf{1.463 \times 10^8}$

Tip: Common wrong answers to part c) are 146.3 × 10⁶ (which isn't in standard form as $A > 10$) and 1.463 × 10⁶ (which is too small).

Exercise 1

Write the following numbers in standard form.

Q1 a) 250   b) 1100   c) 48 000   d) 5 900 000
    e) 2 750 000 f) 8560   g) 808 080   h) 930 078

Q2 a) 0.0025   b) 0.0067   c) 0.0303   d) 0.000056
    e) 0.375   f) 0.07070   g) 0.00000000021 h) 0.0005002

Q3 a) 0.00567×10^9 b) 95.32×10^2 c) 0.034×10^{-4} d) $845\,000 \times 10^{-3}$

Example 2

Write the following standard form numbers as ordinary numbers.

a) **3.5×10^3**   b) **4.67×10^{-5}**

a) 1. The power is positive so the number will be big. $3.5 \times 10^3 = \mathbf{3\,5\,0\,0.0}$
    2. The decimal point moves 3 places.

b) 1. The power is negative so the number will be small.
    2. The decimal point moves 5 places. $4.67 \times 10^{-5} = \mathbf{0.0\,0\,0\,0\,4\,6\,7}$

Exercise 2

Q1 Write the following as ordinary numbers.

a) 3×10^6 b) 9.4×10^4 c) 8.8×10^5 d) 4.09×10^3

e) 1.989×10^8 f) 6.69×10^1 g) 7.20×10^0 h) 3.56×10^{-6}

i) 8.88×10^{-5} j) 1.9×10^{-8} k) 6.69×10^{-1} l) 7.05×10^{-6}

Multiplying and Dividing in Standard Form

Learning Objective — Spec Ref N9:
Multiply and divide numbers in standard form.

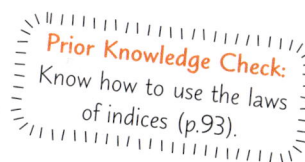

Prior Knowledge Check:
Know how to use the laws of indices (p.93).

To **multiply** or **divide** numbers in standard form, first **rearrange** the calculation so that the **front numbers** and the **powers of 10** are **together** — for example, rewrite $(4 \times 10^6) \times (8 \times 10^2)$ as $(4 \times 8) \times (10^6 \times 10^2)$. Then multiply or divide the front numbers and use the **laws of indices** (see p.93) to multiply or divide the powers of 10. Finally, make sure your answer is still in **standard form** — if not, use the method from p.96.

Example 3

Calculate $(2.4 \times 10^7) \times (5.2 \times 10^3)$. Give your answer in standard form.

1. Rearrange to put the front numbers and powers of 10 together. $(2.4 \times 5.2) \times (10^7 \times 10^3)$
2. Multiply the front numbers and use the laws of indices. $= 12.48 \times 10^{7+3}$
3. 12.48 isn't between 1 and 10 so this isn't in standard form. $= 12.48 \times 10^{10}$
    Convert 12.48 to standard form. $= 1.248 \times 10 \times 10^{10}$
4. Add the indices again to get the answer in standard form. $= \mathbf{1.248 \times 10^{11}}$

Example 4

Calculate $(9.6 \times 10^7) \div (1.2 \times 10^4)$. Give your answer in standard form.

1. Rewrite as a fraction.
2. Separate the front numbers and powers of 10. $\dfrac{9.6 \times 10^7}{1.2 \times 10^4} = \dfrac{9.6}{1.2} \times \dfrac{10^7}{10^4}$
3. Simplify the two fractions. $= 8 \times 10^{7-4} = \mathbf{8 \times 10^3}$

Q1 Calculate the following. Give your answers in standard form.

a) $(3 \times 10^7) \times (4 \times 10^{-4})$

b) $(7 \times 10^9) \times (9 \times 10^{-4})$

c) $(2 \times 10^5) \times (3.27 \times 10^2)$

d) $(3.4 \times 10^{-4}) \times (3 \times 10^2)$

e) $(2 \times 10^{-5}) \times (8.734 \times 10^5)$

f) $(1.2 \times 10^4) \times (5.3 \times 10^6)$

Q2 Calculate the following. Give your answers in standard form.

a) $(3.6 \times 10^7) \div (1.2 \times 10^4)$

b) $(8.4 \times 10^4) \div (7 \times 10^8)$

c) $(1.8 \times 10^{-4}) \div (1.2 \times 10^8)$

d) $(4.8 \times 10^3) \div (1.2 \times 10^{-2})$

e) $(8.1 \times 10^{-1}) \div (0.9 \times 10^{-2})$

f) $(13.2 \times 10^5) \div (1.2 \times 10^4)$

Adding and Subtracting in Standard Form

Learning Objective — Spec Ref N9:
Add and subtract numbers in standard form.

To **add** or **subtract** numbers in standard form, you first need to make the powers of 10 **the same**.
Do this by **multiplying** the **smaller power** of 10 by an appropriate power of 10 —
but make sure to **balance** it out by **dividing** the **front number** by the **same** power of 10.
Then add or subtract the **front numbers** and make sure your answer is still in **standard form** at the end.

Example 5

Calculate: **a) $(3.7 \times 10^4) + (2.2 \times 10^3)$**   **b) $(1.1 \times 10^3) - (9.2 \times 10^2)$**

Give your answers in standard form.

a) 1. The powers of 10 don't match.
   To change 10^3 into 10^4, you
   need to multiply it by 10.
   But then you need to divide
   2.2 by 10 to balance it out.

2. Add the front numbers.

$(3.7 \times 10^4) + (2.2 \times 10^3)$
$= (3.7 \times 10^4) + (0.22 \times 10^4)$
$= (3.7 + 0.22) \times 10^4$
$= \mathbf{3.92 \times 10^4}$

b) 1. Change 9.2×10^2 so that both
   numbers are multiplied by 10^3.

2. Subtract the front numbers.

3. Convert into standard form.

$(1.1 \times 10^3) - (9.2 \times 10^2)$
$= (1.1 \times 10^3) - (0.92 \times 10^3)$
$= (1.1 - 0.92) \times 10^3$
$= 0.18 \times 10^3 = \mathbf{1.8 \times 10^2}$

Tip: Remember, just because a number ends in $\times 10^n$ doesn't mean it's in standard form. The number in front must be between 1 and 10.

Calculate the following, giving your answers in standard form.

Q1 a) $(5.0 \times 10^3) + (3.0 \times 10^2)$

b) $(1.8 \times 10^5) + (3.2 \times 10^3)$

c) $(6.2 \times 10^{-2}) + (4.9 \times 10^{-1})$

d) $(6.9 \times 10^{-4}) + (3.8 \times 10^{-5})$

e) $(3.7 \times 10^{-1}) + (1.1 \times 10^0)$

f) $(5.5 \times 10^7) + (5.5 \times 10^8)$

Q2 a) $(5.2 \times 10^4) - (3.3 \times 10^3)$

b) $(7.2 \times 10^{-3}) - (1.5 \times 10^{-4})$

c) $(6.5 \times 10^2) - (3 \times 10^{-1})$

d) $(8.4 \times 10^2) - (6.3 \times 10^0)$

e) $(8.4 \times 10^4) - (8.3 \times 10^2)$

f) $(28.4 \times 10^{-1}) - (9.3 \times 10^{-2})$

7.4 Surds

Surds are expressions with irrational square roots in them (irrational numbers are ones that you can't write as fractions, e.g. π or $\sqrt{2}$). If a question involving surds asks for an exact answer, you have to leave the surds in.

Multiplying and Dividing Surds

Learning Objective — Spec Ref N8/A4:
Use multiplication and division to simplify surds.

Prior Knowledge Check:
Be able to find the factors of a number. See p.8.

To **multiply** or **divide** two surds, combine them into a **single surd** using the rules below.

$$\sqrt{a} \times \sqrt{b} = \sqrt{a \times b}$$

e.g. $\sqrt{2} \times \sqrt{3} = \sqrt{2 \times 3} = \sqrt{6}$. Also, $(\sqrt{b})^2 = \sqrt{b} \times \sqrt{b} = \sqrt{b \times b} = \sqrt{b^2} = b$

$$\sqrt{a} \div \sqrt{b} = \frac{\sqrt{a}}{\sqrt{b}} = \sqrt{\frac{a}{b}}$$

e.g. $\sqrt{8} \div \sqrt{2} = \frac{\sqrt{8}}{\sqrt{2}} = \sqrt{\frac{8}{2}} = \sqrt{4} = 2$

This is useful for **simplifying** expressions containing surds — the aim is to make the number under the root as **small** as possible or **get rid** of the root completely. To do this, split the number up into two **factors**, one of which should be a **square number**. You can then take the **square root** of this square number to simplify. When you're done, you'll be left with an expression of the form $a\sqrt{b}$ where a and b are **integers** and b is as small as possible.

Example 1

Simplify $\sqrt{72}$.

1. Break 72 down into factors — one of them needs to be a square number.

2. Write as two roots multiplied together.

3. Evaluate $\sqrt{36}$.

$\sqrt{72} = \sqrt{36 \times 2}$

$= \sqrt{36} \times \sqrt{2}$

$= \mathbf{6\sqrt{2}}$

Tip: You could have done $\sqrt{72} = \sqrt{9} \times \sqrt{8}$ $= 3\sqrt{8}$ — but then you'd also have to simplify $\sqrt{8}$ using the same method.

Example 2

Find $\sqrt{5} \times \sqrt{15}$. Simplify your answer.

1. Use the rule $\sqrt{a} \times \sqrt{b} = \sqrt{a \times b}$.

2. Now find factors of 75 so you can simplify — remember that one of the factors needs to be a square number.

$\sqrt{5} \times \sqrt{15} = \sqrt{5 \times 15} = \sqrt{75}$

$= \sqrt{25 \times 3}$

$= \sqrt{25} \times \sqrt{3}$

$= \mathbf{5\sqrt{3}}$

Tip: You could also have done $\sqrt{5} \times \sqrt{15}$ $= \sqrt{5} \times \sqrt{5 \times 3}$ $= \sqrt{5} \times \sqrt{5} \times \sqrt{3}$ $= 5 \times \sqrt{3} = 5\sqrt{3}$

Exercise 1

Q1 Simplify:

a) $\sqrt{12}$ b) $\sqrt{20}$ c) $\sqrt{50}$ d) $\sqrt{32}$

e) $\sqrt{108}$ f) $\sqrt{300}$ g) $\sqrt{98}$ h) $\sqrt{192}$

Q2 Rewrite the following in the form $a\sqrt{b}$, where a and b are integers. Simplify your answers where possible.

a) $\sqrt{2} \times \sqrt{24}$ b) $\sqrt{3} \times \sqrt{12}$ c) $\sqrt{3} \times \sqrt{24}$

d) $\sqrt{2} \times \sqrt{10}$ e) $\sqrt{40} \times \sqrt{2}$ f) $\sqrt{3} \times \sqrt{60}$

g) $\sqrt{7} \times \sqrt{35}$ h) $\sqrt{50} \times \sqrt{10}$ i) $\sqrt{8} \times \sqrt{24}$

Example 3

Find $\sqrt{40} \div \sqrt{10}$.

1. Use the rule $\sqrt{a} \div \sqrt{b} = \sqrt{\dfrac{a}{b}}$.
2. Do the division inside the square root.
3. Simplify.

$$\sqrt{40} \div \sqrt{10} = \sqrt{\frac{40}{10}}$$
$$= \sqrt{4}$$
$$= 2$$

Tip: Don't forget to simplify at the end by evaluating any roots of square numbers.

Example 4

Simplify: a) $\sqrt{\dfrac{1}{4}}$ b) $\sqrt{\dfrac{49}{125}}$

a) Rewrite as two roots, then evaluate $\sqrt{1}$ and $\sqrt{4}$.

$$\sqrt{\frac{1}{4}} = \frac{\sqrt{1}}{\sqrt{4}} = \frac{1}{2}$$

b) 1. Use the rule $\sqrt{\dfrac{a}{b}} = \dfrac{\sqrt{a}}{\sqrt{b}}$.

2. Simplify the surds in the numerator and denominator separately.

$$\sqrt{\frac{49}{125}} = \frac{\sqrt{49}}{\sqrt{125}}$$
$$= \frac{7}{\sqrt{25 \times 5}}$$
$$= \frac{7}{5\sqrt{5}}$$

Tip: You'll usually have to simplify a surd further so that it doesn't have any roots in the denominator. This is covered on page 102.

Exercise 2

Q1 Calculate the exact values of the following. Simplify your answers where possible.

a) $\sqrt{90} \div \sqrt{10}$ b) $\sqrt{72} \div \sqrt{2}$ c) $\sqrt{200} \div \sqrt{8}$ d) $\sqrt{243} \div \sqrt{3}$

e) $\sqrt{294} \div \sqrt{6}$ f) $\sqrt{80} \div \sqrt{10}$ g) $\sqrt{120} \div \sqrt{10}$ h) $\sqrt{180} \div \sqrt{3}$

i) $\sqrt{180} \div \sqrt{9}$ j) $\sqrt{96} \div \sqrt{6}$ k) $\sqrt{484} \div \sqrt{22}$ l) $\sqrt{210} \div \sqrt{35}$

Q2 Simplify the following as far as possible.

a) $\sqrt{\dfrac{1}{9}}$ b) $\sqrt{\dfrac{4}{25}}$ c) $\sqrt{\dfrac{49}{121}}$

d) $\sqrt{\dfrac{100}{64}}$ e) $\sqrt{\dfrac{18}{200}}$ f) $\sqrt{\dfrac{2}{25}}$

g) $\sqrt{\dfrac{108}{147}}$ h) $\sqrt{\dfrac{27}{64}}$ i) $\sqrt{\dfrac{98}{121}}$

Adding and Subtracting Surds

Learning Objective — Spec Ref N8/A4:
Add and subtract surds.

Prior Knowledge Check:
Know how to simplify surds (p.99) and collect like terms (p.75).

You can simplify expressions containing surds by **collecting like terms** — but you can **only** add or subtract terms where the number under the root is **the same**. So you can do $2\sqrt{2} + 3\sqrt{2} = 5\sqrt{2}$ but $\sqrt{2} + \sqrt{3}$ can't be simplified — $\sqrt{a} + \sqrt{b}$ **doesn't equal** $\sqrt{a+b}$.
You'll probably have to **simplify** individual terms first to make the surd parts match.

Example 5

Simplify $\sqrt{12} + 2\sqrt{27}$.

1. Break 12 and 27 down into factors and use the rule $\sqrt{a} \times \sqrt{b} = \sqrt{a \times b}$.

2. Simplify by taking roots of any square numbers.

3. The surds are the same, so collect like terms.

$$\sqrt{12} + 2\sqrt{27} = \sqrt{4 \times 3} + 2\sqrt{9 \times 3}$$
$$= \sqrt{4} \times \sqrt{3} + 2 \times \sqrt{9} \times \sqrt{3}$$
$$= 2\sqrt{3} + 2 \times 3 \times \sqrt{3} = 2\sqrt{3} + 6\sqrt{3}$$
$$= \mathbf{8\sqrt{3}}$$

Exercise 3

Simplify the following as far as possible.

Q1 a) $2\sqrt{3} + 3\sqrt{3}$
    d) $2\sqrt{32} + 3\sqrt{2}$

    b) $7\sqrt{7} - 3\sqrt{7}$
    e) $2\sqrt{27} - 3\sqrt{3}$

    c) $2\sqrt{3} + 3\sqrt{7}$
    f) $5\sqrt{7} + 3\sqrt{28}$

Q2 a) $2\sqrt{125} - 3\sqrt{80}$

    b) $\sqrt{108} + 2\sqrt{300}$

    c) $5\sqrt{294} - 3\sqrt{216}$

Multiplying Brackets Using Surds

Learning Objective — Spec Ref N8/A4:
Expand brackets involving surds.

Prior Knowledge Check:
Be able to expand brackets.
See p.77-79.

Multiply out brackets with surds in them in the same way as you multiply out brackets with **variables**. After expanding, **simplify** the surds that remain if possible. Here are two **special cases** to keep in mind:

- $(a + \sqrt{b})^2 = a^2 + 2a\sqrt{b} + \sqrt{b}^2 = \mathbf{a^2 + 2a\sqrt{b} + b}$
- $(a + \sqrt{b})(a - \sqrt{b}) = a^2 - a\sqrt{b} + a\sqrt{b} - \sqrt{b}^2 = \mathbf{a^2 - b}$ — this is the **difference of two squares** (page 84).

Example 6

Expand and simplify $(3 - \sqrt{7})^2$.

1. Write as two sets of brackets and expand.

2. Simplify $\sqrt{7} \times \sqrt{7}$ (p.99).

3. Collect like terms.

$$(3 - \sqrt{7})^2 = (3 - \sqrt{7})(3 - \sqrt{7})$$
$$= (3 \times 3) + (3 \times -\sqrt{7})$$
$$+ (-\sqrt{7} \times 3) + (-\sqrt{7} \times -\sqrt{7})$$
$$= 9 - 3\sqrt{7} - 3\sqrt{7} + \sqrt{7 \times 7}$$
$$= 9 - 3\sqrt{7} - 3\sqrt{7} + 7 = \mathbf{16 - 6\sqrt{7}}$$

Tip: You could just use the first of the special cases above to expand these brackets.

Example 7

Expand and simplify $(1 + \sqrt{3})(2 - \sqrt{8})$.

1. Expand the brackets first.

2. Simplify the surds — here you can simplify $\sqrt{8}$ and $\sqrt{24}$.

$$(1 + \sqrt{3})(2 - \sqrt{8})$$
$$= (1 \times 2) + (1 \times -\sqrt{8}) + (\sqrt{3} \times 2) + (\sqrt{3} \times -\sqrt{8})$$
$$= 2 - \sqrt{8} + 2\sqrt{3} - \sqrt{24}$$
$$= 2 - \sqrt{4 \times 2} + 2\sqrt{3} - \sqrt{4 \times 6}$$
$$= \mathbf{2 - 2\sqrt{2} + 2\sqrt{3} - 2\sqrt{6}}$$

Exercise 4

Expand these brackets and simplify where possible.

Q1   a) $(2 + \sqrt{3})^2$ b) $(1 + \sqrt{2})(1 - \sqrt{2})$ c) $(5 - \sqrt{2})^2$

       d) $(3 - 3\sqrt{2})(3 - \sqrt{2})$ e) $(5 + \sqrt{3})(3 + \sqrt{3})$ f) $(7 + 2\sqrt{2})(7 - 2\sqrt{2})$

Q2   a) $(2 + \sqrt{6})(4 + \sqrt{3})$ b) $(4 - \sqrt{7})(5 - \sqrt{2})$ c) $(1 - 2\sqrt{10})(6 - \sqrt{15})$

Rationalising the Denominator

Learning Objective — Spec Ref N8/A4:
Rationalise the denominator.

'**Rationalising the denominator**' means 'getting rid of surds from the bottom of a fraction'. For the simplest type, you do this by **multiplying** the top and bottom of the fraction by the surd.

$$\frac{a}{\sqrt{b}} = \frac{a}{\sqrt{b}} \times \frac{\sqrt{b}}{\sqrt{b}} = \frac{a\sqrt{b}}{\sqrt{b} \times \sqrt{b}} = \frac{a\sqrt{b}}{b}$$

Example 8

Rationalise the denominators of: a) $\dfrac{5}{2\sqrt{15}}$ b) $\dfrac{2}{\sqrt{8}}$

a) 1. Multiply by $\dfrac{\sqrt{15}}{\sqrt{15}}$ to eliminate $\sqrt{15}$ from the denominator. Remember that $\sqrt{15} \times \sqrt{15} = 15$.

$$\frac{5}{2\sqrt{15}} = \frac{5}{2\sqrt{15}} \times \frac{\sqrt{15}}{\sqrt{15}}$$
$$= \frac{5\sqrt{15}}{2\sqrt{15} \times \sqrt{15}} = \frac{5\sqrt{15}}{2 \times 15}$$

    2. Simplify the fraction.

$$= \frac{5\sqrt{15}}{30} = \frac{\sqrt{15}}{6}$$

b) 1. Multiply by $\dfrac{\sqrt{8}}{\sqrt{8}}$.

$$\frac{2}{\sqrt{8}} = \frac{2}{\sqrt{8}} \times \frac{\sqrt{8}}{\sqrt{8}} = \frac{2\sqrt{8}}{\sqrt{8} \times \sqrt{8}}$$

    2. Simplify the surd.

$$= \frac{2\sqrt{4 \times 2}}{8} = \frac{2 \times 2\sqrt{2}}{8}$$

    3. Simplify the fraction.

$$= \frac{4\sqrt{2}}{8} = \frac{\sqrt{2}}{2}$$

> **Tip:** You could also simplify the surd while it's still on the denominator and then multiply top and bottom by the surd that remains.

Q1 Rationalise the denominators of the following fractions. Simplify your answers as far as possible.

a) $\dfrac{6}{\sqrt{6}}$

b) $\dfrac{8}{\sqrt{8}}$

c) $\dfrac{5}{\sqrt{5}}$

d) $\dfrac{1}{\sqrt{3}}$

e) $\dfrac{15}{\sqrt{5}}$

f) $\dfrac{9}{\sqrt{3}}$

g) $\dfrac{7}{\sqrt{12}}$

h) $\dfrac{12}{\sqrt{1000}}$

Q2 Rationalise the denominators of the following fractions. Simplify your answers as far as possible.

a) $\dfrac{1}{5\sqrt{5}}$

b) $\dfrac{1}{3\sqrt{3}}$

c) $\dfrac{3}{4\sqrt{8}}$

d) $\dfrac{3}{2\sqrt{5}}$

e) $\dfrac{2}{7\sqrt{3}}$

f) $\dfrac{1}{6\sqrt{12}}$

g) $\dfrac{10}{7\sqrt{5}}$

h) $\dfrac{5}{9\sqrt{10}}$

If the denominator is the **sum** or **difference** of an integer and a surd (e.g. $1 + \sqrt{2}$), use the **difference of two squares** (see p.101) to eliminate the surd: $(a + \sqrt{b})(a - \sqrt{b}) = a^2 - b$. So if the denominator is $a + \sqrt{b}$, multiply by $a - \sqrt{b}$. If it's $a - \sqrt{b}$ then multiply by $a + \sqrt{b}$.

Example 9

Rationalise the denominator of $\dfrac{2 + 2\sqrt{2}}{1 - \sqrt{2}}$.

1. Multiply top and bottom by $(1 + \sqrt{2})$ to get rid of the surd in the denominator.

$$\frac{2 + 2\sqrt{2}}{1 - \sqrt{2}} = \frac{(2 + 2\sqrt{2})(1 + \sqrt{2})}{(1 - \sqrt{2})(1 + \sqrt{2})}$$

2. Expand the brackets in the numerator and denominator

$$= \frac{2 + 2\sqrt{2} + 2\sqrt{2} + 4}{1 + \sqrt{2} - \sqrt{2} - 2}$$

3. Simplify any remaining surds and the fraction.

$$= \frac{6 + 4\sqrt{2}}{-1} = -6 - 4\sqrt{2}$$

Tip: Always multiply by the numbers in the denominator but change the sign of the surd.

Exercise 6

Q1 Rationalise the denominators of the following fractions. Simplify your answers as far as possible.

a) $\dfrac{1}{2 + \sqrt{2}}$

b) $\dfrac{5}{1 - \sqrt{7}}$

c) $\dfrac{10}{5 + \sqrt{11}}$

d) $\dfrac{9}{12 - 3\sqrt{17}}$

Q2 Rewrite the following as fractions with rational denominators in their simplest form.

a) $\dfrac{\sqrt{2}}{2 + 3\sqrt{2}}$

b) $\dfrac{1 + \sqrt{2}}{1 - \sqrt{2}}$

c) $\dfrac{2 + \sqrt{3}}{1 - \sqrt{3}}$

d) $\dfrac{1 - \sqrt{5}}{2 - \sqrt{5}}$

e) $\dfrac{1 + 2\sqrt{2}}{1 - 2\sqrt{2}}$

f) $\dfrac{7 + 8\sqrt{2}}{9 + 5\sqrt{2}}$

Q3 Show that $\dfrac{1}{1 - \dfrac{1}{\sqrt{2}}}$ can be written as $2 + \sqrt{2}$.

Q4 Show that $\dfrac{1}{1 + \dfrac{1}{\sqrt{3}}}$ can be written as $\dfrac{3 - \sqrt{3}}{2}$.

Q1 Simplify the following. Leave your answers in index form.

    a) $7^6 \times 7^9$ b) $d^{-4} \div d^6$ c) $(4^8)^3$

    d) $9^{-2} \times \sqrt[4]{9}$ e) $(c^{10} \div c^2)^{\frac{1}{4}}$ f) $2^4 \times \dfrac{1}{\sqrt[3]{2}} \times 2^{-\frac{1}{2}}$

Q2 Simplify the following.

    a) $(27m)^{\frac{1}{3}}$ b) $(y^4 z^3)^{-\frac{3}{4}}$ c) $\left(\dfrac{b^9}{64c^3}\right)^{\frac{2}{3}}$ d) $\sqrt[4]{u^2} \times (2u)^{-2}$

Q3 Write the following as powers of 2.

    a) 8   b) $\sqrt{2}$ c) $8\sqrt{2}$ d) $\dfrac{1}{8\sqrt{2}}$

Q4 Write the following expressions.

    a) $3 \times \sqrt[3]{3}$ as a power of 3   b) $16\sqrt{4}$ as a power of 4   c) 5 as a power of 25

    d) 2 as a power of 8   e) $\dfrac{\sqrt{10}}{1000}$ as a power of 10   f) $\dfrac{81}{\sqrt[3]{9}}$ as a power of 3

Q5 Albert measured the length of his favourite hair each day for three days. On the first day it grew 3.92×10^{-4} m, on the second day it grew 3.77×10^{-4} m, and on the third day it grew 4.09×10^{-4} m. By how much in metres did the hair grow in total over the three days? Give your answer in standard form.

Q6 The Hollywood film 'The Return of Dr Arzt' cost $\$3.45 \times 10^8$ to make. It made a total of $\$8.9 \times 10^7$ at the box office. What was the loss made by the film in standard form?

Q7 To 3 significant figures, the mass of the Earth is 5.97×10^{24} kg, and the mass of the Sun is 3.33×10^5 times the mass of the Earth. What is the mass of the Sun in kg? Give your answer in standard form to 3 significant figures.

Q8 Write the following as simply as possible.

    a) $\sqrt{96} + 5\sqrt{18}$ b) $\dfrac{\sqrt{48} + \sqrt{363}}{5}$ c) $\sqrt{8} - \dfrac{\sqrt{36}}{2}$

    d) $-\sqrt{98} - \sqrt{2} \times \sqrt{162}$ e) $12\sqrt{99} \div \sqrt{176}$ f) $\dfrac{\sqrt{88}}{\sqrt{32}} + \sqrt{1100}$

Q9 Write the following in the form $a + b\sqrt{c}$ where a, b and c are integers.

    a) $\dfrac{121\sqrt{7}}{-\sqrt{11}}$ b) $\dfrac{60}{6 + \sqrt{6}}$ c) $\dfrac{7 + \sqrt{2}}{\sqrt{8} - 3}$

Q1 Simplify:

a) $2x^3 \times 4x^4$

[1 mark]

b) $(3y^2)^4$

[2 marks]

c) $5z^0$

[1 mark]

Q2 Showing every step of your working, prove that $\left(\dfrac{4}{9}\right)^{-\frac{3}{2}} = 3\dfrac{3}{8}$.

[3 marks]

Q3 A square has a side length of $(3 + 2\sqrt{5})$ cm. Work out the area of the square in cm^2, giving your answer in the form $a + b\sqrt{5}$ where a and b are integers.

[2 marks]

Q4 A smartphone has a 2.4 gigahertz (GHz) processor which means it can do 2400 million calculations in a second.

a) Write 2400 million in standard form.

[1 mark]

b) Work out how many calculations the smartphone can do in a minute, giving your answer in standard form.

[2 marks]

c) To download a particular app to this phone will require it to do 1.8×10^{11} calculations. Work out how long in seconds this will take, giving your answer as an ordinary number.

[2 marks]

Q5 A rectangle has a width of $(5 - \sqrt{10})$ cm and an area of $\sqrt{360}$ cm^2. Find the length of the rectangle, giving your answer in the form $a + b\sqrt{10}$ where a and b are integers.

[4 marks]

Q6 Earth is approximately 4.54×10^9 years old. Humans are thought to have evolved around 2.5×10^5 years ago. For what percentage of the age of the Earth have humans been present? Give your answer as an ordinary number to 3 significant figures.

[4 marks]

Q7 $x = y \times 10^z$ where $4 < y < 10$
Find an expression for x^2 in standard form.

[3 marks]

8.1 Writing Formulas

A formula is a mathematical set of instructions that you can use for working something out.
For example, s = 4t + 3 is a formula for s — it tells you how to find s, given the value of t.
You can write formulas to help solve real-life problems mathematically.

Learning Objective — Spec Ref A3/A21:
Use given information to write a formula.

Prior Knowledge Check:
Be able to simplify algebraic
expressions. See p.75-76.

To write a **formula**, you turn **information** about how different **quantities** relate to each other into **mathematical operations**. **Letters** represent quantities — they're known as **variables**. If a variable is **multiplied** by a number or another variable, you can write them without the × sign, e.g. $4 \times a = 4a$. For example, to find a formula for the amount someone is **paid**, *P*, if you know they're paid **£9 an hour** and work for *h* **hours**, **multiply** the number of hours they work by £9. So as a formula, this is $P = 9h$.

Example 1

The cost (*C*) of hiring a bike is £5 per hour plus a fixed cost of £25.
Write a formula for the cost of hiring a bike for *h* hours.

1. Multiply the number of hours (*h*) by the cost per hour.

2. Add on the fixed cost.

Cost (in £) for *h* hours = $5h$

So $C = 5h + 25$

Tip: The fixed cost doesn't depend on *h* so it can just be added on at the end. Make sure the units are consistent.

Exercise 1

Q1 Claudia owns *f* films. Barry owns twice as many films as Claudia.
   a) How many films does Barry own?
   b) How many films do Claudia and Barry own in total?
   c) How many films would they own in total if they each gave away 3 of their films?

Q2 Alf has £18 in the bank. He gets a job and is paid £8 for every hour he works. Assuming he spends nothing, write a formula for the amount of money (£*M*) Alf will have after he has worked for *h* hours.

Q3 The instructions for cooking a goose are to cook for 50 minutes per kg, plus 25 minutes. Write a formula to find the time taken (*t* minutes) to cook a goose weighing *n* kg.

Q4 Write a formula for the cost (£*C*) of having *t* trees cut down if it costs *p* pounds per tree plus a fixed amount of £30.

Q5 A sequence of shapes is made out of matchsticks.
The first shape in the sequence is made from 4 matchsticks.
Each subsequent shape in the sequence is made by adding 3 matchsticks to the previous shape.
Write a formula for the number of matchsticks (*M*) needed to make the *n*th shape in the sequence.

8.2 Substituting into a Formula

Once you have the formula for a problem, you can substitute in values to find a solution.

Learning Objective — Spec Ref A2:
Substitute values into a given formula.

Prior Knowledge Check:
Be able to use BODMAS. See p.2.

You can **evaluate** a formula by replacing the **letters** in the formula with actual **values**. This is called **substitution**. Here's the method to follow:

- Write the formula out.
- Write it out again, substituting numbers for letters.
- Work out the calculation — using the correct order of operations (BODMAS).

Example 1

Use the formula $v = u + at$ to find v if $u = 2.6$, $a = -18.3$ and $t = 4.9$.

1. Write the formula out. $v = u + at$
2. Replace each letter with its value. $v = 2.6 + (-18.3 \times 4.9)$
3. Work out the calculation step by step — $v = 2.6 + (-89.67)$
   you do multiplication before addition. $v = -87.07$

Tip: This is the formula for the final velocity (v) of an object, where u is the initial velocity, a is its constant acceleration and t is time.

Exercise 1

Q1 If $x = 4$ and $y = 3$, find z when:
    a) $z = x + 2$ b) $z = y - 1$ c) $z = x + y$
    d) $z = 3y$ e) $z = 3y - 2$ f) $z = 6x - y$

Q2 If $a = -4$ and $b = -3$, find c when:
    a) $c = a - 4$ b) $c = 4b$ c) $c = 6b - a$
    d) $c = b^3$ e) $c = -\dfrac{4b}{2a}$ f) $c = 5a - b^2$

Q3 If $r = \dfrac{3}{4}$ and $s = -\dfrac{1}{3}$, find q when:
    a) $q = 4r$ b) $q = -2s$ c) $q = rs$
    d) $q = \dfrac{s}{r}$ e) $q = r + s$ f) $q = 4r + s$

In Questions 4 and 5, give all rounded answers to 2 decimal places.

Q4 Use the formula $v = u + at$ to find v if:
    a) $u = 3$, $a = 7$ and $t = 5$ b) $u = 12$, $a = 17$ and $t = 15$
    c) $u = 2.3$, $a = 4.1$ and $t = 3.4$ d) $u = 5.25$, $a = 9.81$ and $t = 4.39$
    e) $u = 3$, $a = -10$ and $t = 5.6$ f) $u = -34$, $a = -1.37$ and $t = 63.25$

Q5 If $x = 12$, $y = 2.5$ and $z = -0.25$, find w if:
    a) $w = x + 2y - 4z$ b) $w = -3x + y^3 - (2z)^2$ c) $w = 0.5x - yz$
    d) $w = -2x^3 + y^2z$ e) $w = -\dfrac{12}{x} + \dfrac{y}{z}$ f) $w = \dfrac{x^2 + 3y - 8z}{2y^2}$

Wordy problems work in the same way — you write out the formula and substitute in the values. Be careful with **units** — check if the units in the **question** match the units used in the **formula**, and **convert** them if needed (see Section 22).

Example 2

Theo decides to play a game of crazy golf.

The cost of hiring crazy golf equipment, £*C*, is a fixed price of £3 plus 8p for every minute of use. For *g* minutes of crazy golf this can be written as the formula *C* = 0.08*g* + 3.

Theo plays for 2 hours and 30 minutes. How much will hiring the equipment cost him?

1. The formula is for time in minutes, so convert the time in hours to minutes.

2 hours = 2 × 60 = 120 minutes
120 + 30 = 150 minutes
So *g* = 150 minutes

2. Write out the formula.

$C = 0.08g + 3$

3. Replace each letter with its value.

$C = 0.08(150) + 3$

4. Work out the calculation.

$C = 12 + 3 = 15$

So it will cost Theo **£15** to hire the equipment.

Tip: Make sure you give the final answer in the correct units and link it back to the context.

Exercise 2

Q1 The formula for working out the average speed (*s*, in metres per second) of a moving object is $s = \dfrac{d}{t}$, where *d* is the distance travelled (in metres) and *t* is the time taken (in seconds). Find the speed (in metres per second, to 2 d.p.) of each of the following:

a) a runner who travels 800 metres in 110 seconds

b) a cheetah that travels 400 metres in 14 seconds

c) a car that travels 1 km in 1 minute

d) a plane that travels 640 km in 1 hour

Q2 Use the formula $c = \dfrac{5}{9}(f - 32)$ to convert the following temperatures in degrees Fahrenheit (*f*) to degrees Celsius (*c*).

a) 212 °F   b) 68 °F   c) −40 °F   d) 98.6 °F

Q3 The sum (*S*) of the numbers 1 + 2 + 3 + ... + *n* is given by the formula $S = \dfrac{1}{2}n(n + 1)$. Work out the sum for each of the following.

a) 1 + 2 + 3 + ... + 10   b) 1 + 2 + 3 + ... + 100   c) 1 + 2 + 3 + ... + 1000

Q4 Find the volumes (*V* cm³) of the cylinders below to 2 decimal places, using the formula on the left.

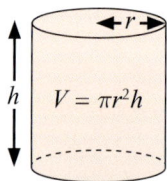

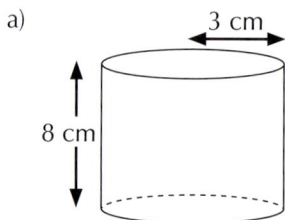

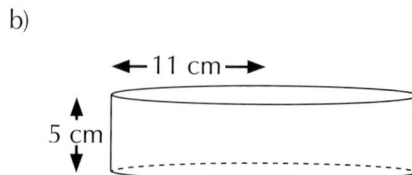

a)

3 cm

8 cm

b)

11 cm

5 cm

h $V = \pi r^2 h$ *r*

8.3 Rearranging Formulas

Rearranging formulas means making a different letter the subject, e.g. getting '$y = \frac{x}{2}$' from '$x = 2y$'.

Learning Objective — Spec Ref A5:
Rearrange formulas to change the subject.

Prior Knowledge Check:
Be able to simplify expressions (see p.75-76), expand brackets (see p.77) and factorise expressions (see p.80-81).

To **rearrange** a formula to make a **different letter** the **subject**, perform **inverse operations** one by one until that letter is **on its own** on one side of the '=' sign. You might have to **expand brackets**, **factorise** and **collect like terms** together to get the subject on its own.

Tip: An inverse operation does the opposite of (or 'undoes') the original operation — e.g. the inverse of +8 is –8.

You're aiming to end up with something in the form '**Ax = B**' (where x is the **subject term** and A and B are **numbers**, **letters** or a mix of both). You then **divide** both sides by A to get '$x = ...$'.

If you ended up with '$Ax^2 = B$', you'd need to take **square roots** after dividing by A. If you do this, remember that there's a **negative root** as well as a **positive root**, so you'll need a ± sign.

Example 1

Make x the subject of the formula $y = 4 + dx$.

1. You need to make x the subject, which means you need to get x on its own.

   $y = 4 + dx$

2. To get the dx term on its own, subtract 4 from both sides.

   $y - 4 = dx$

3. You now have the form '$Ax = B$', where $A = d$ and $B = y - 4$. To get x on its own, divide both sides by d.

   $\frac{y-4}{d} = x$

4. Write as '$x = $'.

   $x = \frac{y-4}{d}$

Tip: Subtraction is the inverse of addition, and division is the inverse of multiplication.

Example 2

Make y the subject of the formula $w = \frac{1-y}{2}$.

1. To make y the subject, you need to get it on its own.

   $w = \frac{1-y}{2}$

2. Multiply both sides by 2 (the denominator of the fraction) to get rid of the fraction.

   $2w = 1 - y$

3. Add y to both sides (so it's positive).

   $2w + y = 1$

4. Subtract $2w$ from each side to get y on its own.

   $y = 1 - 2w$

Example 3

Make r the subject of $V = \frac{4}{3}\pi r^3$.

1. To make r the subject, you need to get it on its own.

   $V = \frac{4}{3}\pi r^3$

2. Multiply both sides by 3 to get rid of the fraction.

   $3V = 4\pi r^3$

3. Divide both sides by 4π.

   $\frac{3V}{4\pi} = r^3$

4. Take the cube root of each side and write as '$r = $'.

   $\sqrt[3]{\frac{3V}{4\pi}} = r \Rightarrow r = \sqrt[3]{\frac{3V}{4\pi}}$

Tip: This is the formula for the volume of a sphere of radius r — see p.368.

If the subject appears **more than once**, you're going to have to do some **factorising** (see p.80).

Example 4

Make a the subject of the formula $x(a + 1) = 3(1 - 2a)$.

1. Multiply out the brackets.

$$x(a + 1) = 3(1 - 2a)$$
$$ax + x = 3 - 6a$$

2. Collect all the a terms on one side and the non-a terms on the other.

$$ax + 6a = 3 - x$$

3. Factorise the left-hand side to get it into the form to '$Aa = B$'.

$$a(x + 6) = 3 - x$$

4. Divide by $(x + 6)$ to get a on its own.

$$a = \frac{3 - x}{x + 6}$$

Exercise 1

Q1 Make x the subject of each of the following formulas.

a) $y = x + 2$
b) $2z = 3r + x$
c) $y = 4x$
d) $k = 2(1 + 2x)$
e) $v = \frac{2}{3}x - 2$
f) $y + 1 = \frac{x - 1}{3}$

Q2 Consider the formula $w = \frac{1}{1 + y}$.

a) Multiply both sides of the formula by $1 + y$.
b) Hence make y the subject of the formula.

Q3 Make y the subject of the following formulas.

a) $w = \frac{3}{2y}$
b) $z + 2 = \frac{2}{1 - y}$
c) $uv = \frac{1}{1 - 2y}$
d) $a + b = \frac{2}{4 - 3y}$

Q4 Consider the formula $2k = 12 - \sqrt{w - 2}$.

a) Make $\sqrt{w - 2}$ the subject of the formula.

b) By first squaring both sides of your answer to part a), make w the subject of the formula.

Q5 Make w the subject of the following formulas.

a) $a = \sqrt{w}$
b) $x = 1 + \sqrt{w}$
c) $y = \sqrt{w - 2}$
d) $f - 3 = 2\sqrt{w}$
e) $j = \sqrt{3 + 4w}$
f) $a = \sqrt{1 - 2w}$

Q6 Consider the formula $t = 1 - 3(z + 1)^2$.

a) Make $(z + 1)^2$ the subject of the formula.

b) By first square rooting both sides of your answer to part a), make z the subject of the formula.

Q7 Make z the subject of the following formulas.

a) $x = 1 + z^2$
b) $2t = 3 - z^2$
c) $xy = 1 - 4z^2$
d) $t + 2 = 3(z - 2)^2$
e) $g = 4 - (2z + 3)^2$
f) $r = 4 - 2(5 - 3z)^2$

Q8 Make a the subject of the following formulas.

a) $x(a + b) = a - 1$
b) $x - ab = c - ad$
c) $c = \frac{1 + a}{1 - 2a}$
d) $2e = \frac{2 + 3a}{a}$

Review Exercise

Q1 Melanie has half as many sweets as Jane.

    a) If Jane has j sweets, write a formula to calculate the number of sweets (m) that Melanie has.

    b) Find m when $j = 24$.

Q2 To book a swimming pool for a party, there is a fixed charge of £30 plus a fee of £1.25 for each person who attends.

    a) Write a formula to calculate the hire cost (£C) for n people.

    b) Calculate C when $n = 32$.

Q3 To hire skates at the park there is a fixed charge of £5, plus a charge of £1.70 for each half-hour.

    a) Write a formula to calculate the cost (£C) for h half-hour periods.

    b) Calculate the cost of hiring skates for two and a half hours.

    c) Rearrange your formula to make h the subject.

    d) Asher spends £15.20 on hiring skates. How long was he skating for?

Q4 The surface area (A cm^2) of the shape on the right is given approximately by the formula $A = 21.5d^2$.

    a) Rearrange the formula to make d the subject.

    b) Find d if $A = 55$ cm^2. Give your answer to 2 s.f.

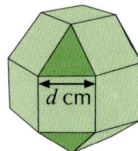

Q5 The time in minutes (T) taken to cook a joint of beef is given by $T = 35w + 25$, where w is the weight of the joint in kg.

    a) How long would it take to cook a 1.5 kg joint?

    b) Make w the subject of the formula.

    c) What weight of beef needs to be cooked for 207 minutes?

Q6 For each of the following formulas, (i) make x the subject, and (ii) find x when $y = -1$.

    a) $-2 + y = \dfrac{3}{4-x}$ b) $y = \dfrac{1}{\sqrt{1-x}}$ c) $2(1-x) = y(3+x)$

    d) $y = \dfrac{2-3x}{1+2x}$ e) $y = 8 - \dfrac{1}{\sqrt{x}}$ f) $2y - 1 = 3\sqrt{2-x}$

Q7 Consider the formula $s = \left(\dfrac{u+v}{2}\right)t$. By rearranging the formula where necessary, find the value of:

    a) s when $u = 2.3$, $v = 1.7$ and $t = 4$.   b) t when $s = 3.3$, $u = 1$ and $v = 2$.

    c) u when $s = 4.5$, $t = 6$ and $v = 7$.   d) v when $s = 0.5$, $t = 0.25$ and $u = 3$.

Q8 Consider the formula $x = \dfrac{1 + \sqrt{y+3}}{2-z}$.

    a) Find the value of x when $y = 1$ and $z = -1$.

    b) By first rearranging the formula, find the value of z when $y = 6$ and $x = -2$.

Exam-Style Questions

Q1 Olivia is on holiday in Las Vegas. She sees a TV weather forecast which reports that today's maximum temperature will be 104° Fahrenheit. The formula $C = \frac{5}{9}(F - 32)$ can be used to convert temperatures in Fahrenheit (F) to Celsius (C).
Work out today's forecast maximum temperature in degrees Celsius.

[2 marks]

Q2 Florence has some matchsticks. The number of matchsticks (m) needed to make a pattern of h hexagons is given by the formula $m = 5h + 1$.
a) How many matchsticks will Florence need to make a pattern of 6 hexagons?

[1 mark]

b) (i) Rearrange the formula to make h the subject.

[2 marks]

(ii) How many hexagons will be in the pattern made with 36 matchsticks?

[1 mark]

Q3 The minimum velocity required for a rocket to leave a planet can be found using the formula $V = \sqrt{\frac{2GM}{r}}$. Make M the subject of this formula. Show your working.

[3 marks]

Q4 You are given the formula $g = \frac{8}{5}h + 17$.
a) Rearrange the formula to make h the subject.

[2 marks]

b) Find h if:
(i) $g = 209$

[1 mark]

(ii) $g = -15$

[1 mark]

Q5 A quarterly gas bill has a fixed charge of £7.50 plus 8p for every unit of gas used.

a) Write the formula to calculate £C, the cost for n units of gas.

[2 marks]

b) José uses 760 units of gas. How much will he have to pay?

[1 mark]

c) Rearrange your formula from part a) to make n the subject.

[2 marks]

d) Anna's gas bill is £39.50. How many units of gas did she use?

[1 mark]

9.1 Solving Equations

Solving equations is one of the most important bits of maths. Solving means finding the value of an unknown letter (usually x) that satisfies the equation, which you do by rearranging it until it's in the form 'x = number'.

Learning Objective — Spec Ref A17:
Be able to solve algebraic equations.

Prior Knowledge Check:
Be able to rearrange algebraic expressions. See Section 6.

To solve an equation, you're going to have to do some rearranging. To rearrange, you always have to **do the same thing to both sides** of the equation — so if you add 7 to one side, you have to add 7 to the other side as well. If you **multiply** or **divide** by a number, you have to multiply or divide **everything** on both sides. Remember, you want to end up with '**x = number**' — so keep this in mind when rearranging.

Example 1

Solve: a) $15 - 2x = 7$ b) $\frac{x}{3} + 2 = -3$

a) 1. Add $2x$ to both sides (to make the coefficient of x positive).

    2. Subtract 7 from both sides.

    3. Divide both sides by 2.

$$15 - 2x = 7$$
$$15 = 7 + 2x$$
$$8 = 2x$$
$$x = 4$$

b) 1. Start by subtracting 2 from both sides. Be careful with the negative numbers here.

    2. Multiply both sides by 3 to get rid of the fraction.

$$\frac{x}{3} + 2 = -3$$
$$\frac{x}{3} = -5$$
$$x = -15$$

Tip: If it helps, write out the working in full:
$$15 - 2x = 7$$
$$15 - 2x + 2x = 7 + 2x$$
$$15 = 7 + 2x$$
$$15 - 7 = 7 + 2x - 7$$
$$8 = 2x$$
$$8 \div 2 = 2x \div 2$$
$$4 = x$$

Exercise 1

Q1 Solve each of the following equations.

    a) $x + 9 = 12$

    b) $x - 7.3 = 1.6$

    c) $12 - x = 9$

    d) $9x = 54$

    e) $-5x = 50$

    f) $40x = -32$

Q2 Find the value of x by solving the following equations.

    a) $\frac{x}{3} = 2$

    b) $\frac{x}{2} = 3.2$

    c) $-\frac{x}{0.2} = 3.2$

    d) $\frac{2x}{5} = 6$

Q3 Solve each of the following equations.

    a) $8x + 10 = 66$

    b) $1.8x - 8 = -62$

    c) $8 - 7x = 22$

    d) $\frac{x}{2} - 1 = 2$

    e) $15x + 12 = 72$

    f) $1.5x - 3 = -24$

    g) $17 - 10x = 107$

    h) $-\frac{2x}{3} - \frac{3}{4} = \frac{1}{4}$

    i) $-\frac{3x}{5} + \frac{1}{3} = \frac{2}{3}$

If the equation has **brackets** in it, **expand** the brackets (see p.77-79) before solving the equation.

see p.77-79

Example 2

Solve the equation $8(x + 2) = 36$.

1. Expand the brackets by multiplying them out.
2. Subtract 16 from both sides.
3. Divide both sides by 8.

$$8(x + 2) = 36$$
$$8x + 16 = 36$$
$$8x = 20$$
$$x = 2.5$$

Tip: You could also start by dividing both sides by 8 to get $x + 2 = (36 \div 8) = 4.5$.

Exercise 2

Q1 Solve each of the following equations by expanding the brackets.

a) $7(x + 4) = 63$
b) $13(x - 4) = -91$
c) $18(x - 3) = -180$
d) $2.5(x + 4) = 30$
e) $3.5(x + 6) = 63$
f) $4.5(x + 3) = 72$
g) $315 = 21(6 - x)$
h) $171 = 4.5(8 - x)$

Q2 Multiply both sides of the equation $\dfrac{1}{x - 2} = 3$ by $(x - 2)$, and hence solve the equation $\dfrac{1}{x - 2} = 3$.

Q3 Solve each of the following equations.

a) $\dfrac{1}{x} = 2$
b) $\dfrac{2}{x} = 5$
c) $\dfrac{12}{x - 2} = 4$
d) $\dfrac{3}{1 - 2x} = 2$

If you have x-terms on **both sides** of the equation, rearrange so you have all the **x-terms** on **one side** of the equation, and all the **numbers** on the **other side**. You can then **collect like terms** (see p.75) and solve.

see p.75

Example 3

Solve the equation $5(x + 2) = 3(x + 6)$.

1. Multiply out the brackets.
2. Rearrange so that all the x-terms are on one side and all the numbers are on the other side.
3. Divide by 2 to find the value of x.

$$5(x + 2) = 3(x + 6)$$
$$5x + 10 = 3x + 18$$
$$5x - 3x = 18 - 10$$
$$2x = 8$$
$$x = 4$$

Exercise 3

Q1 Solve each of the following equations.

a) $6x - 4 = 2x + 16$
b) $17x - 2 = 7x + 8$
c) $6x - 12 = 51 - 3x$
d) $5x - 13 = 87 - 5x$
e) $10x - 18 = 11.4 - 4x$
f) $4x + 9 = 6 - x$

Q2 Solve each of the following equations.

a) $3(x + 2) = x + 14$
b) $5(x + 3) = 2x + 57$
c) $7(x - 7) = 2(x - 2)$
d) $5(x - 4) = 3(x + 8)$
e) $4(x - 3) = 3(x - 8)$
f) $11(x - 2) = 3(x + 6)$

Q3 Solve each of the following equations.

a) $7\left(2x + \dfrac{1}{7}\right) = 8\left(3x - \dfrac{1}{2}\right)$
b) $7(x - 1) = 4(6.2 - 2x)$
c) $-3(x - 3) = 8(0.7 - x)$

If you have **fractions** on **both sides** of the equation, you can **cross-multiply** to save time. All you do is multiply the **numerator** of each fraction by the **denominator** of the other.

$\dfrac{a}{b} \diagdown= \diagup \dfrac{c}{d}$ becomes $a \times d = c \times b$

Example 4

Solve the equation $\dfrac{x-2}{2} = \dfrac{6-x}{6}$.

1. Cross-multiply. This is the same as multiplying both sides by 2 and by 6.

$\dfrac{x-2}{2} \diagdown= \diagup \dfrac{6-x}{6}$

$6(x-2) = 2(6-x) \Rightarrow 6x - 12 = 12 - 2x$

2. Expand the brackets and solve for x.

$\Rightarrow 8x = 24 \Rightarrow x = 3$

Exercise 4

Q1 Solve the following equations.

a) $\dfrac{x}{4} = 1 - x$

b) $\dfrac{x}{3} = 8 - x$

c) $\dfrac{x}{5} = 11 - 2x$

d) $\dfrac{x}{3} = 2(x - 5)$

e) $\dfrac{x}{2} = 4(x - 7)$

f) $\dfrac{x}{5} = 2(x + 9)$

Q2 Solve the following equations.

a) $\dfrac{x+4}{2} = \dfrac{x+10}{3}$

b) $\dfrac{x+2}{2} = \dfrac{x+4}{6}$

c) $\dfrac{x-2}{3} = \dfrac{x+4}{5}$

d) $\dfrac{x-6}{5} = \dfrac{x+3}{8}$

e) $\dfrac{x-2}{4} = \dfrac{15-2x}{3}$

f) $\dfrac{x-4}{6} = \dfrac{12-3x}{2}$

If you get an equation with an x^2 in it, you need to rearrange until you have 'x^2 = number', then take **square roots**. Be careful — whenever you take the square root of a number, you get **two solutions** as the answer can be **positive or negative**. See p.91 and p.99-103 for more about square roots and surds.

Example 5

Solve: a) $3x^2 = 75$ b) $4x^2 + 5 = 33$

a) 1. Divide both sides by 3.

2. Take the square root of both sides, giving the two possible answers.

$3x^2 = 75$
$x^2 = 25$
$x = \pm 5$

b) 1. Rearrange the equation until you end up with 'x^2 = number'.

2. Take the square root. 7 isn't a square number, so the answer is a surd.

$4x^2 + 5 = 33$
$4x^2 = 28$
$x^2 = 7$
$x = \pm\sqrt{7}$

Tip: The $\pm$ symbol means 'plus or minus'. It's just a quick way of writing '$x = 5$ or $x = -5$'.

Exercise 5

Q1 Find the two values of x that satisfy each of the following equations.

a) $x^2 = 16$

b) $x^2 + 10 = 35$

c) $3x^2 = 27$

d) $2x^2 + 1 = 99$

e) $\dfrac{3x^2}{10} = 1.2$

f) $\dfrac{x^2+2}{x^2-4} = \dfrac{11}{10}$

Q2 Solve each of the following equations.

a) $2x^2 = 4$

b) $x^2 + 7 = 13$

c) $3x^2 + 1 = 40$

9.2 Forming Equations from Word Problems

In this section, you'll see how to turn words into maths in order to solve problems with real-life contexts.

Learning Objectives — Spec Ref A21:
- Be able to set up an algebraic equation for a given situation.
- Interpret the solution of an algebraic equation in context.

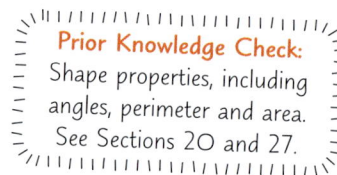

Prior Knowledge Check:
Shape properties, including angles, perimeter and area.
See Sections 20 and 27.

If you're given a **wordy problem**, you can use the information given in the question to **write** an equation.

If you're given information about one or more **unknown quantities**, call one quantity x. Then, if someone has "3 more", this would be $(x + 3)$, if someone has "3 times as many", it would be $3x$, etc. If you know the **total**, add up your different expressions, then set this equal to the total. If you're told the **difference**, you should subtract.

Once you've written your equation, you can **solve** it to find the value you need. You might need to **interpret** your answer in the **context** of the question in order to find some **other values**, too.

Example 1

The sum of three consecutive numbers is 63. What are the numbers?

1. Call the first number x, then the other two numbers are $(x + 1)$ and $(x + 2)$.

2. Form an equation in x and solve it.

3. Don't forget to find the other two numbers once you've found x.

$$x + (x + 1) + (x + 2) = 63$$
$$3x + 3 = 63$$
$$3x = 60$$
$$x = 20$$

So the numbers are **20**, **21** and **22**.

Tip: There's more about writing things like consecutive numbers algebraically coming up on p.119.

Exercise 1

Q1 I think of a number. I double it, and then add 3. The result equals 19.

   a) Write the above description in the form of an equation.

   b) Solve your equation to find the number I was thinking of.

Q2 I think of a number. I divide it by 3, and then subtract 11. The result equals –2. What number was I thinking of?

Q3 The sum of four consecutive numbers is 42. What are the numbers?

Q4 Anna, Bill and Christie are swapping football stickers. Bill has 3 more stickers than Anna. Christie has twice as many stickers as Anna. The three of them have 83 stickers in total. How many stickers does each person have?

Q5 Deb, Eduardo and Fiz are raising money for charity. Eduardo has raised £6 more than Deb. Fiz has raised three times as much as Eduardo. The three of them have raised £106.50 in total. How much did each of them raise?

Q6 Stacey is three years older than Macy. Tracy is twice as old as Stacey. The three of them have a combined age of 41. How old is each person?

You might have to use **properties of shapes** to form an equation which you can then use to find side lengths, angles, area or perimeter — take a look at Sections 20 and 27 for more.

Example 2

Use the triangle to write an equation involving x.
Solve your equation to find x.

1. The angles in a triangle always add up to $180°$ — use this to form an equation.

$x + 2x + 60° = 180°$
$3x + 60° = 180°$
$3x = 120°$

2. Solve the equation.

$x = \mathbf{40°}$

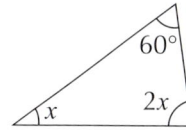

Example 3

A rectangle has sides of length $5x$ cm and $(2x + 1)$ cm. The perimeter of the rectangle is 44 cm. Find the lengths of the sides of the rectangle.

1. Form an equation. Remember to add the length of each side twice to find the perimeter.

$5x + 5x + (2x + 1) + (2x + 1) = 44$
$14x + 2 = 44$
$14x = 42$

2. Solve to find x.

$x = 3$

3. Use x to find the side lengths of the rectangle.

So the rectangle has sides of length:
$5 \times 3 = \mathbf{15}$ **cm** and $(2 \times 3) + 1 = \mathbf{7}$ **cm**.

Exercise 2

Q1   For each shape below, write an equation and solve it to find the value of x.

a)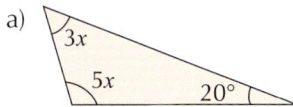

b)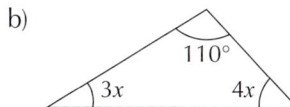

c)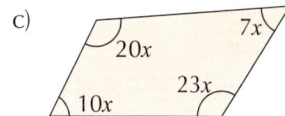

d)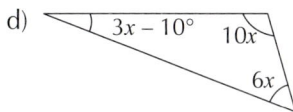

e)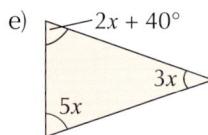

f)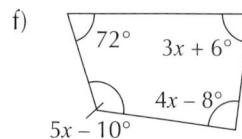

Q2   A triangle has angles of size x, $2x$ and $(70° - x)$. Find the value of x.

Q3   For each shape below, (i) find the value of x, and (ii) find the area of the shape.

a)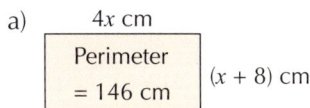

b)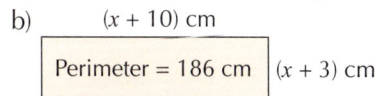

Q4   A rectangle has sides of length $(4 - x)$ cm and $(3x - 2)$ cm.
The perimeter of the rectangle is 8.8 cm. Find the area of the rectangle.

Q5   The triangle and rectangle shown below have the same area. Find the perimeter of each shape.

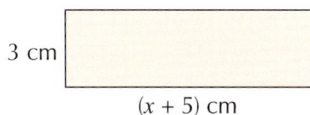

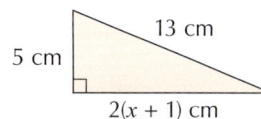

9.3 Identities

Identities look a bit like equations with an extra line, but there's an important difference — equations are only true for a particular value (or values) of x, but identities are always true, no matter what.

> **Learning Objective — Spec Ref A6:**
> Understand the difference between equations and identities.

An **equation** is a way of showing that two expressions are equal for some particular values of an unknown.
Identities are like equations, but are **always true**, for **any value** of the unknown.
Identities have the symbol '$\equiv$' instead of '$=$'.

E.g. $x - 1 = 2$ is an equation — it's only true when $x = 3$.
$\quad\quad x + 1 \equiv 1 + x$ is an identity — it's always true, whatever the value of x.

In identity questions, you should **rearrange** the expressions on **either side** to see if they're the **same**.
You **don't** need to take things to the other side, like you would if you were solving an equation.

Example 1

In which of the following equations could you replace the '$=$' sign with '$\equiv$'?
(i) $6 + 4x = x + 3$ **(ii) $x(x - 1) = -(x - x^2)$**

1. You can rearrange equation (i) to give $3x = -3$.
    This has only one solution ($x = -1$), so: **$6 + 4x = x + 3$ isn't an identity.**

2. If you expand the brackets in (ii) you get $x^2 - x = -x + x^2$.
    Both sides are the same, so: **$x(x - 1) \equiv -(x - x^2)$ is an identity.**

Example 2

Find the value of k if $(x + 2)(x - 3) \equiv x^2 - x + k$.

1. Expand the brackets on the left hand side.

$$(x + 2)(x - 3) \equiv x^2 - x + k$$
$$x^2 + 2x - 3x - 6 \equiv x^2 - x + k$$
$$x^2 - x - 6 \equiv x^2 - x + k$$

2. There's an x^2 and a $-x$ on both sides already,
    so to make both sides identical, k must be -6.   **$k = -6$**

Exercise 1

Q1 For each of the following, state whether or not you could you replace the box with the symbol '$\equiv$'.

a) $x - 1 \,\square\, 0$

b) $x^2 - 3 \,\square\, 3 - x^2$

c) $3(x + 2) - x \,\square\, 2(x + 3)$

d) $x^2 + 2x + 1 \,\square\, (x + 1)^2$

e) $4(2 - x) \,\square\, 2(4 - 2x)$

f) $4x^2 - x \,\square\, 2(x^2 - 2x)$

Q2 Find the value of a if:

a) $2(x + 5) \equiv 2x + 1 + a$

b) $ax + 3 \equiv 5x + 2 - (x - 1)$

c) $(x + 4)(x - 1) \equiv x^2 + ax - 4$

d) $(x + 2)^2 \equiv x^2 + 4x + a$

e) $4 - x^2 \equiv (a + x)(a - x)$

f) $(2x - 1)(3 - x) \equiv ax^2 + 7x - 3$

Q3 Prove that: a) $(x + 5)^2 + 3(x - 1)^2 \equiv 4(x^2 + x + 7)$ b) $3(x + 2)^2 - (x - 4)^2 \equiv 2(x^2 + 10x - 2)$

9.4 Proof

Proof questions might seem a bit confusing — it's not always obvious where to start and where you want to end up. Don't worry though, there are some handy tricks you can use to help you to get started.

Proof

Learning Objective — Spec Ref A6:
Be able to prove that mathematical statements are true.

Proof is all about showing that something is **true**. For example, you can **prove** that two expressions are identical (like on the previous page) by **rearranging** one into the other.
You can use these facts to make proof questions much easier:

- Any **even number** can be written as $2n$ — i.e. as "2 × an integer".
- Any **odd number** can be written as $2n + 1$ — i.e. as "(2 × an integer) + 1".
- **Consecutive numbers** can be written as n, $(n + 1)$, $(n + 2)$, etc. **Consecutive even** numbers can be written as $2n$, $(2n + 2)$, $(2n + 4)$, etc. and **consecutive odd** numbers as $(2n + 1)$, $(2n + 3)$, $(2n + 5)$, etc.
- The **sum**, **difference** and **product** of integers is **always** an integer.

Proof questions can cover any topic, from algebra to geometry to statistics.

Example 1

Prove that the sum of any three odd numbers is odd.

1. Take three odd numbers. $2a + 1$, $2b + 1$ and $2c + 1$, where a, b and c are integers.

2. Add them together and rearrange into the form $2n + 1$ (where n is an integer).

$$2a + 1 + 2b + 1 + 2c + 1 = 2a + 2b + 2c + 2 + 1$$
$$= 2(a + b + c + 1) + 1 = 2n + 1 \text{ where } n = (a + b + c + 1) \text{ is an integer}$$

So the sum of three odd numbers is **odd**.

Tip: The question doesn't say the numbers are consecutive, so you need to use a different letter for each number.

Example 2

Prove that the sum of the exterior angles of a triangle is 360°.

1. Draw a quick sketch.
    Call the interior angles a, b and c.

    The exterior angles are: $180° - a$, $180° - b$ and $180° - c$
    So their sum is: $(180° - a) + (180° - b) + (180° - c) = 540° - (a + b + c)$

2. The angles in a triangle add up to 180°. $= 540° - 180° = \mathbf{360°}$

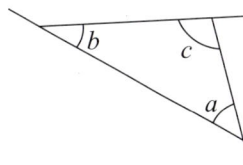

Exercise 1

Q1   Prove that the sum of three consecutive integers is a multiple of 3.

Q2   Prove that the difference between 8^{12} and 12^7 is a multiple of 4.

Q3 $2x + a = 7(x - 2a) - 5$. Prove that if a is odd then x is even.

Q4   Prove that the sum of two consecutive square numbers is always odd.

Q5   The range of a set of positive numbers is 8. Each number in the set is multiplied by 3.
     Show that the new range is 24.

Q6   Prove that, for any integer n, n^2 is larger than the product of the two integers either side of n.

Q7   Triangle numbers are formed by the expression $\frac{1}{2}n(n + 1)$.
     Prove that the sum of two consecutive triangle numbers is a square number.

Q8   A data set with 5 values has a mean of 12. Each value in the data set is increased by 1.
     Show that the new mean is 13.

Disproof by Counter Example

Learning Objective — Spec Ref A6:
Find a counter example to show that a statement isn't true.

To show that a statement is **false**, you can just find an **example** where it **doesn't work**. This is called a
counter example. There are usually many counter examples you could give, but you only need to find one.

Example 3

Disprove the following statement by finding a counter example:
"The difference between two consecutive square numbers is always prime."

Try consecutive square numbers until you find a pair that doesn't work:

   1 and 4 — difference = 3 (prime)
   4 and 9 — difference = 5 (prime)
   9 and 16 — difference = 7 (prime)
   16 and 25 — difference = 9 (NOT prime) so **the statement is false**.

Tip: You don't have
to go through loads of
examples if you can
spot one that's wrong
straightaway — you
could go straight
to 16 and 25.

Example 4

Paz says, "If $x > y$, then $x^2 > y^2$". Is she correct? Explain your answer.

Try some values for x and y: $x = 2$, $y = 1$: $x > y$ and $x^2 = 4 > 1 = y^2$
                              $x = 5$, $y = 2$: $x > y$ and $x^2 = 25 > 4 = y^2$
                              $x = -1$, $y = -2$: $x > y$ but $x^2 = 1 < 4 = y^2$

So **Paz is wrong** as the statement does not hold for all values of x and y.

Tip: When looking
for a counter example,
try different types of
number, e.g. positive
and negative.

Exercise 2

Disprove the following statements by finding a suitable counter example.

Q1   "The sum of three consecutive integers is always bigger than each individual number."

Q2   "The difference between any two prime numbers is always an even number."

Q3   "The sum of two square numbers is never a square number."

Q4   "If $x > 0$, then $x^2 \geq x$."

Q5   "If the median of a data set is equal to the mode, then the mean is also the same."

Q6   "If you draw three points on a grid, joining them up with straight lines always forms a triangle."

9.5 Iterative Methods

Iterative methods give you another way to find solutions to equations. They're most useful when an equation is too hard to solve in the normal way (i.e. when the equation involves high powers of x).

Change of Sign Methods

Learning Objective — Spec Ref A20:
Use change of sign iterative methods to find approximate solutions to equations.

Prior Knowledge Check:
Be able to round to a given degree of accuracy. See Section 2.

Iterative methods can be used to find approximate solutions to equations. The idea is that by repeating steps of the method, you'll get **close enough** to the solution to know it to a **given degree of accuracy**. For example, if you can show that a solution is between 1.62 and 1.63, then you know that the solution will be 1.6 when rounded to 1 d.p. (as all numbers between 1.62 and 1.63 round to 1.6).

Iterative methods for solving equations often involve looking for a **change of sign**. If you have an equation of the form 'f(x) = 0', f(x) will **change from positive to negative** (or vice versa) as x goes past a solution. So if there's a change of sign between two values, then there must be a **solution** between those values. For example, if f(1) is positive, but f(2) is negative, then there is a solution to f(x) = 0 **somewhere between 1 and 2**, as shown in the diagram on the right.

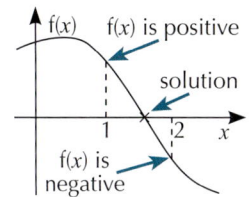

Method 1

To find an approximate solution, **go through values in order** until there's a change of sign. Each time you find a change of sign, **increase** the number of **decimal places** until you know the solution to the required degree of accuracy.

For example, in the diagram above, you know that there's a change of sign between $x = 1$ and $x = 2$, so **add a decimal place** and start testing more values **in order**, i.e. find f(1.1), f(1.2), f(1.3), etc. If you find that f(x) changes sign between $x = 1.6$ and $x = 1.7$, then the solution must be **between 1.6 and 1.7**. Then add another decimal place and start testing f(1.61), f(1.62), f(1.63) etc.

Repeat this method until you know the solution to the level of accuracy you need. You must always test values with **one more decimal place** than the final answer requires — e.g. if you want an answer correct to 2 d.p., you need to test values with 3 d.p. so that you know whether to **round up or down**.

Example 1

The equation $x^3 + 1 = 5x$ has a solution between 2 and 3.
By considering values in this interval, find this solution to 1 decimal place.

1. Rearrange the equation into the form 'f(x) = 0': $x^3 - 5x + 1 = 0$

2. Try the values of x between 2 and 3 with 1 d.p. (2.1, 2.2, etc.) in order until there's a sign change.

   f(2.1) is negative and f(2.2) is positive.
   So there is a solution between 2.1 and 2.2.

3. Now you need to decide whether the answer should round up to 2.2 or down to 2.1. Try the values of x between 2.1 and 2.2 with 2 d.p. in order until there's a sign change.

x	$x^3 - 5x + 1$	
2.0	–1	Negative
2.1	–0.239	Negative
2.2	0.648	Positive
2.11	–0.156069	Negative
2.12	–0.071872	Negative
2.13	0.013597	Positive

   f(2.12) is negative and f(2.13) is positive, so $2.12 < x < 2.13$.
   All the values in this interval round down to 2.1, so the solution is $x = 2.1$ (to 1 d.p.).

Exercise 1

Use Method 1 to answer all the questions in this Exercise.

Q1 A solution to the equation $x^2 + x - 10 = 0$ lies between 2 and 3.

    a) Evaluate $x^2 + x - 10$ for $x = 2$, $x = 2.5$ and $x = 3$.

    b) State whether the solution to $x^2 + x - 10 = 0$ is greater than or less than 2.5.

    c) Work out whether the solution is greater than or less than the following:

        (i) 2.6   (ii) 2.7   (iii) 2.8

    d) Find the solution to $x^2 + x - 10 = 0$ correct to 1 d.p.

Q2 A solution to $x^2 + 2x = 30$ lies between 4.5 and 4.6. Copy and complete the table on the right, adding extra rows as required, to find this solution to 1 d.p.

x	$x^2 + 2x - 30$	
4.5	−0.75	Negative
4.6		
4.51		
4.52		

Q3 The following equations have solutions between 5 and 6. Find the solutions correct to 1 d.p.

    a) $x^3 + 5x = 170$ b) $x^3 - 3x = 133$

Q4 The solution to the equation $2^x = 20$ lies between 4 and 5. Find x correct to 1 decimal place.

Q5 The rectangle on the right has an area of 100 cm².

    a) Write down an equation in x.

    b) Given that $7 < x < 8$, find x correct to 1 decimal place.

    c) Find the lengths of the sides of the rectangle, correct to 1 decimal place.

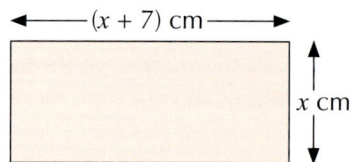

(figure: rectangle labelled $(x + 7)$ cm along the top and x cm along the side)

Method 2

This method also involves finding a change of sign — but rather than testing the values in order, this time you try a value roughly in the **middle** of the interval where there's a sign change. Depending on whether this value is positive or negative, you can use it as the new upper or lower limit, reducing the size of the interval by about half.

For example, if f(1) is positive and f(2) is negative, start by finding **f(1.5)**. If f(1.5) is **positive**, then you know that there's a change of sign between 1.5 and 2, so you've narrowed the interval down to $1.5 < x < 2$. Next, you can try **either 1.7 or 1.8** (they're both roughly in the middle of the interval). Use your **judgement** here — if f(1.5) is **closer to 0** than f(2) is, it might be better to try 1.7. Keep going until you've narrowed the interval down to two values that are **0.1** apart (e.g. $1.6 < x < 1.7$). At this point, you can start testing values with **2 decimal places.**

Once you've narrowed it down to two possible final answers, you need to test **one more value** to see whether to **round up or down**. E.g. if the question asks for the solution to **2 d.p.** and you know that $1.62 < x < 1.63$, then the final answer will be **either 1.62 or 1.63**. Find **f(1.625)** — all values between 1.62 and 1.625 **round down** to 1.62, and values between 1.625 and 1.63 **round up** to 1.63.

This method is usually **quicker** than Method 1.

Example 2

The equation $x^4 + 4x - 6 = 0$ has two solutions, one positive and one negative.
a) Show that the positive solution lies between 1 and 2.
b) Use an iterative method to find the positive solution correct to 1 d.p.

a) Find the value of $x^4 + 4x - 6$ at $x = 1$ and $x = 2$.

$f(1) = 1 + 4 - 6 = -1$ (negative)
$f(2) = 16 + 8 - 6 = 18$ (positive)

There's a change of sign between 1 and 2, so **there is a solution in this interval**.

b) 1. You know that the solution lies between 1 and 2, so try a value in the middle of the interval.

$x = 1.5 \Rightarrow f(x) = 5.0625$

2. $f(1.5)$ is positive and $f(1)$ is negative, from part a):

$f(1.5)$ is positive, so $1 < x < 1.5$

3. Repeat this until you've narrowed it down to 2 possible answers.

$f(1.2) = 0.8736$ is positive, so $1 < x < 1.2$
$f(1.1) = -0.1359$ is negative, so $1.1 < x < 1.2$

> **Tip:** Here, we've tried $f(1.2)$ rather than $f(1.3)$ because $f(1)$ is closer to 0 than $f(1.5)$ is.

4. So the solution to 1 d.p. is either 1.1 or 1.2. Split the interval in half to see what it rounds to.

$f(1.15) = 0.3490...$ is positive, so $1.1 < x < 1.15$
All values in this interval round down, so the solution is $x = 1.1$ (to 1 d.p.)

Exercise 2

Use Method 2 to answer all the questions in this Exercise.

Q1 A solution to $x^2 + x = 35$ lies between 5.4 and 5.5. Copy and complete the table on the right, adding extra rows as required, to find this solution correct to 2 d.p.

x	$x^2 + x - 35$	
5.4	−0.44	Negative
5.5		
5.45		

Q2 Each of the following equations has a solution between 0 and 10. Find this solution correct to 2 d.p.
a) $x^2 + x = 23$ b) $x^2 + 2x = 17$ c) $x^2 + 5x = 62$

Q3 Find, to 3 d.p., the solution to the equation $x^2 + x = 48$ that lies between 6.4 and 6.5.

Q4 The equation $x^3 + 4x = 21$ has a solution between 2.2 and 2.3. Find this solution, correct to 3 decimal places.

Q5 A cannonball is fired upwards from the ground. After x seconds, its height (h) in metres is given by the formula $h = 100x - 5x^2$. It travels upwards for 10 seconds until it reaches a height of 500 m, then falls back towards the ground.

a) How many seconds does it take the ball to reach a height of 200 m as it rises? Give your answer correct to 1 decimal place.

b) After how many seconds does the ball first reach a height of 400 metres? Give your answer correct to 1 decimal place.

c) As it falls back down to the ground, it passes a height of 200 metres for a second time. Find the number of seconds after being fired that it passes this height. Give your answer correct to 1 decimal place.

Recursive Iteration

Recursive iteration is when you put a starting value into an **iteration formula**, complete the calculation, then **put the result back into** the iteration formula. Each iteration gets you **closer** to the solution, so you can **keep going** until you've got the answer to the degree of accuracy you need. You can assume that you've reached the final answer when **two consecutive results** are **the same** when rounded to the required number of decimal places.

You'll be **given** the formula to use in the question, sometimes in the form of an 'iteration machine' — these just explain the method that you should follow, step by step.

Example 3

Use the iteration machine below, with $x_0 = 2$, to find a solution to the equation $x^3 + 2x - 7 = 0$ to 1 d.p.

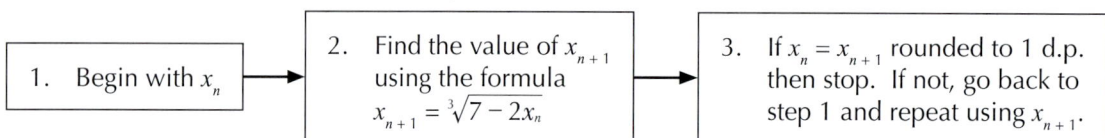

1. Begin with x_n	→	2. Find the value of x_{n+1} using the formula $x_{n+1} = \sqrt[3]{7 - 2x_n}$	→	3. If $x_n = x_{n+1}$ rounded to 1 d.p. then stop. If not, go back to step 1 and repeat using x_{n+1}.

1. Substitute x_0 into the formula to find x_1. $x_1 = \sqrt[3]{7 - 2(2)}$ $x_1 = 1.44224957...$

2. Put x_1 back into the formula to find x_2. $x_2 = \sqrt[3]{7 - 2(1.44224957...)}$ $x_2 = 1.602535155...$

3. Repeat until two consecutive terms round to the same number to 1 d.p. $x_3 = \sqrt[3]{7 - 2(1.602535155...)}$ $x_3 = 1.559796392...$

x_2 and x_3 both round to 1.6 to 1 d.p. so $x = \mathbf{1.6}$ is a solution.

Exercise 3

Q1 Using the following iteration machine, find a solution to the equation $-2x^2 + 3x + 1 = 0$ to 1 d.p. Use the starting value $x_0 = 1$.

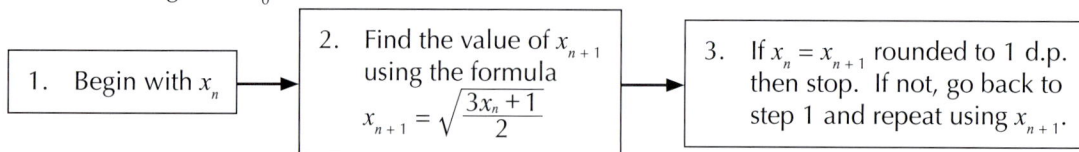

1. Begin with x_n	→	2. Find the value of x_{n+1} using the formula $x_{n+1} = \sqrt{\dfrac{3x_n + 1}{2}}$	→	3. If $x_n = x_{n+1}$ rounded to 1 d.p. then stop. If not, go back to step 1 and repeat using x_{n+1}.

Q2 Use the iteration machine below to find a solution to the equation $x^3 + 4x - 2 = 0$ to 2 d.p. Use the starting value $x_0 = 0$.

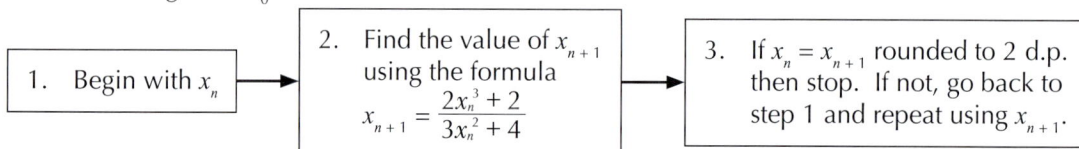

1. Begin with x_n	→	2. Find the value of x_{n+1} using the formula $x_{n+1} = \dfrac{2x_n^3 + 2}{3x_n^2 + 4}$	→	3. If $x_n = x_{n+1}$ rounded to 2 d.p. then stop. If not, go back to step 1 and repeat using x_{n+1}.

Q3 Use the iteration formula $x_{n+1} = \dfrac{2x_n^2 + 11}{4x_n}$ to find a solution to $2x^2 = 11$ to 3 d.p. Use the starting value $x_0 = 3$.

Q4 Using the iteration $x_{n+1} = \sqrt[3]{\dfrac{7 - 2x_n}{2}}$, find a solution to $2x^3 + 2x - 7 = 0$ to 3 d.p. Use $x_0 = 1$ as the starting value.

Review Exercise

Q1 Solve the following equations.

a) $\dfrac{x+8}{3} = 4$

b) $13 - 3.5x = 34$

c) $7(x - 3) = 3(x - 6)$

d) $12(x - 3) = 4(6 + 2x)$

e) $\dfrac{x-2}{5} = \dfrac{9-x}{3}$

f) $\dfrac{2x}{5} = 18 - 2x$

Q2 I think of a number. I subtract 4, then divide by 5. The result equals 15.

a) Write this information in the form of an equation.

b) Solve the equation to find the number I was thinking of.

Q3 Find the value of x in each of the following.

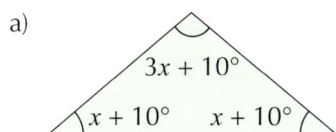

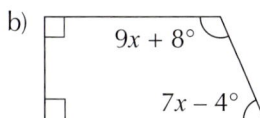

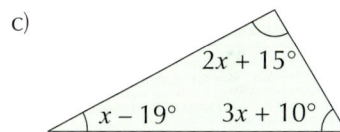

a)

$3x + 10°$

$x + 10°$ $x + 10°$

b)

$9x + 8°$

$7x - 4°$

c)

$2x + 15°$

$x - 19°$ $3x + 10°$

Q4 For each of the following, state whether you could replace the box with the symbol '$\equiv$'.

a) $4x \;\square\; 10$

b) $2(3x + 6) \;\square\; 3(2x + 4)$

c) $3(2 - 3x) + 2 \;\square\; 7x$

Q5 Find the value of a if:

a) $3(x + a) \equiv 12 + 3x$

b) $4(1 - 2x) \equiv 2(a - 4x)$

c) $3(x^2 - 2) \equiv a(6x^2 - 12)$

Q6 a) Prove that the sum of 5 consecutive integers is always a multiple of 5.

b) Mark says "If a and b are prime numbers, then $(2ab - 1)$ is always a prime number." Show that Mark is wrong.

Q7 Each of the following equations has a solution for $0 < x < 10$.
Use an iterative method to find this solution, correct to 1 d.p.

a) $x^2 + 4x = 100$

b) $x^4 = 30 - 10x$

c) $x^4 + 2x^2 + 5x = 20$

Q8 The area of this rectangle is 17 cm².
Given that $x < 3$, use an iterative method
to find the value of x, correct to 1 d.p.

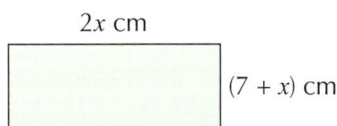

$2x$ cm

$(7 + x)$ cm

Q9 Using the following iteration machine, find a solution to the equation $4x^3 + 2x^2 - 5 = 0$ to 2 d.p.
Use the starting value $x_0 = 1$.

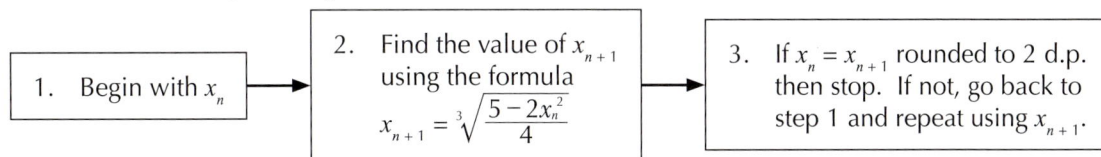

1. Begin with x_n

2. Find the value of x_{n+1} using the formula
$$x_{n+1} = \sqrt[3]{\dfrac{5 - 2x_n^2}{4}}$$

3. If $x_n = x_{n+1}$ rounded to 2 d.p. then stop. If not, go back to step 1 and repeat using x_{n+1}.

Exam-Style Questions

Q1 Solve the following equations:

a) $\dfrac{2(x+3)}{13} = 1$

[2 marks]

b) $\dfrac{x-10}{10} = \dfrac{10-x}{3}$

[2 marks]

c) $3(x^2 + 7) = 4(x^2 - 1)$

[2 marks]

Q2 The area of this rectangle is 12 cm². Find the perimeter of the rectangle.

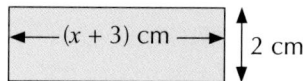

$\longleftarrow (x + 3) \text{ cm} \longrightarrow$ 2 cm

[3 marks]

Q3 a) Will says "The product of two different square numbers is always even." Show that Will is not correct.

[1 mark]

b) Veronica says "When two prime numbers which are both bigger than 2 are added, the answer is always even." Show that Veronica is correct.

[3 marks]

Q4 The triangle and rectangle shown have the same area. The perimeter of the triangle is 12 cm.

a) Find the area of the triangle.

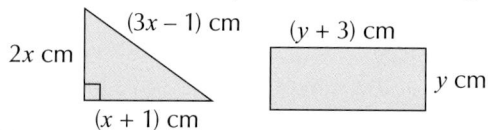

$2x$ cm $(3x - 1)$ cm $(y + 3)$ cm y cm

$(x + 1)$ cm

[3 marks]

b) Use an iterative method to find the value of y, correct to 1 decimal place.

[4 marks]

Q5 The equation $x^2 - 4x + 1 = 0$ has two solutions.

a) Show that one of these solutions lies in the interval $3 < x < 4$.

[2 marks]

b) Use the iteration $x_{n+1} = \sqrt{4x_n - 1}$ to find this solution to 1 decimal place. Use a starting value of $x_0 = 3$.

[3 marks]

Q6 Lol, Maddie and Norm took part in a javelin competition. Lol threw x m. The distance that Maddie threw is the square of the distance that Lol threw. Norm threw 30 m. The total distance of their three throws put together is 128 m.

a) Use the information given to write an equation in x.

[1 mark]

b) The value of x lies in the interval $0 < x < 10$. Use an iterative method to find the distance that Lol threw, correct to the nearest cm.

[4 marks]

10.1 Direct Proportion

Proportion is all about how quantities change in relation to one another. You might have seen proportion before in Section 4 — here you'll see how to tackle these questions algebraically.

Direct Proportion

Learning Objective — Spec Ref R10:
Understand and use direct proportion.

Two quantities are **directly proportional** if they are always in the **same ratio** — so if one quantity **doubles**, the other also **doubles**, if one **triples**, the other also **triples**, and so on.

A **proportional relationship** can be written as a **proportionality statement**: '$y \propto x$', which is read as 'y is proportional to x' or 'as x varies, y varies directly'.

You can write this as an equation: $y = kx$, where k is the **constant of proportionality** (you just replace '$\propto$' with '$= k$' to make the equation).

The graph of a **proportional relationship** is a **straight line** through the **origin**. The **gradient** of the graph (see p.180) is equal to the constant of proportionality.

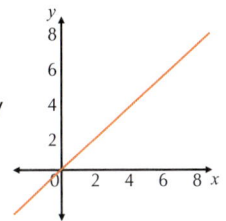

Example 1

y is directly proportional to x. Fill in the gaps in the table.

x	3	5	10	12	
y			25		100

1. Write the proportionality statement and make it into an equation:

    $y \propto x$, so $y = kx$

2. The table shows that when $x = 10$, $y = 25$. Use this to find k.

    $25 = k \times 10$
    $k = 25 \div 10 = 2.5$
    So $y = 2.5x$

    Tip: Rearrange the equation to find the missing x-value (use $x = y \div 2.5$).

3. Use the equation to complete the table:

x	3	5	10	12	$100 \div 2.5 = \mathbf{40}$
y	$2.5 \times 3 = \mathbf{7.5}$	$2.5 \times 5 = \mathbf{12.5}$	25	$2.5 \times 12 = \mathbf{30}$	100

Example 2

m is directly proportional to e. Given that $m = 72$ when $e = 6$, calculate the value of e when $m = 36$.

1. Write the proportionality statement and make it into an equation.

    $m \propto e$, so $m = ke$

2. Use the given values to find k.

    $72 = k \times 6$
    so $k = 72 \div 6 = 12$

3. Put the value of k into the equation $m = ke$.

    $m = 12e$

4. Substitute $m = 36$ into the equation and solve for e.

    $36 = 12e$
    $e = 36 \div 12 = \mathbf{3}$

Q1 In each of the following tables, y is directly proportional to x.
Use this information to fill in the gaps in each table.

a)

x	22	33
y	2	

b)

x	24	
y	18	24

c)

x	2	7	10	21	
y			15		36

d)

x	−4	0		12
y	−14		21	

e)

x		8
y	12	9

f)

x	−27	78		
y		104	272	980

Q2 j is directly proportional to h. When $j = 15$, $h = 5$. What is the value of j when $h = 40$?

Q3 r is directly proportional to t. When $r = 9$, $t = 6$. What is the value of r when $t = 7.5$?

Q4 p is directly proportional to q. When $p = 11$, $q = 3$. What is the value of q when $p = 82.5$?

Q5 Given that $b = 142$ when $s = 16$ and that b is directly proportional to s, find the value of:
a) b when $s = 18$,   b) s when $b = 200$.

Other Types of Direct Proportion

Learning Objective — Spec Ref R10:
Understand and use other types of direct proportion.

There are also other types of direct proportion, where one quantity **varies proportionally** with a **function** of another quantity (rather than with the quantity itself). These relationships can also be written as **proportionality statements**.

Here are some **examples** of other types of **direct proportion**:

- If y is directly proportional to the **square** of x then $y \propto x^2$, so $y = kx^2$.

- If y is directly proportional to the **cube** of x then $y \propto x^3$, so $y = kx^3$.

- If y is directly proportional to the **square root** of x then $y \propto \sqrt{x}$, so $y = k\sqrt{x}$.

Tip: A graph showing direct proportion will always go through the origin.

Example 3

p is directly proportional to the square of q.

a) **Given that $p = 125$ when $q = 5$, find p when $q = 7$.**

1. Write the proportionality statement and make it into an equation.

$p \propto q^2$, so $p = kq^2$

2. Use the given values to find k.

$125 = k \times 5^2 = 25k$, so $k = 125 \div 25 = 5$

3. Substitute $k = 5$ into $p = kq^2$

$p = 5q^2$

4. Substitute in $q = 7$ and solve for p.

$p = 5 \times 7^2 = \mathbf{245}$

b) Hence find q when $p = 3125$.

Substitute $p = 3125$ into the
equation and solve for q.

$3125 = 5q^2$
$q^2 = 3125 \div 5 = 625$
$q = \pm\sqrt{625} = \pm25$

Example 4

p is directly proportional to the cube of q.

a) Given that $p = 81$ when $q = 3$, find p when $q = 5$.

  1. Write the proportionality statement
      and make it into an equation.

  2. Use the given values to find k.

  3. Substitute $k = 3$ into $p = kq^3$

  4. Substitute in $q = 5$ and solve for p.

$p \propto q^3$, so $p = kq^3$
$81 = k \times 3^3 = 27k$,
so $k = 81 \div 27 = 3$

$p = 3q^3$

$p = 3 \times 5^3 = 375$

b) Hence find q when $p = 3000$.

Substitute $p = 3000$ into the
equation and solve for q.

$3000 = 3q^3$
$q^3 = 3000 \div 3 = 1000$
$q = \sqrt[3]{1000} = 10$

Exercise 2

Q1   In each of the following cases, y is directly proportional to the square of x.

      a) If $y = 64$ when $x = 2$, find y when $x = 5$.

      b) If $y = 539$ when $x = 7$, find x when $y = 1331$.

Q2 y is directly proportional to the square root of x.
      Complete the table.

x	1	9	16	
y		84		560

Q3 f is directly proportional to the square of g. It is found that $g = 100$ when $f = 200$.

      a) Find f when $g = 61.5$.

      b) Given that $g > 0$, find the exact value of g when $f = 14$.

Q4   The time taken for a ball to drop from its maximum height is directly proportional
      to the square root of the distance fallen. Given that a ball takes 3 seconds
      to drop 34.1 m, find the time taken for the ball to drop 15 m.

Q5   The volume of a sphere is directly proportional to the cube of its radius.
      The volume of a sphere of radius 12 cm is 2304π cm³.

      a) Find the constant of proportionality in terms of π.
          Use this to write an equation for the volume of a sphere in terms of its radius.

      b) Find the volume of a sphere of radius 21 cm, in terms of π.

      c) Find the radius of a sphere of volume 1000 cm³, correct to 1 decimal place.

10.2 Inverse Proportion

If one quantity decreases as the other increases, you might be dealing with inverse proportion.

Inverse Proportion

Learning Objective — Spec Ref R10:
Understand and use inverse proportion.

If two quantities are **inversely proportional**, as one **increases** the other **decreases** proportionally — so if one quantity **doubles**, the other is **halved**. The **product** of the two quantities remains **constant**.

An **inverse proportional relationship** can be written as a **proportionality statement**:
'$y \propto \frac{1}{x}$' which is read as either '**y is inversely proportional to x**' or '**y varies inversely with x**'.

As with **direct proportion**, you can write this as an **equation** using a **constant of proportionality**: $y = \frac{k}{x}$

A **graph** showing an inverse proportion relationship will be a **reciprocal graph** (see p.200).

Example 1

y is inversely proportional to x. Fill in the gaps in the table.

x	1		5	10
y		100		20

1. Write the proportionality statement and make it into an equation.

$y \propto \frac{1}{x}$, so $y = \frac{k}{x}$

2. The table shows that when $x = 10$, $y = 20$. Use this to find k.

$20 = \frac{k}{10}$

$k = 20 \times 10 = 200$

So $y = \frac{200}{x}$

Tip: Rearrange the equation to find the missing x-value (use $x = \frac{200}{y}$).

3. Use the equation to complete the table:

x	1	$200 \div 100 = \mathbf{2}$	5	10
y	$200 \div 1 = \mathbf{200}$	100	$200 \div 5 = \mathbf{40}$	20

Example 2

a) y is inversely proportional to x, and x = 4 when y = 15. Find y when x = 10.

1. Write the proportionality statement and make it into an equation. $\quad y \propto \frac{1}{x}$, so $y = \frac{k}{x}$

2. Use the given values to find k. $\qquad\qquad\qquad\qquad 15 = k \div 4$, so $k = 15 \times 4 = 60$

3. Put $k = 60$ into the equation. $\qquad\qquad\qquad\qquad y = \frac{60}{x}$

4. Substitute $x = 10$ into the equation and solve for y. $\quad y = \frac{60}{x} = \frac{60}{10} = \mathbf{6}$

b) Sketch the graph of the relationship.

The curve never touches or crosses the x- or y-axes. Both axes are asymptotes to the curve.

The graph is a reciprocal graph.

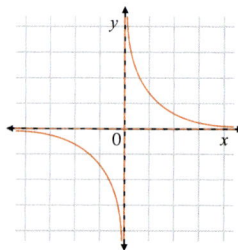

Tip: An asymptote is a straight line that a curve gets closer and closer to but never touches.

Q1 In each of the following tables, y is inversely proportional to x.
Use this information to fill in the gaps in each table.

a)
x	12	
y	15	12

b)
x	11	22
y	4	

c)
x	1	3	6	20	
y			15		270

Q2 p is inversely proportional to q. When $p = 7$, $q = 4$. What is the value of p when $q = 56$?

Q3 s is inversely proportional to t. When $s = 5$, $t = 16$. What is the value of t when $s = 48$?

Q4 Given that w is inversely proportional to z and $w = 15$ when $z = 4$,

a) find z when $w = 25$,

b) explain what happens to z when w is doubled.

Other Types of Inverse Proportion

Learning Objective — Spec Ref R10:
Understand and use other types of inverse proportion.

There are other types of inverse proportion where a quantity is **inversely proportional** to a **function** of another. As for normal inverse proportion, you can also write these as proportionality statements.

Here are some **examples** of other types of **inverse proportion**:

- If y is inversely proportional to the **square** of x then $y \propto \dfrac{1}{x^2}$, so $y = \dfrac{k}{x^2}$.

- If y is inversely proportional to the **cube** of x then $y \propto \dfrac{1}{x^3}$, so $y = \dfrac{k}{x^3}$.

- If y is inversely proportional to the **square root** of x then $y \propto \dfrac{1}{\sqrt{x}}$, so $y = \dfrac{k}{\sqrt{x}}$.

Example 3

y is inversely proportional to the cube of x, and when $x = 4$, $y = 10$.
Find the value of x when $y = 50$.

1. Write the proportionality statement and make it into an equation. $y \propto \dfrac{1}{x^3}$, so $y = \dfrac{k}{x^3}$

2. Use the given values to find k.

$$10 = \frac{k}{4^3} = \frac{k}{64}$$

$$\text{so } k = 10 \times 64 = 640$$

3. Substitute $k = 640$ into the equation for y.

$$y = \frac{640}{x^3}$$

4. Substitute $y = 50$ into the equation and solve for x.

$$50 = \frac{640}{x^3}$$

$$x^3 = 640 \div 50 = 12.8$$

$$x = \sqrt[3]{12.8} = \mathbf{2.34} \text{ (3 s.f.)}$$

Example 4

y is inversely proportional to the square root of x, and when $x = 4$, $y = 4$.
Find the value of x when $y = 2$, given that k is a positive integer.

1. Write the proportionality statement and make it into an equation.
$y \propto \dfrac{1}{\sqrt{x}}$, so $y = \dfrac{k}{\sqrt{x}}$

2. Use the given values to find k.
$4 = k \div \sqrt{4} = k \div 2$
so $k = 4 \times 2 = 8$

3. Substitute $k = 8$ into the equation for y.
$y = \dfrac{8}{\sqrt{x}}$

4. Substitute $y = 2$ into the equation and solve for x.
$2 = \dfrac{8}{\sqrt{x}}$

$\sqrt{x} = 8 \div 2 = 4$
$x = 4^2 = \mathbf{16}$

Exercise 2

Q1 Complete these tables.

a) y is inversely proportional to the square of x.

x	2	5		0.4
y	8		2	

b) y is inversely proportional to the square root of x and k is positive.

x		9	100	
y	6	8		$\dfrac{1}{3}$

Q2 h is inversely proportional to the cube of f. It is known that $h = 12.5$ when $f = 2$.
Find the value of h when $f = 5$.

Q3 a is inversely proportional to the square of c, and when $c = 6$, $a = 3$.
Find the two possible values of c when $a = 12$.

Q4 The air pressure from an electric pump is inversely proportional to the square of the radius of the tube to the pump. A tube with radius 10 mm creates 20 units of air pressure.

a) How much pressure will a tube of radius 15 mm create?

b) If an air-bed is to be pumped up using a maximum of 30 units of air pressure, what radius of tube should be used to achieve the quickest fill?

Q5 b is inversely proportional to the square of c.
When $c = 1$, $b = 64$. Find values of b and c such that $b = c$.

Q6 The quantities u and v are related by the equation $v = \dfrac{k}{u^2}$.

a) Decide which of the following statements are true and which are false:

(i) u is proportional to the square of v. (ii) v multiplied by the square of u is equal to a constant.

(iii) If you double v, you halve u. (iv) If you double u, you divide v by 4.

b) If $k = 900$ and u and v are both positive integers, find at least 3 sets of possible values for u and v.

Review Exercise

Q1 Copy and complete the table on the right if:

a) y is directly proportional to x,

b) y is inversely proportional to x.

x	2	3	9	
y		8		100

Q2 p is inversely proportional to the cube of g and when $g = 1.5$, $p = 10$.

a) Find p when $g = 2.1$.

b) Find g when $p = 15$.

Q3 A person's reach with an upstretched arm is roughly proportional to their height. On average, statistics show that a person can reach 1.3 times their height.

a) Write down both a proportionality statement and an equation for this situation.

b) Would you expect a person of height 1.75 m to be able to touch a ceiling 2.5 m high? Show working to justify your answer.

Q4 Match each of the following statements to one of the sketch graphs below.

A: y is directly proportional to x

B: y is directly proportional to x^2

C: y is directly proportional to the cube of x

D: y is directly proportional to $\sqrt{x}$

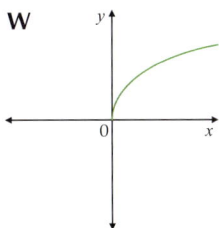

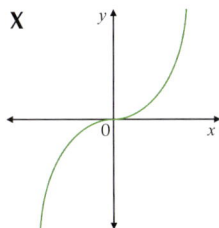

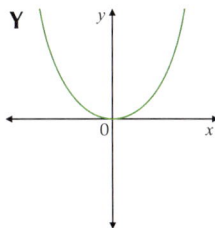

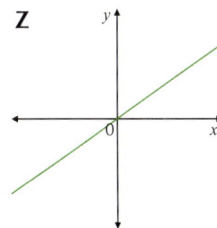

Q5 a) Copy and complete the table on the right if:

(i) y is inversely proportional to the square of x

(ii) y is inversely proportional to the cube of x

x	−6	−4	−2	2	4	6
y				6		

b) Using your tables, sketch graphs of these two types of inverse proportionality.

Q6 Coulomb's inverse square law states that the force of attraction or repulsion between two point charges is directly proportional to the product of the sizes of the charges and inversely proportional to the square of the distance between them.

This can be written as: $F = k \dfrac{Q_1 \times Q_2}{d^2}$, where Q_1 and Q_2 represent the sizes of the point charges, d is the distance between the point charges, and k is a constant.

If the force is found to be 10^6 units when the sizes of two point charges, Q_1 and Q_2, are each equal to 10^{-2} units and the points are 3 metres apart, find the force when these same point charges are 8 metres apart.

Exam-Style Questions

Q1 b is directly proportional to the square root of d.
Given that $b = 5$ when $d = 2.2$, find b (to 3 s.f.) when $d = 0.5$.

[1 mark]

Q2 y is inversely proportional to the square of x.

x		10		30
y	9000	90	40	

Fill in the missing values in the table, given that all values of x are positive.

[3 marks]

Q3   When a mass is attached to the bottom of a 20 cm long vertically held spring,
the length that the spring extends is directly proportional to the magnitude of the mass.
When a mass of 15 g is attached, the spring has a length of 20.9 cm.
Work out how long the spring will be if an additional mass of 10 g is also attached.

[3 marks]

Q4   The maximum possible air pressure, p, from a mountain bike pump
is inversely proportional to the square of the diameter, d, of the pump's cylinder.
If a maximum pressure of 8.5 bars is possible from a pump whose cylinder
has a diameter of 32 mm, find a formula for p in terms of d.

[2 marks]

Q5   In the table of values, $q \propto p^n$.

p	1	4
q	8	0.5

Find the value of n.

[3 marks]

Q6 Z is directly proportional to x^3. Z is also directly proportional to y^2. When $y = 8$, $x = 5$.
Find the value of x when y is 27.

[4 marks]

11.1 Solving Quadratic Equations by Factorising

Quadratic equations are generally given in the form $ax^2 + bx + c = 0$. To solve them, first make sure they are in this form (with 0 on one side) and then use one of the methods in this section. As quadratics contain a squared term there can be two solutions.

Learning Objective — Spec Ref A18:
Solve quadratic equations by factorising.

Prior Knowledge Check:
Be able to factorise quadratic expressions (see p.82-84).

If you know that the **product** of two numbers is equal to 0 (i.e. $p \times q = 0$) then one of the **factors** must be equal to 0 — either $p = 0$ or $q = 0$. You can use this principle to help you to solve a quadratic equation.

Start by **factorising** the quadratic (p.82-84) so that it's in the form $(x \pm m)(x \pm n) = 0$. Then one of the factors must be 0 — either $x \pm m = 0$ or $x \pm n = 0$. Solve each of these for x to give $x = \pm m$ or $x = \pm n$.

If the coefficient of x^2 isn't **1** then the factorisation will be of the form $(px \pm q)(rx \pm s) = 0$. The method to find x is still **the same** though — set **each factor** equal to 0 ($px \pm q = 0$ or $rx \pm s = 0$) and **solve for x.**

There are a couple of **special cases**. Firstly, if a quadratic factorises to the form $(px \pm q)^2$, then there's only **one solution** (found by solving $px \pm q = 0$). Secondly, if a quadratic factorises to the form $rx(px \pm q)$, you'll get two solutions — one from solving $px \pm q = 0$ and one from solving $rx = 0$, i.e. $x = 0$.

Example 1

a) **Solve the equation $x^2 - 3x + 2 = 0$.**

1.	Factorise the left-hand side.	$(x - 1)(x - 2) = 0$
2.	Set each factor equal to 0.	$x - 1 = 0$ or $x - 2 = 0$
3.	Solve to find the two possible values of x.	$x = 1$ or $x = 2$

b) **Solve the equation $8x^2 + 6x - 9 = 0$.**

1.	Factorise the left-hand side.	$(2x + 3)(4x - 3) = 0$
2.	Set each factor equal to 0.	$2x + 3 = 0$ or $4x - 3 = 0$
3.	Solve to find the two possible values of x.	$2x = -3$ or $4x = 3$
		$x = -\dfrac{3}{2}$ or $x = \dfrac{3}{4}$

Exercise 1

Q1 Find the possible values of x for each of the following.

a) $x(x + 8) = 0$ b) $(x - 5)(x - 1) = 0$ c) $(x + 2)(x + 6) = 0$ d) $(x - 9)(x + 7) = 0$

Q2 Solve the following quadratic equations by factorising.

a) $x^2 - 3x = 0$ b) $x^2 + 12x = 0$ c) $x^2 + 3x + 2 = 0$ d) $x^2 - 2x + 1 = 0$

e) $x^2 + 4x + 4 = 0$ f) $x^2 + 3x - 4 = 0$ g) $x^2 - 3x - 4 = 0$ h) $x^2 + 5x + 4 = 0$

i) $x^2 - 5x + 6 = 0$ j) $x^2 + 8x + 12 = 0$ k) $x^2 - 2x - 24 = 0$ l) $x^2 - 15x + 36 = 0$

Q3 Find the possible values of x for each of the following.

a) $x(2x - 3) = 0$ b) $(x - 2)(3x - 1) = 0$ c) $(3x + 4)(2x + 5) = 0$ d) $(4x - 7)(5x + 2) = 0$

Q4 Solve the following equations by factorising.

a) $3x^2 + 5x = 0$ b) $2x^2 + x - 3 = 0$ c) $5x^2 + 3x - 2 = 0$ d) $3x^2 - 11x + 6 = 0$

e) $4x^2 + 17x + 4 = 0$ f) $6x^2 + x - 22 = 0$ g) $4x^2 - 20x + 25 = 0$ h) $9x^2 - 12x + 4 = 0$

It's important that one side of the equation is **0** so it's in the form $ax^2 + bx + c = 0$ **before** you factorise (b or c might be zero). If not, you'll have to **rearrange** into this form by expanding any brackets, getting rid of any fractions, etc. Only then should you factorise your quadratic.

> **Tip:** An equation like $(x + 3)(x + 1) = 1$ isn't factorised as it's equal to 1, so you need to expand and rearrange first.

Example 2

a) Solve the equation $12x^2 - 8x = 15$.

1. Rearrange the equation so it's in the form $ax^2 + bx + c = 0$. $12x^2 - 8x - 15 = 0$
2. Factorise the left-hand side of the equation. $(2x - 3)(6x + 5) = 0$
3. Set each factor equal to 0. $2x - 3 = 0$ or $6x + 5 = 0$
4. Solve for each possible value of x. $x = \dfrac{3}{2}$ or $x = -\dfrac{5}{6}$

b) Solve the equation $x = \dfrac{-(x + 1)}{x - 3}$.

1. Get rid of the fraction by multiplying both sides by $x - 3$. $x(x - 3) = -(x + 1)$
2. Expand the brackets. $x^2 - 3x = -x - 1$
3. Rearrange the equation to get 0 on one side. $x^2 - 2x + 1 = 0$
4. Factorise the quadratic and solve for x. $(x - 1)^2 = 0$ so $x = 1$

Exercise 2

Rearrange the following equations, then solve them by factorising.

Q1 a) $x^2 = x$ b) $x^2 + 2x = 3$ c) $10x - x^2 = 21$ d) $x^2 = 6x - 8$

e) $8x - x^2 = 12$ f) $3x^2 = 6x + 9$ g) $x^2 + 21x = 11 - x^2$ h) $4x^2 + 4x = 3$

i) $6x^2 + x = 1$ j) $6x^2 = 7x - 2$ k) $4x^2 + 1 = 4x$ l) $9x^2 + 25 = 30x$

Q2 a) $x(x - 2) = 8$ b) $x(x + 2) = 35$ c) $(x + 3)(x + 9) + 9 = 0$

d) $(x - 6)(x - 8) + 1 = 0$ e) $(x + 3)(x + 1) = 4x + 7$ f) $(2x + 1)(x - 1) = -16x - 8$

g) $(3x + 4)^2 = 7(3x + 4)$ h) $(3x - 4)(x - 2) - 5 = 0$ i) $2(4x + 1)(x - 1) + 3 = 0$

Q3 a) $x + 1 = \dfrac{6}{x}$ b) $x - 2 = \dfrac{4}{x + 1}$ c) $x + 2 = \dfrac{28}{x - 1}$

d) $2x + 1 = \dfrac{10}{4 - x}$ e) $3x - 1 = \dfrac{4}{2x + 1}$ f) $6x - 1 = \dfrac{4}{4x + 1}$

11.2 Completing the Square

Not all quadratics can be factorised easily. One way you can solve them is by 'completing the square'.

Completing the Square

Learning Objective — Spec Ref A18:
Complete the square for a quadratic equation.

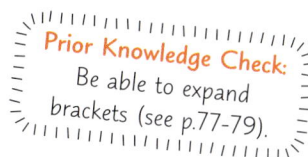

Prior Knowledge Check:
Be able to expand brackets (see p.77-79).

To write the quadratic $x^2 + bx + c$ in '**completed square form**', you need to make it look like this: $(x + p)^2 + q$. To get from the general form to the completed square form, follow this method:

- First, write out the **squared bracket**. The value of p is always half of b, i.e. $p = \frac{b}{2}$, so it's $\left(x + \frac{b}{2}\right)^2$.

- To find q, **expand the squared bracket** — you'll get $x^2 + bx + \left(\frac{b}{2}\right)^2$.

- Now **compare** this with the general form of the quadratic — the x^2 and x terms are the same, so add or subtract the **difference** between c and $\left(\frac{b}{2}\right)^2$ to make the **constant terms** the same.

To get from the completed square form to the general form just **expand the brackets** and **collect like terms**.

Example 1

a) Write $x^2 + 6x - 5$ in completed square form.

1. $b = 6$, so $p = \frac{b}{2} = 3$. The squared bracket is:

2. Expand the brackets.

3. The constant term from the brackets is $+9$, but the constant you want is $c = -5$, so subtract 14 from the squared bracket.

$(x + 3)^2$
$= x^2 + 6x + 9$

$x^2 + 6x - 5 = x^2 + 6x + 9 - 14$
$= (x + 3)^2 - 14$

b) Write $x^2 + 7x + 14$ in completed square form.

1. $b = 7$, so $p = \frac{b}{2} = \frac{7}{2}$. The squared bracket is:

2. Expand the brackets.

3. The constant from the brackets is $+\frac{49}{4} = 12\frac{1}{4}$, but the constant you want is $c = 14$, so add $1\frac{3}{4} = \frac{7}{4}$ to the squared bracket.

$\left(x + \frac{7}{2}\right)^2$
$= x^2 + 7x + \frac{49}{4}$

$x^2 + 7x + 14 = x^2 + 7x + \frac{49}{4} + \frac{7}{4}$
$= \left(x + \frac{7}{2}\right)^2 + \frac{7}{4}$

Exercise 1

Q1 Find the value of q in each of the following equations.

a) $x^2 - 4x + 7 = (x - 2)^2 + q$ b) $x^2 + 2x - 9 = (x + 1)^2 + q$ c) $x^2 + 4x + 2 = (x + 2)^2 + q$

Q2 Write the following quadratics in completed square form.

a) $x^2 + 2x + 6$ b) $x^2 - 2x + 4$ c) $x^2 - 2x - 10$ d) $x^2 - 12x + 100$

e) $x^2 + 12x + 44$ f) $x^2 + 14x$ g) $x^2 - 20x - 200$ h) $x^2 + 20x - 150$

Q3 Rewrite the quadratics below in the form $(x + p)^2 + q$.

a) $x^2 + 3x + 1$ b) $x^2 + 3x - 1$ c) $x^2 - 3x + 1$ d) $x^2 + 5x + 12$

e) $x^2 + 5x + 3$ f) $x^2 - 5x + 20$ g) $x^2 + 7x + 10$ h) $x^2 - 9x - 25$

Section 11 Quadratic Equations 137

If you have a quadratic of the form $ax^2 + bx + c$ ($a \neq 1$), the method is pretty much the same, but the squared bracket is different. The completed square form in this case looks like $a(x + p)^2 + q$.

This time $p = \dfrac{b}{2a}$, so when you expand the squared bracket, you get $a\left(x^2 + 2\dfrac{b}{2a}x + \left(\dfrac{b}{2a}\right)^2\right) = ax^2 + bx + \dfrac{b^2}{4a}$.

The x^2 and x terms are the same, so just **compare** the **constant** terms like before to find q.

Example 2

Complete the square for the quadratic $3x^2 + 12x + 7$.

1. $b = 12$ and $a = 3$ so the bracket is:
$$3\left(x + \frac{12}{2 \times 3}\right)^2 = 3(x + 2)^2$$

2. Expand the brackets — remember to multiply everything by the 3 at the front.
$$= 3(x^2 + 4x + 4)$$
$$= 3x^2 + 12x + 12$$

3. This quadratic has +12 as the constant but you want +7, so you need to subtract 5.
$$3x^2 + 12x + 7$$
$$= 3x^2 + 12x + 12 - 5$$
$$= \mathbf{3(x + 2)^2 - 5}$$

> **Tip:** You could just remember that $p = \dfrac{b}{2a}$ and $q = c - \dfrac{b^2}{4a}$.
>
> When $a = 1$, the values of p and q in completed square form are $p = \dfrac{b}{2}$ and $q = c - \left(\dfrac{b}{2}\right)^2$.

Exercise 2

Q1 Find the value of q in each of the following equations.

a) $2x^2 + x = 2\left(x + \dfrac{1}{4}\right)^2 + q$

b) $2x^2 + 20x + 500 = 2(x + 5)^2 + q$

c) $3x^2 + 4x + 25 = 3\left(x + \dfrac{2}{3}\right)^2 + q$

d) $4x^2 - 7x - 1 = 4\left(x - \dfrac{7}{8}\right)^2 + q$

Q2 Find the values of p and q in each of the following equations.

a) $2x^2 - 12x + 9 = 2(x + p)^2 + q$

b) $3x^2 - 5x - 1 = 3(x + p)^2 + q$

Q3 Write the following quadratics in completed square form.

a) $2x^2 + 8x + 81$

b) $3x^2 + 8x + 10$

c) $2x^2 - 2x + 3$

d) $5x^2 + 5x - 1$

e) $2x^2 - x + 1$

f) $3x^2 + 18x + 90$

g) $4x^2 + 13x + 9$

h) $2x^2 + 11x$

i) $-2x^2 + 10x - 2$

Solving Quadratics by Completing the Square

Learning Objective — Spec Ref A18:
Solve quadratic equations by completing the square.

> **Prior Knowledge Check:**
> Be able to simplify surds (see p.99).

You can **solve** quadratics, including ones that **can't** easily be factorised, by **completing the square**. As with factorising, you should **rearrange** the equation into the form $\mathbf{ax^2 + bx + c = 0}$ before anything else.

Put $ax^2 + bx + c = 0$ into completed square form $(a(x + p)^2 + q)$ and then

rearrange to make x the subject: $x = -p \pm \sqrt{-\dfrac{q}{a}}$. You can't take the square

root of a negative number, so the quadratic only has solutions when $-\dfrac{q}{a} \geq 0$.

> **Tip:** If $a \neq 1$ and you have 0 on one side, you could divide everything by a to potentially make completing the square simpler.

You might end up with a **surd** (see page 99) in your answer. If the question asks you to give an **exact answer**, leave the surd in — don't be tempted to use your calculator to find the decimal answer.

Example 3

a) Find the exact solutions to the equation $x^2 - 4x + 1 = 0$ by completing the square.

1. $b = -4$, so $p = \frac{b}{2} = -2$. Expand $(x - 2)^2$.

$$(x - 2)^2 = x^2 - 4x + 4$$

2. Complete the square — you've got the constant +4 but need +1, so subtract 3.

$$x^2 - 4x + 1 = (x - 2)^2 - 3$$

3. Use the completed square to rewrite and solve the original equation.

$$x^2 - 4x + 1 = 0$$
$$(x - 2)^2 - 3 = 0$$
$$(x - 2)^2 = 3$$

4. Don't forget there's a positive and negative square root.

$$x - 2 = \pm\sqrt{3}$$

5. For the exact solutions, leave in surd form.

$$x = 2 + \sqrt{3} \text{ or } x = 2 - \sqrt{3}$$

b) Solve the equation $2x^2 - 8x + 3 = 0$ by completing the square. Give your answers to 2 decimal places.

1. $b = -8$ and $a = 2$. The squared bracket is:

$$2\left(x + \frac{-8}{2 \times 2}\right)^2 = 2(x - 2)^2$$

2. Expand the brackets.

$$= 2x^2 - 8x + 8$$

3. The quadratic from the brackets has +8 but you want +3, so subtract 5 to complete the square.

$$2x^2 - 8x + 3 = 2x^2 - 8x + 8 - 5$$
$$= 2(x - 2)^2 - 5$$

4. Use the completed square form to rewrite and solve the original equation. Get the squared bracket on its own and then take the square root of both sides (don't forget the ±). Then get x on its own.

$$2x^2 - 8x + 3 = 0,$$
$$\text{so } 2(x - 2)^2 - 5 = 0$$
$$2(x - 2)^2 = 5$$
$$(x - 2)^2 = \frac{5}{2} \Rightarrow x - 2 = \pm\sqrt{\frac{5}{2}}$$
$$\text{so } x = 2 + \sqrt{\frac{5}{2}} \text{ or } x = 2 - \sqrt{\frac{5}{2}}$$

5. You're asked for the answers to 2 d.p., so use your calculator to evaluate the surds.

$$x = 3.58 \text{ (2 d.p.) or } x = 0.42 \text{ (2 d.p.)}$$

Exercise 3

Q1 Find the exact solutions of the following equations by completing the square.

a) $x^2 - 2x - 4 = 0$ b) $x^2 + 4x + 3 = 0$ c) $x^2 + 6x - 4 = 0$

d) $x^2 + 8x + 4 = 0$ e) $x^2 - x - 1 = 0$ f) $x^2 - 11x + 25 = 0$

Q2 Solve the following equations by completing the square. Give your answers to 2 decimal places.

a) $x^2 + 6x + 4 = 0$ b) $x^2 - 2x - 5 = 0$ c) $x^2 + 6x - 3 = 0$

d) $x^2 + 8x + 8 = 0$ e) $x^2 - x - 10 = 0$ f) $x^2 - 5x + 3 = 0$

Q3 Find the exact solutions of the following equations by completing the square.

a) $3x^2 + 2x - 2 = 0$ b) $5x^2 + 2x = 10$ c) $4x^2 - 6x - 1 = 0$

d) $2x^2 = 12x - 5$ e) $3x^2 = 10 - 5x$ f) $10x^2 + 7x - 1 = 0$

Q4 Solve the following equations by completing the square. Give your answers to 2 decimal places.

a) $2x^2 + 2x - 3 = 0$ b) $3x^2 + 2x - 7 = 0$ c) $4x^2 + 8x = 11$

d) $2x^2 - 16x - 19 = 0$ e) $6x^2 = 1 - 3x$ f) $3x^2 + 9x - 7 = 0$

11.3 The Quadratic Formula

There's one final method for solving quadratic equations — and it might be the most convenient one yet.

Learning Objectives — Spec Ref A18:
- Use the quadratic formula to solve quadratic equations.
- Recognise when a quadratic equation has no solutions.

Prior Knowledge Check:
Know how to simplify surds (p.99) and substitute into a formula (p.107).

The **quadratic formula** is a quick way to work out all the possible solutions to a quadratic equation. To use it, first make sure the equation is in the form $ax^2 + bx + c = 0$ and then **substitute** a, b and c into this formula:

$$x = \frac{-b \pm \sqrt{b^2 - 4ac}}{2a}$$

To **derive** the quadratic formula, solve $ax^2 + bx + c = 0$ by '**completing the square**':

$$ax^2 + bx + c = 0 \Rightarrow a\left(x + \frac{b}{2a}\right)^2 + c - \frac{b^2}{4a} = 0 \Rightarrow a\left(x + \frac{b}{2a}\right)^2 = \frac{b^2}{4a} - c$$

$$\Rightarrow a\left(x + \frac{b}{2a}\right)^2 = \frac{b^2 - 4ac}{4a} \Rightarrow \left(x + \frac{b}{2a}\right)^2 = \frac{b^2 - 4ac}{4a^2} \Rightarrow x + \frac{b}{2a} = \pm\sqrt{\frac{b^2 - 4ac}{4a^2}}$$

$$\Rightarrow x = -\frac{b}{2a} \pm \frac{\sqrt{b^2 - 4ac}}{2a} = \frac{-b \pm \sqrt{b^2 - 4ac}}{2a}, \text{ which is the \textbf{quadratic formula}.}$$

The formula might seem pretty simple to use, but **watch out** for these things:

- **Minus signs** can cause confusion — if b is negative then $-b$ will be positive, and if one of a or c is negative then $-4ac$ will be positive.
- Divide **everything** on top by $2a$, not just the square root.
- Don't forget the $\pm$ **sign** — otherwise you won't get **both solutions**.

Tip: If the question mentions exact answers or decimal places then the quadratic probably won't factorise easily, so use the formula (although you could use completing the square in these cases too).

Example 1

a) Solve the equation $x^2 - 5x + 3 = 0$, giving your answers to 2 d.p.

1. The equation is already in the form $ax^2 + bx + c = 0$ so just write down the values of a, b and c.

   $a = 1, b = -5, c = 3$

   Tip: Be careful if any of a, b or c are negative. Here, $b = -5$, not 5.

2. Substitute these values into the formula.

   $$x = \frac{-(-5) \pm \sqrt{(-5)^2 - 4 \times 1 \times 3}}{2 \times 1} = \frac{5 \pm \sqrt{25 - 12}}{2} = \frac{5 \pm \sqrt{13}}{2}$$

3. Use your calculator to get the decimal values.

   $$x = \frac{5 + \sqrt{13}}{2} = 4.30277... = \textbf{4.30} \text{ (2 d.p.)}$$

   $$\text{or } x = \frac{5 - \sqrt{13}}{2} = 0.69722... = \textbf{0.70} \text{ (2 d.p.)}$$

b) Find the exact solutions to the equation $(x + 1)(x - 2) = -7x$.

1. Rearrange the equation into the form $ax^2 + bx + c = 0$ — you'll have to expand the brackets first.

   $x^2 - 2x + x - 2 = -7x$
   $x^2 + 6x - 2 = 0$

2. Write down the values of a, b and c.

   $a = 1, b = 6, c = -2$

3. Substitute the values into the formula.

   $$x = \frac{-6 \pm \sqrt{6^2 - 4 \times 1 \times (-2)}}{2 \times 1}$$

4. You want the exact solutions so leave the surds in the answer. You might have to simplify them (see p.99) — in this case, $\sqrt{44} = \sqrt{4 \times 11} = 2\sqrt{11}$. Then you can simplify the fraction as well.

   $$= \frac{-6 \pm \sqrt{36 + 8}}{2} = \frac{-6 \pm \sqrt{44}}{2}$$

   $$x = \frac{-6 + 2\sqrt{11}}{2} \quad \text{or} \quad x = \frac{-6 - 2\sqrt{11}}{2}$$

   $$= -3 + \sqrt{11} \qquad\qquad = -3 - \sqrt{11}$$

Exercise 1

Q1 Find the exact solutions to the following equations using the quadratic formula.

a) $x^2 - 3x + 1 = 0$

b) $x^2 - 2x - 12 = 0$

c) $4x^2 - 3x - 8 = 0$

d) $x^2 + x - 1 = 0$

e) $x^2 - 8x - 5 = 0$

f) $3x^2 + 6x - 5 = 0$

g) $x^2 - 5x - 3 = 0$

h) $8x + 13 - 2x^2 = 0$

i) $x^2 + 3 - 7x = 0$

Q2 Use the quadratic formula to solve the following equations. Give your answers to 2 decimal places.

a) $x^2 + 3x + 1 = 0$

b) $3x^2 + 2x - 2 = 0$

c) $x^2 - 3x - 3 = 0$

d) $5x + x^2 - 4 = 0$

e) $-x^2 - 8x + 11 = 0$

f) $x^2 - 6 - 7x = 0$

g) $x^2 + 6x - 2 = 0$

h) $x^2 + 4x - 1 = 0$

i) $3 + 2x^2 + 8x = 0$

Q3 By using the quadratic formula, find the exact values of x for which the following equations hold.

a) $x^2 + 3x = 6$

b) $x^2 - 5x + 11 = 2x + 3$

c) $2x^2 - 7x + 14 = x^2 + 7$

d) $(x + 1)^2 = 12$

e) $5x + 12 = (x + 1)(x + 7)$

f) $x^2 = 3(x + 3)$

g) $x(x - 4) = 2(x - 2)$

h) $2(x + 5) = (x + 2)^2$

i) $(2x - 1)(x + 2) = 4x^2 - 9$

Since you can't take the square root of a negative number, if $b^2 - 4ac$ (the bit **under the square root** in the formula) is **negative**, there are **no real solutions** to the quadratic. This means the graph of the quadratic **never crosses** the x-axis. If $b^2 - 4ac$ is **positive**, you get **two solutions** and the graph crosses the axis **twice**. If it's **zero**, you only get **one solution** and the graph **just touches** the axis. See page 221 for more on the relationship between quadratics and their graphs.

Example 2

How many times does the graph of $y = 2x^2 + 2x + 5$ cross the x-axis?

1. Where the graph crosses the x-axis, y is equal to 0.

   $2x^2 + 2x + 5 = 0$

2. Write down the values of a, b and c.

   $a = 2, b = 2, c = 5$

3. Substitute these values into the part of the formula under the square root.

   $b^2 - 4ac = 2^2 - 4 \times 2 \times 5 = 4 - 40 = -36$

4. This value is negative, so there aren't any solutions to $2x^2 + 2x + 5 = 0$.

   Since you can't take the square root of -36, there are no solutions and so the graph **never crosses** the x-axis.

Exercise 2

Q1 Determine how many solutions each of these equations has. You don't need to find the solutions.

a) $3x^2 - x - 1 = 0$

b) $2x^2 - 2x + 3 = 0$

c) $3x^2 + 6x + 3 = 0$

Q2 Each of the graphs (A, B and C) on the right is the graph of one of the equations from Q1. Match up each of the graphs with its equation.

Q3 The three sides of a triangle have lengths x^2 cm, $2x$ cm and $(4 - 2x^2)$ cm. Set up and use an appropriate equation to determine whether it is possible for this triangle to have a perimeter of 6 cm.

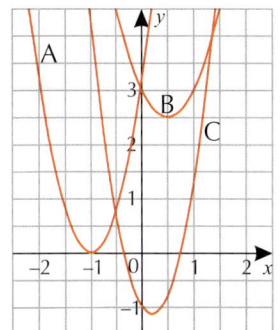

Review Exercise

Q1 Solve the following quadratic equations by factorising.

    a) $x^2 - 12x + 20 = 0$ b) $x^2 + x - 20 = 0$ c) $x^2 - 5x - 50 = 0$

    d) $x^2 + 14x + 48 = 0$ e) $3x^2 - 4x - 15 = 0$ f) $(3x + 2)(4x + 1) = 3$

Q2 Use completing the square to find the exact values of x that satisfy these equations.

    a) $x^2 - 18x - 3 = 0$ b) $x^2 - 5x + 6 = 0$ c) $x(x + 4) - 5 = 0$

    d) $(2x + 1)(x - 2) = 2$ e) $\dfrac{1}{3x + 1} = \dfrac{x}{2x^2 + 3}$ f) $\dfrac{1}{2x - 1} + \dfrac{1}{5x - 3} = 1$

Q3 Find any possible solutions to the following equations by using the quadratic formula. Give your solutions to two decimal places.

    a) $x^2 - 21x + 27 = 0$ b) $2x^2 - x - 18 = 0$ c) $5x^2 = x - 7$

    d) $\dfrac{1}{x + 1} = \dfrac{x + 1}{x - 2}$ e) $\dfrac{x}{2x^2 + 1} = \dfrac{1}{x - 2}$ f) $\dfrac{1}{x + 1} + \dfrac{1}{x - 2} = 5$

Q4 In each of the following cases, let x be the unknown number. For each one, set up and solve an equation to find all possible values of x. Give your answers to 2 d.p. where appropriate.

    a) I think of a number, add one and square the answer. The result is 9.

    b) I think of a number, square it and add 3. The result is 147.

    c) The square of a number plus the original number is 22.

    d) 25 subtract the square of a number makes 9.

    e) A number is doubled, then squared. The result is 36.

    f) I think of a number, add 5, then square the answer. The result is 100.

Q5 The dimensions of two rectangles are given below. Use the fact that the area of a rectangle is its width multiplied by its length to find x in each case.

    a) Width $= x$ m, length $= (x + 3)$ m, area $= 28$ m^2.

    b) Width $= x$ m, length $= (x + 7)$ m, area $= 30$ m^2.

Q6 A fence is erected adjoining a wall, as shown in the diagram. The rectangle enclosed by the fence and the wall has width x m, length y m and area A m^2. The total length of fence used is 20 m.

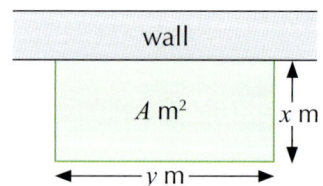

    a) If $A = 50$, show that $x(20 - 2x) = 50$.

    b) Solve the equation $x(20 - 2x) = 50$.
       Hence find the dimensions of the rectangle if its area is 50 m^2.

Q7 Use Pythagoras' theorem to find the value of x in the triangles on the right. Give your answers to 2 decimal places where appropriate.

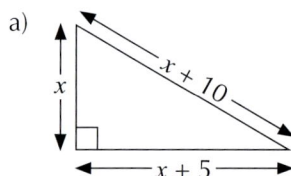

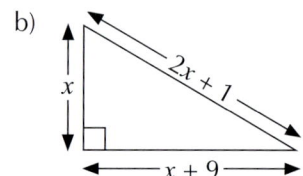

Exam-Style Questions

Q1   a) Find the values of m and n such that $2x^2 - 13x + 21 \equiv (2x + m)(x + n)$.

[1 mark]

b) Hence, find all the possible values of x such that $2x^2 - 13x + 21 = 0$.

[1 mark]

Q2   a) Write the quadratic $x^2 + 8x + 12$ in the form $(x + a)^2 + b$.

[2 marks]

b) Hence, or otherwise, solve the equation $x^2 + 8x + 12 = 0$.

[2 marks]

Q3   Solve $\dfrac{1}{x} - \dfrac{x}{2} = 2$. Give your solutions exactly and fully simplify them.

[4 marks]

Q4   Hannah uses the quadratic formula to solve the equation $3x^2 - 9x + 5 = 1$.
Her working is below.

$$a = 3 \quad b = 9 \quad c = 5$$
$$x = \frac{-9 + \sqrt{9^2 - 4 \times 3 \times 5}}{2 \times 3}$$
$$\text{So } x = \frac{-9 + \sqrt{21}}{6}$$

Hannah has made three mistakes. For each mistake, briefly explain
what she has done wrong and what she should have done instead.

[3 marks]

Q5   The nth term of a number sequence is given by the expression $n^2 - 3n - 60$.
The nth term of a different number sequence is given by the expression $2n + 6$.
Work out the value of n for which the terms of both sequences are equal.

[3 marks]

Q6   The time t, in minutes, that it takes for a train to reach its maximum speed
after leaving a station is modelled by the equation:

$$\frac{3t}{2} + \frac{7t + 2}{3} = t^2$$

Use algebra to solve the equation to find the value of t.

[4 marks]

12.1 Simultaneous Linear Equations

In simultaneous equations you'll be given a pair of equations with two unknowns (e.g. two equations which contain both x and y). The solution will be a pair of values for the unknowns that make both equations true.

Learning Objectives — Spec Ref A19/A21:
- Solve two linear equations simultaneously.
- Form simultaneous equations from real-life situations.

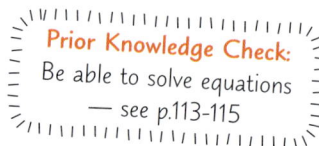

Prior Knowledge Check:
Be able to solve equations
— see p.113-115

There are **two** main algebraic methods for solving **simultaneous equations** — elimination and substitution.

- For the **elimination method**, you **add** the two equations together (or **subtract** one from the other) so that one variable is **eliminated**. To follow this method, you might need to **rearrange** both equations into the form $ax + by = c$ first.

- For the **substitution method**, you **rearrange** one equation to give one **variable** in terms of the **other** (e.g. $2x - y = -3$ would become $y = 2x + 3$). You then **substitute** this expression into the other equation, so only one variable is left (e.g. you would replace y with $2x + 3$, then collect like terms).

Tip: Both methods involve getting rid of one variable first — you'll be left with an equation in terms of the other variable, which you can then solve.

The **final step** in both methods is to **substitute** the value you've just worked out for one variable into either equation to find the value of the other variable.

The **solutions** are where **graphs** of the equations would **cross** (see p.219).

Example 1 — Elimination Method

Solve the simultaneous equations:
(1) $11 - x = y$
(2) $3y = x - 7$

1. Rearrange the equations so that they're both in the form $ax + by = c$.

   (1) $x + y = 11$
   (2) $x - 3y = 7$

2. You've got $+x$ in both equations so subtract equation (2) from equation (1) to eliminate x.

   $$\begin{array}{r} x + y = 11 \\ -(x - 3y = 7) \\ \hline 4y = 4 \\ y = 1 \end{array}$$

3. Solve the resulting equation for y.

4. Put $y = 1$ into one of the original equations and solve for x.

   $x + 1 = 11$
   $x = 10$

5. Use the other equation to check the answer.

   $x - 3y = 10 - 3(1) = 7$ ✔
   So $x = 10, \; y = 1$

Tip: At step 2, if the coefficients of the variable you're eliminating have the same sign (both +ve or both −ve), subtract one equation from the other. If the coefficients have opposite signs (one +ve and one −ve), add the two equations.

Example 2 — Substitution Method

Solve the simultaneous equations: (1) $7x - y = 16$
                                          (2) $y - 2 = x$

1. You've got y in both equations so rearrange $y = x + 2$
    equation (2) to get y on its own.

2. Substitute $y = x + 2$ into equation (1). $7x - (x + 2) = 16$

3. Solve for x. $6x - 2 = 16$
                                                     $6x = 18$
                                                     $x = 3$

Tip: You could have used the elimination method here — you'd add the equations together to eliminate y.

4. Put $x = 3$ into one of the original $y - 2 = 3$
    equations and solve for y. $y = 5$

5. Use the other equation to check the answer. $7x - y = 7(3) - 5 = 16$ ✔
                                                    So **$x = 3$, $y = 5$**

Exercise 1

Solve each of the following pairs of simultaneous equations.

Q1   a) $x + 3y = 13$
         $x - y = 5$

     b) $2x - y = 7$
         $4x + y = 23$

     c) $x + 2y = 6$
         $x + y = 2$

     d) $3x - 2y = 16$
         $2x + 2y = 14$

     e) $x - y = 8$
         $x + 2y = -7$

     f) $2x + 4y = 16$
         $3x + 4y = 24$

     g) $4x - y = -1$
         $4x - 3y = -7$

     h) $3x + y = 11$
         $6x - y = -8$

     i) $6x + y = 9$
         $2x - y = 7$

If **neither** variable has the same coefficient, you'll have to **multiply**
one or both equations to make one set of coefficients match.
For example, if you had the equations $2x + y = 4$ and $3x + 2y = 7$, you'd
multiply the first equation by 2 to make the y-coefficients match.
You then add or subtract using the **elimination method** to find the solutions.

Tip: Pick the variable that needs the least amount of work to match up its coefficients in each equation.

Example 3

Solve the simultaneous equations: (1) $5x - 4y = 23$
                                          (2) $2x + 6y = -25$

1. Multiply equation (1) by 3 and equation
    (2) by 2 to match the y-coefficients.

    $3 \times (1)$: $15x - 12y = 69$
    $2 \times (2)$: $\underline{+ \; 4x + 12y = -50}$
                     $19x \qquad\quad = 19$

2. Add the resulting equations to eliminate
    y and solve the equation for x.

                     $x \qquad\qquad = 1$

Tip: You can label the new equations as (3) and (4) to help you keep track of things. Then Step 2 in this example is (3) + (4).

3. Put $x = 1$ into one of the original
    equations and solve for y.

    $5(1) - 4y = 23 \Rightarrow 5 - 4y = 23$
    $\Rightarrow -4y = 18 \Rightarrow y = -4.5$

4. Use the other equation to
    check the answer.

    $2x + 6y = 2(1) + 6(-4.5) = 2 - 27 = -25$ ✔
    So **$x = 1$, $y = -4.5$**

Exercise 2

Solve each of the following pairs of simultaneous equations.

Q1 a) $3x + 2y = 16$
        $2x + y = 9$

    b) $4x + 3y = 16$
        $5x - y = 1$

    c) $4x - y = 22$
        $3x + 4y = 26$

    d) $2x + 3y = 10$
        $x - y = 5$

    e) $4x - 2y = 8$
        $x - 3y = -3$

    f) $3e - 5r = 17$
        $9e + 2r = -17$

Q2 a) $3x - 2y = 8$
        $5x - 3y = 14$

    b) $4p + 3q = 17$
        $3p - 4q = 19$

    c) $4u + 7v = 15$
        $5u - 2v = 8$

    d) $2c + 6d = 19$
        $3c + 8d = 28$

    e) $3r - 4s = -22$
        $8r + 3s = -4$

    f) $3m + 5n = 14$
        $7m + 2n = 23$

You could also be given **real-life contexts** and have to set up simultaneous equations for yourself.

Example 4

Sue buys 4 dining chairs and 1 table for £142. Ken buys 6 of the same chairs and 2 of the same tables for £254. What is the price of one chair? What is the price of one table? (PROBLEM SOLVING)

1. Choose some variables, then write the question as two simultaneous equations.

    Cost of one chair = £c. Cost of one table = £t.
    (1) $4c + \ t = 142$
    (2) $6c + 2t = 254$

2. Multiply equation (1) by 2 to give the same coefficients of t and label it (3). Subtract equation (2) from (3).

    $2 \times (1) = (3)$: $8c + 2t = 284$
    (2) $\qquad\qquad\ - (6c + 2t = 254)$
    $\qquad\qquad\qquad\quad 2c \qquad = \quad 30$

3. Solve the resulting equation for c.

    $\qquad\qquad\qquad\quad c \quad = \quad 15$

4. Put $c = 15$ into one of the original equations and solve for t.

    $4(15) + t = 142$
    $60 + t = 142$
    $t = 82$

5. Use the other equation to check the answer, then write it in the context of the original question.

    $6c + 2t = 6(15) + 2(82) = 90 + 164 = 254$ ✔
    Chairs cost **£15** each. Tables cost **£82** each.

Exercise 3

(PROBLEM SOLVING)

Q1 The sum of two numbers, x and y, is 58, and the difference between them is 22. Given that x is greater than y, use simultaneous equations to find both numbers.

Q2 A grandfather with 7 grandchildren bought 4 sherbet dips and 3 chocolate bars for £1.91 last week and 3 sherbet dips and 4 chocolate bars for £1.73 the week before. Calculate the price of each item.

Q3 Three friends have just finished a computer game. At the end of the game, Zoe, with 7 yellow aliens and 5 blue spiders scored 85 points; James, with 6 yellow aliens and 11 blue spiders scored 93 points. Hal had 8 yellow aliens and 1 blue spider. How many points did Hal score?

Q4 The lengths of the sides of an equilateral triangle are $3(x + y)$ cm, $(5x + 2y - 1)$ cm and $4x + 4 + y$ cm. Find the side length of the triangle.

12.2 Simultaneous Linear and Quadratic Equations

A set of simultaneous equations including a quadratic term could have more than one pair of solutions.

Learning Objective — Spec Ref A19:
Solve one linear and one quadratic equation simultaneously.

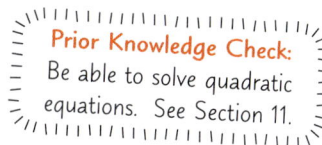

> **Prior Knowledge Check:**
> Be able to solve quadratic
> equations. See Section 11.

To solve simultaneous equations with one **linear equation** and one **quadratic** equation, use the **substitution method**.

- First **rearrange** the **linear equation** to get it in terms of **one variable** (e.g. $2x - y = -3$ would become $y = 2x + 3$).

- **Substitute** this into the **quadratic equation** — you'll be left with a quadratic equation in one variable.

- **Solve** the resulting quadratic equation — you'll probably get **2 solutions**.

- **Substitute** each of these values into one of the original equations to find the values of the other variable — the **linear equation** is often the easiest to use.

Tip: If there's a variable with a coefficient of 1, it's usually best to make that the subject so it can be easily substituted into the other equation.

The **solutions** to these pairs of simultaneous equations correspond to the **points on a graph** where a straight line and a quadratic curve **cross**. If the straight line crosses the curve **twice** there will be **2 pairs** of solutions to the simultaneous equations. The line could also cross the curve only **once** or **not at all**.

Example 1

Solve the simultaneous equations: (1) $x + y = 5$
(2) $y = x^2 - 3x - 30$

1. Rearrange equation (1) to get y on its own. $y = 5 - x$

2. Substitute $y = 5 - x$ into $y = x^2 - 3x - 30$. $5 - x = x^2 - 3x - 30$

3. Rearrange to get zero on the right hand side. $x^2 - 3x + x - 30 - 5 = 0$
$x^2 - 2x - 35 = 0$

4. Solve the quadratic. $(x + 5)(x - 7) = 0$
$x + 5 = 0$ or $x - 7 = 0$
so $x = -5$ or $x = 7$

5. Substitute into equation (1) to find a y-value for each value of x.   If $x = -5$, then $-5 + y = 5$, so $y = 10$.
If $x = 7$, then $7 + y = 5$, so $y = -2$.

6. Write the solutions in pairs. $x = -5, y = 10$ and $x = 7, y = 2$

Tip: If you'd rearranged (1) to make x the subject, you'd have to do a bit more work at step 2.

Exercise 1

Q1 Find the solutions to each of the following pairs of simultaneous equations.

a) $y = x^2 - 4x + 8$
$y = 2x$

b) $y = x^2 - x - 1$
$3x = 2 - y$

c) $y = x^2 - 4x - 28$
$y = 3x + 2$

d) $y = x^2 + 3x - 2$
$y + x = 3$

e) $y = x^2 - 4x + 2$
$y = 2x - 6$

f) $y = x^2 - 2x - 3$
$y = 3x + 11$

g) $y = x^2 - x - 5$
$y = 2x + 5$

h) $y = x^2 - 4x + 8$
$4 = y - x$

i) $y = 2x^2 + x - 2$
$y = 8x - 5$

Q2 The line $2y = x + 3$ and the curve $y = x^2 - 2x - 2$ cross at points M and N. Find the coordinates of M and N by solving the simultaneous equations.

Q3 Find where the line $y = 4 - 3x$ crosses the curve $y = 6x^2 + 10x - 1$ by solving the equations simultaneously.

Q4 Use simultaneous equations to find the coordinates where the line $y = 5x$ meets the curve $y = x^2 + 3x + 1$. What can you say about the line and the curve?

Q5 Solve these equations simultaneously: $x - 4y = 2$ and $y^2 + xy = 0$

Q6 Solve these pairs of simultaneous equations.

a) $x + y = 7$
$x^2 - xy = 4$

b) $x + y = 5$
$x + xy + 2y^2 = 2$

c) $x + y = 2$
$y^2 - x = 0$

d) $x - y = 4$
$x^2 + y = 2$

e) $x + y = 4$
$x^2 + 3xy = 16$

f) $4y + x = 10$
$xy + x = -8$

Example 2

Solve the simultaneous equations: (1) $x^2 + y^2 = 10$
(2) $2x + y = 5$

1. Rearrange equation (2) to get y on its own. $y = 5 - 2x$

    Tip: Equation (1) is the equation of a circle (see p.203).

2. Substitute $y = 5 - 2x$ into $x^2 + y^2 = 10$.

    $x^2 + (5 - 2x)^2 = 10$
    $x^2 + 25 - 20x + 4x^2 = 10$

3. Rearrange to get zero on the right hand side.

    $5x^2 - 20x + 15 = 0$
    $x^2 - 4x + 3 = 0$

4. Solve the quadratic.

    $(x - 1)(x - 3) = 0$
    $x - 1 = 0$ or $x - 3 = 0$
    so $x = 1$ or $x = 3$

5. Substitute into equation (2) to find a y-value for each value of x.

    If $x = 1$, then $2(1) + y = 5$, so $y = 3$
    If $x = 3$, then $2(3) + y = 5$, so $y = -1$

6. Write the solutions in pairs.

    $x = 1, y = 3$ and $x = 3, y = -1$

Exercise 2

Q1 Solve these pairs of simultaneous equations.

a) $2x + y = 3$
$y^2 - x^2 = 0$

b) $3x + y = 4$
$x^2 + 3xy + y^2 = -16$

c) $x - y = -4$
$x^2 + y^2 - x = 20$

Q2 The equations $x - y = -3$ and $3x^2 + 7x + y^2 = 21$ are plotted on a graph.

a) Show that at the points of intersection, $4x^2 + 13x - 12 = 0$.

b) Find the exact values of x when $4x^2 + 13x - 12 = 0$.

c) Find the exact coordinates of the points of intersection.

Q3 Find the exact coordinates where the graphs of the following pairs of equations intersect.

a) $x = 3y + 4$
$x^2 + y^2 = 34$

b) $2x + 2y = 1$
$x^2 + y^2 = 1$

c) $\sqrt{5}y - x = 6$
$x^2 + y^2 = 36$

Review Exercise

Q1 Solve each of the following pairs of simultaneous equations.

a) $e + 2f = 7$
$6e + 2f = 10$

b) $3g + h = 5\frac{1}{5}$
$3g - 2h = -6\frac{4}{5}$

c) $4j - 2i = 8$
$4j + 7i = -37$

Q2 Solve each of the following pairs of simultaneous equations.

a) $5x - 3y = 12$
$2x - y = 5$

b) $2x - y = 11$
$-4x - 7y = 5$

c) $5k + 3l = 4$
$3k + 2l = 3$

d) $2c + 6d = 19$
$3c + 8d = 28$

e) $3r - 4s = -22$
$8r + 3s = -4$

f) $3e - 5f = 8\frac{1}{2}$
$7e - 3f = 15\frac{1}{2}$

Q3 A teacher with a back problem is concerned about the weight of the books she carries home. She knows that 2 textbooks and 30 exercise books weigh a total of 6.9 kg and that 1 textbook and 20 exercise books weigh a total of 4.2 kg. The doctor has suggested she does not carry over 5 kg at a time. ⟨PROBLEM SOLVING⟩

a) Calculate the mass of one exercise book and the mass of one textbook.

b) Can she carry 1 textbook and 25 exercise books?

c) If she needs to carry 2 textbooks, how many exercise books could she carry?

Q4 An interior designer recently spent £1359.55 buying 25 kettles and 20 toasters for a new development. She spent £641.79 on 12 kettles and 9 toasters for a smaller group of new houses. For her current assignment she needs 80 kettles and 56 toasters. If the price of each toaster and each kettle is the same in all three cases, how much should she allow for this cost on her current assignment? ⟨PROBLEM SOLVING⟩

Q5 Find the solution to each of these pairs of simultaneous equations.

a) $y = 2x^2 + 9x + 30$
$y = 9 - 8x$

b) $y = 4x^2 - 5x + 2$
$y = 2x - 1$

c) $y = 7x^2 - 12x - 1$
$y = 8x + 2$

d) $5 = y - 2x$
$2x^2 - y + 3 = x$

e) $3y^2 = x - 8y$
$x + 3y = 4$

f) $3y - x = 2$
$y^2 + xy + x = 1$

g) $x - y + 5x^2 = 1$
$y + 2x = 1$

h) $4y - x = 2$
$x^2 - 3xy = 88$

i) $x + y = 0$
$x^3 - 3x^2y + y^2 = 0$

Q6 As shown on the graph, the line $x + y = 3$ and the circle $x^2 + y^2 = 17$ cross at two points. Find the coordinates for the points where the line meets the circle.

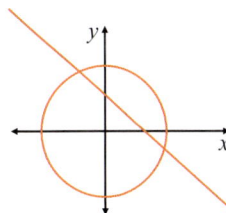

Exam-Style Questions

Q1 Solve the simultaneous equations.

$$7m + 2n = 23$$
$$3m + 5n = 14$$

[3 marks]

Q2 Solve the simultaneous equations.

$$2c + 4v = 580$$
$$3c + 2v = 542$$

[3 marks]

Q3 The equation of line A is $y = 2x - 2$, and the equation of line B is $5y = 20 - 2x$.
Find the coordinates of the point where lines A and B cross.

[3 marks]

Q4 3 kg of organic apples and 2 kg of organic pears cost £19.80,
while 2 kg of these apples and 3 kg of these pears cost £20.70.
Work out the price of 1 kg of the apples and the price of 1 kg of the pears.

[4 marks]

Q5 Find the coordinates of the points where the curve $y = x^2$ and the line $y = 6x - 8$ intersect.

[4 marks]

Q6 A rectangle has an area of 35 cm³ and a perimeter of 24 cm².
Use simultaneous equations to find the lengths of sides x and y, given that $x > y$.

[5 marks]

Q7 Solve the simultaneous equations.

$$y = 2x + 2$$
$$x^2 + y^2 = 8$$

[5 marks]

13.1 Solving Inequalities

Inequality symbols can be used to compare numbers (e.g. 3 < 5 and 10 > 2) or to compare algebraic expressions (e.g. 5x + 4 < 2x − 1). They can also be used to describe a range of values (e.g. x < 4 or −3 > y).

Simple Inequalities

Learning Objectives — Spec Ref N1/A22:
- Solve simple linear inequalities.
- Represent inequalities on a number line and using set notation.

> **Prior Knowledge Check:**
> Be able to solve linear equations (see p.113-115) and use set notation (see p.236-237).

Inequalities are written using the following symbols:

> greater than	< less than	≥ greater than or equal to	≤ less than or equal to

Greater than (>) and less than (<) are called **strict inequalities**.

The rules for **solving inequalities** are very similar to the rules for **solving equations**. The only extra rule you need to know is if you're **multiplying** or **dividing** by a **negative number** then the inequality sign '**flips over**'. For example, to solve $-2x > 6$ you need to divide both sides by -2 and 'flip' the inequality sign to give $x < -3$.

The **solution** to an inequality will usually be an inequality with x on one side and a number on the other. You can also write the solution using **set notation** (see p.236-237) — e.g. if the solution is $x < -3$, you can write it as $\{x : x < -3\}$.

Example 1

a) **Solve the inequality $2x + 4 < 8$.**

Just treat the inequality like an equation.
Subtract 4 from both sides then divide by 2.

$$2x + 4 < 8$$
$$2x < 4$$
$$\boldsymbol{x < 2}$$

b) **Solve $-\frac{x}{3} \geq -5$, giving your answer using set notation.**

1. You need to multiply by -3 to leave x on its own. Remember to 'flip' the inequality sign as you're multiplying by a negative number.

$$-\frac{x}{3} \geq -5$$
$$x \leq -5 \times -3$$
$$x \leq 15$$

2. Don't forget to write your answer using set notation.

$$\{x : x \leq 15\}$$

> **Tip:** To see why the sign 'flips', rearrange the inequality to give $5 \geq \frac{x}{3}$, then multiply by 3 to leave $15 \geq x$ (which is the same as $x \leq 15$).

Exercise 1

Q1 Solve the following inequalities.

a) $3x \geq 36$

b) $-96 \leq -12x$

c) $-\frac{x}{3} < -28$

d) $11 \leq -\frac{x}{7}$

e) $4x + 11 < 23$

f) $5x + 3 \leq 43$

g) $-3x - 7 \geq -1$

h) $65 < 7x - 12$

Q2 Solve these inequalities, giving your answers using set notation.

a) $4x < 16$

b) $-33 \leq 11x$

c) $\frac{x}{6} > -3$

d) $-\frac{x}{5} > -1$

e) $2x + 15 < 21$

f) $-5x - 8 \geq 12$

g) $\frac{x}{3} - 13 \geq -1$

h) $44 < 8x + 16$

Q3 Solve the following inequalities.

a) $\frac{x+2}{3} < 1$

b) $\frac{x+4}{5} \geq 2$

c) $\frac{x-8}{2} > 7$

d) $\frac{x-8}{4} \leq 0.5$

e) $\frac{x}{4} - 2.5 \geq 1$

f) $1 < \frac{x+5.5}{2}$

g) $-1 > \frac{x-3.1}{8}$

h) $-\frac{x}{3} + 1.3 \leq 3.3$

Q4 Solve the following inequalities, giving your answers using set notation.

a) $4x + 2 < 2x - 2$

b) $3x + 5 \leq 4 + x$

c) $3x - 3 \geq -1 + x$

d) $6 - x < 7x - 2$

e) $\frac{x}{2} - 5 \geq 3 - \frac{x}{2}$

f) $1 - 2x < \frac{x+3}{2}$

g) $2x + 4 > \frac{2x-3}{8}$

h) $\frac{x}{4} + \frac{3}{2} \leq \frac{1}{4} - x$

You can also represent the solution to a simple inequality on a **number line**, using a **circle** to show the **boundary value** and an **arrow** to show which values are part of the solution.

Use an **open circle** (O) to represent **strict inequalities** (i.e < or >) and a **solid circle** (●) to represent **non-strict inequalities** (i.e ≤ or ≥). For example:

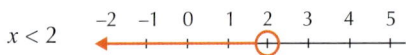

$x < 2$

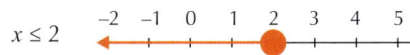

$x \leq 2$

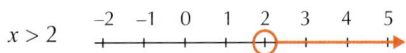

$x > 2$

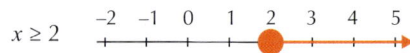

$x \geq 2$

Example 2

Solve the following inequalities. Show the solutions on a number line.

a) $x + 5 < 9$

1. Solve the inequality in the usual way — subtract 5 from both sides.

2. It's a strict inequality so use an open circle to show that the number is not included.

$x + 5 < 9$
$x < 9 - 5$
$\boldsymbol{x < 4}$

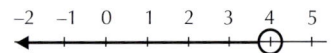

b) $1 + 2x \leq 4x - 5$

1. Solve the inequality in the usual way.

2. It's a non-strict inequality so use a solid circle to show that the number is included.

$1 + 2x \leq 4x - 5$
$1 + 5 \leq 4x - 2x$
$6 \leq 2x$
$\boldsymbol{3 \leq x}$ **(or $x \geq 3$)**

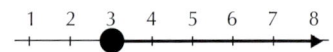

Exercise 2

For Questions 1-3, solve each inequality and show the solution on a number line.

Q1 a) $x + 9 > 14$

b) $x + 3 \leq 12$

c) $x - 2 \geq 14$

d) $x - 7 < 19$

e) $18 < x + 2$

f) $12 \leq x - 4$

g) $1 > x - 17$

h) $31 \geq x + 30$

Q2   a) $3x \geq 9$ b) $5x < 25$ c) $4x < -16$ d) $9x > -72$

      e) $\frac{x}{2} \geq 3$ f) $2 > \frac{x}{5}$ g) $\frac{x}{3} < 8$ h) $\frac{x}{7} \leq 5$

Q3   a) $4x + 3 > x + 15$ b) $x + 2 \leq -3x + 14$ c) $5x + 20 > 7x - 25$

Compound Inequalities

Learning Objectives — Spec Ref N1/A22:
- Solve compound linear inequalities.
- Represent compound inequalities on a number line and using set notation.

A **compound inequality** combines multiple inequalities into one. For example, $3 < x \leq 9$ means that $x > 3$ **and** $x \leq 9$ — so if x is an integer, the solutions are 4, 5, 6, 7, 8 and 9.

To solve a compound inequality, you can just **split it up** into two simple inequalities and solve each one separately. Then **combine** your solutions back into one inequality at the end.

Just like for simple inequalities, you can give a solution using **set notation** or on a **number line** — this time the solution will be shown by two circles with a line between them. For example, if the solution is $-2 < x \leq 1$, then in set notation you write $\{x : -2 < x \leq 1\}$ and on a number line it would be:

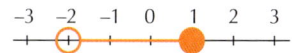

Example 3

Solve the inequality $-4 < 2x + 2 < 6$. Show your solution on a number line.

1. Write down two separate inequalities. $-4 < 2x + 2$ and $2x + 2 < 6$

2. Solve the inequalities separately.

     ① $-4 < 2x + 2$ ② $2x + 2 < 6$
         $-6 < 2x$ $2x < 4$
         $-3 < x$ $x < 2$

3. Combine the inequalities into one and draw the number line — use open circles as both inequalities are strict.

     So $-3 < x < 2$

Tip: You can solve without splitting it up — just do the same thing to each part of the inequality. For example:
$-4 < 2x + 2 < 6$
$-6 < 2x < 4$
$-3 < x < 2$

Exercise 3

Q1   Solve the following inequalities. Show each solution on a number line.

     a) $7 < x + 3 \leq 15$ b) $2 \leq x - 4 \leq 12$ c) $-1 \leq x + 5 \leq 4$ d) $21 \leq x - 16 \leq 44$

Q2   Solve the following inequalities, giving your answers using set notation.

     a) $16 < 4x < 28$ b) $32 < 2x \leq 42$ c) $27 < 4.5x \leq 72$ d) $-22.5 \leq 7.5x < 30$

Q3   Solve the following inequalities:

     a) $17 < 6x + 5 < 29$ b) $8 < 3x - 4 \leq 26$ c) $-42 < 7x + 7 \leq 91$

     d) $-2 < 2x + 3 < 5$ e) $5.1 \leq -x + 2.5 < 9.7$ f) $-5.6 < -x - 6.8 < 12.9$

Q4   Find the integer solutions that satisfy both of the inequalities.

     a) $24 > 8x$ and $9x \geq -18$ b) $-3 > 2x + 5$ and $-4x \leq 32$ c) $9 - x > 8x + 9$ and $2x + 9 > 3$

13.2 Quadratic Inequalities

An inequality where the highest power of x is x^2 is called a quadratic inequality.

Learning Objective — Spec Ref A22:
Solve quadratic inequalities, representing the solution on a number line and using set notation.

Prior Knowledge Check:
Be able to solve quadratic equations (see Section 11) and sketch quadratic graphs (see p.194-197).

In order to solve a **quadratic inequality** you need to sketch a **quadratic graph**. To do this, first solve the equivalent **quadratic equation**. **Replace** the inequality sign with an **equals sign**, then solve the quadratic equation using a method from Section 11 — the solutions are the **boundary values** of the inequality. Next, rearrange the inequality so you have $ax^2 + bx + c$ on one side and **0** on the other. **Sketch the graph** of $y = ax^2 + bx + c$ — the shape of the graph helps you to determine where the quadratic inequality is **satisfied**.

Tip: The solutions of the quadratic equation are needed because they tell you where the graph crosses the x-axis — i.e. where it's positive and where it's negative.

- If the inequality is satisfied **between** the two boundary values then your solution will be a **compound inequality**, e.g. $-5 < x < 5$ or $-1 \leq x \leq 3$.

- If the inequality is satisfied **outside** the two boundary values then your solution will be a **pair** of linear inequalities, e.g. $x \leq -1$ or $x \geq 3$.

Just like for linear inequalities, you can give your solution using **set notation** or on a **number line**.

Example 1

Solve the inequality $x^2 > 4$. Show your solution on a number line.

1. Rewrite the inequality with an equals sign.

2. Take the square root of both sides to solve the equation.

$x^2 = 4$
$x = \pm\sqrt{4}$
$x = -2$ or $x = 2$

3. $x^2 > 4 \Rightarrow x^2 - 4 > 0$ so sketch the graph of $y = x^2 - 4$. It's u-shaped and crosses the x-axis at $x = -2$ and $x = 2$.

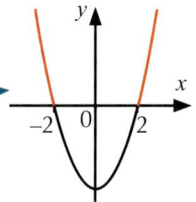

4. Now, $x^2 - 4$ is greater than zero when the graph is above the x-axis (the orange part of the graph).

So $x < -2$ or $x > 2$

5. Show both simple inequalities on the same number line — use open circles as you have strict inequalities.

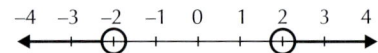

Example 2

Find the integer solutions to the inequality $x + x^2 \leq 6$. Give your answer using set notation.

1. Rewrite the inequality with an equals sign.

2. Rearrange the equation and solve it by factorisation.

$x + x^2 = 6$
$x^2 + x - 6 = 0$
$(x + 3)(x - 2) = 0$
$x = -3$ or $x = 2$

3. $x + x^2 \leq 6 \Rightarrow x^2 + x - 6 \leq 0$ so sketch the graph of $y = x^2 + x - 6$. It is u-shaped and crosses the x-axis at $x = 2$ and $x = -3$.

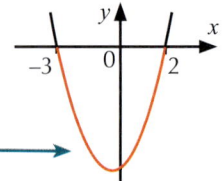

4. $x^2 + x - 6$ is less than or equal to 0 below the x-axis (the orange part of the graph) so the solution is a compound inequality.

So $-3 \leq x \leq 2$

5. Write the integer solutions to this inequality using set notation.

The set of solutions is $\{-3, -2, -1, 0, 1, 2\}$.

Tip: It's not a strict inequality so remember to include the boundary values in the solution.

Q1 Consider the inequality $x^2 > 16$.

    a) Rearrange the inequality into the form $f(x) > 0$, where $f(x)$ is a quadratic expression.

    b) (i) Factorise $f(x)$.

       (ii) Write down the x-coordinates of the points where the graph of $y = f(x)$ crosses the x-axis.

    c) Hence solve the inequality $x^2 > 16$.

Q2 Solve each quadratic inequality and show the solution on a number line.

    a) $x^2 < 4$ b) $x^2 \leq 9$ c) $x^2 > 25$ d) $x^2 \geq 36$

    e) $x^2 < 1$ f) $x^2 \leq 49$ g) $x^2 > 64$ h) $x^2 < 100$

Q3 Solve each quadratic inequality, giving your solution using set notation.

    a) $x^2 < \dfrac{1}{4}$ b) $x^2 \geq \dfrac{1}{25}$ c) $x^2 \leq \dfrac{1}{121}$ d) $x^2 > \dfrac{1}{36}$

    e) $x^2 < \dfrac{4}{9}$ f) $x^2 \geq \dfrac{25}{49}$ g) $x^2 \leq \dfrac{9}{16}$ h) $x^2 < \dfrac{16}{169}$

Q4 Find the integer solutions to these inequalities. Give your answers using set notation.

    a) $2x^2 < 18$ b) $3x^2 \leq 75$ c) $5x^2 < 80$ d) $2x^2 \leq 72$

Q5 Consider the inequality $4x \leq 12 - x^2$.

    a) Rearrange the inequality into the form $g(x) \leq 0$, where $g(x)$ is a quadratic expression.

    b) (i) Factorise $g(x)$.

       (ii) Write down the x-coordinates of the points where the graph of $y = g(x)$ crosses the x-axis.

    c) Hence solve the inequality $4x \leq 12 - x^2$.

Q6 Solve each of these inequalities.

    a) $x^2 + x - 2 < 0$ b) $x^2 - x - 2 \leq 0$ c) $x^2 - 8x + 15 > 0$

    d) $x^2 + 6x + 5 \leq 0$ e) $x^2 - x - 12 \geq 0$ f) $x^2 - 6x - 7 < 0$

    g) $x^2 - 7x + 12 \leq 0$ h) $x^2 + 10x + 24 \geq 0$ i) $x^2 - 6x - 16 < 0$

    j) $x^2 + 2x - 15 < 0$ k) $x^2 - 10x - 11 \leq 0$ l) $x^2 + 11x + 18 < 0$

Q7 Solve each quadratic inequality and show the solution on a number line.

    a) $x^2 - 2x > 48$ b) $x^2 - 3x \leq 10$ c) $x^2 + 20 < 9x$ d) $x^2 + 18 \leq 9x$

    e) $5x \geq 36 - x^2$ f) $x^2 < 9x + 22$ g) $x^2 < 6x + 27$ h) $32 < 12x - x^2$

Q8 Solve each quadratic inequality, giving your solution using set notation.

    a) $x^2 - 4x > 0$ b) $x^2 + 3x \leq 0$ c) $x^2 - 5x < 0$ d) $x^2 + 8x \geq 0$

    e) $x^2 > 12x$ f) $x^2 \leq 2x$ g) $x^2 \geq 9x$ h) $3x \leq -x^2$

13.3 Graphing Inequalities

Another way that you can represent inequalities is by using a graph. Graphs can be used to show the solutions to inequalities in one variable (e.g. $x \geq -2$) and in two variables (e.g. $y < 2x + 5$).

Graphing Inequalities

Learning Objective — Spec Ref A22:
Be able to show inequalities on a graph.

Prior Knowledge Check:
Be able to draw a straight-line graph.
See Section 15.

A linear inequality is shown on a graph using a **straight line** — the set of solutions satisfying the inequality will be the **entire region** on **one side** of the line.

To draw the straight line, replace the inequality sign with an **equals sign** and use one of the methods from Section 15 to draw the line. If you have a **strict inequality** (e.g. < or >) then you should draw a **dashed line** to show that the values on the line are **not included**. If you have a **non-strict inequality** (e.g. ≤ or ≥) then draw a **solid line** to show that the values on the line are **included**.

Once you've drawn the straight line, you need to decide which side of the line **satisfies** the inequality, then **shade** and **label** it. Choose any point on either side of the line. If that point satisfies the inequality (i.e. makes it true), it lies on the required side. If it doesn't, it's the other side you want.

Tip: You can also shade the region that you don't want and then label the region you do want.

Example 1

On a graph, use shading to show the region R that satisfies the inequality $y \leq x + 2$.

1. Draw the line $y = x + 2$ — it's a straight line with a gradient of 1 that crosses the y-axis at $(0, 2)$. It's not a strict inequality so draw a solid line to show that the line is included.

2. Choose a point that's not on the line and see if its coordinates satisfy the inequality. E.g. $(0, 0)$ satisfies the inequality because $0 \leq 0 + 2$ is true. Shade and label the region which includes $(0, 0)$ — below the line.

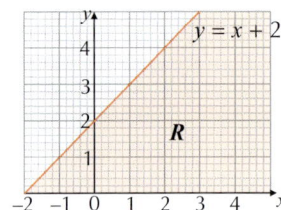

Exercise 1

In Questions 1-2, use shading and an (x, y)-coordinate grid to show the region R that satisfies the inequality.

Q1　a) $x \leq 2$　　　b) $x < -1$　　　c) $y < 2$　　　d) $y \leq -1$

Q2　a) $y < x + 1$　　b) $y > x - 3$　　c) $y \geq x + 3$　　d) $y \leq 5 - x$

　　e) $y \leq 2x$　　　f) $y > 2x + 3$　　g) $y < 3x - 1$　　h) $y > 2x - 5$

　　i) $y < \frac{1}{2}x + 2$　　j) $y \leq \frac{1}{2}x - 3$　　k) $y < \frac{x-1}{2}$　　l) $y > \frac{3x+5}{2}$

Q3　a) Rearrange the inequality $2x > 6 - y$ into the form '$y > ...$'
　　b) Hence draw a graph and show, by shading, the region R that satisfies the inequality $2x > 6 - y$.

Q4　Draw a graph and use shading to show the region R that satisfies each inequality.
　　a) $x + y \leq 5$　　b) $x + y > 0$　　c) $y - x > 4$　　d) $x - y \leq 3$
　　e) $2x - y \geq 8$　　f) $x + 2y \geq 4$　　g) $2x + 3y < 6$　　h) $3x + 2y \geq 6$

Example 2

On a graph, use shading to show the region R that satisfies the inequalities $y < x$, $x < 3$ and $y \geq 4 - x$.

1. Draw each of the lines $y = x$, $x = 3$ and $y = 4 - x$.

2. $y = x$ should be a dotted line as $y < x$. Point (1, 0) satisfies the inequality (0 < 1) so you want the region below $y = x$.

3. $x = 3$ should be a dotted line as $x < 3$. Point (0, 0) satisfies the inequality (0 < 3) so you want the region to the left of $x = 3$.

4. $y = 4 - x$ should be a solid line as $y \geq 4 - x$. Point (0, 0) doesn't satisfy the inequality ($0 \geq 4 - 0$) so you want the region above $y = 4 - x$.

5. The shaded region R satisfies all three inequalities.

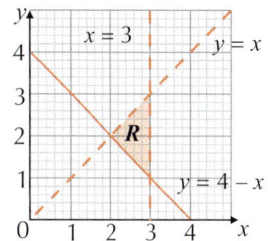

Exercise 2

In Questions 1-6, draw a graph and use shading to show the region R that satisfies the inequalities.

Q1 a) $x > 1$ and $y \leq 2$ b) $x \leq 4$ and $y \leq 3$ c) $x \geq 0$ and $y > 1$

Q2 a) $1 < x < 4$ b) $-2 < x \leq 1$ c) $-4 < y < 2$

Q3 a) $-2 < x < 0$ and $y > 1$ b) $2 < x \leq 6$ and $y \leq 5$ c) $x \geq 0$ and $-6 < y < -2$
    d) $x < -1$ and $-1 \leq y \leq 4$ e) $1 \leq x < 4$ and $2 < y < 5$ f) $-2 < x \leq 3$ and $2 \leq y \leq 6$

Q4 a) $y > 1$ and $y + x < 4$ b) $x < 2$ and $x + y \geq 1$ c) $y > -1$ and $y \leq x - 2$
    d) $x > -4$ and $y \geq x + 3$ e) $y \geq 0$ and $x < 6 - y$ f) $x < 3$ and $3 \leq x - y$

Q5 a) $x > -2$, $y < 5$ and $y > x + 4$ b) $x \geq -1$, $y < 4$ and $y < x + 3$ c) $x \leq 6$, $y > -1$ and $6 < 2x + 2y$
    d) $y > x - 2$ and $x + y \leq 4$ e) $x + y \geq -4$ and $y \leq x + 5$ f) $y \geq x - 3$ and $4 < x + y$

Q6 a) $x > -5$, $y \geq x - 3$ and $x + y < 7$ b) $x < 6$, $x + y > -5$ and $y \leq 2x + 1$
    c) $y < 2x - 4$, $y > x - 4$ and $x + y \leq 8$ d) $y < 3x - 4$, $4y \geq x - 12$ and $x + y \leq 6$

Finding Inequalities from Graphs

Learning Objective — Spec Ref A22:
Be able to describe a region of a graph using inequalities.

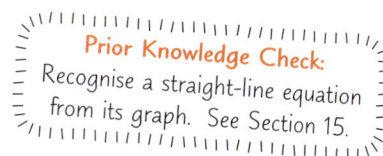

> **Prior Knowledge Check:**
> Recognise a straight-line equation from its graph. See Section 15.

To find the inequalities that define a region on a graph, first **find the equations** of the boundary lines using one of the methods covered in Section 15. Then decide which **inequality sign** to use.

- Look at the type of boundary line. If it's **dashed**, the inequality sign will be **< or >**. If it's a **solid** line, it'll be either **≤ or ≥**.

- Choose a **point** in the region you're trying to define and put the **coordinates** into the equation for each boundary line. Use the result to decide which way round the **inequality sign** should be.

E.g. if you have the equation $y = 4x + 1$ and the point (0, 0) is in the region you want, you get $0 = 1$ — so you should use < or ≤ (depending on whether it is a **dotted** or **dashed line**) to make the statement true.

Example 3

Find the inequalities which are represented by the shaded region shown.

1. The horizontal line is $y = 1$. The point (3, 2) is in the shaded region, and $2 > 1$. The line is dashed, so this represents the inequality $y > 1$.

2. The vertical line is $x = 4$. The point (3, 2) is in the shaded region, and $3 < 4$. The line is dashed, so this represents the inequality $x < 4$.

3. The diagonal line is $y = x$. The point (3, 2) is in the region and $2 < 3$ (i.e. $y < x$). The line is solid, so this represents the inequality $y \leq x$.

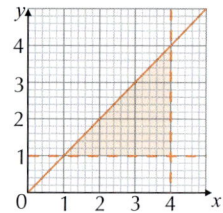

$y > 1$, $x < 4$ and $y \leq x$

Exercise 3

Q1 Write down the inequality represented by the shaded region in each of the following graphs.

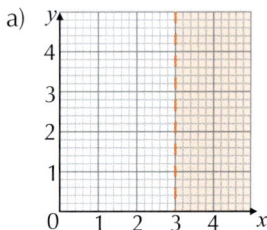

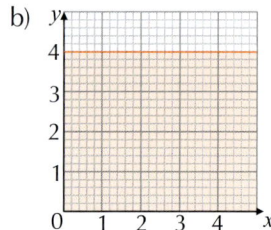

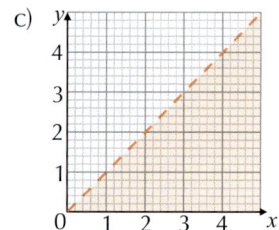

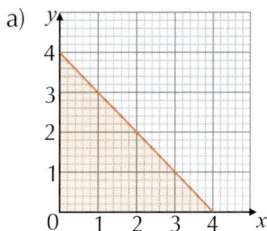

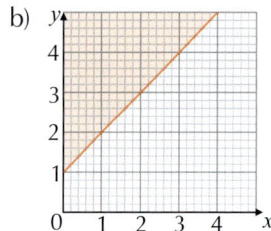

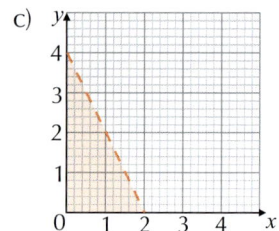

a)

b)

c)

Q2 By first finding the equation of the line shown, write down the inequality represented by the shaded region in each of the following graphs.

a)

b)

c)

Q3 Find the inequalities which are represented by the shaded region in each of the graphs.

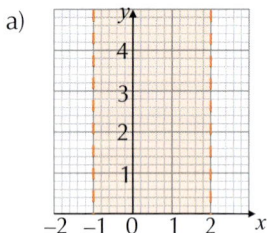

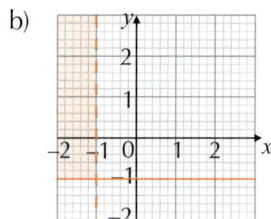

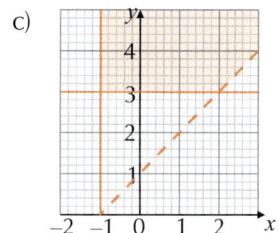

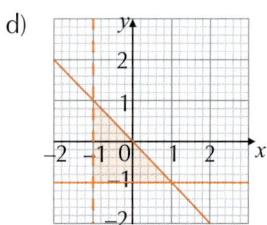

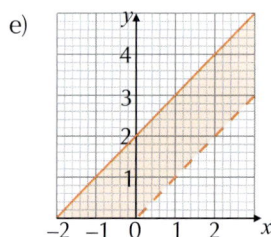

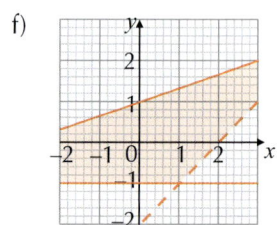

a)

b)

c)

d)

e)

f)

Review Exercise

Q1 Solve the following inequalities. Give your answers:

(i) using set notation. (ii) on a number line.

a) $-1 > \dfrac{x}{8}$ b) $\dfrac{x}{6} \le 0.5$ c) $\dfrac{x}{1.1} \ge 10$ d) $\dfrac{x}{0.2} < -3.2$

e) $2x + 16 \ge -8$ f) $-5 < -9x - 14$ g) $8x - 3.4 < 12.6$ h) $-4x + 2.6 \le 26.6$

Q2 Find the integer solutions to the following compound inequalities.
Give your answers using set notation.

a) $-7 < x - 6 < 4$ b) $-4 \le 2x + 2 \le 10$ c) $-3 < 6x + 3 < 27$

Q3 Solve each of these quadratic inequalities.

a) $x^2 < 81$ b) $x^2 \ge 81$ c) $3x^2 > 48$

d) $x^2 + 7x + 12 > 0$ e) $x^2 - 4x < 5$ f) $-x^2 > 8x - 20$

Q4 Draw a graph and show, by shading, the region R that satisfies each inequality.

a) $x \ge -5$ b) $y > -4$ c) $y \ge 2x + 3$

d) $y < -x - 2$ e) $2x + 5y > 10$ f) $5 < -10x - y$

g) $x - 2y - 6 \le 0$ h) $-4 \le x < -1$ i) $3 < y \le 6$

Q5 Draw a graph and show, by shading, the region R that satisfies the inequalities.

a) $-5 \le y \le -1$ and $y < -x$ b) $-5 \le x < -2$ and $x + 1 < y$

c) $y < 6$, $2y > x - 4$ and $5 < x + y$ d) $y \ge -4$, $y < 2x - 2$ and $x + 2y \le 8$

Q6 Find the inequalities which are represented by the shaded region in each of the graphs.

a)
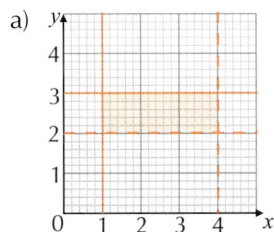

b)
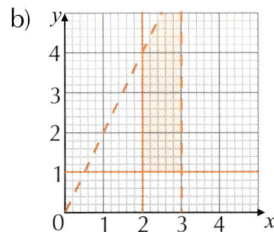

c)
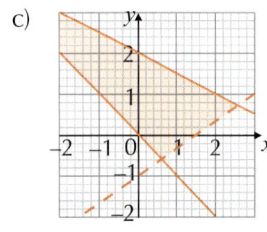

Q7 A delivery company is buying x motorbikes and y vans. Motorbikes cost £8000 and vans cost £16 000. They have £80 000 to spend on buying vehicles, and must buy at least 7 vehicles, including at least 1 van.

a) (i) Write down 4 inequalities which the company must satisfy.

(ii) Represent these inequalities graphically.

b) List the different possible combinations of motorbikes and vans that are available to the company.

Q1 Solve $4(x + 3) > 2(x - 3)$

[2 marks]

Q2 Elisa has drawn the following diagram to show the inequality $-2 < x \leq 3$.

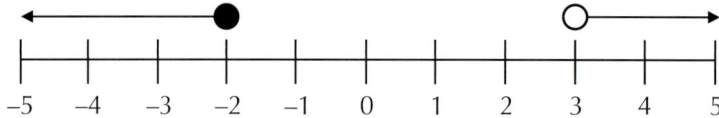

Identify the problems with the inequality Elisa has drawn.

[2 marks]

Q3 Solve $6x^2 + 9x - 15 \leq 0$

[3 marks]

Q4 A charity sells a range of wristbands which all cost the same amount.
Eve buys four wristbands and also makes a £2 donation to the charity.
Faisal buys six wristbands and makes a £1 donation.
Faisal spends less than £13 in total and he spends more than Eve.
If x is the cost of the wristband in £, use inequalities to work out the range
of possible values for x, giving your answer in the form $a < x < b$.

[4 marks]

Q5 A decorator is buying paint from his supplier — undercoat, which costs £12 per tin, and matt emulsion, which costs £24 per tin. He needs to buy at least 5 tins of undercoat and at least as many tins of matt emulsion as undercoat. He has £216 to spend in total. Let x be the number of tins of undercoat he buys and y be the number of tins of matt emulsion he buys.

a) Write down 3 inequalities that the decorator must satisfy when buying his paint.
Simplify the inequalities if possible.

[3 marks]

b) Represent these inequalities graphically.

[3 marks]

c) Use your graph to list all the possible combinations of tins of paint he can buy.

[2 marks]

14.1 Term to Term Rules

A sequence is a list of numbers or shapes which follows a particular rule. Each number or shape in the sequence is called a term. Term to term rules tell you how to get from one term to the next.

Number Sequences

Learning Objectives — Spec Ref A23/A24:
- Find rules for number sequences and use them to find terms in a sequence.
- Recognise and use sequences of square, cube and triangular numbers.

There are different types of number sequences which follow different rules to get from one term to the next. You'll come across the following types: **arithmetic**, **geometric**, **Fibonacci-type** and **quadratic** sequences, and sequences of **triangular**, **square** and **cube** numbers.

Arithmetic Sequences

Arithmetic sequences (also known as **linear** sequences or arithmetic progressions) are ones where the terms **increase** or **decrease** by the **same value** each time — known as the **common difference**. The sequences below are both arithmetic:

1, 7, 13, 19, 25
+6 +6 +6 +6

The rule is 'add 6 to the previous term'.

75.1, 74.5, 73.9, 73.3, 72.7
−0.6 −0.6 −0.6 −0.6

The rule is 'subtract 0.6 from the previous term'.

Geometric Sequences

Geometric sequences (or geometric progressions) are ones where consecutive terms are found by **multiplying** by the **same value** each time, known as the **common ratio**.

2, 8, 32, 128
×4 ×4 ×4

The rule is 'multiply the previous term by 4'. The common ratio is 4.

100, 50, 25, 12.5
÷2 ÷2 ÷2

The rule is 'divide the previous term by 2'. The common ratio is 0.5 since this is what you're multiplying by.

Tip: Terms in a geometric sequence are of the form ar^n, where n is integer (≥ 0) and r is the common ratio.

Example 1

Consider the sequence that starts 2, 6, 18, 54...

a) **Describe the rule for finding the next term in the sequence, and name this type of sequence.**

The difference between terms changes each time, so the rule for finding the next term involves multiplication.

$2 \xrightarrow{\times 3} 6 \xrightarrow{\times 3} 18 \xrightarrow{\times 3} 54...$

Consecutive terms are found by multiplying by the common ratio (which is $6 \div 2 = 3$, as you've just found), so...

So the rule is:
Multiply the previous term by 3.

It's a **geometric sequence**.

b) **Write down the next three terms in the sequence.**

Starting with 54, keep multiplying by 3.

×3 ×3 ×3
54, **162**, **486**, **1458**

Q1 Write down the first 5 terms of the sequence with:

a) first term = 14; term to term rule 'multiply the previous term by 4, then subtract 1'.

b) first term = 11; term to term rule 'multiply the previous term by –2, then add 1'.

Q2 The first four terms of a sequence are 3, 6, 12, 24.

a) Write down what you multiply each term in the sequence by to find the next term.

b) Write down the next three terms in the sequence.

Q3 For each of the following: (i)   explain the rule for finding the next term in the sequence,
                                     (ii)   name the type of sequence,
                                       (iii) find the next three terms in the sequence.

a) 3, 5, 7, 9...   b) 4, 12, 36, 108...   c) 192, 96, 48, 24...

d) 0, –4, –8, –12...   e) 16, 4, 1, $\frac{1}{4}$...   f) 1, –2, 4, –8...

Q4 Copy the following sequences and fill in the blanks.

a) 7, 13, 19, 25, ☐, 37   b) 9, 5, ☐, –3, –7, –11   c) ☐, 0.8, 3.2, 12.8, ☐

Q5 The first four terms of a sequence are –5, –2, 1, 4. Work out the 54th term of the sequence. (PROBLEM SOLVING)

Q6 The first four terms of a sequence are 1, $2\sqrt{3}$, 12, $24\sqrt{3}$.
Find the 7th term of the sequence.

Fibonacci-Type Sequences

For **Fibonacci-type** sequences, the rule is always '**add together** the **two previous terms**'.

2,   2,   4,   6,   10,   16,   26
2 + 2   2 + 4   4 + 6   6 + 10   10 + 16

Quadratic Sequences

In a **quadratic** sequence, the **difference** between terms **changes** by the **same amount** each time.

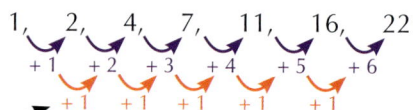

1,   2,   4,   7,   11,   16,   22
+ 1   + 2   + 3   + 4   + 5   + 6
+ 1   + 1   + 1   + 1   + 1

Sequences of Triangular, Square and Cube Numbers

Some sequences are harder to spot by their term to term rules, but you might recognise them as the **square numbers** (1, 4, 9...), **cube** numbers (1, 8, 27...) or **triangular numbers** (shown below):

To make the sequence of triangular numbers, start at 1 and then add 2, then 3, then 4, then 5 etc. to each new term.

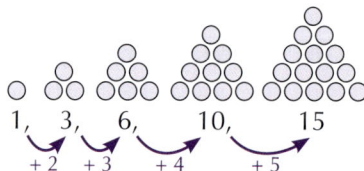

1,   3,   6,   10,   15
+ 2   + 3   + 4   + 5

Tip: These types of sequences are often linked to patterns in shapes, such as the areas or volumes of a sequence of squares or cubes. There's more about shape sequences on p.165.

Other Sequences

Other sequences might have term to term rules that use a **combination** of different types of rules. You'll need to try and spot any **patterns** in the **differences** or **ratios** between the terms. If each term in a sequence is **greater** than the one before, it's an **ascending** sequence, and if each term is **less** than the one before, it's a **descending** sequence. Sequences can be **finite** (i.e. they stop) or **infinite** (never-ending).

a) Find the next three terms in the sequence 4, 5, 7, 10...

1. Try finding the difference between neighbouring terms.

   4   5   7   10
      +1   +2   +3

2. Here, the difference is increasing by 1 each time. (This means it's a quadratic sequence.)

3. Use this to find the next three terms in the sequence. Start with 10, and add 4, then add 5, then add 6.

   10   **14**   **19**   **25**
      +4   +5   +6

b) The first four terms in a sequence are 8, 27, 64 and 125. What is the 9th term?

1. All the numbers in the sequence are cube numbers. So the 9th term will also be a cube number.

2. The first term is 2^3, the second is 3^3, and so on.

3. So the 9th term will be 10^3. $10^3 = 10 \times 10 \times 10 = \mathbf{1000}$

Once you're familiar with the different ways that sequences can be generated, you can solve **algebraic** sequence problems. If you need to, look back at pages 113-115 for a recap on solving equations.

The rule for finding the next term in a sequence is 'multiply the previous term by 3, then add on x'. The 1st term is 2 and the 4th term is 119. Find the value of x.

1. Use the rule to write down an expression for each term until you reach the 4th term.

   1st term $= 2$
   2nd term $= (3 \times 2) + x = 6 + x$
   3rd term $= 3(6 + x) + x = 18 + 3x + x = 18 + 4x$
   4th term $= 3(18 + 4x) + x = 54 + 12x + x = 54 + 13x$

2. You know that the 4th term is 119, so you can write down an equation for the 4th term. Then just solve the equation to find x.

   $54 + 13x = 119$
   $13x = 119 - 54 = 65$
   $x = 65 \div 13 = \mathbf{5}$

Exercise 2

Q1   For each sequence below, find:
     (i) the difference between consecutive terms for the first 5 terms,
     (ii) the next three terms in the sequence.

     a) 3, 4, 6, 9, 13...   b) 20, 18, 15, 11, 6...   c) 1, 2, 0, 3, –1...

Q2   The sequence 1, 1, 2, 3, 5... is a Fibonacci-type sequence, usually referred to as 'the Fibonacci sequence'. Find the next three terms in the sequence.

Q3   The first 5 terms of a sequence are 1, 1, 2, 6, 24.

     a) For the first 5 terms, find the number you multiply by each time to get the next term.

     b) Find the next two terms in the sequence.

Q4   Find the 10th term in the sequence of triangular numbers.

Q5 A sequence begins 0, 3, 8, 15, 24...

    a) What type of sequence is this?

    b) How does it relate to the sequence of square numbers?

    c) Find the 100th term in the sequence. (PROBLEM SOLVING)

Q6 Find b for each of the sequences described below.

    a) 1st term = 1, 4th term = 43; term to term rule = multiply the previous term by 4, then subtract b.

    b) 1st term = 9, 3rd term = –135; term to term rule = multiply the previous term by b, then subtract 54.

Q7 The first three terms of a quadratic sequence are 5, 5 + x and 7 + 2x. (PROBLEM SOLVING)
Find a factorised expression, in terms of x, for the 6th term in the sequence.

Using Sequence Notation

Sequences and their term to term rules can be given in **algebraic notation**.

- A term can be written x_n or u_n, where n tells you the **position** of the term in the sequence — so x_1 is the first term, x_2 is the second term, etc. Sometimes the first term is called x_0 (instead of x_1), in which case x_1 is the second term, etc.

- The term to term rule can then be given as a **formula** for x_{n+1} in terms of x_n.
In other words, it tells you how to get from one term (x_n) to the next (x_{n+1}).
For example, the formula for the rule 'add two to the previous term' is $x_{n+1} = x_n + 2$.

- You're often given a **starting value** (e.g. x_1) and from there you can work out x_2, x_3 etc.

Example 4

The rule for finding the next term in a sequence is $x_{n+1} = 2x_n + 3$ where $x_1 = 4$.
What is the value of x_5?

1. Substitute x_1 into the formula for x_n to get x_2. $x_2 = 2x_1 + 3 = (2 \times 4) + 3 = 11$

2. Substitute x_2 into the formula for x_n to get x_3. $x_3 = 2x_2 + 3 = (2 \times 11) + 3 = 25$

3. Substitute x_3 to get x_4... $x_4 = 2x_3 + 3 = (2 \times 25) + 3 = 53$

4. ...and then x_4 to get x_5. $x_5 = 2x_4 + 3 = (2 \times 53) + 3 = \textbf{109}$

Exercise 3

Q1 In each of the following, use the sequence rules and the values of x_1 to find the value of x_6.

    a) $x_{n+1} = x_n + 5$ where $x_1 = 3$ b) $x_{n+1} = x_n - 7$ where $x_1 = 20$ c) $x_{n+1} = 3x_n$ where $x_1 = 2$

    d) $x_{n+1} = 2x_n + 3$ where $x_1 = 3$ e) $x_{n+1} = \frac{1}{2}x_n + 2$ where $x_1 = 4$ f) $x_{n+1} = \frac{1}{2}x_n + 8$ where $x_1 = 8$

Q2 A sequence is generated using the rule $x_{n+1} = 2x_n - 6$ where $x_1 = 8$. Find the following:

    a) x_4 b) x_6 c) $x_3 + x_5$

Q3 A sequence with the rule $u_{n+1} = 3u_n - 1$ has term $u_5 = 14$. Find the exact value of u_1. (PROBLEM SOLVING)

Shape Sequences

Learning Objective — Spec Ref A23/A24:
Find rules for patterns of shapes and use them to find patterns in a sequence.

Patterns of **shapes** can form **sequences**. Just as with number sequences, you need to find the **rule** to get from one pattern to the next in the sequence. You can then use that rule to find other patterns.

Example 5

The matchstick shapes on the right form the first three patterns in a sequence.

a) **Draw the fourth and fifth patterns in the sequence.**

1. First work out how to get from one pattern to the next. You have to add 3 matchsticks to add an extra square to the previous pattern.

+ 3 matchsticks   + 3 matchsticks

2. The fourth pattern is the next one in the sequence, so add 3 matchsticks to the third pattern to get the fourth...

+ 3 matchsticks

3. ...and add another 3 matchsticks to that to get the fifth.

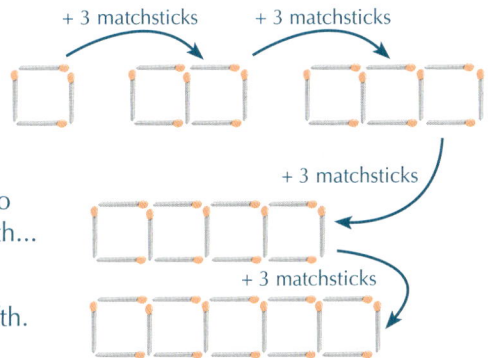

+ 3 matchsticks

b) **How many matchsticks are needed to make the eighth pattern in the sequence?**

1. You don't need to draw all the patterns — just work with the numbers of matchsticks.

   The sequence of numbers of matchsticks is 4, 7, 10, 13, 16...

2. There are 16 matchsticks in the 5th pattern. You need to add on 3 lots of 3 matchsticks to get the 8th pattern.

   $16 + 3 + 3 + 3 = $ **25 matchsticks**

Example 6

The black and white circles on the right form the first 3 patterns in a sequence. Find the number of black circles and the number of white circles in the 9th pattern.

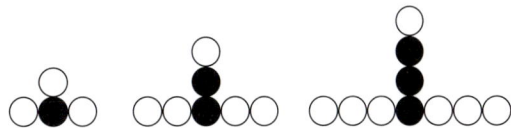

1. Write the numbers of each colour of circle in each pattern as two number sequences.

   Black circles: 1, 2, 3
   White circles: 3, 5, 7

   Tip: It can be easier to pick up on patterns in the numbers than patterns in the shapes.

2. Find the term to term rules for each sequence.

   Rule for black circles: +1 each time.
   Rule for white circles: +2 each time.

3. Apply these rules 6 times, starting at the 3rd term, to get the 9th term.
   Black: add on 6 lots of 1 to the 3rd term.
   White: add on 6 lots of 2 to the 3rd term.

   $3 + (6 \times 1) = $ **9 black circles**
   $7 + (6 \times 2) = $ **19 white circles**

Q1   The first three patterns of some sequences are shown below. For each of the sequences:
      (i)   Explain the rule for making the next pattern.
      (ii) Draw the fourth and fifth patterns in the sequence.
      (iii) Find the number of matches needed to make the sixth pattern.

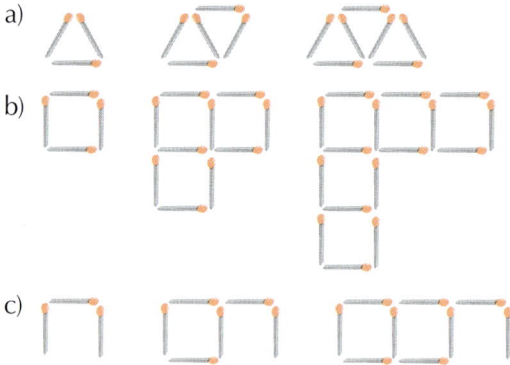

      a)

      b)

      c)

Q2   For each of the sequences below:
      (i)   Draw the next three patterns in the sequence.
      (ii) Explain the rules for generating the number of circles of each colour in the next pattern.
      (iii) Work out how many green circles there are in the 7th pattern.
      (iv) Work out how many orange circles there are in the 10th pattern.

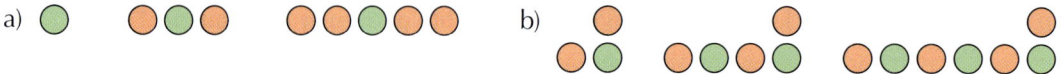

      a)   b)

Q3   For each of the sequences below:
      (i)   Explain the rules for generating the number of circles in the next pattern.
      (ii) Name the special number sequence that the number of circles follows.
      (iii) Draw the 6th pattern in the sequence.

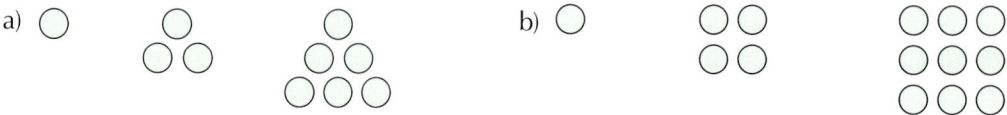

      a)   b)

Q4   The designs shown on the right form the
      first three patterns in a sequence.

      a) Draw the next pattern in the sequence.
      b) Work out the number of white squares in the
          6th pattern in the sequence.
      c) Work out the number of grey squares in the pattern which has 100 squares in total.

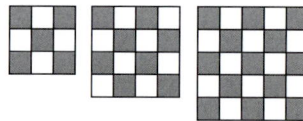

Q5   The matchstick shapes below form the first two patterns in a shape sequence.

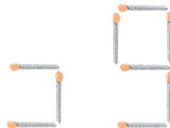

      Describe two different ways that the sequence could be continued,
      and draw the next pattern in the sequence in each case.

14.2 Using the nth Term

Using a term to term rule is fine for the first few terms, but it can be a chore if you want the hundredth term. Luckily, there's a way to get from the position, n, to the term itself by using a formula for the 'nth term'.

Learning Objective — Spec Ref A23:
Use the nth term to find terms in a sequence.

Prior Knowledge Check:
Be able to substitute values into a given formula (p.107) and solve equations (p.113-115).

You can work out the value of a term in a sequence by using its **position** (n) in the sequence, and the **nth term rule** for that sequence. The 1st term (x_1) has $n = 1$, the 2nd term (x_2) has $n = 2$, the 10th term (x_{10}) has $n = 10$, and so on. The nth term formula tells you what to do to n to get the value of that term. For example, if the nth term was $2n + 1$, you'd multiply n by 2 then add 1 to get the term.

Example 1

The nth term of a sequence is $7n - 1$. Find the first four terms of the sequence.

To find the 1st, 2nd, 3rd and 4th terms of the sequence, substitute the values $n = 1$, $n = 2$, $n = 3$ and $n = 4$ into the formula.

$(7 \times 1) - 1 = 6$
$(7 \times 2) - 1 = 13$
$(7 \times 3) - 1 = 20$
$(7 \times 4) - 1 = 27$

So the first four terms are **6**, **13**, **20** and **27**.

Exercise 1

Q1 Find the first five terms of a sequence if the nth term is given by:
   a) $n + 5$ b) $4n - 2$ c) $10 - n$ d) $10n - 8$

Q2 Find the first five terms of a sequence if the nth term is given by:
   a) $n^2 + 1$ b) $2n^2 + 1$ c) $3n^2 - 1$ d) $n(n - 1)$

Q3 The nth term of a sequence is $100 - 3n$. Find the value of:
   a) the 3rd term   b) the 10th term   c) the 30th term   d) the 40th term

Q4 The nth term of a sequence can be found using the formula $x_n = 35 + 5n$. Find the value of:
   a) x_1 b) x_5 c) x_{10} d) x_{100}

Q5 Each of the following gives the nth term of a different sequence.
   For each sequence, find: (i) the 5th term,   (ii) the 10th term,   (iii) the 100th term.
   a) $2n + 3$ b) $4n + 12$ c) $30 - 3n$ d) $-20 + 2n$

Q6 Each of the following rules generates a different sequence.
   For each sequence, find: (i) x_4,   (ii) x_{10},   (iii) x_{20}.
   a) $x_n = \frac{1}{2}n + 30$ b) $x_n = 2n^2 + 8$ c) $x_n = 5n^3 - 50$ d) $x_n = 3n - n^2 + 100$

Q7 Each of the following gives the nth term of a different sequence.
   For each sequence, find: (i) the 2nd term,   (ii) the 5th term,   (iii) the 20th term.
   a) $2n^2 + 3$ b) $3(n^2 + 2)$ c) $n(n + 1)$ d) $n^3 + 2$

If you know the value of a term in a sequence, but not its position, set up and solve an **equation** using the nth term rule to find the position, n.

Example 2

The nth term of a sequence is $4n + 5$.

a) **Which term has the value 41?**

1. Make the nth term equal to 41.	$4n + 5 = 41$
2. Solve the equation to find n.	$4n = 36$
	$n = 9$
	So 41 is the **9th term**.

b) **Which is the first term in this sequence to have a value greater than 100?**

1. Find the value of n which would give a value of 100. As before, set up and solve an equation for n.

$$4n + 5 = 100$$
$$4n = 95$$
$$n = 23.75$$

2. As long as the sequence is increasing, the first term that will give a value over 100 will be the next whole number value of n.

So the first term with a value greater than 100 must be the **24th term**.

Tip: If you're confident using inequalities (see Section 13) you could solve $4n + 5 > 100$ to get the same answer.

3. Check your answer by working out some terms in the sequence.

Check: 23rd term = $(4 \times 23) + 5 = 97$ (<100)
24th term = $(4 \times 24) + 5 = 101$ (>100) ✔

Exercise 2

Q1 a) The nth term of a sequence is $7n + 4$. Which term of the sequence has the value 53?

b) The nth term of a sequence is $5n - 8$. Which term of the sequence has the value 37?

Q2 The nth term of a sequence is $17 - 2n$. Which term has the value: a) 9?   b) −3?

Q3 The nth term of a sequence is $n^2 - 1$. Which term has the value: a) 8?   b) 99?

Q4 The nth term of a sequence is $4n - 10$.
Which term in the sequence is the first to have a value greater than 50?

Q5 The formula for the number of matches in the nth pattern of this 'matchstick sequence' is $2n + 1$.
Which pattern in the sequence is made using:
a) 17 matches?   b) 55 matches?

Q6 The formula for the number of circles in the nth triangle in the sequence shown on the right is $\dfrac{n(n+1)}{2}$.

a) Find the number of circles needed to make the 58th triangle.

b) Which term in the sequence will be the first term to be made from over 200 circles?

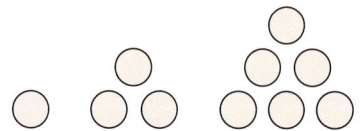

Q7 The nth term of a sequence is $-n(n + 2)$.
a) Which term has the value −48?
b) Which is the first term to have a value less than −20?

You can use the nth term to **check** if a given value is a term in that sequence. As before, set up and solve an equation to find the value of n for a given (suspected) term. The term is **only** part of the sequence if n is a **whole number**. If not, then you know the value is **not** part of the sequence.

Example 3

A sequence has nth term $3n + 2$.

a) **Is 37 a term in the sequence?**

1. Form an equation by setting the formula for the nth term equal to 37.

    $3n + 2 = 37$

2. Solve your equation to find n.

    $3n = 35$
    $n = 11.666...$

    > **Tip:** Don't just answer 'yes' or 'no' — you need to show whether or not n is a whole number.

3. Since n is not a whole number, 37 is not a term in the sequence.

    So 37 is **not a term in the sequence**.

b) **Find the term in the sequence that is closest to 37.**

1. From part a) you know that to get 37, $n = 11.666...$ So the term closest to 37 in the sequence will be where n is the closest whole number to 11.666...

    The closest whole number to 11.666... is 12.

2. Substitute $n = 12$ into the nth term formula.

    $3 \times 12 + 2 = 38$
    The term in the sequence closest to 37 is **38**.

Exercise 3

Q1 Show that 80 is a term in the sequence with nth term equal to $3n - 1$.

Q2 A sequence has nth term $21 - 2n$.

a) Show that -1 is a term in this sequence.

b) Write down the position of -1 in the sequence.

Q3 a) Find the first four terms of the sequence with nth term $6(n + 1)$.

b) Determine whether or not 64 is a term in this sequence.

c) If 64 is a term in the sequence, write down its position.
   If not, find the number closest to 64 that is a term in the sequence.

Q4 A sequence has nth term $17 + 3n$.
Determine whether or not each of the following is a term in this sequence.

a) 52   b) 98   c) 105

d) 248   e) 996   f) $20n$

14.3 Finding the nth Term

You've seen how useful nth terms can be for working out the values of terms in a sequence. It's even more useful to be able to find the nth term for yourself. How you go about this depends on the type of sequence — there are different methods for arithmetic and quadratic sequences.

Arithmetic Sequences

Learning Objective — Spec Ref A25:
Find the nth term of an arithmetic sequence.

Prior Knowledge Check:
Be able to substitute values into a given formula (p.107), solve equations (p.113-115) and solve simultaneous equations (p.144-146).

As you saw on page 161, **arithmetic (linear) sequences** are ones where the terms increase or decrease by the same value each time (the **common difference**, d).

> The nth term of an **arithmetic** sequence is given by $dn + c$.

To find c, look at the value produced by $d \times n$ and work out what you need to **add** or **subtract** to get from dn to the corresponding term in the sequence.

Example 1

a) **Find the nth term of the sequence 45, 42, 39, 36...**

1. First, find the term to term rule to work out what type of sequence it is.

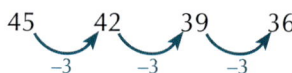

   45 42 39 36
   -3 -3 -3

   Tip: If the terms are decreasing then d will be negative.

2. It's an arithmetic sequence, so find the common difference, d, between the terms.

   It decreases by the same amount (3) each time, so it's an arithmetic sequence. Common difference $d = -3$

3. Work out what you need to add or subtract to get from $d \times n$ ($= -3n$) to the term in the sequence.

   $-3n$: -3 -6 -9 -12
   $+48$ $+48$ $+48$ $+48$
   Term: 45   42   39   36

4. To get from $-3n$ to each term in the sequence, you have to add 48, so $c = +48$.

   So the nth term is **$-3n + 48$**.

5. Check that it works by substituting in a value of n.

   Check: 2nd term ($n = 2$) is
   $(-3 \times 2) + 48 = 42$ ✔

b) **Find the 100th term in the sequence.**

1. You've just worked out the nth term formula, so you can now use it to find the 100th term.

   The nth term is $-3n + 48$.

2. Substitute $n = 100$ into $-3n + 48$.

   $(-3 \times 100) + 48 = -300 + 48 = $ **-252**

Exercise 1

Q1 Find the formula for the nth term of each of the following sequences.

   a) 7, 13, 19, 25...   b) 4, 8, 12, 16...   c) 41, 81, 121, 161...   d) −9, −5, −1, 3...

Q2 For each of the following sequences: (i) Find the nth term formula. (ii) Find the 70th term.

   a) 40, 37, 34, 31...   b) 78, 69, 60, 51...   c) 100, 92, 84, 76...   d) −10, −25, −40, −55...

Q3   a) Find an expression for the nth term of sequence A, which starts 4, 7, 10, 13...

        b) Find an expression for the nth term of sequence B, which starts 5, 8, 11, 14...

        c) (i) How does each term in sequence B compare with the corresponding term in sequence A?

             (ii) What do you notice about the formulas giving the nth term for these two sequences?

Q4   a) Find an expression for the nth term of the sequence 13, 9, 5, 1...

        b) Use your answer to part a) to write down an expression for the nth term (PROBLEM SOLVING) of the sequence 11, 7, 3, –1...

Q5   In the shape sequence on the right, the horizontal and vertical lines (PROBLEM SOLVING) between each pair of dots are 1 unit long. Find:

        a) the number of dots in the nth pattern in the sequence.

        b) the area of the nth pattern in the sequence.

        c) the number of dots and area of the 23rd pattern in the sequence.

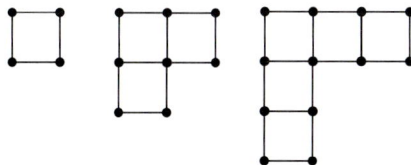

Example 2

Find the nth term of the sequence $\frac{1}{3}, \frac{2}{5}, \frac{3}{7}, \frac{4}{9}$ ...

Tip: When you have a sequence of fractions, always check for separate rules in the top and bottom numbers.

1. Find the nth term for the numerator and denominator of the fractions separately.

Numerator:
$$1 \quad 2 \quad 3 \quad 4$$
$$+1 \quad +1 \quad +1$$

2. The sequence in the numerator is easy to see — the nth term is just n.

nth term of the numerator = n

3. Find the sequence in the denominator in the same way as any other arithmetic sequence. Find the common difference, d, and compare the sequence with dn to find c.

Denominator:
$$3 \quad 5 \quad 7 \quad 9$$
$$+2 \quad +2 \quad +2$$
Common difference $d = 2$

$2n$:   2   4   6   8
          +1 +1 +1 +1
Term: 3   5   7   9
So nth term of the denominator = $2n + 1$

4. Combine the nth terms for the numerator and denominator to find the nth term of the sequence.

nth term of the sequence = $\dfrac{n}{2n + 1}$

Exercise 2

Q1   Find the nth term of each of the following sequences.

      a) $\dfrac{1}{2}, \dfrac{1}{4}, \dfrac{1}{6}, \dfrac{1}{8}$ ...   b) $\dfrac{1}{3}, \dfrac{1}{6}, \dfrac{1}{9}, \dfrac{1}{12}$ ...   c) $\dfrac{5}{2}, \dfrac{6}{3}, \dfrac{7}{4}, \dfrac{8}{5}$ ...

Q2   Find the nth term of each of the following sequences.

      a) $\dfrac{1}{5}, \dfrac{4}{10}, \dfrac{7}{15}, \dfrac{10}{20}$ ...   b) $\dfrac{5}{4}, \dfrac{4}{3}, \dfrac{3}{2}, \dfrac{2}{1}$ ...   c) $\dfrac{5}{9}, \dfrac{8}{8}, \dfrac{11}{7}, \dfrac{14}{6}$ ...

Q3   Find the 100th term of the following sequence: $\dfrac{1}{30}, \dfrac{-1}{40}, \dfrac{-3}{50}, \dfrac{-5}{60}$ ... (PROBLEM SOLVING)

There's another way of finding the **nth term** of an **arithmetic** sequence that uses the **first term** of the sequence, a, as well as the **common difference**, d:

> The nth term of an **arithmetic** sequence is $a + (n - 1)d$.

If you're given the first few terms of a sequence, **identify** a and d, put them in the **formula**, then **simplify**.

Example 3

Find the nth term of the arithmetic sequence that begins –2, 5, 12, 19...

1. It's an arithmetic sequence, so find the value of the first term (a) and the common difference (d).

   $$-2 \quad 5 \quad 12 \quad 19$$
   $$+7 \quad +7 \quad +7$$

   First term: $a = -2$
   Difference: $d = +7$

2. Substitute a and d into the formula $a + (n - 1)d$.

   $$a + (n - 1)d = -2 + (n - 1) \times 7$$

3. Simplify the expression.

   $$= -2 + 7n - 7 = 7n - 9$$

4. Write down your formula.

   So the nth term is **$7n - 9$**.

5. Check that it works.

   Check: 2nd term ($n = 2$) is $(7 \times 2) - 9 = 5$ ✔

If you're given just two terms and their positions in a sequence, you can still work out the nth term of the sequence using the formula $a + (n - 1)d$. You can either use **simultaneous equations** (as shown in Example 4 below), or take a more **problem-solving** approach (as shown in Example 5 on the next page).

Example 4

The 2nd term of an arithmetic sequence is 7 and the 5th term is 19.

a) Find the values of a and d.

1. Write equations for the 2nd and 5th terms using nth term $= a + (n - 1)d$.

   2nd term ($n = 2$): $a + (2 - 1)d = 7$
   $a + d = 7$ ①

   5th term ($n = 5$): $a + (5 - 1)d = 19$
   $a + 4d = 19$ ②

2. Solve the simultaneous equations. Subtract equation ① from equation ② to eliminate a.

   $a + 4d = 19$
   $-(a + d = 7)$
   $\overline{3d = 12}$

3. Solve the equation for d.

   $d = 4$

4. Put $d = 4$ into one of the original equations and solve for a.

   Using equation ①: $a + 4 = 7$
   $a = 3$

   So $a = $ **3** and $d = $ **4**.

5. Put these values into equation ② to check.

   Check: $a + 4d = 3 + 4 \times 4 = 19$ ✔

b) Find the nth term of the sequence.

1. Substitute the values of a and d you found above into the formula $a + (n - 1)d$.

   $$a + (n - 1)d = 3 + (n - 1) \times 4$$
   $$= 3 + 4n - 4$$

2. Simplify the expression.

   $$= 4n - 1$$

3. Write down your formula.

   So the nth term is **$4n - 1$**.

4. Check it works for a term that you know.

   Check: 2nd term ($n = 2$) is $(4 \times 2) - 1 = 7$ ✔

Example 5

a) Find a and d for an arithmetic sequence where the 1st term is 11 and the 6th term is 51.

1. You know a straight away — it's the 1st term. $a = 11$

2. Write an equation for the 6th term using nth term $= a + (n - 1)d$.

3. Solve the equation to find d.

$$11 + (6 - 1)d = 51$$
$$11 + 5d = 51$$
$$5d = 51 - 11 = 40$$
$$d = 40 \div 5 = 8$$

So $a = \mathbf{11}$ and $d = \mathbf{8}$.

b) Find a and d for an arithmetic sequence where the 3rd term is 8 and the 4th term is 5.

1. d is the difference between the 3rd and 4th terms, because they are consecutive.

$$d = \text{4th term} - \text{3rd term} = 5 - 8 = -3$$

2. Write an equation for one of the terms using nth term $= a + (n - 1)d$.

$$a + (3 - 1) \times -3 = 8$$
$$a + 2 \times -3 = 8$$
$$a - 6 = 8$$
$$a = 8 + 6 = 14$$

3. Solve the equation to find a.

So $a = \mathbf{14}$ and $d = \mathbf{-3}$.

Exercise 3

Q1 For each of the following arithmetic sequences, find:
(i) a and d, (ii) the nth term of the sequence.

a) 1st term = 3, 4th term = 9

b) 5th term = 17, 6th term = 21

c) 4th term = −8, 5th term = 2

Q2 In each of the following arithmetic sequences, you are given two of the terms in the form x_n. For each sequence, find the nth term of the sequence.

a) $x_3 = 15$, $x_6 = 30$ b) $x_2 = 12$, $x_5 = 33$ c) $x_4 = 25$, $x_7 = 49$

d) $x_2 = 13$, $x_6 = -7$ e) $x_3 = -3$, $x_6 = 24$ f) $x_6 = 61$, $x_9 = 97$

Q3 Given that the third term of an arithmetic sequence is 4 and the twelfth term is 1, find:

a) a and d

b) the nth term of the sequence

Q4 The 61st term of an arithmetic sequence is 546 and the 81st term is 726. Is 42 a term in this sequence?

Quadratic Sequences

Learning Objective — Spec Ref A25:
Find the nth term of a quadratic sequence.

Prior Knowledge Check:
Be able to find the nth term of an arithmetic sequence. See p.170.

In **quadratic sequences**, the difference between terms changes by the same amount each time (see p.162). It's this **constant**, **second difference** that allows you to find the **nth term** of a quadratic sequence.

> The nth term of a **quadratic sequence** is $an^2 + bn + c$
> where a, b and c are constants and $a \neq 0$.

There are more steps involved in working out the nth term than for an arithmetic sequence. You need to:

- Find the **constant second difference**.
- **Divide** it by **2** to get a.
- **Subtract an^2** from each term in the sequence.
- Find the **nth term** of the remaining **linear** (arithmetic) sequence — this is $bn + c$.

Example 6

a) Find the nth term of the sequence that begins 1, 8, 19, 34, 53...

1. Find the (first) difference between each pair of terms. It's not constant, so the sequence isn't arithmetic.

2. Work out the (second) difference between the differences. It's constant (+4), so the sequence must be quadratic.

$$1 \quad 8 \quad 19 \quad 34 \quad 53$$
$$+7 \quad +11 \quad +15 \quad +19$$
$$+4 \quad +4 \quad +4$$

3. Divide the constant difference by 2 to get a, the coefficient of the n^2 term.

$a = 4 \div 2 = 2$
So the rule contains a $2n^2$ term.

4. For each term in the sequence, subtract the corresponding $2n^2$ term. This leaves a linear sequence.

Term:	1	8	19	34	53
$2n^2$:	2	8	18	32	50
Term $- 2n^2$:	-1	0	1	2	3

5. Work out the rule for the remaining linear sequence (using either the method on p.172 as shown here, or p.170) and add it to the $2n^2$ term.

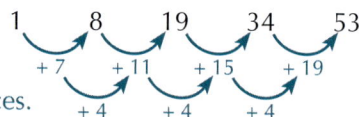

The rule for the linear sequence is
$-1 + (n - 1) \times 1 = n - 2$.
So the rule for the sequence is $\mathbf{2n^2 + n - 2}$.

b) Find the 10th term of the sequence above.

1. Once you've found the nth term, substitute $n = 10$ into the formula.

10th term $(n = 10) = 2 \times 10^2 + 10 - 2$
$= 200 + 10 - 2 = 208$

2. This gives you the value of the 10th term.

So the 10th term is **208**.

Exercise 4

Q1 For each of the following quadratic sequences, find: (i) the nth term, (ii) the 10th term.

a) 1, 5, 11, 19, 29...

b) 5, 12, 21, 32, 45...

c) 5, 12, 23, 38, 57...

d) 7, 20, 39, 64, 95...

e) 6, 24, 52, 90, 138...

f) 10, 23, 44, 73, 110...

Review Exercise

Q1 For each of the following: (i) Explain the rule for finding the next term in the sequence.
(ii) Find the next three terms in the sequence.

a) 1, 2, 4, 8... b) 4, 7, 10, 13... c) 5, 3, 1, –1...

d) 1, 1.5, 2, 2.5... e) 0.01, 0.1, 1, 10... f) –2, –6, –18, –54...

Q2 Copy the following sequences and fill in the blanks.

a) ☐, –4, –16, –64, ☐, –1024 b) –72, ☐, –18, –9, ☐, –2.25 c) –63, –55, ☐, ☐, ☐, –23

Q3 The matchstick shapes on the right form the first three patterns in a sequence. Find the number of matchsticks in the 10th pattern.

Q4 Find the first five terms of a sequence if the nth term is given by:

a) $3n + 2$ b) $5n - 1$ c) $3 - 4n$ d) $-7 - 3n$

Q5 Each of the following rules generates a different sequence.
For each sequence, find: (i) x_4 (ii) x_{10} (iii) x_{20}

a) $x_n = 8n + 4$ b) $x_n = 450 - 5n$ c) $x_n = 3n^2$ d) $x_n = n^2 + n$

Q6 Each of the following gives the nth term of a different sequence.
For each sequence, find: (i) the 2nd term (ii) the 5th term (iii) the 20th term

a) $2n^2$ b) $4n^2 - 5$ c) $\frac{1}{2}n^2 + 20$ d) $400 - n^2$

Q7 For each of the following sequences:
(i) Find the formula for the nth term. (ii) Find the value of the 70th term.

a) 10, 8, 6, 4... b) 70, 60, 50, 40... c) 60, 55, 50, 45... d) 6, 3, 0, –3...

Q8 A sequence starts –5, –1, 3, 7. Determine whether each of the following is a term in this sequence. For those that are, state the position of the term in the sequence.

a) 43 b) 138 c) 384 d) 879

Q9 Find the nth term of the following sequence: $\frac{16}{3}, \frac{36}{7}, \frac{56}{11}, \frac{76}{15}$ ...

Q10 Given that the second term of an arithmetic sequence is –7 and the fifteenth term is 32, find the nth term of the sequence.

Q11 For each quadratic sequence below:
(i) Find the formula for the nth term. (ii) Find the value of the 10th term.

a) 7, 8, 10, 13, 17... b) 5, 7, 11, 17, 25... c) 3, 5, 9, 15, 23...

Exam-Style Questions

Q1 An arithmetic sequence starts with the value –5.
The 19th term in the sequence is –95.

a) Find an expression in terms of n for the nth term of the sequence.

[2 marks]

b) Hence explain why –113 is not a term in the sequence.

[1 mark]

Q2 A sequence is made using the following formula: $\boxed{u_{n+1} = 2u_n^2 - 7, \text{ for } n \geq 1}$
The second term in the sequence is $u_2 = -6.28$.

a) Find u_3.

[2 marks]

b) Find the value of u_1 given that it is positive.

[2 marks]

Q3 In the sequence on the right, the horizontal and vertical lines between each pair of dots are 2 cm long.

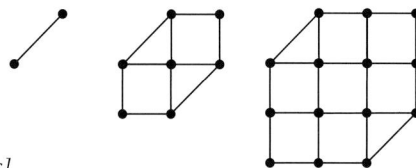

a) Find the number of dots in the
nth pattern in the sequence.

[3 marks]

b) Find the area of the 100th pattern in the sequence.

[5 marks]

Q4 The first five terms of a number sequence are 7, 14, 33, 70 and 131.

a) Is the sequence geometric? Explain your answer.

[1 mark]

b) Is the sequence quadratic? Explain your answer.

[1 mark]

Q5 A Fibonacci-type number sequence is made by following the rules below:

Rule	Example
1. Choose any two numbers.	3, 8
2. To get the next number, add the previous two.	3 + 8 = 11
3. Repeat rule 2 to produce further terms.	3, 8, 11, 19, ...

A Fibonacci-type sequence has a fourth term of 22 and a sixth term of 57.
Find the first two numbers of the sequence.

[4 marks]

15.1 Straight-Line Graphs

This is a gentle introduction to straight-line graphs (or linear graphs) — it starts off with horizontal and vertical lines, before moving on to drawing other straight-line graphs from a given equation.

Horizontal and Vertical Lines

Learning Objectives — Spec Ref A9:
- Recognise horizontal and vertical lines from their equations.
- Be able to draw horizontal and vertical lines given their equations.

All **horizontal** lines have the equation $y = a$ (where a is a number),
since every point on the same horizontal line has the same y-coordinate (a).
To **draw** the line $y = a$, draw a horizontal line that passes through a on the **y-axis**.

All **vertical** lines have the equation $x = b$ (where b is a number),
since every point on the same vertical line has the same x-coordinate (b).
To **draw** the line $x = b$, draw a vertical line that passes through b on the **x-axis**.

The equation of the **x-axis** is $y = 0$ and the equation of the **y-axis** is $x = 0$.

Example 1

Write down the equations of the lines marked A-D.

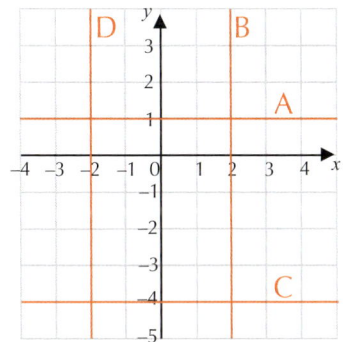

1. Every point on the line marked A has y-coordinate 1.   A is the line $y = 1$

2. Every point on the line marked B has x-coordinate 2.   B is the line $x = 2$

3. Every point on the line marked C has y-coordinate –4.   C is the line $y = -4$

4. Every point on the line marked D has x-coordinate –2.   D is the line $x = -2$

Exercise 1

Q1 Write down the equations of each of the lines labelled A to E on the right.

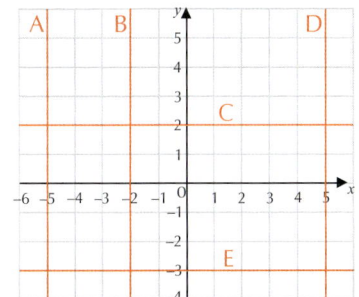

Q2 Draw a set of coordinate axes and plot the graphs with the following equations.

     a) $y = 3$ b) $y = -6$ c) $y = -1$

     d) $x = 2$ e) $x = 4$ f) $x = -6$

Q3 Write down the equation of each of the following.

    a) The line which is parallel to the x-axis, and which passes through the point (4, 8).

    b) The line which is parallel to the y-axis, and which passes through the point (–2, –6).

    c) The line which is parallel to the line $x = 4$, and which passes through the point (1, 1).

    d) The line which is parallel to the line $y = -5$, and which passes through the point (0, 6).

Q4 Write down the coordinates of the points where the following pairs of lines intersect.

    a) $x = 8$ and $y = -11$ b) $x = -5$ and $y = -13$ c) $x = -\dfrac{6}{11}$ and $y = -500$

Other Straight-Line Graphs

Learning Objective — Spec Ref A9:
Use the equation of a straight line to draw its graph.

The equation of a straight line which **isn't** horizontal or vertical contains **both x and y** — e.g. $y = 2x + 4$. If an equation **only** contains x and y terms (e.g. $y = 5x$), then the line passes through the **origin (0, 0)**.

There are a couple of different methods you can use to draw these straight-line graphs:

- Make a **table of values** — find the values of y for different values of x, plot the points and join with a straight line. You only have to plot two points to be able to sketch the graph — but it's often useful to plot more than two, in case one of the points you plot is incorrect.

- Find the **value of x when $y = 0$** and the **value of y when $x = 0$**. Plot these two points and join with a straight line. Both points should lie on the axes. **Extend** your line to cover the range of x-values required (usually specified in the question).

Example 2

a) **Complete the table to show the value of $y = 2x + 1$ for values of x from 0 to 5.**

x	0	1	2	3	4	5
y						

    Use the equation $y = 2x + 1$ to find the y-value corresponding to each value of x.

x	0	1	2	3	4	5
y	$2 \times 0 + 1 = \mathbf{1}$	$2 \times 1 + 1 = \mathbf{3}$	$2 \times 2 + 1 = \mathbf{5}$	$2 \times 3 + 1 = \mathbf{7}$	$2 \times 4 + 1 = \mathbf{9}$	$2 \times 5 + 1 = \mathbf{11}$

b) **Plot the points from the table, and hence draw the graph of $y = 2x + 1$ for values of x from 0 to 5.**

    1. Use your table to find the coordinates to plot — just read off the x- and y-values from each column.

       The points to plot are (0, 1), (1, 3), (2, 5), (3, 7), (4, 9) and (5, 11).

    2. Plot each point on the grid, then join them up with a straight line.

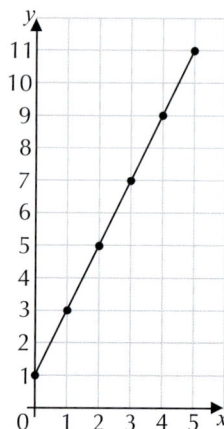

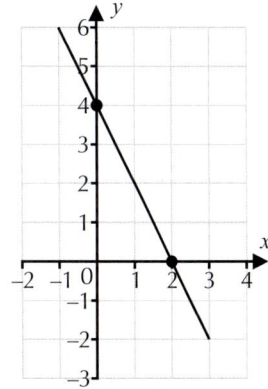

Example 3

Draw the graph of $y = 4 - 2x$ for $-1 \leq x \leq 3$.

1. Put $x = 0$ into the equation to find the value of y — this is where it crosses the y-axis.

   When $x = 0$, $y = 4 - 2(0) = $ **4**.

2. Put $y = 0$ into the equation to find the value of x — this is where it crosses the x-axis.

   When $y = 0$, $0 = 4 - 2x$
   $$2x = 4$$
   $$x = 2$$

   So the graph crosses the axes at **(0, 4)** and **(2, 0)**.

3. Mark the points (0, 4) and (2, 0) on your graph and draw a straight line passing through them. Make sure you extend it to cover the whole range of x-values asked for in the question.

Exercise 2

Q1 a) Copy and complete this table to show the value of $y = 2x$ and the coordinates of points on the line $y = 2x$ for values of x from -2 to 2.

x	-2	-1	0	1	2
y					
Coordinates					

   b) Draw a set of axes with x-values from -5 to 5 and y-values from -10 to 10.
   Plot the coordinates from your table.

   c) Join up the points to draw the graph with equation $y = 2x$ for values of x from -2 to 2.

   d) Use a ruler to extend your line to show the graph of $y = 2x$ for values of x from -5 to 5.

   e) Use your graph to fill in the missing coordinates of these points on the line:

   (i) $(4, \boxed{})$ (ii) $(-3, \boxed{})$ (iii) $(\boxed{}, -10)$

Q2 a) Copy and complete the table to show the value of $y = 8 - x$ and the coordinates of points on the line $y = 8 - x$ for values of x from 0 to 4.

x	0	1	2	3	4
y					
Coordinates					

   b) Draw a set of axes with x-values from -5 to 5 and y-values from 0 to 13.
   Plot the coordinates from your table.

   c) Join up the points to draw the graph of $y = 8 - x$ for values of x from 0 to 4.
   Use a ruler to extend your line to show the graph of $y = 8 - x$ for values of x from -5 to 5.

Q3 For each of the following equations draw a graph for values of x from -5 to 5.

   a) $y = -4x$ b) $y = \frac{x}{2}$ c) $y = 2x + 5$

   d) $y = 4 - x$ e) $y = 8 - 3x$ f) $y = \frac{x}{4} + 1$

Q4 Draw a graph of the following equations for the given range of x-values.

   a) $y = x + 7$ for $-7 \leq x \leq 0$ b) $y = -2x + 8$ for $0 \leq x \leq 5$

   c) $y = 1.5x$ for $-5 \leq x \leq 5$ d) $y = 0.5x + 2$ for $-2 \leq x \leq 4$

15.2 Gradients

The gradient of a straight line tells you how steep it is. The gradient comes in handy for lots of things (as you'll see later in this section), but first you need to know how to find it.

> **Learning Objective — Spec Ref A10:**
> Be able to find the gradient of a straight line.

To find the gradient of a line, divide the '**vertical distance**' (the change in the y-coordinates) between two points on the line by the '**horizontal distance**' (the change in the x-coordinates) between those points.

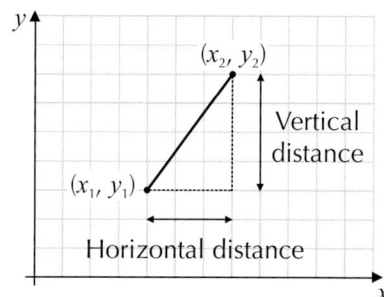

$$\text{Gradient} = \frac{\textbf{Vertical distance}}{\textbf{Horizontal distance}} = \frac{\textbf{Change in } y}{\textbf{Change in } x} = \frac{y_2 - y_1}{x_2 - x_1}$$

Make sure you subtract the x-coordinates and y-coordinates in the **same order** — i.e. if you do $y_2 - y_1$ on the numerator, you must do $x_2 - x_1$ on the denominator.

A line sloping **upwards** from left to right has a **positive gradient**.
A line sloping **downwards** from left to right has a **negative gradient**.

Regardless of the type of graph, the gradient always means '**y-axis units per x-axis units**'.

Example 1

Find the gradient of the line that passes through:
a) points $P(-3, 1)$ and $Q(4, 5)$ b) points $R(-3, 3)$ and $S(2, 1)$

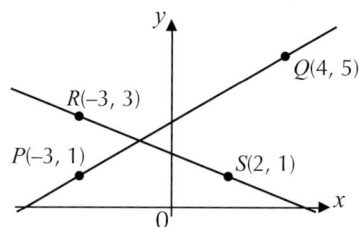

1. Call the coordinates of P (x_P, y_P), the coordinates of Q (x_Q, y_Q), the coordinates of R (x_R, y_R) and the coordinates of S (x_S, y_S).

2. Use the formula for the gradient.

3. The line through P and Q slopes upwards from left to right, so you should get a positive answer.

4. The line through R and S slopes downward from left to right, so you should get a negative answer.

a) Gradient of $PQ = \dfrac{y_Q - y_P}{x_Q - x_P} = \dfrac{5 - 1}{4 - (-3)} = \dfrac{4}{7}$

b) Gradient of $RS = \dfrac{y_R - y_S}{x_R - x_S} = \dfrac{3 - 1}{(-3) - 2} = -\dfrac{2}{5}$

Exercise 1

Q1 Points $P(x_1, y_1)$ and $Q(x_2, y_2)$ are plotted on this graph.

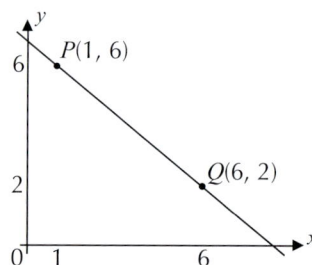

   a) Without doing any calculations, state whether the gradient of the line containing P and Q is positive or negative.

   b) Calculate the vertical distance $y_2 - y_1$ between P and Q.

   c) Calculate the horizontal distance $x_2 - x_1$ between P and Q.

   d) Find the gradient of the line containing P and Q.

Q2 Use the points shown to find the gradient of each of the following lines.

a)

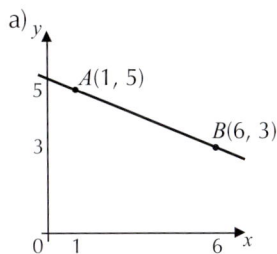

b)

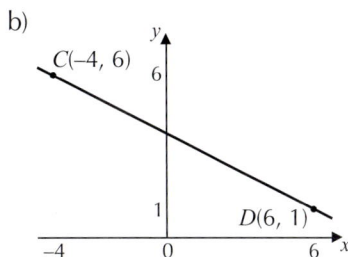

c)

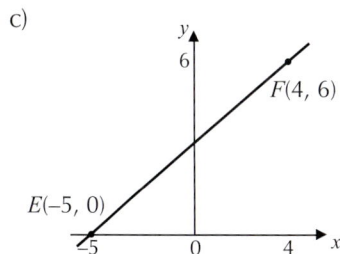

Q3 For each line shown below: (i) Use the axes to find the coordinates of each of the marked points.
(ii) Find the gradient of each of the lines.

a)

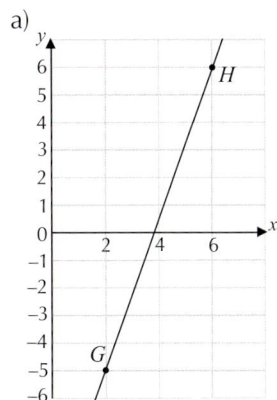

b)

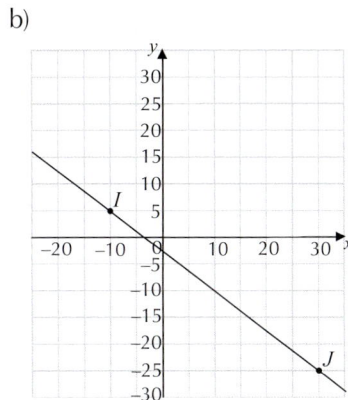

c)

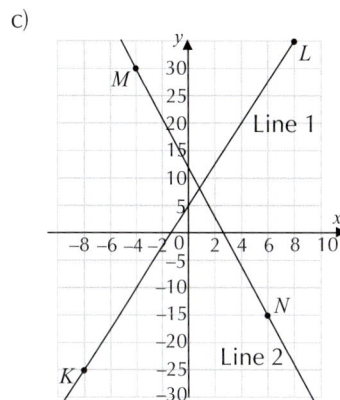

Q4 a) Plot the points $U(-1, 2)$ and $V(2, 5)$ on a grid.

b) Find the gradient of the line joining points U and V.

Q5 a) Find the difference between the y-coordinates of the points $Y(2, 0)$ and $Z(-4, -3)$.

b) Find the difference between the x-coordinates of Y and Z.

c) Hence find the gradient of the line containing points Y and Z.

Q6 Find the gradients of the lines joining the following points.

a) $A(0, 4)$, $B(2, 10)$

b) $C(1, 3)$, $D(5, 11)$

c) $E(-1, 3)$, $F(5, 7)$

d) $G(-3, -2)$, $H(1, -5)$

e) $I(5, -2)$, $J(1, 1)$

f) $K(-4, -3)$, $L(-8, -6)$

Q7 Find the gradients of lines A-D on the graph on the right.

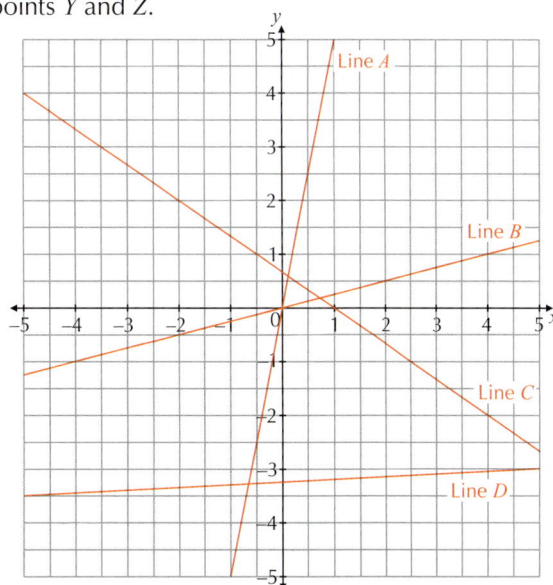

15.3 Equations of Straight-Line Graphs

There are two key facts that help you find the equation of a straight line — the gradient and where it crosses the y-axis (the y-intercept). Armed with these two bits of information, there's nothing you can't do.

Gradients and y-intercepts

Learning Objective — Spec Ref A10:
Find the gradient and y-intercept of a line given its equation.

The equation of a straight line can be written in the form $y = mx + c$.
E.g. for $y = 3x + 5$, $m = 3$ and $c = 5$.

When written in this form:

▪ m is the **gradient** of the line,

▪ c tells you the **y-intercept** — the point where the line crosses the y-axis.

If you're given an equation that isn't in $y = mx + c$ form, rearrange it into this format so that you can read off the values of m and c. For example:

Equation	$y = mx + c$ form	m	c
$y = 2 + 3x$	$y = 3x + 2$	3	2
$x - y = 0$	$y = x + 0$	1	0
$4x - 3 = 5y$	$y = \frac{4}{5}x - \frac{3}{5}$	$\frac{4}{5}$	$-\frac{3}{5}$

Make sure you don't mix up m and c when you get something like $y = 5 + 2x$. Remember, m is the number in front of the x and c is the number on its own. Watch out for **minus signs** too — both m and c can be **negative** (e.g. $y = -2x - 5$), so you have to include the minus sign when you state the gradient and y-intercept.

Example 1

Write down the gradient and the coordinates of the y-intercept of $y = 2x + 1$.

The equation is already in the form $y = mx + c$, so you just need to read the values for the gradient and y-intercept from the equation.

$y = ②x + ①$
$\quad m \qquad c$

gradient = **2**
y-intercept = **(0, 1)**

Tip: The question asks for the coordinates of the y-intercept, so don't forget the x-coordinate (which is 0).

Example 2

Find the gradient and the coordinates of the y-intercept of $2x + 3y = 12$.

1. Rearrange the equation into the form $y = mx + c$.

2. Write down the values for the gradient and y-intercept. Notice that m is negative, so the line slopes downwards from left to right.

$2x + 3y = 12 \quad \searrow -2x$
$3y = -2x + 12 \quad \searrow \div 3$
$y = \left(-\frac{2}{3}\right)x + ④$
$\quad m \qquad\qquad c$

gradient = $-\frac{2}{3}$
y-intercept = **(0, 4)**

Q1   Write down the gradient and the coordinates of the y-intercept for each of the following graphs.

a) $y = 2x - 4$ b) $y = 5x - 11$ c) $y = -3x + 7$ d) $y = 4x$

e) $y = \frac{1}{2}x - 1$ f) $y = -x - \frac{1}{2}$ g) $y = 3 - x$ h) $y = 3$

Q2   Match the graphs to the correct equation from the box.

$y = x + 2$

$y = \frac{7}{3}x - 1$

$y = -x + 6$

$y = 3x$

$y = -\frac{1}{3}x + 4$

$y = \frac{1}{3}x + 2$

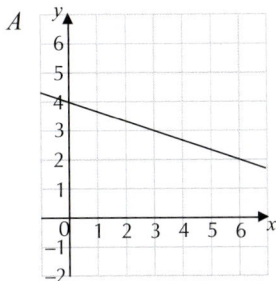

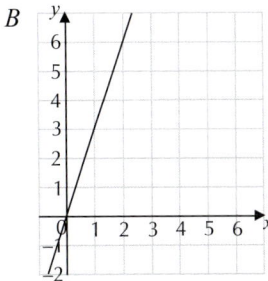

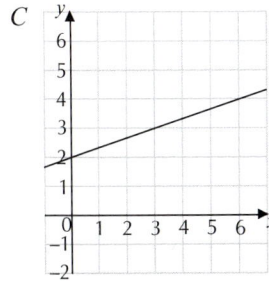

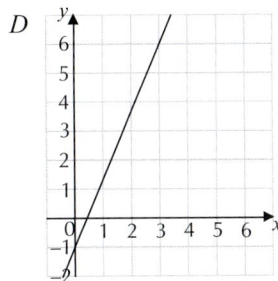

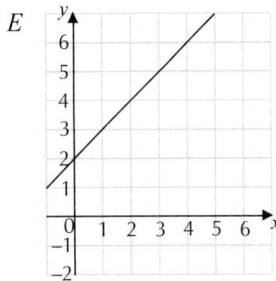

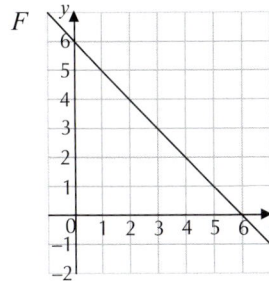

Q3   Find the gradient and the coordinates of the y-intercept for each of the following graphs.

a) $3y = 9 - 3x$ b) $y - 5 = 7x$ c) $y + x = 8$ d) $x = 6 + 2y$

e) $3x + y = 1$ f) $3y - 6x = 15$ g) $4x = 5y - 5$ h) $8x - 2y = 14$

i) $5x + 4y = -3$ j) $4y - 6x + 8 = 0$ k) $6x - 3y + 1 = 0$ l) $\frac{1}{2} = -4x - 2y$

Finding the Equation of a Straight Line

Learning Objective — Spec Ref A9:
Find the equation of a straight line in the form $y = mx + c$.

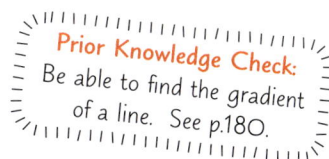

Prior Knowledge Check:
Be able to find the gradient
of a line. See p.180.

You can find the equation of a line using its **gradient** and **one point** on the line.
First, substitute the values of the **gradient** (m) and the **coordinates** of the known point (x, y) into $y = mx + c$.
You'll be left with an equation where c is the only unknown, so **solve** this equation to find the value of c.
Finally, put your values of m and c into $y = mx + c$ to give the **equation of the line**.

If you only know **two points** on the line, you can calculate the gradient of the line using the method
on p.180. Then follow the method above (using either of the two points) to find the **equation of the line**.

Example 3

Find the equation of the straight line that passes through the points $A(-3, -4)$ and $B(-1, 2)$.

1. Write down the equation for a straight line.

   $y = mx + c$ (m = gradient, c = y-intercept)

2. Find the gradient (m) of the line.

   gradient $(m) = \dfrac{y_2 - y_1}{x_2 - x_1} = \dfrac{2 - (-4)}{-1 - (-3)} = \dfrac{6}{2} = 3$

   So the equation of the line must be $y = 3x + c$.

3. Substitute the value for the gradient and the x and y values for one of the points into $y = mx + c$, then solve to find c.

   At point B, $x = -1$ and $y = 2$.
   $2 = 3 \times (-1) + c$
   $2 = -3 + c \Rightarrow c = 5$

4. Finally, rewrite the equation using your values of m and c.

   So the equation of the line is $\mathbf{y = 3x + 5}$

Exercise 2

Q1 Find the equations of the following lines based on the information given.

a) gradient = 8, passes through (0, 2)

b) gradient = –1, passes through (0, 7)

c) gradient = 3, passes through (1, 10)

d) gradient = $\frac{1}{2}$, passes through (4, –5)

e) gradient = –7, passes through (2, –4)

f) gradient = 5, passes through (–3, –7)

Q2 Find the equations of the lines passing through the following points.

a) (3, 7) and (5, 11)

b) (5, 1) and (2, –5)

c) (4, 1) and (–3, –6)

d) (–2, 1) and (1, 7)

e) (2, 8) and (–1, –1)

f) (–3, 2) and (–2, 5)

g) (–7, 8) and (–1, 2)

h) (2, –1) and (4, –9)

i) (3, 4) and (–5, –8)

Q3 Find the equations of the lines A to H shown below. Write all your answers in the form $y = mx + c$.

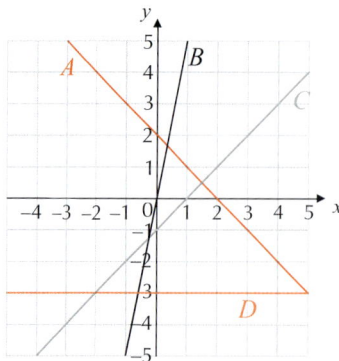

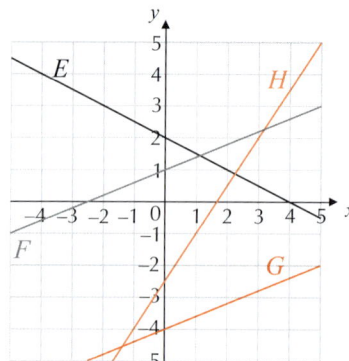

Q4 Find the equations for the lines shown on the graph on the right.
Give your answers in the form $y = mx + c$.

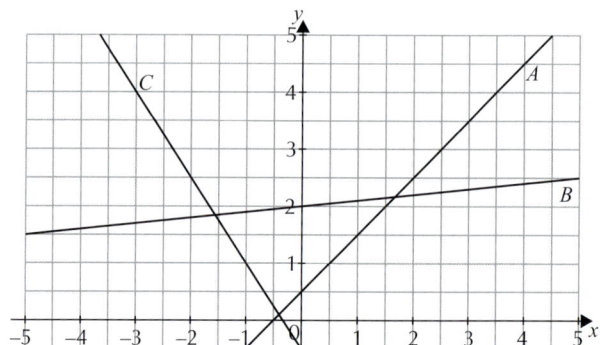

15.4 Parallel and Perpendicular Lines

I told you gradients were going to come in handy — now you're going to use them to identify parallel and perpendicular lines. Remember, parallel lines are always the same distance apart, and never meet.

Parallel Lines

Learning Objectives — Spec Ref A9:
- Identify parallel lines from their equations.
- Be able to find the equation of a parallel line.

Prior Knowledge Check:
Know how to identify the gradient of a line and be able to find the equation of a line. See p.180-184.

Lines that are **parallel** have the **same gradient** — so their equations (in $y = mx + c$ form) all have the same value of *m*.

For example, the lines $y = 3x$, $y = 3x - 2$ and $y = 3x + 4$ are all parallel.

To **check** if two lines are parallel, **rearrange** their equations so that they're both in $y = mx + c$ form, then **compare** the values of *m*.

If you have two parallel lines, *A* and *B*, and you know the **equation** of line *A* and **one point** on line *B*, you can find the equation of line *B*. First, find the **gradient** of line *A* (you can just read off this value if it's in $y = mx + c$ form). You know that line *B* has the **same gradient** as line *A* (as they're parallel), and you know the **coordinates** of one point on line *B*, so now you can use the method from p.183 to find the equation of line *B*.

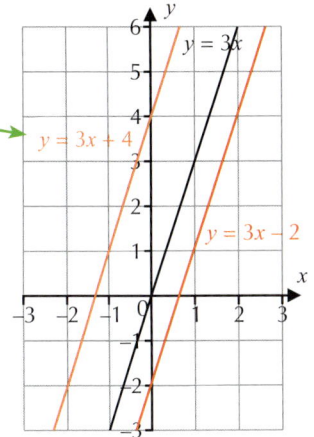

Example 1

Which of the following lines is parallel to the line $2x + y = 5$?
A: $y = 3 - 2x$ **B:** $x + y = 5$ **C:** $y - 2x = 6$

1. Rearrange the equation into the form $y = mx + c$ to find its gradient.

$$2x + y = 5$$
$$y = 5 - 2x$$
$$y = -2x + 5, \text{ so the gradient } (m) = -2.$$

2. Rearrange the other equations in the same way. Any that have $m = -2$ will be parallel to $2x + y = 5$.

A: $y = 3 - 2x$ B: $x + y = 5$ C: $y - 2x = 6$
     $y = -2x + 3$ $y = -x + 5$ $y = 2x + 6$
     $m = -2$ $m = -1$ $m = 2$

So **line A** is parallel to $2x + y = 5$.

Example 2

Find the equation of line *L*, which passes through the point (5, 8) and is parallel to $y = 3x + 2$.

1. Find the gradient of the line $y = 3x + 2$.

2. The lines are parallel, so line *L* will have the same gradient.

Equation of a straight line: $y = mx + c$
The gradient of $y = 3x + 2$ is 3.
So the equation of line *L* must be $y = 3x + c$.

3. Substitute the values for *x* and *y* at the point (5, 8) into the equation. Solve to find *c* and hence the equation of line *L*.

At (5, 8), $x = 5$ and $y = 8$:
$$8 = 3(5) + c$$
$$8 = 15 + c$$
$$c = -7, \text{ so the equation of line } L \text{ is } y = 3x - 7.$$

Q1   Write down the equations of three lines that are parallel to: a) $y = 5x - 1$ b) $x + y = 7$

Q2   Work out which of the following are the equations of lines parallel to: a) $y = 2x - 1$ b) $2x - 3y = 0$
   A: $y - 2x = 4$ B: $2y = 2x + 5$ C: $2x - y = 2$
   D: $2x + y + 7 = 0$ E: $3y + 2x = 2$ F: $6x - 9y = -2$

Q3   Which of the lines listed below are parallel
   to the line shown in the diagram?
   A: $y + 3x = 2$ B: $3y = 7 - x$
   C: $y = 4 - 3x$ D: $x - 3y = 8$
   E: $y = 3 - \dfrac{1}{3}x$ F: $6y = -2x$

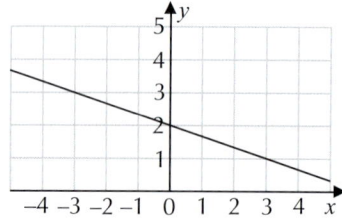

Q4   For each of the following, find the equation of the line which is parallel to the given line
   and passes through the given point.   Give your answers in the form $y = mx + c$.
   a) $y = 5x - 7$, $(1, 8)$ b) $y = 2x$, $(-1, 5)$ c) $y = \dfrac{1}{2}x + 3$, $(6, -7)$
   d) $y = 8x - 1$, $(-3, -5)$ e) $2y = 6x + 3$, $(-3, 4)$ f) $y = 7 - 9x$, $(1, -11)$
   g) $x + y = 4$, $(8, 8)$ h) $2x + y = 12$, $(-4, 0)$ i) $x + 3y + 1 = 0$, $(-9, 9)$

Perpendicular Lines

Learning Objectives — Spec Ref A9:
- Identify perpendicular lines from their equations.
- Be able to find the equation of a perpendicular line.

Prior Knowledge Check:
Know how to identify the gradient
of a line and be able to find the
equation of a line. See p.180-184.

Lines that are **perpendicular** cross at a **right angle**.

If the gradient of a line is m, then the gradient of a line perpendicular to it is $-\dfrac{1}{m}$.

The product of the gradients of two perpendicular lines is **–1**: $m \times -\dfrac{1}{m} = -1$

For example, the lines $y = 3x + 1$ and $y = -\dfrac{1}{3}x + 1$ are perpendicular
because $3 \times -\dfrac{1}{3} = -1$.

To **check** if two lines are perpendicular, **rearrange** their equations
so that they're both in $y = mx + c$ form, then see if their values
of m **multiply** to give **–1**.

If you have two perpendicular lines, P and Q, and you know
the **equation** of line P and **one point** on line Q, you can find
the equation of line Q.   First, find the **gradient** of line P, then
find the gradient of line Q (by doing $-1 \div$ gradient of line P).
You know the **coordinates** of one point on line Q and its gradient,
so now use the method from p.183 to find the equation of line Q.

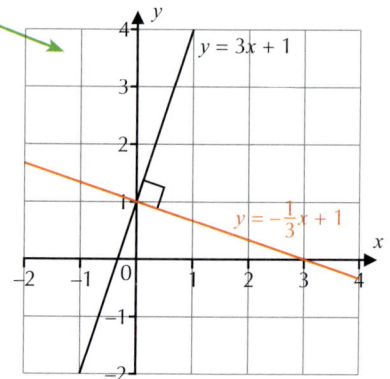

Example 3

Find the equation of line _B_, which is perpendicular to $y = 5x - 2$ and passes through the point (10, 4).

1. Use the gradient of $y = 5x - 2$ to find the gradient of line _B_.

The line $y = 5x - 2$ has gradient $m = 5$.

So the gradient of line _B_ is $-\dfrac{1}{m} = -\dfrac{1}{5}$.

2. Substitute this gradient into the equation for a straight line along with the values for _x_ and _y_ at the point given.

The equation of line _B_ is $y = -\dfrac{1}{5}x + c$.

At the point (10, 4), $x = 10$ and $y = 4$

$4 = -\dfrac{1}{5}(10) + c$

3. Solve this equation to find _c_ and hence the equation of line _B_.

$4 = -2 + c$

$c = 6$

So the equation of line _B_ is $\mathbf{y = -\dfrac{1}{5}x + 6}$.

Exercise 2

Q1 Find the gradient of a line which is perpendicular to a line with gradient:

a) 6 b) −3 c) $-\dfrac{1}{4}$ d) 12

e) −7 f) $\dfrac{2}{3}$ g) −2 h) 1.5

i) 0.3 j) −4.5 k) $-\dfrac{4}{3}$ l) $3\dfrac{1}{2}$

Q2 Write down the equation of any line which is perpendicular to:

a) $y = 2x + 3$ b) $y = -3x + 11$ c) $y = 5 - 6x$

d) $2y = 5x + 1$ e) $x + y = 2$ f) $5x - 10y = 4$

Q3 Match the following equations into pairs of perpendicular lines.

A: $y = 3x - 6$ B: $y = 2x - 3$ C: $8 - x = 3y$

D: $4x - 6y = 3$ E: $y + 3x = 2$ F: $2y - 3x = 6$

G: $x + 2y = 8$ H: $4y + 8x = 6$ I: $8y - 4x = 3$

J: $3y - 4 - x = 0$ K: $4x + 6y - 3 = 0$ L: $2y = 8 - 3x$

Q4 For each of the following, find the equation of the line which is perpendicular to the given line and passes through the given point. Give your answers in the form $y = mx + c$.

a) $y = -3x + 1$, (9, 8) b) $y = \dfrac{1}{2}x - 5$, (3, −4)

c) $y = \dfrac{1}{4}x - 7$, (1, −9) d) $y = \dfrac{4}{3}x + 15$, (12, −1)

e) $y = 8 - 2.5x$, (15, 2) f) $x + y = 8$, (3, 0)

g) $2y = 6x - 1$, (−6, 1) h) $3y + 8x = 1$, (8, 7)

i) $x + 2y = 6$, (1, 9) j) $x - 5y - 11 = 0$, (−2, 8)

15.5 Line Segments

A line segment is just part of a line between two end points — for example, line segment AB is the line joining points AB, rather than the line that passes through A and B and goes on forever in both directions.

Midpoint of a Line Segment

Learning Objective — Spec Ref G11:
Be able to find the midpoint of a line segment.

The **midpoint** of a line segment is **halfway** between the end points.

The **x-coordinate** of the midpoint is the **average** of the **x-coordinates** of the end points — so **add** the x-coordinates of the end points together and **divide by 2**.
The **y-coordinate** of the midpoint is the **average** of the **y-coordinates** of the end points — so **add** the y-coordinates of the end points together and **divide by 2**.

Example 1

Find the midpoint of the line segment *AB*, shown on the right.

1. Write down the coordinates of the end points *A* and *B*.

    $A(-3, -2)$ and $B(1, 4)$

2. Find the average of the x-coordinates by adding them together and dividing by 2. Find the average of the y-coordinates in the same way.

    $\left(\dfrac{-3+1}{2}, \dfrac{-2+4}{2}\right) = (-1, 1)$

    The midpoint has coordinates **(–1, 1)**.

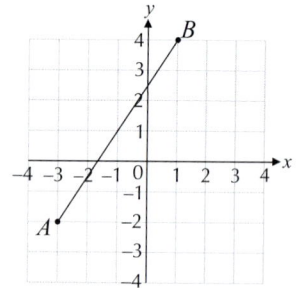

Exercise 1

Q1   Find the coordinates of the midpoint of the line segment *AB*, where *A* and *B* have coordinates:

    a) $A(8, 0)$, $B(4, 6)$ b) $A(-2, 3)$, $B(6, 5)$ c) $A(4, -7)$, $B(-2, 1)$

    d) $A(-3, 0)$, $B(9, -2)$ e) $A(-6, -2)$, $B(-4, 6)$ f) $A(-1, 3)$, $B(-1, -7)$

    g) $A(-\frac{1}{2}, 4)$, $B(\frac{1}{2}, -3)$ h) $A(2p, q)$, $B(6p, 7q)$ i) $A(8p, 2q)$, $B(2p, 14q)$

Q2   Find the midpoint of each side of the following triangles.

    a)

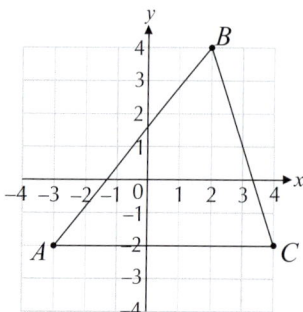

    b)

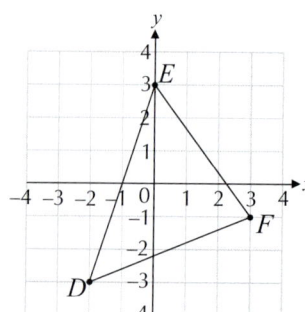

Q3   Point A has the coordinates $A(1, 8)$. The midpoint, M, of the line segment AB
      has the coordinates $M(5, 3)$. Find the coordinates of B.

Q4   The coordinates of the endpoint, C, and midpoint, M,
      of the line segment CD are $C(6, -7)$ and $M(2, -1)$.
      Find the coordinates of point D.

Q5   Use the diagram on the right to find the midpoints
      of the following line segments.

      a) AF b) AC c) DF

      d) BE e) BF f) CE

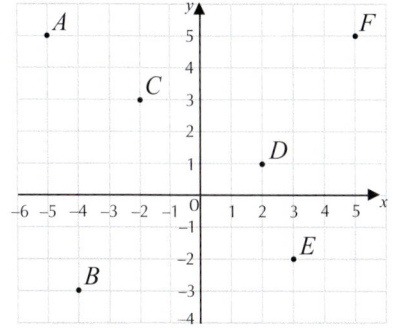

Length of a Line Segment

Learning Objective — Spec Ref G11:
Be able to find the length of a line segment.

Prior Knowledge Check:
Be able to use Pythagoras' theorem. See p.318.

To find the **length** of a line segment (or the **distance** between two points), use **Pythagoras' theorem**:

1.   Create a **right-angled triangle** using the line segment as the **hypotenuse**.

2.   Work out the lengths of the two **shorter sides** — find the **difference** between the x-coordinates
      of the end points to find the length of the **horizontal** side, then find the **difference**
      between the y-coordinates of the end points to find the length of the **vertical** side.

3.   Put these values into Pythagoras' theorem as a and b: $a^2 + b^2 = h^2$ and solve to find h.

Example 2

Calculate the length of the line segment AB. Give your answer to three significant figures.

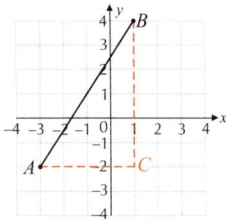

1.   Think of the line segment as the
      hypotenuse of a right-angled triangle ABC.

2.   Calculate the length of AC and BC.

3.   Calculate the length of AB using
      Pythagoras' theorem.

Length of $AC = 1 - (-3) = 4$
Length of $BC = 4 - (-2) = 6$
$AB = \sqrt{AC^2 + BC^2}$
$AB = \sqrt{4^2 + 6^2}$
$AB = \sqrt{52} = \mathbf{7.21}$ (to 3 s.f.)

Exercise 2

Q1   Find the length of the line segments with the following end point coordinates.
      Give your answers to 3 significant figures.

      a) $(5, 9)$ and $(1, 6)$ b) $(15, 3)$ and $(11, 8)$ c) $(5, 4)$ and $(4, 1)$

      d) $(3, 7)$ and $(3, 14)$ e) $(-1, 9)$ and $(9, -3)$ f) $(9, -4)$ and $(-1, 12)$

      g) $(1, -2)$ and $(8, 2)$ h) $(-3, 7)$ and $(-2, -3)$ i) $(-1, -1)$, $(-5, 9)$

      j) $(2, 4)$ and $(-1, -4)$ k) $(0, -1)$ and $(4, 8)$ l) $(-2, -1)$ and $(11, 8)$

Q2 Find the length of each side of the shapes below. Give your answers to 3 significant figures.

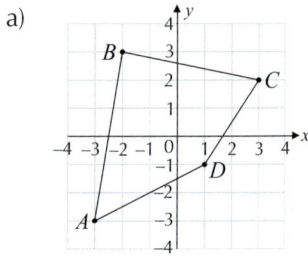

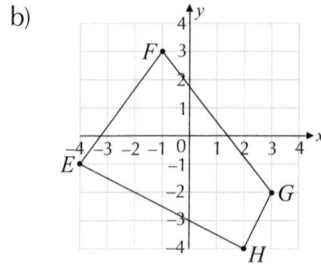

a)

b)

Coordinates and Ratio

Learning Objective — Spec Ref G11:
Use ratios to find coordinates.

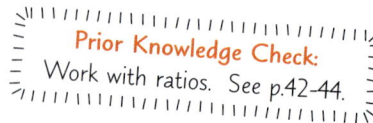

Prior Knowledge Check:
Work with ratios. See p.42-44.

Ratios can be used to express where a **point** is on a **line**. For example, if point R is a point on the line PQ such that $PR : RQ = 2 : 3$, then R is $\frac{2}{2+3} = \frac{2}{5}$ of the way from P to Q.

You can use ratios to find the **coordinates** of a point by treating the x- and y-coordinates **separately**.

Example 3

Points A, B and C lie on a straight line. Point A has coordinates (–3, 4) and point B has coordinates (3, 1). Point C lies on line segment AB such that $AC : CB = 1 : 2$. Find the coordinates of point C.

1. Find the difference between the x- and y-coordinates of A and B.

   x difference: $3 - (-3) = 6$
   y difference: $1 - 4 = -3$

2. Use the given ratio to see how far along AB point C is.

   C is $\frac{1}{1+2} = \frac{1}{3}$ of the way along AB, so

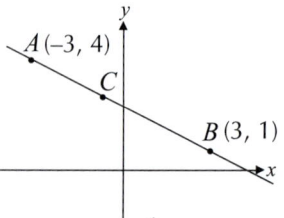

   x: $\frac{1}{3} \times 6 = 2$
   y: $\frac{1}{3} \times -3 = -1$

3. Add the results to the coordinates of A to get C.

   x-coordinate: $-3 + 2 = -1$
   y-coordinate: $4 + -1 = 3$

   Point C lies at **(–1, 3)**.

Exercise 3

Q1 Point C lies on the line segment AB. Find the coordinates of C given that:
   a) $A(3, -3)$, $B(6, 6)$ $AC:CB = 1:2$
   b) $A(-3, 5)$, $B(9, 1)$ $AC:CB = 3:1$
   c) $A(0, 4)$, $B(10, -1)$ $AC:CB = 2:3$
   d) $A(-20, 1)$, $B(8, -13)$ $AC:CB = 3:4$

Q2 Each set of three points below lies on a straight line. Use the points to find the specified ratios.
   a) Find $AB:BC$ given $A(0, 0)$, $B(2, 2)$ and $C(6, 6)$.
   b) Find $DE:EF$ given $D(1, 0)$, $E(-3, 4)$ and $F(-4, 5)$.
   c) Find $GH:HI$ given $G(-1, -2)$, $H(5, 2)$ and $I(14, 8)$.

Q3 Point T lies on the line segment SU. Find the coordinates of U given that:
   a) $S(6, 2)$, $T(12, -4)$ $ST:TU = 3:2$
   b) $S(-2, -4)$, $T(18, 11)$ $ST:TU = 5:4$

Review Exercise

Q1 Draw the following lines.

    a) $y = -2$ b) $x = 4$ c) $y = 4x - 7$

    d) $y = -5x + 1$ e) $2x + y = 3$ f) $x = 2y - 5$

Q2 Find the gradient of the line joining the following points.

    a) (2, 3) and (4, 7)   b) (1, 4) and (3, 2)   c) (3, 1) and (5, 4)

    d) (–1, –2) and (2, –4)   e) (4, –3) and (8, –1)   f) (3, –4) and (–5, –2)

Q3 Match each of the equations with the graphs *A-E*.

$y = 2x$

$y + x = 0$

$y - x = 0$

$y = -0.5x$

$2y = x$

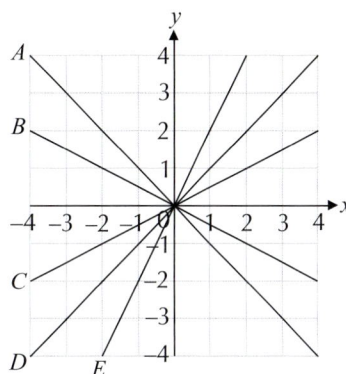

Q4 a) The line through points *A*(–1, 4) and *B*(2, *a*) has a gradient of 5. Find *a*.

    b) Points *C*(2, 7) and *D*(*b*, –2) lie on a line with a gradient of –3. Find *b*.

Q5 Find the gradient and the coordinates of the *y*-intercept of the following lines.

    a) $y = 5x - 9$ b) $y = 11 - 2x$ c) $y = -8 + 3x$

    d) $6 = 2x + 3y$ e) $x = 4y - 7$ f) $2y + 9 = 8x$

Q6 Match the equations below and the graphs shown on the right.

$x + 2y = 6$

$x = -4y$

$y = 2x - 6$

$y = 4(x - 1)$

$4x + y + 4 = 0$

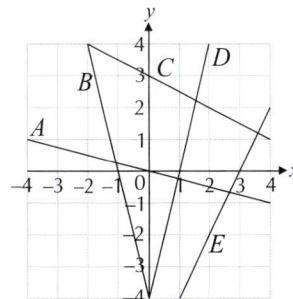

Q7 Find the equations of the lines through the following pairs of points.

    a) (1, 2) and (0, 6)   b) (8, 7) and (0, –9)   c) (4, 5) and (6, 6)

    d) (5, –8) and (–1, 10)   e) (2, –9) and (5, –15)   f) (2, 0) and (–6, –2)

Q8 Find the equations of the lines shown on the graph on the right.

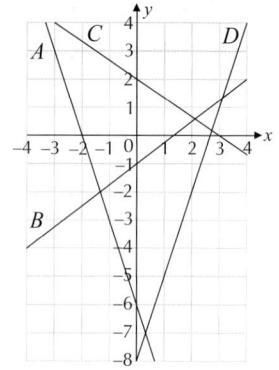

Q9 Find the equation of the line parallel to the given line that passes through the given point.

a) $y = 10 - 7x$, (1, 11)

b) $y + 2 = 9x$, (2, −16)

c) $3x + 2y = 23$, (8, 5)

d) $x - 11 = 3y$, (18, 1)

Q10 Find the equation of the line perpendicular to the given line and passing through the given point.

a) $y = 4x + 1$, (12, 9)

b) $y = 5 - 2x$, (14, 8)

c) $y + 16 = 3x$, (−6, −1)

d) $x + 3y = 17$, (−4, 2)

e) $3x + 2y = 8$, (9, −1)

f) $x - 3 = 5y$, (−2, 2)

Q11 Decide whether each of the following lines are parallel to the line $y = \frac{1}{2}x + 8$, perpendicular to it, or neither.

A: $y = 3 - 2x$ B: $8y - 4x = 5$ C: $y - 2x = 4$ D: $3x - 6y = 1$

E: $4y + 2x = 8$ F: $y = 2(x - 4)$ G: $2y = x - 7$ H: $y + 2x = 8$

Q12 Find the gradient, length and midpoint of each side of the following shapes. Give your answers to 3 significant figures where appropriate.

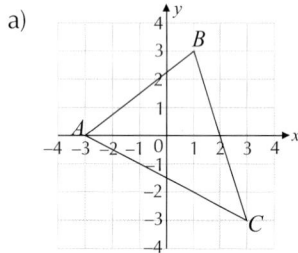

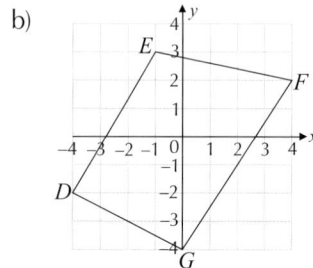

a)

b)

Q13 Find the total perimeter of the following shapes. Give your answers to 3 significant figures.

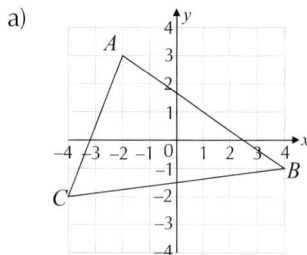

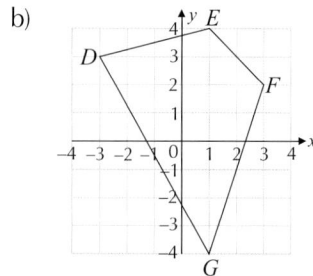

a)

b)

Q14 Jonah plots the points $A(-6, -2)$, $B(p, q)$ and $C(10, 6)$ on the line segment AC.

a) Given that $AB:BC = 1:3$, find the values of p and q.

b) Calculate the length of the line segment AB. Give your answer to one decimal place.

c) Jonah draws another point, D, on the line segment AC such that $AB:DC = 4:1$. What is the length of the line segment DC?

Exam-Style Questions

Q1 Line A has equation $5x + 2y - 8 = 0$.
Line B is parallel to line A but has a y-intercept which is triple that of line A.
Find the equation of line B.

[3 marks]

Q2 Line L passes through the points P and Q, as shown below.

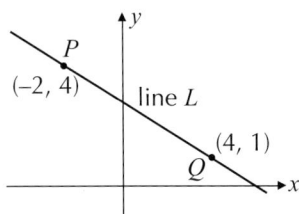

a) Find the equation of line L.

[3 marks]

b) Lines L and M are perpendicular and intersect at the point $(2, 2)$. Find the equation of line M.

[3 marks]

Q3 A straight line passes through the points $(9, 110)$ and $(-5, -100)$.
Does the point $(33, 450)$ lie on the line? Justify your answer.

[4 marks]

Q4 The diagram below shows lines AB and CD.
Line CD intersects AB at the midpoint, M, of line AB.

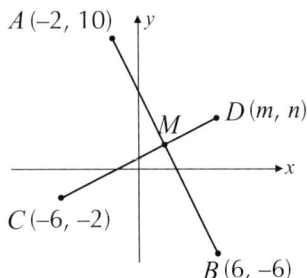

a) Find the midpoint, M, of line AB.

[2 marks]

b) Given that $CM:MD = 2:1$, find the length of the line segment CD. Give your answer to one decimal place.

[3 marks]

Q5 A triangle has vertices $A(1, 5)$, $B(3, -1)$ and $C(6, 0)$.
Is triangle ABC right-angled? Justify your answer.

[4 marks]

16.1 Quadratic Graphs

Quadratics are covered in Sections 6 and 11, so have a look back if you're not familiar with them.
You can plot accurate graphs of quadratics by finding coordinates (as you would with most types of graph).
You can also sketch them once you know how their equations relate to the shape and features of the graph.

Drawing Quadratic Graphs

Learning Objective — Spec Ref A12:
Be able to recognise, draw and read off quadratic graphs.

Quadratic functions always have x^2 as the highest power of x. The **graphs** of quadratic functions are always the same shape (called a **parabola**) and are **symmetrical** about their lowest (or highest) point.

- If the coefficient of x^2 is **positive** (e.g. $y = 2x^2 - 3x - 1$), the parabola is **u-shaped**.

- If the coefficient of x^2 is **negative** (e.g. $y = -2x^2 + 3x + 1$), the parabola is **n-shaped**.

To **draw the graph** of a quadratic function $y = ax^2 + bx + c$, calculate y-values for a set of x-values in a **table**, plot the pairs of values as **coordinates** on a set of axes, then draw a **smooth curve** through the points. You can then **read off** the graph to estimate the value of x for a given value of y.

Example 1

Draw the graph of $y = x^2 - 3$ and use it to estimate x when $y = 2$.

1. Use a table to find the coordinates of points on the graph.

x	−4	−3	−2	−1	0	1	2	3	4
x^2	16	9	4	1	0	1	4	9	16
$x^2 - 3$	13	6	1	−2	−3	−2	1	6	13

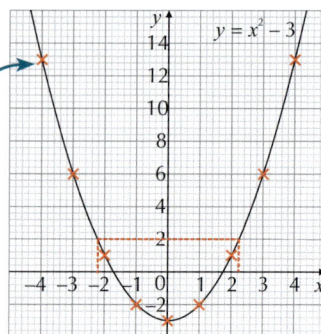

So the coordinates are (−4, 13), (−3, 6), etc.

2. Plot the coordinates on a suitable set of axes.
   Join the points with a smooth curve (<u>not</u> straight lines).

3. Draw a horizontal line at $y = 2$ and read off the values of x: $x = \pm 2.2$ (1 d.p.)

Exercise 1

Q1   For the following quadratic equations, copy and complete the table and draw each graph.

a) $y = 6 - x^2$

x	−4	−3	−2	−1	0	1	2	3	4
x^2			4			1		9	
$6 - x^2$			2			5		−3	

b) $y = 2x^2$

x	−4	−3	−2	−1	0	1	2	3	4
$2x^2$		18				2			

Q2 Draw the graph of $y = x^2 + 5$ for values of x between -4 and 4.

    a) Use your graph to find the value of y when: (i) $x = 2.5$ (ii) $x = -0.5$

    b) Use your graph to find the values of x when: (i) $y = 6.5$ (ii) $y = 10$

Q3 Draw the graph of $y = 4 - x^2$ for values of x between -4 and 4.
Write down the values of x where the graph crosses the x-axis.

Q4 Draw the graph of $y = 3x^2 - 11$ for values of x between -4 and 4.
Estimate the values of x where the graph crosses the x-axis.

Example 2

Draw the graph of $y = x^2 + 3x - 2$.

1. Add extra rows to the table to make it easier to work out the y-values.

x	-4	-3	-2	-1	0	1	2	3	4
x^2	16	9	4	1	0	1	4	9	16
$+3x$	-12	-9	-6	-3	0	3	6	9	12
-2	-2	-2	-2	-2	-2	-2	-2	-2	-2
$x^2 + 3x - 2$	2	-2	-4	-4	-2	2	8	16	26

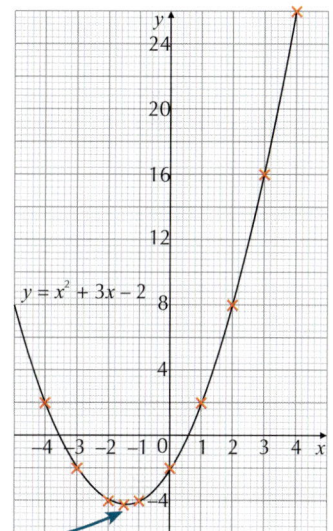

2. The table doesn't tell you the lowest point on the curve, so you need to find one more point before you can draw the graph. Quadratic graphs are always symmetrical, so the x-coordinate of the lowest point on the curve is halfway between the two lowest points from the table (or any pair of points with the same y-coordinate).

So the lowest point of the graph is halfway between $x = -2$ and $x = -1$, when $x = -1.5$ and $y = (-1.5)^2 + (3 \times -1.5) - 2 = -4.25$.

3. Plot the points and join with a smooth curve.

Exercise 2

Q1 Copy and complete the table and draw the graph of $y = 2x^2 + 3x - 7$.

x	-4	-3	-2	-1	0	1	2	3	4
$2x^2$		18						18	
$+3x$		-9						9	
-7		-7						-7	
$2x^2 + 3x - 7$		2						20	

Q2 Draw the graph of $y = x^2 - 5x + 3$ for values of x between -3 and 6.

    a) Use your graph to find the value of y when: (i) $x = -1.5$ (ii) $x = 1.5$

    b) Use your graph to find the values of x when: (i) $y = 8$ (ii) $y = -2$

Q3 Draw the graph of $y = 11 - 2x^2$ for values of x between -4 and 4.

    a) Use your graph to find the value of y when: (i) $x = -2.5$ (ii) $x = 1.25$

    b) Use your graph to find the values of x when: (i) $y = 0$ (ii) $y = 11$

Sketching Quadratic Graphs Using Factorising

Learning Objective — Spec Ref A11/A12:
Be able to sketch quadratic graphs by factorising the equation.

Prior Knowledge Check:
Be able to factorise and solve
quadratic equations. See p.135-136.

Quadratic graphs are always the same shape, which means you can **sketch** them without a table of coordinates. Draw a u-shaped or n-shaped **parabola** in roughly the correct position and label these **important points**:

- The **y-intercept** is the value of the **constant term** in the equation, i.e. $(0, c)$ for $y = ax^2 + bx + c$, because this is the value of y when $x = 0$.

- The **x-intercepts** can be found by **factorising** the equation. A quadratic $y = (x + p)(x + q)$ crosses the x-axis at $(-p, 0)$ and $(-q, 0)$, because $x = -p$ and $x = -q$ are the solutions to $(x + p)(x + q) = 0$.

- The **turning point** is the **lowest point** (for a u-shaped graph) or **highest point** (for an n-shaped graph) on the curve. Due to the symmetry of the graph, the **x-coordinate** of the turning point is always **halfway** between the **x-intercepts**. Put this value into the equation to find the y-coordinate.

Tip: The turning point lies on the vertical line of symmetry.

Example 3

Sketch the graph of $y = x^2 + x - 6$, and label the turning point and intercepts with their coordinates.

1. Find the y-intercept by putting $x = 0$ in the equation (or just look at the constant term).

   When $x = 0$, $y = 0^2 + 0 - 6 = -6$, so the y-intercept is at $(0, -6)$.

2. Factorise and solve the equation $x^2 + x - 6 = 0$ to find the x-intercepts.

   $x^2 + x - 6 = 0 \Rightarrow (x + 3)(x - 2) = 0 \Rightarrow x = -3$ and $x = 2$, so the x-intercepts are $(-3, 0)$ and $(2, 0)$.

3. Use symmetry to find the turning point. The x-coordinate is halfway between the x-intercepts. Find the y-coordinate by putting the x-coordinate into the equation.

   x-coordinate $= (2 + -3) \div 2 = -0.5$
   y-coordinate $= (-0.5)^2 + (-0.5) - 6 = -6.25$
   So the turning point has coordinates $(-0.5, -6.25)$.

4. Use all this information to sketch and label the graph. The x^2 term is positive so the graph is u-shaped. This means that the turning point will be the lowest point on the graph.

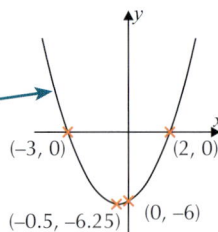

Tip: Sketches don't have to be completely accurate — they just need the correct general shape, and key points labelled and in roughly the correct positions.

Exercise 3

Q1 For each of the following equations, find the coordinates of: (i) the intercepts, (ii) the turning point.

a) $y = (x - 1)(x + 1)$

b) $y = (x + 7)(x + 1)$

c) $y = x^2 + 16x + 60$

Q2 Sketch the graphs of these equations. Label the turning points and intercepts with their coordinates.

a) $y = x^2 - 4$

b) $y = x^2 - 4x - 12$

c) $y = x^2 + 12x + 32$

d) $y = x^2 + x - 20$

e) $y = -x^2 - 2x + 3$

f) $y = -x^2 - 14x - 49$

g) $y = 2x^2 + 4x - 16$

h) $y = 5x^2 - 6x - 8$

i) $y = -2x^2 - x + 6$

Sketching Quadratic Graphs by Completing the Square

Learning Objective — Spec Ref A11/A12:
Be able to sketch quadratic graphs by completing the square.

Prior Knowledge Check:
Be able to complete the square for a quadratic equation. See p.137-139.

You can use the **completed square** form of a quadratic equation to easily find the coordinates of the **turning point** of its graph. For equations of the form $y = (x + p)^2 + q$, the turning point can be found where the expression in **brackets** is **zero**, i.e. where $x = -p$ and $y = q$, at $(-p, q)$.

This also tells you the **number of roots**. For a **positive** quadratic, if the minimum point lies **above** the x-axis (i.e. has a **positive** value of q), the graph won't cross the x-axis and so there are **no real roots**. (The same is true for a **negative** quadratic when the maximum point lies **below** the x-axis with a **negative** value of q.) If the turning point lies **on the x-axis** there is **one repeated root**. Otherwise there are **two real roots**.

Example 4

Find the coordinates of the turning point of the graph with equation $y = 4 + 2x - x^2$.

1. Complete the square.

    $y = 4 + 2x - x^2 = -(x^2 - 2x - 4)$
    $= -[(x-1)^2 - 1 - 4] = 5 - (x-1)^2$

2. Find the x- and y-values when the expression in brackets is zero.

    $(x - 1) = 0$, so $x = 1$, $y = 5 - 0 = 5$.
    So the turning point is at **(1, 5)**.

Tip: The x^2 term is negative, so the turning point is the highest point on the n-shaped graph.

Example 5

Sketch the graph of $y = x^2 + 8x - 5$. Label the y-intercept and turning point.

1. Complete the square.

    $y = (x + 4)^2 - 16 - 5 = (x + 4)^2 - 21$

2. The turning point occurs when the brackets of the completed square are equal to 0.

    When $(x + 4) = 0$, $x = -4$, $y = 0 - 21 = -21$.
    So the turning point is at $(-4, -21)$.

3. Substitute $x = 0$ into $y = x^2 + 8x - 5$ to find the y-intercept.

    $y = 0^2 + (8 \times 0) - 5 = -5$, so the y-intercept is $(0, -5)$.

4. Sketch the graph through these two points and label their coordinates.
    The x^2 term is positive, so the graph is u-shaped, and the turning point is the lowest (minimum) point on the graph.

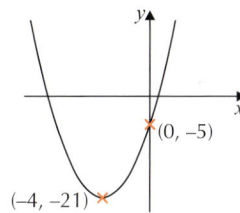

Tip: Here, the turning point lies below the x-axis, so the graph crosses the axis twice and there are 2 real roots, at $(x + 4)^2 - 21 = 0$
$\Rightarrow x = -4 \pm \sqrt{21}$.

Exercise 4

Q1   For each of the following equations, find the coordinates of: (i) the y-intercept, (ii) the turning point.

    a) $y = (x + 5)^2 - 9$ b) $y = (x - 3)^2 - 30$ c) $y = (x - 4)^2 - 13$

Q2   By completing the square, sketch the graphs of the following equations.
    Label the y-intercept and turning point of each graph.

    a) $y = x^2 - 6x - 5$ b) $y = x^2 - 4x + 2$ c) $y = x^2 + 8x - 6$

    d) $y = x^2 + 2x + 8$ e) $y = x^2 - x + 10$ f) $y = -x^2 + 10x - 6$

16.2 Cubic Graphs

Just like quadratic graphs, graphs of cubic equations have their own distinctive shape. You can plot them by calculating coordinates as usual, or sketch them using their general shape and any intercepts.

> **Learning Objective — Spec Ref A12:**
> Be able to recognise, draw and sketch graphs of cubic functions.

Cubic functions have x^3 as the **highest power** of x — they have equations of the form $y = ax^3 + bx^2 + cx + d$, where $a \neq 0$. **Cubic graphs** all have the same basic shape — a curve with a '**wiggle**' in the middle.

- If the coefficient of x^3 is **positive**, the curve goes **up** from the **bottom left**.

- If the coefficient of x^3 is **negative**, the curve goes **down** from the **top left**.

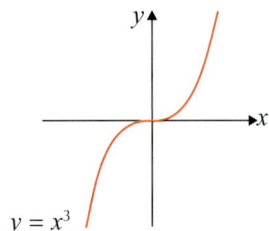

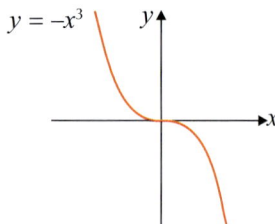

$y = x^3$

$y = -x^3$

> **Tip:** Some cubic graphs will have more of a 'wiggle' depending on the values of b and c — like in Example 1 below.

To **draw the graph** of a cubic function $y = f(x)$, find y-values using x-values in a **table**, then **plot** the coordinates and draw a **smooth curve** through the points. You can then **read off** the graph as usual.

Example 1

a) Draw the graph of $y = x^3 + 3x^2 + 1$.

1. Calculate y-values for a set of x-values in a table. Add extra rows to the table to work out the y-values step by step.

x	-4	-3	-2	-1	0	1	2
x^3	-64	-27	-8	-1	0	1	8
$+3x^2$	48	27	12	3	0	3	12
$+1$	1	1	1	1	1	1	1
$x^3 + 3x^2 + 1$	-15	1	5	3	1	5	21

2. Plot the coordinates and join the points with a smooth curve. The x^3 term is positive so the curve goes up from the bottom left.

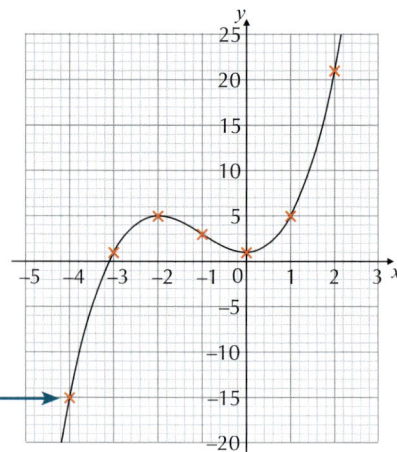

b) Find the coordinates of the x-intercept(s) of $y = x^3 + 3x^2 + 1$.

1. Read off the graph the coordinates where the curve cuts the x-axis.

   It cuts the x-axis once at $x = -3.1$ to 1 d.p., so the x-intercept is at **(−3.1, 0)** (1 d.p.).

2. You can check that this value of x fits with the equation.

   At the x-intercept, $x = -3.1$, so $y = (-3.1)^3 + (3 \times (-3.1)^2) + 1 = 0.0$ (1 d.p.) ✔

Exercise 1

Q1 For the following cubic equations, copy and complete the table and draw each graph.

a) $y = x^3 + 5$

x	−3	−2	−1	0	1	2	3
x^3	−27					8	
$x^3 + 5$	−22					13	

b) $y = 3x^3 - 4x^2 + 2x - 8$

x	−3	−2	−1	0	1	2	3
$3x^3$	−81				3		
$-4x^2$	−36				−4		
$+2x$	−6				2		
-8	−8				−8		
$3x^3 - 4x^2 + 2x - 8$	−131				−7		

c) $y = 5 - x^3$

x	−3	−2	−1	0	1	2	3
$5 - x^3$		13					−22

Q2 Draw the graph of $y = x^3 + 3$ for values of x between −3 and 3.
Use your graph to estimate the value of y when $x = -2.5$.

Q3 Draw the graph of $y = x^3 - 6x^2 + 12x - 5$ for values of x between −1 and 5.
Use your graph to estimate the value of x when $y = 0$.

To **sketch** a cubic graph, first look at the sign of the x^3 term (**positive** or **negative**) to decide on the shape. Find the **y-intercept** by putting **$x = 0$** in the equation. If the equation is in **factorised** form (i.e. $y = (x + p)(x + q)(x + r)$), the solutions to **$y = 0$** are the **x-intercepts** ($x = -p, -q$ and $-r$). You can then draw the '**wiggle**' through these points. If one of the factors is **repeated** (e.g. $y = (x + p)^2(x + q)$), the graph will just **touch** the x-axis at $x = -p$, not cross it.

Example 2

Sketch the cubic graph $y = x(x + 2)(1 - x)$ and label the coordinates of the intercepts.

1. Set $x = 0$ to find the y-intercept. $y = 0 \times (0 + 2) \times (1 - 0) = 0$, so the y-intercept is at (0, 0).

2. Solve $y = 0$ to find any x-intercepts. $x(x + 2)(1 - x) = 0 \Rightarrow x = 0$, $x = -2$ or $x = 1$.
   So the x-intercepts are at (0, 0), (−2, 0) and (1, 0).

3. Expand the brackets to see whether it's a positive or negative cubic. $y = x(x - x^2 + 2 - 2x) = -x^3 - x^2 + 2x$
   The coefficient of x^3 is negative.

4. Sketch a smooth curve with a negative cubic shape (going down from the top left) through the intercepts. Label the coordinates of the intercepts.

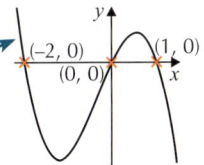

Exercise 2

Q1 For the following cubic functions, find the coordinates of the x- and y-intercepts of their graphs.
   a) $y = x(x - 1)(x + 2)$ b) $y = (x + 1)(x + 3)(1 - x)$ c) $y = x^2(x + 1)$ d) $y = (x - 10)^3$

Q2 Sketch graphs of each function in Q1, labelling the coordinates of the x- and y-intercepts.

16.3 Reciprocal and Exponential Graphs

In mathematics, the reciprocal of a number is 'one divided by it' — reciprocal functions always involve 'one divided by a function'. Exponential functions involve something 'to the power of x'.

Reciprocal Graphs

> **Learning Objective — Spec Ref A12:**
> Be able to recognise, draw and sketch graphs of reciprocal functions.

The equations $y = \dfrac{1}{x}$ and $y = -\dfrac{1}{x}$ are **reciprocal functions** — they can be used to show **inverse proportion** (see p.130). They are **undefined** when $x = 0$ and $y = 0$. This gives their **graphs** their distinctive shape, with horizontal and vertical **asymptotes** at the **axes**. Asymptotes are lines the graph gets very close to, but **never touches**.

The graph of $y = \dfrac{1}{x}$ lies in the bottom left and top right quadrants.

The graph of $y = -\dfrac{1}{x}$ lies in the top left and bottom right quadrants.

Both graphs have asymptotes at $x = 0$ and $y = 0$, and **lines of symmetry** at $y = x$ and $y = -x$.

All **general** reciprocal graphs of the form $y = \dfrac{1}{x-a} + b$ have the **same basic shape** as $y = \pm\dfrac{1}{x}$, but with different asymptotes. They're just **translations** of the graph of $y = \pm\dfrac{1}{x}$ (see p.206).

To **sketch** a reciprocal graph, find the **asymptotes**, then draw the curve **between them** in the right position.

A reciprocal graph will have asymptotes where the function is **undefined**.

- The function is undefined at $x = a$, as this is the value of x that makes the denominator of the fraction 0, so the graph has a **vertical asymptote** with equation $x = a$.

- As $\dfrac{1}{x-a}$ is never equal to 0, the function is also undefined at $y = b$, so the graph has a **horizontal asymptote** with equation $y = b$.

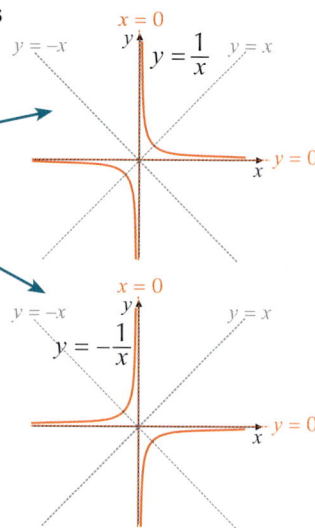

> **Tip:** You can also draw a more accurate reciprocal graph using a table of coordinates.

> **Example 1**
>
> **By finding the equations of the asymptotes, sketch the graph of $y = \dfrac{1}{x} - 2$.**
>
> 1. Find the values of x and y where the function is undefined.
>   When $x = 0$, $\dfrac{1}{x}$ is undefined, so the vertical asymptote is at $x = 0$. Draw this as a dotted line on the axes.
>
> 2. As $\dfrac{1}{x}$ can never equal 0, the function is undefined when $y = -2$. So the horizontal asymptote is at $y = -2$. Draw this as a dotted line on the axes.
>
> 3. The coefficient of the reciprocal term is positive, so draw a sketch with the same shape as $y = \dfrac{1}{x}$ positioned between the two asymptotes.
>
> 4. Check the graph does what you'd expect:
>   When x is very small, $\dfrac{1}{x}$ is very large and when x is large, $\dfrac{1}{x}$ is very small.

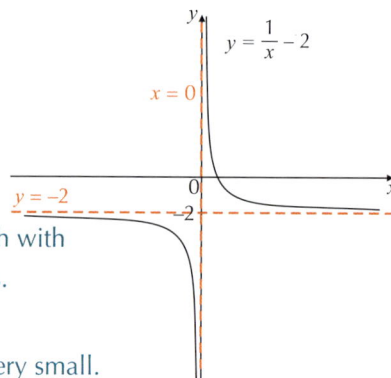

Q1 Copy and complete each table and draw graphs for the following reciprocal functions. Give any rounded numbers to 2 decimal places.

a) $y = \dfrac{4}{x}$

x	-5	-4	-3	-2	-1	-0.5	-0.1	0.1	0.5	1	2	3	4	5
$\dfrac{4}{x}$	-0.8			-2					8					

b) $y = \dfrac{1}{x} + 3$

x	-5	-4	-3	-2	-1	-0.5	-0.1	0.1	0.5	1	2	3	4	5
$\dfrac{1}{x}$			-0.33			-2					0.5			
$+3$			3			3					3			
$\dfrac{1}{x}+3$			2.67			1					3.5			

Q2 a) For what value of x is the expression $y = \dfrac{1}{x-2}$ undefined?

b) For what value of y is the expression $y = \dfrac{1}{x-2}$ undefined?

c) Use your answers to a) and b) to give the equations of the asymptotes of the graph $y = \dfrac{1}{x-2}$.

d) Sketch the graph of $y = \dfrac{1}{x-2}$, and give the coordinates of the y-intercept of the graph.

Q3 Find the equations of the asymptotes of the following functions, and use these to sketch their graphs. Find the coordinates of any x- and y-intercepts.

a) $y = \dfrac{1}{x+5}$

b) $y = \dfrac{1}{2x} - 1$

c) $y = \dfrac{1}{3-x} + 10$

Exponential Graphs

Learning Objective — Spec Ref A12:
Be able to recognise, draw and sketch graphs of exponential functions.

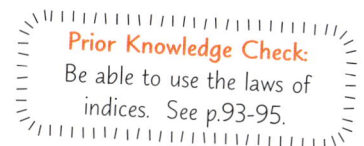

Prior Knowledge Check:
Be able to use the laws of indices. See p.93-95.

Exponential functions have the form $y = a^x$, where $a > 0$. **Graphs** of $y = a^x$ always have the same curved shape and lie above the x-axis (i.e. y is positive for all values of x). They have a **horizontal asymptote** at $y = 0$, and a y-**intercept** at $(0, 1)$ (as $a^0 = 1$ for any value of a). The bigger the value of a, the **steeper** the graph. **Sketches** of these functions will look like one of the graphs below:

When $a > 1$, as x increases, a^x quickly gets very large. As x gets more negative, a^x gets smaller and smaller but never reaches 0, so the x-axis is an **asymptote**.

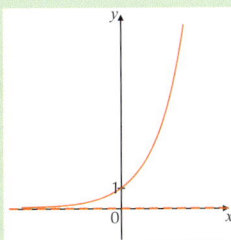

When $0 < a < 1$, as x increases, a^x gets smaller and smaller but never reaches 0, so the x-axis is an **asymptote**. As x gets more negative, a^x quickly gets very large.

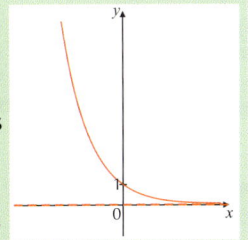

The graph of an exponential function of the form $y = a^x + b$ will have the same shape, but the **asymptote** will be at $y = b$, and the y-**intercept** will be at $(0, 1 + b)$.

Tip: This is a vertical translation of $y = a^x$ by $+b$ (see p.206).

Example 2

On the same axes, sketch the graphs of $y = 4^x$, $y = 4^{-x}$, $y = 6^x$ and $y = 6^x + 4$.

1. For $y = 4^x$, draw an exponential curve that increases as x increases, has an asymptote at $y = 0$, and a y-intercept at $(0, 1)$.

2. Get $y = 4^{-x}$ in the form $y = a^x$ using power laws:
$$y = 4^{-x} = \frac{1}{4^x} = \left(\frac{1}{4}\right)^x$$
This also has an asymptote at $y = 0$ and a y-intercept at $(0, 1)$, but $a < 1$, so draw an exponential curve that decreases as x increases. It will be the reflection of $y = 4^x$ in the y-axis.

3. The graph of $y = 6^x$ is the same as $y = 4^x$ but steeper.

4. The graph of $y = 6^x + 4$ has the same shape as $y = 6^x$, but with a different asymptote and intercepts:

As x gets more negative, 6^x approaches 0, and so y approaches $0 + 4 = 4$.
So the asymptote is at $y = 4$. The y-intercept is at $x = 0 \Rightarrow y = 6^0 + 4 = 1 + 4 = 5$.

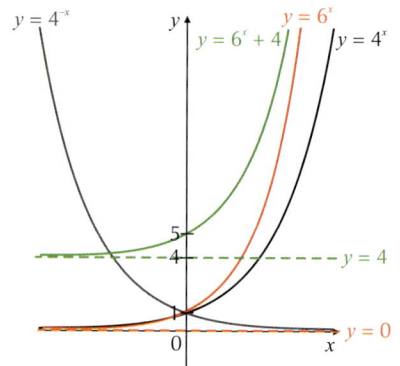

You can also **draw** a more accurate graph using a **table of coordinates**, as usual.
You'll need to draw in the **horizontal asymptote** as you would for a sketch.

Exercise 2

Q1 Copy and complete each table and draw graphs for the following exponential functions.
Give any rounded numbers to 2 decimal places.

a) $y = 3^x$

x	−3	−2	−1	0	1	2	3
3^x		0.11					27

b) $y = 2^{-x}$

x	−3	−2	−1	0	1	2	3
2^{-x}		4		1			

Q2 For each of the following graphs, write down the equation of the asymptote and the point where each graph crosses the y-axis. Use your answers to help you sketch the graphs.

a) $y = 5^x$

b) $y = 2^x - 1$

c) $y = 10^{-x}$

d) $y = 0.1^x$

e) $y = 0.5^x + 3$

f) $y = 10 - 3^x$ (PROBLEM SOLVING)

Q3 £100 is invested in a bank account that pays 5% interest per year.
The amount of money in the account (in pounds) after x years is equal to 100×1.05^x.
Sketch a graph of $y = 100 \times 1.05^x$ for values of x from 0 to 10, to show how the amount of money in the account changes over 10 years.

Q4 The exponential graph on the right shows how the value of a car in pounds, V, changes with t, the car's age in years. (PROBLEM SOLVING)

a) If $V = k \times 0.8^t$, use the graph to find the value of the constant k.

b) What does the value of k mean in the context of the question?

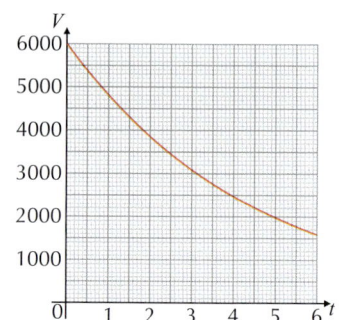

16.4 Circle Graphs

Circles can be described by an equation and plotted on coordinate axes. The equation of a circle tells you the fixed distance of the curve from its centre — also known as the radius.

> **Learning Objective — Spec Ref A16:**
> Be able to draw graphs of circles from their equation.

The equation of a circle with **radius r** and **centre (0, 0)** is $x^2 + y^2 = r^2$.

To **draw a circle** from its equation, take the **square root** to find the value of r, then draw the circle through $\pm r$ on both axes using a pair of **compasses**, with the **centre at (0, 0)**.

> **Tip:** Tangents to circle graphs are covered on page 224.

To **write** the **equation** of a given circle centred at the origin, put its **radius** into the equation given above.

Example 1

Sketch the graph of the equation $x^2 + y^2 + 1 = 17$.

1. Rearrange the equation to get it into the form $x^2 + y^2 = r^2$.

    $x^2 + y^2 + 1 = 17$
    $\Rightarrow x^2 + y^2 = 16$

2. Compare it with $x^2 + y^2 = r^2$ to find r^2.

    $r^2 = 16$

3. Take square roots to find r.

    $r = \sqrt{16} = 4$

4. Use a pair of compasses to draw the circle with a centre at (0, 0) and a radius of 4.

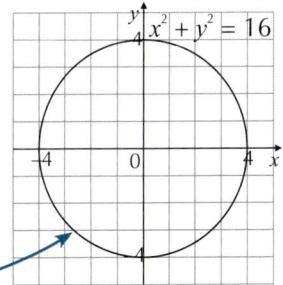

Exercise 1

Q1 Sketch the graphs with the following equations. Label the coordinates of the x- and y-intercepts.

a) $x^2 + y^2 = 25$

b) $x^2 + y^2 = 1$

c) $x^2 + y^2 = 6.25$

d) $y^2 = 4 - x^2$

e) $2x^2 + 2y^2 = 18$

f) $x^2 + 0.04 = 0.2 - y^2$

Q2 Write down the radius and equation of each of the following circles.

a)

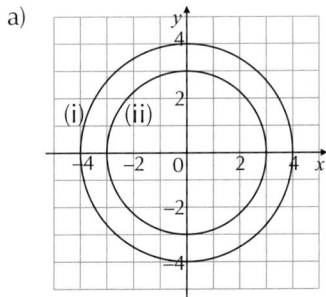

b)

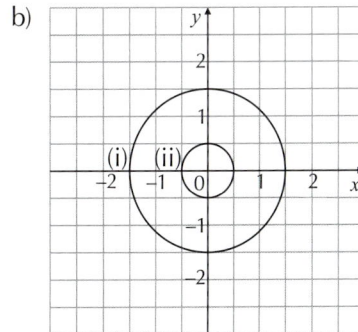

Q3 A circle with centre (0, 0) and diameter 10 is drawn on the same axes as the graph of $y + x = 1$. Draw both graphs and find the coordinates of their points of intersection.

16.5 Trigonometric Graphs

When you see the words sine, cosine and tangent, you'll probably think of right-angled triangles — but if you plot a graph of sin x, cos x or tan x against the angle x, you'll get a very distinctive and useful curve.

Learning Objective — Spec Ref A12:
Be able to recognise and sketch graphs of trigonometric functions.

Equations of the form $y = \sin x$, $y = \cos x$ and $y = \tan x$ (where x is an angle) are known as **trigonometric functions**.

The **graphs** of these trigonometric functions are **periodic** — they all have **patterns** that **repeat**. Once you know the **key features** of a pattern, such as the **intercepts**, **asymptotes** and coordinates of **maximum and minimum points**, you can **sketch the graph** for any range of angles by:

- **Plotting** the **key points** in a pattern and **joining** them with a **smooth curve**,

- Extending as far as necessary in either direction by **repeating** the pattern.

You can also **draw** accurate graphs using a **table of coordinates**, and **read off** other values from the graph.

Sine and Cosine Graphs

Sine and cosine have similar graphs — often described as a **sine 'wave'** $\wedge\!\!\!\vee$ and a **cos 'bucket'** $\vee$. Each function takes values **between –1 and 1**. They both **repeat every 360°** — they have a **period** of 360°.

Graph of $y = \sin x$:

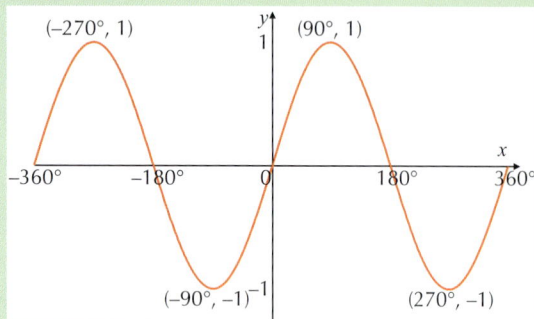

Graph of $y = \cos x$:

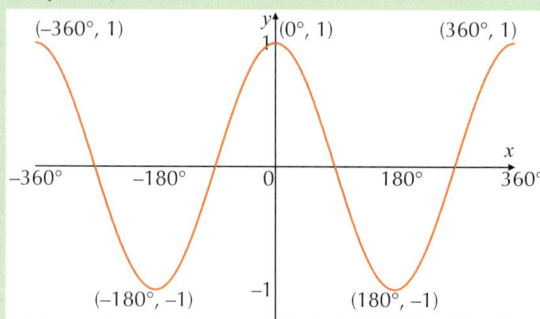

- Crosses **y-axis** at (0°, 0)
- Crosses **x-axis** at (0°, 0), (180°, 0), (360°, 0)...
- **Maximum points** at (–270°, 1), (90°, 1)...
- **Minimum points** at (–90°, –1), (270°, –1)...

- Crosses **y-axis** at (0°, 1)
- Crosses **x-axis** at (–90°, 0), (90°, 0), (270°, 0)...
- **Maximum points** at (–360°, 1), (0°, 1), (360°, 1)...
- **Minimum points** at (–180°, –1), (180°, –1)...

Tangent Graph

The graph of tangent has a **period of 180°** — the pattern repeats every 180°.

Graph of $y = \tan x$:

- **Vertical asymptotes** at $x = –90°, 90°, 270°$...
  So tan x is **undefined** at these points.
- Crosses **y-axis** at (0°, 0).
- Crosses **x-axis** at (0°, 0), (180°, 0), (360°, 0)...
- y can take **any value** from $-\infty$ to ∞.

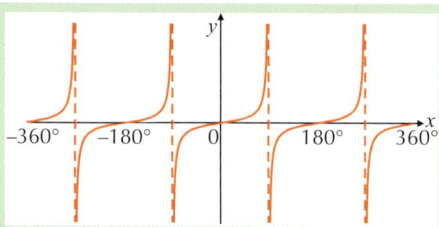

Example 1

Given that one solution to sin x = –0.75 is x = 229°
(to the nearest degree), use the graph of y = sin x on
the right to find all the values of x to the nearest degree
for which sin x = –0.75 in the range –360° ≤ x ≤ 360°.

1. Draw a horizontal line at y = –0.75.
2. The values of x where this line intersects
    the graph are the solutions to sin x = –0.75.
3. There are four solutions in the range. One of them
    was given in the question, so use this value and the
    symmetry of the graph to find the other solutions.

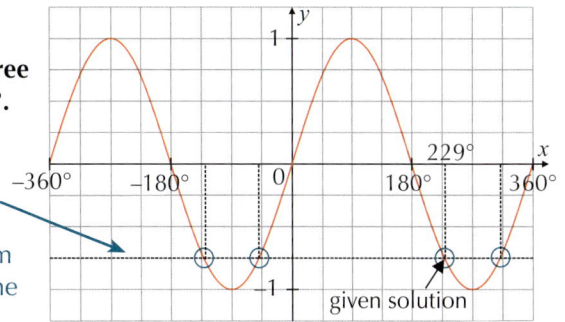

4. The graph between 180° and 360° has a line of
    symmetry at x = 270°, so the other positive solution will
    be the same distance from 360° as 229° is from 180°.

    229° – 180° = 49° away from 180°
    So the other positive solution is at
    360° – 49° = **311°**

5. Using the same method, the negative solutions
    will be 49° away from 0° and –180°.

    The negative solutions are
    0° – 49° = **–49°** and –180° + 49° = **–131°**

Exercise 1

Q1   Sketch graphs of the following equations for values of x between 0° and 720°.
     Label the position of the y-intercept, x-intercepts and any turning points and asymptotes.

a) y = sin x b) y = cos x c) y = tan x

In Questions 2-3, give rounded numbers to 2 d.p., and draw graphs for values of x between 0° and 360°.

Q2   a) Copy and complete the table and draw the graph of y = 2 sin x.

x	0°	30°	45°	60°	90°	120°	135°	150°	180°	210°	225°	240°	270°	300°	315°	330°	360°
sin x		0.5				0.87			0						–0.71		
2 sin x		1				1.73			0						–1.41		

b) Use your graph to find, to the nearest degree, the values of x between 0° and 360° when:

(i) 2 sin x = 1.2   (ii) 2 sin x = –0.8

Q3   a) Copy and complete the table and draw the graph of y = sin 2x.

x	0°	30°	45°	60°	90°	120°	135°	150°	180°	210°	225°	240°	270°	300°	315°	330°	360°
2x		60°			180°					420°				600°			
sin 2x		0.87			0					0.87				–0.87			

b) Describe the connection between the graph of y = sin 2x and the graph of y = sin x.

Q4   The graph of y = 2 + cos x is shown on the right. One solution to
     2 + cos x = 1.75 is x = 104° to the nearest degree. Use the graph
     to find all the solutions of 2 + cos x = 1.75 in the following intervals:

a) 0° ≤ x ≤ 360°   b) 360° ≤ x ≤ 720°

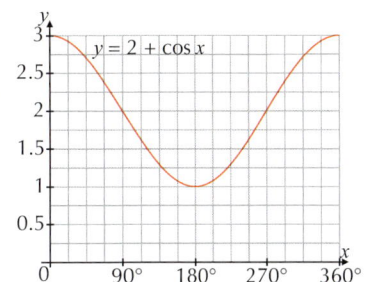

Q5   a) Sketch the graph of y = tan (x + 180°)
         for values of x between –180° and 180°.

b) Explain how the graph is related to the graph of y = tan x.

16.6 Transforming Graphs

The graph of a function can be transformed to give a new graph that is a translation or a reflection of the original. The equation of the function is also altered (depending on the type of transformation).

Translations

Learning Objective — Spec Ref A13:
Be able to translate graphs horizontally and vertically.

Prior Knowledge Check:
Be able to use function notation (see p.228) and vector notation (see p.339).

Translations can be either **horizontal** (i.e. left or right) or **vertical** (i.e. up or down). When a graph is **translated**, the **shape** of the graph **stays the same** but it **moves** (or slides) on the coordinate grid. If the original graph has equation $y = f(x)$, the **equation** and **coordinates** of the translated graph **change** as follows:

- A **horizontal** translation of a **units right** has equation $y = f(x - a)$. This can also be described using a translation vector: $\begin{pmatrix} a \\ 0 \end{pmatrix}$. After a horizontal translation, the **x-coordinates** of all points increase by a.

- A **vertical** translation of a **units up** has equation $y = f(x) + a$. This can also be described using a translation vector: $\begin{pmatrix} 0 \\ a \end{pmatrix}$. After a vertical translation, the **y-coordinates** of all points increase by a.

Example 1

The diagram below shows the graph of $y = x^3$. Draw the graph of: a) $y = x^3 + 2$, b) $y = (x + 2)^3$

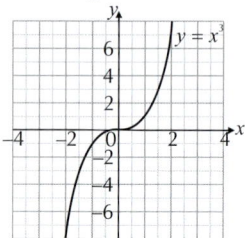

a)
1. $y = x^3 + 2$ is a transformation of the form $y = f(x) + a$, where $a = 2$.
2. So translate the graph of $y = x^3$ by 2 units up, i.e. by the vector $\begin{pmatrix} 0 \\ 2 \end{pmatrix}$.
3. Add 2 to the y-coordinates of any key points (such as the y-intercept) and label them.

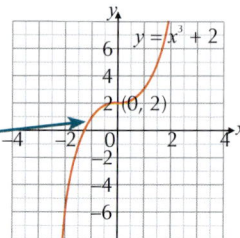

b)
1. $y = (x + 2)^3$ is a transformation of the form $y = f(x - a)$, where $a = -2$.
2. So translate the graph of $y = x^3$ by 2 units left (or −2 units right), i.e. by the vector $\begin{pmatrix} -2 \\ 0 \end{pmatrix}$.
3. Add −2 to the x-coordinates of any key points (such as the x-intercept) and label them.

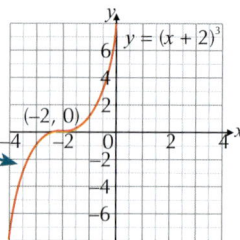

Tip: Take care with the +/− signs for horizontal translations. '+ a' in the brackets is a translation of a units left, or −a units right.

Exercise 1

Q1 For each of the functions below:
(i) Describe the translation, stating the translation vector,
(ii) Find the coordinates of the translated point that had coordinates (0, 0) on the graph of $y = f(x)$.
a) $y = f(x) + 3$ b) $y = f(x - 1)$ c) $y = f(x + 2)$ d) $y = f(x) - 6$

Q2 Each of these functions is a translation of the function $y = x^2$.
For each function, describe the translation and sketch the graph.
a) $y = x^2 + 1$ b) $y = x^2 - 2$ c) $y = (x - 4)^2$ d) $y = (x + 1)^2$

Q3 These graphs are translations of the graph $y = x^2$. Find the equation of each graph.

a)

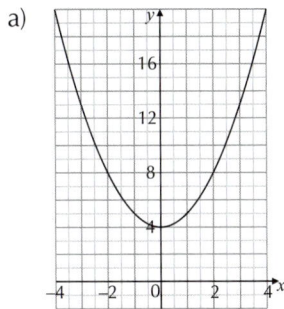

b)

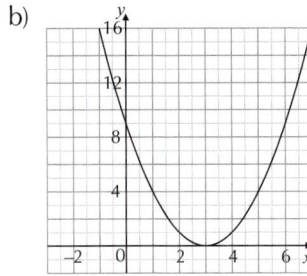

c)

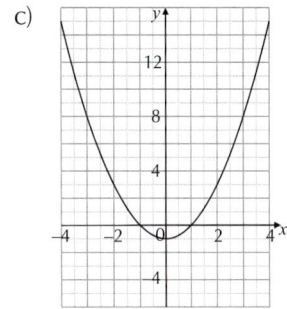

Q4 Each of the following functions is a translation of the function $y = \sin x$.
For each one, describe the translation and sketch the graph for $0° \leq x \leq 360°$.

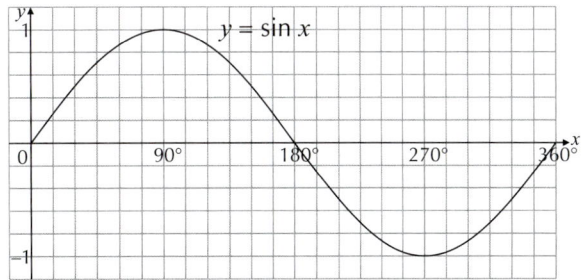

a) $y = (\sin x) + 1$

b) $y = (\sin x) - 2$

c) $y = \sin (x + 60°)$

d) $y = \sin (x - 90°)$

e) $y = \sin (x + 180°)$

f) $y = \sin (x - 360°)$

You can also perform a **combination** of translations on a graph. Just do them **one at a time**.

Example 2

The diagram below shows the graph of $y = x^2$. Draw the graph of $y = (x + 4)^2 - 3$.

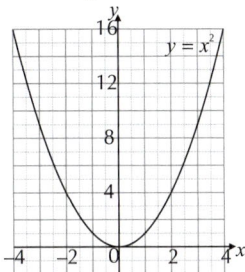

1. $y = (x + 4)^2$ is a transformation of the form $y = f(x - a)$, where $a = -4$. So it's a translation of 4 units left from $y = x^2$.

2. $y = (x + 4)^2 - 3$ is a transformation of the form $y = f(x) + a$, where $a = -3$. So it's a translation of 3 units down from $y = (x + 4)^2$.

3. In vector form, this is an overall translation of $\begin{pmatrix} -4 \\ -3 \end{pmatrix}$.

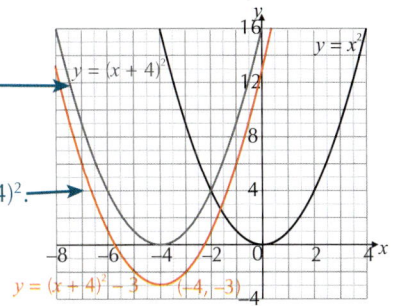

4. Label any key points, such as the turning point. Subtract 4 from the x-coordinate and subtract 3 from the y-coordinate.

Exercise 2

Q1 For each of the functions below:
  (i) Describe the translation, stating the translation vector,
  (ii) Find the coordinates of the translated point that had coordinates $(0, 0)$ on the graph of $y = f(x)$.
  a) $y = f(x - 5) + 6$
  b) $y = f(x + 6) - 4$
  c) $y = f(x - 3) + 2$
  d) $y = f(x + 1) + 9$

Q2 Each of these functions is a translation of the function $y = x^3$.
For each function, describe the translation and sketch the graph.
  a) $y = (x - 5)^3 + 4$
  b) $y = (x + 2)^3 - 3$
  c) $y = (x - 4)^3 + 6.5$
  d) $y = (x + 2.5)^3 - 3$

Q3 These graphs are translations of the graph $y = x^2$. Find the equation of each graph.

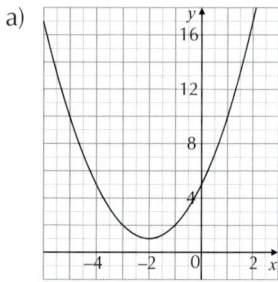

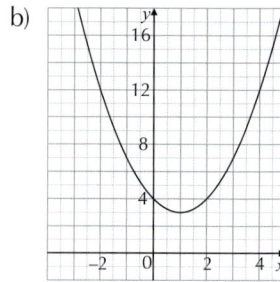

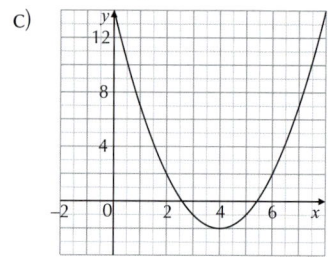

a)

b)

c)

Q4 For each of the following, give another equation that produces the same graph.

a) $y = \tan x$

b) $y = \cos (x + 180°)$

c) $y = \sin (x + 45°)$

PROBLEM SOLVING

Reflections

Learning Objective — Spec Ref A13:
Be able to reflect graphs in the x- or y-axis.

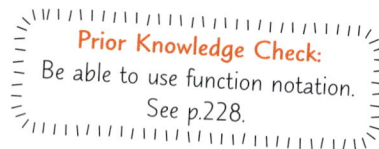

Prior Knowledge Check:
Be able to use function notation.
See p.228.

Graphs of functions can be **reflected** in the x- or y-axis. If the original graph has equation $y = f(x)$, the **equation** and **coordinates** of the reflected graph **change** as follows:

- A reflection in the **x-axis** has equation $y = -f(x)$. After a reflection in the x-axis, the **y-coordinates** of all the points are **multiplied by –1**.

- A reflection in the **y-axis** has equation $y = f(-x)$. After a reflection in the y-axis, the **x-coordinates** of all the points are **multiplied by –1**.

Tip: If a point doesn't change during the transformation, it's called an **invariant point**.

Example 3

The diagram below shows the graph of $y = 3^x$.

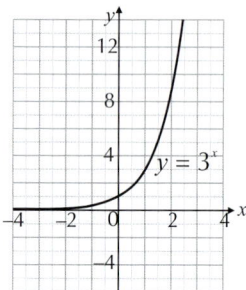

a) Draw the graph of $y = -(3^x)$.

1. $y = -(3^x)$ is a transformation of the form $y = -f(x)$, which is a reflection in the x-axis.

2. The y-intercept has moved from (0, 1) to (0, –1).

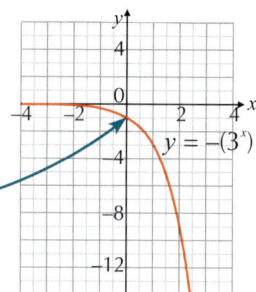

b) Find the equation of the transformation of $y = 3^x$ shown on the graph below.

1. The y-intercept is (0, –1) — the same as the graph of $y = -(3^x)$ above. All the other points are reflections of $y = -(3^x)$ in the y-axis.

2. $y = f(-x)$ represents a reflection in the y-axis, so apply this to $y = -(3^x)$ to find the equation of the transformed graph:

$$f(x) = -(3^x) \text{ so } y = f(-x) \Rightarrow y = -(3^{-x})$$

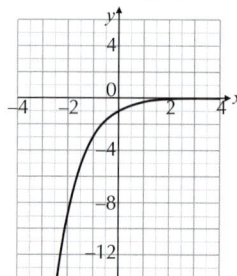

Tip: It can help to look at a couple of key points, such as intercepts or turning points, and see where they have moved to after the transformation.

Example 4

The diagram on the right shows the graph of
$y = \sin(x - 45°)$ for $-180° \leq x \leq 180°$.

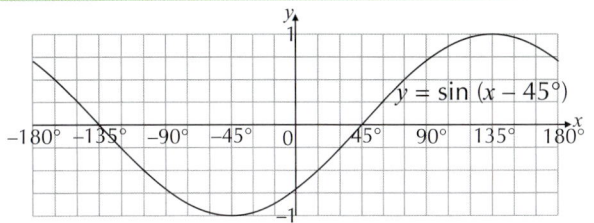

Use this graph to draw graphs
of the following equations:

a) $y = \sin(-x - 45°)$

1. $y = \sin(-x - 45°)$ is a transformation of
   the form $y = f(-x)$, which is a reflection
   of $y = \sin(x - 45°)$ in the y-axis.

2. The y-intercept will stay the same, but the
   turning points at $(-45°, -1)$ and $(135°, 1)$
   will move to $(45°, -1)$ and $(-135°, 1)$:

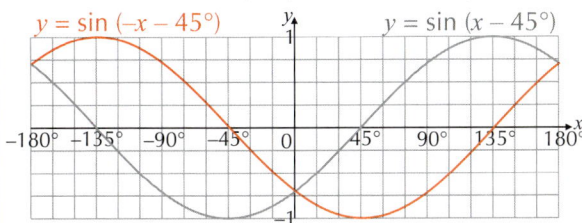

b) $y = -\sin(x - 45°)$

1. $y = -\sin(x - 45°)$ is a transformation of
   the form $y = -f(x)$, which is a reflection
   of $y = \sin(x - 45°)$ in the x-axis.

2. The x-intercepts will stay the same, but the
   turning points at $(-45°, -1)$ and $(135°, 1)$
   will move to $(-45°, 1)$ and $(135°, -1)$:

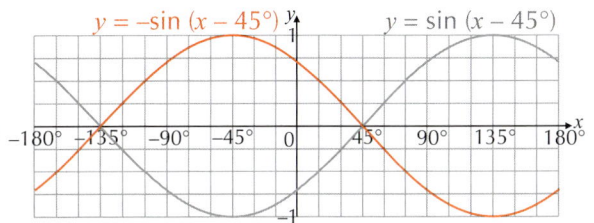

Exercise 3

Q1 The graph on the right shows the function $y = f(x)$.
The graphs of the following functions are reflections of $y = f(x)$.
In each case, describe the reflection and sketch the graph,
labelling the new coordinates of the turning points.

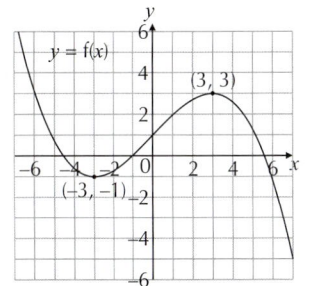

a) $y = f(-x)$

b) $y = -f(x)$

Q2 Sketch the graphs of each of the following pairs of functions.
Use a single set of axes for each pair of graphs.

a) $y = \sin x$, $y = -\sin x$

b) $y = \cos x$, $y = -\cos x$

c) $y = x^2$, $y = -x^2$

d) $y = \dfrac{1}{x}$, $y = -\dfrac{1}{x}$

e) $y = x - 2$, $y = -x - 2$

f) $y = 2^x$, $y = 2^{-x}$

Q3 Sketch the graphs of $y = x^3$, $y = -x^3$ and $y = (-x)^3$. Explain why two of the graphs are the same.

Q4 The graph of $y = (-x - 2)^2$ is obtained by applying a translation then a reflection to the graph of $y = x^2$.

a) Sketch the graph of $y = x^2$.

b) Apply a translation to the graph of $y = x^2$ to give the graph of $y = (x - 2)^2$.

c) Apply a reflection to your graph from part b) to give the graph of $y = (-x - 2)^2$.

Q5 For the following pairs of functions, describe the transformations that
transform the graph of the first function to the graph of the second.

a) $y = x^2$, $y = -(x - 1)^2$

b) $y = x^2$, $y = (-x - 3)^2$

c) $y = \cos x$, $y = -\cos(x + 90°)$

d) $y = \sin x$, $y = \sin(-x) + 1$

Q1 A stone is dropped from the top of a 55 m-high tower. The distance in metres, h, between the stone and the ground after t seconds is given by the formula $h = 55 - 5t^2$.

a) Draw a graph of the height of the stone for $h \geq 0$ and values of t between 0 and 5 seconds.

b) Use your graph to estimate how long it takes the stone to fall to a height of 20 m above the ground.

c) Find the exact time it takes for the stone to reach the ground.

Q2 By completing the square, calculate the position of the y-intercept and turning point of the graph of $y = x^2 - x - 1$.

Q3 Copy and complete the table below and draw a graph of the equation $y = x^3 - 3x^2 - x + 3$.

x	-3	-2	-1	0	1	2	3	4
y		-15					0	

Q4 Match each of these equations to one of the graphs below.

(i) $y = x^3$ (ii) $y = \dfrac{1}{x}$ (iii) $y = 2^x - 1$ (iv) $x^2 + y^2 = 1$

A   B   C   D

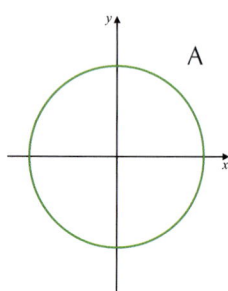

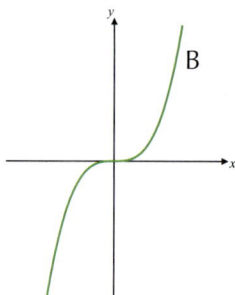

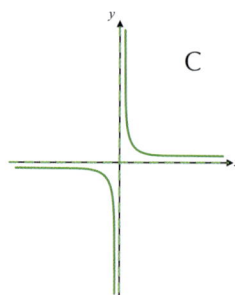

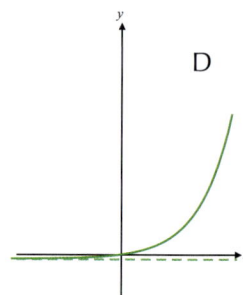

Q5 Susan uses a credit card to buy a computer that costs £500. The credit card company charges 3% interest per month. If she pays nothing back, the amount in pounds Susan will owe after t months is given by the formula $b = 500 \times (1.03)^t$.

a) Explain what the numbers 500 and 1.03 represent in the formula.

b) Draw a graph to show how this debt will increase during one year if Susan doesn't repay any money.

c) Use your graph to estimate how long the interest on the credit card will take to reach £100.

Q6 Point P has coordinates $(3, 4)$.

a) Use Pythagoras' theorem to find the distance of P from the origin.

b) Use your answer to a) to write the equation of the circle with centre $(0, 0)$ that passes through point P.

Q7 The graph of $y = \cos x$ is shown on the right.

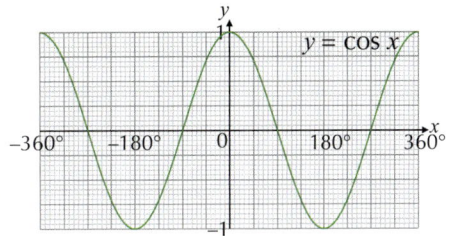

a) Use the graph to find the values of x between $-360°$ and $360°$ for which:

(i) $\cos x = 0.25$ (ii) $\cos x = -0.25$

b) If $\cos 50° = 0.643$, use the graph to write down all the values of x between $-360°$ and $360°$ for which:

(i) $\cos x = 0.643$ (ii) $\cos x = -0.643$

Q8 The graph of $y = x^2$ has a minimum point at $(0, 0)$.

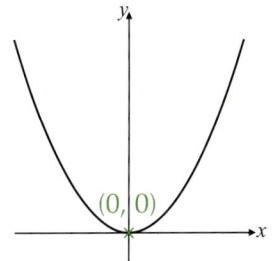

a) (i) The graph of $y = x^2$ is transformed to give the graph of $y = x^2 + 3$. Describe what this transformation does to the coordinates of each point on the original graph.

(ii) What are the coordinates of the minimum point of the graph of $y = x^2 + 3$?

b) Write down the coordinates of the minimum point of:

(i) $y = (x - 1)^2$ (ii) $y = (x + 3)^2 - 2$

Q9 A turning point is a place where the gradient of a graph changes from positive to negative, or from negative to positive. The graph of $y = f(x)$ shown on the right has two turning points, $(0, 4)$ and $(2, 0)$. Use the graph of $f(x)$ to sketch the graphs of the following functions. On each graph, mark the coordinates of the two turning points.

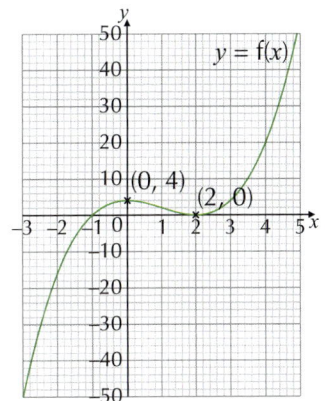

a) $f(x) - 1$ b) $f(x + 2)$

c) $f(-x)$ d) $-f(x)$

e) $f(x + 1) - 2$ f) $f(x - 2) - 3$

g) $-f(x + 1)$ h) $f(-x) - 1$

Q10 Even functions are functions with graphs that have the y-axis as a line of symmetry. $y = x^2$ is an example of an even function.

Odd functions are functions with graphs that have rotational symmetry of order 2 about the origin. $y = x^3$ is an example of an odd function.

Sketch the following graphs and say whether the functions are even, odd or neither.

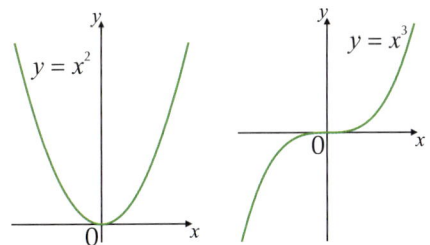

a) $y = x^2 + 5$ b) $y = -x^3$ c) $y = \sin x$ d) $y = (x - 3)^2$

e) $y = \cos x$ f) $y = x^3 - 4$ g) $y = \tan x$ h) $y = \dfrac{1}{x}$

Q1 Sketch graphs of the following equations.
Label the coordinates of any intersections with the axes.

a) $y = x^3$

[1 mark]

b) $y = 1 - x^2$

[3 marks]

Q2 A function $f(x)$ has the graph $y = f(x)$ as shown below left. The graph of $y = f(x)$ can be transformed to make the graph of $y = f(x) + 2$, as shown below right.

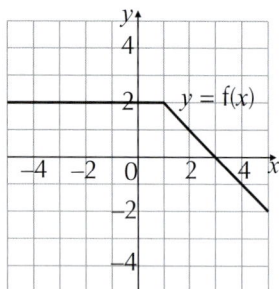

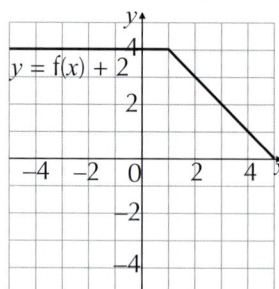

The following two graphs are two other transformations of $y = f(x)$.
For each graph, write its equation in function notation.

a)

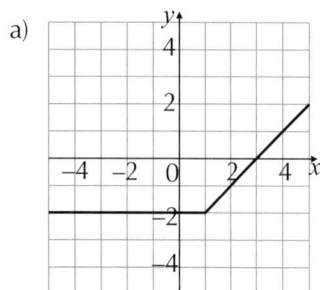

[1 mark]

b)

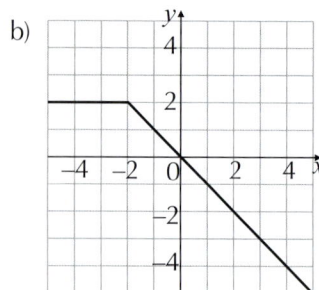

[1 mark]

Q3 A circle with equation $x^2 + y^2 = 9$ is translated 2 units in the positive y-direction.

a) Sketch the translated graph, labelling the y-intercepts.

[2 marks]

b) Find the exact coordinates of the points where the translated graph crosses the x-axis.

[2 marks]

Q4 The table below shows pairs of values satisfying the equation $y = \dfrac{4}{x + 0.5}$.

x	0	0.5	1.5	3.5	7.5
y	p	4	q	1	r

a) Work out the values of p, q and r.

[1 mark]

b) Use your values for p, q, r and the other values in the table to plot the graph of $y = \dfrac{4}{x + 0.5}$ for $0 \leq x \leq 7.5$ on suitable axes.

[2 marks]

c) By plotting the graph of $y = 6 - x$ on the same set of axes, explain how the graphs show that the equation $\dfrac{4}{x + 0.5} = 6 - x$ has 2 solutions.

[2 marks]

Q5 The graph of $y = 4^x - 8$ is sketched on the axes below.

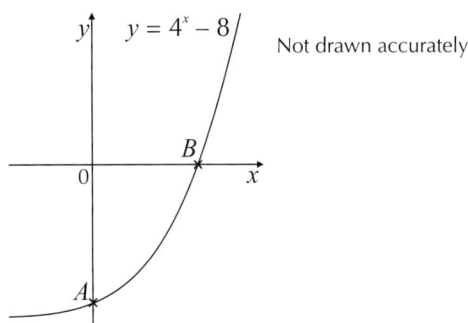

The graph crosses the y-axis at point A and the x-axis at point B.

a) Work out the coordinates of point A.

[1 mark]

b) Work out the coordinates of point B.

[2 marks]

Q6 Point P is on the graph of $y = \tan x°$ and has an x-coordinate of 60.

a) State the exact y-coordinate of P, giving your answer as a surd.

[1 mark]

The graph is transformed by the translation $\begin{pmatrix} 20 \\ 0 \end{pmatrix}$.

b) Find the equation of the transformed graph.

[1 mark]

17.1 Interpreting Real-Life Graphs

Sometimes, graphs show something more interesting than just how y changes with x. They can be used to illustrate motion (such as in distance-time graphs), unit conversions (such as changing between temperatures in °C and °F), and many other connections between real-life quantities.

Learning Objective — Spec Ref A14:
Understand and interpret graphs that represent real-life situations.

Prior Knowledge Check:
Be able to read values off the graphs of straight lines (Section 15) and other functions (Section 16).

Real-life graphs show how one thing changes in relation to another. When **describing** real-life graphs, look at the following **features** and **interpret** them in the **context** of the graph.

- The **direction** of the graph — i.e. is the variable on the vertical axis **increasing** or **decreasing** as the other increases?

- The **gradient** (steepness) — this shows the **rate of change** of one variable with the other. E.g. on a distance-time graph, a **steep gradient** represents a **high speed** because a large distance (on the vertical axis) is being covered in a short time (on the horizontal axis).

- The **change** in **direction** or **gradient** — e.g. a distance-time graph that is initially steep and then levels off shows that an object is moving fast at first then slowing down over time.

Tip: See page 180 for how to find the gradient of a straight line and page 223 for working it out for curves.

Example 1

The graph shows the temperature of an oven as it heats up. Describe how the temperature of the oven changes during the first 10 minutes shown on the graph.

1. Look at the direction of the graph — temperature is increasing with time.

2. The gradient of the graph is steep initially and quite flat towards the end.

3. The graph doesn't change direction (it keeps increasing) but the gradient decreases over time.

4. Relate these features to the context:

   The temperature of the oven **rises** for the entire 10 minutes. This rise is **rapid at first** but then **becomes slower** as the oven heats up.

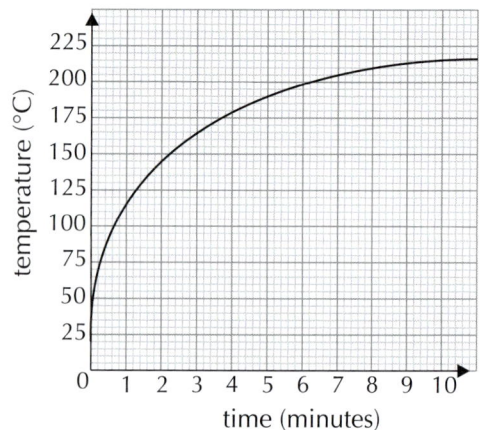

To **read off** values from a graph:

- Draw a **straight line** from **one axis** to the **graph**.

- Draw **another** straight line at a **right angle** to the first from the **graph** to the **other axis**.

- **Read off** the **value** from this axis, including any **units**.

Example 2

The conversion graph shows the connection between temperatures in °C and °F.

a) Convert a temperature of 10 °C to °F.

b) On a distant planet, the average daytime temperature is 80 °F higher than the average night-time temperature. If the night-time temperature is 33 °F, what is the daytime temperature in °C?

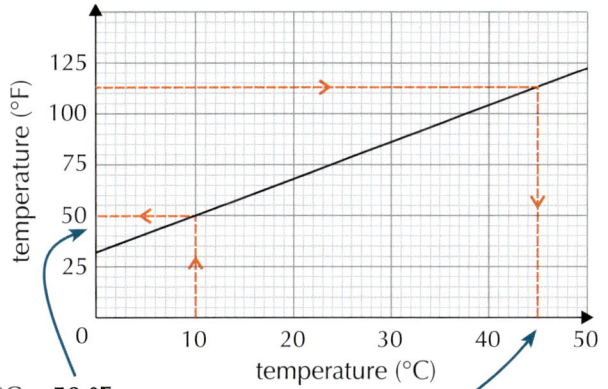

a) Read up from 10 °C and then across.   10 °C = **50 °F**

b) First work out the daytime temperature in °F. Then use the graph to convert this into °C.

$$33 \text{ °F} + 80 \text{ °F} = 113 \text{ °F}$$
$$= \textbf{45 °C}$$

Exercise 1

Q1   Each statement below describes one of the graphs on the right. Match each statement to the correct graph.

   a) The temperature rose quickly, and then fell again gradually.

   b) The number of people who needed hospital treatment stayed at the same level all year.

   c) The cost of gold went up more and more quickly.

   d) The temperature fell overnight, but then climbed quickly again the next morning.

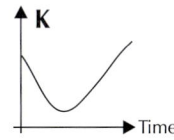

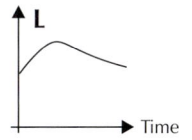

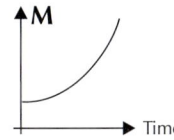

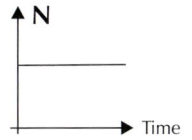

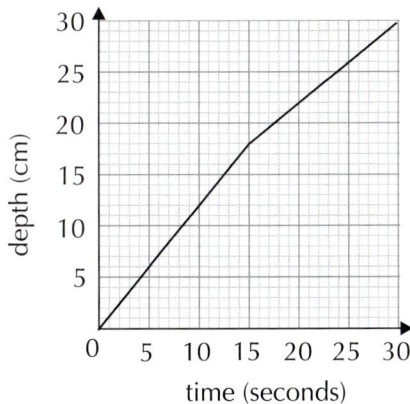

Q2   A vase that is 30 cm tall is filled using a tap flowing at a steady rate. The graph on the left shows how the depth of water varies in the vase over time.

   a) Describe how the depth of water in the vase changes over time.

   b) How long did it take for the water depth to be half the height of the vase?

   c) Which of the diagrams A-D shown below best matches the shape of the vase?

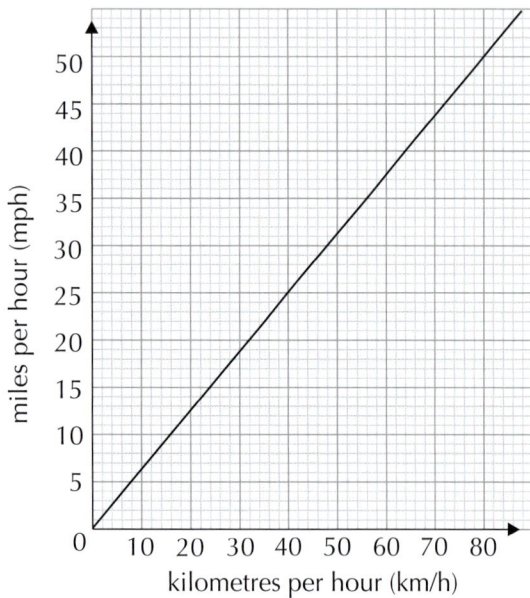

Q3   The graph on the left can be used to convert between speeds in kilometres per hour and miles per hour.

a) Convert 38 km/h into miles per hour, to the nearest 1 mph.

b) (i) Convert 25 mph into km/h.

   (ii) Use your answer to convert 75 mph into km/h.

c) The speed limit on a particular road is 30 mph. A driver travels at 52 km/h. By how many miles per hour is the driver breaking the speed limit?

d) The maximum speed limit in the UK is 70 mph. The maximum speed limit in Spain is 120 km/h. Which country has the greater speed limit, and by how much?

Q4   The graph shows the depth of water in a harbour between 08:00 and 20:00.

a) Describe how the depth of water changed over this time period.

b) At approximately what time was the depth of water the greatest?

c) What was the minimum depth of water during this period?

d) At approximately what times was the water 3 m deep?

e) Mike's boat floats when the depth of the water is 1.6 m or over. Estimate the amount of time that his boat was not floating during this period.

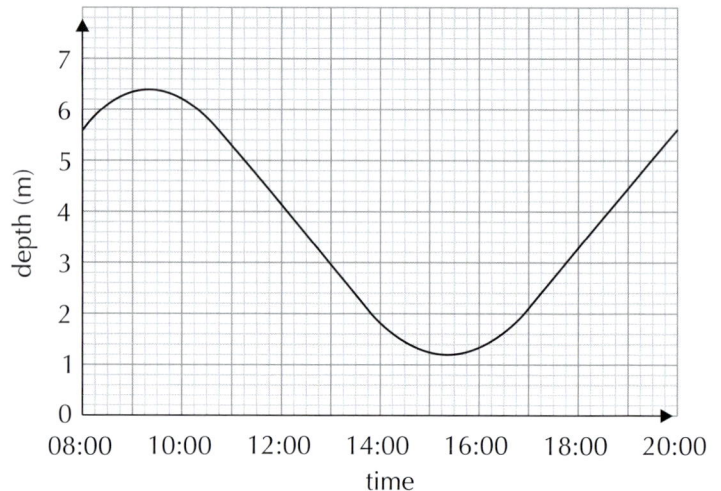

Q5   The graph on the left shows the temperature in two ovens as they warm up.

a) Which oven reaches 100 °C more quickly?

b) Which oven reaches a higher maximum temperature?

c) How long does it take Oven 2 to reach its maximum temperature?

d) (i) After how many seconds are the two ovens at the same temperature?

   (ii) What is the temperature at this time?

e) Calculate the rate at which the temperature of Oven 1 changes in the first 3 minutes after being switched on.

PROBLEM SOLVING

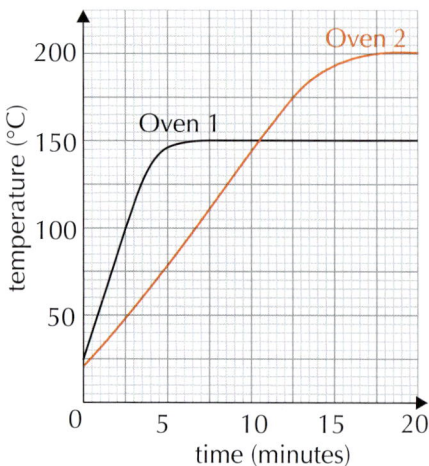

17.2 Drawing Real-Life Graphs

Now you know how to interpret real-life graphs, it's time to have a go at drawing them.

Learning Objective — Spec Ref A14:
Draw graphs that represent real-life situations.

To **draw** a graph of the connection between two **real-life variables**, you'll need a **table of values**. Decide which of the variables should go on the horizontal axis and which on the vertical axis. Generally, the one that **depends** on the other should go on the **vertical axis**. Then **plot** the pairs of values as **coordinates** and carefully draw a **smooth line or curve** through them.

Example 1

A plumber charges customers a standard fee of £40, plus £30 per hour for all work carried out.

a) **Draw a graph to show how the plumber's fee varies with the amount of time the job takes.**

b) **Use the graph to estimate the amount of time a job costing £250 would have taken.**

a) 1. Make a table of values showing the fee for different numbers of hours.
       A 1-hour job will cost £40 + £30 = £70.
       A 2-hour job will cost £40 + (2 × £30) = £100 etc.

Time (hours)	1	2	3	4	5
Fee (£)	70	100	130	160	190

2. Draw your axes. The cost of the job depends on the time it takes to complete so the fee goes on the vertical axis and time on the horizontal axis. Make sure to label them and choose a scale that makes the graph easy to read.

3. Plot the values and join the points to draw the graph. For each extra hour, the fee increases by £30, so this is a straight-line graph.

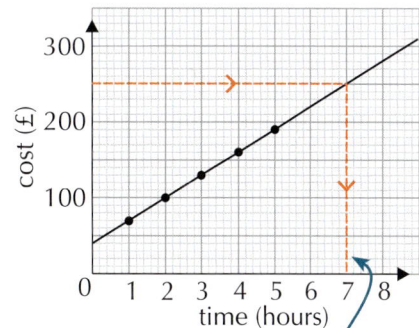

b) Draw a straight line from £250 on the vertical axis and then down to the horizontal axis to find the correct time...   So it would have taken **7 hours**.

Example 2

A different plumber charges less per hour for longer jobs. Some of his fees are shown in the table. Draw a graph to illustrate this plumber's fees. Join your points with a smooth curve.

Time (hours)	1	2	3	4	5	6
Fee per hour (£)	80	78	73	62	50	43

1. Draw the axes — again, the cost goes on the vertical axis and time on the horizontal axis. The cost per hour values are between £43 and £80, so you can cut some of the vertical axis from the bottom of the scale.

2. Plot the values in the table as coordinates.

3. Draw a smooth curve through the points. There shouldn't be any sudden changes of direction or sharp kinks.

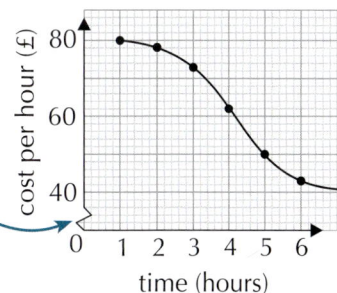

Q1 The instructions for cooking different weights of chicken are as follows:

'Cook for 35 mins per kg, plus an extra 25 minutes.'

a) Copy and complete the table below to show the cooking times for different weights of chicken.

Weight (kg)	1	2	3	4	5
Time (minutes)					

b) Use the values from your table to draw a graph showing the cooking times for different weights of chicken.

c) A chicken cooks in 110 minutes. What is the weight of the chicken?

Q2 The table below shows how the fuel efficiency of a car in miles per gallon (mpg) varies with the speed of the car in miles per hour (mph).

Speed (mph)	55	60	65	70	75	80
Fuel Efficiency (mpg)	32.3	30.7	28.9	27.0	24.9	22.7

a) Plot the points from the table on a graph and join them up with a smooth curve.

b) Use your graph to predict the fuel efficiency of the car when it is travelling at 73 mph.

Q3 Helena is a baby girl. A health visitor records the weight of Helena every two months. The measurements are shown in the table below.

Age (months)	0	2	4	6	8	10	12	14	16
Weight (kg)	3.2	4.6	5.9	7.0	7.9	8.7	9.3	9.8	10.2

a) Draw a graph to show this information. Join your points with a smooth curve.

b) Keira is 9 months old and has a weight of 9.1 kg.
   Use your graph to estimate how much heavier Keira is than Helena was at the same age.

Q4 The number of bacterial cells, N, in a sample after d days is $N = 5000 \times 1.2^d$.

a) Copy and complete the table to show the number of bacterial cells in the sample for the values of d given.

d	0	1	2	3	4
N					

b) Draw a graph to show this information.

c) Use your graph to estimate how many days it takes for there to be 9000 bacterial cells.

Q5 Alfred sells high-tech 'stealth fabric' on his market stall.
The cost of the fabric is £80 per metre for the first 3 metres, then £50 per metre after that.

a) Draw a graph showing how the cost of the stealth fabric varies with the amount purchased.

b) Use your graph to find how much it would cost to buy 6.5 metres of stealth fabric.

c) Bruce bought some fabric to make a stealth cape.
   If he was charged £480, how much fabric did he buy?

Q6 The conical flask shown on the right is filled from a steadily running tap. Sketch a graph to show the depth of water in the flask t seconds after it has started to be filled.

17.3 Solving Simultaneous Equations Graphically

You first met simultaneous equations in Section 12, where you found the solutions algebraically. You can also solve them using their graphs.

Learning Objective — Spec Ref A19:
Solve simultaneous equations using graphs.

Prior Knowledge Check:
Be able to draw the graphs of straight lines (p.177-179) and quadratics (p.194-195).

If you have two equations and you draw their corresponding graphs, the points where the graphs **intersect** will be the **solutions** to **both** equations. So, to solve a pair of **simultaneous equations**, draw their graphs and read off the **intersection points**.

Example 1

Solve the following simultaneous equations graphically: $x + y = 8$ and $y = 2x + 2$.

1. Both equations are straight lines — they can be written in the form $y = mx + c$. Find three pairs of x- and y-values for each equation and plot them.

   $x + y = 8$

x	0	4	8
y	8	4	0

   $y = 2x + 2$

x	−1	0	1
y	0	2	4

2. Draw a straight line through each set of points.

3. Read off the x- and y-values of the point where the graphs intersect. These values are the solution to the pair of equations.

   The graphs cross at (2, 6), so the solution is
   $x = 2$ **and** $y = 6$.

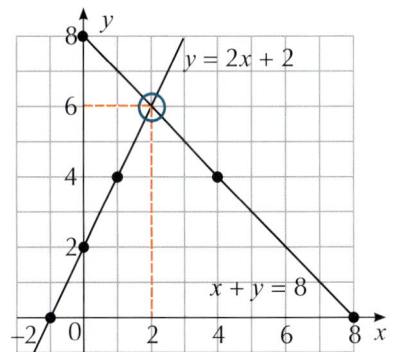

Exercise 1

Q1 The graphs of $y = \frac{1}{2}x + 4$ and $y = -\frac{1}{4}x + 7$ are shown on the right.

Use the graphs to find the solution to the simultaneous equations $y = \frac{1}{2}x + 4$ and $y = -\frac{1}{4}x + 7$.

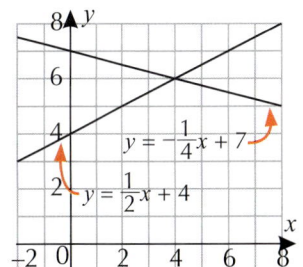

Q2 Solve the following simultaneous equations by drawing graphs.

a) $x + y = 7$ and $y = x - 3$

b) $x + y = 8$ and $y = x$

c) $x + y = 9$ and $y = 2x$

d) $x + y = 1$ and $y = 2x - 7$

e) $x + \frac{1}{2}y = -4$ and $y = 2x$

f) $y = 2x + 3$ and $y = 4x + 2$

g) $y = 2x - 5$ and $2x + y = 1$

h) $y = \frac{1}{4}x$ and $x + 2y = 3$

i) $y = \frac{1}{2}x$ and $x + 2y = 6$

j) $y = 2x$ and $y = 4x + 3$

Q3 a) Draw the graphs of $y = x + 3$ and $y = x - 2$.

b) Explain how this shows that the simultaneous equations $y = x + 3$ and $y = x - 2$ have no solutions.

The **same method** is used for simultaneous equations where one or more of the equations is **non-linear**. If one equation is a **quadratic** and the other is linear, draw the quadratic **curve** and the straight **line** and see where they **intersect**. There could be **up to two points** of intersection which give different **solutions** to the pair of equations.

Example 2

a) **Draw the graphs of $y = \frac{1}{2}x - 1$ and $y = x^2 - 5x - 4$ for $-1 \leq x \leq 7$.**

b) **Solve the simultaneous equations $y = \frac{1}{2}x - 1$ and $y = x^2 - 5x - 4$.**

a) Work out pairs of x- and y-values for each equation and plot them on the same axes. You'll need three pairs for a straight line but more for a quadratic.

$y = \frac{1}{2}x - 1$

x	0	2	4
y	−1	0	1

$y = x^2 - 5x - 4$

x	−1	0	1	2	3	4	5	6	7
y	2	−4	−8	−10	−10	−8	−4	2	10

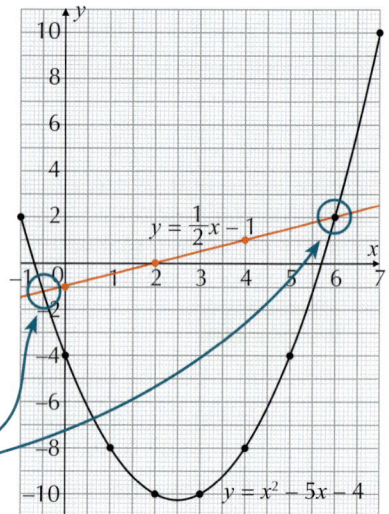

b) The points where the graphs cross are the solutions to the simultaneous equations.

The graphs intersect at (6, 2) and (−0.5, −1.25).
So the solutions are $x = 6$, $y = 2$ and $x = -0.5$, $y = -1.25$.

Tip: There are two points of intersection so there are two solutions.

Exercise 2

Q1 Each of the following diagrams shows the graphs of two equations.
Use the graphs to solve the equations simultaneously.

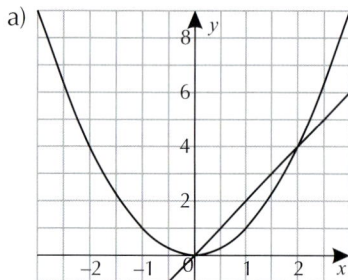

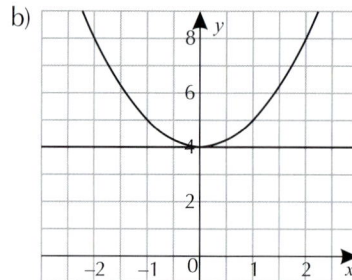

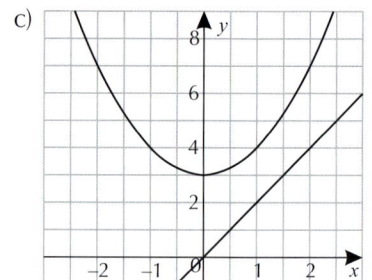

a)

b)

c)

Q2 For each of the following (i) Draw the graphs for the x-value range given.

(ii) Use these graphs to solve the pair of simultaneous equations.

a) $y = x - 1$ and $y = x^2 - 3$ $-4 \leq x \leq 4$

b) $2x + y = 8$ and $y = x^2$ $-4 \leq x \leq 4$

c) $y = x + 2$ and $y = x^2 + x + 1$ $-4 \leq x \leq 4$

d) $x + y = 3$ and $y = x^2 - 3x$ $-3 \leq x \leq 5$

e) $x + y = 1$ and $y = x^2 - 1$ $-4 \leq x \leq 4$

f) $y = x + 6$ and $y = x^2 + 5x + 1$ $-6 \leq x \leq 3$

g) $y = \frac{1}{2}x$ and $y = x^2 - 3x - 2$ $-3 \leq x \leq 5$

h) $y = \frac{1}{2}x + 1$ and $y = \frac{1}{2}x^2 - 5$ $-6 \leq x \leq 6$

17.4 Solving Quadratics Graphically

Quadratic graphs of the form $y = ax^2 + bx + c$ can cross the x-axis in up to two places. These points are the solutions to the quadratic equation $ax^2 + bx + c = 0$ — also known as the 'roots' of the quadratic.

Learning Objective — Spec Ref A18:
Solve quadratic equations using graphs.

Prior Knowledge Check:
Be able to draw the graphs of straight lines (see p.177-179) and quadratics (see p.194-195).

The **solutions** (or **roots**) of a quadratic equation $ax^2 + bx + c = 0$ are the **x-intercepts** of the graph of $y = ax^2 + bx + c$, i.e. the intersection of $y = ax^2 + bx + c$ and $y = 0$. Similarly, the solutions of $ax^2 + bx + c = k$ are the x-values of the intersection points of $y = ax^2 + bx + c$ and **$y = k$**. You can **rearrange** quadratic equations into either of these forms to make it easier to **plot graphs** or read off solutions from graphs that you **already have**.

Example 1

The graph on the right shows $y = x^2 + 2x - 3$ for $-5 \leq x \leq 3$.

a) Use the graph to solve: (i) $x^2 + 2x - 3 = 0$ (ii) $x^2 + 2x - 11 = 0$

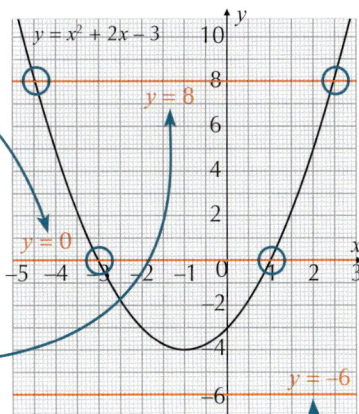

   (i) The roots of a quadratic are at the points where the graph meets the x-axis.   The graph cuts the x-axis twice so there are two solutions: **$x = -3$ and $x = 1$**

   (ii) 1. Rearrange $x^2 + 2x - 11 = 0$ so that $x^2 + 2x - 3$ is on the left-hand side and a constant k is on the right-hand side.

$$x^2 + 2x - 11 = 0$$
$$\Rightarrow x^2 + 2x - 11 + 11 - 3 = 11 - 3$$
$$\Rightarrow x^2 + 2x - 3 = 8$$

       2. Draw the graph of $y = k$. The solutions are where this intersects the curve.   The line $y = 8$ intersects the curve at **$x = 2.5$** and **$x = -4.5$** (to 1 d.p.)

b) Explain how the graph shows that the equation $x^2 + 2x + 3 = 0$ has no solutions.

Rearrange $x^2 + 2x + 3 = 0$ to get $x^2 + 2x - 3$ on the left-hand side and a constant k on the right-hand side. Draw the graph of $y = k$ — this time it never meets the curve.

$x^2 + 2x + 3 = 0 \Rightarrow x^2 + 2x - 3 = -6$

The line $y = -6$ **does not cross** the graph of $y = x^2 + 2x - 3$, so $x^2 + 2x + 3 = 0$ has **no solutions**.

Exercise 1

Q1   a) Draw the graph of $y = x^2 + 2x$ for $-5 \leq x \leq 3$.

       b) Use your graph to solve the following equations to 1 decimal place (where appropriate).

          (i) $x^2 + 2x = 0$ (ii) $x^2 + 2x = 10$ (iii) $x^2 + 2x = 7$

Q2   a) Draw the graph of $y = x^2 - 3x + 1$ for $-2 \leq x \leq 5$.

       b) Use your graph to solve the following equations to 1 decimal place (where appropriate).

          (i) $x^2 - 3x + 1 = 0$ (ii) $x^2 - 3x + 1 = 3$ (iii) $x^2 - 3x + 1 = -0.5$

Q3 a) Draw the graph of $y = x^2 + 4x - 7$ for $-7 \le x \le 3$.

b) Use the graph to find the roots of the following equations to 1 decimal place.

(i) $x^2 + 4x - 7 = 0$ (ii) $x^2 + 4x - 10 = 0$ (iii) $x^2 + 4x - 3 = 0$ (iv) $x^2 + 4x + 2 = 0$

Q4 a) Draw the graph of $y = 6 + x - x^2$ for $-4 \le x \le 5$.

b) Use the graph to find the roots of the following equations to 1 decimal place.

(i) $5 + x - x^2 = 0$ (ii) $16 + x - x^2 = 0$ (iii) $3 + x - x^2 = 0$ (iv) $10 + x - x^2 = 0$

Given a graph of $y = ax^2 + bx + c$, you can also solve $ax^2 + ex + d = 0$ (i.e. where both the **x** and **constant** terms are different). Rearrange to get **$ax^2 + bx + c$** on one side and **$kx + l$** on the other. Plot the graph of $y = kx + l$ and find the solutions where it meets with $y = ax^2 + bx + c$.

Example 2

The graph of $y = x^2 + 5x$ is plotted on the right. By drawing a suitable straight line, use the graph to find the solutions to $x^2 + 6x + 5 = 0$.

1. Rearrange $x^2 + 6x + 5 = 0$ so it has $x^2 + 5x$ on one side and $kx + l$ on the other.

$$x^2 + 6x + 5 = 0$$
$$\Rightarrow x^2 + 6x + 5 - 6x - 5 + 5x$$
$$= -6x - 5 + 5x$$
$$\Rightarrow x^2 + 5x = -x - 5$$

2. Plot $y = kx + l$ (this is the 'suitable straight line'). The solutions to $x^2 + 6x + 5 = 0$ are found where this intersects $y = x^2 + 5x$.

The graph of $y = -x - 5$ intersects $y = x^2 + 5x$ at $x = -1$ and $x = -5$.

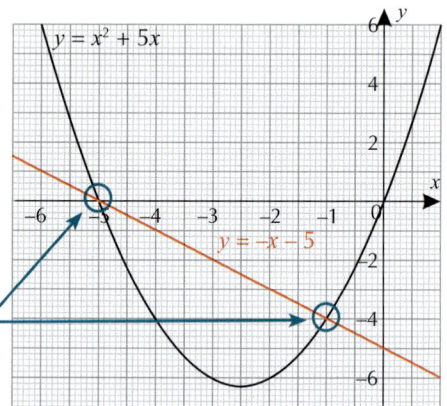

Exercise 2

Q1 The diagram on the right shows the graphs of $y = x^2 + 3x$, $y = x$, $y = 2x + 2$ and $y = -x$.

Use the diagram to solve the following equations graphically:

a) $x^2 + 4x = 0$ b) $x^2 + x - 2 = 0$ c) $x^2 + 2x = 0$

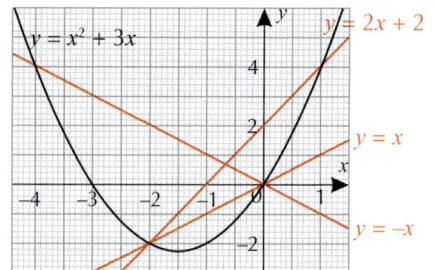

Q2 a) Draw the graph of $y = x^2 - 2x$ for $-3 \le x \le 5$.

b) Rearrange $x^2 - 4x - 1 = 0$ into the form $x^2 - 2x = mx + c$.

c) By drawing a suitable straight line, use your graph to find the solutions to the quadratic equation $x^2 - 4x - 1 = 0$ to 1 decimal place.

Q3 a) Draw the graph of $y = x^2 + 2x + 1$ for $-5 \le x \le 3$.

b) Use your graph to find the solutions to the quadratic equation $x^2 + 3x + 1 = 0$ to 1 decimal place.

Q4 a) Draw the graph of $y = 4x - x^2$ for $-2 \le x \le 5$.

b) Use your graph to find the solutions to the quadratic equation $5x - x^2 - 5 = 0$ to 1 decimal place.

17.5 Gradients of Curves

Working out the gradient of a curve can be tricky as it's not constant — it changes as you go along the curve.

Gradients of Curves

Learning Objective — Spec Ref A15:
Find the gradient of a curve at a given point.

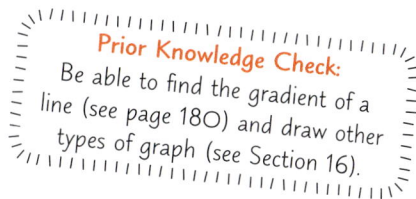

> **Prior Knowledge Check:**
> Be able to find the gradient of a line (see page 180) and draw other types of graph (see Section 16).

A straight line has the **same gradient** everywhere — the steepness/slope never changes. But the gradient of a **curve** is **different** at different points on the curve. For example, a **quadratic graph** gets **less steep** as you move towards the turning point and then **steeper** as you move away from it.

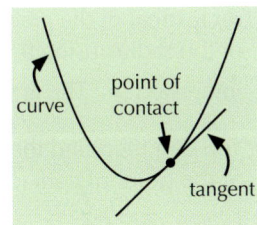

To find the **gradient at a point** on a curve, you can find the gradient of the **tangent** at this point. A tangent is a **straight line** that **just touches** the curve — at this point of contact the **gradient** of the line and the curve are **the same**.

Example 1

The graph of $y = 2x^2 - 5x - 1$ is shown here.
Find the gradient of the curve at $x = 2$.

1. Draw a tangent to the curve at $x = 2$. It should be a line that just touches the curve with the same slope as the curve.

2. Find the gradient of this line using two points, (x_1, y_1) and (x_2, y_2), on the line.

   Using the points $(1, -6)$ and $(3, 0)$ the gradient of the tangent is

   $$\frac{y_2 - y_1}{x_2 - x_1} = \frac{0 - (-6)}{3 - 1} = \frac{6}{2} = 3.$$ So the gradient of the curve at $x = 2$ is also **3**.

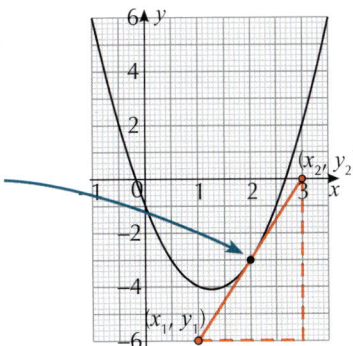

> **Tip:** The gradient of a straight line is $\dfrac{\text{change in } y}{\text{change in } x}$ (see p.180).

Exercise 1

Q1 The graph $y = x^2$ is shown on the right.
   a) Draw tangents to find the gradient at: (i) $x = 4$ (ii) $x = -4$
   b) What do you notice about the gradients at these two points?

Q2 a) Copy and complete the table below for $y = \dfrac{1}{x}$.

x	0.1	0.2	0.5	1	1.5	2	3
y							

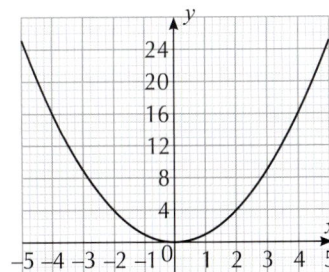

   b) Draw the graph of $y = \dfrac{1}{x}$ for $0 < x \le 3$.
   c) Find the gradient of the graph at: (i) $x = 2$ (ii) $x = 0.2$ (iii) $x = 0.6$

Q3 a) Draw the graph of $y = x^3 + x^2 - 6x$ for $-4 \le x \le 3$.
   b) Find the gradient of the graph at: (i) $x = -1$ (ii) $x = 2$
   c) Use your graph to estimate the x-values at which the gradient of the graph is zero.

Tangents to a Circle

Learning Objective — Spec Ref A16:
Find the equation of a tangent to a circle.

A curve with equation $x^2 + y^2 = r^2$ is a **circle** centred at the **origin** with radius **r**.
The **tangent to a circle** is the same as the tangent to any other curve — a **straight line** that just **touches** the circle at a point and has the **same gradient** as the circle at that point.

The tangent to a circle at a point is **perpendicular** to the **radius** of the circle **at that point**. So the tangent's **gradient** is the **negative reciprocal** of the gradient of the radius. Find the gradient, m, of the radius using the **centre** of the circle and the point you're interested in. The gradient of the tangent at that point is $-\dfrac{1}{m}$.

Tip: See page 186 for more on perpendicular lines and their gradients.

You can then find the **equation** of the tangent using the gradient and the coordinates of the point of contact with the circle in the form $y = mx + c$ (see page 183).

Example 2

The graph of $x^2 + y^2 = 25$ is shown on the right.
Find the equation of the tangent at the point $(-3, 4)$.

1. Find the gradient of the radius of the circle between $(-3, 4)$ and $(0, 0)$.

    Gradient of radius $= \dfrac{4-0}{-3-0} = -\dfrac{4}{3}$

2. The tangent is perpendicular to the radius, so its gradient is $-1 \div$ gradient of radius.

    Gradient of tangent $= -1 \div -\dfrac{4}{3}$
    $= \dfrac{3}{4}$

3. Find the equation of the tangent by substituting $(-3, 4)$ into $y = mx + c$, where m is the gradient $\dfrac{3}{4}$.

    $4 = \dfrac{3}{4} \times -3 + c \Rightarrow c = 4 + \dfrac{9}{4} = \dfrac{25}{4}$

    So the equation of the tangent is $y = \dfrac{3}{4}x + \dfrac{25}{4}$.

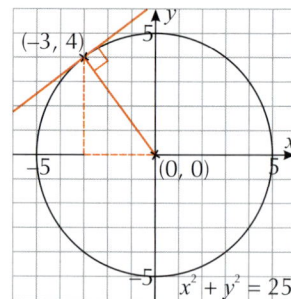

Exercise 2

Q1
a) Find the radius of the circle with equation $x^2 + y^2 = 169$ and hence sketch the graph of the circle.
b) Find the gradient of the radius between the centre of the circle and the point $(5, 12)$ as a fraction.
c) Hence, write down the gradient of the tangent to the circle at the point $(5, 12)$.
d) Find the equation of this tangent.

Q2
a) Sketch the graph of $x^2 + y^2 = 100$.
b) Find the equations of the tangents to the graph at these points:
    (i) $(6, 8)$ (ii) $(8, 6)$ (iii) $(-8, -6)$ (iv) $(-6, 8)$

Q3
a) Sketch the graph of $x^2 + y^2 = 625$.
b) Find the equations of the tangents to the graph at these points:
    (i) $(15, 20)$ (ii) $(-7, 24)$ (iii) $(-15, -20)$ (iv) $(24, -7)$

Q4
a) Find the equation of the tangent to $x^2 + y^2 = 16$ at the point $(-2\sqrt{2}, 2\sqrt{2})$.
b) Find the equation of the tangent to $x^2 + y^2 = 16$ at the point $(2\sqrt{2}, -2\sqrt{2})$.
c) What do you notice about the two tangents from parts a) and b)?

Review Exercise

Q1 A scientist is conducting an experiment.
The graph on the right shows the temperature
of the experiment after t seconds.

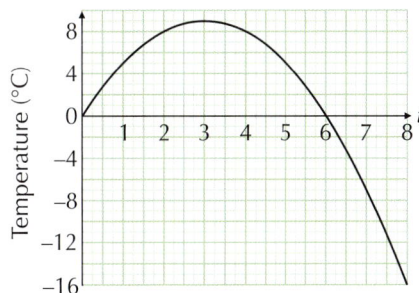

   a) Give a brief description of how the temperature
changes in the first 8 seconds of the experiment.

   b) State the maximum temperature that it reaches.

   c) A temperature of 8 °C was recorded twice.
At what times was the temperature 8 °C?

Q2 A farm allows people to pick their own Brussels sprouts.
They charge 60p per kilogram of sprouts picked, plus an admin fee of £2.40 per customer.

   a) Draw a graph showing how the total cost per customer varies with
the mass of sprouts picked for the range 0 kg to 8 kg.

   b) Use your graph to find the cost of picking 4.5 kg of sprouts.

   c) What mass of sprouts did a customer pick if she was charged £6.60?

Q3 a) Draw the graph of $x + y = 5$.

   b) On the same axes, draw the graph of $y = x - 3$.

   c) Use your graphs to solve the simultaneous equations $x + y = 5$ and $y = x - 3$.

Q4 a) Draw the graph of $y = 2x^2 - 3x - 1$ for $-2 \leq x \leq 4$.

   b) Use the graph to solve the following equations to 1 decimal place (where appropriate).

      (i) $2x^2 - 3x - 1 = 0$ (ii) $2x^2 - 3x - 5 = 0$ (iii) $2x^2 - 3x - 10 = 0$ (iv) $2x^2 - 3x = 0$

Q5 a) Draw the graph of $y = x^2 + 3x - 7$ for $-6 \leq x \leq 3$.

   b) By drawing a suitable straight line, use your graph to find
the solutions to the quadratic equation $x^2 - 1 = 0$.

Q6 The graph of $y = x^3 - 2x^2 - 8x$ is shown on the right.
Find an estimate for the gradient of this curve when:

   a) $x = -3$ b) $x = 2$

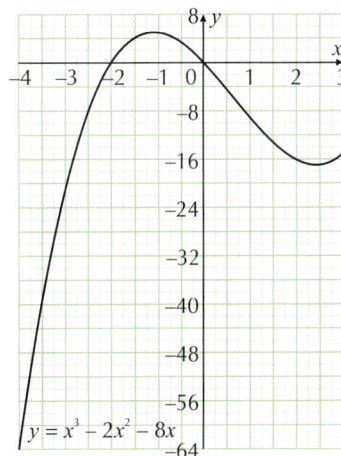

Q7 A circle is described by the equation $x^2 + y^2 = r^2$.
The line T is a tangent to the circle. Given that T meets
the y-axis when $y = 5\sqrt{2}$ and the x-axis when $x = 10\sqrt{2}$,
find the value of r, the radius of the circle.

Exam-Style Questions

Q1 A car is driven along a straight track to test its acceleration.
The distance in metres, x, that the car has travelled after t seconds is given by $x = 2t^2 + t$.

a) By calculating values of x at $t = 0$, 1, 2, 3, 4 and 5, draw a graph of x against t for $0 \leq t \leq 5$.

[2 marks]

b) Use your graph to find the distance the car will have travelled after 3.5 seconds.

[1 mark]

c) Describe how the speed of the car changes with time.

[2 marks]

Q2 The graph below shows the relationship between the total amount in pounds (£), a, that a salesperson is paid in a year and the number of people, p, that they have signed up to a mobile phone contract in that year.

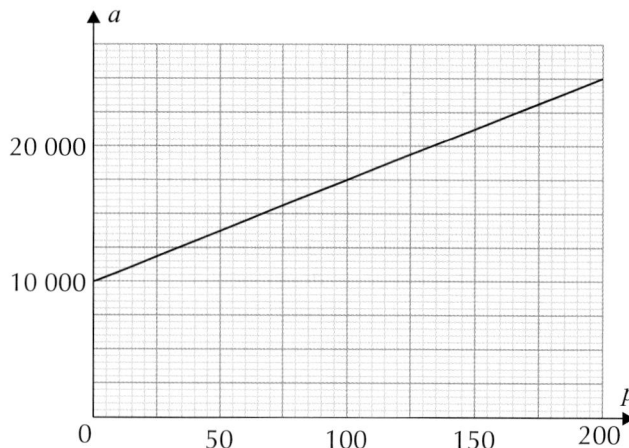

a) How much would the salesperson earn in a year in which they signed up 100 people?

[1 mark]

b) Obtain a formula for a in terms of p.

[3 marks]

Q3 The diagram on the right shows the cross-section of a container used to store chemicals. It is made from a cylinder joined to part of a sphere. The container has a total depth of d, as shown.

A laboratory technician completely fills the container with water from a tap which is running at a constant rate.
On a set of axes, sketch a graph to show how the depth of water in the container changes with time.

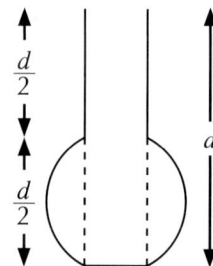

[3 marks]

Q4 Five years ago, two crows were put on a deserted island. The number of crows on the island (C) is given by the equation $C = 2^{x+1}$, where x is the number of years since the crows were first put on the island. The graph of C is shown below.

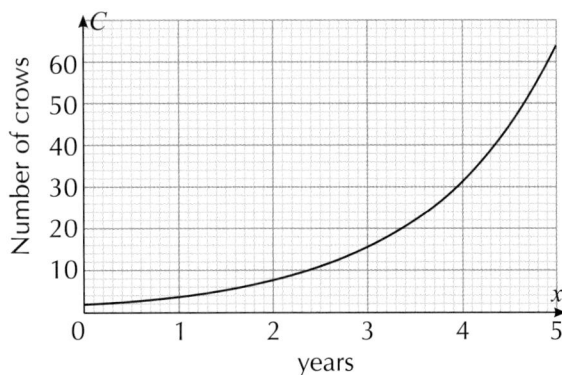

a) Use the graph to find, to 1 decimal place, how many years it took for there to be 20 crows on the island.

[1 mark]

b) Find an estimate for the rate of change of the number of crows when they have been on the island for 4.5 years.

[2 marks]

The number of a particular tree species on the island, T, is given by $T = 5x + 10$.

c) By plotting an appropriate straight line, find how long it will take for the number of trees of this species to equal to the number of crows. Give your answer to 1 decimal place.

[2 marks]

Q5 Two tangents are drawn to the circle with equation $x^2 + y^2 = 25$.
The two tangents touch the circle at (4, 3) and (–4, 3).
Find the coordinates of the point where the two tangents intersect.

[5 marks]

Q6 Hamish has drawn the graph of $y = x^2 + 7x - 3$. He intends to use the graph by plotting straight lines onto it so that the x-coordinates of the intersections will be solutions to the equations he wants to solve. Find the equations of the straight lines he should plot in order to solve:

a) $x^2 + 7x - 6 = 0$

[1 mark]

b) $x^2 + 5x + 8 = 0$

[2 marks]

18.1 Evaluating Functions

Don't be put off functions by the slightly weird notation — they're actually a really simple concept. Functions are just mathematical rules that can be applied to input values to 'map' them to output values.

Learning Objectives — Spec Ref A7:
- Understand and use function notation.
- Be able to evaluate functions.

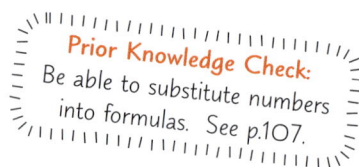

Prior Knowledge Check:
Be able to substitute numbers into formulas. See p.107.

A **function** is a rule that turns one number (the **input**) into another number (the **output**). For example:

- The function $f(x) = x + 3$ is a rule that **adds 3** to any input value x.

- The same function could also be written as f: $x \rightarrow x + 3$
  You read this as 'the function f **maps** the value of x to the value $x + 3$'.

To **evaluate** a function for a particular value of x, **substitute** this value into the expression for $f(x)$. So for function f above, $f(6) = 6 + 3 = 9$, $f(-2) = (-2) + 3 = 1$, and so on.

Tip: You can show functions on graphs (see Sections 15-16).

Each value on the x-axis is 'mapped' to a value on the y-axis. The y-axis is often labelled $y = f(x)$ to show this.

Example 1

For the function $f(x) = 30 - 2x^2$, find: a) f(4),   b) f(-2)

a) To find f(4), substitute $x = 4$ into $30 - 2x^2$. $f(4) = 30 - 2 \times 4^2$
$$= 30 - 2 \times 16 = 30 - 32 = -2$$

b) Substitute $x = -2$ into $30 - 2x^2$. $f(-2) = 30 - 2 \times (-2)^2$
Be careful with the minus signs here. $$= 30 - 2 \times 4 = 30 - 8 = 22$$

Functions can be represented by letters other than f — you might see $g(x)$, $h(x)$ etc.

Example 2

For the function $g(x) = \dfrac{2x + 1}{x + 4}$, find g(3).

To find g(3), substitute $x = 3$ into the function.
Make sure you replace every x with 3.

$$g(3) = \frac{2(3) + 1}{3 + 4} = \frac{6 + 1}{7} = \frac{7}{7} = 1$$

Exercise 1

Q1   a) $f(x) = x - 5$ Work out: (i) f(11)   (ii) f(36)   (iii) f(4)

b) $f(x) = 4x$ Work out: (i) f(0)   (ii) f(7)   (iii) f(1.5)

c) $g(x) = 5x - 7$ Calculate: (i) g(3)   (ii) g(6)   (iii) g(-1)

d) f: $x \rightarrow 11 - 2x$ Work out: (i) f(3)   (ii) f(5.5)   (iii) f(-5)

Q2 a) $f(x) = x^2 - 4x$ Work out: (i) $f(7)$ (ii) $f(-3)$ (iii) $f(1.5)$

b) $g: x \rightarrow 20 - 3x^2$ Work out: (i) $g(2)$ (ii) $g(3)$ (iii) $g(-1)$

c) $h(x) = 2x^2 - x + 4$ Calculate: (i) $h(5)$ (ii) $h(-2)$ (iii) $h(8)$

d) $f(t) = (4t - 3)^2$ Calculate: (i) $f(2)$ (ii) $f(0)$ (iii) $f(0.75)$

Q3 a) $g(x) = \dfrac{20}{x}$ Work out: (i) $g(10)$ (ii) $g(1.25)$ (iii) $g(-0.5)$

b) $g: x \rightarrow \dfrac{18}{x^2 + 2}$ Work out: (i) $g(0)$ (ii) $g(2)$ (iii) $g(-4)$

c) $h(x) = \sqrt{\dfrac{x}{3x + 1}}$ Work out: (i) $h(0)$ (ii) $h(1)$ (iii) $h(16)$

d) $f(x) = 8^x - 1$ Work out: (i) $f(0)$ (ii) $f\left(\dfrac{1}{3}\right)$ (iii) $f\left(-\dfrac{2}{3}\right)$

Example 3

$f(x) = 3x + 4$. **Find the value of x for which $f(x) = 22$.**

Set up and solve an equation in x: $22 = 3x + 4$
$18 = 3x$
$x = 6$

Exercise 2

Q1 For each function below, find the value of x which produces the given output value.

a) $f(x) = 2x - 5$, $f(x) = 35$ b) $g(x) = 8 - 3x$, $g(x) = -10$ c) $h(x) = 9x + 12$, $h(x) = 93$

Q2 For each function below, find the value of x which produces the given output value.

a) $f(x) = x^2$, $f(x) = 225$ b) $g(x) = \sqrt{2x - 1}$, $g(x) = 3$ c) $h(x) = \dfrac{18}{x + 1}$, $h(x) = 2$

Example 4

For the function $f(x) = 30 - 2x^2$, find an expression for $f(2t)$.

To find $f(2t)$, replace x with $2t$ and expand. $f(2t) = 30 - 2 \times (2t)^2$
$= 30 - 2 \times 4t^2$
$= 30 - 8t^2$

Tip: This will come in handy for finding composite functions (see next page).

Exercise 3

Q1 a) $f: x \rightarrow 2x + 1$. Find expressions for: (i) $f(k)$ (ii) $f(2m)$ (iii) $f(3w - 1)$

b) $f(x) = 2x^2 + 3x$. Find expressions for: (i) $f(u)$ (ii) $f(3a)$ (iii) $f(t^2)$

c) $f(x) = \dfrac{4x - 1}{x + 4}$. Find expressions for: (i) $f(t)$ (ii) $f(-x)$ (iii) $f(2x)$

d) $f(x) = \sqrt{5x - 1}$. Find expressions for: (i) $f(w)$ (ii) $f(3x)$ (iii) $f(1 - 2x)$

18.2 Composite Functions

You need to be completely happy with function notation before carrying on. Next up, you'll see what happens when you combine two (or more) functions into one new function — called a composite function.

Learning Objective — Spec Ref A7:
Find composite functions.

If f(x) and g(x) are two functions, then the **combined** function gf(x) is called a **composite function**.

gf(x) means 'put x into function f, then put the answer into function g' — you always do the function **closest** to x first. To find the composite function gf(x), write it as g(f(x)), replace f(x) with the expression it represents, then put this into g.

You'll usually find that fg(x) ≠ gf(x) — in other words, applying function g first, then f is **not** the same as applying f then g, so the order in which you combine the functions matters.

Example 1

If f(x) = 3x − 2 and g(x) = x^2 + 1, then: **a) Calculate gf(3)**   **b) Find fg(x)**

a) Calculate the value of f(3), then use it as the input in g to find gf(3).

$$gf(3) = g(f(3))$$
$$= g(3 \times 3 - 2)$$
$$= g(7) = 7^2 + 1 = \mathbf{50}$$

Tip: Here, gf(x) = $9x^2$ − 12x + 5 — which is not the same as fg(x).

b) Replace x with g(x) in the expression for f(x), then collect like terms to tidy it up.

$$fg(x) = f(g(x))$$
$$= f(x^2 + 1)$$
$$= 3(x^2 + 1) - 2$$
$$= 3x^2 + 3 - 2 = \mathbf{3x^2 + 1}$$

Exercise 1

Q1 f(x) = 2x, g(x) = x + 2   a) Find g(1)   b) Use your answer to a) to find fg(1)

c) Find f(5)   d) Use your answer to c) to find gf(5)

Q2 a) f(x) = x − 5,   g(x) = 4x   Find: (i) fg(7)   (ii) gf(8)   (iii) gf(x)

b) f(x) = 2x − 1,   g(x) = 3x + 1   Find: (i) gf(0)   (ii) fg(2)   (iii) fg(x)

c) f(x) = 5x − 4,   g(x) = $\frac{x}{4}$ Find: (i) fg(4)   (ii) gf(4)   (iii) gf(x)

d) f(x) = 4x + 3,   g(x) = 5x   Find: (i) fg(3)   (ii) ff(2)   (iii) gf(x)

e) f(x) = 2x + 1,   g(x) = 11 − x   Find: (i) gf(3)   (ii) gg(4)   (iii) fg(x)

f) f(x) = $\frac{1}{x}$,   g(x) = 3x − 7   Find: (i) fg(3)   (ii) fg(x)   (iii) gf(x)

Q3 a) f(x) = 10 − x,   g(x) = x^2 Find: (i) gf(8)   (ii) ff(2)   (iii) gf(x)

b) f(x) = $\sqrt{x}$,   g(x) = 2x + 1   Find: (i) ff(16)   (ii) fg(40)   (iii) gf(x)

c) f(x) = x^2 + 5,   g(x) = 3x − 4   Find: (i) gf(−1)   (ii) gf(x)   (iii) gg(x)

d) f(x) = $\sqrt{3x + 1}$, g(x) = 12 − 2x   Find: (i) gf(8)   (ii) gg(x)   (iii) fg(x)

Example 2

If $f(x) = x^2$ and $g(x) = 7 - 2x$, find fgf(x).

Find gf(x) first, then use that as the input for f.

$$fgf(x) = f(gf(x)) = f(g(x^2))$$
$$= f(7 - 2x^2)$$
$$= (7 - 2x^2)^2$$

Tip: Do it one step at a time to avoid mistakes.

Exercise 2

Q1   a) $f(x) = \dfrac{1}{3x + 2}$, $g(x) = 2x - 5$, $h(x) = x^2$ Find: (i) hgf(–1)   (ii) fg(x)   (iii) hf(x)

b) $f(x) = x^3$, $g(x) = 6 - x$, $h(x) = 10 - 2x$ Find: (i) fgh(4)   (ii) gh(x)   (iii) fh(x)

c) $f(x) = 2x + 4$, $g(x) = \dfrac{1}{3x + 1}$, $h(x) = \sqrt{x}$ Find: (i) fh(25)   (ii) fg(0.5)   (iii) hgf(x)

d) $f(x) = x^2 + x$, $g(x) = \dfrac{1}{2x + 3}$, $h(x) = \dfrac{1}{x}$ Find: (i) ff(3)   (ii) gf(–1)   (iii) hgf(x)

Example 3

Let $f(x) = 4x - 7$ and $g(x) = \dfrac{1}{3x + 2}$. Solve the equation fg(x) = 1.

1. First find fg(x).

$$fg(x) = f\left(\frac{1}{3x + 2}\right) = 4\left(\frac{1}{3x + 2}\right) - 7 = \frac{4}{3x + 2} - 7$$

2. Now let fg(x) = 1 and solve for x.

$$\frac{4}{3x + 2} - 7 = 1$$
$$\frac{4}{3x + 2} = 8$$
$$8(3x + 2) = 4$$
$$24x + 16 = 4$$
$$24x = -12$$
$$x = -0.5$$

Tip: Flick back to Section 9 if you need a reminder about solving equations.

Exercise 3

Q1 $f(x) = 9 - x$, $g(x) = 3x$

a) (i) Find fg(x)   (ii) Hence solve fg(x) = –12

b) (i) Find gf(x)   (ii) Hence solve gf(x) = 6

Q2 $f(x) = x + 5$, $g(x) = \frac{x}{2}$. Solve the following equations.

a) fg(x) = 14   b) gf(x) = 8   c) gg(x) = 11

Q3 $f(x) = 6 - x$, $g(x) = 2x^2 - 1$. Solve the following equations.

a) fg(x) = 7   b) gf(x) = 31   c) gg(x) = 97

18.3 Inverse Functions

An inverse operation does the opposite of the original operation — so the inverse of +9 is −9, the inverse of ÷5 is ×5 and so on. Knowing how to use inverse operations will be very helpful in this next topic.

Learning Objective — Spec Ref A7:
Find inverse functions.

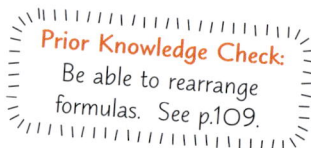

Prior Knowledge Check:
Be able to rearrange
formulas. See p.109.

An **inverse function** reverses the effect of a function. The inverse of f(x) is written **f⁻¹(x)**. So if the function f multiplies by 10, the inverse function f⁻¹(x) would divide by 10 — i.e. f(x) = 10x, so f⁻¹(x) = $\frac{x}{10}$.

If you apply a function to x, then apply its inverse, you just end up with x. This can be written as f⁻¹f(x) = x. This works the other way around as well — applying the inverse, then the function, will get you back to x (ff⁻¹(x) = x). Combining these rules gives **ff⁻¹(x) = f⁻¹f(x) = x**.

There are two methods you can use to find an inverse function — the first one simply involves undoing the operations of the original function, one at a time.

Example 1

Find the inverse of the function g(x) = $\frac{7x + 1}{2}$.

1. Write g(x) out as a function machine.

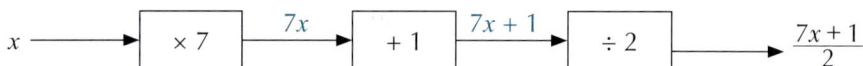

$$x \longrightarrow \boxed{\times 7} \xrightarrow{7x} \boxed{+ 1} \xrightarrow{7x + 1} \boxed{\div 2} \longrightarrow \frac{7x + 1}{2}$$

2. Reverse each step in turn to get the inverse.

$$\frac{2x - 1}{7} \longleftarrow \boxed{\div 7} \xleftarrow{2x - 1} \boxed{- 1} \xleftarrow{2x} \boxed{\times 2} \longleftarrow x$$

So g⁻¹(x) = $\frac{2x - 1}{7}$

3. You can check your answer to make sure that the inverse function does reverse the original function.

$$g(1) = \frac{7(1) + 1}{2} = \frac{8}{2} = 4 \text{ and } g^{-1}(4) = \frac{2(4) - 1}{7} = \frac{8 - 1}{7} = \frac{7}{7} = 1$$

This gives the starting number (1), so the inverse looks to be correct.

Tip: Make sure you reverse the **order** of the operations — so the last operation of the original function becomes the first operation to undo in the inverse function.

Exercise 1

In Questions 1-3, find the inverse of each function.

Q1 a) f(x) = x + 4   b) f(x) = x − 3   c) f(x) = x − 7   d) f(x) = x + 1

     e) f(x) = 8x f) g(x) = 2x g) g(x) = $\frac{x}{3}$ h) f(x) = $\frac{x}{6}$

Q2 a) f(x) = 4x + 3   b) f(t) = 2t − 9   c) g(x) = 3x − 5   d) f(x) = 8x + 11

     e) f(x) = $\frac{x}{5}$ − 7   f) g(x) = $\frac{x}{8}$ + 1   g) f(t) = $\frac{t - 3}{2}$ h) h(x) = $\frac{x + 15}{4}$

Q3 a) f(x) = $\frac{2x + 6}{5}$ b) g(x) = $\frac{3x - 1}{4}$ c) f(x) = x^2 − 3   d) g(x) = (2x + 7)²

For more complicated functions, use the following method:

1. Write out the equation $x = f(y)$
   ($f(y)$ is just $f(x)$ with the x's replaced by y's).

2. **Rearrange** the equation to make y the subject.

3. Finally, **replace** y with $f^{-1}(x)$.

Example 2

Find the inverse of the function $f(x) = \dfrac{\sqrt{4x-1}}{3}$.

1. Write out the equation $x = f(y)$.

$$x = \frac{\sqrt{4y-1}}{3}$$

2. Rearrange the equation to make y the subject.

$$3x = \sqrt{4y-1}$$
$$9x^2 = 4y - 1$$
$$9x^2 + 1 = 4y$$
$$y = \frac{9x^2 + 1}{4}$$

Tip: Don't rush this step — it's easy to make a mistake.

3. Replace y with $f^{-1}(x)$.

$$f^{-1}(x) = \frac{9x^2 + 1}{4}$$

4. Check your inverse function by trying it out on a number.

$$f(2.5) = \frac{\sqrt{4(2.5)-1}}{3} = \frac{\sqrt{10-1}}{3} = \frac{\sqrt{9}}{3} = \frac{3}{3} = 1$$

and $f^{-1}(1) = \dfrac{9(1)^2 + 1}{4} = \dfrac{9+1}{4} = \dfrac{10}{4} = 2.5$

This gives you the starting number (2.5), so it looks like the inverse is correct.

Exercise 2

In Questions 1-5, find the inverse of each function.

Q1 a) $f(x) = \dfrac{x}{5} - 8$ b) $f(x) = \dfrac{3x+1}{5}$ c) $g(x) = \dfrac{2x}{5} - 7$

Q2 a) $f(x) = 7 - 3x$ b) $f(x) = \dfrac{9-7x}{4}$ c) $h(x) = \dfrac{1-6x}{9}$

Q3 a) $g(x) = 4x^2 + 1$ b) $g(x) = (3x-1)^2$ c) $h(x) = \dfrac{(1-2x)^2}{5}$

Q4 a) $f(x) = 6\sqrt{x} + 1$ b) $g(x) = \sqrt{19-2x}$ c) $f(x) = 25 - 4\sqrt{x}$

Q5 a) $f(x) = \dfrac{4}{x} - 7$ b) $g(x) = 8 - \dfrac{2}{\sqrt{x}}$ c) $f(x) = \sqrt{\dfrac{2}{x-1}}$

Q6 The inverse of the function $g(x) = \dfrac{1-2x}{3x+5}$ can be found by rearranging the equation $x = \dfrac{1-2y}{3y+5}$.

    a) Show that, if $x = \dfrac{1-2y}{3y+5}$, then $3xy + 2y = 1 - 5x$.

    b) Factorise the expression $3xy + 2y$, and hence find $g^{-1}(x)$.

Q7 Find the inverse of each of the following functions:

    a) $f(x) = \dfrac{2+3x}{x-2}$ b) $g(x) = \dfrac{4x+1}{2x-7}$ c) $f(x) = \dfrac{x}{3x+2}$

Q1 $f(x) = 5x - 7$ and $g(x) = 2x + 5$

    a) Calculate f(8).   b) Calculate fg(1).   c) Solve $f(x) = g(x)$.   d) Find $g^{-1}(x)$.

Q2 $f(x) = 4x + 1$ and $g(x) = 2x + 5$

    a) Solve $f(x) = 25$ b) Find fg(x)   c) Find $f^{-1}(x)$

Q3 $f(x) = \sqrt{x + 1}$ and $g(x) = 4x$.

    a) Find gf(x).   b) Find $f^{-1}(x)$.

Q4 $f(x) = 2x - 3$

    a) Solve $f(x) = 5$.   b) Find $f^{-1}(x)$.

    c) Solve the equation $f(x) = f^{-1}(x)$ d) Find ff(x)

Q5 $f(x) = 2x - 3$ and $g(x) = 18 - 3x$

    a) Solve $f(x) = g(x)$.   b) Find gf(x).   c) Find $g^{-1}(x)$.

Q6 $f(x) = \dfrac{1}{x} - 5$ and $g(x) = \dfrac{1}{x + 5}$

    a) Calculate g(–3).   b) Find fg(x). How are functions f and g related?

Q7 $f(x) = 2x^2 + 5$ and $g(x) = 3x - 1$

    a) Work out gg(4).   b) Find $g^{-1}(x)$.   c) Solve $fg^{-1}(x) = 55$.

Q8 $f(x) = \dfrac{x}{x - 3}$ and $g(x) = x^2 + 3$

    a) Calculate f(5).   b) Find $f^{-1}(x)$.   c) Find fg(x).

Q9 An object is dropped off the top of a tower. The distance, s metres, the object has travelled t seconds after being released is given by the formula $s = f(t)$, where $f(t) = 5t^2$.

    a) Calculate f(4).   b) Solve the equation $f(t) = 12.8$.

Q10 The function $f(t) = \dfrac{9t}{5} + 32$ can be used to convert a temperature from °C to °F.

    a) Work out f(20).   b) Solve the equation $f(t) = 60.8$.

    c) Find $f^{-1}(t)$.   d) Explain what your answers to parts a)-c) represent.

Q11 Jill uses the following formula to estimate the temperature T (in degrees Fahrenheit) at height h (in thousands of feet) above sea level: $T = f(h)$ where $f(h) = 60 - \dfrac{7}{2}h$.

    a) Calculate f(4).

    b) Find the temperature at a height of 7000 feet above sea level.

    c) Work out $f^{-1}(32)$. Explain what this answer tells you in the context of this question.

Exam-Style Questions

Q1 f and g are two functions, where $f(x) = 6x + 5$ and $g(x) = \dfrac{x+3}{2}$.

    a) Evaluate g(11).

[1 mark]

    b) Find fg(x).

[2 marks]

    c) Find an expression for $g^{-1}(x)$.

[3 marks]

Q2 f and g are two functions, where $f(x) = 2x^2$ and $g(x) = 3x - 5$.
Find fg(x), giving your answer in the form $ax^2 + bx + c$.

[3 marks]

Q3 A function f(x) is defined such that:

$$f(x) = \frac{2x+3}{5x-4}, \quad x \neq \frac{4}{5}$$

Find an expression for the inverse function, $f^{-1}(x)$.

[4 marks]

Q4 $f(x) = \dfrac{4}{5x-15}$ and $g(x) = 2\sqrt{x}$

    a) Solve f(x) = 0.2.

[2 marks]

    b) Work out fg(x).

[2 marks]

Q5 $f(x) = x^2 + 6$, $g(x) = 3x + 4$.
Solve the equation fg(x) = gf(x).

PROBLEM SOLVING

[6 marks]

19.1 Sets

A set is a collection of things — it can be considered as an object in its own right.
Set notation can look quite complicated, but don't worry — it's not as scary as it first appears.

Learning Objectives — Spec Ref P6:
- Understand and use set notation.
- Be able to list elements of sets.

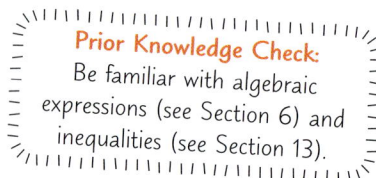

Prior Knowledge Check:
Be familiar with algebraic expressions (see Section 6) and inequalities (see Section 13).

A **set** is a group of items or numbers. Sets are written in pairs of **curly brackets { }**.

- Each item in a set is called an **element** or **member** of the set.

- You can describe a set by **listing every element** in that set (e.g. {2, 4, 6})
  or by **giving a rule** that all elements must follow (e.g. {all red objects}).

- You usually use a **capital letter** to represent the set (e.g. A), so **A = {all red objects}** means 'A is the set of all red objects'. **Lower case letters** represent elements within the set (e.g. *a* might be 'a tomato').

Here's some of the **set notation** you'll need to be familiar with:

A = {...}	A is the set of ...
$x \in A$	x is an element of A
$y \notin A$	y is not an element of A
n(A)	the number of elements in A

Tip: To find n(A), just count the number of elements in the set.

Example 1

a) **List the members of set A, where A = {days of the week with the letter u in their name}.**
b) **Write down the number of elements in set A.**
c) **Is Monday a member of the set?**

a) The days that contain the letter u are Tuesday, Thursday, Saturday and Sunday.

    A = {Tuesday, Thursday, Saturday, Sunday}

b) There are 4 elements in set A.

    n(A) = 4

c) Monday doesn't have the letter u in it, so it's not a member of set A.

    Monday $\notin$ A

Exercise 1

Q1 List the elements of the following sets:
    a) A = {months of the year with fewer than 31 days}
    b) B = {months of the year with fewer than 4 letters in their name}
    c) C = {months of the year with the letter a in their name}

Q2 List the elements of the following sets:
    a) A = {even numbers between 11 and 25}   b) B = {prime numbers less than 20}
    c) C = {square numbers less than 200}   d) D = {factors of 30}

There are two special types of set: the **empty set** and the **universal set**.

∅ or { }	the empty set
ξ	the universal set

Tip: ξ is the lowercase Greek letter xi.

- The **empty set** is a set that has **no members**. E.g. if set B = {square numbers between 50 and 60}, then **B = ∅** because there are no square numbers between 50 and 60.

- The **universal set** is the group of things **under consideration** — the group that members of a set are **selected** from. The universal set is often a set of **integers**, e.g. ξ = {positive integers between 1 and 20}.

- Some sets are described using an **algebraic expression** — e.g. A = {$x : x + 2 < 5$} is read as 'A is the set of numbers x, such that $x + 2$ is less than 5'. To list the elements of a set like this, you need to know the **universal set**. E.g. for A = {$x : x + 2 < 5$}, if ξ = {positive integers}, the elements in set A would be {1, 2}.

Example 2

List the members of A = {$x : x < 15$} if ξ = {$x : x$ is a square number}.

1. Firstly, work out what the symbols and numbers mean:
   A is the set of numbers x, such that x is less than 15.
   ξ is the universal set — it contains all square numbers.

   Tip: The colon just means 'such that'.

2. Elements of A can only come from the universal set ξ, so here A = {square numbers less than 15}.

   A = {1, 4, 9}

Exercise 2

Q1 a) If ξ = {$x : x$ is a positive integer, $x \leq 10$}, list all the elements of:
   (i)   A = {$x : x$ is odd}
   (ii) B = {$x : x$ is a factor of 16}
   (iii) C = {$x : x$ is a square number}
   (iv) D = {$x : x$ is a factor of 30}

   b) If ξ = {$x : x$ is an integer, $20 \leq x \leq 30$}, list all the elements of:
   (i)   A = {$x : x$ is odd}
   (ii) B = {$x : x$ is even}
   (iii) C = {$x : x$ is prime}
   (iv) D = {$x : x$ is a multiple of 3}

Q2 For all the sets in this question, ξ = {$x : x$ is a positive integer, $x < 30$}
   A = {prime numbers}   B = {square numbers}   C = {cube numbers}   D = {multiples of 4}
   a) List the members of A, B, C and D.
   b) Find   (i) n(A)   (ii) n(B)   (iii) n(C)   (iv) n(D)
   c) Find   (i) n(E) where E = {$x : x \in$ both A and B}
              (ii) n(F) where F = {$x : x \in$ B but not D}
              (iii) n(G) where G = {$x : x \notin$ A, $x \notin$ B, $x \notin$ C, $x \notin$ D}

Q3 A = {1, 2, 3, 4}   B = {0, 2, 4}
   a) List the elements of the following sets:
   (i)   C = {$x : x \in$ either A or B or both}
   (ii) D = {$x : x \in$ A but not B}
   (iii) E = {$x : x = a - b, a \in$ A, $b \in$ B, $x < 1$}
   b) Each element of the following sets is a pair of coordinates. List the elements of each set.
   (i)   F = {$(a, b) : a \in$ A, $b \in$ B, $a + b < 5$}
   (ii) G = {$(a, b) : a \in$ A, $b \in$ B, $a \times b > 6$}

19.2 Venn Diagrams

Venn diagrams can be used to display sets. They're great at showing the overlap between sets.

Representing Sets Using Venn Diagrams

Learning Objective — Spec Ref P6:
Use Venn diagrams to show sets.

Venn diagrams use **circles** to represent sets — the **space inside** the circle represents everything in the set. Each **circle** is labelled with a **letter** — this tells you **which set** the circle represents. The **number inside a circle** can tell you either the **number of members** of that set or **actual elements** of the set.

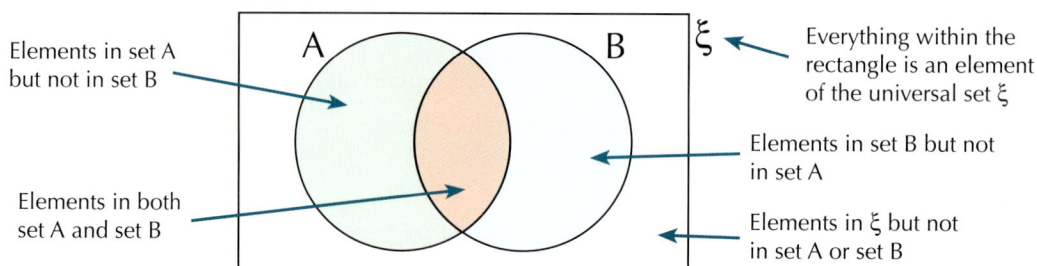

Elements in set A but not in set B

Elements in both set A and set B

A

B

ξ

Everything within the rectangle is an element of the universal set ξ

Elements in set B but not in set A

Elements in ξ but not in set A or set B

Example 1

Given that ξ = {positive integers less than or equal to 12}, draw a Venn diagram to show the elements in the sets A = {$x : x$ is a multiple of 3} and B = {$x : x$ is a factor of 30}.

1. Write out sets A and B.
   A = {3, 6, 9, 12}
   B = {1, 2, 3, 5, 6, 10}

2. Look for the elements that appear in both sets. Here, 3 and 6 are in both A and B, so they go in the overlap between the circles.

3. The other elements of each set go in the circles for A and B, but not in the overlap.

4. The elements of ξ that aren't in either set go outside the circles.

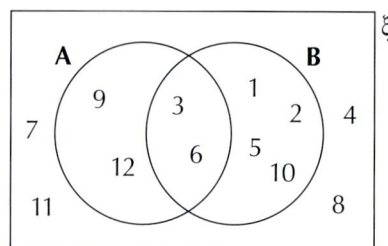

Exercise 1

Q1 Draw Venn diagrams to show the elements in the following pairs of sets, where ξ = {$x : x$ is a positive integer, $x \leq 10$} in each case.
   a) A = {1, 3, 5}   B = {1, 3, 7}
   b) A = {2, 3, 4, 5}   B = {1, 3, 5, 7, 9}
   c) A = {2, 6, 10}   B = {1, 3, 6, 9}

Q2 Draw Venn diagrams to show the elements in the following pairs of sets, where ξ = {$x : x$ is an integer, $20 \leq x \leq 30$} in each case.
   a) A = {$x : x$ is odd}   B = {$x : x$ is a multiple of 5}
   b) A = {$x : x$ is a multiple of 3} B = {$x : x$ is a multiple of 4}
   c) A = {$x : x$ is even}   B = {$x : x$ is a factor of 100}

Q3 The Venn diagram on the right represents sets A, B and C. (PROBLEM SOLVING)

a) List the elements of set A.

b) Which elements are in both set A and set C?

c) What is the value of n(B)?

d) List the elements which are in both set A and set B, but not in set C.

e) Which elements of set C are not also in set B?

f) List the elements which are neither in set A nor set B.

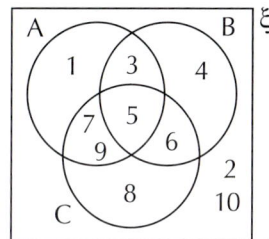

Solving Problems with Venn Diagrams

Learning Objective — Spec Ref P6:
Use Venn diagrams to solve problems.

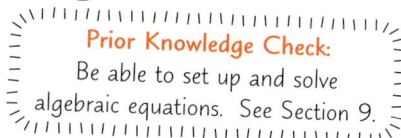

Prior Knowledge Check:
Be able to set up and solve algebraic equations. See Section 9.

When a Venn diagram is labelled with the **number of elements**, it can be used to solve problems involving the numbers of elements in sets. With this information you can **set up equations** and **solve** them to find missing values.

Tip: n(ξ) is the total number of elements in the whole Venn diagram.

Example 2

A and B are sets. n(A) = 24, n(B) = 16 and n(ξ) = 32. 4 elements of ξ are neither in A nor B.

a) Draw a Venn diagram to show sets A and B. Label each part of the diagram with a number or expression representing the number of elements in that part.

b) Find the number of elements that are in both set A and set B.

a) 1. Call the number of elements that are in both A and B x, then write this in the overlap between the circles.

2. You can now write the number of elements in A only as n(A) − x = 24 − x, and the number of elements in B only as n(B) − x = 16 − x. Label the two circles with these expressions.

3. 4 elements are in neither set, so this goes outside the circles.

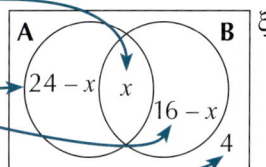

b) Use n(ξ) = 32 to write and solve an equation for x.

$$n(\xi) = (24 − x) + x + (16 − x) + 4$$
$$32 = 44 − x \Rightarrow x = 12$$

So **12 elements** are in both A and B.

Exercise 2

Q1 Below are the Venn diagrams a)-c). They are labelled with the number of elements in each part of the diagram.

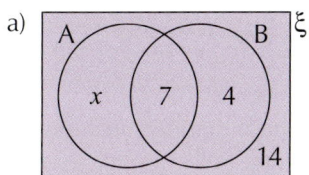

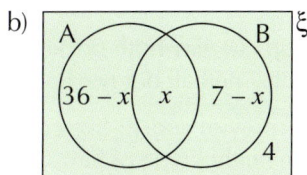

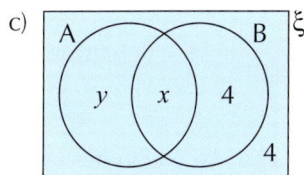

a)

b)

c)

a) For a), n(ξ) = 30. Find: (i) x (ii) n(A)   (iii) n(not B)

b) If the number of elements in A or B or both is 40 in b), find: (i) x (ii) n(A)   (iii) n(ξ)

c) For c), n(A) = 50 and n(B) = 40. Find: (i) x and y (ii) n(ξ)   (iii) n(A or B or both)

Q2   Draw a Venn diagram to represent each situation below, and use it to answer the questions.

a) Set A contains 26 elements in total. 18 of the elements of set A are not in set B.
10 of the elements of set B are not in set A. 12 elements of ξ are neither in A nor B. Find:
(i) n(B)   (ii) n(ξ)

b) 9 of the elements of set A are not in set B. 17 of the elements of set B are not in set A.
4 elements are in both set A and set B. n(ξ) = 40.
Find the number of elements of ξ that are: (i) not in A   (ii) not in B

c) 3 elements are in both set A and set B. 11 elements of ξ are not in A or B. The number of
elements in B but not A is 4. n(ξ) = 25. Find the number of elements:
(i) in B   (ii) not in B   (iii) in A   (iv) in A but not B

d) 37 of the elements of set B are not in set A. If the number of elements that are in both set A
and set B is x, n(A) = 34 + x. 4 elements of ξ are neither in A nor B. n(ξ) = 88.
Find the total number of elements in: (i) both A and B   (ii) B

Example 3

In a class of 30 students, 5 study both geography and history,
15 study history but not geography and 6 study neither subject.

a) Draw a Venn diagram to show this information.

b) Find how many students in total study geography.

a) 1.   As 5 students study both subjects, this goes in the overlap.
      2.   15 study history only, so this goes in the history circle.
      3.   6 students study neither, so this goes outside the circles.
      4.   Label the number of students who study geography only x.

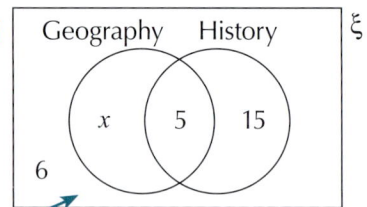

b) Use the Venn diagram to set up an   n(ξ) = 30 $\Rightarrow$ 30 = 6 + 5 + 15 + x $\Rightarrow$ x = 30 − 26 = 4
equation and solve it, using the fact   4 students study geography only,
that there are 30 students in total.   so 4 + 5 = **9 students** in total study geography.

Exercise 3

For each question, draw a Venn diagram to represent the situation and use it to answer the question.

Q1   In a class of pupils, 6 play the guitar and piano, 8 play the guitar only, 4 play the piano only
and 12 play neither instrument. How many pupils are in the class?

Q2   36 children were asked about their pets. 8 owned geese only, 12 owned ducks only
and 12 owned ducks and geese. How many owned neither geese nor ducks?

Q3   Of the 60 members at a sports club, 30 play both hockey and netball, 6 play hockey but not netball
and 14 play neither. How many play netball but not hockey?

Q4   In a group of friends, 8 like horror movies and 12 like comedies. These numbers include 6 people
who like both types of film. 3 like neither type of film. How many are in the group?

Q5   Of the 55 pupils entered for English and maths tests, everybody passed at least one of English
and maths. Including those who passed both subjects, 44 passed maths and 50 passed English.
How many passed exactly one of English and maths?

19.3 Unions and Intersections

Combinations of sets can be described using the terms 'union' and 'intersection' — they represent different regions on a Venn diagram.

Learning Objective — Spec Ref P6:
Find unions and intersections of sets.

The **union** of two sets contains **all the members** that are in **either set**. On a Venn diagram, the union is **everything inside the circles**. You write 'the union of set A and set B' as $A \cup B$.

The **intersection** of two sets **only** contains objects that are members of **both sets**. This is **everything in the overlap** between the circles on a Venn diagram. You write 'the intersection of set A and set B' as $A \cap B$.

Tip: Read $A \cup B$ as "A union B" and $A \cap B$ as "A intersection B".

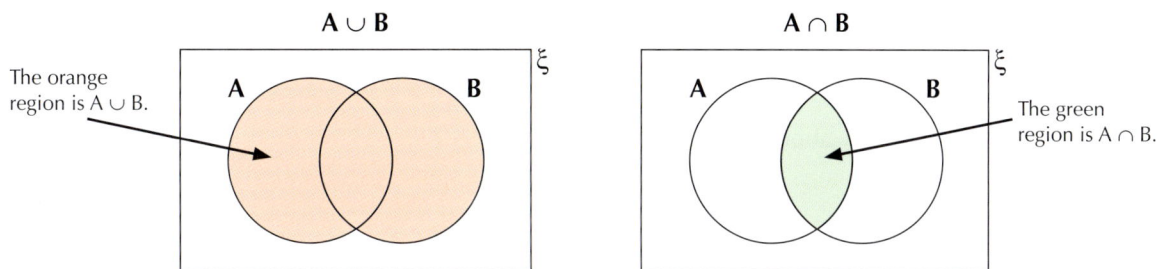

$A \cup B$

The orange region is $A \cup B$.

$A \cap B$

The green region is $A \cap B$.

For example, consider the sets $A = \{1, 2, 3\}$ and $B = \{3, 4, 5, 6\}$. $A \cup B = \{1, 2, 3, 4, 5, 6\}$ as these are all the elements in set A or set B or both, and $A \cap B = \{3\}$ as this is the only element that is in both sets.

Example 1

For the sets represented on the Venn diagram on the right, list the elements of: a) $A \cap B$ b) $A \cup B$

a) $A \cap B$ means "everything that's a member of both A and B". List the elements in the part of the diagram where the circles overlap. $A \cap B = \{0, 8, 19\}$

b) $A \cup B$ means "everything that's in A or B or both", so list everything that's in either circle. $A \cup B = \{-7, 0, 1, 5, 8, 15, 19\}$

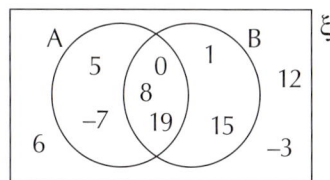

Exercise 1

Q1   For each of the following, list the elements of: (i) $A \cap B$ (ii) $A \cup B$

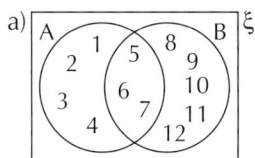

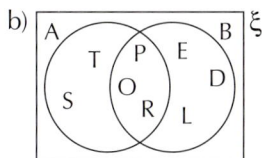

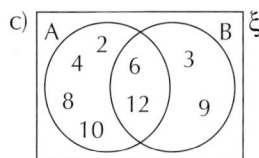

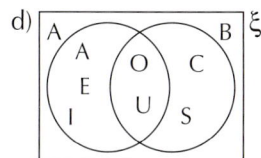

a)

b)

c)

d)

Q2   Describe each of the following:
     a) $A \cap B$ when A = {blue cars} and B = {four-wheel-drive cars}
     b) $A \cup B$ when A = {boys' names} and B = {girls' names}
     c) $A \cap B$ when A = {towns in France} and B = {seaside towns}
     d) $A \cup B$ when A = {countries in Europe} and B = {countries not in Europe}
     e) $A \cap B$ when A = {right-handed people} and B = {people with fair hair}

Q3   For each of the following sets of numbers, list the elements of: (i) $A \cap B$ (ii) $A \cup B$
     a)   A = {positive even numbers less than 30}   B = {positive integers less than 20}
     b)   A = {positive even numbers less than 20}   B = {positive odd numbers less than 20}
     c)   A = {square numbers less than 70)   B = {cube numbers less than 70}
     d)   A = {prime numbers less than 10}   B = {multiples of 3 less than 10}

Q4   Find $A \cap B$ for each of the following, where the universal set is the set of all real numbers.
     a) $A = \{x : 0 < x < 50\}$, $B = \{x : 30 < x < 100\}$ b) $A = \{x : 20 < x \le 30\}$, $B = \{x : 30 \le x < 100\}$
     c) $A = \{x : x \le 100\}$, $B = \{x : x \le 50\}$ d) $A = \{x : x < 50\}$, $B = \{x : x > 60\}$

Q5   a)   Assuming neither A nor B are empty, draw diagrams to show the following:
       (i) $A \cap B = \varnothing$ (ii) $A \cap B = A$ (iii) $A \cup B = B$
     b)   What can you say about A and B if $A \cap B = A \cup B$?

Three Sets on a Venn Diagram

You can use Venn diagrams to show **three sets** in a similar way — just add an **extra circle** to represent the third set. For example:

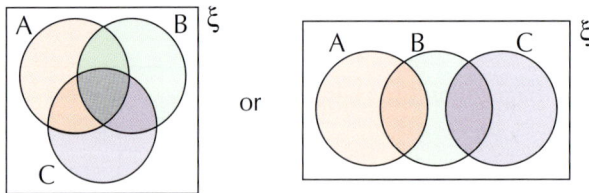

or

Tip: In the right-hand diagram, $A \cap C = \varnothing$.

You can have **unions** and **intersections** between all three sets, which might use **brackets** in the notation. To show unions and intersections, **shade** the relevant regions on **separate diagrams** first, then **compare** the diagrams. For **unions**, you want the area that's shaded on **any** diagram, and for **intersections**, you want the area that's shaded on **all** of them.

▪ $(A \cap B) \cup C$ means 'in both A and B, or in C'.

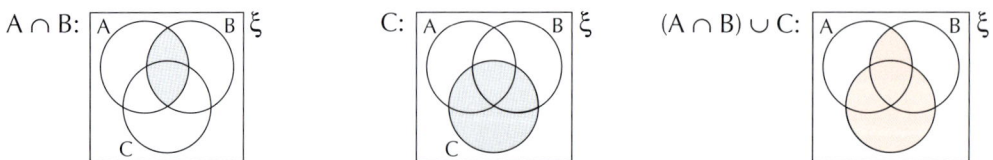

▪ $A \cap (B \cup C)$ means 'in A, and in either B or C or both'.

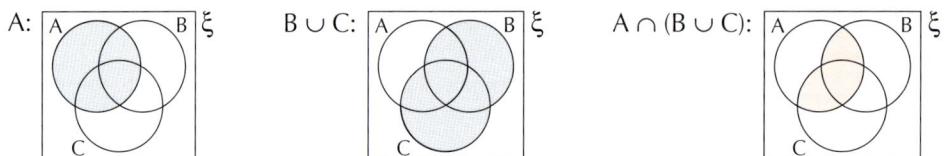

Example 2

The Venn diagram below is labelled with the number of elements in each region. Find n(A ∪ (B ∩ C)).

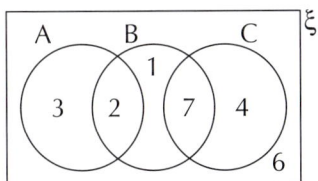

ξ
1. A ∪ (B ∩ C) means 'either in A or in both B and C'. Shade the area that represents this.

2. Find the total number of elements in the shaded regions.

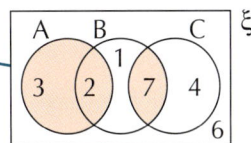

$$n(A \cup (B \cap C)) = (3 + 2) + 7 = \mathbf{12}$$

Exercise 2

Q1 For the sets A, B and C shown on the right, list the elements of:

a) A ∩ C

b) A ∩ B

c) A ∩ B ∩ C

d) B ∩ C

e) (A ∩ C) ∪ (B ∩ C)

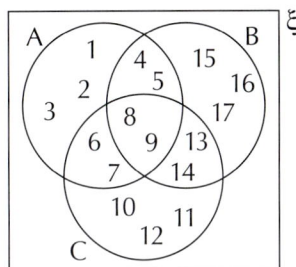

Q2 On a day in winter, members of a class arrived at school wearing hats (H), scarves (S) and gloves (G).

a) Describe what each of the shaded areas represents.

(i) 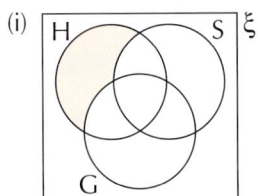 (ii) 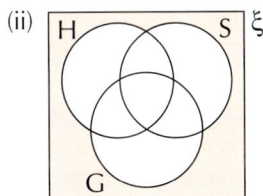 (iii)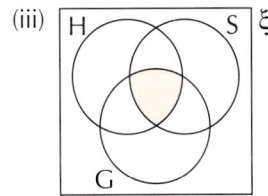

b) Draw diagrams to show the following groups of students.

(i) Those wearing scarves only.

(ii) All those wearing gloves.

(iii) Those wearing exactly two of hat, scarf and gloves.

Q3 The Venn diagram on the right is labelled with the number of elements in each region. Find:

a) n(A)

b) n(A ∩ B)

c) n(B ∪ C)

d) n((A ∪ B) ∩ C)

e) n(A ∪ (B ∩ C))

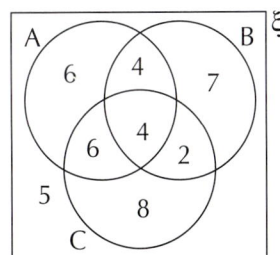

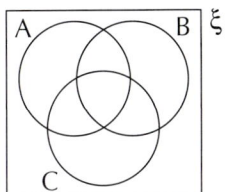

Q4 a) On separate copies of the Venn diagram on the left, shade the following regions.

(i) (A ∪ C) ∩ B

(ii) (A ∩ B) ∪ (B ∩ C)

b) What do your diagrams show about (A ∪ C) ∩ B and (A ∩ B) ∪ (B ∩ C)?

c) Use diagrams to show that A ∪ (B ∩ C) = (A ∪ B) ∩ (A ∪ C).

19.4 Complement of a Set

The complement of a set refers to everything outside the set. It's another way of writing 'the elements that are not in set A'.

> **Learning Objective — Spec Ref P6:**
> Find the complement of a set.

The **complement** of a set A, written as **A'** and read as "A complement", is all the members of the **universal set** that **aren't** in set A. For example, if the universal set is ξ = {integers between 1 and 10 inclusive} and A = {2, 4, 6, 8, 10}, then **A' = {1, 3, 5, 7, 9}**.

- On a Venn diagram, the complement of a set is **everything outside the circle** representing that set (the shaded region on this diagram).

- If you know n(ξ) and n(A), then **n(A') = n(ξ) – n(A)**.

The white region is Set A.

The green region is A'.

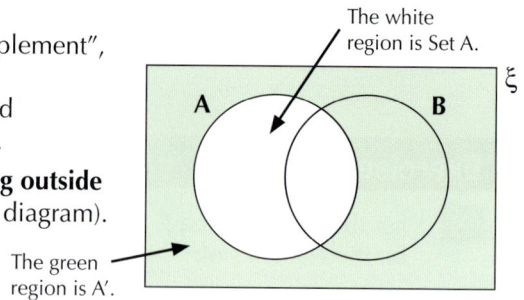

Example 1

For the sets represented on the Venn diagram on the right, list the elements of: a) A'   b) (A $\cup$ B)'

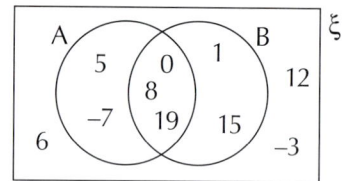

a) A' is everything that's not in A, so list everything that's outside circle A.   **A' = {–3, 1, 6, 12, 15}**

b) (A $\cup$ B)' is everything that's not in A $\cup$ B, so list everything that's outside the circles.   **(A $\cup$ B)' = {–3, 6, 12}**

Example 2

The Venn diagram on the right shows sets A and B. Given that n(A $\cap$ B)' = 15 and n(A) = 12, find: a) x b) y c) n(A' $\cup$ B')

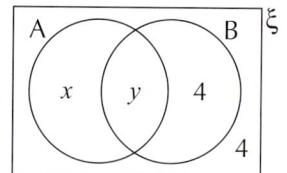

a) (A $\cap$ B)' means everything that's not in A $\cap$ B — i.e. everything that's not in the intersection.

$$n(A \cap B)' = 4 + 4 + x$$
$$15 = 8 + x$$
$$x = 15 - 8 = 7$$

b) n(A) = $x + y$, and, from part a), $x = 7$.

$$n(A) = 12 = 7 + y$$
$$y = 12 - 7 = 5$$

c) (A' $\cup$ B') means everything that's not in set A together with everything that's not in set B.

$$n(A' \cup B') = 4 + 4 + 7 = 15$$

Tip: Notice that n(A' $\cup$ B') = n(A $\cap$ B)'.

Exercise 1

Q1   For each of the following, describe or list the elements of A', the complement of set A.

   a) ξ = {polygons with fewer than 5 sides}   A = {quadrilaterals}
   b) ξ = {months of the year with fewer than 30 days}   A = {February}
   c) ξ = {factors of 18}   A = {multiples of 2}
   d) ξ = {books in a library}   A = {paperback books}
   e) ξ = {cars}   A = {cars with an automatic gearbox}

Q2   For each of the following, list the elements of B′, the complement of set B.

a) ξ = {1, 2, 3, 4, 5, 6, 7, 8, 9, 10}   B = {even numbers}

b) ξ = {prime numbers}   B = {odd numbers}

c) ξ = {$x : x$ is an even number, $0 < x \le 30$}   B = {factors of 100}

d) ξ = {factors of 120}   B = {$x : x < 20$}

Q3   A = {multiples of 3} and B = {multiples of 4}, where the universal set is ξ = {$x : x$ is an integer, $1 \le x \le 20$}.

a) List the elements of A $\cup$ B   b) List the elements of (A $\cup$ B)′

Q4   a) Shade Venn diagrams to show the following sets: (i) (A $\cap$ B)′   (ii) A $\cap$ B′   (iii) A $\cup$ B′

b) Name the sets shaded in each of the following diagrams.

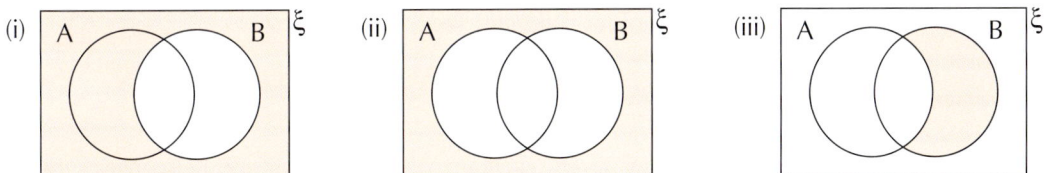

Q5   A group of people were asked to try two new biscuits. If J is the set of people who liked the Chocolate Jamborees and F is the set of people who liked the Apricot Fringits, what do the following sets represent?

a) F′   b) (J $\cup$ F)′   c) J′ $\cup$ F

Q6   For the Venn diagram shown on the right, list the members of the following sets.

a) A′ $\cap$ B

b) (A $\cup$ B $\cup$ C)′

c) (A $\cup$ B) $\cap$ C′

d) A $\cap$ B′ $\cap$ C

e) (B $\cup$ C)′

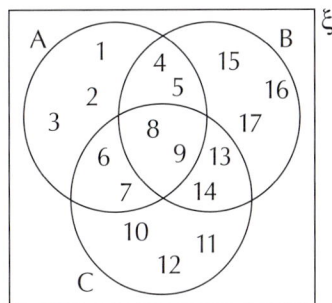

Q7   Given that n(B′) = 20, find:

a) x

b) n(A $\cup$ B′)

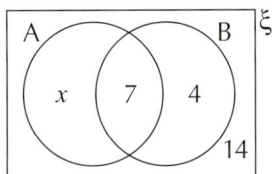

Draw Venn diagrams to help solve Questions 8-10.

Q8   Given that n(L $\cap$ M) = 12, n(L $\cup$ M)′ = 8, n(L′) = 10 and n(M′) = 14, find:

a) n(L)   b) n(ξ)   c) n(L $\cap$ M)′   d) n(L $\cup$ M′)

Q9 ξ = {$x : x$ is an integer, $11 \le x \le 24$} is the universal set for the sets A = {11, 14, 16, 20, 21, 24}, B = {11, 13, 16, 19, 22, 24} and C = {13, 14, 15, 16, 21, 22, 23, 24}. List the elements of each of the following:

a) A′ $\cap$ B   b) A $\cap$ B′ $\cap$ C′   c) B $\cap$ (A′ $\cup$ C′)   d) (A $\cap$ B′) $\cup$ C

Q10   If A = {multiples of 2}, B = {multiples of 5} and C = {multiples of 3}, where ξ = {$x : x$ is an integer, $1 \le x \le 30$}, list the elements of:

a) (A $\cap$ C) $\cup$ B′   b) A′ $\cap$ B′ $\cap$ C   c) (A′ $\cap$ C′) $\cup$ B

Review Exercise

Q1 For the sets shown in the diagram:

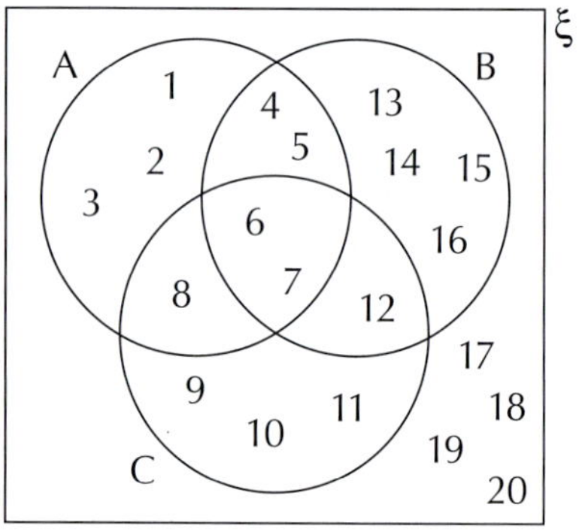

a) State whether each of the following is true or false. For those that are false, change the right hand side of the expression to make a correct statement.

    (i) $n(A) = 7$ (ii) $n(A \cap B) = 4$

    (iii) $20 \in B'$ (iv) $C \cap B' = \{9, 10, 11\}$

b) List the members of each of the following sets.

    (i) $A \cap C$ (ii) $A \cap B \cap C$ (iii) $(B \cap C)$

    (iv) $A' \cap C$ (v) $(A \cup B)'$ (vi) $(A \cap C) \cup (B \cap C)$

Q2 For each of the following, draw a Venn diagram to show the sets and help answer the questions.

a) ξ = {positive integers less than 20} T = {multiples of 2} F = {multiples of 5}

    Find: (i) $n(T \cap F)$ (ii) $n(T \cup F)'$

b) ξ = {positive integers less than 20} T = {multiples of 2} S = {square numbers}

    Find: (i) $T \cap S$ (ii) $n(T')$

Q3 Name the sets corresponding to the shaded area in each of the following diagrams.

a) 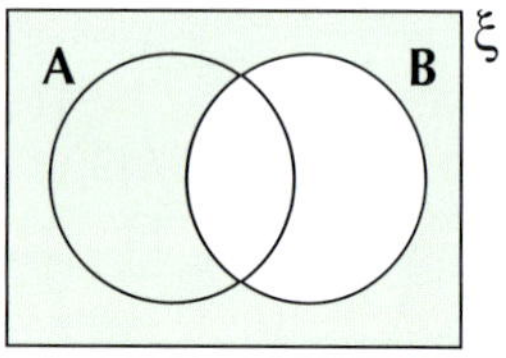   b) 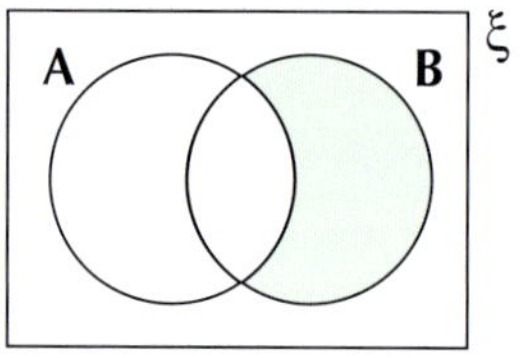   c)

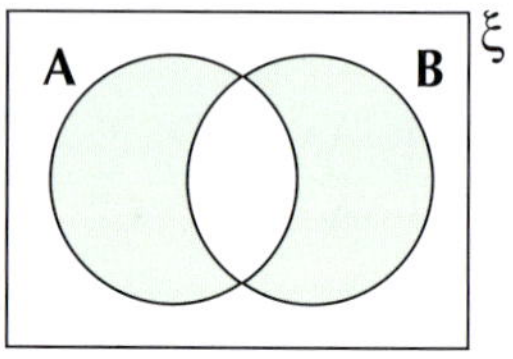

Q4 50 families were asked about the children they had. 30 of those asked said they had girls in the family while 38 said they had boys. 23 families had both boys and girls.

a) Draw a Venn diagram to represent this.

b) How many families had no children?

Q5 On separate copies of the Venn diagram on the right, shade the areas corresponding to the following sets:

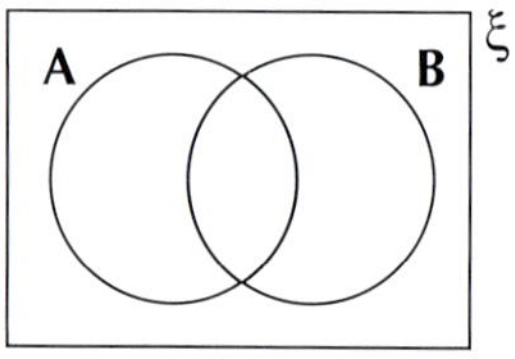

a) $A' \cup B$

b) $(A \cup B)'$

c) $A' \cap B'$

Q6 A group of 100 people were asked whether they had in their pockets any of three items. 72 had all three of keys, crayons and a magic ring. Including those with all three items, 74 had both keys and crayons, 80 had both keys and a magic ring and 78 had both crayons and a magic ring. Of those with exactly one item, 3 had just keys, 5 just crayons and 3 just a magic ring.

a) Show the results of this survey on a Venn diagram.

b) Of those asked, how many had: (i) none of the items,   (ii) exactly two of the items?

Exam-Style Questions

Q1 In a class of 36 students, 28 are taking GCSE French and 18 are taking GCSE Spanish. 13 take both subjects.

a) Represent this information on a Venn diagram.

[1 mark]

b) How many in the class take neither subject?

[1 mark]

Q2 A company produces 120 different products. 32 of the products are made using grommets but not widgets. 20 of the products are made using widgets but not grommets. 46 of the products require neither widgets nor grommets. How many products are made using widgets?

[2 marks]

Q3 100 members of a health spa were asked whether they used the gym, pool or sauna. The replies were as follows:

A total of 52 used the gym, 30 used the pool and 65 used the sauna. 17 used the gym and the pool, 18 used the pool and the sauna, 30 used the gym and sauna. 15 used all three.

a) Draw a Venn diagram showing this information.

[2 marks]

b) Find how many people: (i) don't use the gym, the pool or the sauna,

[1 mark]

(ii) use only one of the gym, pool and sauna,

[1 mark]

(iii) use at least two out of the gym, pool and sauna.

[1 mark]

Q4 A and B are sets such that $n(\xi) = 20$, $n(A' \cap B') = 2$, $n(A') = 5$, $n(A \cap B) = x$ and $n(A \cap B') = y$. Use a Venn diagram to prove that $y = 15 - x$.

[3 marks]

Q5 If ξ = {positive integers less than or equal to 20}, X = {$x : 10 < x < 20$}, Y = {prime numbers}, Z = {multiples of 3}, find:

a) $X \cap Y$

[1 mark]

b) $(X' \cap Y) \cup Z$

[2 marks]

20.1 Angles on Lines and Around Points

When you have a cluster of angles along a straight line or around a single point, you can often find out the size of any unknown angles by using a simple fact: they always add up to a particular number.

Learning Objectives — Spec Ref G3:
- Find angles that lie on a straight line.
- Find angles around a point.

Prior Knowledge Check:
Be able to write and solve equations — see p.113-117.

Angles on a Straight Line

A group of angles on a **straight line** always **add up to 180°**:

$$a + b + c = 180°$$

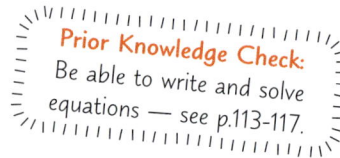

Tip: Angles aren't usually drawn accurately, so you need to know how to work out the size of them geometrically — you can't just measure them with a protractor.

The equation might have a different number of **variables** in it if there are a different number of angles on the line. To find missing angles on a straight line, use your **equation** and **rearrange** it so that the angle you want to find is the **subject**.

Example 1

Find the size of x in the diagram on the right.

1. The angles lie on a straight line and so must add up to 180°.
   $$10° + 115° + 5x = 180°$$
2. Rearrange the equation and solve it for x.
   $$5x = 180° - 10° - 115° = 55°$$
   $$x = 55° \div 5 = \textbf{11°}$$

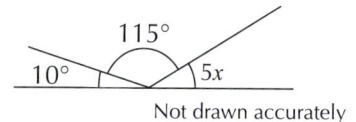

Not drawn accurately

Exercise 1

Q1 Find the value of each letter in the diagrams below. None of the angles are drawn accurately.

a)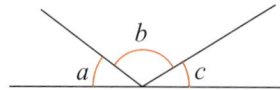
123° 29° a

b) 28° 35° 55° b

c) 57° c 57° c

Q2 Find the value of each letter in the diagrams below. None of the angles are drawn accurately.

a)
$3d$ d

b)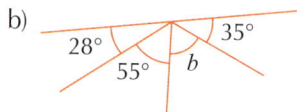
$e + 50°$ e 30°

c)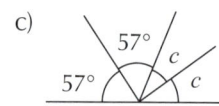
$f + 27°$ f $f - 12°$

Q3 Explain why the line *AOB* on the right cannot be a straight line.

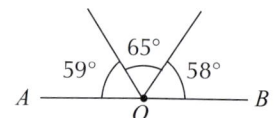
65° 59° 58° A O B

Angles Around a Point

A set of angles around a **point** always **add up to 360°**:

$$a + b + c + d = 360°$$

The equation might have a different number of **variables** in it if there are a different number of angles around the point. Use the **equation** to find any missing angles just like on the previous page.

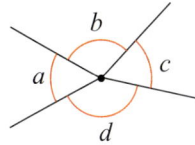

Tip: You could draw a circle through all the angles round a point, and you know there are 360° in a circle.

Example 2

Find the value of y in the diagram on the right.

1. The angles are around a point so they must add up to 360°.

$$y + 2y + 80° + 40° + 90° = 360°$$

2. Rearrange the equation and solve it for y.

$$3y = 360° - 80° - 40° - 90°$$
$$= 150°$$
$$y = 50°$$

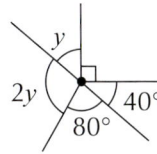

Not drawn accurately

Tip: The little square in the top-right of the diagram means that the angle is a right angle, i.e. it's 90°.

Exercise 2

Q1 Find the value of each letter in the diagrams below. None of the angles are drawn accurately.

a)
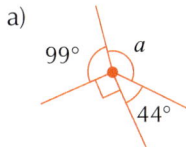
99° a 44°

b)
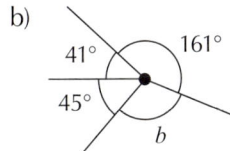
41°   161°   45° b

c)
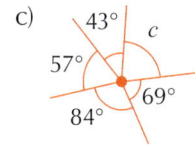
43° c 57°   69°   84°

d)
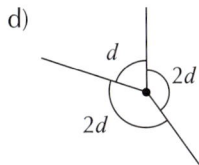
d $2d$ $2d$

e)
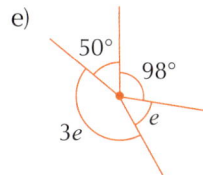
50°   98° $3e$ e

f)
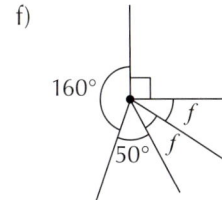
160° f 50° f

g)
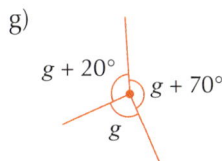
$g + 20°$ $g + 70°$ g

h)
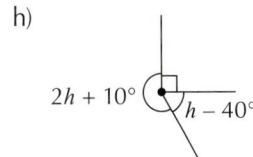
$2h + 10°$ $h - 40°$

i)
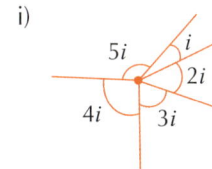
$5i$ i $2i$ $4i$ $3i$

Q2

a) All of the angles in the diagram on the right are larger than 1°. Bernie claims that $m < 307°$. Show that Bernie is right.

b) Given that $x = 40°$, find the value of m.

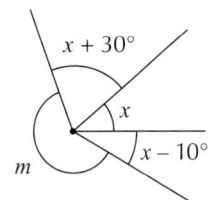
$x + 30°$ x $x - 10°$ m

20.2 Parallel Lines

When a line crosses a pair of parallel lines, it produces special types of angles.

Vertically Opposite and Alternate Angles

Learning Objective — Spec Ref G3:
Find vertically opposite angles and alternate angles.

When **two lines intersect**, it produces **two pairs** of angles. The angles opposite one another are known as **vertically opposite angles**, and vertically opposite angles are always **equal**. In the diagram on the right, a and a are vertically opposite, as are b and b — so $a = a$ and $b = b$. Also, because the two **distinct** angles (i.e. a and b) form a straight line, they add up to **180°** — so $a + b = 180°$.

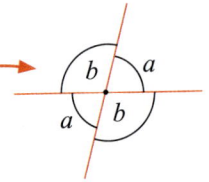

$a + b = 180°$

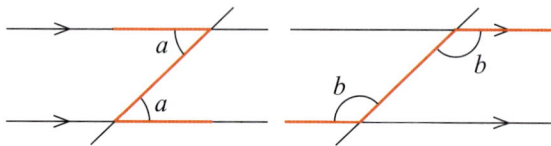

When a straight line crosses two **parallel** lines, it forms two pairs of **alternate angles** (in a sort of **Z-shape**). Alternate angles are always **equal**.

Example 1

Find the values of a, b, c and d in the diagram below.

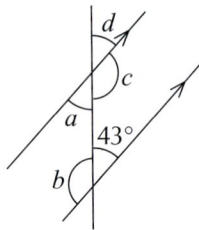

1. a and 43° are alternate angles, so they are equal.

   $a = 43°$

2. b and the angle marked 43° lie on a straight line, so they add up to 180°.

   $43° + b = 180°$
   $b = 180° - 43° = \mathbf{137°}$

3. c and b are alternate angles, so they are equal.

   $c = b = \mathbf{137°}$

4. a and d are vertically opposite angles, so they are equal.

   $d = a = \mathbf{43°}$

Tip: The arrows on a pair of lines indicate those lines are parallel.

Exercise 1

Find the value of each letter in the diagrams below.

Q1   a)
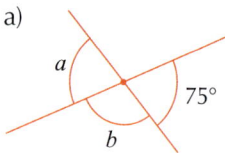

b)
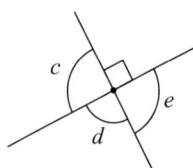

c)
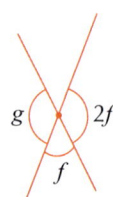

Q2   a)
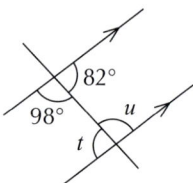

b)
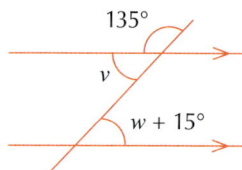

c)
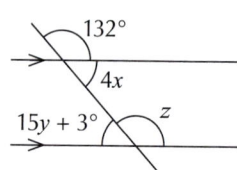

Corresponding and Allied Angles

Learning Objective — Spec Ref G3:
Find corresponding and allied angles.

Corresponding angles form an **F-shape**. Corresponding angles are always **equal**.

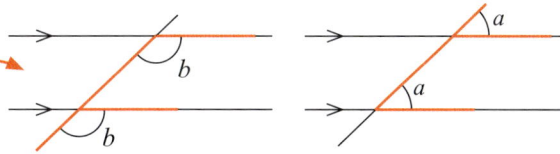

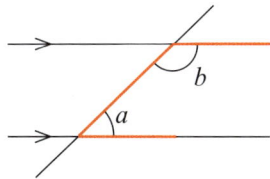

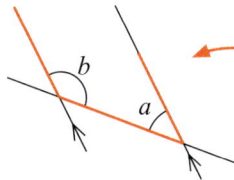

Allied angles form a **C-** or **U-shape**. They always **add up to 180°**.

$$a + b = 180°$$

Tip: Always state which rules you're using when solving geometry problems — e.g. say "because these are allied angles" or "as the angles all lie on a straight line". Make sure you use the proper terms — don't describe them as "angles in a Z-shape".

Example 2

Find the values of *a*, *b* and *c* shown in the diagram below.

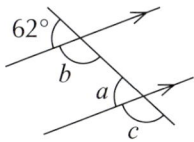

1. *a* and 62° are corresponding angles, so they are equal. $a = 62°$

2. *b* and the angle marked 62° lie on a straight line, so they add up to 180°. (Or you could say *a* and *b* are allied angles, so they add up to 180°.)

   $62° + b = 180°$
   $b = 180° - 62° = 118°$

3. *c* and *b* are corresponding angles, so they're equal. $c = b = 118°$

Exercise 2

Q1 Find the value of each letter in the diagrams below.

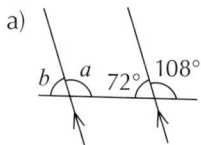

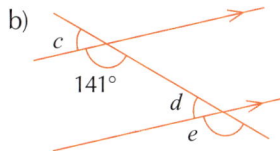

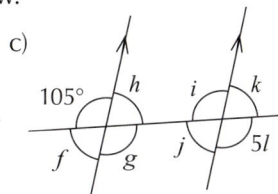

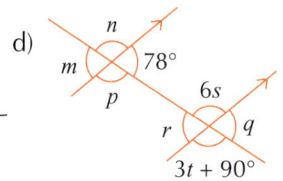

a) b a 72° 108°

b) c 141° d e

c)   105° h i k f g j $5l$

d) n m 78° p $6s$ r q $3t + 90°$

Q2 Two wooden posts stand vertically on sloped ground. The first post makes an angle of 99° with the downward slope, as shown. Find the angle that the second post makes with the upward slope, labelled *y* on the diagram.   99° y

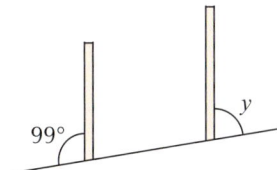

Q3 Decide whether the pairs of orange lines below are parallel. Give reasons for your answers.

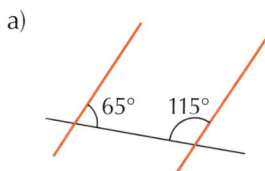

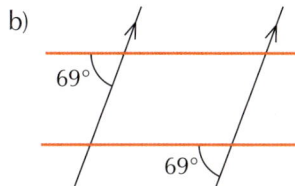

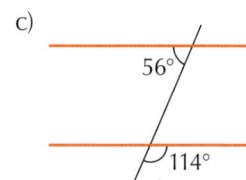

a)   65° 115°

b)   69° 69°

c)   56° 114°

20.3 Triangles

Triangles are perhaps the simplest of the 2D shapes because they have so few sides — three. But they still come in plenty of different forms depending on the lengths of their sides and the sizes of their angles.

Learning Objectives — Spec Ref G3/G4:
- Know the properties of different types of triangles.
- Know and be able to prove that the angles in a triangle sum to 180°.
- Be able to find missing angles in triangles.

There are **different** types of triangles that you need to be familiar with. Make sure you know the defining features of each type.

An **equilateral** triangle has 3 equal sides and 3 equal angles (each of 60°).

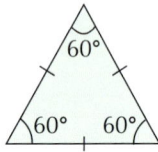

An **isosceles** triangle has 2 equal sides and 2 equal angles.

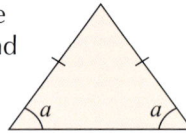

Tip: The little dashes on the sides of a shape mean those sides are of equal length.

The sides and angles of a **scalene** triangle are all different.

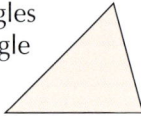

A **right-angled** triangle has 1 right angle (90°).

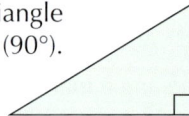

Tip: An isosceles right-angled triangle has one 90° angle and two 45° angles.

You might occasionally see triangles described by the size of their angles:

An **acute-angled** triangle has 3 acute angles (less than 90°).

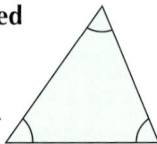

An **obtuse-angled** triangle has 1 obtuse angle (between 90° and 180°).

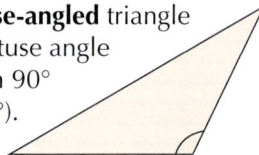

For any triangle, the angles inside (a, b and c) **add up to 180°**. You can use this to set up an **equation** that you can **solve** to find missing angles.

$$a + b + c = 180°$$

To prove this rule, draw parallel lines at the top and base of the triangle, as shown on the right. Then use the fact that **alternate** angles are equal from page 250.

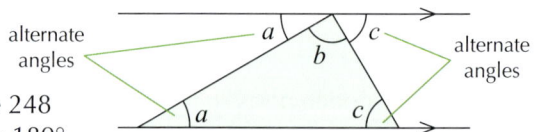

Now, a, b and c lie on a straight line and you saw on page 248 that this means they must add up to 180° — so $a + b + c = 180°$.

Example 1

a) Find the value of x in the triangle below.

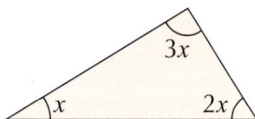

1. The angles in the triangle must add up to 180°, so set up an equation in x.

2. Solve your equation to find x.

$$x + 2x + 3x = 180°$$
$$6x = 180°$$
$$x = 180° \div 6 = \mathbf{30°}$$

b) **Find the value of y in the isosceles triangle below.**

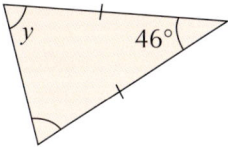

1. The triangle is isosceles so the unmarked angle must be equal to y.

2. All three angles must sum to 180°, so form an equation in y.

3. Solve your equation to find y.

$y + y + 46° = 180°$
$2y = 134°$
$y = 134° \div 2 = \mathbf{67°}$

Exercise 1

In Questions 1-4, the angles aren't drawn accurately, so don't try to measure them.

Q1 Find the missing angles marked with letters.

a)
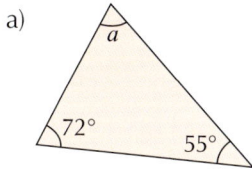

b)
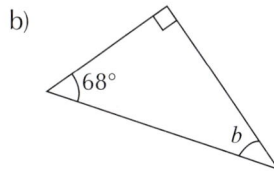

c)
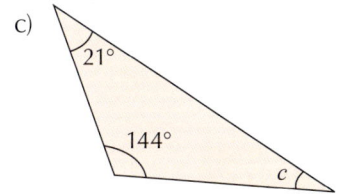

d)
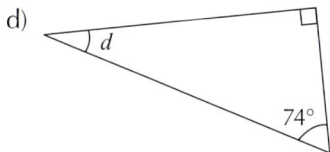

e)
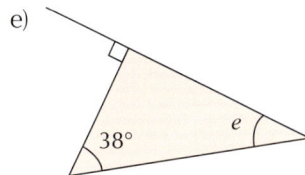

f)
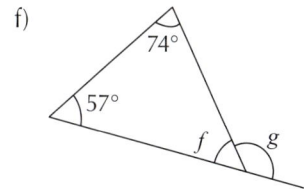

Q2 Find the values of the letters shown in the following diagrams.

a)
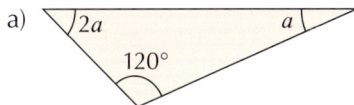

b)
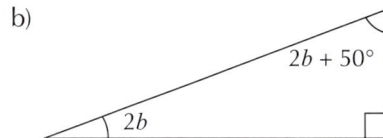

c)
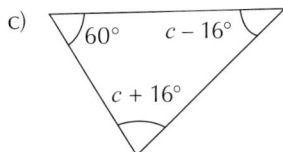

d)
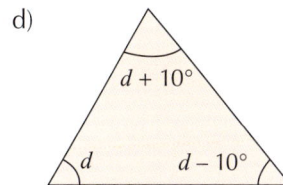

Q3 Find the values of the letters shown in these isosceles triangles.

a)
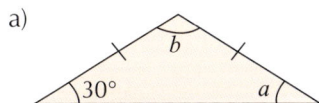

b)
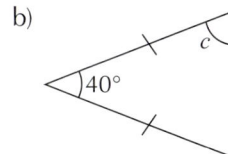

c)
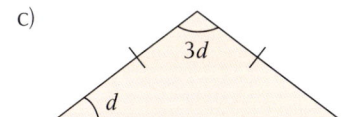

Q4 a) (i) Find the value of x.
      (ii) Find the value of y.

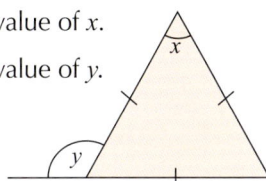

b) (i) Find the value of p.
   (ii) Find the value of q.

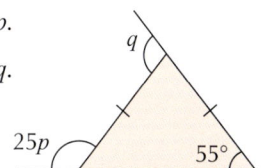

20.4 Quadrilaterals

Quadrilaterals are shapes that have four sides. You're probably familiar with squares and rectangles, two of the simplest quadrilaterals, but the next few pages will introduce you to a whole host of other types.

Quadrilaterals

Learning Objectives — Spec Ref G3:
- Know that the angles in a quadrilateral sum to 360°.
- Be able to find missing angles in quadrilaterals.

The angles in a quadrilateral always **add up to 360°:**

$$a + b + c + d = 360°$$

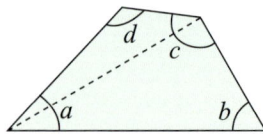

To prove this, **split** the quadrilateral into **two triangles** as shown by the dashed line on the left. Each triangle has angles that add up to **180°** and so the angles in **both triangles**, which is the same as the angles in the quadrilateral, add up to 180° + 180° = **360°**.

Example 1

Find the missing angle x in the quadrilateral below.

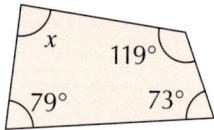

1. The angles in a quadrilateral add up to 360°. Use this to write an equation involving x.

$$79° + 73° + 119° + x = 360°$$
$$271° + x = 360°$$

2. Then solve your equation to find the value of x.

$$x = 360° - 271° = \textbf{89°}$$

Exercise 1

Q1 Find the values of the letters in the following quadrilaterals. They're not drawn accurately, so don't try to measure them.

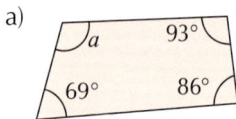

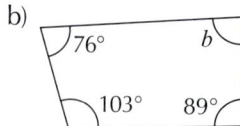

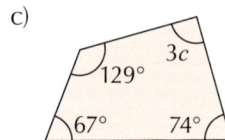

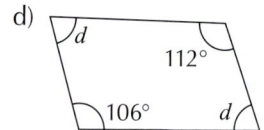

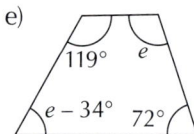

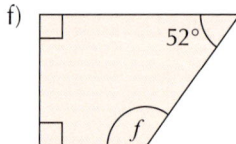

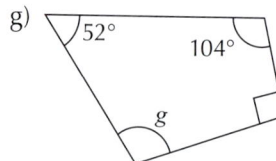

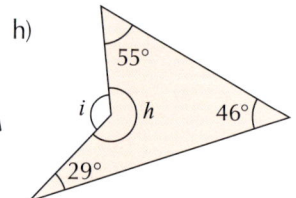

a) a, 93°, 69°, 86°

b) 76°, b, 103°, 89°

c) $3c$, 129°, 67°, 74°

d) d, 112°, 106°, d

e) 119°, e, $e − 34°$, 72°

f) 52°, f

g) 52°, 104°, g

h) 55°, i, h, 46°, 29°

Q2 Find the values of the letters in the quadrilaterals below.

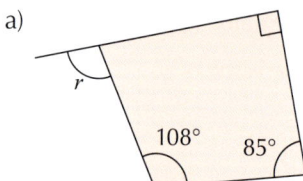

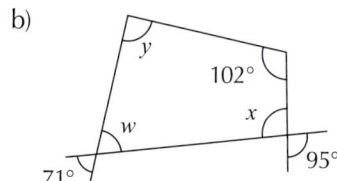

a) r, 108°, 85°

b) y, 102°, w, x, 71°, 95°

Squares, Rectangles, Parallelograms and Rhombuses

Learning Objective — Spec Ref G4:
Know the properties of squares, rectangles, parallelograms and rhombuses.

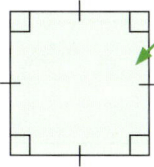

A **square** is a quadrilateral with 4 equal sides and 4 angles of 90°.

A **rectangle** is a quadrilateral with 4 angles of 90° and opposite sides of the same length.

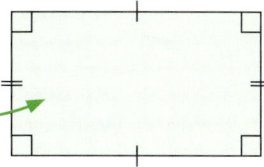

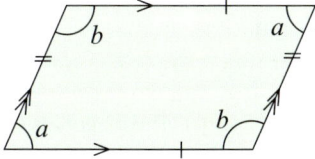

A **parallelogram** is a quadrilateral with 2 pairs of equal, parallel sides.

A **rhombus** is a parallelogram where all the sides are the same length. The **diagonals** of a rhombus **bisect** the angles (i.e. cut them in half) and cross at a **right angle**.

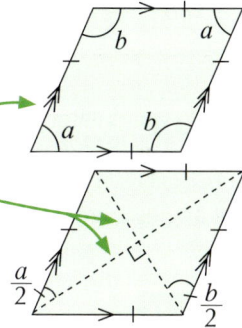

Opposite angles in parallelograms and rhombuses are **equal** and **neighbouring angles** always add up to **180°**: $a + b = 180°$
This is because the **parallel lines** that make up the sides of these shapes mean a and b are **allied angles** (page 251).

Example 2

Find the size of the angles marked with letters in the rhombus below.

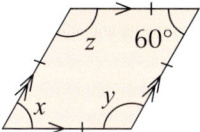

1. Opposite angles in a rhombus are equal. $x = 60°$

2. Neighbouring angles in a rhombus add up to 180°. Use this fact to find angle y.

   $60° + y = 180°$
   $y = 180° - 60° = 120°$

3. Opposite angles in a rhombus are equal, so z is the same size as y. $z = 120°$

Exercise 2

Q1 Calculate the values of the letters in these quadrilaterals.

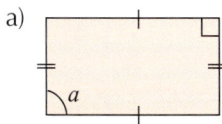

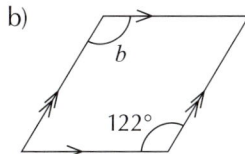

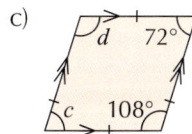

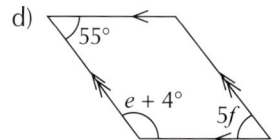

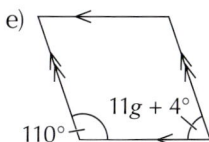

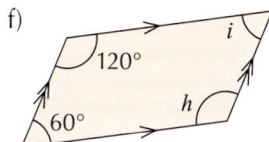

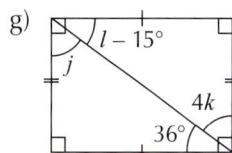

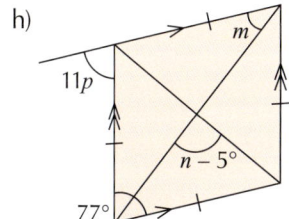

a) a

b) b, $122°$

c) d $72°$, c $108°$

d) $55°$, $e + 4°$, $5f$

e) $11g + 4°$, $110°$

f) $120°$, $60°$, i, h

g) $l - 15°$, j, $4k$, $36°$

h) m, $11p$, $n - 5°$, $77°$

Trapeziums and Kites

Learning Objective — Spec Ref G4:
Know the properties of trapeziums and kites.

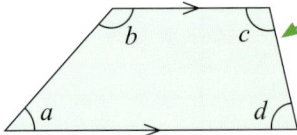

A **trapezium** is a quadrilateral with 1 pair of parallel sides.

An **isosceles trapezium** is a trapezium with 2 pairs of equal angles and 2 sides of the same length.

Because of the **allied angles** created by the parallel sides, pairs of angles **add up to 180°** as shown.

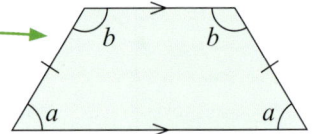

$a + b = 180°$ $c + d = 180°$

$a + b = 180°$

A **kite** is a quadrilateral with 2 pairs of equal sides and 1 pair of equal angles in opposite corners as shown on the diagram (the other pair of angles aren't generally equal).

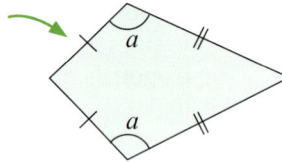

Tip: The diagonals of a kite always cross at a right angle.

Example 3

Find the size of the angles marked with letters in the isosceles trapezium below.

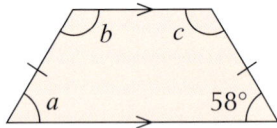

1. This is an isosceles trapezium, so a is the same size as the 58° angle. $a = 58°$

2. Angle c and the angle of 58° must add up to 180°.
   $c + 58° = 180°$
   $c = 180° - 58° = $ **122°**

3. It's an isosceles trapezium, so b is the same size as c.
   $b = $ **122°**

Exercise 3

Q1 Find the value of each letter in the trapeziums below.

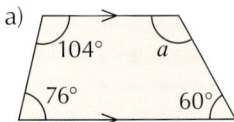

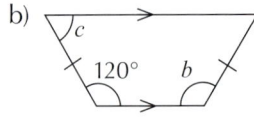

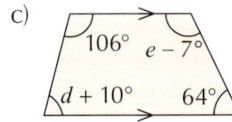

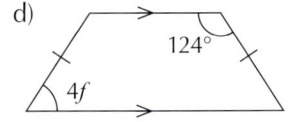

a) 104° a 76° 60°

b) c 120° b

c) 106° $e - 7°$ $d + 10°$ 64°

d) 124° $4f$

Q2 Find the value of each letter in the kites below.

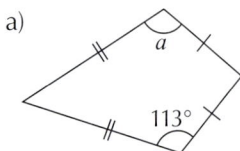

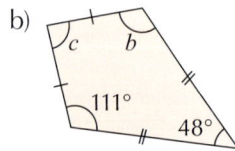

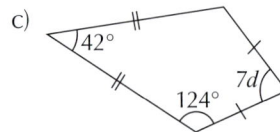

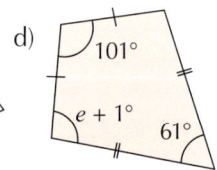

a) a 113°

b) c b 111° 48°

c) 42° 124° $7d$

d) 101° $e + 1°$ 61°

Q3 An isosceles trapezium has two angles of 53°. Find the size of the other two angles.

Q4 A kite has exactly one angle of 50° and exactly one angle of 90°. Find the size of the other two angles.

20.5 Polygons

You've met lots of polygons before — they're just 2D shapes with straight sides. Use the formulas on these pages to work out the number of sides and the size of the angles in polygon problems.

Interior Angles

Learning Objectives — Spec Ref G1/G3:
- Know the names of different types of polygons.
- Find the sum of interior angles in polygons.

A **polygon** is a 2D shape whose sides are all **straight** — the triangles and quadrilaterals on the previous pages were three- and four-sided polygons. The box on the right shows the names of some other polygons — their names depend on the **number of sides** they have.

pentagon = **5** sides	**oct**agon = **8** sides
hexagon = **6** sides	**non**agon = **9** sides
heptagon = **7** sides	**dec**agon = **10** sides

A **regular polygon** is one where the **sides** are all the **same length** and the **angles** are all **equal**. An **equilateral triangle** is a regular triangle and a **square** is a regular quadrilateral.

Interior Angles

The **interior angles** of a polygon are the angles inside each vertex (corner). The interior angles of a regular pentagon are shown on the left.

The **sum** of a polygon's interior angles (S) and its **number of sides** (n) are related by this formula: $S = (n - 2) \times 180°$

Tip: You can prove the formula by splitting the polygon into $n - 2$ triangles (see example below). The case $n = 4$ was done on p.254.

By **rearranging** this formula, you can use it to find the number of sides (n) of a given polygon or the size of missing angles.

Example 1

a) A pentagon has four interior angles of 100°. Find the size of the fifth angle.

1. A pentagon has 5 sides, so substitute $n = 5$ into the formula for the sum of interior angles.

$$S = (n - 2) \times 180°$$
$$= (5 - 2) \times 180° = 540°$$

2. Write an equation for the size of the missing angle, x, and solve it to find x.

$$100° + 100° + 100° + 100° + x = 540°$$
$$400° + x = 540°$$
$$x = 540° - 400° = \mathbf{140°}$$

b) Calculate the size of an interior angle of a regular pentagon.

All the angles will be the same size if the pentagon is regular. So divide the sum of the angles by the number of angles.

The sum of the angles is 540° (from part a)).

So one angle will be 540° ÷ 5 = **108°**.

c) By splitting a pentagon into triangles, prove the formula for the sum of its interior angles.

1. Split the pentagon into 5 − 2 = 3 triangles.

2. Use the fact that the sum of the angles in a triangle is 180°.

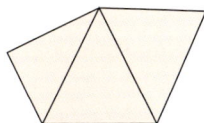

The angles in each triangle add up to 180°. There are three triangles so the angles in the whole shape add up to 180° + 180° + 180° = **540°**.

Q1 For each of the shapes below, determine whether or not it is a polygon and, if so, state its name.

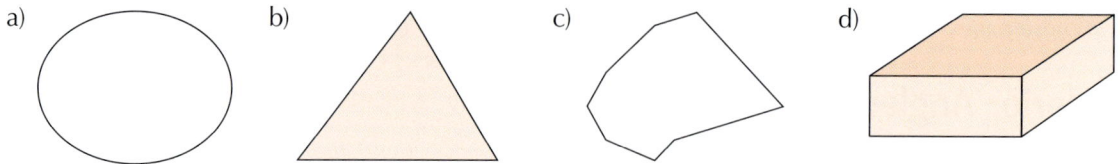

a)   b)   c)   d)

Q2 Find the sum of the interior angles of a polygon with:

a) 6 sides   b) 10 sides   c) 12 sides   d) 20 sides

Q3 For each of the following shapes: (i) Find the sum of the interior angles.

                                           (ii) Find the size of the missing angle.

a)   b)   c)

a) $112°$, $41°$, a, $89°$

b) $101°$, $107°$, b, $85°$

c) $93°$, c, $104°$, $159°$, $121°$, $102°$, $230°$, $150°$, $91°$

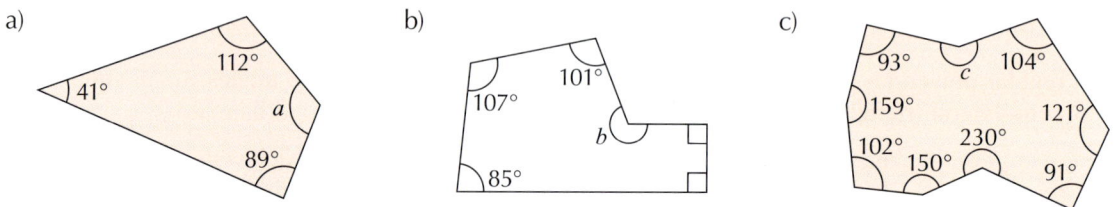

Q4 Find the size of each of the interior angles in the following shapes.

a) Regular octagon   b) Regular nonagon   c) Regular decagon

Q5 a) Seven angles of an octagon are $130°$. Find the size of the eighth angle.

     b) Is this a regular octagon? Give a reason for your answer.

Q6 a) The interior angles of a shape add up to $2520°$. Find the number of sides of this shape.

     b) Half of the angles in this shape are of size $95°$ and the other half are of size x. Find the value of x.

Q7 Find the number of sides of a regular polygon with interior angles of: a) $60°$ b) $150°$

Q8 Find the values of x, y and z in the polygons below.

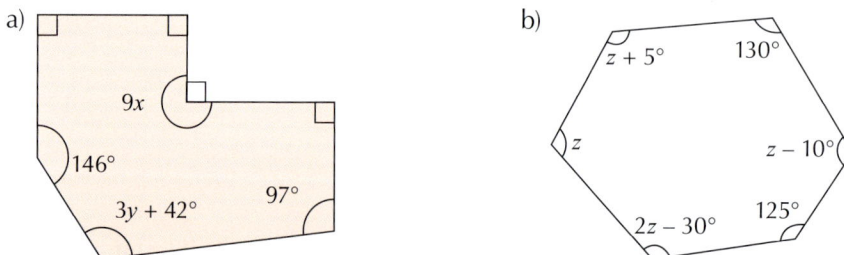

a)   b)

a) $9x$, $146°$, $3y + 42°$, $97°$

b) $z + 5°$, $130°$, z, $z - 10°$, $2z - 30°$, $125°$

Q9 Draw a decagon. By dividing it into triangles, show that the sum of the interior angles of the decagon is $1440°$.

Exterior Angles

Learning Objective — Spec Ref G3:
Find interior and exterior angles of regular and irregular polygons.

An **exterior angle** of a polygon is an angle between a **side**
and a **line** that extends out from one of the **neighbouring sides**.
For example, the exterior angles of a regular pentagon are marked on the right.

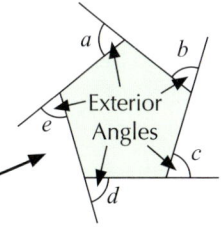

For any polygon, the exterior angles always
add up to 360°. In the case of the pentagon, this is: $a + b + c + d + e = 360°$

Since the exterior angle and the neighbouring interior angle lie on a
straight line, they must **add up to 180°** (see page 248). So you
can use this formula to find an interior angle given the exterior angle:

$$\text{Interior angle} = 180° - \text{Exterior angle}$$

For a **regular** polygon, the interior
angles are all equal, and this means the
exterior angles are all equal too. Then
the formula for the size of an exterior
angle for a regular n-sided polygon is: $\text{Exterior angle} = \dfrac{360°}{n}$

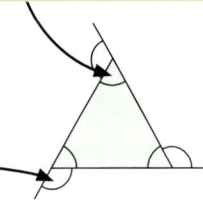

Example 2

Find the size of each of the exterior angles of a regular hexagon.

A hexagon has 6 sides.
The hexagon is regular so put $n = 6$
into the exterior angle formula.

$360° \div 6 = 60°$

So each exterior angle is **60°**.

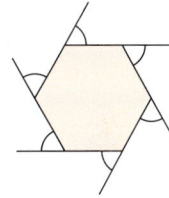

Example 3

A regular polygon has exterior angles of 30°.
How many sides does the polygon have?

1. It's a regular polygon so put 30°
    into the exterior angle formula.

    $30° = \dfrac{360°}{n}$

2. Solve the equation for n.

    $n = 360° \div 30° = 12$
    So the regular polygon has **12 sides**.

Tip: A 12-sided
polygon is called a
dodecagon.

Example 4

Prove that the exterior angle of a triangle is equal to the sum of the two non-adjacent interior angles.

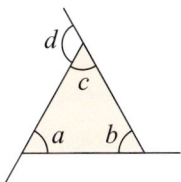

1. You need to show that $a + b = d$.

2. The exterior angle and interior
    angle add up to 180°.

3. The angles in the triangle
    also add up to 180°.

4. Rearrange the equation to get the result.

$d + c = 180° \implies c = 180° - d$

$a + b + c = 180°$

$a + b + (180° - d) = 180°$

$a + b = d$

Q1 Find the size of each of the exterior angles of the following polygons.

a) regular octagon   b) regular nonagon   c) regular heptagon

Q2 Find the size of the angles marked by letters in these diagrams.

a)

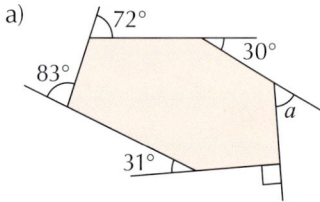

b)

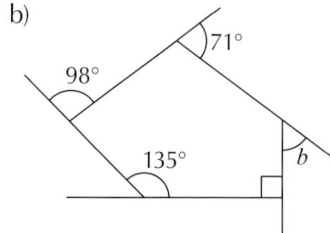

c)

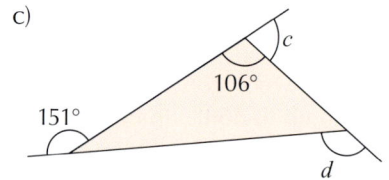

Q3 Find the size of the unknown exterior angle in a shape whose other exterior angles are:

a) 100°, 68°, 84° and 55°

b) 30°, 68°, 45°, 52°, 75° and 50°

c) 42°, 51°, 60°, 49°, 88° and 35°

d) 19°, 36°, 28°, 57°, 101°, 57° and 22°

Q4 A regular polygon has exterior angles of 45°.

a) How many sides does the polygon have? What is the name of this polygon?

b) Sketch the polygon.

c) What is the size of each of the polygon's interior angles?

d) What is the sum of the polygon's interior angles?

Q5 The exterior angles of some regular polygons are given below. For each exterior angle, find:

(i) the number of sides the polygon has,

(ii) the size of each of the polygon's interior angles,

(iii) the sum of the polygon's interior angles.

a) 40°   b) 120°   c) 3°   d) 4.8°

Q6 Find the values of the letters in the following diagrams.

a)

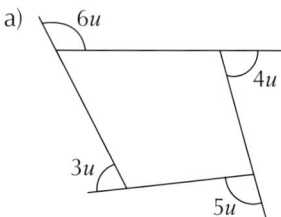

b)

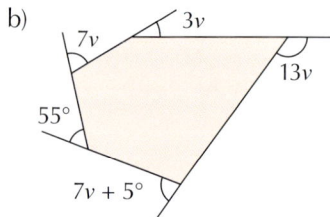

c)

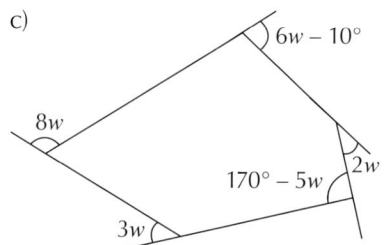

d)

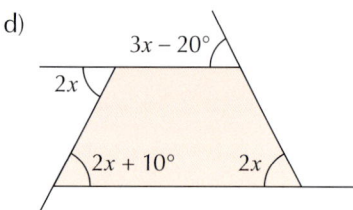

e)

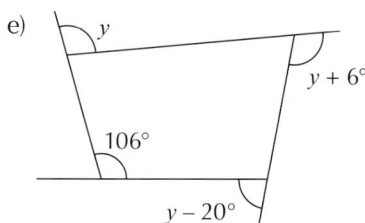

f)

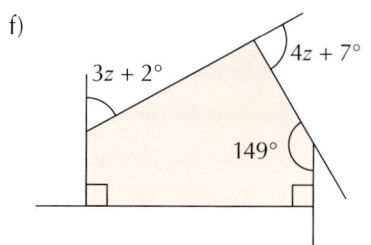

20.6 Symmetry

Symmetry is when a shape can be reflected or rotated and still look exactly the same afterwards. For instance, if you rotate a square by 90° then each of the vertices is in a different place but the square looks unchanged.

Learning Objective — Spec Ref G1:
Recognise lines of symmetry and rotational symmetry in 2D shapes.

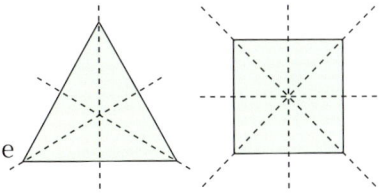

A **line of symmetry** on a shape is a mirror line. Each side of the line of symmetry is a **reflection** of the other. For example, an equilateral triangle has three lines of symmetry and a square has four (shown on the right).

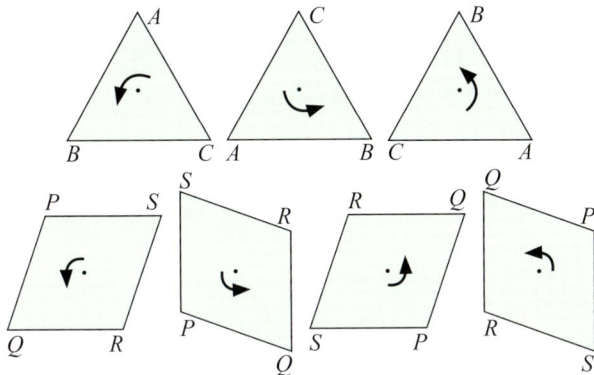

The **order of rotational symmetry** of a shape is the number of positions you can **rotate** (turn) the shape into so that it looks **exactly the same**. For example, an equilateral triangle has rotational symmetry of order 3 and a rhombus has order 2 (shown left). If a shape has **no rotational symmetry** then the order of rotational symmetry is **1**.

A **regular n-sided polygon** has n lines of symmetry and **rotational symmetry of order n**.

Different **quadrilaterals** have different symmetries. A **square** is regular so it has **4** lines of symmetry and rotational symmetry of order **4**. **Rectangles** and **rhombuses** have **2** lines of symmetry and rotational symmetry of order **2**, **kites** and **isosceles trapeziums** have **1** line of symmetry and rotational symmetry of order **1**, and **parallelograms** have **0** lines of symmetry and rotational symmetry of order **2**.

See pages 377 and 380 for how to perform reflections and rotations on a coordinate grid.

Exercise 1

Q1 For each of the shapes below, state: (i) the number of lines of symmetry,
(ii) the order of rotational symmetry.

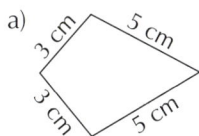

a)

b)

c)

d)

e) a rhombus

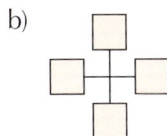

f) an isosceles trapezium

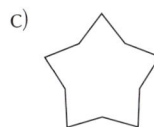

g) a regular pentagon

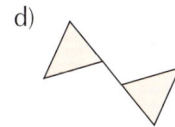

h) a regular decagon

Q2 a) Copy the diagram below, then shade two more squares to make a pattern with no lines of symmetry and rotational symmetry of order 2.

b) Copy the diagram below, then shade four more squares to make a pattern with 4 lines of symmetry and rotational symmetry of order 4.

Q1 Find the value of each letter. The diagrams aren't drawn accurately.

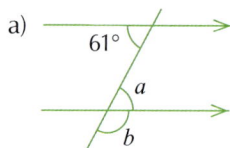

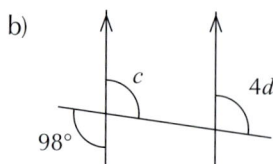

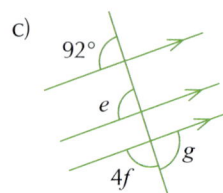

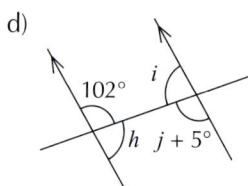

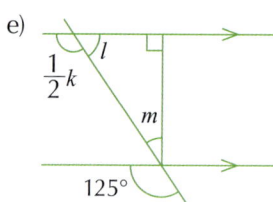

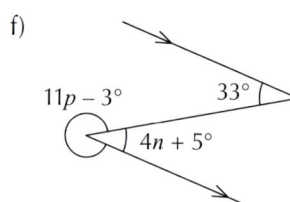

a)

61°

a

b

b)

c

4*d*

98°

c)

92°

e

g

4*f*

d)

102°

i

h   *j* + 5°

e)

$\frac{1}{2}k$

l

m

125°

f)

11*p* – 3°

33°

4*n* + 5°

Q2 Look at the system of lines on the right.

a) (i) Find the value of *x*.

(ii) Hence find the value *y*.

b) Are the lines *AB* and *CD* parallel to each other?
Explain your answer.

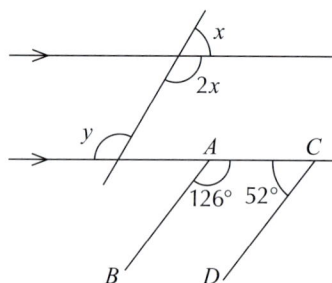

x

2*x*

y

A   *C*

126° 52°

B   *D*

Q3 Find the value of each letter. The diagrams aren't drawn accurately.

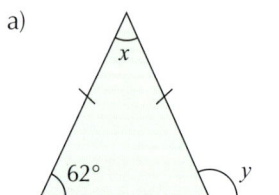

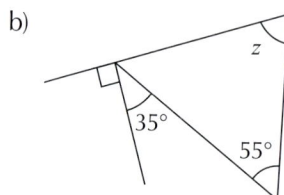

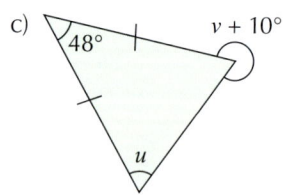

a)

x

62°   *y*

b)

z

35°

55°

c)

48°   *v* + 10°

u

Q4 a) Find angle *p* in the diagram below.

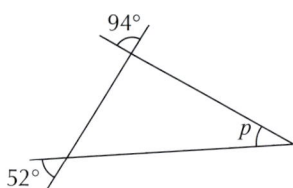

94°

p

52°

b) Find *x* and *y* in the diagram below.

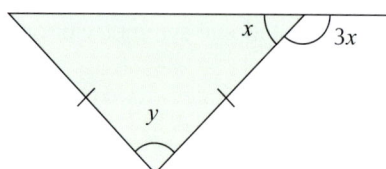

x   3*x*

y

Q5 Find the value of each letter. The diagrams aren't drawn accurately.

a)

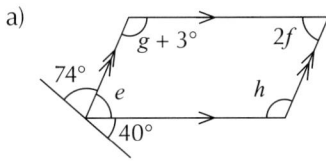

b)

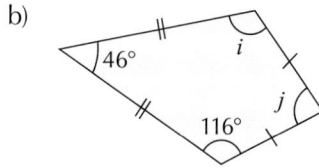

c)

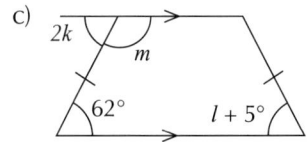

Q6 Write down all the different types of quadrilaterals which satisfy each of the following properties.

a) 4 equal sides

b) 4 angles of 90°

c) exactly 1 pair of equal angles

d) 2 pairs of parallel sides

e) at least 1 pair of parallel sides

f) exactly 1 pair of parallel sides

Q7 Three vertices of a quadrilateral, A, B and C, are plotted on the right. Give the coordinates of the fourth vertex, D, that would make the quadrilateral:

a) a parallelogram

b) a kite

c) an isosceles trapezium

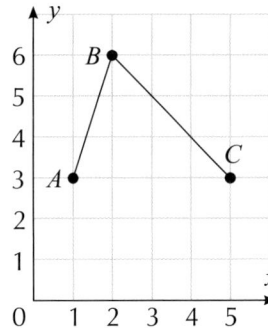

Q8

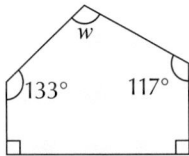

a) (i) What is the name of the polygon on the left.

(ii) Is it regular? Explain your answer.

b) Calculate the sum of the interior angles of the polygon.

c) Use your answer to find the size of angle w.

Q9 Find the size of the exterior angles of a regular polygon with:

a) 10 sides

b) 12 sides

c) 15 sides

d) 25 sides

Q10 Copy and complete the table.

	equilateral triangle	parallelogram	isosceles trapezium	regular nonagon
No. of sides				
Lines of symmetry				
Order of rotational symmetry				
Sum of interior angles				

Exam-Style Questions

Q1 The shape below is made up of a regular polygon and a parallelogram.

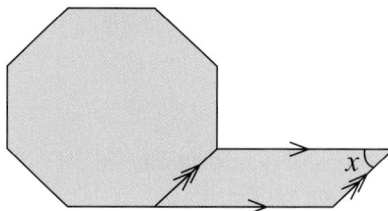

Calculate the size of the angle x.

[2 marks]

Q2 Look at the diagram below. Show that $x = 60° - a$.

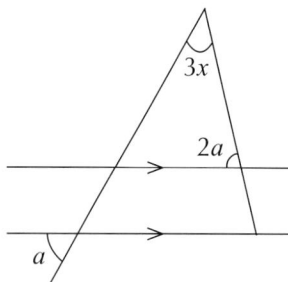

[3 marks]

Q3 Kendra wants to draw a star with 12 equal length sides.
She calculates that each of the acute interior angles needs to be 40°.

Calculate the size of each reflex interior angle.

[3 marks]

Q4 a) Using an appropriate calculation, find the sum of the interior angles of a heptagon.

[2 marks]

b) Prove that the sum of the interior angles of a triangle is 180°.

[3 marks]

c) Hence prove that your answer to part a) is correct.

[2 marks]

Q5 Use the angle properties of parallel lines to prove that the interior angles of any trapezium sum to 360°. Use the diagram below to help you.

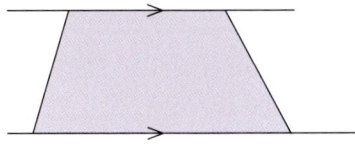

[2 marks]

Q6 A rhombus *ABCD* has diagonals which meet at *X*, as shown. The size of angle *DAX* is 22°.

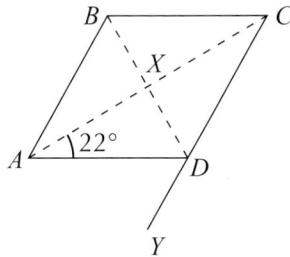

Write down the size of:

a) angle *AXD*

[1 mark]

b) angle *ADX*

[1 mark]

Side *CD* is extended to a point *Y*.

c) Work out the size of angle *ADY*, giving reasons with your working.

[3 marks]

Q7 Two regular polygons have a common side *BD*, as shown in the diagram. A regular polygon can be made which has sides *AB* and *BC*. Work out the number of sides of this polygon.

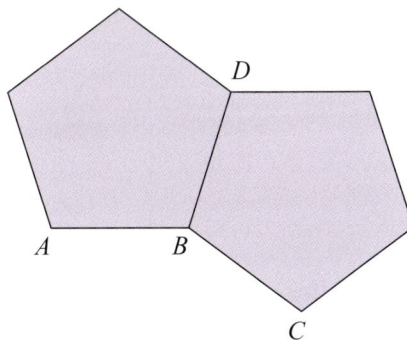

[4 marks]

21.1 Circle Theorems 1

Circle geometry is all about using the properties of circles and lines to find missing angles. Before jumping into the first circle theorems, you'll need to know what all the parts of a circle are called.

Learning Objectives — Spec Ref G9/G10:
- Know the different parts of a circle.
- Be able to apply circle theorems involving radii, tangents, semicircles and chords.

Parts of a Circle

Circumference: the distance around the outside of a circle.

Radius: a line from the centre of a circle to the edge.
The circle's centre is the same distance from all points on the edge.

Diameter: a line from one side of a circle to the other through the centre.
The diameter is **twice** the radius: $d = 2r$

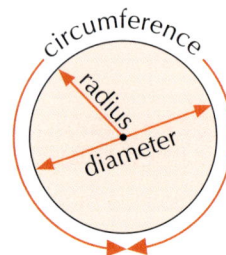

Tangent: a straight line that just touches the circle.

Arc: a part of the circumference.

Chord: a line between two points on the edge of the circle.

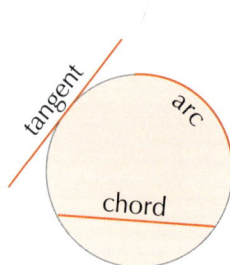

Sector: an area of a circle like a "slice of pie".

Segment: an area of a circle between an arc and a chord.

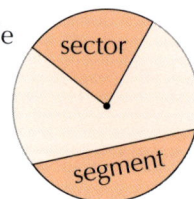

Circle Theorems

There are **nine circle theorems** (or **rules**) to learn in total — here are the first two:

Rule 1: A triangle formed by **two radii** is **isosceles**.

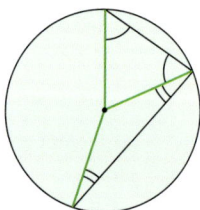

Rule 2: A **tangent** to a circle makes a **right angle with the radius** at that point.

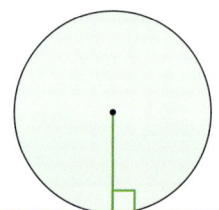

Tip: Rule 2 is used on page 224 to help you find the equation of the tangent to a circle graph at a particular point.

This rule is quite straightforward — all radii are the **same length**, so if two sides of a triangle are radii then it's **isosceles**.

This rule is a bit trickier to get your head around. Draw some **circles**, **tangents** and **radii**, then measure the angle between the tangent and radius to convince yourself that it's true.

Example 1

Find the missing angle x in the diagram on the right.

You have a tangent and a radius so you'll need to use rule 2.

Tangent and radius make an angle of 90°, so $x = 90° - 49° = $ **41°**

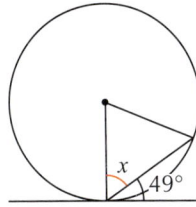

Tip: In circle geometry, it's a good idea to give reasons with your calculations so they're easier to follow.

Example 2

Find the missing angle y in the diagram on the right.

Two sides of the triangle are radii, so use rule 1.

The triangle is isosceles, so both angles at the edge of the circle are the same size. So $y = (180° - 98°) \div 2 = $ **41°**.

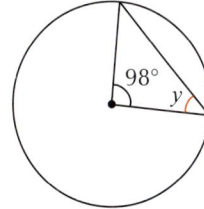

Exercise 1

Q1 Find the value of each letter in the diagrams below.

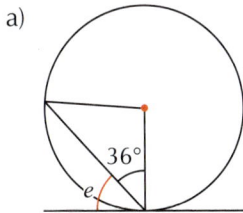

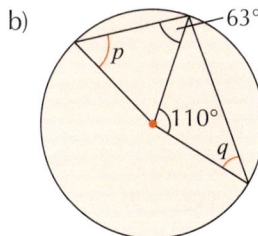

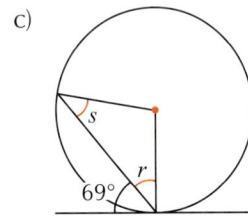

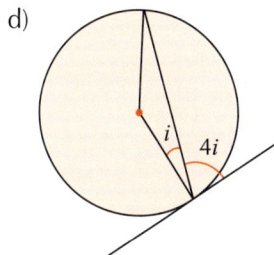

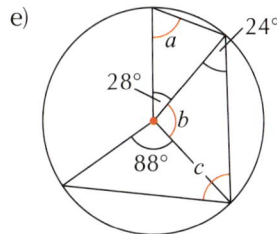

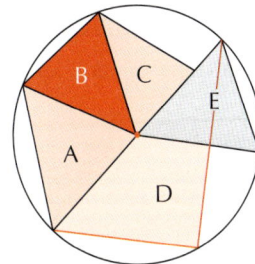

a)

b)

c)

d)

e)

Q2 In the diagram on the right, which of triangles A-E can you be certain are isosceles? Give a reason for your answer.

Rule 3: The **angle in a semicircle** is a **right angle**.

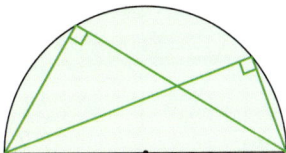

To prove this rule, draw a **radius** going to the angle at the circumference to create two **isosceles** triangles.
The angle at the circumference is $x + y$.
Angles in a triangle add up to 180°,
so $x + y + (x + y) = 180°$
$\Rightarrow 2x + 2y = 180°$
$\Rightarrow x + y = $ **90°**

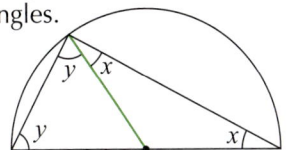

Example 3

Find the missing angle y in the diagram on the right.

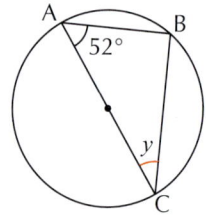

1.   You need to find the angle in a semicircle so use rule 3.

   By the 'angles in a semicircle' rule, ∠ABC = 90°.

2.   You know two angles in a triangle so you can find the third.

   Angles in a triangle add up to 180°, so y = 180° − 90° − 52° = **38°**.

Rule 4: A **diameter bisects** a **chord** at right angles.

Bisecting a chord means splitting it in **exactly in half.**

Any chord which is a **perpendicular bisector** of another chord must be a **diameter**.

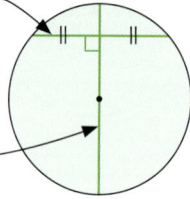

To prove this rule, start with a **diameter** that **bisects a chord**. Draw **two radii** going to either end of the chord. Look at triangles ABC and ADC in the diagram — AC is a common side, AB = AD (both radii) and BC = CD (the chord is bisected). So the triangles are **congruent** (see p.394). ∠ACB and ∠ACD must be **equal** and they lie on a **straight line**, so ∠ACB = ∠ACD = 180° ÷ 2 = **90°**.

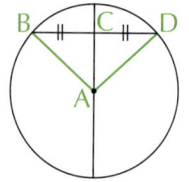

Example 4

On the diagram on the right, BD = DE. Find the missing angle l.

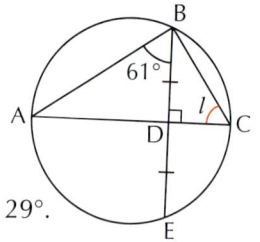

1.   You have a perpendicular bisector of a chord so use rule 4.

   AC is a perpendicular bisector of BE, so AC is a diameter.

2.   As AC is a diameter, you can use rule 3.

   By the 'angles in a semicircle' rule, ∠ABC = 90°, so ∠DBC = 90° − 61° = 29°.

3.   You know two angles in triangle BDC, so you can work out angle l.

   l = 180° − 90° − 29° = **61°**.

Exercise 2

Q1   Find the value of each letter in the diagrams below.

a)
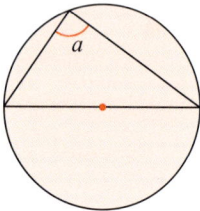

b)
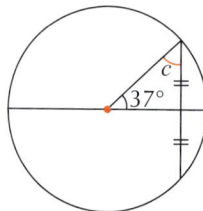

c)
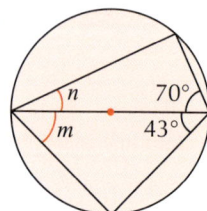

d)
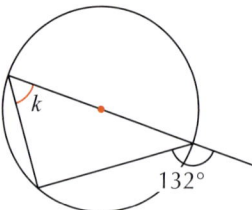

e)
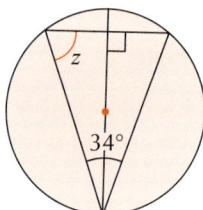

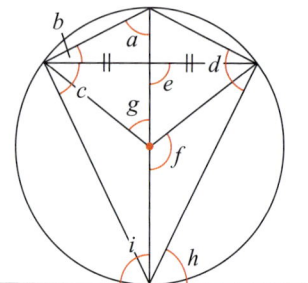

Q2   In the diagram on the right, which of angles a-i must be right angles? Explain your answer.

21.2 Circle Theorems 2

Circle geometry is all about learning the rules and then trying to spot which one to apply in different situations. Make sure you know the rules on the previous pages before having a go at these ones.

Learning Objectives — Spec Ref G10:
- Be able to find angles at the centre or circumference of a circle.
- Be able to find angles in the same segment.
- Be able to find angles in a cyclic quadrilateral.

Rule 5: The angle subtended **at the centre** of a circle is **double** the angle subtended **at the circumference** by the same arc.

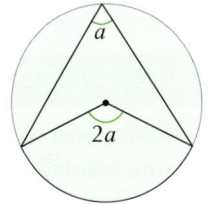

An angle **subtended by an arc** is the angle made where two lines from the ends of the arc meet. The subtended angle is **inside** the shape formed by the arc and the lines.

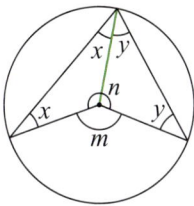

To prove this rule, draw a **radius** going to the angle at the circumference to create **two isosceles triangles** (see rule 1). The angle at the circumference of the circle is $x + y$.
Angles in a quadrilateral add up to 360°,
so $n = 360° - x - y - (x + y) = 360° - 2(x + y)$.
Angles around a point add up to 360°,
so angle $m = 360° - n = 360° - (360° - 2(x + y)) = \mathbf{2(x + y)}$.
So angle m is **double** the angle at the circumference.

Example 1

Find the missing angles a and b in the diagram on the right.

1. Angle a and the 250° angle form the angles around a point.
   Angles at a point add up to 360°, so $a = 360° - 250° = \mathbf{110°}$.
2. Now use rule 5 to find angle b.
   The angle at the centre is double the angle at the edge, so $b = \dfrac{a}{2} = \dfrac{110°}{2} = \mathbf{55°}$.

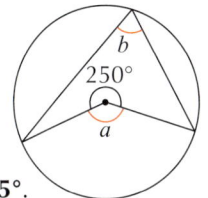

Exercise 1

Q1 Find the value of each letter in the diagrams below.

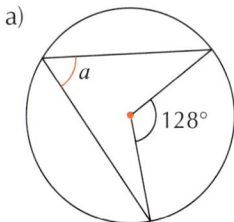

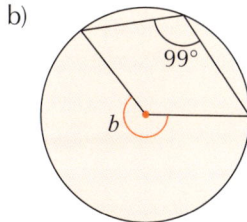

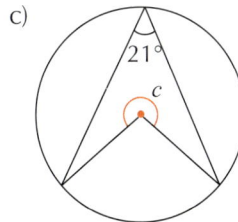

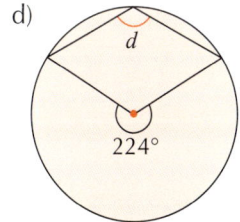

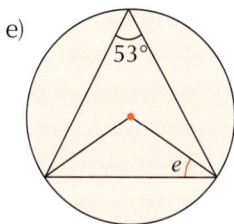

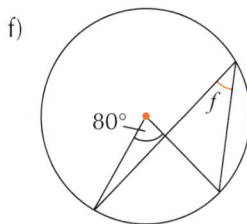

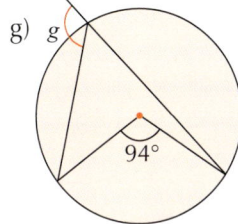

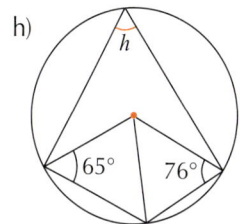

a)

b)

c)

d)

e)

f)

g)

h)

Q1 diagrams contain labels: a) a, 128°; b) 99°, b; c) 21°, c; d) d, 224°; e) 53°, e; f) 80°, f; g) g, 94°; h) h, 65°, 76°.

Q2   Find the value of each letter in the diagrams below.

a)
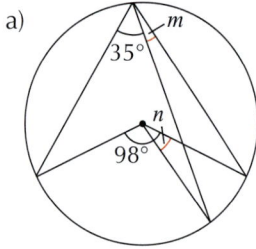

b)
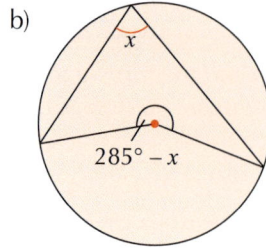

c)
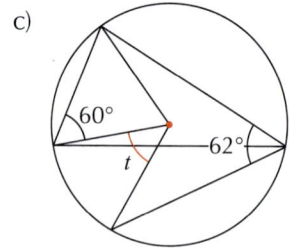

Rule 6: Angles subtended by an arc in the **same segment** are **equal**.

Angles must be in the **same segment** — if you drew an angle subtended in the other segment it **wouldn't** be equal to b.

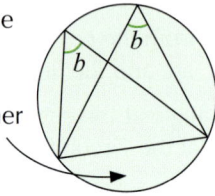

To prove this rule you can just use rule 5. Let m and n be angles subtended by an arc in the **same segment**. Draw angle x subtended at the **centre** of the circle from the **same arc** as m and n.
Now using rule 5: $m = \frac{x}{2}$ and $n = \frac{x}{2}$ so $\boldsymbol{m = n}$.

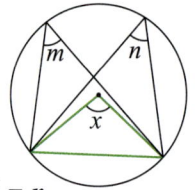

Example 2

Find the missing angles a, b and c in the diagram on the right.

1. Angles in a triangle add up to 180°.

    $a = 180° - 86° - 39° = \mathbf{55°}$

2. Look for angles in the same segment.
    a and b are in the same segment.
    The 39° angle and c are in the same segment.

    Angles in the same segment are equal, so $b = a = \mathbf{55°}$
    and $c = \angle PSQ = \mathbf{39°}$.

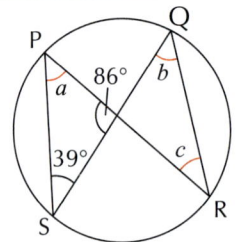

Exercise 2

Q1   Find the value of each letter in the diagrams below.

a)
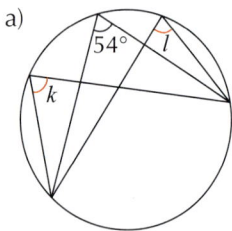

b)
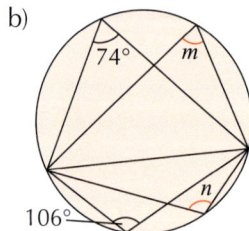

c)
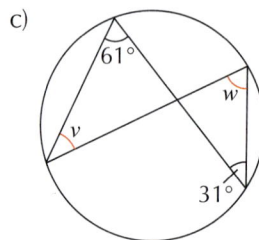

d)
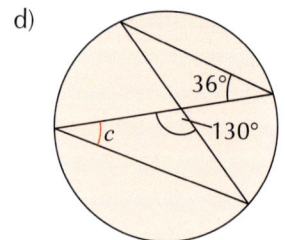

e)
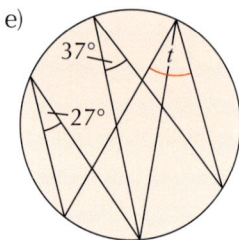

f)
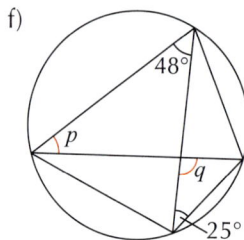

g)
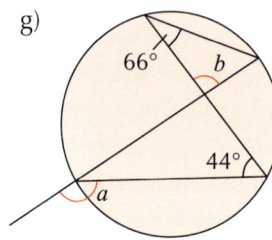

h)
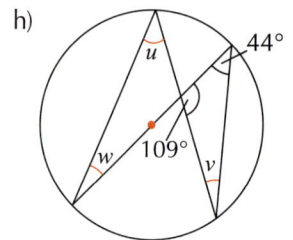

Q2   Find the value of each letter in the diagrams below.

a)

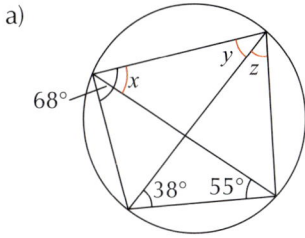

b)

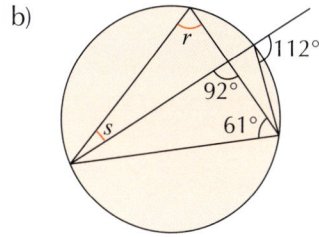

Rule 7: **Opposite angles** in a **cyclic quadrilateral** sum to **180°**.

A **cyclic quadrilateral** is any quadrilateral which can be drawn inside a circle with **all four vertices** touching the circumference.

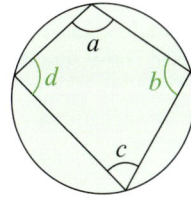

$a + c = 180°$
$b + d = 180°$

To prove this rule, draw **two radii** going to **opposite corners** of the quadrilateral as shown in the diagram on the right. Now you have **two sets of angles** at the centre and circumference subtended by the same arc, so use rule 5. If the angles at the circumference are x and y, the angles in the centre are **twice as big**, so they are $2x$ and $2y$. The angles at the **centre** add up to 360° as they are around a point, so $2x + 2y = 360° \Rightarrow x + y = 180°$

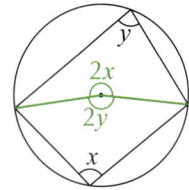

Example 3

Find the missing angles x and y.

It's a cyclic quadrilateral so use rule 7:

Opposite angles in a cyclic quadrilateral add up to 180°.

$x = 180° - 84° = \mathbf{96°}$ $y = 180° - 111° = \mathbf{69°}$

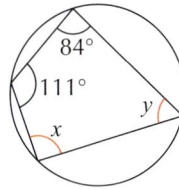

Tip: Keep an eye out for cyclic quadrilaterals — they can be hard to spot if there are other lines on the diagram.

Exercise 3

Q1   Find the value of each letter in the diagrams below.

a)

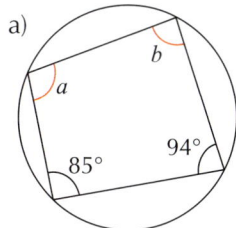

b)

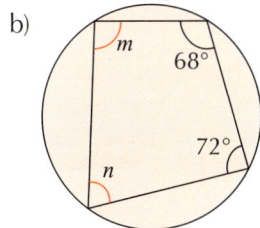

c)

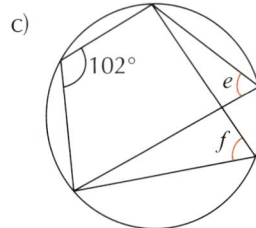

d)

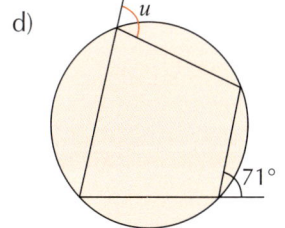

e)

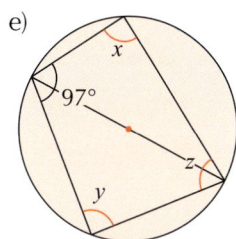

f)

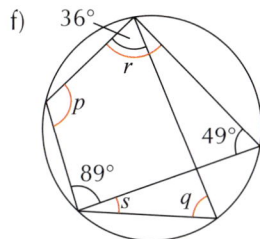

g)

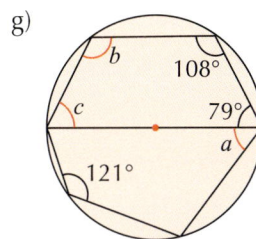

h)

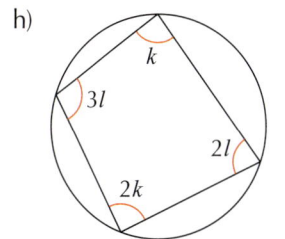

21.3 Circle Theorems 3

Here are the final two circle theorems that you need. Remember that circle geometry often involves using several rules together to find unknown angles, so watch out for places where you can use the other rules.

Learning Objectives — Spec Ref G10:
- Know that two tangents drawn from the same point are equal in length.
- Be able to use the alternate segment theorem.

Rule 8: Two **tangents** to a circle drawn from a single point outside the circle are the **same length**.

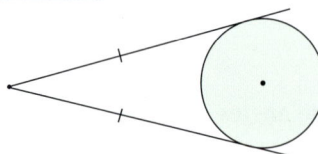

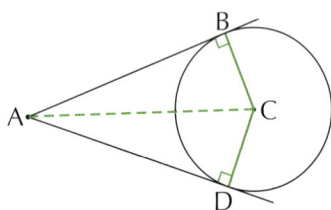

To prove this rule, draw **two radii** going to the points where the **tangents touch the circle**. Then draw another line from the **centre** of the circle to the point **outside** the circle as shown in the diagram on the left. ∠ABC = ∠ADC = **90°** because a tangent meets a radius at 90° (rule 2), so ABC and ADC are **right-angled triangles**. Line AC is the **hypotenuse** of both triangles and BC = DC because they are both **radii**. Condition RHS holds (see p.394), so the triangles are **congruent** and **AB = AD**.

Example 1

Find the size of the missing angle x.

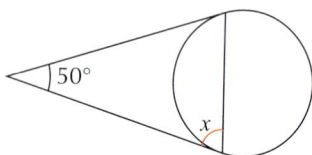

1. The two longer edges can be extended to form tangents.

   The tangents and the chord form an isosceles triangle as tangents from the same point are the same length.

2. The two missing angles are equal as the triangle is isosceles.

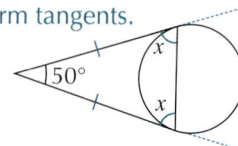

   $x = (180° - 50°) \div 2 = \mathbf{65°}$

Exercise 1

Q1 Find the value of each letter in the diagrams below.

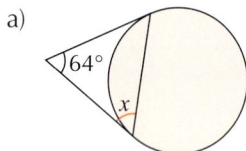

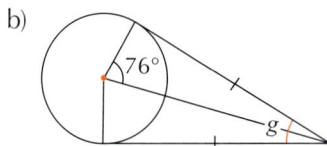

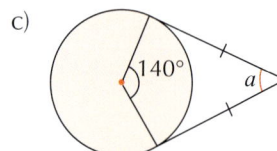

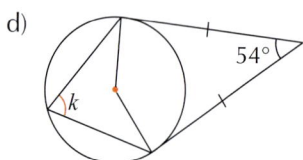

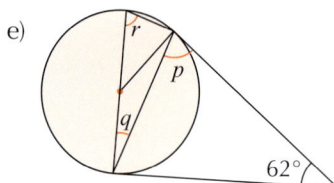

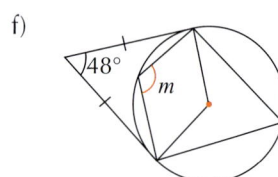

a)

b)

c)

d)

e)

f)

Rule 9: The angle between **a tangent and a chord** is equal to the angle subtended from the ends of the chord in the **alternate segment**.

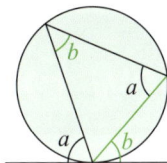

The **alternate segment** to an angle between a tangent and a chord is the segment on the other side of the chord.

This circle rule is called the **alternate segment theorem**.

To prove it, start with a **triangle** inside a circle with a **tangent** at one of the points of the triangle, as shown in the diagram on the right.

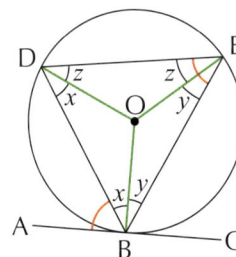

Draw a **radius** to **each corner** of the triangle to create **three isosceles triangles**. Now, aim to show that ∠ABD and ∠BED are **equal**.
Using rule 2 you know that ∠ABO = 90° so **∠ABD = 90° − x**
The angles in triangle BDE add up to 180°, so
$x + x + y + y + z + z = 180°$ ⇒ $2(x + y + z) = 180°$
⇒ $x + y + z = 90°$ ⇒ $y + z = 90° − x$ ⇒ **∠BED = 90° − x**
So **∠ABD = ∠BED**. (You can follow the same method to show **∠CBE = ∠BDE**.)

Example 2

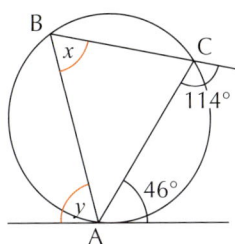

Find the size of angles x and y.

1. You can use the alternate segment theorem to find x.

    x is in the alternate segment to the 46° angle, so x = **46°**.

2. ∠ACB is in the alternate segment to y.

    Using angles on a straight line, y = ∠ACB = 180° − 114° = **66°**.

Exercise 2

Q1 According to the alternate segment theorem, which angle in the diagram on the right is:

a) equal to angle a?

b) equal to angle c?

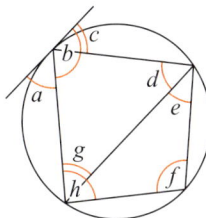

Q2 Find the value of each letter in the diagrams below.

a)

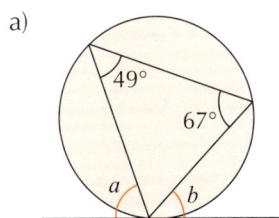

b)

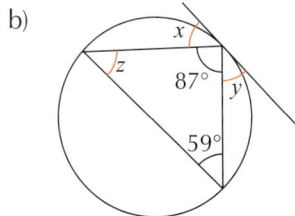

c)

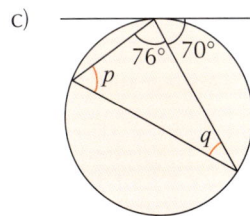

d)

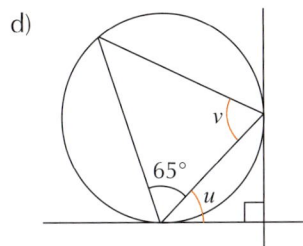

e)

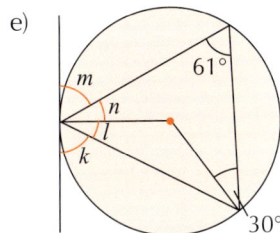

f)

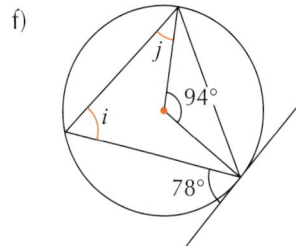

Q1 Find the value of each letter in the diagrams below. Explain your reasoning in each case.

a)

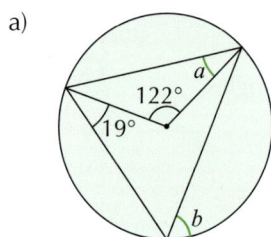

b)

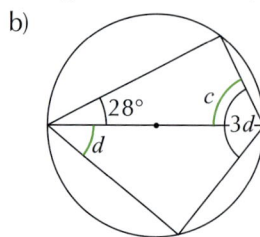

c)

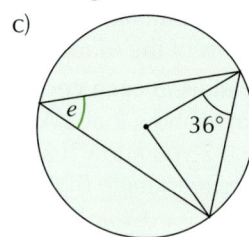

d)

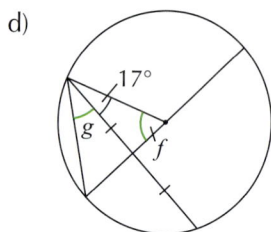

e)

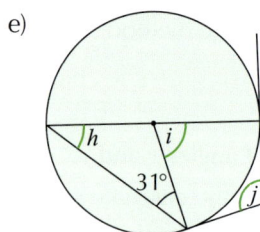

f)

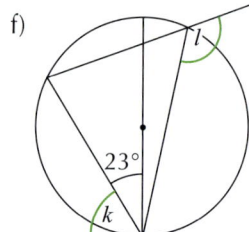

g)

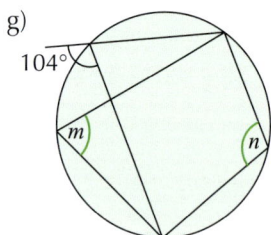

h)

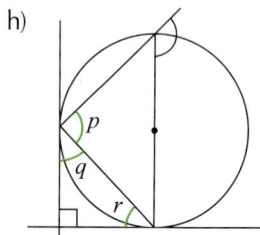

i)

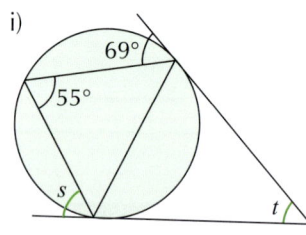

Q2 Look at the diagram on the right.

a) What is the size of ∠BDE?

b) How does the answer to part a) prove that AD is not a diameter of the circle?

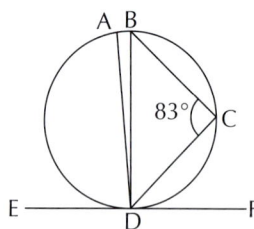

Q3 Joel wants to prove that N is not the centre of the circle shown on the right. Answer the questions below, explaining your answer in each case.

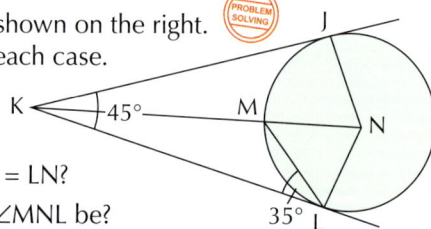

a) If N were the centre of the circle, what would the sizes of ∠KLN and ∠MLN be?

b) (i) If N were the centre of the circle, why would MN = LN?
   (ii) If MN = LN, what would the sizes of ∠LMN and ∠MNL be?

c) (i) If N were the centre of the circle, why would triangles KJN and KLN be congruent?
   (ii) If triangles KJN and KLN were congruent, what would the size of ∠JNL be?

Joel says that N can't be the centre of the circle, because if it is, then ∠JKL is not 45°.

d) If N were the centre of the circle and ∠KLM = 35°, what would the size of ∠JKL be?

Q1 The diagram shows a circle with centre O. The points A, B and C are on the circumference such that AB = AC. DE is a tangent to the circle at C.

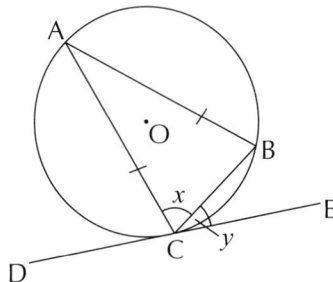

Given that angle ACB = x and BCE = y, prove that $y = 180° - 2x$.
You must give a reason for every statement that you make.

[3 marks]

Q2 In the diagram, a circle with centre O is intersected twice by each of two lines which meet at E.

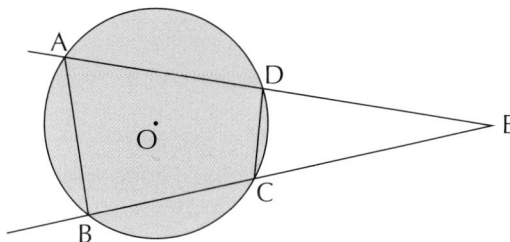

Prove that the triangles AEB and DEC are similar.

[3 marks]

Q3 The diagram shows a semicircle with centre O and diameter AC.

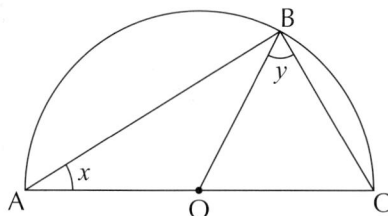

Using angle OAB = x and angle OBC = y, prove that angle ABC = 90°.

[3 marks]

22.1 Metric Units — Length, Mass and Volume

When comparing, adding or subtracting measurements, it's much easier to work with them if they all have the same units — so being able to convert between units is a really useful skill.

Learning Objective — Spec Ref N13/R1/G14:
Convert between different metric units for length, mass and volume.

The **metric units** for length, mass and volume are as follows.

- **Length** is measured in **millimetres** (mm), **centimetres** (cm), **metres** (m) and **kilometres** (km).

- **Mass** is measured in **milligrams** (mg), **grams** (g), **kilograms** (kg) and **tonnes**.

- **Volume** is measured in **millilitres** (ml), **litres** (l) and **cubic centimetres** (cm³).

Tip: Metric units increase by powers of 10 to make larger numbers easier to deal with.

To **convert** between different units of length, mass or volume, you **multiply** or **divide** by a **conversion factor**. The most commonly used conversion factors are shown below.

Length	Mass	Volume
1 cm = 10 mm	1 g = 1000 mg	1 litre (l) = 1000 ml
1 m = 100 cm	1 kg = 1000 g	1 ml = 1 cm³
1 km = 1000 m	1 tonne = 1000 kg	

When converting from small units to **bigger units** (e.g. cm to m), you **divide** by the conversion factor.
When converting from big units to **smaller units**, you **multiply** by the conversion factor.
Always **check** your answers to make sure they seem **reasonable** — e.g. if you converted an elephant's height from m to cm and got 0.025 cm, you'd know you'd gone wrong. You'd expect **more** small units than big units — there are 100 cm in 1 m, so you'd expect there to be more centimetres than metres.

Example 1

What is 0.035 km in cm?

1. Convert km to m — a smaller unit, so multiply by the conversion factor.
   1 km = 1000 m
   0.035 km = 0.035 × 1000 = 35 m

2. Then convert to cm — a smaller unit, so multiply by the conversion factor.
   1 m = 100 cm
   35 m = 35 × 100 = **3500 cm**

Tip: You could convert km into cm in one step by multiplying by 100 × 1000 = 100 000.

Example 2

Find the total of 0.2 tonnes, 31.8 kg and 1700 g. Give your answer in kg.

1. Convert tonnes to kg — a smaller unit, so multiply by the conversion factor.
   1 tonne = 1000 kg
   0.2 tonnes = 0.2 × 1000 = 200 kg

2. Convert g to kg — a bigger unit, so divide by the conversion factor.
   1 kg = 1000 g
   1700 g = 1700 ÷ 1000 = 1.7 kg

3. The masses are now in kg so add them together.
   200 kg + 31.8 kg + 1.7 kg = **233.5 kg**

Example 3

Liam buys a 2.5 litre bottle of lemonade and pours out 175 ml. How much is left in the bottle? Give your answer in cm³.

1. Convert litres to ml — a smaller unit, so multiply by the conversion factor.

    1 litre = 1000 ml
    2.5 litres = 2.5 × 1000 = 2500 ml

2. Subtract 175 ml from 2500 ml.

    2500 ml – 175 ml = 2325 ml

3. Swap ml for cm³, as 1 cm³ = 1 ml.

    2325 ml = **2325 cm³**

Exercise 1

Q1 Convert each measurement into the units given.

   a) 3000 kg into tonnes   b) 0.4 g into mg   c) 123 ml into litres

   d) 5116 g into kg   e) 12.6 kg into tonnes f) 2.7165 m into cm

Q2 Convert each measurement into the units given.

   a) 0.15 kg into mg   b) 1532 g into tonnes   c) 1005 cm into km

   d) 3023 mg into kg   e) 3 mm into km   f) 49 tonnes to g

Q3 Hafsa is having a party for 32 guests. Her glasses have a capacity of 400 ml each. If she wants everyone to have a glass of juice, how many 2 litre bottles of juice should she buy?

Q4 A go-kart has a 5 litre petrol tank. It uses 10 ml of petrol per lap of a 400 m track.

   a) If James fills up the tank, how many laps of the track can he do?

   b) How many km can James travel on each full tank of petrol?

Q5 Sharon, Leuan and Elsie get into a cable car while skiing. Sharon weighs 55.2 kg, Leuan weighs 78.1 kg and Elsie weighs 65.9 kg. Their skis weigh 10 000 g a pair. The cable car is unsafe when carrying a mass of over half a tonne. Will Sharon, Leuan and Elsie be safe?

Q6 Milly runs a 1500 m fun run, a 50 m sprint and a 13.2 km race. How many km has she run in total?

Q7 Find the difference in the amounts of liquid held by the two containers shown on the right. Give your answer in cm³.

Q8 A reservoir contains 600 000 litres of water. During a period of heavy rain, the volume of water in the reservoir increases by 750 000 ml every day. The reservoir can only hold 800 000 litres of water. If the rain continues at this rate, calculate the value of N, the number of whole days that will pass before the reservoir overflows.

Q9 A lasagne recipe requires 0.7 kg of minced beef, 400 g of tomato sauce, 300 g of cheese sauce, 0.2 kg of lasagne sheets and 2500 mg of herbs and spices.

   a) How many kg do these ingredients weigh in total?

   b) 0.2 kg of ingredients are needed for each person. How many people can be fed with this recipe?

22.2 Metric Units — Area and Volume

When converting between units of area and volume, you don't just multiply by the conversion factor.

> **Learning Objective — Spec Ref R1/G14:**
> Convert between different metric units for area and volume.

The **area** of a shape is found by multiplying **two lengths** — so the units for area are the units for length **squared** (e.g. m × m = m²). To **convert** between different units of area, you need to multiply or divide by the **square** of the 'length' conversion factor. For example, to convert 1 m² to cm², you need to multiply by 100 × 100 = 10 000 — you're multiplying by the conversion factor **twice**, once for each dimension.

The **volume** of an object is found by multiplying **three lengths** — so the units for volume are the units for length **cubed** (e.g. m × m × m = m³). To **convert** between different units of volume, you need to multiply or divide by the **cube** of the 'length' conversion factor. For example, to convert 1 m³ to cm³, you need to multiply by 100 × 100 × 100 = 1 000 000 — you're multiplying by the conversion factor **three times**, once for each dimension.

Here are some commonly used **conversion factors** for area and volume:

Area	Volume
1 cm² = 10² mm² = 100 mm²	1 cm³ = 10³ mm³ = 1000 mm³
1 m² = 100² cm² = 10 000 cm²	1 m³ = 100³ cm³ = 1 000 000 cm³
1 km² = 1000² m² = 1 000 000 m²	1 km³ = 1000³ m³ = 1 × 10⁹ m³

> **Tip:** Use standard form (p.96) if the numbers are very large or very small — e.g. 1 × 10⁹ = 1 000 000 000.

As on p.276, you **divide** by the conversion factor when converting from small units to **bigger units** (e.g. cm² to m²) and you **multiply** by the conversion factor when converting from big units to **smaller units**.

Example 1

Convert an area of 0.06 m² to cm².

1. Work out the conversion factor from cm² to m².

   1 m = 100 cm,
   so 1 m² = 100 × 100 = 10 000 cm²

2. You're converting to a smaller unit, so multiply by the conversion factor.

   0.06 m² = 0.06 × 10 000 = **600 cm²**

Example 2

Convert a volume of 382 000 cm³ to m³.

1. Work out the conversion factor from m³ to cm³.

   1 m = 100 cm
   so 1 m³ = 100 × 100 × 100 = 1 000 000 cm³

2. You're converting to a bigger unit, so divide by the conversion factor.

   382 000 cm³ = 382 000 ÷ 1 000 000 = **0.382 m³**

Example 3

A cuboid has a width of 55 mm and a height of 40 mm.

a) What is the area of the front face of the cuboid in cm²?

1.	Work out the area in mm².	$55 \times 40 = 2200$ mm²
2.	Work out the conversion factor from mm² to cm².	1 cm = 10 mm so 1 cm² = 10 × 10 = 100 mm²
3.	You're converting to a bigger unit, so divide by the conversion factor.	2200 mm² = 2200 ÷ 100 = **22 cm²**

> **Tip:** You could also have converted the dimensions into cm first.

b) If the cuboid has a length of 6.8 cm, what is its volume in mm³?

1.	Work out the volume in cm³.	$22 \times 6.8 = 149.6$ cm³
2.	Work out the conversion factor from cm³ to mm³.	1 cm = 10 mm so 1 cm³ = 10 × 10 × 10 = 1000 mm³
3.	You're converting to a smaller unit, so multiply by the conversion factor.	149.6 cm³ = 149.6 × 1000 = **149 600 mm³**

Exercise 1

Q1 Convert each of these measurements into the units given.

a) 84 mm² into cm²   b) 1750 cm² into m²   c) 29 000 mm² into cm²

d) 0.001 km³ into m³   e) 15 cm³ into mm³   f) 0.2 m³ into cm³

g) 3 150 000 m² into km²   h) 8500 mm² into cm²   i) 1700 cm² into m²

j) 0.435 km³ into m³   k) 6.7 km³ into m³   l) 0.00045 cm³ into mm³

Q2 Sandeesh wants to carpet two rectangular floors. One of the floors measures 1.7 m by 3 m, while the other is 670 cm by 420 cm. How many square metres of carpet will she need?

Q3 25 cm³ of squash must be diluted with 0.5 litres of water to make one glass.

a) How many glasses can you make from a 1 litre bottle of squash?

b) What is the total volume of one glass in mm³?

Q4 A swimming pool is 3 m deep and has a base with area 375 m².

a) Find the volume of the pool in cm³.   b) How many litres of water can the pool hold?

Q5 A brand of coffee powder is sold in cuboid packets with dimensions 20.7 cm by 25.5 cm by 10 cm.

a) A volume of 0.003 m³ of coffee powder has already been used. What volume (in m³) is left?

b) Find the total surface area of the packet of coffee powder. Give your answer in mm².

Q6 Convert each of these measurements into the units given.

a) 1.2 m² into mm²   b) 0.001 km² into cm²   c) 50 million mm² into m²

d) 3 million mm³ into km³   e) 0.0006 m³ into mm³   f) 999 cm³ into km³

22.3 Metric and Imperial Units

Unlike metric units, imperial units are not based on powers of 10 — e.g. there are 16 ounces in a pound and 14 pounds in a stone.

Learning Objective — Spec Ref N13/R1/G14:
Convert between metric and imperial units.

The **imperial units** for length, mass and volume are as follows:

- **Length** is measured in **inches** (in), **feet** (ft), **yards** and **miles**. There are **12 inches** in 1 foot, **3 feet** in 1 yard and **1760 yards** in 1 mile.

- **Mass** is measured in **ounces** (oz), **pounds** (lb) and **stones**. There are **16 ounces** in 1 pound and **14 pounds** in a stone.

- **Volume** is measured in **pints** and **gallons**. There are **8 pints** in 1 gallon.

Tip: As 1 foot = 12 in, the conversion factor from feet to inches is 12.

To write small imperial units as a **mixture of big and small units** (e.g. writing inches as feet and inches), you **divide** by the conversion factor and keep the **remainder** in the smaller units.
To write a **mixture** of big and small units in **smaller units** (e.g. writing feet and inches as inches), you **multiply** the big unit by the conversion factor and **add on** the remaining small units.

There are **approximate conversion factors** to switch between metric and imperial units — e.g. there are approximately 2.5 cm in 1 inch. The symbol '≈' means 'approximately equal to'. To **convert** between metric and imperial units, just **multiply** or **divide** by the conversion factors below.

Length
1 inch ≈ 2.5 cm
1 foot ≈ 30 cm
1 yard ≈ 90 cm
1 mile ≈ 1.6 km

Mass
1 ounce ≈ 28 g
1 pound ≈ 450 g
1 stone ≈ 6400 g
1 kg ≈ 2.2 pounds

Volume
1 pint ≈ 0.57 litres
1 gallon ≈ 4.5 litres

Tip: You'll be given metric to imperial conversion factors in an exam if they're needed. Don't be surprised if they're not exactly the same as these ones.

Example 1

Convert 65 cm into feet and inches.

1. Divide by the conversion factor for cm to inches.
   1 inch ≈ 2.5 cm,
   so 65 cm ≈ 65 ÷ 2.5 = 26 inches

2. Convert 26 inches into feet and inches.
   1 foot = 12 inches,
   26 ÷ 12 = 2 remainder 2

3. Keep the remainder in inches.
   So 65 cm ≈ **2 feet 2 inches**

Tip: 1 inch is bigger than 1 cm so you divide by the conversion factor.

Example 2

Convert 6 pounds and 4 ounces into kilograms.

1. Write the whole mass using the same unit.
   1 pound = 16 ounces,
   so 6 pounds and 4 ounces = (6 × 16) + 4 = 100 ounces

2. Convert this into g using the conversion factor for ounces to g.
   1 ounce ≈ 28 g,
   100 ounces ≈ 100 × 28 = 2800 g

3. Then convert the result from g to kg.
   2800 g = 2800 ÷ 1000 = **2.8 kg**

Convert 11 400 ml into both pints and gallons.

1. Convert into litres.

   1 litre = 1000 ml,
   so 11 400 ml = 11 400 ÷ 1000 = 11.4 litres

   Tip: A pint is less than a litre, so you'd expect more pints than litres.

2. Divide by the conversion factor for litres to pints.

   1 pint ≈ 0.57 litres,
   so 11.4 litres ≈ 11.4 ÷ 0.57 = **20 pints**

3. Convert into gallons

   1 gallon = 8 pints,
   so 20 pints = 20 ÷ 8 = **2.5 gallons**

Exercise 1

In these questions, use the conversion factors given on the previous page.

Q1 Convert each of the following measurements into the units given.

   a) 3 ft 7 in to inches
   b) 12 ft 5 in to inches
   c) 5 lb 2 oz to ounces
   d) 280 in to feet and inches
   e) 1001 in to feet and inches
   f) 72 oz to lb and oz
   g) 70 lb to stones and pounds
   h) 200 oz to lb and oz
   i) 5.5 yards to feet and inches
   j) 4.75 ft to feet and inches
   k) 2.5 stone to lb and oz
   l) 8.25 stone to lb and oz

Q2 Convert each of the following masses into pounds and ounces.

   a) 1904 g
   b) 840 g
   c) 2688 g
   d) 4.9 kg

Q3 Convert each of the following into feet and inches.

   a) 2 m
   b) 52.5 cm
   c) 1.5 km
   d) 50 mm

Q4 State which is the greater amount in each of the following pairs.

   a) 10 feet or 3.5 m
   b) 1 stone or 7 kg
   c) 10 miles or 12 km
   d) 15 pints or 9 litres
   e) 3 lb or 1.5 kg
   f) 5 stone or 31 kg
   g) 160 stone or 1 tonne
   h) 2 gallons or 10 litres
   i) 16 lb or 7 kg

Q5 A ride at a theme park states you must be 140 cm or over to ride. Maddie is 4 feet 5 inches and Lily is 4 feet 9 inches. Who can go on the ride — Maddie, Lily, both or neither of them?

Q6 A running track is 400 m. How many laps of the track make one mile?

Q7 Jamie and Oliver are cooking. They need 1 pound 12 ounces of meat for their recipe. They buy 750 g of meat in the supermarket. Will this be enough?

Q8 A large box of juice holds 3 litres.

   a) How many whole pint jugs can be filled from the box?

   b) Approximately how many litres of juice will be left in the box after filling the jugs?

Q9 Marion is on holiday in Spain. Her car measures speed in mph but the road signs are in km/h. The speed limit is 90 km/h. What is this in mph? Give your answer to the nearest mph.

22.4 Estimating in Real Life

When it's not possible to measure something, you might have to estimate its size instead.

> **Learning Objective — Spec Ref N14:**
> Estimate the dimensions of real-life objects by comparison.

You can **estimate** the **dimensions** (e.g. length, mass or volume) of something by **comparing** it with something else you already know the size of. An estimate does not have to be completely accurate, but make sure your answer seems **realistic** and is given in **suitable units**.

Example 1

Estimate the height of this lamp post.

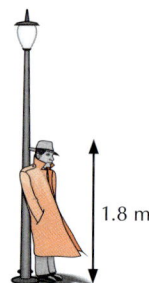

1. Estimate the height of the man.

   Average height of a man ≈ 1.8 m.

2. Compare the height of the lamp post and the man.

   The lamp post is roughly twice the height of the man.

3. Multiply the height of the man by 2.

   Height of the lamp post ≈ 2 × 1.8 m
   = **3.6 m**

1.8 m

Example 2

Give a sensible metric unit for measuring the height of a room.

1. Most rooms are taller than an average person but of the same order of magnitude — at roughly 2 to 3 metres.

2. So it makes sense to measure the height of a room in the same metric unit as a person's height.

   A sensible unit is **metres.**

> **Tip:** Feet would be an acceptable answer if you were asked for an imperial unit.

Exercise 1

Q1 For each of the following, suggest a sensible unit of measurement, using
   (i) metric units   (ii) imperial units.

   a) length of a pencil
   b) mass of a tomato
   c) length of an ant
   d) mass of a bus
   e) volume of a teacup
   f) distance between cities

Q2 Estimate each of the following, using sensible metric units.

   a) height of your bedroom
   b) length of an average car
   c) height of a football goal
   d) arm span of an average adult
   e) diameter of a football
   f) volume of a typical bath tub (PROBLEM SOLVING)

Q1 Convert each measurement into the units given.

a) 10 kg to g

b) 14 cm to mm

c) 4.6 litres to cm^3

d) 22 g to kg

e) 150 ml to litres

f) 6900 cm to m

Q2 Convert each measurement into the units given.

a) 0.006 m to mm

b) 0.57 kg to mg

c) 1.2 km to cm

d) 8 000 000 mg to kg

e) 12 cm to km

f) 1101 g to tonnes

Q3 Convert each of these areas into the units given.

a) 10 cm^2 to m^2

b) 18 km^2 to m^2

c) 8.6 m^2 to mm^2

d) 673 000 000 cm^2 into km^2

e) 60 500 mm^2 into m^2

f) 0.000005 km^2 into mm^2

Q4 Convert each of these volumes into the units given.

a) 0.005 m^3 to cm^3

b) 69 mm^3 to cm^3

c) 720 cm^3 to m^3

d) 19 cm^3 into m^3

e) 17 440 mm^3 into m^3

f) 0.00345 km^3 into mm^3

Q5 The volume of a beachball is 49 900 cm^3. What is the volume of the beachball in m^3?

For Q6-8, use the approximate conversion factors given on page 280.

Q6 Complete each of these calculations. Give each of your answers to two significant figures.

a) 681 cm + 12.4 yards ≈ ⬚ m

b) 16.49 km + 21.5 miles ≈ ⬚ km

c) 3 kg + 1 stone + 1.5 kg ≈ ⬚ kg

d) 100 cm + 2 yards + 15 feet ≈ ⬚ cm

e) 3 l + 4.5 pints + 250 ml ≈ ⬚ ml

f) 7 m + 8 inches + 35 mm ≈ ⬚ cm

Q7 Jakov's mother asks him to buy 4 pints of milk. The shop only sells milk in 1 litre, 2 litre and 4 litre cartons. Which of these amounts is closest to what his mother asked for?

Q8 The weights of 6 people getting into a lift are shown below. The lift has a weight limit of 0.5 tonnes. Will the total weight of the 6 people exceed this limit?

10 stone, 8 stone, 14 stone 1 pound, 9 stone 9 pounds, 12 stone, 17 stone

Q9 Estimate the length and height of the bus. Give your answer in sensible metric units.

Exam-Style Questions

Q1 Estimate the height and length of this dinosaur in metric units by comparing it with a chicken.

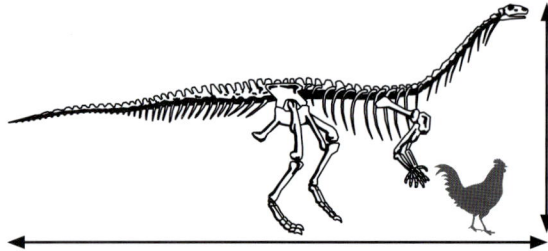

[2 marks]

Q2 Jack buys 1.2 kg of flour. How many pizzas can he make if each pizza needs 300 g of flour?

[1 mark]

Q3 What is the volume of a cube with sides of 150 cm? Give your answer in m^3.

[2 marks]

Q4 Two nature reserves are approximately rectangular, measuring 2.9 km by 3.3 km and 2700 m by 4100 m.
Which nature reserve has the largest area?

[2 marks]

Q5 Elsie adds 16 inches of ribbon to the hats she makes. Ribbon cost £1.50 per metre.
How many hats could Elsie make with £33 worth of ribbon? Use 1 inch ≈ 2.5 cm.

[3 marks]

Q6 A cup of tea contains 0.25 litres of liquid, of which 25 cm^3 is milk and the rest is water.
How much water is needed, in ml, for 5 cups of tea?

[3 marks]

Q7 Lotte's car travels 55 miles per gallon of petrol. Her car contains 45 litres of petrol.

a) How many whole miles can she travel? Use the approximation 1 gallon ≈ 4.5 litres.

[2 marks]

b) If petrol costs £5 per gallon, how much did the petrol in her car cost?

[1 mark]

23.1 Compound Measures

Compound measures are made up of two or more other measurements. For example, speed is a compound measure because it's a combination of the two measurements distance and time.

Speed, Distance and Time

Learning Objective — Spec Ref N13/R1/R11:
Know and use the formula linking speed, distance and time.

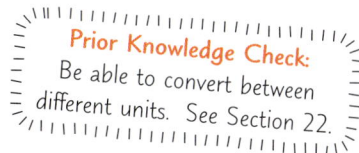

Prior Knowledge Check:
Be able to convert between different units. See Section 22.

The **speed** of an object is the **total distance** it travels divided by the **time** it takes to travel this distance. The units for speed are **distance per unit time**, e.g. **km per hour** or **metres per second**. The **formula** to calculate the (average) speed of an object given the distance and time is:

$$\text{Speed} = \frac{\text{Distance}}{\text{Time}}$$

The formula gives the **average speed** of the journey — the actual speed is likely to change throughout.

Example 1

A car travels 81 km in 45 minutes. What is the average speed of the car in km/h?

1. Convert the time to hours. $45 \text{ minutes} = 45 \div 60 = 0.75 \text{ hours}$

2. Substitute the distance and time into the formula. The units of speed are a combination of the units of distance (km) and time (hours).

$$\text{Speed} = \frac{\text{Distance}}{\text{Time}}$$

$$= \frac{81 \text{ km}}{0.75 \text{ hours}} = \textbf{108 km/h}$$

Tip: Convert the distance and time into the relevant units before doing the calculation.

Exercise 1

Q1   Find the average speed of the following:
  a) a plane flying 1800 miles in 4.5 hours
  b) a lift travelling 100 m in 80 seconds
  c) a cyclist travelling 34 km in 1.6 hours
  d) an escalator step moving 15 m in 24 seconds

Q2   Find the average speed of the following in km/h:
  a) a train travelling 300 000 m in 2.5 hours
  b) a river flowing 5.25 km in 45 minutes
  c) a fish swimming 0.5 km in 12 minutes
  d) a balloon rising 700 m in 3 minutes

Q3   A spacecraft travels 232 900 miles to the moon in 13.7 hours.
    Calculate the average speed of the spacecraft.

Q4   A bobsleigh covers 1.4 km in 65 seconds. Find its average speed in metres per second.
    Give your answer to 2 significant figures.

Q5   A tortoise walks 98 cm in 8 minutes. Find the tortoise's average speed in m/s to 1 significant figure.

You can **rearrange** the speed formula to give formulas for **distance** and **time**:

$$\text{Distance} = \text{Speed} \times \text{Time}$$

$$\text{Time} = \frac{\text{Distance}}{\text{Speed}}$$

You can use the **formula triangle** given below to help you remember all three of the formulas. To use the formula triangle, **cover up** the measurement that you want to find and **write down** the two measurements that are left.

- To find **speed**, cover up **S** to leave $\frac{D}{T}$
- To find **distance**, cover up **D** to leave **S × T**
- To find **time**, cover up **T** to leave $\frac{D}{S}$

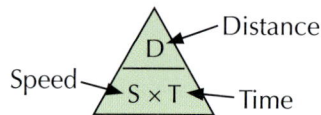

Distance

Speed

Time

Remember to always **check the units** before doing a calculation. If you have **speed** in **mph** and **time** in **minutes**, then doing speed × time won't give you the correct answer.

Example 2

A man sprints for 25 seconds with an average speed of 8 m/s. How far does he sprint?

1. Write down the formula for distance.

   Distance = Speed × Time

2. Check units — you have speed in m/s and time in seconds, so your answer will be in metres.

   Speed = 8 m/s
   Time = 25 seconds

3. Substitute the numbers into the formula.

   Distance = 8 m/s × 25 seconds = **200 m**

Example 3

A car travels 85 miles at an average speed of 34 mph. How many minutes does the journey take?

1. Write down the formula for time.

   $$\text{Time} = \frac{\text{Distance}}{\text{Speed}}$$

2. Check units — you have speed in mph and distance in miles, so your answer will be in hours.

   Distance = 85 miles
   Speed = 34 mph

3. Substitute the numbers into the formula.

   $$\text{Time} = \frac{85 \text{ miles}}{34 \text{ mph}} = 2.5 \text{ hours}$$

4. Finally, convert hours to minutes.

   2.5 hours = 2.5 × 60 = **150 minutes**

Exercise 2

Q1   For each of the following, use the speed and time given to calculate the distance travelled.
   a) speed = 98 km/h, time = 3.5 hours
   b) speed = 25 mph, time = 2.7 hours
   c) speed = 15 m/s, time = 9 minutes
   d) speed = 72 mph, time = 171 minutes

Q2   For each of the following, use the speed and distance given to calculate the time taken.
   a) speed = 2.5 km/h, distance = 4 km
   b) speed = 5 m/s, distance = 9.3 m
   c) speed = 9 m/s, distance = 61.2 km
   d) speed = 8 cm/s, distance = 1.96 m

Q3   A dart is thrown with speed 15 m/s and hits a dartboard 2.4 m away. For how long is the dart in the air?

Q4   A flight to Spain takes 2 hours and 15 minutes. If the plane travels at an average speed of 480 mph, how far does the plane travel?

Q5   A girl skates at an average speed of 7.5 mph. How far does she skate in 75 minutes?

Q6   A train travels at 78 km/h for 5.6 km. How long does the journey take to the nearest minute?

Density, Mass and Volume

Learning Objective — Spec Ref N13/R11:
Know and use the formula linking density, mass and volume.

Density is another compound measure — it's the **mass per unit volume** of a substance and is usually measured in **kg/m³** or **g/cm³**. Different substances have **different densities** — for example, gold has a **higher density** than ice. The **formulas** that connect density, mass and volume are:

$$\text{Density} = \frac{\text{Mass}}{\text{Volume}} \qquad \text{Volume} = \frac{\text{Mass}}{\text{Density}} \qquad \text{Mass} = \text{Density} \times \text{Volume}$$

The **formula triangle** on the right gives a summary of all the formulas. Remember to **check the units** of the measurements that you're putting into a formula — that way you'll know the units of the measurement that comes out.

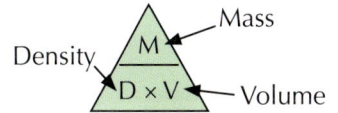

Density $\qquad$ Mass $\qquad$ $D \times V$ $\qquad$ Volume

Example 4

A 1840 kg concrete block has a volume of 0.8 m³. Calculate the density of the concrete block.

1. Write down the formula for density.

   $$\text{Density} = \frac{\text{Mass}}{\text{Volume}}$$

2. Check units — the mass is in kg and the volume is in m³, so your answer will be in kg/m³.

   Mass = 1840 kg, Volume = 0.8 m³

3. Substitute the numbers into the formula.

   $$\text{Density} = \frac{1840 \text{ kg}}{0.8 \text{ m}^3} = \textbf{2300 kg/m}^3$$

Example 5

The mass of a bathtub filled with water is 225 kg. Water has a density of 1000 kg/m³. If the empty bathtub has a mass of 45 kg, what is the volume of water in the bathtub?

1. Calculate the mass of the water on its own.

   225 kg – 45 kg = 180 kg

2. Write down the formula for volume.

   $$\text{Volume} = \frac{\text{Mass}}{\text{Density}}$$

3. Check units — the mass is in kg and the density is in kg/m³, so your answer will be in m³.

   Mass = 180 kg, Density = 1000 kg/m³

4. Substitute the numbers into the formula.

   $$\text{Volume} = \frac{180 \text{ kg}}{1000 \text{ kg/m}^3} = \textbf{0.18 m}^3$$

Exercise 3

Q1 For each of the following, use the mass and volume given to calculate the density.
   a) mass = 200 kg, volume = 540 m³
   b) mass = 23 kg, volume = 0.5 m³
   c) mass = 1088 kg, volume = 1.6 m³
   d) mass = 2498 g, volume = 0.25 cm³

Q2 For each of the following, use the density and mass given to calculate the volume.
   a) density = 8 kg/m³, mass = 1 kg
   b) density = 1510 kg/m³, mass = 3926 kg
   c) density = 90 g/cm³, mass = 540 g
   d) density = 240 kg/m³, mass = 14.4 kg

Q3 A limestone statue has a volume of 0.4 m³. Limestone has a density of 2610 kg/m³. Calculate the mass of the statue.

Pressure, Force and Area

Learning Objective — Spec Ref N13/R11:
Know and use the formula linking pressure, force and area.

Prior Knowledge Check: Be familiar with areas of shapes. See Section 27.

Like speed and density, **pressure** is a compound measure — it's the **force of an object per unit area**. Pressure is usually measured in **N/m²** (also known as pascals, **Pa**), or **N/cm²**, where **N** is **Newtons**, the unit of force. The **formulas** that connect pressure, force and area are:

$$\text{Pressure} = \frac{\text{Force}}{\text{Area}} \qquad \text{Area} = \frac{\text{Force}}{\text{Pressure}} \qquad \text{Force} = \text{Pressure} \times \text{Area}$$

The **formula triangle** on the right gives a summary of these formulas. ⟶

Force / Pressure / Area — F / P × A

Example 6

An object is resting with its base on horizontal ground. The area of the object's base is 20 cm² and the object weighs 60 N. What pressure is the object exerting on the ground?

1. Write down the formula for pressure. $\quad \text{Pressure} = \frac{\text{Force}}{\text{Area}}$

2. Check units — you have force in N and area in cm², so your answer will be in N/cm².
   Force = 60 N
   Area = 20 cm²

   Tip: Weight is the force of the object on the ground due to gravity.

3. Substitute the numbers into the formula. $\quad \text{Pressure} = \frac{60\,\text{N}}{20\,\text{cm}^2} = \textbf{3 N/cm}^2$

Example 7

A laptop with a base of 0.07 m² is resting on a desk and exerting a pressure of 330 Pa. How much does the laptop weigh?

1. Write down the formula for force. $\quad \text{Force} = \text{Pressure} \times \text{Area}$

2. Check units — you have area in m² and pressure in Pa (which is N/m²), so your answer will be in N.
   Pressure = 330 Pa
   Area = 0.07 m²

3. Substitute the numbers into the formula. $\quad \text{Force} = 330\,\text{Pa} \times 0.07\,\text{m}^2 = \textbf{23.1 N}$

Exercise 4

Q1 For each of the following, calculate the missing measure.
   a) Pressure = ? Pa, Area = 4 m², Force = 4800 N
   b) Pressure = ? N/cm², Area = 80 cm², Force = 640 N
   c) Pressure = 180 N/m², Area = ? m², Force = 540 N
   d) Pressure = 36 N/cm², Area = 30 cm², Force = ? N

Q2 A cube of metal with a volume of 512 cm³ is resting with one of its faces on horizontal ground. The cube has a weight of 1792 N. What pressure is the cube exerting on the ground? *(PROBLEM SOLVING)*

Q3 A cylinder with a height of 80 cm is resting with one of its circular faces on horizontal ground. The weight of the cylinder is 560 N and it exerts a pressure of 70 000 N/m² on the ground. What is the volume of the cylinder in cm³? *(PROBLEM SOLVING)*

Converting Compound Measures

Prior Knowledge Check:
Know how to convert between different units. See Section 22.

Learning Objective — Spec Ref R1/R11:
Convert units for compound measures.

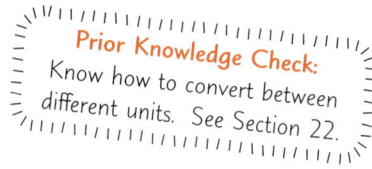

To **convert** the units of a compound measure, you can convert the **individual units** that make up compound measure separately. For example, to convert **km/h to m/s** you would need to do **two separate conversions** — first convert **km** to **metres** (to give m per hour) and then convert **hours** to **seconds** (to give m/s). It doesn't matter which unit you convert first, but **always check your conversions** to make sure your answers are sensible.

If you were converting from, say, **km/h to mph** you would just need to do the conversion from **km** to **miles** because the unit of time (hours) is the **same**.

Tip: Other common compound measures include rates of pay (e.g. £ per hour) and prices per unit mass/volume (e.g. £ per kg or pence per ml).

Example 8

Using the conversion 1 mile ≈ 1.6 km, what is 22 m/s in mph?

1. Use the conversion factor for metres to miles. There'll be fewer miles per second than m/s, so divide.

   1 mile ≈ 1.6 km = 1600 metres
   22 m/s = $\frac{22}{1600}$ = 0.01375 miles per second

2. Work out the conversion factor for seconds to hours. There'll be more mph than miles per second, so multiply.

   1 hour = 60 mins = 3600 seconds
   0.01375 miles/s = 0.01375 × 3600 = **49.5 mph**

Example 9

What is 20 kg/m³ in g/cm³?

1. Work out the conversion factor for kg to grams. There'll be more g/m³ than kg/m³, so multiply.

   1 kg = 1000 g
   20 kg/m³ = 20 × 1000 = 20 000 g/m³

2. Work out the conversion factor for m³ to cm³. There'll be fewer g/cm³ than g/m³, so divide.

   1 m³ = 1 000 000 cm³ (= 100³)
   20 000 g/m³ = $\frac{20\ 000}{1\ 000\ 000}$ = **0.02 g/cm³**

Exercise 5

Q1 For each of the following, convert the compound units:

   a) £5 per gram to £ per kg
   b) £36/hour to £ per minute
   c) 62 m/s to metres per hour
   d) 54 km/h to m/s
   e) 3000 N/cm² to N/m²
   f) 156 m per hour to cm/s
   g) 830 kg/m³ to g/cm³
   h) £12/kg to pence per gram
   i) 29.25 mph to m/s

Q2 An object has a density of 97 g/cm³. What is the density of the object in kg/m³?

Q3 Barney is paid 12p for every minute he works. What is his rate of pay in pounds per hour?

Q4 A train travels from Edinburgh to Newcastle at 40 m/s in 1.5 hours and from Newcastle to London at 46 m/s in 3 hours. What is the average speed in mph? *(PROBLEM SOLVING)*

Q5 A car travels at x km/h for the first half of a journey and then at y km/h for the second half. Write an expression for the average speed of the whole journey in m/s. *(PROBLEM SOLVING)*

23.2 Distance-Time Graphs

Distance-time graphs are used to show the journey of an object over a given period of time.

Learning Objective — Spec Ref A15:
Draw and interpret distance-time graphs.

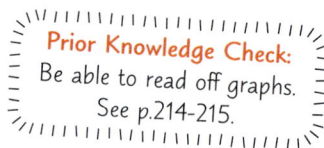

Prior Knowledge Check:
Be able to read off graphs.
See p.214-215.

A **distance-time** graph shows how far an object has travelled in a particular period of time.

- When the graph is **going up**, the object is **moving away**.
  When the graph is **going down**, the object is **coming back**.

- The **gradient** of a distance-time graph shows the **speed** of the object (see p.285). The **steeper** the graph, the **faster** the object is moving.

- A **straight** line shows the object is moving at a **constant speed**.
  A **horizontal** line means the object is **stationary**.

- A **curved graph** shows that the **speed** of the object is **changing**. If the curve is getting **steeper**, the object is **accelerating**. If the curve is getting **flatter**, the object is **decelerating**.

Tip: Be careful when reading distance-time graphs — they show distance from a point, not always total distance travelled.

Example 1

The graph on the right represents a bus journey.

a) How far does the bus travel before it stops?

The bus stops when the graph is a horizontal line. Find the first point where the line becomes horizontal and read the distance off the vertical axis. **4.5 miles**

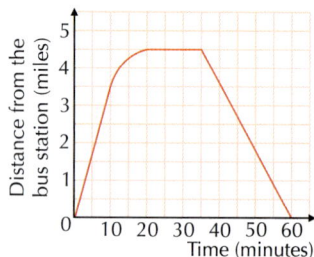

b) Describe the motion of the bus during the journey.

1. At first the graph is a straight line upwards, so the bus is moving at a constant speed away from its starting point.

2. As the graph curves, the gradient decreases — the bus is slowing down.

3. When the graph is flat, the gradient is 0 so the bus has no speed.

4. From 35 minutes, the graph is a straight line again, this time downwards — so the bus must be coming back to its starting point.

Between 0 and 10 minutes the bus was travelling at a **constant speed.**

Between 10 and 20 minutes the bus was **decelerating** (slowing down).

Between 20 and 35 minutes the bus was **stationary.**

Between 35 and 60 minutes the bus travels at a **constant speed** in the opposite direction to before.

Exercise 1

Q1 Harry walks 3 km in 50 minutes at a constant speed to his friend's house. He stays there for 1 hour. He then walks at a constant speed back towards home for 30 minutes until he gets to the shop, 1 km from home. He stays at the shop for 10 minutes before walking home at a constant speed, which takes a further 15 minutes.

Draw a graph to represent Harry's journey.

Q2 The graph on the right shows a family's car journey.
The family left home at 8:00 am.

a) (i) How long did the family travel for before stopping?
   (ii) How far had they travelled when they stopped?

b) How long did the family stay at their destination
   before setting off home?

c) (i) What time did they start the journey back home?
   (ii) How long did the journey home take?

d) Without doing any calculations, state whether the family travelled at a greater speed
   on the way to their destination or on the way back. Explain your answer.

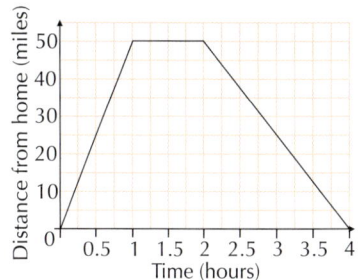

Q3 Describe the journey represented by each of these distance-time graphs.

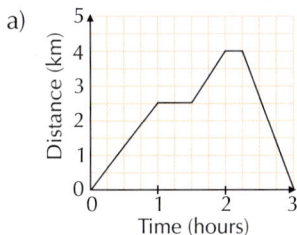

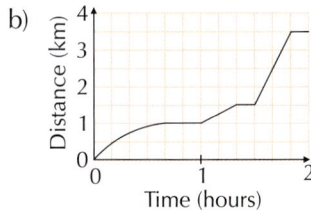

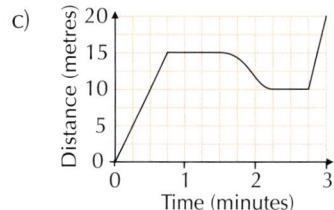

a)

b)

c)

Finding Speed from a Distance-Time Graph

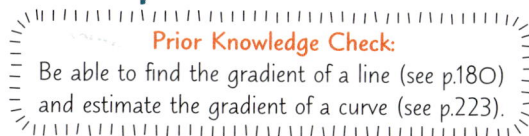

Learning Objective — Spec Ref A15/R14:
Find speed from a distance-time graph.

Prior Knowledge Check:
Be able to find the gradient of a line (see p.180)
and estimate the gradient of a curve (see p.223).

The **gradient** of a distance-time graph shows the **speed** of the object.
You can work out the speed at any stage of the journey by dividing the
distance travelled by the **time taken**. To calculate the **average speed** across
the whole journey, divide the **total distance travelled** by the **total time taken**.

If the graph is **curved**, you can **estimate** the speed at any point by drawing
a **tangent** to the curve at that point and measuring its **gradient** (see p.223).

Tip: If the graph goes
up and down, you'll
have to add up the
distance travelled at
each stage to find the
total distance travelled.

Example 2

**The graph on the right represents a train journey
from Clumpton Station to Hillybrook Station.**

a) **Find the speed of the train (in km/h) at 9:15.**

The speed is constant from
9:00 until 9:30, so use the
graph to find the distance
travelled in this time.

Distance = 75 km
Time = 30 mins = 0.5 hours

$$\text{Speed} = \frac{\text{Distance}}{\text{Time}} = \frac{75 \text{ km}}{0.5 \text{ hours}} = \textbf{150 km/h}$$

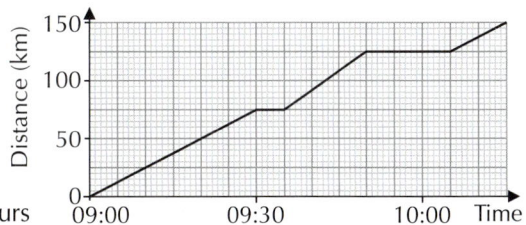

b) **Find the average speed of the train (in km/h) for this journey.**

1. First work out the total distance
   travelled and the total time taken.

   Distance = 150 − 0 = 150 km
   Time = 09:00 to 10:15 = 1 hour 15 mins = 1.25 hours

2. Then use the formula for speed.

   $$\text{Speed} = \frac{\text{Distance}}{\text{Time}} = \frac{150 \text{ km}}{1.25 \text{ hours}} = \textbf{120 km/h}$$

Q1   The graph on the right shows a commuter's journey to work.
     His journey consists of two stages of travelling,
     separated by a break of 30 minutes.

     a) What was his speed (in km/h) during
         the first stage of his journey?

     b) What was his speed (in km/h) during
         the second stage of his journey?

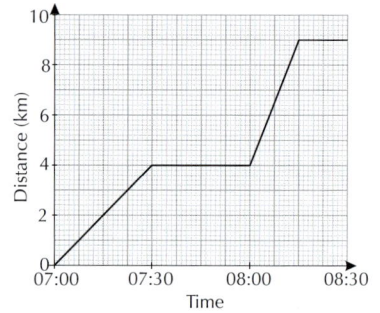

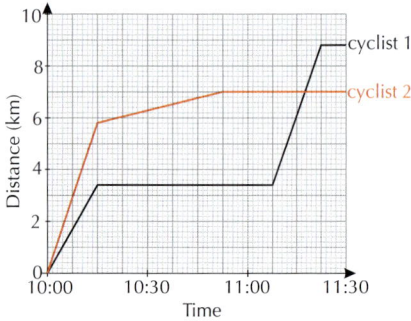

Q2   The graph on the left shows two cyclists' journeys.

     a) Which cyclist had the highest maximum speed?

     b) Find the difference in the speeds of the cyclists
         during the first 15 minutes of their journeys.

Q3   Chay travels to Clapham by taking a bus and a train. His journey is shown in the graph below.

     a) Find the average speed Chay travelled:
         (i) by bus   (ii) by train

     b) Chay spends 30 minutes in Clapham
         before he returns home by taxi at an
         average speed of 60 km/h.

         (i) Find the time it took Chay
              to travel home by taxi.

         (ii) Copy the graph and extend the line
              to show Chay's time in Clapham
              and the taxi ride home.

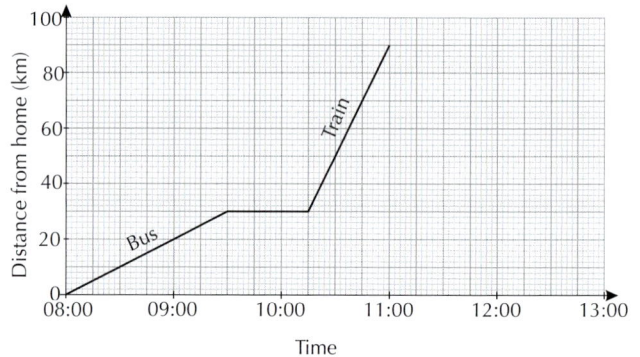

Q4   Draw a distance-time graph to show each of the following journeys:

     a) Ash catches a train at 10:00, then travels for 280 miles at a constant speed of 80 mph.

     b) Corey sets off from home at 09:00 and drives 90 miles at a constant speed of 60 mph.
         After 45 minutes at his destination, he drives back home at a constant speed of 40 mph.

Q5   Estimate the speed of the object represented by each of
     these distance-time graphs at time = 1 hour.

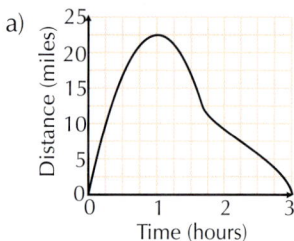

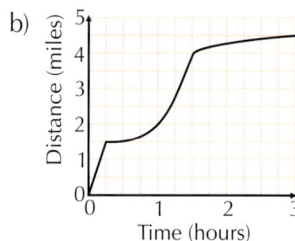

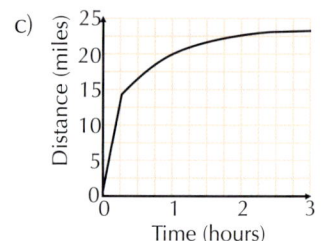

     a)   b)   c)

23.3 Velocity-Time Graphs

Velocity is speed measured in a particular direction. For example, objects with velocities of 5 m/s and −5 m/s are travelling at the same speed but in opposite directions. A velocity-time graph shows the speed and direction of an object over a period of time.

Calculating Acceleration from a Velocity-Time Graph

Learning Objectives — Spec Ref A15/R14:
- Draw and interpret velocity-time graphs.
- Find acceleration from a velocity-time graph.

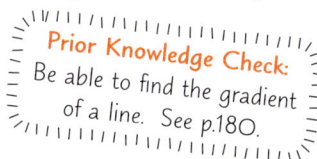

Prior Knowledge Check:
Be able to find the gradient of a line. See p.180.

Acceleration is the **rate of change** of **velocity** over time, i.e. the rate at which an object is speeding up or slowing down. The **units** for acceleration are a combination of the units for velocity and time, e.g. **m/s²**. The **formula** for acceleration is:

$$\text{Acceleration} = \frac{\text{Change in Velocity}}{\text{Change in Time}}$$

The **gradient** of a velocity-time graph shows the **acceleration** (or **deceleration**) of an object — a **positive slope** is acceleration while a **negative slope** is deceleration. The **steeper** the graph, the **greater** the acceleration (or deceleration) of the object.

Tip: A negative acceleration is the same as a deceleration. E.g. an acceleration of −5 m/s² is the same as a deceleration of 5 m/s².

A **straight** line shows that an object has a **constant acceleration**, a **horizontal** line means there is **no acceleration** so the object is moving at a **constant velocity** and a **curved** line means the rate of acceleration is **changing**.

To **estimate** the acceleration from the gradient of a curved graph, draw a **tangent** to the line at the point at which you want to find the acceleration and work out the gradient of the tangent (see p.223).

Example 1

The graph shows the velocity of a taxi during a journey. Calculate the acceleration of the taxi:

a) 6 seconds into the journey

At 6 seconds, the graph is horizontal. This means there is no acceleration, i.e. velocity is constant.

Acceleration at 6 seconds = **0 m/s²**

b) 10 seconds into the journey

1. At 10 seconds, the graph is straight and sloping down so the acceleration is constant and negative — i.e. the taxi is decelerating.

2. Use the acceleration formula. From 7 seconds to 11 seconds the velocity changed from 12 m/s to 0 m/s.

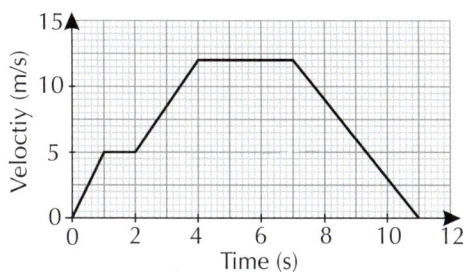

$$\text{Acceleration} = \frac{\text{Change in Velocity}}{\text{Change in Time}}$$

Change in Velocity = 0 − 12 m/s = −12 m/s
Change in Time = 11 − 7 = 4 seconds

$$\text{Acceleration} = \frac{-12 \text{ m/s}}{4 \text{ seconds}} = \textbf{−3 m/s}^2$$

Q1 Calculate the acceleration shown by each of the graphs below.

a)

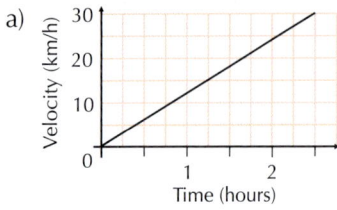

b)

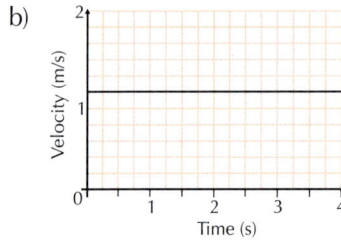

c)

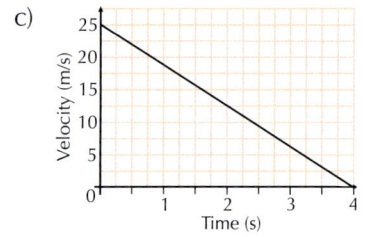

Q2 The diagram shows the velocity of a train travelling between two stations.

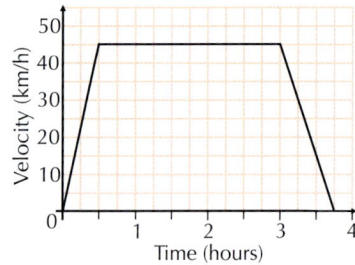

a) What is the maximum velocity of the train during the journey?

b) Calculate the initial acceleration of the train.

c) Calculate the deceleration of the train at the end of its journey.

Example 2

Look at the velocity-time graph for a tram on the right.

a) Describe the acceleration of the tram in the first second of the journey.

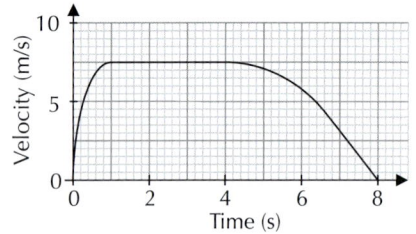

The gradient starts off steep and then decreases. This means the tram's acceleration also decreases.

The acceleration of the tram **decreases** over time.

b) Find the acceleration of the tram at 6 seconds.

1. Draw a tangent to the curve at 6 seconds.

2. The tangent goes through points (4, 9.5) and (8, 2). Work out the gradient of the line — the velocity is decreasing, so it'll be a negative acceleration.

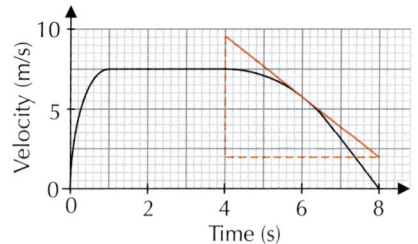

Gradient of tangent $= \dfrac{2-9.5}{8-4} = \dfrac{-7.5}{4} = $ **-1.875 m/s²**

Exercise 2

Q1 Look at the velocity-time graph on the right.

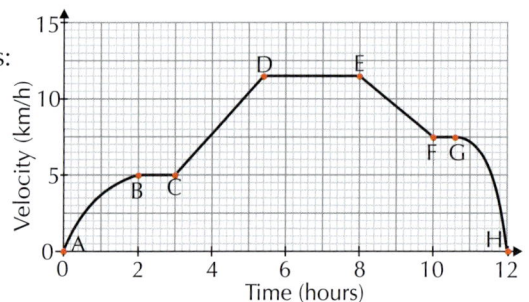

a) Describe the velocity of the object between points:

(i) A and B   (ii) D and E   (iii) G and H

b) Describe the acceleration of the object between points:

(i) A and B   (ii) D and E   (iii) G and H

c) Calculate the acceleration between points:

(i) C and D   (ii) E and F

d) Estimate the acceleration of the object after: (i) 1 hour   (ii) 11 hours

Calculating Distance from a Velocity-Time Graph

Prior Knowledge Check:
Find the area of 2D shapes. See p.349-352.

The **area** under a velocity-time graph is equal to the **distance** travelled. When a graph only has **straight lines**, you can **calculate** the **exact distance travelled** by splitting the area under the graph into triangles, rectangles and trapeziums. Then find the **area** of each shape and **add them up**.

Example 3

The graph on the right shows the velocity of a runner. Calculate the total distance the runner travelled.

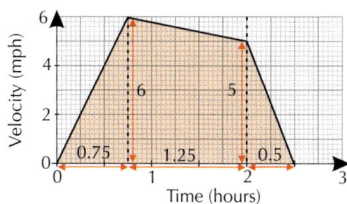
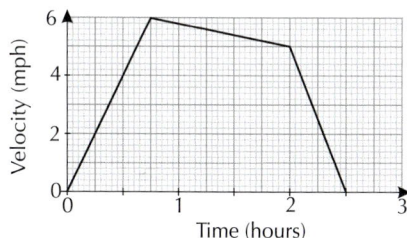

1. Split the area into two triangles and a trapezium and find the dimensions of the shapes using the axes.

2. Find the area of each section of the graph. Then add them up to find the total area.

    Area of 1st triangle = $\frac{1}{2}(6 \times 0.75) = 2.25$

    Area of trapezium = $\frac{1}{2}(6 + 5) \times 1.25 = 6.875$

    Area of 2nd triangle = $\frac{1}{2}(5 \times 0.5) = 1.25$

    Total area = $2.25 + 6.875 + 1.25 = 10.375$

3. The units on the graph are mph and hours so the total distance will be in miles.

    So the total distance travelled is **10.375 miles**.

Exercise 3

Q1 Calculate the distance travelled for each of the following graphs.

a)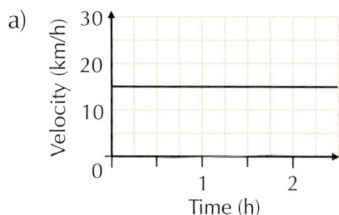

b)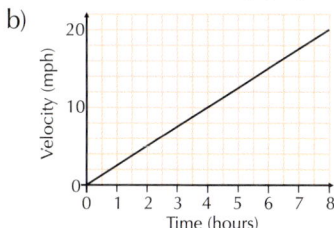

c)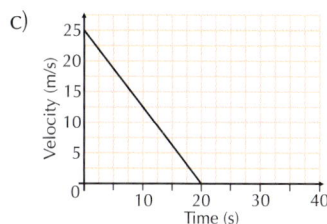

Q2 The diagram shows the velocity of a train for the first 60 seconds of a journey.

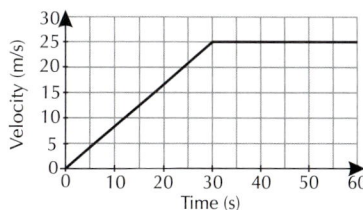

a) Calculate the total distance travelled during this part of the train's journey.

b) Calculate the average velocity of the train.

Q3 A car accelerates for 8 seconds, causing its velocity to increase from 0 m/s to 48 m/s. Then it travels at a constant velocity of 48 m/s for 20 seconds, before decelerating to 23 m/s over 5 seconds.

a) Draw a velocity-time graph to show the car's velocity during this part of the journey.

b) Calculate both the acceleration and deceleration of the car.

c) Calculate the distance travelled by the car.

When a velocity-time graph is **curved**, you can **estimate** the **total distance** travelled by **roughly** splitting the area under the graph into triangles, rectangles and trapeziums. There are often **different ways** to split the graph up and you'll get a **different estimate** depending on how you do it. A common way to split graphs up is into strips of **equal width**, but this doesn't always give the most accurate estimate.

Tip: The greater the number of sections you split the graph into, the more accurate your estimate is likely to be.

If the shapes cover **less than** the whole area under the graph, the distance will be an **underestimate**.
If the shapes cover **more than** the whole area under the graph, the distance will be an **overestimate**.

Example 4

The graph on the right shows the velocity of a ball after it is kicked into the middle of a field. Estimate the total distance the ball travelled.

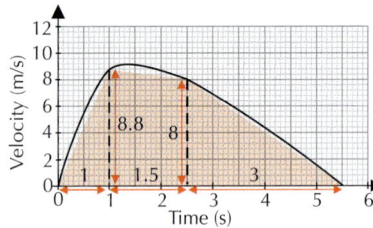

1. Roughly split the area under the curve into simple shapes. Here the area has been split into two triangles and a trapezium. Then use the axes to workout the dimensions of the shapes.

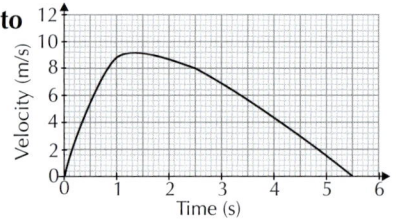

2. Find the area of each section of the graph. Then add them up to find the total area. All of the shapes are under the graph so the answer will be an underestimate of the total area.

Area of 1st triangle $= \frac{1}{2}(8.8 \times 1) = 4.4$

Area of trapezium $= \frac{1}{2}(8.8 + 8) \times 1.5 = 12.6$

Area of 2nd triangle $= \frac{1}{2}(8 \times 3) = 12$

Total area $= 4.4 + 12.6 + 12 = 29$

3. The units on the graph are m/s and seconds so the total distance will be in metres.

So an estimate for the total distance the ball travelled is **29 m**.

Exercise 4

Q1 The diagram shows the velocity of an object over 60 seconds. It has been split up into 20 second intervals.

a) Estimate the distance travelled during the first 20 seconds of the journey.

b) Estimate the distance travelled during the last 20 seconds of the journey.

c) By splitting the middle 20 seconds into two intervals of 10 seconds each, estimate the total distance the object travelled in 60 seconds.

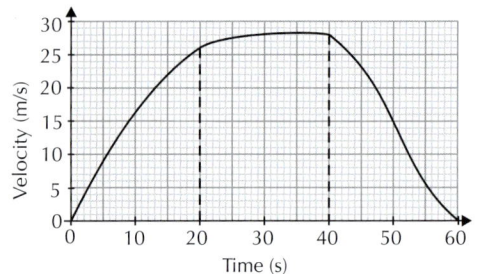

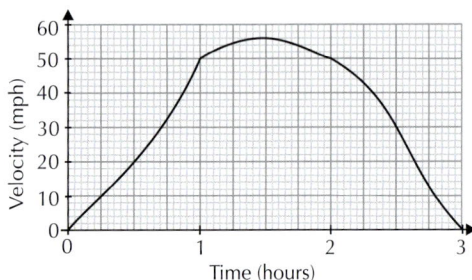

Q2 The graph on the left shows the velocity of an object during a 3 hour journey.

a) Estimate the total distance travelled by the object.

b) Estimate the average velocity of the object over the 3 hour journey.

Review Exercise

Q1 It takes a high-speed train 25 minutes to travel 240 km. Calculate the average speed of the train in kilometres per hour to 3 significant figures.

Q2 If a skydiver falls at a terminal velocity of 120 mph for 16 seconds, how many miles do they fall? Give your answer to 2 significant figures.

Q3 For each of the following, use the formula for density, mass and volume to find the missing value.

a) mass = 642 kg, volume = 0.05 m^3

b) mass = 0.06 kg, volume = 0.025 m^3

c) density = 42 kg/m^3, volume = 6.2 m^3

d) density = 120 g/cm^3, mass = 4.8 g

Q4 A cereal box is resting with its base on a horizontal surface. The weight of the box is 1.5 N and it exerts a pressure of 150 Pa on the surface. What is the area of the base of the box in cm^2?

Q5

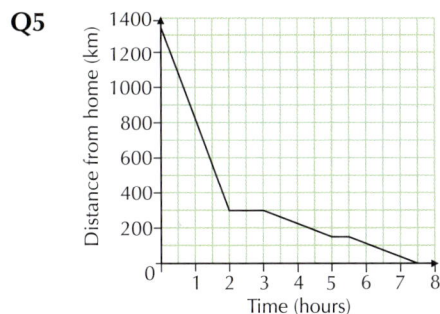

Elsie and Aggie travel home from Vienna to York. They take a flight from Vienna to London, go through passport control, then drive from London to York, stopping once for a break.

a) How long was the flight?

b) How long was their break during the driving section of the journey?

c) How far did they drive?

Q6 Find the speed of the object represented by each of the following distance-time graphs.

a)

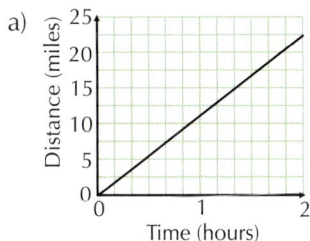

b)

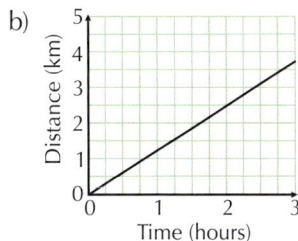

c)

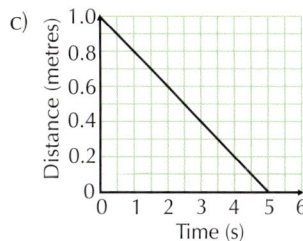

Q7 The velocity-time graph on the right shows the velocity of two ostriches as they run.

a) How far did each ostrich run?

b) Calculate the average velocity of Bert.

c) Work out the initial acceleration of Ernie.

d) Without any further calculation, state which ostrich had the greatest initial acceleration. Explain your answer.

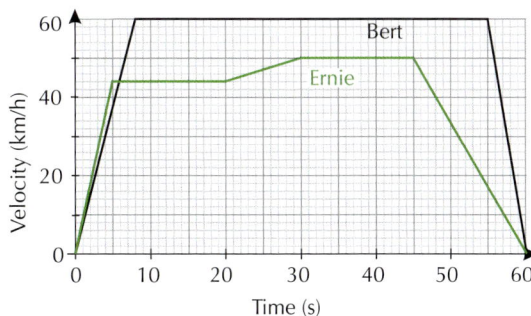

Exam-Style Questions

Q1 36 cm³ of copper and 4 cm³ of tin are melted down and mixed to make a bronze medal. The densities of copper and tin are 9 g/cm³ and 7 g/cm³ respectively. Work out the mass of the medal.

[2 marks]

Q2 A skip exerts a pressure of 625 Newtons per square metre (N/m²) over a ground area of 40 000 cm². Calculate the force in Newtons (N) exerted by the skip on to the ground.

[3 marks]

Q3 Two cyclists, Michael and Nigel, plan to meet for a coffee. Michael starts his ride from Keswick and Nigel from Lorton. Nigel's journey is shown by the graph on the right.

Michael cycles at a constant speed that is half that of Nigel's. He arrives at the meeting point 2 minutes earlier than Nigel, where he waits. What time did Michael set off?

PROBLEM SOLVING

[4 marks]

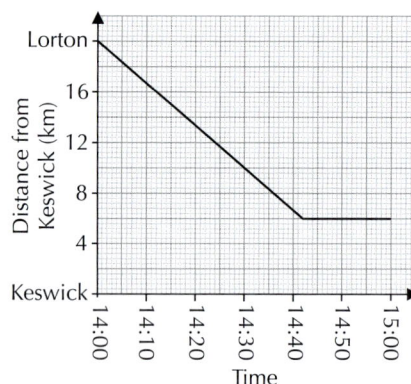

Q4 The density of the metal mercury is 13.6 g/cm³. Work out the density of mercury in kg/m³.

[3 marks]

Q5 The graph on the right shows the velocity of a bus in m/s for the first minute of a journey.

a) Explain what the gradient of the graph represents.

[1 mark]

b) By using three strips of equal width, estimate the distance travelled by the bus during this minute.

[3 marks]

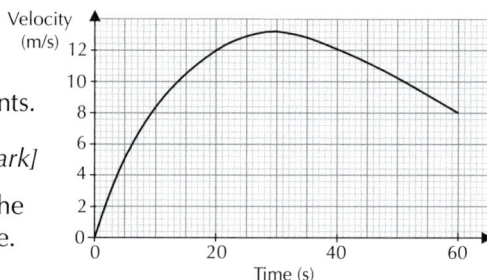

c) Is your answer to part b) an underestimate or an overestimate for the actual distance travelled? You must give a reason for your answer.

[1 mark]

24.1 Scale Drawings

It's important for things like maps to show distances accurately, but scaled down to a manageable size. Drawing diagrams to the correct scale is all about ratios and converting between different units of length.

Learning Objectives — Spec Ref R2/G15:
- Interpret and construct scale drawings.
- Work out real-life distances from maps and vice versa.

Prior Knowledge Check:
Use ratio notation and apply ratios to problems with real contexts (Section 4). Convert between metric units of length (p.276).

The **scale** on a map or plan tells you the **relationship** between the distances on the **map** and in **real life**.

For example, a map scale of 1 cm:100 m means that 1 cm on the **map** represents an **actual distance** of 100 m. A scale **without units** (e.g. 1:100) means you can use **any units** as long as they're the **same** — e.g. 1 cm:100 cm, 1 mm:100 mm, etc.

To work out what a map distance **represents in real life**, **multiply** the map distance by the real-life distance **represented by 1 cm** on the map — so for a map with scale 1 cm:10 km, a map distance of 2.5 cm would be 2.5 × 10 = 25 km in real life. To convert **real-life distances** to **map distances**, you **divide** by this value — so 50 km in real life would be represented by 50 ÷ 10 = 5 cm on a map with scale 1 cm:10 km.

Example 1

The diagram shows a plan of a square garden, drawn to a scale of 1 cm:5 m.

a) **With the help of a ruler, find the actual perimeter of the garden.**

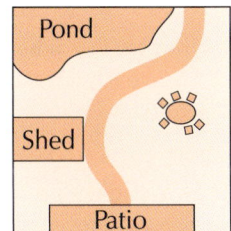

1. Measure a side of the garden with your ruler. Both the length and width should be the same, since the diagram is a square.

   Length = width = 3 cm

2. Use the scale to work out the actual length of one side.

   1 cm represents 5 m, so
   3 cm represents 3 × 5 = 15 m

3. Each length is 15 m, so use this to work out the perimeter.

   Perimeter = 4 × 15 m = **60 m**

b) **Two of the trees in the garden are 2.5 m apart. How far apart should they be drawn on the plan?**

Use the scale to work out the length on the plan that represents an actual distance of 2.5 m.

5 m is shown as 1 cm, so
2.5 m is shown as 2.5 ÷ 5 = **0.5 cm**

Example 2

A plan of the grounds of a palace has a scale of 1:4000.

a) **Represent this scale in the form 1 cm:n m.**

1. Write down the scale using centimetres.

   1 cm:4000 cm

2. Convert the right-hand side to metres by dividing by 100.

   1 cm:(4000 ÷ 100) m
   1 cm:40 m

Tip: You could also do:
   1 m:4000 m
   100 cm:4000 m
   1 cm:40 m

b) On the plan, one of the lakes is 11.5 cm wide. Calculate the actual width of the lake.

Use the scale to work out what
11.5 cm represents in real life.

1 cm represents 40 m, so
11.5 cm represents 11.5 × 40 = **460 m**

Exercise 1

Q1 The scale on a map of Europe is 1 cm : 50 km. Find the distance used on the map to represent:

a) 150 km

b) 600 km

c) 1000 km

d) 25 km

e) 10 km

f) 15 km

Q2 The floor plan of a house is drawn to a scale of 1 cm : 2 m.
Find the actual dimensions of the rooms if they are shown on the plan as:

a) 2.7 cm by 1.5 cm

b) 3.2 cm by 2.2 cm

c) 1.85 cm by 1.4 cm

d) 0.9 cm by 1.35 cm

Q3 A bridge of length 0.8 km is to be drawn on a map with scale 1 cm : 0.5 km.
What length will the bridge appear on the map?

Q4 A set of toy furniture is made to a scale of 1 : 40. Find the dimensions of the actual furniture
when the toys have the following measurements.

a) Width of bed: 3.5 cm

b) Length of table: 3.2 cm

c) Height of chair: 2.4 cm

Q5 A road of length 6.7 km is to be drawn on a map. The scale of the map is 1 : 250 000.
How long will the road be on the map in centimetres?

Q6 A model railway uses a scale of 1 : 500.
Use the actual measurements given below to find measurements for the model in centimetres.

a) Length of footbridge: 100 m

b) Height of signal box: 6 m

Q7 The diagram on the right shows
a plan for a kitchen surface.
The scale is 1 mm : 3 cm.

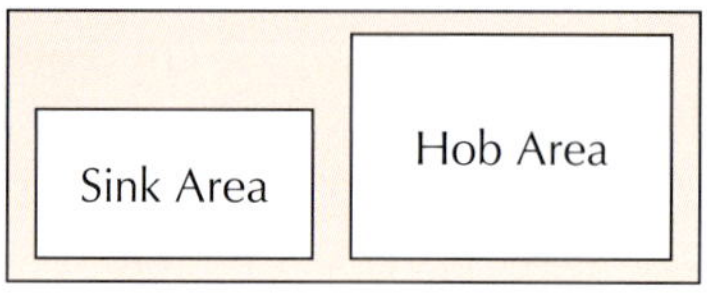

Use the plan to find the actual dimensions of: a) the sink area,   b) the hob area.

Q8 A path of length 4.5 km is shown on a map as a line of length 3 cm.
Express the scale in the form 1 cm : n km.

Q9 The plan for a school has a scale of 1 : 1500.

a) Express this scale in the form 1 cm : n m.

b) The school playground is 60 m in length. Find the corresponding length on the plan.

Q10 On the plans for a house, 30 cm represents the length of a garden with actual length 18 m.

a) Find the map scale in the form 1 : n.

b) On the plan, the width of the garden is 12 cm. What is its actual width in metres?

c) The lounge has a width of 4.5 m. What is the corresponding length on the plan?

If you know the **measurements** of something, you can draw an **accurate plan** using a given scale by first **converting** all the distances to lengths on the plan, then using your **ruler** to draw the lines to the required lengths.

The diagram shows a rough sketch of a garden.
Use the scale 1 : 400 to draw an accurate plan of the garden.

1. Write down the scale in cm.   1 cm : 400 cm

2. Change the right-hand side to metres by dividing by 100.   1 cm : 4 m

3. Use the scale to work out the lengths on the plan, then use these lengths to draw your plan.

    4 m is shown as 1 cm, so:
    12 m is shown as 12 ÷ 4 = 3 cm
    8 m is shown as 8 ÷ 4 = 2 cm
    2 m is shown as 2 ÷ 4 = 0.5 cm
    3 m is shown as 3 ÷ 4 = 0.75 cm

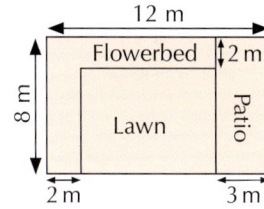

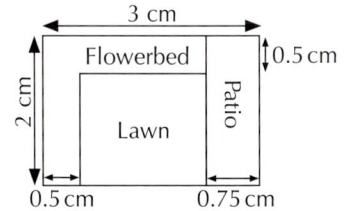

Exercise 2

Q1 Using a scale of 1 : 20, draw an accurate plan of the kitchen shown below.

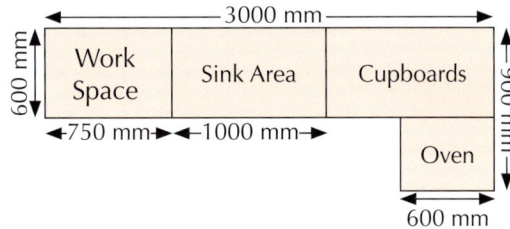

Q2 Use the scale 1 : 25 to draw a scale drawing for the house extension shown in the rough sketch below.

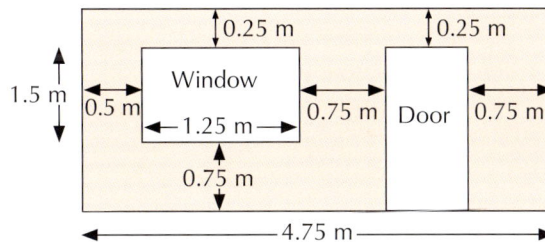

Q3 The diagram on the right shows a sketch of a park lake.

   a) Draw an accurate plan of the lake using the scale 1 cm : 3 m.

   b) There is a duck house at the intersection of *AC* and *BD*. Find the actual distance between the duck house and point *B*.

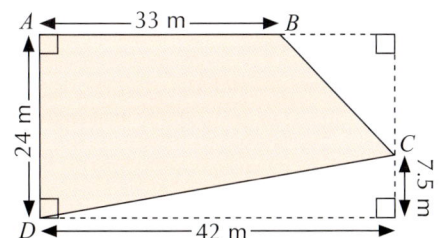

24.2 Bearings

Bearings are often used in navigation to describe which direction something is in, relative to a north line.

Learning Objective — Spec Ref G15:
Understand and use bearings.

Prior Knowledge Check:
Use properties of angles
and lines. See p.248-251.

A **bearing** tells you the direction of one point from another. Bearings are given as **three-figure angles**, measured **clockwise** from the **north line**. Draw the north line at the point you're finding the angle 'from', then go clockwise to draw the angle you want. To work out a bearing, use the information you're given along with **properties of angles** (angles **around a point**, on a **straight line**, and on **parallel lines** — see p.248-251).

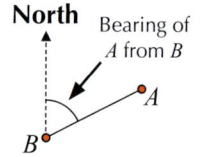

Example 1

a) **Find the bearing of B from A.**

b) **Find the bearing of C from A.**

1. Find the clockwise angle from the north line.

2. Give the bearing as a three-figure number.

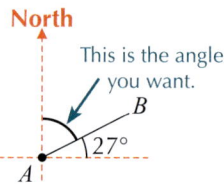

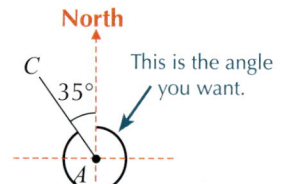

a) $90° - 27° = 63°$, so the bearing of B from A is **063°**.

b) $360° - 35° = 325°$, so the bearing of C from A is **325°**.

Exercise 1

Q1 Find the bearing of B from A in the following diagrams.

a)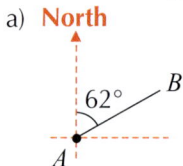

b)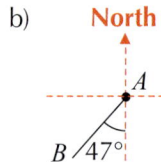

c)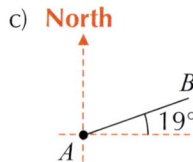

d)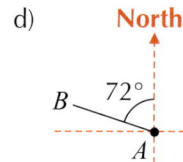

Q2 Find the size of angle x in each of the following diagrams using the information given.

a)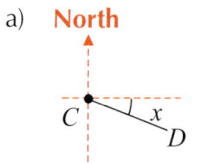

Bearing of D from C is 111°

b)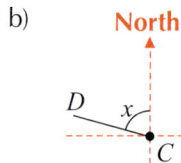

Bearing of D from C is 285°

c)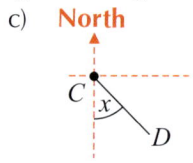

Bearing of D from C is 135°

d)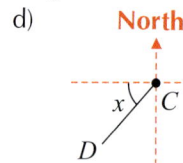

Bearing of D from C is 222°

Q3 Liverchester is 100 km due south of Manpool. King's Hill is 100 km due east of Liverchester.
a) Sketch the layout of the three locations.   b) Find the bearing of King's Hill from Liverchester.
c) Find the bearings from King's Hill of: (i) Liverchester   (ii) Manpool

Q4 Mark a point O and draw in a north line. Use a protractor to help you draw the points a) to d) with the following bearings from O.
a) 040°   b) 321°   c) 163°   d) 283°

Since **north lines** are always **parallel**, you can use the properties of **alternate**, **allied** and **corresponding** angles (see p.250-251) to work out the bearing of A from B, **given** the bearing of B from A. If you're not given a **diagram**, it's always a good idea to **sketch** one to help you to see what to do.

You might notice that the bearing of A from B is always either **180° more** or **180° less** than the bearing of B from A. If the given bearing is **less than 180°**, then you can just **add** 180° to get the other bearing, and if the given bearing is **more than 180°**, then **subtract** 180° to find the one you want.

Example 2

The bearing of X from Y is 244°. Find the bearing of Y from X.

1. Draw a diagram showing what you know. Label the angle you're trying to find.

2. Find the alternate angle to the one you're looking for.
    $244° - 180° = 64°$

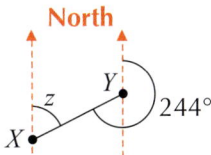

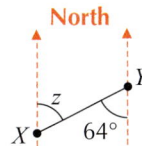

Tip: There are other methods you could use — for example, doing $360° - 244°$ to find the allied angle to the one you're looking for.

3. Alternate angles are equal, so these two are the same. Make sure you give your answer as a three-figure bearing.

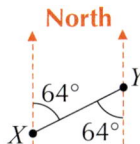

So the bearing of Y from X is **064°**.

Exercise 2

Q1 Find the bearing of A from B in the following diagrams.

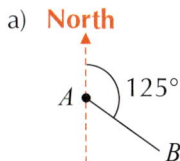

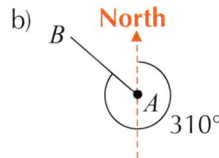

a)

b)

Q2 The bearing of H from G is 023°. Find the bearing of G from H.

Q3 The bearing of K from J is 101°. Find the bearing of J from K.

Q4 Find the bearing of N from M, given that the bearing of M from N is:
a) 200°
b) 310°
c) 080°
d) 117°
e) 015°
f) 099°

Q5 a) Measure the angle x in the diagram on the right.
b) Write down the bearing of R from S.

Q6 The point Q lies due west of point P. Find the bearing of:
a) Q from P
b) P from Q.

Q7 The point Z lies southeast of the point Y. Find the bearing of:
a) Z from Y
b) Y from Z

Q8 In the diagram on the right, find the bearing of:
a) V from U
b) U from V

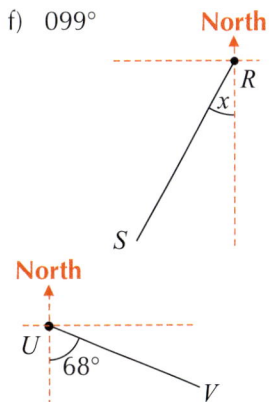

Bearings can also be used to draw **scale diagrams** (see method on p.301).
Use your **protractor** to draw any bearings and make sure any **north lines** are vertical.

Example 3

The points P and Q are 75 km apart. Q lies on a bearing of 055° from P.
Use the scale 1 cm : 25 km to draw an accurate scale diagram of P and Q.

1. Draw P and a north line. Use your protractor to measure the required angle clockwise from north.

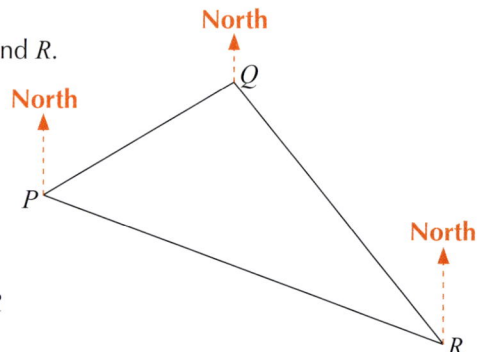

2. Use the scale to work out the distance between the two points.

25 km is shown by 1 cm, so 75 km is shown by
75 ÷ 25 = 3 cm

3. Draw Q the correct distance and direction from P.

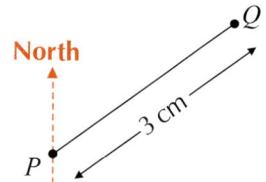

Exercise 3

Q1 A pilot flies from Blissville to Struttcastle, then on to Granich. This journey is shown by the scale diagram on the right, which is drawn using a scale of 1 cm : 100 km.

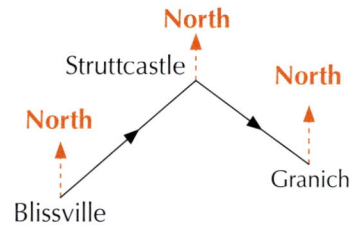

a) Find the distance and bearings from:

(i) Blissville to Struttcastle   (ii) Struttcastle to Granich

b) The pilot returns directly from Granich to Blissville. Find the actual distance travelled in this stage.

Q2 Paul's house is 150 km from Tirana on a bearing of 048°.
Draw an accurate scale diagram of Paul's house and Tirana using the scale 1 cm : 30 km.

Q3 Paradise City lies 540 km from Pretty Grimville on a bearing of 125°.
Draw an accurate scale diagram of the two locations using the scale 1 cm : 90 km.

Q4 A pilot flies 2000 km from Budarid to Madpest on a bearing of 242°.
Draw an accurate scale diagram of the journey using the scale 1 : 100 000 000.

Q5 Use the scale 1 : 22 000 000 to draw an accurate scale diagram of an 880 km journey from Budarid to Blissville on a bearing of 263°.

Q6 The scale drawing on the right shows three cities, P, Q and R. The scale of the diagram is 1 : 10 000 000.

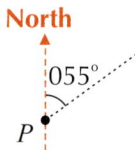

a) Use the diagram to find the following actual distances in km.

(i) PQ (ii) QR (iii) PR

b) Use a protractor to find the following bearings.

(i) Q from P (ii) R from Q (iii) P from R

24.3 Constructions

Constructing a shape means drawing it using just a pencil, a ruler, a pair of compasses and a protractor.
You can also use your ruler and compasses to draw perpendicular lines, split angles in half, and much more.

Constructing Triangles

Learning Objective — Spec Ref G1/G2:
Construct triangles using a ruler, a pair of compasses and a protractor.

To construct a triangle, you need to know **three** pieces of information about it — these could be **lengths** of the sides or **angles** between them. If you're given a **side length**, use your **ruler** to measure the length and draw an arc with your **compasses** if necessary. Measure any **angles** that you're given with your **protractor**.

When constructing shapes and lines, you should **always** leave **compass arcs** and other **construction lines** on your finished drawing to show that you've used the right method.

Example 1

Draw triangle *ABC*, where *AB* is 3 cm, *BC* is 2.5 cm and *AC* is 2 cm.

1. Draw and label side *AB*.

$A \times$————— 3 cm —————$\times B$

2. Set your compasses to 2.5 cm. Draw an arc 2.5 cm from *B*.

3. Now set your compasses to 2 cm. Draw an arc 2 cm from *A*.

4. *C* is where your arcs cross — so draw lines to finish the triangle.

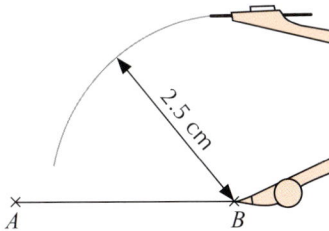
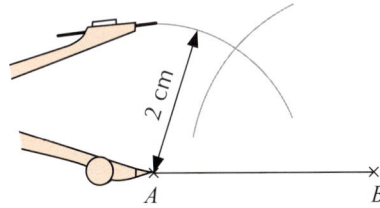
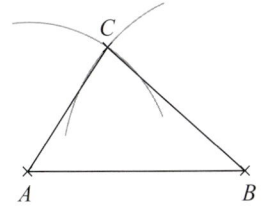

Example 2

Accurately construct the triangle *ABC* shown in the rough sketch on the right.

1. Start by drawing side *AB*.

$A \times$—— 2 cm ——$\times B$

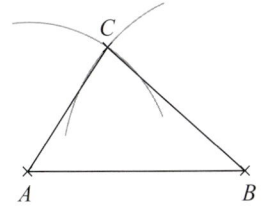

2. Measure an angle of 30° at *A* with a protractor. Mark the angle with a dot.

3. Draw a faint line from *A* through the dot.

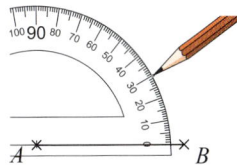
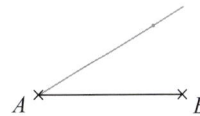

4. Do the same for the 40° angle at *B*.

5. *C* is where these lines cross, so draw darker lines to complete the triangle.

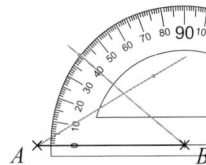
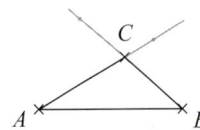

When you're given any of the following **three pieces of information,**
there's **only one triangle** you can draw (ignoring reflections and rotations).

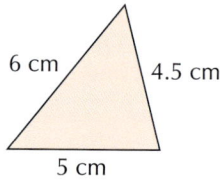

SSS — 3 sides.

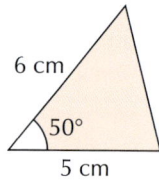

SAS — 2 sides and the angle between them.

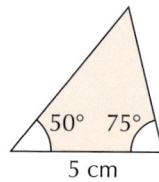

ASA — 2 angles and the side between them.

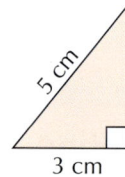

RHS — a right angle, the hypotenuse and another side.

If you're given 2 angles and a side that **isn't between them** (AAS), you can use the fact that angles in a triangle add up to 180° (see p.252) to work out the **third angle** and turn it into an **ASA** triangle.

However, if you're given two sides and an **angle** that **isn't between them** (SSA), then there are often **two possible triangles** you could draw, as shown in the diagram on the right.

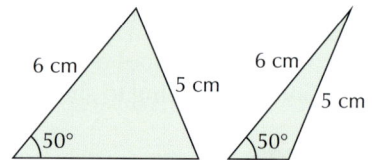

Three angles isn't enough information to draw a triangle because you only know its **shape** — you don't know how big it is.

Both triangles have a 6 cm side and a 5 cm side with a 50° angle opposite it.

Exercise 1

Q1 Draw the following triangles accurately.

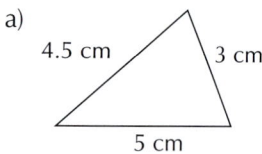

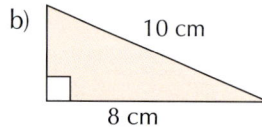

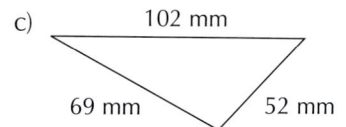

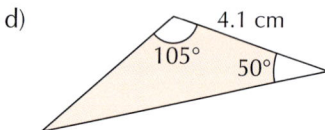

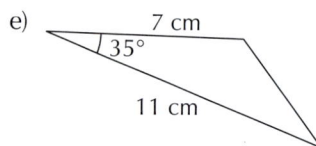

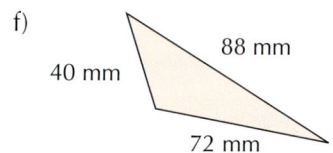

a) 4.5 cm, 3 cm, 5 cm

b) 10 cm, 8 cm

c) 102 mm, 69 mm, 52 mm

d) 4.1 cm, 105°, 50°

e) 7 cm, 35°, 11 cm

f) 88 mm, 40 mm, 72 mm

Q2 Draw each of the triangles *ABC* described below.

a) *AB* is 4 cm, *AC* is 9 cm, angle *BAC* is 50°

b) Angle *BAC* is 40°, angle *ABC* is 25°, *AB* is 7 cm

c) *AB* is 9 cm, *BC* is 9 cm, *AC* is 5 cm

d) Angle *BAC* is 90°, *AB* is 35 mm, *BC* is 61 mm

e) *AB* is 4.6 cm, *BC* is 5.4 cm, *AC* is 8.4 cm

f) *AB* is 4.4 cm, angle *ACB* is 32°, angle *ABC* is 16°

Q3 Draw an isosceles triangle with:

a) one side of length 7 cm and two sides of length 5 cm.

b) two angles of 75° and a side of length 45 mm between them.

c) two sides of length 7.1 cm and an angle of 22° between them.

Q4 In triangle *ABC*, *AB* is 10 cm, *AC* is 6 cm and angle *ABC* is 30°. Use an accurate drawing to find the two possible lengths of *BC*.

Constructing a Perpendicular Bisector

Learning Objective — Spec Ref G2:
Be able to construct a perpendicular bisector.

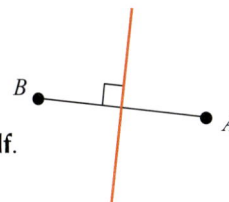

The **perpendicular bisector** of a line AB is at **right angles** to the line, and cuts it in **half**.

All points on the perpendicular bisector are **equally far** from both A and B (this is important when drawing loci — see page 313). You can use this fact to draw perpendicular bisectors **without measuring** the distances — just use your compasses to find two points that are the **same distance** from A and B, then the perpendicular bisector will **pass through both** of these points.

Example 3

Draw a line AB which is 3 cm long and construct its perpendicular bisector.

1. Draw AB.

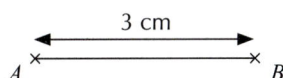

2. Place the compass point at A, with the radius set at more than half of the length AB. Draw two arcs as shown.

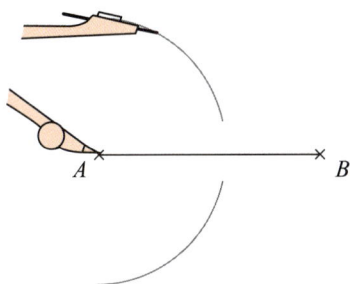

3. Keep the radius the same and put the compass point at B. Draw two more arcs.

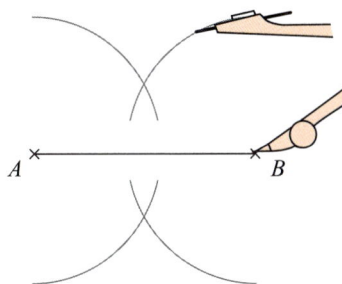

4. Use a ruler to draw a straight line through the points where the arcs meet. This is the perpendicular bisector.

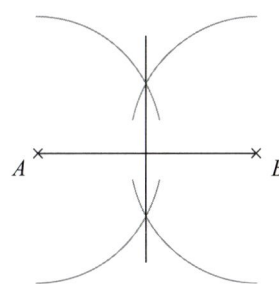

Exercise 2

Q1   Draw the following lines and construct their perpendicular bisectors using only a ruler and a pair of compasses.

a) A horizontal line PQ 5 cm long.   b) A vertical line XY 9 cm long.   c) A line AB 7 cm long.

Q2   a) Draw a line AB 6 cm long. Construct the perpendicular bisector of AB.

b) Draw the rhombus $ACBD$ with diagonals 6 cm and 8 cm.

Q3   a) Draw a circle with radius 5 cm and draw any two chords. Label your chords AB and CD.

b) Construct the perpendicular bisector of chord AB.

c) Construct the perpendicular bisector of chord CD.

d) Where do the two perpendicular bisectors meet?

Constructing an Angle Bisector

Learning Objective — Spec Ref G2:
Be able to construct an angle bisector.

An **angle bisector** is the line that cuts an angle in half.
All points on the angle bisector are the **same distance**
from each of the two lines that enclose the angle (this
is also useful for drawing loci — see page 313). Just like the perpendicular bisector, you can use your
ruler and compasses to construct an angle bisector **without measuring** the angles with a protractor.

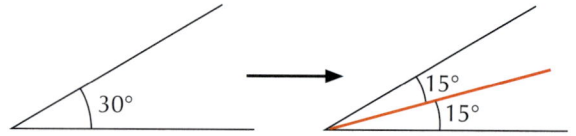

Example 4

Draw an angle of 60° using a protractor, then construct the angle bisector
using only a ruler and compasses.

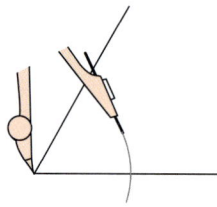

1. Place the point of the   ...and draw arcs   ...using the same radius.
    compasses on the angle...   crossing both lines...

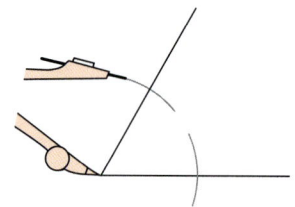

2. Now place the point of the compasses where   3. Draw the angle bisector
    your arcs cross the lines and, from each   through the point
    point, draw a new arc (using the same radius).   where the arcs cross.

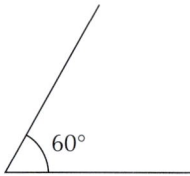

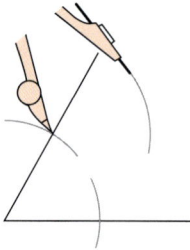

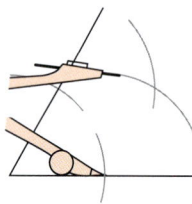

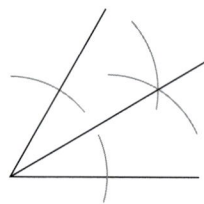

Tip: You can check
your answer by
measuring the angles
with a protractor — they
should both be 30°.

Exercise 3

Q1   Draw each angle using a protractor, then construct its angle bisector using a ruler and compasses.

    a) 100°   b) 140°   c) 96°   d) 44°

    e) 50°   f) 70°   g) 20°   h) 65°

Q2   Draw a triangle and construct the bisectors of each of the angles.
     What do you notice about these bisectors?

Q3   a) Draw an angle ABC of 110°, with $AB = BC = 5$ cm. Construct the bisector of angle ABC.

    b) Mark point D on your drawing, where D is the point on the angle bisector with $BD = 8$ cm.
        What kind of quadrilateral is $ABCD$?

Constructing a Perpendicular from a Point to a Line

perpendicular from X to AB

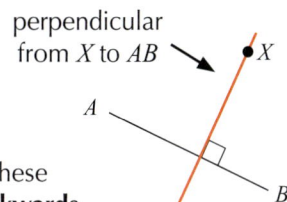

The **perpendicular** from a point to a line is the **shortest path** between them.
It should **pass through** the point, and meet the line at **90°**. Constructing one of these
is similar to constructing a **perpendicular bisector**, except you need to work **backwards**
— first, you find two points **on the line**, then you use them to find the perpendicular.

Example 5

Construct the perpendicular from the point X to the line AB
using only a ruler and a pair of compasses.

A —————— B

1. Draw an arc centred on X cutting the line twice (you may need to extend the line).

2. Draw an arc centred on one of the points where your arc meets the line.

3. Do the same for the other point, keeping the radius the same.

4. Draw the perpendicular to where the arcs cross.

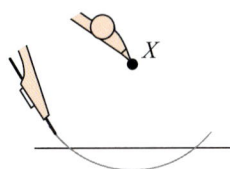
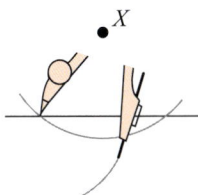
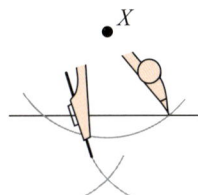
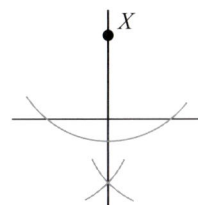

Exercise 4

Q1 Use a ruler to draw any triangle. Label the corners of the triangle X, Y and Z. Construct the perpendicular from X to the line YZ.

Q2 Draw three points that do not lie on a straight line. Label your points P, Q and R.
Draw a straight line passing through points P and Q.
Construct the shortest possible line from R to the line PQ.

Q3 a) On squared paper, draw axes with x-values from 0 to 10 and y-values from 0 to 10.

　　 b) Plot the points $A(1, 2)$, $B(9, 1)$ and $C(6, 8)$, and draw the line AB.

　　 c) Construct the perpendicular from point C to the line AB.

Q4 Draw any triangle. Construct a perpendicular from each of the triangle's corners to the opposite side. What do you notice about these lines?

Q5 a) Construct triangle DEF, where DE = 5 cm, DF = 6 cm and angle FDE = 55°.

　　 b) Construct the perpendicular from F to DE.
　　　 Label the point where the perpendicular meets DE as point G.

　　 c) Measure the length FG. Use your result to work out the area of the triangle.

Constructing 60° and 30° Angles

Learning Objective — Spec Ref G2:
Be able to construct 60° and 30° angles.

To construct a **60° angle** you can take advantage of the fact that all the interior angles of an **equilateral triangle** are 60°. Start by drawing a line, then set your compasses to match its **length** and draw an arc from **each end** of the line. The point where the arcs cross will form an **equilateral triangle** with the **end points** of the line.

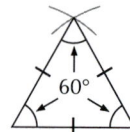

To construct a **30° angle**, first construct a 60° angle. You can then construct the **angle bisector** (see page 308) to get **two** angles of 30°.

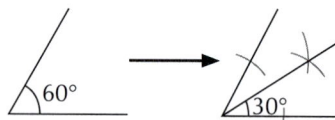

Example 6

Draw a line AB and construct an angle of 30° at A.

1. First construct an angle of 60°. Place the compass point on A and draw a long arc that crosses the line AB.

2. Place the compass point where the arc meets the line and draw another arc of the same radius.

3. Draw a straight line through A and the point where your arcs cross. The angle will be 60°.

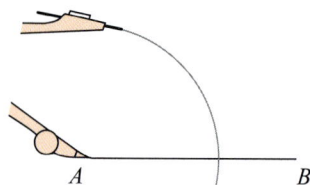

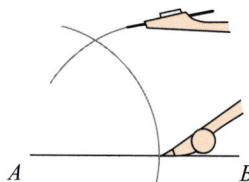

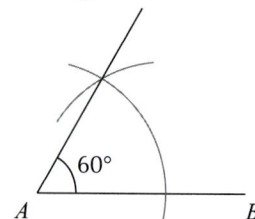

4. Now bisect this angle. You can use the arc from step 1 — so place the compass point at the points where it crosses each line and draw an arc from each of the same radius.

5. Finally, draw a straight line through A and the point where your arcs cross to get a 30° angle.

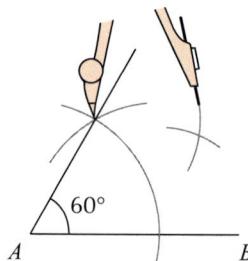

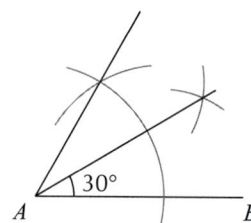

Exercise 5

Q1 Draw a line AB measuring 5 cm. Construct an angle of 60° at A.

Q2 Draw an equilateral triangle with sides measuring 6 cm.

Q3 Draw a line AB measuring 6 cm. Construct an angle of 30° at A.

Q4 Construct the triangle ABC where AB = 7 cm, angle CAB = 60° and angle CBA = 30° using only a ruler and compasses.

Q5 Construct an isosceles triangle PQR where PQ = 8 cm and the angles RPQ and RQP are both 30°, using only a ruler and compasses.

Constructing 90° and 45° Angles

Learning Objective — Spec Ref G2:
Be able to construct 90° and 45° angles.

Constructing a **90° angle** is a bit like constructing a **perpendicular from a point to a line** (see p.309) except now the point is **on the line**.

If you need to construct an angle of 45°, just construct the **angle bisector** (p.308) of the 90° angle.

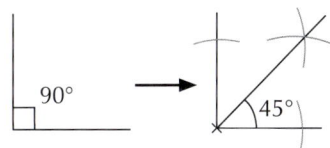

Example 7

Construct an angle of 45° using only compasses and a ruler.

1. Draw a straight line and mark the point where you want to form the angle.

2. Draw arcs of the same radius either side of the point.

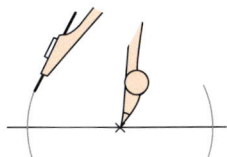

3. Increase the radius of your compasses, and draw two arcs of the same radius — one arc centred on each of the intersections.

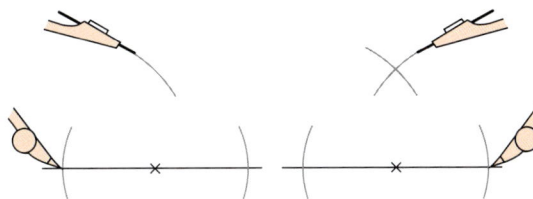

4. Draw a straight line to complete the right angle.

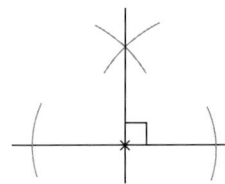

5. Now bisect the right angle to get an angle of 45°.

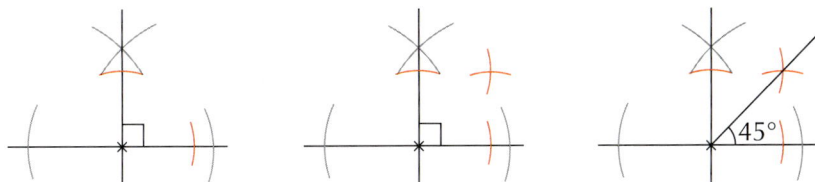

Exercise 6

Q1   Draw a straight line and mark a point roughly halfway along it. Label the point X.
Construct a right angle at X using only a ruler and compasses.

Q2   Using ruler and compasses only, construct a rectangle with sides of length 5 cm and 7 cm.

Q3   Draw a straight line, and mark a point roughly halfway along it. Label the point X.
Construct an angle of 45° at X using only a ruler and compasses.

Q4   Construct an isosceles triangle ABC where $AB = 8$ cm and the angles CAB and CBA are both 45°, using only a ruler and compasses.

Constructing Parallel Lines

To construct a line that is **parallel** to another, passing
through a given point, the first step is to construct
the **perpendicular from the point to the line** (see p.309). Once you've done that,
you just need to construct a **right angle** at the point using the method on the previous page.

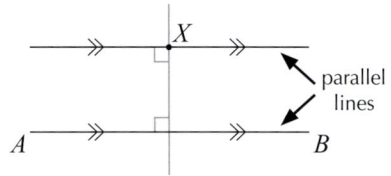

Example 8

Construct a line parallel to AB through the point P.

1. Construct the line perpendicular
   to AB passing through P (p.309).

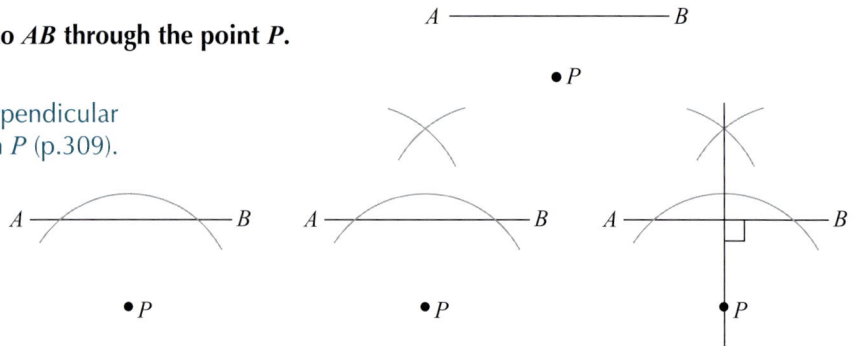

2. Construct a right angle (p.311) to this line at point P.
   This will be parallel to AB.

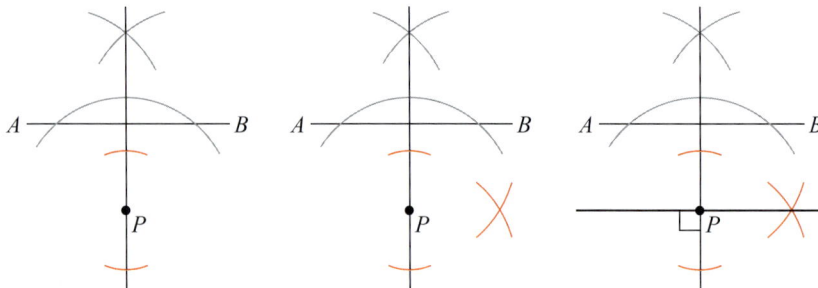

> **Tip:** You could also
> construct the parallel
> line through P by
> constructing a right
> angle anywhere on line
> AB, then constructing
> the perpendicular from
> P to this line.

Exercise 7

Q1 Draw a horizontal line AB and mark a point P approximately 4 cm from your line.
Construct a line parallel to AB through the point P.

Q2 Copy the two straight lines shown on the right.
By adding two lines, one parallel to each of these, construct a parallelogram.

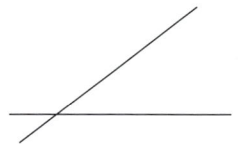

Q3 a) Use only a ruler and compasses to construct the quadrilateral $ABCD$ such that:
  ▪ $AB = 10$ cm
  ▪ angle $ABC = 60°$ and $BC = 3$ cm
  ▪ angle $BAD = 30°$
  ▪ AB and CD are parallel

  b) What type of quadrilateral is $ABCD$?

24.4 Loci

One of the most common uses for constructions is to find all the points that are a given distance from a shape or the same distance from two shapes. These sets of points are called loci (pronounced low-kai).

Learning Objectives — Spec Ref G2:
- Construct the locus of points that are a given distance from a point or a line.
- Construct the locus of points that are equidistant from two points or two lines.
- Solve problems involving loci.

A **locus** (plural **loci**) is a **set of points** which satisfy a particular condition. The types of loci you need to know are the sets of points that are a **fixed distance away** from a point or a line (or another kind of shape), and the sets of points that are **equidistant** (i.e. the **same distance**) from two points or two lines.

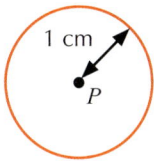

The locus of points that are a fixed distance, e.g. 1 cm, from a **point** P is a **circle** with radius 1 cm centred on P. To construct this, set your **compasses** to the given distance and draw a circle around the point.

The locus of points that are a fixed distance from a **line** AB is a 'sausage shape'. To construct this, use your compasses to draw the ends, which are **semicircles**, then join them up with your ruler.

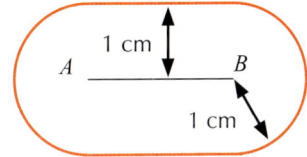

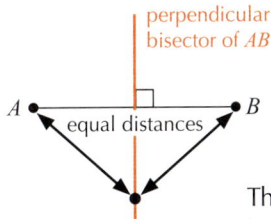

The locus of points equidistant from **two points** A and B is the **perpendicular bisector** of AB (see page 307).

The locus of points equidistant from **two lines** is their **angle bisector** (see page 308).

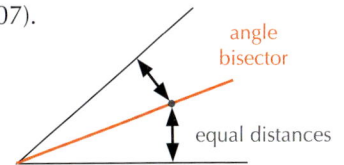

Example 1

The line AB is 2.6 cm long. Construct the locus of points that are 5 mm from AB.

1. Draw AB, then set your compasses to 5 mm and draw arcs around each end of the line. Make sure each arc is slightly more than a semicircle.

2. Using your ruler, join the tops and bottoms of the arcs with straight lines.

3. Mark the locus, leaving your construction lines on the diagram.

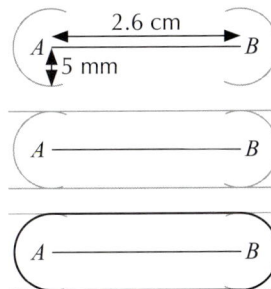

Exercise 1

Q1 Draw a line, AB, that is 7 cm long. Construct the locus of all the points 2 cm from the line.

Q2 a) Mark a point X on your page. Draw the locus of all points which are 3 cm from X.
    b) Shade the locus of all points on the page which are less than 3 cm from X.

Q3   Mark two points A and B that are 6 cm apart.
Construct the locus of all points which are equidistant from A and B.

Q4   Draw two lines that meet at an angle of 50°.
Construct the locus of all points which are equidistant from the two lines.

Q5   Draw a line AB 6 cm long. Draw the locus of all points which are 3 cm from AB.

Q6   a) Draw axes on squared paper with x- and y-values from 0 to 10.
Plot the points $P(2, 7)$ and $Q(10, 3)$.

   b) Construct the locus of points which are equidistant from P and Q.

You might have to draw **more than one locus** to find the region that satisfies **multiple conditions**.

Example 2

Shade the locus of points inside rectangle $ABCD$ that are:

- **more than 2 cm from point A,**
- **less than 1 cm from side CD.**

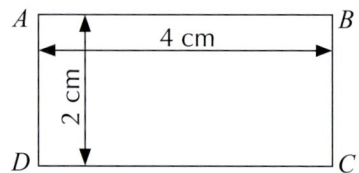

1. Use your compasses to construct an arc of radius 2 cm around A. The first condition means that points within this arc are excluded.

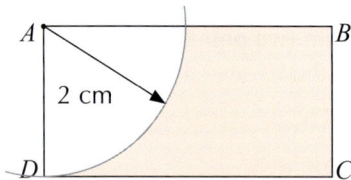

2. Use your ruler to draw a line 1 cm from CD. The second condition means that points on or above this line are also excluded.

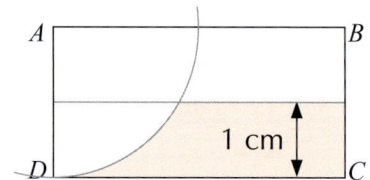

Exercise 2

Q1   a) Mark points P and Q that are 5 cm apart.

   b) Draw the locus of points which are 3 cm from P.

   c) Draw the locus of points which are 4 cm from Q.

   d) Show clearly which points are both 3 cm from P and 4 cm from Q.

Q2   Draw a line RS which is 5 cm long. Construct the locus of points
that are equidistant from R and S and less than 4 cm from R.

Q3   a) Construct a triangle with sides 4 cm, 5 cm and 6 cm.

   b) Draw the locus of all points which are exactly 1 cm from any of the triangle's sides.

Q4   a) Construct an isosceles triangle DEF with $DE = EF = 5$ cm and $DF = 3$ cm.

   b) Draw the locus of points which are equidistant from D and F and less than 2 cm from E.

Q5   Lines AB and CD are both 6 cm long. AB and CD are perpendicular
to each other and cross at the midpoint of each line, M.

   a) Construct lines AB and CD.

   b) Draw the locus of points which are less than 2 cm from AB and CD.

Loci can also be used to solve **real-life problems**, particularly on **scale diagrams** (see p.299-301).

(see p.299-301)

Example 3

The diagram shows a plan of a greenhouse, drawn at a scale of 1:400. The greenhouse has a path through the middle modelled by a straight line, and four sprinklers, shown as dots on the plan.

The sprinklers can water plants within a 4 m radius. The gardener can water anything up to 6 m away from the path using a hosepipe.

Shade the area on the diagram that can be watered.

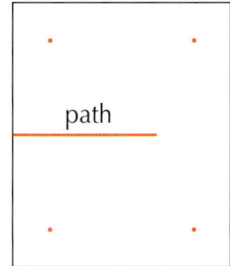

1. At a scale of 1:400, 1 cm:400 cm = 4 m. So draw arcs of radius 1 cm around each sprinkler, and shade the region inside.

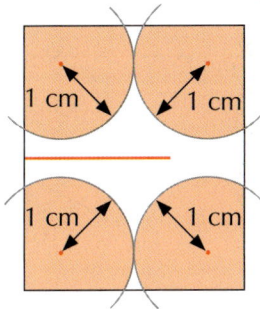

2. 6 m in real life is 6 ÷ 4 = 1.5 cm on the diagram. Construct the locus of points that are 1.5 cm away from the path and shade the region inside.

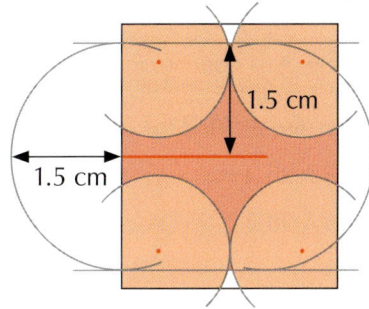

Exercise 3

Q1 A ship sails so that it is always the same distance from a port P and a lighthouse L. The lighthouse and the port are 3 km apart.

   a) Draw a scale diagram showing the port and lighthouse. Use a scale of 1 cm:1 km.

   b) Show the path of the ship on your diagram.

Q2 Two walls of a field meet at an angle of 80°. A bonfire has to be the same distance from each wall and 3 m from the corner. Copy the diagram on the right, then use a ruler and pair of compasses to show the position of the fire.

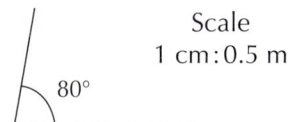

Scale
1 cm:0.5 m

Q3 Two camels set off at the same time from towns A and B, located 50 miles apart in the desert.

   a) Draw a scale diagram showing towns A and B. Use a scale of 1 cm:10 miles.

   b) If a camel can walk up to 40 miles in a day, show on your diagram the region where the camels could possibly meet each other after walking for one day.

Q4 A walled rectangular yard has length 4 m and width 2 m. A dog is secured by a lead of length 1 m to a post in a corner of the yard.

   a) Show on an accurate scale drawing the area in which the dog can move. Use the scale 1 cm:1 m.

   b) The post is replaced with a 3 m rail mounted horizontally along one of the long walls, with one end in the corner as shown. If the end of the lead attached to the rail is free to slide, show the area in which the dog can move now.

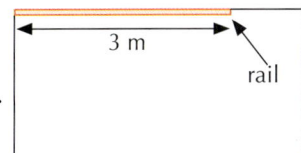

3 m

rail

Review Exercise

Q1 The map shows tourist attractions in a city.
Find the actual distances between:

a) the museum and the cathedral,

b) the art gallery and the theatre,

c) the cathedral and the theatre.

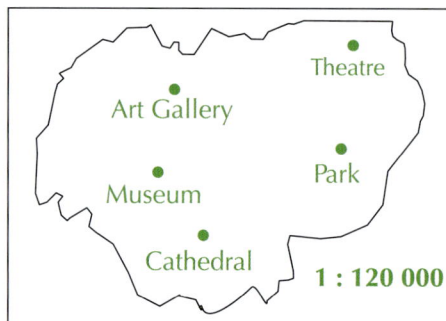

Q2 A ship, *A*, is communicating with four other nearby ships, *B*, *C*, *D* and *E*. The diagram shows the positions of each ship and some angles between them and *A*. Find the bearing of:

a) *B* from *A*   b) *C* from *A*   c) *D* from *A*   d) *E* from *A*

e) *A* from *B*   f) *A* from *C*   g) *A* from *D*   h) *A* from *E*

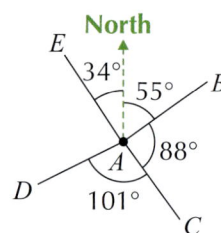

Q3 Using a scale of 1 cm : 10 miles, draw a map showing the relative positions of the three places described on the right.
(All distances are 'as the crow flies'.)

- High Cross is 54 miles due west of Low Cross.
- Very Cross is 48 miles from Low Cross and 27 miles from High Cross
- Very Cross lies to the south of High Cross and Low Cross.

Q4 Draw a line *AB* that is 8 cm long.

a) Construct an angle of 60° at *A*.

b) Complete a construction of a rhombus *ABCD* with sides of length 8 cm.

Q5 a) Construct an equilateral triangle *DEF* with sides of length 5.8 cm.

b) Construct a line that is parallel to side *DE* and passes through point *F*.

Q6 Construct the triangle *ABC* with *AB* = 7.4 cm, angle *CAB* = 60° and angle *ABC* = 45°, using only a ruler and a pair of compasses.

Q7 Construct an angle of 15° using a ruler and a pair of compasses only.

Q8 Some students are doing a treasure hunt. They know the treasure is:
- located in a square region *ABCD*, which measures 10 m × 10 m
- the same distance from *AB* as from *AD*
- 7 m from corner *C*.

Draw a scale diagram to show the location of the treasure. Use a scale of 1 cm : 1 m.

Exam-Style Questions

Q1 Many American cities have a road design of square grids. The diagram shows part of such a design which is a rectangle made up of 12 congruent squares. 3 locations in this city are labelled as the points *A*, *B* and *C*.

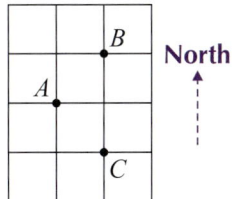

a) Write down the three figure bearing of *B* from *A*.

[1 mark]

b) Work out the three figure bearing of *A* from *C*.

[1 mark]

Q2 On a 1 : 25 000 scale map, the length of a reservoir is 3.8 cm.
Work out the actual length of the reservoir, giving your answer in metres.

[2 marks]

Q3 Accurately construct a triangle with sides of length 37 mm, 52 mm, and 60 mm.

[3 marks]

Q4 A regular hexagon can be made up of 6 identical equilateral triangles, as shown in the diagram.

Construct a regular hexagon with sides of length 3 cm using a ruler and compasses.

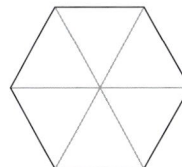

[3 marks]

Q5 A farmer decides to put a scarecrow in a field to protect her crop of wheat. The field is shown in the diagram on the right.

She positions the scarecrow so that it is an equal from each hedge and 40 m from the gate.

Copy the diagram and use a ruler and a pair of compasses to accurately show the two possible positions of the scarecrow.

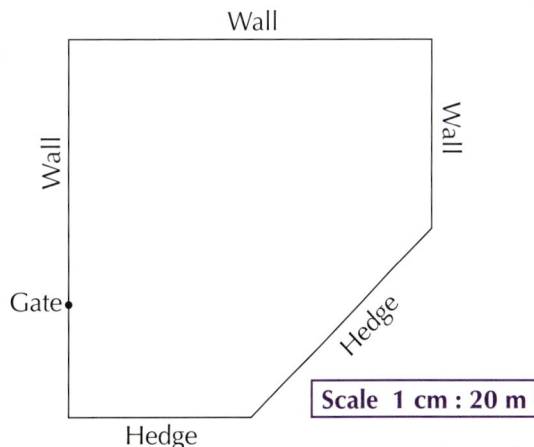

Scale 1 cm : 20 m

[4 marks]

25.1 Pythagoras' Theorem

Pythagoras' theorem can be applied to all right-angled triangles. You use it to find the length of one side if you know the lengths of the other two sides.

Prior Knowledge Check:
Be able to square numbers and find square roots (see p.91) and be familiar with surds (see p.99).

Learning Objective — Spec Ref G6/G20:
Use Pythagoras' theorem to find missing lengths in right-angled triangles.

The lengths of the sides in a **right-angled triangle** always follow the rule: $h^2 = a^2 + b^2$
h is the **hypotenuse** — this is the **longest** side, which is always **opposite** the right angle. a and b are the **shorter** sides.

This is **Pythagoras' theorem** and it is used to find lengths in right-angled triangles.

To find the length of the hypotenuse in a right-angled triangle:

- **Square** the lengths of sides a and b.

- **Add** together the squared lengths, a^2 and b^2, to get h^2.

- Take the **square root** of h^2 to find the hypotenuse, h.

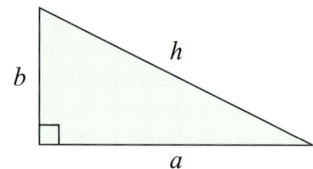

Example 1

Find the exact length of x on the triangle shown.

1. Substitute the values from the diagram into the formula.
$$h^2 = a^2 + b^2$$
$$x^2 = 6^2 + 4^2$$

2. Add together the squared lengths to get x^2.
$$x^2 = 36 + 16 = 52$$

3. Work out the square root to find x.
$$x = \sqrt{52} = \sqrt{4 \times 13}$$
$$= 2\sqrt{13} \text{ cm}$$

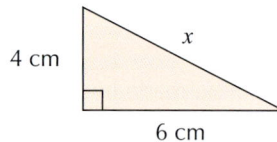

Tip: 'Exact length' means you should give your answer as a surd — simplified if possible (see p.99).

You can also use Pythagoras' theorem to find one of the **shorter sides** if you know the hypotenuse and one of the other sides. To do this, **substitute** in the values and then **rearrange** the formula to make the unknown length the subject.

Example 2

Find the length of b on the triangle shown on the right.

1. Substitute the values from the diagram into the formula.
$$h^2 = a^2 + b^2$$
$$25^2 = 24^2 + b^2$$

2. Rearrange to make b^2 the subject.
$$b^2 = 25^2 - 24^2 = 625 - 576 = 49$$

3. Work out the square root to find b.
$$b = \sqrt{49} = 7 \text{ m}$$

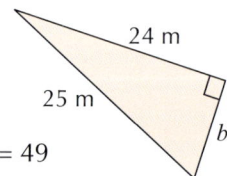

Exercise 1

Q1 Find the exact length of the hypotenuse in each of the triangles below.

a)
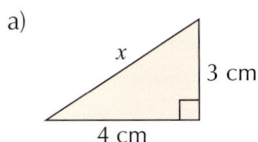

b)
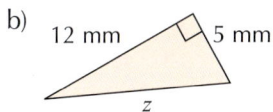

c)
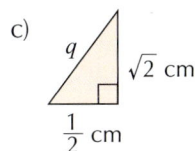

d)
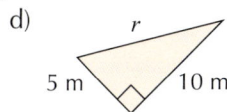

Q2 Find the length of the hypotenuse in each of the triangles below.
Give your answers correct to 2 decimal places where appropriate.

a)
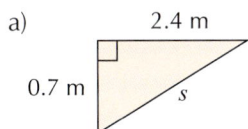

b)
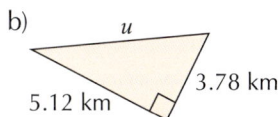

c)
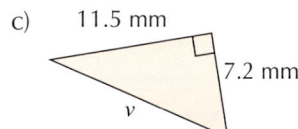

d)
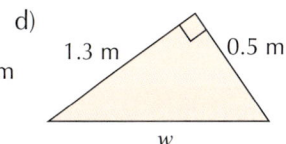

Q3 Find the exact lengths of the unknown sides in these triangles.

a)
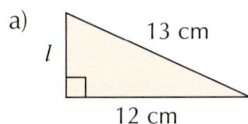

b)
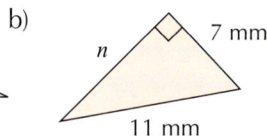

c)
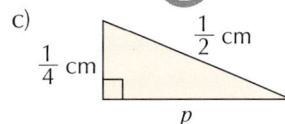

d)
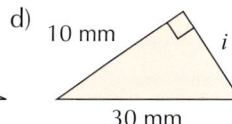

Q4 Find the lengths of the unknown sides in these triangles. Give your answers correct to 2 d.p.

a)
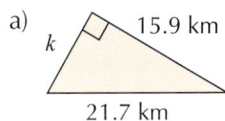

b)
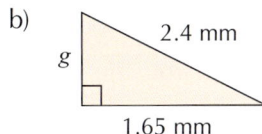

c)
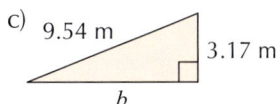

d)
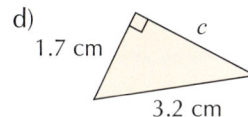

Pythagoras' theorem can be used in lots of other situations too —
you just have to look for ways to **create** a right-angled triangle. For example:

- **Splitting** an **equilateral** or **isosceles triangle** in half can create two identical right-angled triangles.
- The **straight line** between two pairs of **coordinates** forms the hypotenuse of a right-angled triangle with sides equal to the difference in the x-coordinates and the difference in the y-coordinates.

Pythagoras' theorem can also be applied to **real life situations**. The formula is used in the **same way**, you just have to link your answer back to the **context** of the situation.

Example 3

Find the exact distance between points A and B on the grid.

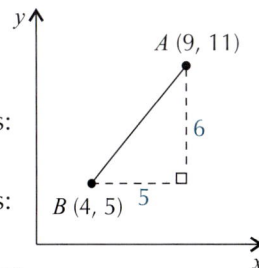

1. Create a right-angled triangle with hypotenuse AB.

2. Find the length of the horizontal side by working out the difference in the x-coordinates.
   Difference in x-coordinates: $9 - 4 = 5$

3. Find the length of the vertical side by working out the difference in the y-coordinates.
   Difference in y-coordinates: $11 - 5 = 6$

4. Substitute the values into the formula.
   $AB^2 = 5^2 + 6^2 = 25 + 36 = 61$

5. Work out the square root to find the distance. Give your answer in surd form.
   $AB = \sqrt{61}$

Section 25 Pythagoras and Trigonometry

319

Example 4

**A TV has a height of 40 cm and width of w cm.
Its diagonal measures 82 cm. Will the TV fit in a box 75 cm wide?**

Tip: Sketch a diagram to help you understand the question, if needed.

1. The diagonal and sides form a right-angled triangle, so substitute the values into the formula.

$h^2 = a^2 + b^2$
$82^2 = 40^2 + w^2$

2. Rearrange to make w^2 the subject.

$w^2 = 82^2 - 40^2 = 6724 - 1600$
$\qquad = 5124$

3. Work out the square root to find w.

$w = \sqrt{5124} = 71.58$ cm (2 d.p.)

4. Use your answer to draw a conclusion.

$71.58 < 75$ so **yes, the TV will fit in the box.**

Exercise 2

Unless told otherwise, give your answers correct to 2 decimal places where appropriate.

Q1 An equilateral triangle has sides of length 10 cm. Find the perpendicular distance from a vertex to its opposite side.

Q2 The triangle JKL is drawn inside a circle centred on O, as shown. JK and KL have lengths 4.9 cm and 6.8 cm respectively.

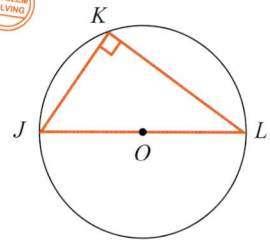

    a) Find the length of JL.   b) Find the radius of the circle.

Q3 A kite gets stuck at the top of a vertical tree. The kite's 15 m string is taut, and its other end is held on the ground, 8.5 m from the base of the tree. Find the height of the tree.

Q4 Newtown is 88 km northwest of Oldtown. Bigton is 142 km from Newtown, and lies northeast of Oldtown. What is the distance from Bigton to Oldtown, to the nearest kilometre?

Q5 A triangle has sides measuring 1.5 m, 2 m and 2.5 m. Show that this is a right-angled triangle.

Q6 A boat is rowed 200 m east and then 150 m south. If it had been rowed to the same point in a straight line instead, how much shorter would the journey have been?

Q7 The points A, B and C are shown on the right. Find the exact lengths of the line segments between the following pairs of points.

    a) A and B b) B and C c) A and C

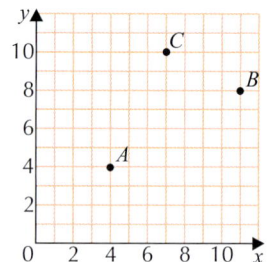

Q8 Find the exact distance between points P and Q with coordinates (11, 1) and (17, 19) respectively.

Q9 Kevin wants to set up a 20 m slide from the top of his 5.95 metre-high tower.

    a) How far from the base of the tower should Kevin anchor the slide?

    b) A safety inspector shortens the slide and anchors it 1.5 m closer to the tower. What is the new length of the slide?

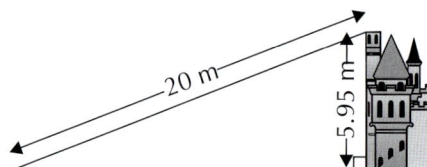

25.2 Pythagoras' Theorem in 3D

You can use Pythagoras' theorem to find dimensions within 3D shapes too.

Learning Objective — Spec Ref G20:
Use Pythagoras' theorem to find lengths in 3D shapes.

For a **cuboid** of length a, width b and height c, the length of the **longest diagonal**, d, can be found using the formula: $\boxed{d^2 = a^2 + b^2 + c^2}$

This formula comes from using the 2D Pythagoras' theorem **twice**. In the diagram to the right, a, b and e make a **right-angled triangle** (e is the **diagonal** of one of the cuboid's faces and forms the **hypotenuse**), so $e^2 = a^2 + b^2$. Side c makes a right-angled triangle with d and e, so $d^2 = e^2 + c^2 = a^2 + b^2 + c^2$.

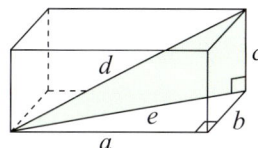

Example 1

Find the exact length *AG* in the cuboid shown on the right.

1. Use Pythagoras on the triangle *ACD* to find the length *AC*.

    $AC^2 = 10^2 + 7^2 = 149$
    $AC = \sqrt{149}$ cm

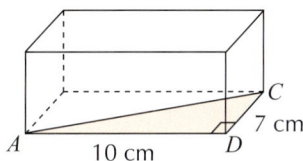

2. Now use Pythagoras again on the triangle *AGC* to find the length *AG*. Give your answer in surd form.

    $AG^2 = (\sqrt{149})^2 + 4^2$
    $\quad\;\; = 149 + 16 = 165$
    $AG = \sqrt{165}$ **cm**

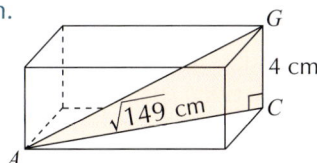

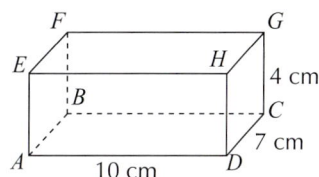

Tip: This example showed the step by step working, but you could have done it using the cuboid formula — $d^2 = 10^2 + 7^2 + 4^2 = 165$.

Example 2

A square-based pyramid *ABCDE* is shown on the right. Point *O* is the midpoint of the base *ABCD* and point *E* lies directly above *O*. Find the exact length *OE*.

1. Use Pythagoras on the triangle *ACD* to find the length *AC*.

    $AC^2 = 6^2 + 6^2 = 72$
    $AC = \sqrt{72} = 6\sqrt{2}$ m

2. *O* is the midpoint of *AC*, so halve *AC* to get *AO*.

    $AO = 6\sqrt{2} \div 2 = 3\sqrt{2}$ m

3. Use Pythagoras on the triangle *AOE* to find the length *OE*. Give your answer in surd form.

    $AE^2 = AO^2 + OE^2$
    $OE^2 = 10^2 - (3\sqrt{2})^2$
    $\quad\;\; = 100 - 18 = 82$
    $OE = \sqrt{82}$ **m**

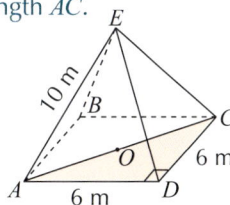

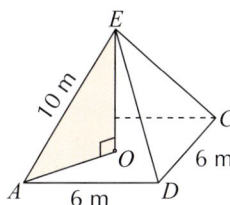

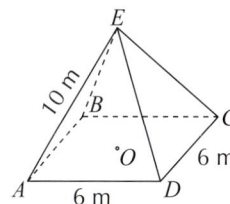

Tip: You have to do this example in stages — you're not finding the diagonal of a cuboid, so you can't use the formula.

Q1 The cuboid *ABCDEFGH* is shown on the right.

a) By considering the triangle *ABD*, find the exact length *BD*.

b) By considering the triangle *BFD*, find the exact length *FD*.

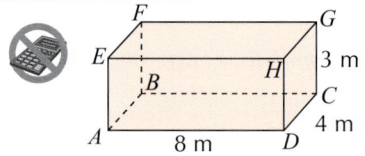

Q2 The cuboid *PQRSTUVW* is shown on the right.

a) Find the exact length *PR*.

b) Find the exact length *RT*.

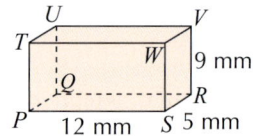

Q3 A cylinder of length 25 cm and radius 4.5 cm is shown on the right. *X* and *Y* are points on opposite edges of the cylinder, such that *XY* is as long as possible. Find the length *XY* to 3 significant figures.

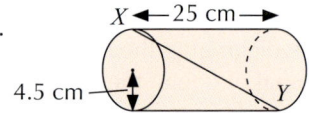

Q4 A cuboid measures 2.5 m × 3.8 m × 9.4 m. Find the length of the diagonal of the cuboid to 2 d.p.

In Questions 5-11, give all answers to 3 significant figures.

Q5 The triangular prism *PQRSTU* is shown on the right.

a) Find the length *QS*.

b) Find the length *ST*.

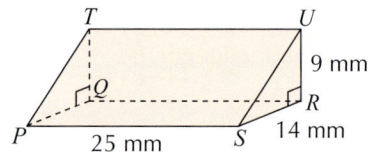

Q6 Find the length of the diagonal of a cube of side 5 m.

Q7 A pencil case in the shape of a cuboid is 16.5 cm long, 4.8 cm wide and 2 cm deep. What is the length of the longest pencil that will fit in the case? Ignore the thickness of the pencil. *(PROBLEM SOLVING)*

Q8 A spaghetti jar is in the shape of a cylinder. The jar has radius 6 cm and height 28 cm. What is the length of the longest stick of dried spaghetti that will fit inside the jar? *(PROBLEM SOLVING)*

Q9 A square-based pyramid has a base of side 4.8 cm and sloped edges all of length 11.2 cm. Find the vertical height of the pyramid from the centre of the base to the highest point.

Q10 A square-based pyramid has a base of side 3.2 m and a vertical height of 9.2 m. Find the length of the sloped edges of the pyramid, given they are all equal.

Q11 The square-based pyramid *JKLMN* is shown on the right. *O* is the centre of the square base, directly below *N*. *P* is the midpoint of *LM*.

a) Find the length *NP*.

b) Find the length *OJ*.

c) Find the length *ON*.

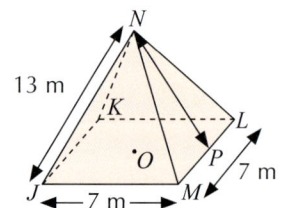

Q12 The square-based pyramid *VWXYZ* is shown on the right. Point *O*, the centre of the square *VWXY*, is directly below *Z*.

a) Find the length *VX*.

b) Hence find the area of the square *VWXY*. *(PROBLEM SOLVING)*

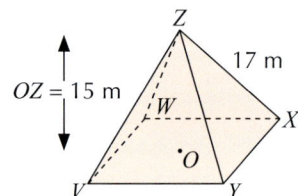

25.3 Trigonometry — Sin, Cos and Tan

Trigonometry allows you to find missing sides and angles in triangles. For right-angled triangles, you'll need to know the sine (sin), cosine (cos) and tangent (tan) formulas.

The Three Formulas

Learning Objective — Spec Ref G20:
Use trigonometry to find missing lengths and angles in right-angled triangles.

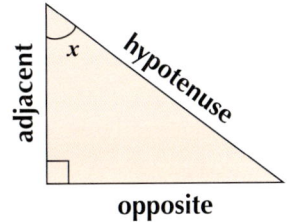

Trigonometry can be used to find lengths or angles in **right-angled triangles**. For a given angle x (as shown on the diagram on the right):

- The side opposite the right angle is the **hypotenuse**.
- The side opposite the given angle is the **opposite**.
- The side between the given angle and the right-angle is the **adjacent**.

The three sides of a right-angled triangle are linked by the following formulas:

Tip: Remember 'SOH CAH TOA' to help you decide which formula you need to use.

$$\sin x = \frac{\text{opp}}{\text{hyp}}, \quad \cos x = \frac{\text{adj}}{\text{hyp}}, \quad \tan x = \frac{\text{opp}}{\text{adj}}$$

If you're given **one angle** and **one side**, you can use trigonometry to find an **unknown side length**. Look at the side you've been **given** and the side you **want to find** to decide which formula to use — e.g. if you had the **hypotenuse** and wanted to find the **adjacent**, you'd use the **cos** formula as it contains these two sides. **Substitute** the values you know into the appropriate formula and **rearrange** it to find the length you want.

Example 1

Find the length of side y.

1. You're given the hypotenuse and asked to find the opposite, so use the formula for $\sin x$ and substitute in the values you know.

$$\sin x = \frac{\text{opp}}{\text{hyp}}$$
$$\sin 30° = \frac{y}{10}$$

2. Rearrange the formula to find y.

$$y = 10 \sin 30°$$

3. Input '10 sin 30' into your calculator and press '=' to find the value of y.

$$y = 5 \text{ m}$$

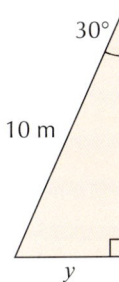

Tip: You might find it helpful to start by labelling the sides O (opposite), A (adjacent) and H (hypotenuse).

Example 2

Find the height of the isosceles triangle shown. Give your answer correct to 3 significant figures.

1. Create a right-angled triangle by splitting the triangle in half. Divide the angle by 2 to find the angle in your right-angled triangle.

$$58 \div 2 = 29°$$

2. Now you have the hypotenuse, and you need to find the adjacent, so use the formula for $\cos x$.

$$\cos x = \frac{\text{adj}}{\text{hyp}} \Rightarrow \cos 29° = \frac{y}{4}$$

3. Rearrange the formula to find y.

$$y = 4 \cos 29°$$

4. Use your calculator to find the value of y.

$$y = 3.498... = 3.50 \text{ cm} \text{ (3 s.f.)}$$

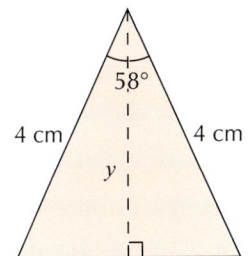

Example 3

Find the length of side y. Give your answer correct to 3 significant figures.

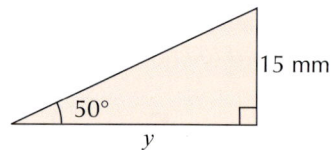

1. You're given the opposite and asked to find the adjacent, so use the formula for tan x.

$$\tan x = \frac{\text{opp}}{\text{adj}}$$

2. Rearrange the formula to find y — this time, y is on the bottom of the fraction, so the rearrangement is slightly different.

$$\tan 50° = \frac{15}{y}$$

$$y \times \tan 50° = 15 \implies y = 15 \div \tan 50°$$

3. Use your calculator to find the value of y.

$$y = 12.586... = \mathbf{12.6 \text{ mm}} \text{ (3 s.f.)}$$

Exercise 1

Q1 Find the lengths of the sides marked with letters below. Give your answers to 3 significant figures.

a)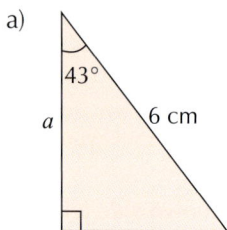

b)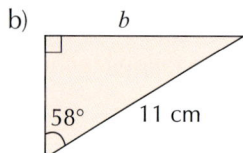

c)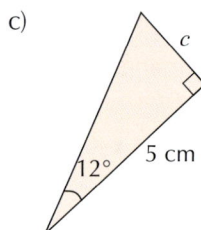

d)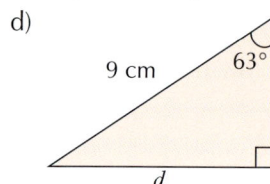

e)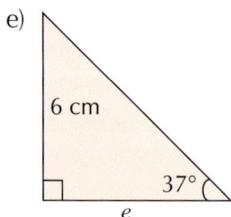

f)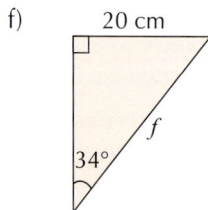

g)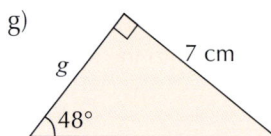

h)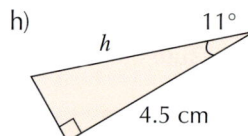

Q2 Find the lengths marked with letters in the triangles below. Give your answers to 3 significant figures.

a)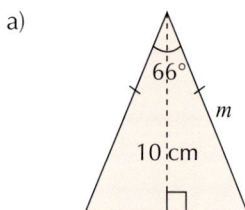

b)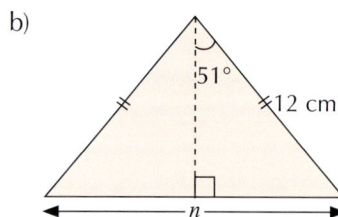

You can also use the trig formulas to find an **angle** if you know two side lengths. You have to use the **inverse functions** of sin, cos and tan (written **sin⁻¹**, **cos⁻¹** and **tan⁻¹**), which return an **angle**. To find an angle, work out which formula you need from the sides you're given as before, then **substitute** in the known values — this will give you a **fraction** on the right-hand side, e.g. $\sin x = \frac{1}{2}$. Take the **inverse trig function** of the fraction to get the angle — so here you'd do $x = \sin^{-1}\left(\frac{1}{2}\right) = 30°$.

Tip: Inverse trig functions are usually found on a calculator by pressing 'shift' or '2nd' before pressing sin, cos or tan.

Example 4

Find the size of angle x. Give your answer correct to 1 decimal place.

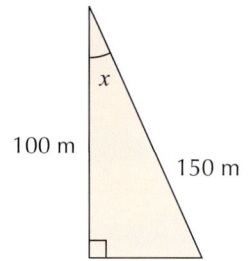

1. You're given the adjacent and the hypotenuse, so use the formula for cos x.

$$\cos x = \frac{\text{adj}}{\text{hyp}} = \frac{100}{150}$$

2. Take the inverse of cos to find the angle.

$$x = \cos^{-1}\left(\frac{100}{150}\right)$$

3. Input 'cos⁻¹(100 ÷ 150)' into your calculator and press '=' to find the value of x.

$$x = 48.189... = \mathbf{48.2°} \text{ (1 d.p.)}$$

Example 5

Find the size of angle x. Give your answer correct to 1 decimal place.

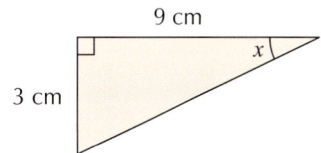

1. You're given the opposite and the adjacent, so use the formula for tan x.

$$\tan x = \frac{\text{opp}}{\text{adj}} = \frac{3}{9}$$

2. Take the inverse of tan to find the angle.

$$x = \tan^{-1}\left(\frac{3}{9}\right)$$

3. Use your calculator to find the value of x.

$$x = 18.434... = \mathbf{18.4°} \text{ (1 d.p.)}$$

Trigonometry can be used to work out angles of **depression** and **elevation**.

The **angle of depression** is the angle between a **horizontal line** and the **line of sight** of an observer at the same level **looking down** — e.g. you could measure the angle of depression of someone looking down from a window.

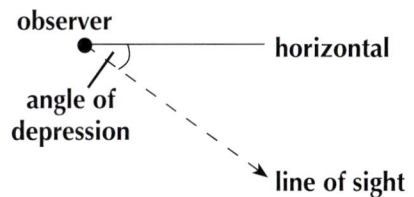

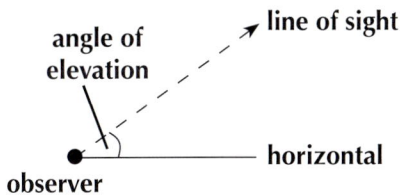

The **angle of elevation** is the angle between a **horizontal line** and the **line of sight** of an observer at the same level **looking up** — e.g. the angle made looking up at a hovering helicopter.

For problems like this, use the information given to **draw** a right-angled triangle and use the formulas in the **same way** as usual. Remember to relate your answer to the **original context** of the problem.

Example 6

Liz holds one end of a 7 m paper chain out of her window. Phil stands in the garden below holding the other end to its full extent. Phil's end of the paper chain is 6 m vertically below Liz's end. Find the size of the angle of depression from Liz to Phil. Give your answer to 1 decimal place.

1. Use the information to draw a right-angled triangle — the angle of depression is the angle below the horizontal.

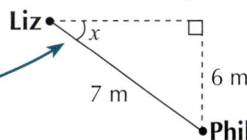

2. You're given the hypotenuse and the opposite, so use the formula for sin x.

$$\sin x = \frac{\text{opp}}{\text{hyp}} = \frac{6}{7}$$

3. Take the inverse of sin to find the angle.

$$x = \sin^{-1}\left(\frac{6}{7}\right)$$

4. Use your calculator to find the value of x.

$$x = 58.997... = \mathbf{59.0°} \text{ (1 d.p.)}$$

Tip: The angle of elevation from Phil to Liz is also 59.0° because they are allied angles (see p.251).

Q1 Find the size of the angles marked with letters. Give your answers to 1 decimal place.

a)

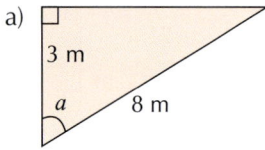

3 m, 8 m, a

b)

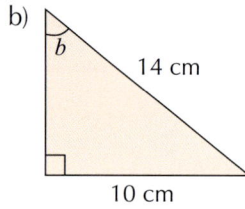

b, 14 cm, 10 cm

c)

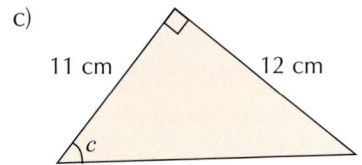

11 cm, 12 cm, c

d)

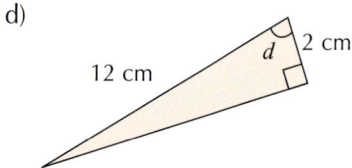

12 cm, d, 2 cm

e)

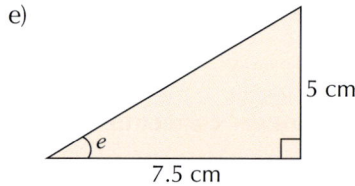

e, 5 cm, 7.5 cm

f)

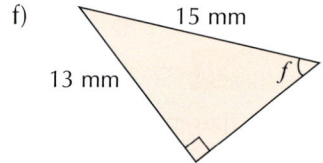

15 mm, 13 mm, f

Q2 Melissa is building slides for an adventure playground.

a) The first slide she builds has an 8 m high vertical ladder and a slide of length 24 m. Find m, the slide's angle of elevation, to 1 d.p.

b) A second slide has a 4 m vertical ladder, and the base of the slide reaches the ground 5.5 m from the base of the ladder as shown. Find q, the angle of depression at the top of the second slide, to 1 d.p.

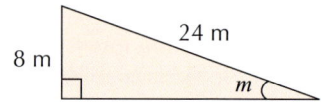

8 m, 24 m, m

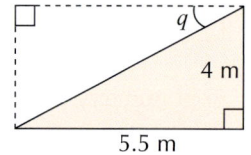

q, 4 m, 5.5 m

Q3 Find the angles marked with letters in the following diagrams. Give your answers correct to 1 d.p.

a)

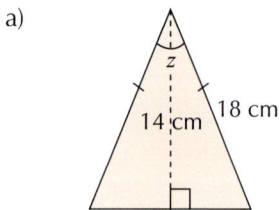

z, 14 cm, 18 cm

b)

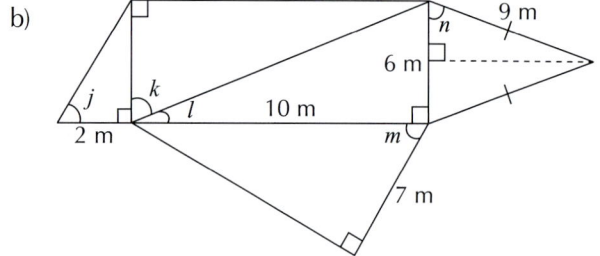

j, k, l, 2 m, 10 m, m, 6 m, n, 9 m, 7 m

Common Trig Values

Learning Objective — Spec Ref G21:
Know and be able to find trig values for common angles.

Prior Knowledge Check:
Be able to simplify surds and rationalise denominators — see p.99-103.

The sin, cos and tan of some angles have **exact values**. You need to either **remember** the values or know how to work them out **without a calculator**. The first of these angles is **45°**.

Start by drawing a **square** with sides of length **1**. Then **split** the square down its **diagonal** to create two **right-angled triangles**. By Pythagoras' theorem, the length of the hypotenuse is $\sqrt{1^2 + 1^2} = \sqrt{2}$. Since the triangle was formed by **bisecting** 90° angles, the two acute angles in the triangle are 90° ÷ 2 = **45°**. Then use the trig formulas to find the sin, cos and tan of 45° — e.g. $\sin 45° = \dfrac{\text{opp}}{\text{hyp}} = \dfrac{1}{\sqrt{2}}$.

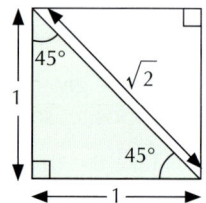

$$\sin 45° = \frac{1}{\sqrt{2}} \qquad \cos 45° = \frac{1}{\sqrt{2}} \qquad \tan 45° = 1$$

To find sin, cos and tan of **30°** and **60°**, start with an **equilateral triangle** with sides of length **2**. The interior angles are **60°** (see p.252). **Split** the triangle in half using a **perpendicular** line from one vertex to the opposite side, leaving two **right-angled triangles** with acute angles of **60°** and $60° \div 2 = 30°$. By Pythagoras' theorem, the perpendicular height is $\sqrt{2^2 - 1^2} = \sqrt{3}$. Then use the trig formulas to get the following results.

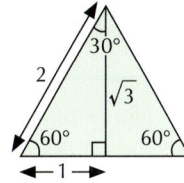

$$\sin 60° = \frac{\sqrt{3}}{2} \qquad \cos 60° = \frac{1}{2} \qquad \tan 60° = \sqrt{3} \qquad\qquad \sin 30° = \frac{1}{2} \qquad \cos 30° = \frac{\sqrt{3}}{2} \qquad \tan 30° = \frac{1}{\sqrt{3}}$$

You **can't** use triangles for **0°** or **90°** but you do need to know their trig values:

$$\sin 0° = 0 \qquad \cos 0° = 1 \qquad \tan 0° = 0$$
$$\sin 90° = 1 \qquad \cos 90° = 0 \qquad \tan 90° = \text{undefined}$$

Tip: The values for 0° and 90° come from the sine, cosine and tangent graphs — see p.204.

Example 7

Without using a calculator, find the exact length of side y on the triangle on the right.

1. You're given the hypotenuse and want to find the adjacent, so use the formula for $\cos x$.

$$\cos x = \frac{\text{adj}}{\text{hyp}}$$

2. Substitute in the values you know and rearrange the formula to make y the subject.

$$\cos 30° = \frac{y}{650}$$
$$y = 650 \times \cos 30°$$

3. Replace $\cos 30°$ with its exact trig value and work through the formula to find y.

$$= 650 \times \frac{\sqrt{3}}{2} = \mathbf{325\sqrt{3}\ mm}$$

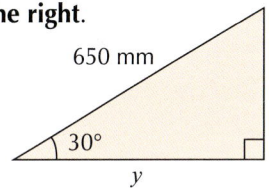

Exercise 3

Q1 Find the size of the angles marked with letters.

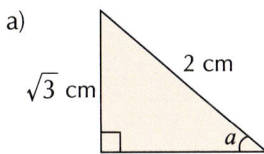

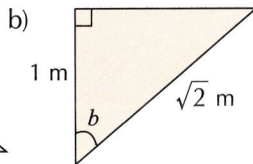

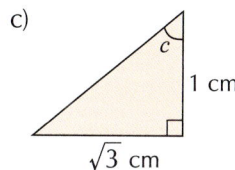

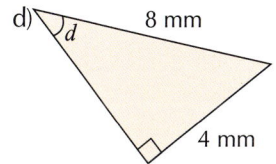

a)

b)

c)

d)

Q2 Find the exact length of the sides marked with letters.

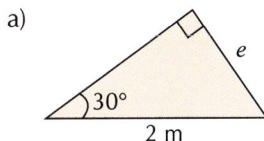

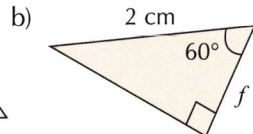

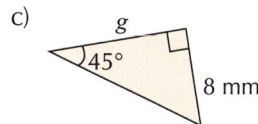

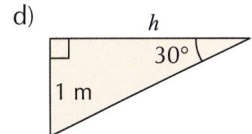

a)

b)

c)

d)

Q3 Show that:

a) $\tan 45° + \sin 60° = \dfrac{2 + \sqrt{3}}{2}$ b) $\sin 45° + \cos 45° = \sqrt{2}$ c) $\tan 30° + \tan 60° = \dfrac{4\sqrt{3}}{3}$ **PROBLEM SOLVING**

Q4 Triangle ABC is isosceles. $AC = 7\sqrt{2}$ cm and angle $ABC = 90°$. **PROBLEM SOLVING** What is the exact length of side AB?

Q5 Triangle DEF is an equilateral triangle with a perpendicular height of 4 mm. What is the exact side length of the triangle? Give your answer in its simplest form.

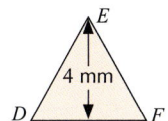

25.4 The Sine and Cosine Rules

You can still use trigonometry to find unknown sides and angles in triangles that aren't right-angled — you just need a few more formulas. And there's also a formula to learn for the area of a triangle too.

The Sine Rule

Learning Objective — Spec Ref G22:
Use the sine rule to find sides and angles in any triangle.

To use trigonometry in triangles that **aren't** right-angled, you must first **label** the sides and angles properly, like in the diagram on the right — side a must always be **opposite** angle A etc. You use **lower case letters** for the **sides** and **upper case letters** for the **angles**.

You can then use the **sine rule**, which is: $\dfrac{a}{\sin A} = \dfrac{b}{\sin B} = \dfrac{c}{\sin C}$

Tip: You will only need to consider 2 at a time — e.g. $\dfrac{a}{\sin A} = \dfrac{b}{\sin B}$.

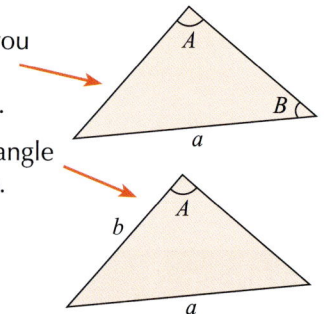

You can also flip the fractions to write it in this form: $\dfrac{\sin A}{a} = \dfrac{\sin B}{b} = \dfrac{\sin C}{c}$

You can use the sine rule if:

- **Two angles and one side** are known (like in the triangle on the right). If you know two angles, you can always find the third by subtracting them from 180°. Then you can use the sine rule to find **either** of the unknown sides.

- **One angle, the opposite side and one other side** are known (like in the triangle below right). Here you can find the angle **opposite** the other known side.

To use the sine rule, **substitute** in the values you know and **rearrange** to find the side or angle you want. You'll probably have to use the **inverse sine function** (see p.324) if you're looking for an angle.

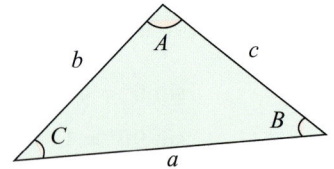

Example 1

Find the length of side x. Give your answer to 3 significant figures.

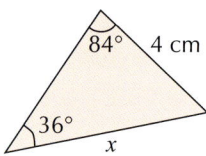

1. Substitute the values from the diagram into the sine rule.

   $$\dfrac{a}{\sin A} = \dfrac{b}{\sin B} \Rightarrow \dfrac{4}{\sin 36°} = \dfrac{x}{\sin 84°}$$

2. Rearrange to make x the subject. $x = \dfrac{4 \sin 84°}{\sin 36°}$

3. Use your calculator to find the value of x.

   $x = \textbf{6.77 cm}$ (3 s.f.)

   Tip: Label the triangle using a, A, b and B if you need to.

Example 2

Find the size of the acute angle y below. Give your answer to 1 decimal place.

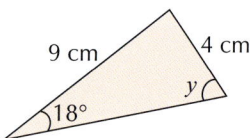

1. Substitute in the values from the diagram. The second version of the formula works best here, but the first would also work.

   $$\dfrac{\sin A}{a} = \dfrac{\sin B}{b} \Rightarrow \dfrac{\sin y}{9} = \dfrac{\sin 18°}{4}$$

2. Rearrange the equation.

   $$\sin y = \dfrac{9 \sin 18°}{4} \Rightarrow y = \sin^{-1}\left(\dfrac{9 \sin 18°}{4}\right)$$

3. Use the inverse function to find y.

   $y = \textbf{44.1°}$ (1 d.p.)

Q1 Find the lengths of the sides marked with letters below. Give your answers to 3 significant figures.

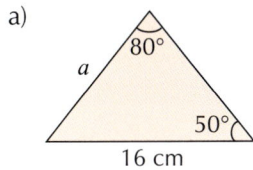

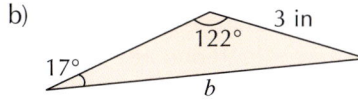

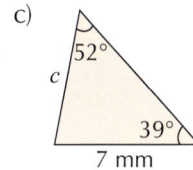

a)

$80°$
a
$50°$
16 cm

b)

$122°$ 3 in
$17°$
b

c)

$52°$
c
$39°$
7 mm

Q2 For each triangle below, find the acute angle marked with a letter. Give your answers to 1 d.p.

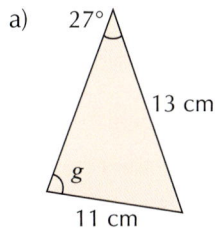

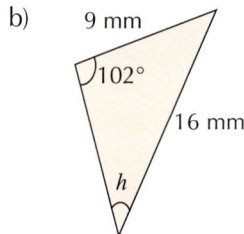

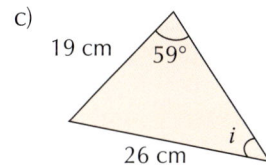

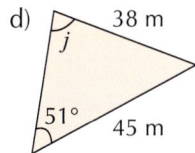

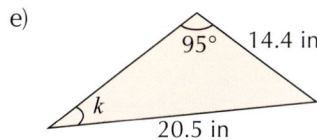

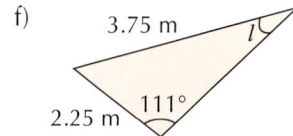

a)

$27°$
13 cm
g
11 cm

b)

9 mm
$102°$
16 mm
h

c)

19 cm $59°$
26 cm
i

d)

38 m
j
$51°$
45 m

e)

$95°$ 14.4 in
k
20.5 in

f)

3.75 m
l
2.25 m $111°$

Q3 The triangle XYZ is such that angle $YXZ = 55°$, angle $XYZ = 40°$ and length $YZ = 83$ m. Find:

a) length XZ, to 3 s.f.
b) length XY, to 3 s.f.

Q4 A triangular piece of metal PQR is such that angle $RPQ = 61°$, length $QR = 13.1$ mm and length $PQ = 7.2$ mm. Find the size of the acute angle PQR, correct to 1 decimal place.

Q5 Point B is 13 km north of point A. Point C lies 19 km from point B, on a bearing of 052° from A. Find the bearing of C from B, to the nearest degree. (PROBLEM SOLVING)

The Cosine Rule

Learning Objective — Spec Ref G22:
Use the cosine rule to find sides and angles in any triangle.

The **cosine rule** can also be used to find unknown angles and sides.

For any triangle labelled as shown in the diagram, the cosine rule is:

$$a^2 = b^2 + c^2 - 2bc\cos A$$

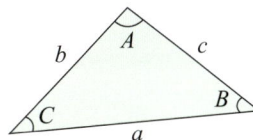

Tip: Whether you use the sine or cosine rule depends on the sides and angles you're given.

This can be rearranged to give $\cos A = \dfrac{b^2 + c^2 - a^2}{2bc}$

This version is useful when you're trying to find an **angle**.

You can use the cosine rule if:

- **Two sides and the angle between them** are known (like in the triangle on the right). Here, you can find the **third side**.

- **All three sides** are known (like in the triangle below right). In this case, you can use the cosine rule to find **any angle**.

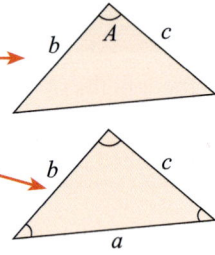

To use the cosine rule, **substitute** in the values you know and **rearrange** to find the side or angle you want. You'll probably have to use the **inverse cosine function** if you're looking for an angle (see p.324).

Example 3

Find the length of side x. Give your answer to 3 significant figures.

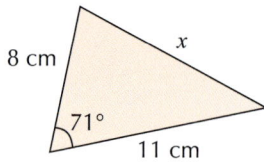

1. Substitute the values from the diagram into the cosine rule — x is side a since it's opposite the known angle (A).

   $a^2 = b^2 + c^2 - 2bc \cos A$

   $x^2 = 8^2 + 11^2 - (2 \times 8 \times 11) \cos 71°$

2. Work it through to find x^2.

   $x^2 = 127.70...$

3. Take the square root to find the value of x.

   $x = \sqrt{127.70...} = \textbf{11.3 cm}$ (3 s.f.)

Example 4

Find the size of angle y in the triangle on the right. Give your answer to 1 decimal place.

1. Use the second version of the cosine rule to find the angle.

   $\cos A = \dfrac{b^2 + c^2 - a^2}{2bc}$

2. Substitute the values from the diagram into the formula — make the side opposite the angle a.

   $\cos y = \dfrac{5.5^2 + 4^2 - 2.5^2}{2 \times 5.5 \times 4}$

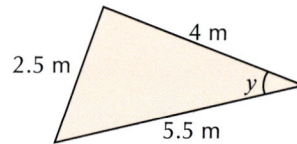

   Tip: Be careful when substituting the values into the formula — the 2.5 m side is side a (and y is angle A).

3. Use the inverse cos function to find the value of y.

   $y = \cos^{-1}\left(\dfrac{5.5^2 + 4^2 - 2.5^2}{2 \times 5.5 \times 4}\right)$

   $= \textbf{24.6°}$ (1 d.p.)

Exercise 2

Q1 Use the cosine rule to find the lengths of the sides marked with letters. Give your answers to 3 s.f.

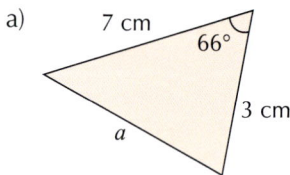

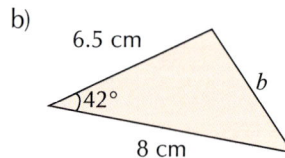

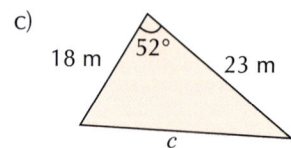

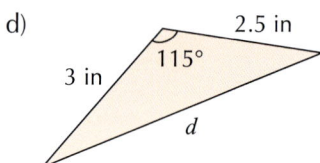

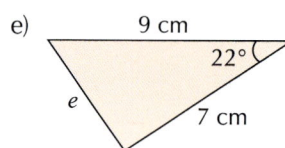

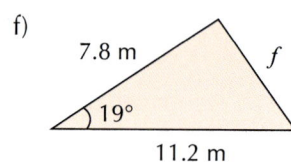

a) 7 cm, 66°, 3 cm, a

b) 6.5 cm, 42°, 8 cm, b

c) 18 m, 52°, 23 m, c

d) 3 in, 115°, 2.5 in, d

e) 9 cm, 22°, 7 cm, e

f) 7.8 m, 19°, 11.2 m, f

Q2 Use the cosine rule to find the sizes of the angles marked with letters.
Give your answers to 1 d.p.

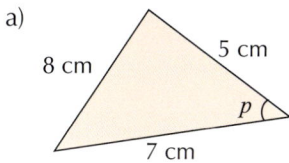

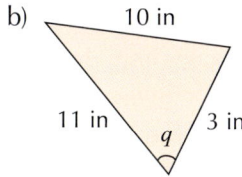

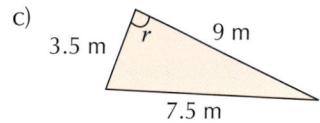

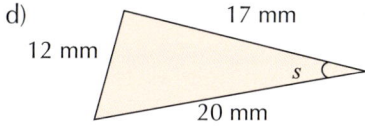

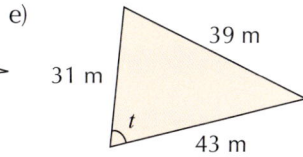

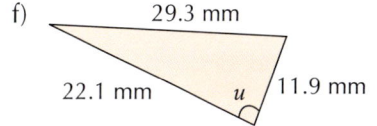

a)

8 cm / 5 cm / 7 cm with angle p

b)

10 in / 11 in / 3 in with angle q

c)

3.5 m / 9 m / 7.5 m with angle r

d)

12 mm / 17 mm / 20 mm with angle s

e)

39 m / 31 m / 43 m with angle t

f)

29.3 mm / 22.1 mm / 11.9 mm with angle u

Q3 A triangular sign XYZ is such that $XY = 67$ cm, $YZ = 78$ cm and $XZ = 99$ cm.
Find the size of the angle XYZ, correct to 1 d.p.

Q4 Find the exact values of the letters in each of these triangles.

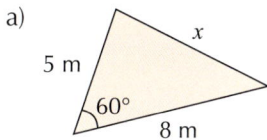

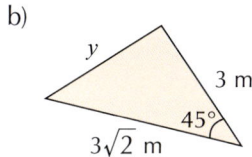

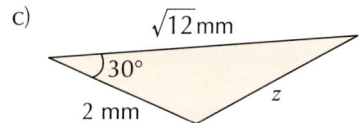

a)

5 m / x / 8 m with 60° angle

b)

y / 3 m / $3\sqrt{2}$ m with 45° angle

c)

$\sqrt{12}$ mm / 2 mm with 30° angle and z

Q5 Village B is 12 miles north of village A. Village C is 37 miles north-east of village A. Find the direct distance between village B and village C, correct to 3 s.f.

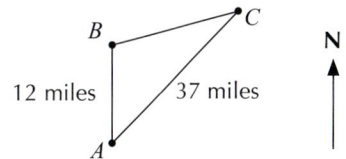

(PROBLEM SOLVING)

Q6 Two ramblers set off walking from point P.
The first rambler walks for 2 km on a bearing of 025° to point A.
The second rambler walks for 3 km on a bearing of 108° to point B.
Find the direct distance between A and B, correct to 3 s.f.

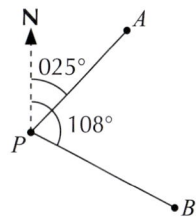

(PROBLEM SOLVING)

Area of a Triangle

Learning Objective — Spec Ref G23:
Find the area of a triangle using the sine function.

To find the **area** of a triangle, use the formula: $\boxed{\textbf{Area} = \frac{1}{2}\,ab\sin C}$

To use this formula, you need to know **two sides and the angle between them**. If you don't, you might have to use the sine and cosine rules first to find the information you need.

You can also use the formula to find the **size of an angle** or **length of a side** when you know the area — just **substitute** in the values you know and **rearrange** the formula to make the angle or side you want the subject.

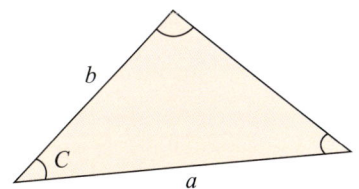

Tip: If you know the base length and height of a triangle, you can use the simpler formula on p.349 to find the area.

Example 5

Find the area of this triangle. Give your answer to 3 significant figures.

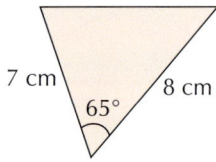

1. Substitute the values from the diagram into the formula.

2. Use your calculator to find the area. Make sure you use the correct units.

$\text{Area} = \frac{1}{2}ab\sin C$

$= \frac{1}{2} \times 7 \times 8 \times \sin 65°$

$= \mathbf{25.4 \ cm^2}$ (3 s.f.)

Example 6

A segment is formed in a sector with angle 60° in a circle of radius 3 cm. Find the area of the segment. Give your answer to 3 significant figures.

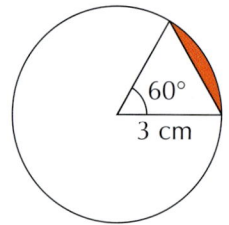

1. Work out the area of the sector.

$\text{Area of sector} = \frac{60}{360} \times \pi \times 3^2 = \frac{3\pi}{2} \ cm^2$

2. The triangle is isosceles (two sides are radii) so you know two sides and the angle between them. Use this information to work out the area.

$\text{Area of triangle} = \frac{1}{2} \times 3 \times 3 \times \sin 60°$

$= \frac{9\sqrt{3}}{4} \ cm^2$

Tip: See p.354 for the formula for the area of a sector.

3. Subtract the area of the triangle from the area of the sector to give the area of the segment.

$\text{Area of segment} = \frac{3\pi}{2} - \frac{9\sqrt{3}}{4}$

$= \mathbf{0.815 \ cm^2}$ (3 s.f.)

Exercise 3

Give all your answers to the questions in this exercise to 3 significant figures.

Q1 Find the area of each of the following triangles.

a)

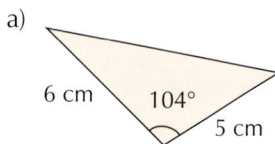

b)

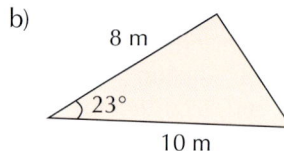

c)

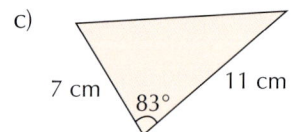

d)

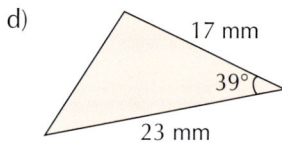

e)

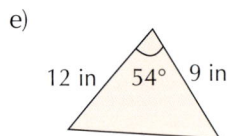

f)

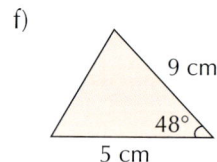

Q2 Find the area of the segment formed in a sector with angle 67° in a circle of radius 4.5 cm.

Q3 A field in the shape of an equilateral triangle has sides of length 32 m. Find the area of the field.

Q4 For the triangle on the right, use the cosine rule to calculate x and hence find the area of the triangle.

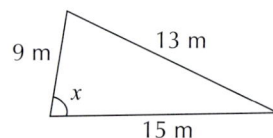

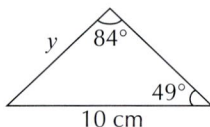

Q5 For the triangle on the left, use the sine rule to calculate y and hence find the area of the triangle.

25.5 Trigonometry in 3D

Like Pythagoras' theorem, trigonometry can also be used in 3D to find lengths and angles.

Learning Objective — Spec Ref G20/G22:
Use trigonometry to find lengths and angles in 3D shapes.

You can use **trigonometry** to find lengths and angles in **3D shapes** by creating triangles within them. You can use sin, cos and tan for **right-angled triangles** and the **sine** or **cosine rules** for other triangles. You might need to use **Pythagoras' theorem** too — see p.321. The formulas are used in the same way as for 2D shapes, but you might have to use them **multiple** times to find what you're looking for. It's often a good idea to **sketch** the triangles as you go along to keep track of what you're doing.

Example 1

Find the angle *BDF* in the cuboid shown on the right.

1. Use Pythagoras' theorem on the triangle *ABD* to find the length *BD*.

    $$BD^2 = AD^2 + AB^2$$
    $$= 6^2 + 7^2 = 85$$
    $$BD = \sqrt{85} \text{ cm}$$

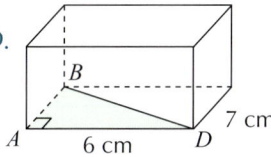

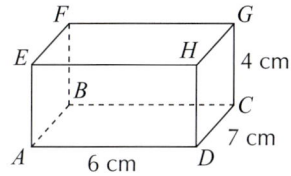

2. *BDF* forms a right-angled triangle and you know the opposite and adjacent sides to angle *BDF*, so use the tan formula.

    $$\tan BDF = \frac{\text{opp}}{\text{adj}} = \frac{FB}{BD} = \frac{4}{\sqrt{85}}$$
    $$BDF = \tan^{-1}\left(\frac{4}{\sqrt{85}}\right)$$
    $$= \textbf{23.5°} \text{ (1 d.p.)}$$

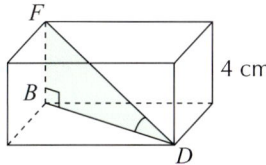

Tip: There may be alternative triangles you can use to find the angle you want.

Exercise 1

Q1 The cube shown has sides of length 3 m. Find:
   a) the exact length *AF*
   b) the exact length *FC*
   c) the angle *AHC*, to 1 d.p.

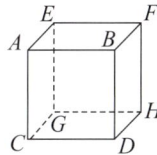

Q2 *ABCDE* is a square-based pyramid. The centre of the base is *O*, and *E* lies directly above *O*. Find:
   a) the angle *BCE*, to 1 d.p
   b) the angle *AEB*, to 1 d.p.
   c) the exact vertical height *EO*
   d) the angle *AEO*, to 1 d.p.

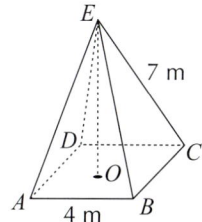

Q3 For the triangular prism shown, find:
   a) the angle *EDF*, to 1 d.p.
   b) the exact length *DC*
   c) the angle *DCE*, to 1 d.p.

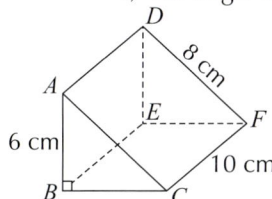

Q4 For the cuboid shown, find:
  a) the exact length AH
  b) the angle EDG, to 1 d.p.

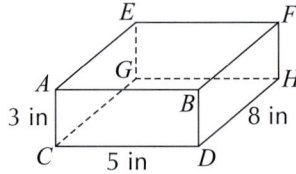

3 in 5 in 8 in

Q5 For the triangular prism shown, where M is the midpoint of AC, find:
  a) the perpendicular height BM, to 3 s.f.
  b) the length EM, to 3 s.f.

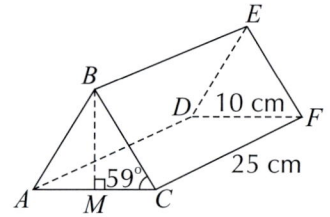

10 cm, 25 cm, 59°

Example 2

Find the size of angle AFH in the cuboid shown on the right.

1. Use Pythagoras' theorem to find the lengths AF, AH and FH.

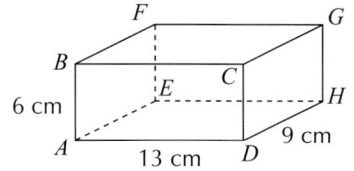

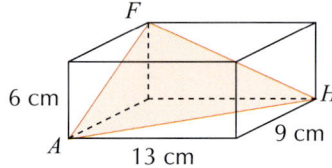

$$AF^2 = 6^2 + 9^2 = 117 \Rightarrow AF = \sqrt{117}$$
$$AH^2 = 13^2 + 9^2 = 250 \Rightarrow AH = \sqrt{250}$$
$$FH^2 = 6^2 + 13^2 = 205 \Rightarrow FH = \sqrt{205}$$

2. AFH isn't a right-angled triangle but you know all 3 sides so use the cosine rule.

$$\cos AFH = \frac{AF^2 + FH^2 - AH^2}{2 \times AF \times FH} = \frac{117 + 205 - 250}{2 \times \sqrt{117} \times \sqrt{205}}$$

3. Take the inverse of cos to find the angle.

$$AFH = \cos^{-1}\left(\frac{117 + 205 - 250}{2 \times \sqrt{117} + \sqrt{205}}\right) = \textbf{76.6°} \text{ (1 d.p.)}$$

Exercise 2

Q1 The diagram shows the triangular prism $IJKLMN$.

  a) Use the cosine rule to find the size of angle JIK, correct to 1 d.p.

  b) Hence find the area of triangle IJK, correct to 1 d.p.

  c) Hence find the volume of the triangular prism $IJKLMN$, to the nearest m³.

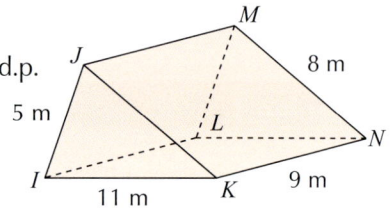

8 m, 5 m, 11 m, 9 m

Q2 The diagram shows the triangle CEH drawn inside the cuboid $ABCDEFGH$.

  a) Find the exact length CE.

  b) Find the exact length CH.

  c) Find the exact length EH.

  d) Hence use the cosine rule to find the size of angle ECH, correct to 1 d.p.

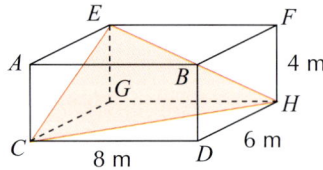

4 m, 6 m, 8 m

Q3 The diagram shows the cuboid $PQRSTUVW$.

  a) Use Pythagoras and the cosine rule to find the size of angle PSU, correct to 1 d.p.

  b) Hence find the area of triangle PSU, correct to 1 d.p.

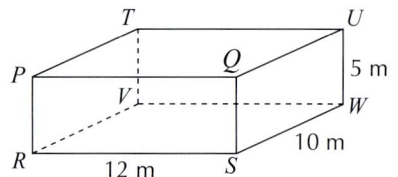

5 m, 10 m, 12 m

Q1 Find the value of the letter in each triangle below. Give your answers to 3 s.f. where necessary.

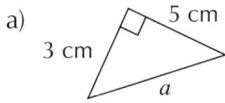

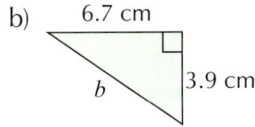

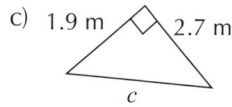

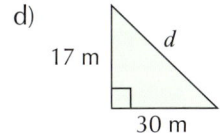

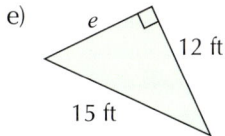

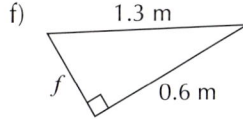

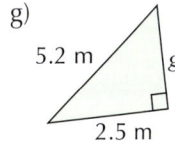

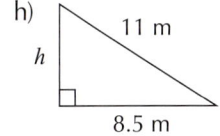

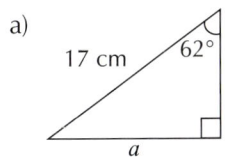

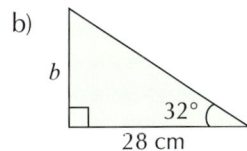

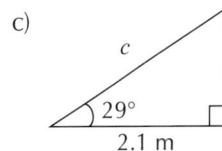

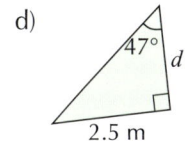

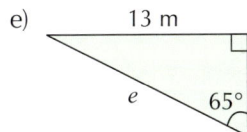

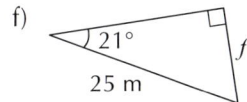

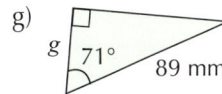

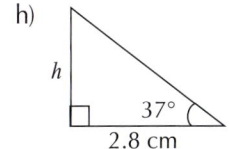

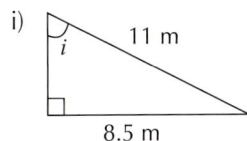

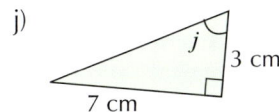

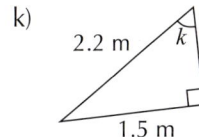

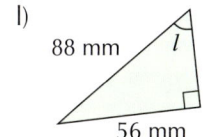

a)
5 cm
3 cm
a

b)
6.7 cm
b
3.9 cm

c)
1.9 m
2.7 m
c

d)
17 m
d
30 m

e)
e
12 ft
15 ft

f)
1.3 m
f
0.6 m

g)
5.2 m
g
2.5 m

h)
11 m
h
8.5 m

Q2 The diagram on the right shows a shape made up of the square-based pyramid *EFGHI* and the cuboid *ABCDEFGH*. *P* is the centre of the square *EFGH* and *O* is the centre of *ABCD*. *P* lies on the vertical line *OI*.

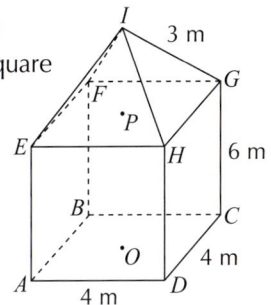

a) (i) Find the exact length *PH*.
(ii) Find the length *PI*.
(iii) Hence find the length *OI*.
b) (i) Find the exact length *OA*.
(ii) Find the length *AI* to 3 s.f.

I 3 m
F *G*
P
E *H* 6 m
B *C*
O 4 m
A 4 m *D*

Q3 A fly is tied to one end of a piece of string. The other end of the string is attached to point *O* on a flat, horizontal surface. The string becomes taut when the fly is 19 cm from *O* in the *x*-direction, 13 cm from *O* in the *y*-direction, and 9 cm from *O* in the *z*-direction. Find the length of the piece of string to 3 s.f.

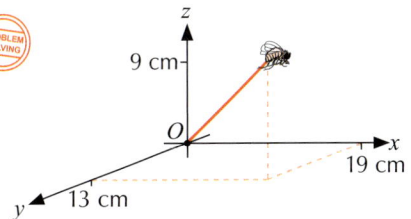

PROBLEM SOLVING

z
9 cm
O
x
19 cm
y 13 cm

Q4 Find the value of the letter in each triangle below. Give your answers to 3 s.f.

a)
17 cm
62°
a

b)
b
32°
28 cm

c)
c
29°
2.1 m

d)
47°
d
2.5 m

e)
13 m
e
65°

f)
21°
25 m
f

g)
g
71°
89 mm

h)
h
37°
2.8 cm

i)
i
11 m
8.5 m

j)
j
3 cm
7 cm

k)
2.2 m
k
1.5 m

l)
88 mm
l
56 mm

Q5 The diagrams below show the cross-sections of two houses. Each has a vertical line of symmetry. Find the vertical height of each house.

a)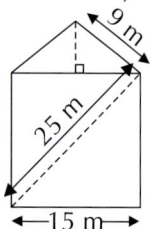

b)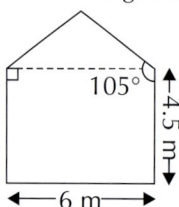

Q6 Points *P*, *Q* and *R* are plotted on a grid of 1 cm squares.
P has coordinates (1, 3), *Q* has coordinates (5, 4) and *R* has coordinates (7, 1).

a) (i) Find the exact distance *PQ*.   (ii) Find the exact distance *PR*.

b) Find the bearing of *Q* from *P* to the nearest whole number.

Q7 Using the sine and/or cosine rules, find the angles *x*, *y* and *z* in the triangle shown on the right, correct to 1 d.p.

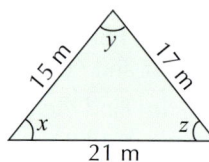

Q8 Leo has a triangular piece of fabric with dimensions as shown.

a) Find the size of angle *p*, to 3 s.f.

b) Hence find the area of the fabric, to 3 s.f.

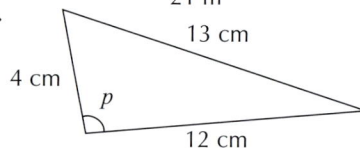

Q9 Find the exact areas of the following triangles.

a)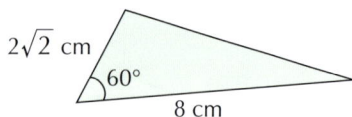

b)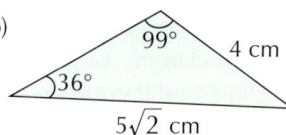

Q10 A wooden crate in the shape of a cuboid is shown on the right.

a) (i) Find the exact length *ED*.
   (ii) Find the angle *FDH*, to 1 d.p.
   (iii) Find the angle *CHD*, to 1 d.p.

b) Will a 10 ft metal pole fit in the crate?

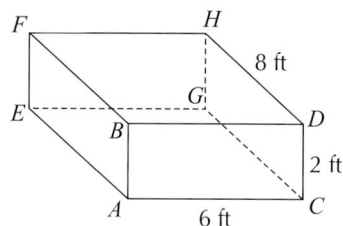

Q11 The diagram on the right shows a shape made up of the square-based pyramid *EFGHI* and the cuboid *ABCDEFGH*. *P* is the centre of the square *EFGH* and *O* is the centre of *ABCD*. *P* lies on the vertical line *OI*.

a) Find the angle *OAI*, to 1 d.p.

b) Find the angle *OAP*, to 1 d.p.

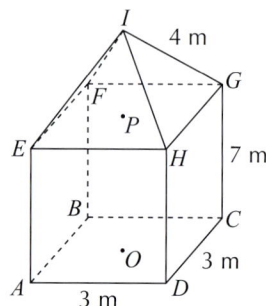

Q1 A ladder leans against the side of a vertical tower. It reaches a window 20 m above the ground. The base of the ladder is placed 8 m away from the bottom of the tower.
What is the angle of elevation? Give your answer correct to 1 decimal place.

[2 marks]

Q2 A taut zip line of length 45 m goes between platforms A and B. Platform A is 80 m above the horizontal ground, and platform B is 65 m above the ground, as shown below.

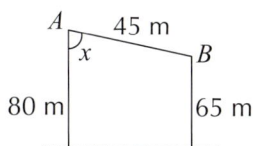

Find x, the angle the wire forms with the vertical at platform A.
Give your answer correct to 1 decimal place.

[2 marks]

Q3 The points P and R have coordinates $P(1, 3)$ and $R(7, 8)$.
Find the distance between P and R, correct to 3 significant figures.

[2 marks]

Q4 The shape shown below is made up of two right-angled triangles.
Find the size of angle x.

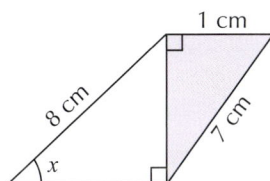

[3 marks]

Q5 Rahim wants to put a gold ribbon around the edges of the kite shown below.

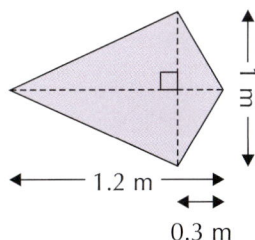

Ribbon is sold in lengths of 10 cm. What length of ribbon should he buy?

[3 marks]

Q6 The pyramid $ABCDE$ has a square base of side length 2.15 m. Point E is directly above the centre of the square $ABCD$. The edge ED is 3.85 m long. Work out the angle between ED and the base, giving your answer to 1 decimal place.

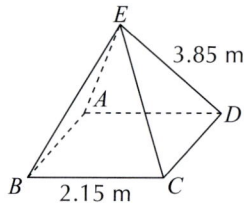

[3 marks]

Q7 An aeroplane takes off from a flat runway and flies for 675 m in a straight line, after which time it is 250 m vertically above the horizontal ground. Work out the angle of elevation of the plane, to 3 significant figures.

[2 marks]

Q8 Bianca (B) and Caleb (C) are standing 25 m apart in a straight line directly facing a vertical tower block, as shown in the diagram. The ground is horizontal. From where Bianca is standing, the angle of elevation from the ground to the top of the tower block is 28°. From where Caleb is standing, the angle of elevation from the ground to the top of the tower block is 47°.

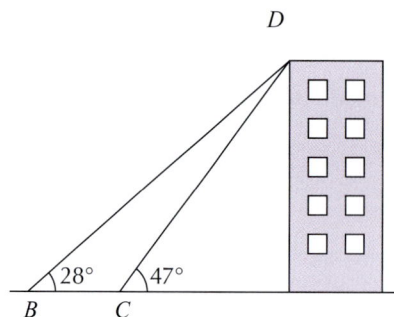

Work out the height of the tower block, giving your answer to three significant figures.

[4 marks]

Q9 The diagram shows a triangle ABC where angle $ACB = 60°$, $AC = x + 5$ cm and $BC = x$ cm. The area of the triangle is $\sqrt{108}$ cm^2.

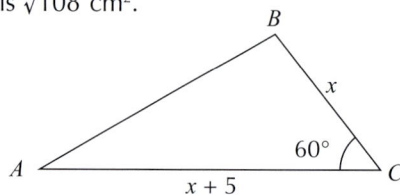

Find the length of AB.

[6 marks]

26.1 Vectors and Scalars

Time for something completely different now — vectors. You can think of them as straight lines from one point in space to another.

Vectors and Scalars

Learning Objectives — Spec Ref G25:
- Understand and use vector notation.
- Be able to multiply vectors by scalars.

A **vector** has **magnitude** (size or length) and **direction**.

There are various ways to represent a vector — it can be:

- written as a **column vector** (positive numbers mean right or up, negative numbers mean left or down).

- shown on a diagram by an **arrow**.

$$\begin{pmatrix} 3 \\ 2 \end{pmatrix}$$

x-component (horizontal): 3 units right

y-component (vertical): 2 units up

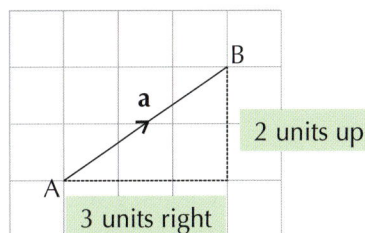

2 units up

3 units right

Vectors can be written using their **end points**, so $\overrightarrow{AB}$ means the vector from **A to B** and $\overrightarrow{BA}$ means the vector from **B to A**. $\overrightarrow{AB}$ and $\overrightarrow{BA}$ are **different** vectors — they have the **same magnitude** but **different directions**. Vectors can also be written using **bold** letters (**a**) or **underlined** letters (a or a).

Two vectors are considered to be **equal** if they have the **same magnitude** and **direction**. E.g. vectors **m** and **n** are equal vectors, even though they have different start and end points.

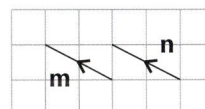

Example 1

On a grid, draw the vector m = $\begin{pmatrix} -3 \\ 5 \end{pmatrix}$.

1. The positive and negative directions in column vectors are the same as they are for coordinates.

2. So –3 in the *x*-component means '3 units left', and 5 in the *y*-component means '5 units up'.

3. Make sure that you label the vector **m**, and add an arrow to show its direction.

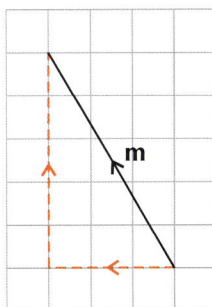

Tip: It doesn't matter where the vector is drawn on the grid. You can choose any starting point — as long as the end point is 3 units to the left and 5 units up.

Exercise 1

Q1 On a grid, draw arrows to represent the following vectors.

a) $\begin{pmatrix} 1 \\ 4 \end{pmatrix}$
b) $\begin{pmatrix} 3 \\ 5 \end{pmatrix}$
c) $\begin{pmatrix} -2 \\ 4 \end{pmatrix}$
d) $\begin{pmatrix} 0 \\ 5 \end{pmatrix}$

e) $\begin{pmatrix} -3 \\ -5 \end{pmatrix}$
f) $\begin{pmatrix} 3 \\ 0 \end{pmatrix}$
g) $\begin{pmatrix} -3 \\ -3 \end{pmatrix}$
h) $\begin{pmatrix} 0 \\ -3 \end{pmatrix}$

Q2 Write down the column vectors represented by these arrows:

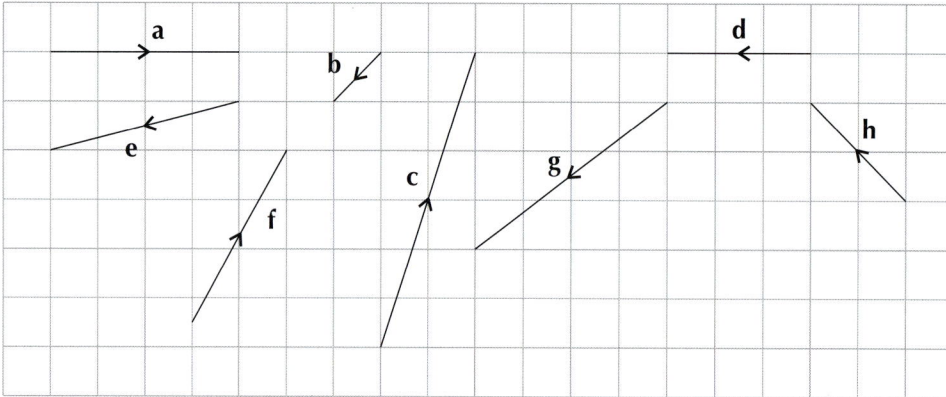

A **scalar** is just a number — scalars have a magnitude (size), but **no** direction. A vector can be **multiplied** by a scalar to give another vector. The resulting vector is **parallel** to the original vector.

E.g. $3 \times \begin{pmatrix} 1 \\ 2 \end{pmatrix} = \begin{pmatrix} 3 \\ 6 \end{pmatrix}$ so the vectors $\begin{pmatrix} 1 \\ 2 \end{pmatrix}$ and $\begin{pmatrix} 3 \\ 6 \end{pmatrix}$ are **parallel**.

If the scalar is **negative**, the direction of the vector is **reversed**.

Example 2

If vector $p = \begin{pmatrix} 2 \\ -3 \end{pmatrix}$, write the following as column vectors: a) 2p   b) $\frac{1}{2}$p   c) −p

Multiply a vector by a scalar by multiplying the x-component and the y-component separately.

a) $2\mathbf{p} = \begin{pmatrix} 2 \times 2 \\ 2 \times -3 \end{pmatrix} = \begin{pmatrix} 4 \\ -6 \end{pmatrix}$

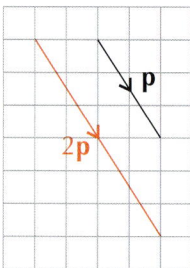

b) $\frac{1}{2}\mathbf{p} = \begin{pmatrix} \frac{1}{2} \times 2 \\ \frac{1}{2} \times -3 \end{pmatrix} = \begin{pmatrix} 1 \\ -\frac{3}{2} \end{pmatrix}$

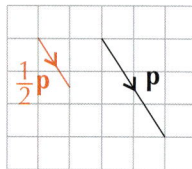

c) $-\mathbf{p} = \begin{pmatrix} -1 \times 2 \\ -1 \times -3 \end{pmatrix} = \begin{pmatrix} -2 \\ 3 \end{pmatrix}$

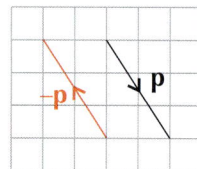

The direction has stayed the same but the magnitude has doubled.

The direction has stayed the same but the magnitude has halved.

The direction has reversed but the magnitude has stayed the same.

Q1 If $\mathbf{q} = \begin{pmatrix} -1 \\ 3 \end{pmatrix}$, find and draw the following vectors.

    a) $3\mathbf{q}$ b) $5\mathbf{q}$ c) $1.5\mathbf{q}$ d) $-2\mathbf{q}$

Q2

$$\mathbf{a} = \begin{pmatrix} 4 \\ -2 \end{pmatrix} \qquad \mathbf{b} = \begin{pmatrix} -1 \\ 4 \end{pmatrix} \qquad \mathbf{c} = \begin{pmatrix} 3 \\ 12 \end{pmatrix} \qquad \mathbf{d} = \begin{pmatrix} 8 \\ -4 \end{pmatrix}$$

$$\mathbf{e} = \begin{pmatrix} 1 \\ 4 \end{pmatrix} \qquad \mathbf{f} = \begin{pmatrix} 0 \\ 3 \end{pmatrix} \qquad \mathbf{g} = \begin{pmatrix} 3 \\ -12 \end{pmatrix} \qquad \mathbf{h} = \begin{pmatrix} 6 \\ 0 \end{pmatrix}$$

From the list of vectors above:

a) Which vector is equal to $2\mathbf{a}$?   b) Which vector is equal to $-3\mathbf{b}$?

c) Which vector is parallel to $\mathbf{e}$?   d) Which two vectors are perpendicular?

Adding and Subtracting Vectors

Learning Objective — Spec Ref G25:
Be able to add and subtract vectors.

To **add** or **subtract** column vectors, you add or subtract the x-components and y-components separately. The sum of two vectors is called the **resultant vector**.

Vectors can also be added by drawing them in a chain, nose-to-tail. The resultant vector goes in a **straight line** from the **start** to the **end** of the chain of vectors. When you add two vectors, it doesn't matter which comes first, i.e. $\mathbf{a} + \mathbf{b} = \mathbf{b} + \mathbf{a}$. Be careful when subtracting though — just like with ordinary numbers $\mathbf{a} - \mathbf{b} = -\mathbf{b} + \mathbf{a}$, not $\mathbf{b} - \mathbf{a}$.

Example 3

If vector $\mathbf{d} = \begin{pmatrix} 3 \\ 1 \end{pmatrix}$, vector $\mathbf{e} = \begin{pmatrix} -2 \\ 0 \end{pmatrix}$ and vector $\mathbf{f} = \begin{pmatrix} 1 \\ -3 \end{pmatrix}$, write the following as column vectors:

a) $\mathbf{d} + \mathbf{e}$ b) $\mathbf{f} - \mathbf{e}$

    Add and subtract the x-components and the y-components separately.

$$\begin{pmatrix} 3 \\ 1 \end{pmatrix} + \begin{pmatrix} -2 \\ 0 \end{pmatrix} = \begin{pmatrix} 3 + (-2) \\ 1 + 0 \end{pmatrix} = \begin{pmatrix} 1 \\ 1 \end{pmatrix} \qquad\qquad \begin{pmatrix} 1 \\ -3 \end{pmatrix} - \begin{pmatrix} -2 \\ 0 \end{pmatrix} = \begin{pmatrix} 1 - (-2) \\ -3 - 0 \end{pmatrix} = \begin{pmatrix} 3 \\ -3 \end{pmatrix}$$

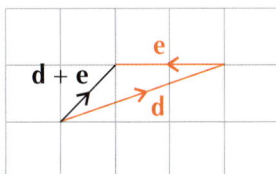

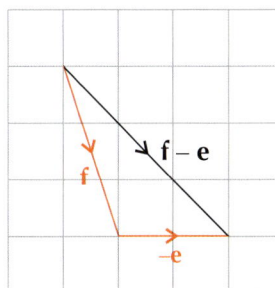

Tip: Subtracting a vector is the same as adding the reverse vector. In part b) the diagram shows that $\mathbf{f} - \mathbf{e}$ is the same as $\mathbf{f} + (-\mathbf{e})$.

Example 4

If vector $p = \begin{pmatrix} 4 \\ 2 \end{pmatrix}$, vector $q = \begin{pmatrix} -3 \\ 1 \end{pmatrix}$ and vector $r = \begin{pmatrix} 1 \\ 2 \end{pmatrix}$, write $2q + p - 3r$ as a column vector.

1. Start by working out the values of the vectors $2q$ and $3r$.

$$2q = \begin{pmatrix} 2 \times -3 \\ 2 \times 1 \end{pmatrix} = \begin{pmatrix} -6 \\ 2 \end{pmatrix}$$

$$3r = \begin{pmatrix} 3 \times 1 \\ 3 \times 2 \end{pmatrix} = \begin{pmatrix} 3 \\ 6 \end{pmatrix}$$

Tip: Always remember to multiply, add and subtract the x-components and the y-components of the vectors separately.

2. Now use vector addition and subtraction to work out the final answer.

$$2q + p - 3r = \begin{pmatrix} -6 \\ 2 \end{pmatrix} + \begin{pmatrix} 4 \\ 2 \end{pmatrix} - \begin{pmatrix} 3 \\ 6 \end{pmatrix}$$

$$= \begin{pmatrix} -6 + 4 - 3 \\ 2 + 2 - 6 \end{pmatrix} = \begin{pmatrix} -5 \\ -2 \end{pmatrix}$$

Exercise 3

Q1 Write the answers to the following calculations as column vectors.
For each expression, draw arrows to represent the two given vectors and the resultant vector.

a) $\begin{pmatrix} 5 \\ 2 \end{pmatrix} + \begin{pmatrix} 3 \\ 4 \end{pmatrix}$

b) $\begin{pmatrix} 4 \\ -1 \end{pmatrix} + \begin{pmatrix} 1 \\ 6 \end{pmatrix}$

c) $\begin{pmatrix} 0 \\ 5 \end{pmatrix} + \begin{pmatrix} -3 \\ 4 \end{pmatrix}$

d) $\begin{pmatrix} 7 \\ 6 \end{pmatrix} - \begin{pmatrix} 3 \\ 4 \end{pmatrix}$

e) $\begin{pmatrix} 5 \\ -1 \end{pmatrix} - \begin{pmatrix} 1 \\ 3 \end{pmatrix}$

f) $\begin{pmatrix} 2 \\ -1 \end{pmatrix} - \begin{pmatrix} -2 \\ 2 \end{pmatrix}$

Q2 If $a = \begin{pmatrix} 2 \\ 3 \end{pmatrix}$, $b = \begin{pmatrix} 0 \\ -2 \end{pmatrix}$ and $c = \begin{pmatrix} -1 \\ 4 \end{pmatrix}$, work out:

a) $b + c$

b) $c - a$

c) $2c + a$

d) $a + b - c$

e) $5b + 4c$

f) $4a - b + 3c$

Q3 If $p = \begin{pmatrix} 5 \\ 0 \end{pmatrix}$, $q = \begin{pmatrix} -3 \\ -1 \end{pmatrix}$ and $r = \begin{pmatrix} 2 \\ 7 \end{pmatrix}$, work out:

a) $p + r$

b) $p - q$

c) $3q + r$

d) $3p + 2r$

e) $p + r - q$

f) $2p - 3q + r$

Q4 $u = \begin{pmatrix} 6 \\ -2 \end{pmatrix}$, $v = \begin{pmatrix} -2 \\ 3 \end{pmatrix}$ and $w = \begin{pmatrix} 1 \\ 2 \end{pmatrix}$

a) Work out $u + 2v$.

b) Draw the vectors $u + 2v$ and w on a grid.

c) What do you notice about the directions of the vectors $u + 2v$ and w?

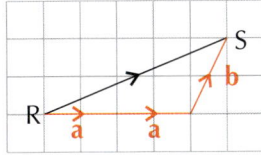

Example 5

**The grid on the right shows vectors a, b, $\vec{RS}$ and $\vec{TU}$.
Describe the following in terms of vectors a and b.**

a) $\vec{RS}$

1. Find a route from R to S using just vectors **a** and **b**.

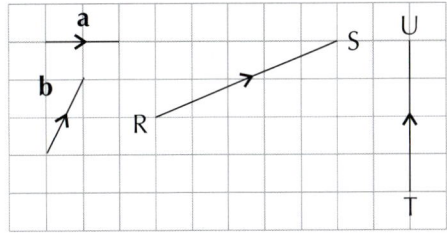

2. So two lots of vector **a** and then vector **b** takes you from R to S:

$$\vec{RS} = 2\mathbf{a} + \mathbf{b}$$

3. Check the answer by doing the vector addition. Work out the component form of the vectors by counting the horizontal and vertical distances.

Check: $2\mathbf{a} + \mathbf{b} = 2\begin{pmatrix}2\\0\end{pmatrix} + \begin{pmatrix}1\\2\end{pmatrix}$

$$= \begin{pmatrix}4\\0\end{pmatrix} + \begin{pmatrix}1\\2\end{pmatrix} = \begin{pmatrix}5\\2\end{pmatrix} = \vec{RS} \ ✔$$

b) $\vec{TU}$

1. Find a route from T to U using just vectors **a** and **b**.

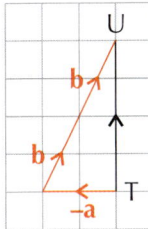

2. The reverse of vector **a** and two lots of vector **b** take you from T to U:

$$\vec{TU} = -\mathbf{a} + 2\mathbf{b}$$

3. Check the answer by doing the vector addition.

Check: $-\mathbf{a} + 2\mathbf{b} = -\begin{pmatrix}2\\0\end{pmatrix} + 2\begin{pmatrix}1\\2\end{pmatrix}$

$$= \begin{pmatrix}-2\\0\end{pmatrix} + \begin{pmatrix}2\\4\end{pmatrix} = \begin{pmatrix}0\\4\end{pmatrix} = \vec{TU} \ ✔$$

Tip: If you're struggling to spot a route between the two points using a diagram, try writing out all the column vectors. Then use them to find a combination of **a** and **b** that gives the vector you're looking for.

Exercise 4

Q1 The grid below shows vectors **a**, **b** and points C-Z.

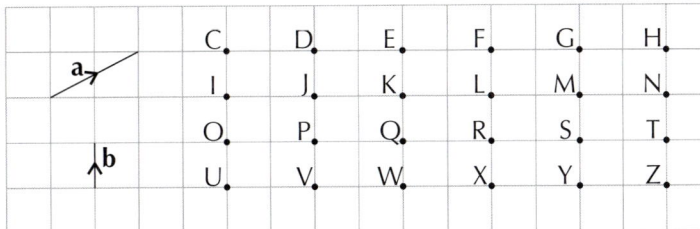

Write the following vectors in terms of **a** and **b**.

a) $\vec{WH}$ b) $\vec{ZH}$ c) $\vec{HR}$ d) $\vec{FX}$

e) $\vec{UD}$ f) $\vec{DU}$ g) $\vec{EM}$ h) $\vec{TP}$

i) $\vec{YJ}$ j) $\vec{CF}$ k) $\vec{FZ}$ l) $\vec{CZ}$

26.2 Vector Geometry

Vectors can be used in geometry to describe the lines that make up shapes. You can use the vectors along with properties of the shape to solve problems and prove other geometrical properties.

Learning Objective — Spec Ref G25:
Use vectors to solve geometry problems.

Prior Knowledge Check:
Understand and use vector notation (p.339), shape properties (Section 20) and ratios (Section 4).

The first step to any vector geometry problem is working out what information you already **know** and what you need to **find**. Always work with a **diagram** — if you're not given one, **draw your own**. Then it'll be easier to see which vectors you need to add, subtract or multiply to get to the solution.

Example 1

In triangle OAB, $\overrightarrow{OA}$ = a and $\overrightarrow{OB}$ = b. M is the midpoint of OB.
Write down $\overrightarrow{AM}$, in terms of vectors a and b:

1. To get from A to M, go from A to O, then from O to M.
   $$\overrightarrow{AM} = \overrightarrow{AO} + \overrightarrow{OM}$$

2. To get from A to O, you go backwards along the vector **a**.
   $$\overrightarrow{AO} = -a$$

3. To get from O to M, you go halfway along the vector **b**.
   $$\overrightarrow{OM} = \tfrac{1}{2}b$$

4. Add the vectors together to find $\overrightarrow{AM}$.
   $$\overrightarrow{AM} = -a + \tfrac{1}{2}b$$

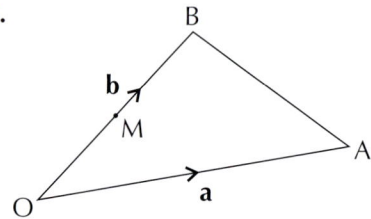

Ratios and **fractions** may be used to describe the position of a point on a line. E.g. if point Z lies on the line XY such that XZ:ZY = 3:5, you know that point Z is $\frac{3}{3+5} = \frac{3}{8}$ of the way from X to Y. If you know the vector of the **whole line**, you can use the ratio or fraction to work out the vector for **part of the line**.

To show that three points lie on a **straight line** you need to show that two vectors between the points are **scalar multiples** of each other. E.g. XYZ is a straight line if $\overrightarrow{XY}$ is a scalar multiple of $\overrightarrow{XZ}$ or $\overrightarrow{YZ}$.

Example 2

ABCD is a parallelogram, $\overrightarrow{AB}$ = r and $\overrightarrow{AD}$ = s. M is the midpoint of BC.
Point T lies on BD such that BT:BD = 1:3.

a) Write down $\overrightarrow{AT}$ in terms of vectors r and s.

1. To get from A to T, go from A to B, then from B to T.
   $$\overrightarrow{AT} = \overrightarrow{AB} + \overrightarrow{BT}$$

2. As BT:BD = 1:3, BT must be $\frac{1}{3}$ of the length of BD.
   $$\overrightarrow{BT} = \tfrac{1}{3}\overrightarrow{BD}$$

3. To get from B to D, go from B to A, then from A to D.
   $$\overrightarrow{BD} = -r + s$$

4. Multiply $\overrightarrow{BD}$ by $\frac{1}{3}$ to find $\overrightarrow{BT}$.
   $$\overrightarrow{BT} = -\tfrac{1}{3}r + \tfrac{1}{3}s$$

5. Add $\overrightarrow{AB}$ and $\overrightarrow{BT}$ to find $\overrightarrow{AT}$.
   $$\overrightarrow{AT} = r - \tfrac{1}{3}r + \tfrac{1}{3}s = \tfrac{2}{3}r + \tfrac{1}{3}s$$

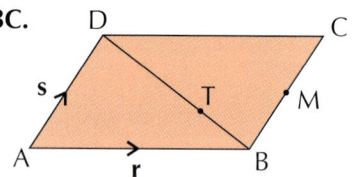

Tip: Be careful when you're using ratios — the ratio given in this question is the ratio of part of the line to the whole line. It's different from the ratio shown in the theory above.

b) Show that ATM is a straight line.

1. First work out $\overrightarrow{AM}$ in terms of **r** and **s**. To get from A to M, go from A to B, then from B to M.

$$\overrightarrow{AM} = \overrightarrow{AB} + \overrightarrow{BM}$$

$$\overrightarrow{BM} = \tfrac{1}{2}\mathbf{s}$$

$$\overrightarrow{AM} = \mathbf{r} + \tfrac{1}{2}\mathbf{s}$$

> **Tip:** Since ABCD is a parallelogram, $\overrightarrow{BC} = \overrightarrow{AD} = \mathbf{s}$.

2. To show that ATM is a straight line you need to show that $\overrightarrow{AT}$ is a scalar multiple of $\overrightarrow{AM}$.

$$\tfrac{2}{3}\overrightarrow{AM} = \tfrac{2}{3}(\mathbf{r} + \tfrac{1}{2}\mathbf{s}) = \tfrac{2}{3}\mathbf{r} + \tfrac{1}{3}\mathbf{s} = \overrightarrow{AT}$$

$\overrightarrow{AT}$ is a scalar multiple of $\overrightarrow{AM}$ so ATM is a straight line.

Exercise 1

Q1 ABCD is a trapezium. $\overrightarrow{AB} = 4\mathbf{p}$, $\overrightarrow{AD} = \mathbf{q}$ and $\overrightarrow{DC} = \mathbf{p}$. Write down, in terms of **p** and **q**:

a) $\overrightarrow{CA}$

b) $\overrightarrow{CB}$

c) $\overrightarrow{BD}$

Not drawn to scale

Q2 In the diagram on the right, $\overrightarrow{OA} = \mathbf{a}$ and $\overrightarrow{OB} = \mathbf{b}$. Point C is added such that $\overrightarrow{OC} = \mathbf{a} + \mathbf{b}$.

a) What type of shape is OACB?

b) Write down, in terms of **a** and **b**:

   (i) $\overrightarrow{CO}$ (ii) $\overrightarrow{AB}$

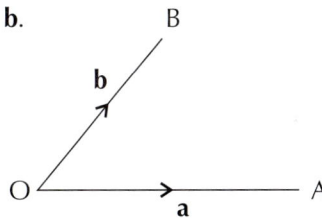

Q3 In the triangle OBD, $\overrightarrow{OA} = \mathbf{a}$ is $\tfrac{1}{4}$ of the length of $\overrightarrow{OB}$, $\overrightarrow{OC} = \mathbf{c}$ is $\tfrac{1}{3}$ of the length of $\overrightarrow{OD}$ and E is the midpoint of BD. Write down, in terms of **a** and **c**:

a) $\overrightarrow{OB}$ b) $\overrightarrow{OD}$ c) $\overrightarrow{AB}$

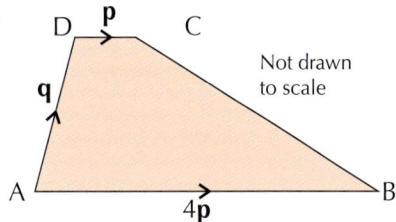

d) $\overrightarrow{BA}$ e) $\overrightarrow{AC}$ f) $\overrightarrow{OE}$

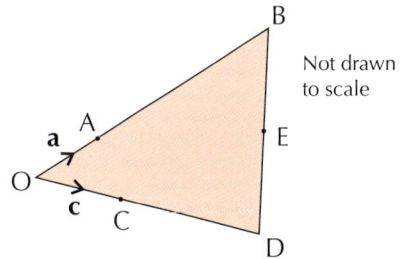

Not drawn to scale

Q4 WXYZ is a parallelogram. $\overrightarrow{WX} = \mathbf{u}$ and $\overrightarrow{WZ} = \mathbf{v}$. M is the midpoint of WY.

a) Write down, in terms of **u** and **v**:

   (i) $\overrightarrow{WY}$ (ii) $\overrightarrow{WM}$

   (iii) $\overrightarrow{XW}$ (iv) $\overrightarrow{XZ}$

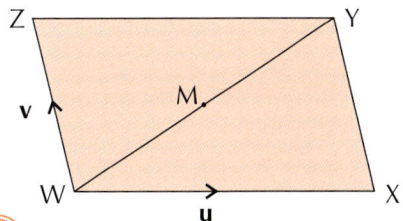

b) Show using vectors that M is the midpoint of XZ.

PROBLEM SOLVING

Q5   ABCDEF is a regular hexagon. M is the centre of the hexagon.
$\vec{AB}$ = **p** and $\vec{BC}$ = **q** and $\vec{CD}$ = **r**.

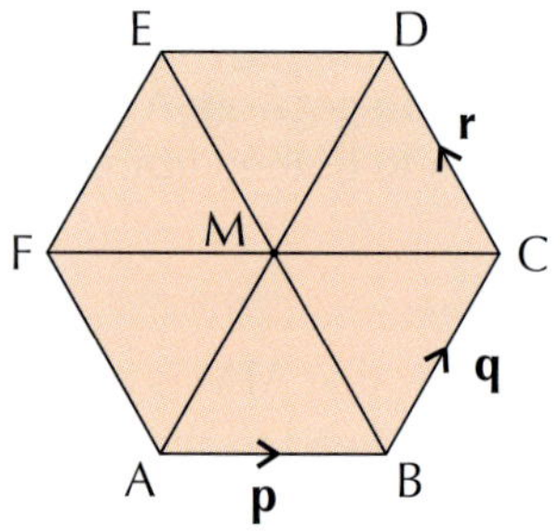

a) Write down, in terms of **p** and **q**:

   (i) $\vec{DE}$ (ii) $\vec{AC}$

b) Using vectors, show that $\vec{FD}$ = $\vec{AC}$.

c) Write down, in terms of **q** and **r**:

   (i) $\vec{AM}$ (ii) $\vec{MB}$

d) Write **p** in terms of **q** and **r**.

Q6   KLMN is a rhombus. $\vec{KN}$ = **a** and $\vec{KL}$ = **b**.
Point S lies on NL such that NS:SL = 2:3.
Point T lies on NM such that $\vec{NT}$ = $\frac{2}{3}\vec{NM}$.

Show that KST is a straight line.

Not drawn to scale

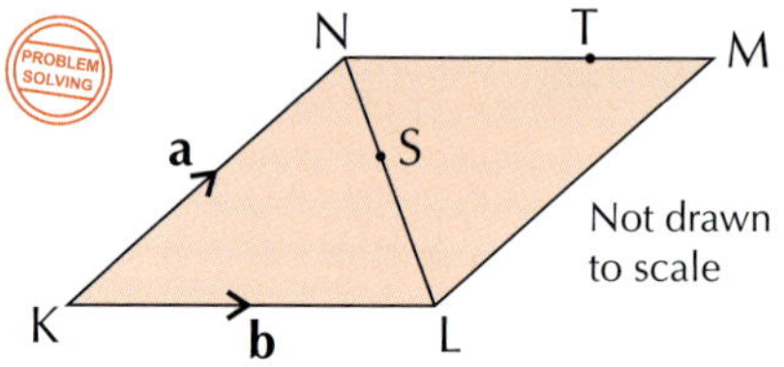

Q7   In the diagram, $\vec{SU}$ = **a**, $\vec{TV}$ = **b**, $\vec{SW}$ = 2$\vec{WU}$ and 3$\vec{TW}$ = 2$\vec{WV}$.

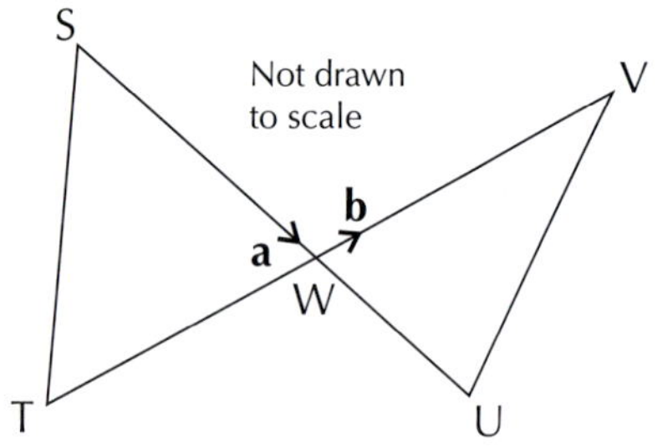

a) Write:

   (i) $\vec{WU}$ in terms of $\vec{SW}$

   (ii) $\vec{SU}$ in terms of $\vec{SW}$

   (iii) $\vec{SW}$ in terms of **a**

b) Write:

   (i) $\vec{WV}$ in terms of $\vec{TW}$

   (ii) $\vec{TV}$ in terms of $\vec{WV}$

   (iii) $\vec{VW}$ in terms of **b**

c) Write, in terms of **a** and **b**:

   (i) $\vec{ST}$ (ii) $\vec{UV}$

Not drawn to scale

Q8 $\vec{GE}$ = **m**, $\vec{GF}$ = **n**, $\vec{GA}$ = 5$\vec{GE}$ and $\vec{GB}$ = 3$\vec{GF}$.
B is the midpoint of AD, and C is the midpoint of BD.

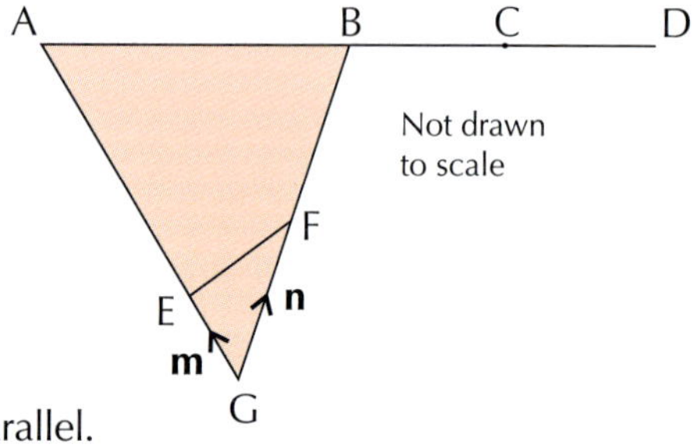

a) Write down, in terms of **m** and **n**:

   (i) $\vec{GA}$ (ii) $\vec{GB}$

   (iii) $\vec{AB}$ (iv) $\vec{BC}$

Not drawn to scale

If E, F and C were on a straight line, $\vec{EF}$ and $\vec{FC}$ would be parallel.

b) Find $\vec{EF}$ and $\vec{FC}$ in terms of **m** and **n**, and use your answers to
    show that E, F and C do not lie on a straight line.

Q1 Match the following column vectors to the correct vectors in the diagram below.

$\begin{pmatrix} 1 \\ 4 \end{pmatrix}$ $\begin{pmatrix} -2 \\ -4 \end{pmatrix}$

$\begin{pmatrix} -4 \\ -1 \end{pmatrix}$ $\begin{pmatrix} 4 \\ -2 \end{pmatrix}$

$\begin{pmatrix} 0 \\ -5 \end{pmatrix}$ $\begin{pmatrix} 3 \\ 4 \end{pmatrix}$

$\begin{pmatrix} 4 \\ -1 \end{pmatrix}$ $\begin{pmatrix} 1 \\ -4 \end{pmatrix}$

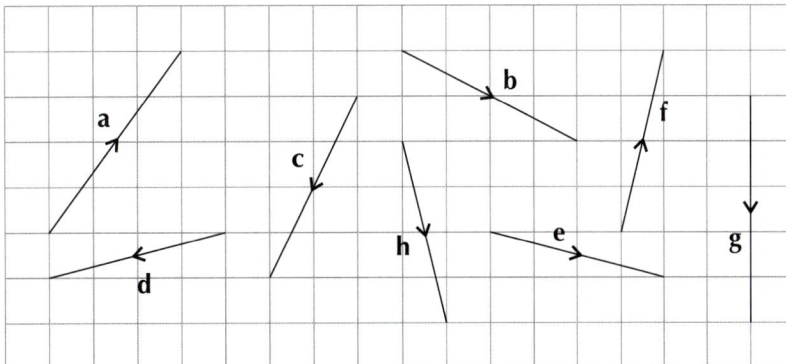

Q2 If $\mathbf{p} = \begin{pmatrix} 4 \\ -3 \end{pmatrix}$, $\mathbf{q} = \begin{pmatrix} 0 \\ 2 \end{pmatrix}$ and $\mathbf{r} = \begin{pmatrix} -1 \\ 5 \end{pmatrix}$, find:

a) $3\mathbf{p}$ b) $2\mathbf{q} + \mathbf{r}$ c) $\mathbf{r} - 2\mathbf{p}$ d) $\mathbf{p} + 5\mathbf{r} - 3\mathbf{q}$

Q3 $\mathbf{a} = \begin{pmatrix} 6 \\ -4 \end{pmatrix}$ $\mathbf{b} = \begin{pmatrix} -2 \\ 3 \end{pmatrix}$ $\mathbf{c} = \begin{pmatrix} 1 \\ 2 \end{pmatrix}$ $\mathbf{d} = \begin{pmatrix} 4 \\ -2 \end{pmatrix}$ $\mathbf{e} = \begin{pmatrix} 3 \\ 6 \end{pmatrix}$ $\mathbf{f} = \begin{pmatrix} 4 \\ -6 \end{pmatrix}$

a) Draw the vector $\mathbf{a} + \mathbf{b}$ and write down the resultant vector as a column vector.

b) Draw the vector $\mathbf{e} - \mathbf{d}$ and write down the resultant vector as a column vector.

c) Which two vectors above can be added to give the resultant vector $\begin{pmatrix} 1 \\ 9 \end{pmatrix}$?

d) Which vector above is parallel to $\mathbf{c}$?

Q4 The diagram on the right shows pentagon ABCDE. $\overrightarrow{BD} = \mathbf{a}$, $\overrightarrow{DE} = \mathbf{a} + 2\mathbf{b}$, $\overrightarrow{EA} = 2\mathbf{b}$ and $\overrightarrow{BC} = -\mathbf{a} - \frac{5}{2}\mathbf{b}$.
Write down, in terms of $\mathbf{a}$ and $\mathbf{b}$:

a) $\overrightarrow{BE}$ b) $\overrightarrow{AB}$

c) $\overrightarrow{CD}$ d) $\overrightarrow{AC}$

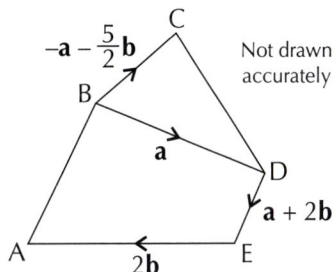

Not drawn accurately

Q5 PQRS is a trapezium, where X is the midpoint of RS. $\overrightarrow{PS} = \mathbf{m}$, $\overrightarrow{PQ} = \mathbf{n}$ and $2\overrightarrow{PS} = 5\overrightarrow{QR}$.

Find $\overrightarrow{PX}$ in terms of $\mathbf{m}$ and $\mathbf{n}$.

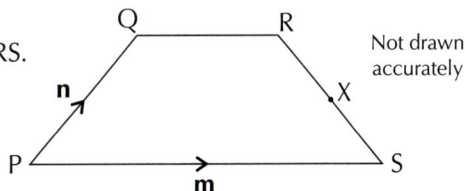

Not drawn accurately

Exam-Style Questions

Q1 **s** and **t** are column vectors such that $\mathbf{s} = \begin{pmatrix} 4 \\ 1 \end{pmatrix}$ and $\mathbf{t} = \begin{pmatrix} -5 \\ 2 \end{pmatrix}$. Work out the value of:

a) 5**s**

[1 mark]

b) 2**s** − **t**

[2 marks]

Q2 **a** and **b** are column vectors such that $\mathbf{a} = \begin{pmatrix} 5p \\ 4q \end{pmatrix}$ and $\mathbf{b} = \begin{pmatrix} -q \\ 0 \end{pmatrix}$ where p and q are integers.

If $2\mathbf{a} - 3\mathbf{b} = \begin{pmatrix} 18 \\ -32 \end{pmatrix}$, find the values of p and q.

[4 marks]

Q3 The diagram shows parallelogram PQRS.

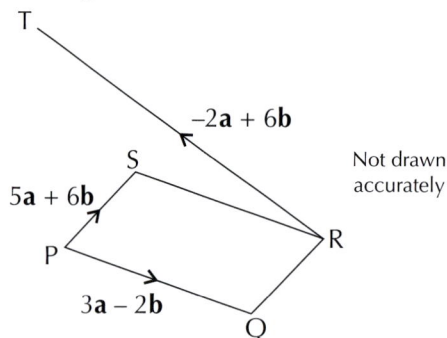

Not drawn accurately

$\overrightarrow{PQ} = 3\mathbf{a} - 2\mathbf{b}$, $\overrightarrow{PS} = 5\mathbf{a} + 6\mathbf{b}$ and T is a point such that $\overrightarrow{RT} = -2\mathbf{a} + 6\mathbf{b}$.

Prove that QST is a straight line.

[4 marks]

Q4 The diagram shows a triangle ODB.

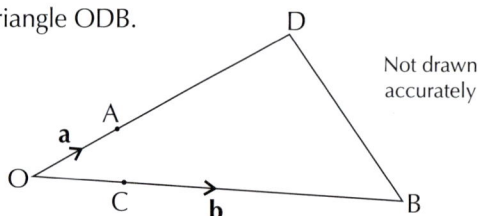

Not drawn accurately

A is a point on OD such that $\overrightarrow{OA} = \mathbf{a}$ and $\overrightarrow{AD} = 3\overrightarrow{OA}$.
$\overrightarrow{OB} = \mathbf{b}$ and C is a point on OB such that OC : OB = 1 : 4.

Prove that ADBC is a trapezium.

[5 marks]

27.1 Triangles and Quadrilaterals

You should already be familiar with area and perimeter, but for GCSE you'll need to work them out for all sorts of shapes, including composite shapes. We'll start with the basics — squares, rectangles and triangles.

Triangles, Squares and Rectangles

Learning Objective — Spec Ref G16/G17:
Find the area and perimeter of rectangles, triangles and composite shapes.

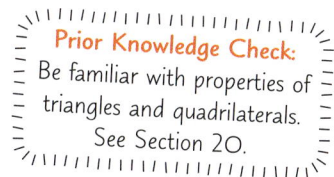
Prior Knowledge Check:
Be familiar with properties of triangles and quadrilaterals. See Section 20.

Perimeter (P) is the distance around the outside of a shape.

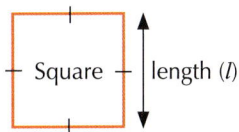

$$P = 4l$$

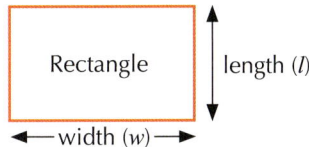

$$P = 2l + 2w$$

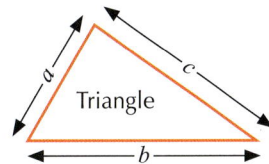

$$P = a + b + c$$

Area (A) is the amount of space inside a shape.

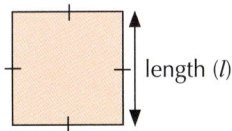

Area = (side length)²

$$A = l^2$$

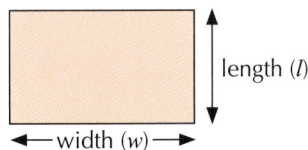

Area = length × width

$$A = lw$$

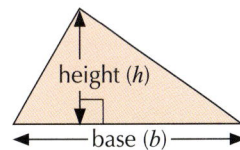

Area = $\frac{1}{2}$ × base × perpendicular height

$$A = \frac{1}{2}bh$$

Exercise 1

Q1 For each shape below, find: (i) its perimeter, (ii) its area.

a)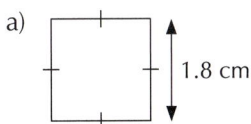

b)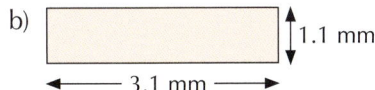

c)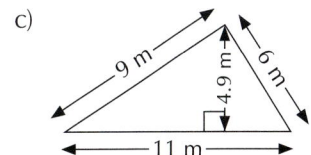

Q2 For each shape described below, find: (i) its perimeter, (ii) its area.

a) a square with sides of length 4 cm

b) a rectangle of width 6 m and length 8 m

c) a rectangle 23 mm long and 15 mm wide

d) a square with 17 m sides

e) a rectangle 22.2 m long and 4.3 m wide

f) a rectangle of length 9 mm and width 2.4 mm

Q3 For each of the triangles below, find: (i) its perimeter, (ii) its area.

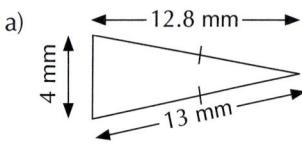

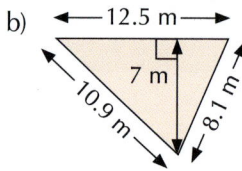

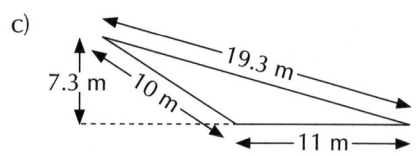

a)

12.8 mm
4 mm
13 mm

b)

12.5 m
7 m
10.9 m
8.1 m

c)

19.3 m
7.3 m
10 m
11 m

You might also have to find the area or perimeter of a **composite shape** (a shape that can be split up into two or more basic shapes). To find the **perimeter**, use the lengths you're given to find any **missing** ones, then **add** all the side lengths up. For the **area**, **split the shape up** into smaller pieces, then work out the area of each piece separately and add them together.

Example 1

For the shape on the right, find: a) its perimeter, b) its area.

7 cm
10 cm
11 cm
4 cm

a) 1. Label the missing sides, and find their lengths.

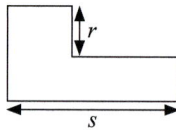

$r = 10 - 4 = 6$ cm, $s = 7 + 11 = 18$ cm

2. Add up the lengths of all the sides. Perimeter $= 7 + 6 + 11 + 4 + 18 + 10 = $ **56 cm**

b) 1. Split the shape into rectangles A and B, and find their areas.

Area of rectangle A $= 10 \times 7$
$= 70$ cm²
Area of rectangle B $= 11 \times 4$
$= 44$ cm²

2. Add these to find the total area.

Total area of shape $= 70 + 44$
$= $ **114 cm²**

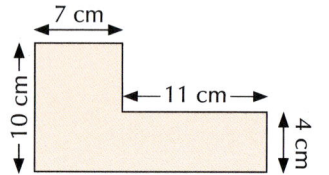

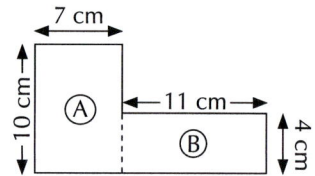

7 cm
10 cm
11 cm
4 cm
(A)
(B)

Exercise 2

Q1 For each shape below, find: (i) its perimeter, (ii) its area.

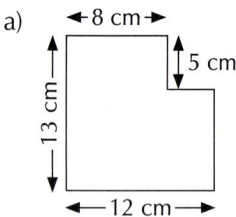

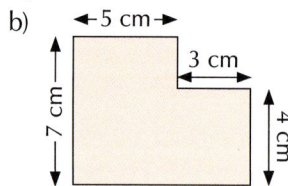

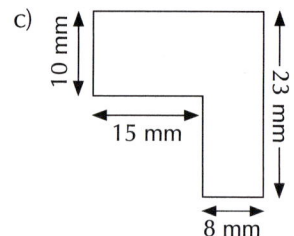

a)

8 cm
5 cm
13 cm
12 cm

b)

5 cm
3 cm
7 cm
4 cm

c)

10 mm
15 mm
23 mm
8 mm

Q2 Find the area of each shape below.

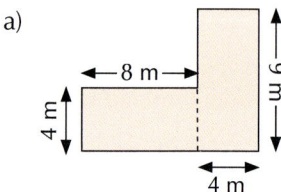

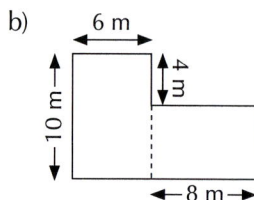

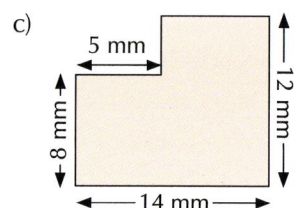

a)

8 m
9 m
4 m
4 m

b)

6 m
4 m
10 m
8 m

c)

5 mm
12 mm
8 mm
14 mm

Q3 Find the areas of the shapes below. The dashed lines show lines of symmetry.

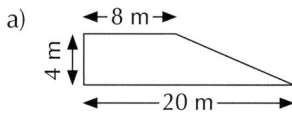

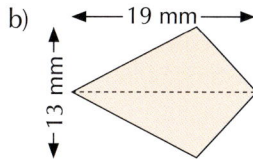

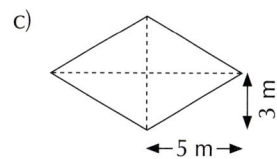

a)

8 m
4 m
20 m

b)

19 mm
13 mm

c)

5 m
3 m

Parallelograms and Trapeziums

Learning Objective — Spec Ref G16/G17:
Find the area and perimeter of parallelograms, trapeziums and composite shapes.

You also need to know the area formulas for **parallelograms** and **trapeziums** — remember that trapeziums always have **one pair** of parallel sides, and parallelograms have **two pairs** of parallel sides (see p.255-256).

The **area** of a **parallelogram** is given by the formula: $A = bh$

Here, h is the **perpendicular height** — it's measured at **right angles** to the base.

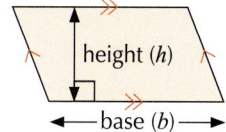

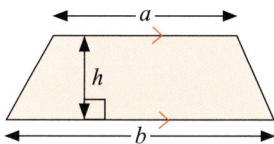

height (h)
base (b)

a
h
b

The **area** of a **trapezium** is given by the formula: $A = \frac{1}{2}(a + b) \times h$

Example 2

Find the area of:
a) the parallelogram,   b) the trapezium.

Just put the numbers into the formulas.

a) $A = bh = 8 \times 3 = $ **24 cm²**

b) $A = \frac{1}{2}(a + b) \times h = \frac{1}{2}(6 + 8) \times 3 = \frac{1}{2} \times 14 \times 3 = $ **21 cm²**

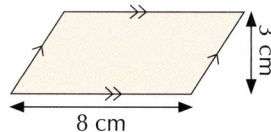

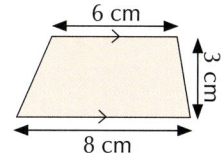

3 cm
8 cm

6 cm
3 cm
8 cm

Exercise 3

Q1 Find the area of each shape below.

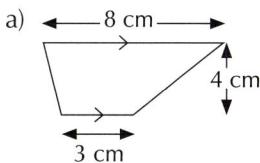

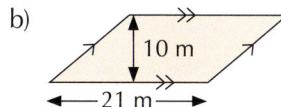

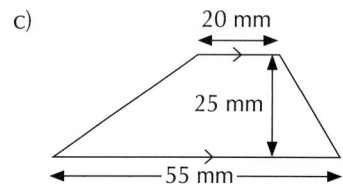

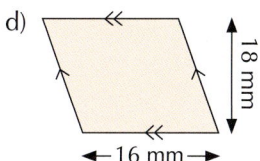

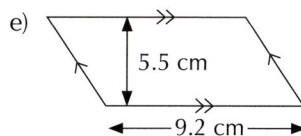

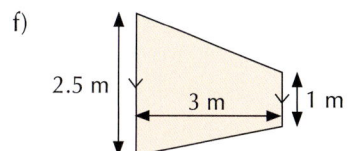

a)

8 cm
4 cm
3 cm

b)

10 m
21 m

c)

20 mm
25 mm
55 mm

d)

18 mm
16 mm

e)

5.5 cm
9.2 cm

f)

2.5 m
3 m
1 m

Composite shapes made up of parallelograms and trapeziums work exactly the same as ones made of rectangles and triangles. When finding perimeters of these shapes, remember that **opposite sides** on a parallelogram are the **same length** — this can be helpful for finding missing lengths. If a big shape has a smaller shape cut out of it, **subtract** the area of the **smaller** shape from the area of the bigger one.

Example 3

For the composite shape on the right, find:
a) its area,   b) its perimeter.

a) 1. Split the shape up into two parallelograms — call them A and B. Find the area of each one.

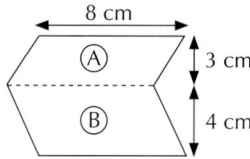

Area of A = bh = 8 × 3 = 24 cm²

Area of B = bh = 8 × 4 = 32 cm²

2. Add these areas to find the total area.   Total area = 24 + 32 = **56 cm²**

b) Opposite sides of a parallelogram are the same length.

P = 8 + 3.3 + 4.4 + 8 + 4.4 + 3.3
  = **31.4 cm**

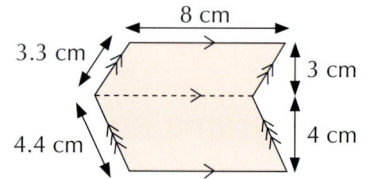

Tip: Make sure you don't accidentally add the dotted line — it's not part of the perimeter.

Exercise 4

Q1 Find each shaded area below.

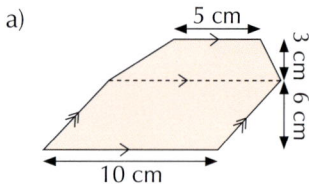

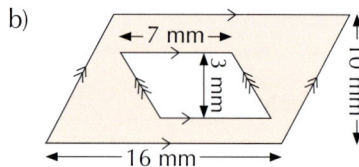

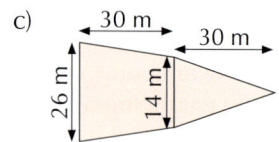

a)

b)

c)

Q2 For each shape below, find: (i) the area, (ii) the perimeter. The dotted lines show lines of symmetry.

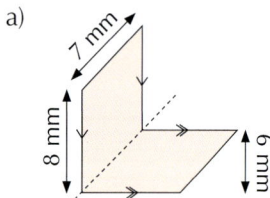

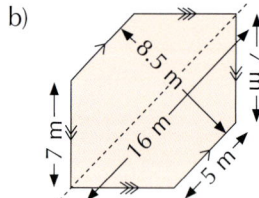

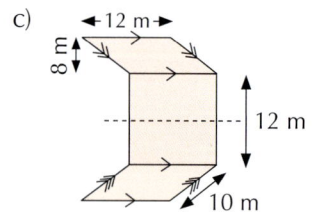

a)

b)

c)

Q3 The picture shows part of a tiled wall. All the tiles are identical parallelograms.
Find the area of one tile.

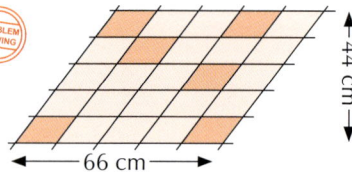

PROBLEM SOLVING

Q4 The flag shown is in the shape of a trapezium. The coloured strips along the top and bottom edges are identical parallelograms.
a) Find the total area of the flag.
b) Find the total area of the coloured strips.

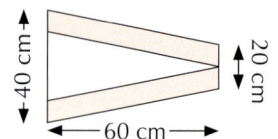

27.2 Circles and Sectors

There are more formulas you need to know on the next few pages, this time featuring your good friend π. Make sure you know where the π button lives on your calculator — you'll need it a lot for circle questions.

Circumference of a Circle

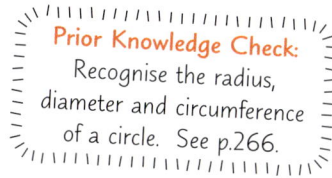
Prior Knowledge Check:
Recognise the radius, diameter and circumference of a circle. See p.266.

Learning Objective — Spec Ref G17:
Find the circumference of a circle.

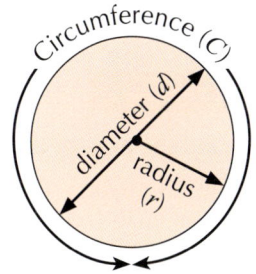

You met the **radius**, **diameter** and **circumference** of a circle on p.266. You can find the circumference (C) of a circle from its diameter (d) or its radius (r) using the formula:

$$C = \pi d = 2\pi r$$

(since $d = 2r$)

Example 1

A circle has a radius of 13 m. Find its circumference to 1 d.p.

You're given r, so use the '$C = 2\pi r$' version of the formula.

$C = 2\pi r = 2 \times \pi \times 13$
$= 81.681... = \mathbf{81.7 \text{ m}}$ (1 d.p.)

Tip: If you're asked for an **exact** answer, leave it in terms of π: $C = 26\pi$.

Example 2

The shape on the right consists of a semicircle on top of a rectangle. Find the perimeter of the shape.

1. Find the curved length. This is half the circumference of a circle with diameter 6 cm.

2. Find the total length of the straight sides and add the two parts.

Curved length $= \pi d \div 2 = \pi \times 6 \div 2$
$= 9.424... \text{ cm}$

Total of straight sides $= 4 + 6 + 4$
$= 14 \text{ cm}$

Total length $= 9.424... + 14 = 23.424... = \mathbf{23.4 \text{ cm}}$ (1 d.p.)

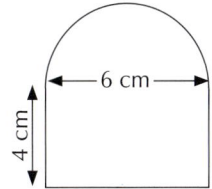

Exercise 1

Q1 Find the circumference of each circle. Give your answers to 1 decimal place.

a)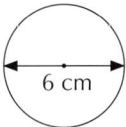
6 cm

b)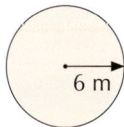
6 m

c)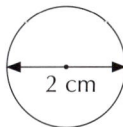
2 cm

d)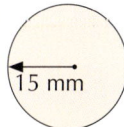
15 mm

Q2 Find, to 1 d.p., the circumference of the circles with the diameter (d) or radius (r) given below.

a) $d = 4$ cm
b) $r = 11$ m
c) $r = 0.1$ km
d) $d = 6.3$ mm

Q3 Find the perimeter of each shape below. Give your answers to 1 decimal place.

a)

←4 cm→

b)

2 m

c)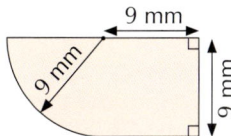
9 mm

9 mm

9 mm

d)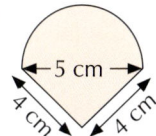
←5 cm→
4 cm   4 cm

Area of a Circle

The **area** (A) of a circle with radius r is given by the formula: $A = \pi r^2$
If you're given the **diameter** of the circle, you need to **divide it by 2** before you can use the formula.

Example 3

Find the area of the circle shown. Give your answer to 1 decimal place.

Substitute the values into the
formula and work out the answer.

$$A = \pi r^2 = \pi \times 3^2 = 28.274...$$
$$= \textbf{28.3 cm}^2 \text{ (1 d.p.)}$$

(3 cm)

Exercise 2

In the following questions, give your answers to 1 decimal place.

Q1   Find the area of each circle.

a) 2 cm   b) 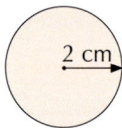 10 mm   c) 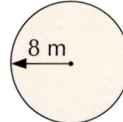 8 m   d) 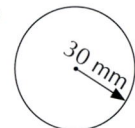 30 mm

Q2   Find, to 1 decimal place, the areas of the circles with the diameter (d) or radius (r) given below.
    a) $r = 6$ mm   b) $r = 5$ cm   c) $r = 4$ m
    d) $d = 8.5$ mm   e) $d = 3.5$ m   f) $d = 1.2$ mm

Q3   Find the area of each shape below to 1 decimal place.

a) 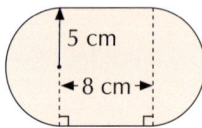   b) 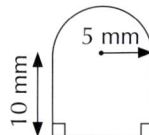   c)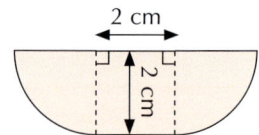

Arcs and Sectors of Circles

Learning Objectives — Spec Ref G17:
- Find the area of a sector of a circle.
- Find the length of an arc of a circle.

Prior Knowledge Check:
Recognise arcs and sectors
of circles. See p.266.

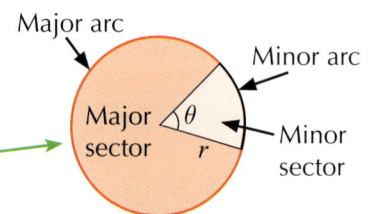

You saw on p.266 that a **sector** is a 'slice' of a circle and an **arc**
is a part of the circumference. If the angle in a sector is **less than 180°**
then it's a **minor** sector, and if the angle is **more than 180°** then
it's a **major** sector.

You can use the following formulas
to find an arc length and a sector area:

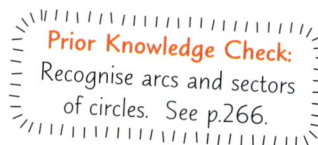

$$\text{Length of arc} = \frac{\theta}{360°} \times \begin{array}{c}\text{circumference}\\\text{of circle}\end{array} = \frac{\theta}{360°} \times 2\pi r$$

$$\text{Area of sector} = \frac{\theta}{360°} \times \text{area of circle} = \frac{\theta}{360°} \times \pi r^2$$

Example 4

For the circle shown, calculate the exact area of the minor sector and the exact length of the minor arc.

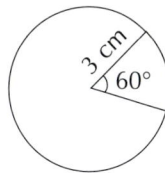

Substitute the values into the formulas and work out the answers.

Area of sector $= \dfrac{60°}{360°} \times \pi \times 3^2$

$= \dfrac{1}{6} \times 9\pi = \dfrac{3}{2}\pi$ **cm²**

Length of arc $= \dfrac{60°}{360°} \times (2 \times \pi \times 3) = \dfrac{1}{6} \times 6\pi = \boldsymbol{\pi}$ **cm**

Tip: The question asks for exact solutions, so leave them in terms of π.

Example 5

The shape shown is a sector of a circle with radius 9 cm. The area of the shape is 27π cm². Find the exact perimeter of the shape.

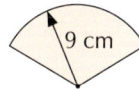

1. You're given the sector area, so use the formula to find the sector angle.

2. Find the length of the arc and add the two straight sides to find the perimeter.

$27\pi = \dfrac{\theta}{360°} \times \pi \times 9^2 \implies \theta = \dfrac{360° \times 27\pi}{81\pi} = 120°$

Arc length $= \dfrac{120°}{360°} \times (2 \times \pi \times 9) = 6\pi$ cm

Perimeter $= 9 + 9 + 6\pi = \boldsymbol{(18 + 6\pi)}$ **cm**

Exercise 3

Q1 For each circle below, find the exact length of the minor arc and exact area of the minor sector.

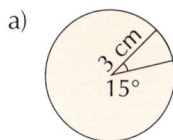

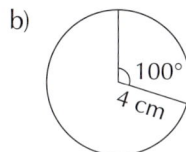

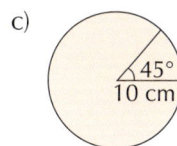

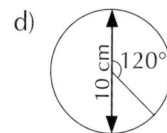

a) 3 cm, 15° b) 100°, 4 cm c) 45°, 10 cm d) 10 cm, 120°

Q2 Find the exact length of the major arc and the exact area of the major sector for the following circles.

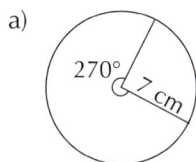

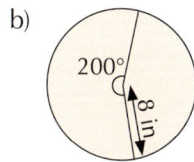

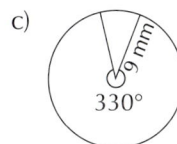

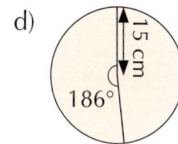

a) 270°, 7 cm b) 200°, 8 in c) 9 mm, 330° d) 15 cm, 186°

Q3 For the circles below, find the length of the major arc and the area of the major sector to 2 d.p.

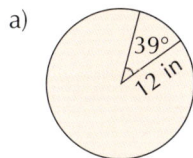

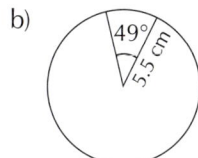

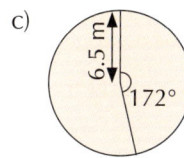

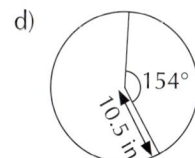

a) 39°, 12 in b) 49°, 5.5 cm c) 6.5 m, 172° d) 154°, 10.5 in

Q4 A sector of a circle has an arc length of 15π cm. The radius of the circle is 12 cm. Find the angle of the minor sector.

Q5 A circular park has an area of 400π m². A children's playground is a sector of the circle and has an area of 80π m². Calculate:
a) the radius of the park,   b) the sector angle of the children's playground,
c) the perimeter of the children's playground, to 1 decimal place.

Review Exercise

Q1 Barbie has a rectangular lawn that is 23.5 m long by 17.3 m wide.
She is going to mow the lawn and then put a fence around the outside.

a) What area will Barbie have to mow (to the nearest m²)?

b) How long will the fence need to be?

Q2 Find the perimeter and area of the triangles below:

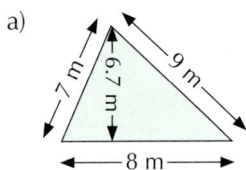

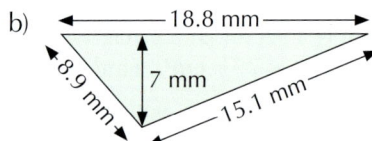

a)

7 m 6.7 m 9 m 8 m

b)

18.8 mm 8.9 mm 7 mm 15.1 mm

Q3 Calculate the area of the following composite shapes:

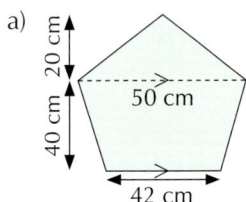

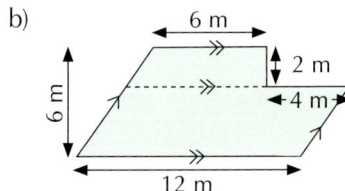

a)

20 cm 40 cm 50 cm 42 cm

b)

6 m 6 m 2 m 4 m 12 m

Q4 Homer and Ned are comparing the flower beds in their gardens.

a) Homer's flower bed is circular with a diameter of 2.7 m.
Calculate the circumference and area of Homer's flower bed to 1 d.p.

b) Ned's flower bed is in the shape of a semicircle. Its area is 19.6 m².
What is the length of its straight edge? Give your answer to 2 d.p. *(PROBLEM SOLVING)*

Q5 Timi has a rectangular sheet of metal that is 50 cm long and 80 cm wide. She cuts a quarter circle of radius 15 cm from each corner. Calculate, to 1 d.p., the remaining area of metal.

Q6 a) Find the length of the major arc and the area of the major sector for a circle with radius 13 cm and minor sector angle 14°. Give your answers to 2 d.p.

b) Find the exact length of the minor arc and the exact area of the minor sector for a circle with diameter 10 m and major sector angle 320°.

Q7 Calculate the total shaded area in the shapes below, giving your answers to 1 decimal place.

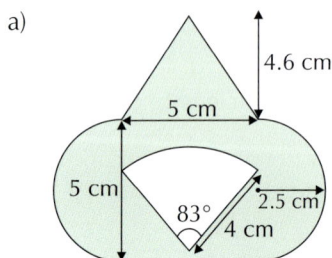

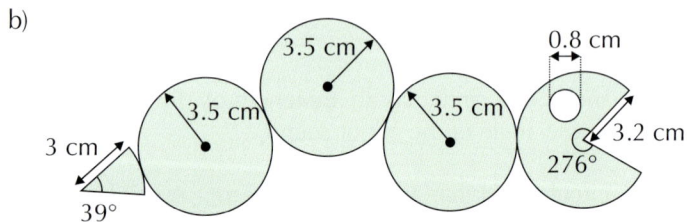

a)

4.6 cm 5 cm 5 cm 83° 2.5 cm 4 cm

b)

3.5 cm 3.5 cm 0.8 cm 3 cm 3.5 cm 3.5 cm 3.2 cm 276° 39°

Exam-Style Questions

Q1 A rectangular floor measures 9 m by 7.5 m.
It is to be tiled using square tiles with sides of length 0.5 m.

a) What is the area of: (i) the floor, (ii) one of the tiles?

[2 marks]

b) How many tiles are needed to cover the floor?

[1 mark]

Q2 Alec has built a semicircular sheep pen. A wall forms the straight side of the semicircle
and Alec used wooden fencing to construct the rest of the pen. The radius of the semicircle
is 16 m. How many metres of fencing has Alec used?

[2 marks]

Q3 The shape shown in the diagram has a line of symmetry,
marked by the dotted line. Find the area of the shape.

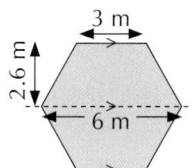

[2 marks]

Q4 Ali bakes a rectangular cake, as shown.

a) What length of ribbon is needed to wrap
once around the outside of the cake,
as shown in the diagram to the right?

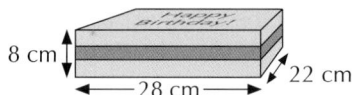

[2 marks]

b) If the top and the four sides are to be iced, what area of icing will be needed?

[2 marks]

c) Tyler bakes a circular cake, and cuts it into 10 segments,
one of which is shown. What length of ribbon is needed to wrap
once around the segment? Give your answer to 1 decimal place.

[3 marks]

Q5 A circular pond of diameter 4 m is surrounded by a path 1 m wide,
as shown in the diagram. What is the area of the path, to 1 d.p.?

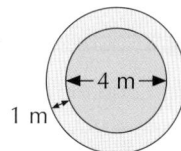

[3 marks]

Q6 A groundsman uses a roller to help dry a flat 22.5 m long
cricket pitch. He pushes the roller in a straight line.
If it takes exactly 17 revolutions of the roller to cover
one length of the pitch, work out the radius of the roller in cm.
Give your answer to an appropriate degree of accuracy.

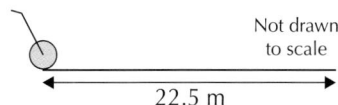

[3 marks]

28.1 Plans, Elevations and Isometric Drawings

It can be difficult to draw 3D shapes accurately on paper — sometimes it's clearer to draw 2D plans and elevations of the 3D shape to show what it looks like from different sides.

Plans and Elevations

Learning Objective — Spec Ref G13:
Draw plans and elevations of 3D shapes.

Plans and **elevations**, also known as **projections**, are **2D representations** of **3D objects** viewed from particular directions. There are **three** different projections:

- **Plan** — the 2D view looking **vertically downwards** on the 3D object.

- **Front elevation** — the 2D view looking **horizontally** from the **front** of the 3D object.

- **Side elevation** — the 2D view looking **horizontally** from the **side** of the 3D object.

Tip: The directions for the front and side elevations are usually indicated by arrows.

For example, the plan of the triangular prism below would be a **rectangle**, the front elevation would be a **triangle** and the side elevation would be another **rectangle**, as shown.

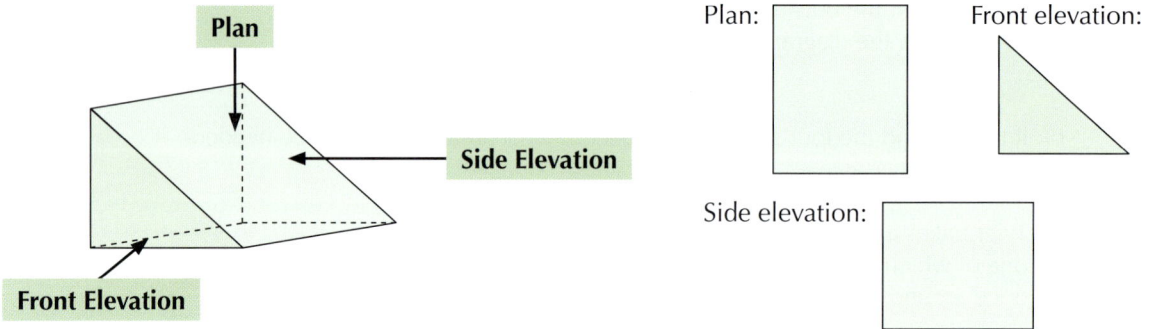

Example 1

Draw the plan and elevations of this 3D shape.

1. Viewed from above, the shape has 4 squares arranged in an L-shape. From this view, you can't tell there's a cube on top of the base layer.

2. Viewed from the front, the shape has 3 squares. You can't see the change in depth.

3. Viewed from the side, the shape has 4 squares in a sideways L-shape. Again, you can't see a change in depth from this elevation.

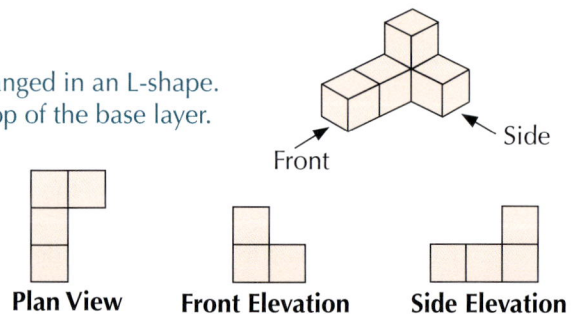

Plan View   Front Elevation   Side Elevation

Example 2

Draw the plan and elevations of the prism on the right.
Label your drawings with their dimensions.

1. Viewed from above, the prism is just a rectangle. So, for the plan, draw a 7 cm × 2 cm rectangle.

2. For the front elevation, you can see the trapezium face, so draw a trapezium with the dimensions given.

3. Viewed from the side, you can see the change of height due to the trapezium. To show this, first draw a 4 cm × 7 cm rectangle. Then, add a straight line 3 cm from the bottom to show the different heights of the top of the trapezium.

Plan View

7 cm
2 cm

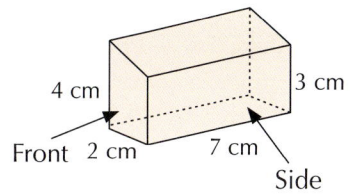

Front Elevation

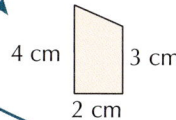

4 cm   3 cm
2 cm

Side Elevation

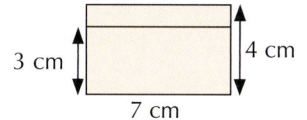

3 cm   4 cm
7 cm

Exercise 1

Q1 For each of the following, draw: (i) the plan view
(ii) the front and side elevations, using the directions shown in a)

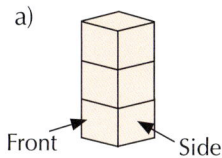

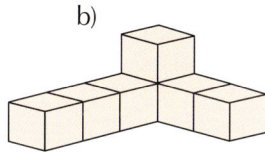

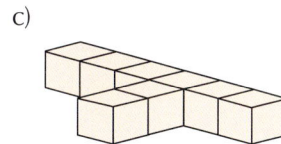

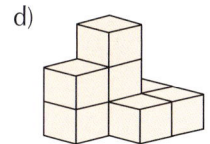

a) b) c) d)

Front Side

Q2 For each of the following, draw the plan, front elevation and side elevation from the directions indicated in part a). Label your diagrams with their dimensions.

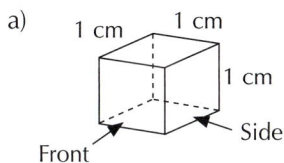

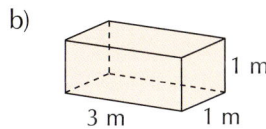

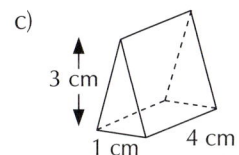

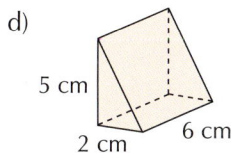

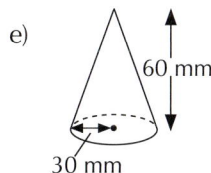

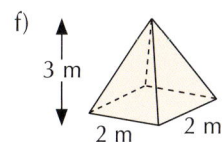

a) 1 cm, 1 cm, 1 cm, Front, Side
b) 1 m, 3 m, 1 m
c) 3 cm, 1 cm, 4 cm
d) 5 cm, 2 cm, 6 cm
e) 60 mm, 30 mm
f) 3 m, 2 m, 2 m

Q3 Draw the plan, front elevation and side elevation for each of the following. Label your diagrams with their dimensions.

a) A cube of side 2 cm

b) A cylinder of radius 3 cm and height 4 cm

c) A 4 cm long prism whose cross-section is an isosceles triangle of height 5 cm and base 3 cm

d) A pyramid of height 5 cm whose base is a square of side 3 cm

Isometric Drawings

Learning Objective — Spec Ref G13:
Draw 3D shapes on isometric paper.

Isometric drawings are drawn on a grid of **dots** or **lines** arranged in a pattern of equilateral triangles. To draw a 3D shape on isometric paper, vertical lines on the shape are shown by **vertical lines** on the isometric paper and horizontal lines are shown by **diagonal lines** on the isometric paper. Each space between the dots in the vertical or diagonal directions represents **one unit** (e.g. 1 cm or 1 m).

Example 3

Draw the triangular prism shown on the right on isometric paper.

1. Join the dots with vertical lines for the vertical lines in the 3D shape and with diagonal lines for the horizontal lines.

2. Build up the drawing as shown below, using the dots to show the dimensions.

3. The triangle has a vertical height of 3 cm so draw a vertical line 3 spaces long.

4. It has a horizontal width of 4 cm so draw a diagonal line 4 spaces long.

5. The depth of the object is 3 cm so draw a diagonal line 3 spaces long.

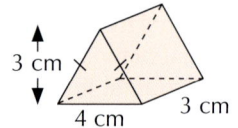

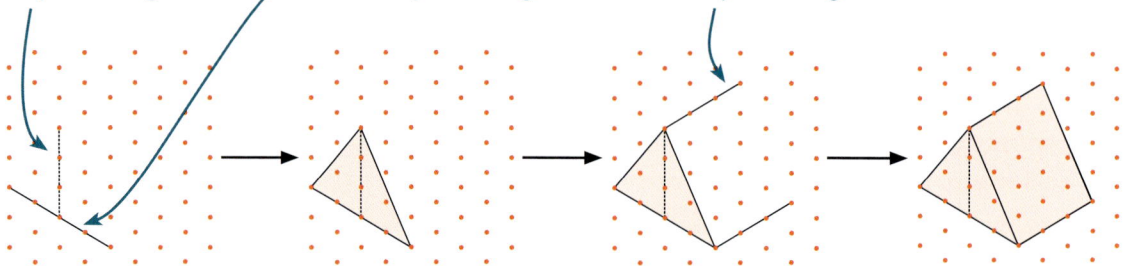

Exercise 2

Q1   Draw the following 3D objects on isometric paper.

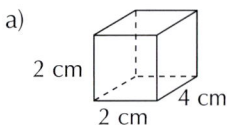

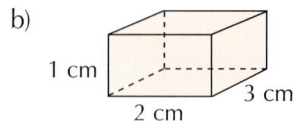

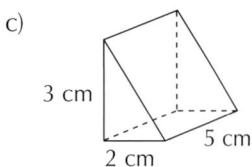

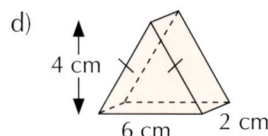

a) 2 cm, 2 cm, 4 cm

b) 1 cm, 2 cm, 3 cm

c) 3 cm, 2 cm, 5 cm

d) 4 cm, 6 cm, 2 cm

Q2   Draw the following 3D objects on isometric paper.

     a) A cube with 2 cm edges

     b) A 2 cm × 2 cm × 3 cm cuboid

     c) A 2 cm long prism whose cross-section is an isosceles triangle with base 4 cm and height 4 cm

     d) A 3 cm long prism whose cross-section is an isosceles triangle with base 2 cm and height 4 cm

You can also use **projections** to draw shapes on isometric paper. It's often helpful to **picture** or **sketch the shape** first, then use the dimensions to draw it **accurately**.

Example 4

The diagram on the right shows the plan and front and side elevations of a 3D object. Draw the object on isometric paper.

1. Try to picture the shape. It's a prism with the front elevation as its cross-section, which you can see since the plan view and side elevation have the same height all the way across.

2. Draw the cross-section on the isometric paper first, using the dots for the dimensions.

3. Use the side elevation and plan view to complete the drawing.

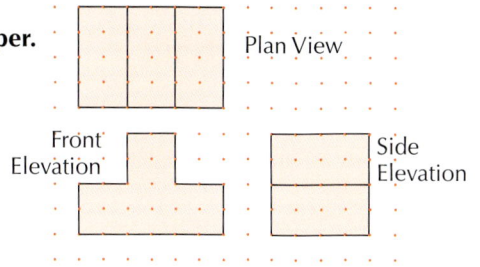

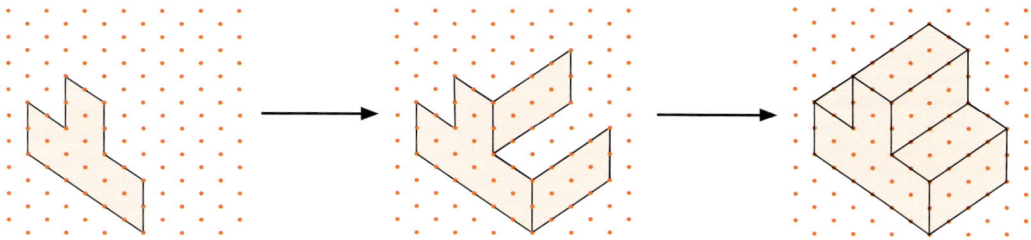

Exercise 3

Q1 The following diagrams show the plan and front and side elevations of different 3D objects.

For each of the objects: (i) use the projections to sketch the object and label the dimensions,
(ii) draw the object accurately on isometric paper.

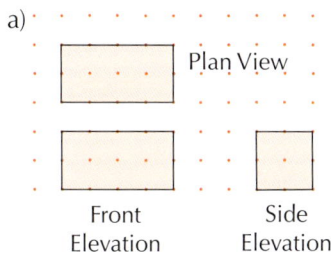

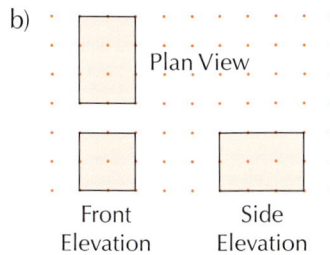

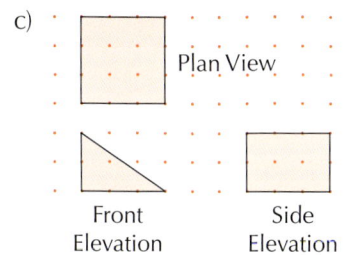

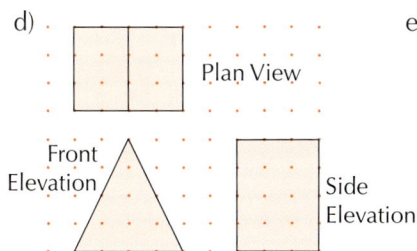

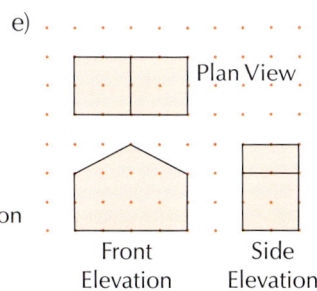

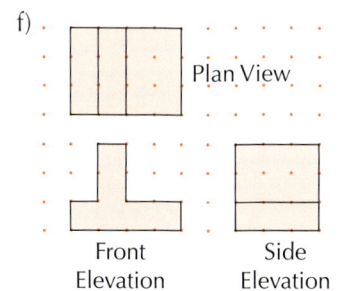

28.2 Volume

3D shapes have vertices (corners), edges and faces (sides). The volume of a 3D shape is the amount of space inside the shape — it's measured in cubic units, e.g. cm^3 or m^3.

Volume of a Cuboid

> **Learning Objective — Spec Ref G16:**
> Calculate the volume of cubes and cuboids.

A **cuboid** is a 3D shape with **six rectangular faces**.
The **volume** of a cuboid is found using the following formula:

> **Volume = Length × Width × Height**

A **cube** is a special type of cuboid — all six of the
faces are **squares**. This means all of the edges have
the **same length**, so the volume is given by:

> **Volume = Length × Length × Length = (Length)³**

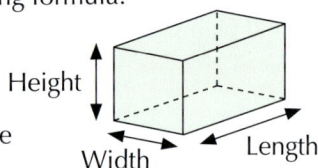

Tip: The units for volume are cubed because you're multiplying three dimensions — here, length, width, height. Make sure all three are in the same units first.

Example 1

Find the volume of the cuboid shown in the diagram.

1. Write down the formula for volume.   Volume = length × width × height

2. Substitute the values into the formula.   Volume = 6 cm × 4 cm × 8 cm

3. Calculate the volume —   **= 192 cm³**
   don't forget the units.

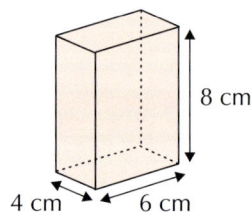

Exercise 1

Q1 Find the volumes of the following cuboids.

a)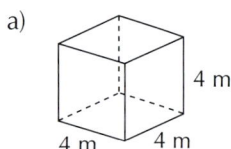

b)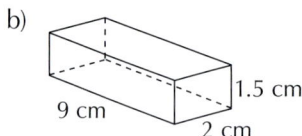

c)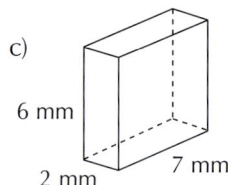

Q2 Find the volume of a cube whose edges are 3.2 mm long.

Q3 Will 3.5 m³ of sand fit in a cuboid-shaped box with dimensions 1.7 m × 1.8 m × 0.9 m?

Q4 A matchbox is 5 cm long and 3 cm wide. The volume of the matchbox is 18 cm³. What is its height?

Q5 A cuboid has a height of 9.2 mm and a width of 1.15 cm. Its volume is 793.5 mm³.
What is its length?

Q6 A bath can be modelled as a cuboid with dimensions 1.5 m × 0.5 m × 0.6 m.
a) What is the maximum volume of water that the bath will hold?
b) Find the volume of water needed to fill the bath to a height of 0.3 m.
c) Find the height of the water in the bath if the volume of water in the bath is 0.3 m³.

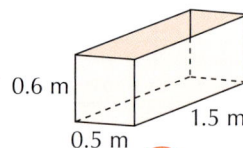

Volume of a Prism

Learning Objective — Spec Ref G16:
Calculate the volume of prisms, including cylinders.

A **prism** is a 3D shape which has a **constant cross-section**. This means that if you **slice** the shape anywhere along its length **parallel** to the faces at the end of the length, the new face you produce is **exactly the same** as those faces at the end. For example, a **triangular prism** has a **triangle** as its constant cross-section and a **cylinder** has a **circle**. The **volume** of a prism is given by the following formula:

$$\text{Volume} = \text{Area of Cross-Section} \times \text{Length}$$

The area of the cross-section for a triangular prism is the area of a triangle ($\frac{1}{2} \times$ **base** $\times$ **height**), and for a cylinder it's the area of a circle (πr^2).

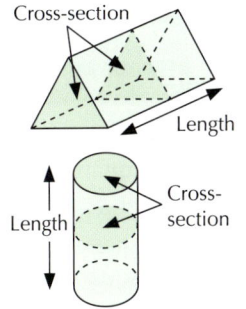

Example 2

Find the volume of each of the shapes shown on the right.

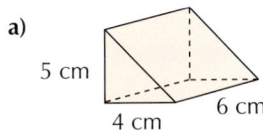

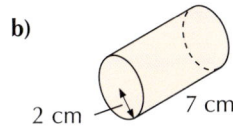

a) 5 cm, 4 cm, 6 cm

b) 2 cm, 7 cm

a) 1. Work out the area of the cross-section. Here it's a triangle, so calculate the area of the triangle.

Area of cross-section = area of triangle
$= \frac{1}{2} \times$ base $\times$ height $= \frac{1}{2} \times 4 \times 5 = 10$ cm²

2. Multiply the cross-sectional area by the length of the prism.

Volume = area of cross-section × length
$= 10 \times 6 = \mathbf{60\ cm^3}$

b) 1. Work out the area of the cross-section. Here it's a circle, so use area $= \pi r^2$.

Area of cross-section = area of circle
$= \pi r^2 = \pi \times 2^2 = 4\pi$ cm²

2. Multiply the cross-sectional area by the length of the prism.

Volume = area of cross-section × length
$= 4\pi \times 7 = \mathbf{88.0\ cm^3}$ (1 d.p.)

Exercise 2

Q1 Find the volumes of the following prisms.

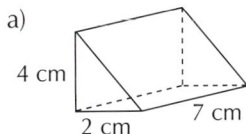

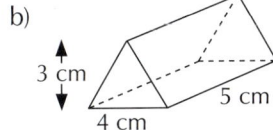

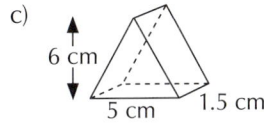

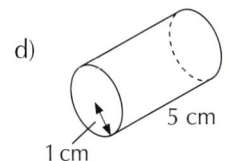

a) 4 cm, 2 cm, 7 cm

b) 3 cm, 4 cm, 5 cm

c) 6 cm, 5 cm, 1.5 cm

d) 1 cm, 5 cm

Q2 Find the volumes of the following prisms.
a) a triangular prism of base 4.2 m, vertical height 1.3 m and length 3.1 m
b) a cylinder of radius 4 m and length 18 m
c) a cylinder of radius 6 mm and length 2.5 mm
d) a prism with parallelogram cross-section of base 3 m and vertical height 4.2 m, and length 1.5 m

Q3 The triangular prism shown on the left has a volume of 936 cm³. Find x, the length of the prism.

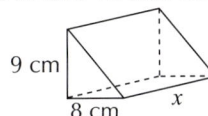

9 cm, 8 cm, x

Q4 Find the values of the missing letters in the following prisms.

a)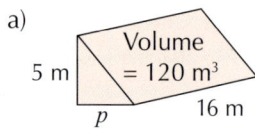
5 m
Volume = 120 m³
16 m
p

b)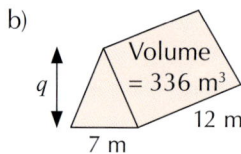
q
Volume = 336 m³
7 m
12 m

c)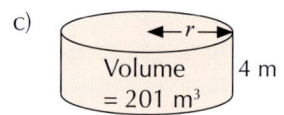
←r→
Volume = 201 m³
4 m

Example 3

The cross-section of a jukebox (shown on the right) is made up of a square and a semicircle. Find the volume of the jukebox.

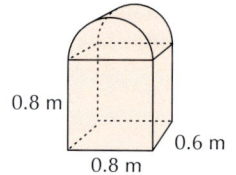
0.8 m
0.8 m
0.6 m

1. The cross-section is a composite shape, so split the shape up into a square and a semicircle. Find each area and then add them together.

Area of square = 0.8 × 0.8 = 0.64 m²

Area of semicircle
$= \frac{1}{2}\pi r^2 = \frac{1}{2}\pi \times 0.4^2 = 0.2513...$ m²

Area of cross-section = 0.64 + 0.2513... = 0.8913... m²

2. Multiply the area of the cross-section by the length.

Volume = area of cross-section × length
= 0.8913... × 0.6 = **0.53 m³** (2 d.p.)

Example 4

A cylindrical tank of radius 28 cm and height 110 cm is half full of water. Find the volume of water in the tank. Give your answer in litres, correct to 1 d.p.

1. Start by working out the capacity of the tank — i.e. its volume.

Area of cross-section = $\pi r^2 = \pi \times 28^2 = 2463.008...$ cm²
Volume of tank = 2463.008... × 110 = 270 930.950... cm³

2. Since the tank is only half full, the volume of water in the tank is half of its total volume.

Volume of water in the tank = 270 930.950... ÷ 2
= 135 465.475... cm³

3. Convert to the unit required. There are 1000 cm³ in 1 litre. (See Section 22 for more on converting between units.)

135 465.475 cm³ = 135 465.475 ÷ 1000
= 135.465... litres
= **135.5 litres** (1 d.p.)

Exercise 3

Q1 By first calculating their cross-sectional areas, find the volumes of the following prisms.

a)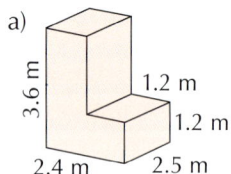
3.6 m
1.2 m
1.2 m
2.4 m
2.5 m

b)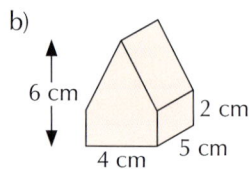
6 cm
2 cm
4 cm
5 cm

c)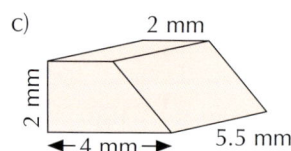
2 mm
2 mm
←4 mm→
5.5 mm

d)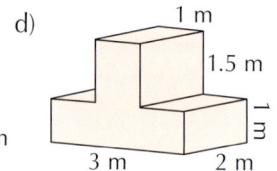
1 m
1.5 m
1 m
3 m
2 m

Use the conversion 1 litre = 1000 cm³ to answer the following questions.

Q2 A cube-shaped box with edges of length 15 cm is filled with water from a glass. The glass has a capacity of 0.125 litres. How many glasses of water will it take to fill the box?

Q3 90 litres of water is pumped into a cylindrical paddling pool, filling it to a depth of 3.5 cm. Find the radius of the paddling pool to 1 decimal place. (PROBLEM SOLVING)

28.3 Nets and Surface Area

Nets are 2D representations of 3D objects. They can be useful for working out the surface area of 3D shapes.

Nets

Learning Objective — Spec Ref G17:
Draw the net of a 3D shape.

A **net** is a **2D drawing** of a 3D shape that can
be folded up to make the 3D shape. To draw
the net, imagine '**unfolding**' the 3D shape so
that **all the faces** of the shape are laid out flat.
Some examples of nets are shown on the right.

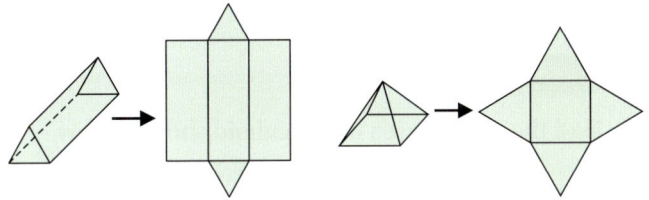

Example 1

Draw a net for the cylinder shown on the right. Label the net with its dimensions.

1. The tube can be 'unfolded' to give a rectangle. Its width
   will be 6 cm but you need to calculate its length, l.

2. The length is the same as the
   circumference of the circular ends of the
   cylinder, so use $C = 2\pi r$ (p.353).

   $l = 2\pi r = 2 \times \pi \times 1$
   $= 6.28$ cm (2 d.p.)

3. Draw the rectangle and add on
   the circular ends to the sides.

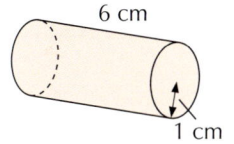

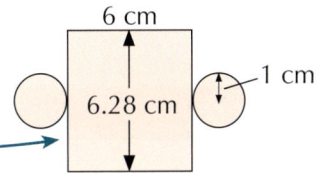

Exercise 1

Q1 Draw a net of each of the following objects. Label each net with its dimensions.

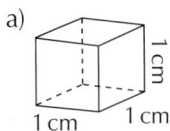

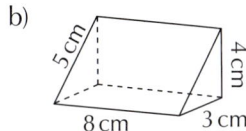

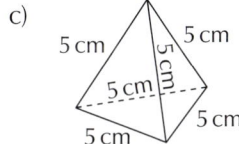

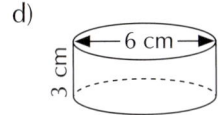

a)

b)

c)

d)

Q2 Draw a net of each of the following objects. Label each net with its dimensions.
a) a cube with 2 cm edges
b) a 1.5 cm × 2 cm × 2.5 cm cuboid
c) a triangular-based pyramid with 3.5 cm edges
d) a cylinder of height 4 cm and radius 2.5 cm
e) a prism of length 3 cm whose cross-section is an equilateral triangle with 2 cm sides
f) a pyramid with square base with 5 cm sides and slanted edge of length 4 cm

Q3 Which of the nets *A*, *B* or *C* is the net of the triangular prism shown below?

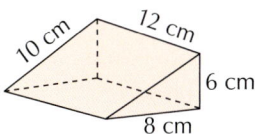

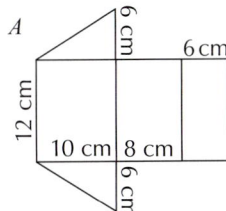

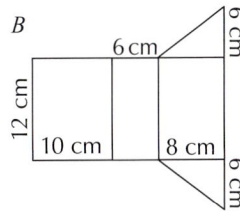

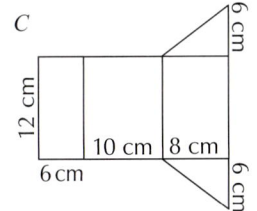

Surface Area

Learning Objective — Spec Ref G17:
Find the surface area of a 3D shape.

Prior Knowledge Check:
Be able to find the area of
2D shapes — see Section 27.

The **surface area** of a 3D shape is the **total area** of all of the **faces** of the shape added together. To find the surface area of a **cuboid** or a **prism**, it's often useful to sketch the **net** of the shape first to make sure that you include **all** of the faces of the shape.

Tip: Spheres and cones have special formulas that give their surface areas. See p.368-369.

Example 2

Find the surface area of the cuboid shown below.

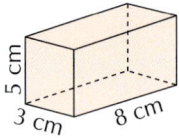

1. Find the area of each face of the shape. Sketch the net first. →

2. You know that a cuboid has six faces, so make sure that each is accounted for.

   2 faces of area $8 \times 5 = 40$ cm²
   2 faces of area $8 \times 3 = 24$ cm²
   2 faces of area $5 \times 3 = 15$ cm²

3. Add together the area of each face to get the surface area.

   Total surface area
   $= (2 \times 40) + (2 \times 24) + (2 \times 15)$
   $= 80 + 48 + 30 = \mathbf{158}$ **cm²**

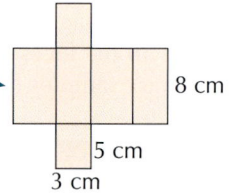

Exercise 2

Q1 Find the surface area of the following prisms.

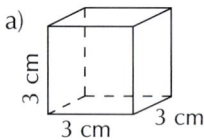

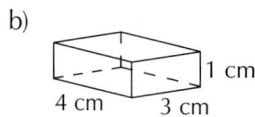

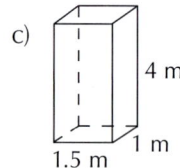

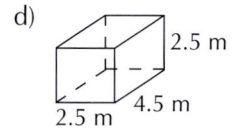

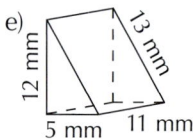

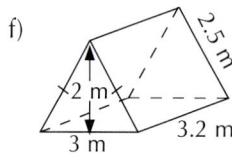

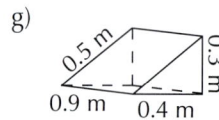

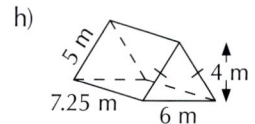

a) 3 cm, 3 cm, 3 cm

b) 1 cm, 4 cm, 3 cm

c) 4 m, 1 m, 1.5 m

d) 2.5 m, 2.5 m, 4.5 m

e) 12 mm, 13 mm, 5 mm, 11 mm

f) 2.5 m, 2 m, 3 m, 3.2 m

g) 0.5 m, 0.3 m, 0.9 m, 0.4 m

h) 5 m, 7.25 m, 4 m, 6 m

Q2 Find the surface area of the following nets and name the shape that the net will make.

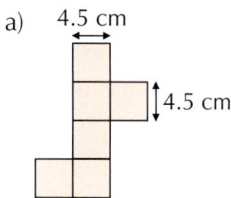

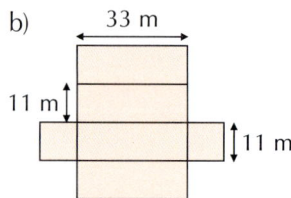

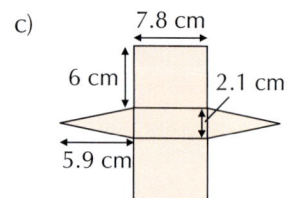

a) 4.5 cm, 4.5 cm

b) 33 m, 11 m, 11 m

c) 7.8 cm, 6 cm, 2.1 cm, 5.9 cm

Q3 Find the surface area of the following prisms.

a) a cube with edges of length 5 m

b) a cube with edges of length 6 mm

c) a 1.5 m × 2 m × 6 m cuboid

d) a 7.5 m × 0.5 m × 8 m cuboid

e) an isosceles triangular prism of height 3 m, slant edges 5 m, base 8 m and length 2.5 m

Q4 The shape shown on the right is made up of a triangular prism (PROBLEM SOLVING) and a cube. The bottom face of the prism is attached to the top face of the cube. Find the surface area of the shape.

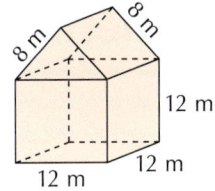

8 m   8 m
12 m
12 m   12 m

Example 3

By considering its net, find the surface area of a cylinder of radius 2 cm and length 8 cm.

1. Sketch the net of the cylinder.

2. Find the length, l, of the rectangle — $l = 2\pi r = 2\pi \times 2$
    it's equal to the circumference of the circle. $= 12.566...$ cm

3. Find the area   Area of each circle $= \pi r^2 = \pi \times 2^2 = 12.566...$ cm^2
    of each face.   Area of rectangle $= 8 \times 12.566... = 100.530...$ cm^2

4. Add the individual areas together.   Total surface area $= (2 \times 12.566...) + 100.530...$
    Remember, there are two circular faces. $= \mathbf{125.7}$ **cm^2** (1 d.p.)

2 cm

l 8 cm

Exercise 3

Q1 Find the surface area of the following cylinders. Give your answers correct to 2 d.p.

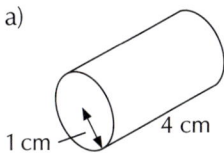

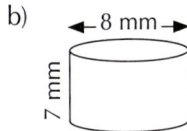

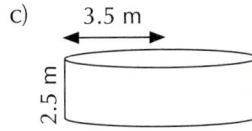

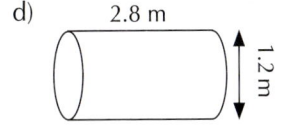

a)   b) ← 8 mm →   c) 3.5 m   d) 2.8 m

1 cm   4 cm   7 mm   2.5 m   1.2 m

Q2 Find the surface area of the cylinders with the following dimensions.
Give your answers correct to 2 d.p.
a) radius = 2 m, length = 7 m   b) radius = 7.5 mm, length = 2.5 mm
c) radius = 12.2 cm, length = 9.9 cm   d) diameter = 22.1 m, length = 11.1 m

Q3 A cylindrical metal pipe has radius 2.2 m and length 7.1 m. The ends of the pipe are open.
a) Find the curved surface area of the outside of the pipe to 2 d.p.
b) A system of pipes consists of 9 of the pipes described above. What area of metal is required to build the system of pipes? Give your answer correct to 2 d.p.

Q4 Ian is painting all the outside surfaces of his cylindrical gas tank. The tank has radius 0.8 m and length 3 m. 1 litre of Ian's paint will cover an area of 14 m^2.
To 2 decimal places, how many litres of paint will Ian need?

Q5 Find the exact surface area of the shapes below.
a)   6 m
9 m
6 m
12 m   5 m

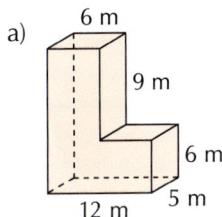

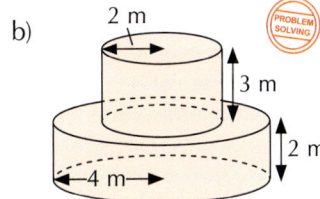

b) 2 m   (PROBLEM SOLVING)
3 m
2 m
← 4 m →

28.4 Spheres, Cones and Pyramids

Spheres, cones and pyramids are 3D shapes which are more complicated than cuboids and prisms.
They each have their own formulas for their surface area and their volume.

Spheres

Learning Objective — Spec Ref G17:
Find the surface area and volume of a sphere.

A **sphere** has one curved face, no vertices and no edges. To find the surface area and volume of a sphere, you need to know its **radius** — the distance from the **centre** of the sphere to **any point on its surface**. For a sphere with radius r, the formulas for **surface area** and **volume** are as follows:

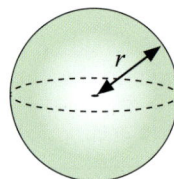

$$\text{Surface area} = 4\pi r^2 \qquad \text{Volume} = \frac{4}{3}\pi r^3$$

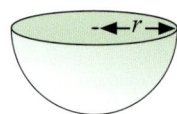

A **hemisphere** is **half** of a sphere. Its volume is just **half the volume** of a sphere with the same radius: $\quad V = \frac{2}{3}\pi r^3$

The surface area is **half of the surface area** of the sphere, **plus** the area of the **flat circular face**: $\quad$ **Surface area** $= 2\pi r^2 + \pi r^2 = 3\pi r^2$

Example 1

Find the exact surface area and volume of a sphere with radius 6 cm.

1. Substitute r into the formula for the surface area and work it through.

   Surface area $= 4\pi r^2 = 4 \times \pi \times 6^2$
   $= \mathbf{144\pi}$ **cm²**

2. Do the same with the formula for the volume.

   Volume $= \frac{4}{3}\pi r^3 = \frac{4}{3} \times \pi \times 6^3$
   $= \mathbf{288\pi}$ **cm³**

Tip: 'Exact' means you should leave your answer in terms of π.

Exercise 1

Q1 For each of the following spheres with the given radius, r, find:
    (i) the exact surface area   (ii) the exact volume
    a) $r = 5$ cm   b) $r = 4$ cm   c) $r = 2.5$ m   d) $r = 10$ mm

Q2 Find the radius of a sphere with surface area 265.9 cm². Give your answer correct to 1 d.p.

Q3 Find the radius of a sphere with volume 24 429 cm³. Give your answer correct to 1 d.p.

Q4 The surface area of a sphere is 2463 mm². Find the volume of the sphere, correct to 1 d.p.

Q5 The volume of a sphere is 6044 m³. Find the surface area of the sphere, correct to 1 d.p.

Q6 The 3D object shown on the right is a hemisphere.
    a) Find the volume of the hemisphere, to 1 d.p.
    b) (i) Find the area of the curved surface of the hemisphere, to 1 d.p.
        (ii) Hence find the total surface area of the hemisphere, to 1 d.p.

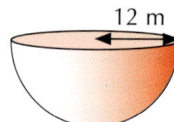

12 m

Cones

Learning Objective — Spec Ref G17:
Find the surface area and volume of cones and frustums.

A **cone** is a 3D shape that has a **circular base** and goes up to a **point** at the top. To find the **volume** and **surface area** of a cone, you need to know its base **radius** r, **perpendicular height** h and **slant height** l. The **volume** of a cone is given by:

$$\text{Volume} = \frac{1}{3}\pi r^2 h$$

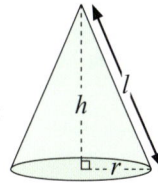

A cone has **two faces** — the curved, sloping face and the circular base. The formula for the **surface area** comes from **adding** up the areas of these faces:

$$\text{Surface area} = \pi r l + \pi r^2$$

Tip: The perpendicular height goes from the centre of the base to the point. The slant height goes from the edge of the base to the point. Don't mix them up.

Example 2

Find the exact surface area and volume of the cone shown on the right.

1. Substitute $r = 3$ and $l = 5$ into the formula for the surface area.

   Surface area $= \pi r l + \pi r^2 = (\pi \times 3 \times 5) + (\pi \times 3^2)$
   $= 15\pi + 9\pi = \mathbf{24\pi \ cm^2}$

2. Substitute $r = 3$ and $h = 4$ into the formula for the volume.

   Volume $= \frac{1}{3}\pi r^2 h = \frac{1}{3} \times \pi \times 3^2 \times 4$
   $= \mathbf{12\pi \ cm^3}$

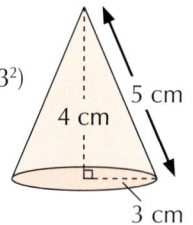

Exercise 2

Q1 Find (i) the exact surface area, and (ii) the exact volume of the cones with the given properties.
   a) $r = 5$ m, $h = 12$ m, $l = 13$ m
   b) $r = 7$ cm, $h = 24$ cm, $l = 25$ cm
   c) $r = 15$ m, $h = 8$ m, $l = 17$ m
   d) $r = 30$ mm, $h = 5.5$ mm, $l = 30.5$ mm

Q2 a) Find the exact surface area of the cone on the right.
   b) (i) Use Pythagoras' theorem to find h, the perpendicular height of the cone.
      (ii) Find the exact volume of the cone.

Q3 A cone has base radius 56 cm and perpendicular height 33 cm.
   a) Find the exact volume of the cone.
   b) (i) Use Pythagoras' theorem to find the slant height of the cone.
      (ii) Find the exact surface area of the cone.

Q4 A cone has perpendicular height 6 mm and volume 39.27 mm³.
   a) Find the base radius of the cone, to 1 d.p.
   b) (i) Use your answer to part a) to find the slant height, l, of the cone.
      (ii) Find the surface area of the cone, to 1 d.p.

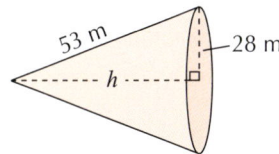

Q5 A cone has vertical height 20 cm and volume 9236.28 cm³. (PROBLEM SOLVING) Find the surface area of the cone, to 1 d.p.

Frustums

A **frustum** of a cone is the 3D shape left once you **chop** the top bit off a cone **parallel** to its circular base. The smaller, removed cone is always **similar** to the larger, original cone — see Section 30 for more on similarity. To find the **volume** of a frustum, find the volume of the **original cone** and take away the volume of the **removed cone** (the top bit).

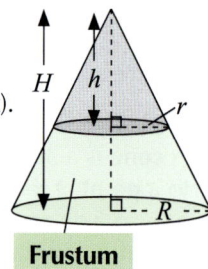

$$\begin{array}{c}\text{Volume of} \\ \text{frustum}\end{array} = \begin{array}{c}\text{Volume of} \\ \text{original cone}\end{array} - \begin{array}{c}\text{Volume of} \\ \text{removed cone}\end{array} = \tfrac{1}{3}\pi R^2 H - \tfrac{1}{3}\pi r^2 h$$

Frustum

Example 3

Find the exact volume of the frustum shown on the right.

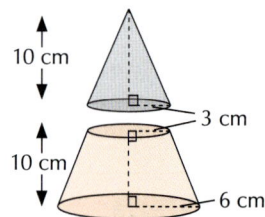

1. Find the volume of the original cone.

   Volume of original cone $= \tfrac{1}{3}\pi R^2 H$

   $= \tfrac{1}{3}\pi \times 6^2 \times (10 + 10)$

   $= \tfrac{1}{3}\pi \times 36 \times 20 = 240\pi$

2. Find the volume of the removed cone.

   Volume of removed cone $= \tfrac{1}{3}\pi r^2 h = \tfrac{1}{3}\pi \times 3^2 \times 10 = \tfrac{1}{3}\pi \times 9 \times 10 = 30\pi$

3. Subtract the volume of the removed cone from the volume of the original cone.

   Volume of frustum = vol. of original cone – vol. of removed cone
   $= 240\pi - 30\pi$
   $= \mathbf{210\pi\ cm^3}$

Exercise 3

Q1 A cone of perpendicular height 4 cm is removed from a cone of perpendicular height 16 cm and base radius 12 cm to leave the frustum shown.

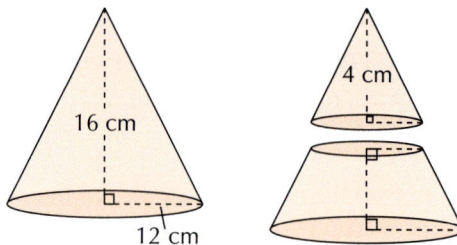

   a) By considering the ratio of the heights of the cones, find the base radius of the smaller cone.

   b) Find the exact volume of the larger cone.

   c) Find the exact volume of the smaller cone.

   d) Hence find the exact volume of the frustum.

Q2 Find the exact volume of the frustum shown. ➡

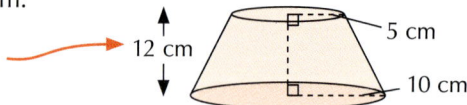

Pyramids

Learning Objective — Spec Ref G17:
Find the surface area and volume of a pyramid.

A **pyramid** is a 3D shape that has a **polygon base** (see page 257) which rises to a **point**. A cone is a bit like a pyramid with a circular base.

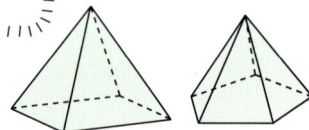

The **volume** of a pyramid can be found by using this formula: **Volume = $\tfrac{1}{3}$ × base area × height**

The **surface area** of a pyramid can be found by adding up the surface area of each of its **faces** (it might be useful to sketch the **net**). There'll be the **base** (which has, say, *n* edges), plus the *n* **triangular faces** on the sides. You might have to use **Pythagoras' theorem** (see p.318) to find the **height** of the triangular faces.

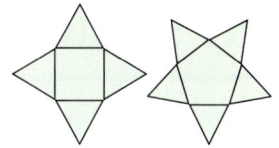

Example 4

Find the volume of the rectangular-based pyramid shown.

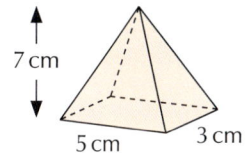

1. Write down the formula for volume of a pyramid.

   $$\text{Volume} = \frac{1}{3} \times \text{base area} \times \text{height}$$

2. Substitute the values into the formula — since the base is a rectangle, base area = length × width.

   $$= \frac{1}{3} \times (5 \times 3) \times 7$$

   $$= \textbf{35 cm}^3$$

Example 5

Find the surface area of the regular tetrahedron shown here. Give your answer correct to 3 significant figures.

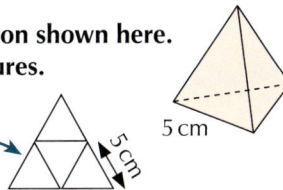

Tip: A tetrahedron is a triangular-based pyramid. 'Regular' means all the triangles are equilateral.

1. Each face is an equilateral triangle with 5 cm sides. Sketch a net to help.

2. Use the formula area = $\frac{1}{2}ab \sin C$ to find the area of each face. All angles are 60° for equilateral triangles.

   Area of one face $= \frac{1}{2} \times 5 \times 5 \times \sin 60°$

   $= 10.825...$

3. There are four identical faces, so multiply this area by four.

   Total surface area $= 4 \times 10.825...$

   $= \textbf{43.3 cm}^2$ (3 s.f.)

Exercise 4

Q1 A hexagon-based pyramid is 15 cm tall. The area of its base is 18 cm². Calculate its volume.

Q2 Just to be controversial, Pharaoh Tim has decided he wants a pentagon-based pyramid. The area of the pentagon is 27 m² and the perpendicular height of the pyramid is 12 m. What is the volume of Tim's pyramid?

In Questions 3-5, give all answers to 1 decimal place.

Q3 Find the volume of a pyramid of height 10 cm with rectangular base with dimensions 4 cm × 7 cm.

Q4 Find the surface area of a regular tetrahedron of side 12 mm.

Q5 A square-based pyramid has height 11.5 cm and volume 736 cm³. Find the side length of its base.

Q6 The diagram on the right shows the square-based pyramid *ABCDE*. The side length of the base is 96 m and the pyramid's height is 55 m. *O* is the centre of the square base, directly below *E*. *M* is the midpoint of *AD*.

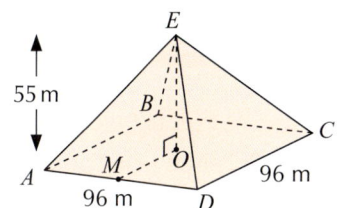

   a) Find the volume of the pyramid.

   b) Use Pythagoras' theorem to find the length *EM*.

   c) Hence find: (i) the area of triangle *ADE*   (ii) the surface area of the pyramid.

28.5 Rates of Flow

Volume can be used to work out the rate of flow of a substance into a 3D container.

Learning Objective — Spec Ref G17:
Calculate rates of flow into and out of 3D shapes.

Prior Knowledge Check:
Be able to convert between different units (Section 22)
and between different compound measures (p.289).

The **rate of flow** tells you **how quickly** a liquid is moving **into**, **out of** or **through** a certain **space**. To work out the rate of flow, you need to know the **total volume** of the space and the **time** it would take to **completely fill** (or empty) the space. The volume **divided** by the time gives the rate of flow.

$$\text{Rate of Flow} = \frac{\text{Volume of container}}{\text{Total time taken}}$$

Rates of flow are measured in **volume per unit time**, e.g. litres per second or m³ per hour (sometimes written as litres/s or m³/hr).

Example 1

Water flows into this cuboid-shaped tank at a rate of 150 cm³ per second. How long does it take to fill the tank?

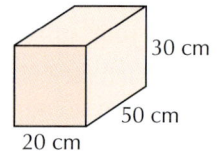

1. Calculate the volume of the cuboid.

   Volume of cuboid = height × length × width
   = 20 × 30 × 50 = 30 000 cm³

2. Rearrange the formula and divide the volume by the rate of flow to get the time taken.

   Time to fill tank = $\frac{30\,000}{150}$ = **200 seconds**

30 cm, 50 cm, 20 cm

Example 2

A spherical water fountain has a radius of 40 cm. Calculate the rate of flow of the water, in litres per second, if it takes 15 minutes for a completely full fountain to empty.

1. Calculate the volume of the fountain in cm³.

   Volume of sphere = $\frac{4}{3}\pi r^3 = \frac{4}{3}\pi \times 40^3$ = 268 082.5... cm³

2. Convert the volume into litres (1 litre = 1000 cm³) and the time into seconds (1 minute = 60 seconds).

   268 082.5... cm³ = 268 082.5... ÷ 1000 = 268.0825... litres
   15 minutes = 15 × 60 = 900 seconds

3. Divide the volume by the time taken to find the rate of flow.

   Rate of flow = $\frac{\text{volume of fountain}}{\text{total time taken}} = \frac{268.0825...}{900}$

   = **0.298 litres per minute** (3 s.f.)

Exercise 1

Q1 Work out the average rates of flow when tanks with these volumes are filled in the given time.
   a) Volume = 600 cm³, Time = 15 seconds
   b) Volume = 150 litres, Time = 8 hours

Q2 A stream flows at a rate of 3 litres per second. Given that 1 ml is 1 cm³, convert this rate of flow into:
   a) cm³ per minute
   b) m³ per minute
   c) m³ per day
   d) litres per day

Q3 Grain is being poured into an empty cylindrical silo with diameter 9 m and height 20 m. The grain is flowing at a rate of 12 m³/minute. How long will it take to half fill the silo, to the nearest minute?

Q4 A spherical mould with radius 60 cm is used to make concrete balls.
   Concrete flows into the mould at a rate of 2π litres/s. How long does it take to completely fill the mould?

28.6 Symmetry of 3D Shapes

While 2D objects have lines of symmetry, 3D objects have planes of symmetry.

Learning Objective — Spec Ref G12:
Work out the number of planes of symmetry in a 3D shape.

Prior Knowledge Check:
Be able to find lines of symmetry in 2D shapes — see p.261.

A **plane of symmetry** cuts a solid into **two identical halves**.

- The number of planes of symmetry of a **prism** is always **one greater** than the number of **lines of symmetry** of its cross-section. This is because a prism also has a plane of symmetry across the prism **parallel** to the cross-section. The only exception to this is a **cube**, which has **nine** planes of symmetry.
- The number of planes of symmetry of a **pyramid** is equal to the number of lines of symmetry of its **base**. The only exception is a **regular tetrahedron**, which has **six** planes of symmetry.

Example 1

How many planes of symmetry does an isosceles triangular prism have?

1. Work out how many lines of symmetry the cross-section has.

   An isosceles triangle has 1 line of symmetry.

2. Add one more to this figure for the plane of symmetry that is parallel to the cross-section.

   1 + 1 = 2, so an isosceles triangular prism has **2 planes of symmetry**.

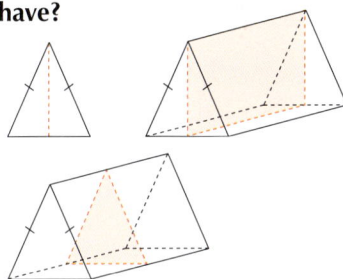

Example 2

How many planes of symmetry does a rectangular-based pyramid have?

Work out how many lines of symmetry the base has.

A rectangle has 2 lines of symmetry, so a rectangle-based pyramid has **2 planes of symmetry**.

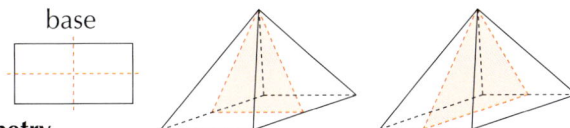

base

Exercise 1

Q1 Write down the number of planes of symmetry of the prisms with the following cross-sections.
   a) regular pentagon
   b) regular octagon
   c) scalene triangle

Q2 Draw the planes of symmetry of the prisms with the following cross-sections.
   a) equilateral triangle
   b) rectangle
   c) regular hexagon

Q3 Draw all the planes of symmetry of a cube.

Q4 Draw a prism with 5 planes of symmetry.

Q5 Write down the number of planes of symmetry of the pyramids with the following bases.
   a) regular pentagon
   b) regular octagon
   c) square

Q6 Draw all the planes of symmetry of a regular tetrahedron.

Review Exercise

Q1 Draw the following prisms on isometric paper.

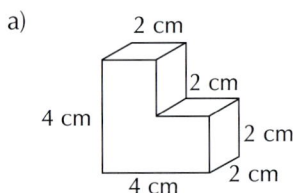

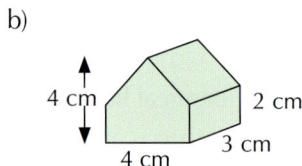

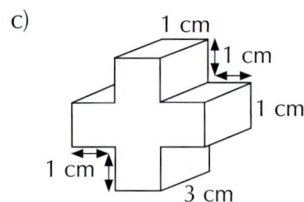

a)

b)

c)

Q2 A cube has a total surface area of 54 cm².

a) Find the area of one face of the cube.

b) Find the length of the edges of the cube.

c) Find the cube's volume.

Q3 A tank is in the shape of a cuboid of length 140 cm and width 85 cm. It is filled with water to a depth of 65 cm. There is room for another 297.5 litres of water in the tank. What is the height of the tank?

Q4 A cylindrical disc has a radius of 4 cm and height of 0.5 cm.

a) Find the surface area of the disc, to 2 d.p.

b) Find the surface area of a stack of 10 discs, to 2 d.p., assuming there are no gaps between each disc.

c) Find the volume of the stack of 10 discs, to 2 d.p.

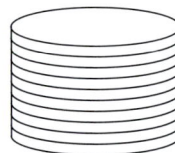

Q5 Beans are sold in cylindrical tins of diameter 7.4 cm and height 11 cm.

a) Find the volume of one tin, to 2 d.p.

The tins are stored in boxes which hold 12 tins in three rows of 4, as shown.

b) Find the dimensions of the box.

c) Find the volume of the box.

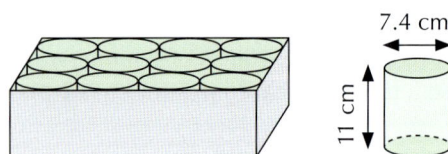

d) Calculate the volume of the box that is not taken up by tins when it is fully packed, to 2 d.p.

Q6 Toilet paper is sold in cylindrical rolls of diameter 12 cm and height 11 cm. The card tube at the centre of the roll is 5 cm in diameter. Find the following, giving your answers to 2 decimal places.

a) The total volume of one roll of toilet paper, including the card tube.

b) The volume of the card tube.   c) The volume of the paper.

Each rectangular sheet of paper measures 11 cm × 13 cm and is 0.03 cm thick.

d) Find the volume of one sheet of paper.

e) Hence find the number of sheets of paper in one roll, to the nearest sheet.

Q7

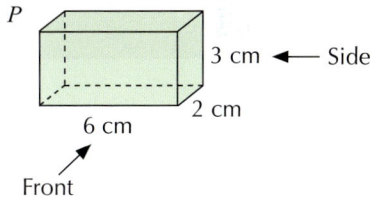

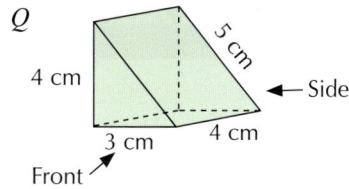

For each of the shapes P and Q above:

a) Draw a net of the shape.

b) Draw the plan, and front and side elevations of the shape from the directions indicated.

c) Draw the shape on isometric paper.

d) Calculate the shape's surface area.

e) Calculate the shape's volume.

f) Calculate how long it would take to fill the shape with water flowing at a rate of 2 cm³ per second.

g) Write down the number of planes of symmetry of the shape.

Q8 The value of the surface area of a sphere (in m²) is equal to the value of the volume of the sphere (in m³). What is the sphere's radius?

Q9 A cone of diameter 25 mm has volume 7854 mm³.

a) Find the perpendicular height of the cone to 1 d.p.

b) (i) Find the slant height of the cone to 1 d.p.

   (ii) Hence, find the surface area of the cone to 1 d.p.

The cone is enclosed in a cuboid, as shown.

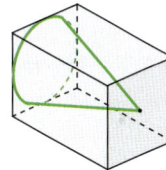

c) Find the volume of the cuboid to the nearest mm³.

d) Find the volume of the empty space between the cone and the cuboid to the nearest mm³.

Q10 The diagram on the right shows the square-based pyramid $ABCDE$. The pyramid's vertical height is 12 m and its volume is 400 m³. O is the centre of the square base, directly below E. M is the midpoint of AD.

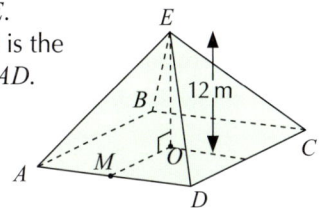

a) Find the side length of the square base.

b) Find the length EM.

c) Find the surface area of the pyramid.

d) How many planes of symmetry does the pyramid have?

e) Sand enters the pyramid at a rate of 100 litres per minute. Calculate how long it would take to completely fill the pyramid, give your answer in hours and minutes.

Q11 A giant novelty pencil is made up of a cone, a cylinder and a hemisphere, as shown.

a) Find the exact volume of the pencil.

b) (i) Find the exact slant height of the cone.

   (ii) Hence find the surface area of the pencil to one decimal place.

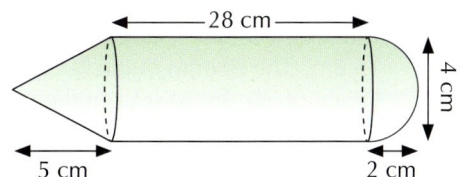

Q1 The diagram shows the front and side elevations of a solid prism drawn accurately on a centimetre square grid.

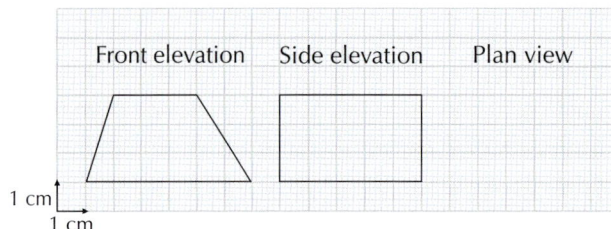

Front elevation   Side elevation   Plan view

1 cm
1 cm

a) On graph paper, accurately draw the plan view of the prism.

[2 marks]

b) Work out the volume of the prism.

[3 marks]

Q2 An industrial oil storage tank is a cylinder with a radius of 5 m and a height of 10 m as shown in the diagram.

Water is being poured into it at a constant rate of 7 cubic metres per minute. The tank has a leak which causes it to lose water at a constant rate of 2 cubic metres per minute. Work out how long it will take for the water to reach halfway up the tank, giving your answer in minutes as a multiple of π.

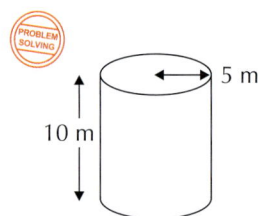

5 m
10 m

[4 marks]

Q3 Spherical tennis balls are sold in cylindrical tubes with lids. There are 4 balls in each tube. The balls fit snugly inside the tubes. Show that four balls take up $\frac{2}{3}$ of the space inside a tube.

[5 marks]

Q4

15 cm
30 cm
Not drawn to scale
12 cm

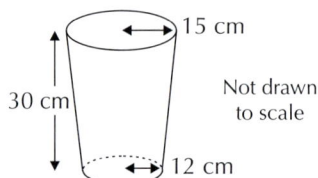

A bucket is in the shape of an upturned frustum. The top of the bucket is a circle of radius 15 cm. The bottom of the bucket is a circle of radius 12 cm. The bucket has a perpendicular height of 30 cm. Work out the capacity of the bucket, giving your answer in litres to 3 s.f.

[5 marks]

29.1 Reflections

Transformations can be used to move and resize shapes on a coordinate grid. In this section, you'll meet four different transformations — the first of which is reflection. To reflect a shape, you draw its mirror image.

Learning Objectives — Spec Ref G7:
- Reflect a shape in a given line.
- Describe the reflection that transforms a shape.

To **reflect** a shape, first reflect the **vertices** of the shape in the line of symmetry (also known as the **mirror line**). The **reflected points** (called the **image points**) should be the **same distance** from the line as the original points but on the **other side** of it. Then **join up** the image points to create the reflected shape. The reflected shape will be the **same size** and **shape** as the original — so the shapes are **congruent** (see p.394).

A reflection in the **y-axis** will send a point (x, y) to $(-x, y)$ and a reflection in the **x-axis** will send a point (x, y) to $(x, -y)$.

An **invariant point** is one that **doesn't move** under a transformation. For reflections, all the points on the **mirror line** are invariant.

To **describe** a reflection, you just need to find the **equation** of the mirror line.

Tip: Mirror lines of the form $x = a$ are vertical and mirror lines of the form $y = a$ are horizontal. See page 177 for more on lines parallel to the axes.

Example 1

Reflect the shape $ABCDE$ in the y-axis. Label the image points A_1, B_1, C_1, D_1 and E_1 with their coordinates.

1. Reflect the shape — one vertex at a time.

2. Each image point should be the same distance from the y-axis as the original point. E.g. C is 2 units to the left of the y-axis so its image C_1 should be 2 units to the right of the y-axis.

3. Write down the coordinates of each of the image points. Each point (x, y) becomes $(-x, y)$.

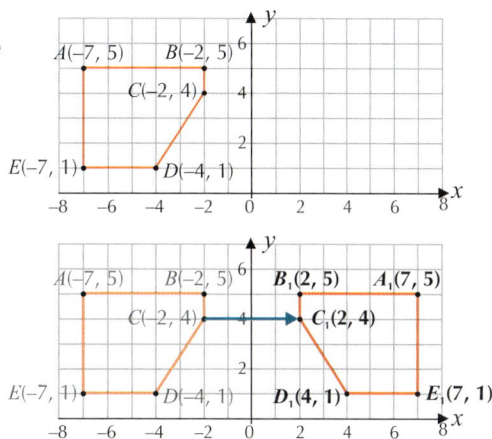

Exercise 1

Q1 Copy the diagram on the right and reflect the shape in the y-axis. Label the image points A_1, B_1, C_1, D_1, E_1 with their coordinates.

Q2 The following points are reflected in the y-axis. Find the coordinates of the image points.
a) $(4, 5)$ b) $(7, 2)$ c) $(-1, 3)$ d) $(-3, -1)$

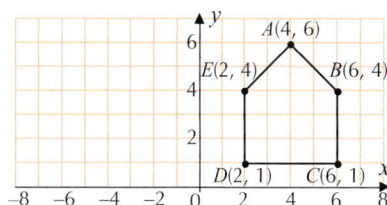

Q3 Copy each of the diagrams below and reflect the shapes in the y-axis.

a)

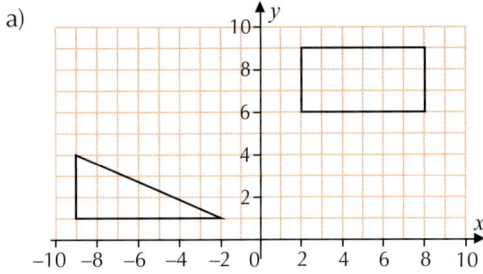

b)

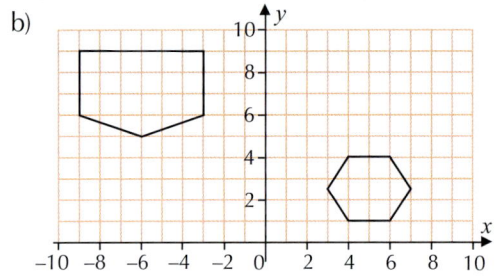

Q4 The following points are reflected in the x-axis. Find the coordinates of the image points.

a) (1, 2)   b) (3, 0)   c) (−2, 4)   d) (−1, −3)   e) (−2, −2)

Q5 Copy each of the diagrams below and reflect the shapes in the x-axis.

a)

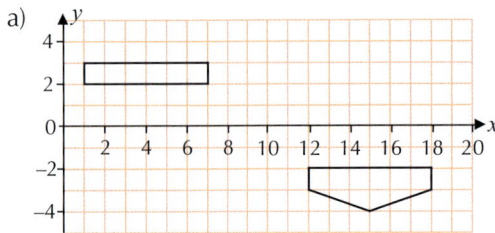

b)

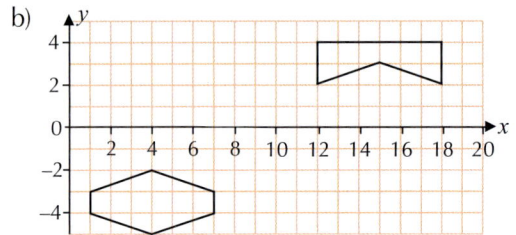

Q6 Copy the diagram shown on the right.

a) Reflect shape A in the line $x = 2$. Label the image A_1.

b) Reflect shape B in the line $x = -1$. Label the image B_1.

c) Reflect shape C in the line $x = 1$. Label the image C_1.

d) Shape B_1 is the reflection of shape A in which mirror line?

e) Give the coordinates of the two invariant points on Shape B when it is reflected in the line $x = -7$.

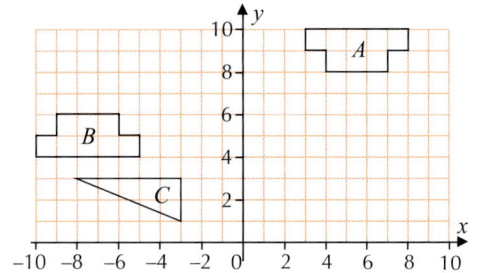

Q7 Copy the diagram shown on the right.

a) Reflect shape A in the line $y = 6$. Label the image A_1.

b) Reflect shape B in the line $y = 4$. Label the image B_1.

c) Reflect shape C in the line $y = 5$. Label the image C_1.

d) Reflect shape D in the line $y = 3$. Label the image D_1.

e) Shape C_1 is the reflection of shape B in which mirror line?

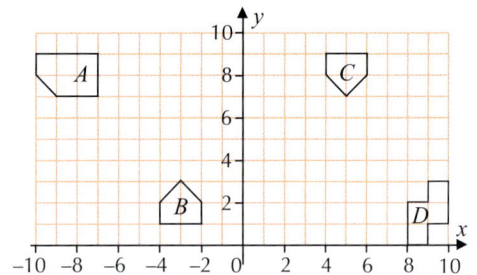

Q8 a) Describe the reflection that would take triangle X to the triangle Y.

b) The image of triangle Z under a single reflection is either triangle X or triangle Y. Determine which and describe the reflection.

c) Which points of triangle X are invariant if it is reflected in the line $y = 5$?

PROBLEM SOLVING

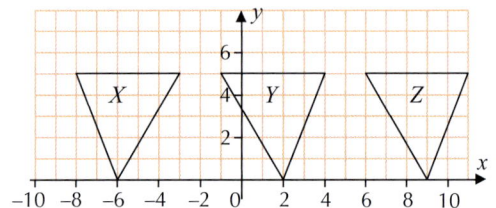

The mirror line isn't always one of the axes or a line parallel to an axis — it could be a **diagonal** line such as $y = x$ or $y = -x$. A reflection in $y = x$ sends (x, y) to **(y, x)** and a reflection in $y = -x$ sends (x, y) to **$(-y, -x)$**. Follow the same method of **reflecting each vertex** and then joining up the **image points**. Make sure the image points are the **same distance** from the mirror line as the original ones — you measure this distance **perpendicular** to the mirror line.

Example 2

Reflect the shape in the line $y = x$ on the grid below.

1. Reflect the shape one vertex at a time.

2. Join the image points. They should be the same perpendicular distance from the line $y = x$ as the original points.

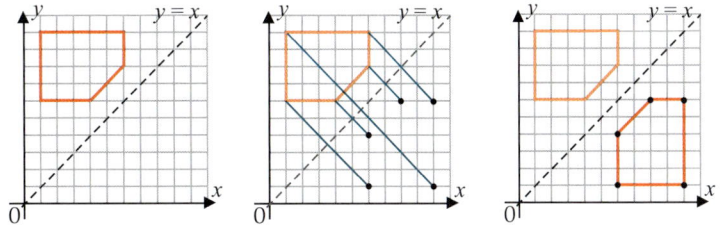

Exercise 2

Q1 The points below are reflected in the line $y = x$. Find the coordinates of their reflections.

a) (1, 2)   b) (3, 0)   c) (–2, 4)   d) (–1, –3)   e) (–2, –2)

Q2 Copy each of the diagrams below and reflect the shapes in the line $y = x$.

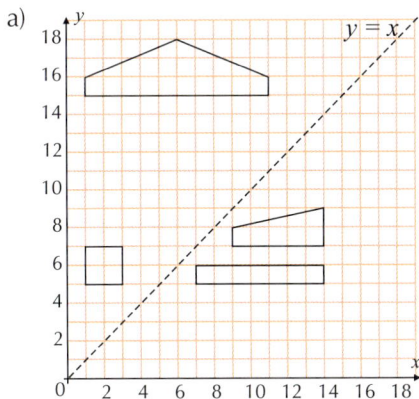

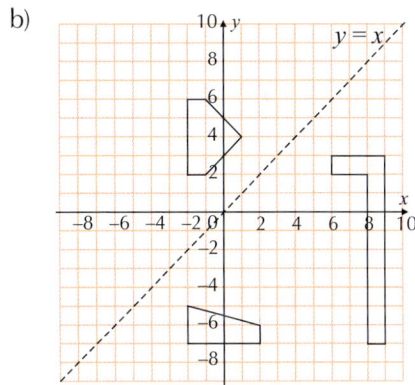

a)

b)

Questions 3 and 4 refer to the diagram on the right.

Q3 Describe the reflection that would map:

a) Shape A onto shape B.

b) Shape C onto shape A.

Q4 a) Copy the diagram and reflect each of the shapes A, B and C in line K. Label them A_1, B_1 and C_1 respectively.

b) Give the points of invariance on shapes A, B or C when they are reflected in line L.

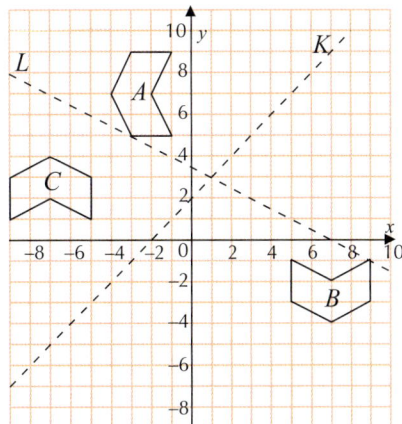

29.2 Rotations

The second transformation to learn about is rotation. Rotations spin everything around a fixed point.

> **Learning Objectives — Spec Ref G7:**
> - Rotate a shape around a given point.
> - Describe the rotation that transforms a shape.

When an object is **rotated** about a point, its size and shape stay the same — so the new shape is **congruent** to the original shape. Also, the **distance** of each vertex from the centre of rotation doesn't change. To describe a rotation, you need to give **three** pieces of information:

    (i) the **centre** of rotation   (ii) the **direction** of rotation   (iii) the **angle** of rotation

The **centre** of rotation can be **any point** — e.g. the origin (0, 0) or (5, 1). The **direction** of rotation will be either **clockwise** or **anticlockwise** and the **angle** might be given in **degrees** or as a **fraction of a turn** (e.g. 90° or a quarter-turn). A rotation of **180°** is the same in both directions so you don't need a direction.

If a point on the shape lies on the **centre of rotation**, it will be an **invariant point** under any rotation.

Example 1

Rotate the shape below 90° clockwise about point P.

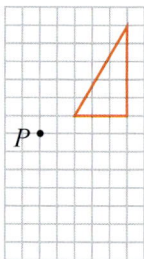

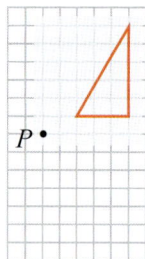

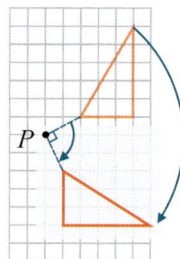

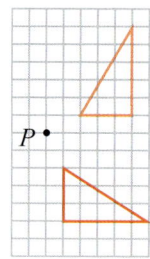

1. Draw the shape on a piece of tracing paper. (Or imagine a drawing of it.)

2. Rotate the tracing paper 90° clockwise about *P*. ('About *P*' means *P* doesn't move.)

3. Draw the image in its new position.

Exercise 1

Q1 a) Copy the diagrams below, then rotate the shapes 180° about *P*.

    (i)   (ii)

    b) Copy the diagrams below, then rotate the shapes 90° clockwise about *P*.

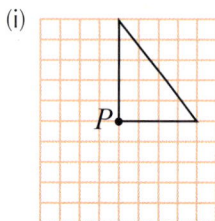

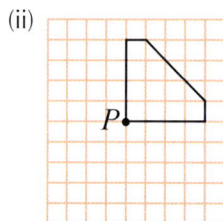

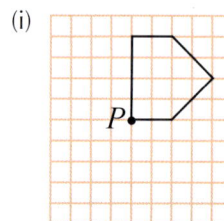

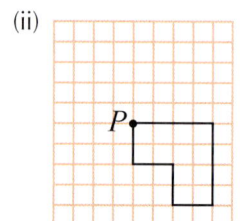

    (i)   (ii)

Q2 Find the new position of the points with the coordinates below if they are rotated anticlockwise about the origin by: (i) 90°   (ii) 180°   (iii) 270°

    a) (1, 0)   b) (0, 2)   c) (3, 1)

Q3 Copy the diagram on the right, then:

a) Rotate A 90° clockwise about the origin.

b) Rotate B 270° anticlockwise about the origin.

Q4 The triangle ABC has vertices $A(-2, 1)$, $B(-2, 6)$ and $C(4, 1)$.

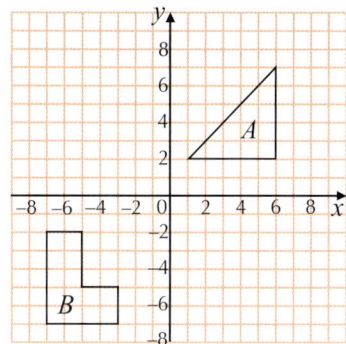

a) Draw the triangle on a pair of axes with x-values from -4 to 8 and y-values from -6 to 6.

b) Rotate the triangle a quarter-turn clockwise about $(5, 4)$. Label the image $A_1B_1C_1$.

c) Write down the coordinates of A_1, B_1 and C_1.

Q5 a) Copy the diagram below, then:

(i) Rotate A 90° clockwise about $(-8, 5)$.

(ii) Rotate B 90° clockwise about $(1, 4)$.

(iii) Rotate C 90° clockwise about $(8, -4)$.

(iv) Rotate D 180° about $(-2, -5)$.

b) Copy the diagram below, then:

(i) Rotate A 180° about $(-5, 6)$.

(ii) Rotate B 90° anticlockwise about $(5, 9)$.

(iii) Rotate C 180° about $(-3, -5)$.

(iv) Rotate D 180° about $(3, -1)$.

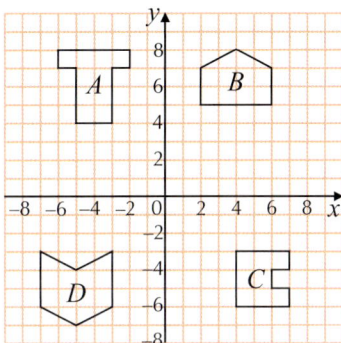

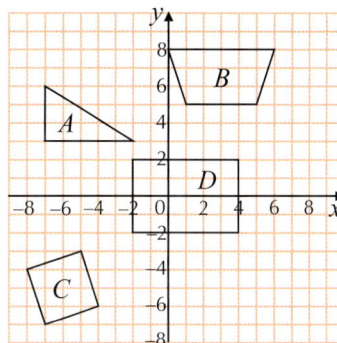

Q6 The triangle DEF has coordinates $D(-2, -2)$, $E(-2, 5)$ and $F(3, 5)$.
DEF is rotated a half-turn about $(2, 0)$ to create the image $D_1E_1F_1$.
Find the coordinates of D_1, E_1 and F_1

Example 2

Describe fully the rotation that transforms shape A to shape B.

1. The shape looks like it has been rotated clockwise by 90°.

2. Trace shape A using tracing paper. Put your pencil on different centres of rotation and turn the tracing paper 90° clockwise until you find a centre that takes shape A onto shape B.

3. To fully describe the rotation, you need to write down the centre, direction and angle of rotation.

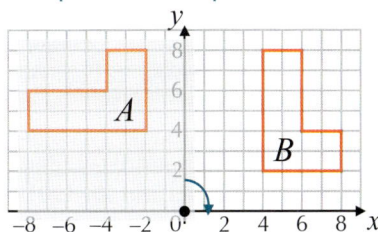

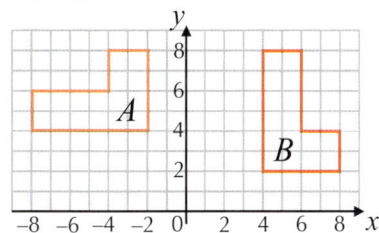

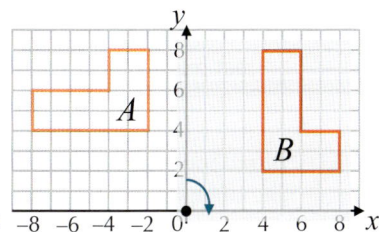

So A is transformed to B by a rotation of **90° clockwise** (or **270° anticlockwise**) about **the origin (0, 0)**.

Q1 **a)**

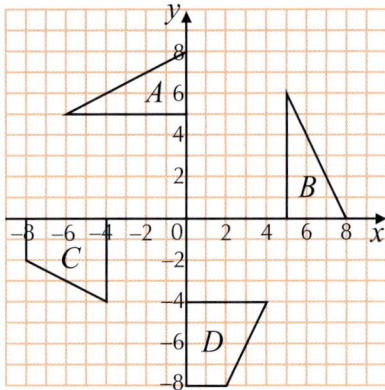

(i) Describe fully the transformation that maps shape *A* to shape *B*.

(ii) Describe fully the transformation that maps shape *C* to shape *D*.

b)

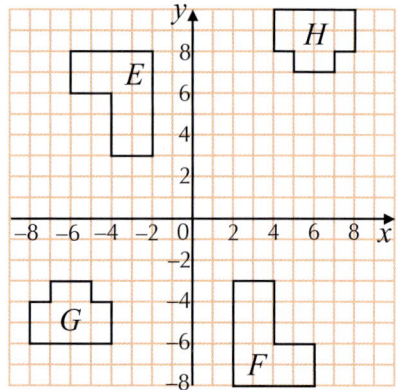

(i) Describe fully the transformation that maps shape *E* to shape *F*.

(ii) Describe fully the transformation that maps shape *G* to shape *H*.

c)

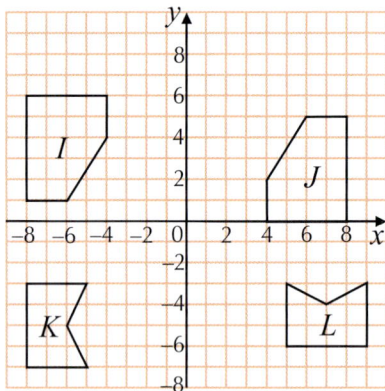

(i) Describe fully the transformation that maps shape *I* to shape *J*.

(ii) Describe fully the transformation that maps shape *K* to shape *L*.

d)

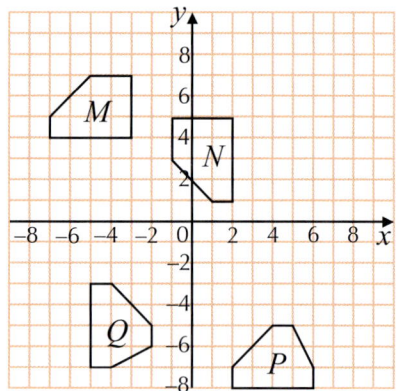

(i) Describe fully the transformation that maps shape *M* to shape *N*.

(ii) Describe fully the transformation that maps shape *P* to shape *Q*.

Q2 **a)** Describe fully the transformation that maps shape *R* to shape *S*.

b) Describe fully the transformation that maps shape *R* to shape *T*.

c) Describe fully the transformation that maps shape *S* to shape *T*.

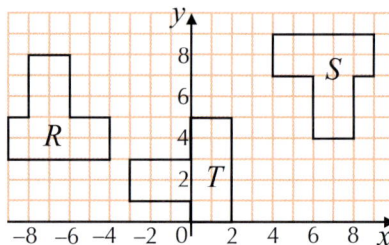

Q3 The triangle *UVW* has vertices *U*(1, 1), *V*(3, 5) and *W*(–1, 3).
The triangle *XYZ* has vertices *X*(–2, 4), *Y*(–6, 6) and *Z*(–4, 2).

a) Draw the two triangles on a pair of axes.

b) Describe fully the transformation that maps *UVW* to *XYZ*.

29.3 Translations

Translations are simple — they just slide a shape up/down and left/right.

Learning Objectives — Spec Ref G7/G24:
- Translate a shape on a coordinate grid.
- Find the vector that describes a translation.

Prior Knowledge Check:
Be familiar with column vectors (see p.339).

To **translate** an object, you need to know the **distance** that it moves and in which **direction** — these are broken down into **horizontal** and **vertical** movements.

- For horizontal movements, **positive** numbers move the shape **right** and **negative** numbers move it **left**.

- For vertical movements, **positive** numbers move the shape **up** and **negative** numbers move it **down**.

You can use **column vectors** to represent this information. For example:

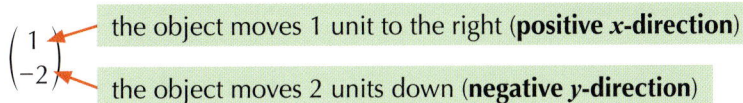

$\begin{pmatrix} 1 \\ -2 \end{pmatrix}$ — the object moves 1 unit to the right (**positive *x*-direction**)

— the object moves 2 units down (**negative *y*-direction**)

An object and its image after a translation are **congruent** — so it doesn't change shape or size. Since a translation moves every point on the shape, there are **no invariant points**.

Example 1

Translate the shape on the axes below by the vector $\begin{pmatrix} 5 \\ -3 \end{pmatrix}$.

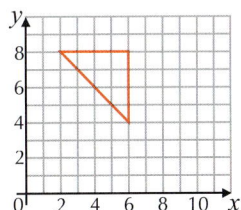

$\begin{pmatrix} 5 \\ -3 \end{pmatrix}$ is a translation of: (i) 5 units to the right,
(ii) 3 units down.

Move each vertex 5 units right and 3 units down and then join them up to create the translated shape.

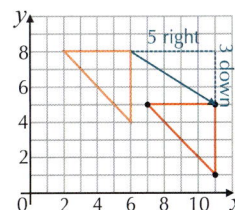

Exercise 1

Q1 Copy the diagrams below, then translate each shape by the vector written next to it.

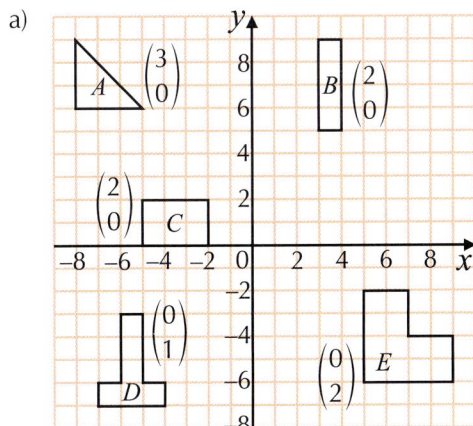

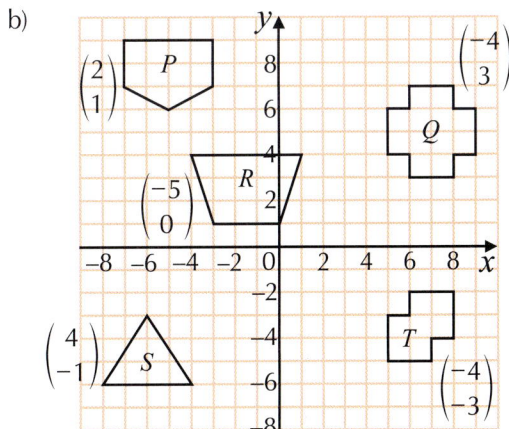

a)

b)

Q2 a) Copy the diagram on the right, then translate the triangle ABC by the vector $\begin{pmatrix} -10 \\ -1 \end{pmatrix}$. Label the image $A_1B_1C_1$.

b) Label A_1, B_1 and C_1 with their coordinates.

c) Describe a rule connecting the coordinates of A, B and C and the coordinates of A_1, B_1 and C_1.

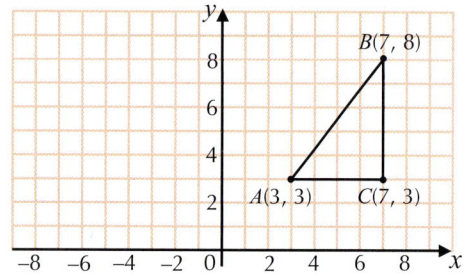

Q3 The triangle DEF has vertices $D(1, 1)$, $E(3, -2)$ and $F(4, 0)$. After the translation $\begin{pmatrix} -3 \\ 2 \end{pmatrix}$, the image of DEF is $D_1E_1F_1$. Find the coordinates of D_1, E_1 and F_1.

PROBLEM SOLVING

Example 2

Describes the transformation that maps shape A onto shape B.

1. The shape hasn't been rotated or reflected and hasn't changed size, so this must be a translation.

2. Choose a pair of corresponding vertices on the two shapes and count how many units horizontally and vertically the point on A has moved to become the point on B.

3. Write the translation as a vector.

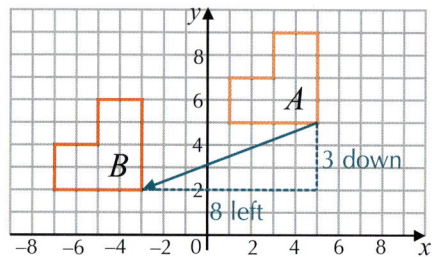

A has moved 8 units to the left and 3 units down so the transformation is a **translation described by the vector** $\begin{pmatrix} -8 \\ -3 \end{pmatrix}$.

Exercise 2

Q1 a)

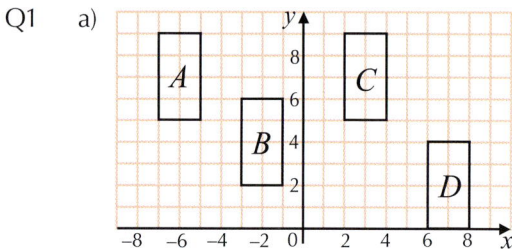

Give the vector that describes each of the following translations.

(i) A onto B

(ii) A onto C

(iii) C onto B

(iv) C onto D

(v) D onto A

(vi) D onto B

b)

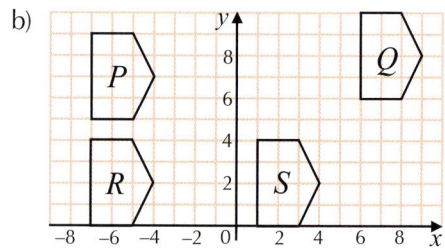

Give the vector that describes each of the following translations.

(i) P onto R

(ii) R onto S

(iii) P onto Q

(iv) S onto R

(v) Q onto R

(vi) S onto P

Q2 The triangle DEF has vertices $D(-3, -2)$, $E(1, -1)$ and $F(0, 2)$. The triangle GHI has vertices $G(0, 2)$, $H(4, 3)$ and $I(3, 6)$. Give the vector that describes the translation that maps DEF onto GHI.

Q3 Shape W is the image of shape Z after the translation $\begin{pmatrix} 1 \\ -4 \end{pmatrix}$.

Write the translation that maps shape W onto shape Z as a vector.

PROBLEM SOLVING

29.4 Enlargements

Enlargements can be trickier than the other transformations. They depend on a scale factor (which can be positive or negative, and sometimes a fraction) and a centre of enlargement.

Learning Objectives — Spec Ref G7:
- Enlarge a shape on a coordinate grid with a given centre of enlargement.
- Enlarge a shape using positive, fractional and negative scale factors.
- Describe an enlargement.

When an object is **enlarged**, its shape stays the same, but its **size changes** — so the image of a shape after an enlargement is **similar** to the original shape (see page 396 for more on similar shapes).

The **scale factor** of an enlargement tells you **how many times longer** the sides of the new shape are **compared** to the old shape. It also tells you **how much further** the points on the new shape are from the **centre of enlargement** than the points on the old shape. For example, enlarging by a **scale factor of 2** makes each side **twice as long** and each point **twice as far** from the centre of enlargement.

To enlarge a shape by a **positive** scale factor, follow this method:

- **Draw lines** from the centre of enlargement to **each vertex** of the shape.

- **Extend** each line depending on the scale factor (e.g. if the scale factor is 3 the line needs to be 3 times as long). Mark the vertices of the new shape at the ends of these extended lines.

- **Join up** the new vertices to create the enlarged shape.

If a point on the shape lies on the **centre of enlargement**, it will be the only **invariant point** under an enlargement.

Example 1

Enlarge the triangle on the axes on the right by scale factor 2 with centre of enlargement (2, 2).

1. Draw a line from (2, 2) through each vertex of the shape.

2. The scale factor is 2, so extend the lines so they are twice as long and mark the new vertices at the ends.

3. Join up the points to create the enlarged triangle. Each side of the triangle should now be twice as long.

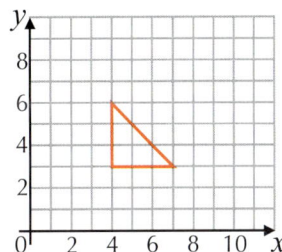

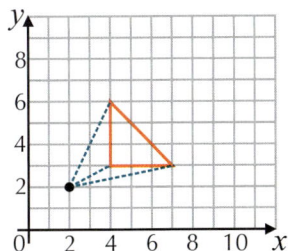

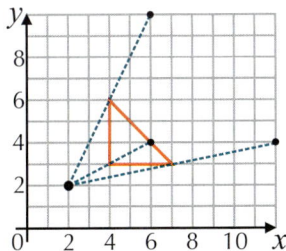

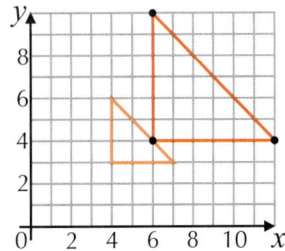

Q1 Copy the diagrams below and enlarge each shape by scale factor 2 with centre of enlargement (0, 0).

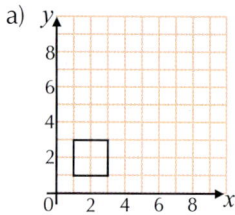

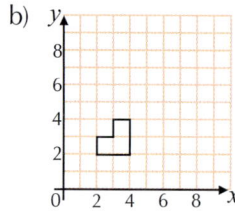

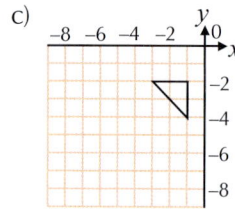

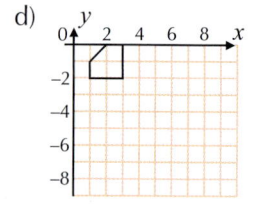

a)

b)

c)

d)

Q2 Copy the diagram on the right.

a) Enlarge *A* by scale factor 2 with centre of enlargement (−8, 9).

b) Enlarge *B* by scale factor 2 with centre of enlargement (9, 9).

c) Enlarge *C* by scale factor 3 with centre of enlargement (9, −8).

d) Enlarge *D* by scale factor 4 with centre of enlargement (−8, −8).

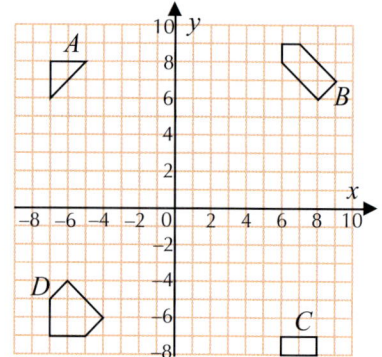

Q3 The triangle *PQR* has corners at *P*(1, 1), *Q*(1, 4) and *R*(4, 2).

a) Draw *PQR* on a pair of axes with *x*- and *y*-values from −1 to 9.

b) Enlarge *PQR* by scale factor 2 with centre of enlargement (−1, 1).

Q4 a) Copy the diagram below, then enlarge each shape by scale factor 2, using the centre of enlargement marked inside the shape.

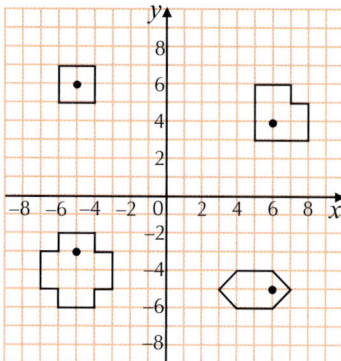

b) Copy the diagram below, then enlarge each shape by scale factor 2, using the centre of enlargement marked on the shape's vertex.

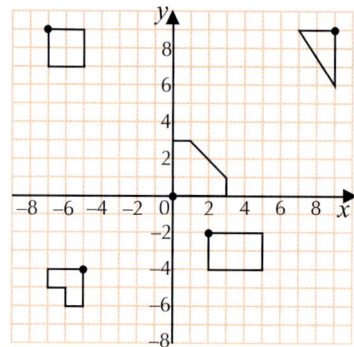

Q5 The shape *WXYZ* has corners at *W*(0, 0), *X*(1, 3), *Y*(3, 3) and *Z*(4, 0).

a) Draw *WXYZ* on a pair of axes with *x*-values from −5 to 9 and *y*-values from −5 to 6.

b) Enlarge *WXYZ* by scale factor 3 with centre of enlargement (2, 2).

Q6 The shape *KLMN* has corners at *K*(3, 4), *L*(3, 6), *M*(5, 6) and *N*(5, 5).

a) Draw *KLMN* on a pair of axes with *x*-values from 0 to 10 and *y*-values from 0 to 7.

b) Enlarge *KLMN* by scale factor 3 with centre of enlargement (3, 6). Label the shape $K_1L_1M_1N_1$.

c) Enlarge *KLMN* by scale factor 2 with centre of enlargement (5, 6). Label the shape $K_2L_2M_2N_2$.

If the scale factor is a **fraction** then the enlargement will actually **shrink** the shape. E.g. a scale factor of $\frac{1}{2}$ makes all the sides of the new shape **half as long**, and each point on the new shape will be **half as far** from the centre of enlargement.

Just like before, **draw lines** from the **centre of enlargement** to **each vertex**. Then mark the vertices of the **new shape** a **fraction** of the way along these lines (depending on the **scale factor**) and join them up. E.g. a scale factor of $\frac{1}{4}$ means the vertices of the new shape are a quarter of the way along the lines.

Example 2

Enlarge the shape on the axes below by scale factor $\frac{1}{2}$ with centre of enlargement (2, 7).

1. Draw lines from (2, 7) to each vertex.

2. Mark the vertices of the new shape half as far from the centre of enlargement as the original vertices. Join them up to create the enlarged shape.

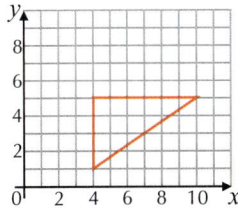

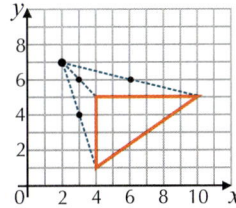

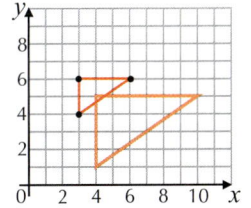

Exercise 2

Q1 Copy the diagrams below. Enlarge each shape by scale factor $\frac{1}{2}$ with centre of enlargement (0, 0).

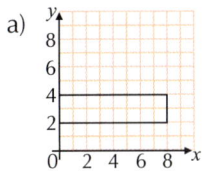

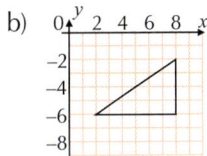

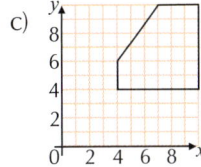

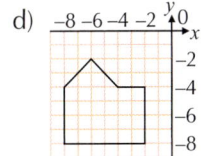

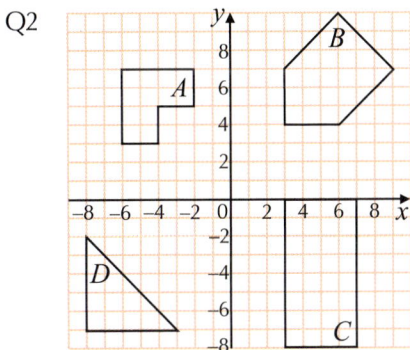

a)

b)

c)

d)

Q2

Copy the diagram on the left.

a) Enlarge A by scale factor $\frac{1}{2}$ with centre of enlargement (–8, 9).

b) Enlarge B by scale factor $\frac{1}{3}$ with centre of enlargement (0, 1).

c) Enlarge C by scale factor $\frac{1}{4}$ with centre of enlargement (–1, –8).

d) Enlarge D by scale factor $\frac{1}{5}$ with centre of enlargement (2, 3).

Q3 Copy the diagram on the right.

a) Enlarge A by scale factor $\frac{1}{2}$ with centre of enlargement (–4, 5).

b) Enlarge B by scale factor $\frac{1}{3}$ with centre of enlargement (7, 7).

c) Enlarge C by scale factor $\frac{1}{4}$ with centre of enlargement (6, –4).

d) Enlarge D by scale factor $\frac{1}{5}$ with centre of enlargement (–3, –2).

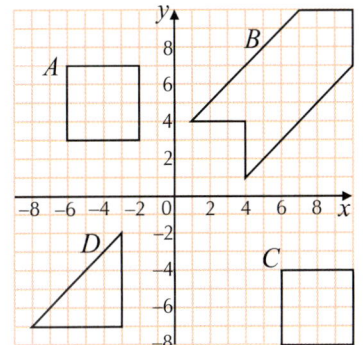

If you have a **negative** scale factor (i.e. –*k*), the sides of the enlarged shape are *k* **times longer**. Each point moves *k* **times further away** from the centre of enlargement but in the **opposite direction**.

To enlarge a shape by scale factor –*k*, **draw lines** from **each vertex** of the shape to the centre of enlargement. **Extend** each line *k* **times further** out of the **other side** of the centre of enlargement. The vertices of the new shape are at the ends of these lines — draw them in and **join them up** to create the new shape.

Tip: Enlarging by a scale factor of –*k* is the same as enlarging by a scale factor of *k* and then rotating 180° about the centre of enlargement.

Example 3

Enlarge the shape on the axes below by scale factor –2 with centre of enlargement (8, 7).

1. Draw lines from each vertex of the shape to (8, 7). The scale factor is –2, so extend the lines twice as far out of the other side of the centre of enlargement.

2. Mark the new vertices and join them up to create the enlarged shape — it's twice as big but on the opposite side of the centre of enlargement.

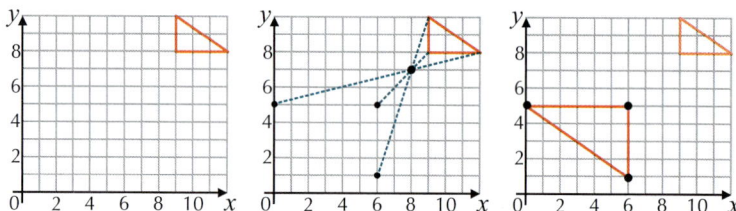

Exercise 3

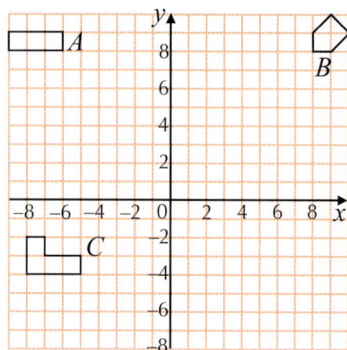

Q1 Copy the diagram on the left.

a) Enlarge *A* by scale factor –2 with centre of enlargement (–6, 7).

b) Enlarge *B* by scale factor –3 with centre of enlargement (7, 7).

c) Enlarge *C* by scale factor –2 with centre of enlargement (–2, –4).

Q2 Copy the diagram on the right.

a) Enlarge *A* by scale factor $-\frac{1}{2}$ with centre of enlargement (–4, 4).

b) Enlarge *B* by scale factor $-\frac{1}{3}$ with centre of enlargement (6, 4).

c) Enlarge *C* by scale factor $-\frac{1}{2}$ with centre of enlargement (4, –4).

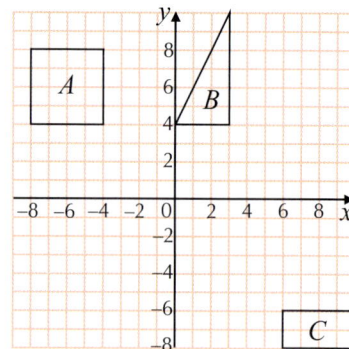

Q3 The shape *TUVW* has corners at (1, 0), (4, 6), (2, 5) and (0, 2).

a) Draw *TUVW* on a pair of axes with *x*-values from –4 to 4 and *y*-values from –6 to 6.

b) Enlarge *TUVW* by scale factor –1, centred at the origin, and label it $T_1U_1V_1W_1$.

c) What other single transformation would map *TUVW* to $T_1U_1V_1W_1$? (PROBLEM SOLVING)

To describe an enlargement, you need to give the **scale factor** and the **centre of enlargement**.

To find the **scale factor**, take the length of **any side** on the new shape and the length of its corresponding side on the old shape and use the **formula**:

$$\text{scale factor} = \frac{\text{new length}}{\text{old length}}$$

For the **centre of enlargement**, **draw** and **extend lines** that go through corresponding vertices of both shapes and see where they all **intersect**. If you're dealing with a **negative scale factor**, the centre of enlargement will lie **between** the two shapes.

Example 4

Describe the enlargement that maps shape X onto shape Y.

1. Pick any side on shape Y and its corresponding side on shape X (it's best to pick a horizontal or vertical side). Measure their lengths and put them into the formula to work out the scale factor.

2. Draw and extend lines from each vertex on shape Y through the corresponding vertex on shape X. The point where these lines meet is the centre of enlargement.

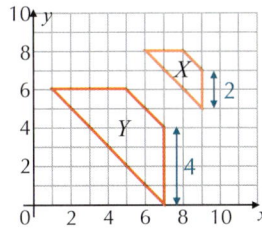

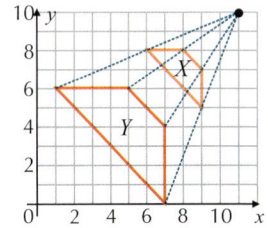

An enlargement by scale factor $\frac{4}{2} = 2$, centre **(11, 10)**.

Exercise 4

Q1 For each of the following, describe fully the transformation that maps shape A onto shape B.

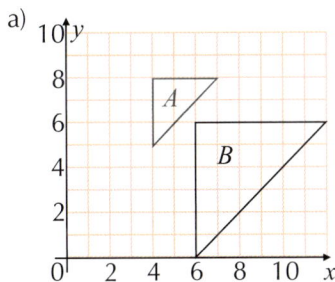

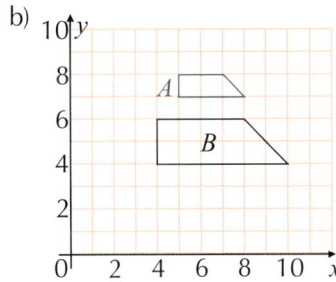

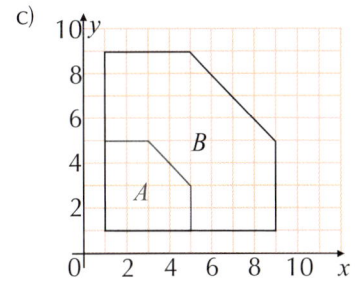

a)

b)

c)

Q2 For each of the following, describe fully the transformation that maps:

(i) shape A onto shape B,

(ii) shape B onto shape A.

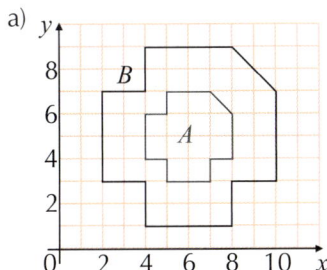

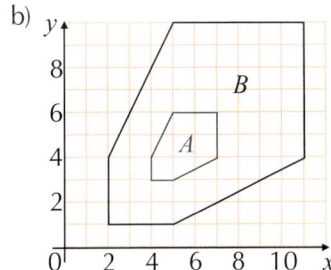

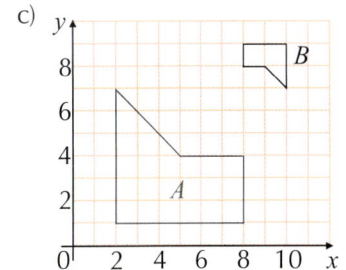

a)

b)

c)

29.5 Combinations of Transformations

Once you've got your head around the transformations covered in this section, the next step is doing them one after another — and figuring out how you could do them as a single transformation.

Learning Objectives — Spec Ref G7/G8:
- Apply combinations of transformations to shapes.
- Describe a combination of transformations as a single transformation.
- Recognise points of invariance under a combination of transformations.

Prior Knowledge Check:
Be familiar with rotation, reflection and translation (see p.377-384).

Doing a **combination** of transformations is as simple as doing **one after the other**, and you can often describe a combination using a **single** transformation. Just make sure you **specify** all the **details** when describing the single transformation — for a **reflection** you need the **equation of the mirror line**, for a **rotation** you need the **centre**, **angle** and **direction of rotation** and for a **translation** you need a **vector**.

Example 1

a) **Rotate triangle ABC on the axes below 180° about the point (1, 1), then translate the image by $\begin{pmatrix} -2 \\ -2 \end{pmatrix}$. Label the final image $A_1B_1C_1$.**

b) **Describe the single rotation that transforms triangle ABC onto the image $A_1B_1C_1$.**

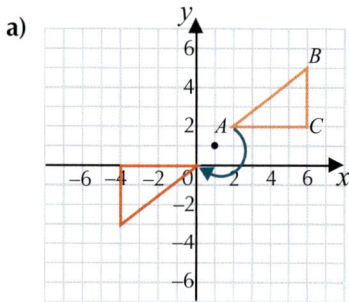

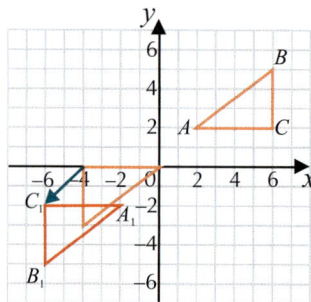

a)

Rotate ABC 180° about (1, 1)...   ... then translate by $\begin{pmatrix} -2 \\ -2 \end{pmatrix}$.

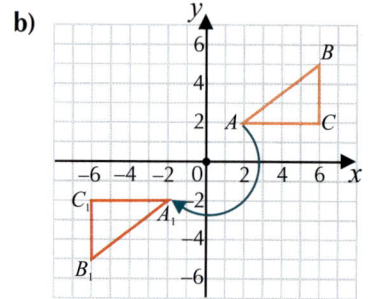

b)

A **rotation** of **180°** about the **origin** maps ABC onto $A_1B_1C_1$.

Exercise 1

Q1 Copy the diagram on the right.

a) (i) Rotate shape P 180° about (4, 5).

   (ii) Translate the image by $\begin{pmatrix} 2 \\ 2 \end{pmatrix}$. Label the final image P_1.

b) Rotate shape P 180° about (5, 6). Label the final image P_2.

c) (i) Rotate shape Q 180° about (4, 5).

   (ii) Translate the image by $\begin{pmatrix} 2 \\ 2 \end{pmatrix}$. Label the final image Q_1.

d) Rotate shape Q 180° about (5, 6). Label the final image Q_2.

e) What do you notice about the images P_1 and P_2 and about the images Q_1 and Q_2?

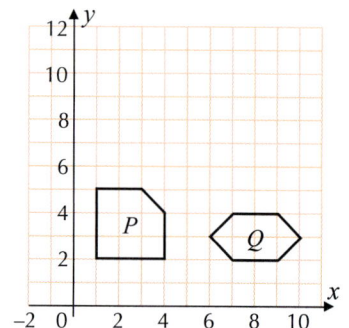

Q2 Triangle ABC has its corners at $A(2, 1)$, $B(6, 4)$ and $C(6, 1)$.

    a) Draw triangle ABC on a pair of axes, where both the x- and y-axes are labelled from -6 to 6.

    b) Reflect ABC in the y-axis. Label the image $A_1B_1C_1$.

    c) Reflect the image $A_1B_1C_1$ in the x-axis. Label the image $A_2B_2C_2$.

    d) Find a single rotation that transforms triangle ABC onto the image $A_2B_2C_2$.

Q3 Draw triangle PQR with corners at $P(2, 3)$, $Q(4, 3)$ and $R(4, 4)$ on a pair of axes with x- and y-values from -6 to 6.

    a) (i) Rotate PQR 90° clockwise about the point $(1, 3)$. Label the image $P_1Q_1R_1$.

       (ii) Write down any points of PQR that are invariant under this transformation.

    b) (i) Translate $P_1Q_1R_1$ by $\begin{pmatrix} 1 \\ 1 \end{pmatrix}$. Label the new image $P_2Q_2R_2$.

       (ii) Write down any points of $P_1Q_1R_1$ that are invariant under this transformation.

    c) (i) Describe a single transformation that maps triangle PQR onto the image $P_2Q_2R_2$.

       (ii) Write down any points of PQR that are invariant under this transformation.

Q4 Triangle WXY has its corners at $W(-5, -5)$, $X(-4, -2)$ and $Y(-2, -4)$.

    a) Draw triangle WXY on a pair of axes, where both the x- and y-axes are labelled from -6 to 6.

    b) Reflect WXY in the line $y = x$. Label the image $W_1X_1Y_1$.

    c) Reflect the image $W_1X_1Y_1$ in the y-axis. Label the image $W_2X_2Y_2$.

    d) Find a single transformation that maps WXY onto the image $W_2X_2Y_2$.

Q5 Copy the diagram on the right.

    a) Rotate triangle ABC 180° about $(0, 0)$. Label the image $A_1B_1C_1$.

    b) $A_1B_1C_1$ is translated such that the point B on the triangle ABC is invariant under the combination of the rotation followed by the translation. Describe this translation.

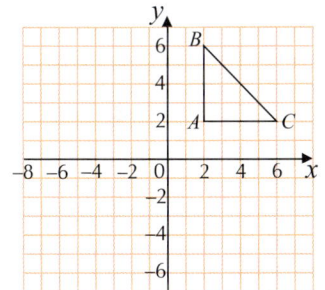

Q6 Shape $WXYZ$ has its corners at $W(-6, 2)$, $X(-3, 3)$, $Y(-2, 6)$ and $Z(-2, 2)$.

    a) Draw $WXYZ$ on a pair of axes, where both the x- and y-axes are labelled from -6 to 6.

    b) Reflect $WXYZ$ in the y-axis. Label the image $W_1X_1Y_1Z_1$.

    c) Rotate the image $W_1X_1Y_1Z_1$ 90° clockwise about $(0, 0)$. Label the image $W_2X_2Y_2Z_2$.

    d) Find a single transformation that maps $WXYZ$ onto $W_2X_2Y_2Z_2$.

Q7 Copy the diagram on the right.

    a) Translate shape $ABCDE$ by $\begin{pmatrix} -2 \\ -2 \end{pmatrix}$. Label the image $A_1B_1C_1D_1E_1$.

    b) The shape $A_1B_1C_1D_1E_1$ is rotated to become $A_2B_2C_2D_2E_2$. Under the combination of the translation from a) and this rotation, point C is invariant and point D maps to point B_2. Describe the rotation.

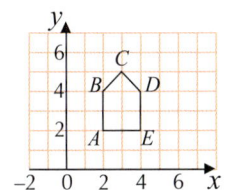

Review Exercise

Q1 a) Copy the grid and shape A on the right and then:
   (i) Reflect shape A in the x-axis and label it A_1.
   (ii) Reflect shape A in the line $x = 3$ and label it A_2.

   b) For each reflection in part a), write down the coordinates of any points of A that are invariant.

   c) Describe the reflection that maps shape A to shape U.

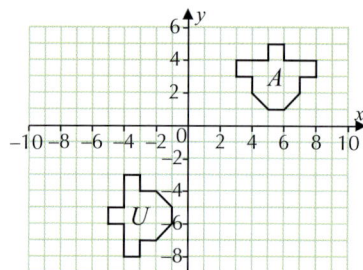

Q2 a) Copy the grid and shape B on the right and then:
   (i) Rotate shape B 180° about the origin and label it B_1.
   (ii) Rotate shape B 90° clockwise about (6, 1) and label it B_2.

   b) For each rotation in part a), write down the coordinates of any points of B that are invariant.

   c) Describe the rotation that maps shape B to shape V.

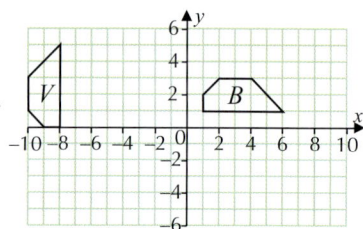

Q3 a) Copy the grid and shape C on the right. Translate shape C by:

   (i) $\begin{pmatrix} 5 \\ 2 \end{pmatrix}$ and label it C_1,   (ii) $\begin{pmatrix} -10 \\ 0 \end{pmatrix}$ and label it C_2.

   b) For each translation in part a), write down the coordinates of any points of C that are invariant.

   c) Describe the translation that maps shape C to shape W.

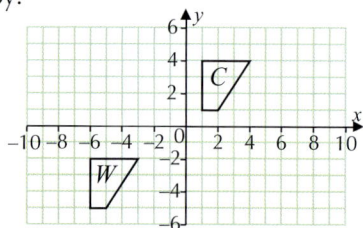

Q4 a) Copy the grid and shape D on the right and then:
   (i) Enlarge shape D by scale factor 2, with centre of enlargement the origin, and label it D_1.

   (ii) Enlarge shape D by scale factor $\frac{1}{2}$, with centre of enlargement (−9, 3), and label it D_2.

   (iii) Enlarge shape D by scale factor −3, with centre of enlargement (3, 1), and label it D_3.

   b) For each enlargement in part a), write down the coordinates of any points of D that are invariant.

   c) Describe the enlargement that maps shape D to shape X.

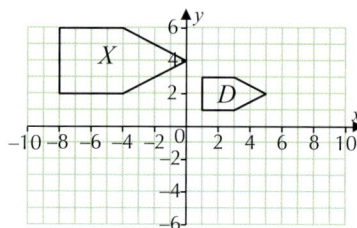

Q5 By considering triangle XYZ with vertices $X(2, 2)$, $Y(4, 4)$ and $Z(6, 2)$, find the single transformation equivalent to a reflection in the line $y = 2$, followed by a reflection in the line $x = 2$.

Q6 By considering shape $ABCD$ with corners at $A(2, -2)$, $B(4, -2)$, $C(5, -5)$ and $D(2, -4)$, find the single transformation equivalent to a rotation of 90° anticlockwise about (2, −2), followed by a reflection in the y-axis, followed by a translation by $\begin{pmatrix} 0 \\ 4 \end{pmatrix}$.

Exam-Style Questions

Q1 a) Describe fully the single transformation equivalent to

- a reflection in the line $x = 0$, followed by
- a reflection in the line $y = -x$.

Use the grid to help you.

[3 marks]

b) State the coordinates of any points of invariance under this transformation.

[1 mark]

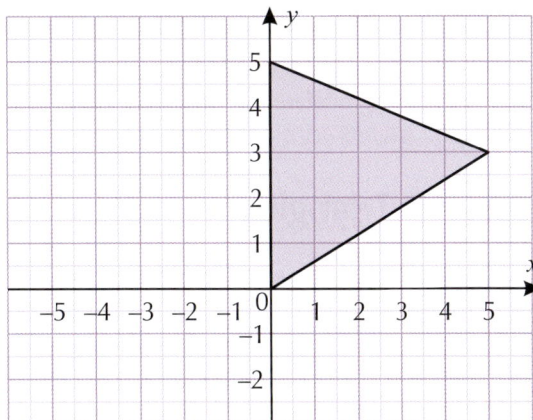

Q2 A triangle with vertices $A(0, 0)$, $B(3, 0)$ and $C(3, 6)$ is reflected in the line $y = k$ and then translated 10 units in the positive y-direction. Point C is invariant under this combination of transformations.

Find the value of k.

[2 marks]

Q3 A shape P is transformed by a translation of $\begin{pmatrix} 1 \\ -4 \end{pmatrix}$ to produce an image Q.

Shape Q is then transformed by an enlargement of scale factor 2 with centre of enlargement $(-2, 3)$ to give shape R. Shape R is shown on the diagram below.

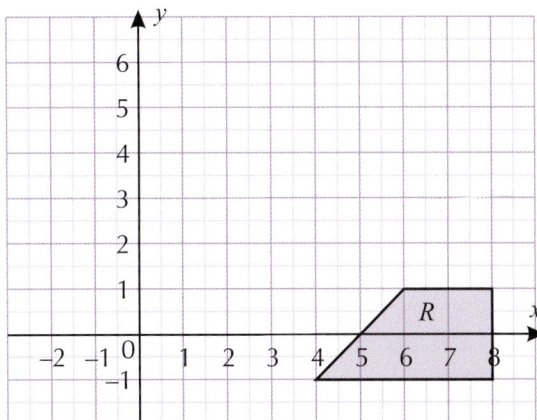

Copy the diagram and draw shape P. Show all your working.

[3 marks]

30.1 Congruence and Similarity

Congruent shapes are exactly the same shape and size. Similar shapes are exactly the same shape but different sizes. Translated, rotated and reflected shapes are congruent to the original shapes, while enlarged shapes are similar to the original shapes.

Congruent Triangles

Learning Objectives — Spec Ref G5/G6:
- Know the congruence conditions for triangles.
- Be able to prove that two triangles are congruent.

In congruent shapes, all **side lengths** and **angles** on one shape are **identical** to the side lengths and angles on the other. However, you don't need to know **all** the side lengths and angles to show that two **triangles** are congruent — just show that they satisfy any **one** of the following 'congruence conditions'.

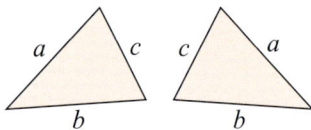

Side, Side, Side:
The three sides on one triangle are the same as the three sides on the other triangle.

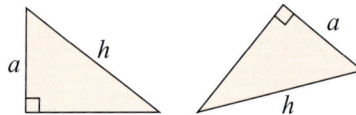

Right angle, Hypotenuse, Side:
Both triangles have a right angle, both triangles have the same hypotenuse and one other side is the same.

Tip: If any of these conditions are true then you can use trigonometry to show that the other sides and angles in the triangles must also be the same.

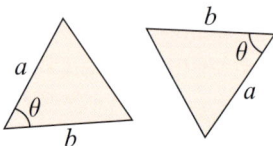

Side, Angle, Side:
Two sides and the angle between them on one triangle are the same as two sides and the angle between them on the other triangle.

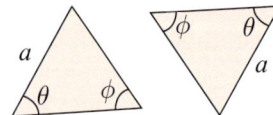

Angle, Angle, Side:
Two angles and an opposite side on one triangle are the same as two angles and the corresponding opposite side on the other triangle.

Be very careful when using 'Side, Angle, Side' and 'Angle, Angle, Side' — you have to have the **correct combination** of sides and angles.

Example 1

Are these two triangles congruent? Give a reason for your answer.

Look at the conditions listed above...

Two of the sides and the angle between them are the same on both triangles.

Condition SAS holds, so the triangles **are congruent**.

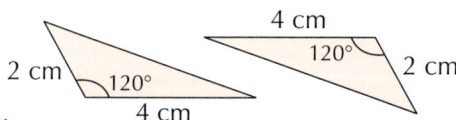

Tip: The conditions are often abbreviated to SAS, SSS, AAS and RHS.

Q1 The following pairs of triangles are congruent. Find the values marked with letters.

a)
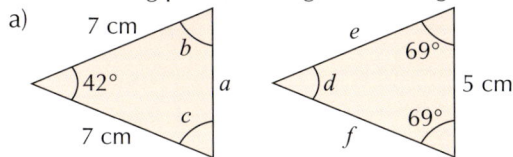

b)
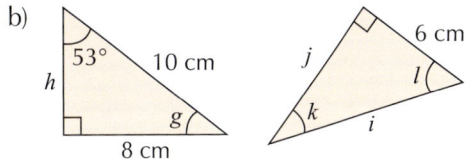

Q2 For each of the following, decide whether the two triangles are congruent.
In each case explain either how you know they are congruent, or why they must not be.

a)

b)
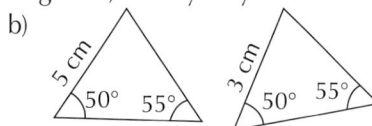

c)
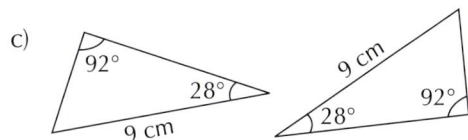

d)
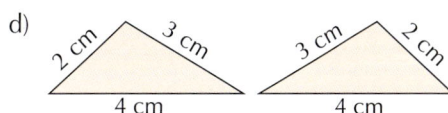

e)
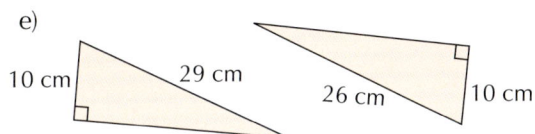

f)
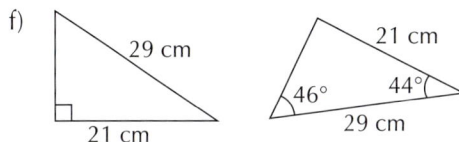

If it's not obvious which of the conditions apply, you might have to figure out some of the side lengths or angles for yourself. Properties of **parallel lines** and **polygons**, **circle theorems** and **Pythagoras' theorem** will all come in handy here — have a look at Sections 20, 21 and 25 if you need a reminder.

Example 2

ABCD **is a cyclic quadrilateral within the circle with centre *O*.**
a) **Prove that triangles *AOD* and *BOC* are congruent.**

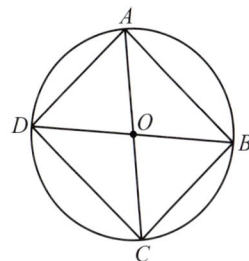

Use properties of vertically opposite angles and circles to match sides and angles in the two triangles.

Vertically opposite angles are equal, so angle *AOD* = angle *BOC*.

Sides *OA*, *OD*, *OB* and *OC* are all radii, so *OA* = *OC* and *OD* = *OB*.

Two sides and the angle between them are the same, which means condition SAS holds, so the triangles **are congruent**.

b) **Hence prove that triangles *ABD* and *BCD* are congruent.**

This time, you need to use your result from part a) and circle theorems.

Triangles *ABD* and *BCD* lie in a semicircle so are right-angled, which means angle *BAD* = angle *BCD* = 90°.

Tip: *BD* is a diameter as it passes through the centre of the circle.

Side *BD* is the hypotenuse and is common to both triangles.

From part a), triangles *AOD* and *BOC* are congruent, so *AD* = *BC*.

Both triangles have a right angle, the same hypotenuse and one other side the same, which means condition RHS holds, so the triangles **are congruent**.

Q1 Show that the two triangles on the right are congruent.

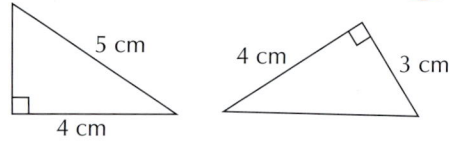

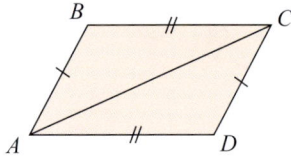

5 cm

4 cm

4 cm

3 cm

Q2 *ABCD* is a parallelogram. Prove that triangles *ABC* and *ADC* are congruent.

B *C*

A *D*

C

B *D*

A *E*

Q3 *ABCDE* is a regular pentagon. Prove that triangles *ABC* and *CDE* are congruent.

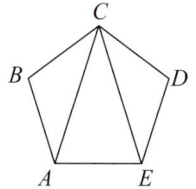

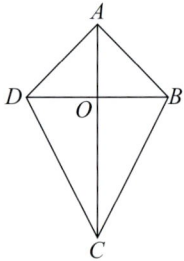

A

D *O* *B*

C

Q4 *ABCD* is a kite and *O* is the point where the diagonals of the kite intersect. Prove that *BOC* and *DOC* are congruent triangles.

A

B

O

D

C

Q5 *AC* and *BD* are diameters of the circle with centre *O*. Prove that triangles *ABC* and *DCB* are congruent.

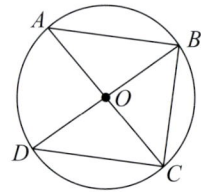

Similar Shapes

Learning Objectives — Spec Ref G6:
- Know the conditions for similar triangles.
- Be able to prove that two triangles are similar.
- Use scale factors to find missing sides and angles.

Prior Knowledge Check:
Properties of triangles and other 2D shapes (see Section 20) and Pythagoras' theorem (Section 25).

Similar Triangles

Similar shapes have the **same angles** as each other, but the **side lengths** are **different** — they are enlarged by the same scale factor. Two **triangles** are similar if they satisfy **any one** of the following conditions.

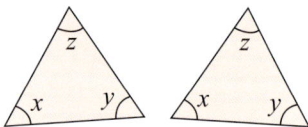

z z

x y x y

a c $2a$ $2c$

b $2b$

a $2a$

x x

b $2b$

All the angles on one triangle are the same as the angles on the other triangle.

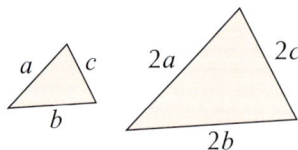

All the sides on one triangle are in the same ratio as the corresponding sides on the other triangle.

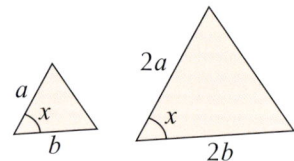

Two sides on one triangle are in the same ratio as the corresponding sides on the other triangle, and the angle between is the same on both triangles.

You might have to work out some **side lengths** or **angles** for yourself before you can decide if two triangles are similar. For example, if you're told **two** angles in one triangle, you can use the fact that the angles in a triangle **add up to 180°** to find the missing angle, then compare it to the other triangle.

To find missing side lengths, you might have to use properties of **isosceles triangles** or other **2D shapes**. You might even have to use **Pythagoras' theorem** or **trigonometry** if it's a right-angled triangle.

Example 3

Are these two triangles similar? Give a reason for your answer.

1. Start by finding the missing angles in each triangle.

   Triangle A: missing angle = $180° - 88° - 54° = 38°$
   Triangle B: missing angle = $180° - 38° - 54° = 88°$

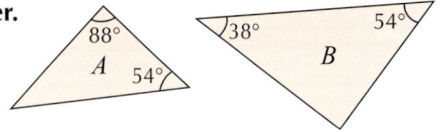

2. Look at the conditions for similarity and see if any of them are true for these triangles.

   All the angles in triangle A are the same as the angles in triangle B.

   So the triangles **are similar**.

Exercise 3

Q1 For each of the following, decide whether the two triangles are similar. In each case explain either how you know they are similar, or why they must not be.

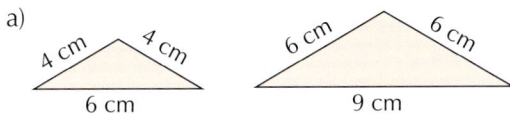

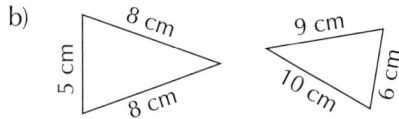

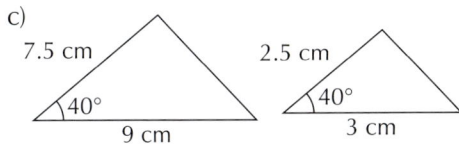

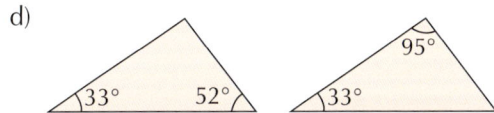

a)

b)

c)

d)

Q2 Show that triangles A and B are similar.

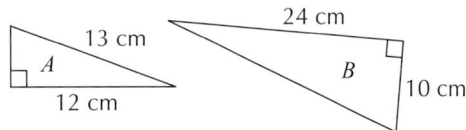

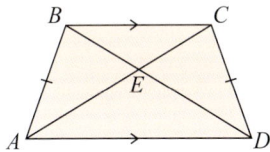

Q3 $ABCD$ is an isosceles trapezium. Prove that triangles BEC and AED are similar.

Q4 ABC is an isosceles triangle.

a) Prove that triangles ABC and DBE are similar.

b) Find another pair of triangles that are similar but not congruent. Explain your answer.

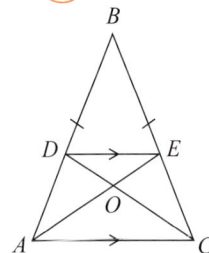

Scale Factors

If you **know** that two shapes are **similar**, you can use that fact to find **missing sides** and **angles** — just remember that similar shapes have **all angles** the same, and sides in the **same ratio**.

All you have to do to find missing sides is find the **scale factor** — the number you have to multiply all the side lengths in one shape by to get the side lengths in the other shape. To find the scale factor to get you from shape A to a similar shape B, use this **formula**:

$$\text{scale factor} = \frac{\text{side length of shape } B}{\text{side length of shape } A}$$

Example 4

Triangles *ABC* and *DEF* are similar. Find the lengths *DE* and *DF*.

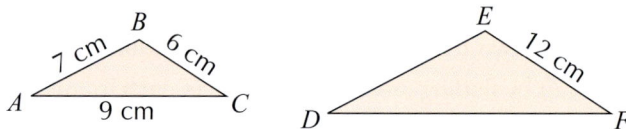

1. Find the scale factor that gets you from *ABC* to *DEF*.

2. Use this scale factor to find the missing side lengths.

scale factor $= \dfrac{EF}{BC} = \dfrac{12}{6} = 2$

So $DE = 2AB = 2 \times 7 = \textbf{14 cm}$
and $DF = 2AC = 2 \times 9 = \textbf{18 cm}$

Tip: You could describe the sides as being in the ratio $1:2$ — see p.400.

Example 5

Trapeziums *PQRS* and T*UVW* are similar. Find angle *x*.

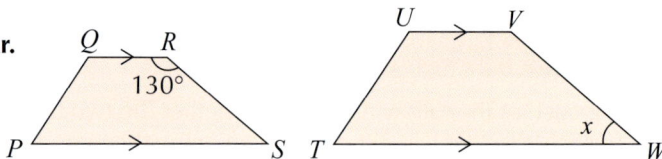

1. The trapeziums are similar, so they will have the same angles.

2. Use the properties of trapeziums and parallel lines to find the equivalent angle on *PQRS*.

3. Calculate the size of this angle.

$x =$ angle *RSP*, as equivalent angles in similar shapes are equal.

Sides *PS* and *QR* are parallel, so angles *QRS* and *RSP* are allied angles, which means they add up to 180°.

So angle $RSP = 180° - 130° = 50°$, which means $x = \textbf{50°}$.

Exercise 4

Q1 Triangles *PQR* and *XYZ* are similar.

a) Find the scale factor that takes you from *PQR* to *XYZ*.

b) Find the length *YZ*.

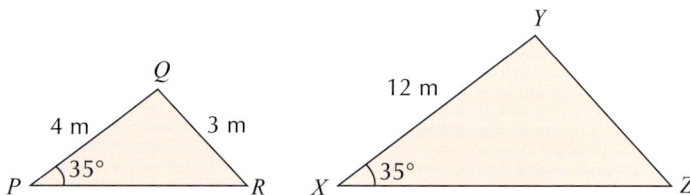

Q2 The diagram on the right shows two similar triangles, *QRU* and *QST*. Find the following lengths.

a) *ST*   b) *QT*   c) *UT*

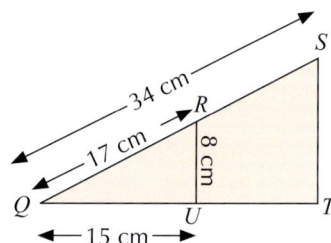

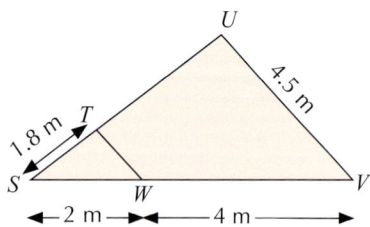

Q3 The diagram on the left shows two similar triangles, *STW* and *SUV*. Find the following lengths.

a) *TW*   b) *SU*   c) *TU*

Q4 Each pair of shapes shown below are similar.

a) Find *x*.

b) Find *y* and *z*.

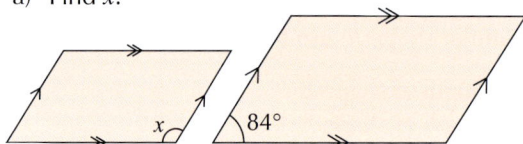

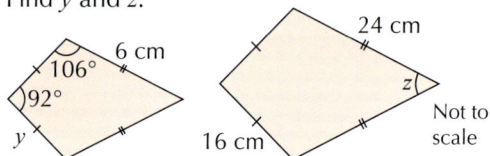

Not to scale

30.2 Areas and Volumes of Similar Shapes

You can use the scale factor of the side lengths to help you find the areas and volumes of similar shapes. Be careful though — you don't just multiply the area or volume of the smaller shape by this scale factor to find the area or volume of the bigger shape. There's a bit more to it than that...

Areas of Similar Shapes

Learning Objectives — Spec Ref G19/R12:
- Find the areas and surface areas of similar shapes.
- Use ratios to solve similarity problems.

Prior Knowledge Check:
Find areas of 2D shapes and surface areas of 3D shapes — see Sections 27 and 28.
Use ratio notation — see p.42.

For two **similar** shapes, where one has sides that are **twice as long** as the sides of the other, the **area** of the larger shape will be 2^2 **times** (i.e. **four** times) the area of the smaller shape. For example, consider a square with side length 3 cm — it has an area of $3^2 = 9$ cm^2. If that square was enlarged by scale factor 5, the larger square would have an area of $9 \times 5^2 = 225$ cm^2.

In general, for an enlargement of **scale factor** n, the area of the new shape will be n^2 **times** the area of the original shape.

To find the scale factor (n) from the areas, use the formula:

$$n^2 = \frac{\text{area of enlarged shape}}{\text{area of original shape}}$$

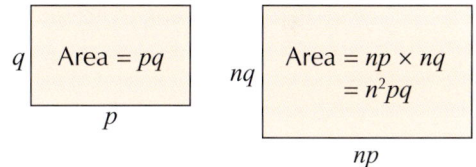

q | Area $= pq$
p

nq | Area $= np \times nq$ $= n^2pq$
np

Example 1

Rectangles A and B are similar. Find the area of rectangle B.

Not to scale

9 cm | A 36 cm^2

45 cm | B

1. First find the scale factor by looking at corresponding sides on each shape.

Scale factor (n) $= \dfrac{\text{length of } B}{\text{length of } A} = \dfrac{45}{9} = 5$

2. Multiply the area of A by the scale factor squared to find the area of B.

Area of $B = 36 \times 5^2 = 36 \times 25 = \textbf{900 cm}^2$

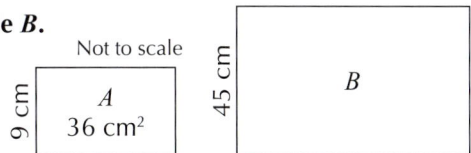

Example 2

Triangles A and B are similar. The area of triangle A is 10 cm^2 and the area of triangle B is 250 cm^2.

Not to scale

A 10 cm^2

B 250 cm^2

12.5 cm

a) **What is the scale factor of the enlargement?**

1. Use the formula above. $n^2 = \dfrac{\text{enlarged area}}{\text{original area}} = \dfrac{250}{10} = 25$

2. Take the square root to find the scale factor of the enlargement. $n = \sqrt{25} = 5$, so the scale factor is **5**

b) **The base of triangle B is 12.5 cm. What is the length of the base of triangle A?**

Divide the length of the base of triangle B by the scale factor you found in part a).

base of triangle A: $12.5 \div 5 = \textbf{2.5 cm}$

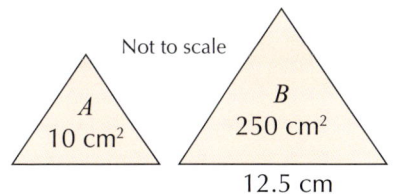

Tip: You divide by the scale factor in part b) because you're going from the larger shape to the smaller shape.

Exercise 1

Q1 Work out the area of the square formed when a square with side length 10 cm is enlarged by a scale factor of 4.

Q2 Triangles A and B are similar. Calculate the area of B.

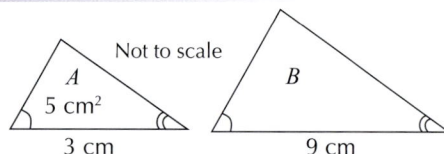

Not to scale

A 5 cm² 3 cm

B 9 cm

Q3 The shapes on the right are similar. Work out the base length of shape Q.

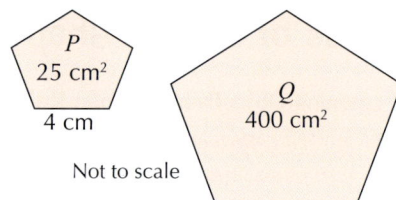

P 25 cm² 4 cm

Q 400 cm²

Not to scale

Q4 The rectangles below are similar. Calculate the area of shape A.

Not to scale

2 cm A

B 125 cm² 5 cm

Q5 A triangle has perimeter 12 cm and area 6 cm². It is enlarged by a scale factor of 3 to produce a similar triangle. What is the perimeter and area of the new triangle?

Q6 Find the dimensions of rectangle B, given that the rectangles are similar.

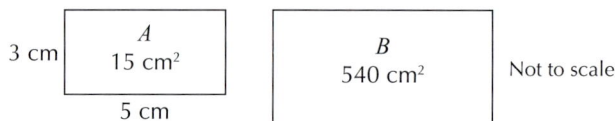

3 cm A 15 cm² 5 cm

B 540 cm² Not to scale

Using Ratios

If the ratio of the side lengths is $a:b$, the ratio of the areas is $a^2:b^2$.
So if the ratio of the sides is $1:2$, the ratio of the areas is $1^2:2^2 = 1:4$.

Example 3

The ratio of the areas of two similar shapes is $49:121$. One side of the smaller shape measures 2.8 cm. Find the length of the corresponding side on the larger shape.

1. First find the ratio of the side lengths by taking the square root of each side of the area ratio.

   $a^2:b^2 = 49:121$
   so $a:b = \sqrt{49}:\sqrt{121} = 7:11$

2. This means that the side of the larger shape is $\frac{11}{7}$ times as big as the corresponding side of the smaller shape — so multiply 2.8 by $\frac{11}{7}$ to find the length you need.

   $2.8 \times \frac{11}{7} = \textbf{4.4 cm}$

Exercise 2

Q1 Squares A and B have side lengths given by the ratio $2:3$. Square A has sides of length 8 cm.
   a) Find the length of one of the sides of B.
   b) Find the area of B.
   c) Find the ratio of the area of A to the area of B.

Q2 The ratio of the areas of two similar shapes is $2.25:6.25$. Find the ratios of their side lengths.

Q3 The ratio of the sides of two similar shapes is $4:5$. The area of the smaller shape is 20 cm². Find the area of the larger shape.

Q4 Stuart is drawing a scale model of his workshop. He uses a scale of 1 cm : 50 cm. The area of the bench on his drawing is 3 cm². What is the area of his bench in real life?

The rules about enlargements and areas of 2D shapes also apply to **surface areas** of **3D shapes**.

Example 4

Shapes A and B on the right are similar triangular prisms.
The surface area of shape A is 35 cm².
Find the surface area of shape B.

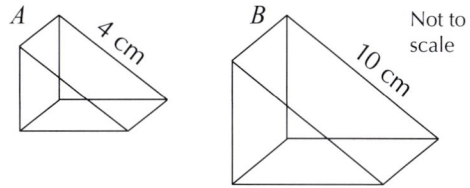

Not to scale

1. Compare corresponding side lengths to find the scale factor of enlargement, n.

    10 cm ÷ 4 cm = 2.5, so the scale factor is $n = 2.5$

2. Multiply the surface area of shape A by n^2 to find the surface area of shape B.

    Surface area of $B = 35 \times 2.5^2 = 35 \times 6.25 = $ **218.75 cm²**

Exercise 3

Q1 A and B are similar prisms. B has a surface area of 1440 in².
     Calculate the surface area of A.

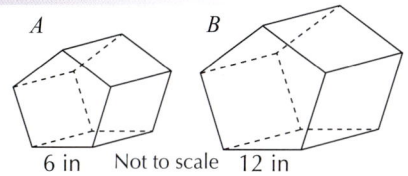

6 in   Not to scale   12 in

Q2   A sphere has surface area 400π mm². What is the surface area
     of a similar sphere with a diameter three times as large?

Q3   The 3D solid P has a surface area of 60 cm². Q, a similar 3D solid, has a surface area of 1500 cm².
     If one side of shape P measures 3 cm, how long is the corresponding side of shape Q?

Q4   Cylinder A has surface area 63π cm².
     Find the surface area of the similar cylinder B.

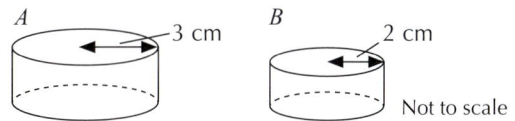

Not to scale

Q5   The vertical height of pyramid A is 5 cm.
     Find the vertical height of the similar pyramid B.

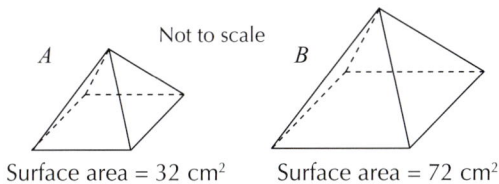

Surface area = 32 cm²   Surface area = 72 cm²

Q6   Similar shapes P, Q and R have surface areas in the
     ratio 4 : 9 : 12.25. Find the ratio of their side lengths.

Volumes of Similar Shapes

Learning Objectives — Spec Ref G19/R12:
- Find the volumes of similar shapes.
- Use ratios to solve similarity problems.

Prior Knowledge Check:
Find volumes of 3D shapes — see Section 28.
Use ratio notation — see p.42.

For two **similar 3D shapes**, when the side lengths are **doubled**, the volume is
multiplied by a factor of 2^3 (= **8**). Consider a cube with sides of length 5 cm
— it has a volume of $5^3 = 125$ cm³. If that cube was
enlarged by scale factor 2, the larger cube would
have a volume of $125 \times 2^3 = 1000$ cm³.

In general, if the sides increase by a scale factor of n,
the volume increases by a scale factor of n^3.

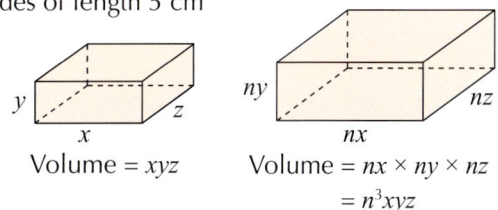

Volume = xyz

Volume = $nx \times ny \times nz$
       = n^3xyz

To find the scale factor (n) from the volumes, use the formula: $n^3 = \dfrac{\text{volume of enlarged shape}}{\text{volume of original shape}}$

For two similar shapes with side lengths in the ratio $a:b$, their volumes will be in the ratio $a^3:b^3$.

Example 5

Triangular prisms A and B are similar. Find the volume of B.

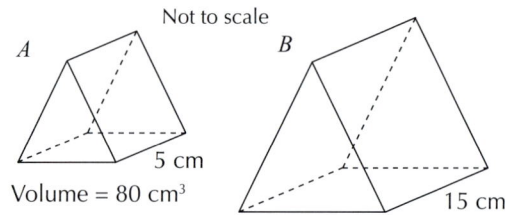

1. Find the scale factor by dividing the side length of B by the corresponding side length of A.
   Scale factor = $15 \div 5 = 3$

2. Multiply the volume of A by the scale factor cubed.
   Volume of B = volume of $A \times 3^3 = 80 \times 27 = \mathbf{2160\ cm^3}$

Exercise 4

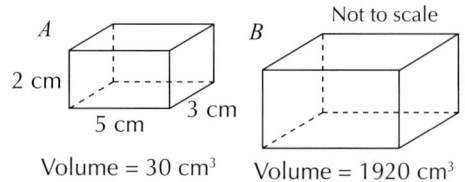

Q1 Cuboids A and B, shown on the right, are similar. Find the dimensions of B.

Q2 A 3D solid of volume 18 m³ is enlarged by scale factor 5. What is the volume of the new solid?

Q3 Triangular prisms A and B, shown on the right, are similar. Find the volume of A.

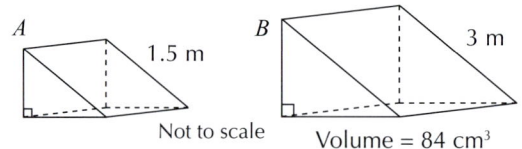

Q4 A cube has sides of length 3 cm. Find the side length of a similar cube whose volume is 3.375 times as big.

Q5 Cone A on the right has a volume of 60π cm³. Find the volume of cone B, given that the shapes are similar.

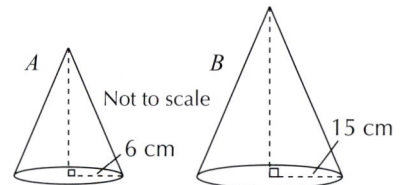

Q6 Two similar solids have side lengths in the ratio $2:5$.
   a) What is the ratio of their volumes?
   b) The smaller shape has a volume of 100 mm³. What is the volume of the larger shape?

Q7 A pyramid has volume 32 cm³ and vertical height 8 cm. A similar pyramid has volume 16 384 cm³. What is its vertical height?

Q8 The radius of the planet Uranus is approximately 4 times the size of the radius of Earth. What does this tell you about the volumes of these planets?

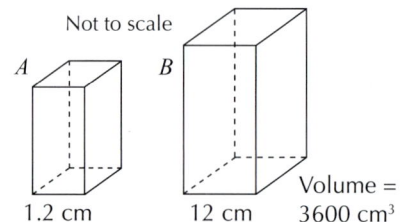

Q9 The rectangular prisms on the right are similar. Calculate the volume of shape A.

Q10 Two similar solids have volumes of 20 m³ and 1280 m³. James says that the surface area of the larger solid is 16 times the surface area of the smaller solid. Claire says that the surface area is 4 times larger. Who is correct?

Review Exercise

Q1 For each of the following, decide whether the two triangles are congruent, similar or neither. In each case, explain your answer.

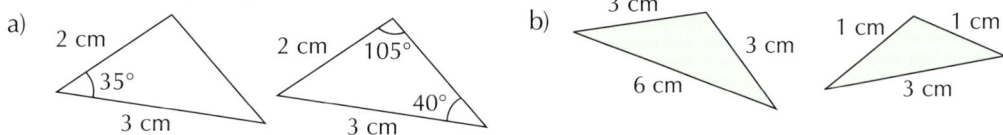

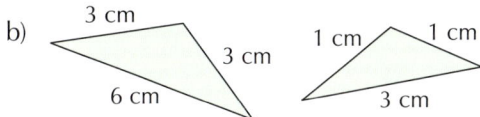

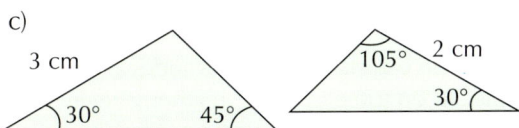

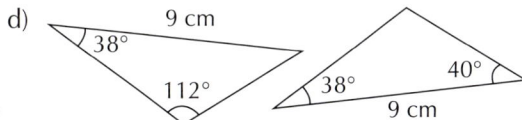

a)

2 cm / 35° / 3 cm   2 cm / 105° / 40° / 3 cm

b)

3 cm / 3 cm / 6 cm   1 cm / 1 cm / 3 cm

c)

3 cm / 30° / 45°   105° / 2 cm / 30°

d)

38° / 9 cm / 112°   38° / 40° / 9 cm

Q2 *JKN* and *JLM* on the right are two similar triangles.

a) Find length *KN*.

b) Find length *KL*.

36 cm / K / 15 cm / L
J / N / M
18 cm / 9 cm

Not to scale

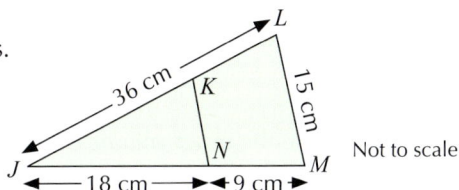

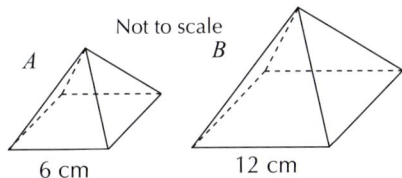

Q3 The square-based pyramid *A* on the left has surface area 96 cm^2 and volume 48 cm^3.

a) Find the surface area of the similar pyramid *B*.

b) Find the volume of *B*.

Not to scale

A / B
6 cm / 12 cm

Q4 a) Two similar solids have side lengths in the ratio $3:5$.

(i) Find the ratio of their surface areas.   (ii) Find the ratio of their volumes.

b) Two similar solids have surface areas in the ratio $49:81$.

(i) Find the ratio of their side lengths.   (ii) Find the ratio of their volumes.

Q5 Vicky has two similar jewellery boxes in the shape of rectangular prisms.
The larger box has dimensions of 9 in × 15 in × 12 in and a volume of 1620 in^3.
The smaller box has a volume of 60 in^3. Find the dimensions of the smaller box.

Q6 A solid has surface area 22 cm^2 and volume 18 cm^3.
A similar solid has sides that are 1.5 times as long.

a) Calculate its surface area.

b) Calculate its volume.

Q7 Mike has an unfortunate habit of falling down wells.
The well he usually falls down has a volume of 1.25π m^3.
His friends need a ladder that's at least 5 m long to rescue him.
One day he falls down a well that is mathematically similar
to his usual well. Given that this well has a volume of 10π m^3,
how long does the ladder need to be this time?

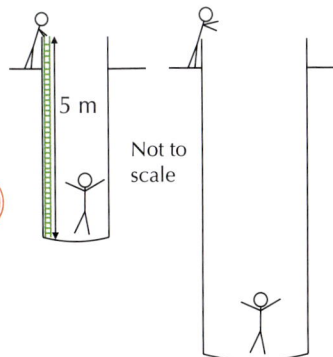

5 m

Not to scale

PROBLEM SOLVING

Exam-Style Questions

Q1 The diagram on the right shows two similar triangles, *ABC* and *ADE*.

a) Find length *BC*.

[2 marks]

b) Find angle *ACB*.

[1 mark]

Not to scale

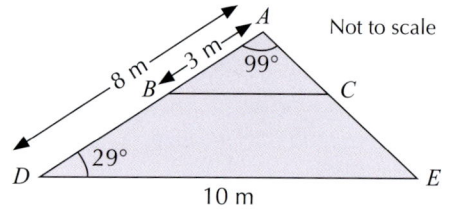

Q2 Josephine is making a two-tiered wedding cake. It consists of a small cylindrical cake with diameter 16 cm and height 6 cm placed on top of a larger, mathematically similar cake. The area of the base of the larger cake is 144π cm².

a) Calculate the diameter of the larger cake.

[2 marks]

b) Calculate the exact volume of the larger cake.

[3 marks]

Q3 The cross-sectional area of a hexagonal prism is 18 cm². The volume of the prism is 270 cm³. A larger similar prism has a cross-sectional area of 162 cm². What is the volume of the larger prism?

[3 marks]

Q4 In the diagram, *ABC* is a triangle in which angle *CAB* = angle *CBA*. *D* and *E* are points on *AC* and *BC* respectively such that *AD* = *BE*. Prove that the triangles *AEC* and *BDC* are congruent.

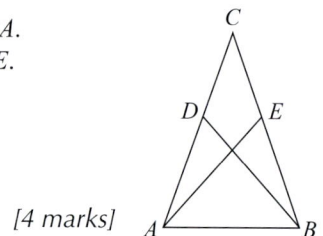

[4 marks]

Q5 An octagonal prism has a surface area of 50 mm² and a volume of 88 mm³. Another octagonal prism has a surface area of 450 mm² and a volume of 792 mm³. Are these two shapes similar? Explain your answer.

[3 marks]

Q6 A shop sells bags of bird food in two different sizes. The bags are mathematically similar. One bag has a height of 3 cm and costs £1.10, while the other has a height of 6 cm and costs £2.50. Is the larger bag better value for money? Explain your answer.

[3 marks]

31.1 Using Different Types of Data

To test a hypothesis or conduct an investigation, you first need to plan what data you need, how you're going to collect it and how you're going to analyse it. Data can come in different types depending on who first gathered it and what sort of values/format it takes. There are benefits to each type of data and it's important to choose the most suitable type for your investigation.

Primary and Secondary Data

Learning Objective — Spec Ref S1:
Know the difference between primary and secondary data.

Primary data is data you **collect yourself**, e.g. by doing a survey or experiment. It's beneficial to use primary data since you can **control** what information is gathered, as well as how **accurate** it is.

Secondary data is data that has been **collected by someone else**. You can get secondary data from things like newspapers or the internet. You're relying on someone else to collect the data — but secondary data is often **cheap** and **quick** to obtain. This is especially useful if you need a **large** amount of information or you require data that would be hard for you to personally **gather**.

Example 1

Jon wants to test whether a six-sided dice has an equal chance of landing on each of its sides. Explain how he could collect data for the test. Will his data be primary or secondary data?

1. Jon needs to do an experiment.   He could roll the dice lots of times and record how many times it lands on each side using a tally chart.

2. He collects the data himself, so...   His data will be **primary** data.

Exercise 1

For each investigation below: a) Describe what data is needed and give a suitable method for collecting it.
b) Say whether the data will be primary or secondary data.

Q1   Nikita wants to know what the girls in her class think about school dinners.

Q2   Dan wants to find the most common colour of car passing his house in a 30-minute interval.

Q3   Anne wants to compare the daily rainfall in London and Manchester last August.

Q4   Rohan wants to test his theory that the boys in his class can throw a ball further than the girls.

Q5   Jim wants to find out how the temperature in his garden at 10 am each morning compares with the temperature recorded by the Met Office for his local area.

Discrete and Continuous Data

Data can be either **qualitative** or **quantitative**. **Qualitative** data is **descriptive**, meaning it uses **words** instead of numbers — e.g. favourite flavours of ice cream ('vanilla', 'chocolate', 'strawberry', etc.) or opinions ('happy', 'neutral', etc.). **Quantitative** data is **numerical** — e.g. the heights of players in a football team or pupils' scores in a Science test. Quantitative data can be split up into two further types:

- **Discrete** data can only take **certain values**. For example, the number of goals scored by a football team — the data must be whole numbers, the team can't score half a goal.

- **Continuous** data can take **any value** in a given **range** — e.g. lengths, heights or weights.

How you **process** and **analyse** your data will depend on the data type. Some of the different types of **graphs** and **diagrams** in Section 33 would be **inappropriate** for showing certain data types (e.g. you shouldn't put qualitative data onto a line graph but you can make a bar chart or pie chart out of it).

Example 2

Say whether the following data is qualitative, discrete quantitative or continuous quantitative.

a) **The hometowns of 100 people.**

This data is in the form of words, so...   It's **qualitative** data.

b) **The weights of the bags of potatoes on sale in a greengrocer's.**

This data is numerical and isn't restricted to certain specific values.   It's **continuous quantitative** data.

Exercise 2

Q1   Say whether the following data is qualitative, discrete quantitative or continuous quantitative.

     a) The number of words in your favourite song.   b) Your favourite food.

     c) The numbers of pets in 20 households.   d) The heights of 100 tomato plants.

     e) The nationalities of the people in a park.   f) The lengths of 30 worms.

     g) The distances of planets from the Sun.   h) The hair colours of 50 people.

Q2   Blair wants to find out what 12 people think about his favourite TV show, 'Quickly Come Financing'.

He first asks 6 people, "How much do you like 'Quickly Come Financing'?", and notes down their responses. He then asks another 6 people the same question, and asks them to give their response as a mark out of 10. His results are shown on the right.

> 'It's alright.'   'It's brilliant!'
>   'OMG, I love it! TJ's so hot!'
>   'It's really annoying, I hate it.'
>   'I shouldn't like it, but I do.'
> 'I liked the first episode, but after that it all just seemed really fake. I don't like it any more.'

a) (i) Does his first set of results consist of qualitative, discrete quantitative or continuous quantitative data?

      (ii) Does his second set of results consist of qualitative, discrete quantitative or continuous quantitative data?

b) Suggest one advantage and one disadvantage of collecting qualitative data instead of quantitative data.

> 2   4   10   8   7   9

31.2 Data Collection

As part of an investigation, you're probably going to collect lots of data, so you need a way of recording and organising it all. Your method of collection should fit the type of data you're trying to collect — certain methods will fit particular data types better.

Data Collection Sheets

Learning Objective — Spec Ref S1:
Design and use data collection sheets.

Tally charts and **frequency tables** can be used to record qualitative or quantitative data — the only thing that changes is the name of the **categories** or **groups**.

- For **qualitative** data, the **names** of each category are used (e.g. if you're asking about favourite colours then the categories might be 'red', 'green, 'blue', ...). It's often useful to have an '**Other**' category so that all the possible options are included without having to list every possible category.

- For **discrete quantitative** data, you can use **individual** values (e.g. 0, 1, 2) if the range is small or **groups** of values if the range is large (e.g. 0-4, 5-9, 10-14). However, you lose some **accuracy** by grouping the data values as you no longer know the actual data values.

- For **continuous quantitative** data, you'll have to group the data — write the groups using **inequalities**.

Questionnaires are especially useful for collecting qualitative data. You can collect large amounts of data through different questions and allow respondents to give more **descriptive responses** to certain questions.

Example 1

An assault course is designed to take about 10 minutes to complete.

a) Design a tally chart that could be used to record the times taken by 100 people.

1. The data is continuous, so write the classes using inequalities.

2. Make use of both the < and ≤ symbols to ensure there are no gaps between classes and that classes don't overlap.

3. Leave the last class open-ended to make sure all possible times are covered.

Time (t mins)	Tally	Frequency
$0 < t \le 5$		
$5 < t \le 10$		
$10 < t \le 15$		
$15 < t \le 20$		
$20 < t$		

Tally column with plenty of space to record the marks

Frequency column for adding up the tally marks

b) Design another tally chart that could be used to record the number of times that the participants have previously attempted the assault course.

1. This time, the data is discrete.

2. Make sure you have clear boundaries between groups — e.g. if you had 1-2 and 2-3 it would be unclear which row would be used for someone who has visited twice previously.

3. Keep the last class open-ended.

No. of attempts	Tally	Frequency
0		
1-2		
3-6		
7 or more		

The groups don't need to be the same width — if certain groups will require closer analysis then make the widths of these smaller.

Q1 Design a tally chart that could be used to record the answers to each of these questions.

a) How many siblings do you have?   b) What transport do you use to get to work?

c) What's your favourite type of fruit?   d) How many days were in the month you were born?

Q2 The ages of 49 guests at a christening are on the right.

a) Copy and complete this tally chart, using groups of 10 years for the rest of the table.

Age (years)	Tally	Frequency
1-9		
10-19		

b) Use your table to find which of these age groups is the most common.

30	56	4	18	19	35	65
79	54	54	45	32	36	39
26	27	1	3	51	56	19
23	9	11	23	45	41	48
23	32	39	43	54	5	77
61	62	78	39	47	52	56
1	2	80	21	48	54	29

Q3 State two things wrong with each of these tally charts and design an improved version of each.

a) Chart for recording the number of times people went to the cinema last year.

No. of cinema trips	Tally	Frequency
1 – 10		
10 – 20		
20 – 30		
30 – 40		
40 or more		

b) Chart for recording the number of people watching a band play at each venue on their tour (max. venue capacity = 25 000).

No. of people	Tally	Frequency
0 – 5000		
6000 – 10 000		
11 000 – 15 000		
16 000 – 20 000		

c) Chart for recording the length of time people can hop on one leg for.

Time (t mins)	Tally	Frequency
$t \leq 5$		
$t \geq 5$		

d) Chart for recording weights of pumpkins.

Weight (w kg)	Tally	Frequency
$w < 3$		
$3 < w < 3.5$		
$3.5 < w < 4$		

Q4 Design a tally chart that could be used to record the following data.

a) The numbers of pairs of socks owned by 50 students.

b) The hand spans of 20 people.

c) The distances that 30 people travel to get to work and back each day.

Q5 Here are questions from Amber and Jay's questionnaires about how much sport people play. Say whose question is better and give two reasons why.

Amber: | How much sport do you play? |

Jay: | How many times a week do you play sport on average? Tick one box.

None ☐   One ☐   Two ☐   Three ☐   More than three ☐

Two-Way Tables

A **two-way table** is a data collection sheet that allows you to record **two** different pieces of information about the **same subject** at once — e.g. for each person, you could collect data on their gender as well as hair colour in one table. One piece of information is covered by the **rows** of the table and the other by the **columns**.

Example 2

Raymond is investigating how fast the players at a tennis club can serve.
He's interested in whether being right-handed or left-handed has any effect on average speed.
Design a two-way table he could use to record the data he needs.

One variable goes down the side and the other goes across the top

Use inequalities for continuous data

	Serve average speed (s mph)				
	$s < 40$	$40 \leq s < 50$	$50 \leq s < 60$	$60 \leq s < 70$	$70 \leq s$
Right-handed					
Left-handed					

Make sure there are no gaps and classes don't overlap

Rows and columns should cover every possible answer

Space for tally marks or frequencies

Exercise 2

Q1 The table below has been designed to record the hair colour and age (in whole years) of 100 adults.
   a) Give three criticisms of the table.   b) Design an improved version of the table.

	Age in whole years					
	0 – 15	15 – 30	30 – 45	45 – 60	60 – 75	75 or older
Blonde						
Light brown						
Dark brown						

Q2 For each of the following, design a two-way table that could be used to record the data described.
   a) The type of music adults and children prefer listening to out of pop, classical and rock.
   b) The average length of time spent doing homework each evening by pupils in each of the school years 7-11. Assume no one spends more than an average of 4 hours an evening on homework.

Q3 For each investigation below, design a two-way table for recording the data.
   a) Hoi Wan is going to ask 50 adults if they prefer cats or dogs.
      She wants to find out if it's true that men prefer dogs and women prefer cats.
   b) Nathan wants to find out whether children watch more TV on average each day than adults.
   c) Chloe is going to ask people how tall they are and how many portions of fruit they eat on average each day.

31.3 Sampling and Bias

It might not be practical or possible to collect data on every individual that you're interested in. If you wanted to know about the wingspan of all the birds in a forest, you wouldn't be able to find and measure every one. Instead, you'll need to collect data from a smaller group — but you have to pick it carefully.

Populations and Samples

Learning Objectives — Spec Ref S1:
- Know the difference between populations and samples.
- Understand the reasons for using a population and for using a sample.

When you're collecting data, the **whole group** of people or things you want to find out about is called the **population**. But populations can be very big and so it is often **impractical** to collect data from **every member**. Usually, it's **quicker**, **cheaper** and **easier** to collect data from a **sample** of the population, rather than the whole thing. A sample is a **smaller group** taken from the population.

Different samples will give different results — the **bigger** the sample, the more reliable it should be. It's also important that your sample **fairly represents** the population so you can apply any **conclusions** to the whole population.

Example 1

Michael and Tina have written questionnaires to find out what students at their college think about public transport. Michael gives the questionnaire to 10 students, and Tina gives the questionnaire to 50 students. Michael concludes that 50% of students are happy with public transport, and Tina concludes that 30% are happy.

a) **Suggest two reasons why Michael and Tina only gave the questionnaire to some of the students.**

Think about the advantages of sampling, for example...

There are fewer copies to print, so **print costs will be lower**.
Also, it will take them **much less time** to collect and analyse the results.

b) **Whose results are likely to be more reliable? Explain your answer.**

Think about the size of the sample...

Tina's results are likely to be more reliable because she has used a **bigger sample**.

Exercise 1

Q1 Jenny wants to know how long it takes the pupils in Year 7 to type out a poem. She plans to time a sample of 30 out of the 216 Year 7 pupils. Give two advantages for Jenny of using a sample.

Q2 Jack, Nikhil and Daisy bake a batch of 200 cupcakes to sell on their market stall. Jack thinks they should taste one cake to check the quality is OK. Nikhil thinks they should taste 10 cakes and Daisy thinks they should taste 50 cakes. Say who you agree with and explain why.

Q3 Melissa and Karen are doing an experiment to see if a coin is fair. They each toss the same coin 100 times and record the number of heads. 52% of Melissa's tosses are heads and 47% of Karen's tosses are heads. Whose result is more reliable? Explain your answer.

Q4   A supermarket chain wants to know what people in a town think of their plan to build a new supermarket there. They've hired a team of researchers to interview a sample of 500 people. Give one reason why they wouldn't want to interview everyone in the town.

Q5   Alfie and Lisa want to find out what people's favourite flavour of ice cream is.
Alfie asks 30 people and finds that most people answer 'chocolate'.
Lisa asks 15 people and finds that most people answer 'strawberry'.
Based on this information, what would you say is the most popular flavour?

Bias and Fairness

Learning Objectives — Spec Ref S1:
- Consider the bias or fairness of a sample.
- Know how to take a random sample.

When you're choosing a sample, it's important to make sure that your selection process is **fair** so that results are more likely to reflect the **whole population**. A sample is **biased** if some members of the population are **more likely** to be included than others. To spot bias, think about **when**, **where** and **how** the sample was taken, and **how many** members are in it — a **small** sample is also likely to be biased.

A **fair** way to select a sample is at **random** — every member of the population has an **equal chance** of being selected. To take a random sample:

- Make or get hold of a **list** of **every member** of the population and assign everyone a **number**.
- Use a computer, calculator or random number generator to create a list of **random numbers**.
- **Match** the random numbers to the assigned numbers of the population to **create the sample**.

Example 2

Adam wants to choose 20 of the 89 people in a sports centre to fill in a questionnaire about their favourite activity at the centre. He goes over to the tennis court and hands the questionnaire to 20 people. Explain whether his sample is fair or biased.

Think about whether some people will have more chance of being included than others.

The sample is **likely** to include more people who like tennis so it will be **biased** in favour of these people.

Tip: When deciding if a sample is biased, see if you can think of any groups that would be excluded.

Exercise 2

Q1   Explain why the following methods of selecting a sample will each result in a biased sample.

a) A library needs to reduce its opening hours. The librarian asks 20 people on a Monday morning whether the library should close on a Monday or a Friday.

b) At a large company, 72% of the employees are women and 28% are men. The manager wants to know what employees think about changing the uniform, so he surveys 20 women and 20 men.

c) A market research company wants to find out about people's working hours. They select 100 home telephone numbers and call them at 2 pm one afternoon.

Q2   The manager of a health club wants to survey a random sample of 40 female members. Explain how she could do this.

Q3   George and Stuart want to find out which football team is most popular at their school. George asks a random selection of 30 pupils and Stuart asks the first 5 pupils he sees at lunchtime. Give two reasons why George's sample is likely to be more representative of the whole school.

Q4   Seema wants to find out about the religious beliefs of people in her street, so she stands outside her house on a Sunday morning and interviews the first 20 people she sees. Explain all the reasons why her sample is likely to be biased and suggest a better method for selecting her sample.

Q5   A group of 300 people are going on a day trip. The organiser wishes to do a survey to help him decide what should be for lunch. 255 of the group are vegetarians. Do you think it would be fair to include the same number of vegetarians as those who are not vegetarians in the sample?

Q6   A water company wants to assess the condition of its underground water pipes along a 200 m stretch of pipe. They decide to take a random sample of five 1 m sections of the pipe. Explain why random sampling might not be the most appropriate method in this case.

Using a Sample to Make Estimates About the Population

Learning Objective — Spec Ref S1:
Know how to infer properties of a population using a sample.

Once you've collected a **fair sample**, you can use it to make **estimates** about the general **population**. E.g. if a **third of the people in the sample** have a certain property, you can estimate that a **third of the population** will have this property.

When you calculate your estimates, you might get a **decimal** answer. If so, you'll need **round** your answer to the nearest **whole number**.

Example 3

There are 10 000 residents in Sheila's town. She wants to find out how satisfied they are with the street furniture in the town so she fairly samples 1000 residents. The results are in the table.

a) **Estimate how many residents in the town are either quite satisfied or very satisfied with the street furniture.**

b) **Could you use this sample to estimate how many people in the country are satisfied with the street furniture in their town?**

Opinion	No. of residents
Very dissatisfied	250
Quite dissatisfied	150
Don't care	216
Quite satisfied	204
Very satisfied	180

a) 1.   Work out the proportion of satisfied residents from the sample.

There are $180 + 204 = 384$ residents satisfied in the sample, so the proportion of residents satisfied in the sample is $\frac{384}{1000} = 0.384$.

2.   Your estimate is the same proportion in the population.

The number of people satisfied in the town is estimated to be $0.384 \times 10\,000 = \textbf{3840}$.

b) Think about whether the sample is still fair or representative.

The sample is **not representative** of the country — it is **biased** in favour of people in one particular town. So you **can't use it** to estimate the opinion of the country.

You can also use sampling to estimate the **size** of a **constant** population.
Call the size of the population N and use this method to find it:

- Take a **sample** of size n from the population and **mark or tag** each one.
  Then **release** the sample back into the population.

- After the sample has **mixed back** into the population,
  collect a **new sample**. It doesn't have to be of the same
  size as the first sample, so let's say it's of size M.

- **Count** the number of **tagged subjects** in the new sample. Let's say
  there are m of them. Assuming that the **proportion** of tagged subjects
  in the new sample $\left(\frac{m}{M}\right)$ is the same as in the population $\left(\frac{n}{N}\right)$, you can
  rearrange $\frac{n}{N} = \frac{m}{M}$ to find N, the size of the population.

Tip: This method can
only be used if the
population doesn't
change between the two
samples being taken.
You also have to assume
that both samples are
perfectly representative
of the population.

Example 4

One morning, a fisherman catches 50 fish from a lake. He places small tags on them and
then releases them back into the lake. That afternoon, he catches 60 fish and finds that 10 of
them are tagged. Make an estimate of the number of fish (N) in the lake.

1. Write down the relevant values (n, M and m) from the question.

   $n = 50$, $M = 60$, $m = 10$

2. Equate the proportion of tagged fish in the population and
   the proportion of tagged fish in the afternoon sample.

   $\dfrac{50}{N} = \dfrac{10}{60}$

3. Rearrange the equation to find N, the number of fish in the lake.

   $N = \dfrac{50 \times 60}{10} = \mathbf{300}$

Exercise 3

Q1 A castle has 450 rooms. The king of the castle believes that many of them have damp
and so hires a dirty rascal to remove it. The king doesn't have time to inspect all the
rooms for damp, so he chooses 20 rooms at random to inspect. Given that seven of these
rooms have damp, estimate how many of the rooms in the entire castle have damp.

Q2 A journalist interviews a sample of 50 out of 650 politicians
to get their views on three policies X, Y and Z.
The results are in the table on the right.

	X	Y	Z
In Favour	34	4	13
Against	13	45	14
Neutral	3	1	23

a) Use the journalist's data to estimate
how many of the 650 politicians:

   (i) are in favour of policy X,   (ii) are neutral about policy Y,   (iii) are against policy Z.

b) What assumption has been made in using this data to make these estimations?

Q3 Laurence aims to estimate the number of newts in his favourite swamp. He finds 110 newts,
marks them harmlessly, then sends them back into the swamp. The next day, Laurence returns to the
swamp and finds 130 newts, of which 44 are marked. Estimate the number of newts in the swamp.

Q4 A conservationist wants to find out how many northern hairy-nosed wombats are living in a forest.
She carefully catches 20 of the wombats and attaches a small band to one of the legs of each.
She then releases them back into the forest. One week later, she returns to the
forest and catches 23 wombats. Three of them have bands on their legs.

a) Estimate the size of the wombat population in the forest.

b) Give two assumptions that you have made.

Review Exercise

Q1 Gemma thinks there is a link between the average number of chocolate bars eaten each week by pupils in her class and how fast they can run 100 metres.

    a) Describe two sets of data Gemma should collect to investigate this link.

    b) Describe suitable methods for collecting the data.

    c) Say whether each set of data is qualitative, discrete quantitative or continuous quantitative.

    d) Say whether each set of data is primary data or secondary data.

Q2 Write down five classes that could form a tally chart for recording each set of data below.

    a) The number of quiz questions, out of a total of 20, answered correctly by some quiz teams.

    b) The weights of 30 bags of apples, where each bag should weigh roughly 200 g.

    c) The volume of tea in 50 cups of tea as they're served in a cafe. Each cup can hold 300 ml.

Q3 For each investigation below, design a data collection sheet to record the data.

    a) Camilla plans to count the number of people in each of 50 cars driving into a car park.

    b) Theo is going to ask 100 people to rate his new hairstyle from 1 to 5.
He's interested in whether children give him higher scores than adults.

    c) Greg is investigating how much pocket money everyone in his class gets.

    d) Simran wants to test her theory that people who play a lot of sport don't watch much TV.

Q4 a) The town council wants to know people's opinions on some new wind turbines.
They decide to take a random sample of 200 of the 6000 residents.
Explain why they may have chosen to take a sample.

    b) An anti-turbine protest group also wants to know people's opinions on the turbines.
They survey four of their friends. Give two reasons why this sample will not be fair.

Q5 A factory makes hundreds of the same component each day.
Describe how a representative sample of 50 components could be
tested each day to make sure the machinery is working properly.

Q6 Eleanor is baking cupcakes for the 75 guests at a wedding. She asks 15 of the guests
what type of cake is their favourite and 6 say it is red velvet. Based on this sample, how
many of the guests at the wedding do you expect would say red velvet is their favourite?

Q7 A shepherd has a very large flock of sheep which he wants to determine the size of.
He herds 20 of the sheep into a pen and places a patch of dye on their wool. He releases
them and returns two days later. He then herds 40 sheep into the pen and finds that 3 of
them have the dye on their wool. Use this information to estimate the size of the flock.

Q1 Christine has the following hypothesis:

"You're more likely to be involved in a car accident on a rainy day."

Christine tests her hypothesis by collecting data on the rainfall in millimetres on 10 consecutive days in her town and the number of reported car accidents in her town on those days.

a) (i) Is the data she collects on the rainfall likely to be discrete or continuous?

   (ii) Is the data she collects on the number of car accidents likely to be primary or secondary?

   [2 marks]

b) It is generally agreed that Christine's hypothesis is true in her town. Comment on whether Christine can use her true hypothesis to draw conclusions about car accidents across the whole country.

   [2 marks]

Q2 Maurice is investigating the number of different coloured cars at a dealership. The dealership has 3000 cars on display. Maurice randomly selects 50 of the cars and takes note of their colour. The results are displayed in the table.

Colour	Black	White	Silver	Blue	Red	Yellow	Other
Number	11	14	12	6	5	0	2

a) (i) State the size of the population that Maurice is interested in.

   (ii) Explain why Maurice chose to use a sample to conduct his investigation.

   [2 marks]

b) Use Maurice's sample to estimate the number of silver cars at the dealership.

   [2 marks]

c) Based on his sample, Maurice claims there are no yellow cars at the dealership. Do you agree? Justify your answer.

   [1 mark]

Q3 Kyros attempts to estimate the number of birds in a bird reserve. Kyros captures 150 birds, attaches a small tag to them and then releases them back into the reserve.

a) Explain why Kyros's estimate might be unreliable if he returns and takes a second sample 30 minutes later.

   [1 mark]

b) Explain why Kyros's estimate might be unreliable if he returns and takes a second sample consisting of 2 birds.

   [1 mark]

c) Kyros returned the next day and captured b birds, 6 of which were tagged. He estimated that there must be 5000 birds in the reserve. Find the value of b.

   [2 marks]

32.1 Averages and Ranges

Averages and ranges are useful ways of summarising a data set — by looking at them you can get a good idea of the data set as a whole without needing to see every piece of data within it.

> **Learning Objectives — Spec Ref S4:**
> - Find the mode, median, mean and range from raw data.
> - Find the mode, median, mean and range from a frequency table.

An **average** is a way of **representing** a whole set of data using a **single value** — it is the **central** or **typical** value within the data set. The **mode**, **median** and **mean** are three common averages that are used.

- The **mode** (or modal value) is the **most common value** — the value that appears in the data set **more often** than any other. Find the mode by counting **how many times** each value appears in the data set.

- The **median** is the **middle value** — the value that is in the **middle** of the data set when all the numbers are put in **ascending order**. For a data set containing n pieces of data, the median is the **value** in position $(n + 1) \div 2$. If n is an **even number**, the median will be **halfway** between two values.

- The **mean** is the **total** of all the values divided by the **number** of values. To find the mean, **add up** all the values in the data set and **divide** by the number of values.

The **range** is the difference between the **largest value** and the **smallest value** in a data set — it tells you how **spread out** the values are within a data set. Data sets with a **small range** are more **consistent** than those with a **large range** — this means there is less variation in the values.

Averages and ranges are useful because they allow us to **compare** one set of data to another.

Example 1

20 customers each gave a restaurant a mark out of 10. Their marks are shown on the right.

3	7	4	8	3	7	5	2	8	9
9	6	1	3	4	5	6	5	7	7

Find: a) the modal mark   b) the median mark
c) the mean mark   d) the range

1. Put the numbers in order first.

From smallest to largest, the numbers are:
1, 2, 3, 3, 3, 4, 4, 5, 5, 5, 6, 6, 7, 7, 7, 7, 8, 8, 9, 9

2. The mode is the most common number.

a) 4 customers gave a mark of 7, so the mode is **7**.

3. There are an even number of values, so the median will be halfway between the two middle numbers.

b) There are 20 values, so the median is in position $(20 + 1) \div 2 = 10.5$, which is halfway between the 10th value = 5 and 11th value = 6.
So the median = $\frac{5+6}{2}$ = **5.5**

4. Divide the total of the values by the number of values to find the mean.

c) Total = $1 + 2 + 3 + 3 + 3 + 4 + 4 + 5 + 5 + 5 + 6$
$+ 6 + 7 + 7 + 7 + 7 + 8 + 8 + 9 + 9 = 109$
Mean = $109 \div 20 = $ **5.45**

5. Subtract the smallest value from the largest.

d) Range = $9 - 1 = $ **8**

Q1 For the following sets of data: (i) Find the mode.   (ii) Find the range.

   a) 6, 9, 2, 7, 7, 6, 5, 9, 6   b) 16, 8, 12, 13, 13, 8, 8, 17   c) 8.2, 8.1, 8.1, 8.2, 8.1, 8.2, 8.2

Q2 Find the median of the following sets of data.

   a) 3, 3, 3, 3, 3, 3, 4, 4, 4, 4, 4   b) 2, 4, 7, 1, 5, 9, 2, 7, 8, 0   c) 5.85, 6.96, 2.04, 7.45, 6.9, 7.8

Q3 The times (to the nearest second) of athletes running the 400 m hurdles are:
78, 78, 84, 81, 90, 79, 84, 78, 95

   a) Find the range of times taken to run the 400 m hurdles.   b) Find the median time.

Q4 Nine students score the following marks on a test: 34, 67, 86, 58, 51, 52, 71, 65, 58
Find the mean score. Give your answer to 3 significant figures.

Q5 Abdul counts the number of crisps in 28 packs. His results are shown below.

   a) Find the modal number of crisps in a pack.

   b) Calculate: (i) the range of the data,

         (ii) the median number of crisps in a pack

         (iii) the mean number of crisps in a pack to 3 s.f.

12	20	21	15	18	20	21
20	15	9	22	16	18	19
20	18	13	15	18	20	17
16	15	16	16	18	21	20

Q6 Look at this data set: 6, 5, 8, 8, 5, ?

   a) If the range was 6, what are the two possible values for the missing number?

   b) If the mean was 7, find the missing value.

(PROBLEM SOLVING)

Q7 The heights (in metres) of a class of students
are given on the right. Which is greater,
the mean or the median height of the students?

1.68	1.45	1.70	1.30	1.72
1.80	1.29	1.40	1.42	1.60
1.65	1.75	1.67	1.69	1.72
1.72	1.63	1.63	1.78	1.70
1.50	1.65	1.40	1.36	1.69

Large data sets can be put into **frequency tables** to make them **more manageable**. The frequency
is the **number of times** a particular value appears within a data set. When **calculating averages**
from frequency tables you treat the data the **same way** as if it had all been written out normally.

Example 2

**This frequency table shows the number of
mobile phones owned by a group of people.**

Number of mobile phones	0	1	2	3	4
Frequency	4	8	5	2	1

a) Find the modal number of mobile phones.

Find the number of mobile phones
with the highest frequency.

Mode = **1 mobile phone**

Tip: The raw data is:
0, 0, 0, 0, 1, 1, 1, 1, 1, 1,
1, 1, 2, 2, 2, 2, 2, 3, 3, 4.

b) What is the median number of mobile phones?

1. Work out the total frequency
   and the position of the median.

Total frequency = 4 + 8 + 5 + 2 + 1 = 20
The median is the position (20 + 1) ÷ 2 = 10.5

2. The data values in the 10th
   and 11th positions are both 1.

It's halfway between the 10th value = 1 and
11th value = 1, so the median = $\frac{1+1}{2}$ = **1 mobile phone**

c) **Find the range for this data.**

Subtract the smallest value from the largest value.

Range = 4 − 0 = **4**

d) **Find the mean number of mobile phones owned by a person.**

1. Work out the total number of mobile phones by multiplying the number of mobile phones by the frequency of each column.

Number of mobile phones	0	1	2	3	4
Frequency	4	8	5	2	1
Phones × frequency	0	8	10	6	4

2. Divide the total number of mobile phones by the total number of people.

Total number of phones = 0 + 8 + 10 + 6 + 4 = **28**

Mean = 28 ÷ 20 = **1.4 mobile phones**

Exercise 2

Q1 The table shows the number of people living in each of 30 houses.

a) Write down the modal number of people per house.

b) Find the median number of people per house.

c) Calculate the mean number of people per house.

d) Work out the range of the data.

Number of people	Frequency
1	4
2	5
3	8
4	10
5	3

Q2 This table shows the number of goals scored one week by 18 teams in the premier division.

a) Write down the modal number of goals.

b) Find the mean number of goals. Give your answer to 3 significant figures.

Number of goals	0	1	2	3	4	5
Number of teams	1	3	4	5	3	2

c) The mean number of goals scored in the same week last year was 2.4. How do these results compare?

Q3 A student wrote down the temperature in his garden in Aberdeen every day at noon during the summer. His results are shown in the table.

a) Find the median noon temperature.

b) Find the mean temperature.

c) The mean midday temperature during the summer in the UK is approximately 18.5 °C. What does this suggest about the temperature in the student's garden?

Temperature (°C)	Frequency
16	18
17	27
18	7
19	15
20	12
21	11

Q4 A survey asked 200 people, 'How many televisions do you own?'. The results are shown in this table.

No. of Televisions	1	2	3	4	5
Frequency	77	p	q	11	3

a) Show that $p + q = 109$.

b) The mean number of televisions is 1.88. Show that $2p + 3q = 240$.

c) Find the number of people who own 2 televisions and 3 televisions.

Measures of Spread — Interquartile Range

Learning Objective — Spec Ref S4:
Find the upper and lower quartiles and the interquartile range for a data set.

The **lower quartile**, the **median** and the **upper quartile** are values that **divide** a data set into **four groups** of **equal size**. The quartiles are positioned **25%**, **50%** and **75%** of the way through the data. For a data set in **ascending order** with n values, you can work out the position of the quartiles using these formulas:

- Lower quartile ($\mathbf{Q_1}$) — $Q_1 = (n + 1) \div 4$

- Median ($\mathbf{Q_2}$) — $Q_2 = (n + 1) \div 2$

- Upper quartile ($\mathbf{Q_3}$) — $Q_3 = 3(n + 1) \div 4$

The lower and upper quartiles can be used to work out the **interquartile range**. This gives you the range of the **middle 50%** of data and can be used to measure how **dispersed** (spread out) the data is.

To find the interquartile range, **subtract** the lower quartile (Q_1) from the upper quartile (Q_3): $IQR = Q_3 - Q_1$

Example 3

The data on the right shows the goals scored by Team A at each hockey game in a season.

Game	1	2	3	4	5	6	7	8	9
Goals scored	2	4	0	3	4	2	2	3	2

a) Calculate the interquartile range for this data.

1. Put the data in ascending order.

   0 2 2 2 2 3 3 4 4

2. Work out the positions of the lower and upper quartiles.

   Q_1 position $= (9 + 1) \div 4 = 2.5$
   Q_3 position $= 3(9 + 1) \div 4 = 7.5$

3. Q_1 will be halfway between the 2nd and 3rd values. Q_3 will be halfway between the 7th and 8th values.

   2nd value = 2, 3rd value = 2, so $Q_1 = 2$
   7th value = 3, 8th value = 4,
   so $Q_3 = (3 + 4) \div 2 = 3.5$

4. Subtract to find the interquartile range.

   $IQR = Q_3 - Q_1 = 3.5 - 2 = \mathbf{1.5}$

b) Team B scores the same total number of goals as Team A, with an interquartile range of 0.5. Which team scored the most consistent number of goals per game?

The smaller the interquartile range, the less spread out and more consistent the data values are.

Team B — their interquartile range is much smaller than for Team A, so their scores were more consistent.

Exercise 3

Q1   Calculate the interquartile range for each of the following data sets.

   a) 8, 9, 9, 9, 10, 10, 12, 15, 16, 17, 19

   b) 80, 70, 34, 21, 21, 56, 75, 89, 84, 20, 17, 45, 87

   c) 1.5, 1.5, 1.3, 1.4, 1.6, 1.8, 1.2

   d) 1, 9, 3, 9, 3, 4, 5, 6, 9, 0, 1, 9, 9, 5, 9, 2, 5

Q2   The table shows the number of spots on 79 ladybirds. Find the interquartile range for the data.

No. of spots	2	3	4	5	6	7	8	9	10
Frequency	3	16	9	18	9	6	9	7	2

Q3 The list on the right shows the exam results for the students in class 3A.

a) Only students with a mark in the upper quartile of the class results passed the exam. What was the pass mark for this exam?

b) Calculate how many people passed the exam.

c) (i) Find the interquartile range for this data.

(ii) Class 3B's results for the same exam have an interquartile range of 20. Comment on how the spread of the results for the two classes differs.

30	36	89	92	76
20	57	89	23	55
56	56	98	35	20
86	38	24	13	90
54	67	67	34	78
72	53	88	24	40
76	20	24		

Choosing the Right Average and Range

Learning Objectives — Spec Ref S4:
- Choose which measure of central tendency or spread to use.
- Understand the advantages and disadvantages of each one.

Averages are a good way to **summarise** and **represent** data, but using the **wrong average** can lead to data being **misinterpreted**. The mode, median and mean averages all have different **advantages** and **disadvantages**, which means that some are better suited to certain data sets than others. For example, some averages are more affected by extreme values (known as **outliers**) than others.

	Advantages	Disadvantages
Mode	It's easy to find. It **doesn't** get distorted by **outliers**. Can be used for **non-numeric** data. It's **always** a value in the data set.	It **doesn't** always exist, or there are **several modes**. Not always a good representation of the data — it **doesn't use all the data values**.
Median	It's easy to find for **ungrouped** data. It **doesn't get distorted** by outliers.	Not always a good representation of the data — it **doesn't use all the data values**. It **isn't always** a value in the data set.
Mean	It is usually the **most representative** average — it uses **all** the data values.	It **can be distorted** by outliers. It **isn't always** a value in the data set.

Exercise 4

Q1 The clothes sizes of 20 women are shown on the left.

a) (i) Find the mode.

(ii) Find the median clothes size.

(iii) Calculate the mean clothes size.

18	4	10	10	16
12	14	14	6	18
14	12	12	8	16
14	12	12	14	14

b) Suggest why the mode might represent this data better than the mean or median.

Q2   The ages of the 30 audience members at a McBeetle concert are shown on the right.

60	12	17	60	14	16	29	15	12	17
60	60	16	17	16	14	13	14	10	16
14	17	19	17	18	19	60	16	19	16

   a) (i) Find the modal age.

      (ii) Work out the median age.

      (iii) Calculate the mean age. Give your answer to 3 significant figures.

   b) A slightly embarrassed 29-year-old says, 'The mean age is the most representative of the ages of the people at the concert.' Do you agree? Explain your answer.

Q3   A company asked 190 people to test their latest wrinkle cream and give it a mark out of 10. The table below shows their results.

Mark	1	2	3	4	5	6	7	8	9	10
Frequency	31	34	35	34	4	6	36	7	2	1

   a) Find the mode, median and mean for the data.
      Give your answers to 3 significant figures where necessary.

   b) (i) The company claims, 'On average, people gave our product 7 out of 10'.
           Which average has it used in its claim?

      (ii) Do you think this average represents the data well? Explain your answer.

The **range** and **interquartile range** also have advantages and disadvantages, but the interquartile range is generally **more reliable** as it isn't affected by **outliers**.

	Advantages	Disadvantages
Range	It's easy to find.	It **can be distorted** by outliers.
Interquartile range	It **doesn't get distorted** by outliers.	Not always a good representation of the data — it **doesn't use all the data values**.

Exercise 5

Q1   Suggest whether the range is a good measure of spread for the following data sets and explain why.

   a) 1, 2, 3, 4, 5, 6, 7, 8, 9, 10, 11   b) 14, 14, 20, 22, 24, 27, 29, 30, 35, 93

   c) 2222, 750, 725, 55, 800, 783, 880   d) 1.8, 1.7, 2.0, 1.0, 1.2, 1.6, 1.5

Q2   Look at the data on the right.

13	28	893	192	89	98	67	45	78
90	34	56	78	20	783	60	33	45
12	67	54	58	101	56	89	708	9
82	76	90	55	89	347	104	43	76
38	13	27	51	92	18	44	87	71

   a) (i) Calculate the range for the data.

      (ii) Calculate the interquartile range.

   b) Which of the measures of spread you calculated in part a) do you think best represents the data? Explain your answer.

32.2 Averages for Grouped Data

Data in grouped frequency tables have to be treated differently as you don't know the actual data values.

> **Learning Objective — Spec Ref S4:**
> Find the mode, median, mean and range for data in a grouped frequency table.

If you're given a data set in the form of a **grouped frequency table** then you don't know the **exact** data values — so you can't find exact values for the averages or range. You can only **identify** the **modal group** and **group containing the median**, and **estimate** the **mean** and the **range**.

- The **modal group** is the group (sometimes called a **class**) that has the **highest frequency**.

- The **group containing the median** is found by working out the **position** of the median in the usual way. You then use the **group frequencies** to identify which group the median falls into.

- The **estimated mean** is found by **multiplying** the **frequencies** by the **midpoints** for each group, **adding** up the results and **dividing** by the **total frequency**. The midpoints of each group are used as **estimate** for all the data values within a group. To find the **midpoint** of each group, add the **lower** and **upper** bounds together and **divide by 2**.

- The **estimated range** is found by **subtracting** the **lower bound** of the smallest group from the **upper bound** of the largest group — this gives you the **largest possible range** for the data set.

Example 1

The table shows a summary of the times taken by 15 people to eat three crackers.

Time (t) in s	$50 \le t < 60$	$60 \le t < 70$	$70 \le t < 80$	$80 \le t < 90$
Frequency	2	3	6	4

a) Write down the modal group.

The modal group is the one with the highest frequency.   Modal group is **$70 \le t < 80$**.

b) Which group contains the median?

Find the position of the median and which group it's in. The $2 + 3 + 1 = 6$th to $2 + 3 + 6 = 11$th values are all in the group $70 \le t < 80$.

There are 15 values, so the median is in the $(15 + 1) \div 2 = 8$th position. The group containing the median is **$70 \le t < 80$**.

c) Find an estimate for the mean.

1. Find the midpoint for each group — add together the upper and lower bounds and divide by 2.

Time (t) in s	$50 \le t < 60$	$60 \le t < 70$	$70 \le t < 80$	$80 \le t < 90$
Frequency	2	3	6	4
Midpoint	$(50 + 60) \div 2 = 55$	$(60 + 70) \div 2 = 65$	$(70 + 80) \div 2 = 75$	$(80 + 90) \div 2 = 85$

2. Now find the mean as before, using the midpoints instead of actual data values.

Midpoint × freq.	$2 \times 55 = 110$	$3 \times 65 = 195$	$6 \times 75 = 450$	$4 \times 85 = 340$

Total frequency $= 110 + 195 + 450 + 340 = 1095$, so the estimated mean $= \dfrac{1095}{15} = $ **73 seconds**.

d) Find an estimate for the range.

Subtract the lower bound of the smallest group from the upper bound of the largest group.

Estimated range $= 90 - 50 = $ **40 seconds**

Q1   The table on the right shows some information about the weights of some tangerines in a supermarket. Find an estimate for the mean tangerine weight.

Weight (w) in grams	Frequency
$0 \leq w < 20$	1
$20 \leq w < 40$	6
$40 \leq w < 60$	9
$60 \leq w < 80$	24

Q2   Troy collected some information about the number of hours students spent watching television over a week. His results are shown in the table.

Time (t) in hours	Frequency
$0 \leq t < 5$	3
$5 \leq t < 10$	8
$10 \leq t < 15$	11
$15 \leq t < 20$	4

     a) Write down the modal group.

     b) Which group contains the median?

     c) Find an estimate for the mean to 3 significant figures.

Q3   This table shows information about the heights of 200 people.

     a) Write down the modal group.

     b) Which group contains the median?

     c) Find an estimate for the mean to 3 significant figures.

     d) Find an estimate for the range.

Height (h) in metres	Frequency
$1.50 \leq h < 1.60$	27
$1.60 \leq h < 1.70$	92
$1.70 \leq h < 1.80$	63
$1.80 \leq h < 1.90$	18

Q4   The table below shows the times taken to deliver pizzas in one week.

Time (t) in minutes	$0 \leq t < 5$	$5 \leq t < 10$	$10 \leq t < 15$	$15 \leq t < 20$	$20 \leq t < 25$	$25 \leq t < 30$
Frequency	40	64	89	82	34	18

     a) Write down the modal group.

     b) Which group contains the median time taken to deliver a pizza?

     c) Estimate the mean time taken to deliver a pizza to 3 significant figures.

     d) Estimate the range of times taken to deliver a pizza.

     e) The pizza company guarantee to deliver your pizza in less than 15 minutes or your pizza is free. What percentage of the pizzas delivered that week were free? Give your answer to 3 s.f.

Q5   The table below shows the number of different species of bird seen by visitors to a nature reserve on one particular day.

No. of bird species (x)	$0 \leq x < 3$	$3 \leq x < 7$	$7 \leq x < 10$	$10 \leq x < 15$	$15 \leq x < 20$	$20 \leq x < 30$
Frequency	2	5	19	16	7	1

     a) Write down the modal group.

     b) Which group contains the median number of different species seen?

     c) Estimate the mean number of different species seen.

     d) Estimate the range of the number of different species seen.

     e) On the next day, the average number of different species seen was 11. How do these results compare to the previous day?

Review Exercise

Q1 A company asks 40 people to taste their new peanut butter and give it a mark out of 10. Their results are shown in the box.

a) (i) Find the modal score.

   (ii) Find the median score.

   (iii) Calculate the mean score.

   (iv) Calculate the range.

5	2	8	9	5	5	7	8	1	2
3	4	6	8	1	5	3	2	4	5
7	7	8	6	2	7	2	8	9	2
5	7	8	9	8	5	7	8	2	6

b) The company want to quote an average score in their next advertising campaign. Which average best represents the data? Explain your answer.

c) Give one reason why the company may choose to use a different average.

Q2 This table shows the shoe sizes of 30 school pupils.

Shoe size	1	2	3	4	5	6	7
Frequency	2	9	4	6	5	0	1

a) Find the mode, median and mean size, rounding to 3 significant figures where necessary.

b) Explain why the mean might not be the best average for this data set.

Q3 A biologist recorded the following weights for 9 hippos in kilograms:

195, 1525, 1340, 245, 1950, 1750, 1600, 1600, 1400

a) Calculate (i) the mode, (ii) the median and (iii) the mean to the nearest kilogram.

b) Calculate (i) the range and (ii) the interquartile range for the weights.

c) The biologist thinks she may have accidentally included the weights of several pygmy hippos in her list. How might this have affected the range and interquartile range?

Q4 A group of 17 pantomime horses were timed as they ran 100 m. The finishing times for the pantomime horses (in seconds) are shown on the right.

15.3	18.9	40.2	20.5	14.0	17.3
17.8	30.5	60.2	56.0	32.1	34.2
22.7	36.2	19.2	41.1	26.4	

17 kittens ran the same course. The mean time it took a kitten to run 100 m was 24.6 seconds. The interquartile range of the kittens' times was 48.6 seconds. Use this data to investigate the hypothesis 'Pantomime horses run faster than kittens'.

Q5 The table shows the number of potatoes some fish and chip shops go through a day.

a) Write down the modal group.

b) Which group contains the median?

c) Find an estimate for the mean.

d) Find an estimate for the range.

Number of potatoes (p)	Frequency
$500 \leq p < 600$	2
$600 \leq p < 700$	14
$700 \leq p < 800$	25
$800 \leq p < 1000$	9

Exam-Style Questions

Q1 Work out the median for the following numbers.

17 83 19 28 29 106

[1 mark]

Q2 Gertrude records the number of eggs laid by hens on her free range farm in a month. Calculate the mean number of eggs laid to 2 decimal places.

No. of eggs laid	24	25	26	27	28	29	30	31
Frequency	7	23	45	109	541	1894	3561	2670

[3 marks]

Q3 The masses of 25 marrows entered into a vegetable growing competition are shown below.

Mass (m kg)	$3.0 \le m < 4.0$	$4.0 \le m < 5.0$	$5.0 \le m < 6.0$	$6.0 \le m < 6.5$
Frequency	11	9	4	1

Using the table, work out the percentage of marrows that were above the mean mass. PROBLEM SOLVING

[3 marks]

Q4 The seven scores given to a competitor in a diving competition were:

7 7 8 8.5 8.5 8.5 9.5

Find the ratio of the interquartile range to the range giving your answer in the form $a : b$, where a and b are both integers.

[3 marks]

Q5 The table shows the number of motorcycles sold each day by a salesman over the last six days.

Number of motorcycles sold	Frequency
0	1
1	3
3	2

The salesman sold the same number of motorcycles each day for the four days prior to these six days. The mean number of motorcycles sold each day over the last ten days was 3.3. Work out the modal number of motorcycles sold each day over the last ten days, explaining why your answer is the mode. PROBLEM SOLVING

[3 marks]

Q6 Four integers have a median and mode of 7. The range of the integers is 10 and the mean is 9. Work out the four integers, giving them in ascending order. PROBLEM SOLVING

[3 marks]

33.1 Tables and Charts

There are many different ways of displaying data, and knowing the best way to display different types makes things much easier when it comes to interpreting it.

Two-Way Tables

Learning Objective — Spec Ref S2:
Display and interpret data in two-way tables.

Prior Knowledge Check:
Be able to write one number as a percentage of another. See p.59.

Two-way tables are used to show the frequencies for **two different variables** — e.g. the hair colour and eye colour of school pupils. The **rows** represent the **categories** for one variable (e.g. 'brown hair', 'blonde hair', etc.) and the **columns** represent the **categories** for the other variable (e.g. 'blue eyes', 'brown eyes', etc.). Each **cell** then shows the number of items in a particular row AND a particular column — e.g. the number of pupils with brown hair AND blue eyes. **Row** and **column totals** show the **total** number of items in each **category** and the bottom-right cell shows the **overall** total.

A complete two-way table can be used to work out the **proportions** of groups with **particular characteristics**. You can find proportions of the **whole group** (e.g. the percentage of pupils with blue eyes and brown hair in the whole school) or proportions within a **category** (e.g. the percentage of blue-eyed pupils that have brown hair). Be careful with which **totals** you use — you need the **overall total** if you're looking at the **whole group**, or the **row/column totals** if you're looking within a **category**.

Example 1

This table shows how students in a class travel to school.

	Walk	Bus	Car	Total
Boys	8	7		19
Girls	6		2	
Total		12		

a) **Complete the table.**

Add frequencies to find row/column totals. Subtract from row/column totals to find other frequencies.

	Walk	Bus	Car	Total
Boys	8	7	19 − 8 − 7 = **4**	19
Girls	6	12 − 7 = **5**	2	6 + 5 + 2 = **13**
Total	8 + 6 = **14**	12	4 + 2 = **6**	19 + 13 = **32**

b) **How many girls take the bus to school?**

Read off the value in the 'Girls' row and the 'Bus' column. → **5 girls** take the bus to school

c) **What percentage of students take the bus to school?**

Divide the total of the 'Bus' column by the overall total.

$$\frac{12}{32} \times 100 = \mathbf{37.5\%}$$

Exercise 1

Q1 This two-way table gives information about the colours of the vehicles in a car park.

a) Copy and complete the table.

b) How many motorbikes were blue?

c) What percentage of vehicles (to 3 s.f.) were:
(i) cars? (ii) vans? (iii) red?

	Red	Black	Blue	White	Total
Cars	8	7	4		22
Vans		2	1	10	
Motorbikes	2	1		2	
Total	12		6		

Q2 A group of schoolchildren were asked if they have been ice skating before. Copy the two-way table below, then use the following information to complete it.

- 20 girls were asked in total.
- Half as many boys as girls were asked.
- 18 of the children have been ice skating.
- One-fifth of the girls haven't been ice skating.

	Have been	Haven't been	Total
Boys			
Girls			
Total			

Q3 This two-way table shows the heights in centimetres of 50 people.

	$h < 160$	$160 \leq h < 170$	$170 \leq h < 180$	$180 \leq h < 190$	$190 \leq h$	Total
Women	4		11		0	24
Men		2	6		5	
Total	4	9				50

a) Copy and complete the table.

b) How many of the people are smaller than 170 cm?

c) What fraction of the women are at least 180 cm tall?

d) Comment on the difference between the heights of the men and the women.

Bar Charts

Bar charts show the **number** (or frequency) of items in **different categories**. They're used for **discrete data** (see p.406). Each bar represents a different category — the bars **shouldn't touch** because the categories are **distinct**.

Multiple sets of data can be displayed on the **same bar chart** — e.g. data for boys and girls or children and adults.

- **Dual bar charts** have two bars per category — one for each data set.

- **Composite bar charts** have single bars split into different sections for each data set.

You can easily read the **mode** (see Section 32) from a bar chart — it's the category with the greatest frequency, so it's shown by the **tallest bar**.

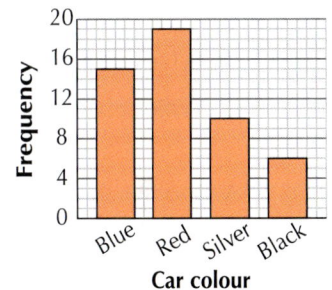

Car colour

Tip: The bars could be replaced by lines, and drawn either vertically or horizontally.

Example 2

Manpreet and Jack recorded how many TV programmes they watched each day for a week. Their results are shown in the table to the right. Draw a dual bar chart to display this information.

Day	M	T	W	T	F	S	S
No. watched by Manpreet	1	2	4	2	3	7	4
No. watched by Jack	2	1	0	2	3	4	0

1. Each day should have two bars — one for Manpreet and one for Jack.

2. The height of each bar is the frequency.

3. Use different colours for Manpreet and Jack's bars — and include a key to show which is which.

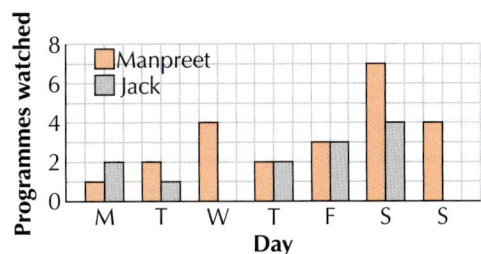

Example 3

The number of cars sold at two car showrooms over a week is shown in the table to the right. Draw a composite bar chart to display this information.

Day	M	T	W	T	F	S	S
No. sold at Showroom 1	3	1	4	2	2	6	3
No. sold at Showroom 2	2	1	1	0	2	5	1

1. Start with Showroom 1 — the height of the orange bar is the frequency for Showroom 1.

2. Add grey bars onto the top of the orange ones. The height of each grey bar should be the frequency for Showroom 2. Make sure you include a key to show which colour is for which showroom.

3. The total height of the composite bar (the orange and grey bars combined) is the total frequency for both showrooms.

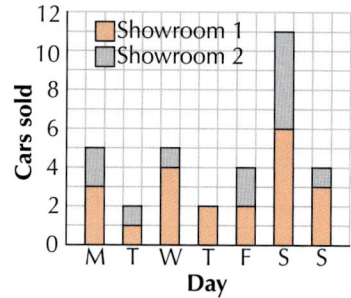

Exercise 2

Q1 The eye colour of 50 students is shown in the table.

Eye colour	Blue	Brown	Green	Other
No. of males	8	7	5	4
No. of females	7	12	5	2

a) Draw a dual bar chart to display the data.

b) Draw a composite bar chart to display the data.

c) Which chart is best for comparing males and females?

d) What is the modal eye colour?

e) Which chart was easier to use to find the mode?

Q2 A group of children and adults were asked to rate a music magazine on a scale from 1 to 5. This dual bar chart shows the results.

a) How many children gave the magazine a score of 1?

b) How many adults rated the magazine?

c) Which score did no children give?

d) Which score was given by twice as many children as adults?

e) Make one comment about the scores given by the adults compared to those given by the children.

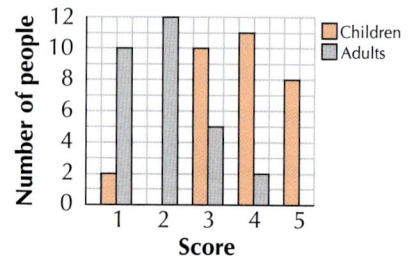

Q3 This composite bar chart shows the ages in whole years of the members of a small gym.

a) How many members are in the age range 26-35 years?

b) How many members does the gym have in total?

c) What is the modal age range for the female members?

d) How many more men aged 26-35 use the gym than women aged 26-35?

e) Which age range has the greatest difference in numbers of men and women?

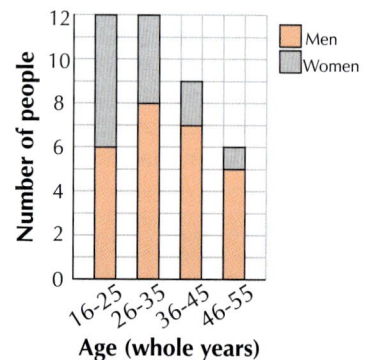

Pie Charts

Like bar charts, **pie charts** show how data is divided into categories, but pie charts show the **proportion** in each category, rather than the actual frequency. The sizes of the **angles** of the sectors are **proportional** to the **frequencies**.

To **draw** a pie chart, you need to find the **angle** that represents a **frequency of 1**. To do this, **divide** 360° (the full circle) by the **total frequency**, which you find by **adding up** the frequencies of each category. **Multiply** this value by the frequency of each category to find the **angle** of the **sector**, then use these angles to draw the pie chart.

Tip: The sum of the sector angles should add up to 360°, so use this to check your working.

Example 4

Jake asked everyone in his class to name their favourite colour. The frequency table on the right shows his results. Draw a pie chart to show his results.

Colour	Red	Green	Blue	Pink
Frequency	12	7	5	6

1. Calculate the total frequency — the total number of people in Jake's class.

   Total frequency = 12 + 7 + 5 + 6 = 30

2. Divide 360° by the total frequency to find the number of degrees needed to represent each person.

   Each person is represented by 360° ÷ 30 = 12°

3. Multiply each frequency by the number of degrees for one person to find each angle (check that the angles add up to 360°).

   Red: 12 × 12° = **144°**
   Green: 7 × 12° = **84°**
   Blue: 5 × 12° = **60°**
   Pink: 6 × 12° = **72°**

   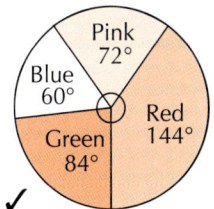

4. Draw the pie chart using the angles you've just calculated to mark out the sectors.

   Check:
   144° + 84° + 60° + 72° = 360° ✓

Exercise 3

Q1 Zofia asked a group of 90 people where they went on holiday last year. Her results are shown in the table. Draw a pie chart to show Zofia's data.

Destination	UK	Europe	USA	Other	Nowhere
Frequency	22	31	8	11	18

Q2 Vicky asked people entering a sports centre what activity they were going to do. 33 were going to play squash, 52 were going to use the gym, 21 were going swimming and 14 had come to play tennis. Draw a pie chart to show this data.

Q3 Pablo surveyed his friends to find out how they travel to school. The table shows his results. Draw a pie chart to represent Pablo's data.

Method of transport	Walking	Bus	Car	Bike
Frequency	36	16	12	16

Q4 Basil used a questionnaire to find out which subject pupils at his school enjoyed the most. His data is shown in the frequency table below. Show this data in a pie chart.

Subject	Maths	Art	PE	English	Science	Other
Frequency	348	297	195	87	108	45

Example 5

A head teacher carried out a survey to find out how pupils travel to school.
The pie chart on the right shows the results of the survey.

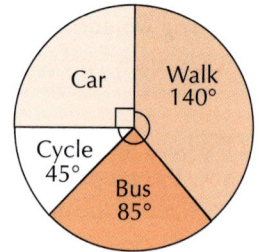

a) **What is the most popular way to travel to school?**

This is the sector with the largest angle.   **Walking** is the most popular.

b) **Which method of transport is twice as common as cycling?**

Cycling is represented by a sector with an
angle of 45°, so look for a sector with an
angle of 2 × 45° = 90° — travelling by car.   **Travelling by car** is twice as common as cycling

c) **280 pupils said they walk to school.**
   How many pupils took part in the survey altogether?

   1. Work out how many pupils
       are represented by 1°.

       140° represents 280 pupils
       So 1° represents 280 ÷ 140 = 2 pupils.

   2. Multiply by 360° to work out how many
       pupils the whole pie chart represents.

       This means 360° represents 360 × 2 = **720 pupils**

d) **Half the children at another school walk to school. Juan says, "A greater number of
   pupils at this other school walk to school." Is Juan correct? Explain your answer.**

   Juan is **not necessarily correct**. Without knowing **how many pupils** are at the other school you
   can't tell whether more or fewer pupils walk to school — you can **only compare the proportions**.

Exercise 4

Q1   Jennifer asked pupils in her school to name their favourite type of pizza.
     The pie chart on the right shows the results.

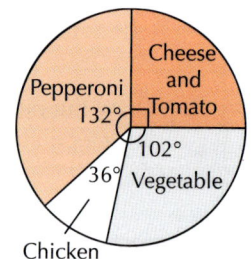

     a) Which was the most popular type of pizza?

     b) What fraction of the pupils said cheese and tomato was their favourite?

     c) Jennifer asked 60 pupils altogether.
         Calculate the number of pupils who said vegetable was their favourite.

Q2   A librarian carried out a survey of the ages of people using the library.
     Chart A on the right shows the results.

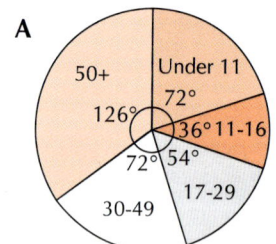

     a) There were 18 people aged 17-29 who took part in the survey.
         How many people took part in the survey altogether?

     b) Use your answer to part a) to calculate
         the number of people in each category.

     Leaflets were handed out to persuade more young people to use the library.
     Another survey was then carried out to find the ages of library users.
     The results are shown in chart B on the right.

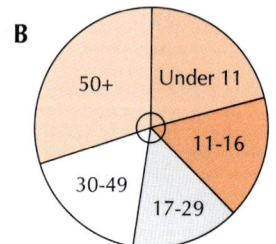

     c) What fraction of the people in the second survey were aged 11-16?

     d) Compare the fraction of people aged 11-16 in the second
         survey with the fraction of people aged 11-16 in the first survey.

     e) Do these pie charts show that there were more people aged 11-16
         in the second survey than in the first survey? Explain your answer.

33.2 Stem and Leaf Diagrams

Stem and leaf diagrams are a bit like horizontal bar charts, but the bars are made out of the actual data. They are useful for showing the spread of data visually, whilst still showing each individual data value.

Learning Objective — Spec Ref S2/S4:
Display and interpret data in stem and leaf diagrams.

Prior Knowledge Check:
Be able to find the mode, median, range and interquartile range. See p.416-419.

In **stem and leaf diagrams**, data values are split up into '**stems**' (their first digits(s)) and **leaves** (the remaining digit). So for the value 25, the stem would be 2 and the leaf would be 5. The leaves are then **ordered** numerically. Stem and leaf diagrams always have a **key** — e.g. '2 | 5 means 25'. Once your data is in a stem and leaf diagram, you can easily find the **mode**, **median**, **range** and **interquartile range**.

Tip: Three figure numbers and decimals can be shown using different keys — e.g. 0 | 3 = 0.3 or 20 | 4 = 204.

Example 1

The marks scored by pupils in a class test are shown here.

56, 52, 82, 65, 76, 82, 57, 63, 69, 73, 58, 81, 73, 52, 73, 71, 67, 59, 63

a) Use this data to draw an ordered stem and leaf diagram.

1. Write down the 'stems' in order in a column — here use the first digit of the marks. The smallest first digit is 5 and the largest is 8, so use 5, 6, 7 and 8.

2. Make a 'leaf' for each data value by writing each second digit next to the correct stem.

```
5 | 6 2 7 8 2 9
6 | 5 3 9 7 3
7 | 6 3 3 3 1
8 | 2 2 1
```

3. Put the leaves in each row in order — from lowest to highest.

4. Remember to add a key.

Key: 5 | 2 means 52 marks

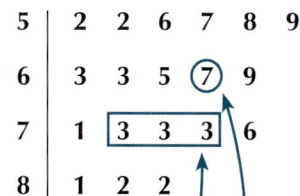

```
5 | 2 2 6 7 8 9
6 | 3 3 5 ⑦ 9
7 | 1 3 3 3 6
8 | 1 2 2
```

b) Use your stem and leaf diagram to find the mode, median and range.

1. Find the mode by looking for the 'leaf' that repeats most often in one of the rows — here, there are three 3s in the third row.

    Mode = **73 marks**

2. There are 19 data values, so the median is the 10th value (p.416).

    Median = **67 marks**

3. Find the range by subtracting the smallest value from the largest.

    Range = 82 – 52 = **30 marks**

Exercise 1

Q1 Draw ordered stem and leaf diagrams for the data sets below, using appropriate keys.

　　a) 41, 48, 51, 54, 59, 65, 74, 80, 86, 89

　　b) 36, 41, 15, 39, 41, 12, 15, 27, 17, 31, 24, 26

　　c) 5.3, 3.1, 4.0, 5.7, 6.0, 5.9, 7.7, 4.4, 3.4

　　d) 205, 203, 232, 211, 236, 234, 240, 203, 221

Q2 Use this diagram showing 11 pupils' test marks to find:

　　a) the mode, median, range and interquartile range,

　　b) the number of pupils who scored between 53 and 63 marks.

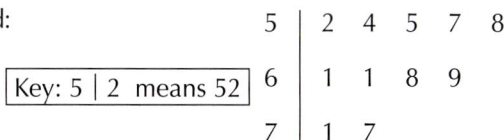

```
5 | 2 4 5 7 8
6 | 1 1 8 9
7 | 1 7
```

Key: 5 | 2 means 52

Back-to-Back Stem and Leaf Diagrams

Learning Objectives — Spec Ref S2/S4:

- Display and interpret data using back-to-back stem and leaf diagrams.
- Compare distributions using back-to-back stem and leaf diagrams.

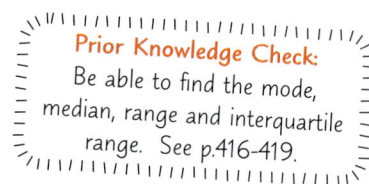

Back-to-back stem and leaf diagrams can be used to display **two sets of data** alongside one other — e.g. they might show data for different genders or age groups. In these diagrams, the stem is in the **centre** and the leaves from the two data sets are placed on **either side** of the stem. This means that one data set has to be read '**backwards**' (the **smaller leaves** are closest to the centre) — so the **key** is really important. It's easy to **compare** the data sets just by looking at the diagram.

Example 2

14 girls and 14 boys completed a puzzle. The times in seconds they took to finish it was recorded. This ordered back-to-back stem and leaf diagram shows the results.

```
        Girls │   Boys
              │ 1 │ 4  4  7  9
  9 8 8 5 5 3 │ 2 │ 7  9
      8 7 4 2 1 │ 3 │ 2  3  8
          4 2 2 │ 4 │ 5  6  6  7  9
```

Key: 2 | 7 for boys means 27
     3 | 2 for girls means 23

a) Find the median times for the girls and the boys.

There are 14 data values for each set of data, so the median is between the 7th and 8th values.

Median for girls = $\dfrac{31+32}{2}$ = **31.5 seconds**

Median for boys = $\dfrac{32+33}{2}$ = **32.5 seconds**

b) Find the range of times for the girls and the boys.

Subtract the first number from the last.

Range for girls = 44 − 23 = **21 seconds**

Range for boys = 49 − 14 = **35 seconds**

c) Use your answers to make two comparisons between the times for the girls and boys.

Compare the medians and the ranges and interpret what these values show about the times.

The **median** for the boys is **higher** than for the girls, suggesting that boys took **longer on average**. The **range** for the girls is **smaller** than for the boys, suggesting the girls' times were **more consistent**.

Exercise 2

Q1 Use these two data sets to draw an ordered back-to-back stem and leaf diagram.

| 18, 48, 38, 29, 41, 28, 33, 24, 12, 37, 32 | 27, 25, 19, 15, 22, 18, 13, 23, 22, 32, 13 |

Q2 The heart rates in beats per minute (bpm) of 15 people at rest and after exercise are shown on the right.

```
          At rest │   After exercise
  8 7 6 4 3 2 2 │ 6 │ 5  8  8  9
    9 8 6 3 2 2 │ 7 │ 4  5  7  7  8
          4 1 │ 8 │ 5  6  7
              │ 9 │ 1  3  7
```

Key: 6 | 5 after exercise means 65 bpm
     2 | 6 at rest means 62 bpm

a) Find the median of each data set.

b) Find the interquartile range for each data set.

c) What conclusions can you draw from your answers?

Q3 The data on the right shows average daily temperatures (in °C) in Dundee and in London during the same 12-day period.

a) Draw a back-to-back stem and leaf diagram to show the data.

b) By calculating the median and range for each set of data, compare the temperatures in the two places.

Dundee	12	19	6	17	23	4
	3	1	15	5	2	3

London	13	21	12	18	24	9
	4	7	15	12	11	16

33.3 Frequency Polygons

Frequency polygons are a visual representation of how the frequency changes throughout a data set.

Learning Objective — Spec Ref S2:
Display and interpret data in frequency polygons.

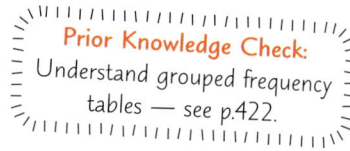

Prior Knowledge Check:
Understand grouped frequency tables — see p.422.

Frequency polygons are used to show the data from a **grouped frequency table**. To draw a frequency polygon, you plot the **frequency** on the **vertical axis** against the **midpoint** of the group on the **horizontal axis** — see p.422 for how to find the midpoint of a group. Frequency polygons are used for **continuous** data (see p.406), so the horizontal axis has a **continuous scale**. You can **compare** the **distributions** of two or more data sets by drawing a frequency polygon for each on the **same axes** and comparing their **shapes**.

Example 1

Amin recorded the speeds of the cars that passed outside his house one day.
The results are shown in the grouped frequency table on the right.
Draw a frequency polygon to represent this information.

Speed (s) in mph	Frequency
$25 \leq s < 30$	1
$30 \leq s < 35$	5
$35 \leq s < 40$	12
$40 \leq s < 45$	16
$45 \leq s < 50$	9
$50 \leq s < 55$	2

1. Find the midpoint of each class by adding the endpoints and dividing by 2.

   E.g. $\frac{25 + 30}{2} = 27.5$ mph

2. For each class, plot the frequency at the midpoint.

   (27.5, 1), (32.5, 5), (37.5, 12), (42.5, 16), (47.5, 9), (52.5, 2)

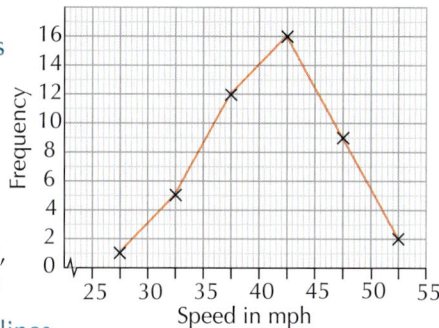

3. Join your points with straight lines.

Exercise 1

Q1 Some students recorded the total floor area of local supermarkets for a project. Their results are shown in the table.

Draw a frequency polygon to show this information.

Floor area (a) in 1000s of square feet	Frequency
$9 \leq a < 13$	3
$13 \leq a < 14$	7
$14 \leq a < 15$	5
$15 \leq a < 16$	2

Q2 Emma recorded the number of hours of sunshine on each day from 1st to 29th February 2016 and each day from 1st to 29th July 2016. She used classes with a width of 1 hour. The frequency polygons below show her results.

a) How many days with 4 or more hours of sunshine were there in February?

b) What is the modal class for February?

c) What is the modal class for July?

d) Use your answers to parts b) and c) to compare the daily hours of sunshine in February and July. Compare the spread of the data for the two months.

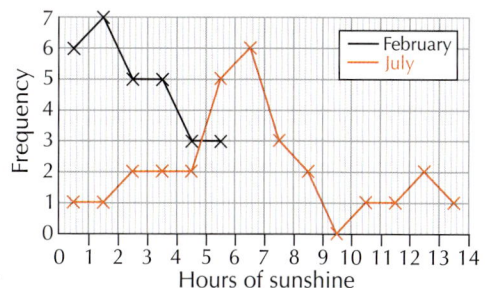

33.4 Histograms

Histograms may look a lot like bar charts, but they're a bit more complicated to draw and interpret.

Histograms

Learning Objective — Spec Ref S3:
Represent data on a histogram.

Prior Knowledge Check:
Understand grouped frequency tables — see p.422.

Histograms are used to show **grouped continuous data** — so the scale on the horizontal axis is **continuous**, and there are **no gaps** between the bars. The vertical axis shows **frequency density** (**not** frequency) — to calculate the frequency density, use the formula:

$$\text{Frequency density} = \frac{\text{Frequency}}{\text{Class width}}$$

This means that the **frequency** of data values is shown by the **area of each bar**, not the height. You can find this by rearranging the formula above:

$$\text{Area} = \text{Frequency} = \text{Frequency density} \times \text{class width}$$

Example 1

Draw a histogram to represent the information about the height, h, of 30 adults shown in the table.

1. Find the class widths for each group.

    E.g. $155 \leq h < 165$: $165 - 155 = 10$
    $165 \leq h < 170$: $170 - 165 = 5$
    $\vdots \quad \vdots \quad \vdots$
    $185 \leq h < 200$: $200 - 185 = 15$

Height (h) in cm	Frequency	Class Width	Frequency Density
$155 \leq h < 165$	4	10	$4 \div 10 = \mathbf{0.4}$
$165 \leq h < 170$	3	5	$3 \div 5 = \mathbf{0.6}$
$170 \leq h < 175$	4	5	$4 \div 5 = \mathbf{0.8}$
$175 \leq h < 180$	9	5	$9 \div 5 = \mathbf{1.8}$
$180 \leq h < 185$	7	5	$7 \div 5 = \mathbf{1.4}$
$185 \leq h < 200$	3	15	$3 \div 15 = \mathbf{0.2}$

2. Add a 'frequency density' column to the table and use the formula to work out the value for each class.

3. Draw axes with a continuous scale for height along the bottom and frequency density up the side.

4. Draw a bar for each class with a height equal to the frequency density. Then check that the area equals the frequency.

    E.g. area of the first bar:

    frequency density × class width = $0.4 \times 10 = 4$ ✓

Exercise 1

Q1 Copy and complete the grouped frequency tables below.

a)

Height (h) in cm	Frequency	Frequency Density
$0 < h \leq 5$	4	
$5 < h \leq 10$	6	
$10 < h \leq 15$	3	
$15 < h \leq 20$	2	

b)

Height (h) in cm	Frequency	Frequency Density
$10 < h \leq 20$	5	
$20 < h \leq 25$	15	
$25 < h \leq 30$	12	
$30 < h \leq 40$	8	

Q2 The tables below show some information about the volume of tea drunk each day by a group of people. Copy and complete each table and use it to draw a histogram to represent the information.

a)

Volume (v) in ml	Frequency	Frequency Density
$0 \leq v < 500$	50	
$500 \leq v < 1000$	75	
$1000 \leq v < 1500$	70	
$1500 \leq v < 2000$	55	

b)

Volume (v) in ml	Frequency	Frequency Density
$0 \leq v < 300$	30	
$300 \leq v < 600$	15	
$600 \leq v < 900$	24	
$900 \leq v < 1500$	42	

Q3 The table on the right shows some information about the masses (m) in kg of 22 pigs on a free range farm. Draw a histogram to represent this information.

Mass (m) in kg	Frequency
$10 \leq m < 20$	5
$20 \leq m < 30$	7
$30 \leq m < 50$	10

Q4 This incomplete table and histogram show information about the length of time (in hours) that some people spend watching the TV programme 'Celebrity Ironing Challenge' each week.

a) Copy the histogram and use the information given to label the vertical axis.

b) Use the information given to fill in the gaps in the table and histogram.

Time (t) in hours	Frequency
$0 \leq t < 1$	
$1 \leq t < 2$	
$2 \leq t < 4$	16
$4 \leq t < 6$	14
$6 \leq t < 10$	12

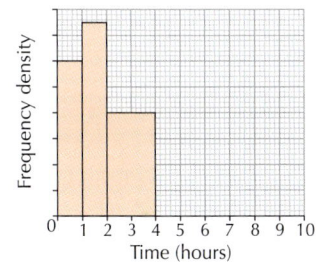

Interpreting Histograms

Learning Objective — Spec Ref S3/S4:
Use histograms to estimate and interpret data.

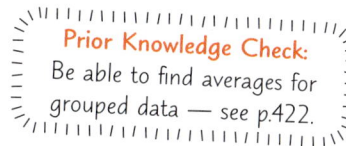

Prior Knowledge Check:
Be able to find averages for grouped data — see p.422.

You can use histograms to **estimate** things about the data — you have to **assume** that the data is **spread evenly** through each interval. For example, the **number of data values** in a particular interval can be found using the **areas of the bars** — so if you wanted to find the number of values between 2 and 2.5 but the class was $2 \leq x < 3$, you just divide the **frequency** (i.e. **area**) of the whole bar by 2. You can estimate values for the **mean** or **median** by working out the **frequency** of each class (class width × frequency density) and following the methods shown for grouped frequency tables on p.422.

Example 2

Some students were asked the length of their journey to university. The histogram on the right shows the information.

a) **Estimate the number of students with a journey of 1.5 km or less.**

1. Work out the frequency of journeys between 0 and 1 km by finding the area of the first bar.

   Frequency = class width × frequency density
   $= 1 \times 2 = 2$

2. Work out the frequency of journeys between 1 and 2 km and divide by 2 to find the frequency of journeys between 1 and 1.5 km.

   Frequency of $1 < j \leq 2$ group
   $= 1 \times 4 = 4$
   Frequency of $1 < j \leq 1.5$
   $\approx 4 \div 2 = 2$

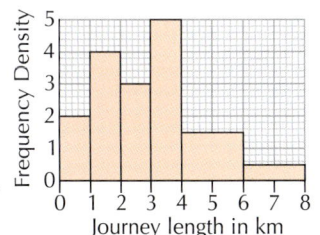

   Tip: You can only estimate here as you don't know the actual data values.

3. Add the values for $0 < j \leq 1$ and $1 < j \leq 1.5$ together.

   $2 + 2 = \textbf{4 students}$

b) Estimate the mean journey length for the students.

1. Draw a grouped frequency table using the classes shown on the histogram.

2. Work out the frequency (class width × frequency density) and midpoint for each class, then multiply them together.

Journey (j) in km	Frequency	Midpoint	Midpoint × frequency
$0 < j \leq 1$	$2 \times 1 = 2$	0.5	1
$1 < j \leq 2$	$4 \times 1 = 4$	1.5	6
$2 < j \leq 3$	$3 \times 1 = 3$	2.5	7.5
$3 < j \leq 4$	$5 \times 1 = 5$	3.5	17.5
$4 < j \leq 6$	$1.5 \times 2 = 3$	5	15
$6 < j \leq 8$	$0.5 \times 2 = 1$	7	7
Totals	18		54

3. Estimate the mean by dividing the total of the 'midpoint × frequency' column by the total frequency.

Estimated mean = 54 ÷ 18
= **3 km**

Tip: You can't tell from the histogram if the inequality signs should be < or ≤ so the classes could be e.g. $0 \leq j < 1$. You just need to make sure there are no overlaps (e.g. you couldn't have $0 < j \leq 1$ and $1 \leq j \leq 2$ since 1 is covered twice).

Exercise 2

Q1 For each of the histograms below, work out the frequency represented by each bar.

a)

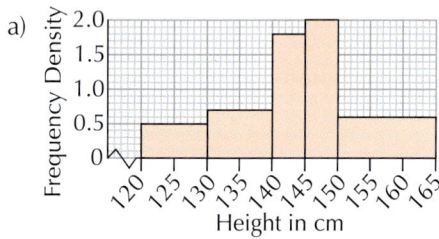

b)

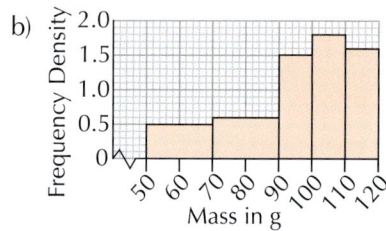

Q2 For the histogram in Q1a) above, estimate the number of data values in the following intervals.

a) 150-155 cm   b) 150-160 cm   c) 145-160 cm   d) 120-126 cm

Q3 The histogram on the right represents the diameters in centimetres of all the cakes made at a bakery one day.

a) Estimate the number of cakes with a diameter of less than 16 cm.

b) Estimate the number of cakes with a diameter in the range 20-30 cm.

c) Work out the total number of cakes made.

d) Calculate an estimate of the mean diameter of the cakes.

e) Explain what assumption you have made for part d) and suggest a problem with this assumption.

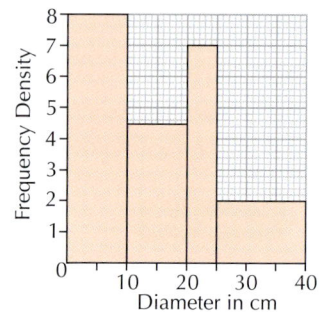

Q4 The histogram on the right represents the heights in centimetres of all the penguins at a wildlife park.

a) Estimate the number of penguins with a height of less than 56 cm.

b) Work out the total number of penguins at the wildlife park.

c) 24 of the penguins have a height of less than H cm. Estimate the value of H. (PROBLEM SOLVING)

d) Use your answers to b) and c) to write down an estimate of the median height. (PROBLEM SOLVING)

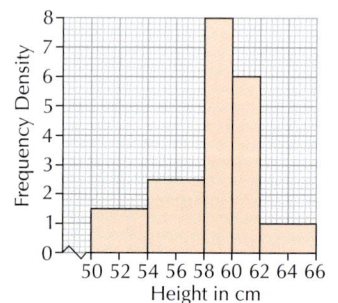

33.5 Cumulative Frequency Diagrams

Cumulative frequency means adding up as you go along — it is the total frequency 'so far' and increases as you go through a data set.

Learning Objectives — Spec Ref S3/S4:
- Display and interpret data on cumulative frequency diagrams.
- Use them to estimate the median, quartiles and the interquartile range.

> **Prior Knowledge Check:**
> Be able to find the median, quartiles and the interquartile range — see p.416 and p.419.

Cumulative frequency is the **running total** of frequencies for a set of data. You find the cumulative frequency by **adding** the frequency of a class to the **sum** of the frequencies of **all the classes** that came before it. For example, if the frequencies were 1, 5, 4, 9..., the cumulative frequencies would be **1**, 1 + 5 = **6**, 6 + 4 = **10**, 10 + 9 = **19**... You can add an **extra column** to your **frequency table** to keep track of the cumulative frequency — the **final entry** in the cumulative frequency column will be the **total frequency** for all the data.

To draw a **cumulative frequency diagram** for grouped data, the cumulative frequency goes on the **vertical axis** and the horizontal axis has a continuous scale. Plot the points at the **upper limit** of each class (e.g. if the class $5 \leq x < 10$ had a cumulative frequency of 15, you'd plot the point (10, 15)). Then draw an **S-shaped curve** (or line) through the points. There are two types of cumulative frequency diagrams:

- **Cumulative frequency curves** — where the points are joined with a smooth curve.

- **Cumulative frequency polygons** — where the points are joined with straight lines instead of a curve.

You can use cumulative frequency diagrams to find the **median** and **interquartile range** of a data set. To find the **median**, **divide** the **total frequency** by 2, draw a horizontal line at this point and read off from the **horizontal axis**. To find the **lower** and **upper quartiles**, divide the total frequency by 4 (and multiply by 3 for the upper quartile), draw horizontal lines at these points and read off from the horizontal axis. You can then calculate the **interquartile range** by **subtracting** the lower quartile from the upper quartile.

> **Tip:** For grouped data, you can only estimate the median and interquartile range because you don't know the actual data values.

Example 1

The table below shows the heights of a set of plants.

a) **Complete the cumulative frequency column for the table.**

Just add up the frequencies, working down the table.

Height, h (cm)	Frequency	Cumulative Frequency
$15 < h \leq 18$	3	**3**
$18 < h \leq 21$	12	3 + 12 = **15**
$21 < h \leq 24$	35	15 + 35 = **50**
$24 < h \leq 27$	26	50 + 26 = **76**
$27 < h \leq 30$	4	76 + 4 = **80**

b) (i) **Draw a cumulative frequency diagram for the data.**

Cumulative frequency goes on the vertical axis and you need a continuous scale between 15 and 30 on the horizontal axis. Plot the cumulative frequency for each group against the upper limit of the group — so you want to plot the points (18, 3), (21, 15), (24, 50), (27, 76) and (30, 80). Then join the points with a smooth curve.

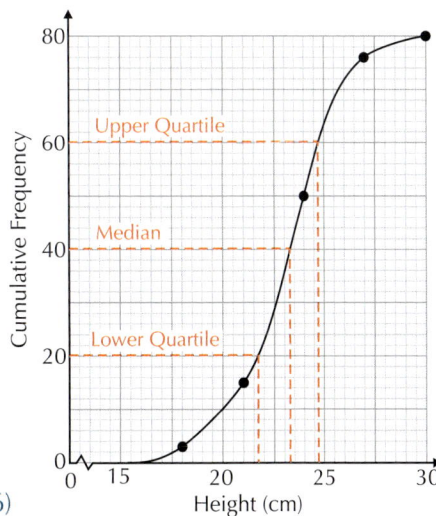

(ii) Estimate the median and interquartile range for the data.

1. Median — read off the plant height that corresponds to a cumulative frequency of 40 (half of 80).

    Median = **23.25 cm**

2. Interquartile range (IQR) — read off the plant height that corresponds to 60 (= ¾ of 80) for the upper quartile and 20 (= ¼ of 80) for the lower quartile. Then find the difference.

    Upper quartile = 24.75
    Lower quartile = 21.75
    IQR = 24.75 − 21.75 = **3 cm**

Exercise 1

Q1 The cumulative frequency graph below shows the monthly earnings for a group of 16-year-olds.

a) How many 16-year-olds took part in the survey?

b) Estimate how many earned less than £20.

c) Estimate how many earned less than £80.

d) Estimate how many earned between £40 and £100.

e) Estimate the median earnings for a 16-year-old in the survey.

f) Estimate the lower quartile, the upper quartile and the interquartile range for this data.

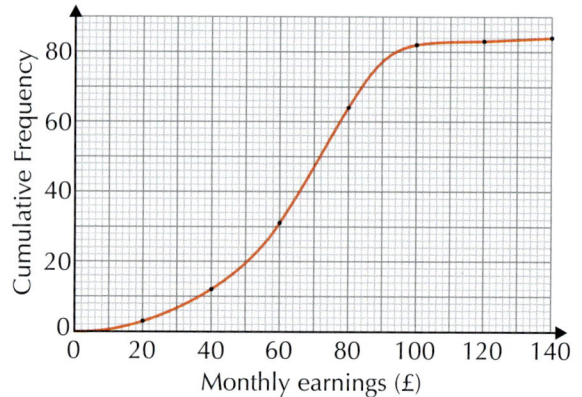

Q2 This table shows Year 11 marks in a Maths test.

a) Copy the table and complete the cumulative frequency column.

b) Draw a cumulative frequency curve for the marks.

c) Estimate the median mark.

d) Estimate the upper and lower quartiles and interquartile range for the data.

e) Pupils who achieved fewer than 45 marks have to resit the test. Estimate how many pupils will resit.

f) Pupils who achieved more than 55 marks will sit the higher tier exam. Estimate how many will be entered for the higher tier.

(PROBLEM SOLVING)

Marks, m	Frequency	Cumulative Frequency
$0 < m \leq 10$	0	
$10 < m \leq 20$	2	
$20 < m \leq 30$	4	
$30 < m \leq 40$	5	
$40 < m \leq 50$	19	
$50 < m \leq 60$	33	
$60 < m \leq 70$	43	
$70 < m \leq 80$	10	
$80 < m \leq 90$	3	
$90 < m \leq 100$	1	

Q3 Some students were asked to pour out a sample of sand that they estimated would have a mass of 25 g. The table shows a summary of the actual masses of their samples.

a) Draw a cumulative frequency curve for this data.

b) Estimate the median and interquartile range.

c) Estimate how many students' samples were within 10 g of the median.

d) Comment on how good the students' estimates were.

Mass, m, in g	Frequency
$m \leq 5$	2
$5 < m \leq 10$	3
$10 < m \leq 15$	4
$15 < m \leq 20$	7
$20 < m \leq 25$	25
$25 < m \leq 30$	51
$30 < m \leq 35$	31
$35 < m \leq 40$	9
$40 < m \leq 45$	5
$45 < m \leq 50$	3

Box Plots

Box plots provide a useful summary of the **distribution of data** in a data set. The **lower quartile**, **median** and **upper quartile** are shown as **vertical lines**, with the lower and upper quartiles forming the **ends** of the **box** — so the **width** of the box is the **interquartile range**. The **maximum** and **minimum** values are also plotted, and joined to the box with **horizontal lines** — you can use these to find the **range**.

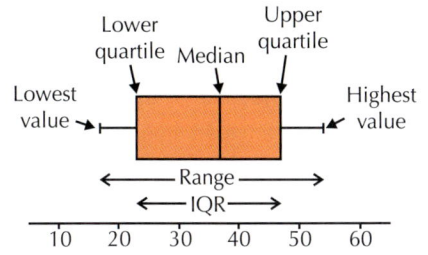

By plotting two box plots on the **same scale**, you can easily **compare** the two distributions.

Example 2

The box plots on the right show the distribution of times taken to answer calls at a call centre over two weeks. The call centre manager says, 'The times taken to answer calls in Week 1 were greater but more consistent than those in Week 2'. Do you agree? Use the data to explain your answer.

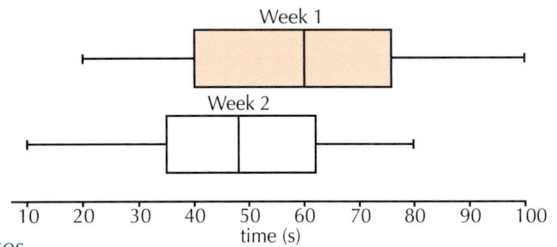

Compare the medians, quartiles and interquartile ranges.

Week 1: $Q_1 = 40$, Median = 60, $Q_3 = 76$, IQR = 36
Week 2: $Q_1 = 35$, Median = 48, $Q_3 = 62$, IQR = 27

All values are greater for Week 1 than the equivalent values for Week 2, so the data supports the first part of the statement. However, the **interquartile range for Week 2 is smaller** than for Week 1, so the times taken to answer calls were more consistent during Week 2.

Tip: You could use the minimum/maximum values and the range instead of the IQR — in this case it'd lead you to the same conclusions.

Exercise 2

Q1 For each of the following box plots, find:

a)

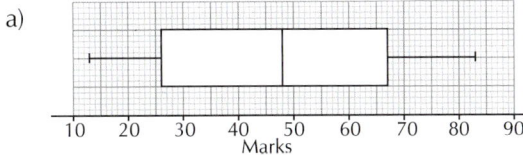

b)

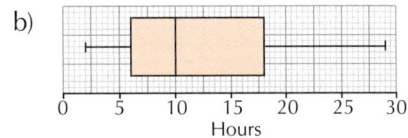

(i) the median,   (ii) the upper quartile,   (iii) the lower quartile,

(iv) the interquartile range,   (v) the lowest value,   (vi) the highest value.

Q2 The box plots on the right show the distributions of the times a group of busybodies and a group of chatterboxes were able to stay silent.

A busybody says, 'The busybodies were definitely better at staying silent than the chatterboxes'. Do you agree? Use the data to support your answer.

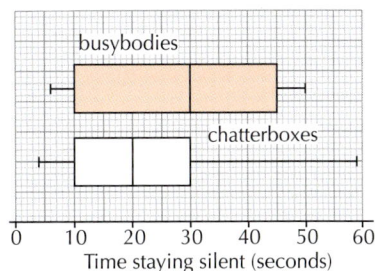

33.6 Time Series

Some line graphs can be used to show how things change over time — these are known as time series graphs. They're useful for spotting repeating patterns or trends in data.

> **Learning Objective — Spec Ref S2:**
> Display and interpret time series on a line graph.

Time series graphs are used to display data collected at **regular intervals** over time — e.g. every day for a week, every month for a year or over several years. As time is a **continuous variable**, time series are plotted on **line graphs** and the plotted coordinates are joined with **straight lines**.

You can use time series to find **trends** in the data over time. Look for:

- **Seasonality** — a basic pattern that is **repeated** regularly over time. For example, the average monthly temperatures will follow a similar pattern year on year, or quarterly sales may follow the same pattern over time. The time taken for a pattern to repeat itself (measured from **peak-to-peak** or **trough-to-trough**) is called the **period**.

> **Tip:** Seasonality doesn't have to match the actual seasons — e.g. tide levels show seasonality over the course of a day.

- An **overall trend** — where the data values **generally** get **smaller** or **larger** over time (ignoring seasonal patterns). For example, the price of a weekly shop may steadily increase or decrease. **Trend lines** can be added to times series to better **illustrate** overall trends.

You can plot **multiple** time series on the same set of axes to **compare** how different data sets change over time — this is called a **comparative time series** graph.

Example 1

The temperature in two countries is recorded once every three months, on the first day of January, April, July and October, for three years. The data for Country X is recorded in the table and the data for Country Y is shown on the time series graph below.

Date	Year 1				Year 2				Year 3			
	Jan	Apr	Jul	Oct	Jan	Apr	Jul	Oct	Jan	Apr	Jul	Oct
Temp. (°C)	6	11	28	12	8	10	29	15	9	13	30	16

a) **Draw the time series for Country X on the same axes.**

   1. Plot the points on the axes. The date is plotted on the horizontal axis and the temperature is plotted on the vertical axis.

   2. Join the plotted points with straight lines and label the line.

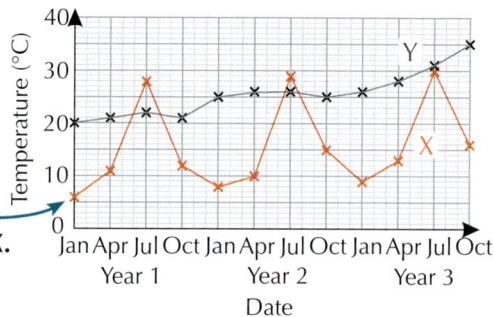

b) **Comment on the trend of the temperatures in Country X.**

The time series for Country X shows seasonality.

There is a **seasonal trend**. Temperatures are **coldest in January** and **warmest in July** every year.

c) **Comment on the trend of the temperatures in Country Y.**

There is an overall trend — the temperatures are generally increasing over time (but there is no seasonal trend).

Temperatures in Country Y are steadily **increasing** with time.

Q1   The table below shows the fastest time in seconds a professional athlete took to run 100 m every day for two weeks leading up to a major competition.

Day	1	2	3	4	5	6	7
Time in seconds	10.8	10.5	10.5	10.4	10.5	10.6	10.5

Day	8	9	10	11	12	13	14
Time in seconds	10.3	10.3	10.2	10.4	10.2	10.1	10.1

Draw a time series graph to show this data and describe the overall trend in the athlete's times.

Q2   The graph shows the average rainfall in London. The average rainfall for Glasgow is shown below.

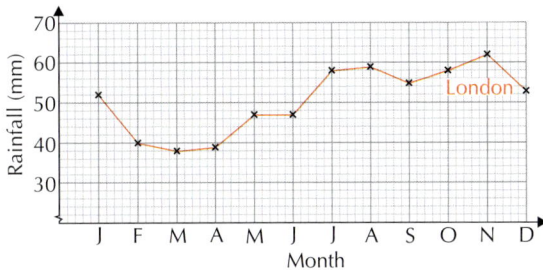

Month	Jan	Feb	Mar	Apr	May	Jun
Rainfall (mm)	111	85	69	67	63	70

Month	Jul	Aug	Sep	Oct	Nov	Dec
Rainfall (mm)	97	93	102	119	106	127

a) Copy the time series graph for London and draw on the data for Glasgow.

b) Write two sentences to compare the rainfall in the two cities.

c) Find the range of each set of data.   PROBLEM SOLVING

Q3   A shop records its sales (in £) of two different brands over an 8-week period in the table below.

Week	1	2	3	4	5	6	7	8
'Impact' brand	520	365	815	960	985	1245	1505	1820
'On Trend' brand	840	795	830	925	960	875	905	965

Draw a comparative time series graph of the data and comment on the sales trends of the two brands.

Q4   A bank converts pounds sterling (£) into US dollars ($). The table below shows the highest conversion rate from £ to $ offered each month in Year 1.
The graph shows the highest conversion rate offered each month for Year 2.

Month (Year 1)	Jan	Feb	Mar	Apr	May	Jun	Jul	Aug	Sep	Oct	Nov	Dec
US dollars ($)	1.45	1.44	1.42	1.47	1.54	1.63	1.64	1.65	1.63	1.62	1.66	1.62

a) Copy the original graph, and add a line showing the data from the table.

b) What was the highest value of £1 in dollars in Year 2?

c) What was the highest value of £1 in dollars in Year 1? When did it occur?

d) A businesswoman always uses this bank to convert her money. She travels to New York in June each year, and changes £500 into dollars each time. Assuming she gets the highest possible rate for June each year, how many more dollars does she get for £500 in Year 1 than in Year 2?

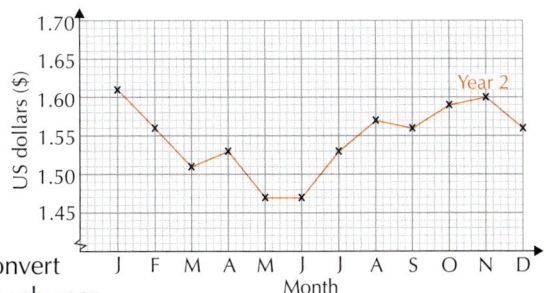

33.7 Scatter Graphs

Scatter graphs are used to show how closely two variables are related to one another — this is known as correlation. A line of best fit can be used to show correlation and estimate values of either variable.

Correlation

Learning Objectives — Spec Ref S6:
- Draw and interpret scatter graphs.
- Recognise and describe correlation.

A **scatter graph** shows **two variables** plotted against each other, e.g. height and weight or temperature and BBQ sales. To **draw** a scatter graph, first decide **which variable** should go on **which axis** — the one that you think **depends** on the other should go on the **vertical** axis. Then **plot** your data as points (x, y), where x is the variable on the horizontal axis and y is the variable on the vertical axis.

If two variables are **related** to each other then they are **correlated**.
Variables can be **positively correlated**, **negatively correlated** or **not correlated** at all.

- **Positive correlation** means that both variables **increase and decrease together** — e.g. the number of ice creams sold and the average daily temperature are likely to be positively correlated, as people buy more ice creams when the weather is hot and fewer ice creams when it is cold. The **points** on the scatter graph will look like a line sloping **upward** from left to right.

- **Negative correlation** means that as one variable **increases**, the other **decreases** — e.g. the number of woolly hats sold and the average daily temperature are likely to be negatively correlated, as people buy fewer woolly hats when the weather is hot and more woolly hats when it is cold. The **points** on the graph will look like a line sloping **downward** from left to right.

- **No correlation** means that there is **no linear relationship** between the variables — e.g. the number of newspapers sold is unlikely to change with the average daily temperature, as people buy newspapers regardless of the temperature. The **points** on the graph will look **randomly scattered**.

If two variables are correlated it **doesn't necessarily** mean that one **causes** the other. There could be a **third factor** affecting both, or it could just be a **coincidence**. For example, the number of people wearing **shorts** and the number of **ice creams** sold are likely to be **positively correlated** — but this doesn't mean that people wearing shorts **cause** ice creams to be sold. Here, both variables are affected by the **weather**.

Example 1

Describe the relationship shown by each of the graphs.

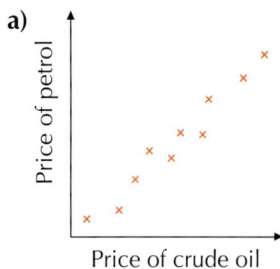

a)

As the price of crude oil increases, the price of petrol also increases, so this is...
Positive correlation

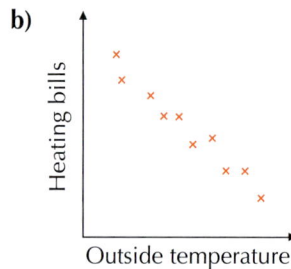

b)

As the temperature outside increases, heating bills decrease, so this is...
Negative correlation

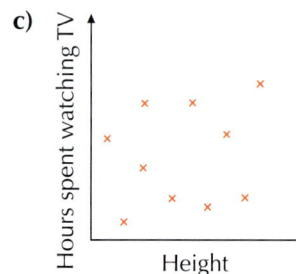

c)

The height of the viewer seems to have no connection to the hours spent watching TV, so this is...
No correlation

Q1 The outside temperature and the number of ice creams sold in a cafe were recorded for six days. The results are shown in the table on the right.

Temp (°C)	28	25	26	21	23	29
Ice creams sold	30	22	27	5	13	33

a) Use the data from the table to plot a scatter graph.

b) Describe the correlation between the outside temperature and the number of ice creams sold.

Q2 Ten children of different ages were asked how many baby teeth they still had.

Age (years)	5	6	8	7	9	7	10	6	8	9
Baby teeth	20	17	11	15	7	17	5	19	13	8

a) Use the data to plot a scatter graph.

b) Describe the relationship between the age of the children and the number of baby teeth they have.

You can describe the **strength** of the correlation as well.
The **closer** the points are to forming a **straight line**, the **stronger** the correlation.

- If most of your points are close to a **straight line**, then you have **strong correlation**.

- If your points are spread **loosely around** a straight line, then you have **moderate correlation**.

- If your points **don't line up** nicely but you can still see that there is a **relationship** between the two variables, then you have **weak correlation**.

Example 2

Describe the strength and type of correlation shown by each of the scatter graphs.

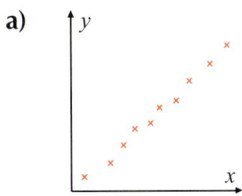

a)

The points form an upward slope fairly close to a straight line, so this is...

Strong positive correlation

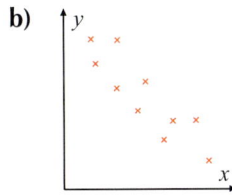

b)

The points form a downward slope loosely around a straight line, so this is...

Moderate negative correlation

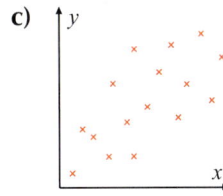

c)

The points form an upward slope, but do not lie close to a straight line, so this is...

Weak positive correlation

Exercise 2

Q1 Describe the strength and type of correlation shown by each of the scatter graphs below.

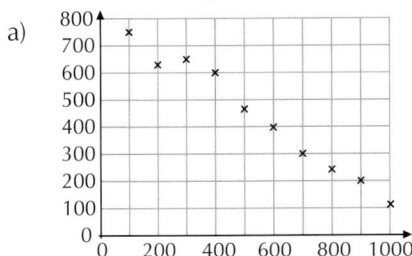

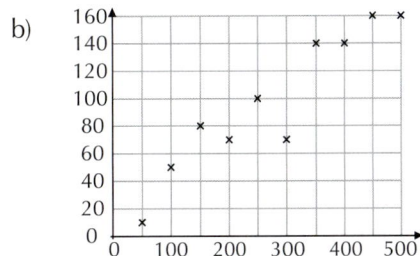

a)

b)

Q2   Jeremy measured the height and shoe size of 10 people.

Height (cm)	165	159	173	186	176	172	181	169	179	194
Shoe size	6	5	8	9	8.5	7	8	6	8	11

a) Use his data to plot a scatter graph of shoe size (on the *y*-axis) against height in cm (on the *x*-axis).

b) Is there evidence to show that height and shoe size are correlated? Explain your answer.

Lines of Best Fit

Learning Objectives — Spec Ref S6:
- Draw a line of best fit and use it to estimate and predict data values.
- Be able to recognise outliers.

If two variables are correlated, then you can draw a **line of best fit** on their scatter graph. This is a **straight line** that passes through the **middle of the points** with a roughly **equal number** on either side.

Outliers are points that **don't fit** the general pattern of the rest of the data. They can **move** your line of best fit away from other values, so are usually **ignored** when drawing your line. Outliers can sometimes be caused by **errors** in the data, but not always — they can just be **unusually high** or **low values**.

You can use a line of best fit to **predict values** for one variable when you know the value of the other. All you have to do is **draw a line** from the value you're given to the line of best fit, then **read off** the value for the variable on the other axis. Using your line to predict values **within** the range of data you have is known as **interpolation**, and should be **fairly reliable**. Using your line to predict values **outside** the range of the data is known as **extrapolation** and can be **unreliable** because you don't know that the pattern continues outside the data range.

Example 3

The scatter graph shows pupils' marks on a Maths test plotted against their marks on an English test.

a) **Draw a line of best fit on the graph.**

b) **Jimmy was ill on the day of the Maths test. If he scored 74 in his English test, predict what his Maths mark would have been.**

c) **Elena was ill on the day of the English test. If she scored 35 on her Maths test, predict what her English result would have been.**

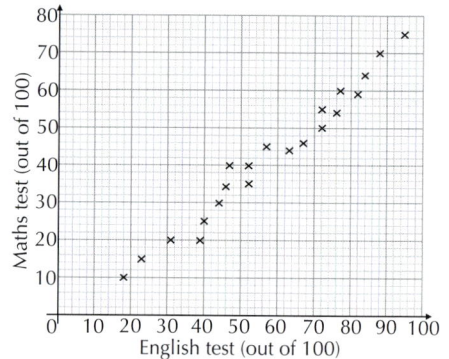

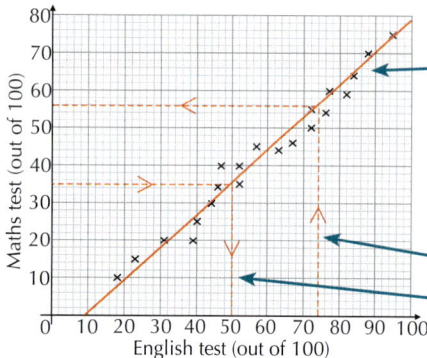

a) Draw a straight line through the middle of the points — there should be about the same number of points on either side of the line.

You can now use the line of best fit to predict Jimmy's and Elena's results:

b) Predicted Maths mark for Jimmy = **56**
c) Predicted English mark for Elena = **50**

Q1 The graph on the right shows the height of a number of trees plotted against the width of their trunks.

a) Describe the strength and type of correlation between the width of the trunks and the height of the trees.

b) Point A does not fit the correlation pattern shown by the rest of the data. Suggest a reason for this.

c) Use the graph to predict the width of a tree's trunk if it is 13 m tall.

d) A tree has grown into power lines, meaning that measuring its height would be dangerous. If the trunk is 100 cm wide, predict the tree's height.

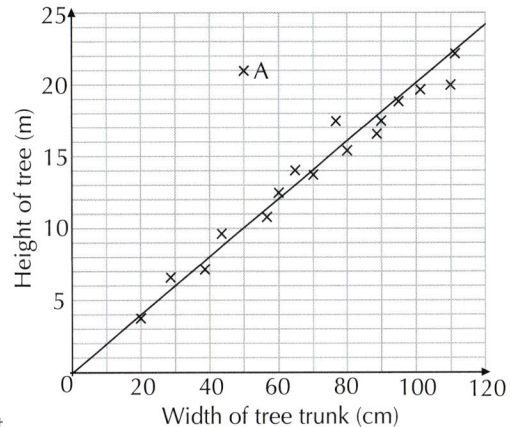

Q2 The outside midday temperature and the number of drinks sold by two vending machines were recorded over a 10-day period. The results are shown in these tables.

Temperature (°C)	14	29	23	19	22	31	33	18	27	21
Drinks sold — Machine 1	6	24	16	13	15	28	31	20	22	14
Drinks sold — Machine 2	7	25	18	15	17	32	35	14	24	17

For each machine:

a) Draw a scatter graph of drinks sold against temperature.

b) Draw a line of best fit for the data.

c) Circle any outliers and suggest a reason for them.

d) Predict the number of drinks that would be sold if the outside midday temperature was 25 °C.

e) Explain why it might not be appropriate to use your line of best fit to estimate the number of drinks sold if the outside midday temperature was 3 °C.

Lucas says, "An increase in temperature causes more vending machine drinks to be sold."

f) Do you agree with Lucas's statement? Explain your answer.

Q3 Craig measures the leg length of 10 members of a running club. He then times how long it takes each member to run 100 m. His results are shown below.

Club member	1	2	3	4	5	6	7	8	9	10
Leg length (in cm)	60	65	75	90	80	69	96	76.5	85	66
Time taken to run 100 m (in s)	16.9	12.8	15.6	13.5	14.3	15.4	13.0	14.8	14.4	16.3

a) Plot a scatter graph of the time taken against leg length.

b) Describe the relationship shown by the scatter graph.

c) Circle any outliers and draw a line of best fit for the data.

d) Predict how long it would take for a club member with a leg length of 87 cm to run 100 m.

e) (i) Estimate how long it would take a club member with a leg length of 100 cm to run 100 m.

(ii) Is your answer to part (i) a reliable estimate? Explain your answer.

33.8 Appropriate Representation of Data

Don't automatically trust statistical diagrams — they can be misleading...

> **Learning Objectives — Spec Ref S2:**
> - Choose an appropriate diagram for a data set.
> - Understand how statistical diagrams can be misleading.

When **choosing** which diagram to use, you need to think about which one will best **represent** your data — for example, although you could use a pie chart to represent continuous grouped data, it would be much more appropriate to show this type of data on a histogram or cumulative frequency diagram. Using the **wrong type of diagram** can make it **difficult** or **impossible** to **interpret** the data and draw **conclusions**.

Data can also be **deliberately** misrepresented to guide you towards a **misleading conclusion**, or to **hide raw data** that presents a negative view. For example, a company could **adjust the scale** on a line graph to make it look like their sales are increasing by more than they actually are. They could also **hide** raw data by using a **pie chart** to show proportions, rather than actual values — for example, if a company had 50 negative reviews out of 1000, using a pie chart to display this data would make the amount of negative reviews seem very small, when there are actually quite a lot.

Example 1

Criticise the diagrams below.

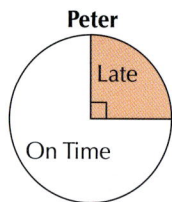

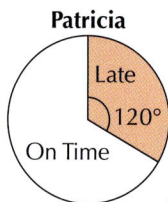

Remember that pie charts show proportion — not actual data values. Here, notice the total number of days is different.

Look at the scale on bar charts to make sure that nothing has been cut off.

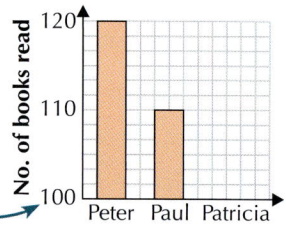

It looks like Patricia is late more than Peter but it's only true that the **proportion** of days she was late is higher — not the actual number of days.

It looks like Paul has read half as many books as Peter, and that Patricia has read none — but that's just because the scale **starts at 100**.

Exercise 1

Q1 This box plot shows information about the temperature inside a fridge, measured at hourly intervals.

Explain whether you think a box plot is a good diagram to use for this data. If not, suggest an alternative diagram.

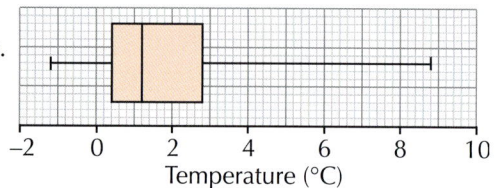

Score (s)	Frequency
$0 < s \le 20$	13
$20 < s \le 30$	35
$30 < s \le 40$	15
$40 < s \le 50$	3
$50 < s \le 55$	20
$55 < s \le 60$	35

Q2 The grouped frequency table to the left shows the scores hit per dart by players during a darts tournament. Would a pie chart be a good way to represent this data? Explain your answer. If not, suggest an alternative diagram.

Review Exercise

Q1 Ten people competed in a quiz. The scores for the first two rounds are shown in the table on the right.

Player	A	B	C	D	E	F	G	H	I	J
Round 1	12	19	6	11	16	15	18	13	12	8
Round 2	9	16	1	8	15	11	13	10	7	4

a) Using this raw data, draw and complete a two-way table to show the frequencies of scores in the following categories for each round: 0-3, 4-7, 8-11, 12-15 and 16-20 points.

b) What percentage of people scored 12 or more in round 1?

c) The questions in one of the rounds were easier than in the other round. Use your two-way table to suggest which round was easier and explain your answer.

Q2 The children at a youth club were asked to name their favourite flavour of ice cream. The dual bar chart on the right shows the results.

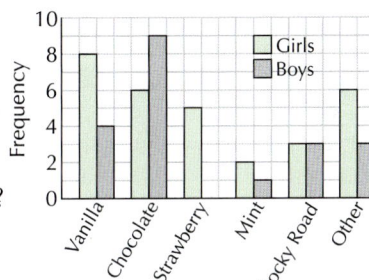

a) How many girls were asked altogether?

b) How many more boys than girls chose chocolate?

c) What is the modal flavour for the girls?

The pie chart on the right shows the same data for the girls.

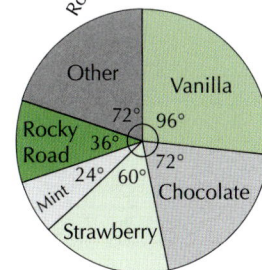

d) Work out the number of degrees used to represent one girl.

e) Draw a pie chart to show the data for the boys.

f) Rocky road was chosen by the same number of girls and boys. Explain why the sectors representing rocky road in each pie chart are different sizes.

Q3 The data below shows the ages of people queuing in a post office at 10 am and at 3 pm.

10 am:	65	48	51	27	29	35	58	51	54	60	59
3 pm:	15	23	32	31	35	22	23	18	27		

a) Draw a back-to-back stem and leaf diagram to show the data.

b) Find and compare the median age of the people queuing at each time.

Q4 The frequency polygon shows information on the heights of members of a gymnastics club. The heights were put into groups with 5 cm intervals.

a) Create and fill in a grouped frequency table using the data from the frequency polygon.

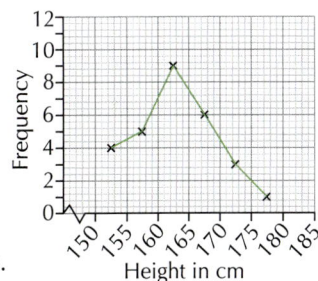

b) What is the modal class for the gymnasts' heights?

c) How many gymnasts are there in the club?

d) Gymnasts who have a height of 160 cm or greater use higher bar apparatus than everyone else. Estimate the percentage of members who use the higher bar apparatus to 1 decimal place.

Q5 The histogram below shows information about the heights of all the members of a netball club.

a) Copy this grouped frequency table and use the histogram to fill in the frequencies.

Height (h) in cm	Frequency
$155 \leq h < 165$	
$165 \leq h < 170$	
$170 \leq h < 175$	
$175 \leq h < 180$	
$180 \leq h < 185$	

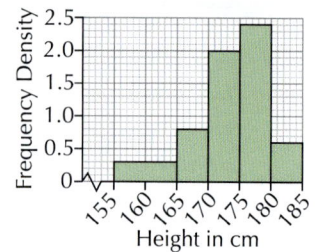

b) Estimate the number of netball players who are taller than 172 cm.

Q6 The cumulative frequency graph on the right shows the distances that 100 school children travel to school.

a) Use the graph to estimate the median.

b) Use the graph to estimate the interquartile range.

c) Children who live more than 4 km away get the school bus. How many children get the school bus?

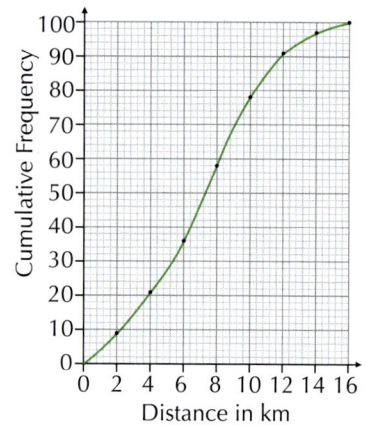

The box plot below shows the distribution of the distances that 100 children at a second school travel to school.

d) Copy the box plot for the second school and draw a box plot for the first school above it. Use 1 km as the shortest distance travelled and 15.5 km as the longest distance travelled.

e) Use your box plots to compare the distances travelled by the children at the two schools.

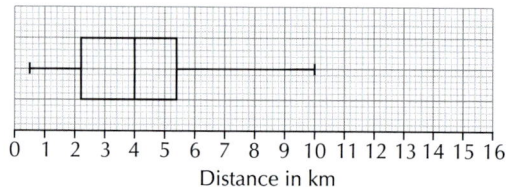

Q7 The table shows the number of components made by a machine each hour for 10 hours.

Hour	1	2	3	4	5	6	7	8	9	10
No. of components	148	151	150	150	149	150	147	142	136	131

a) Draw a time series graph to show this data.

b) The machine is designed to produce 150 components per hour. Draw a line on your graph to show the target number of components for each hour.

c) A factory worker thinks the machine should be checked for faults. Do you agree? Use the data to explain your answer.

Q8 Anton wants to buy a particular model of car. The table below shows the cost of seven of these cars that are for sale, as well as their mileage.

Mileage (×1000)	5	20	10	12	5	25	27
Cost (£)	3500	2000	3000	2500	3900	1000	500

a) Draw a scatter graph to show this data and add a line of best fit.

b) Predict the cost of a car of the same model with a mileage of 15 000 miles.

Exam-Style Questions

Q1 The scatter diagram shows the percentage obtained in a geography exam and the number of lunchtime revision classes attended by 12 students from Years 10 and 11.

a) Compare the correlations between the percentage obtained and the number of revision classes attended by Year 11 and Year 10 students.

[1 mark]

b) Duane was due to sit the exam in Year 11 but was absent. He says:

"I didn't attend any revision classes but I can see from the diagram that I would have got over 20%."

Is Duane correct? Refer to the diagram to explain your answer.

[1 mark]

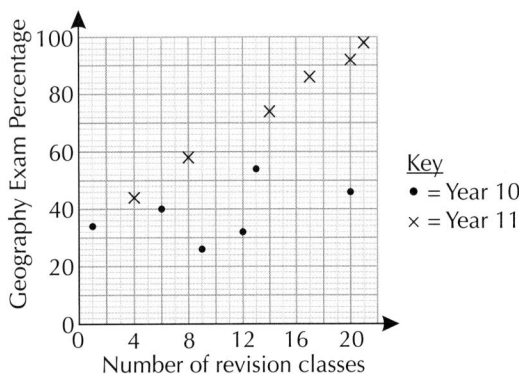

Key
• = Year 10
× = Year 11

Q2 The resting pulse rate in beats per minute (bpm) was measured for a group of athletes. The lowest rate was 44 bpm. 25% of the athletes had a rate of at least 71 bpm. The range was 32 bpm. The interquartile range was 19 bpm. The median was equal to the mean of the lower and upper quartiles. Draw a box plot to show this information.

[3 marks]

Q3 At the end of each month for a year, Midas recorded the price of gold per gram to the nearest £0.10. He displayed the data he recorded in the time series graph below.

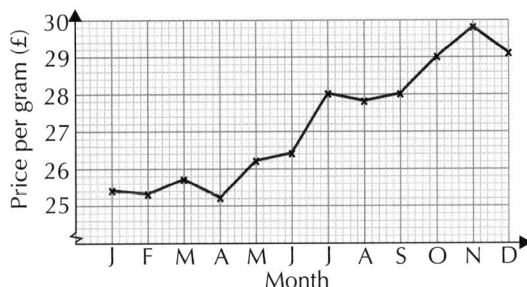

a) Describe the overall trend in gold prices over the course of the year.

[1 mark]

b) Midas owns a 12 kg gold bar. According to the graph, what is the maximum amount of money he could have made from selling the gold bar this year?

[2 marks]

Q4 Participants in a circus skills workshop were asked to juggle three balls for as long as possible. The table on the right shows some information about how long they were able to do this for.

Draw a histogram to represent this information.

[3 marks]

Time (t) in seconds	Frequency
$0 \leq t < 4$	18
$4 \leq t < 8$	12
$8 \leq t < 12$	6
$12 \leq t < 20$	4
$20 \leq t < 30$	1

Q5 Five judges each marked the bands in a competition out of 20. The mean of these five marks is the score that each band received. The scores are represented on the cumulative frequency graph below.

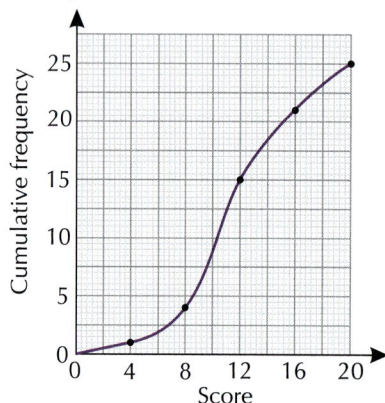

Score	Frequency
$0 < s \leq 4$	1
$4 < s \leq 8$	
$8 < s \leq 12$	
$12 < s \leq 16$	
$16 < s \leq 20$	

a) Copy and complete the frequency table above to show the bands' scores.

[2 marks]

b) Bands that scored over 14 went through to the next round of the competition. Estimate the percentage of bands that went through to the next round.

[2 marks]

Q6 Six classes competed against each other at a sports day. First, second and third places in each event won gold (G), silver (S) and bronze (B) medals. Students in Class 11R won a total of 24 medals, which are shown on the pie chart below.

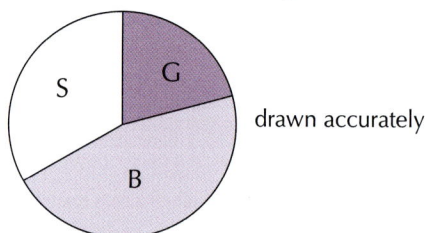

drawn accurately

Students in class 11S won a total of 18 medals. Class 11S won the same number of gold medals as Class 11R and Class 11S won the same proportion of silver medals as Class 11R. If a pie chart was drawn for the medals won by Class 11S, work out the size of the angle of the sector for bronze.

[4 marks]

34.1 Calculating Probabilities

Probability is all about measuring how likely it is for things to happen. Those things could be simple, like rolling a 6 on a dice roll, or they could be more complicated, like rolling a total of 20 over 5 dice rolls.

Calculating Probabilities

Learning Objectives — Spec Ref P2/P3:
- Find the probability of a given event.
- Know that probabilities must be between 0 and 1.

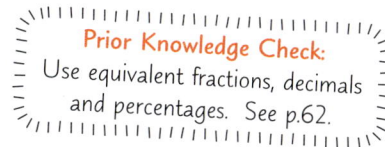

Prior Knowledge Check:
Use equivalent fractions, decimals and percentages. See p.62.

In probability, an **outcome** is the result of an activity (e.g. 'getting tails on a coin flip' or 'rolling 3 on a dice roll') and an **event** is a set of one or more outcomes to which you assign a **probability**. The **probability** of an event is a number that represents how **likely** it is for that event to happen. All probabilities are between **0 and 1**. An **impossible** event has a probability of **0** and an event that's **certain** to happen has a probability of **1**. You can show probabilities on a **scale** (i.e. a **number line** going from 0 to 1).

The probability of an event can be written as **P(event)**, and can be given as a **fraction**, **decimal** or **percentage**. For example, if you flip a fair coin, P(heads) $= \frac{1}{2} = 0.5 = 50\%$.

When **all** the possible outcomes are **equally likely**, you can work out probabilities of events using this formula:

$$P(event) = \frac{\text{number of ways the event can happen}}{\text{total number of possible outcomes}}$$

Example 1

A box contains 20 counters, numbered 1 to 20. If one counter is selected at random, work out the probability that the number is less than 12.

1. All the possible outcomes are equally likely, so use the formula. The total number of possible outcomes is the number of counters.   Total possible outcomes = 20
2. Count the number of outcomes that are less than 12.   11 ways of getting less than 12
3. Put the numbers into the formula.   P(less than 12) $= \frac{11}{20}$

If you're asked for the probability of one event **or** another, you can just **add up** the number of outcomes for each event — as long as both events **can't** happen at the **same time** (see next page).

Example 2

Aidan has 6 apples, 3 bananas, 4 pears and 2 oranges in his fruit bowl. He picks out one piece of fruit at random to eat. What is the probability that it will be an apple or a pear?

1. Work out the total number of possible outcomes (how many pieces of fruit are in the bowl).   Total fruit = 6 + 3 + 4 + 2 = 15
2. Count the number of fruit that are either apples or pears.   Apples or pears = 6 + 4 = 10
3. Use the formula and simplify the fraction.   P(apple or pear) $= \frac{10}{15} = \frac{2}{3}$

Q1 Calculate the probability of rolling a fair, six-sided dice and getting each of the following.

a) 6   b) 7   c) 4 or 5   d) a factor of 6

Q2 A standard pack of 52 playing cards is shuffled and one card is selected at random. Find the probability of selecting each of the following.

a) a club   b) an ace   c) a red card

d) the two of hearts   e) not a spade   f) a 4 or a 5

Q3 A bag contains 16 coloured balls — 2 black, 4 blue, 2 green, 3 red, 2 yellow, 1 orange, 1 brown and 1 purple. If one ball is selected at random, find the probability of getting the following colours.

a) green   b) blue or green   c) not purple   d) red, green or brown

Q4 For each of these questions, draw a copy of the spinner on the right and number the sections in a way that fits the given rule.

a) The probability of getting 2 is $\frac{3}{8}$.   b) The probability of getting 3 is $\frac{1}{2}$.

c) The probability of getting 5 and the probability of getting 6 are both $\frac{1}{4}$.

Q5 Diane has 20 pairs of socks and each pair is different. She picks one sock at random. If she then picks another sock at random from the remaining socks, what is the probability that the 2 socks make a pair?

Q6 A box of identically wrapped chocolates contains 8 caramels, 6 truffles and 4 pralines. Half of each type of chocolate are coated in milk chocolate and half are coated in white chocolate. Chelsea selects a chocolate at random. She doesn't like pralines or white chocolate, but she likes all the others.

a) What is the probability that she gets a white chocolate-coated praline?

b) What is the probability that she gets a chocolate that she likes?

Q7 This table shows information about the books Harry owns. If one of Harry's books is selected at random, what is the probability that it's a:

	Paperback	Hardback
Fiction	14	6
Non-fiction	2	8

a) paperback fiction book?   b) non-fiction book?

Probabilities that Add Up to 1

Learning Objective — Spec Ref P4:
Know that the probabilities of mutually exclusive events sum to 1.

Events that **can't happen at the same time** are called **mutually exclusive** (there's more on mutually exclusive events on p.465). For example, rolling a 1 and rolling a 3 in one dice roll are mutually exclusive events.

The probabilities of **mutually exclusive events** that cover **all possible outcomes** always **add up to 1**.

For any event, there are only two possibilities — it either happens or it doesn't happen. So:

The **probability** that an event **doesn't happen** is equal to **1 minus the probability that it does happen**.

Example 3

The probability that Kamui's train is late is 0.05. Work out the probability that his train is not late.

The train is either late or not late, so
P(train is late) + P(train is not late) = 1.

P(train is not late) = 1 − P(train is late)
= 1 − 0.05 = **0.95**

You can also use the fact that the probabilities for every possible outcome add up to 1 to find **missing probabilities**.

Example 4

A bag contains red, green, blue and white counters. This table shows the probabilities of randomly selecting a red, green or white counter from the bag.

Colour	Red	Green	Blue	White
Probability	0.2	0.1		0.5

Work out the probability of selecting a blue counter.

The probabilities of the 4 colours add up to 1, so the probability of blue is 1 minus the other 3 probabilities.

P(blue) = 1 − (0.2 + 0.1 + 0.5)
= 1 − 0.8 = **0.2**

Exercise 2

Q1 The probability that it will snow in a particular Canadian town on a particular day is $\frac{5}{8}$.
What is the probability that it won't snow there on that day?

Q2 The probability that Clara wins a raffle prize is 25%.
Find the probability that she doesn't win a raffle prize.

Q3 If the probability that Jed doesn't finish a crossword is 0.74, what's the probability he does finish it?

Q4 The probability that Gary wins a tennis match is twice the probability that he loses it.
Work out the probability that he wins a tennis match.

Q5 Everyone taking part in a certain lucky dip wins a prize.
The table on the right shows the probabilities of winning the four possible prizes.

Prize	Lollipop	Pen	Cuddly toy	Gift voucher
Probability	0.4	0.1		0.2

a) Find the missing probability.

b) What's the probability of winning a prize that's not a pen?

Q6 When two football teams play each other the probability that Team A wins is 0.4 and the probability that Team B wins is 0.15. What is the probability that the match is a draw?

Q7 One counter is selected at random from a box containing blue, green and red counters.
The probability that it's a blue counter is 0.5 and the probability that it's a green counter is 0.4.
If there are four red counters in the box, how many counters are there altogether?

Q8 A spinner has three sections, coloured pink, blue and green. The probability it lands on pink is 0.1. If the probability it lands on blue is half the probability it lands on green, find the probability that it lands on green.

34.2 Listing Outcomes

One of the trickiest bits of probability is figuring out what all the possible outcomes are, especially if you've got more than one activity. You'll need to find a way to list or count all the different possible outcomes.

Listing Outcomes

> **Learning Objective — Spec Ref P6/P7:**
> List outcomes of two or more events.

When **two things** are happening at once, e.g. a coin toss and a dice roll, it's much easier to work out probabilities if you **list all the possible outcomes** in a systematic way, so that you don't miss any outcomes. A coin toss has two outcomes (heads 'H' and tails 'T'), and a dice roll has 6 ('1', '2', '3', '4', '5' and '6'), so if both happen together, there will be 12 possible outcomes: 'H and 1', 'T and 1', 'H and 2', 'T and 2', 'H and 3', 'T and 3', etc.

It's often a good idea to use a **sample space diagram** (also called a **possibility diagram**) to write the outcomes in an **ordered** and **logical** way. It can be in the form of a **list**, a **grid** or a **two-way table**.

Example 1

Anne has three tickets for a theme park. She chooses two friends at random to go with her. She chooses one girl from Belinda, Claire and Dee, and one boy from Fred and Greg.

What is the probability that Anne chooses Claire and Fred to go with her?

1. Use a simple two-column table to list the outcomes. Write each girl in turn and fill in all the possibilities for the boys. Each row of the table is a possible outcome.

2. Count the number of rows that Claire and Fred both appear in. Then divide by the total number of rows.

   There's 1 outcome that includes both Claire and Fred, and 6 outcomes in total. So P(Claire and Fred) = $\frac{1}{6}$.

Girls	Boys
Belinda	Fred
Belinda	Greg
Claire	Fred
Claire	Greg
Dee	Fred
Dee	Greg

Exercise 1

Q1 A burger bar offers the meal deal shown on the right. Jana picks one combination of burger and drink at random.

> **Choose 1 burger and 1 drink**
> **Burgers**   **Drinks**
> Hamburger   Cola
> Cheeseburger   Lemonade
> Veggie burger   Coffee

a) What is the probability that she chooses a veggie burger and cola?

b) What is the probability that she chooses at least one of cheeseburger or coffee?

Q2 The fair spinner shown on the right is spun twice.

a) What is the probability of spinning 1 on both spins?

b) What is the probability of getting less than 3 on both spins?

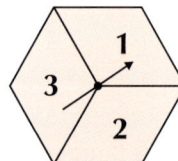

Q3 A fair coin is tossed three times. Work out the probability of getting:

a) three tails

b) no tails

c) one head and two tails

Example 2

Two fair, four-sided dice, one white and one blue, are rolled
and their scores are added together. Both are numbered 1-4.

a) List all the possible total scores.

   1. Draw a two-way table with the outcomes for one dice
      across the top and those for the other dice down the side.

   2. Fill in each square with the score for the row
      plus the score for the column.

	White dice			
	1	2	3	4
Blue dice 1	2	3	4	5
2	3	4	5	6
3	4	5	6	7
4	5	6	7	8

b) What is the probability of scoring a total of 4?

   Count how many times a total of 4
   appears in the table, then divide
   by the total number of outcomes.

   3 of the outcomes are 4, and there are
   16 outcomes in total. So P(total of 4) = $\frac{3}{16}$.

Example 3

A spinner with three equal sections, coloured red, white and blue, is spun twice.
Find the probability of spinning the same colour both times.

1. Draw a two-way table or grid to show all the possible outcomes.

2. Highlight and count the number of ways of spinning the same
   colour both times, then divide by the total number of outcomes.

   There are 3 ways of spinning the same colour, and there
   are 9 outcomes in total. So P(same colour) = $\frac{3}{9}$ = $\frac{1}{3}$.

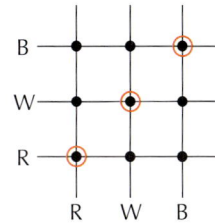

Exercise 2

Q1 Two fair six-sided dice are rolled.

   a) Copy and complete the table on the right to show all the
      possible outcomes when the scores are added together.

	1	2	3	4	5	6
1						
2						
3						
4						
5						
6						

   b) Find the probability of each of the following total scores.

      (i) 6   (ii) 12   (iii) less than 8   (iv) more than 8

Q2 Craig likes to eat curries, but he always finds it difficult to choose what type
    of rice to have. He is equally happy to eat boiled, lemon, pilau or vegetable rice.
    He decides that from now on, he's going to select his rice at random from those four options.

   a) Draw a diagram to show all the possible combinations of rice he could eat with his next two curries.
      Assume he only has one type of rice per curry.

   b) Find the probability that he eats the following types of rice with his next two curries:

      (i) pilau rice both times   (ii) the same type both times   (iii) lemon rice at least once

Q3 Hayley and Asha are playing a game. In each round they both spin a spinner with five equal
    sections labelled 1 to 5, and whoever gets the higher score wins the round. If the winner spins
    a 5, she scores 2 points, and if the winner spins less than 5, she scores 1 point. If both players
    spin the same number, no one wins a point. What is the probability that Hayley wins:

   a) exactly 1 point in the first round?   b) 2 points in the first round?

Product Rule for Counting Outcomes

When you have **lots** of events happening, listing **all** of the outcomes might be **too difficult** or take **too long**. In situations like this, you can use the **product rule** to count the total number of outcomes:

> The number of **possible outcomes** for multiple events is equal to the number of possible outcomes of **each** event **multiplied** together.

Tip: The product rule only works when the outcome of one event doesn't affect the number of outcomes of another event.

Example 4

Gary flips a coin, rolls a standard six-sided dice and draws a card from a standard 52-card deck. How many different possible outcomes are there?

1. Work out what the different events are, and how many possible outcomes each has.

   Flipping the coin = 2 possible outcomes
   Rolling the dice = 6 possible outcomes
   Drawing the card = 52 possible outcomes

2. Use the product rule to find the number of possible combinations.

   Total possible outcomes = $2 \times 6 \times 52 = \mathbf{624}$

Example 5

How many different six-digit numbers are there that only include the digits 4, 8 and 9?

1. Treat each digit of the number as its own event.

   There are 6 'events', each with 3 possible outcomes: 4, 8 or 9.

2. Use the product rule.

   Combinations = $3 \times 3 \times 3 \times 3 \times 3 \times 3$
   $= \mathbf{729}$

Tip: It would take too long to list all 729 outcomes individually.

Exercise 3

Q1 On a menu in a restaurant there are 4 starters, 8 main courses and 3 desserts to choose from.

    a) How many different three-course meals could you choose from?

    b) In a two-course meal one of the meals has to be a main course. How many different two-course meals could you choose from?

Q2 A combination lock has five rotating wheels which can each be set to one of the digits 0-6.

    a) How many different combinations could you set for the lock?

    b) How many different combinations could you set that only include odd digits?

    c) If you randomly choose a combination, what is the probability that all the wheels will be set to odd digits?

Q3 A vending machine contains 6 different flavours of fruit juice. Trish buys 4 cans of fruit juice from the vending machine. She says, "There are 1296 different combinations of fruit juice I could get." Is Trish correct? Explain your answer.

34.3 Probability from Experiments

All the probability so far in this section has been theoretical. In reality, it's difficult to know for sure that the outcomes are equally likely — for example, how can you be certain that a dice isn't more likely to roll a 6 than a 1? The best way to find out is to do experiments and see if the results agree with what you expect.

Relative Frequency

Learning Objective — Spec Ref P1/P5:
Find the relative frequency of an event.

You can **estimate** probabilities using the results of an experiment or what you know has already happened. Your estimate is the **relative frequency** (also called the **experimental probability**), which you work out using this formula:

$$\text{Relative frequency} = \frac{\text{Number of times the result happens}}{\text{Number of times the experiment is done}}$$

The **more times** you do the experiment, the **more accurate** the estimate should be.

Example 1

A biased dice is rolled 100 times. The results are on the right. Estimate the probability of rolling a 1 with this dice.

Score	1	2	3	4	5	6
Frequency	11	14	27	15	17	16

Use the relative frequency as an estimate of the probability — divide the number of times 1 was rolled by the total number of rolls.

1 was rolled 11 times, so $P(1) = \dfrac{11}{100}$

Example 2

Dan caught 64 butterflies in the butterfly house at the zoo. 16 were red, 15 were blue, 17 were green and the rest were purple. After catching each butterfly, he let it go. Estimate the probability that the next butterfly he catches is either blue or green.

1. Find the total number of blue or green butterflies that Dan caught.

   Blue or Green = 15 + 17 = 32

2. Use the relative frequency to estimate the probability.

   $P(\text{Blue or Green}) = \dfrac{32}{64} = \dfrac{1}{2}$

Exercise 1

Q1 A spinner with four sections is spun 100 times. The results are shown in the table below.

a) Estimate the probability of spinning each colour.

b) How could the estimates be made more accurate?

Colour	Red	Green	Yellow	Blue
Frequency	49	34	8	9

Q2 Stacy rolls a six-sided dice 50 times and 2 comes up 13 times.
Jason rolls the same dice 100 times and 2 comes up 18 times.

a) Use Stacy's results to estimate the probability of rolling a 2 on this dice.

b) Use Jason's results to estimate the probability of rolling a 2 on this dice.

c) Explain whose estimate should be more accurate.

Q3 Jamal records the colours of the cars passing his school. Estimate the probability, as a decimal, that the next car passing Jamal's school will be:

Colour	Silver	Black	Red	Blue	Other
Frequency	452	124	237	98	89

    a) silver   b) red   c) not silver, black, red or blue

Q4 Jack and his dad have played tennis against each other 15 times. Jack has won 8 times.
    a) Estimate the probability that Jack will win the next time they play.
    b) Estimate the probability that Jack's dad will win the next time they play.

Q5 Describe how Lilia could estimate the probability that the football team she supports will win a match.

Expected Frequency

> **Learning Objective — Spec Ref P2/P3:**
> Find the expected frequency of an event.

If you know (or have an estimate of) the **probability** of an event, you can work out how many times you would **expect** the event to happen in a given number of experiments. This isn't always going to happen in reality — it's only an **estimate** of roughly how many times you think it will occur.

> **Expected frequency = number of times the experiment is done × probability of the event happening**

Example 3

Ezra rolls a dice 60 times. Estimate the number of times he would expect to roll a 1, if the dice he is rolling is:

a) **biased with P(1) = 0.2**

    Multiply the number of rolls by the probability of rolling a 1.   Expected frequency = 60 × 0.2 = **12 times**

b) **fair, 20-sided and numbered 1-20**

    1. Work out the probability of rolling a 1. $P(1) = \dfrac{1}{20}$

    2. Find the expected frequency.   Expected frequency = $60 \times \dfrac{1}{20}$ = **3 times**

Exercise 2

Q1 The probability that a biased dice lands on 4 is 0.75. How many times would you expect to roll 4 in:
    a) 20 rolls?   b) 60 rolls?   c) 100 rolls?   d) 1000 rolls?

Q2 The spinner on the right has three equal sections. How many times would you expect to spin 'penguin' in:
    a) 60 spins?   b) 300 spins?   c) 480 spins?

Q3 A fair, six-sided dice is rolled 120 times. How many times would you expect to roll:
    a) a 5?   b) a 6?   c) an even number?   d) higher than 1?

Frequency Trees

When you have data relating to **multiple events**, you can use a **frequency tree** to record it.
The **branches** of a frequency tree show the **different possible outcomes** of each event, and the **numbers** at the end of the branches show **how many times** that event or **combination** of events happened.

Example 4

James asks 200 people of various ages to roll a dice. There were 120 people under the age of 20 and 16 of them rolled a six. Of the people aged 20 and over, 18 rolled a six.
Draw a frequency tree to show this information.

1. Draw branches to show the different possibilities.

2. Fill in the bits of the tree given in the question first — the total number of people goes at the start.

3. The numbers at the end of a set of branches should always add up to the number at the start of the branches.

    200 – 120 = 80 people are aged 20 and over
    120 – 16 = 104 people under 20 didn't roll a six
    80 – 18 = 62 people aged 20 and over didn't roll a six

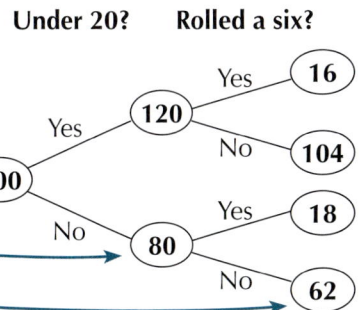

Under 20?   **Rolled a six?**

Yes — 120: Yes — 16, No — 104
No — 80: Yes — 18, No — 62
Start: 200

Example 5

The frequency tree on the right shows weather data for Statston collected on 30 random days of the year. Use the frequency tree to estimate the probability of any day in Statston being:

a) **rainy**   b) **rainy and windy**   c) **windy**

a) Read the number of rainy days from the frequency tree and use the relative frequency to estimate the probability.

There were 12 rainy days
So $P(\text{rainy}) = \frac{12}{30} = \frac{2}{5}$

b) Follow the top branches to find the number of days that were both rainy and windy.

8 days were both rainy and windy
So $P(\text{rainy and windy}) = \frac{8}{30} = \frac{4}{15}$

c) Add up the number of windy days when it did rain and when it didn't rain to find the total number of windy days.

Number of windy days = 8 + 7 = 15
So $P(\text{windy}) = \frac{15}{30} = \frac{1}{2}$

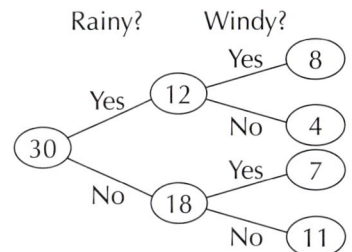

Rainy?   **Windy?**

Yes — 12: Yes — 8, No — 4
No — 18: Yes — 7, No — 11
Start: 30

Exercise 3

Q1 Information on 800 UK residents' hair and eye colour was collected. The data was recorded in the two-way table shown on the right.

a) Show this information on a frequency tree.

b) Find the relative frequency of someone having brown hair and brown eyes.

c) If you collected the same data on another group of 2000 UK residents, how many would you estimate won't have brown eyes or brown hair?

		Has brown hair?	
		Yes	No
Has brown eyes?	Yes	220	105
	No	335	140

Q2 720 truck drivers were asked to do an eye test. 640 said their vision was fine. 180 of the drivers who said their vision was fine failed the eye test. 30 of the drivers who said their vision wasn't fine passed the eye test. Draw and complete a frequency tree to represent this information.

Fair or Biased?

Learning Objective — Spec Ref P2:
Use experimental data to decide if outcomes are fair or biased.

Things like dice and spinners are **fair** if they have the same chance of landing on each side or section. If they're more likely to give some outcomes than others, then they're called **biased**.
To decide whether something is fair or biased, you need to do an **experiment**.
You can then **compare** the experimental results with what you would expect in theory.

For example, if you flip a coin, you expect it to land on heads about half of the time — so if you flipped it 100 times, you'd expect **about** 50 heads. If you actually got 80 heads, you'd be pretty convinced that the coin was **biased**.

Example 6

The table shows the results of 60 rolls of a 6-sided dice.

Score	1	2	3	4	5	6
Frequency	12	3	9	15	9	12

a) **Calculate the relative frequency of each outcome, giving your answers as decimals.**

Divide each frequency by 60 to find the relative frequency for each score.

score 1 = $\frac{12}{60}$ = **0.2**   score 2 = $\frac{3}{60}$ = **0.05**   score 3 = $\frac{9}{60}$ = **0.15**

score 4 = $\frac{15}{60}$ = **0.25**   score 5 = $\frac{9}{60}$ = **0.15**   score 6 = $\frac{12}{60}$ = **0.2**

b) **Do you think the dice is fair or biased? Explain your answer.**

Compare the relative frequencies to the theoretical probability of $\frac{1}{6}$ = 0.166...

The relative frequency of a score of 2 is very different from the theoretical probability, so the experiment suggests that the dice is **biased**.

Exercise 4

Q1 A spinner has four sections: blue, green, white and pink. The table shows the results of 100 spins.

a) Work out the relative frequencies of the four colours.

b) Explain whether you think the spinner is fair or biased.

Colour	Blue	Green	White	Pink
Frequency	22	21	18	39

Q2 A six-sided dice is rolled 120 times and 4 comes up 32 times.

a) How many times would you expect 4 to come up on a fair dice in 120 rolls?

b) Use your answer to part a) to explain whether you think the dice is fair or biased.

Q3 Three friends each toss a coin and record the number of heads they get. The table shows their results.

a) Copy and complete the table.

b) Explain whose results are the most reliable.

c) Explain whether you think the coin is fair or biased.

	Amy	Steve	Hal
Number of tosses	20	60	100
Number of heads	12	33	49
Relative frequency			

Review Exercise

Q1 280 pupils have to choose whether to study history or geography. The numbers choosing each subject are shown in the table.

	History	Geography
Girls	77	56
Boys	63	84

A pupil is chosen at random. Find the probability that the pupil is:

a) a girl studying geography

b) a boy studying geography

c) a girl

d) a boy

e) studying history

f) not studying history

Q2 Amy's CD collection is organised into four categories — groups, male vocal, female vocal and compilations. The table shows the probability that she randomly picks a CD of each category.

Category	Groups	Male vocal	Female vocal	Compilations
Probability	0.45	0.15	0.1	

a) Find the missing probability from the table.

b) Amy picks one CD at random. What is the probability that it isn't a male vocal CD?

c) Given that Amy has 80 CDs in total, how many female vocal CDs does she have?

Q3 How many different 4 digit even numbers are there where the first digit is odd and the second digit is a multiple of 3?

Q4 The probability that a particular component produced in a factory is faulty is 0.002.

a) What is the probability that the component is not faulty?

b) 60 000 of these components are produced in a week.
How many of the components would you expect to be: (i) faulty? (ii) not faulty?

Q5 The two fair spinners shown on the right are both spun once.

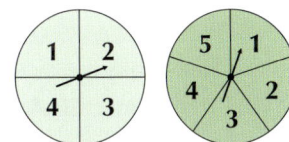

a) Show all the possible outcomes using a sample space diagram.

b) Find the probabilities of spinning the following:

   (i) double 4   (ii) a 2 and a 3   (iii) both numbers the same

   (iv) two different numbers   (v) at least one 5   (vi) no fives

c) If the five-sided spinner is spun 100 times, how many times would you expect to spin 3?

Q6 The results of rolling a four-sided dice 200 times are shown in the table on the right.

Score	1	2	3	4
Frequency	56	34	54	56

a) Work out the relative frequencies of the dice scores.

b) Explain whether these results suggest that the dice is fair or biased.

c) Estimate the probability or rolling a 1 or a 2 with this dice.

Exam-Style Questions

Q1 Ian asked 16 students in one of the Year 11 classes at his school whether they went to the cinema last week. 5 boys said they did. The other 2 boys said they did not. Two thirds of the girls said they did whilst the other girls said they did not.

a) Copy and complete the frequency tree below to show Ian's data.

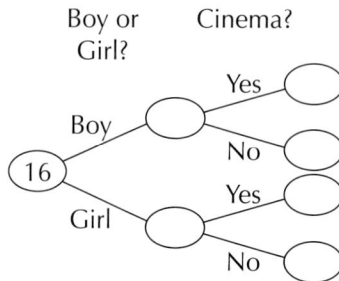

Boy or Girl?   Cinema?

[3 marks]

b) If one of the 16 students is picked at random, work out the probability that they went to the cinema last week.

[2 marks]

Q2 Ellie is playing a game where she has to roll a 6-sided dice, with sides labelled A, B, C, D, E and F. She rolls the dice 4 times in a row to make a 4-letter sequence (the first result is the first letter, the second result is the second letter, etc.).

a) How many different 4-letter sequences can she make?

[2 marks]

b) What is the probability that she gets the sequence 'BEAD'?

[1 mark]

Ellie thinks the dice might not be fair. She writes down the next 10 sequences she makes:

CEAA, ABAA, ACFD, AECE, AFAC, DAEA, DFAE, ADED, AABF, CCAC

c) Copy and complete this table showing the frequency and relative frequency of each letter.

Letter	A	B	C	D	E	F
Frequency						
Relative Frequency						

[3 marks]

d) Do you think the dice is fair or biased? Explain your answer.

[2 marks]

e) If the dice was rolled another 200 times, estimate how many more times it would land on A than B.

[2 marks]

35.1 The AND Rule for Independent Events

One of the most useful probability rules is the AND rule — it lets you work out the probability that one event AND another will both happen. However, it only works when the events don't affect each other.

Learning Objectives — Spec Ref P8:
- Know what it means for events to be independent.
- Use the AND rule to find probabilities.

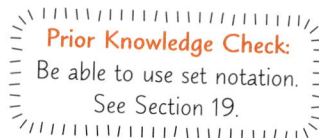

> **Prior Knowledge Check:**
> Be able to use set notation.
> See Section 19.

Events are **independent** if one of them happening has **no effect** on the probability that the others happen.

For example, imagine drawing **two cards** at random from a standard deck of playing cards. If you **put the card back** and shuffle the deck, then the first card drawn **doesn't affect** the second card drawn, so the events are **independent**. However, if you **don't** put the first card back, then the probabilities for the second draw **depend** on the result of the first draw — e.g. the probability of selecting a heart on the second draw is $\frac{12}{51}$ (if the first card **was** a heart) or $\frac{13}{51}$ (if the first card **wasn't** a heart). So the events are **not** independent.

If A and B are **independent** events, then the **AND rule** says that the probability of **both A and B** happening is equal to the probability of A happening **multiplied** by the probability of B happening.

$$P(A \text{ and } B) = P(A) \times P(B) \quad \text{or} \quad P(A \cap B) = P(A) \times P(B) \text{ in set notation.}$$

Example 1

A biased dice has a probability of 0.2 of landing on 6. The dice is rolled twice. What is the probability that: a) **both rolls are a 6?**   b) **neither roll is a 6?**

a)   The first roll has no effect on the second roll, so the events '6 on first roll' and '6 on second roll' are independent.

$P(\text{both rolls are 6}) = P(6 \text{ on first roll}) \times P(6 \text{ on second roll})$
$= 0.2 \times 0.2 = \textbf{0.04}$

b)   1.   On any roll, either a 6 is rolled or a 6 isn't rolled, so $P(6) + P(\text{not } 6) = 1$.

     2.   The events 'not 6 on first roll' and 'not 6 on second roll' are independent.

$P(\text{not } 6) = 1 - P(6) = 1 - 0.2 = 0.8$
$P(\text{neither roll is 6})$
$= P(\text{not 6 on first roll}) \times P(\text{not 6 on second roll})$
$= 0.8 \times 0.8 = \textbf{0.64}$

The AND rule works for **more than 2 independent events** — e.g. $P(A \text{ and } B \text{ and } C) = P(A) \times P(B) \times P(C)$.

Example 2

Jeff, Britta and Annie go to the same library. On any day, the probability that Jeff goes to the library is 0.6, the probability that Britta goes is 0.4 and the probability that Annie goes is 0.75.

a) **In context, describe what assumption needs to be made in order to use the AND rule to find the probability of more than one of Jeff, Britta and Annie visiting library on any given day.**

You need to assume that the probability that each person goes to the library on any day is **independent** of whether or not the others go — i.e. they are not more (or less) likely to go if one of the others goes.

b) Find the probability that Annie and Britta both go to the library on any one day.

Use the AND rule: P(Annie and Britta go to the library) = P(Annie goes) × P(Britta goes)
                                                  = 0.75 × 0.4 = **0.3**

c) Find the probability that all three go to the library on any one day.

The AND rule works for   P(All three go to the library)
3 independent events too.   = P(Jeff goes) × P(Britta goes) × P(Annie goes) = 0.6 × 0.4 × 0.75 = **0.18**

Exercise 1

Q1   Say whether each of these pairs of events are independent or not.

   a) Tossing a coin and getting heads, then tossing the coin again and getting tails.

   b) Selecting a coffee-flavoured chocolate at random from a box of assorted chocolates. Then, after eating the first chocolate, randomly selecting another coffee-flavoured chocolate from the same box.

   c) Rolling a 6 on a dice and randomly selecting a king from a pack of cards.

Q2   A fair coin is tossed and a fair six-sided dice, numbered 1 to 6, is rolled. Find:

   a) P(heads $\cap$ 6)   b) P(heads $\cap$ odd number)   c) P(tails $\cap$ square number)

   d) P(tails $\cap$ prime number)   e) P(heads $\cap$ multiple of 3)   f) P(tails $\cap$ factor of 5)

Q3   Prince is doing a survey in his school. He picks pupils at random to do the survey. The probability of picking a left-handed pupil is 10% and the probability of picking a pupil who wears glasses is 15%. Assuming that the hand they write with and glasses wearing are independent, find the probability that a pupil picked at random:

   a) is right-handed   b) doesn't wear glasses

   c) wears glasses and is left-handed   d) doesn't wear glasses and is right-handed

Q4   A bag contains ten coloured balls. Five of them are red and three of them are blue. A ball is taken from the bag at random, then replaced. A second ball is then selected at random. Find the probability that:

   a) both balls are red   b) neither ball is red

   c) the first ball is red and the second ball is blue

Q5   The probability of Selvi passing some exams is shown in the table. Assuming that her passing any subject is independent of her passing any of the other subjects, find the probability that she:

Subject	Maths	English	Geography	Science
Probability	0.8	0.6	0.3	0.4

   a) passes English and Maths   b) passes Maths and Geography

   c) passes Maths and fails Science

Q6   The probability of randomly selecting an ace from a pack of cards is $\frac{4}{52}$. Len claims that the probability of randomly selecting an ace, then randomly selecting another ace (without replacing the first ace) is $\frac{4}{52} \times \frac{4}{52} = \frac{1}{169}$. Say whether you agree with Len and explain why.

35.2 The OR Rule

The OR rule helps you to work out the probability that either one event OR another will happen.
How it works depends on whether or not it's possible for both events to happen at the same time.

The OR Rule for Mutually Exclusive Events

Learning Objectives — Spec Ref P4/P8:
- Know what it means for events to be mutually exclusive.
- Use the OR rule to find probabilities for mutually exclusive events.

Two events are **mutually exclusive** if they **can't both happen** at the same time (see p.452) — for example, rolling a 2 and a 3 on the same dice roll, or scoring over 70% and less than 40% on the same test.

The **OR rule** (or **addition rule**) says that if events are **mutually exclusive**, then the probability that **at least** one event happens is the **sum** of the probabilities of each event happening.

$$P(A \text{ or } B) = P(A) + P(B) \quad \text{or} \quad P(A \cup B) = P(A) + P(B) \text{ in set notation.}$$

Example 1

A bag contains red, yellow and blue counters. The probabilities of randomly selecting each colour are shown in the table opposite.

Red	Yellow	Blue
0.3	0.5	0.2

Find the probability that a randomly selected counter is red or blue.

The counter can't be both red and blue, so the events 'counter is red' and 'counter is blue' are mutually exclusive.

$$P(\text{red or blue}) = P(\text{red}) + P(\text{blue})$$
$$= 0.3 + 0.2 = \mathbf{0.5}$$

Exercise 1

Q1 A fair spinner has eight sections labelled 1 to 8.
Say whether these pairs of events are mutually exclusive or not.
a) The spinner landing on 6 and the spinner landing on 3.
b) The spinner landing on 2 and the spinner landing on a factor of 6.
c) The spinner landing on a number less than 4 and the spinner landing on a number greater than 3.

Q2 A bag contains some coloured balls. The probability of randomly selecting a pink ball is 0.5, a red ball is 0.4 and an orange ball is 0.1. A ball is picked out at random. Find:
a) $P(\text{pink} \cup \text{orange})$ b) $P(\text{pink} \cup \text{red})$ c) $P(\text{red} \cup \text{orange})$

Q3 Chocolates in a box are wrapped in four different colours of foil — gold, silver, red and blue. The table shows the probabilities of randomly picking a chocolate wrapped in each colour.
Find the probability of picking a chocolate wrapped in:

Colour	Gold	Silver	Red	Blue
Probability	0.4	0.26	0.14	0.2

a) red or gold foil   b) silver or red foil   c) gold or blue foil   d) silver or gold foil

Q4 On sports day, pupils are split into three equal teams — the Eagles, the Falcons and the Ospreys. What is the probability that a pupil picked at random belongs to:
a) the Eagles?   b) the Eagles or the Falcons?   c) the Falcons or the Ospreys?

Q5 Jane is told that in a class of 30 pupils, 4 wear glasses and 10 have blonde hair. She says that the probability of a pupil picked at random from the class having blonde hair or wearing glasses is $\frac{10}{30} + \frac{4}{30} = \frac{14}{30}$. Say whether you agree with Jane and explain why.

Q6 The owner of a cafe records the sandwich fillings chosen by customers who buy a sandwich one lunchtime. The table shows the probabilities that a randomly chosen sandwich buyer chose each of five fillings.

Cheese	Tuna	Salad	Pickle	Mayo
0.54	0.5	0.26	0.22	0.28

Customers can choose any combination of fillings, apart from pickle and mayonnaise together.

a) Find the probability that a randomly selected customer chose pickle or mayonnaise.

b) Explain why the probabilities add up to more than 1.

The General OR Rule

Learning Objective — Spec Ref P8:
Use the OR rule for events that aren't mutually exclusive.

When two events are **not mutually exclusive** (i.e. they **can both happen** at the same time), you need a different version of the OR rule. The **general OR rule** says that for any two events, the probability that **at least one event happens** is equal to the **sum** of the probabilities of **each** event happening **minus** the probability that **both** events happen.

$$P(A \text{ or } B) = P(A) + P(B) - P(A \text{ and } B) \quad \text{or} \quad P(A \cup B) = P(A) + P(B) - P(A \cap B) \text{ in set notation.}$$

To help you visualise this, think about **sets A and B** on a Venn diagram. If there is an **overlap** between sets A and B (i.e. things can be in both set A and set B), then $n(A) + n(B)$ will add things in the overlap **twice**, so you need to **subtract n(A and B)**.

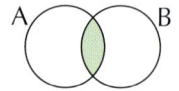

The OR rule for **mutually exclusive** events (see previous page) is actually a **special case** of this rule — if A and B can't happen at the same time, then $P(A \text{ and } B) = 0$, so the OR rule becomes: $P(A \text{ or } B) = P(A) + P(B) - 0 = P(A) + P(B)$.

Example 2

The fair spinner shown on the right is spun once. What is the probability that it lands on a 2 or a shaded sector?

1. The spinner can land on both 2 and a shaded sector at the same time, so the events are not mutually exclusive. Find the probability of the events you're interested in.

    Six sectors on the spinner are numbered 2, so $P(2) = \frac{6}{10}$

    There are five shaded sectors, so $P(\text{shaded}) = \frac{5}{10}$

    There are two shaded sectors numbered 2, so $P(2 \text{ and shaded}) = \frac{2}{10}$

2. Now use the OR rule.

    $P(2 \text{ or shaded}) = P(2) + P(\text{shaded}) - P(2 \text{ and shaded})$
    $$= \frac{6}{10} + \frac{5}{10} - \frac{2}{10} = \mathbf{\frac{9}{10}}$$

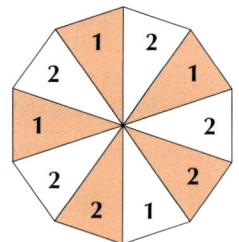

Tip: In this example it would be easy to just count the outcomes, but if you only knew the probabilities you'd have to use the OR rule.

Example 3

The two-way table on the right shows the probabilities for various events.
Find P(X ∪ A).

	A	A′	Totals
X	0.15	0.15	
Y			0.45
Z	0.1		
Totals		0.55	1

1. Use the OR rule to rewrite P(X ∪ A) in terms of P(X), P(A) and P(X ∩ A).

   P(X ∪ A) = P(X) + P(A) − P(X ∩ A)
   P(X ∩ A) = 0.15 from the table.

2. P(X) is the total of the X row of the table.

   P(X) = 0.15 + 0.15 = 0.3

3. P(A) + P(A′) should be 1.

   P(A) = 1 − 0.55 = 0.45

4. Use the formula to find P(X ∪ A).

   P(X ∪ A) = 0.3 + 0.45 − 0.15 = **0.6**

Exercise 2

Q1 A fair 20-sided dice is rolled. What is the probability that it lands on:

   a) a multiple of 3 or an odd number?   b) a factor of 20 or a multiple of 4?

Q2 A card is randomly chosen from a standard pack of 52 playing cards.

   a) What's the probability that it's a red suit or a queen?

   b) What's the probability that it's a club or a picture card?

Q3 The table shows the number of boys and girls in each year group at a school. What is the probability that a randomly chosen pupil is:

	Boys	Girls
Year 9	240	310
Year 10	305	287
Year 11	212	146

   a) in Year 11 or a girl?

   b) a boy or not in Year 9?

Q4 Sumi organises some elements into two sets, A and B.
n(A) = 28, n(B) = 34, n(A ∩ B) = 12, n(ξ) = 100.
Calculate the probability that a randomly chosen element is in set A or set B.

Q5 The frequency tree shows some information about scores achieved by girls and boys on a maths test. Use the frequency tree to find the probability that a randomly chosen student is:

   a) a girl or scored no more than 70% on the maths test.

   b) a boy or scored more than 70% on the maths test.

   c) a boy who scored no more than 70% or a girl who scored more than 70%.

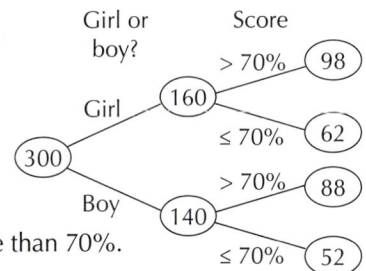

Q6 The Venn diagram on the right shows three events A, B and C.
A and B are independent.
A and C are mutually exclusive.
B and C are mutually exclusive.

Use the Venn diagram to derive an expression for P(A ∪ B ∪ C) in terms of P(A), P(B) and P(C).

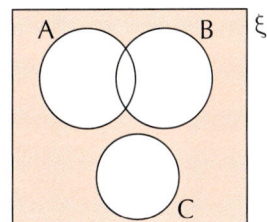

35.3 Using the AND/OR Rules

A lot of tricky probability questions require you to use both the AND rule and the OR rule together.
Don't forget that you need to know that the events are independent before you can use the AND rule.

Learning Objective — Spec Ref P8:
Use the AND and OR rules to solve probability problems.

You can work out more complicated probability questions by **breaking them down** into smaller chunks.

You should always start by figuring out what **events** are happening and whether or not they are **independent** or **mutually exclusive**. Remember that independent events **don't affect each other** and mutually exclusive events **can't both happen** at the same time.

Once you know what your events are, write down what **probabilities** you're looking for, then use the **AND** and **OR rules** as necessary to work them out.

Example 1

A biased dice lands on each number with the probabilities shown in the table opposite. The dice is rolled twice. Find the probability of rolling exactly one 2 and one 3.

1	2	3	4	5	6
0.1	0.15	0.2	0.2	0.3	0.05

1. There are two options for rolling one 2 and one 3.

   (2 and then 3) or (3 and then 2)

2. Find the probability of getting 2 on the first roll and 3 on the second roll. These events are independent, so you can multiply the probabilities.

   P(2 and then 3) = P(2) × P(3)
   = 0.15 × 0.2 = 0.03

3. Find the probability of getting 3 on the first roll and 2 on the second roll.

   P(3 and then 2) = P(3) × P(2)
   = 0.2 × 0.15 = 0.03

4. Find the probability of getting the first option or the second option. These events are mutually exclusive, so you can add the probabilities.

   P((2 and then 3) or (3 and then 2))
   = 0.03 + 0.03 = **0.06**

Example 2

In an online game, you can purchase armour boxes containing one piece of armour. Armour is given a material and a type at random, with the following independent probabilities:

      Material: **P(Leather) = 0.5**   **P(Steel) = 0.35**   **P(Gold) = 0.15**

      Type: **P(Chest) = 0.6**   **P(Leg) = 0.4**

a) Wizards can only use leather or gold armour. Find the probability that an armour box contains a piece of leg armour that a Wizard can use.

1. There are two possibilities — leather leg armour or gold leg armour. The probabilities for each are independent, so use the AND rule.

   P(Leather and legs) = P(Leather) × P(Leg)
   = 0.5 × 0.4 = 0.2

   P(Gold and legs) = P(Gold) × P(Leg)
   = 0.15 × 0.4 = 0.06

2. Now you can use the OR rule.

   P(Leather legs or gold legs) = 0.2 + 0.06 = **0.26**

b) **Fliss wants to get some gold chest armour for her Wizard character, so she buys two boxes. What is the probability that she gets gold chest armour from exactly one box?**

1. First, write down what you're trying to find — use the OR rule to add the results that give exactly 1 gold chest.

    P(one gold chest) = P(gold chest and not gold chest) + P(Not gold chest and gold chest)

2. Find the probability that a box contains gold chest armour using the AND rule.

    P(gold chest) = P(gold) × P(chest) = 0.15 × 0.6 = 0.09

3. Use this to work out the probability that a box doesn't give gold chest armour.

    P(not gold chest) = 1 − P(gold chest) = 1 − 0.09 = 0.91

4. Use the AND rule to find the probabilities that the gold chest comes from each box.

    P(gold chest and not gold chest) = 0.09 × 0.91 = 0.0819
    P(not gold chest and gold chest) = 0.91 × 0.09 = 0.0819

5. Finally add the probabilities of each result.

    P(one gold chest) = 0.0819 + 0.0819 = **0.1638**

Exercise 1

Q1 A fair coin is tossed twice. Work out the probability of getting:

a) 2 heads

b) 2 tails

c) both tosses the same

Q2 All of Justin's shirts are either white or black and all his trousers are either black or grey.
The probability that he chooses a white shirt on any day is 0.8.
The probability that he chooses black trousers on any day is 0.55.
His choice of shirt colour is independent of his choice of trousers colour.

On any given day, find the probability that Justin chooses:

a) a white shirt and black trousers

b) a black shirt and black trousers

c) a black shirt and grey trousers

d) either a black shirt or black trousers, but not both

Q3 A spinner has four sections labelled A, B, C and D.
The probabilities of landing on each section are shown in the table below.

A	B	C	D
0.5	0.15	0.05	0.3

If the spinner is spun twice, find the probability of spinning:

a) A both times

b) B both times

c) not C on either spin

d) C, then not C

e) not C, then C

f) C on exactly one spin

Q4 A class of 30 pupils were asked how they normally travel to school.
The table below shows their responses.

	Car	Bus	Walk	Cycle
Girls	5	6	3	1
Boys	2	7	2	4

One boy and one girl from the class are picked at random. Find the probability that:

a) both travel by bus

b) the girl walks and the boy travels by car

c) exactly one of them cycles

d) at least one of them cycles

35.4 Tree Diagrams

*Tree diagrams can be used to show all the possible results from an experiment
— they're really useful for working out probabilities of combinations of events.*

Learning Objective — Spec Ref P6/P8:
Use tree diagrams to represent events and to find probabilities.

Tree diagrams are similar to **frequency trees** (see p.459), except they show the **probabilities** of the
events rather than the frequency from an experiment. The really useful thing about tree diagrams
is that you can find the probability of specific **results** (e.g. P(A happens and B doesn't happen))
by **multiplying along the branches** that you follow to get to that result. Since branches from the
same point show all the outcomes of a single activity, their probabilities should **add up to 1**.

Example 1

Two fair coins are tossed.

a) **Draw a tree diagram showing all the possible results for the two tosses.**

   1. Draw a set of branches for the first toss.
       You need 1 branch for each of the 2 results.

   2. Draw a set of branches for the second toss.
       Again, there are two possible results.

   3. Write on the probability for each branch.

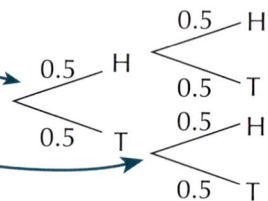

b) **Find the probability of getting 2 heads.**

   Multiply along the branches for heads then heads.   P(H then H) = 0.5×0.5 = **0.25**

c) **Find the probability of getting heads and tails in any order.**

   1. Multiply along the branches for heads then tails.   P(H then T) = 0.5×0.5 = 0.25

   2. Multiply along the branches for tails then heads.   P(T then H) = 0.5×0.5 = 0.25

   3. Both these results give H and T, so add the probabilities.   P(1H and 1T) = $0.25 + 0.25$ = **0.5**

Exercise 1

Q1   Copy and complete the following tree diagrams.

    a) A fair spinner has five equal sections
        — three are red and two are blue.
        The spinner is spun twice.

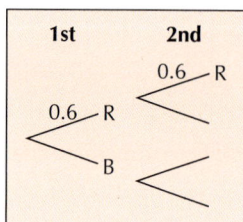

    b) A bag contains ten coloured balls — five red, three
        blue and two green. A ball is selected at random
        and replaced, then a second ball is selected.

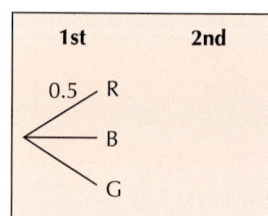

Q2 For each of these questions, draw a tree diagram showing all the possible results. Write the probability on each branch.

a) A biased coin lands on heads with a probability of 0.65. The coin is tossed twice.

b) A fair six-sided spinner has two sections labelled 1, two labelled 2 and two labelled 3. The spinner is spun twice.

c) The probability a football team wins is 0.7, draws is 0.1 and loses is 0.2. The team plays three matches.

Q3 The probability that Freddie beats James at snooker is 0.8. They play two games of snooker.

a) Draw a tree diagram to show all the possible results for the two games.

b) Find the probability that Freddie wins: (i) both games   (ii) exactly one of the games

Q4 On any Saturday, the probability that Alex goes to the cinema is 0.7. Use a tree diagram to work out the probability that Alex either goes to the cinema on all three of the next three Saturdays, or doesn't go to the cinema on any of the next three Saturdays.

Q5 Layla rolls a fair six-sided dice three times. Use a tree diagram to find the probability that she rolls:

a) 'less than 3' on all three rolls   b) 'less than 3' once and '3 or more' twice

If working out the probability of an event happening involves adding up probabilities for **lots of results**, it might be quicker to find the probability of the event **not happening** and then **subtracting** it from 1.

Example 2

Two bags each contain red, green and white balls. This tree diagram shows all the possible results when one ball is picked at random from each bag.

Find the probability that the two balls picked are different colours.

1. The quickest way to do this is to work out the probability that they're the same colour, then subtract this from 1.

2. There are 3 combinations which give the same colour.
    P(same) = P(R, R) + P(G, G) + P(W, W)
              = (0.4 × 0.2) + (0.4 × 0.1) + (0.2 × 0.7) = 0.26

3. P(different colours) = 1 – P(same colour)
    P(different colours) = 1 – 0.26 = **0.74**

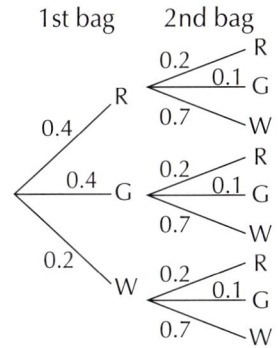

1st bag   2nd bag

R 0.4
   0.2 → R
   0.1 → G
   0.7 → W

G 0.4
   0.2 → R
   0.1 → G
   0.7 → W

W 0.2
   0.2 → R
   0.1 → G
   0.7 → W

Exercise 2

Q1 Sally owns 12 DVDs, four of which are comedies, and Jesse owns 20 DVDs, eight of which are comedies. They each select one of their DVDs at random to watch over the weekend.

a) Draw tree diagram showing the probabilities of each choice being 'comedy' or 'not a comedy'.

b) Find the probability that: (i) neither chooses a comedy   (ii) at least one chooses a comedy

Q2 A fair spinner has six equal sections — three red, two blue and one yellow. It is spun twice. Use a tree diagram to find the probability of spinning:

a) blue then red   b) red once and blue once   c) the same colour twice

d) at least two different colours e) not yellow then not yellow   f) yellow at least once

35.5 Conditional Probability

Earlier, you saw the AND rule for events that were independent. When you're looking at events that are dependent, you need to figure out how they affect each other by working out the conditional probability.

The AND Rule for Dependent Events

> **Learning Objective — Spec Ref P8/P9:**
> Understand conditional probability and use it to find probabilities of dependent events.

Conditional probabilities tell you the probability of an event, **given that** another has happened. They're used when dealing with **dependent events**, where the probabilities **change** depending on what else happens. For example, the probability that it rains tomorrow might depend on whether or not it has rained today. In this case, you might find that P(rain tomorrow, **given that** it rained today) is **higher** than P(rain tomorrow) — **knowing** that it rained today **increases** how likely you think it is to rain tomorrow.

If **A** and **B** are **dependent** events: $P(A \text{ and } B) = P(A) \times P(B \text{ given } A)$

The AND rule for **independent** events from p.463 is a **special case** of this — if A and B are independent, then P(B given A) will be **the same** as P(B), since A happening doesn't affect the probability of B happening. Then the AND rule becomes: $P(A \text{ and } B) = P(A) \times P(B \text{ given } A) = \mathbf{P(A) \times P(B)}$.

Conditional probabilities often come up in situations where objects are selected **without replacement**, where the probabilities for the second selection **depend** on what the first selection was.

Example 1

A box of chocolates contains 10 white chocolates and 10 milk chocolates.

a) **Jane picks a chocolate at random. Find the probability that it's a white chocolate.**

   Use the formula for equally likely outcomes.   20 chocolates in the box, 10 are white chocolates
   $$P(\text{Jane picks white}) = \frac{10}{20} = \frac{1}{2}$$

b) **Given that Jane has picked a white chocolate, what's the probability that Geoff randomly picks a white chocolate?**

   1. Consider the situation where Jane has already picked a white chocolate.   19 chocolates in the box, 9 are white chocolates

   2. Now use the formula for equally likely outcomes again. $P(\text{Geoff picks white given Jane picked white}) = \dfrac{9}{19}$

c) **Find the probability that both Jane and Geoff pick a white chocolate.**

   The events are dependent so use the formula given above and your answers to a) and b).

   P(Jane and Geoff both pick white)
   = P(Jane picks white) × P(Geoff picks white given Jane picked white)
   $$= \frac{10}{20} \times \frac{9}{19} = \frac{90}{380} = \frac{9}{38}$$

   The tree diagram for every possible result is shown on the right — notice that the probability of Geoff picking a white chocolate changes depending on what Jane picked.

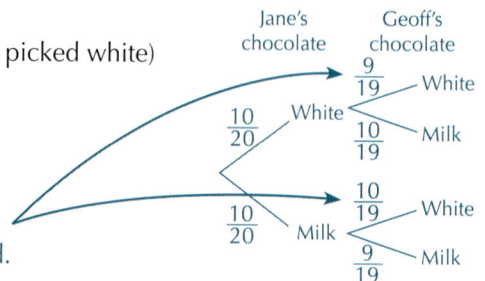

   Jane's chocolate   Geoff's chocolate

   $\frac{10}{20}$ White — $\frac{9}{19}$ White, $\frac{10}{19}$ Milk

   $\frac{10}{20}$ Milk — $\frac{10}{19}$ White, $\frac{9}{19}$ Milk

---

Q1   A bag contains ten balls numbered 1 to 10. Ball 8 is selected at random and not replaced. Find the probability that the next ball selected at random is ball 7.

Q2   In a Year 11 class there are 16 boys and 14 girls. Two names are picked at random. Given that the first student picked is a girl, what is the probability that the second student picked is:
a) a girl?   b) a boy?

Q3   The members of a drama group are choosing their characters for a pantomime. There are nine left to choose — four wizards, three elves and two toadstools. Nobody wants to play the toadstools, so the names of the nine characters are put into a bag so they can be selected at random. John is going to pick first, followed by Kerry.
a) If John picks an elf, what is the probability that Kerry picks:
    (i) an elf?   (ii) a wizard?   (iii)a toadstool?
b) If John picks a toadstool, what is the probability that Kerry also picks a toadstool?

Q4   A club is drawn at random from a standard pack of cards and not replaced. Find the probability that the next card drawn at random:
a) is a club   b) is not a club
c) is a diamond   d) is a black card
e) is a red card   f)   has the same value as the first card

Q5   The probability of rolling a 4 on an unfair dice is 0.1.
The probability that in two rolls you score a total of 10, given that a 4 is rolled first, is 0.4.
What is the probability of rolling a 4 on your first roll and having a total score of 10 after two rolls?

Q6   The probability that Shaun has a stressful day at work is 0.6. If Shaun doesn't have a stressful day at work, then the probability that he eats ice cream in the evening is 0.2. What is the probability that Shaun doesn't have a stressful day and eats ice cream?

Q7   A teacher wants to randomly select two students from his class to be on the school quiz team. There are 12 boys and 8 girls in the class.
a) If the first student chosen is a girl, find the probability that the second one chosen is a girl.
b) If the first student chosen is a boy, find the probability that the second one chosen is a girl.
c) Draw a tree diagram showing all the possible results for the two choices.
d) Find the probability that: (i) two girls are chosen.   (ii) two boys are chosen.

Q8   Three cards are selected at random from ten cards labelled 1 to 10. Find the probability that:
a) all three cards are even numbers   b) one card is odd and the other two are even

Q9   When Tom goes to his favourite Italian restaurant he always orders pizza or pasta.
The probability that he orders pizza is 0.5 if he ate pizza last time, but 0.9 if he ate pasta last time. Given that he ate pizza last time, find the probability that:
a) He orders pizza on each of the next three times he eats there.
b) He doesn't order pizza on any of the next three times he eats there.
c) He orders pizza on two of the next three times he eats there.

Conditional Probability Using Venn Diagrams

Learning Objective — Spec Ref P4/P6/P9:
Use Venn diagrams to find conditional probabilities.

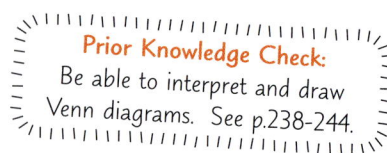

Prior Knowledge Check:
Be able to interpret and draw
Venn diagrams. See p.238-244.

Venn diagrams are really helpful for working out probabilities, particularly **conditional probabilities**. If you've got a **complete** Venn diagram for your events, then it's really easy to find the probability of one event given another:

$$P(A \text{ given } B) = \frac{\text{Number of outcomes in both A and B}}{\text{Number of outcomes in B}} = \frac{n(A \cap B)}{n(B)}$$

You can think of this as the **probability** of getting something from **A** when randomly picking from all the things in **B** (**ignoring** all the things that **aren't in B**).

Example 2

The Venn diagram shows the number of males (M) and the number of people who are drinking coffee (C) at a cafe.

What is the probability that a randomly selected person at the cafe is male, given that they are drinking coffee?

Use the formula — divide the number of people that are both males and drinking coffee by the total number of people drinking coffee.

$$P(M \text{ given } C) = \frac{n(M \text{ and } C)}{n(C)} = \frac{12}{12 + 8} = \frac{12}{20} = \frac{3}{5}$$

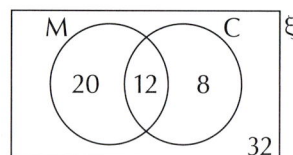

Tip: Remember, n(M and C) or n(M ∩ C) is the number in the bit where M and C overlap.

Exercise 2

Q1 A survey was taken at a hotel in a ski resort which asked 274 people if they could ski and if they could snowboard. The results are shown in the Venn diagram on the right.

What is the probability that a randomly chosen hotel guest:
a) can ski?
b) can snowboard?
c) can ski and can snowboard?
d) can ski given that they can snowboard?
e) can't snowboard given that they can't ski?

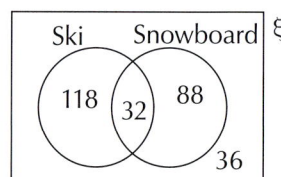

Q2 The Venn diagram below shows the number of members in a drama group who can act (A), dance (D) and sing (S). Each member of the group can do at least one of these.

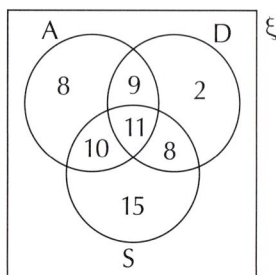

What is the probability that a randomly chosen member:
a) can act given that they can dance?
b) can sing given that they can't dance?
c) can't dance given that they can act?
d) can't act given that they can't sing?
e) can act and dance given that they can sing?
f) can sing and act given that they can't dance?
g) can act or dance given that they can't sing?

Q1 The probability that Colin wins any game of chess is 0.8. Find the probability that he wins:

a) both of his next two games

b) neither of his next two games

c) his next game but loses the one after

d) one of his next two games

Q2 A board game has four categories of question, each split into two difficulty levels. The table shows the probability of randomly selecting a question card of each type.

	Easy	Challenge
TV	0.20	0.10
Music	0.15	0.07
Sport	0.12	0.08
Literature	0.15	x

a) Find the missing probability, x, from the table.

b) Which of the four categories has the most easy questions?

c) Mark randomly selects one card from all the question cards. What is the probability that he selects:

(i) a music question?   (ii) a sport question?   (iii) a music or a sport question?

(iv) a music question or a challenging question?   (v) an easy question or a sport question?

d) Dahlia plays two games of the board game. Each game starts with the first question card being randomly selected from all the cards. Find the probability that both starting cards are:

(i) challenge music questions   (ii) easy literature questions

Q3 Eight friends have to pick three from the group to represent them at a meeting. Five of the friends are in Year 10 and three are in Year 11. If they pick the three representatives at random, find the probability that:

a) all three are in Year 10

b) all three are in Year 11

c) two are in Year 10 and one is in Year 11

d) two are in Year 11 and one is in Year 10

Q4 A box contains ten coloured marbles — five blue, four white and one red. Two marbles are picked at random.

a) Draw a tree diagram showing all the possible results.

b) Work out the probability that:

(i) both are blue   (ii) neither is blue   (iii) exactly one is blue

(iv) at least one is blue   (v) one is red and one is white   (vi) they are different colours

Q5 At a fast food restaurant there are two options for side dishes — fries or salad. The side dish choices of 200 customers are recorded in this Venn diagram. What is the probability that a randomly chosen customer:

a) had fries given that they had salad?

b) didn't have salad given that they had fries?

c) didn't have a side given that they didn't have salad?

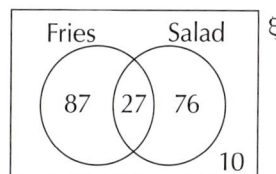

Exam-Style Questions

Q1 30 Sixth Form students were asked if they are studying any of the subjects Art (A), Biology (B) or Chemistry (C) for A-level. The results of this survey are shown in the Venn diagram, where each entry represents the number of students.

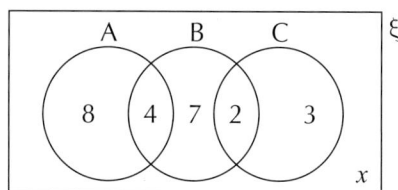

a) Work out the value of x.

[1 mark]

b) Explain clearly why the Venn diagram shows that studying Art and studying Chemistry are "mutually exclusive" events for these 30 students.

[2 marks]

c) Given that a student is studying Biology, work out the probability that they also study Art or Chemistry.

[2 marks]

Q2 Tariq and Udai are researching driving test statistics. The probability of passing a driving test when it is taken for the first time is 0.5. When taken a second time, the probability of passing increases to 0.6. When taken a third time, the probability of passing is 0.8.

a) Work out the probability that Tariq will pass his driving test on his third attempt.

[2 marks]

b) Work out the probability that Tariq and Udai will both pass their driving tests on the third attempt. You may assume that the events are independent of each other.

[1 mark]

Q3 Naasir is playing a board game. He picks two letter tiles at random from a bag. The bag contains the following tiles:

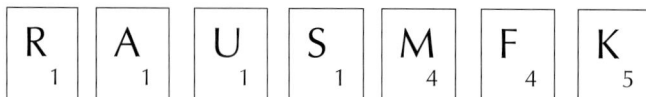

The numbers on each tile represent how many points that tile is worth.

a) Copy and complete the tree diagram below. The first set of branches is drawn for you.

Value of 1st tile   **Value of 2nd tile**

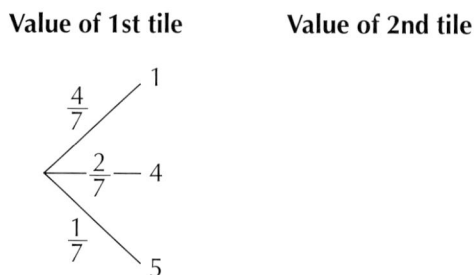

[3 marks]

b) Work out the probability that the two tiles that Naasir picks will be worth less than 6 points in total.

[3 marks]

Mixed Exam-Style Questions

Q1 A brand of breakfast cereal is sold in two different sizes. The 'family' size box contains 750 g. The 'standard' box normally contains 500 g, but as a special offer, these boxes currently contain an extra 25% of cereal. The 'family' box costs £2.25. The special offer 'standard' boxes cost £2.

Work out which size of box is better value for money.
You must show your working.

[4 marks]

Q2 The ratio of boys to girls in a primary school class of 28 children is 3 : 4. Half of the boys and 75% of the girls eat school dinners. The rest eat packed lunches.

Show this information as a frequency tree.

[4 marks]

Q3 Jemima is looking to buy a new tablet computer on the internet. A UK seller offers the model she wants with free delivery. She also sees the same model available from an American seller priced at 432 US dollars ($). Delivery from America will cost $30 and there will also be a 20% import tax on both the cost of the computer and the delivery charge.

The exchange rate is $1 = £0.75.

The price of the computer from the UK seller is £415.
Which seller should Jemima buy the computer from?

[3 marks]

Q4 The diagram below shows a parallelogram with the lengths of three of its sides given in centimetres in terms of x:

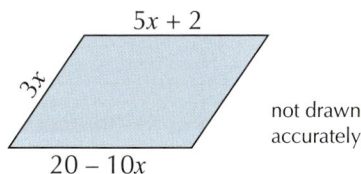

$5x + 2$

$3x$

$20 - 10x$

not drawn accurately

Show that the perimeter of the parallelogram is 23.2 cm.

[4 marks]

The flag of the Asian country Laos is a rectangle with a width to height ratio of $3:2$.
It consists of three horizontal stripes. The top and bottom stripes are the same height
and are half the height of the middle stripe.

In the centre is a circle which has a diameter of $\frac{4}{5}$ of the height of the middle stripe.

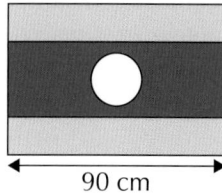

not drawn to scale

90 cm

Work out the area of this circle for a flag of width 90 cm.
Give your answer in cm² as a multiple of π.

[4 marks]

Q6   Andrea has a small city car and Zac has a large sports car.
The efficiency of any car can be measured using the formula:

$$\text{efficiency} = \frac{\text{distance travelled (km)}}{\text{volume of fuel used (litres)}}$$

On a particular trip, Andrea's car travels 81.9 km and uses 2.6 litres of fuel.
The efficiency of Zac's car is two thirds of the efficiency of Andrea's car.

Zac plans to drive a distance of 64.7 km on his next trip. Work out how much fuel he
should expect to use on this trip, giving your answer in litres to 2 significant figures.

[3 marks]

Q7   Erica is deciding between having a mobile phone contract or choosing 'pay as you go'.
She sees two offers from the companies FF and H2:

FF — contract
- £51.99 per month
- 4GB of data free per month then £4.99 per GB
- 5 hours of free calls per month then 7p per minute
- Unlimited texts

H2 — Pay As You Go
Great price on data
Calls cost 8p per minute
Texts costs 4p each

Erica makes the decision based on the figures for her mobile phone use last month
when she used 5 gigabytes (GB) of data, made 7 hours of calls and sent 60 texts.

Work out the maximum price that H2 could charge per GB of data
that would make H2 the better deal. You must show your working.

[5 marks]

Q8 Three interior angles of a pentagon are right angles.
The other two angles are in the ratio $4:11$.

Work out the size of the largest interior angle in the pentagon.

[4 marks]

Q9 A metal shipping container is in the shape of a cuboid with six flat faces.
It has a length of 6 m, a width of 2.6 m and a height of 2.4 m.
The face *ABCD* is on the ground, as shown.

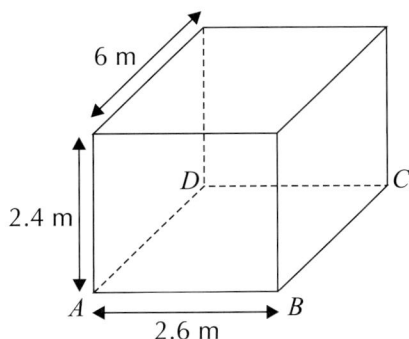

not drawn to scale

A worker is asked to paint the other five faces to protect the container from rust.
She will apply 0.1 litres of paint per square metre.
She is going to paint the four vertical faces first and the top face last.

If she only has 5 litres of paint, work out, to the nearest whole number,
the percentage of the top face she will be able to paint.

[4 marks]

Q10 A teacher randomly chooses three pupils from her class
to take part in a school debating competition.

The teacher chooses one boy and two girls.
There are n boys in the class.
The number of girls in the class is two fewer than the number of boys.

Show that the number of different teams the teacher can
choose is given by the expression $n^3 - 5n^2 + 6n$.

[3 marks]

Q11 The line segment *AB* is such that *A* and *B* have coordinates $(-1, 6)$ and $(3, 4)$ respectively.
Find the equation of the line perpendicular to *AB* which goes through the midpoint of *AB*.

[4 marks]

Q12 Weedkiller is applied to a rectangular lawn at a rate of 20 ml/m².
The length of the lawn is four times its width.

If 8 litres of weedkiller is used in total, calculate the length of the lawn, in metres.

[4 marks]

Q13 When Roger plays tennis against Stan on a grass court, the ratio of Roger's chances
of winning a set to Stan's is $5:2$. When they play on a clay court, this ratio is $4:5$.
They play one set on a grass court and one set on a clay court. The events of
Roger or Stan winning or losing either set are assumed to be independent.

a) Explain what is meant by 'independent' events in probability.

[1 mark]

b) Find the probability that Roger and Stan win one set each.

[4 marks]

Q14 The diagram shows the cross-section of a swimming pool which is in the shape of a trapezium.

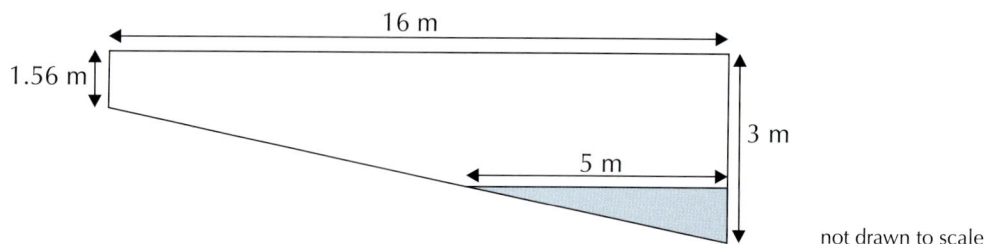

The shallow end has a depth of 1.56 m and the deep end has a depth of 3 m.
The pool is 16 m long. The pool is being filled and the length of the surface
of the water is currently 5 m.

Work out how far the surface of the water is from the top of the pool.

[4 marks]

Q15 Solve the equation: $9^{4x} = 3^{3x+4}$

[3 marks]

Q16 A bag contains 40 sweets. 18 are red and the rest are yellow.

Danesh only likes the red sweets. He eats 9 of the red sweets from the bag and then is given 5 more yellow sweets from a friend. He puts these in the bag.

By how much has the percentage of yellow sweets in the bag increased?

[3 marks]

Q17 For the sequence of triangular numbers, the nth term is given by:

$$u_n = \frac{n(n+1)}{2}, \qquad n > 0.$$

Use this to prove that the sum of two consecutive triangular numbers is a square number.

[3 marks]

Q18 The diagram below shows the cross-section of a garage roof, where ABD is a triangle.
The lengths $AB = 2.5$ m, $AC = 2.3$ m, $CD = 1.3$ m are marked on the diagram.
All lengths are accurate to 1 decimal place.

a) Show that the upper bound for the perpendicular height h of the roof is 1.2 m.

[3 marks]

b) Hence show that the upper bound for the angle BDC is 43.8° to 1 decimal place.
You may use the fact that $\tan x$ increases as x increases.

[3 marks]

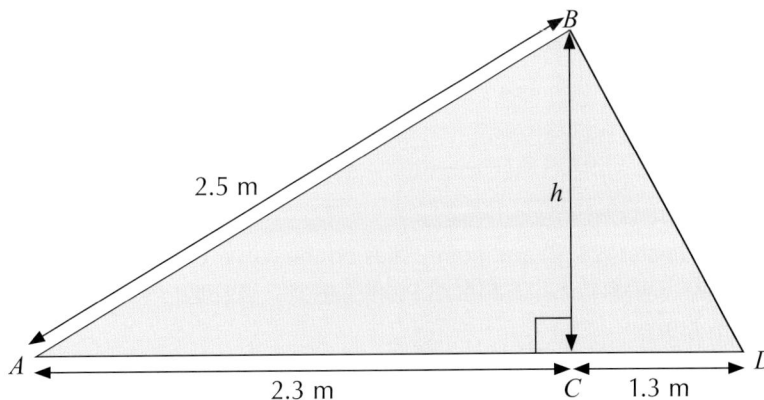

Q19 Eight teams are in a draw for the quarter-finals of a football competition and x of the teams are British. The first two teams are randomly selected. The probability that neither team is British is $\frac{5}{14}$.

Show that $x^2 - 15x + 36 = 0$.

[4 marks]

Q20 A costume jeweller is designing a pendant, as shown in the diagram.

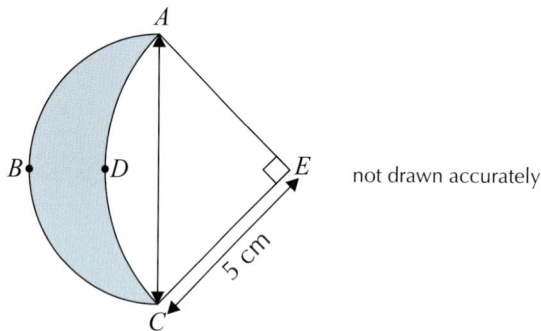

not drawn accurately

$ABCA$ is a semicircle. $ADCEA$ is a sector of a circle of radius 5 cm with centre E and angle $AEC = 90°$. The pendant is the area between the arcs ABC and ADC.

Find the area of the pendant, giving your answer as an improper fraction in its simplest form.

[5 marks]

Q21 Two variables m and h are related by the formula:

$$m^{\frac{1}{4}} h^{-2} = k, \text{ where } k \text{ is a constant.}$$

When $m = 4096$, $h = 2$.

Show that an expression for m in terms of h is $m = ah^8$, where a is a constant to be found.

[4 marks]

Q22 Point P has coordinates $(24, 7)$ and point T lies on the curve $x^2 + y^2 = 400$. The tangent to the curve $x^2 + y^2 = 400$ at point T passes through point P.

Find the length of PT.

[5 marks]

Answers

Section 1 — Arithmetic, Multiples and Factors

1.1 Calculations

Page 2   Exercise 1

1. a) $5 + 1 \times 3 = 5 + 3 = \mathbf{8}$
   b) $11 - 2 \times 5 = 11 - 10 = \mathbf{1}$
   c) $18 - 10 \div 5 = 18 - 2 = \mathbf{16}$
   d) $24 \div 4 + 2 = 6 + 2 = \mathbf{8}$
   e) $35 \div 5 + 2 = 7 + 2 = \mathbf{9}$
   f) $36 - 12 \div 4 = 36 - 3 = \mathbf{33}$
2. a) $2 \times (4 + 10) = 2 \times 14 = \mathbf{28}$
   b) $(7 - 2) \times 3 = 5 \times 3 = \mathbf{15}$
   c) $4 + (48 \div 8) = 4 + 6 = \mathbf{10}$
   d) $56 \div (2 \times 4) = 56 \div 8 = \mathbf{7}$
   e) $(3 + 2) \times (9 - 4) = 5 \times 5 = \mathbf{25}$
   f) $(8 - 7) \times (6 + 5) = 1 \times 11 = \mathbf{11}$
3. a) $2 \times (8 + 4) - 7 = 2 \times 12 - 7 = 24 - 7 = \mathbf{17}$
   b) $100 \div (8 + 3 \times 4) = 100 \div (8 + 12)$
      $= 100 \div 20 = \mathbf{5}$
   c) $7 + (10 - 9 \div 3) = 7 + (10 - 3)$
      $= 7 + 7 = \mathbf{14}$
   d) $20 - (5 \times 3 + 2) = 20 - (15 + 2)$
      $= 20 - 17 = \mathbf{3}$
   e) $48 \div 3 - 7 \times 2 = 16 - 14 = \mathbf{2}$
   f) $36 - (7 + 4 \times 4) = 36 - (7 + 16)$
      $= 36 - 23 = \mathbf{13}$
4. a) $4 - 5 + 2 - 1 = -1 + 2 - 1 = 1 - 1 = \mathbf{0}$
      You'd get the wrong answer if you did $5 + 2 = 7$ first.
   b) $5 \times 4 \div 10 \times 6 = 20 \div 10 \times 6 = 2 \times 6 = \mathbf{12}$

Page 3   Exercise 2

1. a) $\dfrac{16}{4 \times (5 - 3)} = \dfrac{16}{4 \times 2} = \dfrac{16}{8} = \mathbf{2}$
   b) $\dfrac{8 + 2}{15 \div 3} = \dfrac{10}{5} = \mathbf{2}$
   c) $\dfrac{4 \times (7 + 5)}{6 + 3 \times 2} = \dfrac{4 \times 12}{6 + 6} = \dfrac{48}{12} = \mathbf{4}$
   d) $\dfrac{6 + (11 - 8)}{7 - 5} = \dfrac{6 + 3}{2} = \dfrac{\mathbf{9}}{\mathbf{2}}$ or $\mathbf{4\frac{1}{2}}$
2. a) $\dfrac{12 \div (9 - 5)}{25 \div 5} = \dfrac{12 \div 4}{5} = \dfrac{\mathbf{3}}{\mathbf{5}}$
   b) $\dfrac{8 \times 2 \div 4}{5 - 6 + 7} = \dfrac{16 \div 4}{-1 + 7} = \dfrac{4}{6} = \dfrac{\mathbf{2}}{\mathbf{3}}$
   c) $\dfrac{3 \times 3}{21 \div (12 - 5)} = \dfrac{9}{21 \div 7} = \dfrac{9}{3} = \mathbf{3}$
   d) $\dfrac{36 \div (11 - 2)}{8 - 8 \div 2} = \dfrac{36 \div 9}{8 - 4} = \dfrac{4}{4} = \mathbf{1}$

Page 4   Exercise 3

1. a) -1 b) -5 c) 3
   d) -11 e) 6 f) -4
2. a) -10 b) 0 c) 12
   d) 26 e) -60 f) 90

Page 4   Exercise 4

1. a) 5 b) -3 c) 12
   d) -7 e) 48 f) 9
   g) 39 h) -42 i) -68
2. a) $[(-3) \times 7] \div (-21) = (-21) \div (-21) = \mathbf{1}$
   b) $[(-24) \div 8] \div 3 = (-3) \div 3 = \mathbf{-1}$
   c) $[55 \div (-11)] \times (-9) = (-5) \times (-9) = \mathbf{45}$
   d) $[(-63) \div (-9)] \times (-7) = 7 \times (-7) = \mathbf{-49}$
   e) $[35 \div (-7)] \times (-8) = (-5) \times (-8) = \mathbf{40}$
   f) $[(-12) \times 3] \times (-2) = (-36) \times (-2) = \mathbf{72}$
3. a) 2 b) 7 c) -4
   d) 3 e) -12 f) -77

Page 5   Exercise 5

1. a) $\begin{array}{r} 12.74 \\ +\ 7.00 \\ \hline \mathbf{19.74} \end{array}$ b) $\begin{array}{r} 0.\overset{7}{8}\overset{1}{0} \\ -\ 0.03 \\ \hline \mathbf{0.77} \end{array}$ c) $\begin{array}{r} 10.83 \\ +\ 7.40 \\ \hline \mathbf{18.23} \end{array}$

   d) $\begin{array}{r} 91.700 \\ +\ 0.492 \\ \hline \mathbf{92.192} \end{array}$ e) $\begin{array}{r} 6.474 \\ +\ 0.920 \\ \hline \mathbf{7.394} \end{array}$ f) $\begin{array}{r} 16.\overset{2}{3}\overset{1}{0} \\ -\ 5.16 \\ \hline \mathbf{11.14} \end{array}$

   g) $\begin{array}{r} \overset{8}{9}.\overset{1}{2}41 \\ -\ 2.800 \\ \hline \mathbf{6.441} \end{array}$ h) $\begin{array}{r} 2\overset{1}{3}.\overset{9}{0}\overset{9}{0}\overset{1}{0} \\ -\ 18.591 \\ \hline \mathbf{4.409} \end{array}$

2. a) $\begin{array}{r} 7.61 \\ +\ 0.60 \\ \hline 8.21 \end{array}$ b) $\begin{array}{r} 5.43 \\ -\ 2.12 \\ \hline 3.31 \end{array}$

3. $\begin{array}{r} 18.50 \\ +\ 31.99 \\ \hline \mathbf{£50.49} \end{array}$ 4. $\begin{array}{r} 66.49 \\ -\ 15.25 \\ \hline \mathbf{£51.24} \end{array}$

Page 6   Exercise 6

1. a) $31\ 416 \div 10 = \mathbf{3141.6}$
   b) $31\ 416 \div 1000 = \mathbf{31.416}$
   c) $31\ 416 \div 100\ 000 = \mathbf{0.31416}$
   d) $31\ 416 \div 1\ 000\ 000 = \mathbf{0.031416}$
2. a) $\begin{array}{r} 167 \\ \times\quad 8 \\ \hline 1336 \end{array}$, $1336 \div 10 = \mathbf{133.6}$
   b) $\begin{array}{r} 312 \\ \times\quad 6 \\ \hline 1872 \end{array}$, $1872 \div 10 = \mathbf{187.2}$
   c) $\begin{array}{r} 31 \\ \times\ 40 \\ \hline 1240 \end{array}$, $1240 \div 10 = \mathbf{124}$
   d) $7 \times 600 = 4200$, $4200 \div 10 = \mathbf{420}$
   e) $5 \times 4 = 20$, $20 \div 10\ 000 = \mathbf{0.002}$
   f) $8 \times 5 = 40$, $40 \div 1000 = \mathbf{0.04}$
   g) $\begin{array}{r} 21 \\ \times\quad 6 \\ \hline 126 \end{array}$, $126 \div 100 = \mathbf{1.26}$
   h) $16 \times 4 = 64$, $64 \div 1000 = \mathbf{0.064}$
   i) $\begin{array}{r} 52 \\ \times\quad 9 \\ \hline 468 \end{array}$, $468 \div 1000 = \mathbf{0.468}$
   j) $\begin{array}{r} 39 \\ \times\ 83 \\ \hline 117 \\ 3120 \\ \hline 3237 \end{array}$, $3237 \div 100 = \mathbf{32.37}$
   k) $\begin{array}{r} 16 \\ \times\ 33 \\ \hline 48 \\ 480 \\ \hline 528 \end{array}$, $528 \div 1000 = \mathbf{0.528}$
   l) $\begin{array}{r} 64 \\ \times\ 42 \\ \hline 128 \\ 2560 \\ \hline 2688 \end{array}$, $2688 \div 10\ 000 = \mathbf{0.2688}$
3. $176 \times 5 = 880$, $880 \div 100 = \mathbf{8.8\ pints}$

4. $\begin{array}{r} 135 \\ \times\ 92 \\ \hline 270 \\ 12150 \\ \hline 12420 \end{array}$, $12\ 420 \div 1000 = \mathbf{£12.42}$

Page 6   Exercise 7

1. a) $4 \overline{)8.5^{1}2}\ \mathbf{2.1\ 3}$
   b) $4 \overline{)2.^{2}2^{1}1^{1}4^{2}0}\ \mathbf{0.5\ 3\ 5}$
   c) $5 \overline{)8.3^{6}1^{2}2^{2}0}\ \mathbf{1.7\ 2\ 4}$
   d) $6 \overline{)1\ 7.5^{1}1^{3}0}\ \mathbf{2.8\ 5}$
   e) $9 \overline{)0.081}\ \mathbf{0.009}$
   f) $8 \overline{)1\ 2.^{4}0\ 6^{6}0^{4}0}\ \mathbf{1.5\ 0\ 7\ 5}$
2. a) $1.56 \div 0.2 = 15.6 \div 2 = \mathbf{7.8}$
   b) $0.624 \div 0.3 = 6.24 \div 3 = \mathbf{2.08}$
   c) $0.275 \div 0.5 = 2.75 \div 5 = \mathbf{0.55}$
   d) $16.42 \div 0.02 = 1642 \div 2 = \mathbf{821}$
   e) $0.257 \div 0.05 = 25.7 \div 5 = \mathbf{5.14}$
   f) $7.665 \div 0.03 = 766.5 \div 3 = \mathbf{255.5}$
   g) $0.039 \div 0.06 = 3.9 \div 6 = \mathbf{0.65}$
   h) $50.4 \div 0.07 = 5040 \div 7 = \mathbf{720}$
   i) $0.71 \div 0.002 = 710 \div 2 = \mathbf{355}$
   j) $108 \div 0.4 = 1080 \div 4 = \mathbf{270}$
   k) $20.16 \div 0.007 = 20\ 160 \div 7 = \mathbf{2880}$
   l) $1.44 \div 1.2 = 14.4 \div 12 = \mathbf{1.2}$
3. $2.72 \div 0.08 = 272 \div 8 = \mathbf{34}$
4. $6.93 \div 3.5 = 69.3 \div 35 = \mathbf{£1.98}$

1.2 Multiples and Factors

Page 7   Exercise 1

1. a) **9, 18, 27, 36, 45**
   b) **13, 26, 39, 52, 65**
   c) **16, 32, 48, 64, 80**
2. a) **24, 36, 48, 60, 72, 84, 96**
   b) **28, 42, 56, 70, 84**
3. a) **10, 20, 30, 40, 50, 60, 70, 80, 90, 100**
   b) **15, 30, 45, 60, 75, 90, 105**
   c) **30, 60, 90**
4. a) **21, 24, 27, 30, 33**
   b) **20, 24, 28, 32**   c) **24**
5. Multiples of 5: 5, 10, 15, 20, 25, 30, 35, 40
   Multiples of 6: 6, 12, 18, 24, 30, 36
   Common multiple: **30**
6. Multiples of 6: 6, 12, 18, 24, 30, 36, 42, 48, 54, 60, 66, 72, 78, 84, 90, 96
   Multiples of 8: 8, 16, 24, 32, 40, 48, 56, 64, 72, 80, 88, 96
   Multiples of 10: 10, 20, 30, 40, 50, 60, 70, 80, 90, 100
   Common multiples: **None**
7. Multiples of 9: 9, 18, 27, 36, 45, 54, 63, 72, 81, 90, 99
   Multiples of 12: 12, 24, 36, 48, 60, 72, 84, 96
   Multiples of 15: 15, 30, 45, 60, 75, 90
   Common multiples: **None**
8. Multiples of 3: 3, 6, 9, 12, 15, 18, 21, 24, 27, 30, 33, 36...
   Multiples of 6: 6, 12, 18, 24, 30, 36...
   Multiples of 9: 9, 18, 27, 36...
   Common multiples of 3, 6 and 9 are multiples of 18. So the first five common multiples are:
   18, 36, 54, 72, 90

Page 8   Exercise 2

1 a) $1 \times 13 = 13$, so 1 and 13 are both factors
$2 \times — = 13$, so 2 is not a factor
$3 \times — = 13$, so 3 is not a factor
$4 \times — = 13$, so 4 is not a factor
$5 \times — = 13$, so 5 is not a factor
So the factors of 13 are **1** and **13**
You can actually stop checking after 4 as any factor greater than 4 would need to multiply by a number less than 4 to make 13 (since 4 × 4 is greater than 13), and you've already checked 1-3.

b) $1 \times 25 = 25$, so 1 and 25 are both factors
$2 \times — = 25$, so 2 is not a factor
$3 \times — = 25$, so 3 is not a factor
$4 \times — = 25$, so 4 is not a factor
$5 \times 5 = 25$, so 5 is a factor
5 is repeated, so stop
So the factors of 25 are **1, 5** and **25**

Using the same method for c)-h):
c) **1, 2, 3, 4, 6, 8, 12, 24**
d) **1, 5, 7, 35**   **e)** **1, 2, 4, 8, 16, 32**
f) **1, 2, 4, 5, 8, 10, 20, 40**
g) **1, 2, 5, 10, 25, 50** **h)** **1, 7, 49**

2 You need to find the factors of 12:
$1 \times 12 = 12$, $2 \times 6 = 12$, $3 \times 4 = 12$,
so 1, 2, 3, 4, 6 and 12 are factors.
So there are **6** ways to divide the cakes up into equal packets — **1 packet of 12, 2 packets of 6, 3 packets of 4, 4 packets of 3, 6 packets of 2, 12 packets of 1.**

3 You need to find the factors of 100:
$1 \times 100 = 100$, $2 \times 50 = 100$, $3 \times — = 100$,
$4 \times 25 = 100$, $5 \times 20 = 100$, $6 \times — = 100$,
$7 \times — = 100$, $8 \times — = 100$, $9 \times — = 100$,
$10 \times 10 = 100$, so 1, 2, 4, 5, 10, 20, 25, 50 and 100 are factors.
So there are **9** ways to arrange the people into equal groups — **1 group of 100 people, 2 groups of 50 people, 4 groups of 25 people, 5 groups of 20 people, 10 groups of 10 people, 20 groups of 5 people, 25 groups of 4 people, 50 groups of 2 people, 100 groups of 1 person.**

4 a) (i) $1 \times 15 = 15$, $2 \times — = 15$, $3 \times 5 = 15$,
$4 \times — = 15$, so the factors of 15 are **1, 3, 5, 15**
(ii) $1 \times 21 = 21$, $2 \times — = 21$, $3 \times 7 = 21$,
$4 \times — = 21$, $5 \times — = 21$, $6 \times — = 21$,
so the factors of 21 are **1, 3, 7, 21**
b) Common factors appear on both lists, so the common factors of 15 and 21 are **1 and 3**.

5 a) Factors of 15: 1, 3, 5, 15
Factors of 20: 1, 2, 4, 5, 10, 20
Common factors: **1, 5**
b) Factors of 50: 1, 2, 5, 10, 25, 50
Factors of 90: 1, 2, 3, 5, 6, 9, 10, 15, 18, 30, 45, 90
Common factors: **1, 2, 5, 10**
c) Factors of 24: 1, 2, 3, 4, 6, 8, 12, 24
Factors of 32: 1, 2, 4, 8, 16, 32
Common factors: **1, 2, 4, 8**
d) Factors of 64: 1, 2, 4, 8, 16, 32, 64
Factors of 80: 1, 2, 4, 5, 8, 10, 16, 20, 40, 80
Common factors: **1, 2, 4, 8, 16**
e) Factors of 45: 1, 3, 5, 9, 15, 45
Factors of 81: 1, 3, 9, 27, 81
Common factors: **1, 3, 9**
f) Factors of 96: 1, 2, 3, 4, 6, 8, 12, 16, 24, 32, 48, 96
Factors of 108: 1, 2, 3, 4, 6, 9, 12, 18, 27, 36, 54, 108
Common factors: **1, 2, 3, 4, 6, 12**

6 a) Factors of 30: 1, 2, 3, 5, 6, 10, 15, 30
Factors of 45: 1, 3, 5, 9, 15, 45
Factors of 50: 1, 2, 5, 10, 25, 50
Common factors: **1, 5**

b) Factors of 8: 1, 2, 4, 8
Factors of 12: 1, 2, 3, 4, 6, 12
Factors of 20: 1, 2, 4, 5, 10, 20
Common factors: **1, 2, 4**
c) Factors of 9: 1, 3, 9
Factors of 27: 1, 3, 9, 27
Factors of 36: 1, 2, 3, 4, 6, 9, 12, 18, 36
Common factors: **1, 3, 9**
d) Factors of 24: 1, 2, 3, 4, 6, 8, 12, 24
Factors of 48: 1, 2, 3, 4, 6, 8, 12, 16, 24, 48
Factors of 96: 1, 2, 3, 4, 6, 8, 12, 16, 24, 32, 48, 96
Common factors: **1, 2, 3, 4, 6, 8, 12, 24**

1.3 Prime Numbers and Prime Factors

Page 9   Exercise 1
1 a) **15**   **b)** **3, 5**
2 a) **33, 35, 39**
b) $33 = 3 \times 11$, $35 = 5 \times 7$, $39 = 3 \times 13$
3 **5, 47, 59**
4 a) **2, 3, 5, 7**
b) **23, 29, 31, 37, 41, 43, 47**
5 a) (i) **2**   **(ii)** **2 or 7**
(iii) **2 or 17**   **(iv)** **2 or 37**
b) Numbers ending in 4 are even, so they are **divisible by 2**.

Page 11   Exercise 2
1 a) 2×7 **b)** 5×11 **c)** 3×5
d) 3×7 **e)** 5×7 **f)** 3×13
g) 7×11 **h)** 11×11

2 a) (i)

(ii)

(iii)

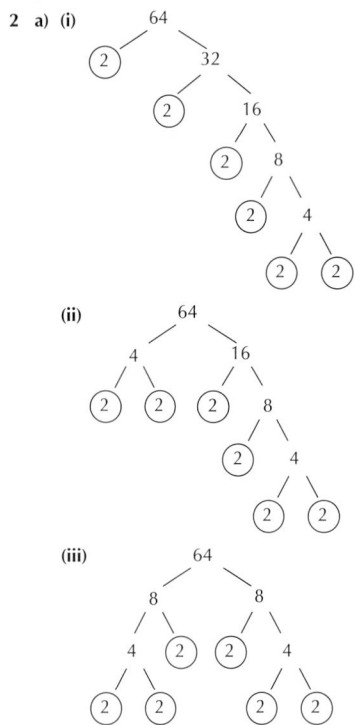

b) 2^6
All the factor trees give the same answer, however you break the number down.

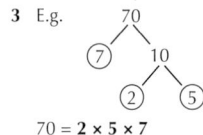

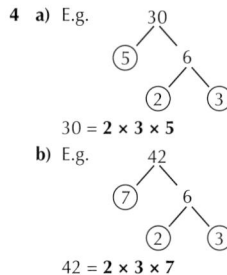

3 E.g.

$70 = 2 \times 5 \times 7$

4 a) E.g.

$30 = 2 \times 3 \times 5$
b) E.g.

$42 = 2 \times 3 \times 7$
Using the same method for c)-h):
c) $66 = 2 \times 3 \times 11$
d) $46 = 2 \times 23$
e) $78 = 2 \times 3 \times 13$
f) $190 = 2 \times 5 \times 19$
g) $210 = 2 \times 3 \times 5 \times 7$
h) $138 = 2 \times 3 \times 23$

5 a)

$44 = 2^2 \times 11$
b)

$48 = 2 \times 2 \times 2 \times 2 \times 3 = 2^4 \times 3$
Using the same method for c)-l):
c) $72 = 2 \times 2 \times 2 \times 3 \times 3 = 2^3 \times 3^2$
d) $90 = 2 \times 3 \times 3 \times 5 = 2 \times 3^2 \times 5$
e) $50 = 2 \times 5 \times 5 = 2 \times 5^2$
f) $28 = 2 \times 2 \times 7 = 2^2 \times 7$
g) $98 = 2 \times 7 \times 7 = 2 \times 7^2$
h) $150 = 2 \times 3 \times 5 \times 5 = 2 \times 3 \times 5^2$
i) $132 = 2 \times 2 \times 3 \times 11 = 2^2 \times 3 \times 11$
j) $168 = 2 \times 2 \times 2 \times 3 \times 7 = 2^3 \times 3 \times 7$
k) $325 = 5 \times 5 \times 13 = 5^2 \times 13$
l) $1000 = 2 \times 2 \times 2 \times 5 \times 5 \times 5 = 2^3 \times 5^3$

6 a) $75 = 3 \times 5 \times 5 = 3 \times 5^2$
b) 3 — this is the smallest number you can multiply by that gives even powers of all the prime factors, giving $3^2 \times 5^2 = 15^2$.

7 a) $250 = 2 \times 5 \times 5 \times 5 = 2 \times 5^3$
To raise all the factors to even powers, multiply by $2 \times 5 = $ **10**
b) $416 = 2 \times 2 \times 2 \times 2 \times 2 \times 13 = 2^5 \times 13$
To raise all the factors to even powers, multiply by $2 \times 13 = $ **26**
c) $756 = 2 \times 2 \times 3 \times 3 \times 3 \times 7 = 2^2 \times 3^3 \times 7$
To raise all the factors to even powers, multiply by $3 \times 7 = $ **21**
d) $1215 = 3 \times 3 \times 3 \times 3 \times 3 \times 5 = 3^5 \times 5$
To raise all the factors to even powers, multiply by $3 \times 5 = $ **15**

1.4 LCM and HCF

Page 12   Exercise 1
1 a) Multiples of 3: 3, 6, 9, 12, 15...
Multiples of 5: 5, 10, 15, 20...
LCM = smallest number in both lists = **15**
b) Multiples of 6: 6, 12, 18, 24, 30...
Multiples of 8: 8, 16, 24, 32...
LCM = smallest number in both lists = **24**
Using the same method for c)-d):
c) **10**   **d)** **42**

2 a) Multiples of 3: 3, 6, 9, 12, 15, 18, 21, 24...
Multiples of 6: 6, 12, 18, 24...
Multiples of 8: 8, 16, 24...
LCM: **24**

b) Multiples of 2: 2, 4, 6, 8 10, 12, 14, 16, 18, 20, 22, 24, 26, 28, 30...
Multiples of 5: 5, 10, 15, 20, 25, 30...
Multiples of 6: 6, 12, 18, 24, 30...
LCM: **30**

Using the same method for c)-d):
c) 36   **d) 70**

3 Mike visits Oscar on day: 4, 8, 12, 16, 20...
Narinda visits Oscar on day: 5, 10, 15, 20...
So they will both visit Oscar on the same day after **20 days.**

4 To arrange exactly in rows of 25, the number of plants must be a multiple 25. Similarly, to arrange exactly in rows of 30, the number of plants must also be a multiple 30. So you need to find a common multiple of 25 and 30. Between 95 and 205, the multiples of 25 are: 100, 125, 150, 175, 200 and the multiples of 30 are: 120, 150, 180.
So there must be **150 plants.**

5 The number of plates must be a multiple of both 40 and 70. Between 240 and 300, the multiples of 40 are: 240 and 280, and the multiples of 70 are: 280.
So there must be **280 plates.**

Page 13   Exercise 2

1 a) Factors of 12: 1, 2, 3, 4, 6, 12
Factors of 20: 1, 2, 4, 5, 10, 20
Common factors: **1, 2, 4**
b) The HCF is the largest factor, so it's **4.**

2 a) Factors of 24: 1, 2, 3, 4, 6, 8, 12, 24
Factors of 32: 1, 2, 4, 8, 16, 32
HCF = highest number in both lists = **8**
b) Factors of 36: 1, 2, 3, 4, 6, 9, 12, 18, 36
Factors of 60: 1, 2, 3, 4, 5, 6, 10, 12, 15, 20, 30, 60
HCF = highest number in both lists = **12**

Using the same method for c)-d):
c) 1   **d) 12**

3 a) Factors of 6: 1, 2, 3, 6
Factors of 8: 1, 2, 4, 8
Factors of 16: 1, 2, 4, 8, 16
HCF = highest number in all lists = **2**
b) Factors of 12: 1, 2, 3, 4, 6, 12
Factors of 15: 1, 3, 5, 15
Factors of 18: 1, 2, 3, 6, 9, 18
HCF = highest number in all lists = **3**

Using the same method for c)-d):
c) 18   **d) 12**

4 a) Each piece of ribbon must be the same length. So the length of the pieces must divide exactly into both 63 and 91 (i.e. it must be a common factor of 63 and 91). Since the pieces must be the greatest possible length, it must be the highest common factor.
Factors of 63 are 1, 3, 7, 9, 21, 63. Factors of 91 are 1, 7, 13, 91. So the length of the pieces will be **7 cm.**
b) The total length of blue ribbon is 63 cm, which means there will be 63 ÷ 7 = **9 pieces.** The total length of the pink ribbon is 91 cm, which means there will be 91 ÷ 7 = **13 pieces.**

5 To create equal-sized piles, the number of counters in a pile must divide into 135 and also into 165, so the pile size is a common factor of 135 and 165.
Factors of 135 are 1, 3, 5, 9, 15, 27, 45, 135.
Factors of 165 are 1, 3, 5, 11, 15, 33, 55, 165.

The common factors of 135 and 165 are 1, 3, 5 and 15. To give the least number of piles, use the HCF. This is 15, so there are 15 counters in each pile.
There will be 135 ÷ 15 = 9 piles of tangerine counters and 165 ÷ 15 = 11 piles of gold counters. So the total number of piles will be 9 + 11 = **20.**

Page 15   Exercise 3

1 a) $120 = 2 \times 2 \times 2 \times 3 \times 5 = \mathbf{2^3 \times 3 \times 5}$, $155 = \mathbf{5 \times 31}$
b) HCF = **5**

2 a) $76 = 2 \times 2 \times 19 = \mathbf{2^2 \times 19}$, $88 = 2 \times 2 \times 2 \times 11 = \mathbf{2^3 \times 11}$
b) LCM = $2^3 \times 11 \times 19 = \mathbf{1672}$

3 a) $60 = 2^2 \times 3 \times 5$, $75 = 3 \times 5^2$
   (i) HCF = $3 \times 5 = \mathbf{15}$
   (ii) LCM = $2^2 \times 3 \times 5^2 = \mathbf{300}$
b) $54 = 2 \times 3^3$, $96 = 2^5 \times 3$
   (i) HCF = $2 \times 3 = \mathbf{6}$
   (ii) LCM = $2^5 \times 3^3 = \mathbf{864}$
c) $108 = 2^2 \times 3^3$, $144 = 2^4 \times 3^2$
   (i) HCF = $2^2 \times 3^2 = \mathbf{36}$
   (ii) LCM = $2^4 \times 3^3 = \mathbf{432}$
d) $200 = 2^3 \times 5^2$, $240 = 2^4 \times 3 \times 5$
   (i) HCF = $2^3 \times 5 = \mathbf{40}$
   (ii) LCM = $2^4 \times 3 \times 5^2 = \mathbf{1200}$
e) $168 = 2^3 \times 3 \times 7$, $196 = 2^2 \times 7^2$
   (i) HCF = $2^2 \times 7 = \mathbf{28}$
   (ii) LCM = $2^3 \times 3 \times 7^2 = \mathbf{1176}$
f) $150 = 2 \times 3 \times 5^2$, $180 = 2^2 \times 3^2 \times 5$
   (i) HCF = $2 \times 3 \times 5 = \mathbf{30}$
   (ii) LCM = $2^2 \times 3^2 \times 5^2 = \mathbf{900}$

4 $93 = 3 \times 31$, $155 = 5 \times 31$
HCF = **31**

5 $316 = 2^2 \times 79$, $408 = 2^3 \times 3 \times 17$
LCM = $2^3 \times 3 \times 17 \times 79 = \mathbf{32\ 232}$

6 The lowest possible number of action figures must be the lowest common multiple of 26 and 12. The prime factorisation of 26 is 2×13 and the prime factorisation of 12 is $2^2 \times 3$. So the LCM = $2^2 \times 3 \times 13$ = **156 action figures.**

7 The number of days until they both go swimming on the same day again is the LCM of 21 and 35. The prime factorisation of 21 is 3×7 and the prime factorisation of 35 is 5×7. So the LCM = $3 \times 5 \times 7 = \mathbf{105\ days}$.

8 a) $30 = \mathbf{2 \times 3 \times 5}$
$140 = \mathbf{2^2 \times 5 \times 7}$
$210 = \mathbf{2 \times 3 \times 5 \times 7}$
b) HCF = $2 \times 5 = \mathbf{10}$

9 a) $121 = \mathbf{11^2}$
$280 = \mathbf{2^3 \times 5 \times 7}$
$550 = \mathbf{2 \times 5^2 \times 11}$
b) LCM = $2^3 \times 5^2 \times 7 \times 11^2 = \mathbf{169\ 400}$

10 a) $65 = 5 \times 13$, $143 = 11 \times 13$, $231 = 3 \times 7 \times 11$
   (i) HCF = **1**
   (ii) LCM = $3 \times 5 \times 7 \times 11 \times 13 = \mathbf{15\ 015}$
b) $175 = 5^2 \times 7$, $245 = 5 \times 7^2$, $1225 = 5^2 \times 7^2$
   (i) HCF = $5 \times 7 = \mathbf{35}$
   (ii) LCM = $5^2 \times 7^2 = \mathbf{1225}$
c) $104 = 2^3 \times 13$, $338 = 2 \times 13^2$, $1078 = 2 \times 7^2 \times 11$
   (i) HCF = **2**
   (ii) LCM = $2^3 \times 7^2 \times 11 \times 13^2 = \mathbf{728\ 728}$

11 a) HCF = $12 = 2^2 \times 3$ and LCM = $120 = 2^3 \times 3 \times 5$
So the prime factorisations of each number must contain at least $2^2 \times 3$ and at most $2^3 \times 3 \times 5$.
So possible pairs are **12 and 120** or $2^3 \times 3 = \mathbf{24}$ and $2^2 \times 3 \times 5 = \mathbf{60}$.

b) HCF = $20 = 2^2 \times 5$ and LCM = $300 = 2^2 \times 3 \times 5^2$
So the prime factorisations of each number must contain at least $2^2 \times 5$ and at most $2^2 \times 3 \times 5^2$.
So possible pairs are **20 and 300** or $2^2 \times 3 \times 5 = \mathbf{60}$ and $2^2 \times 5^2 = \mathbf{100}$.

Page 16   Review Exercise

1 a) $5 \times 6 - 8 \div 2 = 30 - 4 = \mathbf{26}$
b) $18 \div (9 - 12 \div 4) = 18 \div (9 - 3)$
$= 18 \div 6 = \mathbf{3}$

2 $6 - 7 = \mathbf{-1\ °C}$

3 $-4.2 - (1.5 \times -0.3) = -4.2 - (-0.45)$
$= -4.2 + 0.45 = \mathbf{-3.75}$

4
   71.42
 + 11.79
 £83.21
   ₁ ₁

5
11.95   6.59   23.90
 × 2   × 3   + 19.77
23.90 ,   19.77 ,   £43.67
 ₁ ₁   ₁ ₂   ₁ ₁
£50 - £43.67 = **£6.33**

6 One DVD costs:
   £4.55
7)31.³8³5
So 3 DVDs cost:
   455
   × 3
   1365,   1365 ÷ 100 = **£13.65**
   ₁ ₁

7 a) Multiples of 12: 12, 24, 36, 48, 60, 72, 84, 96
Multiples of 16: 16, 32, 48, 64, 80, 96
Common multiples: **48, 96**
b) Multiples of 6: 6, 12, 18, 24, 30, 36, 42, 48, 54, 60, 66, 72, 78, 84, 90, 96
Multiples of 15: 15, 30, 45, 60, 75, 90
Multiples of 20: 20, 40, 60, 80, 100
Common multiples: **60**

8 **8 ways** — 1 pack of 54, 2 packs of 27, 3 packs of 18, 6 packs of 9, 9 packs of 6, 18 packs of 3, 27 packs of 2, 54 packs of 1

9 a) Factors of 12: 1, 2, 3, 4, 6, 12
Factors of 15: 1, 3, 5, 15
Common factors: **1, 3**
b) Factors of 25: 1, 5, 25
Factors of 50: 1, 2, 5, 10, 25, 50
Common factors: **1, 5, 25**
c) Factors of 36: 1, 2, 3, 4, 6, 9, 12, 18, 36
Factors of 48: 1, 2, 3, 4, 6, 8, 12, 16, 24, 48
Common factors: **1, 2, 3, 4, 6, 12**
d) Factors of 64: 1, 2, 4, 8, 16, 32, 64
Factors of 80: 1, 2, 4, 5, 8, 10, 16, 20, 40, 80
Common factors: **1, 2, 4, 8, 16**

10 E.g. **They all end in zero, and are therefore divisible by 10.**

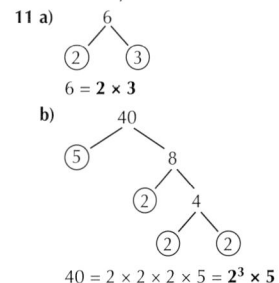

11 a)

$6 = \mathbf{2 \times 3}$

b)

$40 = 2 \times 2 \times 2 \times 5 = \mathbf{2^3 \times 5}$

Using the same method for c)-h):

c) $24 = 2 \times 2 \times 2 \times 3 = 2^3 \times 3$

d) $110 = \textbf{2} \times \textbf{5} \times \textbf{11}$

e) $255 = \textbf{3} \times \textbf{5} \times \textbf{17}$

f) $60 = 2 \times 2 \times 3 \times 5 = \textbf{2}^2 \times \textbf{3} \times \textbf{5}$

g) $360 = 2 \times 2 \times 2 \times 3 \times 3 \times 5 = \textbf{2}^3 \times \textbf{3}^2 \times \textbf{5}$

h) $225 = 3 \times 3 \times 5 \times 5 = \textbf{3}^2 \times \textbf{5}^2$

12 The smallest number of sweets is the lowest common multiple of 5 and 7, which is **35**.

13 a) Factors of 8: 1, 2, 4, 8
Factors of 12: 1, 2, 3, 4, 6, 12
HCF = **4**

b) Factors of 18: 1, 2, 3, 6, 9, 18
Factors of 24: 1, 2, 3, 4, 6, 8, 12, 24
HCF = **6**

c) Factors of 35: 1, 5, 7, 35
Factors of 42: 1, 2, 3, 6, 7, 14, 21, 42
HCF = **7**

d) Factors of 56: 1, 2, 4, 7, 8, 14, 28, 56
Factors of 63: 1, 3, 7, 9, 21, 63
HCF = **7**

14 a) $36 = 2^2 \times 3^2$, $48 = 2^4 \times 3$
(i) HCF = $2^2 \times 3 = \textbf{12}$
(ii) LCM = $2^4 \times 3^2 = \textbf{144}$

b) $210 = 2 \times 3 \times 5 \times 7$, $308 = 2^2 \times 7 \times 11$
(i) HCF = $2 \times 7 = \textbf{14}$
(ii) LCM = $2^2 \times 3 \times 5 \times 7 \times 11 = \textbf{4620}$

c) $126 = 2 \times 3^2 \times 7$, $150 = 2 \times 3 \times 5^2$,
$1029 = 3 \times 7^3$
(i) HCF = **3**
(ii) LCM = $2 \times 3^2 \times 5^2 \times 7^3 = \textbf{154 350}$

Page 17   Exam-Style Questions

1 a) $539 \times 14 = 15\ 092 \div 2 = \textbf{7546}$ *[1 mark]*

b) $5390 \times 0.28 = 15\ 092 \div 10 = \textbf{1509.2}$
[1 mark]

c) $15\ 092 \div 539 = 28$
So $1\ 509\ 200 \div 53.9$
$= (15\ 092 \times 100) \div (539 \div 10)$ *[1 mark]*
$= (15\ 092 \div 539) \times 1000$
$= 28 \times 1000 = \textbf{28 000}$ *[1 mark]*

2 $3.774 \div 0.4 = (3.774 \times 10) \div (0.4 \times 10)$
$= 37.74 \div 4 = 4\overline{)37.^17^14^20}$ $\dfrac{9.\ 4\ 3\ 5}{}$

[3 marks available — 1 mark for converting to a division with a whole-number divisor, 1 mark for carrying out the division, 1 mark for the correct answer]

3 Cost of buying individual packets:
$3 \times 70p = £2.10$
$3 \times 65p = £1.95$ *[1 mark for both]*
$£2.10 + £1.95 = £4.05$ *[1 mark]*
$\dfrac{{}^3 \cancel{4}.{}^9 \cancel{0} 5}{- 3.19}$
$\overline{\textbf{£0.86}}$ *[1 mark]*

4 LCM = $60 = 2^2 \times 3 \times 5$
HCF = $4 = 2^2$ *[1 mark for both]*
So the prime factorisations of each number must contain at least 2^2 and at most $2^2 \times 3 \times 5$.
So the numbers are $2^2 \times 3 = \textbf{12}$ *[1 mark]* and $2^2 \times 5 = \textbf{20}$ *[1 mark]*.

OR

As the LCM is 60, the two numbers must be factors of 60: 1, 2, 3, 4, 5, 6, 10, 12, 15, 20, 30, 60. But as the HCF is 4, both numbers must also be multiples of 4, which leaves 4, 12, 20 and 60 *[1 mark for correct reasoning]*. Since the numbers are not 4 and 60, they must be **12** *[1 mark]* and **20** *[1 mark]*.

---

5 Prime numbers between 50 and 60: 53, 59
[1 mark for both]
Prime numbers between 60 and 70: 61, 67
[1 mark for both]
As the last digit of pq is 9, p must be **59** and q must be **61** *[1 mark for both correct]*.
If you didn't spot the trick with the last digits, you could test 53 and 59 by seeing if they divide exactly into 3599 — you'll find that $3599 \div 59 = 61$.

6 a) E.g.

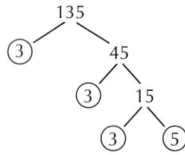

$135 = 3 \times 3 \times 3 \times 5 = \textbf{3}^3 \times \textbf{5}$
[2 marks available — 1 mark for a correct method, 1 mark for the correct factorisation in index form]

b) $60 \times 135 = (2^2 \times 3 \times 5) \times (3^3 \times 5)$
$= 2^2 \times 3^4 \times 5^2$
So $\sqrt{60 \times 135} = \sqrt{2^2 \times 3^4 \times 5^2}$
$= \sqrt{(2 \times 3^2 \times 5)^2}$
$= 2 \times 3^2 \times 5 = \textbf{90}$
[2 marks available — 1 mark for expressing 60×135 as a product of prime factors, 1 mark for taking the square root correctly]

Section 2 — Approximations

2.1 Rounding

Page 18   Exercise 1

1 a) 672.48
(i) The decider is 4 so round down to **672**.
(ii) The decider is 2 so round down to **670**.
(iii) The decider is 7 so round up to **700**.
(iv) The decider is 6 so round up to **1000**.
Using the same method for b)-d):
b) (i) 2536 (ii) 2540 (iii) 2500 (iv) 3000
c) (i) 8499 (ii) 8500 (iii) 8500 (iv) 8000
d) (i) 3823 (ii) 3820 (iii) 3800 (iv) 4000

2 The millions digit is 0 for 390 682 810.
The decider is 6, so round up to
391 000 000 miles.

Page 19   Exercise 2

1 a) 2.6893
(i) For 1 d.p., the decider is 8 so round up to **2.7**.
(ii) For 2 d.p., the decider is 9 so round up to **2.69**.
(iii) For 3 d.p., the decider is 3 so round down to **2.689**.
Using the same method for b)-h):
b) (i) 0.3 (ii) 0.32 (iii) 0.325
c) (i) 5.6 (ii) 5.60 (iii) 5.602
d) (i) 0.1 (ii) 0.05 (iii) 0.053
e) (i) 6.3 (ii) 6.26 (iii) 6.257
f) (i) 0.4 (ii) 0.35 (iii) 0.353
g) (i) 0.1 (ii) 0.08 (iii) 0.080
h) (i) 1.0 (ii) 0.97 (iii) 0.967

2 The decider is 8 so round up to **0.04 kg**.

Page 19   Exercise 3

1 a) (i) To round 46.874 to 1 s.f., the decider is 6 so round up to **50**.
You don't need to add any zeros after the decimal point.
(ii) The decider for 2 s.f. is 8 so round up to **47**.
(iii) The decider for 3 s.f. is 7 so round up to **46.9**.

Using the same method for b)-h):
b) (i) 5000 (ii) 5100 (iii) 5070
c) (i) 40 000 (ii) 36 000 (iii) 35 700
d) (i) 900 000 (ii) 930 000 (iii) 925 000
e) (i) 0.09 (ii) 0.086 (iii) 0.0860
f) (i) 0.1 (ii) 0.11 (iii) 0.107
g) (i) 0.0004 (ii) 0.00042 (iii) 0.000418
h) (i) 30 (ii) 35 (iii) 34.7

2 To round 1236 km/h to 2 s.f., the decider is 3 so round down to **1200 km/h**.

Page 20   Exercise 4

1 a) $102.2 \times 4.2 \approx 100 \times 4 = \textbf{400}$
b) $288.7 \times 7.8 \approx 300 \times 8 = \textbf{2400}$
Using the same method for c)-h):
c) 50   **d) 0.8**   **e) 100**
f) 25   **g) 1000**   **h) 4**

2 a) $101 \times 52 \approx 100 \times 50 = 5000$
The option closest to 5000 is **5252**.
Using the same method for b)-d):
b) 47.42   **c) 1259.50**   **d) 0.255**

3 a) $\dfrac{9.9 \times 285}{18.7 \times 3.2} \approx \dfrac{10 \times 300}{20 \times 3} = \dfrac{3000}{60} = \textbf{50}$
Using the same method for b)-d):
b) 0.3   **c) 7.5**   **d)** $\dfrac{1}{9}$

4 a) $\sqrt{18}$ is nearer $\sqrt{16}$ (=4) than $\sqrt{25}$ (=5), so $\sqrt{18} \approx \textbf{4}$.
Using the same method for b)-d):
b) 8   **c) 7**   **d) 9**

5 a) $\sqrt{11}$ lies between $\sqrt{9}$ and $\sqrt{16}$, but is closer to $\sqrt{11}$, so a sensible estimate is **3.3**.
Using the same method for b)-d):
b) 5.5   **c) 4.5**   **d) 6.1**
For these questions, it's ok if you've gotten slightly different answers for your estimates.

2.2 Upper and Lower Bounds

Page 21   Exercise 1

1 a) The lower bound is $10 - 0.5 = \textbf{9.5}$.
The upper bound is $10 + 0.5 = \textbf{10.5}$.
Using the same method for b)-h):
b) lower bound = **33.5**, upper bound = **34.5**
c) lower bound = **75.5**, upper bound = **76.5**
d) lower bound = **101.5**, upper bound = **102.5**
e) lower bound = **98.5**, upper bound = **99.5**
f) lower bound = **998.5**, upper bound = **999.5**
g) lower bound = **248.5**, upper bound = **249.5**
h) lower bound = **2499.5**, upper bound = **2500.5**

2 a) The lower bound is $645 - 0.5 = \textbf{644.5 kg}$.
The upper bound is $645 + 0.5 = \textbf{645.5 kg}$.
Using the same method for b)-d):
b) lower bound = **254.5 litres**,
upper bound = **255.5 litres**
c) lower bound = **750 g**, upper bound = **850 g**
d) lower bound = **154.5 cm**,
upper bound = **155.5 cm**

3 a) lower bound = $15 - 0.5 = \textbf{£14.50}$
upper bound = $15 + 0.5 = \textbf{£15.50}$
b) lower bound = $320 - 5 = \textbf{£315}$
upper bound = $320 + 5 = \textbf{£325}$
c) lower bound = $76.70 - 0.05 = \textbf{£76.65}$
upper bound = $76.70 + 0.05 = \textbf{£76.75}$
d) lower bound = $600 - 25 = \textbf{£575}$
upper bound = $600 + 25 = \textbf{£625}$

Page 22   Exercise 2

1 a) Use the lower bounds to find the minimum perimeter.
$5 - 0.5 = 4.5$ cm, $6 - 0.5 = 5.5$ cm, so perimeter = $(2 \times 4.5) + (2 \times 5.5) = \textbf{20 cm}$
Use the upper bounds to find the maximum perimeter.
$5 + 0.5 = 5.5$ cm, $6 + 0.5 = 6.5$ cm, so perimeter = $(2 \times 5.5) + (2 \times 6.5) = \textbf{24 cm}$

Using the same method for b)-d):

b) minimum = **14 m**, maximum = **18 m**

c) minimum = **19.5 cm**, maximum = **22.5 cm**

d) minimum = **62.25 cm**,
maximum = **62.75 cm**

2 To find the longest length,
look at the upper bounds.
The upper bound of 2.55 m is 2.555 m.
The upper bound of 3.45 m is 3.455 m.
$(2.555 \times 2) + (3.455 \times 2) = 5.11 + 6.91$
$= \textbf{12.02 m}$

3 Tallest (maximum) = upper bound of 1.3 m
to the nearest 10 cm + upper bound of 5 cm
to the nearest cm = 1.35 m + 5.5 cm
(= 0.055 m). So the tallest he could be is
$1.35 + 0.055 = \textbf{1.405 m}$ (or **140.5 cm**).
Shortest (minimum) = lower bound of 1.3 cm
to the nearest 10 cm + lower bound of 5 cm
to the nearest cm = 1.25 m + 4.5 cm
(= 0.045 m), so $1.25 + 0.045 = \textbf{1.295 m}$
(or **129.5 cm**).

4 a) (i) The minimum gate width is the
lower bound, which is **1.5 m**.

(ii) Maximum length and width are the
upper bounds of the measurements.
These are length = **15.5 m** and
width = **12.5 m**.

(iii) Maximum total length =
maximum garden length – minimum
gate width = 15.5 – 1.5 = **14 m**

(iv) Maximum length of fence for garden
= maximum perimeter – minimum gate
width = $(15.5 \times 2) + (12.5 \times 2) - 1.5$
$= 31 + 25 - 1.5 = \textbf{54.5 m}$

b) Minimum garden length is 14.5 m,
minimum garden width is 11.5 m.
So minimum perimeter of the garden is
$(14.5 \times 2) + (11.5 \times 2) = 29 + 23 = 52$ m.
Maximum gate width is the upper bound
of 2 m, which is 2.5 m. So the minimum
length of fencing is $52 - 2.5 = \textbf{49.5 m}$.

Page 23   Exercise 3

1 The upper bounds are +0.005 m for each
dimension, so the maximum volume
$= 4.005 \times 3.005 \times 1.905 = \textbf{22.93 m}^3$ (2 d.p.).
The lower bounds are –0.005 m off each
dimension, so minimum volume
$= 3.995 \times 2.995 \times 1.895 = \textbf{22.67 m}^3$ (2 d.p.).

2 Maximum time = 13 + 0.5 = 13.5 minutes
13.5 minutes = $\frac{13.5}{60}$ hours = 0.225 hours
So maximum distance = speed × max. time
$= 34 \times 0.225 = \textbf{7.65 km}$

3 Maximum speed = $\frac{\text{maximum distance}}{\text{minimum time}}$
Maximum distance = 500.5 cm
Minimum time = 44.5 mins
44.5 mins = 44.5 × 60 seconds = 2670 s
So max speed = $\frac{500.5}{2670} = \textbf{0.187 cm/s}$ (3 s.f.)

*If you're struggling to get your head around this,
try combinations of upper and lower bounds.
You'll see that $\frac{\text{upper bound}}{\text{lower bound}}$ gives maximum speed.*

4 Maximum volume of paint needed
$= \frac{\text{maximum wall area}}{\text{min. area a tin could cover}} \times$ amount per tin
Maximum wall area is
2.15 m × 5.25 m = 11.2875 m².
Minimum area a tin could cover = 3.45 m².
Maximum volume of paint needed
$= \frac{11.2875}{3.45} \times 0.5 = 1.6358...$
$= \textbf{1.64 litres}$ (2 d.p.)

Page 24   Exercise 4

1 a) To truncate 1.354 to 1 d.p. delete all
the digits after the first decimal place.
So it's **1.3**.

Using the same method for b)-d):

b) 55.7   **c)** 103.6   **d)** 85.9

2 a) To truncate 2738.29109 to 2 d.p. delete all
digits after the second decimal place.
So it's **2738.29**.

Using the same method for b)-d):

b) 1.24   **c)** 17.16   **d)** 100.09

Page 24   Exercise 5

1 a) The lower bound of 15.2 to 1 d.p. is 15.15,
and the upper bound is 15.25.
So the interval is $\textbf{15.15} \le n < \textbf{15.25}$.

Using the same method for b)-d):

b) $37.05 \le p < 37.15$

c) $109.85 \le q < 109.95$

d) $69.95 \le r < 70.05$

2 a) The lower bound is 6.57, and
the upper bound is 6.57999... (≈ 6.58).
So the interval is $6.57 \le s < 6.58$.

Using the same method for b)-d):

b) $25.71 \le t < 25.72$

c) $99.99 \le v < 100.00$

d) $51.00 \le w < 51.01$

3 10 km to the nearest 100 m (0.1 km) is
± 0.05 km. 10 – 0.05 = 9.95,
100 + 0.05 = 10.05. So the inequality
is $\textbf{9.95 km} \le d < \textbf{10.05 km}$.

4 Lower bound for x = 3.2 – 0.05 = 3.15
Lower bound for y = 8.34 – 0.005 = 8.335
Upper bound for x = 3.2 + 0.05 = 3.25
Upper bound for y = 8.34 + 0.005 = 8.345
Minimum: 2x + y = 2(3.15) + 8.335 = 14.635
Maximum: 2x + y = 2(3.25) + 8.345 = 14.845
So the inequality is $\textbf{14.635} \le 2x + y < \textbf{14.845}$.

Page 25   Review Exercise

1 The decider is 4 so round down to **1.2 m**.

2 For the common vole, the decider is 9
so round up to **0.028 kg**.
Using the same method for the other mammals:
Badger = **9.1 kg**
Meerkat = **0.78 kg**
Red Squirrel = **0.20 kg**
Shrew = **0.0061 kg**
Hare = **3.7 kg**

3 To the nearest £, the items cost £1, £9, £14
and £18. Add up the rounded numbers:
1 + 9 + 14 + 18 = 42 = **£42**

4 a) $\frac{64.4 \times 5.6}{17 \times 9.5} \approx \frac{60 \times 6}{20 \times 10} = \frac{360}{200} = \textbf{1.8}$

Using the same method for b)-d):

b) 3   **c)** 20   **d)** 200

5 a) $\sqrt{14}$ lies between $\sqrt{9}$ (=3) and $\sqrt{16}$ (=4).
It's closer to $\sqrt{16}$, so a good estimate is **3.7**.

Using the same method for b)-d):

b) 8.8   **c)** 11.4   **d)** 7.5

*For this question, it's OK if you got slightly
different answers for your estimates.*

6 For 24 minutes to the nearest minute:
lower bound = 24 – 0.5 = **23.5 mins**
upper bound = 24 + 0.5 = **24.5 mins**

7 a) Upper bounds: 5.5 cm and 6.5 cm,
maximum is 5.5 cm × 6.5 cm = **35.75 cm²**
Lower bounds: 4.5 cm and 5.5 cm,
minimum is 4.5 cm × 5.5 cm = **24.75 cm²**

Using the same method for b)-c):

b) maximum = **274.625 cm³**,
minimum = **166.375 cm³**

c) maximum = **89.26 cm³**,
minimum = **83.29 cm³** (2 d.p.)

8 maximum speed = $\frac{\text{max. distance}}{\text{min. time}} = \frac{1505}{259.5}$
$= \textbf{5.80 m/s}$ (2 d.p.)

minimum speed = $\frac{\text{min. distance}}{\text{max. time}} = \frac{1495}{260.5}$
$= \textbf{5.74 m/s}$ (2 d.p.)

9 a) To truncate 78.445 to 1 d.p. delete all
digits after the first decimal place, so **78.4**.

Using the same method for b)-d):

b) 32.5   **c)** 567.8   **d)** 999.9

10 a) Lily: $\textbf{1.40} \le y < \textbf{1.41}$, May: $\textbf{1.43} \le y < \textbf{1.44}$,
Isaac: $\textbf{1.60} \le y < \textbf{1.61}$, Max: $\textbf{1.56} \le y < \textbf{1.57}$,
Daisy: $\textbf{1.28} \le y < \textbf{1.29}$

b) Minimum height difference
= 1.60 – 1.29 = **0.31 m**
Maximum height difference
= 1.61 – 1.28 = **0.33 m**

*For the minimum, take the upper bound of the
shortest child (Daisy) from the lower bound of
the tallest child (Isaac). For the maximum, take
the lower bound of Daisy's height. from the upper
bound of Isaac's height.*

Page 26   Exam-Style Questions

1 The upper bound is **57.5 cm**, and
the lower bound is **56.5 cm**. *[1 mark for both]*

2 a) $l = 6.37 = 6$ cm (1 s.f.)
$a = l^2 \approx 6^2 = \textbf{36 cm}^2$ *[1 mark]*

b) 9.2 cm rounded to 1 s.f. = 9 cm
$V = \frac{1}{3} \times$ base area × perpendicular height
$= \frac{1}{3} \times 36 \times 9 = \textbf{108 cm}^3$
*[2 marks available — 1 mark for rounding,
1 mark for correct answer]*
*Since this is an estimate, you might have gotten
a slightly different final answer — as long as your
estimates are sensible and your working is correct,
you'll get the marks.*

3 a) $628 \approx 600$ to 1 s.f. *[1 mark]*,
$\sqrt{97} \approx \sqrt{100}$ and 9.6 ≈ 10 to 1 s.f. *[1 mark]*
so $\frac{628}{\sqrt{97} + 9.6} \approx \frac{600}{10 + 10} = 30$ *[1 mark]*

b) In the estimation, the numerator is smaller
and the denominator is larger than the
actual values. Both of these will decrease
the overall value, so it is an under-estimate.
[1 mark]

4 7300 to the nearest 100 is ± 50,
so the error interval is $7250 \le r < 7350$
(or $7249 < r < 7350$ or $7250 \le r \le 7349$).
*[2 marks available — 1 mark for correct
numbers, 1 mark for correct signs]*
*The numbers depend on the signs used —
inequalities are covered in Section 13.*

5 a) Maximum perimeter = 3.25 + 7.45 + 4.75
+ 2.85 + 1.55 + 4.65 = **24.5 m**
*[2 marks available — 1 mark for correct
upper bounds, 1 mark for correct answer]*

b) Minimum perimeter = 3.15 + 7.35 + 4.65
+ 2.75 + 1.45 + 4.55 = 23.9 m
interval = **23.9 m** $\le p <$ **24.5 m**
*[2 marks available — 1 mark for correct
minimum perimeter, 1 mark for correct
interval]*

Section 3 — Fractions

3.1 Equivalent Fractions

Page 27   Exercise 1

1 a) $\frac{1}{5} = \frac{1 \times 2}{5 \times 2} = \frac{2}{10} \Rightarrow a = 2$

b) $\frac{1}{4} = \frac{1 \times 3}{4 \times 3} = \frac{3}{12} \Rightarrow b = 3$

c) $\frac{3}{4} = \frac{3 \times 4}{4 \times 4} = \frac{12}{16} \Rightarrow c = 12$

d) $\frac{1}{20} = \frac{1 \times 3}{20 \times 3} = \frac{3}{60} \Rightarrow d = 3$

e) $\frac{1}{5} = \frac{1 \times 5}{5 \times 5} = \frac{5}{25} \Rightarrow e = 25$

f) $\frac{1}{6} = \frac{1 \times 3}{6 \times 3} = \frac{3}{18} \Rightarrow f = 18$

g) $\frac{7}{12} = \frac{7 \times 5}{12 \times 5} = \frac{35}{60} \Rightarrow g = 60$

h) $\frac{9}{10} = \frac{9 \times 9}{10 \times 9} = \frac{81}{90} \Rightarrow h = 90$

2 a) $\frac{5}{15} = \frac{5 \div 5}{15 \div 5} = \frac{1}{3} \Rightarrow a = 3$

b) $\frac{12}{20} = \frac{12 \div 4}{20 \div 4} = \frac{3}{5} \Rightarrow b = 5$

c) $\frac{10}{15} = \frac{10 \div 5}{15 \div 5} = \frac{2}{3} \Rightarrow c = 2$

d) $\frac{9}{42} = \frac{9 \div 3}{42 \div 3} = \frac{3}{14} \Rightarrow d = 3$

e) $\frac{15}{27} = \frac{15 \div 3}{27 \div 3} = \frac{5}{9} \Rightarrow e = 5$

f) $\frac{9}{17} = \frac{9 \times 3}{17 \times 3} = \frac{27}{51} \Rightarrow f = 27$

g) $\frac{55}{80} = \frac{55 \div 5}{80 \div 5} = \frac{11}{16} \Rightarrow g = 16$

h) $\frac{11}{121} = \frac{11 \div 11}{121 \div 11} = \frac{1}{11} \Rightarrow h = 11$

3 $\frac{4}{6} = \frac{4 \times 7}{6 \times 7} = \frac{28}{42}$
28 < 37, so **Dev** got more questions right.

Page 28 Exercise 2
1 a) $\frac{9}{45} = \frac{9 \div 9}{45 \div 9} = \frac{1}{5}$
b) $\frac{15}{36} = \frac{15 \div 3}{36 \div 3} = \frac{5}{12}$
c) $\frac{24}{64} = \frac{24 \div 8}{64 \div 8} = \frac{3}{8}$
d) $\frac{72}{162} = \frac{72 \div 18}{162 \div 18} = \frac{4}{9}$
It's fine if you did any of these in more than one step — as long as your final answer is correct.
2 a) $\frac{21}{35} = \frac{21 \div 7}{35 \div 7} = \frac{3}{5}$
b) $\frac{36}{126} = \frac{36 \div 18}{126 \div 18} = \frac{2}{7}$
c) $\frac{70}{182} = \frac{70 \div 14}{182 \div 14} = \frac{5}{13}$
3 a) $\frac{6}{18} = \frac{6 \div 6}{18 \div 6} = \frac{1}{3}$, $\frac{5}{20} = \frac{5 \div 5}{20 \div 5} = \frac{1}{4}$, $\frac{9}{27} = \frac{9 \div 9}{27 \div 9} = \frac{1}{3}$
So the fraction that is not equivalent to the other two is $\frac{5}{20}$.
Using the same method for b)-d):
b) $\frac{6}{8}$ **c)** $\frac{6}{33}$ **d)** $\frac{24}{40}$
4 a) $\frac{50}{300} = \frac{50 \div 50}{300 \div 50} = \frac{1}{6}$
b) $300 - 50 - 70 = 180$ sheep
$\frac{180}{300} = \frac{180 \div 60}{300 \div 60} = \frac{3}{5}$

3.2 Mixed Numbers
Page 29 Exercise 1
1 a) $1\frac{1}{3} = \frac{3}{3} + \frac{1}{3} = \frac{4}{3} \Rightarrow a = 4$
b) $1\frac{2}{7} = \frac{7}{7} + \frac{2}{7} = \frac{9}{7} \Rightarrow b = 9$
c) $4\frac{1}{2} = \frac{8}{2} + \frac{1}{2} = \frac{9}{2} \Rightarrow c = 9$
d) $3\frac{4}{7} = \frac{21}{7} + \frac{4}{7} = \frac{25}{7} \Rightarrow d = 25$
2 a) $1\frac{4}{5} = \frac{5}{5} + \frac{4}{5} = \frac{9}{5}$
b) $1\frac{5}{12} = \frac{12}{12} + \frac{5}{12} = \frac{17}{12}$
c) $2\frac{9}{10} = \frac{20}{10} + \frac{9}{10} = \frac{29}{10}$
d) $4\frac{3}{4} = \frac{16}{4} + \frac{3}{4} = \frac{19}{4}$
e) $12\frac{2}{5} = \frac{60}{5} + \frac{2}{5} = \frac{62}{5}$
f) $15\frac{5}{7} = \frac{105}{7} + \frac{5}{7} = \frac{110}{7}$
g) $3\frac{1}{9} = \frac{27}{9} + \frac{1}{9} = \frac{28}{9}$
h) $10\frac{3}{10} = \frac{100}{10} + \frac{3}{10} = \frac{103}{10}$
3 a) $\frac{26}{4} = \frac{26 \div 2}{4 \div 2} = \frac{13}{2}$
b) $\frac{26}{4} = \frac{13}{2} = \frac{12+1}{2} = 6\frac{1}{2}$

4 a) $\frac{5}{3} = \frac{3+2}{3} = 1\frac{2}{3}$
b) $\frac{9}{5} = \frac{5+4}{5} = 1\frac{4}{5}$
c) $\frac{17}{10} = \frac{10+7}{10} = 1\frac{7}{10}$
d) $\frac{12}{7} = \frac{7+5}{7} = 1\frac{5}{7}$
e) $\frac{13}{6} = \frac{12+1}{6} = 2\frac{1}{6}$
f) $\frac{18}{12} = \frac{12+6}{12} = 1\frac{6}{12} = 1\frac{1}{2}$
You could also simplify the improper fraction before turning it into a mixed number.
g) $\frac{50}{15} = \frac{45+5}{15} = 3\frac{5}{15} = 3\frac{1}{3}$
h) $\frac{24}{18} = \frac{18+6}{18} = 1\frac{6}{18} = 1\frac{1}{3}$
i) $\frac{13}{11} = \frac{11+2}{11} = 1\frac{2}{11}$
j) $\frac{35}{25} = \frac{25+10}{25} = 1\frac{10}{25} = 1\frac{2}{5}$
k) $\frac{51}{12} = \frac{48+3}{12} = 4\frac{3}{12} = 4\frac{1}{4}$
l) $\frac{98}{8} = \frac{96+2}{8} = 12\frac{2}{8} = 12\frac{1}{4}$
5 a) $\frac{6}{4} = \frac{6 \div 2}{4 \div 2} = \frac{3}{2}$, $\frac{5}{2} = \frac{5}{2}$, $1\frac{1}{2} = \frac{2}{2} + \frac{1}{2} = \frac{3}{2}$
So the fraction that is not equivalent to the other two is $\frac{5}{2}$.
Using the same method for b)-d):
b) $3\frac{1}{2}$ **c)** $\frac{15}{4}$ **d)** $\frac{11}{3}$

3.3 Ordering Fractions
Page 30 Exercise 1
1 a) E.g. $\frac{2}{9} = \frac{2}{9}$, $\frac{1}{3} = \frac{1 \times 3}{3 \times 3} = \frac{3}{9}$
$\frac{3}{9} > \frac{2}{9}$ so $\frac{1}{3}$ is larger.
b) E.g. $\frac{2}{3} = \frac{2 \times 4}{3 \times 4} = \frac{8}{12}$, $\frac{3}{4} = \frac{3 \times 3}{4 \times 3} = \frac{9}{12}$
$\frac{9}{12} > \frac{8}{12}$ so $\frac{3}{4}$ is larger.
c) E.g. $\frac{7}{8} = \frac{7 \times 5}{8 \times 5} = \frac{35}{40}$, $\frac{3}{10} = \frac{3 \times 4}{10 \times 4} = \frac{12}{40}$
$\frac{35}{40} > \frac{12}{40}$ so $\frac{7}{8}$ is larger.
Using the same method for d)-h):
d) E.g. $\frac{2}{5} = \frac{18}{45}$, $\frac{4}{9} = \frac{20}{45}$ so $\frac{4}{9}$ is larger.
e) E.g. $\frac{1}{5} = \frac{4}{20}$, $\frac{7}{10} = \frac{14}{20}$, $\frac{9}{20} = \frac{9}{20}$
so $\frac{7}{10}$ is largest.
f) E.g. $\frac{1}{7} = \frac{6}{42}$, $\frac{4}{21} = \frac{8}{42}$, $\frac{5}{14} = \frac{15}{42}$
so $\frac{5}{14}$ is largest.
g) E.g. $\frac{2}{5} = \frac{24}{60}$, $\frac{5}{12} = \frac{25}{60}$, $\frac{11}{30} = \frac{22}{60}$
so $\frac{5}{12}$ is largest.
h) E.g. $\frac{5}{18} = \frac{100}{360}$, $\frac{7}{24} = \frac{105}{360}$, $\frac{11}{30} = \frac{132}{360}$
so $\frac{11}{30}$ is largest.
You could have used different common denominators for any of the parts in Q1 — the ones used here are the lowest common denominators.
2 a) $\frac{1}{4} = \frac{1 \times 2}{4 \times 2} = \frac{2}{8}$, $\frac{5}{8} = \frac{5}{8}$
So, in order: $\frac{1}{4}$, $\frac{5}{8}$
b) $\frac{5}{6} = \frac{5 \times 2}{6 \times 2} = \frac{10}{12}$, $\frac{3}{4} = \frac{3 \times 3}{4 \times 3} = \frac{9}{12}$
So, in order: $\frac{3}{4}$, $\frac{5}{6}$
c) $\frac{2}{3} = \frac{2 \times 5}{3 \times 5} = \frac{10}{15}$, $\frac{3}{5} = \frac{3 \times 3}{5 \times 3} = \frac{9}{15}$
So, in order: $\frac{3}{5}$, $\frac{2}{3}$
Using the same method for d)-l):
d) $\frac{7}{10}$, $\frac{3}{4}$ **e)** $\frac{7}{16}$, $\frac{1}{2}$, $\frac{5}{8}$
f) $\frac{19}{24}$, $\frac{5}{6}$, $\frac{11}{12}$ **g)** $\frac{5}{12}$, $\frac{4}{9}$, $\frac{2}{3}$

h) $\frac{4}{5}$, $\frac{9}{10}$, $\frac{11}{12}$ **i)** $\frac{7}{9}$, $\frac{4}{5}$, $\frac{13}{15}$
j) $\frac{3}{15}$, $\frac{7}{27}$, $\frac{12}{45}$
The LCM of the denominators is 135.
k) $\frac{5}{16}$, $\frac{7}{20}$, $\frac{9}{25}$
The LCM of the denominators is 400.
l) $\frac{4}{15}$, $\frac{11}{36}$, $\frac{9}{24}$
The LCM of the denominators is 360.

3.4 Adding and Subtracting Fractions
Page 31 Exercise 1
1 a) $\frac{2}{3} - \frac{1}{4} = \frac{8}{12} - \frac{3}{12} = \frac{5}{12}$
b) $\frac{2}{3} + \frac{4}{5} = \frac{10}{15} + \frac{12}{15} = \frac{22}{15} = 1\frac{7}{15}$
Using the same methods for c)-h):
c) $\frac{1}{15}$ **d)** $1\frac{5}{28}$ **e)** $1\frac{32}{99}$
f) $1\frac{29}{63}$ **g)** $\frac{9}{35}$ **h)** $1\frac{43}{80}$
2 a) $\frac{1}{9} + \frac{5}{9} + \frac{11}{18} = \frac{2}{18} + \frac{10}{18} + \frac{11}{18} = \frac{23}{18} = 1\frac{5}{18}$
b) $1 - \frac{2}{10} - \frac{8}{20} = \frac{40}{40} - \frac{8}{40} - \frac{10}{40} = \frac{22}{40} = \frac{11}{20}$
Using the same method for c)-h):
c) $\frac{7}{16}$ **d)** $\frac{3}{7}$ **e)** $1\frac{3}{4}$
f) $\frac{11}{15}$ **g)** $\frac{19}{60}$ **h)** $\frac{23}{42}$
3 $1 - \frac{1}{2} - \frac{1}{5} = \frac{10}{10} - \frac{5}{10} - \frac{2}{10} = \frac{3}{10}$
4 $1 - \frac{2}{9} - \frac{1}{12} - \frac{1}{5} = \frac{180}{180} - \frac{40}{180} - \frac{15}{180} - \frac{36}{180} = \frac{89}{180}$

Page 32 Exercise 2
In Q1 and 2, we've used the method of converting to improper fractions, but in Q3, it's much easier to add the number parts and fraction parts separately. You should get the same answer no matter which method you use.
1 a) $2\frac{3}{8} + \frac{3}{4} = \frac{19}{8} + \frac{6}{8} = \frac{25}{8} = 3\frac{1}{8}$
b) $4\frac{3}{14} + 1\frac{6}{7} = \frac{59}{14} + \frac{26}{14} = \frac{85}{14} = 6\frac{1}{14}$
c) $1\frac{3}{5} + \frac{3}{4} = \frac{8}{5} + \frac{3}{4} = \frac{32}{20} + \frac{15}{20} = \frac{47}{20} = 2\frac{7}{20}$
Using the same method for d)-h) and Q2:
d) $3\frac{7}{24}$ **e)** $5\frac{22}{35}$ **f)** $5\frac{19}{30}$
g) $6\frac{4}{33}$ **h)** $8\frac{59}{60}$
2 a) $1\frac{7}{12}$ **b)** $5\frac{17}{18}$ **c)** $3\frac{1}{28}$
d) $1\frac{28}{45}$ **e)** $\frac{11}{28}$ **f)** $3\frac{11}{72}$
g) $3\frac{17}{30}$ **h)** $\frac{1}{63}$
3 $17\frac{7}{8} + 9\frac{5}{12} + 40\frac{5}{18}$
$= 17 + 9 + 40 + \frac{7}{8} + \frac{5}{12} + \frac{5}{18}$
$= 66 + \frac{63}{72} + \frac{30}{72} + \frac{20}{72} = 66 + \frac{113}{72}$
$= 66 + 1\frac{41}{72} = 67\frac{41}{72}$

3.5 Multiplying and Dividing by Fractions
Page 33 Exercise 1
1 a) $28 \times \frac{3}{4} = (28 \div 4) \times 3 = 7 \times 3 = 21$
b) $\frac{2}{9} \times 36 = 2 \times (36 \div 9) = 2 \times 4 = 8$
Using the same method for c)-h):
c) 18 **d)** 25 **e)** 20
f) 12 **g)** 45 **h)** 56
2 a) $48 \times \frac{2}{7} = \frac{48 \times 2}{7} = \frac{96}{7} = \frac{91+5}{7} = 13\frac{5}{7}$
b) $27 \times \frac{1}{6} = \frac{27}{6} = \frac{24+3}{6} = 4\frac{3}{6} = 4\frac{1}{2}$
Using the same method for c)-h):
c) $21\frac{1}{3}$ **d)** $27\frac{1}{5}$ **e)** $17\frac{7}{9}$
f) $18\frac{3}{4}$ **g)** $19\frac{7}{11}$ **h)** $38\frac{3}{4}$

Page 34 Exercise 2

1 a) $\frac{3}{5} \times \frac{1}{6} = \frac{1}{5} \times \frac{1}{2} = \frac{1 \times 1}{5 \times 2} = \frac{1}{10}$

b) $\frac{5}{6} \times \frac{2}{15} = \frac{1}{6} \times \frac{2}{3} = \frac{1}{3} \times \frac{1}{3} = \frac{1 \times 1}{3 \times 3} = \frac{1}{9}$

Using the same method for c)-f):

c) $\frac{5}{16}$ **d)** $\frac{3}{4}$ **e)** $\frac{1}{4}$ **f)** 1

2 a) $1\frac{5}{6} \times \frac{2}{3} = \frac{11}{6} \times \frac{2}{3} = \frac{11}{3} \times \frac{1}{3} = \frac{11}{9} = 1\frac{2}{9}$

b) $3\frac{3}{4} \times \frac{2}{5} = \frac{15}{4} \times \frac{2}{5} = \frac{3}{2} \times \frac{1}{1} = \frac{3}{2} = 1\frac{1}{2}$

c) $2\frac{1}{7} \times \frac{2}{9} = \frac{15}{7} \times \frac{2}{9} = \frac{5}{7} \times \frac{2}{3} = \frac{10}{21}$

Using the same method for d)-i):

d) $\frac{23}{48}$ **e)** $3\frac{17}{25}$ **f)** $\frac{11}{12}$

g) $7\frac{3}{5}$ **h)** $7\frac{5}{9}$ **i)** 10

Page 35 Exercise 3

1 a) Reciprocal of $7 = \frac{1}{7}$

b) Reciprocal of $\frac{1}{11} = 11$

c) Reciprocal of $\frac{7}{6} = \frac{6}{7}$

d) Reciprocal of $\frac{3}{26} = \frac{26}{3}$

e) $1\frac{11}{12} = \frac{12}{12} + \frac{11}{12} = \frac{23}{12}$ so reciprocal $= \frac{12}{23}$

f) $2\frac{3}{4} = \frac{8}{4} + \frac{3}{4} = \frac{11}{4}$ so reciprocal $= \frac{4}{11}$

g) $5\frac{2}{3} = \frac{15}{3} + \frac{2}{3} = \frac{17}{3}$ so reciprocal $= \frac{3}{17}$

h) $4\frac{2}{7} = \frac{28}{7} + \frac{2}{7} = \frac{30}{7}$ so reciprocal $= \frac{7}{30}$

2 a) $\frac{4}{13} \div \frac{1}{3} = \frac{4}{13} \times \frac{3}{1} = \frac{12}{13}$

b) $\frac{2}{25} \div \frac{1}{5} = \frac{2}{25} \times \frac{5}{1} = \frac{2}{5} \times \frac{1}{1} = \frac{2}{5}$

c) $\frac{2}{5} \div \frac{2}{3} = \frac{2}{5} \times \frac{3}{2} = \frac{1}{5} \times \frac{3}{1} = \frac{3}{5}$

d) $\frac{3}{4} \div \frac{9}{10} = \frac{3}{4} \times \frac{10}{9} = \frac{1}{2} \times \frac{5}{3} = \frac{5}{6}$

e) $\frac{5}{7} \div \frac{11}{14} = \frac{5}{7} \times \frac{14}{11} = \frac{5}{1} \times \frac{2}{11} = \frac{10}{11}$

f) $\frac{2}{5} \div 3 = \frac{2}{5} \times \frac{1}{3} = \frac{2}{15}$

g) $\frac{3}{7} \div 6 = \frac{3}{7} \times \frac{1}{6} = \frac{1}{7} \times \frac{1}{2} = \frac{1}{14}$

h) $7 \div \frac{15}{2} = 7 \times \frac{2}{15} = \frac{7 \times 2}{15} = \frac{14}{15}$

3 a) $2\frac{1}{2} \div \frac{1}{3} = \frac{5}{2} \times \frac{3}{1} = \frac{15}{2} = 7\frac{1}{2}$

b) $1\frac{1}{6} \div \frac{1}{4} = \frac{7}{6} \times \frac{4}{1} = \frac{7}{3} \times \frac{2}{1} = \frac{14}{3} = 4\frac{2}{3}$

c) $2\frac{3}{7} \div 3 = \frac{17}{7} \times \frac{1}{3} = \frac{17}{21}$

Using the same method for d)-h):

d) $\frac{20}{27}$ **e)** $\frac{10}{51}$ **f)** $4\frac{4}{9}$

g) $1\frac{1}{24}$ **h)** $1\frac{27}{50}$

3.6 Fractions and Decimals

Page 36 Exercise 1

1 a) $329 \div 500 = \textbf{0.658}$

b) $2 + (1 \div 8) = 2 + 0.125 = \textbf{2.125}$

c) $2 + (37 \div 100) = 2 + 0.37 = \textbf{2.37}$

d) $4 + (719 \div 1000) = 4 + 0.719 = \textbf{4.719}$

e) $4 \div 9 = 0.44444444... = \mathbf{0.\dot{4}}$

f) $4 \div 15 = 0.26666666... = \mathbf{0.2\dot{6}}$

g) $1234 \div 9999 = 0.12341234... = \mathbf{0.\dot{1}23\dot{4}}$

h) $88 \div 3 = 29.3333333... = \mathbf{29.\dot{3}}$

2 a) $167 \div 287 = 0.58188...$
$87 \div 160 = 0.54375$
$196 \div 360 = 0.54444...$
So, in order: $\frac{87}{160}, \frac{196}{360}, \frac{167}{287}$

b) $96 \div 99 = 0.96969...$
$16 \div 17 = 0.94117...$
$5 \div 6 = 0.83333...$
So, in order: $\frac{5}{6}, \frac{16}{17}, \frac{96}{99}$

---

c) $963 \div 650 = 1.48153...$
$13 \div 9 = 1.44444...$
$77 \div 52 = 1.48076...$
So, in order: $\frac{13}{9}, \frac{77}{52}, \frac{963}{650}$

Page 37 Exercise 2

1 a) $\frac{46}{100} = \textbf{0.46}$

b) $\frac{492}{1000} = \textbf{0.492}$

c) $\frac{9}{30} = \frac{3}{10} = \textbf{0.3}$

d) $\frac{17}{50} = \frac{34}{100} = \textbf{0.34}$

Using the same method for e)-h):

e) $\textbf{0.6}$ **f)** $\textbf{0.88}$

g) $\textbf{0.666}$ **h)** $\textbf{0.615}$

2 a) $16\overline{)1.^{1}0^{10}0^{4}0^{8}0}$ so $\frac{1}{16} = \textbf{0.0625}$

b) $8\overline{)7.^{7}0^{6}0^{4}0}$ so $\frac{7}{8} = \textbf{0.875}$

Using the same method for c)-h):

c) $\textbf{0.008}$ **d)** $\textbf{0.09375}$ **e)** $\textbf{0.65625}$

f) $\textbf{0.4375}$ **g)** $\textbf{0.072}$ **h)** $\textbf{6.625}$

Page 38 Exercise 3

1 a) $\frac{1}{9} = \mathbf{0.\dot{1}}$

b) $\frac{247}{999} = \mathbf{0.\dot{2}4\dot{7}}$

c) $\frac{4}{9} = \mathbf{0.\dot{4}}$

d) $\frac{7}{99} = \frac{07}{99} = \mathbf{0.\dot{0}\dot{7}}$

e) $\frac{4}{999} = \frac{004}{999} = \mathbf{0.\dot{0}0\dot{4}}$

2 a) $\frac{2}{3} = \frac{6}{9} = \mathbf{0.\dot{6}}$

b) $\frac{32}{33} = \frac{96}{99} = \mathbf{0.\dot{9}\dot{6}}$

c) $\frac{80}{111} = \frac{720}{999} = \mathbf{0.\dot{7}2\dot{0}}$

d) $\frac{1}{11} = \frac{9}{99} = \frac{09}{99} = \mathbf{0.\dot{0}\dot{9}}$

e) $\frac{5}{333} = \frac{15}{999} = \frac{015}{999} = \mathbf{0.\dot{0}1\dot{5}}$

3 a) $6\overline{)1.^{1}0^{4}0^{4}0}$ so $\frac{1}{6} = \mathbf{0.1\dot{6}}$

b) $15\overline{)7.^{7}0^{10}0^{10}0^{10}0...}$ so $\frac{7}{15} = \mathbf{0.4\dot{6}}$

c) $45\overline{)1.^{1}0^{10}0^{10}0^{10}0...}$ so $\frac{1}{45} = \mathbf{0.0\dot{2}}$

d) $110\overline{)27.^{27}0^{50}0^{60}0^{50}0^{60}0...}$
so $\frac{27}{110} = \mathbf{0.2\dot{4}\dot{5}}$

Page 39 Exercise 4

1 a) $0.12 = \frac{12}{100} = \frac{3}{25}$

b) $0.084 = \frac{84}{1000} = \frac{21}{250}$

c) $0.375 = \frac{375}{1000} = \frac{3}{8}$

d) $0.7654321 = \frac{7654321}{10000000}$

2 a) $r = 0.\dot{1} = 0.11111...$ $10r = 1.11111...$
$10r - r = 1.11111... - 0.11111...$
$9r = 1 \Rightarrow r = \frac{1}{9}$

b) $r = 0.\dot{3}\dot{4} = 0.343434...$
$100r = 34.343434...$
$100r - r = 34.343434... - 0.343434...$
$99r = 34 \Rightarrow r = \frac{34}{99}$

Using the same method for c)-f):

c) $\frac{18}{99} = \frac{2}{11}$ **d)** $\frac{863}{999}$

e) $\frac{207}{999} = \frac{23}{111}$ **f)** $\frac{7200}{9999} = \frac{800}{1111}$

---

3 a) $r = 0.5444444...$
$10r = 5.4444444...$
$100r = 54.444444...$
$100r - 10r = 54.444444... - 5.4444444...$
$90r = 49 \Rightarrow r = \frac{49}{90}$

b) $r = 0.8\dot{7}\dot{2} = 0.8727272...$
$10r = 8.727272...$
$1000r = 872.727272...$
$1000r - 10r = 872.727272... - 8.727272...$
$990r = 864 \Rightarrow r = \frac{864}{990} = \frac{96}{110} = \frac{48}{55}$

Using the same method for c)-f):

c) $\frac{12}{990} = \frac{2}{165}$ **d)** $\frac{75}{990} = \frac{5}{66}$

e) $\frac{45}{9900} = \frac{1}{220}$ **f)** $\frac{334}{990} = \frac{167}{495}$

Page 40 Review Exercise

1 a) $\frac{21}{35} = \frac{21 \div 7}{35 \div 7} = \frac{3}{5}$

b) $\frac{36}{126} = \frac{36 \div 6}{126 \div 6} = \frac{6}{21} = \frac{6 \div 3}{21 \div 3} = \frac{2}{7}$

c) $\frac{70}{182} = \frac{70 \div 2}{182 \div 2} = \frac{35}{91} = \frac{35 \div 7}{91 \div 7} = \frac{5}{13}$

2 a) $\frac{37}{27} = \frac{27 + 10}{27} = 1\frac{10}{27}$

b) $\frac{89}{5} = \frac{85 + 4}{5} = \frac{(17 \times 5) + 4}{5} = 17\frac{4}{5}$

c) $\frac{230}{11} = \frac{220 + 10}{11} = \frac{(20 \times 11) + 10}{11} = 20\frac{10}{11}$

d) $\frac{135}{19} = \frac{133 + 2}{19} = \frac{(7 \times 19) + 2}{19} = 7\frac{2}{19}$

3 Put the fractions over a common denominator:
$\frac{3}{4} = \frac{45}{60}, \frac{11}{15} = \frac{44}{60}, \frac{7}{10} = \frac{42}{60}, \frac{4}{5} = \frac{48}{60}, \frac{5}{6} = \frac{50}{60}$
So the closest to $\frac{3}{4}$ is $\frac{11}{15}$.

4 a) (i) $\frac{1}{3} + \frac{1}{12} = \frac{4}{12} + \frac{1}{12} = \frac{5}{12}$

(ii) $\frac{1}{2} + \frac{1}{3} + \frac{1}{7} = \frac{21}{42} + \frac{14}{42} + \frac{6}{42} = \frac{41}{42}$

(iii) $\frac{1}{2} + \frac{1}{5} + \frac{1}{20} = \frac{10}{20} + \frac{4}{20} + \frac{1}{20} = \frac{15}{20} = \frac{3}{4}$

b) (i) $\frac{9}{20} - \frac{1}{5} = \frac{9}{20} - \frac{4}{20} = \frac{5}{20} = \frac{1}{4}$
$\Rightarrow a = 4$

(ii) $\frac{11}{18} - \frac{1}{3} - \frac{1}{9} = \frac{11}{18} - \frac{6}{18} - \frac{2}{18} = \frac{3}{18} = \frac{1}{6}$
$\Rightarrow b = 6$

(iii) $\frac{301}{600} - \frac{1}{2} = \frac{301}{600} - \frac{300}{600} = \frac{1}{600}$
$\Rightarrow c = 600$

5 $20 - 7\frac{3}{4} - 5\frac{5}{16} - 2\frac{1}{8} = \frac{20}{1} - \frac{31}{4} - \frac{85}{16} - \frac{17}{8}$
$= \frac{320}{16} - \frac{124}{16} - \frac{85}{16} - \frac{34}{16}$
$= \frac{320 - 124 - 85 - 34}{16}$
$= \frac{77}{16} = 4\frac{13}{16}$ inches

6 $20 \div 1\frac{1}{4} = 20 \div \frac{5}{4} = 20 \times \frac{4}{5}$
$= (20 \div 5) \times 4 = 4 \times 4 = \textbf{16 questions}$

7 a) $\frac{2}{5} \times 100 = 2 \times (100 \div 5) = 2 \times 20 = 40$
40 as a fraction of $50 = \frac{40}{50} = \frac{4}{5}$

b) $\frac{2}{3} \times 90 = 2 \times (90 \div 3) = 2 \times 30 = 60$
60 is $\frac{3}{4}$ of $x \Rightarrow \frac{3}{4} \times x = 60 \Rightarrow x = 60 \div \frac{3}{4}$
$\Rightarrow x = 60 \times \frac{4}{3} = (60 \div 3) \times 4 = 20 \times 4 = \textbf{80}$

c) $\frac{1}{4} \times 64 = 64 \div 4 = 16$
$16 = \frac{1}{7} \times x \Rightarrow x = 16 \div \frac{1}{7} = 16 \times 7 = \textbf{112}$

8 a) $\frac{39}{100} = \textbf{0.39}$
$\frac{7}{20} = \frac{35}{100} = \textbf{0.35}$
$\frac{8}{25} = \frac{32}{100} = \textbf{0.32}$
$\frac{3}{10} = \textbf{0.3}$

b) Order the decimals: $0.3, 0.32, 0.35, 0.39$
So, in order: $\frac{3}{10}, \frac{8}{25}, \frac{7}{20}, \frac{39}{100}$

9 $6 = 2 \times 3$ has a prime factor other than 2 or 5, so $\frac{5}{6}$ is equivalent to a recurring decimal.

5 has no prime factors other than 5, so $\frac{4}{5}$ is equivalent to a terminating decimal.

$9 = 3 \times 3$ has a prime factor other than 2 or 5, so $\frac{2}{9}$ is equivalent to a recurring decimal.

$16 = 2^4$ has no prime factors other than 2, so $\frac{9}{16}$ is equivalent to a terminating decimal.

$40 = 2^3 \times 5$ has no prime factors other than 2 and 5, so $\frac{17}{40}$ is equivalent to a terminating decimal.

So the ones that are equivalent to recurring decimals are $\frac{5}{6}$ and $\frac{2}{9}$.

10 a) E.g. $\frac{24}{112} = \frac{24 \div 8}{112 \div 8} = \frac{3}{14}$. $\frac{3}{14}$ and $\frac{3}{8}$ have the same numerators but different denominators so they are **not equivalent**.

b) E.g. One third of $\frac{1}{15} = \frac{1}{3} \times \frac{1}{15} = \frac{1}{45}$ so $\frac{1}{5}$ is **not** one third of $\frac{1}{15}$.

c) E.g. Halfway between a and b is $(a + b) \div 2$, so halfway between $\frac{1}{2}$ and $\frac{1}{5}$ is:

$\left(\frac{1}{2} + \frac{1}{5}\right) \div 2 = \left(\frac{5}{10} + \frac{2}{10}\right) \div 2 = \frac{7}{10} \div 2 = \frac{7}{20}$

$\frac{1}{4} = \frac{5}{20}$, so $\frac{1}{4}$ is **not** halfway between $\frac{1}{2}$ and $\frac{1}{5}$.

d) E.g. $\frac{8}{9} = 0.\dot{8} = 0.8888...$, $0.\dot{8}\dot{7} = 0.8787...$ So $\frac{8}{9}$ is **greater than** $0.\dot{8}\dot{7}$.

11 a) $\frac{8}{16} = \frac{1}{2} = 0.5$

$\frac{6}{11} = \frac{54}{99} = 0.\dot{5}\dot{4} = 0.545454...$

$0.5\dot{4} = 0.544444...$

$\frac{5}{9} = 0.\dot{5} = 0.555555...$

So, in order: $\frac{8}{16}$, $\mathbf{0.5\dot{4}}$, $\frac{6}{11}$, $\frac{5}{9}$

b) $\frac{51}{100} = 0.51$

$\frac{102}{204} = \frac{1}{2} = 0.5$

$0.4\dot{6} = 0.466666...$

$15\overline{)8.^80^50^50...} \Rightarrow \frac{8}{15} = 0.533333...$

So, in order: $\mathbf{0.4\dot{6}}$, $\frac{102}{204}$, $\frac{51}{100}$, $\frac{8}{15}$

Page 41 Exam-Style Questions

1 $4\frac{1}{2} \div 1\frac{2}{5} = \left(\frac{8}{2} + \frac{1}{2}\right) \div \left(\frac{5}{5} + \frac{2}{5}\right)$

$= \frac{9}{2} \div \frac{7}{5}$ *[1 mark]*

$= \frac{9}{2} \times \frac{5}{7} = \frac{45}{14}$ *[1 mark]*

$= \frac{42 + 3}{14} = 3\frac{3}{14}$ *[1 mark]*

2 $1 - \frac{1}{6} - \frac{3}{24} - \frac{3}{8} = \frac{24}{24} - \frac{4}{24} - \frac{3}{24} - \frac{9}{24}$ *[1 mark]*

$= \frac{8}{24} = \frac{1}{3}$ *[1 mark]*

3 a) Perimeter $= 3\frac{3}{5} + 1\frac{5}{8} + 3\frac{3}{5} + 1\frac{5}{8}$

$= 3 + 1 + 3 + 1 + \frac{3}{5} + \frac{5}{8} + \frac{3}{5} + \frac{5}{8}$

$= 8 + \frac{24}{40} + \frac{25}{40} + \frac{24}{40} + \frac{25}{40}$ *[1 mark]*

$= 8 + \frac{98}{40} = 8 + \frac{49}{20} = 8 + \frac{40 + 9}{20}$

$= 8 + 2 + \frac{9}{20} = 10\frac{9}{20}$ **cm** *[1 mark]*

You could have converted to improper fractions first, then written them over a common denominator. That method is fine, but you end up with pretty big numbers on the numerators.

b) Area $= 3\frac{3}{5} \times 1\frac{5}{8} = \frac{18}{5} \times \frac{13}{8}$ *[1 mark]*

$= \frac{9}{5} \times \frac{13}{4} = \frac{117}{20} = 5\frac{17}{20}$ **cm²** *[1 mark]*

---

4 Let $r = 0.3181818...$
Then $10r = 3.18181818...$
and $1000r = 318.181818...$ *[1 mark]*
$1000r - 10r = 318.1818... - 3.1818...$
$990r = 315$ *[1 mark]*
$\Rightarrow r = \frac{315}{990} = \frac{105}{330} = \frac{35}{110} = \frac{7}{22}$ *[1 mark]*

5 Jess is $1\frac{1}{2}$ m tall, and $\frac{1}{2} > \frac{1}{3}$, so she is tall enough to go on the ride. *[1 mark]*

Eric is $1\frac{1}{2} \times \frac{8}{9} = \frac{3}{2} \times \frac{8}{9} = \frac{24}{18} = \frac{4}{3} = 1\frac{1}{3}$ m tall, so he is tall enough to go on the ride. *[1 mark]*

Xin is $1\frac{1}{2} - \frac{2}{7} = 1\frac{7}{14} - \frac{4}{14} = 1\frac{3}{14}$ m tall.

$\frac{3}{14} = \frac{9}{42}$, but $\frac{1}{3} = \frac{14}{42}$, so $\frac{3}{14}$ is less than $\frac{1}{3}$, i.e. Xin is not tall enough. *[1 mark]*

Abbas is $1.3 = 1\frac{3}{10}$ m tall.

$\frac{3}{10} = \frac{9}{30}$ and $\frac{1}{3} = \frac{10}{30}$, so 1.3 is less than $1\frac{1}{3}$, meaning that Abbas is not tall enough to go on the ride. *[1 mark]* so the only ones who can go on the ride are **Jess** and **Eric**.

6 Let the normal price of the game be £x.

$x \times \frac{3}{4} = 21.99 \Rightarrow x = 21.99 \div \frac{3}{4}$ *[1 mark]*
$= 21.99 \times \frac{4}{3} = (21.99 \div 3) \times 4$ *[1 mark]*
$= 7.33 \times 4 = 29.32$
i.e. the game originally cost **£29.32**. *[1 mark]*
This question is similar to some of the questions in the 'Percentage Increase and Decrease' topic in Section 5 — see p.67-68.

7 E.g. Let $r = 2.\dot{6}$, then $10r = 26.\dot{6}$
$10r - r = 26.\dot{6} - 2.\dot{6}$ *[1 mark]*
$9r = 24 \Rightarrow r = \frac{24}{9} = \frac{8}{3}$ *[1 mark]*
The reciprocal of $\frac{8}{3}$ is $\frac{3}{8}$ *[1 mark]*
Convert this to a decimal using short division:

$$8\overline{)3.^30^60^40} \quad 0.375$$

So the reciprocal of $2.\dot{6}$ is **0.375** *[1 mark]*

Section 4 — Ratio and Proportion

4.1 Ratios

Page 42 Exercise 1

1 a) Divide both sides by 2 (which is the highest common factor of 2 and 6) to get **1:3**.

b) Divide both sides by 10 (which is the highest common factor of 40 and 10) to get **4:1**.

Using the same method for c)-h):

c) ($\div 6$) **4:1**   **d)** ($\div 7$) **1:4**   **e)** ($\div 4$) **4:3**
f) ($\div 12$) **4:3**   **g)** ($\div 16$) **5:2**   **h)** ($\div 11$) **11:3**

2 a) Divide each number in the ratio by 2 (which is the highest common factor of 6, 2 and 4) to get **3:1:2**.

b) Divide each number in the ratio by 3 (which is the highest common factor of 15, 12 and 3) to get **5:4:1**.

Using the same method for c)-d):

c) ($\div 8$) **2:3:10**   **d)** ($\div 7$) **3:7:6**

3 Number of girls = $33 - 18 = 15$. The ratio of boys:girls is 18:15. Divide both numbers by 3 to get **6:5**.

4 Sophia has $42 - 16 = 26$ sweets. The ratio of Sophia's sweets:Isabel's sweets is 26:16. Divide both numbers by 2 to get **13:8**.

Page 43 Exercise 2

1 a) £1 = 100p, so 10p:£1 = 10p:100p. Divide both sides by 10 to get **1:10**.

b) 1 cm = 10 mm, so 20 mm:4 cm = 2 cm:4 cm. Divide both sides by 2 to get **1:2**.

c) 1 week = 7 days, so 2 weeks:7 days = 2 weeks:1 week = **2:1**.

---

d) 1 hour = 60 mins, so 18 mins:1 hour = 18 mins:60 mins. Divide both sides by 6 to get **3:10**.

e) 1 m = 100 cm, so 30 cm:2 m = 30 cm:200 cm. Divide both sides by 10 to get **3:20**.

f) 1 m = 1000 mm, so 1 m:150 mm = 1000 mm:150 mm. Divide both sides by 50 to get **20:3**.

g) 1 year = 12 months, so 6 months:5 years = 6 months:60 months. Divide both sides by 6 to get **1:10**.

h) 1 hour = 60 mins, so 2.5 hours:20 mins = 150 mins:20 mins. Divide both sides by 10 to get **15:2**.

i) 1 m = 100 cm, so 8 cm:1.1 m = 8 cm:110 cm. Divide both sides by 2 to get **4:55**.

j) 1 kg = 1000 g, so 9 g:0.3 kg = 9 g:300 g. Divide both sides by 3 to get **3:100**.

k) 1 km = 1000 m, so 65 m:1.56 km = 65 m:1560 m. Divide both sides by 65 to get **1:24**.
You might not have spotted the highest common factor (65) straight away, but you'll get the same answer if you divide by 5 first, then by 13.

l) 1 kg = 1000 g, so 1.2 kg:480 g = 1200 g:480 g. Divide both sides by 240 to get **5:2**.

2 1 kg = 1000 g. The ratio of the pumpkin's mass:Emma's mass is 6000 g:54 kg = 6 kg:54 kg. Divide both sides by 6 to get **1:9**.

3 1 kg = 1000 g. The ratio of butter:icing sugar is 640 g:1.6 kg = 640 g:1600 g. Divide both sides by 320 to get **2:5**.

Page 44 Exercise 3

1 a) Divide both sides by 2 to get **1:3**.

b) Divide both sides by 30 to get **1:4**.

c) ($\div 8$) **1:3.25**   **d)** ($\div 6$) **1:3.5**
e) ($\div 2$) **1:0.5**   **f)** ($\div 10$) **1:0.3**
g) ($\div 8$) **1:0.625**   **h)** ($\div 5$) **1:1.8**

i) 1 cm = 10 mm, so 10 mm:5 cm = 1 cm:5 cm = **1:5**.

j) 1 hour = 60 mins, so 30 mins:2 hours = 30 mins:120 mins. Divide both sides by 30 to get **1:4**.

k) 1 km = 1000 m, so 90 m:7.2 km = 90 m:7200 m. Divide both sides by 90 to get **1:80**.

l) £1 = 100p, so 50p:£6.25 = 50p:625p. Divide both sides by 50 to get **1:12.5**.

2 1 km = 1000 m, and 1 m = 100 cm, so 1 km = 1000 × 100 = 100000 cm. Map distance:true distance is 12 cm:4.8 km = 12 cm:480000 cm. Divide both sides by 12 to get **1:40 000**.

3 You need to find the ratio of syrup:milk in the form 1:n.

1 litre = 1000 ml, so 125 ml:$2\frac{1}{2}$ litres = 125 ml:2500 ml. Divide both sides by 125 to get 1:20. So the recipe uses **20 ml** of milk for every 1 ml of syrup.

Page 44 Exercise 4

1 Total parts white + brown = 2 + 1 = 3. Of these 1 part in 3 is brown = $\frac{1}{3}$ **brown flour**.

2 Total parts blue + white = 9 + 4 = 13. Of these 9 parts in 13 are blue = $\frac{9}{13}$ **blue**.

3 Total parts home wins + away wins + draws = 7 + 2 + 5 = 14. Of these 7 parts in 14 are home wins = $\frac{7}{14} = \frac{1}{2}$ **home wins**.
Don't forget to simplify the fraction.

4 Total parts spotty + stripy + plain
= 5 + 1 + 4 = 10. Of these 1 part in 10
are stripy = $\frac{1}{10}$ **stripy**.

5 **a)** Total parts purple + red + blue = 3 + 8 + 11
= 22. Of these 11 parts in 22 are blue
= $\frac{11}{22} = \frac{1}{2}$ **blue**.

b) 3 parts in 22 are purple = $\frac{3}{22}$ **purple**.

c) 3 purple + 11 blue = 14 parts aren't red,
out of 22 = $\frac{14}{22} = \frac{7}{11}$ **aren't red**.

Page 45 Exercise 5

1 $\frac{1}{3}$ are red, so 1 − $\frac{1}{3}$ = $\frac{2}{3}$ are green.
So red : green ratio is **1 : 2**.

2 He watched $\frac{3}{10}$, so he didn't watch 1 − $\frac{3}{10}$
= $\frac{7}{10}$. So watched : not watched ratio is **3 : 7**.

3 $\frac{1}{8}$ are red, $\frac{5}{8}$ are blue so 1 − $\left(\frac{1}{8} + \frac{5}{8}\right)$
= 1 − $\frac{6}{8}$ = $\frac{2}{8}$ are green.
Don't simplify the fractions here as you need them
over the same denominator to write as a ratio.
So red : blue : green ratio is **1 : 5 : 2**.

4 Total fraction of wins = $\frac{1}{2} + \frac{1}{4} = \frac{2}{4} + \frac{1}{4} = \frac{3}{4}$.
So losses = 1 − $\frac{3}{4} = \frac{1}{4}$.
So wins : losses ratio is **3 : 1**.

5 $\frac{7}{10}$ choose A, $\frac{1}{5} = \frac{2}{10}$ choose B,
so 1 − $\left(\frac{7}{10} + \frac{2}{10}\right)$ = 1 − $\frac{9}{10} = \frac{1}{10}$ choose C.
So the ratio of A : C is **7 : 1**.

6 **a)** For every 1 lettuce leaf there are $\frac{5}{2}$
tomatoes, so ratio of tomatoes : lettuce is
$\frac{5}{2}$: 1. Multiply both sides by 2 to write in
its simplest form (as integers), so the ratio is
5 : 2.

b) Tomatoes : lettuce = 5 : 2 = 45 : ?
Multiply both sides by 45 ÷ 5 = 9,
so the ratio becomes 45 : (2 × 9) = 45 : 18.
So there are **18 lettuce leaves**.

7 Start by finding a common denominator:
$\frac{1}{3} = \frac{5}{15}$ are red and $\frac{1}{5} = \frac{3}{15}$ are white.
So 1 − $\left(\frac{5}{15} + \frac{3}{15}\right)$ = 1 − $\frac{8}{15} = \frac{7}{15}$ are beige.
So the ratio of red : beige is **5 : 7**.

4.2 Using Ratios

Page 46 Exercise 1

1 **a)** There are 3 snakes for every 8 reptiles = $\frac{3}{8}$.

b) There are 8 − 3 = 5 lizards for every
3 snakes, so snakes : lizards = **3 : 5**.

2 There are 7 − 4 = 3 bourbons for every
4 digestives, so digestives : bourbons = **4 : 3**.

3 There is a total of 15 + 2 = 17 liquid parts for
every 15 water, so water : total = **15 : 17**.

4 There is a total of 53 + 46 = 99 albums for
every 46 sold as CDs, so CDs : total = **46 : 99**.

5 The only pair of whole numbers that have a
sum of 20 and a difference of 10 are 5 and 15,
so Cameron is 15 and Ashley is 5.

a) Ashley : Cameron = 5 : 15. Divide both
sides by 5 to give **1 : 3**.

b) Cameron : total = 15 : 20. Divide both sides
by 5 to give **3 : 4**.
You could also use the simplified ratio in part a)
to find the total parts = 1 + 3 = 4.

Page 48 Exercise 2

1 **a)** Sugar : butter = 2 : 1 = 100 : ?
Multiply both sides by 100 ÷ 2 = 50,
so the ratio becomes 100 : 50. So **50 g
of butter** is needed for 100 g of sugar.

b) Plot the point (100, 50) on the axes and
draw a straight line through this point and
the origin:

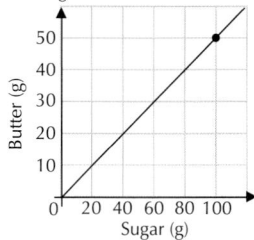

c) Read across from 40 g butter on the vertical
axis. Where it meets the graph, read off
the value for the amount of sugar on the
horizontal axis. **80 g of sugar** is needed for
40 g of butter.

2 Oak : beech = 2 : 9 = 42 : ?
Multiply both sides by 42 ÷ 2 = 21,
so the ratio becomes 42 : (9 × 21) = 42 : 189.
So there are **189 beech trees**.

3 Father : son = 8 : 3 = 48 : ?
Multiply both sides by 48 ÷ 8 = 6,
so the ratio becomes 48 : (3 × 6) = 48 : 18.
So the son is **18 years old**.

4 Cut-out height : actual height = 5 : 6 = 166 : ?
Multiply both sides by 166 ÷ 5 = 33.2,
so the ratio becomes 166 : (6 × 33.2)
= 166 : 199.2. So the footballer is
199.2 cm tall.

5 **a)** Scale the ratio up so you can plot
a sensible point on the graph.
5 : 4 = 500 : 400, so plot the point
(500, 400) on the axes and draw a straight
line through this point and the origin.

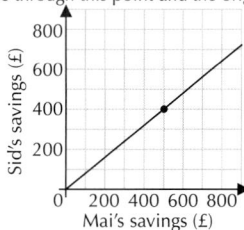

b) Read up from £400 on the horizontal axis.
Where it meets the graph, read off the
amount that Sid gets — **£320**.

c) Mai : Sid = 5 : 4 = ? : 60
Multiply both sides by 60 ÷ 4 = 15,
so the ratio becomes (5 × 15) : 60
= 75 : 60. So they get £75 + £60
= **£135 in total**.
Make sure you answer the question fully —
you're asked for the total, not just Mai's amount.

6 Pineapple : orange = 1 : 3 = 500 : ?
You're only asked about the amount of orange juice,
so you can ignore the part of the ratio for lemonade.
Multiply both sides by 500 ÷ 1 = 500,
so the ratio becomes 500 : (3 × 500)
= 500 : 1500. So **1500 ml of orange juice**
is needed.

7 Olives : courgette : cheese = 8 : 3 : 4 = 24 : ? : ?
Multiply each value by 24 ÷ 8 = 3,
so the ratio becomes 24 : (3 × 3) : (4 × 3)
= 24 : 9 : 12. So she would get **9 slices of
courgette and 12 slices of goat's cheese**.

8 Mo : Liz : Dee = 32 : 33 : 37 = 144 : ? : ?
Multiply each value by 144 ÷ 32 = 4.5, so the
ratio becomes 144 : (33 × 4.5) : (37 × 4.5)
= 144 : 148.5 : 166.5. So their combined
height is 144 + 148.5 + 166.5 = **459 cm**.
You could also have found the ratio of Mo : Total =
32 : (32 + 33 + 37) and scaled it up to find the total
directly, without finding the individual heights.

9 Max : Molly : Maisie = 3 : 7 : 2.
Molly has been waiting 1 h 10 mins = 60 + 10
= 70 mins, so you need the ratio = ? : 70 : ?
Multiply each value by 70 ÷ 7 = 10,
so the ratio becomes (3 × 10) : 70 : (2 × 10)
= 30 : 70 : 20. So **Max has been waiting
30 mins and Maisie has been waiting 20 mins**.

10 **a)** Plot the point (5, 1) on the axes and draw
a straight line through this point and the
origin.

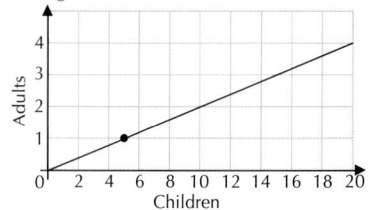

b) Read up from 18 children on the horizontal
axis. Where it meets the graph, read off
the value for the number of adults on the
vertical axis. The number of adults is
between 3 and 4. There must be a whole
number of adults, and since the ratio
should be 5 : 1 or less, round up to **4 adults**.

11 Under-30s : over-30s = 8 : 1 = 100 : ?
Multiply each value by 100 ÷ 8 = 12.5,
so the ratio becomes 100 : 12.5. There must
be a whole number of over-30s, and since the
ratio should be at least 8 : 1, round down to a
maximum of **12 over-30s**.

12 Aubergines : people = 1 : 3 = ? : 10
Multiply each value by 10 ÷ 3 = 3.333...,
so the ratio becomes 3.333... : 10. There must
be a whole number of aubergines, and since
you need enough for at least 10 people, round
up to **4 aubergines**.

13 Reality TV : news = 5 : 1 = 45 : ?
Multiply each value by 45 ÷ 5 = 9,
so the ratio becomes 45 : 9. So she needs to
watch at least **9 minutes of news**.

Page 49 Exercise 3

1 **a)** Fruit : cheese = 3 : 4 = ? : 24
Multiply both sides by 24 ÷ 4 = 6,
so the ratio becomes (3 × 6) : 24 = 18 : 24.
So there are **18 fruit scones**.

b) There are 18 − 6 = 12 fruit scones left, and
still 24 cheese scones. So the new ratio is
12 : 24. Divide both sides by 12 to get **1 : 2**.

2 Fish : crabs = 4 : 1 = 16 : ?
Multiply both sides by 16 ÷ 4 = 4,
so the original ratio becomes 16 : 4.
So there are originally 4 crabs, and 2 are
added to make 4 + 2 = 6 crabs, and 16 fish.
So new ratio is 16 : 6. Divide both sides by 2
to get **8 : 3**.

3 Detached : terraced = 5 : 7 = ? : 56
Multiply both sides by 56 ÷ 7 = 8,
so the original ratio becomes (5 × 8) : 56
= 40 : 56. So there are originally 40 detached
houses, and 12 more are built to make
40 + 12 = 52 detached, and 56 terraced
houses. So new ratio is 52 : 56. Divide both
sides by 4 to get **13 : 14**.

4 English : Maths = 2 : 5 = 48 : ?
Multiply both sides by 48 ÷ 2 = 24,
so the original ratio becomes 48 : (5 × 24)
= 48 : 120. So there are originally 120 Maths
books. At the end of the week there are
48 − 12 = 36 English books, and the new ratio
of English : Maths = 3 : 4 = 36 : ?
Multiply both sides by 36 ÷ 3 = 12,
so the new ratio becomes 36 : (4 × 12)
= 36 : 48. There are now 48 Maths books, so
120 − 48 = **72 Maths books** have been sold.

5 Call the original number of white tokens $2x$, and so the original number of black tokens is $3x$. Adding 8 black and 2 white tokens gives a new ratio of $(3x + 8):(2x + 2)$, which is equivalent to $2:1$. So $(3x + 8)$ is double the value of $(2x + 2)$, which can be written as an equation: $3x + 8 = 2(2x + 2)$. Solve for x: $3x + 8 = 2(2x + 2) \Rightarrow 3x + 8 = 4x + 4$ $\Rightarrow 4x - 3x = 8 - 4 \Rightarrow x = 4$. There were $2x = 2 \times 4 =$ **8 white tokens** to begin with. *There are other ways to solve this problem, but it's always good to practise your algebra skills.*

4.3 Dividing in a Given Ratio

Page 50   Exercise 1

1 a) Number of parts = $2 + 1 = 3$
   Amount for one part = £48 ÷ 3 = £16
   So the shares are $2 \times$ £16 = **£32**, and $1 \times$ £16 = **£16**.
   Don't forget to check that £32 + £16 = £48.

   b) Number of parts = $1 + 3 = 4$
   Amount for one part = £48 ÷ 4 = £12
   So the shares are $1 \times$ £12 = **£12**, and $3 \times$ £12 = **£36**.

   Using the same method for c)-d):
   c) £40 and £8   **d) £28 and £20**

2 a) Number of parts = $2 + 3 + 1 = 6$
   Amount for one part = 72 cm ÷ 6 = 12 cm
   So the shares are 2×12 cm = **24 cm**, 3×12 cm = **36 cm**, and 1×12 cm = **12 cm**.

   b) Number of parts = $2 + 2 + 5 = 9$
   Amount for one part = 72 cm ÷ 9 = 8 cm
   So the shares are 2×8 cm = **16 cm**, 2×8 cm = **16 cm**, and 5×8 cm = **40 cm**.

   Using the same method for c)-d):
   c) 30 cm, 18 cm and 24 cm
   d) 28 cm, 24 cm and 20 cm

3 a) Number of parts = $1 + 4 + 5 = 10$
   Amount for one part = £150 ÷ 10 = £15
   So the shares are $1 \times$ £15 = **£15**, $4 \times$ £15 = **£60**, and $5 \times$ £15 = **£75**.

   b) Number of parts = $15 + 5 + 30 = 50$
   Amount for one part = £150 ÷ 50 = £3
   So the shares are $15 \times$ £3 = **£45**, $5 \times$ £3 = **£15**, and $30 \times$ £3 = **£90**.

   Using the same method for c)-d):
   c) £60, £70 and £20
   d) £65, £55 and £30

4 a) Number of parts = $5 + 6 = 11$
   Amount for one part = £22 ÷ 11 = £2
   So the smallest share (which is the smallest number in the ratio) is $5 \times$ £2 = **£10**.

   b) Number of parts = $22 + 28 = 50$
   Amount for one part = 450 g ÷ 50 = 9 g
   So the smallest share is 22×9 g = **198 g**.

   Using the same method for c)-d):
   c) 10 kg   **d) 500 ml**

5 a) Number of parts = $1 + 3 = 4$
   Amount for one part = £30 ÷ 4 = £7.50
   So the largest share (which is the largest number in the ratio) is $3 \times$ £7.50 = **£22.50**.

   b) Number of parts = $3 + 2 = 5$
   Amount for one part = 36 g ÷ 5 = 7.2 g
   So the largest share is 3×7.2 g = **21.6 g**.

   Using the same method for c)-d):
   c) 75 kg   **d) 15 000 ml**

Page 51   Exercise 2

1 Number of parts = $3 + 5 = 8$
   Amount for one part = 32 ÷ 8 = 4 sandwiches
   So the shares are $3 \times 4 =$ **12 sandwiches**, and $5 \times 4 =$ **20 sandwiches**.

2 Number of parts = $3 + 2 = 5$
   Amount for one part = 30 ÷ 5 = 6 cupcakes
   So Kat gets $3 \times 6 =$ **18 cupcakes**, and Lindsay gets $2 \times 6 =$ **12 cupcakes**.

3 There are 14 parts in total.
   Amount for one part = 112 ÷ 14 = 8 dogs
   So there are $9 \times 8 =$ **72 male dogs**, and $112 - 72 =$ **40 female dogs**.

4 Ratio of Lauren's to Cara's age is $16:14$.
   Number of parts = $16 + 14 = 30$
   Amount for one part = £1200 ÷ 30 = £40
   So Lauren gets $16 \times$ £40 = **£640**, and Cara gets $14 \times$ £40 = **£560**.

5 There are 7 parts in total and 3 parts red paint, so $7 - 3 = 4$ parts yellow paint.
   Amount for one part = 42 litres ÷ 7 = 6 litres
   So $4 \times 6 =$ **24 litres of yellow paint** were used.

6 Number of parts = $2 + 4 + 6 = 12$
   Amount for one part = £150 ÷ 12 = £12.50
   So the smallest share (which is the smallest number in the ratio) is $2 \times$ £12.50 = **£25**.

7 a) Angles in a triangle add up to 180°, so share this in the ratio $2:1:3$.
   Number of parts = $2 + 1 + 3 = 6$
   Amount for one part = 180° ÷ 6 = 30°
   So the angles are $2 \times 30° =$ **60°**, $1 \times 30° =$ **30°**, and $3 \times 30° =$ **90°**.

   b) Angles in a quadrilateral add up to 360°, so share this in the ratio $1:5:2:4$.
   Number of parts = $1 + 5 + 2 + 4 = 12$
   Amount for one part = 360° ÷ 12 = 30°
   So the angles are $1 \times 30° =$ **30°**, $5 \times 30° =$ **150°**, $2 \times 30° =$ **60°**, and $4 \times 30° =$ **120°**.

8 The difference in length between the first and third pieces is 4 parts – 2 parts = 2 parts, and this is equal to 22 cm. If 2 parts = 22 cm, then 1 part = 22 ÷ 2 = 11 cm. The length of the second piece is one part, i.e. **11 cm**.

9 The perimeter is 72 cm, so the length + width is half of this: 72 ÷ 2 = 36 cm. You need to share 36 cm in the ratio $5:1$.
   Number of parts = $5 + 1 = 6$
   Amount for one part = 36 cm ÷ 6 = 6 cm
   So length = 5×6 cm = **30 cm**, and width = 1×6 cm = **6 cm**.

10 Ratio of tomatoes for Pierre:Susanne is $3:5$. The difference in number of parts is $5 - 3 = 2$ parts, which is equal to 16 tomatoes. So 1 part = 16 ÷ 2 = 8 tomatoes. There are $2 + 3 + 5 = 10$ parts in total in the ratio, so there are $10 \times 8 =$ **80 tomatoes in total**.

11 First share £200 in the ratio of Ali's hours to Max's hours = $10:6$.
   Number of parts = $10 + 6 = 16$
   Amount for one part = £200 ÷ 16 = £12.50
   So **Max** gets $6 \times$ £12.50 = **£75**, and Ali and Tim get $10 \times$ £12.50 = £125 to split in the ratio $4:1$. Number of parts = $4 + 1 = 5$.
   Amount for one part = £125 ÷ 5 = £25.
   So **Ali** keeps $4 \times$ £25 = **£100**, and gives **Tim** $1 \times$ £25 = **£25**.

12 If Dan's usual 8 parts = £6, then 1 part = £6 ÷ 8 = £0.75.
   The ratio has $8 + 12 + 16 = 36$ parts in total, which is worth $36 \times$ £0.75 = £27. This is split between Stan and Jan in the ratio $12:16$.
   Number of parts = $12 + 16 = 28$.
   Amount for one part = £27 ÷ 28 = £0.96...
   So, to the nearest penny:
   Stan gets $12 \times$ £0.96... = **£11.57**
   and **Jan** gets $16 \times$ £0.96... = **£15.43**.

Page 52   Exercise 3

1 Ratio of right-handed to left-handed pupils is $7:1$. Number of parts = $7 + 1 = 8$.
   Amount for one part = 600 ÷ 8 = 75 pupils
   So there are $7 \times 75 =$ **525 right-handed pupils**.

2 For every £1 Elsie puts in, Daniel puts in $1\frac{1}{2} \times$ £1 = £1.50, so they share the profits in the ratio $1:1.5$ (Elsie:Daniel).
   Number of parts = $1 + 1.5 = 2.5$
   Amount for one part = £5700 ÷ 2.5 = £2280
   So Daniel gets $1.5 \times$ £2280 = **£3420**.

3 Ratio of passengers on the phone to not on the phone is $2.5:1$.
   Number of parts = $2.5 + 1 = 3.5$
   Amount for one part = 28 ÷ 3.5 = 8 passengers
   So there are $2.5 \times 8 =$ **20 passengers on the phone**.

4 For every 1 g redcurrants, there are 2 g raspberries and 3 g strawberries, so the ratio of redcurrants:raspberries:strawberries is $1:2:3$.
   Number of parts = $1 + 2 + 3 = 6$
   Amount for one part = 450 g ÷ 6 = 75 g
   So there are 1×75 g = **75 g redcurrants**, 2×75 g = **150 g raspberries**, and 3×75 g = **225 g strawberries**.

5 For every 1 part that Jacinta gets, Nicky gets $\frac{1}{2}$ and Charlie gets $\frac{1}{2} \times \frac{1}{2} = \frac{1}{4}$. So the ratio of sweets for Jacinta:Nicky:Charlie is $1:\frac{1}{2}:\frac{1}{4}$, or $4:2:1$.
   Number of parts = $4 + 2 + 1 = 7$
   Amount for one part = 35 ÷ 7 = 5 sweets
   So Jacinta gets $4 \times 5 =$ **20 sweets**, Nicky gets 2×5 sweets = **10 sweets**, and Charlie gets 1×5 sweets = **5 sweets**.
   Check that Nicky gets half as much as Jacinta (10 is half of 20) and Charlie gets half as much as Nicky (5 is half of 10), and that 20 + 10 + 5 = 35 sweets.

4.4 Proportion

Page 53   Exercise 1

1 1 book costs £36 ÷ 8 = £4.50, so 12 books cost £4.50 × 12 = **£54**.

2 1 loaf needs 1.8 kg ÷ 3 = 0.6 kg flour, so 5 loaves need 0.6 kg × 5 = **3 kg flour**.

3 1 jug needs 5 litres ÷ 4 = 1.25 litres of water, so 3 jugs need 1.25 litres × 3 = **3.75 litres of water**.

4 For 1 car he earns £192 ÷ 12 = £16, so for 5 more cars he will earn £16 × 5 = **£80**.

5 £1 can buy $7 \div 84 = \frac{1}{12}$ of a DVD, so £48 can buy $\frac{1}{12} \times 48 =$ **4 DVDs**.

6 a) On 1 litre of petrol, the car can travel 250 km ÷ 35 = 7.142... km, so on 50 litres of petrol it can travel 7.142... km × 50 = 357.142... = **357 km** (to the nearest km).

   b) For 1 km, the car uses 35 litres ÷ 250 = 0.14 litres, so to travel 400 km it would use 0.14 litres × 400 = **56 litres of petrol**.

7 a) For every 1 g of chocolate there are 11 g ÷ 58 = 0.189... g of fat, so in 100 g of chocolate there are 0.189... g × 100 = 18.96... = **19 g of fat** (to the nearest g).

   b) There is 1 g of fat in every 58 g ÷ 11 = 5.272... g of chocolate, so there are 5 g of fat in every 5.272... g × 5 = 26.363... g of chocolate. As this is an upper limit, round down to **26.3 g of chocolate** (1 d.p.).

Page 54   Exercise 2

1 €1 is worth £1 ÷ 1.14 = £0.877...
   a) €10 is worth 10 × £0.877... = **£8.77**
   b) €100 is worth 100 × £0.877... = **£87.72**
   c) €250 is worth 250 × £0.877... = **£219.30**

2 1 lb is approximately equal to 1 kg ÷ 2.2 = 0.454... kg, so 8.25 lbs is approximately equal to 0.454... kg × 8.25 = **3.75 kg**.

3 1 Chinese yuan is worth £100 ÷ 1055 = £0.094..., so 65 Chinese yuan are worth £0.094... × 65 = **£6.16**.

4 a) £1 is worth 1.47 Swiss francs, so £50 is worth $1.47 \times 50 =$ **73.5 Swiss francs**.

b) 1 Swiss franc is worth £1 ÷ 1.50 = £0.666..., so 30 Swiss francs are worth £0.666... × 30 = **£20**.

Always keep the unrounded decimal in your calculator to use again for the second part of your calculation. If you'd rounded part way, you wouldn't have got the exact answer of £20.

Page 55   Exercise 3

1 First change the number of people.
In 5 minutes:
1 person can eat 6 ÷ 3 = 2 hotdogs,
so 5 people can eat 2 × 5 = 10 hotdogs.
Then change the time. 5 people take:
1 minute to eat 10 ÷ 5 = 2 hotdogs,
10 minutes to eat 2 × 10 = **20 hotdogs**.

2 First change the number of people.
In 2 minutes:
1 person can count 850 ÷ 10 = 85 coins,
so 22 people can count 85 × 22 = 1870 coins.
Then change the time. 22 people take:
1 minute to count 1870 ÷ 2 = 935 coins,
5 minutes to count 935 × 5 = **4675 coins**.

3 First change the number of people.
In 90 minutes:
1 person can plant 66 ÷ 2 = 33 bulbs,
so 3 people can plant 33 × 3 = 99 bulbs.
Then change the time. 3 people take:
1 minute to plant 99 ÷ 90 = 1.1 bulbs,
3 h × 60 = 180 minutes to plant
1.1 × 180 = **198 bulbs**.

4 First change the number of people.
In 2 weeks:
1 person can make 8 ÷ 2 = 4 toy soldiers,
so 7 people can make 4 × 7 = 28 toy soldiers.
Then change the time. 7 people take:
1 week to make 28 ÷ 2 = 14 toy soldiers,
3 weeks to make 14 × 3 = **42 toy soldiers**.

Page 55   Exercise 4

1 3 people take 2 h, so 1 person takes 2 h × 3 = 6 h, so 5 people would take 6 h ÷ 5 = 1.2 h = 1 h + (0.2 × 60) mins = **1 hour 12 mins**.

2 4 teachers take 2.5 h, so 1 teacher takes 2.5 h × 4 = 10 h, so 6 teachers would take 10 h ÷ 6 = 1.666... h = 1 h + (0.666... × 60) mins = **1 hour 40 mins**.

3 4 waiters take 20 mins, so 1 waiter takes 20 mins × 4 = 80 mins, so 5 waiters would take 80 mins ÷ 5 = **16 mins**.

4 At a speed of 30 mph it takes 2.25 h, so at 1 mph it would take 2.25 h × 30 = 67.5 h, so at a speed of 45 mph it takes 67.5 h ÷ 45 = 1.5 h = **1 hour 30 mins**.
You could also solve this problem using the formulas for speed, distance and time.

5 a) 5 builders take 62 days, so 1 builder takes 62 days × 5 = 310 days, so 2 builders would take 310 days ÷ 2 = **155 days**.

b) To finish the job in 1 day would take 62 × 5 = 310 builders. So to finish in under 40 days would take at least 310 ÷ 40 = 7.75 builders, which rounds up to **8 builders minimum**.

Page 56   Exercise 5

1 First change the number of people.
For 6 people:
1 person takes 45 × 2 = 90 minutes,
so 5 people would take 90 ÷ 5 = 18 minutes.
Then change the number of rooms.
5 people clean:
1 room in 18 ÷ 6 = 3 minutes, and
20 rooms in 3 × 20 = 60 minutes = **1 hour**.

---

2 First change the number of electricians.
For 2 houses:
1 electrician takes 9 × 2 = 18 days,
so 3 electricians would take 18 ÷ 3 = 6 days.
Then change the number of houses.
3 electricians wire:
1 house in 6 ÷ 2 = 3 days, and
3 houses in 3 × 3 = **9 days**.
You could answer this without doing any calculations if you spotted that, in both cases, there's one electrician for each house, so the time taken will be the same.

3 First change the number of bakers.
For 72 cakes:
1 baker takes 144 × 4 = 576 minutes,
so 6 bakers would take 576 ÷ 6 = 96 minutes.
Then change the number of cakes.
6 bakers ice:
1 cake in 96 ÷ 72 = 1.333... minutes,
96 cakes in 1.333... × 96 = **128 minutes**.

4 First change the number of people.
For 2 driveways:
1 person takes 20 × 3 = 60 minutes,
so 2 people would take 60 ÷ 2 = 30 minutes.
Then change the number of driveways.
2 people shovel:
1 driveway in 30 ÷ 2 = 15 minutes,
5 driveways in 15 × 5 = 75 minutes
= (60 + 15) minutes = **1 hour 15 minutes**.

5 First change the number of plates. In 2 hours:
to paint 1 plate needs 14 ÷ 35 = 0.4 people,
so to paint 60 plates needs 0.4 × 60 = 24 people.
Then change the time. 60 plates can be painted in: 1 hour by 24 × 2 = 48 people, and 2.5 hours by 48 ÷ 2.5 = 19.2 people.
You need at least that amount of people, so round up to **20 people**.

6 a) First change the number of stages.
12 workers complete:
1 stage in 3 ÷ 4 = 0.75 weeks,
and 10 stages in 0.75 × 10 = 7.5 weeks.
Then change the number of workers.
10-stage project can be completed by:
1 worker in 7.5 × 12 = 90 weeks, and
15 workers in 90 ÷ 15 = **6 weeks**.

b) As you found in a), the project takes 90 weeks for 1 worker, so would require 90 workers to complete in 1 week, and 90 ÷ 4 = 22.5 workers to complete in 4 weeks. So the **minimum number of workers is 23**.

Page 57   Review Exercise

1 The ratio of cement : sand is 10 : 25.

a) Divide both sides by 5 to get **2 : 5**.

b) Divide both sides by 10 to get **1 : 2.5**.

c) Divide both sides by 25 to get **0.4 : 1**.

d) For every 2 kg of cement there are 2 + 5 = 7 kg in total, so $\frac{2}{7}$ of the mix must be cement.

2 1 cm = 10 mm, so 120 mm = 12 cm.
Ratio of original : enlarged width is 12 cm : 35 cm = **12 : 35**.

3 a) nails : screws = $\frac{2}{3}$: 1. Multiply each value by 3, so the ratio simplifies to **2 : 3**.

b) There is a total of 2 + 3 = 5 parts for every 3 screws, so screws : total = **3 : 5**.

4 Evie will make a total of 12 + 72 = 84 green cards. Ratio green : purple = 7 : 3 = 84 : ?
Multiply both sides by 84 ÷ 7 = 12, so the ratio becomes 84 : (3 × 12) = 84 : 36.
So she will need to make **36 purple cards**.

---

5 sheep : cows : pigs = 8 : 1 : 3 = 16 : ? : ?
Multiply each value by 16 ÷ 8 = 2, so the ratio becomes 16 : (1 × 2) : (3 × 2) = 16 : 2 : 6. So there are 16 + 2 + 6 = **24 animals in total**.

6 Find the original number of flights to each country by dividing 20 planes in the ratio 4 : 1.
Number of parts = 4 + 1 = 5
Amount for one part = 20 ÷ 5 = 4 flights
So there are usually 4 × 4 = 16 flights to Spain and 1 × 4 = 4 flights to Finland. On the day when 2 extra flights go to Spain, the ratio 3 : 1 = (16 + 2) : ? = 18 : ?. Multiply each value by 18 ÷ 3 = 6 so the ratio becomes 18 : 6. So there are 6 flights to Finland on that day, which is 6 − 4 = **2 additional flights**.

7 a) First share £1440 in the ratio 3 : 5 : 7.
Number of parts = 3 + 5 + 7 = 15
Amount for one part = £1440 ÷ 15 = £96
So **Luiz** gets 3 × £96 = **£288**,
Seth gets 5 × £96 = **£480**, and the remaining amount of 7 × £96 = £672 is split between Fran and Ali in the ratio 4 : 1.
Number of parts = 4 + 1 = 5
Amount for one part = £672 ÷ 5 = £134.40
So **Fran** gets 4 × £134.40 = **£537.60** and **Ali** gets 1 × £134.40 = **£134.40**.

b) Ratio of Fran : Seth = 537.60 : 480, which is not easy to simplify, so work with the fractions of the whole amount.
Fran gets $\frac{4}{5}$ of $\frac{7}{15} = \frac{4 \times 7}{5 \times 15} = \frac{28}{75}$ of the total, and Seth gets $\frac{5}{15} = \frac{25}{75}$ of the total, so the ratio of the numerators is **28 : 25**.

8 Ratio raisins : nuts = 3 : 1, and ratio raisins : oats = 2 : 1. Multiply the first ratio by 2 to get raisins : nuts = 6 : 2, and multiply the second ratio by 3 to get raisins : oats = 6 : 3. The number of parts for raisins is now the same in both, so you can write raisins : nuts : oats = 6 : 2 : 3. Total parts = 6 + 2 + 3 = 11.
2 parts in every 11 are nuts, so $\frac{2}{11} \times 450$ g = 81.818... ≈ **82 g nuts** are needed.

9 a) 1 litre costs £20 ÷ 12.5 = £1.60, so 60 litres costs £1.60 × 60 = **£96**.

b) For £1 I can get 12.5 litres ÷ 20 = 0.625 litres, so for £30 I can get 0.625 litres × 30 = **18.75 litres**.

10 a) £1 = 20.8 rand, so £500 = 20.8 × 500 = **10 400 rand**.

b) Emma had 10400 − 8000 = 2400 rand left over. 1 rand is worth £1 ÷ 19.6 = £0.051..., so 2400 rand is worth £0.051... × 2400 = **£122.45** (to the nearest penny).

11 a) In 1.5 hours they chop down 3 trees, so in 1 hour they chop down 3 ÷ 1.5 = 2 trees, and so in 6 hours they can chop down 2 × 6 = **12 trees**.

b) It takes them 1.5 h ÷ 3 = 0.5 h to chop down 1 tree, so it would take 0.5 h × 8 = **4 hours** to chop down 8 trees.

c) First change the number of lumberjacks.
In 1.5 hours:
1 lumberjack chops 3 ÷ 8 = 0.375 trees, so 6 lumberjacks chop 0.375 × 6 = 2.25 trees.
Then change the time. 6 lumberjacks take:
1 hour to chop 2.25 ÷ 1.5 = 1.5 trees, and 8 hours to chop 1.5 × 8 = **12 trees**.

Page 58   Exam-Style Questions

1 Divide both sides by 0.75 to give
1 : (1.2 ÷ 0.75) *[1 mark]* = 1 : (120 ÷ 75)
= 1 : $\frac{120}{75}$ = 1 : $\frac{8}{5}$ = 1 : $1\frac{3}{5}$ = **1 : 1.6** *[1 mark]*

2 The difference in the number of male and female teachers = 7 − 2 = 5 parts = 40 teachers. So 1 part = 40 ÷ 5 = 8 teachers *[1 mark]*. So the school has a total of (7 + 2) × 8 = **72 teachers** *[1 mark]*

3 The ratio of badgers to foxes is 3 : 1 and the ratio of badgers to hares is 2 : 5 *[1 mark]*. Make the number of parts for badgers equal in both ratios by doubling the first ratio, i.e. badgers : foxes = 6 : 2, and tripling the second ratio, i.e. badgers : hares = 6 : 15. Combining these, for every 6 badgers there are 2 foxes and 15 hares, so the ratio is **6 : 2 : 15** *[1 mark]*.

4 a) To make 1 cupcake, 200 ÷ 25 = 8 g butter is needed *[1 mark]*, so to make 10 cupcakes, 8 g × 10 = **80 g butter is needed** *[1 mark]*.

b) To make 1 cupcake, 280 ÷ 25 = 11.2 g flour is needed *[1 mark]*, so to make 35 cupcakes, 11.2 g × 35 = **392 g flour is needed** *[1 mark]*.

5 8 people take 20 minutes, so 1 person would take 20 minutes × 8 = 160 minutes *[1 mark]*. The remaining 8 − 1 = 7 people would take 160 minutes ÷ 7 *[1 mark]*
= 22.857... minutes
= 22 minutes + (0.857... × 60) seconds
= **22 minutes 51 seconds**, to the nearest second *[1 mark]*.

6 If x × 100 ml of apple juice and y × 100 ml of lemonade is used, then the mixture will contain $(12x + 2.5y)$ grams of sugar per $(x + y)$ × 100 ml, which is equivalent to
$\frac{12x + 2.5y}{x + y}$ g per 100 ml *[1 mark]*.
Set this equal to 6 and simplify: $\frac{12x + 2.5y}{x + y} = 6$
$\Rightarrow 12x + 2.5y = 6x + 6y$ *[1 mark]*
$\Rightarrow 6x = 3.5y \Rightarrow 12x = 7y \Rightarrow x = \frac{7}{12}y$ *[1 mark]*
so the ratio is **7 : 12** *[1 mark]*.

Section 5 — Percentages

5.1 Percentages

Page 59   Exercise 1

1 $\frac{12}{25} = \frac{48}{100} = $ **48%**

2 $\frac{45}{300} = \frac{15}{100} = $ **15%**

3 a) $\frac{11}{25} = \frac{44}{100} = $ **44%**

b) $\frac{33}{50} = \frac{66}{100} = $ **66%**

c) $\frac{3}{20} = \frac{15}{100} = $ **15%**

d) $\frac{100}{400} = \frac{25}{100} = $ **25%**

e) $\frac{48}{32} = \frac{3}{2} = \frac{150}{100} = $ **150%**

f) $\frac{200}{160} = \frac{5}{4} = \frac{125}{100} = $ **125%**

4 a) $\frac{18}{24} = \frac{3}{4} = \frac{75}{100} = $ **75%**

b) 100% − 75% = **25%**

5 39 out of 65 have blonde hair so 65 − 39 = 26 don't have blonde hair.
$\frac{26}{65} = \frac{2}{5} = \frac{40}{100} = $ **40%**

You could also have worked out the percentage with blonde hair and subtracted it from 100%.

Page 60   Exercise 2

1 a) (15 ÷ 24) × 100% = **62.5%**

b) (221 ÷ 260) × 100% = **85%**
Using the same method for c)-f):

c) 132.2%   **d)** 68.8%

e) 38%   **f)** 120.4%

2 (525 ÷ 875) × 100% = **60%**

3 (171 ÷ 180) × 100% = **95%**

4 a) (116.6 ÷ 212) × 100% = **55%**

b) (53.5 ÷ 428) × 100% = **12.5%**

c) (226.8 ÷ 210) × 100% = **108%**

5 (1896.25 ÷ 10 250) × 100% = **18.5%**

Page 61   Exercise 3

1 a) 50% of 24 = 24 ÷ 2 = **12**

b) 25% of 36 = 36 ÷ 4 = **9**

c) 10% of 270 = 270 ÷ 10 = **27**

d) 25% of 20 = 20 ÷ 4 = 5
75% of 20 = 5 × 3 = **15**

e) 10% of 140 = 140 ÷ 10 = 14
5% of 140 = 14 ÷ 2 = **7**

f) 10% of 300 = 300 ÷ 10 = 30
5% of 300 = 30 ÷ 2 = 15
35% of 300 = 30 + 30 + 30 + 15 = **105**

g) 50% of 120 = 120 ÷ 2 = 60
10% of 120 = 120 ÷ 10 = 12
5% of 120 = 12 ÷ 2 = 6
65% of 120 = 60 + 12 + 6 = **78**

h) 10% of 90 = 90 ÷ 10 = 9
130% of 90 = 9 × 13 = **117**

i) 10% of 70 = 70 ÷ 10 = 7
180% of 70 = 7 × 18 = **126**

2 a) 10% of 200 = 200 ÷ 10 = 20
1% of 200 = 200 ÷ 100 = 2
21% of 200 = 20 + 20 + 2 = **42**

b) 1% of 260 = 260 ÷ 100 = 2.6
3% of 260 = 2.6 × 3 = **7.8**

c) 50% of 500 = 500 ÷ 2 = 250
10% of 500 = 500 ÷ 10 = 50
1% of 500 = 500 ÷ 100 = 5
62% of 500 = 250 + 50 + 5 + 5 = **310**

3 50% of 9 m = 9 ÷ 2 = 4.5 m
10% of 9 m = 9 ÷ 10 = 0.9 m
5% of 9 m = 0.9 ÷ 2 = 0.45 m
55% of 9 m = 4.5 + 0.45 = **4.95 m**

Page 61   Exercise 4

1 a) 17% = 17 ÷ 100 = 0.17
0.17 × 200 = **34**

b) 109% = 109 ÷ 100 = 1.09
1.09 × 11 = **11.99**
Using the same method for c)-f):

c) 217.6   **d)** 817.7

e) 485.85   **f)** 874.56

2 12% = 12 ÷ 100 = 0.12
0.12 × 68 = **8.16 kg**

3 31% = 31 ÷ 100 = 0.31
0.31 × 385 = **119.35 km**

4 22% = 22 ÷ 100 = 0.22
0.22 × 57 = £12.54
161% = 161 ÷ 100 = 1.61
1.61 × 8 = £12.88
161% of £8 is larger by
£12.88 − £12.54 = **£0.34 or 34p**.

5 34% = 34 ÷ 100 = 0.34
0.34 × 2.4 = 0.816 litres
It can hold another 2.4 − 0.816 = **1.584 litres**.

5.2 Percentages, Fractions and Decimals

Page 62   Exercise 1

1 a) $\frac{3}{20} = \frac{\mathbf{15}}{\mathbf{100}}$

b) (i) $\frac{15}{100} = $ **15%**

(ii) 15% ÷ 100% = **0.15**

2 a) (i) 75% ÷ 100% = **0.75**

(ii) $0.75 = \frac{75}{100} = \frac{\mathbf{3}}{\mathbf{4}}$
Using the same method for b)-d):

b) (i) 1.3   **(ii)** $\frac{13}{10}$

c) (i) 0.34   **(ii)** $\frac{17}{50}$

d) (i) 0.06   **(ii)** $\frac{3}{50}$

3 a) (i) $\frac{3}{10} = 3 \div 10 = $ **0.3**

(ii) 0.3 × 100% = **30%**

b) (i) $\frac{7}{8} = 7 \div 8 = 8\overline{)7.^706^040} = $ **0.875**

(ii) 0.875 × 100% = **87.5%**
Using the same method for c)-d):

c) (i) $\frac{1}{3} = 1 \div 3 = 3\overline{)1.^10^10^10} = $ **0.3̇**

(ii) 0.3̇ × 100% = **33.3%** (1 d.p.)

d) (i) $\frac{7}{16} = 7 \div 16$
$= 16\overline{)7.^70^{60}12^0\,8^0} = $ **0.4375**

(ii) 0.4375 × 100% = **43.75%**

4 a) (i) 0.35 × 100% = **35%**

(ii) $35\% = \frac{35}{100} = \frac{\mathbf{7}}{\mathbf{20}}$
Using the same method for b)-d):

b) (i) 170%   **(ii)** $\frac{17}{10}$

c) (i) 60%   **(ii)** $\frac{3}{5}$

d) (i) 268%   **(ii)** $\frac{67}{25}$

5 86% ÷ 100% = **0.86**

6 $\frac{3}{5} = \frac{60}{100} = $ **60%**

7 $36\% = \frac{36}{100} = \frac{\mathbf{9}}{\mathbf{25}}$

Page 63   Exercise 2

1 a) 0.35 = 0.35 × 100% = 35%
So **0.35** is larger than 32%.
Using the same method for b)-d):

b) 68%   **c)** 0.4   **d)** 90%

e) $\frac{21}{100} = 0.21$, so $\frac{\mathbf{21}}{\mathbf{100}}$ is larger than 0.2

It's usually better to convert the proportions to decimals or percentages rather than fractions. Fractions are harder to compare as you need to make sure the denominators are the same.

Using the same method for f)-h):

f) $\frac{7}{10}$ **g)** $\frac{3}{4}$ **h)** $\frac{3}{5}$

2 a) 25% = 25 ÷ 100 = 0.25
$\frac{2}{5} = \frac{4}{10} = 0.4$
Putting the decimals in order gives:
0.25, 0.4, 0.42 which is **25%, $\frac{2}{5}$, 0.42**.

b) 45% = 45 ÷ 100 = 0.45
$\frac{1}{2} = 0.5$
Putting the decimals in order gives:
0.45, 0.5, 0.505 which is **45%, $\frac{1}{2}$, 0.505**
Using the same method for c)-h):

c) 0.37, $\frac{3}{8}$, 38%

d) 0.2, 22%, $\frac{2}{9}$

e) 12.5%, 0.13, $\frac{3}{20}$

f) $\frac{9}{40}$, 23%, 0.25

g) 2.5%, $\frac{1}{25}$, 0.4

h) 0.06%, 0.006, $\frac{3}{50}$

3 $\frac{14}{20} = \frac{70}{100} = 70\%$, so **Team X** had a higher proportion of wins.

4 $\frac{11}{20} = \frac{55}{100} = 55\%$, so **Oliver** got more counters into the box.

Page 64   Exercise 3

1 50% prefer pizza
$\frac{1}{5} = \frac{20}{100} = 20\%$ prefer shepherd's pie
So 100% − 50% − 20% = **30%** prefer roast chicken.

2 $\frac{2}{5} = \frac{40}{100} = 40\%$ are red

$0.12 \times 100\% = 12\%$ are white

15% are blue

So $100\% - 40\% - 12\% - 15\% = $ **33%** are black.

3 $\frac{3}{4} = \frac{75}{100} = 75\%$ arrived by train

and 5% walked, so $100\% - 75\% - 5\% = $ **20%** came by car.

4 $\frac{1}{4} = \frac{25}{100} = 25\%$ are leather

$\frac{1}{5} = \frac{20}{100} = 20\%$ are denim

30% are suede

So $100\% - 25\% - 20\% - 30\% = $ **25%** are corduroy.

5 $\frac{3}{8} = 3 \div 8 = 8\overline{)3.{}^30{}^60{}^40}$ = 0.375

$0.375 \times 100\% = 37.5\%$ were sparrows

41.5% were blackbirds. So,

$100\% - 41.5\% - 37.5\% = $ **21%** were robins.

6 $\frac{3}{10} = \frac{30}{100} = 30\%$ of the pupils are girls.

So $100\% - 30\% = 70\%$ are boys.

25% of 30% = 30% $\div$ 4 = 7.5% are girls who are part of a sports club.

$\frac{1}{5}$ of 70% = 70% $\div$ 5 = 14% are boys who are part of a sports club.

So $14\% + 7.5\% = $ **21.5%** of the pupils are part of a sports club.

5.3 Percentage Increase and Decrease

Page 65   Exercise 1

1 a) 10% of 90 = 90 $\div$ 10 = 9

$90 + 9 = $ **99**

Using the same method for b)-d):

b) 75   **c) 140**   **d) 324**

2 a) 60% of 110 = 110 $\times$ 0.6 = 66

$110 + 66 = $ **176**

Using the same method for b)-d):

b) 1032   **c) 203**   **d) 285**

3 a) 10% of 55 = 55 $\div$ 10 = 5.5

$55 - 5.5 = $ **49.5**

Using the same method for b)-d):

b) 12   **c) 17.5**   **d) 48**

4 a) 40% of 125 = 125 $\times$ 0.4 = 50

$125 - 50 = $ **75**

Using the same method for b)-d):

b) 3.3   **c) 67.5**   **d) 442**

5 10% of 380 acres = 380 $\div$ 10 = 38 acres

5% of 380 acres = 38 $\div$ 2 = 19 acres

35% of 380 acres = (3 $\times$ 38) + 19 = 133 acres

So he sold 133 acres, which means he has $380 - 133 = $ **247 acres** left.

6 15% of 2400 = 2400 $\times$ 0.15 = 360

New population = 2400 − 360 = **2040**

7 20% of £485 = 485 $\times$ 0.2 = £97

So £97 of VAT is added, giving a total cost of £485 + £97 = **£582**

8 10% of £400 = 400 $\div$ 10 = £40

5% of £400 = 40 $\div$ 2 = £20

Roberto's weekly wage is £400 − £20 = £380

10% of £350 = 350 $\div$ 10 = £35

Mary's weekly wage is £350 + £35 = £385

£385 − £380 = £5 so **Mary earns more by £5**.

Page 66   Exercise 2

1 a) 11% = 11 $\div$ 100 = 0.11

It's an increase so the multiplier is

$1 + 0.11 = 1.11$

$490 \times 1.11 = $ **543.9**

b) 16% = 16 $\div$ 100 = 0.16

It's an increase so the multiplier is

$1 + 0.16 = 1.16$

$101 \times 1.16 = $ **117.16**

Using the same method for c)-h):

c) 130.35   **d) 143.29**   **e) 177.92**

f) 996.84   **g) 1614.06**   **h) 1346.97**

2 a) 8% = 8 $\div$ 100 = 0.08

It's a decrease so the multiplier is

$1 - 0.08 = 0.92$

$77 \times 0.92 = $ **70.84**

b) 21% = 21 $\div$ 100 = 0.21

It's a decrease so the multiplier is

$1 - 0.21 = 0.79$

$36 \times 0.79 = $ **28.44**

Using the same method for c)-h):

c) 71.34   **d) 57.57**   **e) 117.18**

f) 118.94   **g) 199.95**   **h) 75.87**

3 20% = 20 $\div$ 100 = 0.2

It's an increase so the multiplier is

$1 + 0.2 = 1.2$

50 inches $\times$ 1.2 = **60 inches**

4 3% = 3 $\div$ 100 = 0.03

It's an increase so the multiplier is

$1 + 0.03 = 1.03$

£31 000 $\times$ 1.03 = £31 930

2% = 2 $\div$ 100 = 0.02

It's a decrease so the multiplier is

$1 - 0.02 = 0.98$

£31 930 $\times$ 0.98 = **£31 291.40**

5 2% = 2 $\div$ 100 = 0.02

It's an increase so the multiplier is

$1 + 0.02 = 1.02$

So 24 500 $\times$ 1.02 = 24 990 students go to University A.

9% = 9 $\div$ 100 = 0.09

It's an increase so the multiplier is

$1 + 0.09 = 1.09$

So 22 500 $\times$ 1.09 = 24 525 students go to University B.

24 990 − 24 525 = 465

University A has 465 more students.

6 12% = 12 $\div$ 100 = 0.12

It's an increase so the multiplier is

$1 + 0.12 = 1.12$

152 243 $\times$ 1.12 = 170 512.16

So the population of Barton is 170 512 to the nearest whole person.

8% = 8 $\div$ 100 = 0.08

It's a decrease so the multiplier is

$1 - 0.08 = 0.92$

210 059 $\times$ 0.92 = 193 254.28

So the population of Meristock is 193 254 to the nearest whole person.

193 254 − 170 512 = 22 742

Meristock has a larger population by 22 742.

Page 67   Exercise 3

1 a) 10% of £500 = £500 $\div$ 10 = £50

So in 2 years there will be £50 $\times$ 2 = £100 interest. So there will be £500 + £100 = **£600** in the savings account.

b) 10% of £900 = £900 $\div$ 10 = £90

5% of £900 = £90 $\div$ 2 = £45

So in 6 years there will be £45 $\times$ 6 = £270 interest. So there will be £900 + £270 = **£1170** in the savings account.

2 a) 6% = 6 $\div$ 100 = 0.06

In 1 year he will get 0.06 $\times$ £5500 = £330

So in 3 years he will get £330 $\times$ 3 = £990 interest, which means there will be £5500 + £990 = **£6490** in his account.

b) In 10 years he will get £330 $\times$ 10 = £3300 interest, which means there will be £5500 + £3300 = **£8800** in his account.

3 Raj got £2852 − £2300 = £552 interest.

So he got £552 $\div$ 4 = £138 interest each year.

So the interest rate is $\frac{138}{2300} \times 100\% = $ **6%**.

---

4 Padma got £1632 − £1440 = £192 interest over 9 − 5 = 4 years. So she got £192 $\div$ 4 = £48 interest each year.

In 5 years she would get £48 $\times$ 5 = £240

So her original investment was

£1440 − £240 = **£1200**.

Page 68   Exercise 4

1 $100\% - 50\% = 50\%$

50% = £200

100% = £200 $\times$ 2 = **£400**

2 $100\% - 35\% = 65\%$

65% = £13

1% = £13 $\div$ 65 = £0.20

100% = £0.20 $\times$ 100 = **£20**

3 a) $100\% + 20\% = 120\%$

120% of y = 24

1% of y = 24 $\div$ 120 = 0.2

100% of y = 0.2 $\times$ 100 = **20**

b) $100\% - 40\% = 60\%$

60% of y = 30

1% of y = 30 $\div$ 60 = 0.5

100% of y = 0.5 $\times$ 100 = **50**

c) $100\% - 70\% = 30\%$

30% of y = 99

1% of y = 99 $\div$ 30 = 3.3

100% of y = 3.3 $\times$ 100 = **330**

d) $100\% + 180\% = 280\%$

280% of y = 84

1% of y = 84 $\div$ 280 = 0.3

100% of y = 0.3 $\times$ 100 = **30**

4 $100\% + 10\% = 110\%$

110% = 528

1% = 528 $\div$ 110 = 4.8

100% = 4.8 $\times$ 100 = **480 frogs**

$100\% + 15\% = 115\%$

115% = 621

1% = 621 $\div$ 115 = 5.4

100% = 5.4 $\times$ 100 = **540 newts**

5 $100\% + 20\% = 120\%$

120% = 2268

1% = 2268 $\div$ 120 = 18.9

100% = 18.9 $\times$ 100 = 1890 stamps in Feb

$100\% + 5\% = 105\%$

105% = 1890

1% = 1890 $\div$ 105 = 18

100% = 18 $\times$ 100 = **1800 stamps** in Jan

Page 69   Exercise 5

1 a) $\frac{12 - 10}{10} \times 100\% = $ **20%**

Always remember to divide by the original amount, not the new amount.

b) $\frac{52 - 20}{20} \times 100\% = $ **160%**

2 a) $\frac{10 - 8}{10} \times 100\% = $ **20%**

b) $\frac{25 - 22}{25} \times 100\% = $ **12%**

3 $\frac{72 - 45}{45} \times 100\% = $ **60%**

4 $\frac{100 - 80}{80} \times 100\% = $ **25%**

5 $\frac{80 - 68}{80} \times 100\% = $ **15%**

6 $\frac{50 - 30}{50} \times 100\% = $ **40%**

7 250 rounded to the nearest 100 is 300

$\frac{300 - 250}{250} \times 100\% = $ **20%**

8 a) $\frac{450 - 315}{450} \times 100\% = $ **30%**

b) $\frac{1127 - 980}{980} \times 100\% = $ **15%**

c) In total they cost £450 + £980 = £1430

She sold them for £1127 + £315 = £1442

So the percentage profit was

$\frac{1442 - 1430}{1430} \times 100\% = $ **0.84%** (2 d.p.)

5.4 Compound Percentage Change

Page 70 Exercise 1

1 $£680 \times \left(1 + \frac{2.5}{100}\right)^4 = £750.59$ (2 d.p.)

2 $2025 - 2017 = 8$ years
$66\,000\,000 \times \left(1 + \frac{0.6}{100}\right)^8$
$= 69\,235\,332.35... = \textbf{69\,000\,000}$ (2 s.f.)

Using the same method for Q3-6:

3 **£755.27** (2 d.p.)

4 **319 ants** (to the nearest whole number)

5 **109 cm** (to the nearest cm)

6 **4000 people** (to the nearest hundred)

7 a) $£750 \times \left(1 + \frac{3}{100}\right)^5 = £869.455...$
$£869.455... - £750 = \textbf{£119.46}$ (2 d.p.)

   b) $£50 \times \left(1 + \frac{5.5}{100}\right)^7 = £72.733...$
$£72.733... - £50 = \textbf{£22.73}$ (2 d.p.)

8 $£650 \times \left(1 + \frac{2}{100}\right)^2 = £676.26$
$£676.26 \times \left(1 + \frac{3}{100}\right)^3 = £738.967...$
$£738.967... - £650 = \textbf{£88.97}$ (2 d.p.)
When the compound interest rate changes, it's often best to split the calculation up into stages.

Page 71 Exercise 2

1 $£10\,000 \times \left(1 - \frac{15}{100}\right)^5$
$= £4437.05... = \textbf{£4400}$ (to the nearest £100)

2 $£550 \times \left(1 - \frac{3}{100}\right)^3 = £501.97...$
$£550 - £501.97...$
$= \textbf{£48}$ (to the nearest £)

3 $110\text{ kg} \times \left(1 - \frac{2}{100}\right)^8 = \textbf{94 kg}$ (to the nearest kg)

Using the same method for Q4-5:

4 **£3500** (2 s.f.)

5 **646 Bq** (to the nearest whole number)

6 $£650\,000 \times \left(1 - \frac{2.5}{100}\right)^5 = £572\,712.2...$
$£650\,000 - £572\,712.2...$
$= \textbf{£77\,300}$ (3 s.f.)

7 $£68\,000 \times \left(1 - \frac{20}{100}\right)^2 = £43\,520$
$£43\,520 \times \left(1 - \frac{15}{100}\right)^3 = \textbf{£26\,700}$ (3 s.f.)

8 $£3500 \times \left(1 + \frac{0.75}{100}\right)^2 = £3552.6...$
$£3552.6... \times \left(1 - \frac{1.25}{100}\right)^7 = \textbf{£3250}$ (3 s.f.)

Using the same method for Q9:

9 **£1040** (3 s.f.)

Page 72 Exercise 3

1 Find the first value of n where
$P_n < 500 \div 2 = 250$
$250 = 500 \times \left(1 - \frac{4.9}{100}\right)^n$
Using trial and improvement:
When $n = 10$: $500 \times \left(1 - \frac{4.9}{100}\right)^{10} = 302.534...$
which is too big so $n > 10$.
When $n = 15$: $500 \times \left(1 - \frac{4.9}{100}\right)^{15} = 235.330...$
which is too small so $n < 15$.
When $n = 13$: $500 \times \left(1 - \frac{4.9}{100}\right)^{13} = 260.205...$
which is too big so $n > 13$.
When $n = 14$: $500 \times \left(1 - \frac{4.9}{100}\right)^{14} = 247.455...$
which is too small so $n < 14$.
The population drops below half somewhere between 13 and 14 years, so it will take **14 whole years**.
It might take you a bit longer to find the correct values of n. Remember, you're looking for the smallest value of n that gives a population of less than 250.

Using the same method for Q2-4:

2 **15 days**   3 **4 years**

4 **5 hours**

Page 73 Review Exercise

1 a) $\frac{13}{50} = \frac{26}{100} = \textbf{26\%}$
   Using the same method for b)-c):
   b) **65%**   c) **40%**

2 $(83.10 \div 1385) \times 100\% = \textbf{6\%}$

3 50% of $60 = 60 \div 2 = 30$
10% of $60 = 60 \div 10 = 6$
5% of $60 = 6 \div 2 = 3$
65% of $60 = 30 + 6 + 3 = \textbf{39 games}$

4 a) $3\% = 3 \div 100 = 0.03$
$0.03 \times 210 = \textbf{6.3}$
   Using the same method for b)-c):
   b) **73.8**   c) **40.5**

5 a) $0.5 = 0.5 \times 100\% = 50\%$
$\frac{1}{2} = \frac{50}{100} = 50\%$
So **20%** is not equal to the others.
   Using the same method for b)-c):
   b) **1.25%**   c) **0.22**

6 Kelly scored $(47 \div 68) \times 100\% = 69.117...\%$
Nasir scored $(35 \div 52) \times 100\% = 67.307...\%$
So **Kelly** got the higher percentage score.

7 $27\% = 27 \div 100 = 0.27$
It's a decrease so the multiplier is
$1 - 0.27 = 0.73$
$£356 \times 0.73 = \textbf{£259.88}$

8 10% of $£420\,000 = £420\,000 \div 10 = £42\,000$
5% of $£420\,000 = £42\,000 \div 2 = £21\,000$
15% of $£420\,000 = £42\,000 + £21\,000$
$= £63\,000$
The value of house A is $£420\,000 - £63\,000$
$= £357\,000$
10% of $£340\,000 = £340\,000 \div 10 = £34\,000$
5% of $£340\,000 = £34\,000 \div 2 = £17\,000$
The value of house B is $£340\,000 + £17\,000$
$= £357\,000$ so **there is no difference**.

9 $4\% = 4 \div 100 = 0.04$
$£650 \times 0.04 = £26$
So in 7 years there will be
$£26 \times 7 = £182$ interest. So there will be
$£650 + £182 = \textbf{£832}$ in her savings account.

10 $100\% - 70\% = 30\%$
$30\% = £2.85$
$1\% = £2.85 \div 30 = £0.095$
$100\% = £0.095 \times 100 = \textbf{£9.50}$

11 $\frac{464 - 270}{270} \times 100\% = \textbf{71.9\%}$ (3 s.f.)

12 $£575 \times \left(1 + \frac{7.5}{100}\right)^3 = \textbf{£714.32}$ (2 d.p.)

13 $£2500 \times \left(1 + \frac{0.5}{100}\right)^6 = £2575.943...$
$£2575.943... \times \left(1 - \frac{0.25}{100}\right)^9$
$= \textbf{£2518.56}$ (2 d.p.)

14 Find the first value of n where $P_n < 400$
$400 = 1500 \times \left(1 - \frac{20}{100}\right)^n$
Using trial and improvement:
When $n = 5$: $1500 \times \left(1 - \frac{20}{100}\right)^5 = 491.52$
which is too big so $n > 5$.
When $n = 6$: $1500 \times \left(1 - \frac{20}{100}\right)^6 = 393.216$
which is too small so $n < 6$.
So it's volume will be less than 400 cm³ after
6 whole hours.

Page 74 Exam-Style Questions

1 $100\% - 50\% - 20\% - 5\% = 25\%$ *[1 mark]* are driven to school by their parents.
25% of $1200 = 1200 \div 4$
$= \textbf{300 pupils}$ *[1 mark]*

2 30% of the sofas are leather and 40% of those are black, so you need to find 40% of 30%.
10% of $30\% = 30\% \div 10 = 3\%$
40% of $30\% = 3\% \times 4 = \textbf{12\%}$
[2 marks available — 1 mark for a correct method to find 40% of 30%, 1 mark for the correct answer]
Finding a percentage of a percentage is just the same as finding the percentage of an amount.

3 a) $3.2\% = 3.2 \div 100 = 0.032$
It's an increase so the multiplier is
$1 + 0.032 = 1.032$ *[1 mark]*
$£50 \times 1.032 = \textbf{£51.60}$ *[1 mark]*
   b) $108\% = 108 \div 100 = 1.08$
It's an increase so the multiplier is
$1 + 1.08 = 2.08$ *[1 mark]*
$50\text{Bs} \times 2.08 = \textbf{104Bs}$ *[1 mark]*

4 The amount increased at 5% per annum for
$2017 - 2014 = 3$ years. Put the numbers into the compound growth formula:
$£4\,862\,025 = P_0 \times \left(1 + \frac{5}{100}\right)^3$ *[1 mark]*
$£4\,862\,025 = P_0 \times 1.157625$
$P_0 = £4\,862\,025 \div 1.157625$
$= \textbf{£4\,200\,000}$ *[1 mark]*

5 After 1 year the car's value is
$100\% - 20\% = 80\%$ of the original *[1 mark]*.
After 4 years at 15% depreciation its value is
$80\% \times \left(1 - \frac{15}{100}\right)^4$ *[1 mark]*
$= 41.7605\%$ of the original *[1 mark]*.
So the car has lost
$100\% - 41.7605\% = 58.2395\%$
$= \textbf{58\%}$ (to the nearest whole number) *[1 mark]*

Section 6 — Expressions

6.1 Simplifying Expressions

Page 75 Exercise 1

1 a) $c + c + c + d + d = \textbf{3c + 2d}$
   b) $x + y + x + y + x - y$
$= (x + x + x) + (y + y - y) = \textbf{3x + y}$
   c) $3m + m + 2n = \textbf{4m + 2n}$
   d) $3a + 5b + 8a + 2b = (3a + 8a) + (5b + 2b)$
$= \textbf{11a + 7b}$
   e) $6p + q + p + 3q = (6p + p) + (q + 3q)$
$= \textbf{7p + 4q}$
   Using the same method for f)-i) and Q2-3:
   f) **3b + 3c**   g) **5x + y + 12**
   h) **5m + 2n + 4**   i) **21a + 7b + 7**

2 a) $x^2 + 5x + 5$ b) $2x^2 + 6x + 4$
   c) $3x^2 + x$ d) $7p^2 + 2q$
   e) $7p^2 + pq + 3$ f) $2b^2 + 12b + 7$

3 a) $2x^2 + 2xy + 4ab - cd$
   b) $p^2 + q^2 + 2pq$
   c) $b^2 + 4ab + 3b$
   d) $b^2 + 6abc + 2ab - 3bc + 5b$

Page 76 Exercise 2

1 a) $x \times x \times x \times x = x^3$
   b) $2y \times 3y = 2 \times 3 \times y \times y = \textbf{6y}^2$
   c) $8p \times 2q = 8 \times 2 \times p \times q = \textbf{16pq}$
   Using the same method for d)-h):
   d) **21a²**   e) **15xy**   f) **m⁴**
   g) **48ab**   h) **48p²**

2 a) $p \times pq = p \times p \times q = \textbf{p}^2\textbf{q}$
   b) $4a^2 \times 5a = 4 \times 5 \times a^2 \times a = \textbf{20a}^3$
   c) $4ab \times 2ab = 4 \times 2 \times a \times a \times b \times b = \textbf{8a}^2\textbf{b}^2$
   d) $3i^2 \times 8j^3 = 3 \times 8 \times i^2 \times j^3 = \textbf{24i}^2\textbf{j}^3$
   e) $9n^2m \div 3n^2 = (9 \div 3)(n^2 \div n^2)(m) = \textbf{3m}$
   f) $12a^2 \div 4a = (12 \div 4)(a^2 \div a) = \textbf{3a}$
   g) $16p^3q \div 2p^2 = (16 \div 2)(p^3 \div p^2)(q) = \textbf{8pq}$
   h) $6abc \times 5a^2b^3c^4$
$= 6 \times 5 \times a \times a^2 \times b \times b^3 \times c \times c^4 = \textbf{30a}^3\textbf{b}^4\textbf{c}^5$

3 a) $(2y)^2 = 2y \times 2y = \mathbf{4y^2}$

b) $(r^2)^2 = r^2 \times r^2 = \mathbf{r^4}$

c) $(3c)^3 = 3c \times 3c \times 3c = \mathbf{27c^3}$

d) $(2z^2)^3 = 2z^2 \times 2z^2 \times 2z^2 = \mathbf{8z^6}$

e) $a(2b)^2 = a \times 2b \times 2b = \mathbf{4ab^2}$

f) $5(q^2)^2 = 5 \times q^2 \times q^2 = \mathbf{5q^4}$

g) $(xy)^2 = xy \times xy = \mathbf{x^2y^2}$

h) $i(jk)^2 = i \times jk \times jk = \mathbf{ij^2k^2}$

6.2 Expanding Brackets

Page 77   Exercise 1

1 a) $2(a + 5) = (2 \times a) + (2 \times 5) = \mathbf{2a + 10}$

b) $p(q + 2) = (p \times q) + (p \times 2) = \mathbf{pq + 2p}$

c) $6(5 - r) = (6 \times 5) - (6 \times r) = \mathbf{30 - 6r}$

Using the same method for d)-h):

d) $\mathbf{14t - t^2}$ **e)** $\mathbf{3u + 24v}$

f) $\mathbf{24x + 30y}$ **g)** $\mathbf{3pq - 8p}$

h) $\mathbf{5r - rs}$

2 a) $-(x + 2) = (-1 \times x) + (-1 \times 2) = \mathbf{-x - 2}$

b) $-(n - 11) = (-1 \times n) - (-1 \times 11) = \mathbf{-n + 11}$

c) $-y(4 + y) = (-y \times 4) + (-y \times y) = \mathbf{-4y - y^2}$

Using the same method for d)-h):

d) $\mathbf{-v^2 + 5v}$ **e)** $\mathbf{-30g + 18}$

f) $\mathbf{-56 + 8w}$ **g)** $\mathbf{-8y^2 - 24y}$

h) $\mathbf{-4pu + 28p}$

3 a) $2(z + 3) + 4(z + 2) = 2z + 6 + 4z + 8$
$= \mathbf{6z + 14}$

b) $4(10b - 5) + (b - 2) = 40b - 20 + b - 2$
$= \mathbf{41b - 22}$

c) $11(5x - 3) - (x + 2) = 55x - 33 - x - 2$
$= \mathbf{54x - 35}$

d) $5(2q + 5) - 2(q - 2) = 10q + 25 - 2q + 4$
$= \mathbf{8q + 29}$

e) $4p(3p + 5) - 3(p + 1)$
$= 12p^2 + 20p - 3p - 3 = \mathbf{12p^2 + 17p - 3}$

f) $4(t + 1) - 7t(8t - 11) = 4t + 4 - 56t^2 + 77t$
$= \mathbf{-56t^2 + 81t + 4}$

Page 78   Exercise 2

1 a) $(a + 2)(b + 3) = \mathbf{ab + 3a + 2b + 6}$

b) $(j + 4)(k - 5) = \mathbf{jk - 5j + 4k - 20}$

c) $3(j - 2)(k + 4) = 3 \times (jk + 4j - 2k - 8)$
$= \mathbf{3jk + 12j - 6k - 24}$

Using the same method for d)-h):

d) $\mathbf{xy + 2x + 6y + 12}$

e) $\mathbf{xy - x - 4y + 4}$

f) $\mathbf{8gh + 72g + 40h + 360}$

g) $\mathbf{2wz - 16w - 12z + 96}$

h) $\mathbf{7ab - 56a - 49b + 392}$

2 a) $(x + 8)(x + 3) = x^2 + 3x + 8x + 24$
$= \mathbf{x^2 + 11x + 24}$

b) $(b + 2)(b - 4) = b^2 - 4b + 2b - 8$
$= \mathbf{b^2 - 2b - 8}$

c) $(a - 1)(a + 2) = a^2 + 2a - a - 2$
$= \mathbf{a^2 + a - 2}$

Using the same method for d)-l):

d) $\mathbf{d^2 + 13d + 42}$ **e)** $\mathbf{-c^2 - 2c + 15}$

f) $\mathbf{-y^2 + 14y - 48}$ **g)** $\mathbf{2x^2 + 6x + 4}$

h) $\mathbf{z^2 - 3z - 108}$ **i)** $\mathbf{-3y^2 + 3y + 18}$

j) $\mathbf{4b^2 - 4b - 24}$ **k)** $\mathbf{6x^2 - 36x + 48}$

l) $\mathbf{12a^2 + 12a - 864}$

3 a) $(x + 1)^2 = (x + 1)(x + 1) = x^2 + x + x + 1$
$= \mathbf{x^2 + 2x + 1}$

b) $(x + 4)^2 = (x + 4)(x + 4) = x^2 + 4x + 4x + 16$
$= \mathbf{x^2 + 8x + 16}$

c) $(x - 2)^2 = (x - 2)(x - 2) = x^2 - 2x - 2x + 4$
$= \mathbf{x^2 - 4x + 4}$

Using the same method for d)-l):

d) $x^2 + 10x + 25$ **e)** $x^2 - 6x + 9$

f) $x^2 - 12x + 36$ **g)** $4x^2 + 8x + 4$

h) $2x^2 + 20x + 50$ **i)** $3x^2 - 12x + 12$

j) $2x^2 + 24x + 72$ **k)** $5x^2 - 30x + 45$

l) $2x^2 - 16x + 32$

Page 79   Exercise 3

1 a) $(a + 1)(b + 1)(c + 2)$
$= (a + 1)(bc + 2b + c + 2)$
$= a(bc + 2b + c + 2) + (bc + 2b + c + 2)$
$= \mathbf{abc + 2ab + ac + 2a + bc + 2b + c + 2}$

b) $(m - 5)(n - 1)(p - 3)$
$= (m - 5)(np - 3n - p + 3)$
$= m(np - 3n - p + 3) - 5(np - 3n - p + 3)$
$= \mathbf{mnp - 3mn - mp + 3m}$
$\qquad \mathbf{- 5np + 15n + 5p - 15}$

c) $(-3 + z)(5 - 2a)(b + 3)$
$= (-3 + z)(5b + 15 - 2ab - 6a)$
$= -3(5b + 15 - 2ab - 6a)$
$\qquad + z(5b + 15 - 2ab - 6a)$
$= \mathbf{-15b - 45 + 6ab + 18a}$
$\qquad \mathbf{+ 5bz + 15z - 2abz - 6az}$

Using the same method for d)-i):

d) $\mathbf{xy^2 + 3xy + 2x + 3y^2 + 9y + 6}$

e) $\mathbf{y^3 - 13y^2 + 54y - 72}$

f) $\mathbf{-3z^3 - 5z^2 + 47z - 15}$

g) $\mathbf{2z^3 + 12z^2 + 22z + 12}$

h) $\mathbf{4w^3 - 44w^2 + 152w - 160}$

i) $\mathbf{-18q^3 + 45q^2 + 9q - 54}$

2 a) $(x + 3)^3 = (x + 3)(x + 3)(x + 3)$
$= (x + 3)(x^2 + 3x + 3x + 9)$
$= x(x^2 + 6x + 9) + 3(x^2 + 6x + 9)$
$= x^3 + 6x^2 + 9x + 3x^2 + 18x + 27$
$= \mathbf{x^3 + 9x^2 + 27x + 27}$

b) $(x - 2)^3 = (x - 2)(x - 2)(x - 2)$
$= (x - 2)(x^2 - 2x - 2x + 4)$
$= x(x^2 - 4x + 4) - 2(x^2 - 4x + 4)$
$= x^3 - 4x^2 + 4x - 2x^2 + 8x - 8$
$= \mathbf{x^3 - 6x^2 + 12x - 8}$

c) $(x + 4)^3 = (x + 4)(x + 4)(x + 4)$
$= (x + 4)(x^2 + 4x + 4x + 16)$
$= x(x^2 + 8x + 16) + 4(x^2 + 8x + 16)$
$= x^3 + 8x^2 + 16x + 4x^2 + 32x + 64$
$= \mathbf{x^3 + 12x^2 + 48x + 64}$

Using the same method for d)-i):

d) $\mathbf{-x^3 + 9x^2 - 27x + 27}$

e) $\mathbf{3x^3 - 9x^2 + 9x - 3}$

f) $\mathbf{2x^3 + 30x^2 + 150x + 250}$

g) $\mathbf{8x^3 - 60x^2 + 150x - 125}$

h) $\mathbf{108x^3 - 324x^2 + 324x - 108}$

i) $\mathbf{2x^3 + 12x^2 + 24x + 16}$

6.3 Factorising — Common Factors

Page 80   Exercise 1

1 a) $2a + 10 = (2 \times a) + (2 \times 5) = \mathbf{2(a + 5)}$

b) $3b + 12 = (3 \times b) + (3 \times 4) = \mathbf{3(b + 4)}$

c) $15 + 3y = (3 \times 5) + (3 \times y) = \mathbf{3(5 + y)}$

Using the same method for d)-l):

d) $\mathbf{7(4 + v)}$ **e)** $\mathbf{5(a + 3b)}$

f) $\mathbf{3(3c - 4d)}$ **g)** $\mathbf{3(x + 4y)}$

h) $\mathbf{7(3u - v)}$ **i)** $\mathbf{4(a^2 - 3b)}$

j) $\mathbf{3(c + 5d^2)}$ **k)** $\mathbf{5(c^2 - 5f)}$

l) $\mathbf{6(x - 2y^2)}$

2 a) $3a^2 + 7a = (a \times 3a) + (a \times 7) = \mathbf{a(3a + 7)}$

b) $4b^2 + 19b = (b \times 4b) + (b \times 19) = \mathbf{b(4b + 19)}$

c) $2x^2 + 9x = (x \times 2x) + (x \times 9) = \mathbf{x(2x + 9)}$

Using the same method for d)-l):

d) $\mathbf{x(4x - 9)}$ **e)** $\mathbf{q(21q - 16)}$

f) $\mathbf{y(15 - 7y)}$ **g)** $\mathbf{y(7 + 15y)}$

h) $\mathbf{z(27z + 11)}$ **i)** $\mathbf{d(10d^2 + 27)}$

j) $\mathbf{y^2(4y - 13)}$ **k)** $\mathbf{y^3(11 + 3y)}$

l) $\mathbf{w(22 - 5w^3)}$

Page 81   Exercise 2

1 a) The HCF of 4 and 8 is **4**.

b) The HCF of x^3 and x is **x**.

c) The HCF of y^2 and y^4 is **y^2**.

d) $4x^3y^2 + 8xy^4 = \mathbf{4xy^2(x^2 + 2y^2)}$

2 a) $15a + 10ab = (5a \times 3) + (5a \times 2b)$
$= \mathbf{5a(3 + 2b)}$

b) $12b + 9bc = (3b \times 4) + (3b \times 3c)$
$= \mathbf{3b(4 + 3c)}$

c) $16xy - 4y = (4y \times 4x) - (4y \times 1) = \mathbf{4y(4x - 1)}$

Using the same method for d)-l) and Q3-4:

d) $\mathbf{3x(7 + y)}$ **e)** $\mathbf{6v(4u + 1)}$

f) $\mathbf{5p(2p + 3q)}$ **g)** $\mathbf{6q(2q - 3p)}$

h) $\mathbf{5ab(6b + 5)}$ **i)** $\mathbf{14x(x - 2y^2)}$

j) $\mathbf{2ab(4b + 5a)}$ **k)** $\mathbf{4p(3q - 2p^2)}$

l) $\mathbf{8x^2(3xy - 2)}$

3 a) $\mathbf{x^4(x^2 + 1 - x)}$ **b)** $\mathbf{a^2(8 + 17a^4)}$

c) $\mathbf{6y^4(2y^2 + 1)}$ **d)** $\mathbf{8c(3b^2c^2 - 1)}$

e) $\mathbf{z^2(25 + 13z^4)}$ **f)** $\mathbf{3p^2(4 + 5p^3q^3)}$

g) $\mathbf{9ab(a^3 + 3b^2)}$ **h)** $\mathbf{3(5b^4 - 7a^2 + 6ab)}$

i) $\mathbf{11pq^2(2 - p^2q)}$ **j)** $\mathbf{2xy(8x - 4y + x^2y^2)}$

k) $\mathbf{4x^2y^2(9x^5 + 2y^7)}$ **l)** $\mathbf{x^3(5x + 3y^4 - 25y)}$

4 a) $\mathbf{x^2y^2(13 + 22x^4y + 20x^3y)}$

b) $\mathbf{ab^3(16a^4b^2 - a^3b^2 - 1)}$

c) $\mathbf{7pq(3p^5q - 2q + p^2)}$

d) $\mathbf{xy^3(14y + 13x^4 - 5x^5y)}$

e) $\mathbf{2c^3(8c^3d^5 - 7 + 4d^6)}$

f) $\mathbf{3jk(6 + 7j^2k^5 - 5jk^2)}$

g) $\mathbf{18xy^3(2xy - 4x^4y^4 + 1)}$

h) $\mathbf{a^3b^4(20ab + 4 - 5a^3b^{11})}$

i) $\mathbf{11xy^2(xy + x^2 + 6y^3)}$

6.4 Factorising — Quadratics

Page 83   Exercise 1

1 a) Find pairs of numbers that multiply
to give 6: 1×6, 2×3
To make +7, you need to do +1 + 6, so:
$x^2 + 7x + 6 = \mathbf{(x + 1)(x + 6)}$

b) Find pairs of numbers that multiply
to give 12: 1×12, 2×6, 3×4
To make +7, you need to do +3 + 4, so:
$a^2 + 7a + 12 = \mathbf{(a + 3)(a + 4)}$

Using the same method for c)-f) and Q2:

c) $\mathbf{(x + 1)(x + 7)}$

d) $\mathbf{(z + 2)(z + 6)}$

e) $\mathbf{(x + 1)(x + 4)}$

f) $\mathbf{(v + 3)(v + 3)}$ or $\mathbf{(v + 3)^2}$

2 a) $\mathbf{(x + 1)(x + 3)}$ **b)** $\mathbf{(x - 2)(x - 4)}$

c) $\mathbf{(x - 2)(x - 5)}$ **d)** $\mathbf{(x - 1)(x - 4)}$

e) $\mathbf{(y - 2)(y + 5)}$ **f)** $\mathbf{(x - 2)(x + 4)}$

g) $\mathbf{(s - 3)(s + 6)}$ **h)** $\mathbf{(x + 3)(x - 5)}$

i) $\mathbf{(t + 2)(t - 6)}$

Page 84   Exercise 2

1 a) The only pair of numbers that multiply
to give 2 is 1×2, so write the brackets:
$(x \quad)(2x \quad)$
The only pair of numbers that multiply
to give 1 is 1×1, so the only option is:
$(x \quad 1)(2x \quad 1) \Rightarrow O = x, \ I = 2x$
To make +3x, you need to do $+x + 2x$, so:
$2x^2 + 3x + 1 = \mathbf{(x + 1)(2x + 1)}$

b) The only pair of numbers that multiply
to give 3 is 1×3, so write the brackets:
$(x \quad)(3x \quad)$
The only pair of numbers that multiply
to give 5 is 1×5, so the options are:
$(x \quad 1)(3x \quad 5) \Rightarrow O = 5x, \ I = 3x$
$(x \quad 5)(3x \quad 1) \Rightarrow O = x, \ I = 15x$
To get −16x, you need to do $-x - 15x$, so:
$3x^2 - 16x + 5 = \mathbf{(x - 5)(3x - 1)}$

c) The only pair of numbers that multiply to give 5 is 1×5, so write the brackets:
$(x\quad)(5x\quad)$
List the pairs of numbers that multiply to give 6: 1×6, 2×3
So the options are:
$(x\;\;1)(5x\;\;6) \Rightarrow O = 6x,\; I = 5x$
$(x\;\;6)(5x\;\;1) \Rightarrow O = x,\; I = 30x$
$(x\;\;2)(5x\;\;3) \Rightarrow O = 3x,\; I = 10x$
$(x\;\;3)(5x\;\;2) \Rightarrow O = 2x,\; I = 15x$
To get $-17x$, you need to do $-2x - 15x$, so:
$5x^2 - 17x + 6 = (x - 3)(5x - 2)$

Using the same method for d)-l):
d) $(t - 4)(2t + 3)$ e) $(x - 6)(2x - 1)$
f) $(b - 3)(3b + 2)$ g) $(x + 3)(5x - 3)$
h) $(x - 1)(2x - 1)$ i) $(a + 3)(7a - 2)$
j) $(x - 6)(11x + 4)$ k) $(z + 5)(7z + 3)$
l) $(y - 8)(3y - 2)$

2 a) List the pairs of numbers that multiply to make 6: 1×6, 2×3
So the options for the brackets are:
$(x\quad)(6x\quad)$
$(2x\quad)(3x\quad)$
The only pair of numbers that multiply to give 1 is 1×1, so the only options are:
$(x\;\;1)(6x\;\;1) \Rightarrow O = x,\; I = 6x$
$(2x\;\;1)(3x\;\;1) \Rightarrow O = 2x,\; I = 3x$
To make $+7x$, you need to do $+x + 6x$, so:
$6x^2 + 7x + 1 = (x + 1)(6x + 1)$

b) List the pairs of numbers that multiply to make 6: 1×6, 2×3
So the options for the brackets are:
$(x\quad)(6x\quad)$
$(2x\quad)(3x\quad)$
Again, the numbers that multiply to give 6 are 1×6 or 2×3, so the options are:
$(x\;\;1)(6x\;\;6) \Rightarrow O = 6x,\; I = 6x$
$(x\;\;6)(6x\;\;1) \Rightarrow O = x,\; I = 36x$
$(x\;\;2)(6x\;\;3) \Rightarrow O = 3x,\; I = 12x$
$(x\;\;3)(6x\;\;2) \Rightarrow O = 2x,\; I = 18x$
$(2x\;\;1)(3x\;\;6) \Rightarrow O = 12x,\; I = 3x$
$(2x\;\;6)(3x\;\;1) \Rightarrow O = 2x,\; I = 18x$
$(2x\;\;2)(3x\;\;3) \Rightarrow O = 6x,\; I = 6x$
$(2x\;\;3)(3x\;\;2) \Rightarrow O = 4x,\; I = 9x$
To get $-13x$, you need to do $-4x - 9x$, so:
$6x^2 - 13x + 6 = (2x - 3)(3x - 2)$

You don't need to list every combination — you can stop as soon as you find the one that works.

Using the same method for c)-f):
c) $(3x + 1)(5x - 2)$ d) $(2x - 3)(5x - 2)$
e) $(3u + 1)(4u - 3)$ f) $(2w + 1)(7w + 9)$

Page 84   Exercise 3
1 a) $x^2 - 25 = x^2 - 5^2 = (x + 5)(x - 5)$
b) $x^2 - 9 = x^2 - 3^2 = (x + 3)(x - 3)$
c) $x^2 - 36 = x^2 - 6^2 = (x + 6)(x - 6)$
d) $x^2 - 81 = x^2 - 9^2 = (x + 9)(x - 9)$
e) $x^2 - 64 = x^2 - 8^2 = (x + 8)(x - 8)$
f) $b^2 - 121 = b^2 - 11^2 = (b + 11)(b - 11)$
g) $x^2 - 1 = x^2 - 1^2 = (x + 1)(x - 1)$
h) $c^2 - 4d^2 = c^2 - (2d)^2 = (c + 2d)(c - 2d)$
2 a) $4x^2 - 49 = (2x)^2 - 7^2 = (2x + 7)(2x - 7)$
b) $36x^2 - 4 = (6x)^2 - 2^2 = (6x + 2)(6x - 2)$
c) $9x^2 - 100 = (3x)^2 - 10^2 = (3x + 10)(3x - 10)$
d) $25x^2 - 16 = (5x)^2 - 4^2 = (5x + 4)(5x - 4)$
e) $16z^2 - 1 = (4z)^2 - 1^2 = (4z + 1)(4z - 1)$
f) $27t^2 - 12 = 3(9t^2 - 4) = 3((3t)^2 - 2^2)$
   $= 3(3t + 2)(3t - 2)$
g) $98x^2 - 2 = 2(49x^2 - 1) = 2((7x)^2 - 1^2)$
   $= 2(7x + 1)(7x - 1)$
h) $7p^2 - 175 = 7(p^2 - 25) = 7(p^2 - 5^2)$
   $= 7(p + 5)(p - 5)$
i) $x^2 - 11 = x^2 - (\sqrt{11})^2 = (x + \sqrt{11})(x - \sqrt{11})$
j) $n^2 - 51 = n^2 - (\sqrt{51})^2 = (n + \sqrt{51})(n - \sqrt{51})$

k) $4x^2 - 3 = (2x)^2 - (\sqrt{3})^2 = (2x + \sqrt{3})(2x - \sqrt{3})$
l) $3x^2 - 15 = 3(x^2 - 5) = 3\left(x^2 - (\sqrt{5})^2\right)$
   $= 3(x + \sqrt{5})(x - \sqrt{5})$

6.5 Algebraic Fractions
Page 85   Exercise 1
1 a) $\dfrac{2x}{x^2} = \dfrac{2}{x}$

b) $\dfrac{49x^3}{14x^4} = \dfrac{7x^3(7)}{7x^3(2x)} = \dfrac{7}{2x}$

c) $\dfrac{25s^3t}{5s} = \dfrac{5s \times 5s^2t}{5s} = 5s^2t$

d) $\dfrac{26a^2b^3c^4}{52b^5c} = \dfrac{26b^3c(a^2c^3)}{26b^3c(2b^2)} = \dfrac{a^2c^3}{2b^2}$

2 a) $\dfrac{7x}{5x - x^2} = \dfrac{7x}{x(5 - x)} = \dfrac{7}{5 - x}$

b) $\dfrac{48t - 6t^2}{8s^2t} = \dfrac{2t(24 - 3t)}{2t(4s^2)} = \dfrac{24 - 3t}{4s^2}$

c) $\dfrac{3cd}{8c + 6c^2} = \dfrac{c(3d)}{c(8 + 6c)} = \dfrac{3d}{8 + 6c}$

d) $\dfrac{12}{4a^2b^3 + 8a^7b^9} = \dfrac{4(3)}{4(a^2b^3 + 2a^7b^9)}$
   $= \dfrac{3}{a^2b^3 + 2a^7b^9}$

3 a) $\dfrac{4st + 8s^2}{8t^2 + 16st} = \dfrac{s(4t + 8s)}{2t(4t + 8s)} = \dfrac{s}{2t}$

b) $\dfrac{15xz + 15z}{25xyz - 25yz} = \dfrac{5z(3x + 3)}{5z(5xy - 5y)} = \dfrac{3x + 3}{5xy - 5y}$

c) $\dfrac{6xy - 6x}{3y - 3} = \dfrac{2x(3y - 3)}{(3y - 3)} = 2x$

d) $\dfrac{3a^2b + 5ab^2}{7ab^3} = \dfrac{ab(3a + 5b)}{ab(7b^2)} = \dfrac{3a + 5b}{7b^2}$

4 a) $\dfrac{2x - 8}{x^2 - 5x + 4} = \dfrac{2(x - 4)}{(x - 1)(x - 4)} = \dfrac{2}{x - 1}$

b) $\dfrac{6a - 3}{5 - 10a} = \dfrac{3(2a - 1)}{-5(2a - 1)} = -\dfrac{3}{5}$

c) $\dfrac{x^2 + 7x + 10}{x^2 + 2x - 15} = \dfrac{(x + 2)(x + 5)}{(x - 3)(x + 5)} = \dfrac{x + 2}{x - 3}$

d) $\dfrac{x^2 - 7x + 12}{x^2 - 2x - 8} = \dfrac{(x - 3)(x - 4)}{(x + 2)(x - 4)} = \dfrac{x - 3}{x + 2}$

e) $\dfrac{x^2 + 4x}{x^2 + 7x + 12} = \dfrac{x(x + 4)}{(x + 3)(x + 4)} = \dfrac{x}{x + 3}$

f) $\dfrac{2t^2 + t - 45}{4t^2 - 81} = \dfrac{(t + 5)(2t - 9)}{(2t + 9)(2t - 9)} = \dfrac{t + 5}{2t + 9}$

Page 86   Exercise 2
1 a) $\dfrac{x}{4} + \dfrac{x}{5} = \dfrac{5x}{20} + \dfrac{4x}{20} = \dfrac{9x}{20}$

b) $\dfrac{x}{2} - \dfrac{x}{3} = \dfrac{3x}{6} - \dfrac{2x}{6} = \dfrac{x}{6}$

c) $\dfrac{2b}{7} + \dfrac{b}{6} = \dfrac{12b}{42} + \dfrac{7b}{42} = \dfrac{19b}{42}$

d) $\dfrac{5z}{6} - \dfrac{4z}{9} = \dfrac{15z}{18} - \dfrac{8z}{18} = \dfrac{7z}{18}$

2 a) $\dfrac{x - 2}{5} + \dfrac{x + 1}{3} = \dfrac{3(x - 2)}{15} + \dfrac{5(x + 1)}{15}$
   $= \dfrac{3x - 6 + 5x + 5}{15} = \dfrac{8x - 1}{15}$

b) $\dfrac{2t + 1}{4} + \dfrac{t - 1}{3} = \dfrac{3(2t + 1)}{12} + \dfrac{4(t - 1)}{12}$
   $= \dfrac{6t + 3 + 4t - 4}{12} = \dfrac{10t - 1}{12}$

c) $\dfrac{3x - 1}{4} + \dfrac{2x + 1}{6} = \dfrac{3(3x - 1)}{12} + \dfrac{2(2x + 1)}{12}$
   $= \dfrac{9x - 3 + 4x + 2}{12} = \dfrac{13x - 1}{12}$

d) $\dfrac{c + 2}{c} + \dfrac{c + 1}{2c} = \dfrac{2(c + 2)}{2c} + \dfrac{c + 1}{2c}$
   $= \dfrac{2c + 4 + c + 1}{2c} = \dfrac{3c + 5}{2c}$

3 a) $\dfrac{2}{x + 1} + \dfrac{1}{x - 3} = \dfrac{2(x - 3)}{(x + 1)(x - 3)} + \dfrac{(x + 1)}{(x + 1)(x - 3)}$
   $= \dfrac{2x - 6 + x + 1}{(x + 1)(x - 3)} = \dfrac{3x - 5}{(x + 1)(x - 3)}$

b) $\dfrac{x - 2}{x - 1} - \dfrac{x + 1}{x + 2} = \dfrac{(x + 2)(x - 2)}{(x - 1)(x + 2)} - \dfrac{(x + 1)(x - 1)}{(x - 1)(x + 2)}$
   $= \dfrac{(x^2 - 4) - (x^2 - 1)}{(x - 1)(x + 2)} = \dfrac{-3}{(x - 1)(x + 2)}$

c) $\dfrac{2a - 3}{a + 2} + \dfrac{3a + 2}{a + 3}$
$= \dfrac{(2a - 3)(a + 3)}{(a + 2)(a + 3)} + \dfrac{(3a + 2)(a + 2)}{(a + 2)(a + 3)}$
$= \dfrac{(2a^2 + 6a - 3a - 9) + (3a^2 + 6a + 2a + 4)}{(a + 2)(a + 3)}$
$= \dfrac{5a^2 + 11a - 5}{(a + 2)(a + 3)}$

Using the same method for d)-i):
d) $\dfrac{5x^2 - 6x + 8}{(3x - 2)(2x + 1)}$ e) $\dfrac{6s^2 - 2s - 5}{(3s - 1)(3s + 2)}$
f) $\dfrac{3x^2 - x + 5}{15x}$ g) $\dfrac{3y + 7}{(y + 1)(y + 3)}$
h) $\dfrac{-x^2 - 7x - 5}{(x + 2)(x + 1)}$ i) $\dfrac{x^2 + 15x + 1}{(x - 2)(x + 3)}$

Page 87   Exercise 3
1 a) $\dfrac{x - 2}{(x - 3)(x + 1)} + \dfrac{5}{(x - 3)}$
$= \dfrac{x - 2}{(x - 3)(x + 1)} + \dfrac{5(x + 1)}{(x - 3)(x + 1)}$
$= \dfrac{x - 2 + 5x + 5}{(x - 3)(x + 1)} = \dfrac{6x + 3}{(x - 3)(x + 1)}$

b) $\dfrac{3x}{(x + 1)(x + 2)} + \dfrac{1}{x + 2}$
$= \dfrac{3x}{(x + 1)(x + 2)} + \dfrac{(x + 1)}{(x + 1)(x + 2)}$
$= \dfrac{3x + x + 1}{(x + 1)(x + 2)} = \dfrac{4x + 1}{(x + 1)(x + 2)}$

c) $\dfrac{1}{(x + 4)} - \dfrac{(x - 2)}{(x + 4)(x + 3)}$
$= \dfrac{(x + 3)}{(x + 4)(x + 3)} - \dfrac{(x - 2)}{(x + 4)(x + 3)}$
$= \dfrac{x + 3 - x + 2}{(x + 4)(x + 3)} = \dfrac{5}{(x + 4)(x + 3)}$

Using the same method for Q2-3:
2 a) $\dfrac{11z + 20}{(z + 1)(z + 2)}$ b) $\dfrac{2x^2 + 7x - 3}{(x - 2)(x + 3)}$
c) $\dfrac{4x - 3}{x(x + 4)}$ d) $\dfrac{6}{(a + 3)(a - 3)}$
3 a) $\dfrac{t^2 + t + 1}{t(t + 1)(t + 3)}$ b) $\dfrac{2x^2 + 3x + 6}{(x - 2)(x + 3)(x - 3)}$
c) $\dfrac{4x^2 - 15x}{(x - 1)(x - 3)(x - 4)}$
d) $\dfrac{y^2 - 5y + 2}{(y - 1)(y - 2)(y - 3)}$

Page 88   Exercise 4
1 a) $\dfrac{x}{y} \times \dfrac{3}{x^2} = \dfrac{1}{y} \times \dfrac{3}{x} = \dfrac{3}{xy}$

b) $\dfrac{2a}{4b^2} \times \dfrac{5b}{a^3} = \dfrac{1}{2b} \times \dfrac{5}{a^2} = \dfrac{5}{2a^2b}$

c) $\dfrac{t}{2} \times \dfrac{24st}{6t} = \dfrac{t}{1} \times \dfrac{2s}{1} = 2st$

d) $\dfrac{64xy^2}{9y} \times \dfrac{3x^3}{16x^2y} = \dfrac{4x}{3} \times \dfrac{x}{1} = \dfrac{4x^2}{3}$

e) $\dfrac{1}{x + 4} \times \dfrac{3x}{x + 2} = \dfrac{3x}{(x + 4)(x + 2)}$

f) $\dfrac{6a + b}{12} \times \dfrac{3a}{b + 1} = \dfrac{6a + b}{4} \times \dfrac{a}{b + 1}$
   $= \dfrac{a(6a + b)}{4(b + 1)}$

g) $\dfrac{1}{2z + 5} \times \dfrac{3z}{z - 1} = \dfrac{3z}{(2z + 5)(z - 1)}$

h) $\dfrac{t^2 + 5}{12} \times \dfrac{4t^3}{1 - t} = \dfrac{t^2 + 5}{3} \times \dfrac{t^3}{1 - t} = \dfrac{t^3(t^2 + 5)}{3(1 - t)}$

2 a) $\dfrac{4x}{3y^2} \div \dfrac{2x}{12y^4} = \dfrac{4x}{3y^2} \times \dfrac{12y^4}{2x} = \dfrac{4}{1} \times \dfrac{2y^2}{1} = 8y^2$

b) $\dfrac{18a}{6b^2} \div \dfrac{a}{20b} = \dfrac{18a}{6b^2} \times \dfrac{20b}{a} = \dfrac{3}{b} \times \dfrac{20}{1} = \dfrac{60}{b}$

c) $\dfrac{3x^3y^5}{4x^5y} \div \dfrac{xy}{28} = \dfrac{3x^3y^5}{4x^5y} \times \dfrac{28}{xy}$
   $= \dfrac{3y^3}{x^2} \times \dfrac{7}{x} = \dfrac{21y^3}{x^3}$

d) $\dfrac{ab^2}{15} \div \dfrac{a^2}{5b} = \dfrac{ab^2}{15} \times \dfrac{5b}{a^2} = \dfrac{b^2}{3} \times \dfrac{b}{a} = \dfrac{b^3}{3a}$

e) $\dfrac{y - 2}{3y + 2} \div \dfrac{y - 2}{4y^4} = \dfrac{y - 2}{3y + 2} \times \dfrac{4y^4}{y - 2} = \dfrac{4y^4}{3y + 2}$

f) $\frac{2c+1}{3d^2} \div \frac{cd}{18} = \frac{2c+1}{3d^2} \times \frac{18}{cd}$

$\qquad = \frac{2c+1}{d^2} \times \frac{6}{cd} = \frac{6(2c+1)}{cd^3}$

g) $\frac{1-x^3y^5}{25x^2} \div \frac{y}{5} = \frac{1-x^3y^5}{25x^2} \times \frac{5}{y}$

$\qquad = \frac{1-x^3y^5}{5x^2} \times \frac{1}{y} = \frac{1-x^3y^5}{5x^2y}$

h) $\frac{12t^5}{6t+3t^3} \div \frac{18t^2}{9t} = \frac{12t^5}{3t(2+t^2)} \times \frac{9t}{18t^2}$

$\qquad = \frac{2t^3}{2+t^2}$

3 a) $\frac{x^2-16}{x^2+5x+6} \times \frac{x+3}{x+4}$

$\qquad = \frac{(x+4)(x-4)}{(x+2)(x+3)} \times \frac{(x+3)}{(x+4)} = \frac{x-4}{x+2}$

b) $\frac{x^2+4x+3}{x^2+6x+8} \times \frac{x+4}{2x+6}$

$\qquad = \frac{(x+1)(x+3)}{(x+2)(x+4)} \times \frac{(x+4)}{2(x+3)}$

$\qquad = \frac{(x+1)}{(x+2)} \times \frac{1}{2} = \frac{x+1}{2(x+2)}$

c) $\frac{z^2+3z-10}{z^2+4z+3} \times \frac{z^2+6z+5}{z-2}$

$\qquad = \frac{(z-2)(z+5)}{(z+1)(z+3)} \times \frac{(z+1)(z+5)}{(z-2)}$

$\qquad = \frac{z+5}{z+3} \times \frac{z+5}{1} = \frac{(z+5)^2}{z+3}$

d) $\frac{x^2-4x+3}{x^2+9x+20} \div \frac{x^2-x-6}{x^2+7x+12}$

$\qquad = \frac{x^2-4x+3}{x^2+9x+20} \times \frac{x^2+7x+12}{x^2-x-6}$

$\qquad = \frac{(x-1)(x-3)}{(x+4)(x+5)} \times \frac{(x+3)(x+4)}{(x+2)(x-3)}$

$\qquad = \frac{(x-1)(x+3)}{(x+5)(x+2)}$

e) $\frac{y^2-5y+6}{y^2+y-20} \div \frac{y-2}{3y-12}$

$\qquad = \frac{(y-2)(y-3)}{(y-4)(y+5)} \times \frac{3(y-4)}{(y-2)} = \frac{3(y-3)}{y+5}$

f) $\frac{t^2-9}{t^2+3t+2} \div \frac{t^2+6t+9}{t^2+8t+7}$

$\qquad = \frac{t^2-9}{t^2+3t+2} \times \frac{t^2+8t+7}{t^2+6t+9}$

$\qquad = \frac{(t+3)(t-3)}{(t+1)(t+2)} \times \frac{(t+1)(t+7)}{(t+3)(t+3)}$

$\qquad = \frac{(t-3)(t+7)}{(t+2)(t+3)}$

Page 89   Review Exercise

1 a) $5x$ **b)** $7y-x$ **c)** $5x^2+2x-3$
d) $6a^2$ **e)** $16pq$ **f)** x^3y
g) $11z^3$ **h)** $3g^2h$
i) $x^2(4x)^3 = x^2 \times 4x \times 4x \times 4x = 64x^5$

2 a) $6(x+3)+3(x-4) = 6x+18+3x-12$
$\qquad = 9x+6$

b) $4(a+3)-2(a+2) = 4a+12-2a-4$
$\qquad = 2a+8$

c) $5(p+2)-3(p-4) = 5p+10-3p+12$
$\qquad = 2p+22$

d) $2(2x+3)-3(2x+1) = 4x+6-6x-3$
$\qquad = -2x+3$

e) $x(x+3)+2(x-1) = x^2+3x+2x-2$
$\qquad = x^2+5x-2$

f) $2x(2x+3)+3(x-4) = 4x^2+6x+3x-12$
$\qquad = 4x^2+9x-12$

3 a) $(x-3)(x+4) = x^2+4x-3x-12$
$\qquad = x^2+x-12$

b) $(3x+1)(2x+5) = 6x^2+15x+2x+5$
$\qquad = 6x^2+17x+5$

c) $(x+1)(x-2)(x+3)$
$\qquad = (x+1)(x^2+3x-2x-6)$
$\qquad = x(x^2+x-6)+(x^2+x-6)$
$\qquad = x^3+x^2-6x+x^2+x-6$
$\qquad = x^3+2x^2-5x-6$

d) $-(t-8)(t-1)(t+1)$
$\qquad = (-t+8)(t^2+t-t-1)$
$\qquad = -t(t^2-1)+8(t^2-1)$
$\qquad = -t^3+t+8t^2-8 = -t^3+8t^2+t-8$

e) $(3x-2)^2 = (3x-2)(3x-2)$
$\qquad = 9x^2-6x-6x+4$
$\qquad = 9x^2-12x+4$

f) $(x+2)^3 = (x+2)(x+2)(x+2)$
$\qquad = (x+2)(x^2+2x+2x+4)$
$\qquad = x(x^2+4x+4)+2(x^2+4x+4)$
$\qquad = x^3+4x^2+4x+2x^2+8x+8$
$\qquad = x^3+6x^2+12x+8$

4 a) $4(x-2)$ **b)** $3(2a+1)$
c) $5(t-2)$ **d)** $3x(1+2y)$
e) $4x(2y-3x)$ **f)** $ab(a-2+b)$
g) $4x(4x+3xy-2y^2)$ **h)** $7x(2x^2+xy-y^4)$

5 a) Factors of 8: 1×8, 2×4
$\qquad 2+4 = 6$, so:
$\qquad a^2+6a+8 = (a+2)(a+4)$
Using the same method for b)-e):
b) $(x+1)(x+3)$ **c)** $(z-2)(z-3)$
d) $(x-3)(x+6)$ **e)** $(x+2)(x-5)$
f) Factors of 2: 1×2. Options are:
$\qquad (x\quad 1)(2x\quad 2) \Rightarrow$ O $= 2x$, I $= 2x$
$\qquad (x\quad 2)(2x\quad 1) \Rightarrow$ O $= x$, I $= 4x$
$\qquad +x+4x = 5x$, so:
$\qquad 2x^2+5x+2 = (x+2)(2x+1)$
Using the same method for g)-i):
g) $(m-2)(3m-2)$ **h)** $(x-2)(3x+1)$
i) $(2g+1)(2g+1)$
j) $16a^2-25 = (4a)^2-5^2 = (4a+5)(4a-5)$
k) $4c^2-196 = (2c)^2-14^2 = (2c+14)(2c-14)$
l) $81t^2-121 = (9t)^2-11^2 = (9t+11)(9t-11)$

6 a) $\frac{a}{4}+\frac{a}{8} = \frac{2a}{8}+\frac{a}{8} = \frac{3a}{8}$

b) $\frac{2x}{3}-\frac{2x}{7} = \frac{14x}{21}-\frac{6x}{21} = \frac{8x}{21}$

c) $\frac{y-1}{2}+\frac{y+1}{3} = \frac{3(y-1)}{6}+\frac{2(y+1)}{6}$

$\qquad = \frac{3y-3+2y+2}{6} = \frac{5y-1}{6}$

d) $\frac{2}{x+3}+\frac{4}{x-1} = \frac{2(x-1)}{(x+3)(x-1)}+\frac{4(x+3)}{(x+3)(x-1)}$

$\qquad = \frac{2x-2+4x+12}{(x+3)(x-1)}$

$\qquad = \frac{6x+10}{(x+3)(x-1)}$

e) $\frac{5}{z+2}-\frac{3}{z+3} = \frac{5(z+3)}{(z+2)(z+3)}-\frac{3(z+2)}{(z+2)(z+3)}$

$\qquad = \frac{5z+15-3z-6}{(z+2)(z+3)}$

$\qquad = \frac{2z+9}{(z+2)(z+3)}$

f) $\frac{z+2}{z+3}+\frac{z+1}{z-2} = \frac{(z+2)(z-2)}{(z+3)(z-2)}+\frac{(z+1)(z+3)}{(z+3)(z-2)}$

$\qquad = \frac{z^2-4+z^2+3z+z+3}{(z+3)(z-2)}$

$\qquad = \frac{2z^2+4z-1}{(z+3)(z-2)}$

g) $\frac{3}{x^2+4x+3}+\frac{2}{x^2+x-6}$

$\qquad = \frac{3}{(x+1)(x+3)}+\frac{2}{(x-2)(x+3)}$

$\qquad = \frac{3(x-2)}{(x+1)(x-2)(x+3)}+\frac{2(x+1)}{(x+1)(x-2)(x+3)}$

$\qquad = \frac{3x-6+2x+2}{(x+1)(x-2)(x+3)}$

$\qquad = \frac{5x-4}{(x+1)(x-2)(x+3)}$

h) $\frac{4}{t^2+6t+9}-\frac{3}{t^2+3t}$

$\qquad = \frac{4}{(t+3)(t+3)}-\frac{3}{t(t+3)}$

$\qquad = \frac{4t}{t(t+3)^2}-\frac{3(t+3)}{t(t+3)^2}$

$\qquad = \frac{4t-3t-9}{t(t+3)^2} = \frac{t-9}{t(t+3)^2}$

i) $\frac{x}{x^2-16}+\frac{x-2}{x^2-5x+4}$

$\qquad = \frac{x}{(x+4)(x-4)}+\frac{x-2}{(x-1)(x-4)}$

$\qquad = \frac{x(x-1)}{(x-1)(x+4)(x-4)}+\frac{(x-2)(x+4)}{(x-1)(x+4)(x-4)}$

$\qquad = \frac{x^2-x+x^2+4x-2x-8}{(x-1)(x+4)(x-4)}$

$\qquad = \frac{2x^2+x-8}{(x-1)(x+4)(x-4)}$

7 a) $\frac{x^2-4}{3x-3} \times \frac{9}{2x-4} = \frac{(x+2)(x-2)}{3(x-1)} \times \frac{9}{2(x-2)}$

$\qquad = \frac{x+2}{x-1} \times \frac{3}{2} = \frac{3(x+2)}{2(x-1)}$

b) $\frac{8}{x} \div \frac{6}{x^2} = \frac{8}{x} \times \frac{x^2}{6} = \frac{4}{1} \times \frac{x}{3} = \frac{4x}{3}$

c) $\frac{y-1}{2} \div \frac{x+1}{3} = \frac{y-1}{2} \times \frac{3}{x+1} = \frac{3(y-1)}{2(x+1)}$

d) $\frac{x^2-7x+12}{x^2+3x+2} \times \frac{x+1}{x-3}$

$\qquad = \frac{(x-3)(x-4)}{(x+1)(x+2)} \times \frac{(x+1)}{(x-3)} = \frac{x-4}{x+2}$

e) $\frac{s^2+4s+3}{s^2-16} \div \frac{s+1}{s+4} = \frac{(s+1)(s+3)}{(s+4)(s-4)} \times \frac{(s+4)}{(s+1)}$

$\qquad = \frac{s+3}{s-4}$

f) $\frac{x^2+x-12}{x^2-4} \times \frac{x^2+2x}{3x+12}$

$\qquad = \frac{(x-3)(x+4)}{(x+2)(x-2)} \times \frac{x(x+2)}{3(x+4)}$

$\qquad = \frac{x-3}{x-2} \times \frac{x}{3} = \frac{x(x-3)}{3(x-2)}$

g) $\frac{a^2-7a+10}{a^2+5a+6} \times \frac{a^2+2a-3}{a^2-3a-10}$

$\qquad = \frac{(a-2)(a-5)}{(a+2)(a+3)} \times \frac{(a-1)(a+3)}{(a+2)(a-5)}$

$\qquad = \frac{(a-2)(a-1)}{(a+2)^2}$

h) $\frac{b^2+5b+6}{b^2+6b+5} \div \frac{2b+6}{3b+3}$

$\qquad = \frac{(b+2)(b+3)}{(b+1)(b+5)} \times \frac{3(b+1)}{2(b+3)}$

$\qquad = \frac{b+2}{b+5} \times \frac{3}{2} = \frac{3(b+2)}{2(b+5)}$

i) $\frac{x^2+8x+15}{x^2+4x-12} \div \frac{x^2+4x+3}{x^2+8x+12}$

$\qquad = \frac{(x+3)(x+5)}{(x-2)(x+6)} \times \frac{(x+2)(x+6)}{(x+1)(x+3)}$

$\qquad = \frac{x+5}{x-2} \times \frac{x+2}{x+1} = \frac{(x+5)(x+2)}{(x-2)(x+1)}$

Page 90   Exam-Style Questions

1 a) $x(x^2-4y)+9xy = x^3-4xy+9xy$ *[1 mark]*
$\qquad = x^3+5xy$ *[1 mark]*
b) $(2x-7)^2 = (2x-7)(2x-7)$
$\qquad = 4x^2-14x-14x+49$ *[1 mark]*
$\qquad = 4x^2-28x+49$ *[1 mark]*

2 $x^2-y^2 = (x+y)(x-y)$, so:
$\qquad 145^2-55^2 = (145+55)(145-55)$ *[1 mark]*
$\qquad = 200 \times 90 = \textbf{18 000}$ *[1 mark]*

3 $A = 5x^2+9xy$, $B = 3x(x-y)$
$\qquad A-B = 5x^2+9xy-3x(x-y)$ *[1 mark]*
$\qquad = 5x^2+9xy-(3x^2-3xy)$
$\qquad = 5x^2+9xy-3x^2+3xy$
$\qquad = 2x^2+12xy$ *[1 mark]*
$\qquad = 2x(x+6y)$ *[1 mark]*

4 a) $5a^2-6a = a(5a-6)$ *[1 mark]*
b) $5(2x+3)^2-6(2x+3)$
$\qquad = (2x+3)(5(2x+3)-6)$ *[1 mark]*
$\qquad = (2x+3)(10x+15-6)$
$\qquad = (2x+3)(10x+9)$ *[1 mark]*

5 $6(3x-y)-4(x+5y)$
$\qquad = (18x-6y)-(4x+20y)$ *[1 mark]*
$\qquad = 18x-6y-4x-20y$
$\qquad = 14x-26y$ *[1 mark]*
$\qquad = 2(7x-13y)$
So $a = 2$ and $b = 13$. *[1 mark]*

6 $2x^2 - 4x - 30 = 2(x^2 - 2x - 15)$
$\qquad\qquad\qquad = 2(x + 3)(x - 5)$ *[1 mark]*

$\dfrac{6x^2 + 18x}{2x^2 - 4x - 30} = \dfrac{6x(x + 3)}{2(x + 3)(x - 5)}$ *[1 mark]*

$\qquad\qquad\qquad = \dfrac{3x}{x - 5}$ *[1 mark]*

7 $(x + 3)(x - 2)(x - 2)$
$= (x + 3)(x^2 - 2x - 2x + 4)$
$= (x + 3)(x^2 - 4x + 4)$
$= x(x^2 - 4x + 4) + 3(x^2 - 4x + 4)$
$= x^3 - 4x^2 + 4x + 3x^2 - 12x + 12$
$= \boldsymbol{x^3 - x^2 - 8x + 12}$
[3 marks available — 3 marks for the correct answer, otherwise 1 mark for correctly multiplying two sets of brackets together, 1 mark for attempting to multiply this product by the third set of brackets]

Section 7 — Powers and Roots

7.1 Squares, Cubes and Roots

Page 91 Exercise 1

1 a) (i) $5^2 = 5 \times 5 = \mathbf{25}$
(ii) $5^3 = 5 \times 5 \times 5 = 25 \times 5 = \mathbf{125}$
b) (i) $10^2 = 10 \times 10 = \mathbf{100}$
(ii) $10^3 = 10 \times 10 \times 10 = 100 \times 10 = \mathbf{1000}$
c) (i) $(-2)^2 = -2 \times -2 = \mathbf{4}$
(ii) $(-2)^3 = -2 \times -2 \times -2 = 4 \times -2 = \mathbf{-8}$
d) (i) $0.1^2 = 0.1 \times 0.1 = \mathbf{0.01}$
(ii) $0.1^3 = 0.1 \times 0.1 \times 0.1 = 0.01 \times 0.1$
$\qquad\qquad = \mathbf{0.001}$

2 a) 6 and −6 **b)** 100 and −100
c) −16 has **no square roots** **d)** 9 and −9
3 a) 2 **b)** −4 **c)** 10 **d)** −1
4 a) $\sqrt{4 \times 10^2} = \sqrt{4 \times 100} = \sqrt{400} = \mathbf{20}$
b) $\sqrt[3]{(3 + 5)^2} = \sqrt[3]{8^2} = \sqrt[3]{64} = \mathbf{4}$
c) $\sqrt[3]{3^2 \times 2^3 + 12^2} = \sqrt[3]{9 \times 8 + 144}$
$\qquad\qquad\qquad = \sqrt[3]{72 + 144} = \sqrt[3]{216} = \mathbf{6}$

7.2 Indices and Index Laws

Page 92 Exercise 1

1 a) 2^5 **b)** 7^7 **c)** $3^2 x^3 y^2$
2 a) 10^3 **b)** 10^7 **c)** 10^8
3 a) 81 **b)** 256 **c)** 59 049 **d)** 59 049
e) 768 **f)** 40 **g)** 512 **h)** 537 289
4 a) $\left(\frac{1}{2}\right)^2 = \frac{1^2}{2^2} = \frac{1}{4}$ **b)** $\left(\frac{1}{2}\right)^3 = \frac{1^3}{2^3} = \frac{1}{8}$
c) $\left(\frac{1}{4}\right)^2 = \frac{1^2}{4^2} = \frac{1}{16}$ **d)** $\left(\frac{2}{3}\right)^2 = \frac{2^2}{3^2} = \frac{4}{9}$
e) $\left(\frac{3}{10}\right)^2 = \frac{3^2}{10^2} = \frac{9}{100}$ **f)** $\left(\frac{3}{2}\right)^3 = \frac{3^3}{2^3} = \frac{27}{8}$
g) $\left(\frac{5}{3}\right)^4 = \frac{5^4}{3^4} = \frac{625}{81}$ **h)** $\left(\frac{4}{3}\right)^3 = \frac{4^3}{3^3} = \frac{64}{27}$

Page 93 Exercise 2

1 a) $3^2 \times 3^6 = 3^{2 + 6} = \mathbf{3^8}$
b) $10^7 \div 10^3 = 10^{7 - 3} = \mathbf{10^4}$
c) $a^6 \times a^4 = a^{6 + 4} = \mathbf{a^{10}}$
d) $(4^3)^3 = 4^{3 \times 3} = \mathbf{4^9}$
e) $8^6 \div 8^1 = 8^{6 - 1} = \mathbf{8^5}$
f) $7 \times 7^6 = 7^1 \times 7^6 = 7^{1 + 6} = \mathbf{7^7}$
g) $(c^5)^4 = c^{5 \times 4} = \mathbf{c^{20}}$
h) $\dfrac{b^8}{b^5} = b^8 \div b^5 = b^{8 - 5} = \mathbf{b^3}$
i) $f^{75} \div f^0 = f^{75 - 0} = \mathbf{f^{75}}$
j) $\dfrac{20^{228}}{20^{210}} = 20^{228 - 210} = \mathbf{20^{18}}$
k) $(g^{11})^8 = g^{11 \times 8} = \mathbf{g^{88}}$
l) $(14^7)^d = 14^{7 \times d} = \mathbf{14^{7d}}$
2 a) $8 - 3 = \blacksquare \Rightarrow \blacksquare = \mathbf{5}$
b) $\blacksquare + 10 = 12 \Rightarrow \blacksquare = 12 - 10 = \mathbf{2}$
c) $10 \times 4 = \blacksquare \Rightarrow \blacksquare = \mathbf{40}$

d) $6 \times \blacksquare = 24 \Rightarrow \blacksquare = 24 \div 6 = \mathbf{4}$
e) $\blacksquare \times 10 = 30 \Rightarrow \blacksquare = 30 \div 10 = \mathbf{3}$
f) $7 + \blacksquare = 13 \Rightarrow \blacksquare = 13 - 7 = \mathbf{6}$
g) $\blacksquare - 6 = 7 \Rightarrow \blacksquare = 7 + 6 = \mathbf{13}$
h) $14 - \blacksquare = 7 \Rightarrow \blacksquare = 14 - 7 = \mathbf{7}$
3 a) $3^2 \times 3^5 \times 3^7 = 3^{2 + 5 + 7} = \mathbf{3^{14}}$
b) $5^4 \times 5 \times 5^8 = 5^{4 + 1 + 8} = \mathbf{5^{13}}$
c) $(p^6)^2 \times p^5 = p^{6 \times 2 + 5} = p^{12 + 5} = \mathbf{p^{17}}$
d) $(9^4 \times 9^3)^5 = 9^{(4 + 3) \times 5} = 9^{7 \times 5} = \mathbf{9^{35}}$
e) $7^3 \times 7^5 \div 7^6 = 7^{3 + 5 - 6} = \mathbf{7^2}$
f) $8^3 \div 8^9 \times 8^7 = 8^{3 - 9 + 7} = 8^1 = \mathbf{8}$
g) $(12^8 \div 12^4)^3 = 12^{(8 - 4) \times 3} = 12^{4 \times 3} = \mathbf{12^{12}}$
h) $(q^3)^6 \div q^4 = q^{3 \times 6 - 4} = q^{18 - 4} = \mathbf{q^{14}}$
4 a) $\dfrac{3^4 \times 3^5}{3^6} = \dfrac{3^9}{3^6} = 3^{9 - 6} = \mathbf{3^3}$
b) $\dfrac{s^8 \times s^4}{s^3 \times s^6} = \dfrac{s^{12}}{s^9} = s^{12 - 9} = \mathbf{s^3}$
c) $\left(\dfrac{6^3 \times 6^9}{6^7}\right)^3 = \left(\dfrac{6^{12}}{6^7}\right)^3 = (6^{12 - 7})^3$
$\qquad\qquad\qquad = (6^5)^3 = 6^{5 \times 3} = \mathbf{6^{15}}$
d) $\dfrac{2^5 \times 2^5}{(2^3)^2} = \dfrac{2^{10}}{2^6} = 2^{10 - 6} = \mathbf{2^4}$
e) $\dfrac{5^5 \times 5^5}{5^8 \div 5^3} = \dfrac{5^{10}}{5^5} = 5^{10 - 5} = \mathbf{5^5}$
f) $\dfrac{10^8 \div 10^3}{10^4 \div 10^4} = \dfrac{10^5}{10^0} = 10^{5 - 0} = \mathbf{10^5}$
You could also write $10^0 = 1$ and then use the fact that anything divided by 1 is itself.
g) $\dfrac{(t^6 \div t^3)^4}{t^9 \div t^4} = \dfrac{(t^3)^4}{t^5} = \dfrac{t^{12}}{t^5} = t^{12 - 5} = \mathbf{t^7}$
h) $\dfrac{(8^5)^7 \div 8^{12}}{8^6 \times 8^{10}} = \dfrac{8^{35} \div 8^{12}}{8^{16}} = \dfrac{8^{23}}{8^{16}} = 8^{23 - 16} = \mathbf{8^7}$
5 a) (i) $4 = 2^2$
(ii) $4^5 = (2^2)^5 = 2^{2 \times 5} = \mathbf{2^{10}}$
(iii) $2^3 \times 4^5 = 2^3 \times 2^{10} = 2^{3 + 10} = \mathbf{2^{13}}$
b) (i) $9 \times 3^3 = 3^2 \times 3^3 = 3^{2 + 3} = \mathbf{3^5}$
(ii) $5 \times 25 \times 125 = 5 \times 5^2 \times 5^3$
$\qquad\qquad\qquad = 5^{1 + 2 + 3} = \mathbf{5^6}$
(iii) $16 \times 2^6 = 2^4 \times 2^6 = 2^{4 + 6} = 2^{10}$
$\qquad\qquad = 2^{2 \times 5} = (2^2)^5 = \mathbf{4^5}$

Page 94 Exercise 3

1 a) $4^{-1} = \dfrac{1}{4^1} = \dfrac{1}{4}$ **b)** $2^{-2} = \dfrac{1}{2^2} = \dfrac{1}{4}$
c) $3^{-3} = \dfrac{1}{3^3} = \dfrac{1}{27}$
d) $2 \times 3^{-1} = 2 \times \dfrac{1}{3} = \dfrac{2}{3}$
2 a) 5^{-1} **b)** 11^{-1} **c)** 3^{-2} **d)** 2^{-7}
3 a) $\left(\dfrac{1}{2}\right)^{-1} = \left(\dfrac{2}{1}\right)^1 = \mathbf{2}$
b) $\left(\dfrac{1}{3}\right)^{-2} = \left(\dfrac{3}{1}\right)^2 = 3^2 = \mathbf{9}$
c) $\left(\dfrac{5}{2}\right)^{-3} = \left(\dfrac{2}{5}\right)^3 = \dfrac{2^3}{5^3} = \dfrac{8}{125}$
d) $\left(\dfrac{7}{10}\right)^{-2} = \left(\dfrac{10}{7}\right)^2 = \dfrac{10^2}{7^2} = \dfrac{100}{49}$

Page 95 Exercise 4

1 a) $5^4 \times 5^{-2} = 5^{4 + (-2)} = \mathbf{5^2}$
b) $g^6 \div g^{-6} = g^{6 - (-6)} = \mathbf{g^{12}}$
c) $2^{16} \div \dfrac{1}{2^4} = 2^{16} \div 2^{-4} = 2^{16 - (-4)} = \mathbf{2^{20}}$
d) $k^{10} \times k^{-6} \div k^0 = k^{10 + (-6) - 0} = \mathbf{k^4}$
e) $\left(\dfrac{1}{p^4}\right)^5 = (p^{-4})^5 = p^{-4 \times 5} = \mathbf{p^{-20}}$
f) $\left(\dfrac{l^{-5}}{l^6}\right)^{-3} = (l^{-5 - 6})^{-3} = (l^{-11})^{-3} = l^{-11 \times -3} = \mathbf{l^{33}}$
g) $\dfrac{n^{-4} \times n}{(n^{-3})^6} = \dfrac{n^{-3}}{n^{-18}} = n^{-3 - (-18)} = \mathbf{n^{15}}$
h) $\left(\dfrac{10^7 \times 10^{-11}}{10^9 \div 10^4}\right)^{-5} = \left(\dfrac{10^{-4}}{10^5}\right)^{-5} = (10^{-9})^{-5} = \mathbf{10^{45}}$

2 a) (i) $\dfrac{1}{100}$ **(ii)** $\dfrac{1}{10^2}$ **(iii)** 10^{-2}
b) (i) 10^{-1} **(ii)** 10^{-8} **(iii)** 10^{-4} **(iv)** 10^0
3 a) $3^2 \times 5^{-2} = 3^2 \times \dfrac{1}{5^2} = \dfrac{3^2}{5^2} = \dfrac{9}{25}$
b) $2^{-3} \times 7^1 = \dfrac{1}{2^3} \times 7 = \dfrac{7}{2^3} = \dfrac{7}{8}$
c) $\left(\dfrac{1}{2}\right)^{-2} \times \left(\dfrac{1}{3}\right)^2 = 2^2 \times \dfrac{1}{3^2} = \dfrac{4}{9}$
d) $6^{-4} \div 6^{-2} = 6^{-4 - (-2)} = 6^{-2} = \dfrac{1}{6^2} = \dfrac{1}{36}$
e) $(-9)^2 \times (-5)^{-3} = (-9)^2 \times \dfrac{1}{(-5)^3} = -\dfrac{81}{125}$
f) $8^{-5} \times 8^3 \times 3^3 = 8^{-2} \times 3^3 = \dfrac{3^3}{8^2} = \dfrac{27}{64}$
g) $10^{-5} \div 10^6 \times 10^4 = 10^{-7}$
$\qquad\qquad = \dfrac{1}{10^7} = \dfrac{1}{10\,000\,000}$
h) $\left(\dfrac{3}{4}\right)^{-1} \div \left(\dfrac{1}{2}\right)^{-3} = \dfrac{4}{3} \div 2^3 = \dfrac{4}{3 \times 2^3}$
$\qquad\qquad = \dfrac{4}{3 \times 8} = \dfrac{4}{24} = \dfrac{1}{6}$

Page 95 Exercise 5

1 a) $\sqrt[5]{a}$ **b)** $(\sqrt[5]{a})^3$ **c)** $(\sqrt[5]{a})^2$ **d)** $(\sqrt{a})^5$
2 a) $64^{\frac{1}{2}} = \sqrt{64} = \mathbf{8}$ **b)** $64^{\frac{1}{3}} = \sqrt[3]{64} = \mathbf{4}$
c) $16^{\frac{1}{4}} = \sqrt[4]{16} = \mathbf{2}$
d) $1\,000\,000^{\frac{1}{2}} = \sqrt{1\,000\,000} = \mathbf{1000}$
3 a) $125^{\frac{2}{3}} = (\sqrt[3]{125})^2 = 5^2 = \mathbf{25}$
b) $9^{\frac{3}{2}} = (\sqrt{9})^3 = 3^3 = \mathbf{27}$
c) $1000^{\frac{5}{3}} = (\sqrt[3]{1000})^5 = 10^5 = \mathbf{100\,000}$
d) $8000^{\frac{4}{3}} = (\sqrt[3]{8000})^4 = 20^4 = \mathbf{160\,000}$

7.3 Standard Form

Page 96 Exercise 1

1 a) 2.5×10^2 **b)** 1.1×10^3
c) 4.8×10^4 **d)** 5.9×10^6
e) 2.75×10^6 **f)** 8.56×10^3
g) 8.0808×10^5 **h)** 9.30078×10^5
2 a) 2.5×10^{-3} **b)** 6.7×10^{-3}
c) 3.03×10^{-2} **d)** 5.6×10^{-5}
e) 3.75×10^{-1} **f)** 7.07×10^{-2}
g) 2.1×10^{-10} **h)** 5.002×10^{-4}
3 a) $0.00567 \times 10^9 = (5.67 \times 10^{-3}) \times 10^9$
$\qquad\qquad\qquad = \mathbf{5.67 \times 10^6}$
b) $95.32 \times 10^2 = (9.532 \times 10) \times 10^2$
$\qquad\qquad\qquad = \mathbf{9.532 \times 10^3}$
c) $0.034 \times 10^{-4} = (3.4 \times 10^{-2}) \times 10^{-4}$
$\qquad\qquad\qquad = \mathbf{3.4 \times 10^{-6}}$
d) $845\,000 \times 10^{-3} = (8.45 \times 10^5) \times 10^{-3}$
$\qquad\qquad\qquad = \mathbf{8.45 \times 10^2}$

Page 97 Exercise 2

1 a) 3 000 000 **b)** 94 000
c) 880 000 **d)** 4090
e) 198 900 000 **f)** 66.9
g) 7.20 **h)** 0.00000356
i) 0.0000888 **j)** 0.000000019
k) 0.669 **l)** 0.00000705

Page 98 Exercise 3

1 a) $(3 \times 10^7) \times (4 \times 10^{-4})$
$\qquad = 3 \times 4 \times 10^7 \times 10^{-4} = 12 \times 10^3$
$\qquad = 1.2 \times 10 \times 10^3 = \mathbf{1.2 \times 10^4}$
b) $(7 \times 10^9) \times (9 \times 10^{-4})$
$\qquad = 7 \times 9 \times 10^9 \times 10^{-4} = 63 \times 10^5$
$\qquad = 6.3 \times 10 \times 10^5 = \mathbf{6.3 \times 10^6}$
c) $(2 \times 10^5) \times (3.27 \times 10^2)$
$\qquad = 2 \times 3.27 \times 10^5 \times 10^2 = \mathbf{6.54 \times 10^7}$
d) $(3.4 \times 10^{-4}) \times (3 \times 10^2)$
$\qquad = 3.4 \times 3 \times 10^{-4} \times 10^2 = 10.2 \times 10^{-2}$
$\qquad = 1.02 \times 10 \times 10^{-2} = \mathbf{1.02 \times 10^{-1}}$

e) $(2 \times 10^{-5}) \times (8.734 \times 10^5)$
$= 2 \times 8.734 \times 10^{-5} \times 10^5$
$= 17.468 \times 10^0 = \textbf{1.7468} \times \textbf{10}$

f) $(1.2 \times 10^4) \times (5.3 \times 10^6)$
$= 1.2 \times 5.3 \times 10^4 \times 10^6 = \textbf{6.36} \times \textbf{10}^{\textbf{10}}$

2 a) $(3.6 \times 10^7) \div (1.2 \times 10^4)$
$= \dfrac{3.6 \times 10^7}{1.2 \times 10^4} = \dfrac{3.6}{1.2} \times \dfrac{10^7}{10^4} = \textbf{3} \times \textbf{10}^{\textbf{3}}$

b) $(8.4 \times 10^4) \div (7 \times 10^8)$
$= \dfrac{8.4 \times 10^4}{7 \times 10^8} = \dfrac{8.4}{7} \times \dfrac{10^4}{10^8} = \textbf{1.2} \times \textbf{10}^{\textbf{--4}}$

c) $(1.8 \times 10^{-4}) \div (1.2 \times 10^8)$
$= \dfrac{1.8 \times 10^{-4}}{1.2 \times 10^8} = \dfrac{1.8}{1.2} \times \dfrac{10^{-4}}{10^8} = \textbf{1.5} \times \textbf{10}^{\textbf{--12}}$

d) $(4.8 \times 10^3) \div (1.2 \times 10^{-2})$
$= \dfrac{4.8 \times 10^3}{1.2 \times 10^{-2}} = \dfrac{4.8}{1.2} \times \dfrac{10^3}{10^{-2}} = \textbf{4} \times \textbf{10}^{\textbf{5}}$

e) $(8.1 \times 10^{-1}) \div (0.9 \times 10^{-2})$
$= \dfrac{8.1 \times 10^{-1}}{0.9 \times 10^{-2}} = \dfrac{8.1}{0.9} \times \dfrac{10^{-1}}{10^{-2}} = \textbf{9} \times \textbf{10}$

f) $(13.2 \times 10^5) \div (1.2 \times 10^4)$
$= \dfrac{13.2 \times 10^5}{1.2 \times 10^4} = \dfrac{13.2}{1.2} \times \dfrac{10^5}{10^4} = 11 \times 10$
$= \textbf{1.1} \times \textbf{10}^{\textbf{2}}$

Page 98   Exercise 4

1 a) $(5.0 \times 10^3) + (3.0 \times 10^2)$
$= (5.0 \times 10^3) + (0.3 \times 10^3)$
$= (5.0 + 0.3) \times 10^3 = \textbf{5.3} \times \textbf{10}^{\textbf{3}}$

b) $(1.8 \times 10^5) + (3.2 \times 10^3)$
$= (1.8 \times 10^5) + (0.032 \times 10^5)$
$= (1.8 + 0.032) \times 10^5 = \textbf{1.832} \times \textbf{10}^{\textbf{5}}$

Using the same method for c)-f):
c) $\textbf{5.52} \times \textbf{10}^{\textbf{--1}}$ **d)** $\textbf{7.28} \times \textbf{10}^{\textbf{--4}}$
e) $\textbf{1.47} \times \textbf{10}^{\textbf{0}}$ **f)** $\textbf{6.05} \times \textbf{10}^{\textbf{8}}$

2 a) $(5.2 \times 10^4) - (3.3 \times 10^3)$
$= (5.2 \times 10^4) - (0.33 \times 10^4)$
$= (5.2 - 0.33) \times 10^4 = \textbf{4.87} \times \textbf{10}^{\textbf{4}}$

b) $(7.2 \times 10^{-3}) - (1.5 \times 10^{-4})$
$= (7.2 \times 10^{-3}) - (0.15 \times 10^{-3})$
$= (7.2 - 0.15) \times 10^{-3} = \textbf{7.05} \times \textbf{10}^{\textbf{--3}}$

Using the same method for c)-f):
c) $\textbf{6.497} \times \textbf{10}^{\textbf{2}}$ **d)** $\textbf{8.337} \times \textbf{10}^{\textbf{2}}$
e) $\textbf{8.317} \times \textbf{10}^{\textbf{4}}$ **f)** $\textbf{2.747} \times \textbf{10}^{\textbf{0}}$

7.4 Surds

Page 99   Exercise 1

1 a) $\sqrt{12} = \sqrt{4 \times 3} = \sqrt{4} \times \sqrt{3} = \textbf{2}\sqrt{\textbf{3}}$
b) $\sqrt{20} = \sqrt{4 \times 5} = \sqrt{4} \times \sqrt{5} = \textbf{2}\sqrt{\textbf{5}}$
c) $\sqrt{50} = \sqrt{25 \times 2} = \sqrt{25} \times \sqrt{2} = \textbf{5}\sqrt{\textbf{2}}$
d) $\sqrt{32} = \sqrt{16 \times 2} = \sqrt{16} \times \sqrt{2} = \textbf{4}\sqrt{\textbf{2}}$
e) $\sqrt{108} = \sqrt{36 \times 3} = \sqrt{36} \times \sqrt{3} = \textbf{6}\sqrt{\textbf{3}}$
f) $\sqrt{300} = \sqrt{100 \times 3} = \sqrt{100} \times \sqrt{3} = \textbf{10}\sqrt{\textbf{3}}$
g) $\sqrt{98} = \sqrt{49 \times 2} = \sqrt{49} \times \sqrt{2} = \textbf{7}\sqrt{\textbf{2}}$
h) $\sqrt{192} = \sqrt{64 \times 3} = \sqrt{64} \times \sqrt{3} = \textbf{8}\sqrt{\textbf{3}}$

2 a) $\sqrt{2} \times \sqrt{24} = \sqrt{48} = \sqrt{16 \times 3} = \textbf{4}\sqrt{\textbf{3}}$
b) $\sqrt{3} \times \sqrt{12} = \sqrt{36} = \textbf{6}$
In this case $6\sqrt{1}$ *is simplified to 6 as* $\sqrt{1} = 1$.
c) $\sqrt{3} \times \sqrt{24} = \sqrt{72} = \sqrt{36 \times 2} = \textbf{6}\sqrt{\textbf{2}}$
d) $\sqrt{2} \times \sqrt{10} = \sqrt{20} = \sqrt{4 \times 5} = \textbf{2}\sqrt{\textbf{5}}$
e) $\sqrt{40} \times \sqrt{2} = \sqrt{80} = \sqrt{16 \times 5} = \textbf{4}\sqrt{\textbf{5}}$
f) $\sqrt{3} \times \sqrt{60} = \sqrt{180} = \sqrt{36 \times 5} = \textbf{6}\sqrt{\textbf{5}}$
g) $\sqrt{7} \times \sqrt{35} = \sqrt{245} = \sqrt{49 \times 5} = \textbf{7}\sqrt{\textbf{5}}$
h) $\sqrt{50} \times \sqrt{10} = \sqrt{500} = \sqrt{100 \times 5} = \textbf{10}\sqrt{\textbf{5}}$
i) $\sqrt{8} \times \sqrt{24} = \sqrt{192} = \sqrt{64 \times 3} = = \textbf{8}\sqrt{\textbf{3}}$

Page 100   Exercise 2

1 a) $\sqrt{90} \div \sqrt{10} = \sqrt{90 \div 10} = \sqrt{9} = \textbf{3}$
b) $\sqrt{72} \div \sqrt{2} = \sqrt{72 \div 2} = \sqrt{36} = \textbf{6}$
c) $\sqrt{200} \div \sqrt{8} = \sqrt{200 \div 8} = \sqrt{25} = \textbf{5}$
d) $\sqrt{243} \div \sqrt{3} = \sqrt{243 \div 3} = \sqrt{81} = \textbf{9}$
e) $\sqrt{294} \div \sqrt{6} = \sqrt{294 \div 6} = \sqrt{49} = \textbf{7}$

f) $\sqrt{80} \div \sqrt{10} = \sqrt{80 \div 10} = \sqrt{8} = \textbf{2}\sqrt{\textbf{2}}$
g) $\sqrt{120} \div \sqrt{10} = \sqrt{120 \div 10} = \sqrt{12} = \textbf{2}\sqrt{\textbf{3}}$
h) $\sqrt{180} \div \sqrt{3} = \sqrt{180 \div 3} = \sqrt{60} = \textbf{2}\sqrt{\textbf{15}}$
i) $\sqrt{180} \div \sqrt{9} = \sqrt{180 \div 9} = \sqrt{20} = \textbf{2}\sqrt{\textbf{5}}$
j) $\sqrt{96} \div \sqrt{6} = \sqrt{96 \div 6} = \sqrt{16} = \textbf{4}$
k) $\sqrt{484} \div \sqrt{22} = \sqrt{484 \div 22} = \sqrt{\textbf{22}}$
l) $\sqrt{210} \div \sqrt{35} = \sqrt{210 \div 35} = \sqrt{\textbf{6}}$

2 a) $\sqrt{\dfrac{1}{9}} = \dfrac{\sqrt{1}}{\sqrt{9}} = \dfrac{\textbf{1}}{\textbf{3}}$

b) $\sqrt{\dfrac{4}{25}} = \dfrac{\sqrt{4}}{\sqrt{25}} = \dfrac{\textbf{2}}{\textbf{5}}$

c) $\sqrt{\dfrac{49}{121}} = \dfrac{\sqrt{49}}{\sqrt{121}} = \dfrac{\textbf{7}}{\textbf{11}}$

d) $\sqrt{\dfrac{100}{64}} = \dfrac{\sqrt{100}}{\sqrt{64}} = \dfrac{10}{8} = \dfrac{\textbf{5}}{\textbf{4}}$

e) $\sqrt{\dfrac{18}{200}} = \dfrac{\sqrt{18}}{\sqrt{200}} = \dfrac{3\sqrt{2}}{10\sqrt{2}} = \dfrac{\textbf{3}}{\textbf{10}}$

f) $\sqrt{\dfrac{2}{25}} = \dfrac{\sqrt{2}}{\sqrt{25}} = \dfrac{\sqrt{\textbf{2}}}{\textbf{5}}$

g) $\sqrt{\dfrac{108}{147}} = \dfrac{\sqrt{108}}{\sqrt{147}} = \dfrac{6\sqrt{3}}{7\sqrt{3}} = \dfrac{\textbf{6}}{\textbf{7}}$

h) $\sqrt{\dfrac{27}{64}} = \dfrac{\sqrt{27}}{\sqrt{64}} = \dfrac{\textbf{3}\sqrt{\textbf{3}}}{\textbf{8}}$

i) $\sqrt{\dfrac{98}{121}} = \dfrac{\sqrt{98}}{\sqrt{121}} = \dfrac{\textbf{7}\sqrt{\textbf{2}}}{\textbf{11}}$

Page 101   Exercise 3

1 a) $2\sqrt{3} + 3\sqrt{3} = \textbf{5}\sqrt{\textbf{3}}$
b) $7\sqrt{7} - 3\sqrt{7} = \textbf{4}\sqrt{\textbf{7}}$
c) $2\sqrt{3} + 3\sqrt{7}$
Not all sums of surds can be simplified.
d) $2\sqrt{32} + 3\sqrt{2} = 2 \times 4\sqrt{2} + 3\sqrt{2} = \textbf{11}\sqrt{\textbf{2}}$
e) $2\sqrt{27} - 3\sqrt{3} = 2 \times 3\sqrt{3} - 3\sqrt{3} = \textbf{3}\sqrt{\textbf{3}}$
f) $5\sqrt{7} + 3\sqrt{28} = 5\sqrt{7} + 3 \times 2\sqrt{7} = \textbf{11}\sqrt{\textbf{7}}$

2 a) $2\sqrt{125} - 3\sqrt{80} = 2 \times 5\sqrt{5} - 3 \times 4\sqrt{5} = \textbf{--2}\sqrt{\textbf{5}}$
b) $\sqrt{108} + 2\sqrt{300} = 6\sqrt{3} + 2 \times 10\sqrt{3} = \textbf{26}\sqrt{\textbf{3}}$
c) $5\sqrt{294} - 3\sqrt{216} = 5 \times 7\sqrt{6} - 3 \times 6\sqrt{6} = \textbf{17}\sqrt{\textbf{6}}$

Page 102   Exercise 4

1 a) $(2 + \sqrt{3})^2 = 2^2 + 2 \times 2\sqrt{3} + 3 = \textbf{7} + \textbf{4}\sqrt{\textbf{3}}$
b) $(1 + \sqrt{2})(1 - \sqrt{2}) = 1^2 - 2 = \textbf{--1}$
This is the difference of two squares.
c) $(5 - \sqrt{2})^2 = 5^2 - 2 \times 5\sqrt{2} + 2 = \textbf{27} - \textbf{10}\sqrt{\textbf{2}}$
d) $(3 - 3\sqrt{2})(3 - \sqrt{2}) = 9 - 3\sqrt{2} - 9\sqrt{2} + 6$
$= \textbf{15} - \textbf{12}\sqrt{\textbf{2}}$
e) $(5 + \sqrt{3})(3 + \sqrt{3}) = 15 + 5\sqrt{3} + 3\sqrt{3} + 3$
$= \textbf{18} + \textbf{8}\sqrt{\textbf{3}}$
f) $(7 + 2\sqrt{2})(7 - 2\sqrt{2}) = 7^2 - (2\sqrt{2})^2$
$= 49 - 4 \times 2 = \textbf{41}$
This is also the difference of two squares
— but don't forget you need to square the
coefficient in front of the $\sqrt{2}$.

2 a) $(2 + \sqrt{6})(4 + \sqrt{3}) = 8 + 2\sqrt{3} + 4\sqrt{6} + \sqrt{18}$
$= \textbf{8} + \textbf{2}\sqrt{\textbf{3}} + \textbf{4}\sqrt{\textbf{6}} + \textbf{3}\sqrt{\textbf{2}}$
b) $(4 - \sqrt{7})(5 - \sqrt{2}) = \textbf{20} - \textbf{4}\sqrt{\textbf{2}} - \textbf{5}\sqrt{\textbf{7}} + \sqrt{\textbf{14}}$
c) $(1 - 2\sqrt{10})(6 - \sqrt{15})$
$= 6 - \sqrt{15} - 12\sqrt{10} + 2\sqrt{150}$
$= 6 - \sqrt{15} - 12\sqrt{10} + 2 \times 5\sqrt{6}$
$= \textbf{6} - \sqrt{\textbf{15}} - \textbf{12}\sqrt{\textbf{10}} + \textbf{10}\sqrt{\textbf{6}}$

Page 103   Exercise 5

1 a) $\dfrac{6}{\sqrt{6}} = \dfrac{6\sqrt{6}}{\sqrt{6} \times \sqrt{6}} = \dfrac{6\sqrt{6}}{6} = \sqrt{\textbf{6}}$

b) $\dfrac{8}{\sqrt{8}} = \dfrac{8\sqrt{8}}{\sqrt{8} \times \sqrt{8}} = \dfrac{8\sqrt{8}}{8} = \sqrt{8} = \textbf{2}\sqrt{\textbf{2}}$

c) $\dfrac{5}{\sqrt{5}} = \dfrac{5\sqrt{5}}{\sqrt{5} \times \sqrt{5}} = \dfrac{5\sqrt{5}}{5} = \sqrt{\textbf{5}}$

d) $\dfrac{1}{\sqrt{3}} = \dfrac{\sqrt{3}}{\sqrt{3} \times \sqrt{3}} = \dfrac{\sqrt{\textbf{3}}}{\textbf{3}}$

e) $\dfrac{15}{\sqrt{5}} = \dfrac{15\sqrt{5}}{5} = \textbf{3}\sqrt{\textbf{5}}$

f) $\dfrac{9}{\sqrt{3}} = \dfrac{9\sqrt{3}}{3} = \textbf{3}\sqrt{\textbf{3}}$

g) $\dfrac{7}{\sqrt{12}} = \dfrac{7\sqrt{12}}{12} = \dfrac{7 \times 2\sqrt{3}}{12} = \dfrac{\textbf{7}\sqrt{\textbf{3}}}{\textbf{6}}$

h) $\dfrac{12}{\sqrt{1000}} = \dfrac{12\sqrt{1000}}{1000} = \dfrac{12 \times 10\sqrt{10}}{1000} = \dfrac{\textbf{3}\sqrt{\textbf{10}}}{\textbf{25}}$

2 a) $\dfrac{1}{5\sqrt{5}} = \dfrac{\sqrt{5}}{5\sqrt{5} \times \sqrt{5}} = \dfrac{\sqrt{5}}{5 \times 5} = \dfrac{\sqrt{\textbf{5}}}{\textbf{25}}$

b) $\dfrac{1}{3\sqrt{3}} = \dfrac{\sqrt{3}}{3\sqrt{3} \times \sqrt{3}} = \dfrac{\sqrt{3}}{3 \times 3} = \dfrac{\sqrt{\textbf{3}}}{\textbf{9}}$

c) $\dfrac{3}{4\sqrt{8}} = \dfrac{3\sqrt{8}}{4 \times 8} = \dfrac{3 \times 2\sqrt{2}}{32} = \dfrac{\textbf{3}\sqrt{\textbf{2}}}{\textbf{16}}$

d) $\dfrac{3}{2\sqrt{5}} = \dfrac{3\sqrt{5}}{2 \times 5} = \dfrac{\textbf{3}\sqrt{\textbf{5}}}{\textbf{10}}$

e) $\dfrac{2}{7\sqrt{3}} = \dfrac{2\sqrt{3}}{7 \times 3} = \dfrac{\textbf{2}\sqrt{\textbf{3}}}{\textbf{21}}$

f) $\dfrac{1}{6\sqrt{12}} = \dfrac{\sqrt{12}}{6 \times 12} = \dfrac{2\sqrt{3}}{72} = \dfrac{\sqrt{\textbf{3}}}{\textbf{36}}$

g) $\dfrac{10}{7\sqrt{5}} = \dfrac{10\sqrt{5}}{7 \times 5} = \dfrac{10\sqrt{5}}{35} = \dfrac{\textbf{2}\sqrt{\textbf{5}}}{\textbf{7}}$

h) $\dfrac{5}{9\sqrt{10}} = \dfrac{5\sqrt{10}}{9 \times 10} = \dfrac{5\sqrt{10}}{90} = \dfrac{\sqrt{\textbf{10}}}{\textbf{18}}$

Page 103   Exercise 6

1 a) $\dfrac{1}{2 + \sqrt{2}} = \dfrac{2 - \sqrt{2}}{(2 + \sqrt{2})(2 - \sqrt{2})}$

$= \dfrac{2 - \sqrt{2}}{4 - 2} = \dfrac{\textbf{2} - \sqrt{\textbf{2}}}{\textbf{2}}$

b) $\dfrac{5}{1 - \sqrt{7}} = \dfrac{5(1 + \sqrt{7})}{(1 - \sqrt{7})(1 + \sqrt{7})}$

$= \dfrac{5 + 5\sqrt{7}}{1 - 7} = \dfrac{5 + 5\sqrt{7}}{-6} = -\dfrac{\textbf{5} + \textbf{5}\sqrt{\textbf{7}}}{\textbf{6}}$

c) $\dfrac{10}{5 + \sqrt{11}} = \dfrac{10(5 - \sqrt{11})}{(5 + \sqrt{11})(5 - \sqrt{11})}$

$= \dfrac{50 - 10\sqrt{11}}{25 - 11} = \dfrac{50 - 10\sqrt{11}}{14}$

$= \dfrac{\textbf{25} - \textbf{5}\sqrt{\textbf{11}}}{\textbf{7}}$

d) $\dfrac{9}{12 - 3\sqrt{17}} = \dfrac{3}{4 - \sqrt{17}} = \dfrac{3(4 + \sqrt{17})}{(4 - \sqrt{17})(4 + \sqrt{17})}$

$= \dfrac{12 + 3\sqrt{17}}{16 - 17} = -(\textbf{12} + \textbf{3}\sqrt{\textbf{17}})$

If you can, always simplify a fraction before
rationalising the denominator.

2 a) $\dfrac{\sqrt{2}}{2 + 3\sqrt{2}} = \dfrac{\sqrt{2}(2 - 3\sqrt{2})}{(2 + 3\sqrt{2})(2 - 3\sqrt{2})}$

$= \dfrac{2\sqrt{2} - 3 \times 2}{4 - 9 \times 2} = \dfrac{2\sqrt{2} - 6}{-14}$

$= \dfrac{\textbf{3} - \sqrt{\textbf{2}}}{\textbf{7}}$

b) $\dfrac{1 + \sqrt{2}}{1 - \sqrt{2}} = \dfrac{(1 + \sqrt{2})(1 + \sqrt{2})}{(1 - \sqrt{2})(1 + \sqrt{2})}$

$= \dfrac{1 + \sqrt{2} + \sqrt{2} + 2}{1 - 2} = -(\textbf{3} + \textbf{2}\sqrt{\textbf{2}})$

c) $\dfrac{2 + \sqrt{3}}{1 - \sqrt{3}} = \dfrac{(2 + \sqrt{3})(1 + \sqrt{3})}{(1 - \sqrt{3})(1 + \sqrt{3})}$

$= \dfrac{2 + 2\sqrt{3} + \sqrt{3} + 3}{1 - 3}$

$= \dfrac{5 + 3\sqrt{3}}{-2} = -\dfrac{\textbf{5} + \textbf{3}\sqrt{\textbf{3}}}{\textbf{2}}$

d) $\dfrac{1 - \sqrt{5}}{2 - \sqrt{5}} = \dfrac{(1 - \sqrt{5})(2 + \sqrt{5})}{(2 - \sqrt{5})(2 + \sqrt{5})}$

$= \dfrac{2 + \sqrt{5} - 2\sqrt{5} - 5}{4 - 5} = \textbf{3} + \sqrt{\textbf{5}}$

e)
$$\frac{1+2\sqrt{2}}{1-2\sqrt{2}} = \frac{(1+2\sqrt{2})(1+2\sqrt{2})}{(1-2\sqrt{2})(1+2\sqrt{2})}$$
$$= \frac{1+2\sqrt{2}+2\sqrt{2}+4\times2}{1-4\times2}$$
$$= \frac{9+4\sqrt{2}}{-7} = -\frac{9+4\sqrt{2}}{7}$$

f)
$$\frac{7+8\sqrt{2}}{9+5\sqrt{2}} = \frac{(7+8\sqrt{2})(9-5\sqrt{2})}{(9+5\sqrt{2})(9-5\sqrt{2})}$$
$$= \frac{63-35\sqrt{2}+72\sqrt{2}-40\times2}{81-25\times2}$$
$$= \frac{37\sqrt{2}-17}{31}$$

3 Multiply top and bottom by $\sqrt{2}$ and then rationalise the denominator:
$$\frac{1}{1-\frac{1}{\sqrt{2}}} = \frac{\sqrt{2}}{\sqrt{2}-1} = \frac{\sqrt{2}(-\sqrt{2}-1)}{(\sqrt{2}-1)(-\sqrt{2}-1)}$$
$$= \frac{-2-\sqrt{2}}{-2+1} = 2+\sqrt{2}$$

4 Multiply top and bottom by $\sqrt{3}$ and then rationalise the denominator:
$$\frac{1}{1+\frac{1}{\sqrt{3}}} = \frac{\sqrt{3}}{\sqrt{3}+1} = \frac{\sqrt{3}(-\sqrt{3}+1)}{(\sqrt{3}+1)(-\sqrt{3}+1)}$$
$$= \frac{-3+\sqrt{3}}{-3+1} = \frac{3-\sqrt{3}}{2}$$

Page 104   Review Exercise
1 a) $7^6 \times 7^9 = 7^{6+9} = 7^{15}$
b) $d^{-4} \div d^6 = d^{-4-6} = d^{-10}$
c) $(4^8)^3 = 4^{8\times3} = 4^{24}$
d) $9^{-2} \times \sqrt[4]{9} = 9^{-2} \times 9^{\frac{1}{4}} = 9^{-2+\frac{1}{4}} = 9^{-\frac{7}{4}}$
e) $(c^{10} \div c^2)^{\frac{1}{4}} = (c^8)^{\frac{1}{4}} = c^2$
f) $2^4 \times \frac{1}{\sqrt[3]{2}} \times 2^{-\frac{1}{2}} = 2^4 \times 2^{-\frac{1}{3}} \times 2^{-\frac{1}{2}}$
$$= 2^{4-\frac{1}{3}-\frac{1}{2}} = 2^{\frac{19}{6}}$$

2 a) $(27m)^{\frac{1}{3}} = \sqrt[3]{27m} = \sqrt[3]{27} \times \sqrt[3]{m} = 3m^{\frac{1}{3}}$
b) $(y^4z^3)^{-\frac{3}{4}} = y^{4\times-\frac{3}{4}}z^{3\times-\frac{3}{4}} = y^{-3}z^{-\frac{9}{4}}$
c) $\left(\frac{b^9}{64c^3}\right)^{\frac{2}{3}} = \frac{b^{9\times\frac{2}{3}}}{\sqrt[3]{64}^2\, c^{3\times\frac{2}{3}}} = \frac{b^6}{16c^2}$
d) $\sqrt[4]{u^2} \times (2u)^{-2} = u^{\frac{2}{4}} \times 2^{-2} \times u^{-2}$
$$= \frac{1}{4}u^{\frac{1}{2}-2} = \frac{1}{4u^{\frac{3}{2}}}$$

3 a) 2^3 b) $2^{\frac{1}{2}}$
c) $8\sqrt{2} = 2^3 \times 2^{\frac{1}{2}} = 2^{3+\frac{1}{2}} = 2^{\frac{7}{2}}$
d) $\frac{1}{8\sqrt{2}} = \frac{1}{2^{\frac{7}{2}}} = 2^{-\frac{7}{2}}$

4 a) $3 \times \sqrt[3]{3} = 3 \times 3^{\frac{1}{3}} = 3^{1+\frac{1}{3}} = 3^{\frac{4}{3}}$
b) $16\sqrt[4]{4} = 4^2 \times 4^{\frac{1}{2}} = 4^{2+\frac{1}{2}} = 4^{\frac{5}{2}}$
c) $5 = \sqrt{25} = 25^{\frac{1}{2}}$
d) $2 = \sqrt[3]{8} = 8^{\frac{1}{3}}$
e) $\frac{\sqrt{10}}{1000} = \frac{10^{\frac{1}{2}}}{10^3} = 10^{\frac{1}{2}-3} = 10^{-\frac{5}{2}}$
f) $\frac{81}{\sqrt[3]{9}} = \frac{3^4}{3^{\frac{2}{3}}} = 3^{4-\frac{2}{3}} = 3^{\frac{10}{3}}$

5 $(3.92 \times 10^{-4}) + (3.77 \times 10^{-4}) + (4.09 \times 10^{-4})$
$= (3.92 + 3.77 + 4.09) \times 10^{-4}$
$= 11.78 \times 10^{-4} = 1.178 \times 10^{-3}$ m

6 $(3.45 \times 10^8) - (8.9 \times 10^7)$
$= (3.45 \times 10^8) - (0.89 \times 10^8)$
$= (3.45 - 0.89) \times 10^8 = \mathbf{\$2.56 \times 10^8}$

7 Mass of Sun $= (3.33 \times 10^5) \times$ Mass of Earth
$= (3.33 \times 10^5) \times (5.97 \times 10^{24})$
$= (3.33 \times 5.97) \times (10^5 \times 10^{24})$
$= (19.8801) \times 10^{29}$
$= 1.98801 \times 10^{30}$
$= \mathbf{1.99 \times 10^{30}}$ kg (3 s.f.)

8 a) $\sqrt{96} + 5\sqrt{18} = 4\sqrt{6} + 5 \times 3\sqrt{2}$
$= 4\sqrt{6} + 15\sqrt{2}$
b) $\frac{\sqrt{48}+\sqrt{363}}{5} = \frac{4\sqrt{3}+11\sqrt{3}}{5}$
$= \frac{15\sqrt{3}}{5} = 3\sqrt{3}$
c) $\sqrt{8} - \frac{\sqrt{36}}{2} = 2\sqrt{2} - \frac{6}{2} = 2\sqrt{2} - 3$
d) $-\sqrt{98} - \sqrt{2} \times \sqrt{162} = -7\sqrt{2} - \sqrt{324}$
$= -7\sqrt{2} - 18$
e) $12\sqrt{99} \div \sqrt{176} = 12 \times 3\sqrt{11} \div 4\sqrt{11} = 9$
f) $\frac{\sqrt{88}}{\sqrt{32}} + \sqrt{1100} = \frac{2\sqrt{22}}{4\sqrt{2}} + 10\sqrt{11}$
$= \frac{\sqrt{11}}{2} + 10\sqrt{11} = \frac{21\sqrt{11}}{2}$

9 a) $\frac{121\sqrt{7}}{-\sqrt{11}} = \frac{121\sqrt{7} \times \sqrt{11}}{-\sqrt{11} \times \sqrt{11}} = \frac{121\sqrt{77}}{-11}$
$= -11\sqrt{77}$
b) $\frac{60}{6+\sqrt{6}} = \frac{60(6-\sqrt{6})}{(6+\sqrt{6})(6-\sqrt{6})}$
$= \frac{60(6-\sqrt{6})}{36-6} = \frac{60(6-\sqrt{6})}{30}$
$= 2(6-\sqrt{6}) = 12 - 2\sqrt{6}$
c) $\frac{7+\sqrt{2}}{\sqrt{8}-3} = \frac{(7+\sqrt{2})(-\sqrt{8}-3)}{(\sqrt{8}-3)(-\sqrt{8}-3)}$
$= \frac{-7\sqrt{8}-21-\sqrt{16}-3\sqrt{2}}{-8+9}$
$= \frac{-7\times2\sqrt{2}-21-4-3\sqrt{2}}{1}$
$= -25 - 17\sqrt{2}$

Page 105   Exam-Style Questions
1 a) $2x^3 \times 4x^4 = 2 \times 4 \times x^3 \times x^4 = 8x^7$ [1 mark]
b) $(3y^2)^4 = 3^4 \times (y^2)^4 = 81y^8$
[2 marks available — 1 mark for 81, 1 mark for y^8]
c) $5z^0 = 5 \times 1 = 5$ [1 mark]
2 $\left(\frac{4}{9}\right)^{-\frac{3}{2}} = \left(\frac{9}{4}\right)^{\frac{3}{2}}$ [1 mark]
$= \left(\sqrt{\frac{9}{4}}\right)^3 = \frac{27}{8}$ [1 mark]
$= 3\frac{3}{8}$
[1 mark for final answer with full working and no errors]
Alternatively, you could have interpreted the negative index as $\frac{1}{\left(\frac{4}{9}\right)^{\frac{3}{2}}}$ and then dealt with the fractional index in the denominator.
3 Area $= (3 + 2\sqrt{5})(3 + 2\sqrt{5})$
$= 9 + 6\sqrt{5} + 6\sqrt{5} + 20$ [1 mark]
$= 29 + 12\sqrt{5}$ cm² [1 mark]
4 a) 2400 million = 2 400 000 000
$= 2.4 \times 10^9$ [1 mark]
b) $2.4 \times 10^9 \times 60$ [1 mark]
$= 2.4 \times 6 \times 10^9 \times 10$
$= 14.4 \times 10^{10} = 1.44 \times 10^{11}$ [1 mark]
c) $(1.8 \times 10^{11}) \div (2.4 \times 10^9)$
$= \frac{1.8}{2.4} \times \frac{10^{11}}{10^9}$ [1 mark]
$= 0.75 \times 10^2 = 75$ seconds [1 mark]
5 Length $= \frac{\text{Area}}{\text{Width}} = \frac{\sqrt{360}}{5-\sqrt{10}}$ [1 mark]
$= \frac{\sqrt{36}\times\sqrt{10}}{5-\sqrt{10}} = \frac{6\sqrt{10}}{5-\sqrt{10}}$
Rationalise the denominator:
$\frac{6\sqrt{10}(5+\sqrt{10})}{(5-\sqrt{10})(5+\sqrt{10})}$ [1 mark]
$= \frac{30\sqrt{10}+6\times10}{25-10} = \frac{30\sqrt{10}+60}{15}$ [1 mark]
$= 4 + 2\sqrt{10}$ cm [1 mark]

6 $\frac{\text{Age of Humanity}}{\text{Age of Earth}} = \frac{2.5\times10^5}{4.54\times10^9}$ [1 mark]
$= \frac{2.5}{4.54} \times \frac{10^5}{10^9}$
$= 0.5506... \times 10^{-4}$ [1 mark]
As a percentage, this is $0.5506... \times 10^{-4} \times 100$
$= 0.5506... \times 10^{-2}$
$= \mathbf{0.00551\%}$ (3 s.f.)
[2 marks available — 1 mark for multiplying by 100, 1 mark for the correct final answer]
7 $x^2 = y \times 10^z \times y \times 10^z = y^2 \times 10^{2z}$
[1 mark for 10^{2z}, 1 mark for fully correct expression]
This is not in standard form since $4 < y < 10$ and so $16 < y^2 < 100$. You need to divide y^2 by 10 and then multiply 10^{2z} by 10, giving
$x^2 = \frac{y^2}{10} \times 10^{2z+1}$ [1 mark]

Section 8 — Formulas
8.1 Writing Formulas
Page 106   Exercise 1
1 a) Barry owns twice as many films as Claudia, so he has $2 \times f = 2f$ films
b) In total there are $f + 2f = 3f$ films
c) $3f - 3 - 3 = 3f - 6$ films
2 The amount earned in h hours is $£8 \times h = 8h$. Adding the starting value of £18 gives
$M = 18 + 8h$
3 The time taken to cook n kg at 50 minutes per kg is $50n$. Adding the extra 25 minutes gives $t = 50n + 25$
4 The cost for t trees at £p per tree is $£p \times t = pt$. Adding the fixed value of £30 gives $C = pt + 30$
5 The first shape has 4 matchsticks, so there's a fixed value of 4. Then for every subsequent shape above $n = 1$ there's an extra 3 matchsticks, which means adding $3 \times (n - 1)$ to the original 4. Together this gives
$M = 3(n - 1) + 4 \Rightarrow M = 3n + 1$
You can check this formula works by generating a few terms of the sequence (4, 7, 10...) and putting $n = 1, 2, 3...$ into the formula.

8.2 Substituting into a Formula
Page 107   Exercise 1
1 a) $z = x + 2 = 4 + 2 = 6$
b) $z = y - 1 = 3 - 1 = 2$
c) $z = x + y = 4 + 3 = 7$
d) $z = 3y = 3(3) = 9$
e) $z = 3y - 2 = 3(3) - 2 = 9 - 2 = 7$
f) $z = 6x - y = 6(4) - 3 = 24 - 3 = 21$
2 a) $c = a - 4 = (-4) - 4 = -8$
b) $c = 4b = 4(-3) = -12$
c) $c = 6b - a = 6(-3) - (-4) = -18 + 4 = -14$
d) $c = b^3 = (-3)^3 = -27$
e) $c = -\frac{4b}{2a} = -\frac{4(-3)}{2(-4)} = -\frac{(-12)}{(-8)} = -1.5$
f) $c = 5a - b^2 = 5(-4) - (-3)^2 = (-20) - 9 = -29$
3 a) $q = 4r = 4\left(\frac{3}{4}\right) = 3$
b) $q = -2s = -2\left(-\frac{1}{3}\right) = \frac{2}{3}$
c) $q = rs = \left(\frac{3}{4}\right)\left(-\frac{1}{3}\right) = -\frac{3}{12} = -\frac{1}{4}$
d) $q = \frac{s}{r} = -\frac{1}{3} \div \frac{3}{4} = -\frac{1}{3} \times \frac{4}{3} = -\frac{4}{9}$
e) $q = r + s = \frac{3}{4} + \left(-\frac{1}{3}\right) = \frac{9}{12} - \frac{4}{12} = \frac{5}{12}$
f) $q = 4r + s = 4\left(\frac{3}{4}\right) + \left(-\frac{1}{3}\right) = 3 - \frac{1}{3} = 2\frac{2}{3}$
4 a) $v = u + at = 3 + (7 \times 5) = 3 + 35 = 38$
b) $v = u + at = 12 + (17 \times 15) = 12 + 255 = 267$

Using the same method for c)-f):
c) $v = 16.24$ d) $v = 48.32$ (2 d.p.)
e) $v = -53$ f) $v = -120.65$ (2 d.p.)

5 a) $w = x + 2y - 4z$
$= 12 + 2(2.5) - 4(-0.25) = 18$
b) $w = -3x + y^3 - (2z)^2$
$= -3(12) + 2.5^3 - (2 \times -0.25)^2$
$= -20.625 = -20.63$ (2 d.p.)
c) $w = 0.5x - yz = 0.5(12) - (2.5 \times -0.25)$
$= 6.625 = 6.63$ (2 d.p.)
d) $w = -2x^3 + y^2z$
$= -2(12)^3 + (2.5^2 \times -0.25)$
$= -3457.5625 = -3457.56$ (2 d.p.)
e) $w = -\dfrac{12}{x} + \dfrac{y}{z}$
$= -\dfrac{12}{12} + \dfrac{2.5}{-0.25} = -11$
f) $w = \dfrac{x^2 + 3y - 8z}{2y^2}$
$= \dfrac{12^2 + 3(2.5) - 8(-0.25)}{2(2.5)^2} = 12.28$

Page 108   Exercise 2
1 a) $s = \dfrac{d}{t} = \dfrac{800}{110} = 7.27$ m/s (2 d.p.)
b) $s = \dfrac{d}{t} = \dfrac{400}{14} = 28.57$ m/s (2 d.p.)
c) 1 km = 1000 m
1 minute = 60 seconds
$s = \dfrac{d}{t} = \dfrac{1000}{60} = 16.67$ m/s (2 d.p.)
Don't forget to convert the units here — you need the distance in metres and the time in seconds to get a speed in metres per second.
d) 640 km = 640 × 1000 = 640 000 m
1 hour = 60 minutes = 60 × 60 = 3600 secs
$s = \dfrac{d}{t} = \dfrac{640\,000}{3600} = 177.78$ m/s (2 d.p.)
2 a) $c = \dfrac{5}{9}(f - 32) = \dfrac{5}{9}(212 - 32) = 100$ °C
Using the same method for b)-d):
b) 20 °C   c) -40 °C   d) 37 °C
3 a) $n = 10$, so
$S = \dfrac{1}{2}n(n + 1) = \dfrac{1}{2} \times 10 \times (10 + 1) = 55$
Using the same method for b)-c):
b) $S = 5050$ c) $S = 500\,500$
4 a) $V = \pi r^2 h = \pi \times 3^2 \times 8 = 226.19$ cm³ (2 d.p.)
b) $V = \pi r^2 h = \pi \times 11^2 \times 5$
$= 1900.66$ cm³ (2 d.p.)

8.3 Rearranging Formulas
Page 110   Exercise 1
1 a) $y = x + 2 \Rightarrow x = y - 2$
b) $2z = 3r + x \Rightarrow x = 2z - 3r$
c) $y = 4x \Rightarrow x = \dfrac{y}{4}$
d) $k = 2(1 + 2x) \Rightarrow k = 2 + 4x \Rightarrow k - 2 = 4x$
$\Rightarrow x = \dfrac{k-2}{4}$
e) $v = \dfrac{2}{3}x - 2 \Rightarrow 3v = 2x - 6 \Rightarrow 3v + 6 = 2x$
$\Rightarrow x = \dfrac{3v+6}{2}$ or $x = \dfrac{3}{2}(v + 2)$
f) $y + 1 = \dfrac{x-1}{3} \Rightarrow 3y + 3 = x - 1$
$\Rightarrow x = 3y + 4$

2 a) $w = \dfrac{1}{1+y} \Rightarrow w(1 + y) = \dfrac{1}{1+y}(1 + y)$
$\Rightarrow w(1 + y) = 1$
b) $w(1 + y) = 1 \Rightarrow w + wy = 1$
$\Rightarrow wy = 1 - w \Rightarrow y = \dfrac{1-w}{w}$
You might have rearranged this differently to get $y = \dfrac{1}{w} - 1$.
3 a) $w = \dfrac{3}{2y} \Rightarrow 2wy = 3 \Rightarrow y = \dfrac{3}{2w}$
b) $z + 2 = \dfrac{2}{1-y} \Rightarrow (z + 2)(1 - y) = 2$
$\Rightarrow 1 - y = \dfrac{2}{z+2} \Rightarrow y = 1 - \dfrac{2}{z+2}$

c) $uv = \dfrac{1}{1-2y} \Rightarrow uv(1 - 2y) = 1$
$\Rightarrow 1 - 2y = \dfrac{1}{uv} \Rightarrow 2y = 1 - \dfrac{1}{uv}$
$\Rightarrow y = \dfrac{1}{2} - \dfrac{1}{2uv}$
d) $a + b = \dfrac{2}{4-3y} \Rightarrow (a + b)(4 - 3y) = 2$
$\Rightarrow 4 - 3y = \dfrac{2}{a+b} \Rightarrow 3y = 4 - \dfrac{2}{a+b}$
$\Rightarrow y = \dfrac{4}{3} - \dfrac{2}{3(a+b)}$
You might have ended up with slightly different answers for parts b)-d) if you rearranged differently.

4 a) $2k = 12 - \sqrt{w-2} \Rightarrow 2k + \sqrt{w-2} = 12$
$\Rightarrow \sqrt{w-2} = 12 - 2k$
b) $\sqrt{w-2} = 12 - 2k \Rightarrow w - 2 = (12 - 2k)^2$
$\Rightarrow w - 2 = 144 - 48k + 4k^2$
$\Rightarrow w = 146 - 48k + 4k^2$
5 a) $a = \sqrt{w} \Rightarrow w = a^2$
b) $x = 1 + \sqrt{w} \Rightarrow x - 1 = \sqrt{w}$
$w = (x - 1)^2$ or $w = x^2 - 2x + 1$
Using the same method for c)-f):
c) $w = y^2 + 2$ d) $w = \left(\dfrac{f-3}{2}\right)^2$
e) $w = \dfrac{j^2-3}{4}$ f) $w = \dfrac{1-a^2}{2}$
6 a) $t = 1 - 3(z + 1)^2 \Rightarrow t + 3(z + 1)^2 = 1$
$\Rightarrow 3(z + 1)^2 = 1 - t \Rightarrow (z + 1)^2 = \dfrac{1-t}{3}$
b) $(z + 1)^2 = \dfrac{1-t}{3} \Rightarrow z + 1 = \pm\sqrt{\dfrac{1-t}{3}}$
$\Rightarrow z = \pm\sqrt{\dfrac{1-t}{3}} - 1$
7 a) $x = 1 + z^2 \Rightarrow x - 1 = z^2 \Rightarrow z = \pm\sqrt{x-1}$
b) $2t = 3 - z^2 \Rightarrow 2t + z^2 = 3 \Rightarrow z^2 = 3 - 2t$
$\Rightarrow z = \pm\sqrt{3 - 2t}$
Using the same method for c)-f):
c) $z = \pm\sqrt{\dfrac{1-xy}{2}}$ d) $z = \pm\sqrt{\dfrac{t+2}{3}} + 2$
e) $z = \dfrac{\pm\sqrt{4-g} - 3}{2}$ f) $z = \dfrac{5 - \left(\pm\sqrt{\dfrac{4-r}{2}}\right)}{3}$
8 a) $x(a + b) = a - 1 \Rightarrow ax + bx = a - 1$
$\Rightarrow bx + 1 = a - ax \Rightarrow bx + 1 = a(1 - x)$
$\Rightarrow a = \dfrac{bx+1}{1-x}$
b) $x - ab = c - ad \Rightarrow x - c = ab - ad$
$\Rightarrow x - c = a(b - d) \Rightarrow a = \dfrac{x-c}{b-d}$
If you'd rearranged this a bit differently, you'd end up with $a = \dfrac{c-x}{d-b}$
Using the same method for c)-d):
c) $a = \dfrac{c-1}{1+2c}$ d) $a = \dfrac{2}{2e-3}$

Page 111   Review Exercise
1 a) Melanie has half as many as Jane which can be written as $\dfrac{1}{2}j$, or $\dfrac{j}{2}$ so $m = \dfrac{j}{2}$
b) $m = \dfrac{j}{2} = \dfrac{24}{2} = 12$
2 a) n people costs $1.25n$. Adding the fixed charge of £30 gives $C = 30 + 1.25n$
b) $C = 30 + 1.25n = 30 + 1.25(32) = £70$
3 a) h half-hours at £1.70 per half hour costs $1.7h$. Adding the fixed charge of £5 gives $C = 5 + 1.7h$
b) 2.5 hours ÷ 0.5 = 5 half-hours $\Rightarrow h = 5$
$C = 5 + 1.7h = 5 + 1.7(5) = £13.50$
c) $C = 5 + 1.7h \Rightarrow 1.7h = C - 5$
$\Rightarrow h = \dfrac{C-5}{1.7}$
d) $h = \dfrac{C-5}{1.7} = \dfrac{15.2-5}{1.7} = 6$ half-hours
$= 3$ hours
4 a) $A = 21.5d^2 \Rightarrow \dfrac{A}{21.5} = d^2$
$\Rightarrow d = \pm\sqrt{\dfrac{A}{21.5}}$

b) $d = \sqrt{\dfrac{A}{21.5}} = \sqrt{\dfrac{55}{21.5}} = 1.599...$
$= 1.6$ cm (2 s.f.)
Ignore the negative square root as length has to be positive.
5 a) $T = 35w + 25 = 35(1.5) + 25$
$= 77.5$ minutes or 77 mins 30 sec
b) $T = 35w + 25 \Rightarrow T - 25 = 35w$
$\Rightarrow w = \dfrac{T-25}{35}$
c) $w = \dfrac{T-25}{35} = \dfrac{207-25}{35} = 5.2$ kg
6 a) (i) $-2 + y = \dfrac{3}{4-x}$
$\Rightarrow (4 - x)(-2 + y) = 3$
$\Rightarrow 4 - x = \dfrac{3}{-2+y}$
$\Rightarrow x = 4 - \dfrac{3}{-2+y}$
An alternative answer is $x = \dfrac{11-4y}{2-y}$ *but there are others too.*
(ii) $x = 4 - \dfrac{3}{-2+y} = 4 - \dfrac{3}{-2+(-1)}$
$= 4 - \dfrac{3}{-3} = 4 - (-1) = 5$
b) (i) $y = \dfrac{1}{\sqrt{1-x}} \Rightarrow y\sqrt{1-x} = 1$
$\Rightarrow \sqrt{1-x} = \dfrac{1}{y} \Rightarrow 1 - x = \dfrac{1}{y^2}$
$\Rightarrow x = 1 - \dfrac{1}{y^2}$
(ii) $x = 1 - \dfrac{1}{y^2} = 1 - \dfrac{1}{(-1)^2} = 0$
c) (i) $2(1 - x) = y(3 + x) \Rightarrow 2 - 2x = 3y + xy$
$\Rightarrow 2 - 3y = xy + 2x$
$\Rightarrow 2 - 3y = x(y + 2)$
$\Rightarrow x = \dfrac{2-3y}{y+2}$
(ii) $x = \dfrac{2-3y}{y+2} = \dfrac{2-3(-1)}{(-1)+2} = 5$
d) (i) $y = \dfrac{2-3x}{1+2x} \Rightarrow y(1 + 2x) = 2 - 3x$
$\Rightarrow y + 2xy = 2 - 3x \Rightarrow 2xy + 3x = 2 - y$
$\Rightarrow x(2y + 3) = 2 - y \Rightarrow x = \dfrac{2-y}{2y+3}$
(ii) $x = \dfrac{2-y}{2y+3} = \dfrac{2-(-1)}{2(-1)+3} = 3$
e) (i) $y = 8 - \dfrac{1}{\sqrt{x}} \Rightarrow \dfrac{1}{\sqrt{x}} = 8 - y$
$\Rightarrow \dfrac{1}{x} = (8 - y)^2 \Rightarrow x = \dfrac{1}{(8-y)^2}$
(ii) $x = \dfrac{1}{(8-y)^2} = \dfrac{1}{(8-(-1))^2} = \dfrac{1}{9^2} = \dfrac{1}{81}$
f) (i) $2y - 1 = 3\sqrt{2-x} \Rightarrow \dfrac{2y-1}{3} = \sqrt{2-x}$
$\Rightarrow \left(\dfrac{2y-1}{3}\right)^2 = 2 - x$
$\Rightarrow x = 2 - \left(\dfrac{2y-1}{3}\right)^2$
(ii) $x = 2 - \left(\dfrac{2(-1)-1}{3}\right)^2 = 2 - \left(\dfrac{-3}{3}\right)^2 = 1$
7 a) $s = \left(\dfrac{u+v}{2}\right)t = \left(\dfrac{2.3+1.7}{2}\right)4 = \left(\dfrac{4}{2}\right)4 = 8$
b) $s = \left(\dfrac{u+v}{2}\right)t \Rightarrow 2s = (u + v)t \Rightarrow t = \dfrac{2s}{u+v}$
$t = \dfrac{2(3.3)}{1+2} = 2.2$
c) $s = \left(\dfrac{u+v}{2}\right)t \Rightarrow \dfrac{s}{t} = \dfrac{u+v}{2} \Rightarrow \dfrac{2s}{t} = u + v$
$\Rightarrow u = \dfrac{2s}{t} - v \Rightarrow u = \dfrac{2(4.5)}{6} - 7 = -5.5$
d) $s = \left(\dfrac{u+v}{2}\right)t \Rightarrow \dfrac{s}{t} = \dfrac{u+v}{2} \Rightarrow \dfrac{2s}{t} = u + v$
$\Rightarrow v = \dfrac{2s}{t} - u \Rightarrow v = \dfrac{2(0.5)}{0.25} - 3 = 1$
8 a) $x = \dfrac{1+\sqrt{y+3}}{2-z} = \dfrac{1+\sqrt{1+3}}{2-(-1)}$
$x = \dfrac{1+\sqrt{4}}{3} = \dfrac{1+2}{3} = \dfrac{3}{3} = 1$

b) $x = \dfrac{1+\sqrt{y+3}}{2-z} \Rightarrow x(2-z) = 1 + \sqrt{y+3}$

$2x - xz = 1 + \sqrt{y+3}$

$\Rightarrow xz = 2x - (1 + \sqrt{y+3})$

$\Rightarrow z = \dfrac{2x - (1 + \sqrt{y+3})}{x}$

$\Rightarrow z = \dfrac{2(-2) - (1 + \sqrt{6+3})}{-2} = 4$

Page 112   Exam-Style Questions

1 $C = \dfrac{5}{9}(F - 32) = \dfrac{5}{9}(104 - 32) = 5 \times \dfrac{1}{9} \times 72$

$= 5 \times 8 = $ **40 °C**

[2 marks available — 1 mark for correct substitution, 1 mark for correct answer]

2 a) $m = 5h + 1 = 5(6) + 1 = $ **31** *[1 mark]*

   b) (i) $m = 5h + 1 \Rightarrow m - 1 = 5h$

$\Rightarrow h = \dfrac{m-1}{5}$

[2 marks for the correct final answer, otherwise 1 mark for at least one correct step in the rearrangement]

     (ii) $h = \dfrac{m-1}{5} = \dfrac{36-1}{5} = \dfrac{35}{5}$

$\Rightarrow h = $ **7 hexagons** *[1 mark]*

If you'd got the expression wrong in part (i), you'd still get the mark if you substituted $m = 36$ correctly into your expression.

3 $V = \sqrt{\dfrac{2GM}{r}} \Rightarrow V^2 = \dfrac{2GM}{r} \Rightarrow rV^2 = 2GM$

$\Rightarrow \dfrac{rV^2}{2G} = M \Rightarrow M = \dfrac{rV^2}{2G}$

[3 marks available — 1 mark for squaring both sides, 1 mark for multiplying both sides by r, 1 mark for correct answer]

4 a) $g = \dfrac{8}{5}h + 17 \Rightarrow g - 17 = \dfrac{8}{5}h$

$\Rightarrow h = \dfrac{5}{8}(g - 17)$

[2 marks for the correct final answer, otherwise 1 mark for at least one correct step in the rearrangement]

   b) (i) $h = \dfrac{5}{8}(g - 17) \Rightarrow h = \dfrac{5}{8}(209 - 17)$

$\Rightarrow h = $ **120** *[1 mark]*

     (ii) $h = \dfrac{5}{8}(g - 17) \Rightarrow h = \dfrac{5}{8}(-15 - 17)$

$\Rightarrow h = $ **−20** *[1 mark]*

5 a) The cost in £ of n units at 8p per unit is $0.08n$. Adding the fixed charge of £7.50 gives $C = $ **0.08n + 7.5**

[2 marks available — 1 mark for 0.08n, 1 mark for +7.5]

   b) $C = 0.08n + 7.5 = 0.08(760) + 7.5 = 68.3$

$= $ **£68.30** *[1 mark]*

   c) $C = 0.08n + 7.5 \Rightarrow C - 7.5 = 0.08n$

$\Rightarrow n = \dfrac{C - 7.5}{0.08}$

[2 marks for the correct final answer, otherwise 1 mark for at least one correct step in the rearrangement]

   d) $n = \dfrac{C - 7.5}{0.08} \Rightarrow n = \dfrac{39.5 - 7.5}{0.08}$

$\Rightarrow n = $ **400 units** *[1 mark]*

For parts b) and d) you'd still get the marks if you substituted the values correctly into whatever answers you gave for a) and c).

Section 9 — Equations

9.1 Solving Equations

Page 113   Exercise 1

1 a) $x + 9 = 12 \Rightarrow x = 3$

   b) $x - 7.3 = 1.6 \Rightarrow x = 8.9$

   c) $12 - x = 9 \Rightarrow 12 = 9 + x \Rightarrow x = 3$

   d) $9x = 54 \Rightarrow x = 6$

   e) $-5x = 50 \Rightarrow x = -10$

   f) $40x = -32 \Rightarrow x = -0.8$

---

2 a) $\dfrac{x}{3} = 2 \Rightarrow x = 6$

   b) $\dfrac{x}{2} = 3.2 \Rightarrow x = 6.4$

   c) $-\dfrac{x}{0.2} = 3.2 \Rightarrow x = -0.64$

   d) $\dfrac{2x}{5} = 6 \Rightarrow 2x = 30 \Rightarrow x = 15$

3 a) $8x + 10 = 66 \Rightarrow 8x = 56 \Rightarrow x = 7$

   b) $1.8x - 8 = -62 \Rightarrow 1.8x = -54 \Rightarrow x = -30$

   c) $8 - 7x = 22 \Rightarrow 8 = 22 + 7x \Rightarrow -14 = 7x$

$\Rightarrow x = -2$

   d) $\dfrac{x}{2} - 1 = 2 \Rightarrow \dfrac{x}{2} = 3 \Rightarrow x = 6$

   e) $15x + 12 = 72 \Rightarrow 15x = 60 \Rightarrow x = 4$

   f) $1.5x - 3 = -24 \Rightarrow 1.5x = -21 \Rightarrow x = -14$

   g) $17 - 10x = 107 \Rightarrow 17 = 107 + 10x$

$\Rightarrow -90 = 10x \Rightarrow x = -9$

   h) $-\dfrac{2x}{3} - \dfrac{3}{4} = \dfrac{1}{4} \Rightarrow -\dfrac{2x}{3} = 1 \Rightarrow 2x = -3$

$\Rightarrow x = -\dfrac{3}{2}$

   i) $-\dfrac{3x}{5} + \dfrac{1}{3} = \dfrac{2}{3} \Rightarrow -\dfrac{3x}{5} = \dfrac{1}{3} \Rightarrow 3x = -\dfrac{5}{3}$

$\Rightarrow x = -\dfrac{5}{9}$

Page 114   Exercise 2

1 a) $7(x + 4) = 63 \Rightarrow 7x + 28 = 63 \Rightarrow 7x = 35$

$\Rightarrow x = 5$

   b) $13(x - 4) = -91 \Rightarrow 13x - 52 = -91$

$\Rightarrow 13x = -39 \Rightarrow x = -3$

   c) $18(x - 3) = -180 \Rightarrow 18x - 54 = -180$

$\Rightarrow 18x = -126 \Rightarrow x = -7$

   d) $2.5(x + 4) = 30 \Rightarrow 2.5x + 10 = 30$

$\Rightarrow 2.5x = 20 \Rightarrow x = 8$

   e) $3.5(x + 6) = 63 \Rightarrow 3.5x + 21 = 63$

$\Rightarrow 3.5x = 42 \Rightarrow x = 12$

   f) $4.5(x + 3) = 72 \Rightarrow 4.5x + 13.5 = 72$

$\Rightarrow 4.5x = 58.5 \Rightarrow x = 13$

   g) $315 = 21(6 - x) \Rightarrow 315 = 126 - 21x$

$\Rightarrow 315 + 21x = 126 \Rightarrow 21x = -189$

$\Rightarrow x = -9$

   h) $171 = 4.5(8 - x) \Rightarrow 171 = 36 - 4.5x$

$\Rightarrow 171 + 4.5x = 36 \Rightarrow 4.5x = -135$

$\Rightarrow x = -30$

2 $\dfrac{1}{x-2} = 3 \Rightarrow \dfrac{x-2}{x-2} = 3(x-2) \Rightarrow 1 = 3(x-2)$

$\Rightarrow 1 = 3x - 6 \Rightarrow 7 = 3x \Rightarrow x = \dfrac{7}{3}$

3 a) $\dfrac{1}{x} = 2 \Rightarrow 1 = 2x \Rightarrow x = 0.5$

   b) $\dfrac{2}{x} = 5 \Rightarrow 2 = 5x \Rightarrow x = 0.4$

   c) $\dfrac{12}{x-2} = 4 \Rightarrow 12 = 4(x-2) \Rightarrow 12 = 4x - 8$

$\Rightarrow 20 = 4x \Rightarrow x = 5$

   d) $\dfrac{3}{1-2x} = 2 \Rightarrow 3 = 2(1 - 2x) \Rightarrow 3 = 2 - 4x$

$\Rightarrow 3 + 4x = 2 \Rightarrow 4x = -1 \Rightarrow x = -0.25$

Page 114   Exercise 3

1 a) $6x - 4 = 2x + 16 \Rightarrow 4x = 20 \Rightarrow x = 5$

   b) $17x - 2 = 7x + 8 \Rightarrow 10x = 10 \Rightarrow x = 1$

   c) $6x - 12 = 51 - 3x \Rightarrow 9x = 63 \Rightarrow x = 7$

   d) $5x - 13 = 87 - 5x \Rightarrow 10x = 100$

$\Rightarrow x = 10$

   e) $10x - 18 = 11.4 - 4x \Rightarrow 14x = 29.4$

$\Rightarrow x = 2.1$

   f) $4x + 9 = 6 - x \Rightarrow 5x = -3 \Rightarrow x = -0.6$

2 a) $3(x + 2) = x + 14 \Rightarrow 3x + 6 = x + 14$

$\Rightarrow 2x = 8 \Rightarrow x = 4$

   b) $5(x + 3) = 2x + 57 \Rightarrow 5x + 15 = 2x + 57$

$\Rightarrow 3x = 42 \Rightarrow x = 14$

   c) $7(x - 7) = 2(x - 2) \Rightarrow 7x - 49 = 2x - 4$

$\Rightarrow 5x = 45 \Rightarrow x = 9$

   d) $5(x - 4) = 3(x + 8) \Rightarrow 5x - 20 = 3x + 24$

$\Rightarrow 2x = 44 \Rightarrow x = 22$

   e) $4(x - 3) = 3(x - 8) \Rightarrow 4x - 12 = 3x - 24$

$\Rightarrow x = -12$

   f) $11(x - 2) = 3(x + 6) \Rightarrow 11x - 22 = 3x + 18$

$\Rightarrow 8x = 40 \Rightarrow x = 5$

---

3 a) $7\left(2x + \dfrac{1}{7}\right) = 8\left(3x - \dfrac{1}{2}\right)$

$\Rightarrow 14x + 1 = 24x - 4 \Rightarrow 5 = 10x$

$\Rightarrow x = \dfrac{1}{2}$

   b) $7(x - 1) = 4(6.2 - 2x) \Rightarrow 7x - 7 = 24.8 - 8x$

$\Rightarrow 15x = 31.8 \Rightarrow x = 2.12$

   c) $-3(x - 3) = 8(0.7 - x)$

$\Rightarrow -3x + 9 = 5.6 - 8x \Rightarrow 5x = -3.4$

$\Rightarrow x = -0.68$

Page 115   Exercise 4

1 a) $\dfrac{x}{4} = 1 - x \Rightarrow x = 4(1 - x) \Rightarrow x = 4 - 4x$

$\Rightarrow 5x = 4 \Rightarrow x = 0.8$

   b) $\dfrac{x}{3} = 8 - x \Rightarrow x = 3(8 - x) \Rightarrow x = 24 - 3x$

$\Rightarrow 4x = 24 \Rightarrow x = 6$

   c) $\dfrac{x}{5} = 11 - 2x \Rightarrow x = 5(11 - 2x)$

$\Rightarrow x = 55 - 10x \Rightarrow 11x = 55 \Rightarrow x = 5$

   d) $\dfrac{x}{3} = 2(x - 5) \Rightarrow x = 6(x - 5)$

$\Rightarrow x = 6x - 30 \Rightarrow x + 30 = 6x$

$\Rightarrow 30 = 5x \Rightarrow x = 6$

   e) $\dfrac{x}{2} = 4(x - 7) \Rightarrow x = 8(x - 7)$

$\Rightarrow x = 8x - 56 \Rightarrow x + 56 = 8x$

$\Rightarrow 56 = 7x \Rightarrow x = 8$

   f) $\dfrac{x}{5} = 2(x + 9) \Rightarrow x = 10(x + 9)$

$\Rightarrow x = 10x + 90 \Rightarrow x - 90 = 10x$

$\Rightarrow 9x = -90 \Rightarrow x = -10$

2 a) $\dfrac{x+4}{2} = \dfrac{x+10}{3} \Rightarrow 3(x + 4) = 2(x + 10)$

$\Rightarrow 3x + 12 = 2x + 20 \Rightarrow x = 8$

   b) $\dfrac{x+2}{2} = \dfrac{x+4}{6} \Rightarrow 6(x + 2) = 2(x + 4)$

$\Rightarrow 6x + 12 = 2x + 8 \Rightarrow 4x = -4$

$\Rightarrow x = -1$

   c) $\dfrac{x-2}{3} = \dfrac{x+4}{5} \Rightarrow 5(x - 2) = 3(x + 4)$

$\Rightarrow 5x - 10 = 3x + 12 \Rightarrow 2x = 22$

$\Rightarrow x = 11$

   d) $\dfrac{x-6}{5} = \dfrac{x+3}{8} \Rightarrow 8(x - 6) = 5(x + 3)$

$\Rightarrow 8x - 48 = 5x + 15 \Rightarrow 3x = 63$

$\Rightarrow x = 21$

   e) $\dfrac{x-2}{4} = \dfrac{15-2x}{3} \Rightarrow 3(x - 2) = 4(15 - 2x)$

$\Rightarrow 3x - 6 = 60 - 8x \Rightarrow 11x = 66 \Rightarrow x = 6$

   f) $\dfrac{x-4}{6} = \dfrac{12-3x}{2} \Rightarrow 2(x - 4) = 6(12 - 3x)$

$\Rightarrow 2x - 8 = 72 - 18x \Rightarrow 20x = 80$

$\Rightarrow x = 4$

Page 115   Exercise 5

1 a) $x^2 = 16 \Rightarrow x = \pm 4$

   b) $x^2 + 10 = 35 \Rightarrow x^2 = 25 \Rightarrow x = \pm 5$

   c) $3x^2 = 27 \Rightarrow x^2 = 9 \Rightarrow x = \pm 3$

   d) $2x^2 + 1 = 99 \Rightarrow 2x^2 = 98 \Rightarrow x^2 = 49$

$\Rightarrow x = \pm 7$

   e) $\dfrac{3x^2}{10} = 1.2 \Rightarrow 3x^2 = 12 \Rightarrow x^2 = 4$

$\Rightarrow x = \pm 2$

   f) $\dfrac{x^2+2}{x^2-4} = \dfrac{11}{10} \Rightarrow 10(x^2 + 2) = 11(x^2 - 4)$

$\Rightarrow 10x^2 + 20 = 11x^2 - 44 \Rightarrow 64 = x^2$

$\Rightarrow x = \pm 8$

2 a) $2x^2 = 4 \Rightarrow x^2 = 2 \Rightarrow x = \pm\sqrt{2}$

   b) $x^2 + 7 = 13 \Rightarrow x^2 = 6 \Rightarrow x = \pm\sqrt{6}$

   c) $3x^2 + 1 = 40 \Rightarrow 3x^2 = 39 \Rightarrow x^2 = 13$

$\Rightarrow x = \pm\sqrt{13}$

9.2 Forming Equations from Word Problems

Page 116   Exercise 1

1 a) **2x + 3 = 19**

   b) $2x + 3 = 19 \Rightarrow 2x = 16 \Rightarrow x = 8$

2 $\dfrac{x}{3} - 11 = -2 \Rightarrow \dfrac{x}{3} = 9 \Rightarrow x = 27$

3 $x + (x + 1) + (x + 2) + (x + 3) = 42$

$\Rightarrow 4x + 6 = 42 \Rightarrow 4x = 36 \Rightarrow x = 9$

So the numbers are **9**, **10**, **11** and **12**.

4 If Anna has x stickers, then Bill has $(x + 3)$ stickers and Christie has $2x$ stickers.
$x + (x + 3) + 2x = 83 \Rightarrow 4x + 3 = 83$
$\Rightarrow 4x = 80 \Rightarrow x = 20$
So Anna has **20** stickers, Bill has **23** stickers and Christie has **40** stickers.

5 If Eduardo has raised £x, then Deb has raised £$(x - 6)$ and Fiz has raised £$3x$.
$(x - 6) + x + 3x = 106.5 \Rightarrow 5x - 6 = 106.5$
$\Rightarrow 5x = 112.5 \Rightarrow x = 22.5$
So Deb has raised **£16.50**, Eduardo has raised **£22.50** and Fiz has raised **£67.50**.

6 If Stacey is x years old, then Macy is $(x - 3)$ years old and Tracy is $2x$ years old.
$x + (x - 3) + 2x = 41 \Rightarrow 4x - 3 = 41$
$\Rightarrow 4x = 44 \Rightarrow x = 11$
So Stacey is **11** years old, Macy is **8** years old and Tracy is **22** years old.

Page 117   Exercise 2

1 a) $3x + 5x + 20° = 180°$
$\Rightarrow 8x = 160° \Rightarrow x = 20°$
b) $3x + 4x + 110° = 180°$
$\Rightarrow 7x = 70° \Rightarrow x = 10°$
c) $20x + 7x + 23x + 10x = 360°$
$\Rightarrow 60x = 360° \Rightarrow x = 6°$
d) $6x + 10x + (3x - 10°) = 180°$
$\Rightarrow 19x = 190° \Rightarrow x = 10°$
e) $3x + 5x + (2x + 40°) = 180°$
$\Rightarrow 10x = 140° \Rightarrow x = 14°$
f) $72° + (3x + 6°) + (4x - 8°) + (5x - 10°) = 360°$
$\Rightarrow 12x = 300° \Rightarrow x = 25°$

2 $x + 2x + (70° - x) = 180°$
$\Rightarrow 2x = 110° \Rightarrow x = 55°$

3 a) (i) $4x + 4x + (x + 8) + (x + 8) = 146$
$\Rightarrow 10x + 16 = 146 \Rightarrow 10x = 130$
$\Rightarrow x = 13$
(ii) $4x \times (x + 8) = 52 \times 21 = \textbf{1092 cm}^2$
b) (i) $(x + 3) + (x + 3) + (x + 10) + (x + 10) = 186$
$\Rightarrow 4x + 26 = 186 \Rightarrow 4x = 160$
$\Rightarrow x = 40$
(ii) $(x + 3) \times (x + 10) = 43 \times 50 = \textbf{2150 cm}^2$

4 $(4 - x) + (4 - x) + (3x - 2) + (3x - 2) = 8.8$
$\Rightarrow 4x + 4 = 8.8 \Rightarrow 4x = 4.8 \Rightarrow x = 1.2$
Area $= (4 - x) \times (3x - 2) = (4 - 1.2) \times (3.6 - 2)$
$= 2.8 \times 1.6 = \textbf{4.48 cm}^2$

5 $3 \times (x + 5) = \frac{1}{2} \times 2 \times (x + 1) \times 5$
$\Rightarrow 3(x + 5) = 5(x + 1) \Rightarrow 3x + 15 = 5x + 5$
$\Rightarrow 2x = 10 \Rightarrow x = 5$
Perimeter of the rectangle
$= 3 + 3 + (x + 5) + (x + 5)$
$= 3 + 3 + 10 + 10 = \textbf{26 cm}$
Perimeter of the triangle $= 5 + 13 + 2(x + 1)$
$= 5 + 13 + 12 = \textbf{30 cm}$

9.3 Identities

Page 118   Exercise 1

1 a) $x - 1 = 0$ only when $x = 1$, so **no**, the symbol '$\equiv$' cannot be used.
b) $x^2 - 3 \equiv -(3 - x^2)$, so **no**, the symbol '$\equiv$' cannot be used.
c) $3(x + 2) - x \equiv 3x + 6 - x \equiv 2x + 6$
$\equiv 2(x + 3)$,
so **yes**, the symbol '$\equiv$' can be used.
d) Expanding $(x + 1)^2$ gives $x^2 + 2x + 1$, so **yes**, the symbol '$\equiv$' can be used.
e) $4(2 - x) \equiv 8 - 4x \equiv 2(4 - 2x)$, so **yes**, the symbol '$\equiv$' can be used.
f) $2(x^2 - 2x) \equiv 2x^2 - 4x$, so **no**, the symbol '$\equiv$' cannot be used.

2 a) $2(x + 5) \equiv 2x + 1 + a$
$\Rightarrow 2x + 10 \equiv 2x + 1 + a$
$\Rightarrow 2x + 1 + 9 \equiv 2x + 1 + a$
So $a = 9$.

b) $ax + 3 \equiv 5x + 2 - (x - 1)$
$\Rightarrow ax + 3 \equiv 4x + 3$
So $a = 4$.
c) $(x + 4)(x - 1) \equiv x^2 + ax - 4$
$\Rightarrow x^2 - x + 4x - 4 \equiv x^2 + ax - 4$
$\Rightarrow x^2 + 3x - 4 \equiv x^2 + ax - 4$
So $a = 3$.
d) $(x + 2)^2 \equiv x^2 + 4x + a$
$\Rightarrow x^2 + 2x + 2x + 4 \equiv x^2 + 4x + a$
$\Rightarrow x^2 + 4x + 4 \equiv x^2 + 4x + a$
So $a = 4$.
e) $4 - x^2 \equiv (a + x)(a - x)$
$\Rightarrow 4 - x^2 \equiv a^2 - ax + ax - x^2$
$\Rightarrow 4 - x^2 \equiv a^2 - x^2$
So $a^2 = 4 \Rightarrow a = \pm 2$.
f) $(2x - 1)(3 - x) \equiv ax^2 + 7x - 3$
$\Rightarrow 6x - 2x^2 - 3 + x \equiv ax^2 + 7x - 3$
$\Rightarrow -2x^2 + 7x - 3 \equiv ax^2 + 7x - 3$
So $a = -2$.

3 a) Rearrange the left-hand side until it matches the right-hand side:
$(x + 5)^2 + 3(x - 1)^2$
$\equiv x^2 + 5x + 5x + 25 + 3x^2 - 3x - 3x + 3$
$\equiv 4x^2 + 4x + 28$
$\equiv 4(x^2 + x + 7)$
b) $3(x + 2)^2 - (x - 4)^2$
$\equiv 3x^2 + 6x + 6x + 12 - x^2 + 4x + 4x - 16$
$\equiv 2x^2 + 20x - 4$
$\equiv 2(x^2 + 10x - 2)$

9.4 Proof

Page 119   Exercise 1

1 Take three consecutive integers —
x, $(x + 1)$ and $(x + 2)$. Then their sum is:
$x + (x + 1) + (x + 2) = 3x + 3 = 3(x + 1)$
$= 3n$ where n is an integer.
So the sum of three consecutive integers is a **multiple of 3**.

2 $8^{12} - 12^7 = 8 \times 8^{11} - 12 \times 12^6$
$= 4 \times 2 \times 8^{11} - 4 \times 3 \times 12^6$
$= 4(2 \times 8^{11} - 3 \times 12^6)$
$= 4n$ where n is an integer.
So $8^{12} - 12^7$ is a **multiple of 4**.

3 If a is odd, then $a = (2n + 1)$ where n is an integer. So:
$2x + a = 7(x - 2a) - 5$
$\Rightarrow 2x + (2n + 1) = 7x - 14(2n + 1) - 5$
$\Rightarrow 2x + 2n + 1 = 7x - 28n - 14 - 5$
$\Rightarrow 30x = 5x - 20 \Rightarrow 5x = 30n + 20$
$\Rightarrow x = 6n + 4 = 2(3n + 2)$
So if a is odd, then x is **even**.

4 Take two consecutive square numbers — x^2 and $(x + 1)^2$. Then their sum is:
$x^2 + (x + 1)^2 = x^2 + x^2 + 2x + 1$
$= 2x^2 + 2x + 1 = 2(x^2 + x) + 1$
$= 2n + 1$ where n is an integer.
So the sum of two consecutive square numbers is **odd**.

5 Let the smallest value in the data set be x. Then the largest value is $(x + 8)$, since the range of the data set is 8.
If each number in the set is multiplied by 3, then the smallest value will become $3x$ and the largest value will become $3(x + 8) = 3x + 24$. So, the new range will be $(3x + 24) - 3x = \textbf{24}$.

6 The product of the two integers either side of n is: $(n + 1)(n - 1) = n^2 - n + n - 1$
$= n^2 - 1$
$n^2 > n^2 - 1$ for any integer n.

7 Take two consecutive triangle numbers — $\frac{1}{2}n(n + 1)$ and $\frac{1}{2}[n + 1]([n + 1] + 1) = \frac{1}{2}(n + 1)(n + 2)$.

Then their sum is:
$\frac{1}{2}n(n + 1) + \frac{1}{2}(n + 1)(n + 2)$
$= \frac{1}{2}(n^2 + n + n^2 + 3n + 2)$
$= \frac{1}{2}(2n^2 + 4n + 2) = n^2 + 2n + 1$
$= (n + 1)^2$
So the sum of two consecutive triangle numbers is a **square number**.

8 Let the five values in the data set be a, b, c, d and e. Then since the mean is 12:
$\frac{1}{5}(a + b + c + d + e) = 12$
$\Rightarrow a + b + c + d + e = 60$
If each value in the data set is increased by 1, then the new mean will be:
$\frac{1}{5}((a + 1) + (b + 1) + (c + 1) + (d + 1) + (e + 1))$
$= \frac{1}{5}(a + b + c + d + e + 5)$
$= \frac{1}{5}(60 + 5) = \frac{65}{5} = 13$

Page 120   Exercise 2

1 E.g. $(-3) + (-2) + (-1) = -6$, which is **less than** each individual number.

2 E.g. 2 and 3 are both prime, but $3 - 2 = 1$ which is **odd**.

3 E.g. $3^2 + 4^2 = 9 + 16 = 25 = 5^2$

4 E.g. If $x = 0.5$, then $x^2 = 0.25 < x$.

5 E.g. For the data set $\{1, 1, 2, 2\}$, the median and mode are both 1, but the mean is $\frac{7}{5} = 1.4$.

6 E.g. If you draw three points **in a row**, joining them up forms a straight line rather than a triangle.

9.5 Iterative Methods

Page 122   Exercise 1

1 a) $x = 2 \Rightarrow x^2 + x - 10 = 4 + 2 - 10 = -4$
$x = 2.5 \Rightarrow x^2 + x - 10 = 6.25 + 2.5 - 10$
$= -1.25$
$x = 3 \Rightarrow x^2 + x - 10 = 9 + 3 - 10 = 2$
b) f(2) and f(2.5) are both negative, and f(3) is positive. There's a change of sign between 2.5 and 3, so the solution is **greater than 2.5**.
c) (i) $x = 2.6 \Rightarrow x^2 + x - 10 = 6.76 + 2.6 - 10$
$= -0.64$
There's a change of sign between 2.6 and 3, so the solution is **greater than 2.6**.
(ii) $x = 2.7 \Rightarrow x^2 + x - 10 = 7.29 + 2.7 - 10$
$= -0.01$
There's a change of sign between 2.7 and 3, so the solution is **greater than 2.7**.
(iii) $x = 2.8 \Rightarrow x^2 + x - 10 = 7.84 + 2.8 - 10$
$= 0.64$
There's a change of sign between 2.7 and 2.8, so the solution is **less than 2.8**.
d) f(2.71) = 0.0541, which is positive. The solution is between 2.7 and 2.71, so $x = \textbf{2.7}$ (1 d.p.)

2

x	$x^2 + 2x - 30$	
4.5	−0.75	Negative
4.6	0.36	Positive
4.51	−0.6399	Negative
4.52	−0.5296	Negative
4.53	−0.4191	Negative
4.54	−0.3084	Negative
4.55	−0.1975	Negative
4.56	−0.0864	Negative
4.57	0.0249	Positive

The solution is between 4.56 and 4.57, so $x = \textbf{4.6}$ (1 d.p.)

3 a) $x^3 + 5x = 170 \Rightarrow x^3 + 5x - 170 = 0$

x	$x^3 + 5x - 170$	
5	–20	Negative
6	76	Positive
5.1	–11.849	Negative
5.2	–3.392	Negative
5.3	5.377	Positive
5.21	–2.529239	Negative
5.22	–1.663352	Negative
5.23	–0.794333	Negative
5.24	0.077824	Positive

The solution is between 5.23 and 5.24, so **$x = 5.2$** (1 d.p.)

b) $x^3 - 3x = 133 \Rightarrow x^3 - 3x - 133 = 0$

x	$x^3 - 3x - 133$	
5	–23	Negative
6	65	Positive
5.1	–15.649	Negative
5.2	–7.992	Negative
5.3	–0.023	Negative
5.4	8.264	Positive
5.31	0.791291	Positive

The solution is between 5.3 and 5.31, so **$x = 5.3$** (1 d.p.)

4 $2^x = 20 \Rightarrow 2^x - 20 = 0$

x	$2^x - 20$	
4	–4	Negative
5	12	Positive
4.1	–2.8516...	Negative
4.2	–1.6208...	Negative
4.3	–0.3016...	Negative
4.4	1.1121...	Positive
4.31	–0.1646...	Negative
4.32	–0.0267...	Negative
4.33	0.1122...	Positive

The solution is between 4.32 and 4.33, so **$x = 4.3$** (1 d.p.)

5 a) $x(x + 7) = 100 \Rightarrow x^2 + 7x - 100 = 0$

b)

x	$x^2 + 7x - 100$	
7	–2	Negative
8	20	Positive
7.1	0.11	Positive
7.01	–1.7899	Negative
7.02	–1.5796	Negative
⋮	⋮	⋮
7.09	–0.1019	Negative

The solution is between 7.09 and 7.1, so **$x = 7.1$** (1 d.p.)

c) $x + 7 = 14.1$ (1 d.p.), so the lengths of the sides to 1 d.p. are **14.1 cm and 7.1 cm**.

Page 123   Exercise 2

1 $x^2 + x = 35 \Rightarrow x^2 + x - 35 = 0$

E.g. x	$x^2 + x - 35$	
5.4	–0.44	Negative
5.5	0.75	Positive
5.45	0.1525	Positive
5.43	–0.0851	Negative
5.44	0.0336	Positive
5.435	–0.025775	Negative

The solution is between 5.435 and 5.44, so **$x = 5.44$** (2 d.p.)

2 a) $x^2 + x = 23 \Rightarrow x^2 + x - 23 = 0$

E.g. x	$x^2 + x - 23$	
0	–23	Negative
10	87	Positive
5	7	Positive
3	–11	Negative
4	–3	Negative
4.5	1.75	Positive
4.3	–0.21	Negative
4.4	0.76	Positive
4.35	0.2725	Positive
4.33	0.0789	Positive
4.32	–0.0176	Negative
4.325	0.030625	Positive

The solution is between 4.32 and 4.325, so **$x = 4.32$** (2 d.p.)

b) $x^2 + 2x = 17 \Rightarrow x^2 + 2x - 17 = 0$

x	$x^2 + 2x - 17$	
0	–17	Negative
10	103	Positive
⋮	⋮	⋮
3.24	–0.0224	Negative
3.245	0.020025	Positive

The solution is between 3.24 and 3.245, so **$x = 3.24$** (2 d.p.)

c) $x^2 + 5x = 62 \Rightarrow x^2 + 5x - 62 = 0$

x	$x^2 + 5x - 62$	
0	–62	Negative
10	88	Positive
⋮	⋮	⋮
5.76	–0.0224	Negative
5.765	0.060225	Positive

The solution is between 5.76 and 5.765, so **$x = 5.76$** (2 d.p.)

3 $x^2 + x = 48 \Rightarrow x^2 + x - 48 = 0$

x	$x^2 + x - 48$	
6.4	–0.64	Negative
6.5	0.75	Positive
⋮	⋮	⋮
6.446	–0.003084	Negative
6.4465	0.00386225	Positive

The solution is between 6.446 and 6.4465, so **$x = 6.446$** (3 d.p.)

4 $x^3 + 4x = 21 \Rightarrow x^3 + 4x - 21 = 0$

x	$x^3 + 4x - 21$	
2.2	–1.552	Negative
2.3	0.367	Positive
⋮	⋮	⋮
2.281	–0.00804...	Negative
2.2815	0.00176...	Positive

The solution is between 2.281 and 2.2815, so **$x = 2.281$** (3 d.p.)

5 a) $100x - 5x^2 = 200 \Rightarrow x^2 - 20x + 40 = 0$
As the cannonball is still rising, $x < 10$ seconds.

x	$x^2 - 20x + 40$	
0	40	Positive
10	–60	Negative
⋮	⋮	⋮
2.3	–0.71	Negative
2.25	0.0625	Positive

The solution is between 2.25 and 2.3, so **$x = 2.3$ seconds** (1 d.p.)

b) $100x - 5x^2 = 400 \Rightarrow x^2 - 20x + 80 = 0$
As the cannonball is still rising, $x < 10$ seconds.

x	$x^2 - 20x + 80$	
0	80	Positive
10	–20	Negative
⋮	⋮	⋮
5.5	0.25	Positive
5.55	–0.1975	Negative

The solution is between 5.5 and 5.55, so **$x = 5.5$ seconds** (1 d.p.)

c) $100x - 5x^2 = 200 \Rightarrow x^2 - 20x + 40 = 0$
As the cannonball is falling back down towards the ground, $x > 10$ seconds.

x	$x^2 - 20x + 40$	
10	–60	Negative
20	40	Positive
⋮	⋮	⋮
17.7	–0.71	Negative
17.75	0.0625	Positive

The solution is between 17.7 and 17.75, so **$x = 17.7$ seconds** (1 d.p.)

Page 124   Exercise 3

1 $x_0 = 1$

$x_1 = \sqrt{\dfrac{3(1) + 1}{2}} = 1.414...$

$x_2 = \sqrt{\dfrac{3(1.414...) + 1}{2}} = 1.619...$

$x_3 = \sqrt{\dfrac{3(1.619...) + 1}{2}} = 1.711...$

$x_4 = \sqrt{\dfrac{3(1.711...) + 1}{2}} = 1.751...$

$x_5 = \sqrt{\dfrac{3(1.751...) + 1}{2}} = 1.768...$

x_4 and x_5 both round to **1.8** (1 d.p.)

You can use the 'Ans' button on your calculator to make these questions a lot quicker. Just type in '1 =' (or whatever value you're using for x_0), then type in the iteration formula with 'Ans' instead of x_n. Then, pressing '=' will give you the next iteration without you needing to type anything else.

2 $x_0 = 0$

$x_1 = \dfrac{2(0)^3 + 2}{3(0)^2 + 4} = 0.5$

$x_2 = \dfrac{2(0.5)^3 + 2}{3(0.5)^2 + 4} = 0.4736...$

$x_3 = \dfrac{2(0.473...)^3 + 2}{3(0.473...)^2 + 4} = 0.4734...$

x_2 and x_3 both round to **0.47** (2 d.p.)

3 $x_0 = 3$

$x_1 = \dfrac{2(3)^2 + 11}{4(3)} = 2.41666...$

$x_2 = \dfrac{2(2.41666...)^2 + 11}{4(2.41666...)} = 2.34626...$

$x_3 = \dfrac{2(2.34626...)^2 + 11}{4(2.34626...)} = 2.34520...$

$x_4 = \dfrac{2(2.34520...)^2 + 11}{4(2.34520...)} = 2.34520...$

x_3 and x_4 both round to **2.345** (3 d.p.)

4 $x_0 = 1$

$x_1 = \sqrt[3]{\dfrac{7 - 2(1)}{2}} = 1.35720...$

$x_2 = \sqrt[3]{\dfrac{7 - 2(1.35720...)}{2}} = 1.28921...$

...

$x_6 = 1.30047...$

$x_7 = 1.30049...$

x_6 and x_7 both round to **1.300** (3 d.p.)

1 a) $\frac{x+8}{3} = 4 \Rightarrow x + 8 = 12 \Rightarrow$ **$x = 4$**

b) $13 - 3.5x = 34 \Rightarrow 13 = 34 + 3.5x$
$\Rightarrow 3.5x = -21 \Rightarrow$ **$x = -6$**

c) $7(x - 3) = 3(x - 6) \Rightarrow 7x - 21 = 3x - 18$
$\Rightarrow 4x = 3 \Rightarrow$ **$x = 0.75$**

d) $12(x - 3) = 4(6 + 2x)$
$\Rightarrow 12x - 36 = 24 + 8x \Rightarrow 4x = 60$
$\Rightarrow$ **$x = 15$**

e) $\frac{x-2}{5} = \frac{9-x}{3} \Rightarrow 3(x - 2) = 5(9 - x)$
$\Rightarrow 3x - 6 = 45 - 5x \Rightarrow 8x = 51$
$\Rightarrow$ **$x = 6.375$**

f) $\frac{2x}{5} = 18 - 2x \Rightarrow 2x = 5(18 - 2x)$
$\Rightarrow 2x = 90 - 10x \Rightarrow 12x = 90 \Rightarrow$ **$x = 7.5$**

2 a) $\dfrac{x-4}{5} = 15$

b) $\frac{x-4}{5} = 15 \Rightarrow x - 4 = 75 \Rightarrow$ **$x = 79$**

3 a) $(x + 10°) + (x + 10°) + (3x + 10°) = 180°$
$\Rightarrow 5x + 30° = 180° \Rightarrow 5x = 150°$
$\Rightarrow$ **$x = 30°$**

b) $90° + (9x + 8°) + (7x - 4°) + 90° = 360°$
$\Rightarrow 16x + 184° = 360° \Rightarrow 16x = 176°$
$\Rightarrow$ **$x = 11°$**

c) $(x - 19°) + (2x + 15°) + (3x + 10°) = 180°$
$\Rightarrow 6x + 6° = 180° \Rightarrow 6x = 174°$
$\Rightarrow$ **$x = 29°$**

4 a) $4x = 10 \Rightarrow x = 2.5$, so **no**,
the symbol '$\equiv$' cannot be used.

b) $2(3x + 6) = 6x + 12 = 3(2x + 4)$, so **yes**,
the symbol '$\equiv$' can be used.

c) $3(2 - 3x) + 2 = 7x \Rightarrow 6 - 9x + 2 = 7x$
$\Rightarrow 8 = 16x \Rightarrow x = 0.5$, so **no**,
the symbol '$\equiv$' cannot be used.

5 a) $3(x + a) \equiv 12 + 3x \Rightarrow 3x + 3a \equiv 3x + 12$
$\Rightarrow$ **$a = 4$**.

b) $4(1 - 2x) \equiv 2(a - 4x) \Rightarrow 4 - 8x \equiv 2a - 8x$
$\Rightarrow$ **$a = 2$**.

c) $3(x^2 - 2) \equiv a(6x^2 - 12)$
$\Rightarrow 3x^2 - 6 \equiv 6ax^2 - 12a \Rightarrow$ **$a = 0.5$**.

6 a) Take five consecutive numbers —
x, $(x + 1)$, $(x + 2)$, $(x + 3)$ and $(x + 4)$.
Then their sum is:
$x + (x + 1) + (x + 2) + (x + 3) + (x + 4)$
$= 5x + 1 + 2 + 3 + 4 = 5x + 10 = 5(x + 2)$
So the sum of five consecutive numbers
is a **multiple of 5**.

b) E.g. if $a = 5$ and $b = 7$,
then a and b are prime, but
$2ab - 1 = 70 - 1 = 69 = 3 \times 23$ is **not prime**.

7 a) $x^2 + 4x = 100 \Rightarrow x^2 + 4x - 100 = 0$

x	$x^2 + 4x - 100$	
0	−100	Negative
10	40	Positive
⋮	⋮	⋮
8.2	0.04	Positive
8.15	−0.9775	Negative

The solution is between 8.15 and 8.2,
so **$x = 8.2$** (1 d.p.)

b) $x^4 = 30 - 10x \Rightarrow x^4 + 10x - 30 = 0$

x	$x^4 + 10x - 30$	
0	−30	Negative
10	10 070	Positive
⋮	⋮	⋮
1.8	−1.5024	Negative
1.85	0.21350625	Positive

The solution is between 1.8 and 1.85,
so **$x = 1.8$** (1 d.p.)

c) $x^4 + 2x^2 + 5x = 20$
$\Rightarrow x^4 + 2x^2 + 5x - 20 = 0$

x	$x^4 + 2x^2 + 5x - 20$	
0	−20	Negative
10	10 230	Positive
⋮	⋮	⋮
1.6	−0.3264	Negative
1.65	1.10700625	Positive

The solution is between 1.6 and 1.65,
so **$x = 1.6$** (1 d.p.)

8 $2x(7 + x) = 17 \Rightarrow 2x^2 + 14x - 17 = 0$

x	$2x^2 + 14x - 17$	
0	−17	Negative
3	43	Positive
⋮	⋮	⋮
1.1	0.82	Positive
1.05	−0.095	Negative

The solution is between 1.05 and 1.1,
so **$x = 1.1$** (1 d.p.)

9 $x_0 = 1$

$x_1 = \sqrt[3]{\dfrac{5 - 2(1)^2}{4}} = 0.9085...$

$x_2 = \sqrt[3]{\dfrac{5 - 2(0.9085...)^2}{4}} = 0.9425...$

$x_3 = \sqrt[3]{\dfrac{5 - 2(0.9425...)^2}{4}} = 0.9305...$

$x_4 = \sqrt[3]{\dfrac{5 - 2(0.9305...)^2}{4}} = 0.9348...$

x_3 and x_4 both round to **0.93** (2 d.p.)

1 a) $\frac{2(x+3)}{13} = 1 \Rightarrow 2(x + 3) = 13$
$\Rightarrow 2x + 6 = 13$ *[1 mark]*
$\Rightarrow 2x = 7 \Rightarrow$ **$x = 3.5$** *[1 mark]*

b) $\frac{x-10}{10} = \frac{10-x}{3} \Rightarrow 3(x - 10) = 10(10 - x)$
$\Rightarrow 3x - 30 = 100 - 10x$ *[1 mark]*
$\Rightarrow 13x = 130 \Rightarrow$ **$x = 10$** *[1 mark]*

c) $3(x^2 + 7) = 4(x^2 - 1)$
$\Rightarrow 3x^2 + 21 = 4x^2 - 4$ *[1 mark]*
$\Rightarrow x^2 = 25 \Rightarrow$ **$x = \pm5$** *[1 mark]*

2 Area of the rectangle $= 2(x + 3) = 12$ *[1 mark]*
$2(x + 3) = 12 \Rightarrow 2x + 6 = 12$
$\Rightarrow 2x = 6 \Rightarrow x = 3$ *[1 mark]*
Perimeter $= 2 + 2 + (x + 3) + (x + 3)$
$= 2 + 2 + 6 + 6 =$ **16 cm** *[1 mark]*

3 a) E.g. $1^2 \times 3^2 = 1 \times 9 = 9$ which is **odd**,
so Will's statement is **not correct**. *[1 mark]*

b) 2 is the only even prime, so all prime
numbers bigger than 2 are odd. *[1 mark]*
Take two prime numbers bigger than 2 —
since they are odd, they can be written
as $(2a + 1)$ and $(2b + 1)$, where a and b
are integers. Then their sum is:
$(2a + 1) + (2b + 1) = 2a + 2b + 2$ *[1 mark]*
$= 2(a + b + 1) = 2n$ where n
is an integer, i.e. even.
So the sum of two prime numbers greater
than 2 is always even — Veronica's
statement is **correct**. *[1 mark]*

4 a) $2x + (3x - 1) + (x + 1) = 12$ *[1 mark]*
$\Rightarrow 6x = 12 \Rightarrow x = 2$ cm *[1 mark]*
Area $= \frac{1}{2} \times (x + 1) \times 2x$
$= \frac{1}{2} \times 3 \times 4 =$ **6 cm²** *[1 mark]*

c) $y(y + 3) = 6 \Rightarrow y^2 + 3y - 6 = 0$
E.g.

y	$y^2 + 3y - 6$	
0	−6	Negative
5	34	Positive
2	4	Positive
1	−2	Negative
1.5	0.75	Positive
1.3	−0.41	Negative
1.4	0.16	Positive
1.35	−0.1275	Negative

The solution is between 1.35 and 1.4,
so **$x = 1.4$** (1 d.p.)
*[4 marks available — 1 mark for forming
the correct equation in y, 1 mark for rows
showing a sign change between y = 1
and y = 2, 1 mark for rows showing a sign
change between y = 1.3 and y = 1.4,
1 mark for the correct value of y]*

*We've used the second method (from p.122)
here, but you could use either method
and get full marks.*

5 a) $x = 3 \Rightarrow x^2 - 4x + 1 = -2$
$x = 4 \Rightarrow x^2 - 4x + 1 = 1$ *[1 mark for both]*
Since there is a change of sign
between $x = 3$ and $x = 4$, the equation
**$x^2 - 4x + 1 = 0$ has a solution
in the interval $3 < x < 4$.** *[1 mark]*

b) $x_0 = 3$
$x_1 = \sqrt{4(3) - 1} = 3.3166...$
$x_2 = \sqrt{4(3.3166...) - 1} = 3.5023...$
$x_3 = \sqrt{4(3.5023...) - 1} = 3.6068...$
$x_4 = \sqrt{4(3.6068...) - 1} = 3.6643...$
$x_5 = \sqrt{4(3.6643...) - 1} = 3.6955...$
x_4 and x_5 both round to **3.7** (1 d.p.)
*[3 marks available — 1 mark for carrying
out the iteration correctly, 1 mark for
stopping when at least two consecutive
iterations are the same to 1 d.p.,
1 mark for the correct value of x]*

*You could do more iterations to make sure your
answer is definitely accurate, but it's usually fine
to stop once two iterations round to the same
value to the given degree of accuracy.*

6 a) **$x + x^2 + 30 = 128$** *[1 mark]*

b) $x + x^2 + 30 = 128 \Rightarrow x^2 + x - 98 = 0$
E.g.

x	$x^2 + x - 98$	
0	−98	Negative
10	12	Positive
5	−68	Negative
8	−26	Negative
9	−8	Negative
9.5	1.75	Positive
9.3	−2.21	Negative
9.4	−0.24	Negative
9.45	0.7525	Positive
9.42	0.1564	Positive
9.41	−0.0419	Negative
9.415	0.057225	Positive

The solution is between 9.41 and 9.415,
so **$x = 9.41$ m** to the nearest cm.
*[4 marks available — 1 mark for rows
showing a sign change between x = 9
and x = 10, 1 mark for rows showing
a sign change between x = 9.4 and
x = 9.5, 1 mark for rows showing a sign
change between x = 9.41 and x = 9.42,
1 mark for the correct value of x]*

Section 10 — Direct and Inverse Proportion

10.1 Direct Proportion

Page 128   Exercise 1

1 a) $y = kx \Rightarrow 2 = 22k \Rightarrow k = 2 \div 22 = \frac{1}{11}$

$y = \frac{1}{11} \times 33 = \mathbf{3}$

b) $y = kx \Rightarrow 18 = 24k \Rightarrow k = 18 \div 24 = 0.75$
$24 = 0.75x \Rightarrow x = 24 \div 0.75 = \mathbf{32}$

c) $y = kx \Rightarrow 15 = 10k \Rightarrow k = 15 \div 10 = 1.5$
$y = 1.5 \times 2 = \mathbf{3}$
$y = 1.5 \times 7 = \mathbf{10.5}$
$y = 1.5 \times 21 = \mathbf{31.5}$
$36 = 1.5x \Rightarrow x = 36 \div 1.5 = \mathbf{24}$

d) $y = kx \Rightarrow -14 = -4k$
$\Rightarrow k = -14 \div -4 = 3.5$
$y = 3.5 \times 0 = \mathbf{0}$
$21 = 3.5x \Rightarrow x = 21 \div 3.5 = \mathbf{6}$
$y = 3.5 \times 12 = \mathbf{42}$

e) $y = kx \Rightarrow 9 = 8k \Rightarrow k = \frac{9}{8}$
$12 = \frac{9}{8}x \Rightarrow x = \frac{32}{3}$

f) $y = kx \Rightarrow 104 = 78k \Rightarrow k = 104 \div 78 = \frac{4}{3}$
$y = \frac{4}{3} \times -27 = \mathbf{-36}$
$272 = \frac{4}{3}x \Rightarrow x = 272 \div \frac{4}{3} = \mathbf{204}$
$980 = \frac{4}{3}x \Rightarrow x = 980 \div \frac{4}{3} = \mathbf{735}$

2 $j = kh \Rightarrow 15 = 5k \Rightarrow k = 15 \div 5 = 3$
$j = 3 \times 40 = \mathbf{120}$

3 $r = kt \Rightarrow 9 = 6k \Rightarrow k = 9 \div 6 = 1.5$
$r = 1.5 \times 7.5 = \mathbf{11.25}$

4 $p = kq \Rightarrow 11 = 3k \Rightarrow k = \frac{11}{3}$
$82.5 = \frac{11}{3}q \Rightarrow q = 82.5 \div \frac{11}{3} = \mathbf{22.5}$

5 a) $b = ks \Rightarrow 142 = 16k$
$\Rightarrow k = 142 \div 16 = \frac{71}{8}$
$b = \frac{71}{8} \times 18 = \frac{\mathbf{639}}{\mathbf{4}}$

b) $200 = \frac{71}{8}s \Rightarrow s = 200 \div \frac{71}{8} = \frac{\mathbf{1600}}{\mathbf{71}}$

Page 129   Exercise 2

1 a) $y = kx^2 \Rightarrow 64 = k \times 2^2 = 4k$
$\Rightarrow k = 64 \div 4 = 16$
$y = 16 \times 5^2 = \mathbf{400}$

b) $y = kx^2 \Rightarrow 539 = k \times 7^2 = 49k$
$\Rightarrow k = 539 \div 49 = 11$
$1331 = 11 \times x^2 \Rightarrow x^2 = 1331 \div 11 = 121$
$\Rightarrow x = \pm\sqrt{121} = \mathbf{\pm 11}$

2 $y = k\sqrt{x} \Rightarrow 84 = k \times \sqrt{9} \Rightarrow k = 84 \div 3 = 28$
$y = 28 \times \sqrt{1} = \mathbf{28}$
$y = 28 \times \sqrt{16} = \mathbf{112}$
$560 = 28 \times \sqrt{x} \Rightarrow \sqrt{x} = 560 \div 28 = 20,$
$x = 20^2 = \mathbf{400}$

3 a) $f = kg^2 \Rightarrow 200 = k \times 100^2$
$\Rightarrow k = 200 \div 10\,000 = 0.02$
$f = 0.02 \times 61.5^2 = \mathbf{75.645}$

b) $14 = 0.02 \times g^2 \Rightarrow g^2 = 14 \div 0.02 = 700$
$\Rightarrow g = \sqrt{700} = \sqrt{100} \times \sqrt{7} = \mathbf{10\sqrt{7}}$

4 $y = k\sqrt{x} \Rightarrow 3 = k \times \sqrt{34.1}$
$\Rightarrow k = 3 \div \sqrt{34.1} = 0.513...$
$y = 0.513... \times \sqrt{15} = \mathbf{1.99}$ **seconds** (to 3 s.f.)

5 a) $V = kr^3 \Rightarrow 2304\pi = k \times 12^3$
$\Rightarrow k = 2304\pi \div 1728 = \frac{4}{3}\pi \Rightarrow V = \frac{4}{3}\pi r^3$

b) $V = \frac{4}{3}\pi \times 21^3 = \mathbf{12\,348\pi}$ **cm³**

c) $1000 = \frac{4}{3}\pi \times r^3,$
$r^3 = 1000 \div \frac{4}{3}\pi = 238.73...$
$r = \mathbf{6.2}$ **cm** (to 1 d.p.)

10.2 Inverse Proportion

Page 131   Exercise 1

1 a) $y = \frac{k}{x} \Rightarrow 15 = \frac{k}{12} \Rightarrow k = 15 \times 12 = 180$
$12 = \frac{180}{x} \Rightarrow x = 180 \div 12 = \mathbf{15}$

b) $y = \frac{k}{x} \Rightarrow 4 = \frac{k}{11} \Rightarrow k = 4 \times 11 = 44$
$y = \frac{44}{22} = \mathbf{2}$

c) $y = \frac{k}{x}, 15 = \frac{k}{6} \Rightarrow k = 15 \times 6 = 90$
$y = \frac{90}{1} = \mathbf{90}, y = \frac{90}{3} = \mathbf{30}, y = \frac{90}{20} = \mathbf{4.5}$
$270 = \frac{90}{x} \Rightarrow x = 90 \div 270 = \frac{\mathbf{1}}{\mathbf{3}}$

2 $p = \frac{k}{q} \Rightarrow 7 = \frac{k}{4} \Rightarrow k = 7 \times 4 = 28$
$p = \frac{28}{56} = \mathbf{0.5}$

3 $s = \frac{k}{t} \Rightarrow 5 = \frac{k}{16} \Rightarrow k = 5 \times 16 = 80$
$48 = \frac{80}{t} \Rightarrow t = \frac{80}{48} = \frac{\mathbf{5}}{\mathbf{3}}$

4 a) $w = \frac{k}{z} \Rightarrow 15 = \frac{k}{4} \Rightarrow k = 15 \times 4 = 60$
$25 = \frac{60}{z} \Rightarrow z = 60 \div 25 = \mathbf{2.4}$

b) If w is doubled (or multiplied by 2), then z is halved (or divided by 2).

Page 132   Exercise 2

1 a) $y = \frac{k}{x^2} \Rightarrow 8 = \frac{k}{2^2} \Rightarrow k = 8 \times 4 = 32$
$y = \frac{32}{5^2} = \mathbf{1.28}$
$2 = \frac{32}{x^2} \Rightarrow x^2 = 32 \div 2 = 16 \Rightarrow x = \mathbf{\pm 4}$
$y = \frac{32}{0.4^2} = \mathbf{200}$

b) $y = \frac{k}{\sqrt{x}} \Rightarrow 8 = \frac{k}{\sqrt{9}} \Rightarrow k = 8 \times 3 = 24$
$6 = \frac{24}{\sqrt{x}} \Rightarrow \sqrt{x} = 24 \div 6 = 4 \Rightarrow x = \mathbf{16}$
$y = \frac{24}{\sqrt{100}} = \mathbf{2.4}$
$\frac{1}{3} = \frac{24}{\sqrt{x}} \Rightarrow \sqrt{x} = 24 \div \frac{1}{3} = 72$
$\Rightarrow x = \mathbf{5184}$

2 $h = \frac{k}{f^3} \Rightarrow 12.5 = \frac{k}{2^3} \Rightarrow k = 12.5 \times 8 = 100$
$h = \frac{100}{5^3} = \mathbf{0.8}$

3 $a = \frac{k}{c^2} \Rightarrow 3 = \frac{k}{6^2} \Rightarrow k = 3 \times 36 = 108$
$12 = \frac{108}{x^2} \Rightarrow x^2 = 108 \div 12 = 9 \Rightarrow x = \mathbf{\pm 3}$

4 a) $p = \frac{k}{r^2} \Rightarrow 20 = \frac{k}{10^2}$
$\Rightarrow k = 20 \times 100 = 2000$
$p = \frac{2000}{15^2} = \mathbf{8.89}$ **units** (to 3 s.f.)

b) $30 = \frac{2000}{x^2} \Rightarrow x^2 = 2000 \div 30 = 66.666...$
$\Rightarrow x = \sqrt{66.666...} = \mathbf{8.16}$ **mm** (to 3 s.f.)

5 $b = \frac{k}{c^2} \Rightarrow 64 = \frac{k}{1^2} \Rightarrow k = 64$
If $b = c$, then $b = \frac{64}{b^2}$, $b^3 = 64$, $b = 4$
Therefore $\mathbf{b = c = 4}$

6 a) (i) **false** — u is inversely proportional to the square root of v.
(ii) **true** — $k = v \times u^2$.
(iii) **false** — if you double v, you halve u^2.
(iv) **true** — $(2u)^2 = 4u^2$, so v is divided by 4

b) $v = \frac{k}{u^2} = \frac{900}{u^2}$
e.g. when $u = 1$, $v = \mathbf{900}$
when $u = 2$, $v = \mathbf{225}$
when $u = 3$, $v = \mathbf{100}$,
when $u = 5$, $v = \mathbf{36}$
when $u = 10$, $v = \mathbf{9}$

Page 133   Review Exercise

1 a) $y = kx \Rightarrow 8 = 3k \Rightarrow k = \frac{8}{3}$
$y = \frac{8}{3} \times 2 = \frac{\mathbf{16}}{\mathbf{3}}, y = \frac{8}{3} \times 9 = \mathbf{24}$
$100 = \frac{8}{3}x \Rightarrow x = 100 \div \frac{8}{3} = \mathbf{37.5}$

b) $y = \frac{k}{x} \Rightarrow 8 = \frac{k}{3} \Rightarrow k = 8 \times 3 = 24$
$y = \frac{24}{2} = \mathbf{12}, y = \frac{24}{9} = \frac{\mathbf{8}}{\mathbf{3}}$
$100 = \frac{24}{x} \Rightarrow x = 24 \div 100 = \mathbf{0.24}$

2 a) $p = \frac{k}{g^3} \Rightarrow 10 = \frac{k}{1.5^3} \Rightarrow k = 33.75$
$p = \frac{33.75}{2.1^3} = \mathbf{3.64}$ (to 3 s.f.)

b) $15 = \frac{33.75}{g^3} \Rightarrow g^3 = 33.75 \div 15 = 2.25$
$g = \sqrt[3]{2.25} = \mathbf{1.31}$ (to 3 s.f.)

3 a) E.g. $r \propto h \Rightarrow r = kh \Rightarrow \mathbf{r = 1.3h}$
b) $r = 1.3 \times 1.75 = 2.275$ m
2.275 m < 2.5 m, so **no**, you wouldn't expect a person of height 1.75 m to be able to touch a ceiling 2.5 m high.

4 A and Z — direct proportionality between x and y gives a straight line through the origin.
B and Y — the shape of the curve is that of an x^2 graph.
C and X — the shape of the curve is that of an x^3 graph.
D and W — the graph does not exist for negative values of x.

5 a) (i) $y = \frac{k}{x^2} \Rightarrow 6 = \frac{k}{4} \Rightarrow k = 24$
$y = \frac{24}{36} = \frac{\mathbf{2}}{\mathbf{3}}, y = \frac{24}{16} = \frac{\mathbf{3}}{\mathbf{2}}, y = \frac{24}{4} = \mathbf{6},$
$y = \frac{24}{16} = \frac{\mathbf{3}}{\mathbf{2}}, y = \frac{24}{36} = \frac{\mathbf{2}}{\mathbf{3}}$

(ii) $y = \frac{k}{x^3} \Rightarrow 6 = \frac{k}{8} \Rightarrow k = 48$
$y = \frac{48}{-216} = -\frac{\mathbf{2}}{\mathbf{9}}, y = \frac{48}{-64} = -\frac{\mathbf{3}}{\mathbf{4}},$
$y = \frac{48}{-8} = \mathbf{-6}, y = \frac{48}{64} = \frac{\mathbf{3}}{\mathbf{4}},$
$y = \frac{48}{216} = \frac{\mathbf{2}}{\mathbf{9}}$

b) (i)

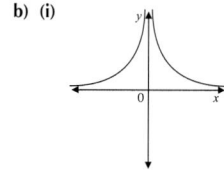

(ii)

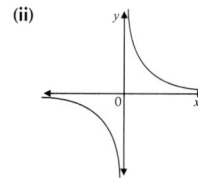

6 $F = k\frac{Q_1 \times Q_2}{d^2} \Rightarrow 10^6 = k\frac{10^{-2} \times 10^{-2}}{3^2}$
$\Rightarrow 10^6 = \frac{10^{-4}}{9}k \Rightarrow k = 10^6 \div \frac{10^{-4}}{9} = 9 \times 10^{10}$
$F = 9 \times 10^{10}\left(\frac{10^{-4}}{8^2}\right) = \mathbf{140\,625}$ **units**

Page 134   Exam-Style Questions

1 $b = k\sqrt{d} \Rightarrow 5 = k\sqrt{2.2}$
$\Rightarrow k = 5 \div \sqrt{2.2} = 3.37...$
$b = 3.37...\sqrt{0.5} = \mathbf{2.38}$ (to 3 s.f.)
[1 mark]

508   Answers

2 $y \propto \dfrac{1}{x^2} \Rightarrow y = \dfrac{k}{x^2} \Rightarrow 90 = \dfrac{k}{10^2}$

$\Rightarrow k = 90 \times 100 = 9000$

$9000 = \dfrac{9000}{x^2} \Rightarrow x^2 = 1 \Rightarrow x = 1$

$40 = \dfrac{9000}{x^2} \Rightarrow x^2 = \dfrac{9000}{40} = 225$

$\Rightarrow x = \sqrt{225} = 15$

$y = \dfrac{9000}{30^2} = \dfrac{9000}{900} = 10$

[3 marks available — 1 mark for finding k, 2 marks for all 3 values correct, or 1 mark for 2 values correct]

3 $E = km \Rightarrow 0.9 = 15k \Rightarrow k = 0.06$

$E = 0.06 \times 25 = 1.5$

$20 + 1.5 = 21.5$ **cm**

[3 marks available — 1 mark for finding the value of k, 1 mark for calculating the new extension for 25 g, 1 mark for the correct answer]

4 $p \propto \dfrac{1}{d^2} \Rightarrow p = \dfrac{k}{d^2}$

$8.5 = \dfrac{k}{32^2} = 8.5 \times 32^2 = 8704$

$p = \dfrac{8704}{d^2}$

[2 marks available — 1 mark for substituting correct values into equation and rearranging to find constant, 1 mark for formula]

5 $q = kp^n \Rightarrow 8 = k \times 1^n \Rightarrow k = 8$

$0.5 = 8 \times 4^n \Rightarrow 4^n = \dfrac{1}{16} = 4^{-2}$

$\Rightarrow n = -2$

[3 marks available — 1 mark for finding constant of proportionality, 1 mark for equation for 4^n, 1 mark for correct answer]

6 $Z = kx^3$ and $Z = jy^2$,

so $kx^3 = jy^2 \Rightarrow x^3 = \dfrac{j}{k}y^2 = Ky^2$,

where $K = \dfrac{j}{k}$

$5^3 = 8^2K \Rightarrow 125 = 64K \Rightarrow K = \dfrac{125}{64}$

$x^3 = \dfrac{125}{64}(27)^2 = 1423.8...$

$x = \sqrt[3]{1423.8...} = 11.25$

[4 marks available — 1 mark for finding equations for Z in terms of x and y, 1 mark for equation for x in terms of y, 1 mark for constant of proportionality, 1 mark for correct answer]

Section 11 — Quadratic Equations

11.1 Solving Quadratic Equations by Factorising

Page 135 Exercise 1

1 a) $x(x + 8) = 0 \Rightarrow x = 0$ or $x + 8 = 0$
$\Rightarrow x = 0$ or $x = -8$
b) $(x - 5)(x - 1) = 0 \Rightarrow x - 5 = 0$ or $x - 1 = 0$
$\Rightarrow x = 5$ or $x = 1$
c) $(x + 2)(x + 6) = 0 \Rightarrow x + 2 = 0$ or $x + 6 = 0$
$\Rightarrow x = -2$ or $x = -6$
d) $(x - 9)(x + 7) = 0 \Rightarrow x - 9 = 0$ or $x + 7 = 0$
$\Rightarrow x = 9$ or $x = -7$

2 a) $x(x - 3) = 0 \Rightarrow x = 0$ or $x = 3$
b) $x(x + 12) = 0 \Rightarrow x = 0$ or $x = -12$
c) $(x + 2)(x + 1) = 0 \Rightarrow x = -2$ or $x = -1$
d) $(x - 1)^2 = 0 \Rightarrow x = 1$
e) $(x + 2)^2 = 0 \Rightarrow x = -2$
f) $(x + 4)(x - 1) = 0 \Rightarrow x = -4$ or $x = 1$
g) $(x - 4)(x + 1) = 0 \Rightarrow x = 4$ or $x = -1$
h) $(x + 4)(x + 1) \Rightarrow x = -4$ or $x = -1$
i) $(x - 3)(x - 2) = 0 \Rightarrow x = 3$ or $x = 2$
j) $(x + 6)(x + 2) = 0 \Rightarrow x = -6$ or $x = -2$
k) $(x - 6)(x + 4) = 0 \Rightarrow x = 6$ or $x = -4$
l) $(x - 12)(x - 3) = 0 \Rightarrow x = 12$ or $x = 3$

3 a) $x(2x - 3) = 0 \Rightarrow x = 0$ or $2x - 3 = 0$
$\Rightarrow x = 0$ or $2x = 3$
$\Rightarrow x = 0$ or $x = \dfrac{3}{2}$
b) $(x - 2)(3x - 1) = 0$
$\Rightarrow x - 2 = 0$ or $3x - 1 = 0$
$\Rightarrow x = 2$ or $3x = 1$
$\Rightarrow x = 2$ or $x = \dfrac{1}{3}$
c) $(3x + 4)(2x + 5) = 0$
$\Rightarrow 3x + 4 = 0$ or $2x + 5 = 0$
$\Rightarrow 3x = -4$ or $2x = -5$
$\Rightarrow x = -\dfrac{4}{3}$ or $x = -\dfrac{5}{2}$
d) $(4x - 7)(5x + 2) = 0$
$\Rightarrow 4x - 7 = 0$ or $5x + 2 = 0$
$\Rightarrow 4x = 7$ or $5x = -2$
$\Rightarrow x = \dfrac{7}{4}$ or $x = -\dfrac{2}{5}$

4 a) $x(3x + 5) = 0 \Rightarrow x = 0$ or $x = -\dfrac{5}{3}$
b) $(2x + 3)(x - 1) = 0 \Rightarrow x = -\dfrac{3}{2}$ or $x = 1$
c) $(5x - 2)(x + 1) = 0 \Rightarrow x = \dfrac{2}{5}$ or $x = -1$
d) $(3x - 2)(x - 3) = 0 \Rightarrow x = \dfrac{2}{3}$ or $x = 3$
e) $(4x + 1)(x + 4) = 0 \Rightarrow x = -\dfrac{1}{4}$ or $x = -4$
f) $(6x - 11)(x + 2) = 0 \Rightarrow x = \dfrac{11}{6}$ or $x = -2$
g) $(2x - 5)^2 = 0 \Rightarrow x = \dfrac{5}{2}$
h) $(3x - 2)^2 = 0 \Rightarrow x = \dfrac{2}{3}$

Page 136 Exercise 2

1 a) $x^2 = x \Rightarrow x^2 - x = 0 \Rightarrow x(x - 1) = 0$
$\Rightarrow x = 0$ or $x = 1$
b) $x^2 + 2x = 3 \Rightarrow x^2 + 2x - 3 = 0$
$\Rightarrow (x + 3)(x - 1) = 0 \Rightarrow x = -3$ or $x = 1$
Using the same method for c)-l):
c) $(x - 7)(x - 3) = 0 \Rightarrow x = 7$ or $x = 3$
d) $(x - 4)(x - 2) = 0 \Rightarrow x = 4$ or $x = 2$
e) $(x - 6)(x - 2) = 0 \Rightarrow x = 6$ or $x = 2$
f) $(3x - 9)(x + 1) = 0 \Rightarrow x = 3$ or $x = -1$
g) $(2x - 1)(x + 11) = 0 \Rightarrow x = \dfrac{1}{2}$ or $x = -11$
h) $(2x + 3)(2x - 1) = 0 \Rightarrow x = -\dfrac{3}{2}$ or $x = \dfrac{1}{2}$
i) $(3x - 1)(2x + 1) = 0 \Rightarrow x = \dfrac{1}{3}$ or $x = -\dfrac{1}{2}$
j) $(3x - 2)(2x - 1) = 0 \Rightarrow x = \dfrac{2}{3}$ or $x = \dfrac{1}{2}$
k) $(2x - 1)^2 = 0 \Rightarrow x = \dfrac{1}{2}$
l) $(3x - 5)^2 = 0 \Rightarrow x = \dfrac{5}{3}$

2 a) $x(x - 2) = 8 \Rightarrow x^2 - 2x = 8$
$\Rightarrow x^2 - 2x - 8 = 0$
$\Rightarrow (x - 4)(x + 2) = 0 \Rightarrow x = 4$ or $x = -2$
b) $x(x + 2) = 35 \Rightarrow x^2 + 2x = 35$
$\Rightarrow x^2 + 2x - 35 = 0$
$\Rightarrow (x + 7)(x - 5) = 0 \Rightarrow x = -7$ or $x = 5$
Using the same method for c)-i):
c) $(x + 6)^2 = 0 \Rightarrow x = -6$
d) $(x - 7)^2 = 0 \Rightarrow x = 7$
e) $(x + 2)(x - 2) = 0 \Rightarrow x = -2$ or $x = 2$
This one's the difference of two squares (see p.84).
f) $(2x + 1)(x + 7) = 0 \Rightarrow x = -\dfrac{1}{2}$ or $x = -7$
g) $(3x + 4)(3x - 3) = 0 \Rightarrow x = -\dfrac{4}{3}$ or $x = 1$
h) $(3x - 1)(x - 3) = 0 \Rightarrow x = \dfrac{1}{3}$ or $x = 3$
i) $(4x - 1)(2x - 1) = 0 \Rightarrow x = \dfrac{1}{4}$ or $x = \dfrac{1}{2}$

3 a) $x + 1 = \dfrac{6}{x} \Rightarrow x(x + 1) = 6 \Rightarrow x^2 + x = 6$
$\Rightarrow x^2 + x - 6 = 0$
$\Rightarrow (x + 3)(x - 2) = 0 \Rightarrow x = -3$ or $x = 2$
b) $x - 2 = \dfrac{4}{x + 1} \Rightarrow (x - 2)(x + 1) = 4$
$\Rightarrow x^2 + x - 2x - 2 = 4$
$\Rightarrow x^2 - x - 6 = 0$
$\Rightarrow (x + 2)(x - 3) = 0 \Rightarrow x = -2$ or $x = 3$

Using the same method for c)-f):
c) $(x + 6)(x - 5) = 0 \Rightarrow x = -6$ or $x = 5$
d) $(2x - 3)(x - 2) = 0 \Rightarrow x = \dfrac{3}{2}$ or $x = 2$
e) $(6x - 5)(x + 1) \Rightarrow x = \dfrac{5}{6}$ or $x = -1$
f) $(12x - 5)(2x + 1) = 0 \Rightarrow x = \dfrac{5}{12}$ or $x = -\dfrac{1}{2}$

11.2 Completing the Square

Page 137 Exercise 1

1 a) Expand the brackets:
$(x - 2)^2 = x^2 - 4x + 4$
This quadratic has 4 as the constant but you want 7, so you need to add 3. Therefore, **$q = 3$**.
b) Expand the brackets:
$(x + 1)^2 = x^2 + 2x + 1$
This quadratic has 1 as the constant but you want –9, so you need to subtract 10. Therefore, **$q = -10$**.
c) Expand the brackets:
$(x + 2)^2 = x^2 + 4x + 4$
This quadratic has 4 as the constant but you want 2, so you need to subtract 2. Therefore, **$q = -2$**.

2 a) $b = 2$ so $\dfrac{b}{2} = 1$
$(x + 1)^2 = x^2 + 2x + 1$
$x^2 + 2x + 6 = x^2 + 2x + 1 + 5$
$= (x + 1)^2 + 5$
b) $b = -2$ so $\dfrac{b}{2} = -1$
$(x - 1)^2 = x^2 - 2x + 1$
$x^2 - 2x + 4 = x^2 - 2x + 1 + 3$
$= (x - 1)^2 + 3$
Using the same method for c)-h):
c) $(x - 1)^2 - 11$ **d)** $(x - 6)^2 + 64$
e) $(x + 6)^2 + 8$ **f)** $(x - 7)^2 - 49$
g) $(x - 10)^2 - 300$ **h)** $(x + 10)^2 - 250$

3 a) $b = 3$ so $\dfrac{b}{2} = \dfrac{3}{2}$
$\left(x + \dfrac{3}{2}\right)^2 = x^2 + 3x + \dfrac{9}{4}$
$x^2 + 3x + 1 = x^2 + 3x + \dfrac{9}{4} - \dfrac{5}{4}$
$= \left(x + \dfrac{3}{2}\right)^2 - \dfrac{5}{4}$
b) $b = 3$ so $\dfrac{b}{2} = \dfrac{3}{2}$
$\left(x + \dfrac{3}{2}\right)^2 = x^2 + 3x + \dfrac{9}{4}$
$x^2 + 3x - 1 = x^2 + 3x + \dfrac{9}{4} - \dfrac{13}{4}$
$= \left(x + \dfrac{3}{2}\right)^2 - \dfrac{13}{4}$
Using the same method for c)-h):
c) $\left(x - \dfrac{3}{2}\right)^2 - \dfrac{5}{4}$ **d)** $\left(x + \dfrac{5}{2}\right)^2 + \dfrac{23}{4}$
e) $\left(x + \dfrac{5}{2}\right)^2 - \dfrac{13}{4}$ **f)** $\left(x - \dfrac{5}{2}\right)^2 + \dfrac{55}{4}$
g) $\left(x + \dfrac{7}{2}\right)^2 - \dfrac{9}{4}$ **h)** $\left(x - \dfrac{9}{2}\right)^2 - \dfrac{181}{4}$

Page 138 Exercise 2

1 a) Expand the brackets:
$2\left(x + \dfrac{1}{4}\right)^2 = 2x^2 + x + \dfrac{1}{8}$
This quadratic has $\dfrac{1}{8}$ as the constant but you want 0, so you need to subtract $\dfrac{1}{8}$. Therefore, **$q = -\dfrac{1}{8}$**.
b) Expand the brackets:
$2(x + 5)^2 = 2x^2 + 20x + 50$
This quadratic has 50 as the constant but you want 500, so you need to add 450. Therefore, **$q = 450$**.

c) Expand the brackets:

$$3\left(x+\tfrac{2}{3}\right)^2 = 3x^2 + 4x + \tfrac{4}{3}$$

This quadratic has $\tfrac{4}{3}$ as the constant but you want 25, so you need to add $23\tfrac{2}{3} = \tfrac{71}{3}$. Therefore, $q = \dfrac{71}{3}$.

d) Expand the brackets:

$$4\left(x-\tfrac{7}{8}\right)^2 = 4x^2 - 7x + \tfrac{49}{16}$$

This quadratic has $\tfrac{49}{16}$ as the constant but you want -1, so you need to subtract $1\tfrac{49}{16} = \tfrac{65}{16}$. Therefore, $q = -\dfrac{65}{16}$.

2 a) $p = \dfrac{b}{2a} = \dfrac{-12}{2\times 2} = -3$

To find q, expand the brackets:
$$2(x-3)^2 = 2x^2 - 12x + 18$$
This quadratic has 18 as the constant but you want 9, so you need to subtract 9. Therefore, $q = -9$.

b) $p = \dfrac{b}{2a} = \dfrac{-5}{2\times 3} = -\dfrac{5}{6}$

To find q, expand the brackets:
$$3\left(x-\tfrac{5}{6}\right)^2 = 3x^2 - 5x + \tfrac{25}{12}$$
This quadratic has $\tfrac{25}{12} = 2\tfrac{1}{12}$ as the constant but you want -1, so you need to subtract $3\tfrac{1}{12} = \tfrac{37}{12}$. Therefore $q = -\dfrac{37}{12}$.

3 a) $a\left(x+\tfrac{b}{2a}\right)^2 = 2\left(x+\tfrac{8}{4}\right)^2 = 2(x+2)^2$
$$= 2(x^2 + 4x + 4) = 2x^2 + 8x + 8$$
$$2x^2 + 8x + 81 = 2x^2 + 8x + 8 + 73$$
$$= 2(x+2)^2 + 73$$

b) $a\left(x+\tfrac{b}{2a}\right)^2 = 3\left(x+\tfrac{8}{6}\right)^2 = 3\left(x+\tfrac{4}{3}\right)^2$
$$= 3\left(x^2 + \tfrac{8}{3}x + \tfrac{16}{9}\right)$$
$$= 3x^2 + 8x + \tfrac{16}{3}$$
$$3x^2 + 8x + 10 = 3x^2 + 8x + \tfrac{16}{3} + \tfrac{14}{3}$$
$$= 3\left(x+\tfrac{4}{3}\right)^2 + \tfrac{14}{3}$$

Using the same method for c)-i):

c) $2\left(x-\tfrac{1}{2}\right)^2 + \dfrac{5}{2}$ **d)** $5\left(x+\tfrac{1}{2}\right)^2 - \dfrac{9}{4}$

e) $2\left(x-\tfrac{1}{4}\right)^2 + \dfrac{7}{8}$ **f)** $3(x+3)^2 + 63$

g) $4\left(x+\tfrac{13}{8}\right)^2 - \dfrac{25}{16}$ **h)** $2\left(x+\tfrac{11}{4}\right)^2 - \dfrac{121}{8}$

i) $-2\left(x-\tfrac{5}{2}\right)^2 + \dfrac{21}{2}$

Here, $a = -2$ so the bracket is
$-2\left(x+\tfrac{10}{2\times -2}\right)^2 = -2\left(x-\tfrac{5}{2}\right)^2.$

Page 139   Exercise 3

1 a) $(x-1)^2 = x^2 - 2x + 1$
$x^2 - 2x - 4 = x^2 - 2x + 1 - 5 = (x-1)^2 - 5$
$x^2 - 2x - 4 = 0 \Rightarrow (x-1)^2 - 5 = 0$
$\Rightarrow (x-1)^2 = 5 \Rightarrow x - 1 = \pm\sqrt{5}$
$\Rightarrow x = 1 \pm \sqrt{5}$

b) $(x+2)^2 = x^2 + 4x + 4$
$x^2 + 4x + 3 = x^2 + 4x + 4 - 1 = (x+2)^2 - 1$
$x^2 + 4x + 3 = 0 \Rightarrow (x+2)^2 - 1 = 0$
$\Rightarrow (x+2)^2 = 1 \Rightarrow x + 2 = \pm 1$
$\Rightarrow x = -2 \pm 1 \Rightarrow x = -1 \text{ or } x = -3$
This quadratic factorises quite easily but the question tells you to complete the square.

Using the same method for c)-f):

c) $x = -3 \pm \sqrt{13}$ **d)** $x = -4 \pm 2\sqrt{3}$

e) $x = \dfrac{1}{2} \pm \dfrac{\sqrt{5}}{2}$ **f)** $x = \dfrac{11}{2} \pm \dfrac{\sqrt{21}}{2}$

2 a) $(x+3)^2 = x^2 + 6x + 9$
$x^2 + 6x + 4 = x^2 + 6x + 9 - 5 = (x+3)^2 - 5$
$x^2 + 6x + 4 = 0 \Rightarrow (x+3)^2 - 5 = 0$
$\Rightarrow (x+3)^2 = 5 \Rightarrow x + 3 = \pm\sqrt{5}$
$\Rightarrow x = -3 \pm \sqrt{5}$
$\Rightarrow x = -0.7639... = -0.76$ (2 d.p.)
or $x = -5.2360... = -5.24$ (2 d.p.)

b) $(x-1)^2 = x^2 - 2x + 1$
$x^2 - 2x - 5 = x^2 - 2x + 1 - 6 = (x-1)^2 - 6$
$x^2 - 2x - 5 = 0 \Rightarrow (x-1)^2 - 6 = 0$
$\Rightarrow (x-1)^2 = 6 \Rightarrow x - 1 = \pm\sqrt{6}$
$\Rightarrow x = 1 \pm \sqrt{6}$
$\Rightarrow x = 3.4494... = 3.45$ (2 d.p.)
or $x = -1.4494... = -1.45$ (2 d.p.)

Using the same method for c)-f):

c) $x = 0.4641... = 0.46$ (2 d.p.)
or $x = -6.4641... = -6.46$ (2 d.p.)

d) $x = -1.1715... = -1.17$ (2 d.p.)
or $x = -6.8284... = -6.83$ (2 d.p.)

e) $x = 3.7015... = 3.70$ (2 d.p.)
or $x = -2.7015... = -2.70$ (2 d.p.)

f) $x = 4.3027... = 4.30$ (2 d.p.)
or $x = 0.6972... = 0.70$ (2 d.p.)

3 a) $a\left(x+\tfrac{b}{2a}\right)^2 = 3\left(x+\tfrac{2}{6}\right)^2 = 3\left(x+\tfrac{1}{3}\right)^2$
$$= 3\left(x^2 + \tfrac{2}{3}x + \tfrac{1}{9}\right)$$
$$= 3x^2 + 2x + \tfrac{1}{3}$$
$$3x^2 + 2x - 2 = 3x^2 + 2x + \tfrac{1}{3} - \tfrac{7}{3}$$
$$= 3\left(x+\tfrac{1}{3}\right)^2 - \tfrac{7}{3}$$
$3x^2 + 2x - 2 = 0 \Rightarrow 3\left(x+\tfrac{1}{3}\right)^2 - \tfrac{7}{3} = 0$
$\Rightarrow \left(x+\tfrac{1}{3}\right)^2 = \tfrac{7}{9} \Rightarrow x + \tfrac{1}{3} = \pm\tfrac{\sqrt{7}}{3}$
$\Rightarrow x = -\dfrac{1}{3} \pm \dfrac{\sqrt{7}}{3}$

b) $a\left(x+\tfrac{b}{2a}\right)^2 = 5\left(x+\tfrac{2}{10}\right)^2 = 5\left(x+\tfrac{1}{5}\right)^2$
$$= 5\left(x^2 + \tfrac{2}{5}x + \tfrac{1}{25}\right)$$
$$= 5x^2 + 2x + \tfrac{1}{5}$$
$$5x^2 + 2x - 10 = 5x^2 + 2x + \tfrac{1}{5} - \tfrac{51}{5}$$
$$= 5\left(x+\tfrac{1}{5}\right)^2 - \tfrac{51}{5}$$
$5x^2 + 2x - 10 = 0 \Rightarrow 5\left(x+\tfrac{1}{5}\right)^2 - \tfrac{51}{5} = 0$
$\Rightarrow \left(x+\tfrac{1}{5}\right)^2 = \tfrac{51}{25} \Rightarrow x + \tfrac{1}{5} = \pm\tfrac{\sqrt{51}}{5}$
$\Rightarrow x = -\dfrac{1}{5} \pm \dfrac{\sqrt{51}}{5}$

Using the same method for c)-f):

c) $x = \dfrac{3}{4} \pm \dfrac{\sqrt{13}}{4}$ **d)** $x = 3 \pm \sqrt{\dfrac{13}{2}}$

e) $x = -\dfrac{5}{6} \pm \dfrac{\sqrt{145}}{6}$ **f)** $x = -\dfrac{7}{20} \pm \dfrac{\sqrt{89}}{20}$

4 a) $a\left(x+\tfrac{b}{2a}\right)^2 = 2\left(x+\tfrac{2}{4}\right)^2 = 2\left(x+\tfrac{1}{2}\right)^2$
$$= 2\left(x^2 + x + \tfrac{1}{4}\right) = 2x^2 + 2x + \tfrac{1}{2}$$
$$2x^2 + 2x - 3 = 2x^2 + 2x + \tfrac{1}{2} - \tfrac{7}{2}$$
$$= 2\left(x+\tfrac{1}{2}\right)^2 - \tfrac{7}{2}$$
$2x^2 + 2x - 3 = 0 \Rightarrow 2\left(x+\tfrac{1}{2}\right)^2 - \tfrac{7}{2} = 0$
$\Rightarrow \left(x+\tfrac{1}{2}\right)^2 = \tfrac{7}{4} \Rightarrow x + \tfrac{1}{2} = \pm\tfrac{\sqrt{7}}{2}$
$\Rightarrow x = -\dfrac{1}{2} \pm \dfrac{\sqrt{7}}{2}$
So $x = 0.8228... = 0.82$ (2 d.p.)
or $x = -1.8228... = -1.82$ (2 d.p.)

b) $a\left(x+\tfrac{b}{2a}\right)^2 = 3\left(x+\tfrac{2}{6}\right)^2 = 3\left(x+\tfrac{1}{3}\right)^2$
$$= 3\left(x^2 + \tfrac{2}{3}x + \tfrac{1}{9}\right)$$
$$= 3x^2 + 2x + \tfrac{1}{3}$$
$$3x^2 + 2x - 7 = 3x^2 + 2x + \tfrac{1}{3} - \tfrac{22}{3}$$
$$= 3\left(x+\tfrac{1}{3}\right)^2 - \tfrac{22}{3}$$
$3x^2 + 2x - 7 = 0 \Rightarrow 3\left(x+\tfrac{1}{3}\right)^2 - \tfrac{22}{3} = 0$
$\Rightarrow \left(x+\tfrac{1}{3}\right)^2 = \tfrac{22}{9} \Rightarrow x + \tfrac{1}{3} = \pm\tfrac{\sqrt{22}}{3}$
$\Rightarrow x = -\dfrac{1}{3} \pm \dfrac{\sqrt{22}}{3}$
So $x = 1.2301... = 1.23$ (2 d.p.)
or $x = -1.8968... = -1.90$ (2 d.p.)

Using the same method for c)-f):

c) $x = 0.9364... = 0.94$ (2 d.p.)
or $x = -2.9365... = -2.94$ (2 d.p.)

d) $x = 9.0497... = 9.05$ (2 d.p.)
or $x = -1.0497... = -1.05$ (2 d.p.)

e) $x = 0.2287... = 0.23$ (2 d.p.)
or $x = -0.7287... = -0.73$ (2 d.p.)

f) $x = 0.6408... = 0.64$ (2 d.p.)
or $x = -3.6408... = -3.64$ (2 d.p.)

11.3 The Quadratic Formula

Page 141   Exercise 1

1 a) $a = 1, b = -3, c = 1$
$$x = \frac{-(-3) \pm \sqrt{(-3)^2 - 4\times 1\times 1}}{2\times 1}$$
$$= \frac{3 \pm \sqrt{9-4}}{2} = \frac{3 \pm \sqrt{5}}{2}$$

b) $a = 1, b = -2, c = -12$
$$x = \frac{-(-2) \pm \sqrt{(-2)^2 - 4\times 1\times(-12)}}{2\times 1}$$
$$= \frac{2 \pm \sqrt{4+48}}{2} = \frac{2 \pm \sqrt{52}}{2}$$
But $\sqrt{52} = \sqrt{4\times 13} = 2\sqrt{13}$ and so
$$x = \frac{2 \pm 2\sqrt{13}}{2} = 1 \pm \sqrt{13}$$

Using the same method for c)-i):

c) $a = 4, b = -3, c = -8$
Then $x = \dfrac{3 \pm \sqrt{137}}{8}$

d) $a = 1, b = 1, c = -1$
Then $x = \dfrac{-1 \pm \sqrt{5}}{2}$

e) $a = 1, b = -8, c = -5$
Then $x = 4 \pm \sqrt{21}$

f) $a = 3, b = 6, c = -5$
Then $x = \dfrac{-3 \pm 2\sqrt{6}}{3}$

g) $a = 1, b = -5, c = -3$
Then $x = \dfrac{5 \pm \sqrt{37}}{2}$

h) $a = -2, b = 8, c = 13$
Then $x = \dfrac{4 \pm \sqrt{42}}{2}$
Make sure you get the right values for a, b and c — the x^2, x and constant terms might not be written in the order you expect.

i) $a = 1, b = -7, c = 3$
Then $x = \dfrac{7 \pm \sqrt{37}}{2}$

2 a) $a = 1, b = 3, c = 1$
$$x = \frac{-3 \pm \sqrt{3^2 - 4\times 1\times 1}}{2\times 1}$$
$$= \frac{-3 \pm \sqrt{9-4}}{2} = \frac{-3 \pm \sqrt{5}}{2}$$
So $x = -0.3819... = -0.38$ (2 d.p.)
or $x = -2.6180... = -2.62$ (2 d.p.)

b) $a = 3, b = 2, c = -2$
$$x = \frac{-2 \pm \sqrt{2^2 - 4\times 3\times(-2)}}{2\times 3}$$
$$= \frac{-2 \pm \sqrt{4+24}}{6} = \frac{-2 \pm \sqrt{28}}{6}$$
So $x = 0.5485... = 0.55$ (2 d.p.)
or $x = -1.2152... = -1.22$ (2 d.p.)

Using the same method for c)-i):

c) $a = 1, b = -3, c = -3$
Then $x = 3.7912... = 3.79$ (2 d.p.)
or $x = -0.7912... = -0.79$ (2 d.p.)

d) $a = 1, b = 5, c = -4$
Then $x = 0.7015... = 0.70$ (2 d.p.)
or $x = -5.7015... = -5.70$ (2 d.p.)

e) $a = -1, b = -8, c = 11$
Then $x = 1.1961... = 1.20$ (2 d.p.)
or $x = -9.1961... = -9.20$ (2 d.p.)

f) $a = 1, b = -7, c = -6$
Then $x = 7.7720... = 7.77$ (2 d.p.)
or $x = -0.7720... = -0.77$ (2 d.p.)

g) $a = 1$, $b = 6$, $c = -2$
Then $x = 0.3166... = \mathbf{0.32}$ (2 d.p.)
or $x = -6.3166... = \mathbf{-6.32}$ (2 d.p.)

h) $a = 1$, $b = 4$, $c = -1$
Then $x = 0.2360... = \mathbf{0.24}$ (2 d.p.)
or $x = -4.2360... = \mathbf{-4.24}$ (2 d.p.)

i) $a = 2$, $b = 8$, $c = 3$
Then $x = -0.4188... = \mathbf{-0.42}$ (2 d.p.)
or $x = -3.5811... = \mathbf{-3.58}$ (2 d.p.)

3 a) $x^2 + 3x = 6 \Rightarrow x^2 + 3x - 6 = 0$
$a = 1$, $b = 3$, $c = -6$
$$x = \frac{-3 \pm \sqrt{3^2 - 4 \times 1 \times (-6)}}{2 \times 1}$$
$$= \frac{-3 \pm \sqrt{9 + 24}}{2} = \frac{-3 \pm \sqrt{33}}{2}$$
So $\boldsymbol{x = \dfrac{-3 \pm \sqrt{33}}{2}}$

b) $x^2 - 5x + 11 = 2x + 3 \Rightarrow x^2 - 7x + 8 = 0$
$a = 1$, $b = -7$, $c = 8$
$$x = \frac{-(-7) \pm \sqrt{(-7)^2 - 4 \times 1 \times 8}}{2 \times 1}$$
$$= \frac{7 \pm \sqrt{49 - 32}}{2} = \frac{7 \pm \sqrt{17}}{2}$$
So $\boldsymbol{x = \dfrac{7 \pm \sqrt{17}}{2}}$

Using the same method for c)-i):

c) $a = 1$, $b = -7$, $c = 7$
Then $\boldsymbol{x = \dfrac{7 \pm \sqrt{21}}{2}}$

d) $a = 1$, $b = 2$, $c = -11$
Then $\boldsymbol{x = -1 + 2\sqrt{3}}$ or $\boldsymbol{x = -1 - 2\sqrt{3}}$

e) $a = 1$, $b = 3$, $c = -5$
Then $\boldsymbol{x = \dfrac{-3 \pm \sqrt{29}}{2}}$

f) $a = 1$, $b = -3$, $c = -9$
Then $\boldsymbol{x = \dfrac{3 \pm 3\sqrt{5}}{2}}$

g) $a = 1$, $b = -6$, $c = 4$
Then $\boldsymbol{x = 3 \pm \sqrt{5}}$

h) $a = 1$, $b = 2$, $c = -6$
Then $\boldsymbol{x = -1 \pm \sqrt{7}}$

i) $a = 2$, $b = -3$, $c = -7$
Then $\boldsymbol{x = \dfrac{3 \pm \sqrt{65}}{4}}$

Page 141   Exercise 2

1 a) $b^2 - 4ac = (-1)^2 - 4 \times 3 \times -1 = 13$
Since this is positive, there will be **two solutions**.

b) $b^2 - 4ac = 2^2 - 4 \times 2 \times 3 = -20$
Since this is negative, there will be **no solutions**.

c) $b^2 - 4ac = 6^2 - 4 \times 3 \times 3 = 0$
So there will be **one solution**.

2 Graph A just touches the x-axis, so there is one solutions, matching the equation from **c)**.
Graph B never crosses the x-axis, so there are no solutions, matching the equation from **b)**.
Graph C crosses the x-axis twice, so there are two solutions, matching the equation from **a)**.

3 The perimeter of the triangle is
$x^2 + 2x + (4 - 2x^2) = -x^2 + 2x + 4$.
If the perimeter equals 6 cm then you get the
equation $-x^2 + 2x + 4 = 6 \Rightarrow -x^2 + 2x - 2 = 0$.
If you try to solve this using the formula,
you'll get $b^2 - 4ac = 2^2 - 4 \times -1 \times -2 = -4$
under the square root. So there are no solutions, i.e. the situation is **impossible**.

Page 142   Review Exercise

1 a) $(x - 10)(x - 2) = 0 \Rightarrow \boldsymbol{x = 10}$ or $\boldsymbol{x = 2}$

b) $(x + 5)(x - 4) = 0 \Rightarrow \boldsymbol{x = -5}$ or $\boldsymbol{x = 4}$

c) $(x - 10)(x + 5) = 0 \Rightarrow \boldsymbol{x = 10}$ or $\boldsymbol{x = -5}$

d) $(x + 8)(x + 6) = 0 \Rightarrow \boldsymbol{x = -8}$ or $\boldsymbol{x = -6}$

e) $(3x + 5)(x - 3) = 0 \Rightarrow \boldsymbol{x = -\dfrac{5}{3}}$ or $\boldsymbol{x = 3}$

---

f) $(3x + 2)(4x + 1) = 3$
$\Rightarrow 12x^2 + 11x + 2 = 3$
$\Rightarrow 12x^2 + 11x - 1 = 0$
$\Rightarrow (12x - 1)(x + 1) = 0$
$\Rightarrow \boldsymbol{x = \dfrac{1}{12}}$ or $\boldsymbol{x = -1}$

2 a) $(x - 9)^2 = x^2 - 18x + 81$
$x^2 - 18x - 3 = x^2 - 18x + 81 - 84$
$\qquad = (x - 9)^2 - 84 = 0$
$\Rightarrow x - 9 = \pm\sqrt{84} = \pm 2\sqrt{21}$
$\Rightarrow \boldsymbol{x = 9 \pm 2\sqrt{21}}$

b) $\left(x - \dfrac{5}{2}\right)^2 = x^2 - 5x + \dfrac{25}{4}$
$x^2 - 5x + 6 = x^2 - 5x + \dfrac{25}{4} - \dfrac{1}{4}$
$\qquad = \left(x - \dfrac{5}{2}\right)^2 - \dfrac{1}{4} = 0$
$\Rightarrow x - \dfrac{5}{2} = \pm\dfrac{1}{2} \Rightarrow \boldsymbol{x = 3, \ x = 2}$

c) $x(x + 4) - 5 = 0 \Rightarrow x^2 + 4x - 5 = 0$
$(x + 2)^2 = x^2 + 4x + 4$
$x^2 + 4x - 5 = x^2 + 4x + 4 - 9$
$\qquad = (x + 2)^2 - 9 = 0$
$\Rightarrow x + 2 = \pm 3 \Rightarrow \boldsymbol{x = 1, \ x = -5}$

d) $(2x + 1)(x - 2) = 2 \Rightarrow 2x^2 - 3x - 2 = 2$
$\Rightarrow 2x^2 - 3x - 4 = 0$
$2\left(x - \dfrac{3}{4}\right)^2 = 2x^2 - 3x + \dfrac{9}{8}$
$2x^2 - 3x - 4 = 2x^2 - 3x + \dfrac{9}{8} - \dfrac{41}{8}$
$\qquad = 2\left(x - \dfrac{3}{4}\right)^2 - \dfrac{41}{8} = 0$
$\Rightarrow x - \dfrac{3}{4} = \dfrac{\pm\sqrt{41}}{4} \Rightarrow \boldsymbol{x = \dfrac{3 \pm \sqrt{41}}{4}}$

e) $\dfrac{1}{3x + 1} = \dfrac{x}{2x^2 + 3} \Rightarrow 2x^2 + 3 = x(3x + 1)$
$\Rightarrow x^2 + x - 3 = 0$
$\left(x + \dfrac{1}{2}\right)^2 = x^2 + x + \dfrac{1}{4}$
$x^2 + x - 3 = x^2 + x + \dfrac{1}{4} - \dfrac{13}{4}$
$\qquad = \left(x + \dfrac{1}{2}\right)^2 - \dfrac{13}{4} = 0$
$x + \dfrac{1}{2} = \dfrac{\pm\sqrt{13}}{2} \Rightarrow \boldsymbol{x = \dfrac{-1 \pm \sqrt{13}}{2}}$

f) $\dfrac{1}{2x - 1} + \dfrac{1}{5x - 3} = 1$
$\Rightarrow \dfrac{5x - 3 + 2x - 1}{(2x - 1)(5x - 3)} = 1$
$\Rightarrow 7x - 4 = (2x - 1)(5x - 3)$
$\Rightarrow 7x - 4 = 10x^2 - 11x + 3$
$\Rightarrow 10x^2 - 18x + 7 = 0$
$10\left(x - \dfrac{9}{10}\right)^2 = 10x^2 - 18x + \dfrac{81}{10}$
$10x^2 - 18x + 7 = 10x^2 - 18x + \dfrac{81}{10} - \dfrac{11}{10}$
$\qquad = 10\left(x - \dfrac{9}{10}\right)^2 - \dfrac{11}{10} = 0$
$\Rightarrow x - \dfrac{9}{10} = \dfrac{\pm\sqrt{11}}{10} \Rightarrow \boldsymbol{x = \dfrac{9 \pm \sqrt{11}}{10}}$

3 a) $a = 1$, $b = -21$, $c = 27$
$$x = \frac{-(-21) \pm \sqrt{(-21)^2 - 4 \times 1 \times 27}}{2 \times 1}$$
$$= \frac{21 \pm \sqrt{333}}{2} = \frac{21 \pm 3\sqrt{37}}{2}$$
So $x = 19.6241... = \mathbf{19.62}$ (2 d.p.)
or $x = 1.3758... = \mathbf{1.38}$ (2 d.p.)

b) $a = 2$, $b = -1$, $c = -18$
$$x = \frac{-(-1) \pm \sqrt{1^2 - 4 \times 2 \times (-18)}}{2 \times 2}$$
$$= \frac{1 \pm \sqrt{145}}{4}$$
So $x = 3.2603... = \mathbf{3.26}$ (2 d.p.)
or $x = -2.7603... = \mathbf{-2.76}$ (2 d.p.)

c) $5x^2 = x - 7 \Rightarrow 5x^2 - x + 7 = 0$
$a = 5$, $b = -1$, $c = 7$
$$x = \frac{-(-1) \pm \sqrt{(-1)^2 - 4 \times 5 \times 7}}{2 \times 5}$$
$$= \frac{1 \pm \sqrt{-139}}{10}$$
So there are **no solutions**.

---

d) $\dfrac{1}{x + 1} = \dfrac{x + 1}{x - 2} \Rightarrow x - 2 = (x + 1)^2$
$\Rightarrow x - 2 = x^2 + 2x + 1 \Rightarrow x^2 + x + 3 = 0$
$a = 1$, $b = 1$, $c = 3$
$$x = \frac{-1 \pm \sqrt{1^2 - 4 \times 1 \times 3}}{2 \times 1} = \frac{-1 \pm \sqrt{-11}}{2}$$
So there are **no solutions**.

e) $\dfrac{x}{2x^2 + 1} = \dfrac{1}{x - 2} \Rightarrow x(x - 2) = 2x^2 + 1$
$\Rightarrow x^2 - 2x = 2x^2 + 1 \Rightarrow x^2 + 2x + 1 = 0$
$a = 1$, $b = 2$, $c = 1$
$$x = \frac{-2 \pm \sqrt{2^2 - 4 \times 1 \times 1}}{2 \times 1} = \frac{-2 \pm \sqrt{0}}{2} = \boldsymbol{-1}$$

f) $\dfrac{1}{x + 1} + \dfrac{1}{x - 2} = 5 \Rightarrow \dfrac{x - 2 + x + 1}{(x + 1)(x - 2)} = 5$
$\Rightarrow 2x - 1 = 5(x + 1)(x - 2)$
$\Rightarrow 2x - 1 = 5x^2 - 5x - 10$
$\Rightarrow 5x^2 - 7x - 9 = 0$
$a = 5$, $b = -7$, $c = -9$
$$x = \frac{-(-7) \pm \sqrt{(-7)^2 - 4 \times 5 \times (-9)}}{2 \times 5}$$
$$= \frac{7 \pm \sqrt{229}}{10}$$
So $x = 2.2132... = \mathbf{2.21}$ (2 d.p.)
or $x = -0.8132... = \mathbf{-0.81}$ (2 d.p.)

4 a) $(x + 1)^2 = 9 \Rightarrow x + 1 = \pm 3 \Rightarrow x = 1 \pm 3$
So $\boldsymbol{x = 2}$ or $\boldsymbol{x = -4}$

b) $x^2 + 3 = 147 \Rightarrow x^2 = 144 \Rightarrow \boldsymbol{x = \pm 12}$

c) $x^2 + x = 22 \Rightarrow x^2 + x - 22 = 0$
$a = 1$, $b = 1$, $c = -22$
$$x = \frac{-1 \pm \sqrt{1^2 - 4 \times 1 \times (-22)}}{2 \times 1} = \frac{-1 \pm \sqrt{89}}{2}$$
So $x = 4.2169... = \mathbf{4.22}$ (2 d.p.)
or $x = -5.2169... = \mathbf{-5.22}$ (2 d.p.)

d) $25 - x^2 = 9 \Rightarrow x^2 = 16 \Rightarrow \boldsymbol{x = \pm 4}$

e) $(2x)^2 = 36 \Rightarrow 2x = \pm 6 \Rightarrow \boldsymbol{x = \pm 3}$

f) $(x + 5)^2 = 100 \Rightarrow x + 5 = \pm 10$
$\Rightarrow x = -5 \pm 10$
So $\boldsymbol{x = 5}$ or $\boldsymbol{x = -15}$

5 a) $x(x + 3) = 28 \Rightarrow x^2 + 3x - 28 = 0$
$\Rightarrow (x + 7)(x - 4) = 0 \Rightarrow x = -7$ or $x = 4$
But x cannot be negative so $\boldsymbol{x = 4}$.

b) $x(x + 7) = 30 \Rightarrow x^2 + 7x - 30 = 0$
$(x + 10)(x - 3) = 0 \Rightarrow x = -10$ or $x = 3$
But x cannot be negative so $\boldsymbol{x = 3}$.

6 a) Length of fence $= 20 = x + y + x$
$\Rightarrow y = 20 - 2x$
Area $= 50 = xy = x(20 - 2x)$
So $\boldsymbol{x(20 - 2x) = 50}$, as required.

b) $x(20 - 2x) = 50 \Rightarrow 2x^2 - 20x + 50 = 0$
$\Rightarrow x^2 - 10x + 25 = 0 \Rightarrow (x - 5)^2 = 0$
So $x = 5$ and then $y = 20 - (2 \times 5) = 10$.
Therefore, the width of the area is **5 m** and the length is **10 m**.

7 a) $x^2 + (x + 5)^2 = (x + 10)^2$
$\Rightarrow x^2 + x^2 + 10x + 25 = x^2 + 20x + 100$
$\Rightarrow x^2 - 10x - 75 = 0$
$\Rightarrow (x + 5)(x - 15) = 0 \Rightarrow x = -5$ or $x = 15$
But x cannot be negative so $\boldsymbol{x = 15}$.

b) $x^2 + (x + 9)^2 = (2x + 1)^2$
$\Rightarrow x^2 + x^2 + 18x + 81 = 4x^2 + 4x + 1$
$\Rightarrow 2x^2 - 14x - 80 = 0$
$\Rightarrow x^2 - 7x - 40 = 0$
$a = 1$, $b = -7$, $c = -40$
$$x = \frac{-(-7) \pm \sqrt{(-7)^2 - 4 \times 1 \times (-40)}}{2 \times 1}$$
$$= \frac{7 \pm \sqrt{209}}{2}$$
So $x = 10.7284...$ or $x = -3.7284...$
But x cannot be negative
so $\boldsymbol{x = 10.73}$ (2 d.p.).

Page 143   Exam-Style Questions

1 a) The only factors of 21 are 1×21 and 3×7
$2x^2 - 13x + 21 = (2x - 7)(x - 3)$
$\boldsymbol{m = -7}$ and $\boldsymbol{n = -3}$ *[1 mark]*

b) $2x - 7 = 0$ or $x - 3 = 0$
$\boldsymbol{x = \dfrac{7}{2}, \ x = 3}$ *[1 mark]*

2 a) $a = 8 \div 2 = 4$ *[1 mark]*
$(x + 4)^2 = x^2 + 8x + 16$
$x^2 + 8x + 12 = x^2 + 8x + 16 - 4$
$= (x + 4)^2 - 4$ *[1 mark]*

b) $(x + 4)^2 - 4 = 0 \Rightarrow (x + 4)^2 = 4$
$\Rightarrow x + 4 = \pm 2 \Rightarrow x = -4 \pm 2$ *[1 mark]*
$x = -2, \ x = -6$ *[1 mark]*

3 $\frac{1}{x} - \frac{x}{2} = 2 \Rightarrow \frac{2 - x^2}{2x} = 2$
$\Rightarrow 2 - x^2 = 4x$ *[1 mark]*
$\Rightarrow x^2 + 4x - 2 = 0$ *[1 mark]*
$a = 1, b = 4, c = -2$

So $x = \dfrac{-4 \pm \sqrt{4^2 - 4 \times 1 \times (-2)}}{2 \times 1}$

$= \dfrac{-4 \pm \sqrt{24}}{2}$ *[1 mark]*

$= \dfrac{-4 \pm 2\sqrt{6}}{2} = -2 \pm \sqrt{6}$ *[1 mark]*

4 Hannah has used $b = 9$ when it should have been $b = -9$ *[1 mark]*. She has used $c = 5$ when it should have been $c = 4$ (the right-hand side of the equation needs to be 0) *[1 mark]*. She has used + instead of ± (so she will only get one solution, not two) *[1 mark]*.

5 $n^2 - 3n - 60 = 2n + 6$
so $n^2 - 3n - 2n - 60 - 6 = 0$
$\Rightarrow n^2 - 5n - 66 = 0$ *[1 mark]*
$\Rightarrow (n + 6)(n - 11) = 0$ *[1 mark]*
So $n + 6 = 0$ or $n - 11 = 0$
$n = -6$ or $n = 11$
But n cannot be negative so $n = 11$ *[1 mark]*

6 $\dfrac{3(3t) + 2(7t + 2)}{6} = t^2$
$\Rightarrow 3(3t) + 2(7t + 2) = 6t^2$ *[1 mark]*
$\Rightarrow 9t + 14t + 4 = 6t^2$
$\Rightarrow 6t^2 - 23t - 4 = 0$ *[1 mark]*
$\Rightarrow (6t + 1)(t - 4) = 0$ *[1 mark]*
So $6t + 1 = 0$ or $t - 4 = 0$
$t = -\frac{1}{6}$ or $t = 4$
But t cannot be negative so $t = 4$ *[1 mark]*

Section 12 — Simultaneous Equations

12.1 Simultaneous Linear Equations

Page 145 Exercise 1

1 a) $x + 3y = 13$
$\underline{- (x - y = 5)}$
$\quad\ 4y = 8$
$\quad\quad y = 2$
$x + 3(2) = 13 \Rightarrow x = 7$
Don't forget to put your values into the other equation to check them.

b) $2x - y = 7$
$\underline{+ (4x + y = 23)}$
$\quad 6x = 30$
$\quad\quad x = 5$
$2(5) - y = 7 \Rightarrow y = 3$

Using the same method for c)-i):
c) $x = -2, y = 4$
d) $x = 6, y = 1$
e) $x = 3, y = -5$
f) $x = 8, y = 0$
g) $x = \frac{1}{2}, y = 3$
h) $x = \frac{1}{3}, y = 10$
i) $x = 2, y = -3$
You could also use the substitution method for all the questions in Exercise 1.

Page 146 Exercise 2

1 a) (1) $3x + 2y = 16$, (2) $2x + y = 9$
(1) $\quad\quad 3x + 2y = 16$
(2) × 2: $\underline{- (4x + 2y = 18)}$
$\quad\quad\quad\quad -x = -2$
$\quad\quad\quad\quad\ \ x = 2$
$2 \times 2 + y = 9 \Rightarrow y = 5$

b) (1) $4x + 3y = 16$, (2) $5x - y = 1$
(1) $\quad\quad 4x + 3y = 16$
(2) × 3: $\underline{+ (15x - 3y = 3)}$
$\quad\quad\quad\quad 19x = 19$
$\quad\quad\quad\quad\ \ x = 1$
$5 \times 1 - y = 1 \Rightarrow y = 4$

Using the same method for c)-f):
c) $x = 6, y = 2$
d) $x = 5, y = 0$
e) $x = 3, y = 2$
f) $e = -1, r = -4$

2 a) (1) $3x - 2y = 8$, (2) $5x - 3y = 14$
(2) × 2: $(10x - 6y = 28)$
(1) × 3: $\underline{- (9x - 6y = 24)}$
$\quad\quad\quad\quad x = 4$
$3(4) - 2y = 8 \Rightarrow 2y = 4 \Rightarrow y = 2$

b) (1) $4p + 3q = 17$, (2) $3p - 4q = 19$
(1) × 4: $(16p + 12q = 68)$
(2) × 3: $\underline{+ (9p - 12q = 57)}$
$\quad\quad\quad\quad 25p = 125$
$\quad\quad\quad\quad\ \ p = 5$
$4(5) + 3q = 17 \Rightarrow 3q = -3 \Rightarrow q = -1$

Using the same method for c)-f):
c) $u = 2, v = 1$
d) $c = 8, d = \frac{1}{2}$
e) $r = -2, s = 4$
f) $m = 3, n = 1$

Page 146 Exercise 3

1 $\quad x + y = 58$
$\quad \underline{+ (x - y = 22)}$
$\quad\quad\ 2x = 80$
$\quad\quad\quad x = 40$
$40 + y = 58 \Rightarrow y = 18$

2 Let d = sherbet dip, c = chocolate bar
(1) $4d + 3c = 1.91$, (2) $3d + 4c = 1.73$
(1) × 4: $(16d + 12c = 7.64)$
(2) × 3: $\underline{- (9d + 12c = 5.19)}$
$\quad\quad\quad\quad 7d = 2.45$
$\quad\quad\quad\quad\ d = 0.35$
$4 \times 0.35 + 3c = 1.91 \Rightarrow 3c = 0.51 \Rightarrow c = 0.17$
So a sherbet dip costs **35p** and a chocolate bar costs **17p**

3 Let y = yellow aliens, b = blue spiders
(1) $7y + 5b = 85$, (2) $6y + 11b = 93$
(1) × 6: $(42y + 30b = 510)$
(2) × 7: $\underline{- (42y + 77b = 651)}$
$\quad\quad\quad\quad -47b = -141$
$\quad\quad\quad\quad\quad\ b = 3$
$7y + 5(3) = 85 \Rightarrow 7y = 70 \Rightarrow y = 10$
So Hal's score = $8(10) + 3 =$ **83 points**.

4 $3(x + y) = 5x + 2y - 1$
$\Rightarrow 3x + 3y - 5x - 2y = -1$
$\Rightarrow 2x - y = 1$ (1)
$3(x + y) = 4x + 4 + y$
$\Rightarrow 3x + 3y = 4x + 4 + y$
$\Rightarrow x - 2y = -4$ (2)
(1) $\quad\quad\quad 2x - y = 1$
(2) × 2: $\underline{- (2x - 4y = -8)}$
$\quad\quad\quad\quad 3y = 9$
$\quad\quad\quad\quad\ \ y = 3$
$2x - 3 = 1 \Rightarrow 2x = 4, x = 2$
$3(x + y) = 3(2 + 3) =$ **15 cm**
Your working might look a bit different if you used different pairs of sides.

12.2 Simultaneous Linear and Quadratic Equations

Page 147 Exercise 1

1 a) Substitute $y = 2x$ into $y = x^2 - 4x + 8$:
$2x = x^2 - 4x + 8$
$x^2 - 6x + 8 = 0$
$(x - 2)(x - 4) = 0$
so $x = 2$ or $x = 4$
$y = 2(2) = 4$ or $y = 2(4) = 8$
$x = 2, y = 4$ and $x = 4, y = 8$

b) $3x = 2 - y \Rightarrow y = 2 - 3x$.
Substitute $y = 2 - 3x$ into $y = x^2 - x - 1$:
$2 - 3x = x^2 - x - 1$
$x^2 + 2x - 3 = 0$
$(x - 1)(x + 3) = 0$
so $x = 1$ or $x = -3$
$y = 2 - 3(1) = -1$ or $y = 2 - 3(-3) = 11$
$x = 1, y = -1$ and $x = -3, y = 11$

Using the same method for c)-i):
c) $x = 10, y = 32$ and $x = -3, y = -7$
d) $x = 1, y = 2$ and $x = -5, y = 8$
e) $x = 4, y = 2$ and $x = 2, y = -2$
f) $x = 7, y = 32$ and $x = -2, y = 5$
g) $x = 5, y = 15$ and $x = -2, y = 1$
h) $x = 4, y = 8$ and $x = 1, y = 5$
i) $x = 0.5, y = -1$ and $x = 3, y = 19$

2 $2y = x + 3 \Rightarrow y = 0.5x + 1.5$
Substitute $y = 0.5x + 1.5$ into $y = x^2 - 2x - 2$:
$0.5x + 1.5 = x^2 - 2x - 2$
$x^2 - 2.5x - 3.5 = 0 \Rightarrow 2x^2 - 5x - 7 = 0$
$(2x - 7)(x + 1) = 0$
so $x = 3.5$ or $x = -1$
$y = 0.5(3.5) + 1.5 = 3.25$
or $y = 0.5(-1) + 1.5 = 1$
N = (3.5, 3.25) and M = (-1, 1)
Make sure you give the solutions as coordinates.

3 Substitute $y = 4 - 3x$ into $y = 6x^2 + 10x - 1$:
$4 - 3x = 6x^2 + 10x - 1$
$6x^2 + 13x - 5 = 0$
$(3x - 1)(2x + 5) = 0$
so $x = \frac{1}{3}$ or $-\frac{5}{2}$
$y = 4 - 3\left(\frac{1}{3}\right) = 3$ or $y = 4 - 3\left(-\frac{5}{2}\right) = \frac{23}{2}$
$\left(\frac{1}{3}, 3\right)$ and $\left(-\frac{5}{2}, \frac{23}{2}\right)$

4 $x^2 + 3x + 1 = 5x$
$x^2 - 2x + 1 = 0$
$(x - 1)(x - 1) = 0$
so $x = 1$ and $y = 5$
The line and curve only meet at one point, so the line is a tangent to the curve.

5 $x - 4y = 2 \Rightarrow x = 2 + 4y$
$y^2 + y(2 + 4y) = 0$
$5y^2 + 2y = 0 \Rightarrow y(5y + 2) = 0$
so $y = 0$ or $y = -0.4$
$x = 2 + 4(0) = 2$ or $x = 2 + 4(-0.4) = 0.4$
$x = 2, y = 0$ and $x = 0.4, y = -0.4$

6 a) $x + y = 7 \Rightarrow y = 7 - x$
$x^2 - x(7 - x) - 4 = 0$
$2x^2 - 7x - 4 = 0$
$(2x + 1)(x - 4) = 0$
so $x = -0.5$ or $x = 4$
$y = 7 - (-0.5) = 7.5$ or $y = 7 - 4 = 3$
$x = -0.5, y = 7.5$ and $x = 4, y = 3$

b) $x + y = 5 \Rightarrow x = 5 - y$
$5 - y + y(5 - y) + 2y^2 = 2$
$2y^2 - y^2 + 5y - y + 5 - 2 = 0$
$y^2 + 4y + 3 = 0$
$(y + 1)(y + 3) = 0$
so $y = -1$ or $y = -3$
$x = 5 - (-1) = 6$ or $x = 5 - (-3) = 8$
$x = 6, y = -1$ and $x = 8, y = -3$

Using the same method for c)-f):
c) $x = 1, y = 1$ and $x = 4, y = -2$
d) $x = 2, y = -2$ and $x = -3, y = -7$
e) $x = 2, y = 2$ and $x = 4, y = 0$
f) $x = 16, y = -1.5$ and $x = -2, y = 3$

Page 148 Exercise 2

1 a) $2x + y = 3 \Rightarrow y = 3 - 2x$
$(3 - 2x)^2 - x^2 = 0$
$3x^2 - 12x + 9 = 0 \Rightarrow x^2 - 4x + 3 = 0$
$(x - 1)(x - 3) = 0$
so $x = 1$ or $x = 3$
$y = 3 - 2(1) = 1$ or $y = 3 - 2(3) = -3$
$x = 1, y = 1$ and $x = 3, y = -3$

b) $3x + y = 4 \implies y = 4 - 3x$
$x^2 + 3x(4 - 3x) + (4 - 3x)^2 = -16$
$x^2 + 12x - 9x^2 + 16 - 24x + 9x^2 + 16 = 0$
$x^2 - 12x + 32 = 0$
$(x - 8)(x - 4) = 0$
so $x = 8$ or $x = 4$
$y = 4 - 3(8) = -20$ or $y = 4 - 3(4) = -8$
$x = 8, y = -20$ and $x = 4, y = -8$

c) $x - y = -4 \implies y = x + 4$
$x^2 + (x + 4)^2 - x = 20$
$x^2 + x^2 + 8x + 16 - x - 20 = 0$
$2x^2 + 7x - 4 = 0$
$(2x - 1)(x + 4) = 0$
so $x = \frac{1}{2}$ or $x = -4$
$y = \frac{1}{2} + 4 = \frac{9}{2}$ or $y = (-4) + 4 = 0$
$x = \frac{1}{2}, y = \frac{9}{2}$ and $x = -4, y = 0$

2 a) $x - y = -3 \implies y = x + 3$
$3x^2 + 7x + (x + 3)^2 = 21$
$3x^2 + 7x + x^2 + 6x + 9 = 21$
$4x^2 + 13x - 12 = 0$

b) $(4x - 3)(x + 4) = 0$
so $x = \frac{3}{4}$ or $x = -4$

c) $y = \frac{3}{4} + 3 = \frac{15}{4}$ or $y = (-4) + 3 = -1$
So they cross at $\left(\frac{3}{4}, \frac{15}{4}\right)$ and $(-4, -1)$

3 a) $x = 3y + 4$
$(3y + 4)^2 + y^2 = 34$
$10y^2 + 24y - 18 = 0$
$5y^2 + 12y - 9 = 0$
$(5y - 3)(y + 3) = 0$
$y = \frac{3}{5}$ or $y = -3$
$x = 3\left(\frac{3}{5}\right) + 4 = \frac{29}{5}$ or $x = 3(-3) + 4 = -5$
$\left(\frac{29}{5}, \frac{3}{5}\right)$ and $(-5, -3)$

b) $y = \frac{1}{2} - x$
$x^2 + (\frac{1}{2} - x)^2 = 1$
$2x^2 - x - 0.75 = 0$
$x = \frac{1 \pm \sqrt{(-1)^2 - 4 \times 2 \times (-0.75)}}{2(2)} = \frac{1 \pm \sqrt{7}}{4}$
$y = \frac{1}{2} - \frac{1 \pm \sqrt{7}}{4} = \frac{2 - (1 \pm \sqrt{7})}{4} = \frac{1 \mp \sqrt{7}}{4}$
$\left(\frac{1 + \sqrt{7}}{4}, \frac{1 - \sqrt{7}}{4}\right)$ and $\left(\frac{1 - \sqrt{7}}{4}, \frac{1 + \sqrt{7}}{4}\right)$

c) $x = \sqrt{5}y - 6$
$(\sqrt{5}y - 6)^2 + y^2 = 36$
$\implies 5y^2 - 12\sqrt{5} + 36 + y^2 = 36$
$\implies 6y^2 - 12\sqrt{5}y = 0$
$\implies 6y(y - 2\sqrt{5}) = 0$
$\implies y = 0$ or $y = 2\sqrt{5}$
$x = \sqrt{5}(0) - 6 = -6$ or $x = \sqrt{5}(2\sqrt{5}) - 6 = 4$
$(-6, 0)$ and $(4, 2\sqrt{5})$

Page 149   Review Exercise

1 a)
$e + 2f = 7$
$\underline{-(6e + 2f = 10)}$
$-5e = -3$
$e = 0.6$
$0.6 + 2f = 7 \implies f = 3.2$
Using the same method for b)-c):

b) $h = 4, g = \frac{2}{5}$

c) $i = -5, j = -\frac{1}{2}$

2 a) (1) $5x - 3y = 12$, (2) $2x - y = 5$
(2) × 3: $6x - 3y = 15$
(1): $\underline{-(5x - 3y = 12)}$
$x = 3$
$2(3) - y = 5 \implies y = 1$

b) (1) $2x - y = 11$, (2) $-4x - 7y = 5$
(1) × 2: $4x - 2y = 22$
(2): $+(-4x - 7y = 5)$
$-9y = 27$
$y = -3$
$2x - (-3) = 11$ fi $x = 4$

Using the same method for c)-f):

c) $k = -1, l = 3$

d) $c = 8, d = 0.5$

e) $r = -2, s = 4$

f) $e = 2, f = -0.5$

3 a) Let t kg = mass of one textbook and
e kg = mass of one exercise book.
$2t + 30e = 6.9$ and $t + 20e = 4.2$
$t = 4.2 - 20e$
$2(4.2 - 20e) + 30e = 6.9 \implies -10e = -1.5$
$\implies e = 0.15$
$t + 20(0.15) = 4.2 \implies t = 4.2 - 3 = 1.2$
An exercise book weighs 150 g and a textbook weighs 1.2 kg.

b) $t + 25e = 1.2 + 25(0.15) = 4.95$
4.95 kg < 5 kg, so **yes.**

c) $5 - 2t = 5 - 2(1.2) = 2.6$
$2.6 \div e = 2.6 \div 0.15 = 17.33...$
so she can carry **17** exercise books.

4 Let £k = cost of one kettle and
£t = cost of one toaster.
(1) $25k + 20t = 1359.55$, (2) $12k + 9t = 641.79$
(1) × 9: $(225k + 180t = 12235.95)$
(2) × 20: $\underline{-(240k + 180t = 12835.8)}$
$-15k = -599.85$
$k = 39.99$
$12(39.99) + 9t = 641.79$
$9t = 161.91 \implies t = 17.99$
$80k + 56t = 80(39.99) + 56(17.99) = $ **£4206.64**

5 a) $2x^2 + 9x + 30 = 9 - 8x$
$2x^2 + 17x + 21 = 0$
$(2x + 3)(x + 7) = 0$
so $x = -1.5$ or $x = -7$
$y = 9 - 8(-1.5) = 21$ or $y = 9 - 8(-7) = 65$
$x = -1.5, y = 21$ and $x = -7, y = 65$

b) $4x^2 - 5x + 2 = 2x - 1$
$4x^2 - 7x + 3 = 0$
$(4x - 3)(x - 1) = 0$
so $x = 0.75$ or $x = 1$
$y = 2(0.75) - 1 = 0.5$ or $y = 2(1) - 1 = 1$
$x = 0.75, y = 0.5$ and $x = 1, y = 1$

Using the same method for c)-i):

c) $x = -\frac{1}{7}, y = \frac{6}{7}$ and $x = 3, y = 26$

d) $x = -0.5, y = 4$ and $x = 2, y = 9$

e) $x = 3, y = \frac{1}{3}$ and $x = 16, y = -4$

f) $x = 0.25, y = 0.75$ and $x = -5, y = -1$

g) $x = 0.4, y = 0.2$ and $x = -1$ and $y = 3$

h) $x = -16, y = -3.5$ and $x = 22, y = 6$

i) $x = 0, y = 0$ and $x = -0.25, y = 0.25$

6 $x + y = 3 \implies y = 3 - x$
$x^2 + (3 - x)^2 = 17$
$x^2 + 9 - 6x + x^2 = 17$
$2x^2 - 6x - 8 = 0 \implies x^2 - 3x - 4 = 0$
$(x + 1)(x - 4) = 0$
so $x = -1$ or $x = 4$
$y = 3 - (-1) = 4$ or $y = 3 - 4 = -1$
$(-1, 4)$ and $(4, -1)$

Page 150   Exam-Style Questions

1 (1) $7m + 2n = 23$, (2) $3m + 5n = 14$
(1) × 5: $35m + 10n = 115$
(2) × 2: $\underline{-(6m + 10n = 28)}$
$29m = 87$
$m = 3$
$3(3) + 5n = 14 \implies n = 1$
[3 marks available — 1 mark for a suitable method, 1 mark for the correct value of m, 1 mark for the correct value of n]

2 (1) $2c + 4v = 580$, (2) $3c + 2v = 542$
(2) × 2: $(6c + 4v = 1084)$
(1) $\underline{-(2c + 4v = 580)}$
$4c = 504$
$c = 126$
$2(126) + 4v = 580 \implies 4v = 328 \implies v = 82$
[3 marks available — 1 mark for a suitable method, 1 mark for the correct value of c, 1 mark for the correct value of v]

---

3 $5(2x - 2) = 20 - 2x \implies 10x - 10 = 20 - 2x$
$12x = 30 \implies x = 2.5$
so $y = 2(2.5) - 2 = 3$
(2.5, 3)
[3 marks available — 1 mark for a suitable method, 1 mark for the correct value of x, 1 mark for the correct value of y]

4 Let a = kg of apples and p = kg of pears
(1) $3a + 2p = 19.8$ and (2) $2a + 3p = 20.7$
(1) × 3: $(9a + 6p = 59.4)$
(2) × 2: $\underline{-(4a + 6p = 41.4)}$
$5a = 18$
$a = 3.6$
$3(3.6) + 2p = 19.8$, so $p = 4.5$
1 kg of apples = £3.60 and 1 kg of pears = £4.50.
[4 marks available — 1 mark for setting up equations, 1 mark for suitable multipliers for equations, 1 mark for each correct answer]

5 $x^2 = 6x - 8$, so $x^2 - 6x + 8 = 0$
$(x - 4)(x - 2) = 0$
so $x = 4$ or $x = 2$
$y = 4^2 = 16$ or $y = 2^2 = 4$
(4, 16) and (2, 4)
[4 marks available — 1 mark for rearranging equations into a quadratic equation, 1 mark for factorising, 1 mark for each pair of coordinates]

6 Area $= xy$, Perimeter $= 2x + 2y$
So $xy = 35$ (1)
and $2x + 2y = 24 \implies 2y = 24 - 2x$
$\implies y = 12 - x$ (2)
Substitute equation (2) into equation (1):
$x(12 - x) = 35 \implies 12x - x^2 = 35$
$x^2 - 12x + 35 = 0$
$(x - 7)(x - 5) = 0$
$x > y$, so $x = 7$ cm and $y = 5$ cm
[5 marks available — 1 mark for equations for area and perimeter, 1 mark for substitution, 1 mark for the quadratic equation, 1 mark for factorising, 1 mark for the correct values of x and y]

7 $x^2 + (2x + 2)^2 = 8$
$5x^2 + 8x - 4 = 0$
$(5x - 2)(x + 2) = 0$
so $x = 0.4$ or $x = -2$
$y = 2(0.4) + 2 = 2.8$ or $y = 2(-2) + 2 = -2$
$x = 0.4, y = 2.8$ and $x = -2, y = -2$
[5 marks available — 1 mark for substitution, 1 mark for the quadratic equation, 1 mark for factorising, 1 mark for each pair of x and y values]

Section 13 — Inequalities

13.1 Solving Inequalities

Page 151   Exercise 1

1 a) $3x \geq 36 \implies x \geq 12$

b) $-96 \leq -12x \implies 8 \geq x$ or $x \leq 8$
Remember to flip the inequality sign when you divide or multiply by a negative number.

c) $-\frac{x}{3} < -28 \implies x > 84$

d) $11 \leq -\frac{x}{7} \implies -77 \geq x$ or $x \leq -77$

e) $4x + 11 < 23 \implies 4x < 12 \implies x < 3$

f) $5x + 3 \leq 43 \implies 5x \leq 40 \implies x \leq 8$

g) $-3x - 7 \geq -1 \implies -3x \geq 6 \implies x \leq -2$

h) $65 < 7x - 12 \implies 77 < 7x$
$\implies$ **11 < x** or **x > 11**

2 a) $4x < 16 \implies x < 4 \implies \{x : x < 4\}$

b) $-33 \leq 11x \implies -3 \leq x$
$\implies \{x : -3 \leq x\}$ or $\{x : x \geq -3\}$

c) $\frac{x}{6} > -3 \implies x > -18 \implies \{x : x > -18\}$

d) $-\frac{x}{5} > -1 \implies x < 5 \implies \{x : x < 5\}$

e) $2x + 15 < 21 \implies 2x < 6 \implies x < 3$
$\implies \{x : x < 3\}$

f) $-5x - 8 \geq 12 \Rightarrow -5x \geq 20 \Rightarrow x \leq -4$
$\Rightarrow \{x : x \leq -4\}$

g) $\frac{x}{3} - 13 \geq -1 \Rightarrow \frac{x}{3} \geq 12 \Rightarrow x \geq 36$
$\Rightarrow \{x : x \geq 36\}$

h) $44 < 8x + 16 \Rightarrow 28 < 8x \Rightarrow 3.5 < x$
$\Rightarrow \{x : 3.5 < x\}$ or $\{x : x > 3.5\}$

3 a) $\frac{x+2}{3} < 1 \Rightarrow x + 2 < 3 \Rightarrow x < 1$

b) $\frac{x+4}{5} \geq 2 \Rightarrow x + 4 \geq 10 \Rightarrow x \geq 6$
Using the same method for c)-h):
c) $x > 22$ **d)** $x \leq 10$ **e)** $x \geq 14$
f) $x > -3.5$ **g)** $x < -4.9$ **h)** $x \geq -6$

4 a) $4x + 2 < 2x - 2 \Rightarrow 2x < -4 \Rightarrow x < -2$
$\Rightarrow \{x : x < -2\}$

b) $3x + 5 \leq 4 + x \Rightarrow 2x \leq -1 \Rightarrow x \leq -0.5$
$\Rightarrow \{x : x \leq -0.5\}$

c) $3x - 3 \geq -1 + x \Rightarrow 2x \geq 2 \Rightarrow x \geq 1$
$\Rightarrow \{x : x \geq 1\}$

d) $6 - x < 7x - 2 \Rightarrow 8 < 8x \Rightarrow 1 < x$
$\Rightarrow \{x : 1 < x\}$ or $\{x : x > 1\}$

e) $\frac{x}{2} - 5 \geq 3 - \frac{x}{2} \Rightarrow x \geq 8 \Rightarrow \{x : x \geq 8\}$

f) $1 - 2x < \frac{x+3}{2} \Rightarrow 2 - 4x < x + 3$
$\Rightarrow -1 < 5x \Rightarrow -0.2 < x$
$\Rightarrow \{x : -0.2 < x\}$ or $\{x : x > -0.2\}$

g) $2x + 4 > \frac{2x-3}{8} \Rightarrow 16x + 32 > 2x - 3$
$\Rightarrow 14x > -35 \Rightarrow x > -2.5$
$\Rightarrow \{x : x > -2.5\}$

h) $\frac{x}{4} + \frac{3}{2} \leq \frac{1}{4} - x \Rightarrow x + 6 \leq 1 - 4x$
$\Rightarrow 5x \leq -5 \Rightarrow x \leq -1 \Rightarrow \{x : x \leq -1\}$

Page 152   Exercise 2

1 a) $x + 9 > 14 \Rightarrow x > 5$

b) $x + 3 \leq 12 \Rightarrow x \leq 9$

Using the same method for c)-h):

c) $x \geq 16$

d) $x < 26$

e) $x > 16$

f) $x \geq 16$

g) $x < 18$

h) $x \leq 1$

2 a) $3x \geq 9 \Rightarrow x \geq 3$

b) $5x \leq 25 \Rightarrow x < 5$

Using the same method for c)-h):

c) $x < -4$

d) $x > -8$

e) $x \geq 6$

f) $x < 10$

g) $x < 24$

h) $x \leq 35$

---

3 a) $4x + 3 > x + 15 \Rightarrow 3x > 12 \Rightarrow x > 4$

Using the same method for b)-c):

b) $x \leq 3$

c) $x < 22.5$

Page 153   Exercise 3

1 a) $7 < x + 3 \leq 15 \Rightarrow 7 < x + 3$ and $x + 3 \leq 15$
$\Rightarrow 4 < x$ and $x \leq 12 \Rightarrow 4 < x \leq 12$

Using the same method for b)-d):

b) $6 \leq x \leq 16$

c) $-6 \leq x \leq -1$

d) $37 \leq x \leq 60$

2 a) $16 < 4x < 28 \Rightarrow 16 < 4x$ and $4x < 28$
$\Rightarrow 4 < x$ and $x < 7 \Rightarrow 4 < x < 7$
$\Rightarrow \{x : 4 < x < 7\}$

Using the same method for b)-d):

b) $\{x : 16 < x \leq 21\}$ **c)** $\{x : 6 < x \leq 16\}$

d) $\{x : -3 \leq x < 4\}$

3 a) $17 < 6x + 5 < 29$
$\Rightarrow 17 < 6x + 5$ and $6x + 5 < 29$
$\Rightarrow 12 < 6x$ and $6x < 24$
$\Rightarrow 2 < x$ and $x < 4 \Rightarrow 2 < x < 4$

b) $8 < 3x - 4 \leq 26$
$\Rightarrow 8 < 3x - 4$ and $3x - 4 \leq 26$
$\Rightarrow 12 < 3x$ and $3x \leq 30$
$\Rightarrow 4 < x$ and $x \leq 10$
$\Rightarrow 4 < x \leq 10$

Using the same method for c)-d):

c) $-7 < x \leq 12$ **d)** $-2.5 < x < 1$

e) $5.1 \leq -x + 2.5 < 9.7$
$\Rightarrow 2.6 \leq -x < 7.2$
$\Rightarrow -2.6 \geq x > -7.2 \Rightarrow -7.2 < x \leq -2.6$

Using the same method for f):

f) $-19.7 < x < -1.2$

4 a) $24 > 8x \Rightarrow 3 > x$
$9x \geq -18 \Rightarrow x \geq -2$
So $-2 \leq x < 3$ which gives integer solutions
-2, -1, 0, 1, 2.

b) $-3 > 2x + 5 \Rightarrow -8 > 2x \Rightarrow -4 > x$
$-4x \leq 32 \Rightarrow x \geq -8$
So $-8 \leq x < -4$ which gives integer
solutions **-8, -7, -6, -5.**

c) $9 - x > 8x + 9 \Rightarrow 0 > 9x \Rightarrow 0 > x$
$2x + 9 > 3 \Rightarrow 2x > -6 \Rightarrow x > -3$
So $-3 < x < 0$ which gives integer solutions
-2, -1.

13.2 Quadratic Inequalities

Page 155   Exercise 1

1 a) $x^2 > 16 \Rightarrow x^2 - 16 > 0$

b) (i) $f(x) = x^2 - 16 = (x + 4)(x - 4)$
(ii) $x = -4$ and $x = 4$

c)

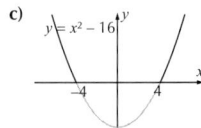

The graph of $x^2 - 16$ is u-shaped so $f(x) > 0$
outside the boundary values, so
$x < -4$ or $x > 4$.

---

2 a) $x^2 = 4 \Rightarrow x = -2$ or $x = 2$
$x^2 < 4 \Rightarrow x^2 - 4 < 0$

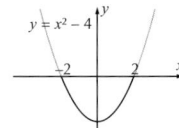

The graph of $x^2 - 4$ is u-shaped so
$x^2 - 4 < 0$ between the boundary values,
so $-2 < x < 2$.

b) $x^2 = 9 \Rightarrow x = -3$ or $x = 3$
$x^2 \leq 9 \Rightarrow x^2 - 9 \leq 0$

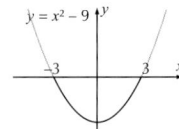

The graph of $x^2 - 9$ is u-shaped so
$x^2 - 9 \leq 0$ between the boundary values,
so $-3 \leq x \leq 3$.

Using the same method for c)-h):

c) $x < -5$ or $x > 5$

d) $x \leq -6$ or $x \geq 6$

e) $-1 < x < 1$

f) $-7 \leq x \leq 7$

g) $x < -8$ or $x > 8$

h) $-10 < x < 10$

3 a) $x^2 = \frac{1}{4} \Rightarrow x = -\frac{1}{2}$ or $x = \frac{1}{2}$
$x^2 < \frac{1}{4} \Rightarrow x^2 - \frac{1}{4} < 0$

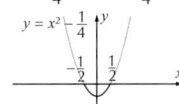

The graph of $x^2 - \frac{1}{4}$ is u-shaped so
$x^2 - \frac{1}{4} < 0$ between the boundary values,
so $-\frac{1}{2} < x < \frac{1}{2} \Rightarrow \{x : -\frac{1}{2} < x < \frac{1}{2}\}$.

b) $x^2 = \frac{1}{25} \Rightarrow x = -\frac{1}{5}$ or $x = \frac{1}{5}$
$x^2 \geq \frac{1}{25} \Rightarrow x^2 - \frac{1}{25} \geq 0$

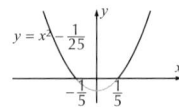

The graph of $x^2 - \frac{1}{25}$ is u-shaped so
$x^2 - \frac{1}{25} \geq 0$ outside the boundary values,
so $x \leq -\frac{1}{5}$ or $x \geq \frac{1}{5}$
$\Rightarrow \{x : x \leq -\frac{1}{5}\} \cup \{x : x \geq \frac{1}{5}\}$
∪ *is the union of the two sets — so anything*
that is in the first set or the second set.

Using the same method for c)-h):

c) $\{x : -\frac{1}{11} \leq x \leq \frac{1}{11}\}$

d) $\{x : x < -\frac{1}{6}\} \cup \{x : x > \frac{1}{6}\}$

e) $\{x : -\frac{2}{3} < x < \frac{2}{3}\}$

f) $\{x : x \le -\frac{5}{7}\} \cup \{x : x \ge \frac{5}{7}\}$

g) $\{x : -\frac{3}{4} \le x \le \frac{3}{4}\}$

h) $\{x : -\frac{4}{13} < x < \frac{4}{13}\}$

4 a) $2x^2 = 18 \Rightarrow x^2 = 9 \Rightarrow x = -3$ or $x = 3$
$2x^2 < 18 \Rightarrow 2x^2 - 18 < 0 \Rightarrow x^2 - 9 < 0$

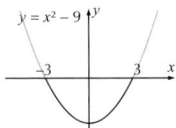

The graph of $x^2 - 9$ is u-shaped so $x^2 - 9 < 0$ between the boundary values, so $-3 < x < 3$. The integer solutions are $\{-2, -1, 0, 1, 2\}$.

Using the same method for b)-d):

b) $-5 \le x \le 5$ so the integer solutions are $\{-5, -4, -3, -2, -1, 0, 1, 2, 3, 4, 5\}$

c) $-4 < x < 4$ so the integer solutions are $\{-3, -2, -1, 0, 1, 2, 3\}$

d) $-6 \le x \le 6$ so the integer solutions are $\{-6, -5, -4, -3, -2, -1, 0, 1, 2, 3, 4, 5, 6\}$

5 a) $4x \le 12 - x^2 \Rightarrow x^2 + 4x - 12 \le 0$

b) (i) $g(x) = x^2 + 4x - 12 = (x + 6)(x - 2)$
(ii) $x = -6$ and $x = 2$

c)

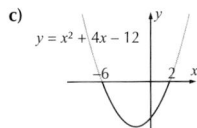

The graph of $x^2 + 4x - 12$ is u-shaped so $x^2 + 4x - 12 \le 0$ between the boundary values, so $-6 \le x \le 2$.

6 a) $x^2 + x - 2 = 0 \Rightarrow (x + 2)(x - 1) = 0$
$\Rightarrow x = -2$ or $x = 1$

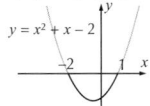

The graph of $x^2 + x - 2$ is u-shaped so $x^2 + x - 2 < 0$ between the boundary values, so $-2 < x < 1$.

b) $x^2 - x - 2 = 0 \Rightarrow (x - 2)(x + 1) = 0$
$\Rightarrow x = 2$ or $x = -1$

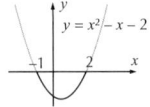

The graph of $x^2 - x - 2$ is u-shaped so $x^2 - x - 2 \le 0$ between the boundary values, so $-1 \le x \le 2$.

Using the same method for c)-l):

c) $x < 3$ or $x > 5$ **d)** $-5 \le x \le -1$
e) $x \le -3$ or $x \ge 4$ **f)** $-1 < x < 7$
g) $3 \le x \le 4$ **h)** $x \le -6$ or $x \ge -4$
i) $-2 < x < 8$ **j)** $-5 < x < 3$
k) $-1 \le x \le 11$ **l)** $-9 < x < -2$

7 a) $x^2 - 2x > 48 \Rightarrow x^2 - 2x - 48 > 0$
$x^2 - 2x - 48 = 0 \Rightarrow (x - 8)(x + 6) = 0$
$\Rightarrow x = 8$ or $x = -6$

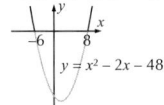

The graph of $x^2 - 2x - 48$ is u-shaped so $x^2 - 2x - 48 > 0$ outside the boundary values, so $x < -6$ or $x > 8$.

b) $x^2 - 3x \le 10 \Rightarrow x^2 - 3x - 10 \le 0$
$x^2 - 3x - 10 = 0 \Rightarrow (x - 5)(x + 2) = 0$
$\Rightarrow x = 5$ or $x = -2$

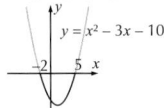

The graph of $x^2 - 3x - 10$ is u-shaped so $x^2 - 3x - 10 \le 0$ between the boundary values, so $-2 \le x \le 5$.

Using the same method for c)-h):

c) $4 < x < 5$

d) $3 \le x \le 6$

e) $x \le -9$ or $x \ge 4$

f) $-2 < x < 11$

g) $-3 < x < 9$

h) $4 < x < 8$

8 a) $x^2 - 4x = 0 \Rightarrow x(x - 4) = 0$
$\Rightarrow x = 0$ or $x = 4$

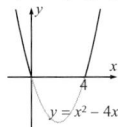

The graph of $x^2 - 4x$ is u-shaped so $x^2 - 4x > 0$ outside the boundary values, so $x < 0$ or $x > 4 \Rightarrow \{x : x < 0\} \cup \{x : x > 4\}$.

b) $x^2 + 3x = 0 \Rightarrow x(x + 3) = 0$
$\Rightarrow x = 0$ or $x = -3$

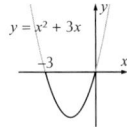

The graph of $x^2 + 3x$ is u-shaped so $x^2 + 3x \le 0$ between the boundary values, so $-3 \le x \le 0 \Rightarrow \{x : -3 \le x \le 0\}$.

Using the same method for c)-h):

c) $\{x : 0 < x < 5\}$
d) $\{x : x \le -8\} \cup \{x : x \ge 0\}$
e) $\{x : x < 0\} \cup \{x : x > 12\}$
f) $\{x : 0 \le x \le 2\}$
g) $\{x : x \le 0\} \cup \{x : x \ge 9\}$
h) $\{x : -3 \le x \le 0\}$

13.3 Graphing Inequalities

Page 156   Exercise 1

1 a) Draw the graph of $x = 2$. The inequality sign is $\le$ so draw a solid line. Using point $(0, 0)$ gives $0 \le 2$ which is true, so $(0, 0)$ is in region R.

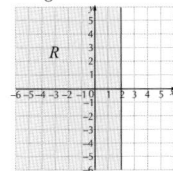

b) Draw the graph of $x = -1$. The inequality sign is $<$ so draw a dashed line. Using point $(0, 0)$ gives $0 < -1$ which is not true, so $(0, 0)$ is not in region R.

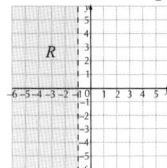

Using the same method for c)-d):

c)   **d)**

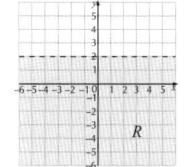 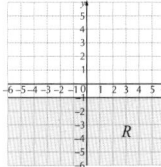

2 a) Draw the graph of $y = x + 1$. It's a straight line with gradient $= 1$ and y-intercept $= 1$. The inequality sign is $<$ so draw a dashed line. Using point $(0, 0)$ gives $0 < 0 + 1$ which is true, so $(0, 0)$ is in region R.

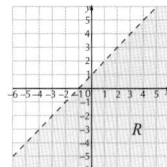

b) Draw the graph of $y = x - 3$. It's a straight line with gradient $= 1$ and y-intercept $= -3$. The inequality sign is $>$ so draw a dashed line. Using point $(0, 0)$ gives $0 > 0 - 3$ which is true, so $(0, 0)$ is in region R.

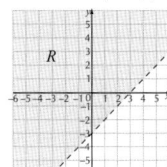

Using the same method for c)-l):

c)   **d)**

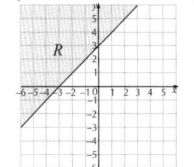 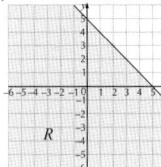

e)   **f)**

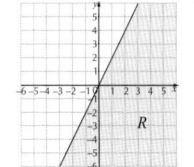 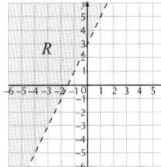

g)   **h)**

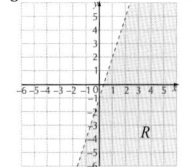 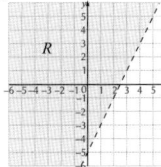

i)   j)

k)   l)

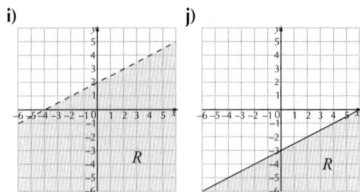

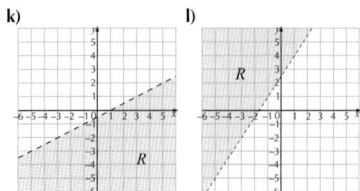

3 **a)** $2x > 6 - y \Rightarrow y > 6 - 2x$

b) Draw the graph of $y = 6 - 2x$. It's a straight line with gradient $= -2$ and y-intercept $= 6$. The inequality sign is $>$ so draw a dashed line. Using point $(0, 0)$ gives $0 > 6 - 2 \times 0$ $\Rightarrow$ $0 > 6$ which is not true, so $(0, 0)$ is not in region R.

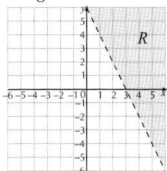

4 **a)** $x + y \le 5 \Rightarrow y \le 5 - x$

Draw the graph of $y = 5 - x$. It's a straight line with gradient $= -1$ and y-intercept $= 5$. The inequality sign is $\le$ so draw a solid line. Using point $(0, 0)$ gives $0 \le 5 - 0$ which is true, so $(0, 0)$ is in region R.

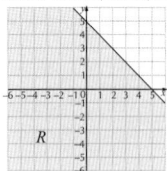

Using the same method for b)-h):

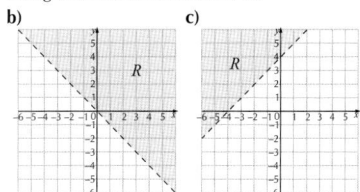

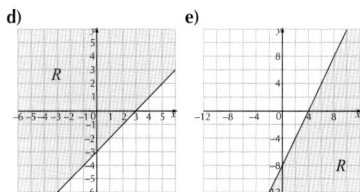

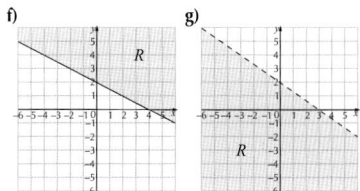

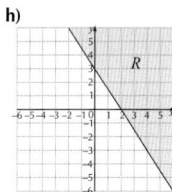

b)   c)

d)   e)

f)   g)

Page 157   Exercise 2

1 **a)** Draw the graphs of $x = 1$ and $y = 2$.
For $x > 1$ the inequality sign is $>$ so draw a dashed line. Using point $(0, 0)$ gives $0 > 1$ which is not true. $(0, 0)$ doesn't satisfy the inequality so region R is to the right of $x = 1$.
For $y \le 2$ the inequality sign is $\le$ so draw a solid line. Using point $(0, 0)$ gives $0 \le 2$ which is true. $(0, 0)$ does satisfy the inequality so region R is below $y = 2$.

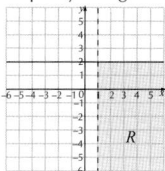

Using the same method for b)-c):

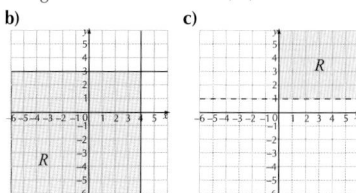

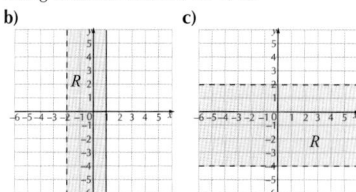

b)   c)

2 **a)** $1 < x < 4 \Rightarrow 1 < x$ and $x < 4$
Draw the graphs of $1 = x$ and $x = 4$.
For $1 < x$ the inequality sign is $<$ so draw a dashed line. For $x < 4$ the inequality sign is $<$ so draw a dashed line. It's a compound inequality so region R will be between the two lines.

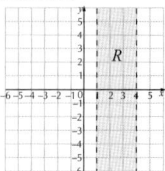

Using the same method for b)-c):

b)   c)

3 **a)** $-2 < x < 0 \Rightarrow -2 < x$ and $x < 0$
Draw the graphs of $-2 = x$, $x = 0$ and $y = 1$.
For $-2 < x$ the inequality sign is $<$ so draw a dashed line. For $x < 0$ the inequality sign is $<$ so draw a dashed line. It's a compound inequality so region R will be between the two lines.
For $y > 1$ the inequality sign is $>$ so draw a dashed line. Using point $(0, 0)$ gives $0 > 1$ which not true. $(0, 0)$ doesn't satisfy the inequality so region R is above $y = 1$.

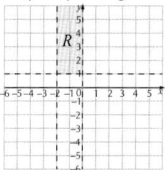

Using the same method for b)-f):

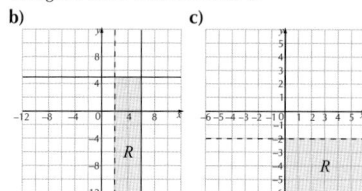

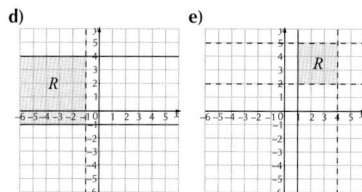

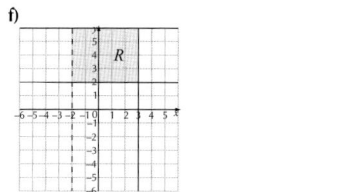

b)   c)

d)   e)

f)

4 **a)** $y + x < 4 \Rightarrow y < 4 - x$
Draw the graphs of $y = 1$ and $y = 4 - x$.
For $y > 1$ the inequality sign is $>$ so draw a dashed line. Using point $(0, 0)$ gives $0 > 1$ which is not true. $(0, 0)$ doesn't satisfy the inequality so region R is above $y = 1$.
For $y < 4 - x$ the inequality sign is $<$ so draw a dashed line. Using point $(0, 0)$ gives $0 < 4 - 0$ which is true. $(0, 0)$ satisfies the inequality so region R is below $y = 4 - x$.

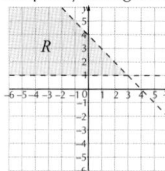

Using the same methods for b)-f) and Q5-6:

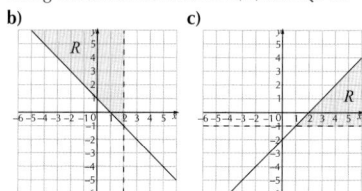

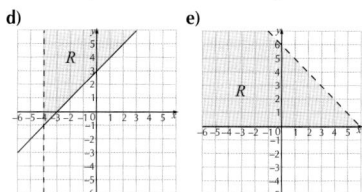

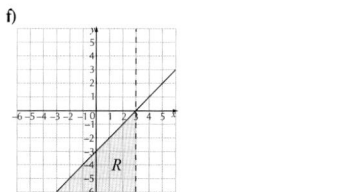

b)   c)

d)   e)

f)

5 a)

b)

c)

d)

e)

f)

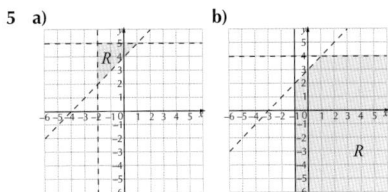

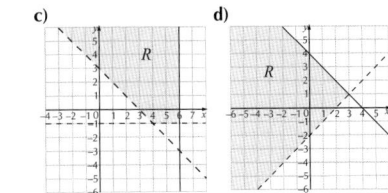

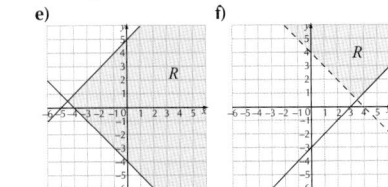

If you prefer, you can shade the region you don't want, and leave R unshaded — as long as you label it. This method is sometimes easier if you're dealing with more than two inequalities.

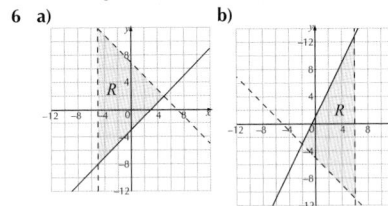

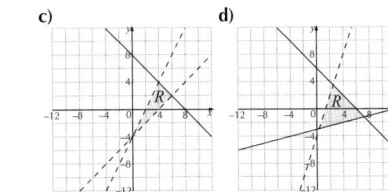

6 a)

b)

c)

d)

Page 158 Exercise 3

1 a) The boundary line has equation $x = 3$.
The line is dashed so it's either $<$ or $>$.
Point $(4, 1)$ lies in the shaded region and $4 > 3$ so the inequality is $x > 3$.
Using the same method for b)-c):

   b) $y \leq 4$　　　c) $y \leq x$

2 a) The boundary line has equation $y = 4 - x$.
$y = 4 - x$ is solid so it's either $\leq$ or $\geq$.
Point $(1, 1)$ lies in the shaded region and $1 \leq 4 - 1$ so the inequality is $y \leq 4 - x$.
Using the same method for b)-c):

   b) $y \geq x + 1$　　　c) $y < 4 - 2x$

3 a) The boundary lines are at $x = -1$ and $x = 2$.
$x = -1$ and $x = 2$ are dashed so they're either $<$ or $>$. The shaded region is between the two lines so it's a compound inequality so $-1 < x < 2$.

   b) The boundary lines are at $x = -1$ and $y = -1$.
$x = -1$ is dashed so it's either $<$ or $>$.
$y = -1$ is solid so it's either $\leq$ or $\geq$.
Point $(-2, 0)$ lies in the shaded region.
$-2 < -1$ so $x < -1$ and $0 \geq -1$ so $y \geq -1$.

   c) The boundary lines are at $x = -1$, $y = 3$ and $y = x + 1$. $x = -1$ and $y = 3$ are solid so they're both either $\leq$ or $\geq$.
$y = x + 1$ is dashed so it's either $<$ or $>$.
Point $(0, 4)$ lies in the shaded region.
$0 \geq -1$ so $x \geq -1$, $4 \geq 3$ so $y \geq 3$ and $4 > 0 + 1$ so $y > x + 1$.

Using the same method for d)-f):

   d) $x > -1$, $y \geq -1$, $y \leq -x$

   e) $y > x$, $y \leq x + 2$

   f) $y \geq -1$, $y > x - 2$, $3y \leq x + 3$

Page 159 Review Exercise

1 a) $-1 > \frac{x}{8} \Rightarrow -8 > x$
   (i) $\{x : -8 > x\}$ or $\{x : x < -8\}$
   (ii)

Using the same method for b)-d):

   b) $x \leq 3$
   (i) $\{x : x \leq 3\}$
   (ii)

   c) $x \geq 11$
   (i) $\{x : x \geq 11\}$
   (ii)

   d) $x < -0.64$
   (i) $\{x : x < -0.64\}$
   (ii)

   e) $2x + 16 \geq -8 \Rightarrow 2x \geq -24 \Rightarrow x \geq -12$
   (i) $\{x : x \geq -12\}$
   (ii)

Using the same method for f)-h):

   f) $x < -1$
   (i) $\{x : x < -1\}$
   (ii)

   g) $x < 2$
   (i) $\{x : x < 2\}$
   (ii)

   h) $x \geq -6$
   (i) $\{x : x \geq -6\}$
   (ii)

2 a) $-7 < x - 6 < 4 \Rightarrow -7 < x - 6$ and $x - 6 < 4$
   $\Rightarrow -1 < x$ and $x < 10 \Rightarrow -1 < x < 10$
   $\Rightarrow \{0, 1, 2, 3, 4, 5, 6, 7, 8, 9\}$

Using the same method for b)-c):

   b) $\{-3, -2, -1, 0, 1, 2, 3, 4\}$

   c) $\{0, 1, 2, 3\}$

3 a) $x^2 = 81 \Rightarrow x = -9$ or $x = 9$
   $x^2 < 81 \Rightarrow x^2 - 81 < 0$
   The graph of $x^2 - 81$ is u-shaped so $x^2 - 81 < 0$ between the boundary values, so $-9 < x < 9$.

   b) $x^2 = 81 \Rightarrow x = -9$ or $x = 9$
   $x^2 \geq 81 \Rightarrow x^2 - 81 \geq 0$
   The graph of $x^2 - 81$ is u-shaped so $x^2 - 81 \geq 0$ outside the boundary values, so $x \leq -9$ or $x \geq 9$.

   c) $3x^2 = 48 \Rightarrow x^2 = 16 \Rightarrow x = -4$ or $x = 4$
   $3x^2 > 48 \Rightarrow 3x^2 - 48 > 0$
   The graph of $3x^2 - 48$ is u-shaped so $3x^2 - 48 > 0$ outside the boundary values, so $x < -4$ or $x > 4$.

   d) $x^2 + 7x + 12 = 0 \Rightarrow (x + 4)(x + 3) = 0$
   $\Rightarrow x = -4$ or $x = -3$
   The graph of $x^2 + 7x + 12$ is u-shaped so $x^2 + 7x + 12 > 0$ outside the boundary values, so $x < -4$ or $x > -3$.

   e) $x^2 - 4x = 5 \Rightarrow x^2 - 4x - 5 = 0$
   $\Rightarrow (x - 5)(x + 1) = 0 \Rightarrow x = 5$ or $x = -1$
   The graph of $x^2 - 4x - 5$ is u-shaped so $x^2 - 4x - 5 < 0$ between the boundary values, so $-1 < x < 5$.

   f) $-x^2 = 8x - 20 \Rightarrow x^2 + 8x - 20 = 0$
   $\Rightarrow (x + 10)(x - 2) = 0 \Rightarrow x = -10$ or $x = 2$
   The graph of $x^2 + 8x - 20$ is u-shaped so $x^2 + 8x - 20 < 0$ between the boundary values, so $-10 < x < 2$.

---

4 a) Draw the graph of $x = -5$. The inequality sign is $\geq$ so draw a solid line. Using point $(0, 0)$ gives $0 \geq -5$ which is true, so $(0, 0)$ is in region R.

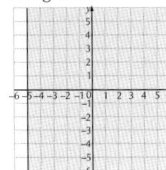

Using the same method for b)-g):

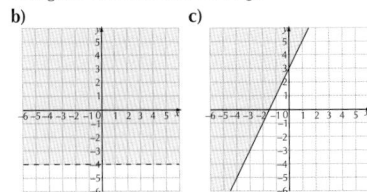

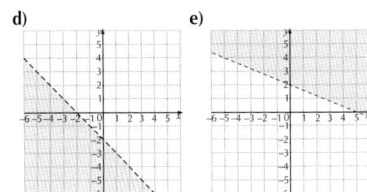

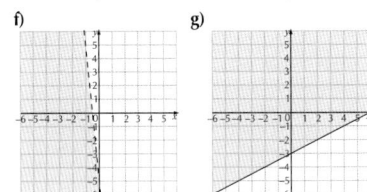

   b)

   c)

   d)

   e)

   f)

   g)

   h) $-4 \leq x \leq -1 \Rightarrow -4 \leq x$ and $x \leq -1$.
Draw the graphs of $-4 = x$ and $x = -1$.
For $-4 \leq x$ the inequality sign is $\leq$ so draw a solid line. For $x < -1$ the inequality sign is $<$ so draw a dashed line. It's a compound inequality so region R will be between the two lines.

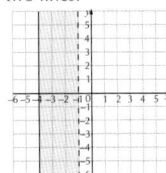

Using the same method for i):

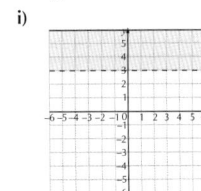

   i)

5 a) $-5 \leq y \leq -1 \Rightarrow -5 \leq y$ and $y \leq -1$.
Draw the graphs of $-5 = y$, $y = -1$ and $y = -x$. For $-5 \leq y$ the inequality sign is $\leq$ so draw a solid line. For $y \leq -1$ the inequality sign is $\leq$ so draw a solid line. It's a compound inequality so region R will be between the two lines.
For $y < -x$ the inequality sign is $<$ so draw a dashed line. Using point $(1, 0)$ gives $0 < -1$ which not true. $(1, 0)$ doesn't satisfy the inequality so region R is below $y = -x$.

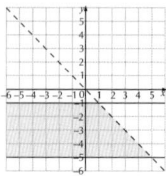

Using the same method for b)-d):

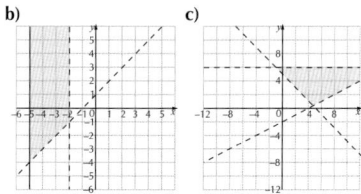

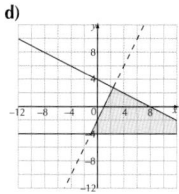

b)

c)

d)

6 a) The boundary lines are at $x = 1$, $x = 4$, $y = 2$ and $y = 3$.
$x = 1$ and $y = 3$ are solid so they're either $\leq$ or $\geq$.
$x = 4$ and $y = 2$ are dashed so they're either $<$ or $>$.
Point $(2, 2.5)$ lies in the shaded region.
$2 \geq 1$ so $x \geq 1$ and $2 < 4$ so $x < 4$.
$2.5 > 2$ so $y > 2$ and $2.5 \leq 3$ so $y \leq 3$.
(You could also give the compound inequalities $1 \leq x < 4$, $2 < y \leq 3$.)

Using the same method for b)-c):

b) $2 \leq x < 3$, $y \geq 1$, $y < 2x$

c) $3y > 2x - 3$, $y \geq -x$, $2y \leq 4 - x$

7 a) (i) The total cost must be less than £80 000 so $8000x + 16\,000y \leq 80\,000$
$\Rightarrow x + 2y \leq 10$
They must buy at least 7 vehicles so $x + y \geq 7$.
They must buy at least 1 van so $y \geq 1$
The number of motorbikes is never negative, so $x \geq 0$.

(ii) Draw the lines $x + 2y = 10$, $x + y = 7$, $y = 1$ and $x = 0$. All inequalities are $\leq$ or $\geq$ so all the lines should be solid. The region satisfying all the inequalities is below $x + 2y = 10$, above $x + y = 7$, above $y = 1$ and to the right of $x = 0$. It is the shaded region on the graph below.

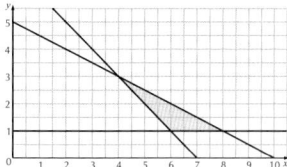

b) The points with whole number coordinates in the shaded region are $(4, 3)$, $(5, 2)$, $(6, 1)$, $(6, 2)$, $(7, 1)$ and $(8, 1)$. So they could buy:
4 motorbikes and 3 vans,
5 motorbikes and 2 vans,
6 motorbikes and 2 vans,
6 motorbikes and 1 van,
7 motorbikes and 1 van,
8 motorbikes and 1 van.

Page 160 Exam-Style Questions

1 $4(x + 3) > 2(x - 3)$
$\Rightarrow 4x + 12 > 2x - 6$ *[1 mark]*
$\Rightarrow 2x > -18 \Rightarrow$ **$x > -9$** *[1 mark]*

2 The circles are the wrong way round.
The open circle should be at -2 and the solid circle should be at 3 *[1 mark]*.
The arrows are pointing the wrong way — the values that satisfy the inequality are between -2 and 3 *[1 mark]*.

3 $6x^2 + 9x - 15 = 0 \Rightarrow 2x^2 + 3x - 5 = 0$
$\Rightarrow (x - 1)(2x + 5) = 0$ *[1 mark]*
$\Rightarrow x - 1 = 0$ or $2x + 5 = 0$
$\Rightarrow x = 1$ or $2x = -5$
$\Rightarrow x = 1$ or $x = -2.5$ *[1 mark]*
The graph of $6x^2 + 9x - 15$ is u-shaped so $6x^2 + 9x - 15 \leq 0$ between the boundary values, so **$-2.5 \leq x \leq 1$** *[1 mark]*.

4 Faisal spends £$(6x + 1)$ and this is less than £13, so $6x + 1 < 13$ *[1 mark]*.
$6x + 1 < 13 \Rightarrow 6x < 12 \Rightarrow x < 2$ *[1 mark]*
Faisal spends more than Eve, who spends £$(4x + 2)$, so $6x + 1 > 4x + 2$ *[1 mark]*
$6x + 1 > 4x + 2 \Rightarrow 2x > 1 \Rightarrow x > 0.5$
So **$0.5 < x < 2$** *[1 mark]*

5 a) He needs at least 5 tins of undercoat, so $x \geq 5$ *[1 mark]*.
He needs at least as many tins of matt emulsion as undercoat, so $y \geq x$ *[1 mark]*.
He has £216 to spend in total, so $12x + 24y \leq 216 \Rightarrow$ **$x + 2y \leq 18$** *[1 mark]*.

b) Draw the lines $x = 5$, $y = x$ and $x + 2y = 18$. All the inequalities are $\leq$ or $\geq$ so all lines should be solid.

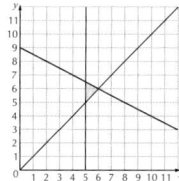

[3 marks available — 2 marks for all three lines correctly drawn, otherwise 1 mark for one or two lines correctly drawn, 1 mark for shading correct region]

c) The region satisfying all the inequalities is to the right of $x = 5$, above $y = x$, and below $x + 2y = 18$. It is the shaded region shown on the graph. The points with whole number coordinates in the shaded region are $(5, 5)$, $(5, 6)$ and $(6, 6)$. So he could get:
5 tins of undercoat and 5 tins of matt emulsion,
5 tins of undercoat and 6 tins of matt emulsion,
6 tins of undercoat and 6 tins of matt emulsion.
[2 marks available — 1 mark for identifying all three correct points with integer coordinates, 1 mark for interpreting these points in terms of tins of paint]

Section 14 — Sequences

14.1 Term to Term Rules

Page 162 Exercise 1

1 a) 1st term: **14**
2nd term: $14 \times 4 - 1 = $ **55**
3rd term: $55 \times 4 - 1 = $ **219**
4th term: $219 \times 4 - 1 = $ **875**
5th term: $875 \times 4 - 1 = $ **3499**

b) 1st term: **11**
2nd term: $11 \times -2 + 1 = $ **-21**
3rd term: $-21 \times -2 + 1 = $ **43**
4th term: $43 \times -2 + 1 = $ **-85**
5th term: $-85 \times -2 + 1 = $ **171**

---

2 a) $6 \div 3 = 2$, $12 \div 6 = 2$, $24 \div 12 = 2$
So you need to multiply by **2** each time.

b) $24 \times 2 = $ **48**, $48 \times 2 = $ **96**, $96 \times 2 = $ **192**

3 a) (i) $5 - 3 = 2$, $7 - 5 = 2$, $9 - 7 = 2$, so the rule is **add 2 to the previous term**.

(ii) The terms are increasing by the same amount each time, so it's **arithmetic**.

(iii) $9 + 2 = $ **11**, $11 + 2 = $ **13**, $13 + 2 = $ **15**

b) (i) $12 \div 4 = 3$, $36 \div 12 = 3$, $108 \div 36 = 3$, so the rule is **multiply the previous term by 3**.

(ii) Consecutive terms are found by multiplying by the same value each time, so it's **geometric**.

(iii) $108 \times 3 = $ **324**, $324 \times 3 = $ **972**, $972 \times 3 = $ **2916**

c) (i) $96 \div 192 = 0.5$, $48 \div 96 = 0.5$, $24 \div 48 = 0.5$
so the rule is **multiply the previous term by 0.5**.
Or 'divide the previous term by 2'.

(ii) Consecutive terms are found by multiplying by the same value each time, so it's **geometric**.

(iii) $24 \times 0.5 = $ **12**, $12 \times 0.5 = $ **6**, $6 \times 0.5 = $ **3**

d) (i) $-4 - 0 = -4$, $-8 - (-4) = -4$, $-12 - (-8) = -4$, so the rule is **subtract 4 from the previous term**.

(ii) The terms are decreasing by the same amount each time, so it's **arithmetic**.

(iii) $-12 - 4 = $ **-16**, $-16 - 4 = $ **-20**, $-20 - 4 = $ **-24**

e) (i) $4 \div 16 = \frac{1}{4}$, $1 \div 4 = \frac{1}{4}$, $\frac{1}{4} \div 1 = \frac{1}{4}$
so the rule is **multiply the previous term by $\frac{1}{4}$**.
Or 'divide the previous term by 4'.

(ii) Consecutive terms are found by multiplying by the same value each time, so it's **geometric**.

(iii) $\frac{1}{4} \times \frac{1}{4} = \frac{1}{16}$, $\frac{1}{16} \times \frac{1}{4} = \frac{1}{64}$, $\frac{1}{64} \times \frac{1}{4} = \frac{1}{256}$

f) (i) $-2 \div 1 = -2$, $4 \div -2 = -2$, $-8 \div 4 = -2$
So the rule is **multiply the previous term by -2**.

(ii) Consecutive terms are found by multiplying by the same value each time, so it's **geometric**.

(iii) $-8 \times -2 = $ **16**, $16 \times -2 = $ **-32**, $-32 \times -2 = $ **64**

4 a) The terms are increasing by 6 each time $(13 - 7 = 6, 19 - 13 = 6...)$. So the missing number is $25 + 6 = $ **31**.

b) The terms are decreasing by 4 each time $(5 - 9 = -4, -7 - (-3) = -4...)$. So the missing number is $5 - 4 = $ **1**.

c) The terms are being multiplied by 4 each time $(3.2 \div 0.8 = 4, 12.8 \div 3.2 = 4)$. So the missing numbers are $0.8 \div 4 = $ **0.2**, and $12.8 \times 4 = $ **51.2**.

5 $-2 - (-5) = 3$, $1 - (-2) = 3$, $4 - 1 = 3$
So the rule is add 3 to the previous term.
The 4th term is 4, so you need to add $54 - 4 = 50$ lots of 3 to the 4th term to get the 54th term. The 54th term is: $4 + (50 \times 3) = 4 + 150 = $ **154**.

6 The terms aren't increasing by a constant amount each time, so the sequence is not arithmetic.
$2\sqrt{3} \div 1 = 2\sqrt{3}$,
$12 \div 2\sqrt{3} = \frac{6}{\sqrt{3}} = \frac{6\sqrt{3}}{\sqrt{3} \times \sqrt{3}} = \frac{6\sqrt{3}}{3} = 2\sqrt{3}$
$24\sqrt{3} \div 12 = 2\sqrt{3}$

So the sequence is geometric, and the rule is multiply the previous term by $2\sqrt{3}$.

To get the 7th term, multiply the 4th term $(24\sqrt{3})$ by $2\sqrt{3}$ three times:

$24\sqrt{3} \times 2\sqrt{3} \times 2\sqrt{3} \times 2\sqrt{3}$
$= (24 \times 2 \times 2 \times 2) \times (\sqrt{3} \times \sqrt{3} \times \sqrt{3} \times \sqrt{3})$
$= 192 \times (3 \times 3) = \textbf{1728}$.

To revise rationalising the denominator, see p.102.

Page 163   Exercise 2

1 a) (i) $4 - 3 = 1$, $6 - 4 = 2$,
$9 - 6 = 3$, $13 - 9 = 4$
So the differences are **+1, +2, +3, +4**...

(ii) Carrying on the sequence rule:
$13 + 5 = \textbf{18}$, $18 + 6 = \textbf{24}$, $24 + 7 = \textbf{31}$

b) (i) $18 - 20 = -2$, $15 - 18 = -3$,
$11 - 15 = -4$, $6 - 11 = -5$
So the differences are **–2, –3, –4, –5**...

(ii) Carrying on the sequence rule:
$6 - 6 = \textbf{0}$, $0 - 7 = \textbf{–7}$, $-7 - 8 = \textbf{–15}$

c) (i) $2 - 1 = 1$, $0 - 2 = -2$,
$3 - 0 = 3$, $-1 - 3 = -4$
So the differences are **+1, –2, +3, –4**...

(ii) Carrying on the sequence rule:
$-1 + 5 = \textbf{4}$, $4 - 6 = \textbf{–2}$, $-2 + 7 = \textbf{5}$

2 $3 + 5 = \textbf{8}$, $5 + 8 = \textbf{13}$, $8 + 13 = \textbf{21}$

3 a) $1 \div 1 = 1$, $2 \div 1 = 2$, $6 \div 2 = 3$, $24 \div 6 = 4$
So the terms are multiplied by **1**, then **2**, then **3**, then **4**.

b) Carrying on the sequence rule:
$24 \times 5 = \textbf{120}$, $120 \times 6 = \textbf{720}$

4 The sequence of triangular numbers starts at 1, then adds 2, then 3, then 4 etc.
So the 10th term is:
$1 + 2 + 3 + 4 + 5 + 6 + 7 + 8 + 9 + 10 = \textbf{55}$.

5 a) You can't multiply 0 by anything to get 3, so the sequence can't be geometric. Look at the differences between each term:
$3 - 0 = 3$, $8 - 3 = 5$,
$15 - 8 = 7$, $24 - 15 = 9$
So the differences are +3, +5, +7, +9...
The differences are increasing by 2 each time. This is a constant value, so the sequence must be **quadratic**.

b) Write the sequence below the sequence of square numbers to compare:

1	4	9	16	25
0	3	8	15	24

Each number in the sequence is **one less than the corresponding term in the square number sequence**.

c) Use your answer to part b):
The 100th square number $= 100^2 = 10\,000$
So the 100th term in the sequence is:
$10\,000 - 1 = \textbf{9999}$.

6 a) 1st term = 1
2nd term = $1 \times 4 - b = 4 - b$
3rd term = $(4 - b) \times 4 - b = 16 - 5b$
4th term = $(16 - 5b) \times 4 - b = 64 - 21b$
The 4th term is 43, so:
$64 - 21b = 43 \Rightarrow 21b = 21 \Rightarrow \textbf{\textit{b}} = \textbf{1}$

b) 1st term = 9
2nd term = $9b - 54$
3rd term = $(9b - 54)b - 54$
$= 9b^2 - 54b - 54 = 9(b^2 - 6b - 6)$
The 3rd term is –135 so:
$9(b^2 - 6b - 6) = -135$
$b^2 - 6b - 6 = -15$
$b^2 - 6b + 9 = 0$
$(b - 3)^2 = 0 \Rightarrow b - 3 = 0 \Rightarrow \textbf{\textit{b}} = \textbf{3}$

7 $(5 + x) - 5 = x$, $(7 + 2x) - (5 + x) = 2 + x$,
The first difference is x, the next difference is $2 + x$, and you've been told it's a quadratic sequence so you know the differences will increase by the same amount each time.

The difference between the differences is +2, so the next difference between terms will be $(2 + x) + 2 = 4 + x$, etc.
4th term = $7 + 2x + (4 + x) = 11 + 3x$
5th term = $11 + 3x + (6 + x) = 17 + 4x$
6th term = $17 + 4x + (8 + x) = 25 + 5x$
$= \textbf{5(5 + \textit{x})}$

Page 164   Exercise 3

1 a) $x_1 = 3$, so:
$x_2 = 3 + 5 = 8$, $x_3 = 8 + 5 = 13$,
$x_4 = 13 + 5 = 18$, $x_5 = 18 + 5 = 23$,
$x_6 = 23 + 5 = \textbf{28}$

b) $x_1 = 20$, so:
$x_2 = 20 - 7 = 13$, $x_3 = 13 - 7 = 6$,
$x_4 = 6 - 7 = -1$, $x_5 = -1 - 7 = -8$,
$x_6 = -8 - 7 = \textbf{–15}$

Using the same method for c)-f):
c) 486   d) 189   e) 4   f) 15.75

2 a) $x_1 = 8$, so:
$x_2 = 2 \times 8 - 6 = 10$,
$x_3 = 2 \times 10 - 6 = 14$,
$x_4 = 2 \times 14 - 6 = \textbf{22}$

b) $x_4 = 22$, so:
$x_5 = 2 \times 22 - 6 = 38$,
$x_6 = 2 \times 38 - 6 = \textbf{70}$

c) $x_3 + x_5 = 14 + 38 = \textbf{52}$
You had already calculated the two terms in parts a) and b).

3 Rearrange the formula:
$u_{n+1} = 3u_n - 1$, so $3u_n = u_{n+1} + 1$
and so $u_n = (u_{n+1} + 1) \div 3$.
Starting from $u_{n+1} = u_5 = 14$:
$u_4 = (14 + 1) \div 3 = 5$
$u_3 = (5 + 1) \div 3 = 2$
$u_2 = (2 + 1) \div 3 = 1$
$u_1 = (1 + 1) \div 3 = \dfrac{\textbf{2}}{\textbf{3}}$
You're asked for the exact value, so leave it as a fraction rather than a rounded decimal.

Page 166   Exercise 4

1 a) (i) E.g. Add 2 matches to the right hand side to form another equilateral triangle.

(ii)

(iii) There are 11 matches in the fifth pattern (from part (ii)) and 2 matches added each time, so $11 + 2 = \textbf{13 matches}$.

b) (i) Add 3 matches to form another square to the right of the top right square and 3 matches to form another square below the bottom left square.

(ii)

(iii) There are 28 matches in the fifth pattern (from part (ii)) and $3 + 3$ matches added each time, so $28 + 3 + 3 = \textbf{34 matches}$.

c) (i) E.g. Add three matches to the left hand side to form a new square.

(ii)

(iii) There are 15 matches in the fifth pattern (from part (ii)) and 3 matches added each time, so $15 + 3 = \textbf{18 matches}$.

2 In the answers to part (i) of this question, the orange circles have been shaded and the green circles are white.

a) (i)

(ii) The number of green circles in each pattern is always 1.
The number of orange circles in each pattern increases by 2 each time.

(iii) There is always 1 green circle, so there will be **1 green circle** in the 7th pattern.

(iv) There are 10 orange circles in the 6th pattern (from part (i)) and 2 added each time, so add on 4 lots of 2 to get the 10th pattern:
$10 + (4 \times 2) = \textbf{18 orange circles}$.

b) (i)

(ii) The number of green circles increases by 1 each time. The number of orange circles increases by 1 each time.

(iii) There are 6 green circles in the 6th pattern (from part (i)) and 1 added each time, so in the 7th pattern there are:
$6 + 1 = \textbf{7 green circles}$.

(iv) There are 7 orange circles in the 6th pattern (from part (i)) and 1 added each time, so add on 4 lots of 1 to get the 10th pattern:
$7 + (4 \times 1) = \textbf{11 orange circles}$.

3 a) (i) E.g. Add a row of circles to the bottom of the triangle, with one more circle in the new row than in the bottom row of the previous shape in the pattern.

(ii) The number of circles is 1, 3, 6... which is the sequence of **triangular numbers**.

(iii) The 6th pattern in the sequence will have a row of 1 circle, then a row of 2 circles, then 3... up to 6 circles at the bottom:

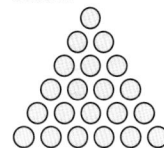

b) (i) E.g. Add an extra row and column of circles to the right and bottom of the shape so that a new 'square' is formed that is wider and taller than the previous square by 1 circle.

(ii) The number of circles is 1, 4, 9... which is the sequence of **square numbers**.

(iii) The 6th pattern in the sequence will have 6 rows and 6 columns of circles:

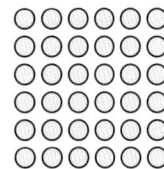

4 a) To make each new term, a row of squares has been added to the bottom, and a column of squares has been added to the right of the previous term, with the colours of each new square being different to any adjacent squares. So the next pattern is:

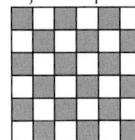

b) The number of white squares in each pattern follows the sequence: 4, 8, 12, 18... There is a different number of white squares being added on each time, so you need to look at the patterns themselves. The 4th pattern is shown in part a), and so the next 2 patterns are:

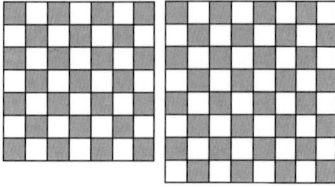

So there are **32 white squares** in the 6th pattern.

c) The total number of squares in each pattern follows the sequence: 9, 16, 25, 36, 49, 64... which is a sequence of square numbers. Continuing the sequence: 7th pattern = 81 (9^2), 8th pattern = 100 (10^2). So you're looking for the number of grey squares in the 8th pattern. You could draw this pattern and count the squares:

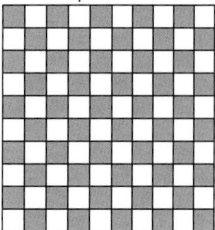

So there are **50 grey squares**.
Or you could investigate the patterns in the numbers by drawing a table:

Pattern	1	2	3	4	5	6
Total	9	16	25	36	49	64
White	4	8	12	18	24	32
Grey	5	8	13	18	25	32

In every even numbered pattern, there is an even number of squares which is split equally between white and grey.
So you can predict that the pattern with 100 squares will be an even numbered pattern (because 100 is even) and there will be $100 \div 2 = \textbf{50 grey squares}$.

5 There are several ways to continue the sequence, e.g.:
The second term has three more matchsticks than the first term, so the pattern could continue by adding another three matchsticks to the top of the second term:

The second term has double the number of matchsticks as the first term, with the extra matchsticks being set out as a 90° rotation anticlockwise on top of the first term. So the pattern could continue by doubling the number of matchsticks in the second term, and arranging them as follows:

The patterns could form a Fibonacci-type sequence, where the previous two terms are added to get the next term, so the next pattern could be:

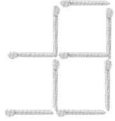

14.2 Using the nth Term

Page 167   Exercise 1

1 a) 1st term ($n = 1$) = $1 + 5 = $ **6**
2nd term ($n = 2$) = $2 + 5 = $ **7**
3rd term ($n = 3$) = $3 + 5 = $ **8**
4th term ($n = 4$) = $4 + 5 = $ **9**
5th term ($n = 5$) = $5 + 5 = $ **10**

b) 1st term ($n = 1$) = $4 \times 1 - 2 = $ **2**
2nd term ($n = 2$) = $4 \times 2 - 2 = $ **6**
3rd term ($n = 3$) = $4 \times 3 - 2 = $ **10**
4th term ($n = 4$) = $4 \times 4 - 2 = $ **14**
5th term ($n = 5$) = $4 \times 5 - 2 = $ **18**

Using the same method for c)-d):
c) 9, 8, 7, 6, 5   **d) 2, 12, 22, 32, 42**

2 a) 1st term ($n = 1$) = $1^2 + 1 = $ **2**
2nd term ($n = 2$) = $2^2 + 1 = $ **5**
3rd term ($n = 3$) = $3^2 + 1 = $ **10**
4th term ($n = 4$) = $4^2 + 1 = $ **17**
5th term ($n = 5$) = $5^2 + 1 = $ **26**

b) 1st term ($n = 1$) = $2 \times 1^2 + 1 = $ **3**
2nd term ($n = 2$) = $2 \times 2^2 + 1 = $ **9**
3rd term ($n = 3$) = $2 \times 3^2 + 1 = $ **19**
4th term ($n = 4$) = $2 \times 4^2 + 1 = $ **33**
5th term ($n = 5$) = $2 \times 5^2 + 1 = $ **51**

Using the same method for c)-d):
c) 2, 11, 26, 47, 74   d) 0, 2, 6, 12, 20

3 a) $n = 3$, so $100 - (3 \times 3) = $ **91**
b) $n = 10$, so $100 - (3 \times 10) = $ **70**
Using the same method for c)-d):
c) 10   d) −20

4 a) $n = 1$, so $35 + (5 \times 1) = $ **40**
b) $n = 5$, so $35 + (5 \times 5) = $ **60**
Using the same method for c)-d):
c) 85   d) 535

5 a) (i) $n = 5$, so $2 \times 5 + 3 = $ **13**
(ii) $n = 10$, so $2 \times 10 + 3 = $ **23**
(iii) $n = 100$, so $2 \times 100 + 3 = $ **203**
Using the same method for b)-d):
b) (i) 32   (ii) 52   (iii) 412
c) (i) 15   (ii) 0   (iii) −270
d) (i) −10   (ii) 0   (iii) 180

6 a) (i) $n = 4$, so $x_4 = \frac{1}{2} \times 4 + 30 = $ **32**
(ii) $n = 10$, so $x_{10} = \frac{1}{2} \times 10 + 30 = $ **35**
(iii) $n = 20$, so $x_{20} = \frac{1}{2} \times 20 + 30 = $ **40**
b) (i) $n = 4$, so $x_4 = 2 \times 4^2 + 8 = $ **40**
(ii) $n = 10$, so $x_{10} = 2 \times 10^2 + 8 = $ **208**
(iii) $n = 20$, so $x_{20} = 2 \times 20^2 + 8 = $ **808**
c) (i) $n = 4$, so $x_4 = 5 \times 4^3 - 50 = $ **270**
(ii) $n = 10$, so $x_{10} = 5 \times 10^3 - 50 = $ **4950**
(iii) $n = 20$, so $x_{20} = 5 \times 20^3 - 50 = $ **39 950**

d) (i) $n = 4$, so $x_4 = (3 \times 4) - 4^2 + 100 = $ **96**
(ii) $n = 10$,
so $x_{10} = (3 \times 10) - 10^2 + 100 = $ **30**
(iii) $n = 20$,
so $x_{20} = (3 \times 20) - 20^2 + 100 = $ **−240**

7 a) (i) $n = 2$, so $2 \times 2^2 + 3 = $ **11**
(ii) $n = 5$, so $2 \times 5^2 + 3 = $ **53**
(iii) $n = 20$, so $2 \times 20^2 + 3 = $ **803**
Using the same method for b)-d):
b) (i) 18   (ii) 81   (iii) 1206
c) (i) 6   (ii) 30   (iii) 420
d) (i) 10   (ii) 127   (iii) 8002

Page 168   Exercise 2

1 a) Solve the equation:
$7n + 4 = 53 \Rightarrow 7n = 49 \Rightarrow n = 7$
So 53 is the **7th** term.
b) Solve the equation:
$5n - 8 = 37 \Rightarrow 5n = 45 \Rightarrow n = 9$
So 37 is the **9th** term.

2 a) Solve the equation:
$17 - 2n = 9 \Rightarrow 2n = 8 \Rightarrow n = 4$
So 9 is the **4th** term.
b) Solve the equation:
$17 - 2n = -3 \Rightarrow 2n = 20 \Rightarrow n = 10$
So −3 is the **10th** term.

3 a) Solve the equation:
$n^2 - 1 = 8 \Rightarrow n^2 = 9 \Rightarrow n = \sqrt{9} = 3$
So 8 is the **3rd** term.
You can ignore the negative square root — the position in a sequence must be a positive integer.
b) Solve the equation:
$n^2 - 1 = 99 \Rightarrow n^2 = 100 \Rightarrow n = \sqrt{100} = 10$
So 99 is the **10th** term.

4 Find the value of n that would give a value of 50 by solving:
$4n - 10 = 50 \Rightarrow 4n = 60 \Rightarrow n = 15$
If the 15th term is equal to 50, then as long as the terms are increasing, the **16th term** is the first to be greater than 50. Check with $n = 16$:
$4 \times 16 - 10 = 54$, which is greater than 50.
You could also have solved the inequality:
$4n - 10 > 50.$

5 a) Solve the equation:
$2n + 1 = 17 \Rightarrow 2n = 16 \Rightarrow n = 8$
So the **8th** pattern uses 17 matches.
b) Solve the equation:
$2n + 1 = 55 \Rightarrow 2n = 54 \Rightarrow n = 27$
So the **27th** pattern uses 54 matches.

6 a) Number of circles in 58th triangle ($n = 58$)
$= \frac{58(58 + 1)}{2} = $ **1711**
b) Find the value of n that would give a value of 200 by solving:
$\frac{n(n + 1)}{2} = 200$
$n(n + 1) = 400$
$n^2 + n = 400$
$n^2 + n - 400 = 0$
Using the quadratic formula:
$n = \frac{-1 \pm \sqrt{1^2 - 4 \times 1 \times -400}}{2 \times 1}$
$n = 19.506...$
You only need to think about positive values of n...
n is not a whole number, so there is no term with exactly 200 circles. But the nearest whole number above this is 20. As the terms are increasing, the **20th term** must be the first to be made from over 200 circles. Check with $n = 20$:
$\frac{20(20 + 1)}{2} = 210$,
which is greater than 200.
You can also check by making sure the 19th term has fewer than 200 circles:
$\frac{19(19 + 1)}{2} = 190$,
which is smaller than 200.

7 a) Solve the equation:
$-n(n + 2) = -48$
$-n^2 - 2n = -48$
$0 = n^2 + 2n - 48$
$0 = (n + 8)(n - 6)$
$n = -8$ or $n = 6$. Since n must be positive,
-48 must be the **6th** term.

b) Find the value of n that would give a value
of -20 by solving:
$-n(n + 2) = -20$
$-n^2 - 2n = -20$
$0 = n^2 + 2n - 20$
Using the quadratic formula:
$n = \dfrac{-2 \pm \sqrt{2^2 - 4 \times 1 \times -20}}{2 \times 1}$
$n = 3.582...$
Just take the positive answer.

n is not a whole number, so -20 is not a
term in the sequence. Try terms around
this value of n:
$n = 3: -3(3 + 2) = -15$
$n = 4: -4(4 + 2) = -24$
The terms are decreasing, so the **4th term**
is the first to have a value less than -20.

Page 169   Exercise 3

1 If 80 is a term, then solving $3n - 1 = 80$ will
give a positive integer value for n:
$3n - 1 = 80 \Rightarrow 3n = 81 \Rightarrow n = 27$
n is a positive integer, so 80 is
a term in the sequence.
It's the 27th term in the sequence.

2 a) $21 - 2n = -1 \Rightarrow 2n = 22 \Rightarrow n = 11$,
which is a positive integer, so -1 is a term
in the sequence.

b) $n = 11$ so -1 is the **11th term**.

3 a) 1st term $(n = 1) = 6(1 + 1) = $ **12**
2nd term $(n = 2) = 6(2 + 1) = $ **18**
3rd term $(n = 3) = 6(3 + 1) = $ **24**
4th term $(n = 4) = 6(4 + 1) = $ **30**

b) If 64 is a term, then solving $6(n + 1) = 64$
will give a positive integer value for n:
$6(n + 1) = 64$
$n + 1 = 64 \div 6 = 10.666...$
$n = 10.666... - 1 = 9.666...$
n is not an integer, so **64 is not a term in
the sequence**.
*Alternatively, if you'd spotted that all the terms in
the sequence were multiples of 6, you could have
said that 64 couldn't be in the sequence as it's
not a multiple of 6.*

c) Try numbers for n around 9.666...
$n = 9: 6(9 + 1) = 60$
$n = 10: 6(10 + 1) = 66$
The closest of these to 64 is **66**.

4 a) If 52 is a term, then solving $17 + 3n = 52$
will give a positive integer value for n:
$17 + 3n = 52 \Rightarrow 3n = 35 \Rightarrow n = 11.666...$
n is not an integer, so **52 is not a term in
the sequence**.

b) If 98 is a term, then solving $17 + 3n = 98$
will give a positive integer value for n:
$17 + 3n = 98 \Rightarrow 3n = 81 \Rightarrow n = 27$
n is a positive integer, so **98 is a term in the
sequence**.

Using the same method for c)-e):

c) **105 is not a term in the sequence**

d) **248 is a term in the sequence**

e) **996 is not a term in the sequence**

f) If $20n$ is a term, then solving $17 + 3n = 20n$
will give a positive integer value for n:
$17 + 3n = 20n \Rightarrow 17 = 17n \Rightarrow n = 1$
n is a positive integer, so **$20n$ is a term in
the sequence**.
It's the first term in the sequence.

14.3 Finding the nth Term

Page 170   Exercise 1

1 a) $13 - 7 = 6$, $19 - 13 = 6$, $25 - 13 = 6$
So common difference $d = +6$.
Compare sequence with $6n$:

$6n$:	6	12	18	24
	↓+1	↓+1	↓+1	↓+1
Term:	7	13	19	25

So $c = 1$, and the nth term is **$6n + 1$**.

b) $8 - 4 = 4$, $12 - 8 = 4$, $16 - 12 = 4$
So common difference $d = +4$.
Compare sequence with $4n$:

$4n$:	4	8	12	16
	↓+0	↓+0	↓+0	↓+0
Term:	4	8	12	16

So $c = 0$, and the nth term is **$4n$**.
*Any sequence that's a 'times table' like this will
have an nth term of this form. So the 3 times
table has the nth term 3n, the 12 times table has
the nth term 12n, etc.*

Using the same method for c)-d):

c) **$40n + 1$**   **d)** **$4n - 13$**

2 a) (i) $37 - 40 = -3$, $34 - 37 = -3$,
$31 - 34 = -3$
So common difference $d = -3$.
Compare sequence with $-3n$:

$-3n$:	-3	-6	-9	-12
	↓+43	↓+43	↓+43	↓+43
Term:	40	37	34	31

So $c = 43$, and the nth term is **$43 - 3n$**.

(ii) Substitute $n = 70$ into $43 - 3n$:
$43 - (3 \times 70) = $ **-167**

b) (i) $69 - 78 = -9$, $60 - 69 = -9$,
$51 - 60 = -9$
So common difference $d = -9$.
Compare sequence with $-9n$:

$-9n$:	-9	-18	-27	-36
	↓+87	↓+87	↓+87	↓+87
Term:	78	69	60	51

So $c = 87$, and the nth term is **$87 - 9n$**.

(ii) Substitute $n = 70$ into $87 - 9n$:
$87 - (9 \times 70) = $ **-543**

Using the same method for c)-d):

c) (i) **$108 - 8n$**   **(ii)** **-452**

d) (i) **$5 - 15n$**   **(ii)** **-1045**

3 Using the same method for Q3a)-b):

a) **$3n + 1$**   **b)** **$3n + 2$**

c) (i) Each term is 1 greater than in A.

(ii) They differ only in the constant — the
coefficient of n is the same in both (+3).

4 a) Using the same method for a):
$17 - 4n$

b) Each term in this sequence is two less than
the sequence in a). So subtract 2 from the
nth term for a): $17 - 4n - 2 = $ **$15 - 4n$**.

5 a) Sequence of number of dots: 4, 8, 12...
This is the 4 times table, so the nth term
is **$4n$**.
*If you didn't spot that it was the 4 times table,
use the method from Q1 to find the nth term.*

b) Sequence of areas (in units²): 1, 3, 5...
So $d = +2$. Compare sequence with $2n$:

$2n$:	2	4	6
	↓−1	↓−1	↓−1
Term:	1	3	5

So $c = -1$, and the nth term is
$2n - 1$ units².

c) Substitute $n = 23$ into each nth term:
Number of dots $= 4 \times 23 = $ **92 dots**
Area $= 2 \times 23 - 1 = $ **45 units²**

Page 171   Exercise 2

1 a) Numerator sequence: 1, 1, 1, 1...
So nth term $= 1$
Denominator sequence: 2, 4, 6, 8...
It's the 2 times table, so nth term $= 2n$
So the combined nth term is $\dfrac{1}{2n}$

b) Numerator sequence: 1, 1, 1, 1...
So nth term $= 1$
Denominator sequence: 3, 6, 9, 12...
It's the 3 times table, so nth term $= 3n$
So the combined nth term is $\dfrac{1}{3n}$

c) Numerator sequence: 5, 6, 7, 8...
So $d = +1$. Compare sequence with n:

n:	1	2	3	4
	↓+4	↓+4	↓+4	↓+4
Term:	5	6	7	8

So $c = 4$, and nth term $= n + 4$
Denominator sequence: 2, 3, 4, 5...
So $d = +1$. Compare sequence with n:

n:	1	2	3	4
	↓+1	↓+1	↓+1	↓+1
Term:	2	3	4	5

So $c = 1$, and nth term $= n + 1$
So the combined nth term is $\dfrac{n + 4}{n + 1}$

2 a) Numerator sequence: 1, 4, 7, 10...
So $d = +3$. Compare sequence with $3n$:

$3n$:	3	6	9	12
	↓−2	↓−2	↓−2	↓−2
Term:	1	4	7	10

So $c = -2$, and nth term $= 3n - 2$
Denominator sequence: 5, 10, 15, 20...
It's the 5 times table, so nth term $= 5n$
So the combined nth term is $\dfrac{3n - 2}{5n}$

Using the same method for b)-c):

b) $\dfrac{6 - n}{5 - n}$ **c)** $\dfrac{3n + 2}{10 - n}$

3 First find the nth term:
Numerator sequence: 1, -1, -3, -5...
So $d = -2$. Compare sequence with $-2n$:

$-2n$:	-2	-4	-6	-8
	↓+3	↓+3	↓+3	↓+3
Term:	1	-1	-3	-5

So $c = 3$, and nth term $= 3 - 2n$
Denominator sequence: 30, 40, 50, 60...
So $d = +10$. Compare sequence with $10n$:

$10n$:	10	20	30	40
	↓+20	↓+20	↓+20	↓+20
Term:	30	40	50	60

So $c = 20$, and nth term $= 10n + 20$
So the combined nth term is $\dfrac{3 - 2n}{10n + 20}$.
Substitute $n = 100$ to get the 100th term
$= \dfrac{3 - 2 \times 100}{10 \times 100 + 20} = -\dfrac{197}{1020}$

Page 173   Exercise 3

*You don't need to use simultaneous equations for Q1,
although you can if you prefer.*

1 a) (i) 1st term $= a = 3$,
4th term $= a + (4 - 1)d = 9$, and $a = 3$:
$3 + 3d = 9 \Rightarrow 3d = 6 \Rightarrow d = 2$
So $a = 3$ and $d = 2$.

(ii) nth term $= a + (n - 1)d$
$= 3 + (n - 1) \times 2 = $ **$2n + 1$**

b) (i) d is the difference between consecutive
terms, so $d = 21 - 17 = 4$.
5th term $= a + (5 - 1)d = 17$, and $d = 4$:
$a + 4 \times 4 = 17 \Rightarrow a = 17 - 16 = 1$
So $a = 1$ and $d = 4$.

(ii) nth term $= a + (n - 1)d$
$= 1 + (n - 1) \times 4 = $ **$4n - 3$**

Using the same method for c):

c) (i) $a = -38$ and $d = 10$

(ii) **$10n - 48$**

2 a) $x_3 = a + (3 - 1)d = 15 \Rightarrow a + 2d = 15$ (1)
$x_6 = a + (6 - 1)d = 30 \Rightarrow a + 5d = 30$ (2)
Subtract (1) from (2):
$5d - 2d = 30 - 15 \Rightarrow 3d = 15 \Rightarrow d = 5$
Put d back into (1):
$a + 2 \times 5 = 15 \Rightarrow a = 15 - 10 = 5$
nth term $= a + (n - 1)d$
$= 5 + (n - 1) \times 5 = $ **$5n$**

b) $x_2 = a + (2 - 1)d = 12 \Rightarrow a + d = 12$ (1)
$x_5 = a + (5 - 1)d = 33 \Rightarrow a + 4d = 33$ (2)
Subtract (1) from (2):
$4d - d = 33 - 12 \Rightarrow 3d = 21 \Rightarrow d = 7$
Put d back into (1):
$a + 7 = 12 \Rightarrow a = 12 - 7 = 5$
nth term $= a + (n - 1)d$
$= 5 + (n - 1) \times 7 = \mathbf{7n - 2}$

Using the same method for c)-f):
c) $\mathbf{8n - 7}$ **d)** $\mathbf{23 - 5n}$
e) $\mathbf{9n - 30}$ **f)** $\mathbf{12n - 11}$

3 a) $x_3 = a + (3 - 1)d = 4 \Rightarrow a + 2d = 4$ (1)
$x_{12} = a + (12 - 1)d = 1 \Rightarrow a + 11d = 1$ (2)
Subtract (1) from (2):
$11d - 2d = 1 - 4$
Be careful with your subtraction here
— it's 1 − 4 not 4 − 1...
$9d = -3 \Rightarrow d = -3 \div 9 = -\dfrac{1}{3}$
Put d back into (1):
$a + 2 \times -\dfrac{1}{3} = 4$
$\Rightarrow a = 4 + \dfrac{2}{3} = \dfrac{14}{3}$
So $a = \dfrac{14}{3}$ and $d = -\dfrac{1}{3}$
b) nth term $= a + (n - 1)d$
$= \dfrac{14}{3} + (n - 1) \times -\dfrac{1}{3} = \dfrac{14}{3} + \dfrac{1}{3} - \dfrac{1}{3}n$
$= 5 - \dfrac{1}{3}n$

4 First, find the nth term of the sequence using simultaneous equations:
$x_{61} = a + (61 - 1)d = 546 \Rightarrow a + 60d = 546\,(1)$
$x_{81} = a + (81 - 1)d = 726 \Rightarrow a + 80d = 726\,(2)$
Subtract (1) from (2):
$80d - 60d = 726 - 546 \Rightarrow 20d = 180 \Rightarrow d = 9$
Put d back into (1):
$a + 60 \times 9 = 546 \Rightarrow a = 546 - 540 = 6$
nth term $= a + (n - 1)d$
$= 6 + (n - 1) \times 9 = 9n - 3$
If 42 is in the sequence, then solving
$9n - 3 = 42$ will give a positive integer n:
$9n - 3 = 42 \Rightarrow 9n = 45 \Rightarrow n = 5$
So **yes, 42 is in the sequence.**
$n = 5$, so 42 is the 5th term.

Page 174   Exercise 4

1 a) (i) First differences between terms are:
$+4, +6, +8, +10...$
So the second difference is $+2$, and the coefficient of n^2 in the nth term formula is $2 \div 2 = 1$.
So the expression for the nth term starts with an n^2 term...
Subtract $n^2 \times 1$ from each term:

Term:	1	5	11	19	29
n^2:	1	4	9	16	25
Term $- n^2$:	0	1	2	3	4

This linear sequence has first term 0 and difference $+1$, so the linear part of the nth term is $0 + (n - 1) \times 1 = n - 1$.
So the full expression for the nth term is $\mathbf{n^2 + n - 1}$.
You might have used a different method to get the linear part of the nth term.
(ii) Substitute $n = 10$ to get the 10th term:
$10^2 + 10 - 1 = \mathbf{109}$
b) (i) First differences between terms are:
$+7, +9, +11, +13...$
So the second difference is $+2$, and the coefficient of n^2 in the nth term formula is $2 \div 2 = 1$.
Subtract $1 \times n^2$ from each term:

Term:	5	12	21	32	45
n^2:	1	4	9	16	25
Term $- n^2$:	4	8	12	16	20

This linear sequence is the 4 times table, so the nth term is $4n$.
So the full expression for the nth term is $\mathbf{n^2 + 4n}$.

(ii) Substitute $n = 10$ to get the 10th term:
$10^2 + 4 \times 10 = \mathbf{140}$
Using the same method for c)-f):
c) (i) $\mathbf{2n^2 + n + 2}$ **(ii)** $\mathbf{212}$
d) (i) $\mathbf{3n^2 + 4n}$ **(ii)** $\mathbf{340}$
e) (i) $\mathbf{5n^2 + 3n - 2}$ **(ii)** $\mathbf{528}$
f) (i) $\mathbf{4n^2 + n + 5}$ **(ii)** $\mathbf{415}$

Page 175   Review Exercise

1 a) (i) $2 \div 1 = 2$, $4 \div 2 = 2$, $8 \div 4 = 2$
So the rule is **multiply the previous term by 2.**
(ii) $8 \times 2 = \mathbf{16}$, $16 \times 2 = \mathbf{32}$,
$32 \times 2 = \mathbf{64}$
b) (i) $7 - 4 = 3$, $10 - 7 = 3$, $13 - 10 = 3$
So the rule is **add 3 to the previous term.**
(ii) $13 + 3 = \mathbf{16}$, $16 + 3 = \mathbf{19}$, $19 + 3 = \mathbf{22}$
Using the same method for c)-f):
c) (i) **Subtract 2 from the previous term.**
(ii) $\mathbf{-3, -5, -7}$
d) (i) **Add 0.5 to the previous term.**
(ii) $\mathbf{3, 3.5, 4}$
e) (i) **Multiply the previous term by 10.**
(ii) $\mathbf{100, 1000, 10\,000}$
f) (i) **Multiply the previous term by 3.**
(ii) $\mathbf{-162, -486, -1458}$

2 a) The terms are being multiplied by 4 each time $(-4 \times 4 = -16, -16 \times 4 = -64...)$.
So the missing numbers are:
$-4 \div 4 = \mathbf{-1}$ and $-64 \times 4 = \mathbf{-256}$
b) The terms are being divided by 2 each time $(-18 \div 2 = -9)$, so the missing numbers are:
$-72 \div 2 = \mathbf{-36}$ and $-9 \div 2 = \mathbf{-4.5}$
You might need to use a bit of guesswork here because you're only given two consecutive numbers. Always check that the rule still works after you've found the missing numbers:
$-36 \div 2 = -18$ and $-4.5 \div 2 = -2.25$.
c) The terms are increasing by 8 each time $(-63 + 8 = -55)$, so the missing numbers are: $-55 + 8 = \mathbf{-47}$, $-47 + 8 = \mathbf{-39}$ and $-39 + 8 = \mathbf{-31}$
Again, check that $-31 + 8 = -23$.

3 Sequence of number of matchsticks: 3, 6, 9...
The term to term rule is 'add three matchsticks each time', so to get to the 10th pattern add on 7 lots of 3 matchsticks to the 3rd pattern:
10th pattern $= 9 + (7 \times 3) = \mathbf{30\ matchsticks}$.
You could also have used the nth term — the sequence is just the 3 times table, so the nth term is 3n, and the 10th term is 3 × 10 = 30.

4 a) 1st term $(n = 1) = 3 \times 1 + 2 = \mathbf{5}$
2nd term $(n = 2) = 3 \times 2 + 2 = \mathbf{8}$
3rd term $(n = 3) = 3 \times 3 + 2 = \mathbf{11}$
4th term $(n = 4) = 3 \times 4 + 2 = \mathbf{14}$
5th term $(n = 5) = 3 \times 5 + 2 = \mathbf{17}$
b) 1st term $(n = 1) = 5 \times 1 - 1 = \mathbf{4}$
2nd term $(n = 2) = 5 \times 2 - 1 = \mathbf{9}$
3rd term $(n = 3) = 5 \times 3 - 1 = \mathbf{14}$
4th term $(n = 4) = 5 \times 4 - 1 = \mathbf{19}$
5th term $(n = 5) = 5 \times 5 - 1 = \mathbf{24}$
Using the same method for c)-d):
c) $\mathbf{-1, -5, -9, -13, -17}$
d) $\mathbf{-10, -13, -16, -19, -22}$

5 a) (i) $n = 4$, so $x_4 = 8 \times 4 + 4 = \mathbf{36}$
(ii) $n = 10$, so $x_{10} = 8 \times 10 + 4 = \mathbf{84}$
(iii) $n = 20$, so $x_{20} = 8 \times 20 + 4 = \mathbf{164}$
Using the same method for b)-d):
b) (i) **430**   **(ii) 400**   **(iii) 350**
c) (i) **48**   **(ii) 300**   **(iii) 1200**
d) (i) **20**   **(ii) 110**   **(iii) 420**

6 a) (i) $n = 2$, so $2 \times 2^2 = \mathbf{8}$
(ii) $n = 5$, so $2 \times 5^2 = \mathbf{50}$
(iii) $n = 20$, so $2 \times 20^2 = \mathbf{800}$
Using the same method for b)-d):
b) (i) **11**   **(ii) 95**   **(iii) 1595**
c) (i) **22**   **(ii) 32.5**   **(iii) 220**
d) (i) **396**   **(ii) 375**   **(iii) 0**

7 a) (i) $8 - 10 = -2$, $6 - 8 = -2$, $4 - 6 = -2$
So common difference $d = -2$.
Compare sequence with $-2n$:

$-2n$:	-2	-4	-6	-8
	↓+12	↓+12	↓+12	↓+12
Term:	10	8	6	4

So $c = 12$, and the nth term is $\mathbf{12 - 2n}$
You might have used the $a + (n - 1)d$ method instead.
(ii) Substitute $n = 70$ into $12 - 2n$:
$12 - (2 \times 70) = \mathbf{-128}$
b) (i) $60 - 70 = -10$, $50 - 60 = -10$,
$40 - 50 = -10$, so $d = -10$.
Compare sequence with $-10n$:

$-10n$:	-10	-20	-30	-40
	↓+80	↓+80	↓+80	↓+80
Term:	70	60	50	40

So $c = 80$, and the nth term is $\mathbf{80 - 10n}$
(ii) Substitute $n = 70$ into $80 - 10n$:
$80 - (10 \times 70) = \mathbf{-620}$
Using the same method for c)-d):
c) (i) $\mathbf{65 - 5n}$ **(ii)** $\mathbf{-285}$
d) (i) $\mathbf{9 - 3n}$ **(ii)** $\mathbf{-201}$

8 First find the nth term.
$-1 - (-5) = 4$, $3 - (-1) = 4$, $7 - 3 = 4$
So $d = +4$. Compare sequence with $4n$:

$4n$:	4	8	12	16
	↓−9	↓−9	↓−9	↓−9
Term:	-5	-1	3	7

So $c = -9$, and the nth term $= 4n - 9$.
You might have used a different method here.
a) If 43 is a term, then solving $4n - 9 = 43$ will give a positive integer value for n:
$4n - 9 = 43 \Rightarrow 4n = 52 \Rightarrow n = 13$
n is a positive integer, so **43 is a term in the sequence**, with $n = 13$.
b) If 138 is a term, then solving $4n - 9 = 138$ will give a positive integer value for n:
$4n - 9 = 138 \Rightarrow 4n = 147 \Rightarrow n = 36.75$
n is not a positive integer, so **138 is not a term in the sequence.**
Using the same method for c)-d):
c) **384 is not a term in the sequence.**
d) **879 is a term in the sequence,** with $n = 222$.

9 Numerator sequence: 16, 36, 56, 76...
So $d = +20$. Compare sequence with $20n$:

$20n$:	20	40	60	80
	↓−4	↓−4	↓−4	↓−4
Term:	16	36	56	76

So $c = -4$, and the nth term $= 20n - 4$
Denominator sequence: 3, 7, 11, 15...
So $d = +4$. Compare sequence with $4n$:

$4n$:	4	8	12	16
	↓−1	↓−1	↓−1	↓−1
Term:	3	7	11	15

So $c = -1$, and the nth term $= 4n - 1$
So the combined nth term is $\dfrac{20n - 4}{4n - 1}$.
You might have used a different method to find each nth term.

10 $x_2 = a + (2 - 1)d = -7 \Rightarrow a + d = -7$ (1)
$x_{15} = a + (15 - 1)d = 32 \Rightarrow a + 14d = 32$ (2)
Subtract (1) from (2):
$14d - d = 32 - (-7) \Rightarrow 13d = 39 \Rightarrow d = 3$
Put d back into (1):
$a + 3 = -7 \Rightarrow a = -7 - 3 = -10$
nth term $= a + (n - 1)d$
$= -10 + (n - 1) \times 3 = \mathbf{3n - 13}$

11 a) (i) First differences between terms are:
+1, +2, +3, +4...
So the second difference is +1, and the coefficient of n^2 in the nth term formula is $1 \div 2 = \frac{1}{2}$.
Subtract $\frac{1}{2} \times n^2$ from each term:

Term:	7	8	10	13	17
$\frac{1}{2}n^2$:		0.5	2	4.5	8	12.5
Term $- \frac{1}{2}n^2$:		6.5	6	5.5	5	4.5

This linear sequence has first term 6.5 and difference –0.5, so the linear part of the nth term is:
$6.5 + (n - 1) \times -0.5 = -\frac{1}{2}n + 7$.
You might have used a different method to get the linear part of the nth term.
So the full expression for the nth term is $\frac{1}{2}n^2 - \frac{1}{2}n + 7$.

(ii) Substitute $n = 10$ into $\frac{1}{2}n^2 - \frac{1}{2}n + 7$:
10th term $= \frac{1}{2} \times 10^2 - \frac{1}{2} \times 10 + 7$
$= 50 - 5 + 7 = \mathbf{52}$

b) (i) First differences between terms are:
+2, +4, +6, +8...
So the second difference is +2, and the coefficient of n^2 in the nth term formula is $2 \div 2 = 1$.
Subtract n^2 from each term:

Term:	5	7	11	17	25
n^2:	1	4	9	16	25
Term $- n^2$:	4	3	2	1	0

This linear sequence has first term 4 and difference –1, so the linear part of the nth term is $4 + (n - 1) \times -1 = 5 - n$
So the full expression for the nth term is $\mathbf{n^2 - n + 5}$.

(ii) Substitute $n = 10$ into $n^2 - n + 5$:
10th term $= 10^2 - 10 + 5 = \mathbf{95}$
Using the same method for c):

c) (i) $\mathbf{n^2 - n + 3}$ **(ii)** **93**

Page 176 Exam-Style Questions

1 a) 1st term $= a = -5$,
19th term $= a + (19 - 1)d = -95$,
and $a = -5$, so $-5 + 18d = -95$
$18d = -95 + 5 = -90$
$d = -90 \div 18 = -5$
[1 mark for finding a and d]
nth term $= a + (n - 1)d$
$= -5 + (n - 1) \times -5 = \mathbf{-5n}$ *[1 mark]*

b) For –113 to be a term in the sequence, the solution to $-5n = -113$ would have to be an integer value of n. Since –113 is not a multiple of 5 (or –5), n will not be an integer, and so –113 is not a term in the sequence *[1 mark]*.

2 a) $u_3 = 2u_2^2 - 7 = 2 \times (-6.28)^2 - 7$ *[1 mark]*
$= \mathbf{71.8768}$ *[1 mark]*

b) $u_2 = 2u_1^2 - 7$ so rearrange to get:
$2u_1^2 = u_2 + 7$
$u_1 = \sqrt{\frac{u_2 + 7}{2}}$ *[1 mark]*
so $u_1 = \sqrt{\frac{-6.28 + 7}{2}} = \sqrt{0.36}$
$= \mathbf{0.6}$ *[1 mark]*

3 a) First, draw the next pattern in the sequence:

Sequence of number of dots: 2, 7, 14, 23...
First differences: +5, +7, +9
So the second difference is +2.

The sequence must be quadratic, and the coefficient of n^2 in the nth term formula is $2 \div 2 = 1$. So there is an n^2 term *[1 mark]*.
Subtract n^2 from each term:

Term:	2	7	14	23
n^2:	1	4	9	16
Term $- n^2$:	1	3	5	7

This linear sequence has first term 1 and difference +2 *[1 mark]*, so the linear part of the nth term is $1 + (n - 1) \times 2 = 2n - 1$ *[1 mark]*.
You might have used a different method.
So the full expression for the nth term is $\mathbf{n^2 + 2n - 1}$.

b) First, find the nth term of the areas.
Each square in the patterns has an area of $2 \times 2 = 4$ cm^2, and each triangle must be half of this, i.e. 2 cm^2.
So the sequence of areas is: 0, 12, 32, 60...
[1 mark for correct areas]
First differences: +12, +20, +28
So the second difference is +8. The sequence must be quadratic, and the coefficient of n^2 in the nth term formula is $8 \div 2 = 4$. So there is a $4n^2$ term *[1 mark]*.
Subtract $4n^2$ from each term:

Term:	0	12	32	60
$4n^2$:	4	16	36	64
Term $- 4n^2$:	–4	–4	–4	–4

The linear part of the nth term is just a constant –4 *[1 mark]*.
So the whole nth term is $4n^2 - 4$, and the area of the 100th pattern (with $n = 100$)
$= 4 \times 100^2 - 4$ *[1 mark]*
$= \mathbf{39\ 996\ cm^2}$ *[1 mark]*
There are other ways to get the correct answer, but as long as you show your working you'll still get all the marks.

4 a) No — the sequence does not have a common ratio / you don't multiply by the same number as you go from one term to the next (e.g. $7 \times 2 = 14$, but $14 \times 2 = 28 \neq 33$) *[1 mark]*.

b) No — the second differences (or the differences between the differences) are not the same (e.g. the difference between the first 4 terms are 7, 19 and 37, but the difference between the first two differences are 12 and 18 respectively, which are not the same) *[1 mark]*.

5 If the first and second terms are x and y then the sequence is:
$x, y, x + y, x + 2y, 2x + 3y, 3x + 5y$ ...
So the 4th term is $x + 2y = 22$ *[1 mark]* (1)
and the 6th term is $3x + 5y = 57$ *[1 mark]* (2)
Solve these two equations simultaneously:
Multiply (1) by 3 to get $3x + 6y = 66$ (3)
Subtract (2) from (3): $6y - 5y = 66 - 57$, which gives $y = \mathbf{9}$ *[1 mark]*.
Substitute in (1): $x + 2 \times 9 = 22$, so $x = 22 - 18 = \mathbf{4}$ *[1 mark]*.

Section 15 — Straight-Line Graphs

15.1 Straight-Line Graphs

Page 177 Exercise 1

1 All points on line A have an x-coordinate of –5, so the equation is $\mathbf{x = -5}$.
All points on line B have an x-coordinate of –2, so the equation is $\mathbf{x = -2}$.
All points on line C have a y-coordinate of 2, so the equation is $\mathbf{y = 2}$.
All points on line D have an x-coordinate of 5, so the equation is $\mathbf{x = 5}$.
All points on line E have a y-coordinate of –3, so the equation is $\mathbf{y = -3}$.

2

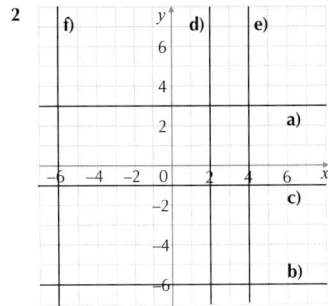

f) d) e) a) c) b)

3 a) $y = 8$ **b)** $x = -2$
c) $x = 1$ **d)** $y = 6$

4 a) $(8, -11)$ **b)** $(-5, -13)$
c) $(-\frac{6}{11}, -500)$

Page 179 Exercise 2

1 a)

x	–2	–1	0	1	2
y	–4	–2	0	2	4
Coord.	(–2, –4)	(–1, –2)	(0, 0)	(1, 2)	(2, 4)

b)-d)

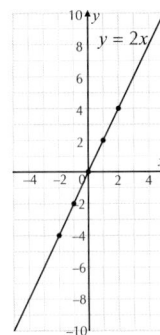
$y = 2x$

e) (i) 8 **(ii)** –6 **(iii)** –5

2 a)

x	0	1	2	3	4
y	8	7	6	5	4
Coord.	(0, 8)	(1, 7)	(2, 6)	(3, 5)	(4, 4)

b)-c)

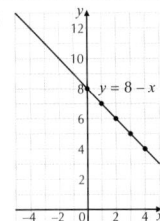
$y = 8 - x$

3 a) Find the coordinates of at least two points on the line: e.g. when $x = -5$, $y = -4(-5) = 20$ and when $x = 0$, $y = -4(0) = 0$. Plot these points, join with a straight line, then extend the line to cover the whole range of x.

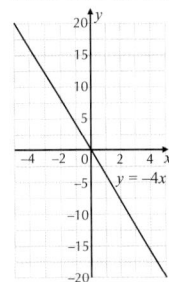
$y = -4x$

b) Find the coordinates of at least two points on the line: e.g. when $x = -2$, $y = \frac{-2}{2} = -1$ and when $x = 4$, $y = \frac{4}{2} = 2$. Plot these points, join with a straight line, then extend the line to cover the whole range of x.

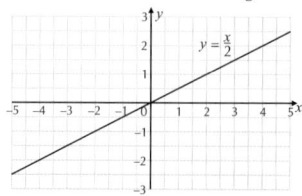

Using the same method for c)-f) and Q4:

c)

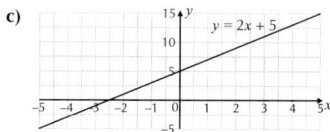

d)

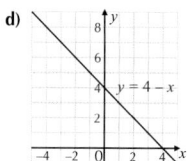

e)

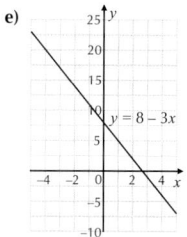

f)

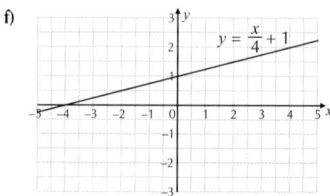

4 a)

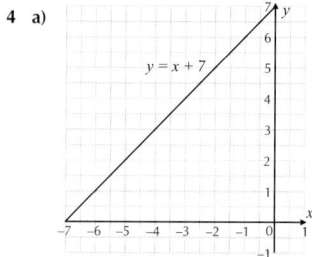

b)

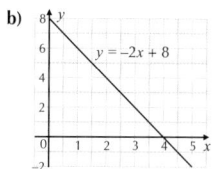

c)

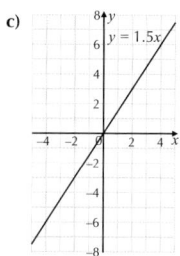

d)

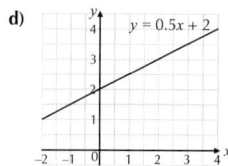

15.2 Gradients

Page 180   Exercise 1

1 a) The line slopes downwards from left to right, so the gradient is **negative**.

b) $y_2 - y_1 = 2 - 6 = \mathbf{-4}$

c) $x_2 - x_1 = 6 - 1 = \mathbf{5}$

d) gradient $= \dfrac{\text{change in } y}{\text{change in } x} = \dfrac{-4}{5} = -\dfrac{4}{5}$

2 a) gradient $= \dfrac{3-5}{6-1} = -\dfrac{2}{5}$

b) gradient $= \dfrac{1-6}{6-(-4)} = \dfrac{-5}{10} = -\dfrac{1}{2}$

c) gradient $= \dfrac{6-0}{4-(-5)} = \dfrac{6}{9} = \dfrac{2}{3}$

3 a) (i) $G = (2, -5)$, $H = (6, 6)$

(ii) gradient $= \dfrac{6-(-5)}{6-2} = \dfrac{11}{4} = \mathbf{2.75}$

b) (i) $I = (-10, 5)$, $J = (30, -25)$

(ii) gradient $= \dfrac{(-25)-5}{30-(-10)} = \dfrac{-30}{40} = -\dfrac{3}{4}$

c) (i) $K = (-8, -25)$, $L = (8, 35)$
$M = (-4, 30)$, $N = (6, -15)$

(ii) Line 1 gradient $= \dfrac{35-(-25)}{8-(-8)}$

$= \dfrac{60}{16} = \dfrac{15}{4} = \mathbf{3.75}$

Line 2 gradient $= \dfrac{(-15)-30}{6-(-4)}$

$= \dfrac{-45}{10} = -\dfrac{9}{2} = \mathbf{-4.5}$

4 a)

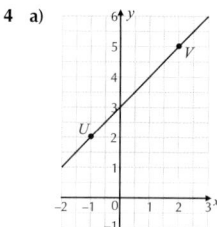

b) gradient $= \dfrac{5-2}{2-(-1)} = \dfrac{3}{3} = \mathbf{1}$

5 a) $0 - (-3) = \mathbf{3}$　　**b)** $2 - (-4) = \mathbf{6}$

c) gradient $= \dfrac{3}{6} = \dfrac{1}{2}$

6 a) gradient $= \dfrac{10-4}{2-0} = \dfrac{6}{2} = \mathbf{3}$

b) gradient $= \dfrac{11-3}{5-1} = \dfrac{8}{4} = \mathbf{2}$

Using the same method for c)-f):

c) $\dfrac{2}{3}$　**d)** $-\dfrac{3}{4}$　**e)** $-\dfrac{3}{4}$　**f)** $\dfrac{3}{4}$

7 Pick 2 points on line A, e.g. $(0, 0)$ and $(1, 5)$
Then gradient $= \dfrac{5-0}{1-0} = \dfrac{5}{1} = \mathbf{5}$
Pick 2 points on line B, e.g. $(0, 0)$ and $(4, 1)$
Then gradient $= \dfrac{1-0}{4-0} = \dfrac{1}{4}$
Using the same method for lines C-D:
gradient of $C = -\dfrac{2}{3}$, gradient of $D = \dfrac{1}{20}$

For all the lines in Q7, you could have picked any two points on the line — you'd end up with the same gradient. It's usually easiest to pick coordinates with integer values where possible, and if the line goes through the origin, that's a good point to use.

15.3 Equations of Straight-Line Graphs

Page 183   Exercise 1

1 For a line of the form $y = mx + c$, the gradient is m and the y-intercept has coordinates $(0, c)$.

a) gradient $= \mathbf{2}$, y-intercept $= \mathbf{(0, -4)}$

b) gradient $= \mathbf{5}$, y-intercept $= \mathbf{(0, -11)}$

c) gradient $= \mathbf{-3}$, y-intercept $= \mathbf{(0, 7)}$

d) gradient $= \mathbf{4}$, y-intercept $= \mathbf{(0, 0)}$

e) gradient $= \dfrac{1}{2}$, y-intercept $= \mathbf{(0, -1)}$

f) gradient $= \mathbf{-1}$, y-intercept $= \left(\mathbf{0, -\dfrac{1}{2}}\right)$

g) gradient $= \mathbf{-1}$, y-intercept $= \mathbf{(0, 3)}$
Be careful here — the mx and c terms are the wrong way round in the given equation.

h) gradient $= \mathbf{0}$, y-intercept $= \mathbf{(0, 3)}$

2 Start by looking at the y-intercepts of the graphs and equations. If two lines have the same y-intercept, look at the gradient.
A: $y = -\dfrac{1}{3}x + 4$　　B: $y = 3x$
C: $y = \dfrac{1}{3}x + 2$　　D: $y = \dfrac{7}{3}x - 1$
E: $y = x + 2$　　F: $y = -x + 6$

3 a) $3y = 9 - 3x \Rightarrow y = -x + 3$
So gradient $= \mathbf{-1}$, y-intercept $= \mathbf{(0, 3)}$

b) $y - 5 = 7x \Rightarrow y = 7x + 5$
So gradient $= \mathbf{7}$, y-intercept $= \mathbf{(0, 5)}$
Using the same method for c)-l):

c) gradient $= \mathbf{-1}$, y-intercept $= \mathbf{(0, 8)}$

d) gradient $= \dfrac{1}{2}$, y-intercept $= \mathbf{(0, -3)}$

e) gradient $= \mathbf{-3}$, y-intercept $= \mathbf{(0, 1)}$

f) gradient $= \mathbf{2}$, y-intercept $= \mathbf{(0, 5)}$

g) gradient $= \dfrac{4}{5}$, y-intercept $= \mathbf{(0, 1)}$

h) gradient $= \mathbf{4}$, y-intercept $= \mathbf{(0, -7)}$

i) gradient $= -\dfrac{5}{4}$, y-intercept $= \left(\mathbf{0, -\dfrac{3}{4}}\right)$

j) gradient $= \dfrac{3}{2}$, y-intercept $= \mathbf{(0, -2)}$

k) gradient $= \mathbf{2}$, y-intercept $= \left(\mathbf{0, \dfrac{1}{3}}\right)$

l) gradient $= \mathbf{-2}$, y-intercept $= \left(\mathbf{0, -\dfrac{1}{4}}\right)$

Page 184   Exercise 2

1 For parts a) and b) you're given the gradient (m) and the y-intercept (in the form $(0, c)$), so put these values into $y = mx + c$:

a) $y = 8x + 2$　　**b)** $y = -x + 7$

For the rest of the questions, use the gradient and the given point to find the value of c:

c) Gradient $= 3$, so $10 = 3(1) + c \Rightarrow c = 7$.
So the equation of the line is $y = 3x + 7$.

d) Gradient $= \dfrac{1}{2}$, so $-5 = \dfrac{1}{2}(4) + c \Rightarrow c = -7$.
So the equation of the line is $y = \dfrac{1}{2}x - 7$.

e) Gradient $= -7$, so $-4 = -7(2) + c \Rightarrow c = 10$.
So the equation of the line is $y = -7x + 10$.

f) Gradient $= 5$, so $-7 = 5(-3) + c \Rightarrow c = 8$.
So the equation of the line is $y = 5x + 8$.

2 a) Gradient $= \dfrac{11-7}{5-3} = \dfrac{4}{2} = 2$, so $7 = 2(3) + c$
$\Rightarrow c = 1$. So the equation of the line is
$y = 2x + 1$.

b) Gradient $= \dfrac{(-5)-1}{2-5} = \dfrac{-6}{-3} = 2$,
so $1 = 2(5) + c \Rightarrow c = -9$.
So the equation of the line is $y = 2x - 9$.
Using the same method for c)-i):

c) $y = x - 3$　　**d)** $y = 2x + 5$

e) $y = 3x + 2$　　**f)** $y = 3x + 11$

g) $y = -x + 1$　　**h)** $y = -4x + 7$

i) $y = \dfrac{3}{2}x - \dfrac{1}{2}$

3 Gradient of line $A = \dfrac{4-0}{(-2)-2} = \dfrac{4}{-4} = -1$,
y-intercept $= 2$, so its equation is $y = -x + 2$.
Gradient of line $B = \dfrac{5-0}{1-0} = \dfrac{5}{1} = 5$.
It goes through the origin so y-intercept $= 0$,
so its equation is $y = 5x$.

Using the same method for lines *C–H*:

line *C*: $y = x - 1$ line *D*: $y = -3$

line *E*: $y = -\frac{1}{2}x + 2$ line *F*: $y = \frac{2}{5}x + 1$

line *G*: $y = \frac{2}{5}x - 4$ line *H*: $y = \frac{3}{2}x - \frac{5}{2}$

4 Using the same method as Q3:
Line *A* has equation $y = x + 0.5$.

Line *B* has equation $y = \frac{1}{10}x + 2$.

Line *C* has equation $y = -\frac{3}{2}x - \frac{1}{2}$.

15.4 Parallel and Perpendicular Lines

Page 186   Exercise 1

1 **a)** Lines parallel to $y = 5x - 1$ have a gradient of 5, e.g. **$y = 5x + 1$, $y = 5x + 2$, $y = 5x + 3$**
You can have any value of c here.

  b) $x + y = 7 \Rightarrow y = -x + 7$. So parallel lines will have a gradient of –1, e.g. **$y = -x + 6$, $y = -x + 5$, $y = -x + 4$**
You could have given these equations in the form $x + y = 6, x + y = 5, x + y = 4$ as the question doesn't ask for y = mx + c form.

2 Rearrange lines A-F into $y = mx + c$ form:

  A: $y = 2x + 4$ B: $y = x + 2.5$

  C: $y = 2x - 2$ D: $y = -2x - 7$

  E: $y = -\frac{2}{3}x + \frac{2}{3}$ F: $y = \frac{2}{3}x + \frac{2}{9}$

  a) $y = 2x - 1$ has a gradient of 2, so lines **A** and **C** are parallel to it.

  b) $2x - 3y = 0 \Rightarrow y = \frac{2}{3}x$. This line has a gradient of $\frac{2}{3}$, so line **F** is parallel to it.

3 The line on the diagram has gradient $\frac{3-1}{(-3)-3} = \frac{2}{-6} = -\frac{1}{3}$.
Rearrange lines A-F into $y = mx + c$ form:

  A: $y = -3x + 2$ B: $y = -\frac{1}{3}x + \frac{7}{3}$

  C: $y = -3x + 4$ D: $y = \frac{1}{3}x - \frac{8}{3}$

  E: $y = -\frac{1}{3}x + 3$ F: $y = -\frac{1}{3}x$

  So lines **B**, **E** and **F** are parallel to the line on the diagram.

4 **a)** gradient = 5, so $8 = 5(1) + c \Rightarrow c = 3$
So the line has equation **$y = 5x + 3$**.

  b) gradient = 2, so $5 = 2(-1) + c \Rightarrow c = 7$
So the line has equation **$y = 2x + 7$**.

Using the same method for c)-i):

  c) $y = \frac{1}{2}x - 10$ **d)** $y = 8x + 19$

  e) $y = 3x + 13$ **f)** $y = -9x - 2$

  g) $y = -x + 16$ **h)** $y = -2x - 8$

  i) $y = -\frac{1}{3}x + 6$

   For parts e)-i), you'll have to rearrange into $y = mx + c$ form first.

Page 187   Exercise 2

1 **a)** $-1 \div 6 = -\frac{1}{6}$ **b)** $-1 \div 3 = \frac{1}{3}$

  c) $-1 \div -\frac{1}{4} = 4$ **d)** $-1 \div 12 = -\frac{1}{12}$

Using the same method for e)-l):

  e) $\frac{1}{7}$ **f)** $-\frac{3}{2}$ **g)** $\frac{1}{2}$ **h)** $-\frac{2}{3}$

  i) $-\frac{10}{3}$ **j)** $\frac{2}{9}$ **k)** $\frac{3}{4}$ **l)** $-\frac{2}{7}$

2 **a)** A line perpendicular to $y = 2x + 3$ will have gradient $-1 \div 2 = -\frac{1}{2}$, e.g. **$y = -\frac{1}{2}x + 3$**.
   For all parts of Q2, you can have any value of c.

  b) A line perpendicular to $y = -3x + 11$ will have gradient $-1 \div -3 = \frac{1}{3}$, e.g. **$y = \frac{1}{3}x + 5$**.

Using the same method for c)-f):

  c) E.g. $y = \frac{1}{6}x + 5$ **d)** E.g. $y = -\frac{2}{5}x + 1$

  e) E.g. $y = x + 2$ **f)** E.g. $y = -2x + 8$

---

3 Find the gradient of each line:

  A: 3   B: 2   C: $-\frac{1}{3}$ D: $\frac{2}{3}$

  E: -3 F: $\frac{3}{2}$ G: $-\frac{1}{2}$ H: -2

  I: $\frac{1}{2}$ J: $\frac{1}{3}$ K: $-\frac{2}{3}$ L: $-\frac{3}{2}$

  The gradients of perpendicular lines multiply to give –1, so the pairs of perpendicular lines are: **A** and **C**, **B** and **G**, **D** and **L**, **E** and **J**, **F** and **K**, **H** and **I**.

4 **a)** Gradient of perpendicular line:
   $-1 \div -3 = \frac{1}{3}$. So $8 = \frac{1}{3}(9) + c \Rightarrow c = 5$.
   So the equation of the line is **$y = \frac{1}{3}x + 5$**.

  b) Gradient of perpendicular line:
   $-1 \div \frac{1}{2} = -2$. So $-4 = -2(3) + c \Rightarrow c = 2$.
   So the equation of the line is **$y = -2x + 2$**.

Using the same method for c)-j):

  c) $y = -4x - 5$ **d)** $y = -\frac{3}{4}x + 8$

  e) $y = \frac{2}{5}x - 4$ **f)** $y = x - 3$

  g) $y = -\frac{1}{3}x - 1$ **h)** $y = \frac{3}{8}x + 4$

  i) $y = 2x + 7$ **j)** $y = -5x - 2$

15.5 Line Segments

Page 188   Exercise 1

1 **a)** Midpoint $= \left(\frac{8+4}{2}, \frac{0+6}{2}\right) = $ **(6, 3)**

  b) Midpoint $= \left(\frac{(-2)+6}{2}, \frac{3+5}{2}\right) = $ **(2, 4)**

  c) Midpoint $= \left(\frac{4+(-2)}{2}, \frac{(-7)+1}{2}\right) = $ **(1, –3)**

Using the same method for d)-i):

  d) **(3, –1)**   **e)** **(–5, 2)**   **f)** **(–1, –2)**

  g) $\left(0, \frac{1}{2}\right)$ **h)** **(4p, 4q)**   **i)** **(5p, 8q)**

2 **a)** Point *A* has coordinates (–3, –2),
   point *B* has coordinates (2, 4),
   point *C* has coordinates (4, –2).
   Midpoint of $AB = \left(\frac{(-3)+2}{2}, \frac{(-2)+4}{2}\right)$
   $= $ **(–0.5, 1)**
   Midpoint of $BC = \left(\frac{2+4}{2}, \frac{4+(-2)}{2}\right) = $ **(3, 1)**
   Midpoint of $CA = \left(\frac{4+(-3)}{2}, \frac{(-2)+(-2)}{2}\right)$
   $= $ **(0.5, –2)**

  b) Point *D* has coordinates (–2, –3),
   point *E* has coordinates (0, 3),
   point *F* has coordinates (3, –1).
   Midpoint of $DE = \left(\frac{(-2)+0}{2}, \frac{(-3)+3}{2}\right)$
   $= $ **(–1, 0)**
   Midpoint of $EF = \left(\frac{0+3}{2}, \frac{3+(-1)}{2}\right)$
   $= $ **(1.5, 1)**
   Midpoint of $FD = \left(\frac{3+(-2)}{2}, \frac{(-1)+(-3)}{2}\right)$
   $= $ **(0.5, –2)**

3 Call the coordinates of *B* (*x*, *y*). Then
   $\frac{1+x}{2} = 5 \Rightarrow x = 9$ and $\frac{8+y}{2} = 3 \Rightarrow y = -2$.
   So the coordinates of *B* are **(9, –2)**.

4 Call the coordinates of *D* (*x*, *y*). Then
   $\frac{6+x}{2} = 2 \Rightarrow x = -2$ and $\frac{(-7)+y}{2} = -1$
   $\Rightarrow y = 5$.
   So the coordinates of *D* are **(–2, 5)**.

5 Find the coordinates of each point: *A*(–5, 5), *B*(–4, –3), *C*(–2, 3), *D*(2, 1), *E*(3, –2), *F*(5, 5).

  a) Midpoint of $AF = \left(\frac{(-5)+5}{2}, \frac{5+5}{2}\right) = $ **(0, 5)**

  b) Midpoint of $AC = \left(\frac{(-5)+(-2)}{2}, \frac{5+3}{2}\right)$
   $= $ **(–3.5, 4)**

Using the same method for c)-f):

  c) Midpoint of $DF = $ **(3.5, 3)**

  d) Midpoint of $BE = $ **(–0.5, –2.5)**

  e) Midpoint of $BF = $ **(0.5, 1)**

  f) Midpoint of $CE = $ **(0.5, 0.5)**

---

1 **a)** Change in *x*-coordinates = $5 - 1 = 4$
   Change in *y*-coordinates = $9 - 6 = 3$
   So length $= \sqrt{4^2 + 3^2} = \sqrt{25} = $ **5**

  b) Change in *x*-coordinates = $15 - 11 = 4$
   Change in *y*-coordinates = $3 - 8 = -5$
   So length $= \sqrt{4^2 + (-5)^2} = \sqrt{41}$
   $= $ **6.40** (3 s.f.)

Using the same method for c)-l):

  c) **3.16** (3 s.f.)   **d)** **7**

  e) **15.6** (3 s.f.)   **f)** **18.9** (3 s.f.)

  g) **8.06** (3 s.f.)   **h)** **10.0** (3 s.f.)

  i) **10.8** (3 s.f.)   **j)** **8.54** (3 s.f.)

  k) **9.85** (3 s.f.)   **l)** **15.8** (3 s.f.)

2 **a)** Find the coordinates of each point:
   A(–3, –3), *B*(–2, 3), *C*(3, 2), *D*(1, –1)
   Length of $AB = \sqrt{((-3)-(-2))^2 + ((-3)-3)^2}$
   $= \sqrt{(-1)^2 + (-6)^2} = \sqrt{37} = $ **6.08** (3 s.f.)
   Length of $BC = \sqrt{((-2)-3)^2 + (3-2)^2}$
   $= \sqrt{(-5)^2 + 1^2} = \sqrt{26} = $ **5.10** (3 s.f.)
   Using the same method for *CD* and *DA*:
   Length of $CD = $ **3.61** (3 s.f.)
   Length of $DA = $ **4.47** (3 s.f.)

  b) Find the coordinates of each point:
   E(–4, –1), *F*(–1, 3), *G*(3, –2), *H*(2, –4)
   Length of $EF = \sqrt{((-4)-(-1))^2 + ((-1)-3)^2}$
   $= \sqrt{(-3)^2 + (-4)^2} = \sqrt{25} = $ **5**
   Length of $FG = \sqrt{((-1)-3)^2 + (3-(-2))^2}$
   $= \sqrt{(-4)^2 + 5^2} = \sqrt{41} = $ **6.40** (3 s.f.)
   Using the same method for *GH* and *HE*:
   Length of $GH = $ **2.24** (3 s.f.)
   Length of $HE = $ **6.71** (3 s.f.)

Page 190   Exercise 3

1 **a)** *x*-difference: $6 - 3 = 3$,
   y-difference: $6 - (-3) = 9$
   C lies $\frac{1}{1+2} = \frac{1}{3}$ of the way along *AB*, so
   x: $\frac{1}{3} \times 3 = 1$ and *y*: $\frac{1}{3} \times 9 = 3$
   x-coordinate of *C*: $3 + 1 = 4$,
   y-coordinate of *C*: $-3 + 3 = 0$,
   so *C* has coordinates **(4, 0)**.

  b) *x*-difference: $9 - (-3) = 12$,
   y-difference: $1 - 5 = -4$
   C lies $\frac{3}{3+1} = \frac{3}{4}$ of the way along *AB*, so
   x: $\frac{3}{4} \times 12 = 9$ and *y*: $\frac{3}{4} \times (-4) = -3$
   x-coordinate of *C*: $(-3) + 9 = 6$,
   y-coordinate of *C*: $5 + (-3) = 2$,
   so *C* has coordinates **(6, 2)**.

Using the same method for c)-d):

  c) **(4, 2)**   **d)** **(–8, –5)**

2 **a)** *x*-difference between *A* and *B*: $2 - 0 = 2$,
   x-difference between *B* and *C*: $6 - 2 = 4$
   So ratio = $2 : 4 = $ **1 : 2**
   You could have used the y-differences here instead — the ratio would be the same.

  b) *x*-difference between *D* and *E*: $(-3) - 1 = -4$,
   x-difference between *E* and *F*:
   $(-4) - (-3) = -1$. So ratio = $-4 : -1 = $ **4 : 1**

  c) *x*-difference between *G* and *H*: $5 - (-1) = 6$,
   x-difference between *H* and *I*: $14 - 5 = 9$
   So ratio = $6 : 9 = $ **2 : 3**

3 **a)** *x*-difference between *S* and *T*: $12 - 6 = 6$,
   y-difference between *S* and *T*: $(-4) - 2 = -6$
   These distances are $\frac{3}{3+2} = \frac{3}{5}$ of the distances from *S* to *U*, so *x*-difference between S and *U* is $(6 \div 3) \times 5 = 10$ and *y*-difference between S and *U* is $(-6 \div 3) \times 5 = -10$.
   So the *x*-coordinate of *U* is $6 + 10 = 16$ and the *y*-coordinate of *U* is $2 + (-10) = -8$.
   So *U* has coordinates **(16, –8)**.

b) x-difference between S and T:
$18 - (-2) = 20$,
y-difference between S and T:
$11 - (-4) = 15$
These distances are $\frac{5}{5+4} = \frac{5}{9}$ of the
distances from S to U, so x-difference
between S and U is $(20 \div 5) \times 9 = 36$
and y-difference between S and U is
$(15 \div 5) \times 9 = 27$.
So the x-coordinate of U is $(-2) + 36 = 34$
and the y-coordinate of U is $(-4) + 27 = 23$.
So U has coordinates **(34, 23)**.

Page 191   Review Exercise

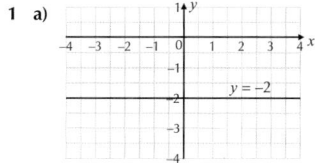

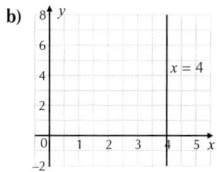

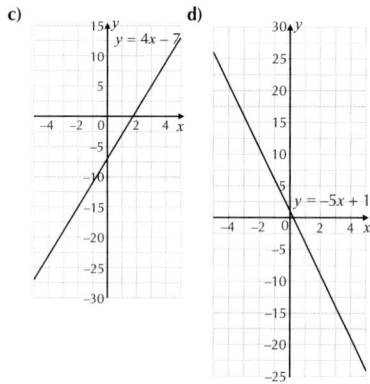

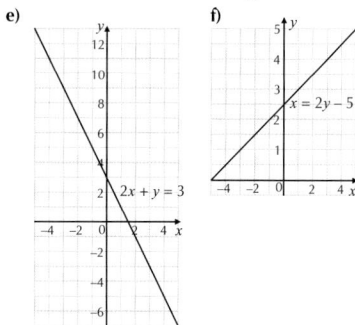

1 a)

b)

c)

d)

e)

f)

2 a) gradient $= \dfrac{\text{change in } y}{\text{change in } x} = \dfrac{7-3}{4-2} = \dfrac{4}{2} = \mathbf{2}$

b) gradient $= \dfrac{\text{change in } y}{\text{change in } x} = \dfrac{2-4}{3-1} = \dfrac{-2}{2} = \mathbf{-1}$

Using the same method for c)-f):

c) $\dfrac{3}{2}$ **d)** $-\dfrac{2}{3}$ **e)** $\dfrac{1}{2}$ **f)** $-\dfrac{1}{4}$

3 All the lines go through the origin, so have
equations of the form $y = mx$. Rearrange all
the equations into this form, then compare the
gradients: $y = 2x$, $y = -x$, $y = x$, $y = -0.5x$ and
$y = 0.5x$.

The line with the steepest positive gradient
is line E, so it has equation $y = 2x$. The next
steepest positive gradient is line D, so it has
equation $y = x$ or $y - x = 0$. The line with the
smallest positive gradient is line C, so it has
equation $y = 0.5x$ or $2y = x$. Line B has a
small negative gradient, so has equation
$y = -0.5x$. Line A has a steeper negative
gradient, so has equation $y = -x$ or $y + x = 0$.
*If you need to, work out the gradients of lines A-E
and compare them to the given equations.*

4 a) $\dfrac{a-4}{2-(-1)} = 5 \Rightarrow \dfrac{a-4}{3} = 5 \Rightarrow a = 19$

b) $\dfrac{(-2)-7}{b-2} = -3 \Rightarrow \dfrac{-9}{b-2} = -3 \Rightarrow b = 5$

5 Read off the values of m and c from the
equation in $y = mx + c$ form:

a) gradient = **5**, y-intercept = **(0, −9)**

b) gradient = **−2**, y-intercept = **(0, 11)**
  *Watch out — the mx and c terms are the wrong
  way round in the given equation.*

c) gradient = **3**, y-intercept = **(0, −8)**

d) $6 = 2x + 3y \Rightarrow y = -\dfrac{2}{3}x + 2$
  gradient $= -\dfrac{2}{3}$, y-intercept = **(0, 2)**

e) $x = 4y - 7 \Rightarrow y = \dfrac{1}{4}x + \dfrac{7}{4}$
  gradient $= \dfrac{1}{4}$, y-intercept $= \left(0, \dfrac{7}{4}\right)$

f) $2y + 9 = 8x \Rightarrow y = 4x - \dfrac{9}{2}$
  gradient = **4**, y-intercept $= \left(0, -\dfrac{9}{2}\right)$

6 Rearrange all equations into $y = mx + c$ form,
then match the y-intercepts or gradients to the
lines on the graph.
$x + 2y = 6 \Rightarrow y = -\dfrac{1}{2}x + 3$,
$x = -4y \Rightarrow y = -\dfrac{1}{4}x$, $y = 2x - 6$,
$y = 4(x - 1) = 4x - 4$
$4x + y + 4 = 0 \Rightarrow y = -4x - 4$
Line A goes through the origin so has equation
$x = -4y$. Line B has a negative gradient
and a y-intercept of -4 so has equation
$4x + y + 4 = 0$. Line C has a negative gradient
and a y-intercept of 3 so has equation
$x + 2y = 6$. Line D has a positive gradient and
a y-intercept of -4 so has equation
$y = 4(x - 1)$. Line E has a gradient of 2 so has
equation $y = 2x - 6$.
*Look at the y-intercept first, then if two
lines have the same intercept, look at their
gradients. If you can't see the y-intercept,
you might have to calculate the gradient.*

7 a) gradient $= \dfrac{6-2}{0-1} = \dfrac{4}{-1} = -4$
  So $2 = -4(1) + c \Rightarrow c = 6$
  So the line has equation $y = -4x + 6$.

b) gradient $= \dfrac{(-9)-7}{0-8} = \dfrac{-16}{-8} = 2$
  So $-9 = 2(0) + c \Rightarrow c = -9$
  So the line has equation $y = 2x - 9$.
  Using the same method for c)-f):

c) $y = \dfrac{1}{2}x + 3$ **d)** $y = -3x + 7$

e) $y = -2x - 5$ **f)** $y = \dfrac{1}{4}x - \dfrac{1}{2}$

8 Line A has a gradient of -3 and a y-intercept of
-6, so has equation $y = -3x - 6$.
Line B has a gradient of $\dfrac{3}{4}$ and a y-intercept
of -1, so has equation $y = \dfrac{3}{4}x - 1$.
Using the same method for lines C-D:
C: $y = -\dfrac{2}{3}x + 2$ D: $y = 3x - 8$
*If you can't just read off the gradient, pick two points
on the line and use the gradient formula.*

9 a) Gradient of parallel line $= -7$,
  so $11 = -7(1) + c \Rightarrow c = 18$, so the
  equation of the line is $y = -7x + 18$.

b) Gradient of parallel line = 9,
  so $-16 = 9(2) + c \Rightarrow c = -34$, so the
  equation of the line is $y = 9x - 34$.

Using the same method for c)-d):

c) $y = -\dfrac{3}{2}x + 17$ **d)** $y = \dfrac{1}{3}x - 5$

10 a) Gradient of perpendicular line $= -1 \div 4$
  $= -\dfrac{1}{4}$, so $9 = -\dfrac{1}{4}(12) + c \Rightarrow c = 12$, so the
  line has equation $y = -\dfrac{1}{4}x + 12$.

b) Gradient of perpendicular line $= -1 \div -2$
  $= \dfrac{1}{2}$, so $8 = \dfrac{1}{2}(14) + c \Rightarrow c = 1$, so the line
  has equation $y = \dfrac{1}{2}x + 1$.
  Using the same method for c)-f):

c) $y = -\dfrac{1}{3}x - 3$ **d)** $y = 3x + 14$

e) $y = \dfrac{2}{3}x - 7$ **f)** $y = -5x - 8$

11 Parallel lines have gradient $\dfrac{1}{2}$ and
perpendicular lines have gradient $-1 \div \dfrac{1}{2} = -2$.

A: Gradient $= -2$ so line is **perpendicular**.

B: Gradient $= \dfrac{1}{2}$ so line is **parallel**.

C: Gradient $= 2$, so line is **neither**.

D: Gradient $= \dfrac{1}{2}$ so line is **parallel**.

E: Gradient $= -\dfrac{1}{2}$, so line is **neither**.

F: Gradient $= 2$, so line is **neither**.

G: Gradient $= \dfrac{1}{2}$ so line is **parallel**.

H: Gradient $= -2$ so line is **perpendicular**.

12 a) Coordinates of $A = (-3, 0)$, coordinates of
  $B = (1, 3)$, coordinates of $C = (3, -3)$
  AB: Gradient $= \dfrac{3-0}{1-(-3)} = \dfrac{3}{4}$,
  length $= \sqrt{(1-(-3))^2 + (3-0)^2} = \sqrt{4^2 + 3^2}$
  $= \sqrt{25} = 5$,
  midpoint $= \left(\dfrac{(-3)+1}{2}, \dfrac{0+3}{2}\right) = \left(-1, \dfrac{3}{2}\right)$
  Using the same methods for BC and CA:
  BC: Gradient $= -3$, length $= \mathbf{6.32}$ (3 s.f.),
    midpoint $= \mathbf{(2, 0)}$
  CA: Gradient $= -\dfrac{1}{2}$, length $= \mathbf{6.71}$ (3 s.f.),
    midpoint $= \left(0, -\dfrac{3}{2}\right)$

b) Coordinates of $D = (-4, -2)$, coordinates of
  $E = (-1, 3)$, coordinates of $F = (4, 2)$,
  coordinates of $G = (0, -4)$
  DE: Gradient $= \dfrac{3-(-2)}{(-1)-(-4)} = \dfrac{5}{3}$,
  length $= \sqrt{((-1)-(-4))^2 + (3-(-2))^2}$
  $= \sqrt{3^2 + 5^2} = \sqrt{34} = 5.83$ (3 s.f.),
  midpoint $= \left(\dfrac{(-4)+(-1)}{2}, \dfrac{(-2)+3}{2}\right)$
  $= \left(-\dfrac{5}{2}, \dfrac{1}{2}\right)$
  Using the same methods for EF, FG
  and GD:
  EF: Gradient $= -\dfrac{1}{5}$, length $= \mathbf{5.10}$ (3 s.f.),
    midpoint $= \left(\dfrac{3}{2}, \dfrac{5}{2}\right)$
  FG: Gradient $= \dfrac{3}{2}$, length $= \mathbf{7.21}$ (3 s.f.),
    midpoint $= \mathbf{(2, -1)}$
  GD: Gradient $= -\dfrac{1}{2}$, length $= \mathbf{4.47}$ (3 s.f.),
    midpoint $= \mathbf{(-2, -3)}$

13 a) Coordinates of $A = (-2, 3)$, coordinates of
  $B = (4, -1)$, coordinates of $C = (-4, -2)$
  Length of $AB = \sqrt{(4-(-2))^2 + ((-1)-3)^2}$
  $= \sqrt{6^2 + (-4)^2} = \sqrt{52}$
  Length of $BC = \sqrt{65}$, length of $CA = \sqrt{29}$
  Perimeter of $ABC = \sqrt{52} + \sqrt{65} + \sqrt{29}$
  $= \mathbf{20.7}$ (3 s.f.)

b) Coordinates of $D = (-3, 3)$, coordinates of
  $E = (1, 4)$, coordinates of $F = (3, 2)$,
  coordinates of $G = (1, -4)$

Length of $DE = \sqrt{(1-(-3))^2 + (4-3)^2}$
$= \sqrt{4^2 + 1^2} = \sqrt{17}$
Length of $EF = \sqrt{8}$, length of $FG = \sqrt{40}$,
length of $GE = \sqrt{65}$
Perimeter of $DEFG =$
$\sqrt{17} + \sqrt{8} + \sqrt{40} + \sqrt{65} = $ **21.3** (3 s.f.)

14 a) x-difference between A and C:
$10 - (-6) = 16$,
y-difference between A and C: $6 - (-2) = 8$
B lies $\frac{1}{1+3} = \frac{1}{4}$ of the way along AC, so
x: $\frac{1}{4} \times 16 = 4$ and y: $\frac{1}{4} \times 8 = 2$
$p = x$-coordinate of B: $(-6) + 4 = $ **−2**,
$q = y$-coordinate of B: $(-2) + 2 = $ **0**.

b) length $= \sqrt{((-2)-(-6))^2 + (0-(-2))^2}$
$= \sqrt{4^2 + 2^2} = \sqrt{20} = $ **4.5** (1 d.p.)

c) length of $DC = \frac{1}{4}$ of length $AB = \frac{1}{4} \times \sqrt{20}$
$= $ **1.1** (1 d.p.)
You could have worked out the coordinates of point D using the ratio, but that would have involved a lot more work.

Page 193   Exam-Style Questions

1 Rewrite the equation of line A in $y = mx + c$ form:
$5x + 2y - 8 = 0 \Rightarrow 2y = -5x + 8$
$\Rightarrow y = -\frac{5}{2}x + 4$
[1 mark for correct value of m, 1 mark for correct value of c].
Line B is parallel to line A so has a gradient of $-\frac{5}{2}$. Its y-intercept is $4 \times 3 = 12$, so the equation of line B is $y = -\frac{5}{2}x + 12$ *[1 mark].*

2 a) First find the gradient of line L:
gradient $(m) = \dfrac{\text{change in } y}{\text{change in } x} = \dfrac{1-4}{4-(-2)}$
$= \dfrac{-3}{6} = -\dfrac{1}{2}$
Find the y-intercept by substituting m and one pair of coordinates into $y = mx + c$:
$1 = -\frac{1}{2}(4) + c \Rightarrow 1 = -2 + c \Rightarrow c = 3$
So the equation of line L is: $y = -\frac{1}{2}x + 3$
[3 marks available — 1 mark for finding the gradient, 1 mark for a correct method to find c, 1 mark for the correct value of c]

b) Gradient of line $M =$
$-1 \div$ gradient of line $L = -1 \div -\frac{1}{2} = 2$
Find the y-intercept by substituting m and $(2, 2)$ into $y = mx + c$:
$2 = 2(2) + c \Rightarrow 2 = 4 + c \Rightarrow c = -2$
So the equation of line M is: $y = 2x - 2$
[3 marks available — 1 mark for finding the gradient, 1 mark for a correct method to find c, 1 mark for the correct value of c]

3 Find the equation of the line passing through the two given points:
gradient $(m) = \dfrac{\text{change in } y}{\text{change in } x} = \dfrac{110-(-100)}{9-(-5)}$
$= \dfrac{210}{14} = 15$
Find the y-intercept by substituting m and one pair of coordinates into $y = mx + c$:
$110 = 15(9) + c \Rightarrow 110 = 135 + c$
$\Rightarrow c = -25$
So the equation of the line is: $y = 15x - 25$
Now put the x-coordinate of the point you're testing into this equation:
$y = 15(33) - 25 = 470 \neq 450$, the y-coordinate of the point.
So (33, 450) does not lie on the line.
[4 marks available — 1 mark for finding the gradient, 1 mark for a correct method to find c, 1 mark for the correct value of c, 1 mark for showing that (33, 450) does not lie on the line]

4 a) Midpoint $= \left(\dfrac{(-2)+6}{2}, \dfrac{10+(-6)}{2}\right)$
$= $ **(2, 2)**
[2 marks available — 1 mark for correct x-coordinate, 1 mark for correct y-coordinate]

b) M lies $\frac{2}{2+1} = \frac{2}{3}$ of the way along CD, so CM will be $\frac{2}{3}$ of the length of CD.
Length $CM = \sqrt{(2-(-6))^2 + (2-(-2))^2}$
$= \sqrt{8^2 + 4^2} = \sqrt{64 + 16} = \sqrt{80}$
$CM = \frac{2}{3}CD$, so $CD = \frac{3}{2}CM = \frac{3}{2} \times \sqrt{80}$
$= 13.416... = $ **13.4** (1 d.p.)
[3 marks available — 1 mark for finding the length of CM, 1 mark for multiplying by 1.5, 1 mark for the correct answer]
You could have also done this by finding the coordinates of M using the given ratio, then working out the length of CM using Pythagoras.

5 Draw a quick sketch of the triangle to help you see what's going on:

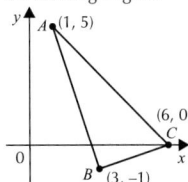

If the triangle is right-angled, angle ABC must be a right angle (as it is opposite the longest side). So lines AB and BC must be perpendicular.
Gradient of $AB = \dfrac{\text{change in } y}{\text{change in } x} = \dfrac{(-1)-5}{3-1}$
$= \dfrac{-6}{2} = -3$
Gradient of $BC = \dfrac{\text{change in } y}{\text{change in } x} = \dfrac{0-(-1)}{6-3}$
$= \dfrac{1}{3}$
Gradient of AB × gradient of $BC = -3 \times \frac{1}{3}$
$= -1$
The product of the gradients is −1, so lines AB and BC are perpendicular, which means **triangle ABC is right-angled**.
[4 marks available — 1 mark for finding the gradient of AB, 1 mark for finding the gradient of BC, 1 mark for showing their product is −1, 1 mark for using this to conclude that the triangle is right-angled]

Section 16 — Other Types of Graph

16.1 Quadratic Graphs

Page 194   Exercise 1

1 a)

x	−4	−3	−2	−1	0	1	2	3	4
x^2	16	9	4	1	0	1	4	9	16
$6-x^2$	−10	−3	2	5	6	5	2	−3	−10

Plot each pair of values (−4, −10), (−3, −3) etc. on a coordinate grid, making sure the axes span the required values, i.e. $-4 \leq x \leq 4$ and $-10 \leq y \leq 6$:

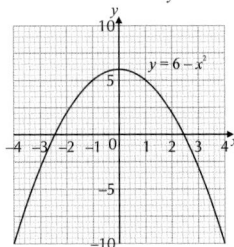

b) Using the same method as part a):

x	−4	−3	−2	−1	0	1	2	3	4
$2x^2$	32	18	8	2	0	2	8	18	32

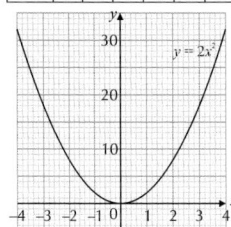

2 Calculate coordinates in a table of values, e.g:

x	−4	−3	−2	−1	0	1	2	3	4
x^2+5	21	14	9	6	5	6	9	14	21

Plot them on a set of suitable axes:

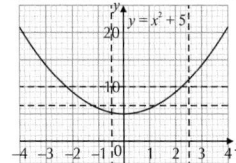

a) Draw vertical lines at the given x-values (as shown on graph above) and read off the y-values where these lines meet the graph.
(i) When $x = 2.5$, $y = $ **11.3** (1 d.p.)
(Accept 11.2 to 11.4)
(ii) When $x = -0.5$, $y = $ **5.3** (1 d.p.)
(Accept 5.2 to 5.4)

b) Draw horizontal lines at the given y-values (as shown on graph above) and read off both x-values where these lines meet the graph.
(i) When $y = 6.5$, $x = $ **1.2** and $x = $ **−1.2**
(both to 1 d.p.)
(Accept 1.1 to 1.3 and −1.3 to −1.1)
(ii) When $y = 10$, $x = $ **2.2** and $x = $ **−2.2**
(both to 1 d.p.)
(Accept 2.1 to 2.3 and −2.3 to −2.1)

3 Calculate coordinates in a table of values, e.g:

x	−4	−3	−2	−1	0	1	2	3	4
$4-x^2$	−12	−5	0	3	4	3	0	−5	−12

Plot them on a set of suitable axes:

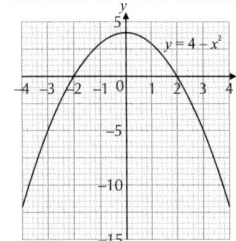

The graph crosses the x-axis at $x = $ **2** and $x = $ **−2**.

4 Calculate coordinates in a table of values, e.g:

x	−4	−3	−2	−1	0	1	2	3	4
$3x^2-11$	37	16	1	−8	−11	−8	1	16	37

Plot them on a set of suitable axes:

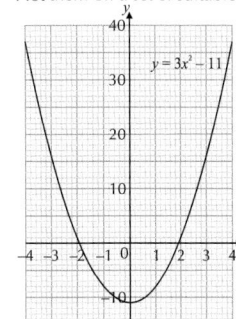

Read off the values where the graph crosses the x-axis: **x = –1.9** and **x = 1.9** (both to 1 d.p.) (Accept –2.0 to –1.8, 1.8 to 2.0)

Page 195   Exercise 2

1

x	–4	–3	–2	–1	0	1	2	3	4
$2x^2$	32	18	8	2	0	2	8	18	32
$+3x$	–12	–9	–6	–3	0	3	6	9	12
-7	–7	–7	–7	–7	–7	–7	–7	–7	–7
$2x^2 + 3x - 7$	13	2	–5	–8	–7	–2	7	20	37

Plot each pair of values (–4, 13), (–3, 2) etc. on a coordinate grid, making sure the axes span the required values, i.e. $-4 \le x \le 4$ and $-8 \le y \le 37$:

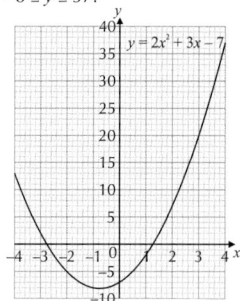

2 Calculate coordinates in a table of values, e.g:

x	–3	–2	–1	0	1	2	3	4	5	6
$x^2 - 5x + 3$	27	17	9	3	–1	–3	–3	–1	3	9

Plot them on a set of suitable axes:

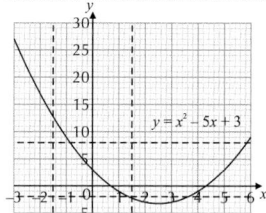

a) Draw vertical lines at the given x-values (as shown on graph above) and read off the y-values where these lines meet the graph.
  (i) When $x = -1.5$, **y = 12.8** (1 d.p.) (Accept 12.7 to 12.9)
  (ii) When $x = 1.5$, **y = –2.3** (1 d.p.) (Accept –2.4 to –2.2)

b) Draw horizontal lines at the given y-values (as shown on graph above) and read off both x-values where these lines meet the graph.
  (i) When $y = 8$,
    x = –0.9 and **x = 5.9** (both to 1 d.p.) (Accept –1.0 to –0.8, 5.8 to 6.0)
  (ii) When $y = -2$,
    x = 1.4 and **x = 3.6** (both to 1 d.p.) (Accept 1.3 to 1.5, 3.5 to 3.7)

3 Using the same method as for question 2:

x	–4	–3	–2	–1	0	1	2	3	4
$11 - 2x^2$	–21	–7	3	9	11	9	3	–7	–21

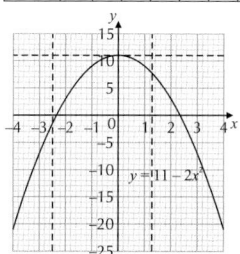

a) **(i)** **y = –1.5** (1 d.p.) (Accept –1.6 to –1.4)
  (ii) **y = 7.9** (1 d.p.) (Accept 7.8 to 8.0)
b) **(i)** **x = –2.3** and **x = 2.3** (both to 1 d.p.) (Accept –2.4 to –2.2, 2.2 to 2.4)
  (ii) **x = 0**

Page 196   Exercise 3

1 a) (i) When $x = 0$, $y = (0 - 1)(0 + 1) = -1$.
  So the y-intercept is **(0, –1)**.
  When $(x - 1)(x + 1) = 0$, $x = 1$ or $x = -1$.
  So the x-intercepts are **(–1, 0)** and **(1, 0)**.
  (ii) The x-coordinate of the turning point is halfway between the x-intercepts i.e. $(-1 + 1) \div 2 = 0$. When $x = 0$, $y = (0 - 1)(0 + 1) = -1$. So the turning point is at **(0, –1)**.
  Here, the turning point is also the y-intercept.

Using the same method for b)-c):
b) (i) The y-intercept is **(0, 7)**.
  The x-intercepts are **(–7, 0)** and **(–1, 0)**.
  (ii) The turning point is at **(–4, –9)**.
c) (i) First factorise: $y = (x + 10)(x + 6)$.
  The y-intercept is **(0, 60)**.
  The x-intercepts are **(–10, 0)** and **(–6, 0)**.
  (ii) The turning point is at **(–8, –4)**.

2 a) When $x = 0$, $y = 0^2 - 4 = -4$.
  So the y-intercept is (0, –4).
  $y = x^2 - 4$ factorises to $y = (x - 2)(x + 2)$.
  When $(x - 2)(x + 2) = 0$, $x = 2$ or $x = -2$.
  So the x-intercepts are (–2, 0) and (2, 0).
  The x-coordinate of the turning point is halfway between the x-intercepts i.e. $(-2 + 2) \div 2 = 0$. When $x = 0$, $y = 0^2 - 4 = -4$. So the turning point is at (0, –4). The x^2 term is positive, so sketch a u-shaped curve passing through the intercepts and turning point:

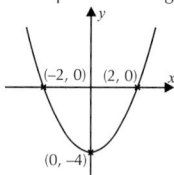

b) When $x = 0$, $y = 0^2 - (4 \times 0) - 12 = -12$.
  So the y-intercept is (0, –12).
  $y = x^2 - 4x - 12 = (x - 6)(x + 2)$.
  When $(x - 6)(x + 2) = 0$, $x = 6$ or $x = -2$.
  So the x-intercepts are (–2, 0) and (6, 0).
  The x-coordinate of the turning point is halfway between the x-intercepts i.e. $(-2 + 6) \div 2 = 2$. When $x = 2$, $y = 2^2 - (4 \times 2) - 12 = -16$. So the turning point is at (2, –16). The x^2 term is positive, so sketch a u-shaped curve passing through the intercepts and turning point:

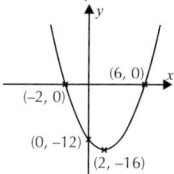

Using the same method for c)-i):
c) $y = x^2 + 12x + 32 = (x + 4)(x + 8)$

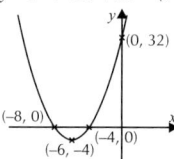

d) $y = x^2 + x - 20 = (x + 5)(x - 4)$

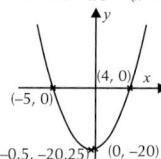

e) $y = -x^2 - 2x + 3 = (x + 3)(1 - x)$
Here, the x^2 term is negative, so it's an n-shaped curve:

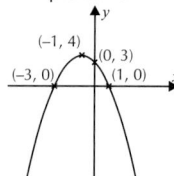

f) $y = -x^2 - 14x - 49 = -(x + 7)^2$
  or $(x + 7)(-x - 7)$

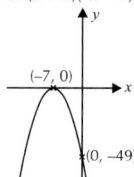

g) $y = 2x^2 + 4x - 16 = 2(x + 4)(x - 2)$

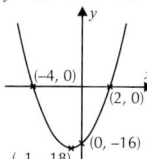

h) $y = 5x^2 - 6x - 8 = (5x + 4)(x - 2)$

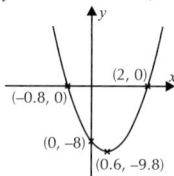

i) $y = -2x^2 - x + 6 = (3 - 2x)(x + 2)$

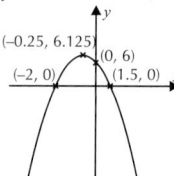

Page 197   Exercise 4

1 a) (i) When $x = 0$, $y = (0 + 5)^2 - 9 = 16$.
  So the y-intercept is **(0, 16)**.
  (ii) The turning point is when $(x + 5) = 0$ $\Rightarrow x = -5$ and $y = 0^2 - 9 = -9$. So the turning point is at **(–5, –9)**.
  You can just read this off from the completed square form of the equation.

Using the same method for b)-c):
b) (i) The y-intercept is **(0, –21)**.
  (ii) The turning point is at **(3, –30)**.
c) (i) The y-intercept is **(0, 3)**.
  (ii) The turning point is at **(4, –13)**.

2 a) When $x = 0$, $y = 0^2 - (6 \times 0) - 5 = -5$.
So the y-intercept is $(0, -5)$.
Completing the square:
$y = x^2 - 6x - 5 = (x - 3)^2 - 14$, so the
turning point is at $(3, -14)$. The x^2 term
is positive, so sketch a u-shaped curve
passing through the coordinates for the
intercept and turning point:

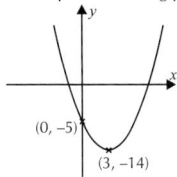

b) When $x = 0$, $y = 0^2 - (4 \times 0) + 2 = 2$.
So the y-intercept is $(0, 2)$.
Completing the square:
$y = x^2 - 4x + 2 = (x - 2)^2 - 2$, so the turning
point is at $(2, -2)$. The x^2 term is positive,
so sketch a u-shaped curve passing through
the coordinates for the intercept and
turning point:

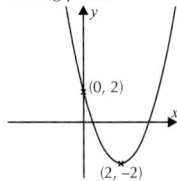

Using the same method for c)-f):

c) $y = x^2 + 8x - 6 = (x + 4)^2 - 22$

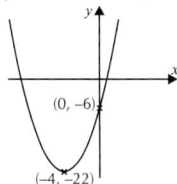

d) $y = x^2 + 2x + 8 = (x + 1)^2 + 7$

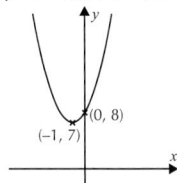

This graph does not cross the x-axis because the turning point is above the x-axis on a u-shaped quadratic.

e) $y = x^2 - x + 10 = \left(x - \frac{1}{2}\right)^2 + 9\frac{3}{4}$

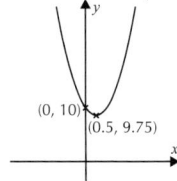

f) $y = -x^2 + 10x - 6 = 19 - (x - 5)^2$
The x^2 term is negative, so it's n-shaped:

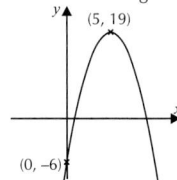

16.2 Cubic Graphs

Page 199   Exercise 1

1 a)

x	-3	-2	-1	0	1	2	3
x^3	-27	-8	-1	0	1	8	27
$x^3 + 5$	-22	-3	4	5	6	13	32

Plot each pair of values $(-3, -22)$, $(-2, -3)$
etc. on a coordinate grid with suitable axes:

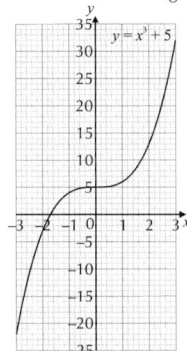

Using the same method for b)-c):

b)

x	-3	-2	-1	0	1	2	3
$3x^3$	-81	-24	-3	0	3	24	81
$-4x^2$	-36	-16	-4	0	-4	-16	-36
$+2x$	-6	-4	-2	0	2	4	6
-8	-8	-8	-8	-8	-8	-8	-8
$3x^3 - 4x^2 + 2x - 8$	-131	-52	-17	-8	-7	4	43

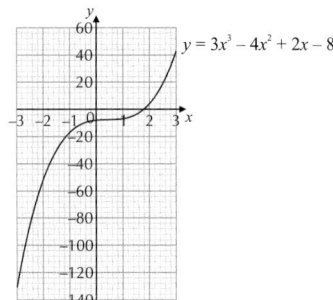

c)

x	-3	-2	-1	0	1	2	3
$5 - x^3$	32	13	6	5	4	-3	-22

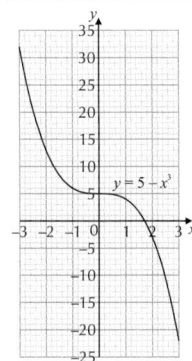

2 Calculate coordinates in a table of values, e.g:

x	-3	-2	-1	0	1	2	3
$x^3 + 3$	-24	-5	2	3	4	11	30

Plot them on a set of suitable axes:

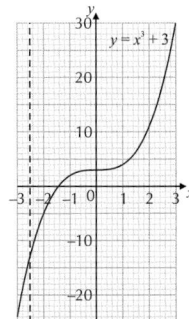

Draw a vertical line at $x = -2.5$ (as shown on
graph) and read off the y-value where it meets
the graph: **$y = -12.6$** (1 d.p.)
(Accept y-values from -12.7 to -12.5)

3 Calculate coordinates in a table of values, e.g:

x	-1	0	1	2	3	4	5
$x^3 - 6x^2 + 12x - 5$	-24	-5	2	3	4	11	30

Plot them on a set of suitable axes:

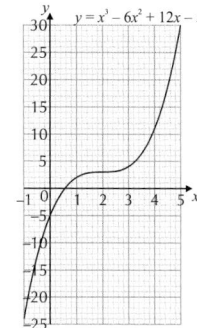

The graph crosses the x-axis at **$x = 0.6$** (1 d.p.)
(Accept x-values from 0.5 to 0.7)

Page 199   Exercise 2

1 a) At $x = 0$, $y = 0 \times (0 - 1) \times (0 + 2) = 0$,
so the y-intercept is at **(0, 0)**.
Solve $x(x - 1)(x + 2) = 0$. Either $x = 0$
or $(x - 1) = 0 \Rightarrow x = 1$ or $(x + 2) = 0$
$\Rightarrow x = -2$. So there are three x-intercepts
at **(0, 0)**, **(1, 0)** and **(-2, 0)**.
These are the roots of the cubic equation.

b) At $x = 0$, $y = (0 + 1) \times (0 + 3) \times (1 - 0) = 3$,
so the y-intercept is at **(0, 3)**.
Solve $(x + 1)(x + 3)(1 - x) = 0$
$\Rightarrow x = -1, -3$ or 1. So there are three
x-intercepts at **(-1, 0)**, **(-3, 0)** and **(1, 0)**.

c) At $x = 0$, $y = 0^2 \times (0 + 1) = 0$, so the
y-intercept is at **(0, 0)**.
Solve $x^2(x + 1) = 0 \Rightarrow x = 0$ or -1.
So there are two x-intercepts at **(0, 0)** and
(-1, 0).

d) At $x = 0$, $y = (0 - 10)^3 = -1000$, so the
y-intercept is at **(0, -1000)**.
Solve $(x - 10)^3 = 0 \Rightarrow x - 10 = 0$
$\Rightarrow x = 10$, so there is one x-intercept at
(10, 0).

2 a) Draw a positive cubic curve through the
intercepts:

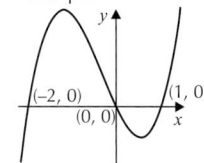

b) Draw a negative cubic curve through the
intercepts:

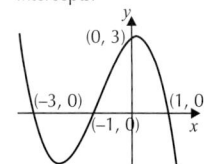

c) Draw a positive cubic curve through the
two x-intercepts:

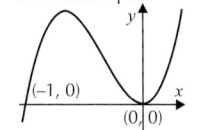

d) Draw a curve the same shape as $y = x^3$ through (0, –1000) and (10, 0):

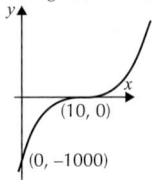

16.3 Reciprocal and Exponential Graphs

Page 201   Exercise 1

1 a)

x	–5	–4	–3	–2	–1	–0.5	–0.1
$\frac{4}{x}$	–0.8	–1	–1.33	–2	–4	–8	–40

x	0.1	0.5	1	2	3	4	5
$\frac{4}{x}$	40	8	4	2	1.33	1	0.8

Plot each pair of values (–5, –0.8), (–4, –1) etc. on a coordinate grid with suitable axes, and draw in the asymptotes. There is a vertical asymptote where the denominator of the fraction is zero, i.e. at $x = 0$.
$\frac{4}{x}$ can never be zero, so there is a horizontal asymptote at $y = 0$.

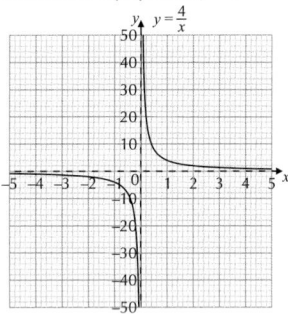

Using the same method for b):

b)

x	–5	–4	–3	–2	–1	–0.5	–0.1
$\frac{1}{x}$	–0.2	–0.25	–0.33	–0.5	–1	–2	–10
+3	3	3	3	3	3	3	3
$\frac{1}{x} + 3$	2.8	2.75	2.67	2.5	2	1	–7

x	0.1	0.5	1	2	3	4	5
$\frac{1}{x}$	10	2	1	0.5	0.33	0.25	0.2
+3	3	3	3	3	3	3	3
$\frac{1}{x} + 3$	13	5	4	3.5	3.33	3.25	3.2

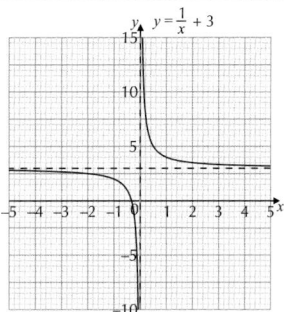

2 a) $\frac{1}{x-2}$ is undefined where the denominator $x - 2 = 0 \Rightarrow x = 2$.

b) $\frac{1}{x-2}$ can never be 0, so the function is undefined when $y = 0$.

c) The vertical asymptote is at $x = 2$ (from part a)). The horizontal asymptote is at $y = 0$ (from part b)).

d) Draw a positive reciprocal graph in between the asymptotes found in part c):

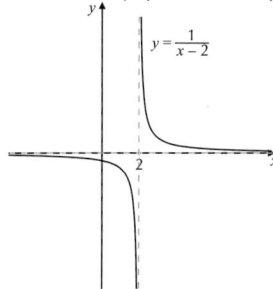

When $x = 0$, $y = \frac{1}{0-2} = -0.5$.
So the y-intercept is at **(0, –0.5)**.

3 a) There is a vertical asymptote where the denominator of the fraction is zero, i.e. at $x = -5$. $\frac{1}{x+5}$ can never be zero, so there is a horizontal asymptote at $y = 0$.
When $x = 0$, $y = \frac{1}{0+5} = 0.2$.
So the y-intercept is at **(0, 0.2)**.

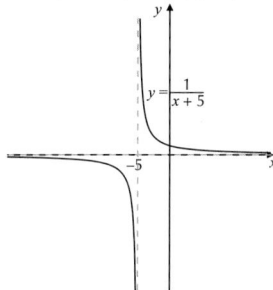

Using the same method for b)-c):

b) Asymptotes at $x = 0$ and $y = -1$. x-intercept at **(0.5, 0)**.

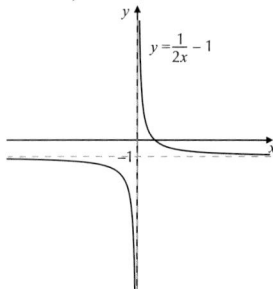

c) Asymptotes at $x = 3$ and $y = 10$. x-intercept at **(3.1, 0)**. y-intercept at **(0, 10.3)** (1 d.p.). The negative sign of the x in the fraction means the shape of the graph is like $y = -\frac{1}{x}$:

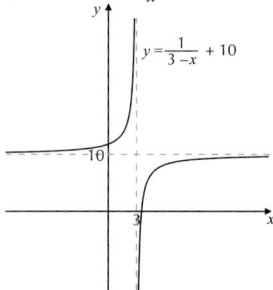

Page 202   Exercise 2

1 a)

x	–3	–2	–1	0	1	2	3
3^x	0.04	0.11	0.33	1	3	9	27

Plot each pair of values (–3, 0.04), (–2, 0.11) etc. on a coordinate grid with suitable axes, and draw in the asymptote. As x gets more positive, $y = 3^x$ gets larger. As x gets more negative, $y = 3^x$ gets closer to zero, so there is a horizontal asymptote at $y = 0$.

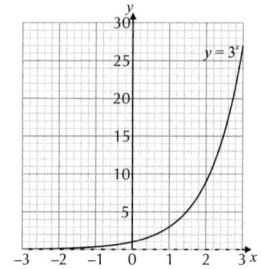

Using the same method for b):

b)

x	–3	–2	–1	0	1	2	3
2^{-x}	8	4	2	1	0.5	0.25	0.13

As x gets more positive, $-x$ gets more negative, and $y = 2^{-x}$ gets closer to zero, so there is a horizontal asymptote at $y = 0$.

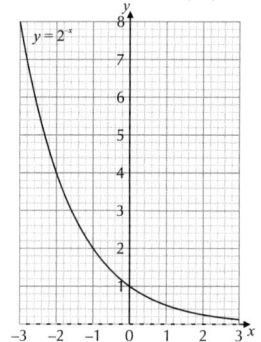

2 a) As x gets more positive, $y = 5^x$ gets larger. As x gets more negative, $y = 5^x$ gets closer to zero, so the asymptote is $y = 0$. When $x = 0$, $y = 5^0 = 1$, so the graph crosses y-axis at $y = 1$.

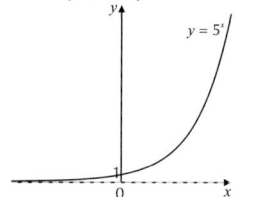

b) As x gets more positive, $y = 2^x - 1$ gets larger. As x gets more negative, 2^x gets closer to zero and $y = 2^x - 1$ gets closer to –1, so the asymptote is $y = -1$. When $x = 0$, $y = 2^0 - 1 = 0$, so the graph crosses y-axis at $y = 0$.

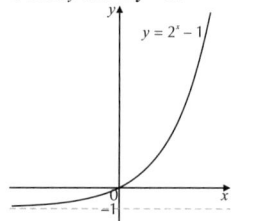

Using the same method for c)-f):

c) Asymptote is $y = 0$.
Graph crosses y-axis at $y = 1$.

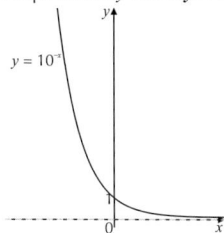
$y = 10^{-x}$

d) Asymptote is $y = 0$.
Graph crosses y-axis at $y = 1$.

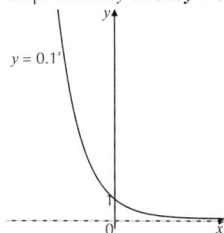
$y = 0.1^x$

This is the same function as in part c).

e) Asymptote is $y = 3$.
Graph crosses y-axis at $y = 4$.

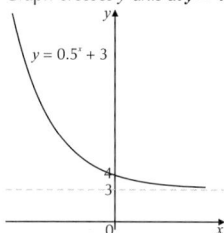
$y = 0.5^x + 3$

f) Asymptote is $y = 10$.
Graph crosses y-axis at $y = 9$.
As x gets more positive, 3^x gets larger and more positive, and so $y = 10 - 3^x$ gets more negative. So the graph looks like this:

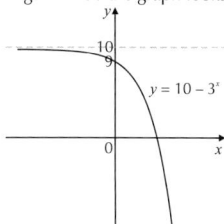
$y = 10 - 3^x$

3 As x gets more positive, $y = 100 \times 1.05^x$ gets larger, up to $100 \times 1.05^{10} = £162.89$ after 10 years. When $x = 0$, $y = 100$, i.e. the initial amount invested, so the graph crosses y-axis at $y = 100$.

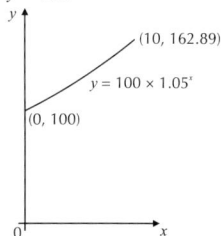
(10, 162.89)
$y = 100 \times 1.05^x$
(0, 100)

You only need to sketch the graph for positive x-values, because the number of years after the initial investment can't be negative.

4 a) When $t = 0$, $V = k \times 0.8^0 = k \times 1 = k$.
So k is the vertical axis intercept of the graph: $k = 6000$.

b) k is the value of V when $t = 0$, so it is the **value of the car when it was brand new**.

16.4 Circle Graphs

Page 203   Exercise 1

1 a) It's a circle with centre (0, 0) and radius $\sqrt{25} = 5$, so it cuts each axis at ±5.

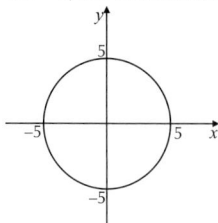

b) It's a circle with centre (0, 0) and radius $\sqrt{1} = 1$, so it cuts each axis at ±1.

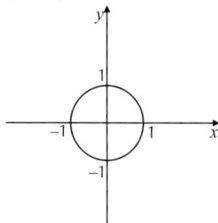

c) It's a circle with centre (0, 0) and radius $\sqrt{6.25} = 2.5$, so it cuts each axis at ±2.5.

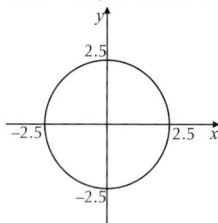

d) Rearrange into the form $x^2 + y^2 = r^2$:
$y^2 = 4 - x^2 \Rightarrow x^2 + y^2 = 4$. So it's a circle with centre (0, 0) and radius $\sqrt{4} = 2$, which cuts each axis at ±2.

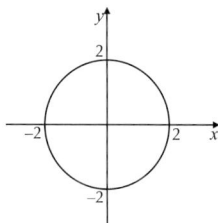

e) $2x^2 + 2y^2 = 18 \Rightarrow x^2 + y^2 = 9$. So it's a circle with centre (0, 0) and radius $\sqrt{9} = 3$, which cuts each axis at ±3.

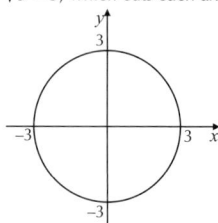

f) $x^2 + 0.04 = 0.2 - y^2 \Rightarrow x^2 + y^2 = 0.16$.
So it's a circle with centre (0, 0) and radius $\sqrt{0.16} = 0.4$, which cuts each axis at ±0.4.

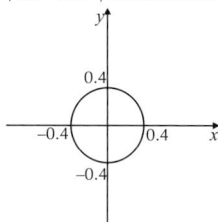

2 a) (i) Radius = **4**, so equation is $x^2 + y^2 = 4^2$
$\Rightarrow x^2 + y^2 = 16$.
   (ii) Radius = **3**, so equation is $x^2 + y^2 = 3^2$
$\Rightarrow x^2 + y^2 = 9$.

b) (i) Radius = **1.5**, so equation is
$x^2 + y^2 = 1.5^2 \Rightarrow x^2 + y^2 = 2.25$.
   (ii) Radius = **0.5**, so equation is
$x^2 + y^2 = 0.5^2 \Rightarrow x^2 + y^2 = 0.25$.

3 The circle has diameter of 10, so radius = $10 \div 2 = 5$. The line $y + x = 1$ has y-intercept at $y + 0 = 1 \Rightarrow y = 1$, and x-intercept at $0 + x = 1 \Rightarrow x = 1$. Use this information to plot both graphs on the same axes:

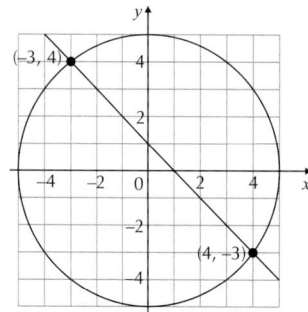
(−3, 4)
(4, −3)

The graphs intersect at **(−3, 4)** and **(4, −3)**.

16.5 Trigonometric Graphs

Page 205   Exercise 1

1 a) $y = \sin x$ has a y-intercept at (0°, 0), x-intercepts at (0°, 0), (180°, 0) and (360°, 0), a maximum point at (90°, 1) and a minimum point at (270°, −1). These all repeat every 360°, so add on 360° to the x-coordinates of turning points and intercepts.

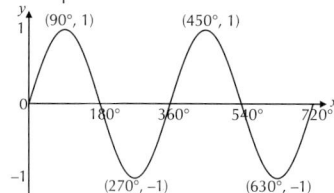
(90°, 1)   (450°, 1)
(270°, −1)   (630°, −1)

b) $y = \cos x$ has a y-intercept at (0°, 1), x-intercepts at (90°, 0) and (270°, 0), a maximum point at (0°, 1) and a minimum point at (180°, −1). These all repeat every 360°, so add on 360° to the x-coordinates of turning points and intercepts.

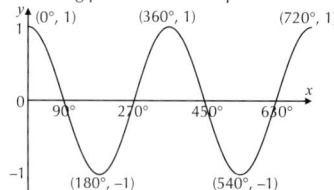
(0°, 1)   (360°, 1)   (720°, 1)
(180°, −1)   (540°, −1)

c) $y = \tan x$ has a y-intercept at (0°, 0), x-intercepts at (0°, 0), (180°, 0) and (360°, 0), and vertical asymptotes at $x = 90°$ and $x = 270°$. These all repeat every 180°, so add on 180° to the x-coordinates of asymptotes and intercepts.

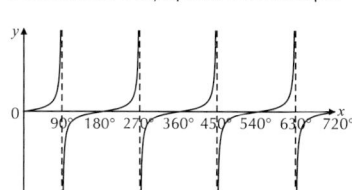

2 a)

x	0°	30°	45°	60°	90°	120°	135°	150°	180°
$\sin x$	0	0.5	0.71	0.87	1	0.87	0.71	0.5	0
$2\sin x$	0	1	1.41	1.73	2	1.73	1.41	1	0

x	210°	225°	240°	270°	300°	315°	330°	360°
$\sin x$	−0.5	−0.71	−0.87	−1	−0.87	−0.71	−0.5	0
$2\sin x$	−1	−1.41	−1.73	−2	−1.73	−1.41	−1	0

Plot each pair of values (0°, 0), (30°, 1) etc. on a coordinate grid with suitable axes:

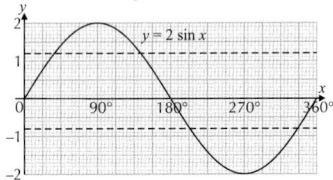

b) Draw horizontal lines at the given y-values (as shown on the graph) and read off the corresponding x-values.

(i) **37°** and **143°**
(Accept 36° to 38°, 142° to 144°)

(ii) **204°** and **336°**
(Accept 203° to 205°, 335° to 337°)

3 a)

x	0°	30°	45°	60°	90°	120°	135°	150°	180°
$2x$	0°	60°	90°	120°	180°	240°	270°	300°	360°
$\sin 2x$	0	0.87	1	0.87	0	−0.87	−1	−0.87	0

x	210°	225°	240°	270°	300°	315°	330°	360°
$2x$	420°	450°	480°	540°	600°	630°	660°	720°
$\sin 2x$	0.87	1	0.87	0	−0.87	−1	−0.87	0

Plot each pair of values (0°, 0), (30°, 0.87) etc. on a coordinate grid with suitable axes:

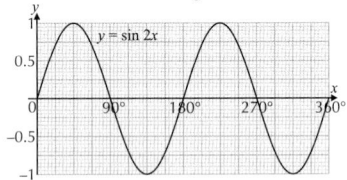

b) E.g. The graphs of $y = \sin x$ and $y = \sin 2x$ have the same shape, but $y = \sin 2x$ repeats twice as often as $y = \sin x$ / it has a period of 180° instead of 360°.

4 a) Draw a horizontal line at $y = 1.75$:

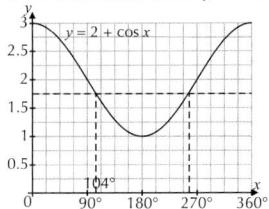

You can see there are 2 solutions — one at $x = 104°$ and the other to be found. The graph has a line of symmetry at $x = 180°$, so the second solution will be 104° away from 360°, so $x = 360° − 104° = $ **256°**.

b) The graph repeats every 360°, so add 360° to each solution from part a):
$x = 104° + 360° = $ **464°** and $256° + 360° = $ **616°**.

---

5 a) $y = \tan x$ has x-intercepts at
$x = 0°, 180°, 360°...$, so $y = \tan (x + 180°)$ has them at $x + 180° = 0°, 180°, 360°...$ i.e. at $x = −180°, 0°, 180°...$
$y = \tan x$ has vertical asymptotes at $x = 90°, 270°...$ so $y = \tan (x + 180°)$ has them at $x + 180° = 90°, 270°...$ i.e. at $x = −90°, 90°...$ Use these features to sketch the graph:

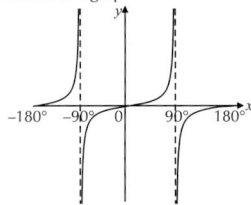

b) It's the same as $y = \tan x$. This is because it's a horizontal translation of $y = \tan x$ by 180°, and since $y = \tan x$ repeats itself every 180°, the translation maps onto the original graph.

16.6 Transforming Graphs

Page 206   Exercise 1

1 a) (i) $y = f(x) + a$ is a translation a units up, so $y = f(x) + 3$ is a translation **3 units up**, or by the vector $\begin{pmatrix} 0 \\ 3 \end{pmatrix}$.

(ii) The y-coordinates increase by 3, so (0, 0) moves to **(0, 3)**.

b) (i) $y = f(x − a)$ is a translation a units right, so $y = f(x − 1)$ is a translation **1 unit right**, or by the vector $\begin{pmatrix} 1 \\ 0 \end{pmatrix}$.

(ii) The x-coordinates increase by 1, so (0, 0) moves to **(1, 0)**.

c) (i) $y = f(x − a)$ is a translation a units right, so $y = f(x + 2)$ is a translation **2 units left**, or by the vector $\begin{pmatrix} -2 \\ 0 \end{pmatrix}$.

(ii) The x-coordinates increase by −2 (i.e. they decrease by 2), so (0, 0) moves to **(−2, 0)**.

d) (i) $y = f(x) + a$ is a translation a units up, so $y = f(x) − 6$ is a translation **6 units down**, or by the vector $\begin{pmatrix} 0 \\ -6 \end{pmatrix}$.

(ii) The y-coordinates increase by −6 (i.e. they decrease by 6), so (0, 0) moves to **(0, −6)**.

2 a) $f(x) = x^2$, so this is $y = f(x) + 1$, which is a translation of **1 unit up**:

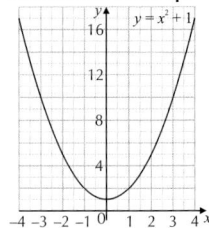

The turning point moves from (0, 0) to (0, 1).

b) This is $y = f(x) − 2$, which is a translation of **2 units down**:

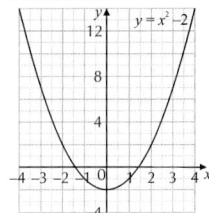

---

c) This is $y = f(x − 4)$, which is a translation of **4 units right**:

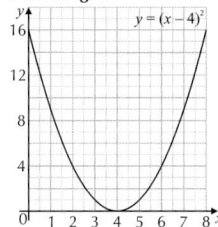

The turning point moves from (0, 0) to (4, 0).

d) This is $y = f(x + 1)$, which is a translation of **1 unit left**:

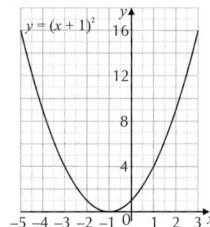

3 a) The turning point has moved from (0, 0) up 4 units to (0, 4). So the equation is $y = f(x) + 4 \Rightarrow y = x^2 + 4$.

b) The turning point has moved from (0, 0) right 3 units to (3, 0). So the equation is $y = f(x − 3) \Rightarrow y = (x − 3)^2$.

c) The turning point has moved from (0, 0) down 1 unit to (0, −1). So the equation is $y = f(x) − 1 \Rightarrow y = x^2 − 1$.

4 a) $f(x) = \sin x$, so this is $y = f(x) + 1$, which is a translation of **1 unit up**:

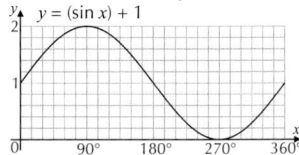

The y-coordinates of the y-intercept and maximum and minimum points have increased by 1.

b) This is $y = f(x) − 2$, which is a translation of **2 units down**:

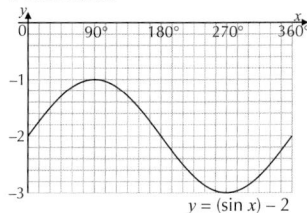

c) This is $y = f(x + 60°)$, which is a translation of **60° left**:

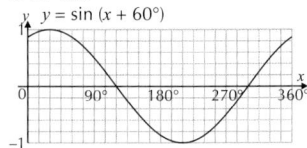

The x-coordinates of the x-intercepts and both turning points have decreased by 60°.

d) This is $y = f(x − 90°)$, which is a translation of **90° right**:

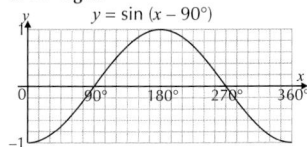

e) This is $y = f(x + 180°)$, which is a translation of **180° left**:

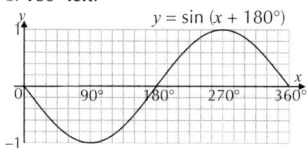
$y = \sin(x + 180°)$

f) This is $y = f(x - 360°)$, which is a translation of **360° right**:

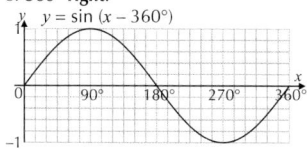
$y = \sin(x - 360°)$

This maps back onto the graph of $y = \sin x$, which repeats itself every 360°.

Page 207   Exercise 2

1 a) (i) The $y = f(x - 5)$ part is a translation of **5 units right**. The $y = f(x) + 6$ part is a translation of **6 units up**. So the combined translation vector is $\begin{pmatrix} 5 \\ 6 \end{pmatrix}$.

(ii) The x-coordinates increase by 5 and the y-coordinates increase by 6, so (0, 0) moves to **(5, 6)**.

b) (i) The $y = f(x + 6)$ part is a translation of **6 units left**. The $y = f(x) - 4$ part is a translation of **4 units down**. So the combined translation vector is $\begin{pmatrix} -6 \\ -4 \end{pmatrix}$.

(ii) The x-coordinates decrease by 6 and the y-coordinates decrease by 4, so (0, 0) moves to **(−6, −4)**.

c) (i) The $y = f(x - 3)$ part is a translation of **3 units right**. The $y = f(x) + 2$ part is a translation of **2 units up**. So the combined translation vector is $\begin{pmatrix} 3 \\ 2 \end{pmatrix}$.

(ii) The x-coordinates increase by 3 and the y-coordinates increase by 2, so (0, 0) moves to **(3, 2)**.

d) (i) The $y = f(x + 1)$ part is a translation of **1 unit left**. The $y = f(x) + 9$ part is a translation of **9 units up**. So the combined translation vector is $\begin{pmatrix} -1 \\ 9 \end{pmatrix}$.

(ii) The x-coordinates decrease by 1 and the y-coordinates increase by 9, so (0, 0) moves to **(−1, 9)**.

2 a) $f(x) = x^3$, so this is $y = f(x - 5) + 4$, which is a translation of **5 units right** and **4 units up**:

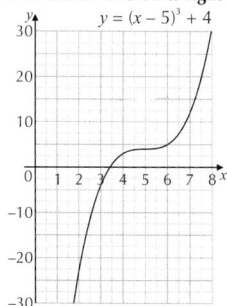
$y = (x - 5)^3 + 4$

The point (0, 0) moves to the point (5, 4).

b) This is $y = f(x + 2) - 3$, which is a translation of **2 units left** and **3 units down**:

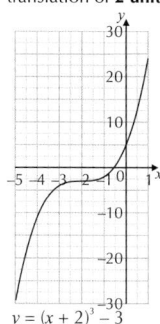
$y = (x + 2)^3 - 3$

c) This is $y = f(x - 4) + 6.5$, which is a translation of **4 units right** and **6.5 units up**:

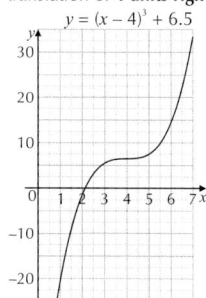
$y = (x - 4)^3 + 6.5$

d) This is $y = f(x + 2.5) - 3$, which is a translation of **2.5 units left** and **3 units down**:

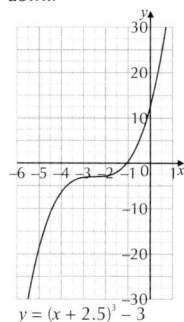
$y = (x + 2.5)^3 - 3$

3 a) The turning point has moved from (0, 0) left 2 units and up 1 unit to (−2, 1). So the equation is $y = f(x + 2) + 1$
$\Rightarrow y = (x + 2)^2 + 1$.

b) The turning point has moved from (0, 0) right 1 unit and up 3 units to (1, 3). So the equation is $y = f(x - 1) + 3$
$\Rightarrow y = (x - 1)^2 + 3$.

c) The turning point has moved from (0, 0) right 4 units and down 2 units to (4, −2). So the equation is $y = f(x - 4) - 2$
$\Rightarrow y = (x - 4)^2 - 2$.

4 a) $y = \tan x$ repeats every 180°, so translating it by a multiple of 180° left or right will produce the same graph. So give any equation of the form $y = \tan(x + 180°n)$, where n is a positive or negative integer, e.g. **$y = \tan(x + 180°)$**.

b) $y = \cos x$, and hence $y = \cos(x + 180°)$, repeats every 360°, so translating it by a multiple of 360° left or right will produce the same graph. So give any equation of the form $y = \cos(x + 180° + 360°n)$, where n is a positive or negative integer, e.g. **$y = \cos(x - 180°)$**.
You could also give this in terms of sin — any answer of the form $y = \sin(x - 90° + 360°n)$ will also be correct.

c) $y = \sin x$, and hence $y = \sin(x + 45°)$, repeats every 360°, so translating it by a multiple of 360° left or right will produce the same graph. So give any equation of the form $y = \sin(x + 45° + 360°n)$, where n is a positive or negative integer, e.g. **$y = \sin(x + 405°)$**.
You could also give this in terms of cos — any answer of the form $y = \cos(x - 45° + 360°n)$ will also be correct.

Page 209   Exercise 3

1 a) $y = f(-x)$ is a **reflection in the y-axis**. The x-coordinates of all points on the graph are multiplied by −1, and the y-coordinates stay the same, so the turning points move to (−3, 3) and (3, −1):

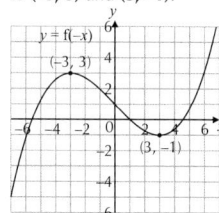
$y = f(-x)$

Reflect the coordinates of key features such as the turning points and intercepts first, then sketch in the reflected graph through these points.

b) $y = -f(x)$ is a **reflection in the x-axis**. The y-coordinates of all points on the graph are multiplied by −1, and the x-coordinates stay the same, so the turning points move to (−3, 1) and (3, −3):

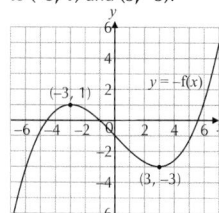
$y = -f(x)$

2 Sketches drawn for Question 2 may vary from those given here if different ranges of x-values have been used.

a) If $f(x) = \sin x$, then $y = -\sin x = -f(x)$, which is a reflection in the x-axis, e.g.

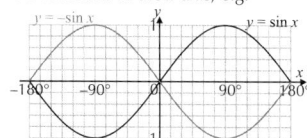
$y = -\sin x$; $y = \sin x$

b) If $f(x) = \cos x$, then $y = -\cos x = -f(x)$, which is a reflection in the x-axis, e.g.

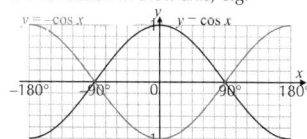
$y = -\cos x$; $y = \cos x$

c) If $f(x) = x^2$, then $y = -x^2 = -f(x)$, which is a reflection in the x-axis, e.g.

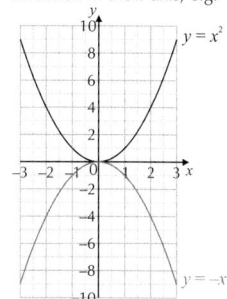
$y = x^2$; $y = -x^2$

1 a)

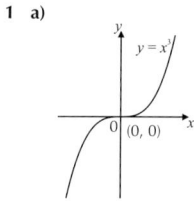

[1 mark]

b) When $x = 0$, $y = 1 - 0 = 1$, so the y-intercept is at (0, 1). When $y = 0$, $x^2 = 1$ $\Rightarrow x = \pm 1$, so the x-intercepts are at (–1, 0) and (1, 0). The x^2 term is negative, so draw an n-shaped curve through the intercepts:

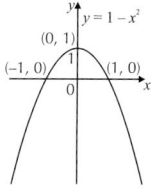

[3 marks available — 1 mark for an n-shaped quadratic graph, 1 mark for the y-intercept labelled correctly, 1 mark for the x-intercepts labelled correctly]

2 a) There has been a reflection in the x-axis so the equation is $y = -f(x)$ *[1 mark]*.

b) There has been a translation of 3 units to the left so the equation is $y = f(x + 3)$ *[1 mark]*.

3 a) The original graph is a circle with centre (0, 0) and radius $\sqrt{9} = 3$, so it cuts each axis at ±3, with y-intercepts at (0, –3) and (0, 3). After a vertical translation of +2, the y-coordinates of the y-intercepts will increase by 2 to (0, –1) and (0, 5):

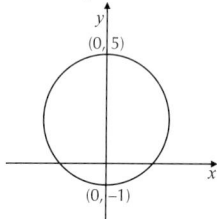

[1 mark for a sketch of circle with radius 3, 1 mark for correct y-intercepts]

b) A translation of 2 units in the positive y-direction is $y = f(x) + 2$. Get the original equation in $y = f(x)$ form:
$x^2 + y^2 = 9 \Rightarrow y^2 = 9 - x^2 \Rightarrow y = \sqrt{9 - x^2}$
So $y = f(x) + 2 = \sqrt{9 - x^2} + 2$ *[1 mark]*.
$y = 0$ at the x-intercepts, so:
$\sqrt{9 - x^2} + 2 = 0 \Rightarrow 9 - x^2 = (-2)^2$
$\Rightarrow x^2 = 5 \Rightarrow x = \pm\sqrt{5}$. So the coordinates are $(-\sqrt{5}, 0)$ and $(\sqrt{5}, 0)$ *[1 mark]*.

4 a) $p = \dfrac{4}{0 + 0.5} = \dfrac{4}{0.5} = 8$
$q = \dfrac{4}{1.5 + 0.5} = \dfrac{4}{2} = 2$
$r = \dfrac{4}{7.5 + 0.5} = \dfrac{4}{8} = 0.5$ *[1 mark for all 3]*

b)

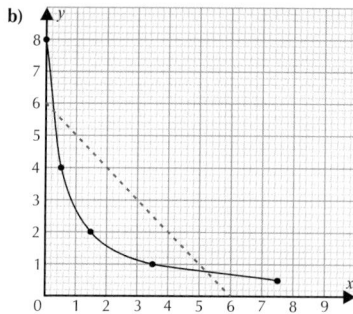

[2 marks available — 1 mark for plotting all five points correctly, 1 mark for a smooth curve joining them]

c) $y = 6 - x$ is a straight line through (0, 6) and (6, 0) (as drawn on the graph in b)). Since the line crosses the curve twice, the equation in the question has **two solutions**.
[2 marks available — 1 mark for drawing the line $y = 6 - x$, 1 mark for a correct statement]
You may have used a table of values or $y = mx + c$ to get $c = 6$ and $m = -1$.

5 a) At A, $x = 0$ so $y = 4^0 - 8 = 1 - 8 = -7$. So A is at **(0, –7)** *[1 mark]*.

b) At B, $y = 0$ so $4^x - 8 = 0 \Rightarrow 4^x = 8$. 4 is 2^2 and 8 is 2^3, so $(2^2)^x = 2^3$ *[1 mark]* $\Rightarrow 2^{2x} = 2^3 \Rightarrow 2x = 3 \Rightarrow x = 1.5$. So B is at **(1.5, 0)** *[1 mark]*.

6 a) The y-coordinate of P is the value of $\tan 60° = \sqrt{3}$ *[1 mark]*
This is one of the common trig angles you need to know.

b) A translation of $\begin{pmatrix} 20 \\ 0 \end{pmatrix}$ (i.e. 20 units right) would produce the graph of $y = f(x - 20)$, which is $y = \tan (x - 20)°$ *[1 mark]*.

3 a) Read up from 38 km/h on the horizontal axis and then across to the vertical axis to get an answer of **24 mph**.

b) (i) Read across from 25 mph on the vertical axis and then down to the horizontal axis to get an answer of **40 km/h**.

  (ii) 75 mph is 3×25 mph. Since speeds in mph and km/h are directly proportional (the graph is a straight line through the origin), the km/h equivalent of 75 mph is 3×40 km/h = **120 km/h**.

c) 52 km/h = 32.5 mph so the driver is $32.5 - 30 = $ **2.5 mph** over the speed limit.

d) From part b)(ii), 120 km/h = 75 mph, which is greater than 70 mph. So the speed limit is greater in **Spain** by $75 - 70 = $ **5 mph**.
Alternatively, you could convert 70 mph into km/h (it's 112 km/h) using the method in part b)(ii) and conclude the speed limit is greater in Spain by 8 km/h.

4

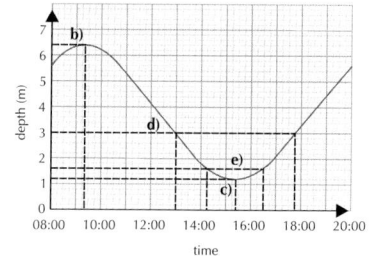

a) The water got deeper for about an hour (since the graph increases), then shallower for about 6 hours (since the graph decreases). Finally it got deeper for about 5 hours (since the graph increases again).

b) The greatest depth corresponds to the highest point on the graph. This occurs at **09:20** (accept 09:15-09:30).

c) The minimum depth corresponds to the lowest point on the graph. At this point, the depth is **1.2 m**.

d) Read across from 3 m. The graph is at this depth twice, once at **12:55** (accept 13:00) and again at **17:45**.

e) The depth of the water is below 1.6 m between 14:15 and 16:30. So his boat is not floating for **2h 15m** (135 minutes). (Accept answers between 2 hours 10 minutes and 2 hours 20 minutes.)

5

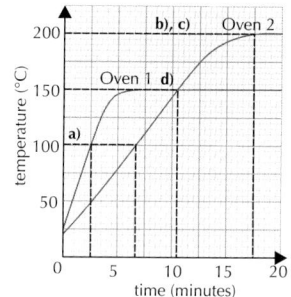

a) Reading across from 100 °C on the vertical axis, the graph for Oven 1 shows it's at this temperature after 2.5 minutes while the graph for Oven 2 shows it's at this temperature after about 7 minutes. So **Oven 1** reaches 100 °C more quickly.

b) The maximum point on the graph for Oven 1 is at 150 °C while the maximum point on the graph for Oven 2 is at 200 °C. So **Oven 2** reaches a higher maximum temperature.

Section 17 — Using Graphs

17.1 Interpreting Real-Life Graphs

Page 215   Exercise 1

1 a) Look for a graph that increases with a steep gradient and then decreases gently. This matches graph **L**.

b) Look for a graph that neither rises nor falls (i.e. is horizontal). This matches graph **N**.

c) Look for a graph that increases and has a gradient which gets more and more steep. This matches graph **M**.

d) Look for a graph that decreases initially but then rises steeply. This matches graph **K**.

2 a) The graph increases throughout the time period so the depth of water **deepens** for the entire 30 seconds. For the first 15 seconds, the gradient is steep and constant (i.e. it's a straight line) so the depth gets deeper **rapidly** and at a **constant rate**. For the next 15 seconds, the gradient is less steep but still constant so the depth gets deeper **less rapidly** but still at a **constant rate**.

b) The vase is 30 cm tall so half the height is 15 cm. Reading across from 15 cm on the vertical axis and then down to the horizontal axis, the answer is **12.5 seconds**.

c) Look for a vase that will fill quickly but at a constant rate initially and then less quickly but still at a constant rate. This matches vase **A**.
Vase B would fill at the same constant rate throughout, vase C would fill more quickly to start with and then slower, and vase D would initially fill at a constant rate and then start filling more quickly.

c) Reading across from 200 °C on the vertical axis to the graph for Oven 2, this corresponds to a time of around **18.5 minutes** (accept 18-19 minutes).

d) The ovens are at the same temperature where the lines intersect.
This point occurs at:
(i) 10.5 minutes = 10.5 × 60
= **630 seconds**
(ii) 150 °C

e) Find the gradient of the graph between 0 and 3 minutes. Choose two points: e.g. (0, 25) and (2.5, 100).
Then the gradient is
$\frac{\text{change in } y}{\text{change in } x} = \frac{100-25}{2.5-0} = 30$ °C/minute
(Accept 28-30 °C/minute)

17.2 Drawing Real-Life Graphs

Page 218   Exercise 1

1 a) You would cook a chicken weighing 1 kg for (35 × 1) + 25 = 60 minutes.
You would cook a chicken weighing 2 kg for (35 × 2) + 25 = 95 minutes.
Using the same method for the remaining weights, the table can be completed as follows.

Weight (kg)	1	2	3	4	5
Time (minutes)	60	95	130	165	200

b) Plot each pair of values on suitable axes and join with a straight line:

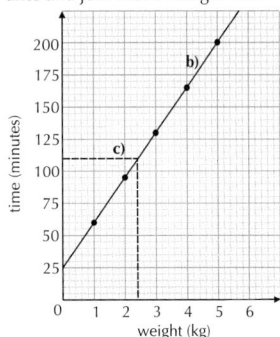

c) Reading across from 110 minutes on the vertical axis and then down to the horizontal axis, the weight is **2.4 kg**. (Accept 2.4-2.5 kg).

2 a)

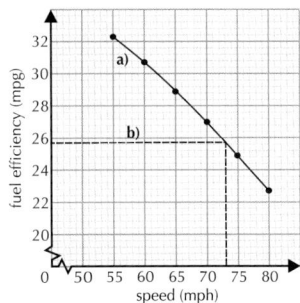

You don't need speeds lower than 55 mph or fuel efficiencies lower than 22.7 mpg so you can remove some of the axes (shown by the squiggles).

b) Reading up from 73 mph on the horizontal axis and then across to the vertical axis, the fuel efficiency is **25.5-25.9 mpg**.

3 a)

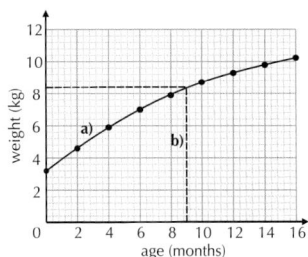

b) Reading up from 9 months on the horizontal axis and across to the vertical axis, the weight is about 8.4 kg. So Keira is 9.1 – 8.4 = **0.7 kg** heavier. (Accept 0.7-0.8 kg)

4 a) For $d = 0$, $N = 5000 × 1.2^0 = 5000$.
For $d = 1$, $N = 5000 × 1.2^1 = 7200$.
Using the same method for the remaining values of d, the table can be completed as follows.

d	0	1	2	3	4
N	5000	6000	7200	8640	10 368

b)

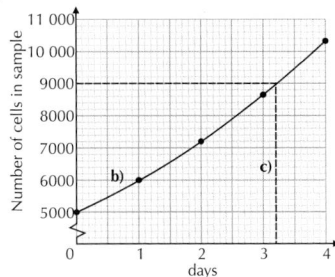

The lowest number of cells is 5000 so you can remove some of the vertical axis below this.

c) Reading across from 9000 on the vertical axis and down to the horizontal axis, it takes **3.2 days**. (Accept 3.1-3.3 days)

5 a) For the first three metres, the cost is £80 × l, where l is the length.
Thereafter, the cost is
£80 × 3 + £50 × (l – 3) = £240 + £50(l – 3).
This gives the table of values below.

Length (l m)	1	2	3	4	5
Cost (£)	80	160	240	290	340

Plot these points and join them with two straight lines to get the graph below.

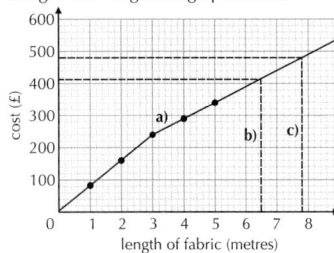

The relationship between length and cost changes after 3 metres, so you get two 'parts' to the graph (the two straight lines). You wouldn't join these points with a smooth curve because that would imply there is a single 'rule' defining the relationship.

b) Reading up from 6.5 metres on the horizontal axis and across to the vertical axis, the cost is **£415**.

c) Reading across from £480 on the vertical axis and down to the horizontal axis, the length is **7.8 metres**.

6 The flask fills slowly initially because it is wide but then will fill faster as it narrows. This corresponds to an **increasing** graph with an **initially gentle gradient** which gets **steeper with time**. At the top, the flask is cylindrical so the flask will fill at a constant rate. This corresponds to a **constant gradient**, i.e. a straight line. So the graph looks something like this:

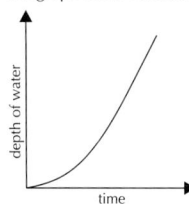

17.3 Solving Simultaneous Equations Graphically

Page 219   Exercise 1

1 The lines intersect at the point (4, 6) so the solution to the simultaneous equations is **$x = 4$ and $y = 6$**.

2 For each part a)-j), draw the graphs of the two straight lines and read off the x- and y-values of the point of intersection to find the solutions.

a)

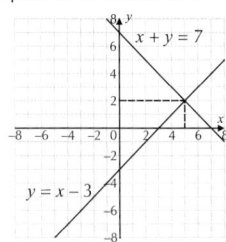

$x = 5$ and $y = 2$

b)

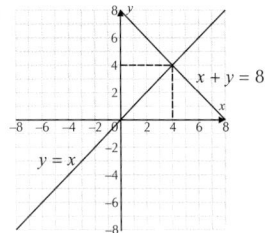

$x = 4$ and $y = 4$

c)

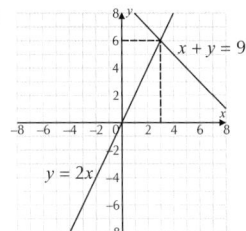

$x = 3$ and $y = 6$

d)

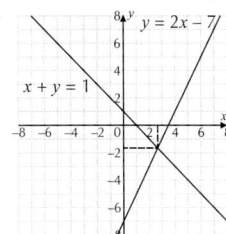

$x = 2\frac{2}{3}$ and $y = -1\frac{2}{3}$

e)

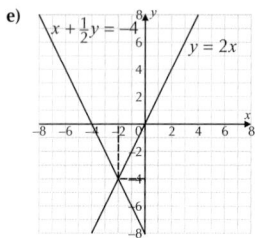

$x + \frac{1}{2}y = -4$
$y = 2x$

$x = -2$ and $y = -4$

f)

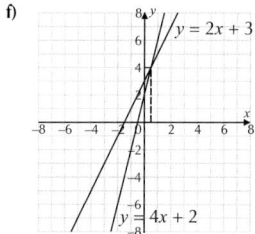

$y = 2x + 3$
$y = 4x + 2$

$x = \frac{1}{2}$ and $y = 4$

g)

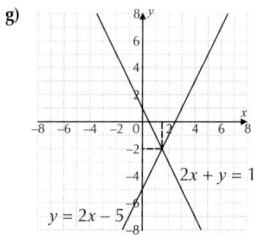

$2x + y = 1$
$y = 2x - 5$

$x = 1\frac{1}{2}$ and $y = -2$

h)

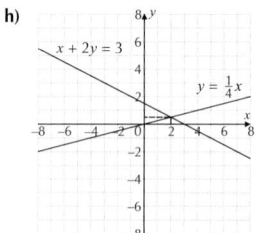

$x + 2y = 3$
$y = \frac{1}{4}x$

$x = 2$ and $y = \frac{1}{2}$

i)

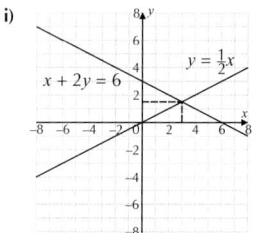

$x + 2y = 6$
$y = \frac{1}{2}x$

$x = 3$ and $y = 1\frac{1}{2}$

j)

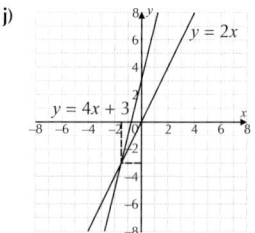

$y = 2x$
$y = 4x + 3$

$x = -1\frac{1}{2}$ and $y = -3$

3 a)

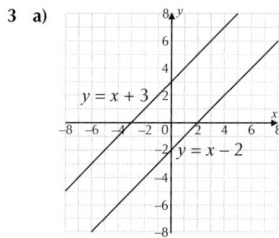

$y = x + 3$
$y = x - 2$

b) The lines are parallel and so do not intersect. This means there are no points where both the x- and the y-values are the same for both lines, and so there are no solutions to the simultaneous equations.

Page 220   Exercise 2

1 a) The graphs intersect at the points $(0, 0)$ and $(2, 4)$ so the solutions are $x = 0, y = 0$ and $x = 2, y = 4$.

b) The graphs meet at $(0, 4)$ so the only solution is $x = 0, y = 4$.

c) The graphs don't intersect so there are **no solutions**.

2 For each part a)-h), draw the graphs of the straight line and the quadratic curve, then read off the x- and y-values of the points of intersection to find the solutions.

a) (i)

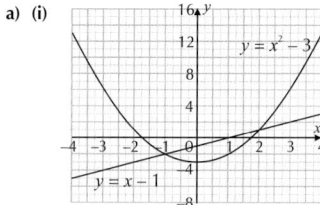

$y = x^2 - 3$
$y = x - 1$

(ii) $x = -1, y = -2$ and $x = 2, y = 1$

b) (i)

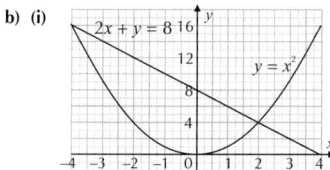

$2x + y = 8$
$y = x^2$

(ii) $x = -4, y = 16$ and $x = 2, y = 4$

c) (i)

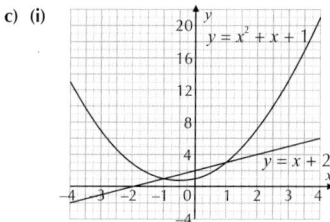

$y = x^2 + x + 1$
$y = x + 2$

(ii) $x = -1, y = 1$ and $x = 1, y = 3$

d) (i)

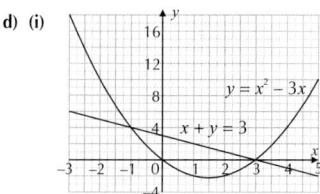

$y = x^2 - 3x$
$x + y = 3$

(ii) $x = -1, y = 4$ and $x = 3, y = 0$

e) (i)

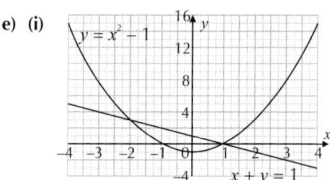

$y = x^2 - 1$
$x + y = 1$

(ii) $x = -2, y = 3$ and $x = 1, y = 0$

f) (i)

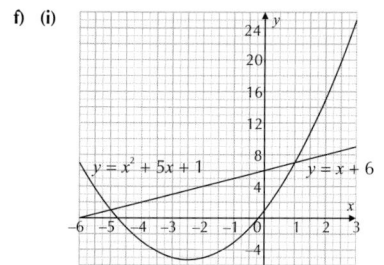

$y = x^2 + 5x + 1$
$y = x + 6$

(ii) $x = -5, y = 1$ and $x = 1, y = 7$

g) (i)

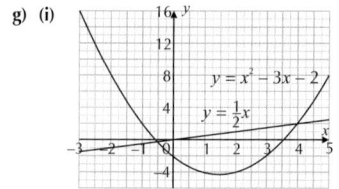

$y = x^2 - 3x - 2$
$y = \frac{1}{2}x$

(ii) $x = -\frac{1}{2}, y = -\frac{1}{4}$ and $x = 4, y = 2$

h) (i)

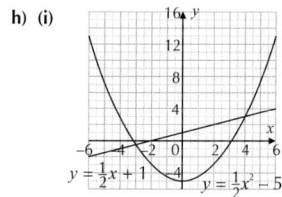

$y = \frac{1}{2}x + 1$
$y = \frac{1}{2}x^2 - 5$

(ii) $x = -3, y = -\frac{1}{2}$ and $x = 4, y = 3$

17.4 Solving Quadratics Graphically

Page 221   Exercise 1

1 a)

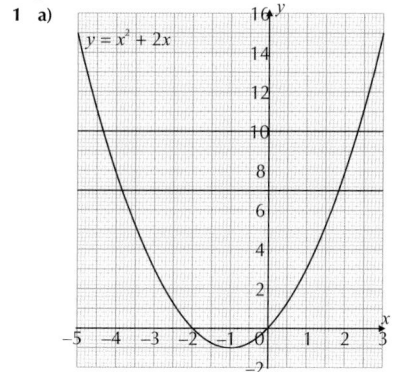

$y = x^2 + 2x$

b) (i) $x^2 + 2x = 0$ has solutions where the graph intersects the x-axis:
$x = -2$ and $x = 0$

(ii) $x^2 + 2x = 10$ has solutions where the graph intersects the line $y = 10$:
$x = -4.3$ and $x = 2.3$ (to 1 d.p.)

(iii) $x^2 + 2x = 7$ has solutions where the graph intersects the line $y = 7$:
$x = -3.8$ and $x = 1.8$ (to 1 d.p.)

2 a)

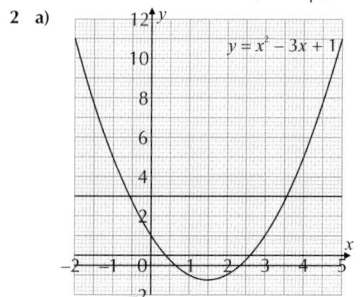

$y = x^2 - 3x + 1$

b) (i) $x^2 - 3x + 1 = 0$ has solutions where the graph intersects the x-axis:
$x = 0.4$ and $x = 2.6$ (to 1 d.p.)

(ii) $x^2 - 3x + 1 = 3$ has solutions where the graph intersects the line $y = 3$:
$x = -0.6$ and $x = 3.6$ (to 1 d.p. — allow −0.5 and 3.5)

(iii) $x^2 - 3x + 1 = -0.5$ has solutions where the graph intersects the line $y = -0.5$:
$x = 0.6$ and $x = 2.4$ (to 1 d.p. — allow 0.7 and 2.3)

3 a)

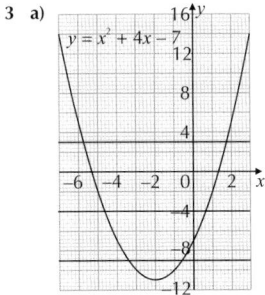

$y = x^2 + 4x - 7$

b) (i) $x^2 + 4x - 7 = 0$ has solutions where the graph intersects the x-axis:
$x = -5.3$ and $x = 1.3$ (to 1 d.p.)

(ii) $x^2 + 4x - 10 = 0 \Rightarrow x^2 + 4x - 7 = 3$, which has solutions where the graph intersects the line $y = 3$:
$x = -5.7$ and $x = 1.7$ (to 1 d.p. — allow −5.8 and 1.8)

(iii) $x^2 + 4x - 3 = 0 \Rightarrow x^2 + 4x - 7 = -4$, which has solutions where the graph intersects the line $y = -4$:
$x = -4.6$ and $x = 0.6$ (to 1 d.p. — allow −4.7 and 0.7)

(iv) $x^2 + 4x + 2 = 0 \Rightarrow x^2 + 4x - 7 = -9$, which has solutions where the graph intersects the line $y = -9$:
$x = -3.4$ and $x = -0.6$ (to 1 d.p.)

4 a)

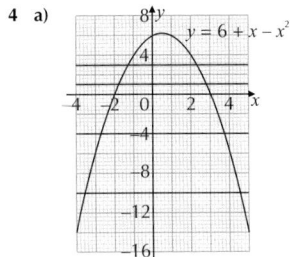

$y = 6 + x - x^2$

b) (i) $5 + x - x^2 = 0$
$\Rightarrow 5 + x - x^2 - 5 + 6 = -5 + 6$
$\Rightarrow 6 + x - x^2 = 1$,
which has solutions where the graph intersects the line $y = 1$:
$x = -1.8$ and $x = 2.8$ (to 1 d.p.)

(ii) $16 + x - x^2 = 0$
$\Rightarrow 16 + x - x^2 - 16 + 6 = -16 + 6$
$\Rightarrow 6 + x - x^2 = -10$,
which has solutions where the graph intersects the line $y = -10$:
$x = -3.5$ and $x = 4.5$ (to 1 d.p.)

(iii) $3 + x - x^2 = 0$
$\Rightarrow 3 + x - x^2 - 3 + 6 = -3 + 6$
$\Rightarrow 6 + x - x^2 = 3$,
which has solutions where the graph intersects the line $y = 3$:
$x = -1.3$ and $x = 2.3$ (to 1 d.p.)

(iv) $10 + x - x^2 = 0$
$\Rightarrow 10 + x - x^2 - 10 + 6 = -10 + 6$
$\Rightarrow 6 + x - x^2 = -4$,
which has solutions where the graph intersects the line $y = -4$:
$x = -2.7$ and $x = 3.7$ (to 1 d.p.)

1 a) Rearrange $x^2 + 4x = 0$ so that one side is $x^2 + 3x$:
$x^2 + 4x = 0 \Rightarrow x^2 + 4x - 4x + 3x = -4x + 3x$
$\Rightarrow x^2 + 3x = -x$
The solutions to $x^2 + 4x = 0$ are found where $y = x^2 + 3x$ and $y = -x$ meet:
$x = 0$ and $x = -4$

b) Rearrange $x^2 + x - 2 = 0$ so that one side is $x^2 + 3x$:
$x^2 + x - 2 = 0$
$\Rightarrow x^2 + x - 2 - x + 2 + 3x = -x + 2 + 3x$
$\Rightarrow x^2 + 3x = 2x + 2$
The solutions to $x^2 + x - 2 = 0$ are found where $y = x^2 + 3x$ and $y = 2x + 2$ meet:
$x = 1$ and $x = -2$

c) Rearrange $x^2 + 2x = 0$ so that one side is $x^2 + 3x$:
$x^2 + 2x = 0 \Rightarrow x^2 + 2x - 2x + 3x = -2x + 3x$
$\Rightarrow x^2 + 3x = x$
The solutions to $x^2 + 2x = 0$ are found where $y = x^2 + 3x$ and $y = x$ meet:
$x = 0$ and $x = -2$
You could have algebraically solved all of the quadratic equations in this question fairly easily by factorising — but the question tells you to solve the equations graphically.

2 a) See below

b) $x^2 - 4x - 1 = 0$
$\Rightarrow x^2 - 4x - 1 + 4x + 1 - 2x = 4x + 1 - 2x$
$\Rightarrow x^2 - 2x = 2x + 1$

c) You already have the graph of $y = x^2 - 2x$ so a suitable straight line would be the right-hand side of part b), i.e. $y = 2x + 1$.

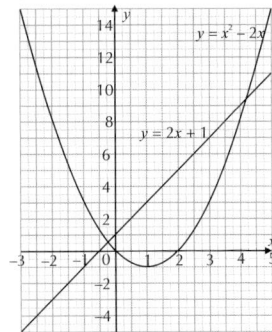

$y = x^2 - 2x$
$y = 2x + 1$

The solutions are the x-values of the points of intersection:
$x = -0.2$ and $x = 4.2$ (to 1 d.p.)

3 a) See below

b) Rearrange $x^2 + 3x + 1 = 0$ so that one side is $x^2 + 2x + 1$:
$x^2 + 3x + 1 = 0$
$\Rightarrow x^2 + 3x + 1 - 3x + 2x = -3x + 2x$
$\Rightarrow x^2 + 2x + 1 = -x$
Sketch the line $y = -x$ on the same set of axes as the quadratic curve.

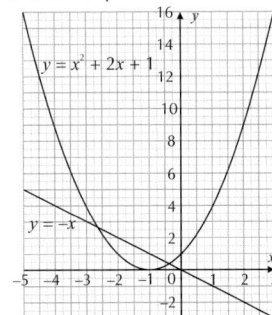

$y = x^2 + 2x + 1$
$y = -x$

The solutions are the x-values of the points of intersection:
$x = -2.6$ and $x = -0.4$ (to 1 d.p.)

4 a) See below

b) Rearrange $5x - x^2 - 5 = 0$ so that one side is $4x - x^2$:
$5x - x^2 - 5 = 0$
$\Rightarrow 5x - x^2 - 5 - 5x + 5 + 4x = -5x + 5 + 4x$
$\Rightarrow 4x - x^2 = -x + 5$
Sketch the line $y = -x + 5$ on the same set of axes as the quadratic curve.

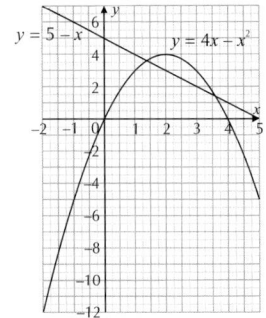

$y = 5 - x$
$y = 4x - x^2$

The solutions are the x-values of the points of intersection:
$x = 1.4$ and $x = 3.6$ (to 1 d.p.)

17.5 Gradients of Curves

It's unlikely you'll be able to draw your graphs as accurately as we have here. Don't worry if your answers don't match up to the ones below, so long as you're following the same method correctly and obtaining estimates that are reasonably close to these ones.

1 a) Draw tangents at the points on the curve where $x = 4$ and $x = -4$, and calculate their gradients.

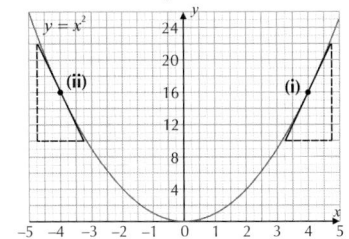

$y = x^2$

(i) E.g. using the points (3.25, 10) and (4.75, 22), the gradient is
$\dfrac{22 - 10}{4.75 - 3.25} = \dfrac{12}{1.5} = \mathbf{8}$.

(ii) E.g. using the points (−3.25, 10) and (−4.75, 22), the gradient is
$\dfrac{22 - 10}{-4.75 - (-3.25)} = \dfrac{12}{-1.5} = \mathbf{-8}$.

b) They are the same apart from their sign.
Since $y = x^2$ is symmetric about the y-axis, the gradient at $-x$ will always be the same as the gradient at x but with the sign changed.

2 a) For $x = 0.1$, $y = \dfrac{1}{0.1} = 10$.

For $x = 0.2$, $y = \dfrac{1}{0.2} = 5$.

Using the same method for the remaining values of x, the table can be completed as follows.

x	0.1	0.2	0.5	1	1.5	2	3
y	10	5	2	1	$\frac{2}{3}$	$\frac{1}{2}$	$\frac{1}{3}$

b) See below

c) Draw tangents at the points on the curve where $x = 2$, $x = 0.2$ and $x = 0.6$, and calculate their gradients.

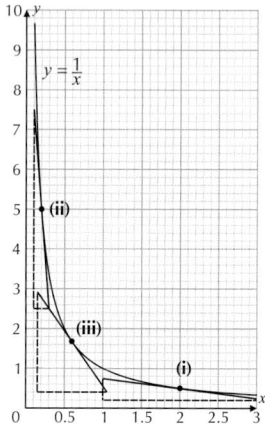

(i) E.g. using the points $(1, 0.7)$ and $(3, 0.2)$, the gradient is
$$\frac{0.2 - 0.7}{3 - 1} = \frac{-0.5}{2} = -\frac{1}{4}.$$

(ii) E.g. using the points $(0.1, 7.5)$ and $(0.3, 2.5)$, the gradient is
$$\frac{7.5 - 2.5}{0.1 - 0.3} = \frac{5}{-0.2} = -25.$$

(iii) E.g. using the points $(0.15, 2.9)$ and $(1.05, 0.4)$, the gradient is
$$\frac{2.9 - 0.4}{0.15 - 1.05} = \frac{2.5}{-0.9} = -\frac{25}{9}.$$

3 a) Use a table of values, e.g:

x	-4	-3	-2	-1	0	1	2	3
y	-24	0	8	6	0	-4	0	18

See below for the graph.

b) Draw tangents at the points on the curve where $x = -1$ and $x = 2$, and calculate their gradients.

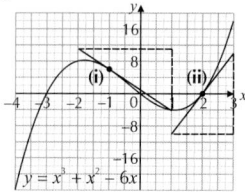

(i) E.g. using the points $(-2, 11)$ and $(1, -4)$, the gradient is
$$\frac{11 - (-4)}{-2 - 1} = \frac{15}{-3} = -5.$$

(ii) E.g. using the points $(1, -10)$ and $(3, 10)$, the gradient is
$$\frac{10 - (-10)}{3 - 1} = \frac{20}{2} = 10.$$

c) The gradient is zero where the tangent to the curve is horizontal, i.e. at the turning points. These occur around $x = -1.8$ (to 1 d.p.) and $x = 1.1$ (to 1 d.p.).

Page 224   Exercise 2

1 a) The radius of the circle is $\sqrt{169} = 13$.

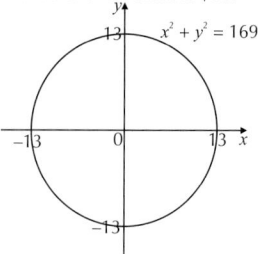

---

b) Using the points $(0, 0)$ and $(5, 12)$, the gradient of the radius is $\frac{12 - 0}{5 - 0} = \frac{12}{5}$.

c) The gradient of the tangent is the negative reciprocal of the gradient of the radius: $-1 \div \frac{12}{5} = -\frac{5}{12}$

d) The equation of the tangent is of the form $y = mx + c$, where the gradient $m = -\frac{5}{12}$. To find c, substitute $(5, 12)$ into the equation:
$$12 = -\frac{5}{12} \times 5 + c \implies c = \frac{169}{12}.$$
So the equation of the tangent is
$$y = -\frac{5}{12}x + \frac{169}{12}.$$

2 a)

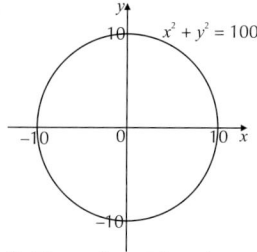

b) (i) The gradient of the radius at $(6, 8)$ is $\frac{8 - 0}{6 - 0} = \frac{4}{3}$.
So the gradient of the tangent is $-\frac{3}{4}$.
Substituting $(6, 8)$ into $y = -\frac{3}{4}x + c$
gives $8 = -\frac{3}{4} \times 6 + c \implies c = \frac{25}{2}$
So the equation of the tangent is
$$y = -\frac{3}{4}x + \frac{25}{2}.$$

(ii) The gradient of the radius at the point $(8, 6)$ is $\frac{6 - 0}{8 - 0} = \frac{3}{4}$.
So the gradient of the tangent is $-\frac{4}{3}$.
Substituting $(8, 6)$ into $y = -\frac{4}{3}x + c$
gives $6 = -\frac{4}{3} \times 8 + c \implies c = \frac{50}{3}$
So the equation of the tangent is
$$y = -\frac{4}{3}x + \frac{50}{3}.$$
Using the same method for (iii)-(iv):

(iii) $y = -\frac{4}{3}x - \frac{50}{3}$ **(iv)** $y = \frac{3}{4}x + \frac{25}{2}$

3 a)

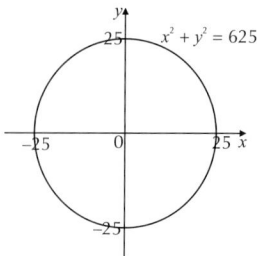

b) (i) The gradient of the radius at $(15, 20)$ is $\frac{20 - 0}{15 - 0} = \frac{4}{3}$.
So the gradient of the tangent is $-\frac{3}{4}$.
Substituting $(15, 20)$ into $y = -\frac{3}{4}x + c$
gives $20 = -\frac{3}{4} \times 15 + c \implies c = \frac{125}{4}$
So the equation of the tangent is
$$y = -\frac{3}{4}x + \frac{125}{4}.$$
Using the same method for (ii)-(iv):

(ii) $y = \frac{7}{24}x + \frac{625}{24}$

(iii) $y = -\frac{3}{4}x - \frac{125}{4}$

(iv) $y = \frac{24}{7}x - \frac{625}{7}$

---

4 a) The gradient of the radius at $(-2\sqrt{2}, 2\sqrt{2})$ is $\frac{2\sqrt{2} - 0}{-2\sqrt{2} - 0} = -1$.
So the gradient of the tangent is 1. Substituting $(-2\sqrt{2}, 2\sqrt{2})$ into $y = x + c$ gives $2\sqrt{2} = -2\sqrt{2} + c \implies c = 4\sqrt{2}$
So the equation of the tangent is
$$y = x + 4\sqrt{2}.$$
Using the same method for b):

b) $y = x - 4\sqrt{2}$

c) E.g. The two tangents have the same gradient so they are parallel.

Page 225   Review Exercise

1 a) The graph initially increases and then decreases so the temperature **rises for the first 3 seconds** and then **decreases for the remaining 5 seconds**. Over the first 3 seconds, the gradient of the graph decreases to zero, so the temperature **rises at a slower rate over time** until it **stops rising**. In the final 5 seconds, the gradient gets steeper and steeper so the temperature **falls at a more rapid rate** as time goes on.

b) The graph is at its highest when $t = 3$. Reading across to the vertical axis, this is at a temperature of **9 °C**.

c) Reading across from 8 °C on the vertical axis and down to the horizontal axis, the times are $t = 2$ **seconds** and $t = 4$ **seconds**.

2 a) Plot a straight line with a vertical axis intercept of £2.40, which increases by £0.60 for every kg of sprouts (i.e. the gradient of the line is 0.6):

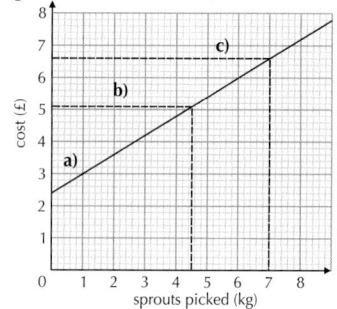

b) Reading up from 4.5 kg on the horizontal axis and across to the vertical axis, the cost is **£5.10**.

c) Reading across from £6.60 on the vertical axis and down to the horizontal axis, the weight is **7 kg**.

3 a), b)

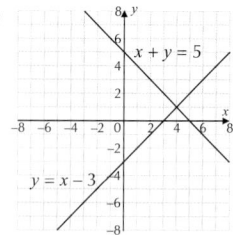

c) The graphs intersect at $(4, 1)$ so the solution to the simultaneous equations is $x = 4$ and $y = 1$.

4 a)

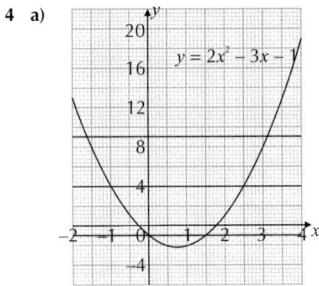

$y = 2x^2 - 3x - 1$

b) (i) The solutions to $2x^2 - 3x - 1 = 0$ occur where the graph crosses the x-axis:
$x = -0.3$ and $x = 1.8$ (to 1 d.p.)

(ii) $2x^2 - 3x - 5 = 0 \Rightarrow 2x^2 - 3x - 1 = 4$
So the solutions to $2x^2 - 3x - 5 = 0$ occur where the graph crosses $y = 4$:
$x = -1$ and $x = 2.5$

(iii) $2x^2 - 3x - 10 = 0 \Rightarrow 2x^2 - 3x - 1 = 9$
So the solutions to $2x^2 - 3x - 10 = 0$ occur where the graph crosses $y = 9$:
$x = -1.6$ and $x = 3.1$ (to 1 d.p.)

(iv) $2x^2 - 3x = 0 \Rightarrow 2x^2 - 3x - 1 = -1$
So the solutions to $2x^2 - 3x = 0$ occur where the graph crosses $y = -1$:
$x = 0$ and $x = 1.5$

5 a)

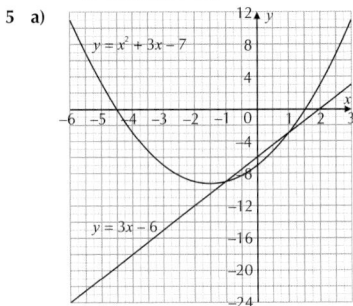

$y = x^2 + 3x - 7$

$y = 3x - 6$

b) $x^2 - 1 = 0$
$\Rightarrow x^2 - 1 + 1 + 3x - 7 = 1 + 3x - 7$
$\Rightarrow x^2 + 3x - 7 = 3x - 6$
So a suitable line to draw would be $y = 3x - 6$ (shown above). Then the solutions to $x^2 - 1 = 0$ occur where the graphs intersect: $x = -1$ and $x = 1$

6 Draw tangents at the points on the curve where $x = -3$ and $x = 2$ and calculate their gradients.

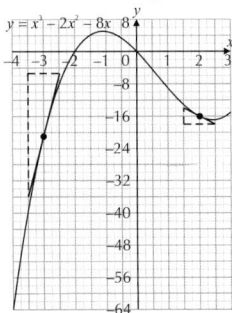

$y = x^3 - 2x^2 - 8x$

a) E.g. use the points $(-3.5, -36)$ and $(-2.5, -5)$ to find the gradient of the tangent:
$\frac{-36 - (-5)}{-3.5 - (-2.5)} = \frac{-31}{-1} = 31$

b) E.g. use the points $(1.5, -14)$ and $(2.5, -18)$ to find the gradient of the tangent:
$\frac{-18 - (-14)}{2.5 - 1.5} = \frac{-4}{1} = -4$

---

7 The equation of the tangent is $y = mx + c$, where m is the gradient and $c = 5\sqrt{2}$ is the y-intercept.
Substituting $(10\sqrt{2}, 0)$ into the equation:
$0 = 10\sqrt{2}m + 5\sqrt{2} \Rightarrow m = -\frac{1}{2}$
The radius between $(0, 0)$ and the point of contact of this tangent is given by the equation
$y = -\frac{1}{m}x$ (since its gradient is the negative reciprocal of the tangent and the y-intercept is the origin). So the radius is given by $y = 2x$.
Find the point of contact by putting the two equations equal to one another:
$-\frac{1}{2}x + 5\sqrt{2} = 2x \Rightarrow x = 2\sqrt{2}$
$\Rightarrow y = 2 \times 2\sqrt{2} = 4\sqrt{2}$
Now substitute this point into the equation of the circle to find r:
$(2\sqrt{2})^2 + (4\sqrt{2})^2 = 8 + 32 = 40$
$\Rightarrow r = \sqrt{40} = 2\sqrt{10}$

Page 226 Exam-Style Questions

1 a) For $t = 0$, $x = 2 \times 0^2 + 0 = 0$.
For $t = 1$, $x = 2 \times 1^2 + 1 = 3$.
Using the same method for the remaining values of t, a table of values can be completed as follows.

t	0	1	2	3	4	5
x	0	3	10	21	36	55

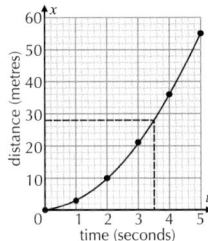

[2 marks available — 1 mark for five points plotted correctly, 1 mark for a smooth curve joining them]

b) Reading up from 3.5 seconds on the horizontal axis and across to the vertical axis, the distance is **28 metres** *[1 mark]*.

c) The graph gets steeper and steeper *[1 mark]* so the speed is increasing with time *[1 mark]*.

2 a) Reading up from 100 people on the horizontal axis and across to the vertical axis, they'd earn **£17 500** *[1 mark]*.

b) The formula will be the equation of the graph in the form $a = mp + c$, where m is the gradient and c is the intercept with the vertical axis.
Find the gradient using two known points, e.g. $(0, 10\,000)$ and $(100, 17\,500)$:
$\frac{17\,500 - 10\,000}{100 - 0}$ *[1 mark]* $= \frac{7500}{100} = 75$.
The intercept is $10\,000$ *[1 mark]* so the formula is $a = 75p + 10\,000$ *[1 mark]*.

3 Until the water reaches the level of the centre of the sphere, the rate of increase of depth will be decreasing. For the rest of the sphere part, this rate will be increasing. The cylinder will fill quickly at a constant rate.

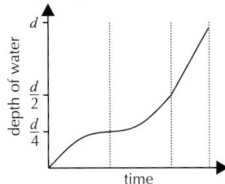

[3 marks available — 1 mark for approximately correct shape for each of the three sections of the graph. The third mark is only obtained if the time taken for the first two sections is equal, and longer than the third]

4

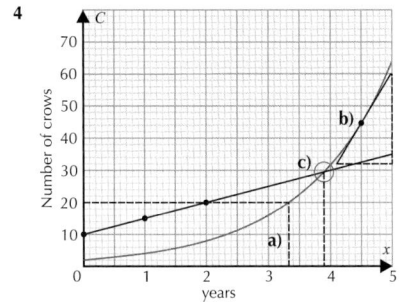

a) Reading across from 20 on the vertical axis and down to the horizontal axis, the answer is **3.3 years** (1 d.p.) *[1 mark]*.

b) The rate of change is given by the gradient of the curve at $x = 4.5$, which is the same as the gradient of the tangent at that point.
Using the points e.g. $(4.1, 32)$ and $(5, 60)$, the gradient is $\frac{60 - 32}{5 - 4.1}$ *[1 mark]* $= \frac{28}{0.9}$
≈ 31 *[1 mark]*
So the rate of change at this point is 31 crows per year. Allow the mark for 29-33 crows per year.

c) Plot the line $T = 5x + 10$. To do so, work out three points on the line — e.g:

x	0	2	4
T	10	20	30

Then the solution to the equation $T = C$ is the x-value of the point of intersection: **3.9 years** (to 1 d.p.).
[2 marks available — 1 mark for correctly plotting the line $T = 5x + 10$, 1 mark for the correct solution]

5 The gradient of the radius at $(4, 3)$ is $\frac{3}{4}$ so the gradient of the tangent is $-\frac{4}{3}$. Then the equation of the tangent is $y = -\frac{4}{3}x + c$.
Find c by substituting $(4, 3)$ into the equation:
$3 = -\frac{16}{3} + c \Rightarrow c = \frac{25}{3}$.
So, the equation of the tangent at $(4, 3)$ is $y = -\frac{4}{3}x + \frac{25}{3}$.
Using the same method, the equation of the tangent at $(-4, 3)$ is $y = \frac{4}{3}x + \frac{25}{3}$.
To find where the two lines intersect, set them equal to one another:
$-\frac{4}{3}x + \frac{25}{3} = \frac{4}{3}x + \frac{25}{3} \Rightarrow \frac{8}{3}x = 0 \Rightarrow x = 0$
Then $y = -\frac{4}{3} \times 0 + \frac{25}{3} = \frac{25}{3}$.
So the point of intersection is $\left(0, \frac{25}{3}\right)$.
[5 marks available — 1 mark for finding the gradient of one of the tangents, 1 mark for finding the equation of one of the tangents, 1 mark for setting both equations equal to one another, 1 mark for the correct x-coordinate, 1 mark for the correct y-coordinate]

6 a) $x^2 + 7x - 6 = 0$
$\Rightarrow x^2 + 7x - 6 + 6 - 3 = 6 - 3$
$\Rightarrow x^2 + 7x - 3 = 3$
so he needs to plot $y = 3$ *[1 mark]*

b) $x^2 + 5x + 8 = 0$
$\Rightarrow x^2 + 5x + 8 - 5x - 8 + 7x - 3$
$ = -5x - 8 + 7x - 3$ *[1 mark]*
$\Rightarrow x^2 + 7x - 3 = 2x - 11$
so he needs to plot $y = 2x - 11$ *[1 mark]*

Section 18 — Functions

18.1 Evaluating Functions

Page 228   Exercise 1

1 **a)** (i) $f(11) = 11 - 5 = \textbf{6}$
   (ii) $f(36) = 36 - 5 = \textbf{31}$
   (iii) $f(4) = 4 - 5 = \textbf{-1}$

   b) (i) $f(0) = 4 \times 0 = \textbf{0}$
   (ii) $f(7) = 4 \times 7 = \textbf{28}$
   (iii) $f(1.5) = 4 \times 1.5 = \textbf{6}$

   c) (i) $g(3) = 5(3) - 7 = \textbf{8}$
   (ii) $g(6) = 5(6) - 7 = \textbf{23}$
   (iii) $g(-1) = 5(-1) - 7 = \textbf{-12}$

   d) (i) $f(3) = 11 - 2(3) = \textbf{5}$
   (ii) $f(5.5) = 11 - 2(5.5) = \textbf{0}$
   (iii) $f(-5) = 11 - 2(-5) = \textbf{21}$

2 **a)** (i) $f(7) = 7^2 - 4(7) = \textbf{21}$
   (ii) $f(-3) = (-3)^2 - 4(-3) = \textbf{21}$
   (iii) $f(1.5) = 1.5^2 - 4(1.5) = \textbf{-3.75}$

   b) (i) $g(2) = 20 - 3(2^2) = \textbf{8}$
   (ii) $g(3) = 20 - 3(3^2) = \textbf{-7}$
   (iii) $g(-1) = 20 - 3((-1)^2) = \textbf{17}$

   c) (i) $h(5) = 2(5^2) - 5 + 4 = \textbf{49}$
   (ii) $h(-2) = 2((-2)^2) - (-2) + 4 = \textbf{14}$
   (iii) $h(8) = 2(8^2) - 8 + 4 = \textbf{124}$

   d) (i) $f(2) = (4(2) - 3)^2 = \textbf{25}$
   (ii) $f(0) = (4(0) - 3)^2 = \textbf{9}$
   (iii) $f(0.75) = (4(0.75) - 3)^2 = \textbf{0}$

3 **a)** (i) $g(10) = \dfrac{20}{10} = \textbf{2}$
   (ii) $g(1.25) = \dfrac{20}{1.25} = \textbf{16}$
   (iii) $g(-0.5) = \dfrac{20}{-0.5} = \textbf{-40}$

   b) (i) $g(0) = \dfrac{18}{0^2 + 2} = \textbf{9}$
   (ii) $g(2) = \dfrac{18}{2^2 + 2} = \textbf{3}$
   (iii) $g(-4) = \dfrac{18}{(-4)^2 + 2} = \textbf{1}$

   c) (i) $h(0) = \sqrt{\dfrac{0}{3(0) + 1}} = \textbf{0}$
   (ii) $h(1) = \sqrt{\dfrac{1}{3(1) + 1}} = \dfrac{\textbf{1}}{\textbf{2}}$
   (iii) $h(16) = \sqrt{\dfrac{16}{3(16) + 1}} = \dfrac{\textbf{4}}{\textbf{7}}$

   d) (i) $f(0) = 8^0 - 1 = \textbf{0}$
   (ii) $f\left(\dfrac{1}{3}\right) = 8^{\frac{1}{3}} - 1 = \textbf{1}$
   (iii) $f\left(-\dfrac{2}{3}\right) = 8^{-\frac{2}{3}} - 1 = -\dfrac{\textbf{3}}{\textbf{4}}$

Page 229   Exercise 2

1 **a)** $2x - 5 = 35 \Rightarrow 2x = 40 \Rightarrow \textbf{\textit{x} = 20}$
   b) $8 - 3x = -10 \Rightarrow -3x = -18 \Rightarrow \textbf{\textit{x} = 6}$
   c) $9x + 12 = 93 \Rightarrow 9x = 81 \Rightarrow \textbf{\textit{x} = 9}$

2 **a)** $x^2 = 225 \Rightarrow \textbf{\textit{x} = ±15}$
   b) $\sqrt{2x - 1} = 3 \Rightarrow 2x - 1 = 9 \Rightarrow \textbf{\textit{x} = 5}$
   c) $\dfrac{18}{x + 1} = 2 \Rightarrow 18 = 2(x + 1)$
   $\Rightarrow 9 = x + 1 \Rightarrow \textbf{\textit{x} = 8}$

Page 229   Exercise 3

1 **a)** (i) $f(k) = \textbf{2\textit{k} + 1}$
   (ii) $f(2m) = 2(2m) + 1 = \textbf{4\textit{m} + 1}$
   (iii) $f(3w - 1) = 2(3w - 1) + 1 = \textbf{6\textit{w} - 1}$

   b) (i) $f(u) = \textbf{2\textit{u}}^2 + \textbf{3\textit{u}}$
   (ii) $f(3a) = 2(3a)^2 + 3(3a) = \textbf{18\textit{a}}^2 + \textbf{9\textit{a}}$
   (iii) $f(t^2) = 2(t^2)^2 + 3(t^2) = \textbf{2\textit{t}}^4 + \textbf{3\textit{t}}^2$

   c) (i) $f(t) = \dfrac{\textbf{4\textit{t} - 1}}{\textbf{\textit{t} + 4}}$
   (ii) $f(-x) = \dfrac{4(-x) - 1}{(-x) + 4} = \dfrac{\textbf{-4\textit{x} - 1}}{\textbf{4 - \textit{x}}}$
   (iii) $f(2x) = \dfrac{4(2x) - 1}{(2x) + 4} = \dfrac{\textbf{8\textit{x} - 1}}{\textbf{2\textit{x} + 4}}$

   d) (i) $f(w) = \sqrt{\textbf{5\textit{w} - 1}}$
   (ii) $f(3x) = \sqrt{5(3x) - 1} = \sqrt{\textbf{15\textit{x} - 1}}$
   (iii) $f(1 - 2x) = \sqrt{5(1 - 2x) - 1} = \sqrt{\textbf{4 - 10\textit{x}}}$

18.2 Composite Functions

Page 230   Exercise 1

1 **a)** $g(1) = 1 + 2 = \textbf{3}$
   b) $fg(1) = f(3) = 2(3) = \textbf{6}$
   c) $f(5) = 2(5) = \textbf{10}$
   d) $gf(5) = g(10) = 10 + 2 = \textbf{12}$

2 **a)** (i) $fg(7) = f(4 \times 7) = f(28) = 28 - 5 = \textbf{23}$
   (ii) $gf(8) = g(8 - 5) = g(3) = 4 \times 3 = \textbf{12}$
   (iii) $gf(x) = g(x - 5) = \textbf{4(\textit{x} - 5)}$ or $\textbf{4\textit{x} - 20}$

   b) (i) $gf(0) = g(2(0) - 1) = g(-1) = 3(-1) + 1$
   $= \textbf{-2}$
   (ii) $fg(2) = f(3(2) + 1) = f(7) = 2(7) - 1 = \textbf{13}$
   (iii) $fg(x) = f(3x + 1) = 2(3x + 1) - 1$
   $= 6x + 2 - 1 = \textbf{6\textit{x} + 1}$

   c) (i) $fg(4) = f\left(\dfrac{4}{4}\right) = f(1) = 5(1) - 4 = \textbf{1}$
   (ii) $gf(4) = g(5(4) - 4) = g(16) = \left(\dfrac{16}{4}\right) = \textbf{4}$
   (iii) $gf(x) = g(5x - 4) = \dfrac{\textbf{5\textit{x} - 4}}{\textbf{4}}$

   d) (i) $fg(3) = f(5 \times 3) = f(15) = 4(15) + 3 = \textbf{63}$
   (ii) $ff(2) = f(4(2) + 3) = f(11) = 4(11) + 3$
   $= \textbf{47}$
   (iii) $gf(x) = g(4x + 3) = 5(4x + 3) = \textbf{20\textit{x} + 15}$

   e) (i) $gf(3) = g(2(3) + 1) = g(7) = 11 - 7 = \textbf{4}$
   (ii) $gg(4) = g(11 - 4) = g(7) = 11 - 7 = \textbf{4}$
   (iii) $fg(x) = f(11 - x) = 2(11 - x) + 1$
   $= \textbf{23 - 2\textit{x}}$

   f) (i) $fg(3) = f(3(3) - 7) = f(2) = \dfrac{\textbf{1}}{\textbf{2}}$
   (ii) $fg(x) = f(3x - 7) = \dfrac{\textbf{1}}{\textbf{3\textit{x} - 7}}$
   (iii) $gf(x) = g\left(\dfrac{1}{x}\right) = 3\left(\dfrac{1}{x}\right) - 7 = \dfrac{\textbf{3}}{\textbf{\textit{x}}} - \textbf{7}$

3 **a)** (i) $gf(8) = g(10 - 8) = g(2) = 2^2 = \textbf{4}$
   (ii) $ff(2) = f(10 - 2) = f(8) = 10 - 8 = \textbf{2}$
   (iii) $fg(x) = g(10 - x) = \textbf{(10 - \textit{x})}^2$
   or $= \textbf{100 - 20\textit{x} + \textit{x}}^2$

   b) (i) $ff(16) = f(\sqrt{16}) = f(4) = \sqrt{4} = \textbf{2}$
   (ii) $fg(40) = f(2(40) + 1) = f(81) = \sqrt{81} = \textbf{9}$
   (iii) $gf(x) = g(\sqrt{x}) = \textbf{2}\sqrt{\textbf{\textit{x}}} + \textbf{1}$

   c) (i) $gf(-1) = g((-1)^2 + 5) = g(6) = 3(6) - 4$
   $= \textbf{14}$
   (ii) $gf(x) = g(x^2 + 5) = 3(x^2 + 5) - 4$
   $= \textbf{3\textit{x}}^2 + \textbf{11}$
   (iii) $gg(x) = g(3x - 4) = 3(3x - 4) - 4$
   $= \textbf{9\textit{x} - 16}$

   d) (i) $gf(8) = g(\sqrt{3(8) + 1}) = g(5) = 12 - 2(5)$
   $= \textbf{2}$
   (ii) $gg(x) = g(12 - 2x) = 12 - 2(12 - 2x)$
   $= \textbf{4\textit{x} - 12}$
   (iii) $fg(x) = f(12 - 2x)$
   $= \sqrt{3(12 - 2x) + 1} = \sqrt{\textbf{37 - 6\textit{x}}}$

Page 231   Exercise 2

1 **a)** (i) $hgf(-1) = hg\left(\dfrac{1}{3(-1) + 2}\right) = hg(-1)$
   $= h(2(-1) - 5) = h(-7) = (-7)^2 = \textbf{49}$
   (ii) $fg(x) = f(2x - 5) = \dfrac{1}{3(2x - 5) + 2}$
   $= \dfrac{\textbf{1}}{\textbf{6\textit{x} - 13}}$
   (iii) $hf(x) = h\left(\dfrac{1}{3x + 2}\right) = \left(\dfrac{1}{3x + 2}\right)^2$
   $= \dfrac{\textbf{1}}{\textbf{(3\textit{x} + 2)}^2}$

   b) (i) $fgh(4) = fg(10 - 2(4)) = fg(2) = f(6 - 2)$
   $= f(4) = 4^3 = \textbf{64}$
   (ii) $gh(x) = g(10 - 2x) = 6 - (10 - 2x)$
   $= \textbf{2\textit{x} - 4}$
   (iii) $fh(x) = f(10 - 2x) = \textbf{(10 - 2\textit{x})}^3$

   c) (i) $fh(25) = f(\sqrt{25}) = f(5) = 2(5) + 4 = \textbf{14}$
   (ii) $fg(0.5) = f\left(\dfrac{1}{3(0.5) + 1}\right) = f(0.4)$
   $= 2(0.4) + 4 = \textbf{4.8}$

(iii) $hgf(x) = hg(2x + 4)$
$= h\left(\dfrac{1}{3(2x + 4) + 1}\right)$
$= h\left(\dfrac{1}{6x + 13}\right) = \sqrt{\dfrac{1}{6x + 13}}$
$= \dfrac{\textbf{1}}{\sqrt{\textbf{6\textit{x} + 13}}}$

   d) (i) $ff(3) = f(3^2 + 3) = f(12) = 12^2 + 12 = \textbf{156}$
   (ii) $gf(-1) = f((-1)^2 + (-1)) = g(0) = \dfrac{1}{2(0) + 3}$
   $= \dfrac{\textbf{1}}{\textbf{3}}$
   (iii) $hgf(x) = hg(x^2 + x) = h\left(\dfrac{1}{2(x^2 + x) + 3}\right)$
   $= \textbf{2\textit{x}}^2 + \textbf{2\textit{x} + 3}$

Page 231   Exercise 3

1 **a)** (i) $fg(x) = f(3x) = \textbf{9 - 3\textit{x}}$
   (ii) $fg(x) = -12 \Rightarrow 9 - 3x = -12$
   $\Rightarrow -3x = -21 \Rightarrow \textbf{\textit{x} = 7}$

   b) (i) $gf(x) = g(9 - x) = 3(9 - x) = \textbf{27 - 3\textit{x}}$
   (ii) $gf(x) = 6 \Rightarrow 27 - 3x = 6 \Rightarrow 21 = 3x$
   $\Rightarrow \textbf{\textit{x} = 7}$

2 **a)** $fg(x) = f\left(\dfrac{x}{2}\right) = \dfrac{x}{2} + 5$, so $fg(x) = 14$
   $\Rightarrow \dfrac{x}{2} + 5 = 14 \Rightarrow \dfrac{x}{2} = 9 \Rightarrow \textbf{\textit{x} = 18}$

   b) $gf(x) = g(x + 5) = \dfrac{x + 5}{2}$, so $gf(x) = 8$
   $\Rightarrow \dfrac{x + 5}{2} = 8 \Rightarrow x + 5 = 16 \Rightarrow \textbf{\textit{x} = 11}$

   c) $gg(x) = g\left(\dfrac{x}{2}\right) = \dfrac{\left(\frac{x}{2}\right)}{2} = \dfrac{x}{4}$, so $gg(x) = 11$
   $\Rightarrow \dfrac{x}{4} = 11 \Rightarrow \textbf{\textit{x} = 44}$

3 **a)** $fg(x) = f(2x^2 - 1) = 6 - (2x^2 - 1) = 7 - 2x^2$
   So $fg(x) = 7 \Rightarrow 7 - 2x^2 = 7 \Rightarrow -2x^2 = 0$
   $\Rightarrow \textbf{\textit{x} = 0}$

   b) $gf(x) = g(6 - x) = 2(6 - x)^2 - 1$
   $= 71 - 24x + 2x^2$
   So $gf(x) = 31 \Rightarrow 71 - 24x + 2x^2 = 31$
   $\Rightarrow 2x^2 - 24x + 40 = 0$
   $\Rightarrow x^2 - 12x + 20 = 0$
   $\Rightarrow (x - 2)(x - 10) = 0 \Rightarrow \textbf{\textit{x} = 2 or \textit{x} = 10}$

   c) $gg(x) = g(2x^2 - 1) = 2(2x^2 - 1)^2 - 1$
   $= 8x^4 - 8x^2 + 1$
   So $gg(x) = 97 \Rightarrow 8x^4 - 8x^2 + 1 = 97$
   $\Rightarrow 8x^4 - 8x^2 = 96 \Rightarrow x^4 - x^2 = 12$
   $\Rightarrow x^4 - x^2 - 12 = 0 \Rightarrow (x^2 + 3)(x^2 - 4) = 0$
   So $x^2 = -3$, which has no real solutions, or
   $x^2 = 4$, so $\textbf{\textit{x} = 2 or \textit{x} = -2}$
   If you're struggling to factorise $x^4 - x^2 - 12 = 0$,
   you can break it down into more steps like this:
   Let $y = x^2$, so it becomes $y^2 - y - 12 = 0$ —
   which is a normal quadratic to factorise. After
   solving to find values for y, replace y with x^2 and
   take square roots to find the value(s) of x.

18.3 Inverse Functions

Page 232   Exercise 1

1 **a)** The reverse of +4 is -4 so $f^{-1}(x) = \textbf{\textit{x} - 4}$
   b) The reverse of -3 is +3 so $f^{-1}(x) = \textbf{\textit{x} + 3}$
   c) The reverse of -7 is +7 so $f^{-1}(x) = \textbf{\textit{x} + 7}$
   d) The reverse of +1 is -1 so $f^{-1}(x) = \textbf{\textit{x} - 1}$
   e) The reverse of ×8 is ÷8 so $f^{-1}(x) = \dfrac{\textbf{\textit{x}}}{\textbf{8}}$
   f) The reverse of ×2 is ÷2 so $g^{-1}(x) = \dfrac{\textbf{\textit{x}}}{\textbf{2}}$
   g) The reverse of ÷3 is ×3 so $g^{-1}(x) = \textbf{3\textit{x}}$
   h) The reverse of ÷6 is ×6 so $f^{-1}(x) = \textbf{6\textit{x}}$

2 **a)** $f(x)$ multiplies by 4 then adds 3, so the
   inverse subtracts 3 then divides by 4:
   $f^{-1}(x) = \dfrac{\textbf{\textit{x} - 3}}{\textbf{4}}$

   b) $f(t)$ multiplies by 2 then subtracts 9, so the
   inverse adds 9 then divides by 2:
   $f^{-1}(t) = \dfrac{\textbf{\textit{t} + 9}}{\textbf{2}}$

   Using the same method for c)-h):
   c) $g^{-1}(x) = \dfrac{\textbf{\textit{x} + 5}}{\textbf{3}}$ **d)** $f^{-1}(x) = \dfrac{\textbf{\textit{x} - 11}}{\textbf{8}}$
   e) $f^{-1}(x) = \textbf{5(\textit{x} + 7)}$ **f)** $g^{-1}(x) = \textbf{8(\textit{x} - 1)}$
   g) $f^{-1}(t) = \textbf{2\textit{t} + 3}$ **h)** $h^{-1}(x) = \textbf{4\textit{x} - 15}$

3 a) To reverse the operations of f, multiply by 5, subtract 6, then divide by 2:
$$f^{-1}(x) = \frac{5x - 6}{2}$$

b) To reverse the operations of g, multiply by 4, add 1, then divide by 3: $g^{-1}(x) = \frac{4x + 1}{3}$

c) To reverse the operations of f, add 3, then take the square root: $f^{-1}(x) = \sqrt{x + 3}$

d) To reverse the operations of g, take the square root, subtract 7, then divide by 2:
$$g^{-1}(x) = \frac{\sqrt{x} - 7}{2}$$
In parts c)-d), you don't need to worry about the negative square roots because functions only ever map to one value.

Page 233   Exercise 2

1 a) Let $x = f(y)$, so $x = \frac{y}{5} - 8 \Rightarrow x + 8 = \frac{y}{5}$
$\Rightarrow 5(x + 8) = y$. So $f^{-1}(x) = 5(x + 8)$

b) Let $x = f(y)$, so $x = \frac{3y + 1}{5} \Rightarrow 5x = 3y + 1$
$\Rightarrow 5x - 1 = 3y \Rightarrow \frac{5x - 1}{3} = y$.
So $f^{-1}(x) = \frac{5x - 1}{3}$

c) Let $x = g(y)$, so $x = \frac{2y}{5} - 7 \Rightarrow x + 7 = \frac{2y}{5}$
$\Rightarrow 5(x + 7) = 2y \Rightarrow \frac{5(x + 7)}{2} = y$.
So $g^{-1}(x) = \frac{5(x + 7)}{2}$

2 a) Let $x = f(y)$, so $x = 7 - 3y \Rightarrow 3y = 7 - x$
$\Rightarrow y = \frac{7 - x}{3}$. So $f^{-1}(x) = \frac{7 - x}{3}$

b) Let $x = f(y)$, so $x = \frac{9 - 7y}{4} \Rightarrow 4x = 9 - 7y$
$\Rightarrow 7y = 9 - 4x \Rightarrow y = \frac{9 - 4x}{7}$.
So $f^{-1}(x) = \frac{9 - 4x}{7}$

c) Let $x = h(y)$, so $x = \frac{1 - 6y}{9} \Rightarrow 9x = 1 - 6y$
$\Rightarrow 6y = 1 - 9x \Rightarrow y = \frac{1 - 9x}{6}$.
So $h^{-1}(x) = \frac{1 - 9x}{6}$

3 a) Let $x = g(y)$, so $x = 4y^2 + 1 \Rightarrow x - 1 = 4y^2$
$\Rightarrow \frac{x - 1}{4} = y^2 \Rightarrow \sqrt{\frac{x - 1}{4}} = y$.
So $g^{-1}(x) = \sqrt{\frac{x - 1}{4}}$

b) Let $x = g(y)$, so $x = (3y - 1)^2 \Rightarrow \sqrt{x} = 3y - 1$
$\Rightarrow \sqrt{x} + 1 = 3y \Rightarrow \frac{\sqrt{x} + 1}{3} = y$.
So $g^{-1}(x) = \frac{\sqrt{x} + 1}{3}$

c) Let $x = h(y)$, so $x = \frac{(1 - 2y)^2}{5}$
$\Rightarrow 5x = (1 - 2y)^2 \Rightarrow \sqrt{5x} = 1 - 2y$
$\Rightarrow 2y = 1 - \sqrt{5x} \Rightarrow y = \frac{1 - \sqrt{5x}}{2}$.
So $h^{-1}(x) = \frac{1 - \sqrt{5x}}{2}$

4 a) Let $x = f(y)$, so $x = 6\sqrt{y} + 1 \Rightarrow x - 1 = 6\sqrt{y}$
$\Rightarrow \frac{x - 1}{6} = \sqrt{y} \Rightarrow \left(\frac{x - 1}{6}\right)^2 = y$.
So $f^{-1}(x) = \left(\frac{x - 1}{6}\right)^2$

b) Let $x = g(y)$, so $x = \sqrt{19 - 2y} \Rightarrow x^2 = 19 - 2y$
$\Rightarrow 2y = 19 - x^2 \Rightarrow y = \frac{19 - x^2}{2}$.
So $g^{-1}(x) = \frac{19 - x^2}{2}$

c) Let $x = f(y)$, so $x = 25 - 4\sqrt{y}$
$\Rightarrow 4\sqrt{y} = 25 - x \Rightarrow \sqrt{y} = \frac{25 - x}{4}$
$\Rightarrow y = \left(\frac{25 - x}{4}\right)^2$. So $f^{-1}(x) = \left(\frac{25 - x}{4}\right)^2$

5 a) Let $x = f(y)$, so $x = \frac{4}{y} - 7 \Rightarrow x + 7 = \frac{4}{y}$
$\Rightarrow y = \frac{4}{x + 7}$. So $f^{-1}(x) = \frac{4}{x + 7}$

b) Let $x = g(y)$, so $x = 8 - \frac{2}{\sqrt{y}} \Rightarrow \frac{2}{\sqrt{y}} = 8 - x$
$\Rightarrow \sqrt{y} = \frac{2}{8 - x} \Rightarrow y = \left(\frac{2}{8 - x}\right)^2$.
So $g^{-1}(x) = \left(\frac{2}{8 - x}\right)^2$

---

c) Let $x = f(y)$, so $x = \sqrt{\frac{2}{y - 1}} \Rightarrow x^2 = \frac{2}{y - 1}$
$\Rightarrow y - 1 = \frac{2}{x^2} \Rightarrow y = \frac{2}{x^2} + 1$.
So $f^{-1}(x) = \frac{2}{x^2} + 1$

6 a) $x = \frac{1 - 2y}{3y + 5} \Rightarrow x(3y + 5) = 1 - 2y$
$\Rightarrow 3xy + 5x = 1 - 2y \Rightarrow 3xy + 2y = 1 - 5x$

b) $3xy + 2y = y(3x + 2)$
So $y(3x + 2) = 1 - 5x \Rightarrow y = \frac{1 - 5x}{3x + 2}$
$\Rightarrow g^{-1}(x) = \frac{1 - 5x}{3x + 2}$

7 a) Let $x = f(y)$, so $x = \frac{2 + 3y}{y - 2}$
$\Rightarrow x(y - 2) = 2 + 3y \Rightarrow xy - 2x = 2 + 3y$
$\Rightarrow y(x - 3) = 2x + 2 \Rightarrow y = \frac{2x + 2}{x - 3}$
So $f^{-1}(x) = \frac{2x + 2}{x - 3}$
Using the same method for b)-c):

b) $g^{-1}(x) = \frac{7x + 1}{2x - 4}$ **c)** $f^{-1}(x) = \frac{2x}{1 - 3x}$
The trick with Q7 is to get all the terms involving y on one side of the equation, then take y out as a factor to get it on its own.

Page 234   Review Exercise

1 a) $f(8) = 5(8) - 7 = 33$

b) $fg(1) = f(2(1) + 5) = f(7) = 5(7) - 7 = 28$

c) $5x - 7 = 2x + 5 \Rightarrow 3x = 12 \Rightarrow x = 4$

d) $g(x)$ multiplies by 2 then adds 5, so the inverse subtracts 5 then divides by 2:
$$g^{-1}(x) = \frac{x - 5}{2}$$

2 a) $4x + 1 = 25 \Rightarrow 4x = 24 \Rightarrow x = 6$

b) $fg(x) = f(2x + 5) = 4(2x + 5) + 1 = 8x + 21$

c) $f(x)$ multiplies by 4 then adds 1, so the inverse subtracts 1 then divides by 4:
$$f^{-1}(x) = \frac{x - 1}{4}$$

3 a) $gf(x) = g(\sqrt{x + 1}) = 4\sqrt{x + 1}$

b) To reverse the operations of f, square then subtract 1: $f^{-1}(x) = x^2 - 1$

4 a) $2x - 3 = 5 \Rightarrow 2x = 8 \Rightarrow x = 4$

b) $f(x)$ multiplies by 2 then subtracts 3, so the inverse adds 3 then divides by 2:
$$f^{-1}(x) = \frac{x + 3}{2}$$

c) $2x - 3 = \frac{x + 3}{2} \Rightarrow 4x - 6 = x + 3$
$\Rightarrow 3x = 9 \Rightarrow x = 3$

d) $ff(x) = f(2x - 3) = 2(2x - 3) - 3 = 4x - 9$

5 a) $2x - 3 = 18 - 3x \Rightarrow 5x = 21 \Rightarrow x = 4.2$

b) $gf(x) = g(2x - 3) = 18 - 3(2x - 3) = 27 - 6x$

c) Let $x = g(y)$, so $x = 18 - 3y \Rightarrow 3y = 18 - x$
$\Rightarrow y = \frac{18 - x}{3}$. So $g^{-1}(x) = \frac{18 - x}{3}$

6 a) $g(-3) = \frac{1}{(-3) + 5} = \frac{1}{2}$

b) $fg(x) = f\left(\frac{1}{x + 5}\right) = \frac{1}{\frac{1}{x + 5}} - 5 = x + 5 - 5 = x$
f and g are **inverses** of each other.

7 a) $gg(4) = g(3(4) - 1) = g(11) = 3(11) - 1 = 32$

b) $g(x)$ multiplies by 3 then subtracts 1, so the inverse adds 1 then divides by 3:
$$g^{-1}(x) = \frac{x + 1}{3}$$

c) $fg^{-1}(x) = f\left(\frac{x + 1}{3}\right) = 2\left(\frac{x + 1}{3}\right)^2 + 5$
$fg^{-1}(x) = 55 \Rightarrow 2\left(\frac{x + 1}{3}\right)^2 + 5 = 55$
$\Rightarrow 2\left(\frac{x + 1}{3}\right)^2 = 50 \Rightarrow \left(\frac{x + 1}{3}\right)^2 = 25$
$\Rightarrow \frac{x + 1}{3} = \pm 5 \Rightarrow x + 1 = \pm 15$
$\Rightarrow x = 14$ or $x = -16$

8 a) $f(5) = \frac{5}{5 - 3} = \frac{5}{2} = 2.5$

b) Let $x = f(y)$, so $x = \frac{y}{y - 3} \Rightarrow x(y - 3) = y$
$\Rightarrow xy - 3x = y \Rightarrow y(x - 1) = 3x$
$\Rightarrow y = \frac{3x}{x - 1}$. So $f^{-1}(x) = \frac{3x}{x - 1}$

---

c) $fg(x) = f(x^2 + 3) = \frac{x^2 + 3}{x^2 + 3 - 3} = \frac{x^2 + 3}{x^2}$
$= 1 + \frac{3}{x^2}$

9 a) $f(4) = 5(4^2) = 80$

b) $5t^2 = 12.8 \Rightarrow t^2 = 2.56 \Rightarrow t = \pm 1.6$, but time cannot be negative, so $t = 1.6$

10 a) $f(20) = \frac{9(20)}{5} + 32 = 36 + 32 = 68$

b) $\frac{9t}{5} + 32 = 60.8 \Rightarrow \frac{9t}{5} = 28.8 \Rightarrow 9t = 144$
$\Rightarrow t = 16$

c) Let $t = f(y)$, so $t = \frac{9y}{5} + 32 \Rightarrow t - 32 = \frac{9y}{5}$
$\Rightarrow 5(t - 32) = 9y \Rightarrow y = \frac{5(t - 32)}{9}$.
So $f^{-1}(t) = \frac{5(t - 32)}{9}$

d) a) 20 °C is equivalent to 68 °F.
   b) 60.8 °F is equivalent to 16 °C.
   c) The inverse function converts temperatures from °F to °C.

11 a) $f(4) = 60 - \frac{7}{2}(4) = 60 - 14 = 46$

b) A height of 7000 ft means $h = 7$, so
$f(7) = 60 - \frac{7}{2}(7) = 60 - 24.5 = 35.5$ °F

c) Let $h = f(y)$, so $h = 60 - \frac{7}{2}y \Rightarrow \frac{7}{2}y = 60 - h$
$\Rightarrow y = \frac{2}{7}(60 - h)$, so $f^{-1}(h) = \frac{2}{7}(60 - h)$
$f^{-1}(32) = \frac{2}{7}(60 - 32) = \frac{2}{7}(28) = 8$.
At 8000 feet, the temperature is 32 °F.

Page 235   Exam-Style Questions

1 a) $g(11) = \frac{11 + 3}{2} = \frac{14}{2} = 7$ *[1 mark]*

b) $fg(x) = f\left(\frac{x + 3}{2}\right) = 6\left(\frac{x + 3}{2}\right) + 5 = 3x + 14$
[2 marks available — 1 mark for substituting the expression for g into f, 1 mark for the correct answer]

c) Let $x = g(y)$, so $x = \frac{y + 3}{2} \Rightarrow 2x = y + 3$
$\Rightarrow 2x - 3 = y$. So $g^{-1}(x) = 2x - 3$
[3 marks available — 1 mark for writing x as a function of y, 1 mark for attempting to rearrange to make y the subject, 1 mark for the correct answer]

2 $fg(x) = f(3x - 5) = 2(3x - 5)^2 = 2(9x^2 - 30x + 25)$
$= 18x^2 - 60x + 50$
[3 marks available — 1 mark for substituting the expression for g into f, 1 mark for expanding the brackets, 1 mark for the correct answer in the specified form]

3 Let $x = f(y)$, so $x = \frac{2y + 3}{5y - 4} \Rightarrow x(5y - 4) = 2y + 3$
$\Rightarrow 5xy - 4x = 2y + 3 \Rightarrow y(5x - 2) = 4x + 3$
$\Rightarrow y = \frac{4x + 3}{5x - 2}$. So $f^{-1}(x) = \frac{4x + 3}{5x - 2}$
[4 marks available — 1 mark for writing x as a function of y, 1 mark for multiplying both sides by 5y − 4, 1 mark for factorising to take out a factor of y, 1 mark for the correct answer]

4 a) $\frac{4}{5x - 15} = 0.2$ *[1 mark]* $\Rightarrow 4 = 0.2(5x - 15)$
$\Rightarrow 4 = x - 3 \Rightarrow x = 7$ *[1 mark]*

b) $fg(x) = f(2\sqrt{x}) = \frac{4}{5(2\sqrt{x}) - 15} = \frac{4}{10\sqrt{x} - 15}$
[2 marks available — 1 mark for substituting the expression for g into f, 1 mark for the correct answer]

5 $fg(x) = f(3x + 4) = (3x + 4)^2 + 6 = 9x^2 + 24x + 22$
$gf(x) = g(x^2 + 6) = 3(x^2 + 6) + 4 = 3x^2 + 22$
So $9x^2 + 24x + 22 = 3x^2 + 22$
$\Rightarrow 6x^2 + 24x = 0 \Rightarrow x^2 + 4x = 0$
$\Rightarrow x(x + 4) = 0 \Rightarrow x = 0$ or $x = -4$
[6 marks available — 1 mark for a correct method to find composite functions, 1 mark for a correct expression for fg(x), 1 mark for a correct expression for gf(x), 1 mark for setting these equal to each other, 1 mark for rearranging to form a quadratic, 1 mark for solving the quadratic to find both solutions]

Section 19 — Sets

19.1 Sets

Page 236 Exercise 1

1 a) **A = {February, April, June, September, November}**
b) **B = {May}**
c) **C = {January, February, March, April, May, August}**

2 a) **A = {12, 14, 16, 18, 20, 22, 24}**
b) **B = {2, 3, 5, 7, 11, 13, 17, 19}**
c) **C = {1, 4, 9, 16, 25, 36, 49, 64, 81, 100, 121, 144, 169, 196}**
d) **D = {1, 2, 3, 5, 6, 10, 15, 30}**

Page 237 Exercise 2

1 a) (i) Set A is all odd numbers that are also positive integers ≤ 10.
So **A = {1, 3, 5, 7, 9}**
(ii) Set B is all factors of 16 that are also positive integers ≤ 10.
So **B = {1, 2, 4, 8}**
Using the same method for (iii)-(iv):
(iii) **C = {1, 4, 9}**
(iv) **D = {1, 2, 3, 5, 6, 10}**

b) (i) Set A is all odd numbers that are also integers between 20 and 30.
So **A = {21, 23, 25, 27, 29}**
Using the same method for (ii)-(iv):
(ii) **B = {20, 22, 24, 26, 28, 30}**
(iii) **C = {23, 29}**
(iv) **D = {21, 24, 27, 30}**

2 a) Set A is all prime numbers that are also positive integers less than 30.
So **A = {2, 3, 5, 7, 11, 13, 17, 19, 23, 29}**
Using the same method for B, C and D:
B = {1, 4, 9, 16, 25}
C = {1, 8, 27}
D = {4, 8, 12, 16, 20, 24, 28}

b) (i) There are 10 elements in A, so **n(A) = 10**
Using the same method for (ii)-(iv):
(ii) **n(B) = 5**
(iii) **n(C) = 3**
(iv) **n(D) = 7**

c) (i) Set E is all elements that belong to both sets. As there are no elements that are in both sets, **n(E) = 0**
(ii) Set F is all elements of B that aren't also in D. F = {1, 9, 25}, so **n(F) = 3**
(iii) Set G is all elements of ξ that are not also in A, B, C or D. G = {6, 10, 14, 15, 18, 21, 22, 26}, so **n(G) = 8**

3 a) (i) Set C contains all elements in A or B, so **C = {0, 1, 2, 3, 4}**
(ii) A = {1, ~~2~~, 3, ~~4~~}, B = {0, 2, 4}.
So **D = {1, 3}**
(iii) E = $\{x : x = a - b, a \in A, b \in B, x < 1\}$
The possibilities for $a - b = x$ are
$1 - 0 = 1$, $1 - 2 = -1$, $1 - 4 = -3$,
$2 - 0 = 2$, $2 - 2 = 0$, $2 - 4 = -2$,
$3 - 0 = 3$, $3 - 2 = 1$, $3 - 4 = -1$,
$4 - 0 = 4$, $4 - 2 = 2$ and $4 - 4 = 0$.
As x has to be < 1, **E = {0, –1, –2, –3}**.

b) (i) Set F has members (a, b) such that a is a member of A, b is a member of B and $a + b < 5$. So **F = {(1, 0), (1, 2), (2, 0), (2, 2), (3, 0), (4, 0)}**
(ii) Set G has members (a, b) such that a is a member of A, b is a member of B and $a \times b > 6$. So **G = {(2, 4), (3, 4), (4, 2), (4, 4)}**

19.2 Venn Diagrams

Page 238 Exercise 1

1 a) Look for elements that are in both A and B. Here, that's only 3 so that goes in the overlap. The other elements of A and B go in the respective circles (but not in the overlap). The numbers 2, 4, 6, 8, 9 and 10 are in ξ but in neither A nor B so they go outside the circles.

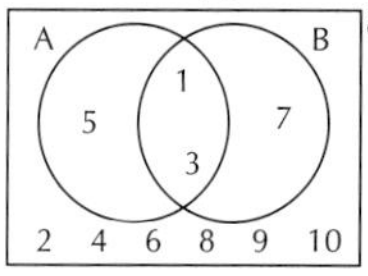

Using the same method for b)-c):

b)
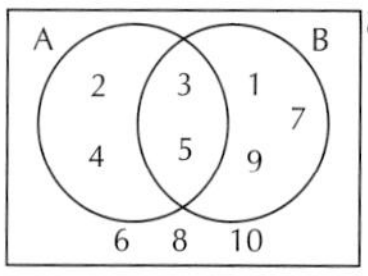

c)
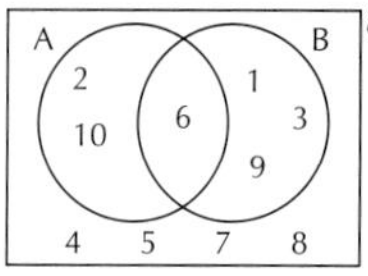

2 a) Write out A and B: A = {21, 23, 25, 27, 29} and B = {20, 25, 30}. Look for elements that are in both A and B. Here, that's only 25 so that goes in the overlap. The other elements of A and B go in the respective circles (but not in the overlap). The numbers 22, 24, 26 and 28 are in ξ but in neither A nor B so they go outside the circles.

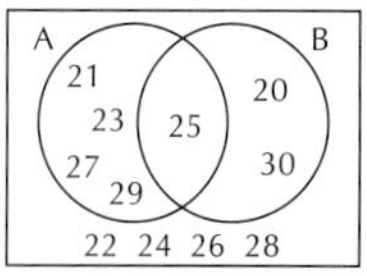

Using the same method for b)-c):

b)
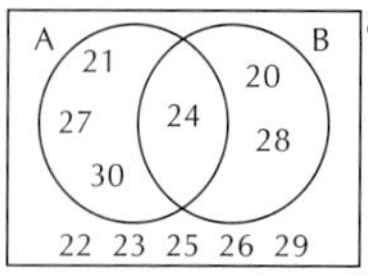

c)
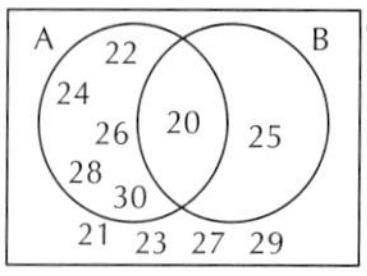

3 a) This is all the elements in the circle for A:
A = {1, 3, 5, 7, 9}
b) The elements in both set A and set C are in the overlap of circles A and C: **{5, 7, 9}**.
c) n(B) is the number of elements in circle B, so **n(B) = 4**
d) The overlap between A and B but not with C just contains the element **{3}**.
e) The elements in the circle for C but not in the circle for B are **{7, 8, 9}**.
f) The elements not in the circles for A or B are the elements only in circle C and the elements outside the circles. These are **{2, 8, 10}**.

Page 239 Exercise 2

1 a) (i) $x = n(\xi) - (7 + 4 + 14) = 30 - 25 =$ **5**
(ii) $n(A) = x + 7 = 5 + 7 =$ **12**
(iii) $n(\text{not } B) = x + 14 = 5 + 14 =$ **19**
You could do $30 - n(B) = 30 - (7 + 4)$ here.

b) (i) $40 = (36 - x) + x + (7 - x)$
$\Rightarrow 40 = 43 - x$, so $\boldsymbol{x = 3}$
(ii) $n(A) = (36 - x) + x =$ **36**
(iii) $n(\xi) = 40 + 4 =$ **44**

c) (i) $x = n(B) - 4 = 40 - 4 =$ **36**
$y = n(A) - x = 50 - 36 =$ **14**
(ii) $n(\xi) = y + x + 4 + 4 = 14 + 36 + 4 + 4$ $=$ **58**
(iii) $n(A \text{ or } B \text{ or both}) = y + x + 4$ $= 14 + 36 + 4 =$ **54**

2 a) n(A) = 26. If you label the number of elements in the overlap as x, then $26 = 18 + x$, so $x = 8$.

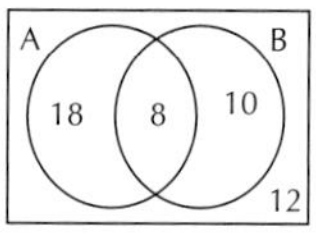

(i) $n(B) = 10 + 8$ $=$ **18**
(ii) $n(\xi) = 18 + 8 + 10 + 12 =$ **48**

b) $n(\text{not in A or B}) = 40 - (9 + 4 + 17)$ $= 40 - 30 = 10$

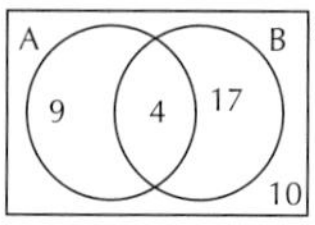

(i) $n(\text{not in A})$ $= 17 + 10 =$ **27**
(ii) $n(\text{not in B})$ $= 9 + 10 =$ **19**

c)
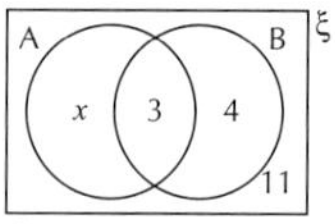

(i) $n(B) = 4 + 3 =$ **7**
(ii) $n(\text{not } B) = 25 - 7 =$ **18**
(iii) $n(A) = 11 - 4 =$ **10**
(iv) $n(\text{A but not B}) = x = 10 - 3 =$ **7**

d)
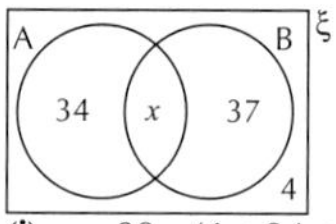

(i) $x = 88 - (4 + 34 + 37) = 88 - 75 =$ **13**
(ii) $n(B) = 37 + 13 =$ **50**

Page 240 Exercise 3

1
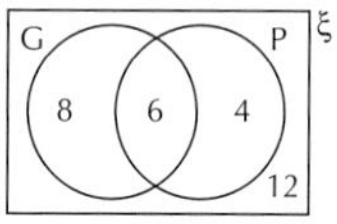

$n(\xi) = 8 + 6 + 4 + 12$ = **30 pupils in the class**

2
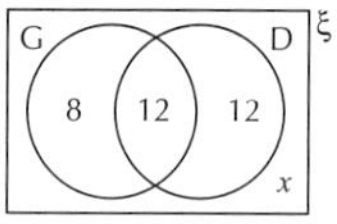

$36 - (8 + 12 + 12)$
$= 36 - 32$
= **4 children don't own geese or ducks**

3
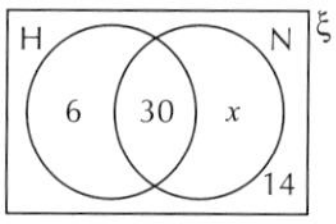

$x = 60 - (6 + 30 + 14)$
$= 60 - 50 =$ **10 people play netball but not hockey**

4
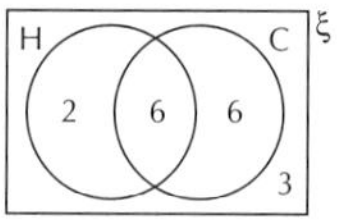

$2 + 6 + 6 + 3 =$ **17 friends in the group**

5
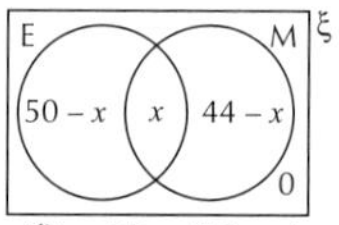

$n(\xi) = 55 = (50 - x) + x + (44 - x) = 94 - x$
$\Rightarrow x = 39$ pupils who passed both subjects.
$55 - 39 =$ **16 pupils passed exactly one of English and maths**

19.3 Unions and Intersections

Page 241 Exercise 1

1 a) (i) Elements in the intersection are {5, 6, 7}
   (ii) Elements in the union are
   {1, 2, 3, 4, 5, 6, 7, 8, 9, 10, 11, 12}
  b) (i) {P, O, R} **(ii)** {T, S, P, O, R, E, D, L}
  c) (i) {6, 12} **(ii)** {2, 3, 4, 6, 8, 9, 10, 12}
  d) (i) {O, U} **(ii)** {A, E, I, O, U, C, S}

2 a) The elements must be a member of both
   sets, so A ∩ B = **{blue four-wheel-drive cars}**
  b) The elements can be a member of set A or
   set B or both, so A ∪ B = **{children's names}**
  Using the same method for c)-e):
  c) A ∩ B = **{seaside towns in France}**
  d) A ∪ B = **{countries of the world}**
  e) A ∩ B = **{right-handed people
         with fair hair}**

3 a) A = {2, 4, 6, 8, 10, 12, 14, 16, 18, 20, 22,
   24, 26, 28}, B = {1, 2, 3, 4, 5, 6, 7, 8, 9,
   10, 11, 12, 13, 14, 15, 16, 17, 18, 19}.
   (i) A ∩ B = **{2, 4, 6, 8, 10, 12, 14, 16, 18}**
   (ii) A ∪ B = **{1, 2, 3, 4, 5, 6, 7, 8, 9, 10, 11,
       12, 13, 14, 15, 16, 17, 18, 19, 20, 22,
       24, 26, 28}**
  Using the same method for b)-d):
  b) (i) ∅ or { }
   (ii) {1, 2, 3, 4, 5, 6, 7, 8, 9, 10, 11, 12, 13,
       14, 15, 16, 17, 18, 19}
  c) (i) {1, 64}
   (ii) {1, 4, 8, 9, 16, 25, 27, 36, 49, 64}
  d) (i) {3} **(ii)** {2, 3, 5, 6, 7, 9}

4 a) A ∩ B is the intersection, so it's all the
   elements that belong to both sets —
   {x : 30 < x < 50}
   *Since the universal set is all real numbers, you
   can't list the elements of A ∩ B. Instead, use
   inequalities to give the range of elements.*
  b) {30} **c)** {x : x ≤ 50} **d)** ∅ or { }

5 a) (i) The intersection is empty so the two sets
   A and B don't overlap.

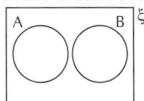

   (ii) The intersection is all of A, so the
   overlap between A and B is all of A
   (i.e. A lies inside B).

   (iii) The union is all of B, so everything in
   either circle is also in the circle for B.

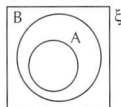

  b) If A ∩ B = A ∪ B, then **A = B**.

Page 243 Exercise 2

1 a) A ∩ C is the intersection of A and C, so the
   elements are {6, 7, 8, 9}
  Using the same method for b)-d):
  b) {4, 5, 8, 9} **c)** {8, 9}
  d) {8, 9, 13, 14}
  e) (A ∩ C) ∪ (B ∩ C) contains the elements
   that are in A and C, or B and C, or both.
   So this is {6, 7, 8, 9, 13, 14}.

2 a) (i) **Class members wearing hats only
   (i.e. not gloves or scarves).**
   (ii) **Class members not wearing hats,
   gloves or scarves.**

**(iii) Class members wearing hats,
   gloves and scarves.**

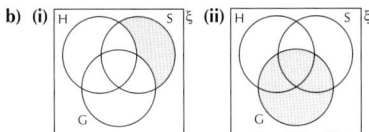

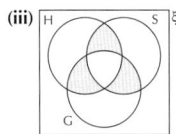

b) (i) **(ii)**

(iii)

3 a) n(A) = 6 + 6 + 4 + 4 = **20**
  b) n(A ∩ B) = 4 + 4 = **8**
  c) n(B ∪ C) = 4 + 7 + 4 + 2 + 6 + 8 = **31**
  d) n((A ∪ B) ∩ C) = 6 + 4 + 2 = **12**
  e) n(A ∪ (B ∩ C)) = 6 + 6 + 4 + 4 + 2 = **22**

4 a) (i) **(ii)**

b) The diagrams show that
   (A ∪ C) ∩ B = (A ∩ B) ∪ (B ∩ C)

c) A: B ∩ C: A ∪ (B ∩ C):

   A ∪ B: A ∪ C: (A∪B)∩(A∪C):

19.4 Complement of a Set

Page 244 Exercise 1

1 a) The complement of A is all the elements in
   ξ that aren't also in A. Polygons with fewer
   than 5 sides that aren't quadrilaterals are
   triangles, so **A' = {triangles}**
  b) There are no other months in the year
   with fewer than 30 days, so **A' = ∅ or { }**
  c) Factors of 18 that aren't also
   multiples of 2 are **A' = {1, 3, 9}**
  d) A' = **{books in the library which aren't
   paperbacks}**
  e) A' = **{cars which don't have an
   automatic gearbox}**

2 a) The complement of B is all the elements in
   ξ that aren't in B. So **B' = {1, 3, 5, 7, 9}**
  b) The only prime number that's
   not odd is 2, so **B' = {2}**
  c) B' is the set of all the even numbers ≤ 30
   and that aren't factors of 100. So **B' = {6,
   8, 12, 14, 16, 18, 22, 24, 26, 28, 30}**
  d) B' is the set of factors of 120 which aren't
   smaller than 20, so **B' = {20, 24, 30, 40,
   60, 120}**

3 a) A = {3, 6, 9, 12, 15, 18},
   B = {4, 8, 12, 16, 20}.
   A ∪ B = {3, 4, 6, 8, 9, 12, 15, 16, 18, 20}
  b) This is all the integers between 1 and 20
   which are not listed in part a):
   **(A ∪ B)' = {1, 2, 5, 7, 10, 11,
       13, 14, 17, 19}**

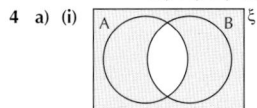

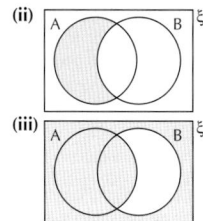

4 a) (i)

(ii)

(iii)

b) (i) B' **(ii)** (A ∪ B)' OR A' ∩ B'
  (iii) A' ∩ B

5 a) F' represents the set of **people who
   didn't like Apricot Fringits.**
  b) (J ∪ F)' represents the set of **people
   who didn't like either biscuit.**
  c) J' ∪ F represents the set of **people who
   liked Apricot Fringits or those who didn't
   like Chocolate Jamborees or both.**

6 a) A' ∩ B is the set of all elements that
   aren't in set A but that are in set B.
   So **A' ∩ B = {13, 14, 15, 16, 17}**
  b) (A ∪ B ∪ C)' is the set of all element not in
   A or B or C. So **(A ∪ B ∪ C)' = ∅ or { }**
  c) (A ∪ B) ∩ C' is the set of all elements of
   set A and set B that are also not in set C.
   So **(A ∪ B) ∩ C' = {1, 2, 3, 4, 5, 15, 16, 17}**
  d) A ∩ B' ∩ C is the set of elements that are
   in set A and not in set B and in set C.
   So **A ∩ B' ∩ C = {6, 7}**
  e) (B ∪ C)' is the set of all elements in neither
   set B nor set C. So **(B ∪ C)' = {1, 2, 3}**

7 a) x = n(B') − 14 = 20 − 14 = **6**
  b) n(A ∪ B') = 6 + 7 + 14 = **27**

8 For the Venn diagram: n(L ∩ M) = 12, so this
   goes in the overlap. n(L ∩ M)' = 8, so this
   goes outside the circles. n(L') = 10,
   so n(M ∩ L') = 10 − 8 = 2. n(M') = 14,
   so n(L ∩ M') = 14 − 8 = 6.

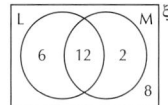

  a) n(L) = 6 + 12 = **18**
  b) n(ξ) = 6 + 12 + 2 + 8 = **28**
  c) n(L ∩ M)' = 6 + 2 + 8 = **16**
  d) n(L ∪ M') = 6 + 12 + 8 = **26**

9 **a)** {13, 19, 22}
  b) {20}
  c) {11, 13, 19, 22}
  d) {13, 14, 15, 16,
       20, 21, 22, 23, 24}

10

  a) {1, 2, 3, 4, 6, 7, 8, 9, 11, 12, 13, 14, 16, 17,
       18, 19, 21, 22, 23, 24, 26, 27, 28, 29, 30}
  b) {3, 9, 21, 27}
  c) {1, 5, 7, 10, 11, 13, 15, 17, 19, 20,
       23, 25, 29, 30}

Page 246 Review Exercise

1 a) (i) Count the number of elements in the
   circle for A. There are 8 so n(A) = 8 and
   the statement is **false.**

(ii) Count the number of elements in the overlap of the circles for A and B. There are 4 so the statement is **true**.

(iii) 20 is not in the circle for B so it is in the complement of B, so the statement is **true**.

(iv) Look at which elements are in C but not in B. These are 8, 9, 10 and 11 so the statement is **false**.

b) (i) {6, 7, 8}   **(ii)** {6, 7}
(iii) {6, 7, 12}   **(iv)** {9, 10, 11, 12}
(v) {9, 10, 11, 17, 18, 19, 20}
(vi) {6, 7, 8, 12}

2 a)

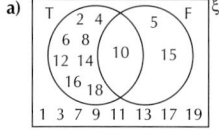

(i) $T \cap F = \{10\}$ so $n(T \cap F) = \mathbf{1}$
(ii) $(T \cup F)' = \{1, 3, 7, 9, 11, 13, 17, 19\}$ so $n(T \cup F)' = \mathbf{8}$

b)

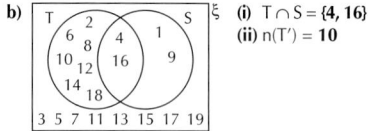

(i) $T \cap S = \mathbf{\{4, 16\}}$
(ii) $n(T') = \mathbf{10}$

3 a) B′   **b)** A′ ∩ B
c) (A′ ∩ B) ∪ (A ∩ B′) OR (A ∪ B) ∩ (A ∩ B)′

4 a) Let G = {families with girls},
B = {families with boys}.
$n(G \cap B') = 30 - 23 = 7$, $n(B \cap G')$
$= 38 - 23 = 15$.

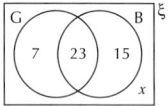

b) $x = 50 - (7 + 23 + 15) = 50 - 45$
= **5 families had no children.**

5 a)

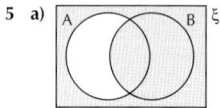

b)

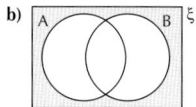

c)

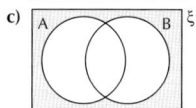

6 a) Let K = {people with keys},
C = {people with crayons},
R = {people with magic rings}
number of people with keys, crayons and magic rings = 72,
number of people with just keys and crayons = 74 − 72 = 2,
number of people with just keys and magic rings = 80 − 72 = 8,
number of people with just crayons and magic rings = 78 − 72 = 6,
number of people with just keys = 3,
number of people with just crayons = 5,
number of people with just magic rings = 3,
number of people with no keys, crayons or magic rings = 100 − 3 − 2 − 5 − 8 − 72
− 6 − 3 = 1

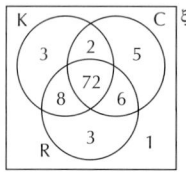

b) (i) $n(K \cup C \cup R)' = \mathbf{1}$
(ii) Exactly 2 items = 2 + 8 + 6 = **16**

Page 247 Exam-Style Questions

1 a) Let F = {students studying French},
S = {students studying Spanish}
$n(F \cap S') = 28 - 13 = 15$,
$n(S \cap F') = 18 - 13 = 5$

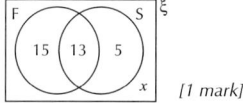

[1 mark]

b) $x = 36 - (15 + 13 + 5) = 36 - 33 = \mathbf{3}$
[1 mark]

2 Sketch a Venn diagram showing the information:

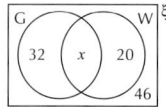

$x = 120 - (32 + 20 + 46) = 120 - 98 = 22$
[1 mark]
$n(W) = 22 + 20 = \mathbf{42\ products}$ are made using widgets *[1 mark]*.

3 a) Let G = {people who use the gym},
P = {people who use the pool},
S = {people who use the sauna}
Number of people who use the gym, the pool and the sauna = 15,
number of people who just use the gym and the pool = 17 − 15 = 2,
number of people who just use the pool and the sauna = 18 − 15 = 3,
number of people who just use the gym and the sauna = 30 − 15 = 15,
number of people who just use the gym
= 52 − 15 − 15 − 2 = 20
number of people who just use the pool
= 30 − 2 − 15 − 3 = 10
number of people who just use the sauna
= 65 − 15 − 15 − 3 = 32

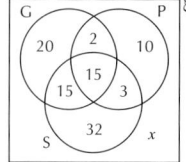

[2 marks available — 1 mark for the correct values in the intersections, 1 mark for correct values in the rest of the circles]

b) (i) Number of people who use none of the facilities = 100 − 20 − 15 − 15 − 2 − 10
− 3 − 32 = **3** *[1 mark]*
(ii) 20 + 10 + 32 = **62** *[1 mark]*
(iii) 2 + 15 + 15 + 3 = **35** *[1 mark]*

4 $n(A' \cap B') = 2$, so 2 goes outside both circles.
$n(A') = 5$, so in the 'B only' region is 5 − 2 = 3.
The intersection is x. $n(A \cap B') = y$, so this goes in the 'A only' region.

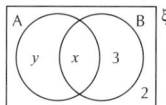

$n(\xi) = 20 = y + x + 3 + 2 = y + x + 5$,
so $y = 15 - x$

[3 marks available — 1 mark for 2 correct regions in the diagram, 1 mark for fully correct diagram, 1 mark for sum of all regions = 20 and writing given result]

5 a) X = {11, 12, 13, 14, 15, 16, 17, 18, 19},
Y = {2, 3, 5, 7, 11, 13, 17, 19}. X ∩ Y is the set of elements that are in both set Y and set X, so X ∩ Y = **{11, 13, 17, 19}**
[1 mark]

b) (X′ ∩ Y) ∪ Z is the set of elements that aren't in set X but are in set Y, together with those in set Z. Z = {3, 6, 9, 12, 15, 18} so (X′ ∩ Y) ∪ Z = **{2, 3, 5, 6, 7, 9, 12, 15, 18}**
[2 marks available — 2 marks for all correct values, otherwise 1 mark for 6 to 8 correct values. Lose 1 mark if incorrect values have been included]

Section 20 — Angles and 2D Shapes

20.1 Angles on Lines and Around Points

Page 248 Exercise 1

1 a) $123° + 29° + a = 180° \Rightarrow a = \mathbf{28°}$
b) $28° + 55° + b + 35° = 180° \Rightarrow b = \mathbf{62°}$
c) $57° + 57° + c + c = 180°$
$\Rightarrow 2c = 66° \Rightarrow c = \mathbf{33°}$
2 a) $d + 3d = 180° \Rightarrow 4d = 180° \Rightarrow d = \mathbf{45°}$
b) $(e + 50°) + e + 30° = 180°$
$\Rightarrow 2e = 100° \Rightarrow e = \mathbf{50°}$
c) $(f + 27°) + f + (f - 12°) = 180°$
$\Rightarrow 3f = 165° \Rightarrow f = \mathbf{55°}$
3 Angle $AOB = 59° + 65° + 58° = 182°$ but angles on a straight line **must add up to 180°** so AOB **cannot be a straight line.**

Page 249 Exercise 2

1 a) $99° + a + 44° + 90° = 360° \Rightarrow a = \mathbf{127°}$
b) $41° + 161° + b + 45° = 360° \Rightarrow b = \mathbf{113°}$
c) $43° + c + 69° + 84° + 57° = 360°$
$\Rightarrow c = \mathbf{107°}$
d) $d + 2d + 2d = 360°$
$\Rightarrow 5d = 360° \Rightarrow d = \mathbf{72°}$
e) $50° + 98° + e + 3e = 360°$
$4e = 212° \Rightarrow e = \mathbf{53°}$
f) $160° + 90° + f + f + 50° = 360°$
$\Rightarrow 2f = 60° \Rightarrow f = \mathbf{30°}$
g) $(g + 20°) + (g + 70°) + g = 360°$
$\Rightarrow 3g = 270° \Rightarrow g = \mathbf{90°}$
h) $(2h + 10°) + 90° + (h - 40°) = 360°$
$\Rightarrow 3h = 300° \Rightarrow h = \mathbf{100°}$
i) $i + 2i + 3i + 4i + 5i = 360°$
$\Rightarrow 15i = 360° \Rightarrow i = \mathbf{24°}$
2 a) If every angle is greater than 1° then
$x - 10° > 1°$ so $x > 11°$.
The angles are around a point so:
$m + (x + 30°) + x + (x - 10°) = 360°$
$\Rightarrow m + 3x = 340° \Rightarrow m = 340° - 3x$
Now, $3x > 33°$ so $m < 340° - 33° = \mathbf{307°}$
b) Using $m + 3x = 340°$:
$m = 340° - 3 \times 40° = \mathbf{220°}$

20.2 Parallel Lines

Page 250 Exercise 1

1 a) a and 75° are vertically opposite so $a = \mathbf{75°}$,
b and 75° lie on a straight line
so $b + 75° = 180° \Rightarrow b = \mathbf{105°}$
b) c and the right angle lie on a straight line so
$c + 90° = 180° \Rightarrow c = \mathbf{90°}$,
d and the right angle are vertically opposite so $d = \mathbf{90°}$,
e and c are vertically opposite
so $e = c = \mathbf{90°}$
c) f and $2f$ lie on a straight line
so $f + 2f = 180° \Rightarrow 3f = 180° \Rightarrow f = \mathbf{60°}$,
g and $2f$ are vertically opposite
so $g = 2f = \mathbf{120°}$

2 a) *t* and 82° are alternate angles so *t* = **82°**,
  u and 98° are alternate angles so *u* = **98°**

b) *v* and 135° lie on a straight line
  so $v + 135° = 180° \Rightarrow$ *v* = **45°**,
  v and (*w* + 15°) are alternate angles
  so $v = 45° = w + 15° \Rightarrow$ *w* = **30°**

c) 4*x* and 132° lie on a straight line
  so $4x + 132° = 180°$
  $\Rightarrow 4x = 48° \Rightarrow$ *x* = **12°**,
  4*x* and (15*y* + 3°) are alternate angles
  so $4x = 48° = 15y + 3°$
  $\Rightarrow 15y = 45° \Rightarrow$ *y* = **3°**,
  z and (15*y* + 3°) lie on a straight line
  so $z + 48° = 180° \Rightarrow$ *z* = **132°**

Page 251   Exercise 2

1 a) *a* and 108° are corresponding angles
  so *a* = **108°**,
  b and 72° are corresponding angles
  so *b* = **72°**
  Alternatively, you could have used the fact that a and 72° are allied angles, and then a = 180° − 72° = 108°.

b) *c* and 141° lie on a straight line
  so $c + 141° = 180° \Rightarrow$ *c* = **39°**,
  c and *d* are corresponding angles
  so *d* = **39°**,
  e and 141° are corresponding angles
  so *e* = **141°**

c) *f* and 105° lie on a straight line
  so $f + 105° = 180° \Rightarrow$ *f* = **75°**,
  g and 105° are vertically opposite
  so *g* = **105°**,
  h and *f* are vertically opposite so *h* = **75°**,
  i and 105° are corresponding angles
  so *i* = **105°**,
  j and *f* are corresponding angles so *j* = **75°**,
  k and *h* are corresponding angles
  so *k* = **75°**,
  5*l* and *i* are vertically opposite
  so $5l = 105° \Rightarrow$ *l* = **21°**
  There are loads of different ways to get to these answers — e.g. you could have found g by using the fact that it's on a straight line with f, or you could have found i by using the fact that i and g are alternate angles.

d) *m* and 78° are vertically opposite
  so *m* = **78°**,
  n and 78° lie on a straight line
  so $n + 78° = 180° \Rightarrow$ *n* = **102°**,
  n and *p* are vertically opposite so *p* = **102°**,
  q and 78° are corresponding angles
  so *q* = **78°**,
  r and *q* are vertically opposite so *r* = **78°**,
  6*s* and *n* are corresponding angles
  so $6s = 102° \Rightarrow$ *s* = **17°**,
  p and (3*t* + 90°) are corresponding angles
  so $3t + 90° = 102° \Rightarrow 3t = 12° \Rightarrow$ *t* = **4°**

2 The angle made by the second post and the
  downward slope is 99° since this and the
  99° shown in the diagram are corresponding
  angles. Then *y* and 99° lie on a straight line so
  $y + 99° = 180° \Rightarrow$ *y* = **81°**.

3 a) If the lines are parallel then the angles
  would be allied, so they should add
  up to 180°. $65° + 115° = 180°$,
  so these lines **are parallel**.

b)

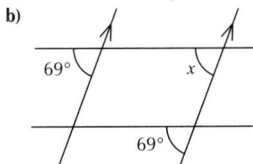

  x = 69° using the known parallel
  lines and corresponding angles with
  the 69° in the top-left. If the other
  pair of lines were parallel then, by
  corresponding angles, *x* would be
  equal to the 69° in the bottom-right —
  which it is. So the lines **are parallel**.

c)

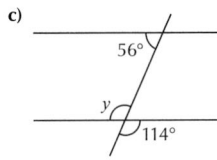

  Using vertically opposite angles, *y* = 114°.
  If the lines were parallel then, by allied
  angles, $56° + y = 180°$.
  But $56° + 114° = 170° \neq 180°$,
  so the lines **are not parallel**.

20.3 Triangles

Page 253   Exercise 1

1 a) $a + 72° + 55° = 180° \Rightarrow$ *a* = **53°**

b) $68° + 90° + b = 180° \Rightarrow$ *b* = **22°**

c) $21° + 144° + c = 180° \Rightarrow$ *c* = **15°**

d) $d + 90° + 74° = 180° \Rightarrow$ *d* = **16°**

e) Angles on a straight line add up to 180° so
  the unmarked angle is $180° − 90° = 90°$.
  Then $90° + 38° + e = 180° \Rightarrow$ *e* = **52°**

f) $57° + 74° + f = 180° \Rightarrow$ *f* = **49°**
  $49° + g = 180° \Rightarrow$ *g* = **131°**

2 a) $2a + a + 120° = 180°$
  $\Rightarrow 3a = 60° \Rightarrow$ *a* = **20°**

b) $2b + (2b + 50°) + 90° = 180°$
  $\Rightarrow 4b = 40° \Rightarrow$ *b* = **10°**

c) $60° + (c − 16°) + (c + 16°) = 180°$
  $2c = 120° \Rightarrow$ *c* = **60°**

d) $d + (d + 10°) + (d − 10°) = 180°$
  $\Rightarrow 3d = 180° \Rightarrow$ *d* = **60°**

3 a) *a* = **30°** since the triangle is isosceles
  $b + 30° + 30° = 180° \Rightarrow$ *b* = **120°**

b) $40° + 2c = 180° \Rightarrow 2c = 140° \Rightarrow$ *c* = **70°**

c) $3d + 2d = 180° \Rightarrow 5d = 180° \Rightarrow$ *d* = **36°**

4 a) (i) The triangle is equilateral so *x* = **60°**.
  (ii) *y* lies on a straight line with an angle of
  60° so $y + 60° = 180° \Rightarrow$ *y* = **120°**

b) (i) The triangle is isosceles so the angle in
  the bottom-left of the triangle is 55°.
  Then 25*p* lies on a straight line with this
  angle so $25p + 55° = 180°$
  $\Rightarrow 25p = 125° \Rightarrow$ *p* = **5°**
  (ii) The angle at the top of the triangle is
  $180° − 2 × 55° = 70°$. Then *q* lies on a
  straight line with this angle so
  $q + 70° = 180° \Rightarrow$ *q* = **110°**.

20.4 Quadrilaterals

Page 254   Exercise 1

1 a) $a + 93° + 69° + 86° = 360° \Rightarrow$ *a* = **112°**

b) $76° + b + 103° + 89° = 360° \Rightarrow$ *b* = **92°**

c) $129° + 3c + 67° + 74° = 360°$
  $\Rightarrow 3c = 90° \Rightarrow$ *c* = **30°**

d) $d + 112° + 106° + d = 360°$
  $\Rightarrow 2d = 142° \Rightarrow$ *d* = **71°**

e) $119° + e + (e − 34°) + 72° = 360°$
  $\Rightarrow 2e = 203° \Rightarrow$ *e* = **101.5°**

f) $90° + 52° + 90° + f = 360° \Rightarrow$ *f* = **128°**

g) $52° + 104° + g + 90° = 360° \Rightarrow$ *g* = **114°**

h) $29° + h + 55° + 46° = 360° \Rightarrow$ *h* = **230°**
  i and *h* are angles around a point so
  $i + 230° = 360° \Rightarrow$ *i* = **130°**

2 a) The missing angle in the quadrilateral is
  $360° − 90° − 108° − 85° = 77°$.
  Then this angle and r lie on a straight line
  so $r + 77° = 180° \Rightarrow$ *r* = **103°**

b) *w* and 71° are vertically opposite
  so *w* = **71°**,
  x and 95° are vertically opposite
  so *x* = **95°**,
  $71° + 95° + 102° + y = 360° \Rightarrow$ *y* = **92°**

Page 255   Exercise 2

1 a) The shape has two pairs of opposite equal
  sides, so it's a parallelogram or rectangle.
  a must therefore equal the angle opposite
  it, so *a* = **90°**.
  *All four angles are actually 90°, so the shape is
  a rectangle.*

b) Opposite angles in a parallelogram
  are equal so *b* = **122°**.

c) Opposite angles in a rhombus are equal so
  c = **72°** and *d* = **108°**.

d) Neighbouring angles sum to 180°
  so $(e + 4°) + 55° = 180° \Rightarrow$ *e* = **121°**.
  Opposite angles in a parallelogram are
  equal so $5f = 55° \Rightarrow$ *f* = **11°**.

e) Neighbouring angles sum to 180°
  so $(11g + 4°) + 110° = 180°$
  $\Rightarrow 11g = 66° \Rightarrow$ *g* = **6°**

f) Opposite angles in a parallelogram are
  equal so *h* = **120°** and *i* = **60°**.

g) Angles in a triangle sum to 180°
  so $90° + 36° + j = 180° \Rightarrow$ *j* = **54°**,
  $36° + 4k = 90° \Rightarrow 4k = 54° \Rightarrow$ *k* = **13.5°**,
  $54° + (l − 15°) = 90° \Rightarrow$ *l* = **51°**
  *You could also find k and l using alternate
  angles.*

h) Opposite angles in a rhombus are equal
  so the angle opposite 77° is 77°.
  Then diagonals in a rhombus bisect the
  angles so $m = 77° ÷ 2 =$ **38.5°**.
  Diagonals in a rhombus meet at a right
  angle so $n − 5° = 90° \Rightarrow$ *n* = **95°**.
  11*p* and 77° are alternate angles
  so $11p = 77° \Rightarrow$ *p* = **7°**.

Page 256   Exercise 3

1 a) $a + 60° = 180° \Rightarrow$ *a* = **120°**

b) This is an isosceles trapezium so *b* = **120°**
  and $c + 120° = 180° \Rightarrow$ *c* = **60°**

c) $(d + 10°) + 106° = 180° \Rightarrow$ *d* = **64°**,
  $(e − 7°) + 64° = 180° \Rightarrow$ *e* = **123°**

d) $124° + 4f = 180° \Rightarrow 4f = 56° \Rightarrow$ *f* = **14°**

2 a) The opposite angles are equal so *a* = **113°**.

b) *b* and 111° are equal so *b* = **111°**.
  Then the angles must sum to 360° so
  $111° + 48° + 111° + c = 360° \Rightarrow$ *c* = **90°**.

c) $124° + 124° + 42° + 7d = 360°$
  $\Rightarrow 7d = 70° \Rightarrow$ *d* = **10°**

d) $(e + 1°) + (e + 1°) + 101° + 61° = 360°$
  $\Rightarrow 2e = 196° \Rightarrow$ *e* = **98°**

3 An isosceles trapezium has two pairs of equal
  angles. Since the pair of angles given are
  equal, the missing pair must also be equal.
  Call them both *x*. Then each one added to 53°
  gives 180°, so $x + 53° = 180° \Rightarrow$ *x* = **127°**

4 A kite has one pair of equal angles. Since
  neither of the angles given are part of this
  pair, the missing angles must be equal.
  Call them both *y*. Then $90° + 50° + 2y = 360°$
  $\Rightarrow 2y = 220° \Rightarrow$ *y* = **110°**

20.5 Polygons

Page 258   Exercise 1

1 a) This shape doesn't have straight sides so it's
  not a polygon.

b) This is a polygon — it's a **triangle**.

c) This is a polygon — it's an **octagon**.

d) This shape isn't 2D so it's **not a polygon**.

2 a) $S = (6 − 2) × 180° =$ **720°**

b) $S = (10 − 2) × 180° =$ **1440°**

c) $S = (12 − 2) × 180° =$ **1800°**

d) $S = (20 − 2) × 180° =$ **3240°**

3 a) (i) $S = (4 − 2) × 180° =$ **360°**
  *This is a quadrilateral so you already know
  that the angles have to sum to 360°.*

(ii) $41° + 112° + 89° + a = 360°$
$\Rightarrow \boldsymbol{a = 118°}$

b) (i) $S = (6 - 2) \times 180° = \boldsymbol{720°}$

(ii) $107° + 101° + b$
$+ 90° + 90° + 85° = 720°$
$\Rightarrow \boldsymbol{b = 247°}$

c) (i) $S = (9 - 2) \times 180° = \boldsymbol{1260°}$

(ii) $93° + c + 104° + 121° + 91°$
$+ 230° + 150° + 102° + 159° = 1260°$
$\Rightarrow \boldsymbol{c = 210°}$

4 a) $S = (8 - 2) \times 180° = 1080°$
But all 8 angles are the same size
since the shape is regular, so one
angle is $1080° \div 8 = \boldsymbol{135°}$.

b) $S = (9 - 2) \times 180° = 1260°$
But all 9 angles are the same size
since the shape is regular, so one
angle is $1260° \div 9 = \boldsymbol{140°}$.

c) $S = (10 - 2) \times 180° = 1440°$
But all 10 angles are the same size
since the shape is regular, so one
angle is $1440° \div 10 = \boldsymbol{144°}$.

5 a) Using the formula, the sum of the
interior angles in an octagon is
$S = (8 - 2) \times 180° = 1080°$.
Also, $S = 7 \times 130° + x$, where x is
the size of the unknown angle
$\Rightarrow 1080° = 910° + x \Rightarrow \boldsymbol{x = 170°}$

b) It is **not** a regular octagon since the angles
aren't all the same size.

6 a) $S = 2520° = (n - 2) \times 180°$
$\Rightarrow n - 2 = 14 \Rightarrow \boldsymbol{n = 16}$

b) 8 of the angles are 95° so these add up
to $8 \times 95° = 760°$. So the other 8 angles
add up to $2520° - 760° = 1760°$.
Then $8x = 1760° \Rightarrow \boldsymbol{x = 220°}$

7 a) $S = n \times 60° = (n - 2) \times 180°$
$\Rightarrow 60n = 180n - 360$
$\Rightarrow 120n = 360 \Rightarrow \boldsymbol{n = 3}$

b) $S = n \times 150° = (n - 2) \times 180°$
$\Rightarrow 150n = 180n - 360$
$\Rightarrow 30n = 360 \Rightarrow \boldsymbol{n = 12}$

8 a) $9x + 90° = 360° \Rightarrow 9x = 270° \Rightarrow \boldsymbol{x = 30°}$
$S = (7 - 2) \times 180° = 900°$
$90° + 90° + 270° + 90°$
$+ 97° + (3y + 42°) + 146° = 900°$
$\Rightarrow 3y = 75° \Rightarrow \boldsymbol{y = 25°}$

b) $S = (6 - 2) \times 180° = 720°$
$z + (z + 5°) + 130° + (z - 10°)$
$+ 125° + (2z - 30°) = 720°$
$\Rightarrow 5z = 500 \Rightarrow \boldsymbol{z = 100°}$

9 Split the decagon into $10 - 2 = 8$ triangles, e.g.

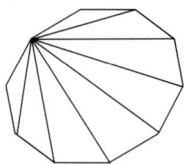

The interior angles of each triangle
sum to 180°. So the interior angles
of all eight triangles, which is the
same as the interior angles of the
decagon, sum to $8 \times 180° = \boldsymbol{1440°}$.

Page 260   Exercise 2

1 a) $360° \div 8 = \boldsymbol{45°}$

b) $360° \div 9 = \boldsymbol{40°}$

c) $360° \div 7 = \boldsymbol{51.4285...°}$
It was fine to use the formula $360° \div n$
in this question because the polygons were
regular — but remember that it doesn't
work for a polygon that isn't regular.

2 a) $90° + 31° + 83° + 72° + 30° + a = 360°$
$\Rightarrow \boldsymbol{a = 54°}$

b) The missing exterior angles are
$180° - 135° = 45°$ and $180° - 90° = 90°$.
$90° + 45° + 98° + 71° + b = 360°$
$\Rightarrow \boldsymbol{b = 56°}$

c) $c = 180° - 106° = \boldsymbol{74°}$
$151° + 74° + d = 360° \Rightarrow \boldsymbol{d = 135°}$

3 a) $360° - 100° - 68° - 84° - 55° = \boldsymbol{53°}$

b) $360° - 30° - 68° - 45° - 52°$
$- 75° - 50° = \boldsymbol{40°}$

c) $360° - 42° - 51° - 60° - 49°$
$- 88° - 35° = \boldsymbol{35°}$

d) $360° - 19° - 36° - 28° - 57°$
$- 101° - 57° - 22° = \boldsymbol{40°}$

4 a) Since the polygon is regular,
$360° \div n = 45° \Rightarrow n = 360° \div 45° = \boldsymbol{8}$
This is an **octagon**.

b)

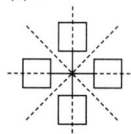

c) Interior angle $= 180° - 45° = \boldsymbol{135°}$

d) $8 \times 135° = \boldsymbol{1080°}$
Alternatively, you could use $(8 - 2) \times 180°$.

5 a) (i) $360° \div n = 40° \Rightarrow n = 360° \div 40° = \boldsymbol{9}$

(ii) $180° - 40° = \boldsymbol{140°}$

(iii) $9 \times 140° = \boldsymbol{1260°}$

b) (i) $360° \div n = 120°$
$\Rightarrow n = 360° \div 120° = \boldsymbol{3}$

(ii) $180° - 120° = \boldsymbol{60°}$

(iii) $3 \times 60° = \boldsymbol{180°}$

c) (i) $360° \div n = 3° \Rightarrow n = 360° \div 3° = \boldsymbol{120}$

(ii) $180° - 3° = \boldsymbol{177°}$

(iii) $120 \times 177° = \boldsymbol{21\,240°}$

d) (i) $360° \div n = 4.8°$
$\Rightarrow n = 360° \div 4.8° = \boldsymbol{75}$

(ii) $180° - 4.8° = \boldsymbol{175.2°}$

(iii) $75 \times 175.2° = \boldsymbol{13\,140°}$

6 a) $3u + 6u + 4u + 5u = 360°$
$\Rightarrow 18u = 360° \Rightarrow \boldsymbol{u = 20°}$

b) $55° + 7v + 3v + 13v + (7v + 5°) = 360°$
$\Rightarrow 30v = 300° \Rightarrow \boldsymbol{v = 10°}$

c) $3w + 8w + (6w - 10°) + 2w$
$+ [180° - (170° - 5w)] = 360°$
$\Rightarrow 24w = 360° \Rightarrow \boldsymbol{w = 15°}$
$170° - 5w$ is the interior angle, so the
exterior angle is $180° - (170° - 5w)$.

d) $2x + (3x - 20°) + [180° - 2x]$
$+ [180° - (2x + 10°)] = 360°$
$\Rightarrow \boldsymbol{x = 30°}$

e) $y + (y + 6°) + (y - 20°)$
$+ [180° - 106°] = 360°$
$\Rightarrow 3y = 300° \Rightarrow \boldsymbol{y = 100°}$

f) $90° + 90° + (3z + 2°) + (4z + 7°)$
$+ [180° - 149°] = 360°$
$\Rightarrow 7z = 140° \Rightarrow \boldsymbol{z = 20°}$

20.6 Symmetry

Page 261   Exercise 1

1 a) (i) **1** — below

(ii) 1

b) (i) **4** — below

(ii) 4

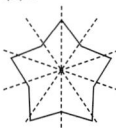

c) (i) **5** — below

(ii) 5

d) (i) 0

(ii) 2

e) (i) **2** — below

(ii) 2

f) (i) **1** — below

(ii) 1

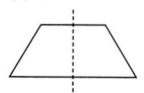

g) (i) **5** – below

(ii) 5

h) (i) **10** — below

(ii) 10

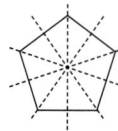

2 a)

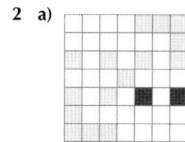

b)

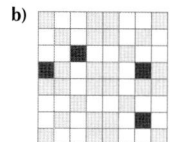

Page 262   Review Exercise

1 a) a and 61° are alternate angles so $\boldsymbol{a = 61°}$,
b and a lie on a straight line
so $b = 180° - 61° = \boldsymbol{119°}$

b) c and 98° are vertically opposite
so $\boldsymbol{c = 98°}$,
$4d$ and c are corresponding angles
so $4d = 98° \Rightarrow \boldsymbol{d = 24.5°}$

c) e and 92° are corresponding angles
so $\boldsymbol{e = 92°}$,
e lies on a straight line with an unmarked
angle of $180° - 92° = 88°$ and
then this is a corresponding angle
with $4f$, so $4f = 88° \Rightarrow \boldsymbol{f = 22°}$,
g and $4f$ lie on a straight line
so $g = 180° - 88° = \boldsymbol{92°}$

d) h and 102° lie on a straight line
so $h = 180° - 102° = \boldsymbol{78°}$,
i and h are alternate angles so $i = h = \boldsymbol{78°}$,
$(j + 5°)$ and 102° are alternate angles
so $j + 5° = 102° \Rightarrow \boldsymbol{j = 97°}$

e) $\frac{1}{2}k$ and 125° are corresponding angles
so $\frac{1}{2}k = 125° \Rightarrow \boldsymbol{k = 250°}$,
l and $\frac{1}{2}k$ lie on a straight line
so $l = 180° - 125° = \boldsymbol{l = 55°}$,
$55° + 90° + m = 180° \Rightarrow \boldsymbol{m = 35°}$

f) $(4n + 5°)$ and 33° are alternate angles
so $4n + 5° = 33° \Rightarrow 4n = 28° \Rightarrow \boldsymbol{n = 7°}$,
$(11p - 3°)$ and $(4n + 5°)$ are angles around
a point so $(11p - 3°) + 33° = 360°$
$\Rightarrow 11p = 330° \Rightarrow \boldsymbol{p = 30°}$

2 a) (i) x and $2x$ lie on a straight line
so $x + 2x = 180°$
$\Rightarrow 3x = 180° \Rightarrow \boldsymbol{x = 60°}$

(ii) $2x$ and y are alternate angles
so $y = 2x = \boldsymbol{120°}$

b) If AB and CD were parallel then the
allied angles BAC and ACD would sum
to 180°. But $126° + 52° = 178° \neq 180°$,
so the lines are **not parallel**.

3 a) $62° + 62° + x = 180° \Rightarrow \boldsymbol{x = 56°}$
$62°$ and y lie on a straight line
so $y = 180° - 62° = \boldsymbol{118°}$

b) The unmarked angle in the triangle is
$180° - 35° - 90° = 55°$.
Then $z = 180° - 55° - 55° = \boldsymbol{70°}$.

c) $48 + u + u = 180°$
$\Rightarrow 2u = 132° \Rightarrow \boldsymbol{u = 66°}$,
$(v + 10°)$ and u are angles around a point so
$(v + 10°) + 66° = 360° \Rightarrow \boldsymbol{v = 284°}$

4 a) Using vertically opposite angles, the angles
of the triangle are p, 94° and 52°.
Then $p + 94° + 52° = 180° \Rightarrow \boldsymbol{p = 34°}$.

b) x and $3x$ lie on a straight line so
$x + 3x = 180° \Rightarrow 4x = 180° \Rightarrow x = 45°$,
the triangle is isosceles so
$45° + 45° + y = 180° \Rightarrow y = 90°$

5 a) $74° + e + 40° = 180° \Rightarrow e = 66°$
$2f = e \Rightarrow f = 33°$
$(g + 3°) + e = 180° \Rightarrow g = 111°$
$h = g + 3° = 114°$

b) $i = 116°$
$116° + 116° + 46° + j = 360° \Rightarrow j = 82°$

c) $2k$ and $62°$ are alternate angles
so $2k = 62° \Rightarrow k = 31°$,
the trapezium is isosceles
so $l + 5° = 62° \Rightarrow l = 57°$
m and $2k$ lie on a straight line
so $m = 180° - 62° = 118°$,

6 a) square, rhombus

b) square, rectangle

c) kite

d) square, rectangle, parallelogram, rhombus

e) square, rectangle, parallelogram, rhombus, trapezium

f) trapezium

7 a) **(4, 0)** so that AB and CD are parallel

b) **(2, 0)** so that AB and AD are the same length and BC and CD are the same length

c) **(2, 2)** so that AD and BC are parallel and AB and CD are the same length

8 a) (i) There are five sides so it's a **pentagon**.
(ii) It's **not regular** since the angles aren't all the same size.

b) $S = (5 - 2) \times 180° = 540°$

c) $w + 133° + 117° + 90° + 90° = 540°$
$\Rightarrow w = 110°$

9 a) $360° \div 10 = 36°$

b) $360° \div 12 = 30°$

c) $360° \div 15 = 24°$

d) $360° \div 25 = 14.4°$

10

	equilateral triangle	parallelogram
No. of sides	3	4
Lines of symmetry	3	0
Order of rotational symmetry	3	2
Sum of interior angles	180°	360°

	isosceles trapezium	regular nonagon
No. of sides	4	9
Lines of symmetry	1	9
Order of rotational symmetry	1	9
Sum of interior angles	360°	1260°

Page 264 Exam-Style Questions

1 The angle marked p in the diagram below is the exterior angle of the regular polygon. So $p = 360° \div 8 = 45°$ *[1 mark]*. Opposite angles in a parallelogram are equal so $x = 45°$ *[1 mark]*.

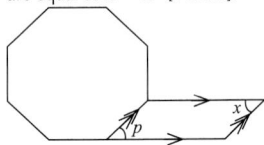

You might have approached this in a slightly different way — if you used a sensible method and got the correct answer, you'd get all the marks.

2 The angle marked p in the diagram below and the angle a are corresponding angles so $p = a$ *[1 mark]*. Then the angle marked q and the angle p are vertically opposite so $q = p = a$ *[1 mark]*. Angles in a triangle add up to $180°$ so $3x = 180° - 2a - a = 180° - 3a$ $\Rightarrow x = (180° - 3a) \div 3 = 60° - a$ *[1 mark]*.

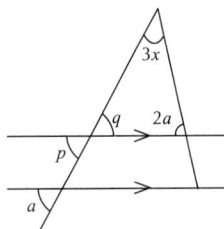

You might have used different methods to answer this question — as long as you explain all your reasoning and reach the correct conclusion, you'd get all the marks.

3 The sum of the interior angles is $(12 - 2) \times 180° = 1800°$ *[1 mark]*. Six of the angles are $40°$ and six are the unknown size, say x. So $6 \times 40° + 6x = 1800°$ *[1 mark]*
$\Rightarrow 6x = 1560° \Rightarrow x = 260°$ *[1 mark]*

4 a) A heptagon has 7 sides, using the formula, the sum of the interior angles is $(7 - 2) \times 180°$ *[1 mark]* $= 900°$ *[1 mark]*

b) Draw parallel lines on top of the triangle and along the base. Label the angles inside the triangle a, b and c, e.g.

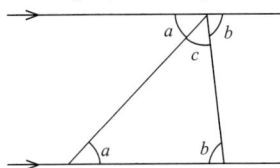

Using alternate angles, there is an angle a and an angle b along the top line. Then angles on a straight line add up to $180°$ so $a + b + c = 180°$. *[3 marks available — 1 mark for considering parallel lines, 1 mark for using a suitable method to collect angles at the top of the triangle, 1 mark for using angles on a straight line to reach the conclusion]*

c) Split the heptagon into 5 triangles, e.g.

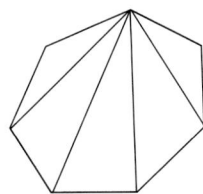

The angles in each triangle add up to $180°$ so the angles in all the triangles, and hence the heptagon, add up to $5 \times 180° = 900°$. *[2 marks available — 1 mark for correctly splitting a heptagon into 5 triangles, 1 mark for using the sum of angles in a triangle to reach the correct conclusion]*

5 Using the labelling in the diagram below:
$b = 180° - a$ as a and b are allied angles,
$d = 180° - c$ as c and d are allied angles.
[1 mark for writing b and d in terms of a and c]
Then the sum of the angles is
$a + b + c + d = a + (180° - a) + c + (180° - c)$
$= 180° + 180° = 360°$ *[1 mark]*

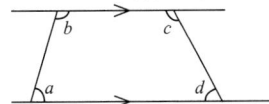

6 a) Diagonals in a rhombus meet at a right angle so $AXD = 90°$ *[1 mark]*

b) Angles in a triangle sum to $180°$ so $ADX = 180° - 22° - 90° = 68°$ *[1 mark]*

c) Diagonals in a rhombus bisect the angles so $CDX = ADX = 68°$. Then angles on a straight line sum to $180°$ so $ADY = 180° - 68° - 68° = 44°$.

[3 marks available — 1 mark for the correct size of CDX, 1 mark for the correct size of ADY, 1 mark for stating the correct reasons] There are different ways to work out this angle — as long as you explain all your reasoning and get the right answer, you'd get all the marks.

7 Since the pentagons are regular the exterior angles are $360° \div 5 = 72°$. So the interior angles are $180° - 72° = 108°$ *[1 mark]*. Since angles around a point sum to $360°$, the angle $ABC = 360° - 108° - 108° = 144°$. *[1 mark]* The exterior angle of this polygon is $180° - 144° = 36°$ *[1 mark]*.

Then using the formula exterior angle $= \frac{360°}{n}$, the number of sides $n = \frac{360°}{36°} = 10$ *[1 mark]*.

There are a couple of different ways of answering this question — just make sure you show all your working (and get the answer right) and you'll get all the marks.

Section 21 — Circle Geometry

21.1 Circle Theorems 1

Page 267 Exercise 1

1 a) A radius meets a tangent at $90°$ so $e + 36° = 90° \Rightarrow e = 90° - 36° = 54°$

b) Both triangles have two sides that are radii, so they are isosceles.
$p = 63°$
$q = (180° - 110°) \div 2 = 35°$

c) A radius meets a tangent at $90°$ so $r + 69° = 90° \Rightarrow r = 90° - 69° = 21°$
The triangle has two sides that are radii, so it is isosceles. Therefore $s = r = 21°$

d) A radius meets a tangent at $90°$ so $i + 4i = 90° \Rightarrow 5i = 90° \Rightarrow i = 18°$

e) All three triangles have two sides that are radii, so they are isosceles.
$a = (180° - 28°) \div 2 = 76°$
$b = 180° - 24° - 24° = 132°$
The missing angles in the triangle with the $88°$ angle are both $(180° - 88°) \div 2 = 46°$.
So $c = 46° + 24° = 70°$

2 **A**, **B** and **E** must all be isosceles, as they each have two sides which are radii.

Page 268 Exercise 2

1 a) The angle in a semicircle is $90°$ so $a = 90°$.

b) A diameter bisects a chord at $90°$ so the unlabelled angle in the triangle is $90°$.
$c = 180° - 90° - 37° = 53°$

c) The angle in a semicircle is $90°$ so the unlabelled angle in both triangles is $90°$.
$m = 180° - 90° - 43° = 47°$
$n = 180° - 90° - 70° = 20°$

d) Angles on a straight line add up to $180°$ so one missing angle in the triangle is $180° - 132° = 48°$
The angle in a semicircle is $90°$ so the other missing angle in the triangle is $90°$.
$k = 180° - 90° - 48° = 42°$

e) The diameter is bisecting the chord as they meet at a right angle. The two triangles are congruent because they share a common side, they both have a side that is half the chord length and the angle between these sides on each triangle is $90°$. The smallest angle in each triangle is $34 \div 2 = 17°$ so $z = 180° - 90° - 17° = 73°$.

2 d is a right angle, because it's the angle in a semicircle.
e is a right angle, because the diameter is a perpendicular chord bisector.
i is a right angle, because a tangent and radius meet at a right angle.

Page 269 Exercise 1

For each of these questions you'll have to use the fact that the angle subtended at the centre of a circle is double the angle subtended at the circumference by the same arc.

1 a) Angle a and 128° are subtended from the same arc so $a = 128° \div 2 = $ **64°**

b) Angle b and 99° are subtended from the same arc so $b = 99° \times 2 = $ **198°**

c) The other angle around the centre and 21° are subtended from the same arc so the missing angle at the centre is $21° \times 2 = 42°$
So $c = 360° - 42° = $ **318°**

d) Angle d and 224° are subtended from the same arc so $d = 224° \div 2 = $ **112°**.

e) The angle at the centre and 53° are subtended from the same arc so the angle at the centre in the small triangle is $53° \times 2 = 106°$.
The triangle containing angle e is isosceles as two sides are radii, so
$e = (180° - 106°) \div 2 = $ **37°**

f) Angle f and 80° are subtended from the same arc so $f = 80° \div 2 = $ **40°**.

g) The 94° angle and the angle on a straight line with g are subtended from the same arc so the angle on the straight line with g is $94° \div 2 = 47°$ and $g = 180° - 47° = $ **133°**.

h) The two smaller triangles have two sides that are radii so they are isosceles.
The angles at the centre of the circle in each small triangle are
$180° - 76° - 76° = 28°$
$180° - 65° - 65° = 50°$
So the total angle at the centre is
$50° + 28° = 78°$
Angle h and 78° are subtended from the same arc so $h = 78° \div 2 = $ **39°**.

2 a) $35° + m$ and the 98° angle are subtended from the same arc, so
$35° + m = 98° \div 2 \Rightarrow 35° + m = 49°$
$\Rightarrow m = 49° - 35° = $ **14°**
m and n are subtended from the same arc, so $n = 14 \times 2 = $ **28°**

b) The other angle around the point is
$360° - (285° - x) = x + 75°$
$x + 75°$ and x are subtended from the same arc, so $x \times 2 = x + 75°$
$\Rightarrow 2x = x + 75° \Rightarrow x = $ **75°**

c) The triangle containing 60° is isosceles as two sides are radii. So the other angle at the circumference is 60° and the angle at the centre of the circle is
$180° - 60° - 60° = 60°$
$60° + t$ and the 62° angle are subtended from the same arc, so
$60° + t = 62° \times 2 \Rightarrow 60° + t = 124°$
$\Rightarrow t = 124° - 60° = $ **64°**

Page 270 Exercise 2

For each of these questions you'll have to use the fact that the angles subtended by an arc in the same segment are equal.

1 a) k, l and the 54° angle are all subtended from an arc in the same segment so
$k = $ **54°**, $l = $ **54°**.

Using the same method for b)-c):

b) $m = $ **74°**, $n = $ **106°**   **c)** $v = $ **31°**, $w = $ **61°**

d) The unlabelled angle in the triangle containing c is also 36°.
$c = 180° - 130° - 36° = $ **14°**

e) t is made up of two smaller angles, one is subtended from an arc in the same segment as 27° and the other is subtended from an arc in the same segment as 37°.
So $t = 27° + 37° = $ **64°**

f) p and the 25° angle are subtended from an arc in the same segment so $p = $ **25°**.
The unlabelled angle in the triangle containing q is also 48°.
$q = 180° - 48° - 25° = $ **107°**
You could also work out the unlabelled angle in the triangle containing p then use vertically opposite angles to find q.

g) The angle on a straight line with a is also 66° so $a = 180° - 66° = $ **114°**
The unlabelled angle in the triangle containing b is also 44° so
$b = 180° - 44° - 66° = $ **70°**

h) u and the 44° angle are subtended from an arc in the same segment so $u = $ **44°**.
$v = 180° - 44° - 109° = $ **27°**
v and w are subtended from an arc in the same segment so $w = $ **27°**.

2 a) x and the 38° angle are subtended from an arc in the same segment so $x = $ **38°**.
y and the 55° angle are subtended from an arc in the same segment so $y = $ **55°**.
The angles making up the 68° angle are x and $68° - x = 68° - 38° = 30°$
z and the 30° angle are subtended from an arc in the same segment so $z = $ **30°**.

b) The angle on a straight line with the 112° angle is $180° - 112° = 68°$.
r and the 68° angle are subtended from an arc in the same segment so $r = $ **68°**.
The missing angle in the same triangle as r and s is on a straight line with the 92° angle so it is $180° - 92° = 88°$.
So $s = 180° - 68° - 88° = $ **24°**

Page 271 Exercise 3

For each of these questions you'll have to use the fact that opposite angles in a cyclic quadrilateral add up to 180°.

1 a) $a + 94° = 180° \Rightarrow a = 180° - 94° = $ **86°**
$b + 85° = 180° \Rightarrow b = 180° - 85° = $ **95°**

Using the same method for b)-c):

b) $m = $ **108°**, $n = $ **112°**

c) $e = $ **78°**, $f = $ **78°**

d) Angle on a straight line with 71° is
$180° - 71° = 109°$
Angle opposite 109° in the cyclic quadrilateral is $180° - 109° = 71°$
u is on a straight line with the 71° angle, so $u = 180° - 71° = $ **109°**

e) x and y are both angles in a semicircle so $x = $ **90°** and $y = $ **90°**.
z is opposite the 97° angle in the cyclic quadrilateral, so $z = 180° - 97° = $ **83°**

f) $p = 180° - 49° = $ **131°**
$q = 180° - p = 180° - 131° = $ **49°**
$r = 180° - 89° = $ **91°**
$s + 89° = 180° - 36° = 144° \Rightarrow s = $ **55°**
You could also find q and s using rule 6 (angles subtended from the same arc in the same segment are equal). q is equal to the 49° angle and s is equal to $r - 36°$.

g) $a = 180° - 121° = $ **59°**
$b = 180° - 79° = $ **101°**
$c = 180° - 108° = $ **72°**

h) $k + 2k = 180° \Rightarrow 3k = 180° \Rightarrow k = $ **60°**
$3l + 2l = 180° \Rightarrow 5l = 180° \Rightarrow l = $ **36°**

21.3 Circle Theorems 3

Page 272 Exercise 1

1 a) The lines can be extended to form tangents so the triangle is isosceles as both tangents come from the same point.
$x = (180° - 64°) \div 2 = $ **58°**

b) A tangent meets a radius at right angles so one missing angle in the triangle is 90° and the other is $180° - 90° - 76° = 14°$.
The two triangles are congruent so
$g = 14° \times 2 = $ **28°**

c) The two tangents meet the radius at 90° so
$a = 360° - 140° - 90° - 90 = $ **40°**

d) The two tangents meet the radius at 90° so the angle in the centre of the circle is
$360° - 54° - 90° - 90 = 126°$
The angle subtended by the same arc at the centre is twice that at the circumference so
$k = 126° \div 2 = $ **63°**.

e) The lines can be extended to form tangents so the triangle formed by the tangents is isosceles as they both come from the same point. So $p = (180° - 62°) \div 2 = $ **59°**.
The other angle in that triangle is also 59° and a tangent meets a radius at 90°,
so $q = 90° - 59° = $ **31°**.
The angle in the semicircle is 90°, so
$r = 180° - 90° - q = 180° - 90° - 31° = $ **59°**
You can also work out the size of angle r using the alternate segment theorem (see p.273).

f) The two tangents meet the radius at 90° so the angle in the centre of the circle is
$360° - 48° - 90° - 90° = 132°$
The other angle around the centre point is
$360° - 132° = 228°$.
The angle subtended by the same arc at the centre is twice that at the circumference so
$m = 228° \div 2 = $ **114°**.

Page 273 Exercise 2

1 a) d is the angle in the alternate segment to a.

b) g is the angle in the alternate segment to c.

2 a) 67° is in the alternate segment to a and 49° is in the alternate segment to b so
$a = $ **67°** and $b = $ **49°**.

b) 59° is in the alternate segment to x so
$x = $ **59°**.
y lies on a straight line with x and 87° so
$y = 180° - 87° - 59° = $ **34°**.
z is in the alternate segment to y so
$z = $ **34°**.

c) p is in the alternate segment to 70° so
$p = $ **70°**.
Angles in a triangle add up to 180° so
$q = 180° - 76° - 70° = $ **34°**

d) The tangents form an isosceles triangle as they both come from the same point. So
$u = (180° - 90°) \div 2 = $ **45°**
Angles on a straight line add up to 180° so the missing angle on the straight line with u and 65° is $180° - 45° - 65° = 70°$.
v is in the alternate segment to 70° so
$v = $ **70°**.

e) 61° is in the alternate segment to k so
$k = $ **61°**
A tangent meets a radius at 90° so
$l = 90° - k = 90° - 61° = $ **29°**
Angle l is in an isosceles triangle as two sides are radii, so the angle at the other side is also 29°.
$30° + 29° = 59°$ is in the alternate segment to m, so $m = $ **59°**.
A tangent meets a radius at 90° so
$n = 90° - m = 90° - 59° = $ **31°**

f) The angle subtended by the same arc at the centre is twice that at the circumference so
$i = 94° \div 2 = $ **47°**.
The triangle containing the 94° angle is isosceles as two of its sides are radii, so the other two angles are
$(180° - 94°) \div 2 = 43°$.
$j + 43°$ is in the alternate segment to 78° so
$j + 43° = 78° \Rightarrow j = 78° - 43° = $ **35°**

1 a) A triangle formed by two radii is isosceles, so $a = (180° - 122°) \div 2 = \mathbf{29°}$.
The other angle in the triangle is also 29°.
So $19° + 29° = 48°$ is in the alternate segment to b, so $b = \mathbf{48°}$.

b) Angles in a semicircle are always 90°, so both triangles shown are right-angled triangles. So $c = 180° - 90° - 28° = \mathbf{62°}$.
Opposite angles in a cyclic quadrilateral sum to 180°, so $28° + d + 3d = 180°$
$\Rightarrow 4d = 152° \Rightarrow d = \mathbf{38°}$.

c) Two radii make an isosceles triangle, so the angle at the centre is
$180° - 36° - 36° = 108°$.
The angle at the centre is twice the angle at the circumference, so $e = 108 \div 2 = \mathbf{54°}$.

d) A diameter which bisects a chord meets it at a right angle so
$f = 180° - 17° - 90° = \mathbf{73°}$.
Two radii make an isosceles triangle, so
$g + 17° = (180° - 73°) \div 2$
$\Rightarrow g + 17° = 53.5° \Rightarrow g = \mathbf{36.5°}$.

e) Two radii make an isosceles triangle, so $h = \mathbf{31°}$, and the other angle in that triangle is $180° - 31° - 31° = 118°$.
Angles on a straight line add up to 180°, so $i = 180° - 118° = \mathbf{62°}$.
Tangents and radii form right angles, and angles in a quadrilateral sum to 360°, so $j = 360° - 90° - 90° - 62° = \mathbf{118°}$.

f) Tangent and radius make a right angle, so $k = 90° - 23° = \mathbf{67°}$.
Angle in alternate segment to k is also 67°, and angles on a straight line sum to 180°, so $l = 180° - 67° = \mathbf{113°}$.

g) Angles on a straight line sum to 180°, so the angle on a straight line with 104° is $180° - 104° = 76°$.
m and the 76° angle are subtended by an arc in the same segment, so $m = \mathbf{76°}$.
Opposite angles in a cyclic quadrilateral sum to 180°, so $n = 180° - 76° = \mathbf{104°}$.

h) Angles in a semicircle are right angles so $p = \mathbf{90°}$.
Tangents from the same point are equal in length so they form an isosceles triangle so $q = r = (180° - 90°) \div 2 = \mathbf{45°}$.

i) Using the alternate segment theorem, one angle in the triangle inside the circle is also 69° and the other is $180° - 55° - 69° = 56°$.
56° is in an alternate segment to s so $s = \mathbf{56°}$.
The angle on a straight line with s and the 69° angle (in the triangle) is
$180° - 69° - 56° = 55°$.
Two tangents from the same point form an isosceles triangle so
$t = 180° - 55° - 55° = \mathbf{70°}$.

2 a) $\angle BCD$ is in the alternate segment to $\angle BDE$, so $\angle BDE = \mathbf{83°}$.

b) $\angle BDE = 83°$, so $\angle ADE$ must be less than 83°. A tangent always makes a right angle with a radius or diameter, so as $\angle ADE \neq 90°$, AD cannot be a diameter.

3 a) If N were the centre of the circle, then the tangent KL and the radius NL would meet at a right angle, so $\angle KLN = \mathbf{90°}$.
So $\angle MLN = 90° - 35° = \mathbf{55°}$.

b) (i) If N were the centre, then MN and LN would both be radii, so they would be the same length.

(ii) If MN = LN, then triangle LMN would be isosceles, so $\angle LMN = \mathbf{55°}$ and $\angle MNL = 180° - 55° - 55° = \mathbf{70°}$.

c) (i) If N were the centre, then JN and LN would both be radii, so they would form congruent right-angled triangles with tangents from the same point.

(ii) If triangles KJN and KLN were congruent, and $\angle JNK = \angle KNL = 70°$, then $\angle JNL = 70° + 70° = \mathbf{140°}$.
If you worked this out using the angles in quadrilateral KJNL you'd get a different answer. $\angle KJN$ and $\angle KLN$ are 90° as a tangent meets a radius at 90°, so $\angle JNL = 360° - 90° - 90° - 45° = 135°$.

d) Angles in a quadrilateral sum to 360°, so if $\angle JNL = 140°$ and $\angle KJN = \angle KLN = 90°$, then $\angle JKL = 360° - 140° - 90° - 90° = \mathbf{40°}$.

1 AB = AC so triangle ABC is isosceles and $\angle ABC = x$ [1 mark].
Using the alternate segment theorem, $\angle ACD = x$ [1 mark].
As points on a straight line add up to 180°, $y = 180° - x - x = 180° - 2x$ [1 mark]
You could also use the alternate segment theorem to show that $\angle CAB = y$, then use the angles in an isosceles triangle to show that $y = 180° - 2x$.

2 To show that AEB and DEC are similar you can show that they have the same angles.
They share an angle so $\angle DEC = \angle AEB$.
ABCD is a cyclic quadrilateral, so
$\angle ABC = 180° - \angle ADC$
$\angle BAD = 180° - \angle BCD$
Angles on a straight line add to 180°, so
$\angle EDC = 180° - \angle ADC = \angle ABC$
$\angle ECD = 180° - \angle BCD = \angle BAD$
Now $\angle ABC = \angle EDC$ and $\angle BAD = \angle ECD$.
All three angles in the triangles are equal, so they are similar.
[3 marks available — 1 mark for showing the triangles have one pair of corresponding angles, 1 mark for showing that the triangles have a second pair of corresponding angles, 1 mark for the correct conclusion]
It's enough to show that any two angles are equal and then state that the third must be equal because angles in a triangle add up to 180°.

3 AOB and BOC are isosceles triangles because they have two sides that are radii, so $\angle ABO = x$ and $\angle BCO = y$ [1 mark]
The angles in triangle ABC add up to 180° so
$x + (x + y) + y = 180°$ [1 mark]
$\Rightarrow 2x + 2y = 180° \Rightarrow x + y = 90°$
So $\angle ABC = x + y = 90°$ [1 mark]

Section 22 — Units, Measuring and Estimating

22.1 Metric Units — Length, Mass and Volume

1 a) 1 tonne = 1000 kg, so conversion factor = 1000
You're converting to a bigger unit, so divide by the conversion factor.
$3000 \text{ kg} = 3000 \div 1000 = \mathbf{3 \text{ tonnes}}$

b) 1 g = 1000 mg, so conversion factor = 1000
You're converting to a smaller unit, so multiply by the conversion factor.
$0.4 \text{ g} = 0.4 \times 1000 = \mathbf{400 \text{ mg}}$
Using the same method for c)-f):

c) **0.123 litres**   **d)** **5.116 kg**

e) **0.0126 tonnes**   **f)** **271.65 cm**

2 a) 1 kg = 1000 g
You're converting to a smaller unit, so multiply by the conversion factor.
$0.15 \text{ kg} = 0.15 \times 1000 = 150 \text{ g}$
1 g = 1000 mg
You're converting to a smaller unit, so multiply by the conversion factor.
$150 \times 1000 = \mathbf{150\,000 \text{ mg}}$
*You could do the two steps all in one go.
1 kg = 1 000 000 mg , so you would multiply by 1 000 000.*

b) 1 kg = 1000 g
You're converting to a bigger unit, so divide by the conversion factor.
$1532 \text{ g} = 1532 \div 1000 = 1.532 \text{ kg}$
1 tonne = 1000 kg
You're converting to a bigger unit, so divide by the conversion factor.
$1.532 \text{ kg} = 1.532 \div 1000$
$= \mathbf{0.001532 \text{ tonnes}}$
Using the same method for c)-f):

c) **0.01005 km**   **d)** **0.003023 kg**

e) **0.000003 km**   **f)** **49 000 000 g**

3 $400 \text{ ml} \times 32 = 12\,800 \text{ ml}$
$12\,800 \text{ ml} = 12\,800 \div 1000 = 12.8 \text{ litres}$
$12.8 \text{ litres} \div 2 \text{ litres/bottle} = 6.4$, so **7 bottles**
You need to round up because 6 bottles would not give enough to fill 32 glasses.

4 a) 5 litres = $5 \times 1000 = 5000$ ml
$5000 \text{ ml} \div 10 \text{ ml/lap} = \mathbf{500 \text{ laps}}$

b) 400 m = $400 \div 1000 = 0.4$ km
$0.4 \text{ km} \times 500 \text{ laps} = \mathbf{200 \text{ km}}$

5 1 pair of skis weighs 10 000 g = 10 kg, so 3 pairs of skis weigh $10 \times 3 = 30$ kg.
$55.2 + 78.1 + 65.9 + 30 = 229.2$ kg
$229.2 \text{ kg} = 229.2 \div 1000 = 0.2292$ tonnes, which is less than half a tonne, so **yes, they will be safe.**

6 $1500 \text{ m} = 1500 \div 1000 = 1.5$ km
50 m = 0.05 km
$1.5 + 0.05 + 13.2 = \mathbf{14.75 \text{ km}}$

7 Beaker = 750 cm³, bottle = 0.55 litres
$0.55 \text{ litres} = 0.55 \times 1000 = 550 \text{ cm}^3$
$750 - 550 = \mathbf{200 \text{ cm}^3}$

8 The reservoir can take $800\,000 - 600\,000 = 200\,000$ litres before it overflows.
$750\,000 \text{ ml} = 750\,000 \div 1000 = 750 \text{ litres}$
$N = 200\,000 \text{ litres} \div 750 \text{ litres/day} = 266.66...$
It'll overflow after 266.66... days, so **266 whole days** would have passed.

9 a) $400 \text{ g} = 400 \div 1000 = 0.4 \text{ kg}$
$300 \text{ g} = 300 \div 1000 = 0.3 \text{ kg}$
$2500 \text{ mg} = 2500 \div 1000 = 2.5 \text{ g}$
$2.5 \text{ g} = 2.5 \div 1000 = 0.0025 \text{ kg}$
$0.7 + 0.4 + 0.3 + 0.2 + 0.0025 = \mathbf{1.6025 \text{ kg}}$

b) $1.6025 \text{ kg} \div 0.2 \text{ kg/person}$
$= 8.0125$, so **8 people**

22.2 Metric Units — Area and Volume

1 a) 1 cm = 10 mm, so 1 cm² = 10^2 = 100 mm²
You're converting to a bigger unit, so divide by the conversion factor.
$84 \text{ mm}^2 = 84 \div 100 = \mathbf{0.84 \text{ cm}^2}$

b) 1 m = 100 cm, so 1 m² = 100^2
$= 10\,000 \text{ cm}^2$
You're converting to a bigger unit, so divide by the conversion factor.
$1750 \text{ cm}^2 = 1750 \div 10\,000 = \mathbf{0.175 \text{ m}^2}$
Using the same method for c)-l):

c) **290 cm²**   **d)** **1 000 000 m³**

e) **15 000 mm³**   **f)** **200 000 cm³**

g) **3.15 km²**   **h)** **85 cm²**

i) **0.17 m²**   **j)** **435 000 000 m³**

k) **6 700 000 000 m³** **l)** **0.45 mm³**

2 Area 1 = $1.7 \times 3 = 5.1 \text{ m}^2$
Area 2 = $670 \times 420 = 281\,400 \text{ cm}^2$
$281\,400 \text{ cm}^2 = 281\,400 \div 100^2 = 28.14 \text{ m}^2$
Total area = $5.1 + 28.14 = \mathbf{33.24 \text{ m}^2}$

3 a) 1 litre = 1000 ml = 1000 cm³
$1000 \div 25 = \mathbf{40 \text{ glasses}}$

b) 0.5 litres = 500 cm³
$500 + 25 = 525 \text{ cm}^3$
$525 \times 1000 = \mathbf{525\,000 \text{ mm}^3}$

4 a) Volume = $3 \times 375 = 1125 \text{ m}^3$
$1125 \text{ m}^3 = 1125 \times 100^3$
$= \mathbf{1\,125\,000\,000 \text{ cm}^3}$

b) 1 litre = 1000 ml = 1000 cm³
So 1 125 000 000 cm³ ÷ 1000
= **1 125 000 litres**

5 a) Volume = 20.7 × 25.5 × 10 = 5278.5 cm³
5278.5 cm³ = 5278.5 ÷ 100³
= 0.0052785 m³
0.0052785 − 0.003 = **0.0022785 m³**

b) Area of face 1: 20.7 × 25.5 = 527.85 cm²
Area of face 2: 20.7 × 10 = 207 cm²
Area of face 3: 25.5 × 10 = 255 cm²
(2 × 527.85) + (2 × 207) + (2 × 255)
= 1979.7 cm²
1979.7 cm² = 1979.7 × 10²
= **197 970 mm²**

6 a) 1 m = 1000 mm,
so 1 m² = 1000² = 1 000 000 mm²
You're converting to a smaller unit, so
multiply by the conversion factor.
1.2 m² = 1.2 × 1 000 000 = **1 200 000 mm²**

b) 1 km = 100 000 cm, so
1 km² = 100 000² = 10 000 000 000 cm²
You're converting to a smaller unit, so
multiply by the conversion factor.
0.001 km² = 0.001 × 10 000 000 000
= **10 000 000 cm²**

Using the same method for c)-f):
c) 50 m²   **d) 0.000000000003 km³**
e) 600 000 mm³
f) 0.000000000000999 km³
You could write these answers in standard form.
For example, 0.000000000000999 km³
= 9.99 × 10⁻¹³ km³.

22.3 Metric and Imperial Units
Page 281   Exercise 1
1 a) 3 ft 7 in = (3 × 12) + 7 = **43 inches**
b) 12 ft 5 in = (12 × 12) + 5 = **149 inches**
c) 5 lb 2 oz = (5 × 16) + 2 = **80 ounces**
d) 280 in = 280 ÷ 12 = 23 remainder 4
= **23 ft 4 in**
e) 1001 in = 1001 ÷ 12 = 83 remainder 5
= **83 ft 5 in**
f) 72 oz = 72 ÷ 16 = 4 remainder 8
= **4 lb 8 oz**
g) 70 lb = 70 ÷ 14 = **5 stone**
h) 200 oz = 200 ÷ 16 = 12 remainder 8
= **12 lb 8 oz**
i) 5.5 yards = 5.5 × 3 = 16.5 ft
= **16 ft 6 in**
j) 4.75 ft = 4 ft + (0.75 × 12) in = **4 ft 9 in**
k) 2.5 stone = 2.5 × 14 = **35 lb**
l) 8.25 stone = 8.25 × 14 = 115.5 lb
115 lb 8 oz

2 a) 1904 g ≈ 1904 ÷ 28 = 68 oz
68 ÷ 16 = 4 remainder 4 = **4 lb 4 oz**
You might have divided by 450 first to get
4.23... lb and then converted to lb and oz.
b) 840 g ≈ 840 ÷ 28 = 30 ounces
30 ÷ 16 = 1 remainder 14 = **1 lb 14 oz**
c) 2688 g ≈ 2688 ÷ 28 = 96 ounces
96 ÷ 16 = **6 lb**
d) 4.9 kg = 4900 g
4900 g ≈ 4900 ÷ 28 = 175 ounces
175 ÷ 16 = 10 remainder 15 = **10 lb 15 oz**
Alternatively, you might have converted 4.9 kg
to 10.78 pounds using 1 kg ≈ 2.2 lb. Then
this is approximately equal to 10 lb 12 oz.
Since these conversions are only
approximations, you sometimes get different
answers using different methods.

3 a) 2 m = 200 cm ≈ 200 ÷ 2.5 = 80 in
80 ÷ 12 = 6 remainder 8 = **6 ft 8 in**
b) 52.5 cm ≈ 52.5 ÷ 2.5 = 21 in
21 ÷ 12 = 1 remainder 9 = **1 ft 9 in**

Using the same method for c)-d):
c) 5000 ft   **d) 2 in**
4 a) 10 feet ≈ 10 × 30 = 300 cm
300 cm = 3 m so **3.5 m > 10 ft**
b) 1 stone ≈ 6400 g
6400 g = 6.4 kg so **7 kg > 1 stone**
Using the same method for c)-i):
c) 10 miles > 12 km
d) 9 litres > 15 pints
e) 1.5 kg > 3 lb
f) 5 stone > 31 kg
g) 160 stone > 1 tonne
h) 10 litres > 2 gallons
i) 16 lb > 7 kg

5 Maddie: 4 ft 5 in = 4 × 12 + 5 = 53 in
53 in ≈ 53 × 2.5 = 132.5 cm
Lily: 4 ft 9 in = 57 in
57 in ≈ 57 × 2.5 = 142.5 cm
Only Lily is tall enough to ride.

6 400 m = 0.4 km
1 mile ≈ 1.6 km
1.6 ÷ 0.4 = **4 laps**

7 1 lb 12 oz = 28 oz
28 oz ≈ 28 × 28 = 784 g, so **no.**

8 a) 1 pt ≈ 0.57 litres
3 ÷ 0.57 = 5.263... pints,
so **5 jugs can be filled.**
b) 0.57 × 5 = 2.85 litres, 3 − 2.85 = **0.15 litres**

9 90 km ≈ 90 ÷ 1.6 = 56.25 miles
So the speed limit is **56 mph**
(to the nearest mph).

22.4 Estimating in Real Life
Page 282   Exercise 1
1 a) (i) cm   **(ii) in**
b) (i) grams   **(ii) oz**
c) (i) mm   **(ii) in**
The length of an ant would be
a fraction of an inch.
d) (i) tonnes   **(ii) stone**
You could also use imperial tons, which
are equivalent to 160 stone.
e) (i) ml, cm³   **(ii) pints**
The volume of a tea cup would
be a fraction of a pint.
f) (i) km   **(ii) miles**
2 a) About 1.5 × the height of
a person so **2-3 m**.
b) Depending on the length of the car, about
2-2.5 × the height of a person so **3.5-5 m**.
c) A bit taller than the height of
a person so **2-2.5 m**.
d) A person's arm span is about the same as
their height so **1.5-2 m**.
e) A football is a bit smaller than a
30 cm ruler so **20-25 cm**.
f) Model the bath as a cuboid. Then it's
a bit shorter and wider than a person,
and about 0.5 m deep. So the volume is
approximately 1.6 × 0.7 × 0.5 = 0.56 m³
= **560 litres.**
This is very accurate for an estimate, you
could model it more approximately as
2 m × 0.5 m × 0.5 m = 0.5 m³ = 500 litres.

Page 283   Review Exercise
1 a) 1 kg = 1000 g, so the conversion factor is
1000. You're converting to a smaller unit,
so multiply by the conversion factor.
10 kg = 10 × 1000 = **10 000 g**
b) 1 cm = 10 mm, so the conversion factor is
10. You're converting to a smaller unit, so
multiply by the conversion factor.
14 cm = 14 × 10 = **140 mm**

Using the same method for c)-f):
c) 4600 cm³   **d) 0.022 kg**
e) 0.15 litres   **f) 69 m**
2 a) 1 m = 1000 mm, so the conversion factor is
1000. You're converting to a smaller unit,
so multiply by the conversion factor.
0.006 m = 0.006 × 1000 = **6 mm**
b) 1 kg = 1000 g and 1 g = 1000 mg
so 1 kg = 1000 × 1000 = 1 000 000 mg
You're converting to a smaller unit, so
multiply by the conversion factor.
0.57 × 1 000 000 = **570 000 mg**
Using the same method for c)-f):
c) 120 000 cm   **d) 8 kg**
e) 0.00012 km   **f) 0.001101 tonnes**
3 a) 1 m = 100 cm,
so 1 m² = 100² = 10 000 cm²
You're converting to a bigger unit,
so divide by the conversion factor.
10 cm² = 10 ÷ 10 000 = **0.001 m²**
b) 1 km = 1000 m,
so 1 km² = 1000² = 1 000 000 m²
You're converting to a smaller unit, so
multiply by the conversion factor.
18 km² = 18 × 1 000 000 = **18 000 000 m²**
Using the same method for c)-f):
c) 8 600 000 mm²   **d) 0.0673 km²**
e) 0.0605 m²   **f) 5 000 000 mm²**
4 a) 1 m = 100 cm,
so 1 m³ = 100³ = 1 000 000 cm³
You're converting to a smaller unit, so
multiply by the conversion factor.
0.005 m³ = 0.005 × 1 000 000 = **5000 cm³**
b) 1 cm = 10 mm,
so 1 cm³ = 10³ = 1000 mm³
You're converting to a bigger unit,
so divide by the conversion factor.
69 mm³ = 69 ÷ 1000 = **0.069 cm³**
Using the same method for c)-f):
c) 0.00072 m³   **d) 0.000019 m³**
e) 0.00001744 m³
f) 3 450 000 000 000 000 mm³
This is 3.45 × 10¹⁵ in standard form.

5 1 m = 100 cm,
so 1 m³ = 100³ = 1 000 000 cm³
You're converting to a bigger unit, so divide
by the conversion factor.
49 900 cm³ = 49 900 ÷ 1 000 000
= **0.0499 m³**

6 a) 1 yard ≈ 90 cm, so conversion factor is 90
12.4 yards ≈ 12.4 × 90 = 1116 cm
681 + 1116 = 1797 cm
1 m = 100 cm
1797 ÷ 100 = 17.97 = **18 m** (2 s.f.)
b) 1 mile ≈ 1.6 km, so conversion factor is 1.6
21.5 miles ≈ 21.5 × 1.6 = 34.4 km
16.49 + 34.4 = 50.89 = **51 km** (2 s.f.)
Using the same method for c)-f):
c) 11 kg (2 s.f.)   **d) 730 cm** (2 s.f.)
e) 5800 ml (2 s.f.)   **f) 720 cm** (2 s.f.)

7 1 pint ≈ 0.57 litres, so the
conversion factor is 0.57
4 pints ≈ 4 × 0.57 = 2.28 litres
So **2 litres** is closest to 4 pints.

8 1 stone = 14 lb so:
10 stone = 140 lb, 8 stone = 112 lb,
14 stone 1 lb = 197 lb, 9 stone 9 lb = 135 lb,
12 stone = 168 lb, 17 stone = 238 lb
140 + 112 + 197 + 135 + 168 + 238 = 990 lb
990 lb ≈ 990 ÷ 2.2 = 450 kg
450 kg ≈ 450 ÷ 1000 = 0.45 < 0.5 tonnes so
no, the limit is not exceeded.

9 The bus is about twice as tall as the man
leaning against it. So if you use the height
of an average man as 1.8 m, the bus will be
3.6 m tall. The bus is about 3 times as long as
it is tall, so the bus will be **10.8 m** long.

1 The dinosaur is about 3 times as tall as the chicken and 7 times as long. So if you take the height and length of an average chicken to be 30 cm, the dinosaur will be around $30 \times 3 = 90$ cm = **0.9 m** tall and $30 \times 7 = 210$ cm = **2.1 m** long.
[2 marks available — 1 mark for estimating the height and length of the chicken, 1 mark for using these measurements to estimate the height and length of the dinosaur in m or cm]

2 1.2 kg = $1.2 \times 1000 = 1200$ kg
$1200 \div 300 =$ **4 pizzas** *[1 mark]*

3 Volume = $150^3 = 3\ 375\ 000$ cm^3 *[1 mark]*
1 m = 100 cm,
so 1 m$^3 = 100^3 = 1\ 000\ 000$ cm^3
You're converting to a bigger unit, so divide by the conversion factor.
$3\ 375\ 000$ cm$^3 = 3\ 375\ 000 \div 1\ 000\ 000$
$= $ **3.375 m^3** *[1 mark]*
You could've converted the side lengths to metres first.

4 Reserve 1: $2.9 \times 3.3 = 9.57$ km^2
1 km = 1000 m, so the conversion factor is 1000. You're converting to a bigger unit, so divide by the conversion factor.
2700 m = $2700 \div 1000 = 2.7$ km
4100 m = 4.1 km
Reserve 2: $2.7 \times 4.1 = 11.07$ km^2
So **Reserve 2 is the largest**.
[2 marks available — 1 mark for the correct conversion(s) into the same units, 1 mark for the correct answer justified by the calculation of both areas]

5 1 inch ≈ 2.5 cm, so conversion factor is 2.5
16 in ≈ $16 \times 2.5 = 40$ cm of ribbon
1 m = 100 cm, so conversion factor is 100
You're converting to a bigger unit, so divide by the conversion factor.
40 cm = $40 \div 100 = 0.4$ m
$0.4 \times 1.5 = $ £0.60 per hat
$33 \div 0.6 = $ **55 hats**
[3 marks available — 1 mark for conversion between inches and cm, 1 mark for conversion between cm and m, 1 mark for the correct answer]

6 1 litre = 1000 cm^3, so the conversion factor is 1000. You're converting to a smaller unit, so multiply by the conversion factor.
0.25 litres = $0.25 \times 1000 = 250$ cm^3
Subtract the volume of milk from cup of tea:
$250 - 25 = 225$ cm^3
Multiply by the number of cups:
$225 \times 5 = 1125$ cm^3
1125 cm$^3 =$ **1125 ml**
[3 marks available — 1 mark for conversion between litres and cm^3, 1 mark for volume of water in one cup, 1 mark for correct answer with correct units]

7 a) 45 litres ≈ $45 \div 4.5 = 10$ gallons
10 gallons $\times$ 55 miles/gallon = **550 miles**
[2 marks available — 1 mark for conversion between litres and gallons, 1 mark for correct answer]

b) 45 litres ≈ 10 gallons (from a))
10 gallons $\times$ £5 = **£50** *[1 mark]*

Section 23 — Compound Measures

23.1 Compound Measures

Page 285 Exercise 1

1 a) speed $= \dfrac{\text{distance}}{\text{time}} = \dfrac{1800}{4.5} = $ **400 mph**
Remember to check units — you're putting in miles and hours so the speed will be in miles per hour (mph).
Using the same method for b)-d):
b) **1.25 m/s**   c) **21.25 km/h**
d) **0.625 m/s**

---

2 a) 300 000 m = 300 km
speed $= \dfrac{300}{2.5} = $ **120 km/h**

b) 45 minutes = $45 \div 60 = 0.75$ hours
speed $= \dfrac{5.25}{0.75} = $ **7 km/h**

Using the same method for c)-d):
c) **2.5 km/h**   d) **14 km/h**

3 speed $= \dfrac{232\,900}{13.7} = $ **17 000 mph**

4 1.4 km = 1400 m
speed $= \dfrac{1400}{65} = 21.5384...$ m/s
$=$ **22 m/s** (2 s.f.)

5 98 cm = 0.98 m
8 minutes = $8 \times 60 = 480$ seconds
speed $= \dfrac{0.98}{480} = 0.00204...$ m/s
$=$ **0.002 m/s** (1 s.f.)

Page 286 Exercise 2

1 a) distance = speed $\times$ time = $98 \times 3.5 = $ **343 km**
b) distance = $25 \times 2.7 = $ **67.5 miles**
c) 9 minutes = $9 \times 60 = 540$ seconds
distance = $15 \times 540 = $ **8100 m**
d) 171 minutes = $171 \div 60 = 2.85$ hours
distance = $72 \times 2.85 = $ **205.2 miles**

2 a) time $= \dfrac{\text{distance}}{\text{speed}} = \dfrac{4}{2.5} = $ **1.6 hours**
b) time $= \dfrac{9.3}{5} = $ **1.86 seconds**
c) 61.2 km = 61 200 m
time $= \dfrac{61\,200}{9} = $ **6800 seconds**
d) 1.96 m = $1.96 \times 100 = 196$ cm
time $= \dfrac{196}{8} = $ **24.5 seconds**

3 time $= \dfrac{2.4}{15} = $ **0.16 seconds**

4 2 hours 15 mins = 2.25 hours
distance = $480 \times 2.25 = $ **1080 miles**

5 75 minutes = $75 \div 60 = 1.25$ hours
distance = $7.5 \times 1.25 = $ **9.375 miles**

6 time $= \dfrac{5.6}{78} = 0.0717...$ hours
$0.0717... \times 60 = 4.3076...$ minutes
$=$ **4 minutes** (to nearest minute)

Page 287 Exercise 3

1 a) density $= \dfrac{\text{mass}}{\text{volume}} = \dfrac{200}{540}$
$=$ **0.37 kg/m^3** (2 d.p.)
Remember to check units — you're putting in kg and m^3 so the density will be in kg/m^3.
Using the same method for b)-d):
b) **46 kg/m^3**   c) **680 kg/m^3**
d) **9992 g/cm^3**

2 a) volume $= \dfrac{\text{mass}}{\text{density}} = \dfrac{1}{8} = $ **0.125 m^3**
Using the same method for b)-d):
b) **2.6 m^3**   c) **6 cm^3**   d) **0.06 m^3**

3 mass = density $\times$ volume = 2610×0.4
$=$ **1044 kg**

Page 288 Exercise 4

1 a) pressure $= \dfrac{\text{force}}{\text{area}} = \dfrac{4800}{4} = $ **1200 Pa**
b) pressure $= \dfrac{640}{80} = $ **8 N/cm^2**
c) area $= \dfrac{\text{force}}{\text{pressure}} = \dfrac{540}{180} = $ **3 m^2**
d) force = pressure $\times$ area = $36 \times 30 = $ **1080 N**

2 Side length of cube = $\sqrt[3]{512} = 8$ cm
Area of cube = $8^2 = 64$ cm^2
Pressure $= \dfrac{1792}{64} = $ **28 N/cm^2**

3 Volume of a cylinder = area of base $\times$ height
Area of base $= \dfrac{\text{force}}{\text{pressure}} = \dfrac{560}{70\,000} = 0.008$ m^2
0.008 m$^2 = 0.008 \times 100^2 = 80$ cm^2
Volume of cylinder = 80 cm$^2 \times 80$ cm
$=$ **6400 cm^3**

Page 289 Exercise 5

1 a) 1 kg = 1000 g
£5 per gram = $5 \times 1000 = $ **£5000 per kg**
A kilogram is larger than a gram so you'd expect the cost per kg to be more than the cost per g — so multiply by the conversion factor.

b) 1 hour = 60 minutes
£36/hour = $36 \div 60 = 0.6$
$=$ **£0.60 per minute**
A minute is smaller than an hour so you'd expect the cost per minute to be less than the cost per hour — so divide by the conversion factor.

c) 1 hour = 60 minutes = 3600 seconds
62 m/s = 62×3600
$=$ **223 200 metres per hour**

d) 1 km = 1000 m
54 km/h = $54 \times 1000 = 54\,000$ m per hour
1 hour = 3600 seconds
54 000 m per hour = $54\,000 \div 3600$
$=$ **15 m/s**

e) 1 m$^2 = 100^2$ cm$^2 = 10\,000$ cm^2
3000 N/cm$^2 = 3000 \times 10\,000$
$=$ **30 000 000 N/m^2**

f) 1 m = 100 cm
156 m per hour = 156×100
$= 15\,600$ cm per hour
1 hour = 3600 seconds
15 600 cm per hour = $\dfrac{15\,600}{3600} = $ **4$\frac{1}{3}$ cm/s**

g) 1 kg = 1000 g
830 kg/m^3 = $830 \times 1000 = 830\,000$ g/m^3
1 m$^3 = 100^3$ cm$^3 = 1\,000\,000$ cm^3
830 000 g/m$^3 = \dfrac{830\,000}{1\,000\,000} = $ **0.83 g/cm^3**

h) £1 = 100p
£12/kg = $12 \times 100 = 1200$p per kg
1 kg = 1000 g
1200p per kg $= \dfrac{1200}{1000} = $ **1.2p per gram**

i) 1 mile ≈ 1.6 km = 1600 m
29.25 mph = 29.25×1600
$= 46\,800$ m per hour
1 hour = 3600 seconds
46 800 m per hour $= \dfrac{46\,800}{3600} = $ **13 m/s**

2 1 kg = 1000 g
97 g/cm$^3 = \dfrac{97}{1000} = 0.097$ kg/cm^3
1 m$^3 = 100^3$ cm$^3 = 1\,000\,000$ cm^3
0.097 kg/cm$^3 = 0.097 \times 1\,000\,000$
$=$ **97 000 kg/m^3**

3 £1 = 100p
12p per minute = $12 \div 100 = $ £0.12 per minute
1 hour = 60 minutes
£0.12 $\times 60 = $ **£7.20 per hour**

4 The journey was $1.5 + 3 = 4.5$ hours in total.
So $\dfrac{1.5}{4.5} = \dfrac{1}{3}$ of the journey averaged 40 m/s
and $\dfrac{3}{4.5} = \dfrac{2}{3}$ of the journey averaged 46 m/s.
The average speed for the whole journey was
$\left(\dfrac{1}{3} \times 40\right) + \left(\dfrac{2}{3} \times 46\right) = 44$ m/s.
1 hour = 60 minutes = 3600 seconds
44 m/s = $44 \times 3600 = 158\,400$ m per hour
1 mile ≈ 1.6 km = 1600 m
158 400 m per hour $= \dfrac{158\,400}{1600} = $ **99 mph**

5 Average speed $= \dfrac{x+y}{2}$ km/h
1 km = 1000 m
$\dfrac{x+y}{2}$ km/h $= \dfrac{x+y}{2} \times 1000$
$= 500(x+y)$ m per hour
1 hour = 60 minutes = 3600 seconds
$500(x+y)$ m per hour $= \dfrac{500(x+y)}{3600}$
$= \dfrac{5(x+y)}{36}$ **m/s**

Page 290   Exercise 1

1 Put distance in km on the y-axis and time in hours on the x-axis. Plot each point where a change in speed occurs, then join the points up. The points are (0h50, 3 km), (1h50, 3 km), (2h20, 1 km), (2h30, 1 km) and (2h45, 0 km)

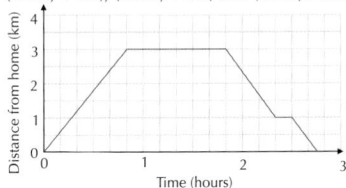

2 a) (i) They stopped at the horizontal part of the graph so they travelled for **1 hour** before stopping.

   (ii) Reading the y-axis at the point where they stopped, they travelled **50 miles**.

   b) The graph is horizontal from 1 hour to 2 hours, so they stayed at their destination for 2 − 1 = **1 hour**.

   c) (i) The graph starts to decline at 2 hours which shows they set off for home then. 8:00 am + 2 hours = **10:00 am**

   (ii) They set off at 2 hours and got back at 4 hours so it took 4 − 2 = **2 hours**.

   d) The family travelled at a greater speed **on the way to their destination** because, e.g. the line for the journey there is **steeper** than for the journey home.
   You could also mention that it took less time to get to their destination than it did to get back.

3 a) Between 0 and 1 hours the object travelled 2.5 km at a constant speed. Between 1 hour and 1 h 30 mins the object was stationary. Between 1 h 30 mins and 2 hours the object travelled 1.5 km at a constant speed. Between 2 hours and 2 h 15 mins the object was stationary. Between 2 h 15 mins and 3 hours the object travelled 4 km in the opposite direction at a constant speed.

   b) For the first 40 minutes the object travelled 1 km while decelerating. Between 40 minutes and 1 hour the object was stationary. Between 1 hour and 1 h 20 mins, the object travelled 0.5 km at a constant speed. Between 1 h 20 mins and 1 h 30 mins the object was stationary. Between 1 h 30 mins and 1 h 50 mins the object travelled 2 km at a constant speed. Between 1 h 50 mins and 2 hours the object was stationary.

   c) In the first 45 seconds the object travelled 15 m at a constant speed. Between 45 seconds and 1 min 30 s the object was stationary. Between 1 min 30 s and 2 mins 15 s the object travelled 10 m in the opposite direction, accelerating for the first 2.5 m then decelerating for the second 2.5 m. Between 2 mins 15 s and 2 mins 45 s the object was stationary. In the last 15 seconds the object travelled 10 m at a constant speed.

Page 292   Exercise 2

1 a) Speed = $\dfrac{\text{distance travelled}}{\text{time taken}}$
   He travelled 4 km in 30 minutes.
   30 minutes = $\dfrac{30}{60}$ = 0.5 hours
   Speed = $\dfrac{4}{0.5}$ = **8 km/h**

   b) Distance travelled = 9 − 4 = 5 km
   It took from 08:00 to 08:15, so
   time = 15 minutes = $\dfrac{15}{60}$ = 0.25 hours
   Speed = $\dfrac{5}{0.25}$ = **20 km/h**

---

2 a) Steepest gradient for Cyclist 1 is near the end of the journey — they travel
   8.8 − 3.4 = 5.4 km in 15 minutes.
   15 minutes = 0.25 hours
   Speed = $\dfrac{5.4}{0.25}$ = 21.6 km/h

   Steepest gradient for Cyclist 2 is at the start of their journey — they travel
   5.8 − 0 = 5.8 km in 15 minutes.
   15 minutes = 0.25 hours
   Speed = $\dfrac{5.8}{0.25}$ = 23.2 km/h
   So **Cyclist 2** has the maximum speed.
   You could also just look at the distances — Cyclist 2 travelled further in the same time so must have had a higher maximum speed.

   b) Cyclist 1 travelled 3.4 km in the first
   15 minutes = 0.25 hours.
   Speed = $\dfrac{3.4}{0.25}$ = 13.6 km/h
   Speed of Cyclist 2 = 23.2 km/h from part a).
   Difference = 23.2 − 13.6 = **9.6 km/h**

3 a) (i) Chay was on the bus between 8:00 and 9:30, so
   time = 1.5 hours, distance = 30 km
   Speed = $\dfrac{30}{1.5}$ = **20 km/h**

   (ii) Chay was on the train between 10:15 and 11:00, so
   time = 0.75 hours,
   distance = 90 − 30 = 60 km
   Speed = $\dfrac{60}{0.75}$ = **80 km/h**

   b) (i) Chay is 90 km from home and he travels at a speed of 60 km/h.
   Time = $\dfrac{\text{distance}}{\text{speed}}$ = $\dfrac{90}{60}$ = **1.5 hours**

   (ii) From 11:00 till 11:30 he is in Clapham and doesn't travel any distance. Then it takes 1 hour 30 minutes to get home so he arrives back home (0 km) at 13:00.

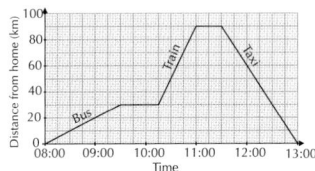

4 a) 280 miles at 80 mph gives a time of
   $\dfrac{\text{distance}}{\text{speed}} = \dfrac{280}{80}$ = 3.5 hours.
   So he arrives at 10:00 + 3.5 hours = 13:30.

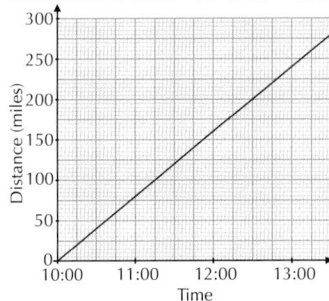

   b) For the first part:
   Time = $\dfrac{90}{60}$ = 1.5 hours.
   So he arrives at his destination at
   9:00 + 1.5 hours = 10:30.
   Corey sets off home at
   10:30 + 45 mins = 11:15.
   For the last part, he travels 90 miles at the speed of 40 mph.
   Time = $\dfrac{90}{40}$ = 2.25 hours.
   Corey arrives home at
   11:15 + 2.25 hours = 13:30

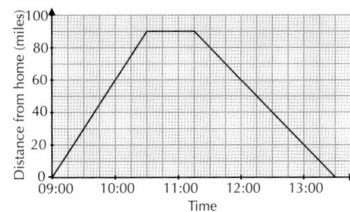

5 Draw tangents to each curve at 1 hour.

   a)
   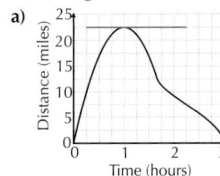
   The tangent is horizontal so the gradient is 0, so speed = **0 mph**.

   b)
   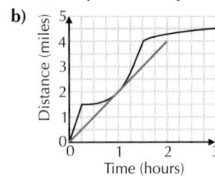
   Tangent passes through (0, 0) and (2, 4), so
   gradient = $\dfrac{4-0}{2-0}$ = **2 mph**.
   (Accept 1.8 mph to 2.2 mph)

   c)
   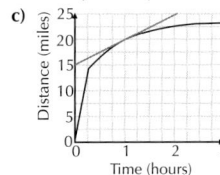
   Tangent passes through (0, 15) and (2, 25), so gradient = $\dfrac{25-15}{2-0}$ = **5 mph**.
   (Accept 4.8 mph to 5.2 mph)
   You can review finding a gradient in Section 17.

23.3 Velocity-Time Graphs

Page 294   Exercise 1

In this exercise, use the following formula:
Acceleration = $\dfrac{\text{change in velocity}}{\text{change in time}}$

1 a) $\dfrac{30-0}{2.5-0}$ = **12 km/h²**
   Always check units — you're putting km/h and hours in so you'll get km/h² out.

   b) The line is horizontal so it's a constant velocity, which means acceleration is **0 m/s²**.

   c) $\dfrac{0-25}{4-0}$ = **−6.25 m/s²**

2 a) Maximum velocity is the highest point on the graph = **45 km/h**

   b) $\dfrac{45-0}{0.5-0}$ = **90 km/h²**

   c) $\dfrac{0-45}{0.75-0}$ = −60 km/h², so the deceleration is **60 km/h²**.

Page 294   Exercise 2

1 a) (i) The velocity **increases from 0 km/h to 5 km/h**.
   (ii) The velocity is **constant at 11.5 km/h**.
   (iii) The velocity **decreases from 7.5 km/h to 0 km/h**.

   b) (i) The object is **accelerating** and the **rate of acceleration is decreasing**.
   (ii) **No acceleration**.
   (iii) The object is **decelerating** and the **rate of deceleration is increasing**.

c) (i) Acceleration $= \frac{11.5 - 5}{5.4 - 3} = \frac{6.5}{2.4}$

$= 2.708\ldots$ km/h^2 = **2.71 km/h^2**

(ii) Acceleration $= \frac{7.5 - 11.5}{10 - 8}$

$= \frac{-4}{2} =$ **−2 km/h^2**

d) Draw tangents at 1 hour and 11 hours.

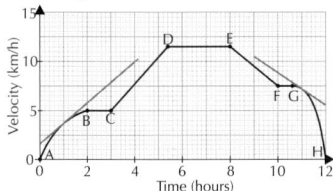

(i) The tangent at 1 hour goes through (0.4, 2.5) and (4, 10), so

gradient $= \frac{10 - 2.5}{4 - 0.4} =$ **2.1 km/h^2** (1 d.p.)

(Accept 1.9 km/h^2 to 2.3 km/h^2)

(ii) The tangent at 11 hours goes through (9, 10.5) and (12, 5.5), so

gradient $= \frac{5.5 - 10.5}{12 - 9}$

$=$ **−1.7 km/h^2** (1 d.p.)

(Accept −1.5 km/h^2 to −1.9 km/h^2)

Page 295   Exercise 3

To find distances you need to work out the area under the graph each time.

1 a) The area under the graph is a rectangle with length 2.5 and width 15.

Distance $= 15 \times 2.5 =$ **37.5 km**

b) The area under the graph is a triangle with base 8 and height 20.

Distance $= \frac{1}{2}(8 \times 20) =$ **80 miles**

c) The area under the graph is a triangle with base 20 and height 25.

Distance $= \frac{1}{2}(20 \times 25) =$ **250 m**

2 a) The area under the graph is a trapezium with height 25 and parallel sides of 60 and 60 − 30 = 30.

Distance $= \frac{1}{2}(60 + 30) \times 25 =$ **1125 m**

You could also have split the area under the curve into a triangle and rectangle and added their areas together.

b) Average velocity $= \frac{\text{total distance}}{\text{total time}}$

$= \frac{1125}{60} =$ **18.75 m/s**

3 a) Draw a straight line starting at 0 m/s and reaching 48 m/s after 8 seconds. Then draw a horizontal line showing a constant velocity of 48 m/s for 20 seconds ending at 28 seconds. Finally, draw another straight line from 48 m/s to 23 m/s over 5 seconds so ending at 33 seconds.

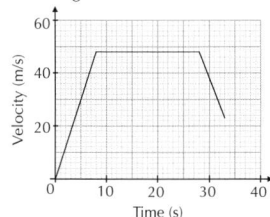

b) Acceleration $= \frac{48 - 0}{8 - 0} =$ **6 m/s^2**

Deceleration $= \frac{48 - 23}{5} =$ **5 m/s^2**

The acceleration would be −5 m/s^2.

c) Area under the graph can be broken into two trapeziums.

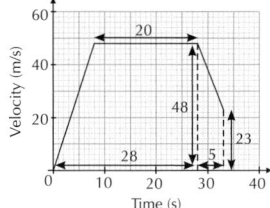

1st trapezium $= \frac{1}{2}(28 + 20) \times 48 = 1152$

2nd trapezium $= \frac{1}{2}(48 + 23) \times 5 = 177.5$

Total distance $= 1152 + 177.5 =$ **1329.5 m**

There are lots of other ways to split the area up but they should all give the same answer.

Page 296   Exercise 4

1 Use this graph to answer the questions.

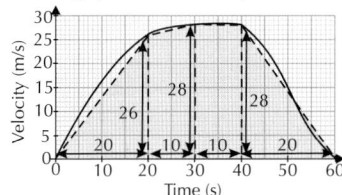

a) The area under the first 20 seconds is a triangle. Distance $= \frac{1}{2}(20 \times 26) =$ **260 m**

b) The area under the last 20 seconds is a triangle. Distance $= \frac{1}{2}(20 \times 28) =$ **280 m**

c) Splitting the middle section into 2 gives a trapezium and a rectangle.

Area of trapezium $= \frac{1}{2}(26 + 28) \times 10 = 270$

Area of rectangle $= 28 \times 10 = 280$

Total distance $= 260 + 270 + 280 + 280$

$=$ **1090 m**

2 You can very roughly estimate the area under the curve using two triangles and a rectangle, as shown below:

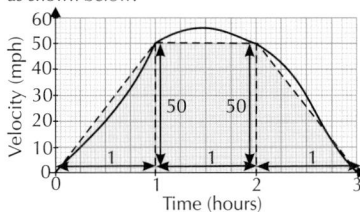

a) Area of each triangle $= \frac{1}{2}(1 \times 50) = 25$

Area of rectangle $= 50 \times 1 = 50$

Total distance $= 25 + 50 + 25 =$ **100 miles**

b) average velocity $= \frac{\text{total distance}}{\text{total time}} = \frac{100}{3}$

$=$ **33 mph** (to the nearest whole number)

Depending on how you split up the area under the graph, you may get a different answer.

Page 297   Review Exercise

1 25 minutes $= \frac{25}{60} = 0.416\ldots$ hours

Speed $= \frac{\text{distance}}{\text{time}} = \frac{240}{0.416\ldots}$

$=$ **576 km/h** (3 s.f.)

2 1 hour = 3600 seconds

16 seconds $= \frac{16}{3600} = 0.004\ldots$ hours

Distance = speed × time $= 120 \times 0.004\ldots$

$= 0.533\ldots =$ **0.53 miles** (2 s.f.)

3 a) Density $= \frac{\text{mass}}{\text{volume}} = \frac{642}{0.05} =$ **12 840 kg/m^3**

b) Density $= \frac{0.06}{0.025} =$ **2.4 kg/m^3**

c) Mass = density × volume $= 42 \times 6.2$

$=$ **260.4 kg**

d) Volume $= \frac{\text{mass}}{\text{density}} = \frac{4.8}{120} =$ **0.04 cm^3**

4 Area $= \frac{\text{force}}{\text{pressure}} = \frac{1.5}{150} = 0.01$ m^2.

1 m = 100 cm

0.01 m^2 = 0.01 × 100^2 = **100 cm^2**

5 a) The flight is the first section of the journey. The first section of the graph lasts for 2 hours, so the flight was **2 hours** long.

b) The driving section is the less steep descent from 3 hours until 7.5 hours. The graph is horizontal between 5 and 5.5 hours, so the break was **30 minutes**.

c) They began driving at 300 km from home and drove until 0 km, so they drove 300 − 0 = **300 km**.

6 The speed is the gradient of the graph.

a) Gradient $= \frac{22.5 - 0}{2 - 0} =$ **11.25 mph**

b) Gradient $= \frac{3.75 - 0}{3 - 0} =$ **1.25 km/h**

c) Gradient $= \frac{0 - 1.0}{5 - 0} = -0.2$

So the speed is **0.2 m/s**.

The gradient is negative so the object is moving back towards you — but an object can't have a negative speed so you take the positive value. (The velocity, on the other hand, could be negative).

7 a) Convert seconds to hours by dividing by 3600 (= 60 × 60).

Bert's distance = area under graph

$= \frac{1}{2}\left(60 \times \frac{8}{3600}\right) + \left(60 \times \frac{47}{3600}\right)$

$+ \frac{1}{2}\left(60 \times \frac{5}{3600}\right)$

$= 0.891\ldots =$ **0.9 km** (1 d.p.)

Ernie's distance = area under graph

$= \frac{1}{2}\left(44 \times \frac{5}{3600}\right) + \left(44 \times \frac{15}{3600}\right)$

$+ \left[\frac{1}{2}(44 + 50) \times \frac{10}{3600}\right]$

$+ \left(50 \times \frac{15}{3600}\right) + \frac{1}{2}\left(50 \times \frac{15}{3600}\right)$

$= 0.656\ldots =$ **0.7 km** (1 d.p.)

b) 60 s $= \frac{60}{3600} = \frac{1}{60}$ hours.

Average velocity $= \frac{\text{total distance}}{\text{total time}}$

$= \frac{0.891\ldots}{\frac{1}{60}} =$ **53.5 km/h** (1 d.p.)

c) acceleration $= \frac{\text{change in velocity}}{\text{change in time}}$

$= \frac{44}{\frac{5}{3600}} =$ **31 680 km/h^2**

d) **Ernie** had the greatest initial acceleration since his velocity-time graph is steeper.

Page 298   Exam-Style Questions

1 Density $= \frac{\text{mass}}{\text{volume}}$

$\Rightarrow$ mass = density × volume

Mass of copper $= 9 \times 36 = 324$ g

Mass of tin $= 4 \times 7 = 28$ g *[1 mark for both]*

Total mass $= 324 + 28 =$ **352 g** *[1 mark]*

2 Convert cm^2 to m^2: 40 000 cm^2

$= 40\,000 \div 100^2 = 4$ m^2 *[1 mark]*

Pressure $= \frac{\text{force}}{\text{area}} \Rightarrow$ force = pressure × area

F $= 625 \times 4$ *[1 mark]* = **2500 N** *[1 mark]*

3 Nigel cycles 14 km in 42 minutes, which is $\frac{42}{60} = 0.7$ hours, so his speed is:

$\frac{\text{distance}}{\text{time}} = \frac{14}{0.7} = 20$ km/h *[1 mark]*

As Michael's speed is half of Nigel's, Michael cycles at 10 km/h *[1 mark]* and so his journey of 6 km will take $\frac{6}{10} = 0.6$ hours

$= 0.6 \times 60$ mins $= 36$ mins *[1 mark]*.

Michael arrives 2 mins earlier than Nigel, so he arrives at 14:40. As it took 36 mins for Michael to cycle there, he must have set off at 14:40 − 36 mins = **14:04** *[1 mark]*.

4 $1 \text{ kg} = 1000 \text{ g}$
$13.6 \text{ g/cm}^3 = 13.6 \div 1000 = 0.0136 \text{ kg/cm}^3$
[1 mark]
$1 \text{ m}^3 = 100^3 \text{ cm}^3 = 1\,000\,000 \text{ cm}^3$ *[1 mark]*
$0.0136 \text{ kg/cm}^3 \times 1\,000\,000 = \textbf{13\,600 kg/m}^3$
[1 mark]
You expect to have fewer kg than grams, so divide by the conversion factor. You also expect to have more kg per m^3 than cm^3, so multiply by the conversion factor.

5 a) The gradient of the velocity-time graph is **the acceleration of the bus** *[1 mark]*.
b) Estimate the area under the curve by drawing a triangle, rectangle and trapezium (as shown below) and adding the areas together.

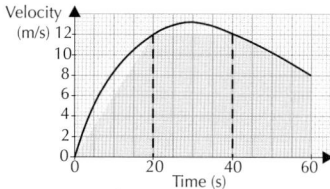

So distance $= \frac{1}{2}(20 \times 12) + (20 \times 12)$
$+ \frac{1}{2}(12 + 8) \times 20 = 120 + 240 + 200$
$= \textbf{560 m}$

[3 marks available – 1 mark for the correct areas of both the triangle and rectangle, 1 mark for the correct area of the trapezium, 1 mark for the correct answer]
c) It is an **underestimate** because the shapes used for the estimate lie **underneath the curve** *[1 mark]*.

Section 24 — Constructions

24.1 Scale Drawings

Page 300   Exercise 1

1 The scale is 1 cm:50 km, so to convert from km in real life to cm on the map, divide by 50.
a) $150 \div 50 = \textbf{3 cm}$
b) $600 \div 50 = \textbf{12 cm}$
Using the same method for c)-f):
c) 20 cm   **d) 0.5 cm**
e) 0.2 cm   **f) 0.3 cm**
2 The scale is 1 cm:2 m, so to convert from cm on the plan to metres in real life, multiply by 2.
a) $2.7 \times 2 = 5.4$ m, $1.5 \times 2 = 3$ m
So the room is: **5.4 m by 3 m**
Using the same method for b)-d):
b) 6.4 m by 4.4 m
c) 3.7 m by 2.8 m
d) 1.8 m by 2.7 m
3 The scale is 1 cm:0.5 km, so to convert from km in real life to cm on the map, divide by 0.5: $0.8 \div 0.5 = \textbf{1.6 cm}$
4 The scale is 1:40, so to convert from the size of the toy furniture to the size of the real furniture, multiply by 40.
a) $3.5 \text{ cm} \times 40 = \textbf{140 cm}$
b) $3.2 \text{ cm} \times 40 = \textbf{128 cm}$
c) $2.4 \text{ cm} \times 40 = \textbf{96 cm}$
5 The scale is $1:250\,000 = 1 \text{ cm}:250\,000 \text{ cm}$
$= 1 \text{ cm}:2500 \text{ m} = 1 \text{ cm}:2.5 \text{ km}$, so to convert from distances in real life to distances on the map, divide by 2.5:
$6.7 \div 2.5 = \textbf{2.68 cm}$
You could also divide by 250 000 to get the answer in km and then convert to cm.
6 The scale is $1:500 = 1 \text{ cm}:500 \text{ cm}$
$= 1 \text{ cm}:5 \text{ m}$, so to convert from actual size in metres to model size in cm, divide by 5:
a) $100 \div 5 = \textbf{20 cm}$

b) $6 \div 5 = \textbf{1.2 cm}$
You could also divide by 500 to get the answers in metres and then convert them to cm.
7 The scale is 1 mm:3 cm, so to convert mm on the diagram to cm in real life, multiply by 3.
a) Sink area measures 19 mm by 10 mm.
$19 \times 3 = 57$ cm, $10 \times 3 = 30$ cm
So sink area is **57 cm by 30 cm**.
b) Hob area measures 22 mm by 15 mm.
$22 \times 3 = 66$ cm, $15 \times 3 = 45$ cm
So hob area is **66 cm by 45 cm**.
8 3 cm:4.5 km ($\div$ 3) **1 cm:1.5 km**
9 a) 1 cm:1500 cm = **1 cm:15 m**
b) To convert metres in real life to cm on the plan, divide by 15:
$60 \div 15 = \textbf{4 cm}$
10 a) 30 cm:18 m ($\div$ 30)
1 cm:0.6 m = 1 cm:60 cm = **1:60**
b) To convert cm on the plan to metres in real life, multiply by 0.6:
$12 \times 0.6 = \textbf{7.2 m}$
c) To convert metres in real life to cm on the plan, divide by 0.6:
$4.5 \div 0.6 = \textbf{7.5 cm}$

Page 301   Exercise 2
For this exercise, use a ruler to make sure that your drawings have the correct measurements.
1 The scale is 1:20, so to convert measurements in real life to measurements on the plan, divide by 20. E.g.
$3000 \text{ mm} \div 20 = 150 \text{ mm}$
$600 \text{ mm} \div 20 = 30 \text{ mm}$
Using the same method for the rest of the measurements:

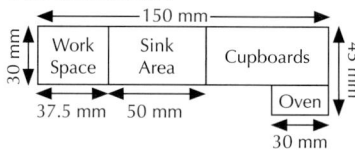

2 The scale is 1:25, so to convert measurements in real life to measurements on the plan, divide by 25. E.g.
$0.25 \text{ m} \div 25 = 0.01 \text{ m} = 1 \text{ cm}$
$1.5 \text{ m} \div 25 = 0.06 \text{ m} = 6 \text{ cm}$
$1.25 \text{ m} \div 25 = 0.05 \text{ m} = 5 \text{ cm}$
Using the same method for the rest of the measurements:

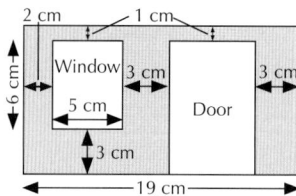

You could also have used the scale 1 cm:0.25 m and divided by 0.25 to change all measurements in metres directly into cm.
3 a) The scale is 1 cm:3 m, so to convert metres in real life to cm on the plan, divide by 3:
$33 \div 3 = 11$ cm, $24 \div 3 = 8$ cm
$42 \div 3 = 14$ cm, $7.5 \div 3 = 2.5$ cm

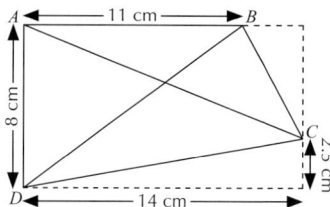

b) The distance between the duck house and *B* measures about 48 mm = 4.8 cm.
To convert from cm on the plan to metres in real life, multiply by 3:
$4.8 \times 3 = \textbf{14.4 m}$
You might have a slightly different measurement for the distance between the duck house and B — any answer between 13.8 m and 14.7 m is fine.

24.2 Bearings

Page 302   Exercise 1

1 a) 062°
b) $47° + 180° = \textbf{227°}$
c) $90° - 19° = \textbf{071°}$
d) $360° - 72° = \textbf{288°}$
2 a) $x = 111° - 90° = \textbf{21°}$
b) $x = 360° - 285° = \textbf{75°}$
c) $x = 180° - 135° = \textbf{45°}$
d) $x = 270° - 222° = \textbf{48°}$
These angles don't need to be three-figure as they're not bearings.
3 a)

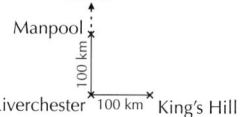

b) King's Hill is due east, so bearing = **090°**.
c) (i) Liverchester is due west, so bearing = **270°**.
(ii) Manpool is 100 km west and 100 km north, i.e. due northwest, so bearing = $270° + 45° = \textbf{315°}$.
North, east, south and west make right angles with each other, so have bearings of 000°, 090°, 180° and 270° respectively. Northeast, southeast, southwest and northwest are 45° between them, so have bearings of 045°, 135°, 225° and 315° respectively.
4 For this question, use a protractor to check that the angles match the diagrams below.

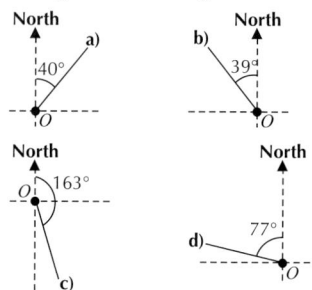

Page 303   Exercise 2
1 a) Using alternate angles:

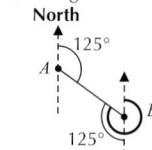

So bearing = $125° + 180° = \textbf{305°}$
You can also use other angle rules, e.g. allied angles, to help you find the missing bearings.
b) Using alternate angles:

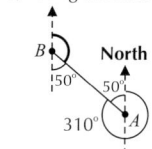

So bearing = $180° - 50° = \textbf{130°}$

You didn't need to draw a diagram for these questions — you can always work out the bearing of B from A by subtracting 180° from the bearing of A from B.

Using the same method for Q2-4:

2 $023° + 180° = \mathbf{203°}$

3 $101° + 180° = \mathbf{281°}$

4 a) $200° - 180° = \mathbf{020°}$
b) $310° - 180° = \mathbf{130°}$
c) $080° + 180° = \mathbf{260°}$
d) $117° + 180° = \mathbf{297°}$
e) $015° + 180° = \mathbf{195°}$
f) $099° + 180° = \mathbf{279°}$
If you're not sure whether to add or subtract 180°, draw a diagram and use angle properties to find the bearing.

5 a) 30°
b) Using alternate angles:

So bearing = $x = \mathbf{030°}$
You might have a slightly different measurement for x — anything between 29° and 31° is fine. Your answer for part b) should match whatever you got for part a).

6 a) Q is due west, so bearing = **270°**
b) P is due east, so bearing = **090°**

7 a) Z is due southeast, so bearing = **135°**
b) Y is due northwest, so bearing = **315°**

8 a) Bearing of V from $U = 180° - 68° = \mathbf{112°}$
b) Bearing of U from $V = 112° + 180° = \mathbf{292°}$

Page 304   Exercise 3

1 The scale is 1 cm:100 km, so to convert from cm on the diagram to km in real life, multiply by 100.
a) (i) The line measures 2.4 cm, so distance = $2.4 \times 100 = \mathbf{240\ km}$
Accept 230-250 km.
Measuring the angle gives a bearing of **050°**.
Accept 049°-051°.
(ii) The line measures 1.9 cm, so distance = $1.9 \times 100 = \mathbf{190\ km}$
Accept 180-200 km.
Measuring the angle gives a bearing of **125°**.
Accept 124°-126°.
b) The distance measures 3.4 cm (1 d.p.), so distance = $3.4 \times 100 = \mathbf{340\ km}$

For Questions 2-5, use a ruler and a protractor to check that your diagrams have the measurements shown.

2 The scale is 1 cm:30 km, so to convert km in real life to cm on the diagram, divide by 30: $150 \div 30 = 5$ cm

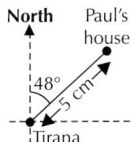

3 The scale is 1 cm:90 km, so to convert km in real life to cm on the diagram, divide by 90: $540 \div 90 = 6$ cm

4 The scale is 1 cm:100 000 000 cm $\Rightarrow$ 1 cm:1 000 000 m $\Rightarrow$ 1 cm:1000 km, so to convert km in real life to cm on the diagram, divide by 1000: $2000 \div 1000 = 2$ cm
$360° - 242° = 118°$

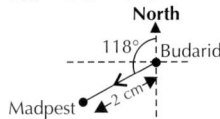

5 The scale is 1 cm:22 000 000 cm $\Rightarrow$ 1 cm:220 000 m $\Rightarrow$ 1 cm:220 km, so to convert km in real life to cm on the diagram, divide by 220: $880 \div 220 = 4$ cm
$360° - 263° = 97°$

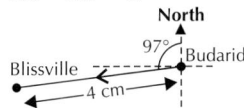

6 a) The scale is 1 cm:10 000 000 cm $\Rightarrow$ 1 cm:100 000 m $\Rightarrow$ 1 cm:100 km, so to convert cm on the diagram to km in real life, multiply by 100:
(i) PQ measures 3 cm, so distance = $3 \times 100 = \mathbf{300\ km}$
Accept 290-310 km.
(ii) QR measures 4.5 cm, so distance = $4.5 \times 100 = \mathbf{450\ km}$
Accept 440-460 km.
(iii) PR measures 5.8 cm, so distance = $5.8 \times 100 = \mathbf{580\ km}$
Accept 570-590 km.
b) (i) Bearing (Q from P) = **060°**
Accept 059°-061°.
(ii) Bearing (R from Q) = **140°**
Accept 139°-141°.
(iii) Bearing (P from R) = **290°**
Accept 289°-291°.

24.3 Constructions

Page 306   Exercise 1

In this exercise, use a ruler and a protractor to make sure that your diagrams have the measurements shown (given rounded to the nearest mm or degree). Allow 1 mm or 1° either side.

1 a) E.g. First, draw the 5 cm side, using a ruler to measure the correct length. Next, set your compasses 4.5 cm and draw an arc from the left-hand end of the 5 cm line. Then, set your compasses to 3 cm and draw an arc from the other end. Mark the point where these arcs cross and join it to the ends of the 5 cm line.

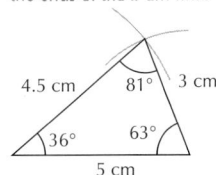

b) E.g. Draw the 8 cm line. Then, set your compasses to 10 cm, and draw an arc from the right-hand end. Next, use your protractor to measure an angle of 90° from the other end, marking it with a dot. Draw a line from the end of the 8 cm line through the dot. Mark the point where this line crosses the arc and join it to the ends of the 8 cm line.

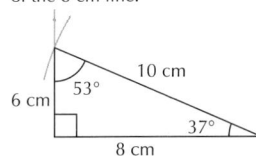

c) Using the same method as part a):

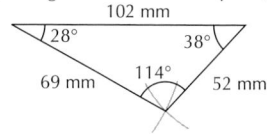

d) E.g. Draw the 4.1 cm line. Measure an angle of 105° from one end and mark it with a dot, then draw a line through the dot. Repeat at the other end of the line with an angle of 50°. Mark the point where these lines cross and join it to each end of the 4.1 cm line.

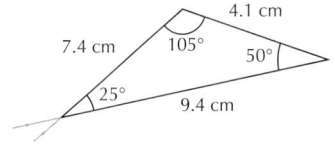

e) E.g. Draw the 11 cm line. Measure an angle of 35° at the left-hand end, marking it with a dot. Draw a line from the end of the line through the dot. Then measure 7 cm along, mark this point, and join it to the other end of the 11 cm line.

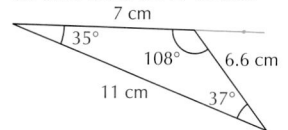

f) Using the same method as part a):

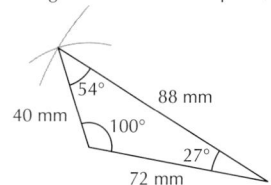

Using the same methods (as appropriate) for Q2-3:

2 a)

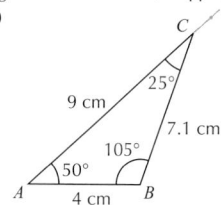

b)

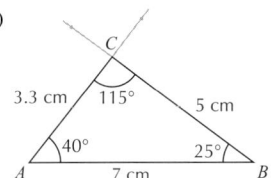

c)

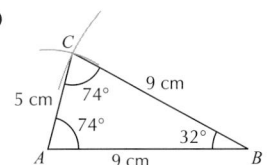

d)

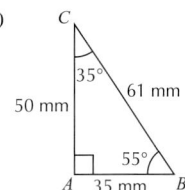

e)

8.4 cm 30° 5.4 cm 114° 36° 4.6 cm C A B

f)

C 32° 6.2 cm 2.3 cm 132° 16° A 4.4 cm B

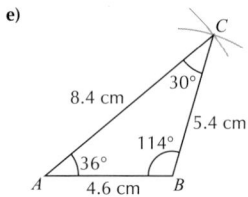

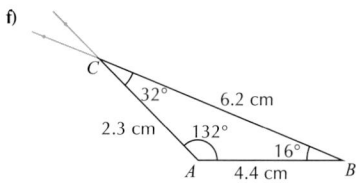

The easiest way to do part f) is to work out angle BAC: 180° − 16° − 32° = 132°. You can then do a normal ASA construction.

3 a)

42° 7 cm 7 cm 69° 69° 5 cm

b)

30° 87 mm 87 mm 75° 75° 45 mm

c)

7.1 cm 79° 2.7 cm 22° 79° 7.1 cm

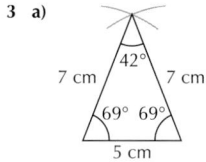

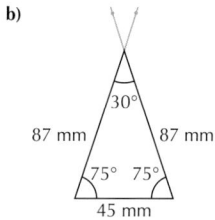

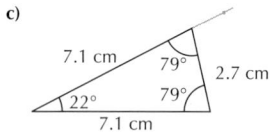

4 E.g. Start by drawing AB with a length of 10 cm. Measure an angle of 30° at B, and draw a line in this direction. Next, set your compasses at 6 cm and draw an arc from A. The arc should cross the line from B in two places — these are the two possible locations of C.

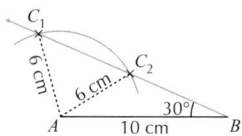

C_1 6 cm 6 cm C_2 30° A 10 cm B

Measure the distance from B to each of these points to find the two possible lengths of BC — **12.0 cm** and **5.3 cm** (to the nearest mm).

Page 307   Exercise 2

In this exercise, mark your answers by checking that the angle between the two lines is 90° and that the perpendicular bisector is half-way between the two points.

1 a) First, draw a horizontal line that is 5 cm long. Next, set your compasses so that they are more than 2.5 cm apart, and draw two arcs from P (one above PQ and one below) and two arcs from Q. Then, draw the line that passes through the points where the arcs cross.

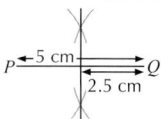

P 5 cm Q 2.5 cm

Using the same method for b)-c) and Q2-3:

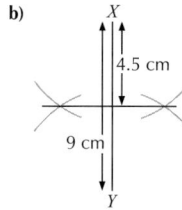

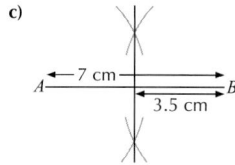

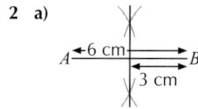

b)

X 4.5 cm 9 cm Y

c)

7 cm A B 3.5 cm

2 a)

A 6 cm B 3 cm

b) E.g. Measure 4 cm along the perpendicular bisector each way, and mark these points C and D. Then you can form a rhombus by joining the points $ACBD$.

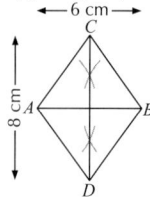

6 cm C 8 cm A B D

Remember, the diagonals of a rhombus cross at right angles — see page 255.

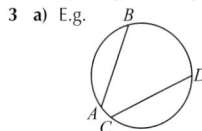

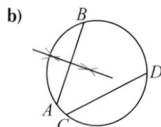

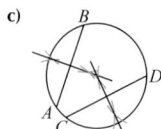

3 a) E.g.

B D A C

b)

B D A C

c)

B D A C

d) The lines meet at the **centre of the circle**.
This comes from rule 4 of the circle theorems (see p.268).

Page 308   Exercise 3

In this exercise, mark your answers by using a protractor to measure the angles and checking that they match the diagrams given. Allow 1° either side.

1 a) Draw the 100° angle using your protractor. Next, place the point of your compasses on the angle and draw arcs crossing both lines. Then, put the point of your compasses on the points where these arcs cross the lines and draw two more arcs using the same radius. Finally, draw a line through the point where these arcs cross.

50° 50°

Using the same method for b)-h) and Q2-3:

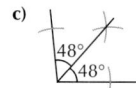

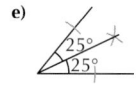

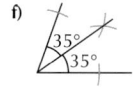

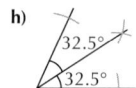

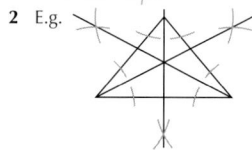

b)

70° 70°

c)

48° 48°

d)

22° 22°

e)

25° 25°

f)

35° 35°

g)

10° 10°

h)

32.5° 32.5°

2 E.g.

The bisectors all **intersect at a single point**.

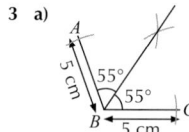

3 a)

A 5 cm 55° 55° B 5 cm C

b) Use your ruler to measure 8 cm along the angle bisector:

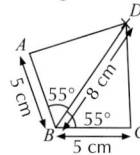

D A 5 cm 55° 8 cm 55° B 5 cm C

The shape is symmetrical along the line BD, so AD and CD are the same length, meaning $ABCD$ is a **kite**.

Page 309   Exercise 4

In this exercise, mark your answers by using a protractor to check that all perpendiculars make an angle of 90° with the line.

1 Draw a triangle XYZ, then draw an arc centred on X that crosses the line YZ twice (you may need to extend the line so that you get two crossing points). Then, draw two arcs of the same radius centred on each of these crossing points. Finally, draw a line from X through the point where these two arcs intersect.
E.g.

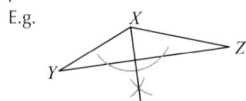

X Y Z

Using the same method for Q2-5:

2 The shortest possible line is perpendicular.
E.g.

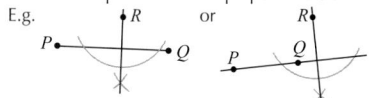

R or R P Q P Q

Your diagram might look different depending on where you placed P, Q and R.

3 a)-c)

4 E.g.

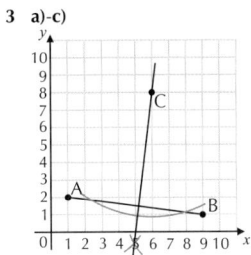

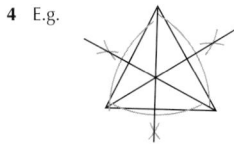

The perpendiculars **intersect at a single point**.

5 a) E.g. Draw DE by using your ruler to measure 5 cm. Next, measure an angle of 55° at D, marking it with a dot. Draw a line from D through this dot. Finally, measure 6 cm from D with your ruler and mark F, joining it to E to complete the triangle.

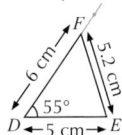

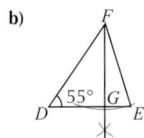

b)

c) Using your ruler, $FG = $ **4.9 cm**.
Accept 4.8-5.0 cm.

Area of a triangle $= \frac{1}{2} \times$ base $\times$ height
$= \frac{1}{2} \times 5 \times 4.9 = $ **12.25 cm²**
Accept 12.0-12.5 cm².

Page 310   Exercise 5

In this exercise, mark your answers by using a ruler and protractor to measure the lengths and angles and checking that they match the diagrams given.

1 Use your ruler to measure a 5 cm line and label it AB. Then, place your compass point at A and draw a long arc that crosses AB. Next, draw an arc of the same radius centred on the point where the first arc crosses AB. Finally, draw a line from A through the point where these arcs cross.

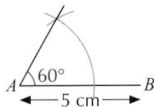

2 Draw a line that is 6 cm long. Then, set your compasses to 6 cm, and draw arcs from each end of this line. Join the point where these arcs cross to each end of the line.

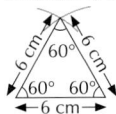

3 Use your ruler to measure a line 6 cm long, and construct a 60° at A using the same method as Q1. Then, use your compasses to draw two arcs of the same radius about the points where the long arc crosses each line. Finally, draw a line from A through the point where these arcs cross.

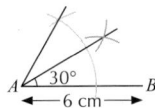

4 E.g. Use your ruler to measure a line 7 cm long and label it AB. Next, set your compasses to 7 cm and draw arcs from both ends of the line. Join A and B to the point where these lines cross to form an equilateral triangle. Then, bisect the 60° angle at B using the method from p.308. Mark the point where the angle bisector crosses the 60° line from A as point C, then join C to A and B to complete the triangle.

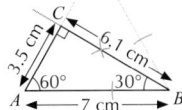

Using the same method for Q5:

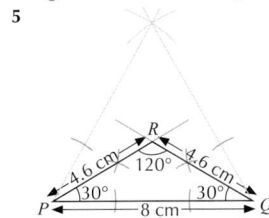

5

Here, you had to construct two 60° angles, then bisect both of them.

Page 311   Exercise 6

In this exercise, mark your answers by using a ruler and protractor to measure the lengths and angles and checking that they match the diagrams given.

1 Place your compass point at X and draw two arcs that cross the line either side of X. Then, increase the radius of your compasses and draw an arc from each intersection. Finally, draw a line through X and the point where these arcs cross.

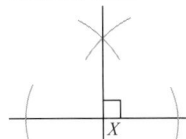

2 E.g. Draw the 7 cm line, mark each end of the line, then extend the line at each end. Next, construct 90° angles at each end of the 7 cm line. Measure 5 cm along each line and mark these points. Finally, join these points to complete the rectangle.

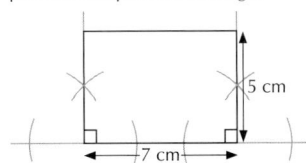

3 Construct a 90° angle using the same method as Q1. Next, place your compass point at X and draw two arcs — one through the original line and one through the perpendicular line. Then, use your compasses to draw two arcs of the same radius about the intersections. Finally, draw a line from X through the point where these arcs cross.

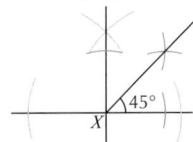

4 Draw a line 8 cm long, label the ends A and B, and extend it at each end. Then, construct 45° angles at A and B using the same method as for Q3. Label the point where the 45° lines intersect as C, then join it to A and B to complete the triangle.

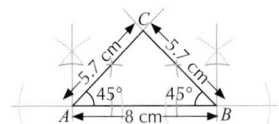

Page 312   Exercise 7

In this exercise, mark your answers by using a protractor to measure any right angles and checking that they are all 90°.

1 Place your compass point at P and draw an arc that crosses AB twice, then draw an arc from each intersection. Draw a line from the point where these arcs cross through P. Next, draw two arcs of the same radius from P that cross this new line, and draw an arc from each intersection. Finally, draw another line through P and the point where these arcs cross.

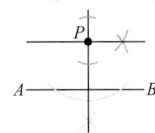

2 First, draw the two lines and draw a point not on the lines. Using the same method as for Q1, construct a line parallel to one of the original lines through this point. Then, draw a new point (or use the same one) and construct a line parallel to the other original line through this point. The shape enclosed by these four lines will be a parallelogram.

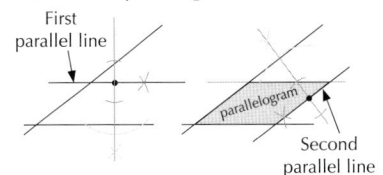

3 a) E.g. First, draw line AB by measuring 10 cm with your ruler. Next, construct a 60° angle at B using the method from page 310 and draw a line 3 cm long in this direction, marking C at the end. Then, construct a 30° angle at A by first constructing a 60° angle and then bisecting it. Finally, use the method from Q1 to construct a line parallel to AB that passes through C — the point where this line and the 30° line from A intersect is point D.

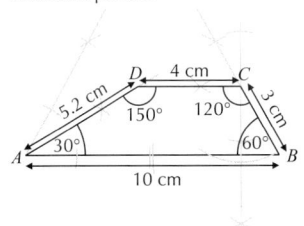

b) $ABCD$ is a **trapezium** (since it has one pair of parallel sides).

Page 313 Exercise 1

For this exercise, use a ruler to check that your drawings have the measurements shown.

1 Use your ruler to draw a 7 cm line. Next, set your compasses to 2 cm and draw a long arc around each end of the line. Finally, use your ruler to join the tops and the bottoms of each arc.

2 a) Set your compasses to 3 cm, place the point on X and draw a **circle**.

   b) Shade the **inside** of the circle.

3 Use your ruler to draw two points 6 cm apart, then construct the **perpendicular bisector** of these points using the method from page 307.

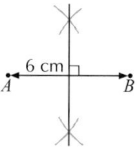

4 Use your protractor to draw two lines that meet at an angle of 50°, then construct the **angle bisector** of these lines using the method from page 308.

5 Using the same method as Q1:

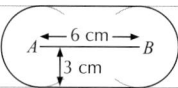

6 a), b) Using the same method as Q3:

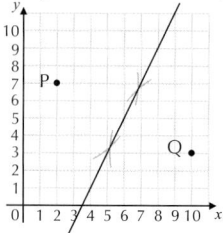

Page 314 Exercise 2

1 a)-d) Draw a circle of radius 3 cm around P, and a circle of radius 4 cm around Q:

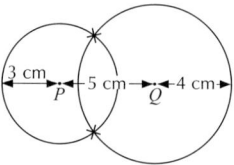

The points marked with crosses are both 3 cm from P and 4 cm from Q.

2 E.g. Use your ruler to draw line RS 5 cm long. Then, set your compasses to 4 cm and draw a circle around R. Next, construct the perpendicular bisector of R and S. The required locus is the set of points that are both on this line and inside the circle:

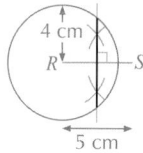

3 a) Using the method from page 305:

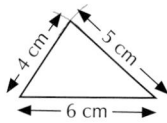

   b) Set your compasses to 1 cm, and draw circles around each corner of the triangle. Then, for each side, join the top and bottom of the circles at each end with straight lines. Finally, mark the required locus as shown below (it has two parts — one inside the original triangle and one outside).

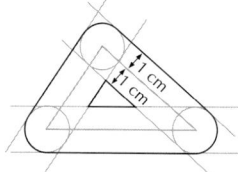

4 a) Using the method from page 305:

   b) E.g. First, use your compasses to draw a circle of radius 2 cm around E. Next, construct the perpendicular bisector of DF using the method from page 307. The required locus is the set of points that are both on this line and inside the circle:

5 a) E.g. First, use your ruler to measure a line 6 cm long. Next, construct the perpendicular bisector of this line using the method from page 307. Finally, measure 3 cm along this line in each direction with your ruler and mark these points.

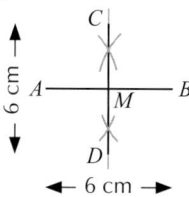

   b) Construct the locus of points that are 2 cm from each line separately. The required area is the square in the middle where these shapes overlap (the shaded area):

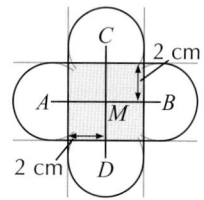

Page 315 Exercise 3

1 a), b) The scale is 1 cm : 1 km, so 3 km in real life will be 3 cm on the diagram. Draw P and L 3 cm apart, then construct the perpendicular bisector using the method from page 307.

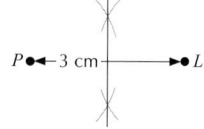

2 Copy the diagram, then construct the angle bisector using the method from page 308. The scale is 1 cm : 0.5 m, so 3 m in real life will be 6 cm on the diagram. Use your ruler to measure 6 cm along the angle bisector and mark the position of the bonfire.

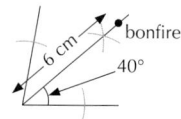

3 a), b) The scale is 1 cm : 10 miles, so 50 miles will be $50 \div 10 = 5$ cm on the diagram. Draw A and B 5 cm apart. 40 miles will be $40 \div 10 = 4$ cm on the diagram, so draw circles of radius 4 cm around each point. The area where these circles overlap is the region where the camels could possibly meet.

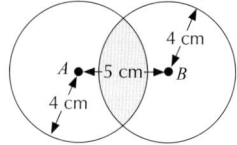

4 a) The scale is 1 cm : 1 m so the length of the yard on the diagram is 4 cm, the width is 2 cm and the dog's lead is 1 cm. Use your ruler to draw a rectangle 2 cm by 4 cm. Then, set your compasses to 1 cm and draw an arc from one of the corners. The required area is the inside of this quarter-circle:

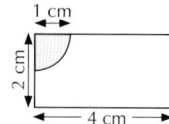

   b) The rail will be 3 cm on the diagram. Use your ruler to measure 3 cm from the corner, then use your compasses to draw an arc of radius 1 cm around this point. Then, draw a horizontal line 1 cm from the top of the rectangle. Finally, shade the required area as shown below:

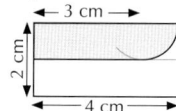

1 1 cm : 120 000 cm = 1 cm : 1200 m
            = 1 cm : 1.2 km

To convert cm on the map to km in real life, multiply by 1.2.

a) Museum to cathedral measures 1 cm, so distance = 1 × 1.2 = **1.2 km**.
You might have a slightly different measurement between the museum and the cathedral — any answer between 1.08 km and 1.32 km is fine.

b) Art gallery to theatre measures 2.5 cm, so distance = 2.5 × 1.2 = **3 km**.
Any answer between 2.88 km and 3.12 km is fine.

c) Cathedral to theatre measures 3.2 cm, so distance = 3.2 × 1.2 = **3.84 km**.
Any answer between 3.72 km and 3.96 km is fine.

2 a) Bearing (*B* from *A*) = **055°**
Don't forget to give your answer a 3-figure bearing.

b) Bearing (*C* from *A*) = 55° + 88° = **143°**

c) Bearing (*D* from *A*) = 143° + 101° = **244°**

d) Bearing (*E* from *A*) = 360° − 34° = **326°**

e) Bearing (*A* from *B*) = 055° + 180° = **235°**

f) Bearing (*A* from *C*) = 143° + 180° = **323°**

g) Bearing (*A* from *D*) = 244° − 180° = **064°**

h) Bearing (*A* from *E*) = 326° − 180° = **146°**

3 The scale is 1 cm : 10 miles, so to convert miles in real life to cm on the map, divide by 10. High Cross is 54 ÷ 10 = 5.4 cm from Low Cross, Very Cross is 48 ÷ 10 = 4.8 cm from Low Cross and 27 ÷ 10 = 2.7 cm from High Cross. Use a ruler to measure a horizontal line 5.4 cm long from High Cross to Low Cross. Next, set your compasses to 4.8 cm and draw an arc from Low Cross (below the line, since Very Cross is to the south). Then set your compasses to 2.7 cm and draw an arc from High Cross. Very Cross is at the point where these arcs intersect.

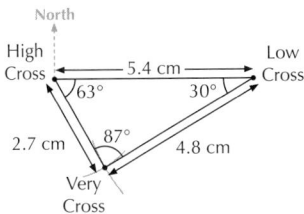

4 a) Use your ruler to draw line *AB* 8 cm long. Then, set your compasses to 8 cm and draw arcs from *A* and *B*. Finally, draw a line from *A* through where these arcs cross.

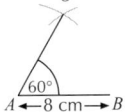

b) E.g. Measure 8 cm along the 60° line and mark this point *D* (this will be the point where the arcs crossed if you used the same method as above). Next, set your compasses to 8 cm and draw arcs from *B* and *D*. Mark the point where these cross as point *C* and join it to *B* and *D* to complete the rhombus.

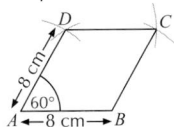

5 a) E.g. Use your ruler to draw line *DE* 5.8 cm long. Then, set your compasses to 5.8 cm and draw arcs from *D* and *E*. Mark point *F* where these arcs intersect.

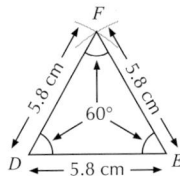

b) Using the method from page 312:

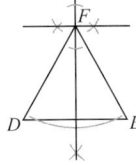

6 E.g. Use your ruler to draw line *AB* 7.4 cm long. Then, construct a 60° angle at *A* using the method from page 310. Next, construct a 90° angle at *B* and bisect it, as shown on page 311. Mark the point where these two lines cross as *C* and join it to *A* and *B*.

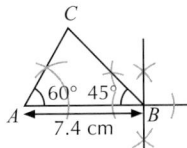

7 Construct an angle of 60° using the method from page 310. Then, bisect this angle twice, using the method from page 308.

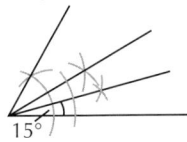

8 The scale is 1 cm : 1 m so on the diagram, region *ABCD* is 10 cm × 10 cm and the treasure is 7 cm from corner *C*.
E.g. First, use your ruler to draw a 10 cm by 10 cm square. Since the treasure is the same distance from *AB* and *AD*, the treasure must lie on the angle bisector of these two lines. Since this is a square, the angle bisector is *AC*, so join *A* and *C* with a straight line. Next, set your compasses to 7 cm and draw an arc around *C*. The treasure is at the point where these lines intersect.

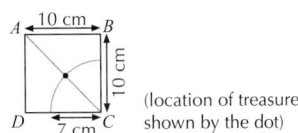

(location of treasure shown by the dot)

1 a) The clockwise angle between a north line from *A* and *AB* is 45° since the grid is made of squares. So the bearing is **045°** *[1 mark]*.

b) The clockwise angle between a north line from *C* and *CA* is 360° − 45° (or 270° + 45°) = **315°** *[1 mark]*.

2 1 cm : 25 000 cm = 1 cm : 250 m *[1 mark]*
So to convert from cm on the map to metres in real life, multiply by 250.
Actual length of reservoir = 3.8 × 250
                    = **950 m** *[1 mark]*

3 E.g. First, use your ruler to draw the 37 mm line. Next, use your compasses to draw a 52 mm arc around one end of the line, and a 60 mm arc around the other end. Join the point where these arcs cross to each end of the line to complete the triangle.

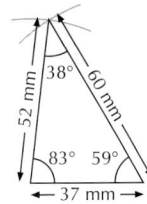

[3 marks available in total — 1 mark for accurately drawing one line, 1 mark for two construction arcs that intersect, 1 mark for a fully correct triangle with required lengths]
Check your measurements are accurate by measuring the other length and angles of your triangle — if they're within 1 mm/1° of the ones given above, your diagram should be accurate.

4 E.g. Draw a horizontal line, and mark a point along it. Then, measure 3 cm along the line from this point in each direction and mark these points. Next, set your compasses to 3 cm and draw arcs from all three of these points, giving 4 intersection points as shown below. Join up the outer points to form a regular hexagon.

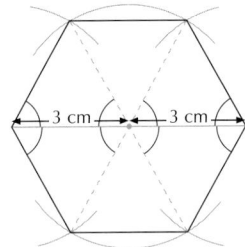

(all the marked angles are 60°)

[3 marks available — 1 mark for correct use of 60° construction to form at least 1 equilateral triangle, 1 mark for attempt to use multiple 60° angles to construct a hexagon, 1 mark for a correct and accurate final diagram]

5 The scale is 1 cm : 20 m, so to convert metres in real life to cm on the diagram, divide by 20. The scarecrow is 40 ÷ 20 = 2 cm from the gate.
E.g. The scarecrow should be the same distance from both hedges, so construct the angle bisector of these two lines using the method from page 308. It should also be 2 cm from the gate, so set your compasses to 2 cm and draw an arc that crosses the angle bisector twice. Mark these two intersection points as the two possible positions of the scarecrow.

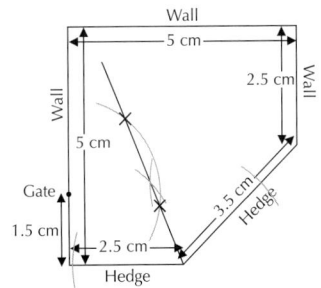

[4 marks available — 1 mark for an attempt to bisect the angle between the hedges, 1 mark for this angle bisected accurately with construction arcs shown (the line should make an angle of 67°-68° with each hedge), 1 mark for an arc or 2 arcs of radius 2 cm from the gate that intersect the line twice, 1 mark for a fully correct answer with both scarecrow positions clearly indicated]

Section 25 — Pythagoras and Trigonometry

25.1 Pythagoras' Theorem

Page 319 Exercise 1

1 a) $x^2 = 3^2 + 4^2 = 9 + 16 = 25$
$x = \sqrt{25} =$ **5 cm**

b) $z^2 = 5^2 + 12^2 = 25 + 144 = 169$
$z = \sqrt{169} =$ **13 mm**

c) $q^2 = \left(\frac{1}{2}\right)^2 + \sqrt{2}^2 = \frac{1}{4} + 2 = \frac{9}{4}$
$q = \sqrt{\frac{9}{4}} = \frac{3}{2}$ **cm** or **1.5 cm**

d) $r^2 = 5^2 + 10^2 = 125$
$r = \sqrt{125} = \sqrt{25 \times 5} = 5\sqrt{5}$ **m**

2 a) $s^2 = 0.7^2 + 2.4^2 = 0.49 + 5.76 = 6.25$
$s = \sqrt{6.25} =$ **2.5 m**

b) $u^2 = 3.78^2 + 5.12^2 = 40.5028$
$u = \sqrt{40.5028} =$ **6.36 km** (2 d.p.)

c) **13.57 mm** (2 d.p.) d) **1.39 m** (2 d.p.)

3 a) $13^2 = 12^2 + l^2 \Rightarrow l^2 = 13^2 - 12^2$
$= 169 - 144 = 25 \Rightarrow l = \sqrt{25} =$ **5 cm**

b) $11^2 = 7^2 + n^2 \Rightarrow n^2 = 11^2 - 7^2 = 121 - 49$
$= 72 \Rightarrow n = \sqrt{72} = \sqrt{36 \times 2} = 6\sqrt{2}$ **mm**

c) $\left(\frac{1}{2}\right)^2 = \left(\frac{1}{4}\right)^2 + p^2 \Rightarrow p^2 = \left(\frac{1}{2}\right)^2 - \left(\frac{1}{4}\right)^2$
$= \frac{1}{4} - \frac{1}{16} = \frac{3}{16}$
$p = \sqrt{\frac{3}{16}} = \frac{\sqrt{3}}{4}$ **cm**

d) $30^2 = 10^2 + i^2 \Rightarrow i^2 = 30^2 - 10^2$
$= 900 - 100 = 800$
$i = \sqrt{800} = \sqrt{400 \times 2} = 20\sqrt{2}$ **mm**

4 a) $21.7^2 = 15.9^2 + k^2 \Rightarrow k^2 = 21.7^2 - 15.9^2$
$= 470.89 - 252.81 = 218.08$
$k = \sqrt{218.08} =$ **14.77 km** (2 d.p.)

b) $2.4^2 = 1.65^2 + g^2 \Rightarrow g^2 = 2.4^2 - 1.65^2$
$= 5.76 - 2.7225 = 3.0375$
$g = \sqrt{3.0375} =$ **1.74 mm** (2 d.p.)

Using the same method for c)-d):
c) **9.00 m** (2 d.p.) d) **2.71 cm** (2 d.p.)

Page 320 Exercise 2

1 The perpendicular from a vertex to the opposite side splits the triangle in half, leaving two identical right-angled triangles, each with a hypotenuse of 10 cm and another side of $10 \div 2 = 5$ cm.
Let b = perpendicular height, then
$10^2 = 5^2 + b^2 \Rightarrow b^2 = 100 - 25 = 75$
$b = \sqrt{75} = 8.6602... =$ **8.66 cm** (2 d.p.)

2 a) $JL^2 = JK^2 + KL^2 = 4.9^2 + 6.8^2$
$= 24.01 + 46.24 = 70.25$
$JL = \sqrt{70.25} =$ **8.38 cm** (2 d.p.)

b) Radius $= 8.381... \div 2 =$ **4.19 cm** (2 d.p.)

3 $15^2 = 8.5^2 + a^2 \Rightarrow a^2 = 225 - 72.25 = 152.75$
$a = \sqrt{152.75} = 12.359...$
So the height of the tree is **12.36 m** (2 d.p.).

4 Sketch a diagram to show the positions of the towns:

$142^2 = 88^2 + OB^2$
$OB^2 = 20\,164 - 7744 = 12\,420$
$OB = \sqrt{12\,420} =$ **111 km** (nearest km)

5 $1.5^2 + 2^2 = 2.25 + 4 = 6.25$ and $2.5^2 = 6.25$
Since $1.5^2 + 2^2 = 2.5^2$, the triangle is right-angled.

6 Let h = length of the straight line
$h^2 = 200^2 + 150^2 = 62\,500$
$h = \sqrt{62\,500} = 250$ m
Subtract the length of the straight line from the original distance: $(200 + 150) - 250 = 100$
So the journey would be **100 m shorter**.

7 a) Horizontal distance $= 11 - 4 = 7$
Vertical distance $= 8 - 4 = 4$
$AB^2 = 7^2 + 4^2 = 65$
$AB = \sqrt{65}$

b) Horizontal distance $= 11 - 7 = 4$
Vertical distance $= 10 - 8 = 2$
$BC^2 = 4^2 + 2^2 = 20$
$BC = \sqrt{20} = 2\sqrt{5}$

c) Horizontal distance $= 7 - 4 = 3$
Vertical distance $= 10 - 4 = 6$
$AC^2 = 3^2 + 6^2 = 45$
$AC = \sqrt{45} = 3\sqrt{5}$

8 Horizontal distance $= 17 - 11 = 6$
Vertical distance $= 19 - 1 = 18$
$h^2 = 6^2 + 18^2 = 36 + 324 = 360$
$h = \sqrt{360} = 6\sqrt{10}$

9 a) $20^2 = 5.95^2 + a^2 \Rightarrow a^2 = 400 - 35.4025$
$\Rightarrow a^2 = 364.5975$
$a = \sqrt{364.5975} = 19.0944...$
So he should anchor the slide **19.09 m** (2 d.p.) from the base of the tower.

b) $h^2 = 5.95^2 + (19.09... - 1.5)^2$
$= 35.4025 + 309.56... = 344.966...$
$h = \sqrt{344.966...} = 18.573...$
So the new length of the slide is **18.57 m** (2 d.p.).

25.2 Pythagoras' Theorem in 3D

Page 322 Exercise 1

1 a) $BD^2 = AB^2 + AD^2 = 4^2 + 8^2 = 80$
$BD = \sqrt{80} = 4\sqrt{5}$ **m**

b) $FD^2 = BD^2 + BF^2 = 80 + 3^2 = 89$
$FD = \sqrt{89}$ **m**

2 a) $PR^2 = PS^2 + RS^2 = 12^2 + 5^2 = 169$
$PR = \sqrt{169} =$ **13 mm**

b) $RT^2 = PR^2 + PT^2 = 169 + 9^2 = 250$
$RT = \sqrt{250} = 5\sqrt{10}$ **mm**

3 Diameter $= 4.5 \times 2 = 9$ cm
$XY^2 = 9^2 + 25^2 = 81 + 625 = 706$
$XY = \sqrt{706} =$ **26.6 cm** (3 s.f.)

4 $d^2 = 2.5^2 + 3.8^2 + 9.4^2 = 6.25 + 14.44 + 88.36$
$= 109.05$
$d = \sqrt{109.05} =$ **10.44 m** (2 d.p.)

5 a) $QS^2 = PQ^2 + PS^2 = 14^2 + 25^2 = 821$
$QS = \sqrt{821} =$ **28.7 mm** (3 s.f.)

b) $ST^2 = QS^2 + QT^2 = 821 + 9^2 = 902$
$ST = \sqrt{902} =$ **30.0 mm** (3 s.f.)

6 $d^2 = 5^2 + 5^2 + 5^2 = 75$
$d = \sqrt{75} =$ **8.66 m** (3 s.f.)

7 $d^2 = 16.5^2 + 4.8^2 + 2^2 = 299.29$
$d = \sqrt{299.29} = 17.3$
So the longest pencil that can fit would be **17.3 cm** long.

8 Diameter $= 6 \times 2 = 12$ cm
$h^2 = 12^2 + 28^2 = 144 + 784 = 928$
$h = \sqrt{928} = 30.463...$
So the longest stick of dried spaghetti that can fit would be **30.5 cm** (3 s.f.).

9 Let d = diagonal of base
$d^2 = 4.8^2 + 4.8^2 = 46.08$
$d = \sqrt{46.08} = 6.788...$ cm
$6.788... \div 2 = 3.39...$
Let V = vertical height of pyramid
$11.2^2 = 3.39...^2 + V^2$
$V^2 = 125.44 - 11.52 = 113.92$
$V = \sqrt{113.92} =$ **10.7 cm** (3 s.f.)

10 Let d = diagonal of base
$d^2 = 3.2^2 + 3.2^2 = 20.48$
$d = \sqrt{20.48} = 4.525...$ m
$4.525... \div 2 = 2.262...$
Let s = length of sloped edge
$s^2 = 2.262...^2 + 9.2^2 = 89.76$
$s = \sqrt{89.76} =$ **9.47 m** (3 s.f.)

11 a) $LN^2 = LP^2 + NP^2$
$13^2 = 3.5^2 + NP^2$
$NP^2 = 13^2 - 3.5^2 = 169 - 12.25 = 156.75$
$NP = \sqrt{156.75} =$ **12.5 m** (3 s.f.)

b) $JL^2 = JM^2 + LM^2 = 7^2 + 7^2 = 98$
$JL = \sqrt{98}$
$OJ = JL \div 2 = \sqrt{98} \div 2 =$ **4.95 m** (3 s.f.)
You could find OJ directly by using the fact that the perpendicular distance from O to each edge of the square is 3.5 m ($OJ^2 = 3.5^2 + 3.5^2$).

c) $NJ^2 = OJ^2 + ON^2$
$ON^2 = NJ^2 - OJ^2 = 13^2 - 4.949...^2$
$= 169 - 24.5 = 144.5$
$ON = \sqrt{144.5} =$ **12.0 m** (3 s.f.)

12 a) $XZ^2 = OX^2 + OZ^2$
$OX^2 = XZ^2 - OZ^2 = 17^2 - 15^2 = 64$
$OX = \sqrt{64} = 8$ m
$VX = OX \times 2 =$ **16 m**

b) $VXWY$ is a square so $VY = XY$, which means area $= VY^2$
Using Pythagoras' theorem:
$VX^2 = VY^2 + XY^2 = 2VY^2$
$VY^2 = VX^2 \div 2 = 16^2 \div 2 =$ **128 m²**

25.3 Trigonometry — Sin, Cos and Tan

Page 324 Exercise 1

1 a) You're given the hypotenuse and want to find the adjacent, so use the formula for cos x.
$\cos 43° = \frac{a}{6}$
$a = 6 \cos 43° =$ **4.39 cm** (3 s.f.)

b) You're given the hypotenuse and want to find the opposite, so use the formula for sin x.
$\sin 58° = \frac{b}{11}$
$b = 11 \sin 58° =$ **9.33 cm** (3 s.f.)

Using the same method for c)-d):
c) **1.06 cm** (3 s.f.) d) **8.02 cm** (3 s.f.)

e) You're given the opposite and want to find the adjacent, so use the formula for tan x.
$\tan 37° = \frac{6}{e} \Rightarrow e \times \tan 37° = 6$
$e = \frac{6}{\tan 37°} =$ **7.96 cm** (3 s.f.)

f) You're given the opposite and want to find the hypotenuse, so use the formula for sin x.
$\sin 34° = \frac{20}{f} \Rightarrow f \times \sin 34° = 20$
$f = \frac{20}{\sin 34°} =$ **35.8 cm** (3 s.f.)

Using the same method for g)-h):
g) **6.30 cm** (3 s.f.) h) **4.58 cm** (3 s.f.)

2 a) Splitting the triangle in half along the dotted line gives a right-angled triangle. You're given the adjacent and asked to find the hypotenuse, so use the formula for cos x. Divide the given angle by 2 to find the angle in the right-angled triangle.
$\cos 33° = \frac{10}{m}$
$m = 10 \div \cos 33° =$ **11.9 cm** (3 s.f.)

b) Splitting the triangle in half along the dotted line gives a right-angled triangle. You're given the hypotenuse and need to find the opposite, so use the formula for sin x.
Let $N = \frac{n}{2}$
$\sin 51° = \frac{N}{12} \Rightarrow N = 12 \sin 51° = 9.325...$
$n = 2N = 18.651... =$ **18.7 cm** (3 s.f.)

Page 326 Exercise 2

1 a) You're given the adjacent and hypotenuse, so use the formula for cos x.

$\cos a = \frac{3}{8}$

$a = \cos^{-1}\left(\frac{3}{8}\right) =$ **68.0°** (1 d.p.)

b) You're given the opposite and hypotenuse, so use the formula for sin x.

$\sin b = \frac{10}{14}$

$b = \sin^{-1}\left(\frac{10}{14}\right) =$ **45.6°** (1 d.p.)

Using the same method for c)-f):

c) **47.5°** (1 d.p.)   **d)** **80.4°** (1 d.p.)

e) **33.7°** (1 d.p.)   **f)** **60.1°** (1 d.p.)

2 a) You're given the opposite and hypotenuse, so use the formula for sin x.

$\sin m = \frac{8}{24} \Rightarrow m = \sin^{-1}\left(\frac{8}{24}\right) = 19.471...$

So the angle of elevation of the slide is **19.5°** (1 d.p.).

b) Using alternate angles, q is the same as the angle opposite the ladder. So you know the opposite and the adjacent so use the formula for tan x.

$\tan q = \frac{4}{5.5}$

$q = \tan^{-1}\left(\frac{4}{5.5}\right) = 36.027...$

So the angle of depression from the top of the slide is **36.0°** (1 d.p.).

3 a) Splitting the triangle in half along the dotted line gives a right-angled triangle.

Let $Z = \frac{z}{2}$

You're given the adjacent and the hypotenuse, so use the formula for cos x.

$\cos Z = \frac{14}{18} \Rightarrow Z = \cos^{-1}\left(\frac{14}{18}\right) = 38.942...$

$z = 2Z =$ **77.9°** (1 d.p.)

b) In this part, use the properties of rectangles and triangles to find the lengths of missing sides.

j: You know the opposite and adjacent, so use the formula for tan x.

$\tan j = \frac{6}{2} \Rightarrow j = \tan^{-1}\left(\frac{6}{2}\right) =$ **71.6°** (1 d.p.)

k: You know the opposite and adjacent, so use the formula for tan x.

$\tan k = \frac{10}{6}$

$k = \tan^{-1}\left(\frac{10}{6}\right) =$ **59.0°** (1 d.p.)

Using the same method for angles l-n:

$l =$ **31.0°**, $m =$ **45.6°**, $n =$ **70.5°**

Page 327 Exercise 3

1 Using the common values:

a) $\sin a = \frac{\sqrt{3}}{2} \Rightarrow a =$ **60°**

b) $\cos b = \frac{1}{\sqrt{2}} \Rightarrow b =$ **45°**

c) $\tan c = \frac{\sqrt{3}}{1} = \sqrt{3} \Rightarrow c =$ **60°**

d) $\sin d = \frac{4}{8} = \frac{1}{2} \Rightarrow d =$ **30°**

2 a) $\sin 30° = \frac{e}{2}$

$\Rightarrow e = 2\sin 30° = 2 \times \frac{1}{2} =$ **1 m**

b) $\cos 60° = \frac{f}{2}$

$\Rightarrow f = 2\cos 60° = 2 \times \frac{1}{2} =$ **1 cm**

c) $\tan 45° = \frac{8}{g}$

$\Rightarrow g = \frac{8}{\tan 45°} = \frac{8}{1} =$ **8 mm**

d) $\tan 30° = \frac{1}{h}$

$\Rightarrow h = \frac{1}{\tan 30°} = \frac{1}{\left(\frac{1}{\sqrt{3}}\right)} = \sqrt{3}$ **m**

3 a) $\tan 45° = 1, \sin 60° = \frac{\sqrt{3}}{2}$

$1 + \frac{\sqrt{3}}{2} = \frac{2}{2} + \frac{\sqrt{3}}{2} = \frac{2+\sqrt{3}}{2}$

---

b) $\sin 45° = \frac{1}{\sqrt{2}}, \cos 45° = \frac{1}{\sqrt{2}}$

$\frac{1}{\sqrt{2}} + \frac{1}{\sqrt{2}} = \frac{2}{\sqrt{2}} = \frac{2\sqrt{2}}{2} = \sqrt{2}$

c) $\tan 30° = \frac{1}{\sqrt{3}}, \tan 60° = \sqrt{3}$

$\frac{1}{\sqrt{3}} + \sqrt{3} = \frac{1}{\sqrt{3}} + \frac{3}{\sqrt{3}} = \frac{4}{\sqrt{3}} = \frac{4\sqrt{3}}{3}$

4

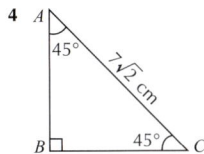

The triangle is right-angled with acute angles of 45°. You're given the hypotenuse and need to find the opposite or adjacent, so use the formula for sin x or cos x. E.g.

$\sin 45° = \frac{AB}{7\sqrt{2}}$

$\Rightarrow AB = 7\sqrt{2} \sin 45° = 7\sqrt{2} \times \frac{1}{\sqrt{2}} =$ **7 cm**

You could also solve this using Pythagoras' theorem.
$AB = BC$ so $AB^2 + AB^2 = 2AB^2 = (7\sqrt{2})^2$.

5 E.g. Angle $EDF = 60°$ since the triangle is equilateral. Then use the formula for sin x:

$\sin 60° = \frac{4}{ED} \Rightarrow \frac{\sqrt{3}}{2} = \frac{4}{ED}$

$\Rightarrow ED (= DF = ED) = \frac{4}{\left(\frac{\sqrt{3}}{2}\right)} = \frac{8}{\sqrt{3}} = \frac{8\sqrt{3}}{3}$ **mm**

25.4 The Sine and Cosine Rules

Page 329 Exercise 1

1 a) $\frac{a}{\sin 50°} = \frac{16}{\sin 80°}$

$a = \frac{16\sin 50°}{\sin 80°} =$ **12.4 cm** (3 s.f.)

b) $\frac{b}{\sin 122°} = \frac{3}{\sin 17°}$

$b = \frac{3\sin 122°}{\sin 17°} =$ **8.70 in** (3 s.f.)

Using the same method for c):

c) **5.59 mm** (3 s.f.)

2 a) $\frac{\sin g}{13} = \frac{\sin 27°}{11}$

$g = \sin^{-1}\left(\frac{13\sin 27°}{11}\right) =$ **32.4°** (1 d.p.)

b) $\frac{\sin h}{9} = \frac{\sin 102°}{16}$

$h = \sin^{-1}\left(\frac{9\sin 102°}{16}\right) =$ **33.4°** (1 d.p.)

Using the same method for c)-f):

c) **38.8°** (1 d.p.)   **d)** **67.0°** (1 d.p.)

e) **44.4°** (1 d.p.)   **f)** **34.1°** (1 d.p.)

3 a)

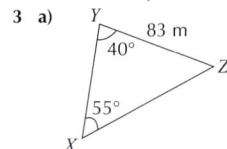

$\frac{YZ}{\sin YXZ} = \frac{XZ}{\sin XYZ} \Rightarrow \frac{83}{\sin 55°} = \frac{XZ}{\sin 40°}$

$XZ = \frac{83\sin 40°}{\sin 55°} =$ **65.1 m** (3 s.f.)

b) $XZY = 180° - (55° + 40°) = 85°$

$\frac{YZ}{\sin YXZ} = \frac{XY}{\sin XZY} \Rightarrow \frac{83}{\sin 55°} = \frac{XY}{\sin 85°}$

$XY = \frac{83\sin 85°}{\sin 55°} =$ **101 m** (3 s.f.)

4

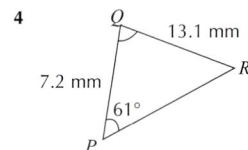

$\frac{\sin RPQ}{QR} = \frac{\sin PRQ}{PQ} \Rightarrow \frac{\sin 61°}{13.1} = \frac{\sin PRQ}{7.2}$

$PRQ = \sin^{-1}\left(\frac{7.2\sin 61°}{13.1}\right) = 28.731...°$

$PQR = 180° - 61° - 28.731...° =$ **90.3°** (3 s.f.)

5

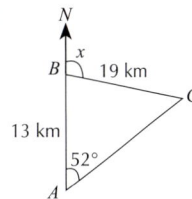

$\frac{\sin 52°}{19} = \frac{\sin ACB}{13} ACB = \sin^{-1}\left(\frac{13\sin 52°}{19}\right) =$

$32.626...°$

$ABC = 180° - 52° - 32.626...° = 95.373...°$

$x = 180° - 95.373...° = 84.626...°$

$= $ **085°** (to the nearest degree)

The answer is a bearing, so you need to give it as three figures.

Page 330 Exercise 2

1 a) $a^2 = b^2 + c^2 - 2bc\cos A$

$a^2 = 7^2 + 3^2 - (2 \times 7 \times 3 \times \cos 66°) = 40.9...$

$a = \sqrt{40.9...} =$ **6.40 cm** (3 s.f.)

b) $b^2 = 6.5^2 + 8^2 - (2 \times 6.5 \times 8 \times \cos 42°)$

$b^2 = 28.9...$

$b = \sqrt{28.9...} =$ **5.38 cm** (3 s.f.)

Using the same method for c)-f):

c) **18.5 m** (3 s.f.)   **d)** **4.65 in** (3 s.f.)

e) **3.63 cm** (3 s.f.)   **f)** **4.59 m** (3 s.f.)

2 a) $\cos A = \frac{b^2 + c^2 - a^2}{2bc}$

$\cos p = \frac{5^2 + 7^2 - 8^2}{2 \times 5 \times 7}$

$p = \cos^{-1}\left(\frac{5^2 + 7^2 - 8^2}{2 \times 5 \times 7}\right) =$ **81.8°** (1 d.p.)

b) $\cos q = \frac{3^2 + 11^2 - 10^2}{2 \times 3 \times 11}$

$q = \cos^{-1}\left(\frac{3^2 + 11^2 - 10^2}{2 \times 3 \times 11}\right) =$ **63.0°** (1 d.p.)

Using the same method for c)-f):

c) **54.0°** (1 d.p.)   **d)** **36.7°** (1 d.p.)

e) **61.1°** (1 d.p.)   **f)** **115.7°** (1 d.p.)

3 $\cos XYZ = \frac{XY^2 + YZ^2 - XZ^2}{2 \times XY \times YZ} = \frac{67^2 + 78^2 - 99^2}{2 \times 67 \times 78}$

$XYZ = \cos^{-1}\left(\frac{67^2 + 78^2 - 99^2}{2 \times 67 \times 78}\right) =$ **85.8°** (3 s.f.)

4 Since this question is non-calculator, you'll have to use the common trig values from pages 326 and 327.

a) $x^2 = 5^2 + 8^2 - (2 \times 5 \times 8 \times \cos 60°)$

$x^2 = 25 + 64 - 80 \times \frac{1}{2} = 89 - 40 = 49$

$x = \sqrt{49} =$ **7 m**

b) $y^2 = 3^2 + (3\sqrt{2})^2 - (2 \times 3 \times 3\sqrt{2} \times \cos 45°)$

$y^2 = 9 + 18 - 18\sqrt{2} \times \frac{1}{\sqrt{2}} = 9 + 18 - 18 = 9$

$y = \sqrt{9} =$ **3 m**

c) $z^2 = 2^2 + (\sqrt{12})^2 - (2 \times 2 \times \sqrt{12} \times \cos 30°)$

$z^2 = 4 + 12 - 4\sqrt{12} \times \frac{\sqrt{3}}{2} = 16 - 12 = 4$

$z = \sqrt{4} =$ **2 mm**

5 $BC^2 = 12^2 + 37^2 - (2 \times 12 \times 37 \times \cos 45°)$

$BC^2 = 885.08...$

$BC = \sqrt{885.08...} =$ **29.8 miles** (3 s.f.)

6 $APB = 108° - 25° = 83°$

$AB^2 = 2^2 + 3^2 - (2 \times 2 \times 3 \times \cos 83°) = 11.5...$

$AB = \sqrt{11.5...} =$ **3.40 km** (3 s.f.)

Page 332 Exercise 3

1 a) Area $= \frac{1}{2} \times 6 \times 5 \times \sin 104°$

$= $ **14.6 cm²** (3 s.f.)

b) Area $= \frac{1}{2} \times 8 \times 10 \times \sin 23°$

$= $ **15.6 m²** (3 s.f.)

Using the same method for c)-f):

c) **38.2 cm²** (3 s.f.)   **d)** **123 mm²** (3 s.f.)

e) **43.7 in²** (3 s.f.)   **f)** **16.7 cm²** (3 s.f.)

2 Area of sector $= \frac{67°}{360°} \times \pi \times 4.5^2 = 11.8...$ cm^2

Area of triangle $= \frac{1}{2} \times 4.5 \times 4.5 \times \sin 67°$
$= 9.32...$ cm^2

Area of sector $= 11.8... - 9.32...$
$= \textbf{2.52 cm}^2$ (3 s.f.)

3 An equilateral triangle has sides of equal length and angles of equal size (60°).

Area $= \frac{1}{2} \times 32 \times 32 \times \sin 60° = \textbf{443 m}^2$ (3 s.f.)

4 $\cos x = \frac{9^2 + 15^2 - 13^2}{2 \times 9 \times 15}$

$x = \cos^{-1}\left(\frac{9^2 + 15^2 - 13^2}{2 \times 9 \times 15}\right) = \textbf{59.5°}$ (3 s.f.)

Area $= \frac{1}{2} \times 9 \times 15 \times \sin 59.50...°$
$= \textbf{58.2 m}^2$ (3 s.f.)

5 $\frac{10}{\sin 84°} = \frac{y}{\sin 49°}$

$y = \frac{10 \sin 49°}{\sin 84°} = 7.5886... = \textbf{7.59 cm}$ (3 s.f.)

Missing angle $= 180° - 84° - 49° = 47°$

Area $= \frac{1}{2} \times 10 \times 7.5886... \times \sin 47°$
$= \textbf{27.7 cm}^2$ (3 s.f.)

25.5 Trigonometry in 3D

Page 333 Exercise 1

1 a) $AF^2 = AB^2 + BF^2 = 3^2 + 3^2 = 18$
$AF = \sqrt{18} = \textbf{3}\sqrt{\textbf{2}}$ **m**

b) $FC^2 = AC^2 + AF^2 = 3^2 + (\sqrt{18})^2 = 27$
$FC = \sqrt{27} = \textbf{3}\sqrt{\textbf{3}}$ **m**

c) $FC = AH = 3\sqrt{3}$
$\sin AHC = \frac{AC}{AH} = \frac{3}{3\sqrt{3}}$
$AHC = \sin^{-1}\left(\frac{3}{3\sqrt{3}}\right) = \textbf{35.3°}$ (1 d.p.)

2 a) Splitting BCE in half with a perpendicular line from E to BC gives a right-angled triangle. Let M = midpoint of BC.
$\cos MCE = \frac{CM}{CE} = \frac{2}{7}$
$BCE = MCE = \cos^{-1}\left(\frac{2}{7}\right) = \textbf{73.4°}$ (1 d.p.)

b) The triangular faces of the pyramid are isosceles, so
$AEB = 180° - (2 \times 73.39...°)$
$= \textbf{33.2°}$ (1 d.p.)

c) $AC^2 = AB^2 + BC^2 = 4^2 + 4^2 = 32$
$AC = \sqrt{32} = 4\sqrt{2}$ m
$AO = 4\sqrt{2} \div 2 = 2\sqrt{2}$ m
$AE = EC = 7$ m
$AE^2 = AO^2 + EO^2 \Rightarrow EO^2 = AE^2 - AO^2$
$EO^2 = 7^2 - (2\sqrt{2})^2 = 41 \Rightarrow EO = \sqrt{41}$ **m**

d) $\cos AEO = \frac{EO}{AE} = \frac{\sqrt{41}}{7}$
$AEO = \cos^{-1}\left(\frac{\sqrt{41}}{7}\right) = \textbf{23.8°}$ (1 d.p.)

Alternatively, you could have used the sin or tan formulas, as you know all three lengths.

3 a) $\cos EDF = \frac{DE}{DF} = \frac{6}{8}$
$EDF = \cos^{-1}\left(\frac{6}{8}\right) = \textbf{41.4°}$ (1 d.p.)

b) $DC^2 = DF^2 + CF^2 = 8^2 + 10^2 = 164$
$DC = \sqrt{164} = \textbf{2}\sqrt{\textbf{41}}$ **cm**

c) $\sin DCE = \frac{DE}{DC} = \frac{6}{2\sqrt{41}}$
$DCE = \sin^{-1}\left(\frac{6}{2\sqrt{41}}\right) = \textbf{27.9°}$ (1 d.p.)

4 a) Using the formula for the longest diagonal in a cuboid:
$AH^2 = AC^2 + CD^2 + DH^2$
$= 3^2 + 5^2 + 8^2 = 98$
$AH = \sqrt{98} = \textbf{7}\sqrt{\textbf{2}}$ **in**

b) $ED = AH = 7\sqrt{2}$ in
$\sin EDG = \frac{EG}{ED} = \frac{3}{7\sqrt{2}}$
$EDG = \sin^{-1}\left(\frac{3}{7\sqrt{2}}\right) = \textbf{17.6°}$ (1 d.p.)

---

5 a) $MC = 10 \div 2 = 5$ cm
$\tan 59° = \frac{BM}{5}$
$BM = 5 \tan 59° = 8.321... = \textbf{8.32 cm}$ (3 s.f.)

b) $EM^2 = BM^2 + BE^2$
$= 8.321...^2 + 25^2 = 694.24...$
$EM = \sqrt{694.24...} = \textbf{26.3 cm}$ (3 s.f.)

Page 334 Exercise 2

1 a) $\cos JIK = \frac{IJ^2 + IK^2 - JK^2}{2 \times IJ \times IK} = \frac{5^2 + 11^2 - 8^2}{2 \times 5 \times 11}$
$JIK = \cos^{-1}\left(\frac{5^2 + 11^2 - 8^2}{2 \times 5 \times 11}\right) = \textbf{41.8°}$ (1 d.p.)

b) Area $= \frac{1}{2} \times 5 \times 11 \times \sin 41.8...° = \textbf{18.3 m}^2$

c) Volume $= 18.3... \times 9$
$= \textbf{165 m}^3$ (to nearest m^3)

2 a) $CE^2 = AC^2 + AE^2 = 4^2 + 6^2 = 52$
$CE = \sqrt{52} = \textbf{2}\sqrt{\textbf{13}}$ **m**

b) $CH^2 = CD^2 + DH^2 = 8^2 + 6^2 = 100$
$CH = \sqrt{100} = \textbf{10 m}$

c) $EH^2 = EF^2 + FH^2 = 8^2 + 4^2 = 80$
$EH = \sqrt{80} = \textbf{4}\sqrt{\textbf{5}}$ **m**

d) $\cos ECH = \frac{CE^2 + CH^2 - EH^2}{2 \times CE \times CH}$
$= \frac{52 + 100 - 80}{2 \times 2\sqrt{13} \times 10}$
$ECH = \cos^{-1}\left(\frac{52 + 100 - 80}{2 \times 2\sqrt{13} \times 10}\right)$
$= \textbf{60.1°}$ (1 d.p.)

3 a) $PU^2 = PQ^2 + QU^2 = 12^2 + 10^2 = 244$
$PU = \sqrt{244} = 2\sqrt{61}$
$PS^2 = PR^2 + RS^2 = 5^2 + 12^2 = 169$
$PS = \sqrt{169} = 13$ m
$SU^2 = SW^2 + UW^2 = 10^2 + 5^2 = 125$
$SU = \sqrt{125} = 5\sqrt{5}$ m
$\cos PSU = \frac{PS^2 + SU^2 - PU^2}{2 \times PS \times SU}$
$= \frac{169 + 125 - 244}{2 \times 13 \times 5\sqrt{5}}$
$PSU = \cos^{-1}\left(\frac{169 + 125 - 244}{2 \times 13 \times 5\sqrt{5}}\right)$
$= \textbf{80.1°}$ (1 d.p.)

b) Area $= \frac{1}{2} \times 13 \times 5\sqrt{5} \times \sin 80.09...°$
$= \textbf{71.6 m}^2$ (1 d.p.)

Page 335 Review Exercise

1 a) $a^2 = 3^2 + 5^2 = 9 + 25 = 34$
$a = \sqrt{34} = \textbf{5.83 cm}$ (3 s.f.)

b) $b^2 = 6.7^2 + 3.9^2 = 60.1$
$b = \sqrt{60.1} = \textbf{7.75 cm}$ (3 s.f.)

Using the same method for c-d):

c) **3.30 m** (3 s.f.)　　**d)** **34.5 m** (3 s.f.)

e) $15^2 = 12^2 + e^2 \Rightarrow e^2 = 15^2 - 12^2 = 81$
$e = \sqrt{81} = \textbf{9 ft}$

f) $1.3^2 = 0.6^2 + f^2 \Rightarrow f^2 = 1.3^2 - 0.6^2 = 1.33$
$f = \sqrt{1.33} = \textbf{1.15 m}$ (3 s.f.)

Using the same method for g)-h):

g) **4.56 m** (3 s.f.)　　**h)** **6.98 m** (3 s.f.)

2 a) (i) $FH^2 = 4^2 + 4^2 = 32$
$FH = \sqrt{32} = 4\sqrt{2}$
$PH = FH \div 2 = \textbf{2}\sqrt{\textbf{2}}$ **m**

(ii) $HI^2 = PI^2 + PH^2$
$PI^2 = HI^2 - PH^2 = 3^2 - (2\sqrt{2})^2 = 1$
$PI = \sqrt{1} = \textbf{1 m}$

(iii) $OI = OP + PI = 6 + 1 = \textbf{7 m}$

b) (i) $OA = PH = \textbf{2}\sqrt{\textbf{2}}$ **m**

(ii) $AI^2 = OA^2 + OI^2 = (2\sqrt{2})^2 + 7^2 = 57$
$AI = \sqrt{57} = \textbf{7.55 m}$ (3 s.f.)

3 You're finding the diagonal of a cuboid 9 cm × 13 cm × 19 cm, so use the cuboid formula.
$d^2 = 9^2 + 13^2 + 19^2 = 611$
$d = \sqrt{611} = \textbf{24.7 cm}$ (3 s.f.)

---

4 a) You're given the hypotenuse and want the opposite side, so use the formula for sin x.
$\sin 62° = \frac{a}{17}$
$a = 17 \sin 62° = \textbf{15.0 cm}$ (3 s.f.)

b) You're given the adjacent and want the opposite, so use the formula for tan x.
$\tan 32° = \frac{b}{28}$
$b = 28 \tan 32° = \textbf{17.5 cm}$ (3 s.f.)

Using the same method for c)-h):

c) **2.40 m** (3 s.f.)　　**d)** **2.33 m** (3 s.f.)

e) **14.3 m** (3 s.f.)　　**f)** **8.96 m** (3 s.f.)

g) **29.0 mm** (3 s.f.)　　**h)** **2.11 cm** (3 s.f.)

i) You're given the opposite and hypotenuse, so use the formula for sin x.
$\sin i = \frac{8.5}{11}$
$i = \sin^{-1}\left(\frac{8.5}{11}\right) = \textbf{50.6°}$ (3 s.f.)

j) You're given the opposite and adjacent, so use the formula for tan x.
$\tan j = \frac{7}{3}$
$j = \tan^{-1}\left(\frac{7}{3}\right) = \textbf{66.8°}$ (3 s.f.)

Using the same method for k)-l):

k) **43.0°** (3 s.f.)　　**l)** **39.5°** (3 s.f.)

5 a) (Height of rectangle)$^2 = 25^2 - 15^2 = 400$
Height of rectangle $= \sqrt{400} = 20$ m
Splitting the triangle along the vertical line of symmetry gives a right-angled triangle with a base length of $15 \div 2 = 7.5$
(Height of triangle)$^2 = 9^2 - 7.5^2 = 24.75$
Height of triangle $= \sqrt{24.75} = 4.97...$ m
Total height $= 20 + 4.97... = \textbf{25.0 m}$ (3 s.f.)

b) Small angle in triangle $= 105° - 90° = 15°$
Splitting the triangle along a vertical line of symmetry gives a right-angled triangle with a base length of $6 \div 2 = 3$. You have the adjacent and want the opposite, so use the formula for tan x.
$\tan 15° = \frac{x}{3} \Rightarrow x = 3 \tan 15° = 0.803...$
Height $= 4.5 + 0.803... = \textbf{5.30 m}$ (3 s.f.)

6 a) (i)

$PQ^2 = (5 - 1)^2 + (4 - 3)^2 = 17$
$PQ = \sqrt{17}$ **cm**
(ii) $PR^2 = (7 - 1)^2 + (1 - 3)^2 = 40$
$PR = \sqrt{40} = \textbf{2}\sqrt{\textbf{10}}$ **cm**

b) Mark the angle you're looking for (x), then form a right-angled triangle using angle x.

$\tan x = \frac{4}{1} \Rightarrow x = \tan^{-1}(4) = 75.96...°$
So the bearing is **076°**.
You could also have used sin x or cos x, since you know that PQ is $\sqrt{17}$ from part a).

7 $\cos x = \frac{Y^2 + Z^2 - X^2}{2 \times Y \times Z} = \frac{15^2 + 21^2 - 17^2}{2 \times 15 \times 21}$
$\Rightarrow x = \cos^{-1}\left(\frac{15^2 + 21^2 - 17^2}{2 \times 15 \times 21}\right) = \textbf{53.2°}$ (1 d.p.)
$\frac{\sin 53.24...°}{17} = \frac{\sin y}{21}$
$\Rightarrow y = \sin^{-1}\left(\frac{21 \sin 53.24...°}{17}\right) = \textbf{81.8°}$ (1 d.p.)
$z = 180° - 53.24...° - 81.77...° = \textbf{45.0°}$ (1 d.p.)
You could have used the sine and cosine rules differently — for example, you could have used the cosine rule to find all 3 angles.

8 a) $\cos p = \frac{4^2 + 12^2 - 13^2}{2 \times 4 \times 12}$
$\Rightarrow p = \cos^{-1}\left(\frac{4^2 + 12^2 - 13^2}{2 \times 4 \times 12}\right) = \textbf{95.4°}$ (3 s.f.)

b) Area $= \frac{1}{2} \times 4 \times 12 \times \sin 95.37...°$
$= \textbf{23.9 cm}^2$ (3 s.f.)

9 a) Area $= \frac{1}{2} \times 8 \times 2\sqrt{2} \times \sin 60°$
$= 4 \times 2\sqrt{2} \times \frac{\sqrt{3}}{2} = \textbf{4}\sqrt{\textbf{6}}$ **cm²**

b) $180° - 99° - 36° = 45°$
Area $= \frac{1}{2} \times 4 \times 5\sqrt{2} \times \sin 45°$
$= 2 \times 5\sqrt{2} \times \frac{1}{\sqrt{2}} = 2 \times 5 = \textbf{10 cm}^2$

10 a) (i) $ED^2 = EA^2 + AC^2 + CD^2$
$= 8^2 + 6^2 + 2^2 = 104$
$ED = \sqrt{104} = \textbf{2}\sqrt{\textbf{26}}$ **ft**

(ii) You're given the opposite and adjacent, so use the formula for tan x.
$\tan FDH = \frac{FH}{DH} = \frac{6}{8}$
$FDH = \tan^{-1}\left(\frac{6}{8}\right) = \textbf{36.9}°$ (1 d.p.)

(iii) You're given the opposite and adjacent, so use the formula for tan x.
$\tan CHD = \frac{CD}{DH} = \frac{2}{8}$
$CHD = \tan^{-1}\left(\frac{2}{8}\right) = \textbf{14.0}°$ (1 d.p.)

b) The diagonal of the crate is ED, which is $2\sqrt{26} \approx 10.2$ ft, so **yes the pole will fit** as 10 ft $< 2\sqrt{26}$ ft.

11 a) $AC^2 = AB^2 + BC^2 = 3^2 + 3^2 = 18$
$AC = \sqrt{18} = 3\sqrt{2}$ m
$EP = AO = \frac{AC}{2} = \frac{3\sqrt{2}}{2}$ m
$EI^2 = EP^2 + IP^2$
$IP^2 = EI^2 - EP^2 = 4^2 - \left(\frac{3\sqrt{2}}{2}\right)^2 = 11.5$
$IP = \sqrt{11.5}$ m
$IO = OP + IP = 7 + \sqrt{11.5} = 10.39...$
You know the opposite and adjacent, so use the formula for tan x.
$\tan OAI = \frac{IO}{AO} = 10.39... \div \frac{3\sqrt{2}}{2}$
$OAI = \tan^{-1}\left(10.39... \div \frac{3\sqrt{2}}{2}\right)$
$= \textbf{78.5}°$ (1 d.p.)

b) You know the opposite and adjacent, so use the formula for tan x.
$\tan OAP = \frac{OP}{AO} = 7 \div \frac{3\sqrt{2}}{2}$
$OAP = \tan^{-1}\left(7 \div \frac{3\sqrt{2}}{2}\right) = \textbf{73.1}°$ (1 d.p.)

Page 337 Exam-Style Questions

1 You're given the opposite and adjacent sides in a right-angled triangle, so use the formula for tan x, where x is the angle of elevation.
$\tan x = \frac{20}{8} \Rightarrow \tan^{-1}\left(\frac{20}{8}\right) = \textbf{68.2}°$ (1 d.p.)
[2 marks available — 1 mark for using tan, 1 mark for the correct answer]
Sketch a diagram if you need to.

2 Drawing a horizontal line from B to the point vertically below A gives a right-angled triangle with a height of $80 - 65 = 15$ m
You have the adjacent and hypotenuse of this triangle, so use cos to find x.
$\cos x = \frac{15}{45} \Rightarrow x = \cos^{-1}\left(\frac{15}{45}\right) = \textbf{70.5}°$ (1 d.p.)
[2 marks available — 1 mark for using cos, 1 mark for the correct answer]

3

R (7, 8)
P (1, 3)
With PR as the hypotenuse, the coordinates form a right-angled triangle with side lengths of $7 - 1 = 6$ and $8 - 3 = 5$.
$PR^2 = 6^2 + 5^2 = 61$
$\Rightarrow PR = \sqrt{61} = \textbf{7.81 units}$ (3 s.f.)
[2 marks available — 1 mark for a correct method, 1 mark for the correct answer]

4 Use Pythagoras's theorem to find the vertical height of both triangles:
$h^2 = a^2 + b^2 \Rightarrow 7^2 = 1^2 + b^2$
$b^2 = 7^2 - 1^2 = 49 - 1 = 48$
$b = \sqrt{48} = 4\sqrt{3}$ cm
You have the opposite and hypotenuse of the unshaded triangle, so use sin.
$\sin x = \frac{4\sqrt{3}}{8}$
$x = \sin^{-1}\left(\frac{4\sqrt{3}}{8}\right) = \sin^{-1}\left(\frac{\sqrt{3}}{2}\right) = \textbf{60}°$
[3 marks available — 1 mark for using Pythagoras to find the vertical height, 1 mark for using sin, 1 mark for the correct answer]
60° is a common angle so you need to know the values of sin 60°, cos 60° and tan 60°.

5 The side lengths of the larger right-angled triangles are $1.2 - 0.3 = 0.9$ m and $1 \div 2 = 0.5$ m
$h^2 = 0.5^2 + 0.9^2 = 1.06$
$h = \sqrt{1.06} = 1.02...$ m
The side lengths of the smaller right-angled triangles are 0.5 m and 0.3 m
$H^2 = 0.5^2 + 0.3^2 = 0.34$
$H = \sqrt{0.34} = 0.58...$ m
Perimeter $= 2h + 2H = 3.22...$ m
so he should buy **3.3 m** or **330 cm** of ribbon.
[3 marks available — 1 mark for a correct method to find the perimeter of the kite, 1 mark for the correct value of at least one of the sides, 1 mark for the correct answer]

6 $BD^2 = AB^2 + AD^2 = 2.15^2 + 2.15^2 = 9.245$
$BD = \sqrt{9.245} = 3.04...$
Let M = midpoint of BD
$MD = 3.04... \div 2 = 1.52...$
You know the adjacent and hypotenuse of triangle EDM, so use the formula for cos x.
$\cos EDM = \frac{MD}{ED} = \frac{1.52...}{3.85}$
$EDM = \cos^{-1}\left(\frac{1.52...}{3.85}\right) = \textbf{66.7}°$ (1 d.p.)
[3 marks available — 1 mark for a correct method to find MD, 1 mark for using the formula for cos EDB, 1 mark for the correct answer]

7

675 m   250 m
The angle required is that between the flight path and the runway. You're given the opposite and hypotenuse, so use the formula for sin x, where x is the angle of elevation.
$\sin x = \frac{250}{675} \Rightarrow x = \sin^{-1}\left(\frac{250}{675}\right) = \textbf{21.7}°$ (3 s.f.)
[2 marks available — 1 mark for using sin, 1 mark for the correct answer]

8 $BCD = 180° - 47° = 133°$
$BDC = 180° - 133° - 28° = 19°$
Using the sine rule, in triangle BCD:
$\frac{BD}{\sin 133°} = \frac{25}{\sin 19°}$
$\Rightarrow BD = \frac{25 \sin 133°}{\sin 19°} = 56.159...$ m
Let A = point at base of tower. In triangle ABD:
You know the hypotenuse and want the opposite, so use the formula for sin x.
$\sin 28° = \frac{AD}{56.159...}$
$AD = 56.159... \times \sin 28° = \textbf{26.4 m}$ (3 s.f.)
[4 marks available — 1 mark for finding BDC using angles on the diagram, 1 mark for using sine rule to find the distance between either person and the top of the tower, 1 mark for using the formula for sin x to find the height of the tower, 1 mark for correct answer]
There are other ways to do this — for example you can find CD then use triangle ACD.

9 Area $= \frac{1}{2}ab \sin C = \frac{1}{2} \times x \times (x + 5) \times \sin 60°$
$= \frac{1}{2} \times (x^2 + 5x) \times \frac{\sqrt{3}}{2} = \frac{\sqrt{3}}{4}(x^2 + 5x)$

Area $= \sqrt{108} = 6\sqrt{3}$, so $6\sqrt{3} = \frac{\sqrt{3}}{4}(x^2 + 5x)$
$\Rightarrow 24 = x^2 + 5x \Rightarrow x^2 + 5x - 24 = 0$
$\Rightarrow (x - 3)(x + 8) = 0$, so $x = 3$ or $x = -8$
x is a length and cannot be negative, so $x = 3$
$BC = 3$ cm and $AC = 3 + 5 = 8$ cm
Use the cosine rule to find AB:
$AB^2 = BC^2 + AC^2 - (2 \times BC \times AC \times \cos ACB)$
$= 3^2 + 8^2 - (2 \times 3 \times 8 \times \cos 60°)$
$= 9 + 64 - (48 \times \frac{1}{2}) = 49$
$AB = \sqrt{49} = \textbf{7 cm}$
[6 marks available — 1 mark for inputting values into the formula for area, 1 mark for rearranging into a quadratic equation, 1 mark for solving the quadratic equation, 1 mark for values for AC and BC, 1 mark for using cosine rule correctly, 1 mark for the correct answer]

Section 26 — Vectors

26.1 Vectors and Scalars

Page 340 Exercise 1

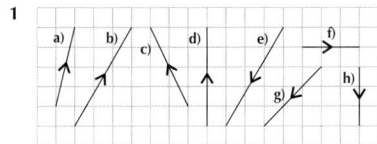

1

2 $a = \begin{pmatrix} 4 \\ 0 \end{pmatrix}$ $b = \begin{pmatrix} -1 \\ -1 \end{pmatrix}$ $c = \begin{pmatrix} 2 \\ 6 \end{pmatrix}$ $d = \begin{pmatrix} -3 \\ 0 \end{pmatrix}$

$e = \begin{pmatrix} -4 \\ -1 \end{pmatrix}$ $f = \begin{pmatrix} 2 \\ 3.5 \end{pmatrix}$ $g = \begin{pmatrix} -4 \\ -3 \end{pmatrix}$ $h = \begin{pmatrix} -2 \\ 2 \end{pmatrix}$

Page 341 Exercise 2

1 a) $3\mathbf{q} = \begin{pmatrix} 3 \times -1 \\ 3 \times 3 \end{pmatrix} = \begin{pmatrix} -3 \\ 9 \end{pmatrix}$

Using the same method for b)-d):
b) $\begin{pmatrix} -5 \\ 15 \end{pmatrix}$ **c)** $\begin{pmatrix} -1.5 \\ 4.5 \end{pmatrix}$ **d)** $\begin{pmatrix} 2 \\ -6 \end{pmatrix}$

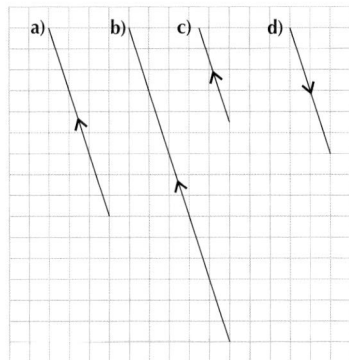

2 a) $2\mathbf{a} = \begin{pmatrix} 2 \times 4 \\ 2 \times -2 \end{pmatrix} = \begin{pmatrix} 8 \\ -4 \end{pmatrix} = \mathbf{d}$

b) $-3\mathbf{b} = \begin{pmatrix} -3 \times -1 \\ -3 \times 4 \end{pmatrix} = \begin{pmatrix} 3 \\ -12 \end{pmatrix} = \mathbf{g}$

c) $\mathbf{c}$ because $3\mathbf{e} = \begin{pmatrix} 3 \times 1 \\ 3 \times 4 \end{pmatrix} = \begin{pmatrix} 3 \\ 12 \end{pmatrix} = \mathbf{c}$

d) $\mathbf{f}$ and $\mathbf{h}$ because $\mathbf{f}$ is vertical and $\mathbf{h}$ is horizontal.

Page 342 Exercise 3

1 a) $\begin{pmatrix} 5 \\ 2 \end{pmatrix} + \begin{pmatrix} 3 \\ 4 \end{pmatrix} = \begin{pmatrix} 5+3 \\ 2+4 \end{pmatrix} = \begin{pmatrix} \textbf{8} \\ \textbf{6} \end{pmatrix}$

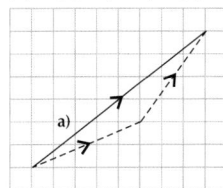

Using the same method for b)-c):

b) $\begin{pmatrix}5\\5\end{pmatrix}$ c) $\begin{pmatrix}-3\\9\end{pmatrix}$

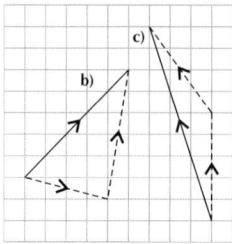

d) $\begin{pmatrix}7\\6\end{pmatrix}-\begin{pmatrix}3\\4\end{pmatrix}=\begin{pmatrix}7-3\\6-4\end{pmatrix}=\begin{pmatrix}4\\2\end{pmatrix}$

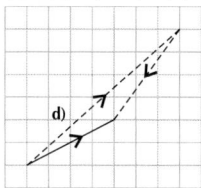

Using the same method for e)-f):

e) $\begin{pmatrix}4\\-4\end{pmatrix}$ f) $\begin{pmatrix}4\\-3\end{pmatrix}$

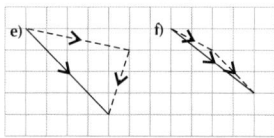

2 a) $\begin{pmatrix}0\\-2\end{pmatrix}+\begin{pmatrix}-1\\4\end{pmatrix}=\begin{pmatrix}0+(-1)\\-2+4\end{pmatrix}=\begin{pmatrix}-1\\2\end{pmatrix}$

b) $\begin{pmatrix}-1\\4\end{pmatrix}-\begin{pmatrix}2\\3\end{pmatrix}=\begin{pmatrix}-1-2\\4-3\end{pmatrix}=\begin{pmatrix}-3\\1\end{pmatrix}$

c) $2\begin{pmatrix}-1\\4\end{pmatrix}+\begin{pmatrix}2\\3\end{pmatrix}=\begin{pmatrix}2(-1)+2\\2(4)+3\end{pmatrix}=\begin{pmatrix}0\\11\end{pmatrix}$

d) $\begin{pmatrix}2\\3\end{pmatrix}+\begin{pmatrix}0\\-2\end{pmatrix}-\begin{pmatrix}-1\\4\end{pmatrix}=\begin{pmatrix}2+0-(-1)\\3+(-2)-4\end{pmatrix}=\begin{pmatrix}3\\-3\end{pmatrix}$

e) $5\begin{pmatrix}0\\-2\end{pmatrix}+4\begin{pmatrix}-1\\4\end{pmatrix}=\begin{pmatrix}5(0)+4(-1)\\5(-2)+4(4)\end{pmatrix}=\begin{pmatrix}-4\\6\end{pmatrix}$

f) $4\begin{pmatrix}2\\3\end{pmatrix}-\begin{pmatrix}0\\-2\end{pmatrix}+3\begin{pmatrix}-1\\4\end{pmatrix}$
$=\begin{pmatrix}4(2)-0+3(-1)\\4(3)-(-2)+3(4)\end{pmatrix}=\begin{pmatrix}5\\26\end{pmatrix}$

Using the same method for Q3:

3 a) $\begin{pmatrix}7\\7\end{pmatrix}$ b) $\begin{pmatrix}8\\1\end{pmatrix}$ c) $\begin{pmatrix}-7\\4\end{pmatrix}$

d) $\begin{pmatrix}19\\14\end{pmatrix}$ e) $\begin{pmatrix}10\\8\end{pmatrix}$ f) $\begin{pmatrix}21\\10\end{pmatrix}$

4 a) $\begin{pmatrix}6\\-2\end{pmatrix}+2\begin{pmatrix}-2\\3\end{pmatrix}=\begin{pmatrix}6+2(-2)\\-2+2(3)\end{pmatrix}=\begin{pmatrix}2\\4\end{pmatrix}$

b)

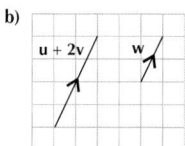

c) The vectors are **parallel** because
$\mathbf{u}+2\mathbf{v}=\begin{pmatrix}2\\4\end{pmatrix}=2\begin{pmatrix}1\\2\end{pmatrix}=2\mathbf{w}$

Page 343   Exercise 4

1 a)

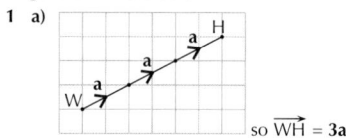

so $\overrightarrow{WH}=3\mathbf{a}$

Using the same method for b)-l):

b) $3\mathbf{b}$ c) $-2\mathbf{a}$ d) $-3\mathbf{b}$
e) $\mathbf{a}+2\mathbf{b}$ f) $-\mathbf{a}-2\mathbf{b}$ g) $2\mathbf{a}-3\mathbf{b}$
h) $-4\mathbf{a}+4\mathbf{b}$ i) $-3\mathbf{a}+5\mathbf{b}$ j) $3\mathbf{a}-3\mathbf{b}$
k) $2\mathbf{a}-5\mathbf{b}$ l) $5\mathbf{a}-8\mathbf{b}$

26.2 Vector Geometry

Page 345   Exercise 1

1 a) $\overrightarrow{CA}=\overrightarrow{CD}+\overrightarrow{DA}=-\mathbf{p}-\mathbf{q}$
b) $\overrightarrow{CB}=\overrightarrow{CA}+\overrightarrow{AB}=(-\mathbf{p}-\mathbf{q})+4\mathbf{p}=3\mathbf{p}-\mathbf{q}$
c) $\overrightarrow{BD}=\overrightarrow{BA}+\overrightarrow{AD}=-4\mathbf{p}+\mathbf{q}$

2 a) Drawing point C gives:

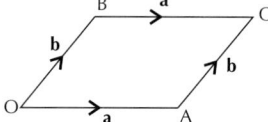

So shape OACB is a **parallelogram**.
b) (i) $\overrightarrow{CO}=\overrightarrow{CB}+\overrightarrow{BO}=-\mathbf{a}-\mathbf{b}$
(ii) $\overrightarrow{AB}=\overrightarrow{AC}+\overrightarrow{CB}=\mathbf{b}-\mathbf{a}$

3 a) $\overrightarrow{OB}=4\overrightarrow{OA}=4\mathbf{a}$
b) $\overrightarrow{OD}=3\overrightarrow{OC}=3\mathbf{c}$
c) $\overrightarrow{AB}=-\overrightarrow{OA}+\overrightarrow{OB}=-\mathbf{a}+4\mathbf{a}=3\mathbf{a}$
d) $\overrightarrow{BA}=-\overrightarrow{AB}=-3\mathbf{a}$
e) $\overrightarrow{AC}=\overrightarrow{AO}+\overrightarrow{OC}=-\mathbf{a}+\mathbf{c}$
f) $\overrightarrow{OE}=\overrightarrow{OD}+\overrightarrow{DE}$
$\overrightarrow{DE}=\frac{1}{2}\overrightarrow{DB}=\frac{1}{2}(\overrightarrow{DO}+\overrightarrow{OB})$
$=\frac{1}{2}(-3\mathbf{c}+4\mathbf{a})=-\frac{3}{2}\mathbf{c}+2\mathbf{a}$
$\overrightarrow{OE}=3\mathbf{c}+(-\frac{3}{2}\mathbf{c}+2\mathbf{a})=2\mathbf{a}+\frac{3}{2}\mathbf{c}$

Using the same method for Q4a):

4 a) (i) $\mathbf{u}+\mathbf{v}$
(ii) $\frac{1}{2}\mathbf{u}+\frac{1}{2}\mathbf{v}$
(iii) $-\mathbf{u}$
(iv) $-\mathbf{u}+\mathbf{v}$
b) If M is the midpoint of XZ,
then $\frac{1}{2}\overrightarrow{XZ}=\overrightarrow{XM}$
$\frac{1}{2}\overrightarrow{XZ}=\frac{1}{2}(-\mathbf{u}+\mathbf{v})=-\frac{1}{2}\mathbf{u}+\frac{1}{2}\mathbf{v}$
$\overrightarrow{XM}=\overrightarrow{XW}+\overrightarrow{WM}=-\mathbf{u}+\frac{1}{2}\mathbf{u}+\frac{1}{2}\mathbf{v}$
$=-\frac{1}{2}\mathbf{u}+\frac{1}{2}\mathbf{v}$
So $\frac{1}{2}\overrightarrow{XZ}$ and $\overrightarrow{XM}$ are equal,
so M is the midpoint of XZ.

5 a) (i) $\overrightarrow{DE}=-\overrightarrow{AB}=-\mathbf{p}$
(ii) $\overrightarrow{AC}=\overrightarrow{AB}+\overrightarrow{BC}=\mathbf{p}+\mathbf{q}$
b) $\overrightarrow{FD}=\overrightarrow{FE}+\overrightarrow{ED}=\mathbf{q}+\mathbf{p}$
$\overrightarrow{AC}=\mathbf{p}+\mathbf{q}$ from a)(ii) so $\overrightarrow{FD}=\overrightarrow{AC}$.
c) (i) $\overrightarrow{AM}=\overrightarrow{BC}=\mathbf{q}$
(ii) $\overrightarrow{MB}=-\overrightarrow{CD}=-\mathbf{r}$
d) $\mathbf{p}=\overrightarrow{AB}=\overrightarrow{AM}+\overrightarrow{MB}=\mathbf{q}-\mathbf{r}$

6 For KST to be a straight line, $\overrightarrow{KT}$ and $\overrightarrow{KS}$ must be scalar multiples of one another. Find $\overrightarrow{KT}$ and $\overrightarrow{KS}$ in terms of $\mathbf{a}$ and $\mathbf{b}$:
$\overrightarrow{KT}=\overrightarrow{KN}+\overrightarrow{NT}=\overrightarrow{KN}+\frac{2}{3}\overrightarrow{NM}=\mathbf{a}+\frac{2}{3}\mathbf{b}$
$\overrightarrow{KS}=\overrightarrow{KN}+\overrightarrow{NS}=\overrightarrow{KN}+\frac{2}{2+3}\overrightarrow{NL}=\overrightarrow{KN}+\frac{2}{5}\overrightarrow{NL}$
Now, $\overrightarrow{NL}=\overrightarrow{NK}+\overrightarrow{KL}=-\mathbf{a}+\mathbf{b}$ so
$\overrightarrow{KS}=\mathbf{a}+\frac{2}{5}(-\mathbf{a}+\mathbf{b})=\frac{3}{5}\mathbf{a}+\frac{2}{5}\mathbf{b}$
Finally, $\frac{3}{5}\overrightarrow{KT}=\frac{3}{5}(\mathbf{a}+\frac{2}{3}\mathbf{b})=\frac{3}{5}\mathbf{a}+\frac{2}{5}\mathbf{b}=\overrightarrow{KS}$
and so $\overrightarrow{KS}$ is a scalar multiple of $\overrightarrow{KT}$.
Hence K, S and T lie on a straight line.

7 a) (i) $\overrightarrow{WU}=\frac{1}{2}\overrightarrow{SW}$
(ii) $\overrightarrow{SU}=\overrightarrow{SW}+\overrightarrow{WU}=\overrightarrow{SW}+\frac{1}{2}\overrightarrow{SW}=\frac{3}{2}\overrightarrow{SW}$
(iii) $\overrightarrow{SU}=\frac{3}{2}\overrightarrow{SW}=\mathbf{a}$ so $\overrightarrow{SW}=\frac{2}{3}\mathbf{a}$

---

b) (i) $\overrightarrow{WV}=\frac{3}{2}\overrightarrow{TW}$

(ii) $\overrightarrow{TV}=\overrightarrow{TW}+\overrightarrow{WV}=\frac{2}{3}\overrightarrow{WV}+\overrightarrow{WV}=\frac{5}{3}\overrightarrow{WV}$

(iii) $\overrightarrow{TV}=\frac{5}{3}\overrightarrow{WV}=\mathbf{b}$ so $\overrightarrow{WV}=\frac{3}{5}\mathbf{b}$
$\overrightarrow{VW}=-\overrightarrow{WV}=-\frac{3}{5}\mathbf{b}$

c) (i) $\overrightarrow{ST}=\overrightarrow{SW}+\overrightarrow{WT}$
$\overrightarrow{SW}=\frac{2}{3}\mathbf{a}$
$\overrightarrow{WT}=\overrightarrow{VT}-\overrightarrow{VW}=-\mathbf{b}-(-\frac{3}{5}\mathbf{b})=-\frac{2}{5}\mathbf{b}$
So $\overrightarrow{ST}=\frac{2}{3}\mathbf{a}-\frac{2}{5}\mathbf{b}$

(ii) $\overrightarrow{UV}=\overrightarrow{UW}+\overrightarrow{WV}$
$\overrightarrow{UW}=\overrightarrow{US}-\overrightarrow{WS}=-\mathbf{a}-(-\frac{2}{3}\mathbf{a})=-\frac{1}{3}\mathbf{a}$
$\overrightarrow{WV}=\frac{3}{5}\mathbf{b}$
So $\overrightarrow{UV}=-\frac{1}{3}\mathbf{a}+\frac{3}{5}\mathbf{b}$

8 a) (i) $\overrightarrow{GA}=5\overrightarrow{GE}=5\mathbf{m}$
(ii) $\overrightarrow{GB}=3\overrightarrow{GF}=3\mathbf{n}$
(iii) $\overrightarrow{AB}=\overrightarrow{AG}+\overrightarrow{GB}=-5\mathbf{m}+3\mathbf{n}$
(iv) $\overrightarrow{BC}=\frac{1}{2}\overrightarrow{AB}=\frac{1}{2}(-5\mathbf{m}+3\mathbf{n})$
$=-\frac{5}{2}\mathbf{m}+\frac{3}{2}\mathbf{n}$
b) $\overrightarrow{EF}=\overrightarrow{EG}+\overrightarrow{GF}=-\mathbf{m}+\mathbf{n}$
$\overrightarrow{FC}=\overrightarrow{FB}+\overrightarrow{BC}$
$\overrightarrow{FB}=\overrightarrow{GB}-\overrightarrow{GF}=3\mathbf{n}-\mathbf{n}=2\mathbf{n}$
$\overrightarrow{FC}=2\mathbf{n}+(-\frac{5}{2}\mathbf{m}+\frac{3}{2}\mathbf{n})=-\frac{5}{2}\mathbf{m}+\frac{7}{2}\mathbf{n}$

But there's no number a such that $\overrightarrow{FC}=a\overrightarrow{EF}$
— e.g multiplying $\overrightarrow{EF}$ by $\frac{5}{2}$ gives
$-\frac{5}{2}\mathbf{m}+\frac{5}{2}\mathbf{n}$, so the coefficient of $\mathbf{m}$
matches $\overrightarrow{FC}$ but the coefficient of $\mathbf{n}$ doesn't.
So $\overrightarrow{EF}$ and $\overrightarrow{FC}$ are not parallel and E, F and
C don't all lie on a straight line.

Page 347   Review Exercise

1 $\mathbf{a}=\begin{pmatrix}3\\4\end{pmatrix}$ $\mathbf{b}=\begin{pmatrix}4\\-2\end{pmatrix}$ $\mathbf{c}=\begin{pmatrix}-2\\-4\end{pmatrix}$

$\mathbf{d}=\begin{pmatrix}-4\\-1\end{pmatrix}$ $\mathbf{e}=\begin{pmatrix}4\\-1\end{pmatrix}$ $\mathbf{f}=\begin{pmatrix}1\\4\end{pmatrix}$

$\mathbf{g}=\begin{pmatrix}0\\-5\end{pmatrix}$ $\mathbf{h}=\begin{pmatrix}1\\-4\end{pmatrix}$

2 a) $3\begin{pmatrix}4\\-3\end{pmatrix}=\begin{pmatrix}3\times4\\3\times-3\end{pmatrix}=\begin{pmatrix}12\\-9\end{pmatrix}$

b) $2\begin{pmatrix}0\\2\end{pmatrix}+\begin{pmatrix}-1\\5\end{pmatrix}=\begin{pmatrix}2(0)-1\\2(2)+5\end{pmatrix}=\begin{pmatrix}-1\\9\end{pmatrix}$

c) $\begin{pmatrix}-1\\5\end{pmatrix}-2\begin{pmatrix}4\\-3\end{pmatrix}=\begin{pmatrix}-1-2(4)\\5-2(-3)\end{pmatrix}=\begin{pmatrix}-9\\11\end{pmatrix}$

d) $\begin{pmatrix}4\\-3\end{pmatrix}+5\begin{pmatrix}-1\\5\end{pmatrix}-3\begin{pmatrix}0\\2\end{pmatrix}$
$=\begin{pmatrix}4-5(1)-3(0)\\-3+5(5)-3(2)\end{pmatrix}=\begin{pmatrix}-1\\16\end{pmatrix}$

3 a)

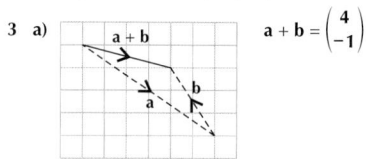

$\mathbf{a}+\mathbf{b}=\begin{pmatrix}4\\-1\end{pmatrix}$

b)

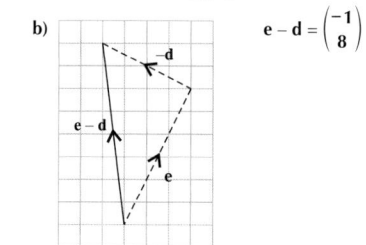

$\mathbf{e}-\mathbf{d}=\begin{pmatrix}-1\\8\end{pmatrix}$

c) Look for two x-components that add to give 1 or two y-components that add to give 9:
$\begin{pmatrix} -2 \\ 3 \end{pmatrix} + \begin{pmatrix} 3 \\ 6 \end{pmatrix} = \begin{pmatrix} 1 \\ 9 \end{pmatrix}$ so the vectors are **b** and **e**.

d) $3\mathbf{c} = 3\begin{pmatrix} 1 \\ 2 \end{pmatrix} = \begin{pmatrix} 3 \times 1 \\ 3 \times 2 \end{pmatrix} = \begin{pmatrix} 3 \\ 6 \end{pmatrix} = \mathbf{e}$
so **e** is parallel to c.

4 a) $\overrightarrow{BE} = \overrightarrow{BD} + \overrightarrow{DE} = \mathbf{a} + (\mathbf{a} + 2\mathbf{b}) = \mathbf{2a + 2b}$
b) $\overrightarrow{AB} = \overrightarrow{AE} + \overrightarrow{ED} + \overrightarrow{DB}$
$= -2\mathbf{b} + (-\mathbf{a} - 2\mathbf{b}) + (-\mathbf{a}) = \mathbf{-2a - 4b}$
c) $\overrightarrow{CD} = \overrightarrow{CB} + \overrightarrow{BD} = (\mathbf{a} + \frac{5}{2}\mathbf{b}) + \mathbf{a} = \mathbf{2a} + \frac{5}{2}\mathbf{b}$
d) $\overrightarrow{AC} = \overrightarrow{AB} + \overrightarrow{BC} = (-2\mathbf{a} - 4\mathbf{b}) + (-\mathbf{a} - \frac{5}{2}\mathbf{b})$
$= \mathbf{-3a} - \frac{13}{2}\mathbf{b}$

5 $\overrightarrow{PX} = \overrightarrow{PS} + \overrightarrow{SX}$
$\overrightarrow{SX} = \frac{1}{2}\overrightarrow{SR}$
$\overrightarrow{SR} = \overrightarrow{SP} + \overrightarrow{PQ} + \overrightarrow{QR}$
$\overrightarrow{QR} = \frac{2}{5}\overrightarrow{PS} = \frac{2}{5}\mathbf{m}$
$\overrightarrow{SR} = -\mathbf{m} + \mathbf{n} + \frac{2}{5}\mathbf{m} = -\frac{3}{5}\mathbf{m} + \mathbf{n}$
$\overrightarrow{SX} = \frac{1}{2}(-\frac{3}{5}\mathbf{m} + \mathbf{n}) = -\frac{3}{10}\mathbf{m} + \frac{1}{2}\mathbf{n}$
$\overrightarrow{PX} = \mathbf{m} + (-\frac{3}{10}\mathbf{m} + \frac{1}{2}\mathbf{n}) = \frac{7}{10}\mathbf{m} + \frac{1}{2}\mathbf{n}$

Page 348   Exam-Style Questions

1 a) $5\mathbf{s} = 5\begin{pmatrix} 4 \\ 1 \end{pmatrix} = \begin{pmatrix} 5 \times 4 \\ 5 \times 1 \end{pmatrix} = \begin{pmatrix} 20 \\ 5 \end{pmatrix}$ *[1 mark]*

b) $2\mathbf{s} - \mathbf{t} = 2\begin{pmatrix} 4 \\ 1 \end{pmatrix} - \begin{pmatrix} -5 \\ 2 \end{pmatrix} = \begin{pmatrix} 8 \\ 2 \end{pmatrix} - \begin{pmatrix} -5 \\ 2 \end{pmatrix} = \begin{pmatrix} 13 \\ 0 \end{pmatrix}$
[2 marks available — 2 marks for the correct answer, otherwise 1 mark for working out 2s correctly]

2 $2\begin{pmatrix} 5p \\ 4q \end{pmatrix} - 3\begin{pmatrix} -q \\ 0 \end{pmatrix} = \begin{pmatrix} 10p \\ 8q \end{pmatrix} - \begin{pmatrix} -3q \\ 0 \end{pmatrix} = \begin{pmatrix} 10p + 3q \\ 8q \end{pmatrix}$
So $\begin{pmatrix} 10p + 3q \\ 8q \end{pmatrix} = \begin{pmatrix} 18 \\ -32 \end{pmatrix}$
$8q = -32$ so $\mathbf{q = -4}$
$10p + 3 \times -4 = 18$
$10p - 12 = 18 \Rightarrow 10p = 30 \Rightarrow \mathbf{p = 3}$
[4 marks available — 1 mark for correctly multiplying 2a or 3b, 1 mark for finding 2a − 3b, 1 mark for the correct value of q, 1 mark for the correct value of p]

3 $\overrightarrow{QS} = \overrightarrow{QP} + \overrightarrow{PS}$
$= -(3\mathbf{a} - 2\mathbf{b}) + (5\mathbf{a} + 6\mathbf{b})$
$= 2\mathbf{a} + 8\mathbf{b}$ *[1 mark]*
$\overrightarrow{QT} = \overrightarrow{QR} + \overrightarrow{RT}$
$= (5\mathbf{a} + 6\mathbf{b}) + (-2\mathbf{a} + 6\mathbf{b})$ *[1 mark]*
$= 3\mathbf{a} + 12\mathbf{b}$ *[1 mark]*
$\frac{2}{3}\overrightarrow{QT} = \frac{2}{3}(3\mathbf{a} + 12\mathbf{b}) = 2\mathbf{a} + 8\mathbf{b} = \overrightarrow{QS}$ so $\overrightarrow{QS}$ and $\overrightarrow{QT}$ are parallel (since the vectors are scalar multiples of each other) so QST is a straight line. *[1 mark]*

4 E.g.
To be a trapezium AC must be parallel to DB.
Since $\overrightarrow{OB} = \mathbf{b}$ and OC : OB is 1 : 4,
$\overrightarrow{OC} = \frac{1}{4}\overrightarrow{OB} = \frac{1}{4}\mathbf{b}$ *[1 mark]*
$\overrightarrow{AC} = \overrightarrow{AO} + \overrightarrow{OC} = -\mathbf{a} + \frac{1}{4}\mathbf{b}$ *[1 mark]*
$\overrightarrow{DB} = \overrightarrow{DO} + \overrightarrow{OB}$
$\overrightarrow{DO} = \overrightarrow{DA} + \overrightarrow{AO} = 3\overrightarrow{AO} + \overrightarrow{AO} = 4\overrightarrow{AO}$
$= -4\mathbf{a}$ *[1 mark]*
$\overrightarrow{DB} = -4\mathbf{a} + \mathbf{b}$ *[1 mark]*
$4\overrightarrow{AC} = 4(-\mathbf{a} + \frac{1}{4}\mathbf{b}) = -4\mathbf{a} + \mathbf{b} = \overrightarrow{DB}$ so AC is parallel to DB (since the vectors are scalar multiples of each other). Therefore ADBC is a trapezium. *[1 mark]*

Section 27 — Perimeter and Area

27.1 Triangles and Quadrilaterals

Page 349   Exercise 1

1 a) (i) $P = 4 \times 1.8 = \mathbf{7.2\ cm}$
   (ii) $A = 1.8^2 = \mathbf{3.24\ cm^2}$
b) (i) $P = (2 \times 1.1) + (2 \times 3.1) = \mathbf{8.4\ mm}$
   (ii) $A = 1.1 \times 3.1 = \mathbf{3.41\ mm^2}$
c) (i) $P = 6 + 9 + 11 = \mathbf{26\ m}$
   (ii) $A = \frac{1}{2} \times 11 \times 4.9 = \mathbf{26.95\ m^2}$

2 a) (i) $P = 4 \times 4 = \mathbf{16\ cm}$
   (ii) $A = 4^2 = \mathbf{16\ cm^2}$
b) (i) $P = (2 \times 6) + (2 \times 8) = \mathbf{28\ m}$
   (ii) $A = 6 \times 8 = \mathbf{48\ m^2}$
c) (i) $P = (2 \times 23) + (2 \times 15) = \mathbf{76\ mm}$
   (ii) $A = 23 \times 15 = \mathbf{345\ mm^2}$
d) (i) $P = 4 \times 17 = \mathbf{68\ m}$
   (ii) $A = 17^2 = \mathbf{289\ m^2}$
e) (i) $P = (2 \times 22.2) + (2 \times 4.3) = \mathbf{53\ m}$
   (ii) $A = 22.2 \times 4.3 = \mathbf{95.46\ m^2}$
f) (i) $P = (2 \times 9) + (2 \times 2.4) = \mathbf{22.8\ mm}$
   (ii) $A = 9 \times 2.4 = \mathbf{21.6\ mm^2}$

3 a) (i) $P = 4 + 13 + 13 = \mathbf{30\ mm}$
   (ii) $A = \frac{1}{2} \times 4 \times 12.8 = \mathbf{25.6\ mm^2}$
b) (i) $P = 8.1 + 10.9 + 12.5 = \mathbf{31.5\ m}$
   (ii) $A = \frac{1}{2} \times 12.5 \times 7 = \mathbf{43.75\ m^2}$
c) (i) $P = 10 + 11 + 19.3 = \mathbf{40.3\ m}$
   (ii) $A = \frac{1}{2} \times 11 \times 7.3 = \mathbf{40.15\ m^2}$

Page 350   Exercise 2

1 a) (i) Missing lengths are: $12 - 8 = 4$ cm and $13 - 5 = 8$ cm.
   $P = 8 + 5 + 4 + 8 + 12 + 13 = \mathbf{50\ cm}$
   (ii) Split horizontally into two rectangles:
   $A = (8 \times 5) + (12 \times 8) = \mathbf{136\ cm^2}$
   You could also split the shape vertically.
b) (i) Missing lengths are: $7 - 4 = 3$ cm and $5 + 3 = 8$ cm.
   $P = 5 + 3 + 3 + 4 + 8 + 7 = \mathbf{30\ cm}$
   (ii) Split vertically into two rectangles:
   $A = (5 \times 7) + (4 \times 3) = \mathbf{47\ cm^2}$
c) (i) Missing lengths are $15 + 8 = 23$ mm and $23 - 10 = 13$ mm.
   $P = 23 + 23 + 8 + 13 + 15 + 10$
   $= \mathbf{92\ mm}$
   (ii) Split vertically into two rectangles:
   $A = (15 \times 10) + (8 \times 23) = \mathbf{334\ mm^2}$

2 a) $A = (8 \times 4) + (4 \times 9) = \mathbf{68\ m^2}$
b) Height of right-hand rectangle
   $= 10 - 4 = 6$ m
   $A = (6 \times 10) + (8 \times 6) = \mathbf{108\ m^2}$
c) Split vertically into two rectangles:
   Width of right-hand rectangle
   $= 14 - 5 = 9$ mm
   $A = (5 \times 8) + (9 \times 12) = \mathbf{148\ mm^2}$

3 a) Width of triangle $= 20 - 8 = 12$ m
   $A = (8 \times 4) + (\frac{1}{2} \times 12 \times 4) = \mathbf{56\ m^2}$
   This is a trapezium so you could use the formula on page 351.
b) Height of each triangle $= 13 \div 2 = 6.5$ mm
   $A = 2 \times \frac{1}{2} \times 19 \times 6.5 = \mathbf{123.5\ mm^2}$
c) Area of each triangle $= \frac{1}{2} \times 5 \times 3 = 7.5$ m²
   $A = 4 \times 7.5 = \mathbf{30\ m^2}$

Page 351   Exercise 3

1 a) $A = \frac{1}{2}(8 + 3) \times 4 = \mathbf{22\ cm^2}$
b) $A = 21 \times 10 = \mathbf{210\ m^2}$
c) $A = \frac{1}{2}(20 + 55) \times 25 = \mathbf{937.5\ mm^2}$
d) $A = 16 \times 18 = \mathbf{288\ mm^2}$
e) $A = 9.2 \times 5.5 = \mathbf{50.6\ cm^2}$
f) $A = \frac{1}{2}(2.5 + 1) \times 3 = \mathbf{5.25\ m^2}$

Page 352   Exercise 4

1 a) $A = (10 \times 6) + \frac{1}{2}(5 + 10) \times 3 = \mathbf{82.5\ cm^2}$
b) $A = (16 \times 10) - (7 \times 3) = \mathbf{139\ mm^2}$
c) $A = \left(\frac{1}{2}(26 + 14) \times 30\right) + \left(\frac{1}{2} \times 14 \times 30\right)$
   $= \mathbf{810\ m^2}$

2 a) (i) $A = 2 \times (8 \times 6) = \mathbf{96\ mm^2}$
   (ii) $P = (4 \times 8) + (2 \times 7) = \mathbf{46\ mm}$
b) (i) Perpendicular height of each trapezium
   $= 8.5 \div 2 = 4.25$ m
   $A = 2 \times \frac{1}{2}(5 + 16) \times 4.25 = \mathbf{89.25\ m^2}$
   (ii) $P = (4 \times 7) + (2 \times 5) = \mathbf{38\ m}$
c) (i) $A = 2 \times (12 \times 8) + 12^2 = \mathbf{336\ m^2}$
   (ii) $P = (4 \times 10) + (4 \times 12) = \mathbf{88\ m}$

3 Area of whole shape $= 66 \times 44 = 2904$ cm²
   Area of one tile $= 2904 \div 25 = \mathbf{116.16\ cm^2}$
   You could also have found the base length and height of a single tile (13.2 cm and 8.8 cm), and multiplied them together.

4 a) $A = \frac{1}{2}(40 + 20) \times 60 = \mathbf{1800\ cm^2}$
b) Area of each strip $= 10 \times 60 = 600$ cm²
   So $A = 600 \times 2 = \mathbf{1200\ cm^2}$

27.2 Circles and Sectors

Page 353   Exercise 1

1 a) $C = \pi \times 6 = \mathbf{18.8\ cm}$ (1 d.p.)
b) $C = 2\pi \times 6 = \mathbf{37.7\ m}$ (1 d.p.)
c) $C = \pi \times 2 = \mathbf{6.3\ cm}$ (1 d.p.)
d) $C = 2\pi \times 15 = \mathbf{94.2\ mm}$ (1 d.p.)

2 a) $C = \pi \times 4 = \mathbf{12.6\ cm}$ (1 d.p.)
b) $C = 2\pi \times 11 = \mathbf{69.1\ m}$ (1 d.p.)
c) $C = 2\pi \times 0.1 = \mathbf{0.6\ km}$ (1 d.p.)
d) $C = \pi \times 6.3 = \mathbf{19.8\ mm}$ (1 d.p.)

3 a) Curved length $= \frac{1}{2} \times \pi \times 4 = 6.28...$ cm
   $P = 6.28... + 4 = \mathbf{10.3\ cm}$ (1 d.p.)
b) Curved length $= \frac{1}{2} \times 2\pi \times 2 = 6.28...$ m
   $P = 6.28... + 2 + 2 = \mathbf{10.3\ m}$ (1 d.p.)
c) Curved length $= \frac{1}{4} \times 2\pi \times 9 = 14.13...$ mm
   $P = 14.13... + (4 \times 9) = \mathbf{50.1\ mm}$ (1 d.p.)
d) Curved length $= \frac{1}{2} \times \pi \times 5 = 7.85...$ cm
   $P = 7.85... + 4 + 4 = \mathbf{15.9\ cm}$ (1 d.p.)

Page 354   Exercise 2

1 a) $A = \pi \times 2^2 = \mathbf{12.6\ cm^2}$ (1 d.p.)
b) $r = 10 \div 2 = 5$ mm
   $A = \pi \times 5^2 = \mathbf{78.5\ mm^2}$ (1 d.p.)
c) $A = \pi \times 8^2 = \mathbf{201.1\ mm^2}$ (1 d.p.)
d) $A = \pi \times 30^2 = \mathbf{2827.4\ mm^2}$ (1 d.p.)

2 a) $A = \pi \times 6^2 = \mathbf{113.1\ mm^2}$ (1 d.p.)
b) $A = \pi \times 5^2 = \mathbf{78.5\ cm^2}$ (1 d.p.)
c) $A = \pi \times 4^2 = \mathbf{50.3\ m^2}$ (1 d.p.)
d) $r = 8.5 \div 2 = 4.25$ m
   $A = \pi \times 4.25^2 = \mathbf{56.7\ mm^2}$ (1 d.p.)
e) $r = 3.5 \div 2 = 1.75$ m
   $A = \pi \times 1.75^2 = \mathbf{9.6\ m^2}$ (1 d.p.)
f) $r = 1.2 \div 2 = 0.6$ mm
   $A = \pi \times 0.6^2 = \mathbf{1.1\ mm^2}$ (1 d.p.)

3 a) Area of circular sections $= \pi \times 5^2$
   $= 78.53...$ cm²
   $A = 78.53... + (10 \times 8) = \mathbf{158.5\ cm^2}$ (1 d.p.)
b) Area of circular section $= \frac{1}{2} \times \pi \times 5^2$
   $= 39.26...$ mm²
   $A = 39.26... + 10^2 = \mathbf{139.3\ mm^2}$ (1 d.p.)
c) Area of circular sections $= \frac{1}{2} \times \pi \times 2^2$
   $= 6.28...$ cm²
   $A = 6.28... + 2^2 = \mathbf{10.3\ cm^2}$ (1 d.p.)

Page 355   Exercise 3

1 a) Arc length $= \frac{15°}{360°} \times 2\pi \times 3 = \frac{\pi}{4}$ **cm**

Sector area $= \frac{15°}{360°} \times \pi \times 3^2 = \frac{3\pi}{8}$ **cm²**

b) Arc length $= \frac{100°}{360°} \times 2\pi \times 4 = \frac{20\pi}{9}$ **cm**

Sector area $= \frac{100°}{360°} \times \pi \times 4^2 = \frac{40\pi}{9}$ **cm²**

Using the same method for c)-d) and Q2:

c) Arc length $\frac{5\pi}{2}$ **cm**, sector area $\frac{25\pi}{2}$ **cm²**

d) Arc length $\frac{10\pi}{3}$ **cm**, sector area $\frac{25\pi}{3}$ **cm²**

Make sure you work out r first in part d).

2 a) Arc length $\frac{21\pi}{2}$ **cm**, sector area $\frac{147\pi}{4}$ **cm²**

b) Arc length $\frac{80\pi}{9}$ **in**, sector area $\frac{320\pi}{9}$ **in²**

c) Arc length $\frac{33\pi}{2}$ **mm**, sector area $\frac{297\pi}{4}$ **mm²**

d) Arc length $\frac{31\pi}{2}$ **cm**, sector area $\frac{465\pi}{4}$ **cm²**

3 a) Major sector angle $= 360 - 39 = 321°$

Arc length $= \frac{321°}{360°} \times 2\pi \times 12 = $ **67.23 in**

Sector area $= \frac{321°}{360°} \times \pi \times 12^2 = $ **403.38 in²**

Using the same method for b)-d):

b) Arc length **29.85 cm**, sector area **82.10 cm²**

c) Arc length **21.33 m**, sector area **69.32 m²**

d) Arc length **37.75 in**, sector area **198.20 in²**

4 Arc length $= \frac{\theta}{360°} \times 2\pi r$

$\Rightarrow 15\pi = \frac{\theta}{360°} \times 2\pi \times 12$

$\Rightarrow 15 = \frac{24\theta}{360°} \Rightarrow 15 = \frac{\theta}{15°} \Rightarrow \theta = 225°$

$\theta > 180°$ so this is the angle of the major sector.
Angle of the minor sector $= 360° - 225° = $ **135°**

5 a) Area $= \pi r^2 \Rightarrow 400\pi = \pi r^2$

$\Rightarrow r^2 = 400 \Rightarrow $ **r = 20 m**

b) Sector area $= \frac{\theta}{360°} \times \pi r^2$

$\Rightarrow 80\pi = \frac{\theta}{360°} \times \pi \times 20^2 \Rightarrow \frac{\theta}{360°} = \frac{1}{5}$

$\Rightarrow$ **$\theta = 72°$**

c) Arc length $= \frac{72°}{360°} \times 2\pi \times 20 = 8\pi$

$P = 8\pi + 20 + 20 = $ **65.1 m** (1 d.p.)

Page 356   Review Exercise

1 a) $23.5 \times 17.3 = 406.55$

$= $ **407 m²** to the nearest m²

b) $(2 \times 23.5) + (2 \times 17.3) = $ **81.6 m**

2 a) $P = 7 + 8 + 9 = $ **24 m**

$A = \frac{1}{2} \times 8 \times 6.7 = $ **26.8 m²**

b) $P = 8.9 + 15.1 + 18.8 = $ **42.8 mm**

$A = \frac{1}{2} \times 18.8 \times 7 = $ **65.8 mm²**

3 a) $A = \left(\frac{1}{2} \times 50 \times 20\right) + \left(\frac{1}{2}(50 + 42) \times 40\right)$

$= $ **2340 cm²**

b) Base of trapezium $= 12 - 4 = 8$ m
Height of parallelogram $= 6 - 2 = 4$ m

$A = \left(\frac{1}{2}(8 + 6) \times 2\right) + (12 \times 4) = $ **62 m²**

4 a) $C = \pi \times 2.7 = 8.482... = $ **8.5 m** (1 d.p.)

$r = 2.7 \div 2 = 1.35$

$A = \pi \times 1.35^2 = 5.725... = $ **5.7 m²** (1 d.p.)

b) $A = \frac{1}{2}(\pi r^2) \Rightarrow \pi r^2 = 19.6 \times 2 = 39.2$ m²

$\Rightarrow r^2 = 39.2 \div \pi$

$\Rightarrow r = \sqrt{12.477...} = 3.532...$ m

Straight length $= 3.532... + 3.532...$
$= 7.064... = $ **7.06 m** (2 d.p.)

5 Area of metal rectangle $= 50 \times 80 = 4000$ cm²

Area of each quarter circle $= \frac{1}{4} \times \pi \times 15^2$
$= 176.71...$ cm²

Remaining area of metal
$= 4000 - (4 \times 176.71...) = $ **3293.1 cm²** (1 d.p.)

6 a) Major sector angle $= 360° - 14° = 346°$

Arc length $= \frac{346°}{360°} \times 2\pi \times 13$
$= $ **78.50 cm** (2 d.p.)

Sector area $= \frac{346°}{360°} \times \pi \times 13^2$
$= $ **510.28 cm²** (2 d.p.)

b) $r = 10 \div 2 = 5$ m

Minor sector angle $= 360° - 320° = 40°$

Arc length $= \frac{40°}{360°} \times \pi \times 10 = \frac{10\pi}{9}$ **m**

Sector area $= \frac{40°}{360°} \times \pi \times 5^2 = \frac{25\pi}{9}$ **m²**

7 a) Area of triangle $= \frac{1}{2} \times 5 \times 4.6 = 11.5$ cm²

Area of circular sections $= \pi \times 2.5^2$
$= 19.63...$ cm²

Area of central square $= 5^2 = 25$ cm²

Area of missing section $= \frac{83°}{360°} \times \pi \times 4^2$
$= 11.58...$ cm²

$A = 11.5 + 19.63... + 25 - 11.58...$
$= $ **44.5 cm²** (1 d.p.)

b) Area of each body segment $= \pi \times 3.5^2$
$= 38.48...$ cm²

Area of tail $= \frac{39°}{360°} \times \pi \times 3^2 = 3.06...$ cm²

Area of head $= \frac{276°}{360°} \times \pi \times 3.2^2$
$= 24.66...$ cm²

Area of eye $= \pi \times (0.8 \div 2)^2 = 0.50...$ cm²

$A = (3 \times 38.48...) + 3.06... + 24.66... - 0.50...$
$= $ **142.7 cm²** (1 d.p.)

Page 357   Exam-Style Questions

1 a) (i) $A = 9 \times 7.5 = $ **67.5 m²** *[1 mark]*

(ii) $A = 0.5^2 = $ **0.25 m²** *[1 mark]*

b) $67.5 \div 0.25 = $ **270 tiles** *[1 mark]*

You could also have worked out the number of tiles in each row and column (18 and 15) and multiplied them together.

2 Curved length $= \frac{1}{2} \times (2\pi \times 16)$ *[1 mark]*

$= 50.265... = $ **50.3 m** (1 d.p.) *[1 mark]*

3 Area of trapezium $= \frac{1}{2}(3 + 6) \times 2.6$

$= 11.7$ m² *[1 mark]*

$A = 2 \times 11.7 = $ **23.4 m²** *[1 mark]*

4 a) Length of ribbon
$= (2 \times 28) + (2 \times 22)$ *[1 mark]*
$= $ **100 cm** *[1 mark]*

b) Area of top $= 22 \times 28 = 616$ cm²
Area of front $= 8 \times 28 = 224$ cm²
Area of side $= 8 \times 22 = 176$ cm²
[1 mark for all three]
Total $= 616 + (2 \times 224) + (2 \times 176)$
$= $ **1416 cm²** *[1 mark]*

c) Arc length $= \frac{36°}{360°} \times 2\pi \times 7$ *[1 mark]*
$= 4.39...$ cm *[1 mark]*

Length of ribbon
$= 4.39... + 7 + 7 = 18.39...$
$= $ **18.4 cm** (1 d.p.) *[1 mark]*

5 The pond is a circle of radius 2 m. Including the path makes a circle of radius $2 + 1 = 3$ m.
Area including path $= \pi \times 3^2$
$= 28.274...$ m² *[1 mark]*
Area of pond $= \pi \times 2^2 = 12.566...$ m² *[1 mark]*
So area of path $= 28.274... - 12.566...$
$= 15.707...$
$= $ **15.7 m²** (1 d.p.) *[1 mark]*

6 Circumference of roller $= 22.5 \div 17$
$= 1.323...$ m *[1 mark]*
$C = 2\pi r \Rightarrow 1.323... = 2\pi \times r$ *[1 mark]*
$\Rightarrow r = 1.323... \div 2\pi = 0.2106...$ m
$= 21.06...$ cm $= $ **21.1 cm** (3 s.f.) *[1 mark]*
3 significant figures is a sensible degree of accuracy here — but 1 d.p. would have been fine too (you'd get the same answer to 1 d.p. as to 3 s.f.).

Section 28 — 3D Shapes

28.1 Plans, Elevations and Isometric Drawings

Page 359   Exercise 1

1 a) (i) The plan view is the view from above, so only one square is seen:

Plan

(ii) The front elevation and side elevation both show three squares on top of each other:

Front   Side

b) (i) Looking vertically from above, you can see an L-shape:

Plan

(ii) From the front and the side, you can only see L-shapes:

Front   Side

Using the same method for c)-d):

c) (i)

Plan

(ii)

Front   Side

d) (i)

Plan

(ii)

Front   Side

2 a) As the shape is a cube, all elevations will be the same:

Plan   Front   Side

b) The plan and front elevation will be rectangles. The side elevation will be a square:

Plan   Front   Side

Using the same method for c)-f):

c)

Plan   Front   Side

d)

Plan   Front   Side

e)

Plan   Front   Side

f)

Plan   Front   Side

3 a)

Plan   Front   Side

b)

Plan   Front   Side

c)

Plan   Front   Side

d)

Plan   Front   Side

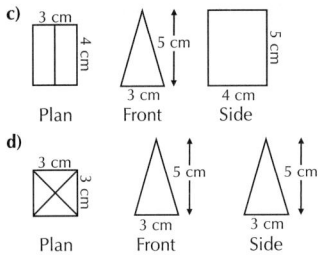

If you're struggling with Q3, try sketching the 3D shapes first. You might get slightly different answers depending on how you imagine each shape to be orientated.

Page 360   Exercise 2

1 Each space between the dots in the vertical and diagonal directions is equal to 1 cm.

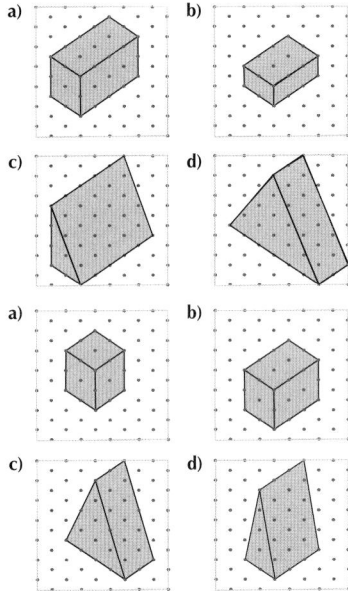

a)   **b)**

c)   **d)**

2 a)   **b)**

c)   **d)**

Page 361   Exercise 3

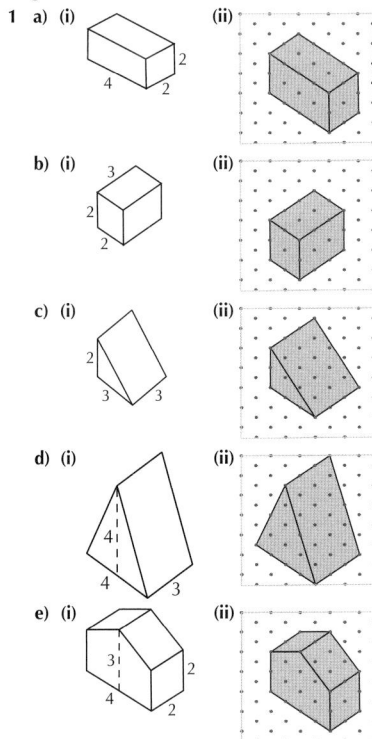

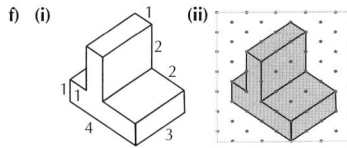

1 a) (i)   **(ii)**

b) (i)   **(ii)**

c) (i)   **(ii)**

d) (i)   **(ii)**

e) (i)   **(ii)**

f) (i)   **(ii)**

28.2 Volume

Page 362   Exercise 1

1 a) Volume = length × width × height
$$= 4 \times 4 \times 4 = \textbf{64 m}^3$$

b) Volume = $2 \times 9 \times 1.5 = \textbf{27 cm}^3$

c) Volume = $7 \times 2 \times 6 = \textbf{84 mm}^3$

2 Volume = $3.2^3 = \textbf{32.768 mm}^3$

3 Volume = $1.7 \times 1.8 \times 0.9 = 2.754 \text{ m}^3$.
The volume of sand is 3.5 m³ so **no, the sand will not fit into the box**.

4 Volume = 18 cm³
Length × width = $5 \times 3 = 15$ cm
So $18 = 15 \times$ height
$\Rightarrow$ height $= \frac{18}{15} = \textbf{1.2 cm}$

5 Volume = 793.5 mm³
1.15 cm = $1.15 \times 10 = 11.5$ mm
Height × width = $9.2 \times 11.5 = 105.8$
So $793.5 = 105.8 \times$ length
$\Rightarrow$ length $= \frac{793.5}{105.8} = \textbf{7.5 mm}$

6 a) Maximum volume = $1.5 \times 0.5 \times 0.6$
$$= \textbf{0.45 m}^3 \textbf{ of water}$$

b) If height = 0.3 m then
volume = $1.5 \times 0.5 \times 0.3 = \textbf{0.225 m}^3$

c) Volume = 0.3 m³
Length × width = $1.5 \times 0.5 = 0.75$
So $0.3 = 0.75 \times$ height
$\Rightarrow$ height $= \frac{0.3}{0.75} = \textbf{0.4 m}$

Page 363   Exercise 2

1 a) V = area of cross-section × length
$$= \frac{1}{2} \times \text{base} \times \text{height} \times \text{length}$$
$$= \frac{1}{2} \times 2 \times 4 \times 7 = \textbf{28 cm}^3$$

b) $V = \frac{1}{2} \times 4 \times 3 \times 5 = \textbf{30 cm}^3$

c) $V = \frac{1}{2} \times 5 \times 6 \times 1.5 = \textbf{22.5 cm}^3$

d) $V = \pi r^2 \times$ length
$$= \pi \times 1^2 \times 5 = \textbf{15.7 cm}^3 \text{ (1 d.p.)}$$

2 a) $V = \frac{1}{2} \times 4.2 \times 1.3 \times 3.1 = \textbf{8.463 m}^3$

b) $V = \pi r^2 \times$ length
$$= \pi \times 4^2 \times 18 = \textbf{904.8 m}^3 \text{ (1 d.p.)}$$

Using the same methods for c)-d):

c) **282.7 mm³** (1 d.p.)

d) **18.9 m³**
The area of a parallelogram is given by base × vertical height.

3 Area of cross-section $= \frac{1}{2} \times 9 \times 8 = 36$ cm²
so volume = $936 = 36x$
$\Rightarrow x = 936 \div 36 = \textbf{26 cm}$

4 a) Area of cross-section $= \frac{1}{2} \times 5 \times p = \frac{5}{2}p$
so volume = $120 = \frac{5}{2}p \times 16$
$\Rightarrow 120 = 40p \Rightarrow p = \textbf{3 m}$

b) Area of cross-section $= \frac{1}{2}(7 \times q) = \frac{7}{2}q$
so volume = $336 = \frac{7}{2}q \times 12$
$\Rightarrow 336 = 42p \Rightarrow p = \textbf{8 m}$

c) Area of cross-section $= \pi r^2$
so volume = $201 = \pi r^2 \times 4$
$\Rightarrow r^2 = \frac{201}{4\pi} = 15.99...$
$\Rightarrow r = \sqrt{15.99...} = \textbf{4.0 m}$ (1 d.p.)

Page 364   Exercise 3

1 a) Split the cross-section into a rectangle and square and add the areas:
Area of rectangle = $3.6 \times (2.4 - 1.2)$
$$= 4.32 \text{ m}^2$$
Area of square = $1.2 \times 1.2 = 1.44$ m²
$\Rightarrow$ Area of cross-section = $4.32 + 1.44$
$$= 5.76 \text{ m}^2$$
So $V = 5.76 \times 2.5 = \textbf{14.4 m}^3$

b) Split the cross-section into a rectangle and triangle and add the areas:
Area of rectangle = $4 \times 2 = 8$ cm²
Area of triangle = $\frac{1}{2} \times 4 \times (6 - 2) = 8$ cm²
$\Rightarrow$ Area of cross-section = $8 + 8 = 16$ cm²
So $V = 16 \times 5 = \textbf{80 cm}^3$

c) Split the cross-section into a square and a triangle and add the areas:
Area of square = $2 \times 2 = 4$ mm²
Area of triangle = $\frac{1}{2} \times 2 \times (4 - 2) = 2$ mm²
$\Rightarrow$ Area of cross-section = $4 + 2 = 6$ mm²
So $V = 6 \times 5.5 = \textbf{33 mm}^3$
Here, you could also have worked out the area of the cross-section using the trapezium formula.

d) Split the cross-section into two rectangles and add the areas:
Area of bottom rectangle = $3 \times 1 = 3$ m²
Area of top rectangle = $1 \times 1.5 = 1.5$ m²
$\Rightarrow$ Area of cross-section = $3 + 1.5 = 4.5$ m²
So $V = 4.5 \times 2 = \textbf{9 m}^3$

2 Volume of cube = $15^3 = 3375$ cm³
Glass capacity = 0.125 litres
$= 0.125 \times 1000 = 125$ cm³
Number of glasses = $\frac{3375}{125} = \textbf{27 glasses}$

3 90 litres = $90 \times 1000 = 90\,000$ cm³
Volume = $90\,000 = \pi r^2 \times$ height = $\pi r^2 \times 3.5$
$\Rightarrow r^2 = \frac{90\,000}{3.5\pi} = 8185.11...$
$\Rightarrow r = \sqrt{8185.11...} = \textbf{90.5 cm}$ (1 d.p.)

28.3 Nets and Surface Area

Page 365   Exercise 1

1 a) There should be six faces, each with 1 cm sides.
E.g.

b) There should be three rectangular faces and two triangular faces, with dimensions as shown.
E.g.

c) There should be four equilateral triangles, each with sides of 5 cm.
E.g.

d) There should be one rectangular face of width 3 cm and length $\pi d = \pi \times 6$ = 18.85 (2 d.p.), and two circular with diameter 6 cm faces attached to either side of the rectangle as shown.

E.g.

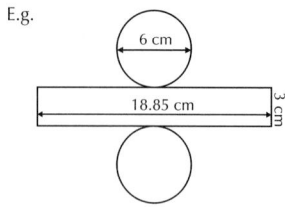

Using the same methods for Q2:

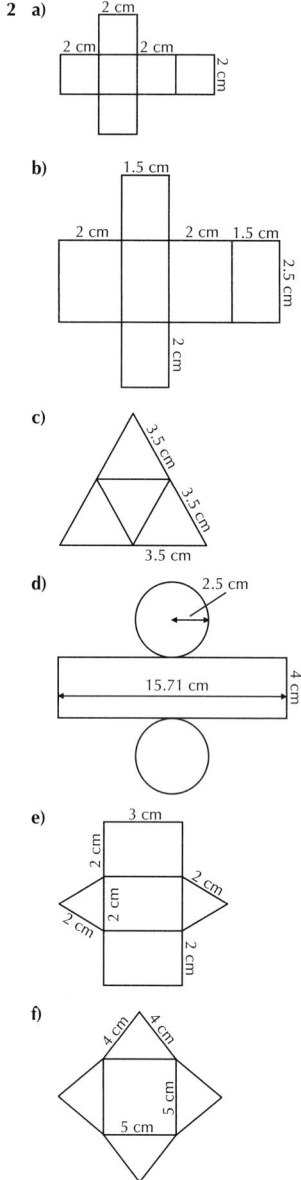

2 a)

b)

c)

d)

e)

f)

There should be one square face and four triangular faces.

3 Net *C* because it's the only net where the edges match correctly when its folded.

Page 366   Exercise 2

1 a) There are 6 faces that are 3 cm × 3 cm.
Surface area = 6(3 × 3) = 6 × 9 = **54 cm²**

b) There are 2 faces that are 4 cm × 3 cm,
2 faces that are 4 cm × 1 cm, and 2 faces
that are 3 cm × 1 cm.
Surface area = 2(4 × 3) + 2(4 × 1) + 2(3 × 1)
= 24 + 8 + 6 = **38 cm²**

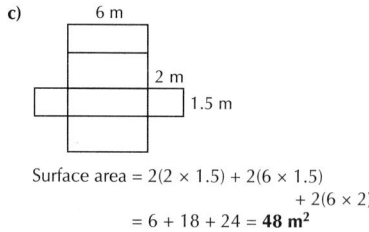

Using the same method for c)-d):

c) 23 m²   **d)** 57.5 m²

---

e) Area of 2 triangles = 2 × $\frac{1}{2}$(5 × 12)
= 60 mm²
Area of base = 5 × 11 = 55 mm²
Area of vertical face = 12 × 11 = 132 mm²
Area of slanting face = 13 × 11 = 143 mm²
Surface area = 60 + 55 + 132 + 143
= **390 mm²**

f) Area of 2 triangles = 2 × $\frac{1}{2}$(3 × 2) = 6 m²
Area of base = 3 × 3.2 = 9.6 m²
Area of the 2 other faces = 2(3.2 × 2.5)
= 16 m²
Surface area = 6 + 9.6 + 16 = **31.6 m²**

Using the same methods for g) and h):

g) 1.2 m²   **h)** 140 m²

2 a) The net will make a **cube** with
edges of length 4.5 cm.
1 square face = 4.5 × 4.5 = 20.25 cm²
Surface area = 6 × 20.25 = **121.5 cm²**

b) The net will make a **cuboid** with width and
height of 11 m and length of 33 m.
Surface area = 4(33 × 11) + 2(11 × 11)
= 1452 + 242 = **1694 m²**

c) The net will make a **triangular prism**.
Surface area
= 2(7.8 × 6) + (7.8 × 2.1) + 2 × $\frac{1}{2}$(5.9 × 2.1)
= 93.6 + 16.38 + 12.39 = **122.37 cm²**

3 a) 1 square face = 5 × 5 = 25 m²
Surface area = 6 squares faces
= 6 × 25 m² = **150 m²**

Using the same method for b):

b) 216 mm²

c)

Surface area = 2(2 × 1.5) + 2(6 × 1.5)
+ 2(6 × 2)
= 6 + 18 + 24 = **48 m²**

Using the same method for d):

d) 135.5 m²

e)

Surface area = 2 × $\frac{1}{2}$(8 × 4) + 2(5 × 2.5)
+ (8 × 2.5)
= **77 m²**

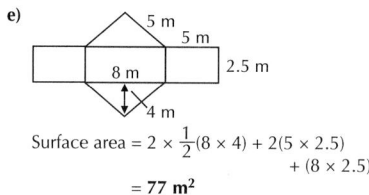

4 Area of 5 square faces = 5 × 12² = 720 m²
*There are only five faces to the cube part of the
shape since what would be the sixth is attached
to the triangular prism inside the shape.*
Area of 2 slanting faces = 2(12 × 8) = 192 m²
Height of triangle can be found using
Pythagoras' theorem: $a^2 + b^2 = c^2$
$\Rightarrow 6^2 + b^2 = 8^2 \Rightarrow b^2 = 28$
$\Rightarrow b = \sqrt{28} = 5.29...$
Area of 2 triangular faces
= 2 × $\frac{1}{2}$(12 × 5.29...) = 63.49... m²
Surface area of shape
= 720 + 192 + 63.49... = **975.5 m²** (1 d.p.)

Page 367   Exercise 3

1 a) Surface area = 2 × area of circle
+ area of curved surface
Area of circle = πr^2 = π × 1² = 3.1415...
Area of curved surface
= circumference of circle × length
= $2\pi r \times l$ = 2π × 1 × 4 = 25.13...
Surface area = (2 × 3.1415...) + 25.13...
= **31.42 cm²** (2 d.p.)

Using the same method for b)-d):

b) 276.46 mm² (2 d.p.)

c) 131.95 m² (2 d.p.)   **d)** 12.82 m² (2 d.p.)

2 a) Surface area = 2 × area of circle
+ area of curved surface
Area of circle = πr^2 = π × 2² = 12.57...
Area of curved surface
= $2\pi r \times l$ = 2π × 2 × 7 = 87.96...
Surface area = (2 × 12.57...) + 87.96...
= **113.10 m²** (2 d.p.)

Using the same method for b)-c):

b) 471.24 mm² (2 d.p.)

c) 1694.07 cm² (2 d.p.)

d) Radius = 22.1 ÷ 2 = 11.05 m
Then using the same method as above,
the surface area is **1537.86 m²** (2 d.p.).

3 a) Curved surface area
= circumference of the circle × length
= $2\pi r \times l$ = 2π × 2.2 × 7.1
= 98.143... = **98.14 m²** (2 d.p.)

b) 9 pipes = 9 × 98.143... = **883.29 m²** (2 d.p.)

4 Surface area of gas tank
= 2 × area of circle + area of curved surface
Area of circle = πr^2 = π × 0.8² = 2.01... m²
Area of curved surface
= $2\pi r \times l$ = 2π × 0.8 × 3 = 15.079...
Surface area = (2 × 2.01...) + 15.079...
= 19.10... m²
Paint needed = 19.10... ÷ 14
= **1.36 litres** (2 d.p.)

5 a)

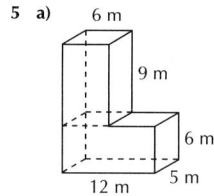

Split the shape along the dotted line.
Surface area of bottom cuboid
= 2(12 × 6) + 2(5 × 6)
+ (12 × 5) + ((12 – 6) × 5)
= 144 + 60 + 60 + 30 = 294 m²
Surface area of top cuboid
= 2(9 × 5) + 2(9 × 6) + (6 × 5)
= 90 + 108 + 30 = 228 m²
Total surface area = 294 + 228 = **522 m²**
*Be careful that you don't include any area where
the two cuboids are joined together. These are
within the shape so not on the surface.*

b) Surface area of top cylinder
= area of circle + area of curved surface
Area of circle = πr^2 = π × 2² = 4π
Area of curved surface = $2\pi r \times l$
= 2π × 2 × 3 = 12π
So surface area of top cylinder
= 4π + 12π = 16π m²
Surface area of bottom cylinder
= area of circle + area of curved surface
+ (area of circle
– area of base of top cylinder)
Area of circle = πr^2 = π × 4² = 16π
Area of curved surface = $2\pi r \times l$
= 2π × 4 × 2 = 16π
Surface area of bottom cylinder
= 16π + 16π + (16π – 4π)
= 44π m²
Surface area of shape = 16π + 44π
= **60π m²**

28.4 Spheres, Cones and Pyramids

Page 368   Exercise 1

1 a) (i) Surface area = $4\pi r^2$ = 4π × 5²
= **100π cm²**

(ii) Volume = $\frac{4}{3}\pi r^3 = \frac{4}{3}$π × 5³
= $\frac{500\pi}{3}$ cm²

Using the same method for b)-d):

b) (i) 64π cm^2 **(ii) $\frac{256\pi}{3}$ cm^3**

c) (i) 25π m^2 **(ii) $\frac{125\pi}{6}$ m^3**

d) (i) 400π mm^2 **(ii) $\frac{4000\pi}{3}$ mm^3**

2 Surface area = $265.9 = 4\pi r^2 \Rightarrow \pi r^2 = 66.475$
$\Rightarrow r^2 = 21.15... \Rightarrow r = \sqrt{21.15...}$
= **4.6 cm** (1 d.p.)

3 Volume = $24\ 429 = \frac{4}{3}\pi r^3 \Rightarrow \pi r^3 = 18\ 321.75$
$\Rightarrow r^3 = 5831.99... \Rightarrow r = \sqrt[3]{5831.99...}$
= **18.0 cm** (1 d.p.)

4 Find the radius:
Surface area = $4\pi r^2 = 2463 \Rightarrow \pi r^2 = 615.75$
$\Rightarrow r^2 = 195.99...$
$\Rightarrow r = \sqrt{195.99...} = 13.99...$ mm
Use the radius to find the volume:
Volume = $\frac{4}{3}\pi r^3 = \frac{4}{3}\pi \times 13.99...^3$
= **11 494.0 mm^3** (1 d.p.)

5 Find the radius:
Volume = $6044 = \frac{4}{3}\pi r^3 \Rightarrow \pi r^3 = 4533$
$\Rightarrow r^3 = 1442.89... \Rightarrow r = \sqrt[3]{1442.89...}$
= 11.30... m
Use the radius to find the surface area:
Surface area = $4\pi r^2 = 4\pi \times 11.30...^2$
= **1604.6 m^2** (1 d.p.)

6 a) Volume of hemisphere
= $\frac{2}{3}\pi r^3 = \frac{2}{3}\pi \times 12^3 = $ **3619.1 m^3** (1 d.p.)

b) (i) Curved surface of hemisphere
= $2\pi r^2 = 2\pi \times 12^2 = $ **904.8 m^2** (1 d.p.)

(ii) Total surface area = area of circle
+ area of curved surface
Area of circle = $\pi r^2 = \pi \times 12^2$
= 452.38...

Total surface area
= 452.38... + 904.77...
= **1357.2 m^2** (1 d.p.)
You could have done this in one step using the formula for the surface area of a hemisphere on p.368.

Page 369 Exercise 2

1 a) (i) Surface area = $\pi r l + \pi r^2$
= $\pi \times 5 \times 13 + \pi \times 5^2$
= $65\pi + 25\pi = $ **90π m^2**

(ii) Volume = $\frac{1}{3}\pi r^2 h = \frac{1}{3}\pi \times 5^2 \times 12$
= $\frac{1}{3}\pi \times 300 = $ **100π m^3**

Using the same method for b)-d):

b) (i) 224π cm^2 **(ii) 392π cm^3**
c) (i) 480π m^2 **(ii) 600π m^3**
d) (i) 1815π mm^2 **(ii) 1650π mm^3**

2 a) Surface area = $\pi r l + \pi r^2$
= $\pi \times 28 \times 53 + \pi \times 28^2$
= **2268π m^2**

b) (i)

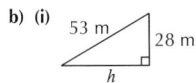

$53^2 = 28^2 + h^2$
$\Rightarrow h^2 = 53^2 - 28^2 = 2025$
$\Rightarrow h = \sqrt{2025} = $ **45 m**

(ii) Volume = $\frac{1}{3}\pi r^2 h = \frac{1}{3}\pi \times 28^2 \times 45$
= $\frac{1}{3}\pi \times 35\ 280 = $ **11 760π m^3**

3 a) Volume = $\frac{1}{3}\pi r^2 h = \frac{1}{3}\pi \times 56^2 \times 33$
= $\frac{1}{3}\pi \times 103\ 488 = $ **34 496π cm^3**

b) (i)

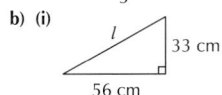

$l^2 = 33^2 + 56^2 = 4225$
$\Rightarrow l = \sqrt{4225} = $ **65 cm**

(ii) Surface area = $\pi r l + \pi r^2$
= $\pi \times 56 \times 65 + \pi \times 56^2$
= $3640\pi + 3136\pi = $ **6776π cm^2**

4 a) Volume = $\frac{1}{3}\pi r^2 h = 39.27$ mm^3
$39.27 = \frac{1}{3}\pi r^2 \times 6 = 2\pi r^2$
$\pi r^2 = 19.635 \Rightarrow r^2 = 6.25...$
$\Rightarrow r = \sqrt{6.25...} = $ **2.5 mm** (1 d.p.)

b) (i)

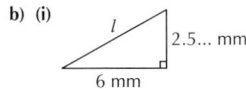

$l^2 = 6^2 + 2.5...^2 = 42.25...$
$\Rightarrow l = \sqrt{42.25...} = $ **6.5 mm** (1 d.p.)

(ii) Surface area = $\pi r l + \pi r^2$
= $\pi \times 2.5... \times 6.5... + \pi \times 2.5...^2$
= **70.7 mm^2** (1 d.p.)

5 Volume = $\frac{1}{3}\pi r^2 h = 9236.28$ cm^3
$9236.28 = \frac{1}{3}\pi r^2 \times 20 = \frac{20}{3}\pi r^2$
$\Rightarrow 1385.442 = \pi r^2 \Rightarrow r^2 = 440.99...$
$\Rightarrow r = \sqrt{440.99...} = 20.99...$ cm
Using Pythagoras' theorem:
$20^2 + 20.99...^2 = l^2$
$\Rightarrow l = \sqrt{840.99...} = 28.99...$ cm
Surface area = $\pi r l + \pi r^2$
= $\pi \times 20.99... \times 28.99... + \pi \times 20.99...^2$
= **3298.7 cm^2** (1 d.p.)

Page 370 Exercise 3

1 a) The heights of the cones have a ratio of
$16:4 = 4:1$, so the base radius will also
be of the ratio $4:1$. So the base of the
smaller cone is $12 \div 4 = $ **3 cm**.

b) Volume = $\frac{1}{3}\pi R^2 H = \frac{1}{3}\pi \times 12^2 \times 16$
= $\frac{1}{3}\pi \times 2304 = $ **768π cm^3**

c) Volume = $\frac{1}{3}\pi r^2 h = \frac{1}{3}\pi \times 3^2 \times 4$
= $\frac{1}{3}\pi \times 36 = $ **12π cm^3**

d) Volume of frustum
= vol. of larger cone – vol. of smaller cone
= $768\pi - 12\pi = $ **756π cm^3**

2 The ratio of the radius of the original
cone to the radius of the removed cone
is $10:5 = 2:1$, so the height of the
original cone will be $12 \times 2 = 24$ cm.
Volume of original cone = $\frac{1}{3}\pi R^2 H$
= $\frac{1}{3}\pi \times 10^2 \times 24 = \frac{1}{3}\pi \times 2400 = 800\pi$
The height of the removed cone will be the
half the height of the original cone as the ratio
is $2:1$ and the cones are similar.
Volume of removed cone = $\frac{1}{3}\pi r^2 h$
= $\frac{1}{3}\pi \times 5^2 \times 12 = \frac{1}{3}\pi \times 300 = 100\pi$
So volume of the frustum
= $800\pi - 100\pi = $ **700π cm^3**
You could have done this in one step using the formula for the volume of a frustum on p.370.

Page 371 Exercise 4

1 Volume of pyramid = $\frac{1}{3} \times$ base area $\times$ height
= $\frac{1}{3} \times 18 \times 15 = $ **90 cm^3**

2 Volume of pyramid = $\frac{1}{3} \times 27 \times 12 = $ **108 m^3**

3 Volume of pyramid = $\frac{1}{3} \times (4 \times 7) \times 10$
= $\frac{1}{3} \times 28 \times 10 = $ **93.3 cm^3** (1 d.p.)

4 Area of one face = $\frac{1}{2}ab \sin C$
= $\frac{1}{2} \times 12 \times 12 \times \sin 60° = 72 \times \sin 60°$
= 62.35... mm^2
Total surface area = $4 \times 62.35...$
= **249.4 mm^2** (1 d.p.)

---

5 Volume = $736 = \frac{1}{3} \times$ base area $\times 11.5$
$\Rightarrow$ base area = 192 cm^2
The base is a square,
so base area = (side length)^2 = 192 cm^2
$\Rightarrow$ side length = $\sqrt{192} = $ **13.9 cm** (1 d.p.)

6 a) Volume = $\frac{1}{3} \times 96^2 \times 55 = $ **168 960 m^3**

b) $OM = \frac{96}{2} = 48$ m
$EM^2 = OM^2 + OE^2$
$\Rightarrow EM^2 = 48^2 + 55^2 = 5329$
$\Rightarrow EM = \sqrt{5329} = $ **73 m**

c) (i) Area of triangle = $\frac{1}{2} \times$ base $\times$ height
= $\frac{1}{2} \times 96 \times 73 = $ **3504 m^2**

(ii) Surface area = base area + 4 triangles
= $96^2 + 4 \times 3504 = = $ **23 232 m^2**

28.5 Rates of Flow
Page 372 Exercise 1

1 a) Rate of flow = $\dfrac{\text{volume of container}}{\text{total time taken}}$
= $\dfrac{600}{15} = $ **40 cm^3/s**

b) Rate of flow = $\dfrac{150}{8} = $ **18.75 litres/hr**

2 a) 1 litre = 1000 ml,
so 3 litres/s = $3 \times 1000 = 3000$ ml/s.
As 1 ml = 1 cm^3, 3000 ml/s = 3000 cm^3/s.
As 1 minute = 60 seconds, 3000 cm^3/s
= $3000 \times 60 = $ **180 000 cm^3/min**

b) 1 m^3 = 1 000 000 cm^3
$\dfrac{180\ 000}{1\ 000\ 000} = $ **0.18 m^3/min**

c) 1 day = 24 hours, 1 hour = 60 minutes,
so 1 day = $24 \times 60 = 1440$ minutes.
0.18 m^3/min = 0.18×1440
= **259.2 m^3/day**

d) 1 m^3 = 1000 litres
259.2 m^3/day = 259.2×1000
= **259 200 litres/day**

3 Radius = diameter $\div$ 2, so $r = 9 \div 2 = 4.5$ m
Volume of silo = $\pi r^2 \times h = \pi \times 4.5^2 \times 20$
= 1272.3... m^3
To half fill the silo, the volume
needs to be $\dfrac{1272.3...}{2} = 636.17...$ m^3
Time taken = $\dfrac{636.17...}{12} = 53.01...$
= **53 minutes** (to the nearest minute)

4 Volume of sphere = $\frac{4}{3}\pi r^3 = \frac{4}{3}\pi \times 60^3$
= 288 000π cm$^3 = \dfrac{288\ 000\pi}{1000} = 288\pi$ litres
Time taken = $\dfrac{288\pi}{2\pi} = $ **144 seconds**

28.6 Symmetry of 3D Shapes
Page 373 Exercise 1

1 a) A pentagon has 5 lines of symmetry so the
prism has **6 planes of symmetry**.

b) An octagon has 8 lines of symmetry so the
prism has **9 planes of symmetry**.

c) A scalene triangle has 0 lines of symmetry
so the prism has **1 plane of symmetry**.

2 a)

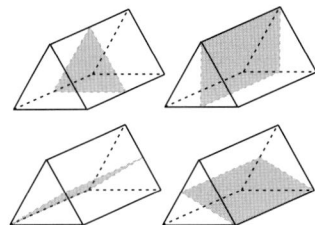

b)

c)

3

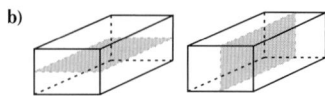

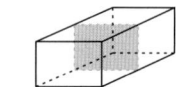

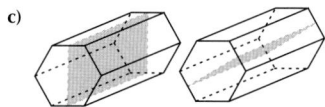

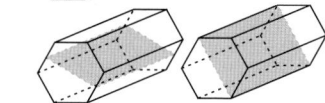

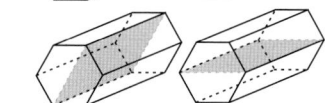

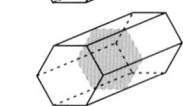

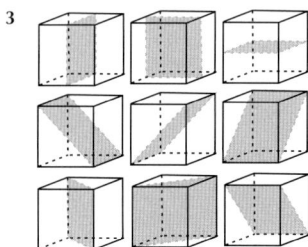

4 Choose a 2D shape that has 4 lines of symmetry and draw the prism of this shape.
E.g.

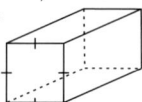

5 a) A pentagon has 5 lines of symmetry, so the pyramid has **5 planes of symmetry**.
b) An octagon has 8 lines of symmetry so the pyramid has **8 planes of symmetry**.
c) A square has 4 lines of symmetry, so the pyramid has **4 planes of symmetry**.

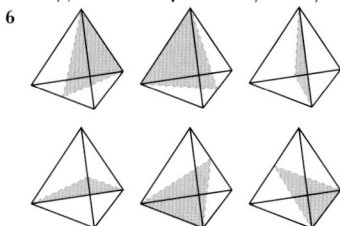

6

Page 374   Review Exercise

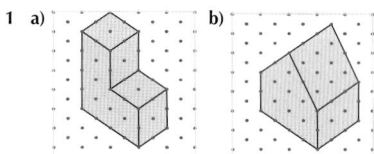

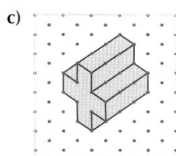

1 a)

b)

c)

---

2 a) A cube has 6 faces, so area of
1 face = 54 ÷ 6 = **9 cm²**
b) Length of edges of the cube = $\sqrt{9}$ = **3 cm**
c) Volume = length³ = 3³ = **27 cm³**

3 Volume of water = 140 × 85 × 65
  = 773 500 cm³
Space left in tank = 297.5 litres = 297 500 cm³
So total volume of tank = 773 500 + 297 500
  = 1 071 000 cm³
Height of tank = 1 071 000 ÷ (140 × 85)
  = **90 cm**

4 a) Surface area = 2 × area of circle
  + area of curved surface
Area of circle = πr^2 = π × 4² = 50.26...
Area of curved surface
  = $2\pi r \times l$ = 2π × 4 × 0.5 = 12.56...
Surface area = (2 × 50.26...) + 12.56...
  = **113.10 cm²** (2 d.p.)
b) Surface area = 10 × area of curved surface
  + 2 × area of circle
  = (10 × 12.56...) + (2 × 50.26...)
  = **226.19 cm²** (2 d.p.)
c) Volume = area of cross-section × length
  = 50.26... × (10 × 0.5)
  = **251.33 cm³** (2 d.p.)

5 a) Radius = diameter ÷ 2 = 7.4 ÷ 2 = 3.7 cm
Volume = area of cross-section × length
  = $\pi r^2 \times l$ = π × 3.7² × 11
  = 473.092... = **473.09 cm³** (2 d.p.)
b) Dimensions = (4 × 7.4) by (3 × 7.4) by 11
  = **29.6 cm by 22.2 cm by 11 cm**
c) Volume = length × width × height
  = 29.6 × 22.2 × 11 = **7228.32 cm³**
d) Volume of box not taken up by tins
  = volume of box − volume of 12 tins
  = 7228.32 − (12 × 473.092...)
  = **1551.21 cm³** (2 d.p.)

6 a) Radius = 12 ÷ 2 = 6 cm
Volume = area of cross-section × length
  = $\pi r^2 \times l$ = π × 6² × 11
  = **1244.07 cm³** (2 d.p.)
b) Radius = 5 ÷ 2 = 2.5 cm
Volume = $\pi r^2 \times l$ = π × 2.5² × 11
  = **215.98 cm³** (2 d.p.)
c) Volume of paper
  = vol. of whole roll − vol. of card tube
  = 1244.07... − 215.98...
  = **1028.09 cm³** (2 d.p.)
d) Volume = length × width × height
  = 11 × 13 × 0.03 = **4.29 cm³**
e) Number of sheets per roll
  = $\dfrac{\text{Volume of paper in roll}}{\text{volume of one sheet}}$ = $\dfrac{1028.09}{4.29}$
  = 239.6... = **240 sheets**

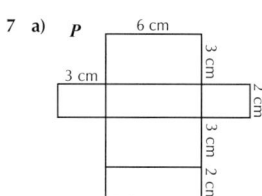

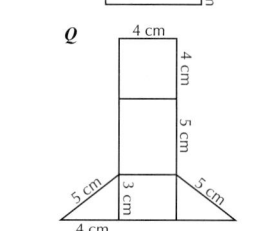

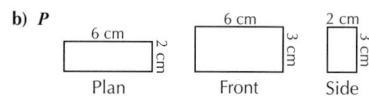

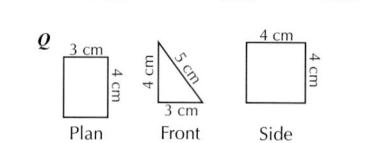

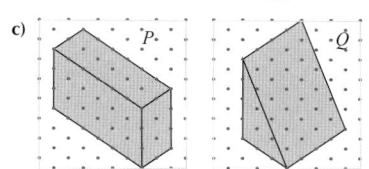

7 a) *P*

  Q

---

b) *P*

  Q

c)

d) *P*: Surface area
  = 2(6 × 3) + 2(2 × 3) + 2(6 × 2)
  = 36 + 12 + 24 = **72 cm²**
Q: Surface area
  = $2 \times \frac{1}{2}(4 \times 3) + (3 \times 4)$
    + (4 × 4) + (5 × 4)
  = 12 + 12 + 16 + 20 = **60 cm²**
e) *P*: volume = length × width × height
  = 6 × 2 × 3 = **36 cm³**
Q: volume = $\frac{1}{2}$(base × height) × length
  = $\frac{1}{2}$(3 × 4) × 4 = **24 cm³**
f) *P*: Time taken = $\dfrac{\text{volume}}{\text{rate}} = \dfrac{36}{2}$
  = **18 seconds**
Q: Time taken = $\dfrac{\text{volume}}{\text{rate}} = \dfrac{24}{2}$
  = **12 seconds**
g) *P*: The rectangle has 2 lines of symmetry,
  so the cuboid has **3 planes of symmetry**.
Q: The triangle has 0 lines of symmetry,
  so the prism has **1 plane of symmetry**.

8 $4\pi r^2 = \frac{4}{3}\pi r^3 \Rightarrow r^2 = \frac{1}{3}r^3 \Rightarrow r^3 - 3r^2 = 0$
$r^2(r - 3) = 0$
$r > 0$, so **r = 3 m**

9 a) Radius = 25 ÷ 2 = 12.5 mm
  Volume = 7854 = $\frac{1}{3}\pi \times 12.5^2 \times h$
  $\Rightarrow h = 48.000... = $ **48.0 mm** (1 d.p.)
b) (i) $l^2 = 12.5^2 + 48.000...^2 = 2460.26...$
  $\Rightarrow l = \sqrt{2460.26...} = $ **49.6 mm** (1 d.p.)
(ii) Surface area = $\pi r l + \pi r^2$
  = π × 12.5 × 49.6... + π × 12.5²
  = **2438.7 mm²** (1 d.p.)
c) Length of cuboid = diameter of cone
  Width of cuboid = height of cone
  Height of cuboid = diameter of cone
  Volume = 25 × 48.000... × 25
  = **30 000 mm³** (to nearest mm³)
d) Volume of empty space
  = volume of cuboid − volume of cone
  = 30 000.0... − 7854
  = **22 146 mm³** (to nearest mm³)

10 a) Volume = 400 = $\frac{1}{3}$ × base area × 12
  $\Rightarrow$ base area = 100 m²
  The base is a square,
  so base area = (side length)² = 100 m²
  $\Rightarrow$ side length = $\sqrt{100}$ = **10 m**
b) $EM^2 = OE^2 + OM^2 = 12^2 + (10 \div 2)^2 = 169$
  $\Rightarrow EM = \sqrt{169} = $ **13 m**
c) Surface area = 4 × area of sides
    + area of base
  Area of one side = $\frac{1}{2}$ × base × height
    = $\frac{1}{2}$ × 10 × 13 = 65 m²
  Surface area = (4 × 65) + 100 = **360 m²**
d) A square has 4 lines of symmetry
  so a square-based pyramid has
  4 planes of symmetry.

---

e) 1 m^3 = 1000 litres

Volume of pyramid = 400 m^3 = 400 × 1000
= 400 000 litres

Time taken = $\dfrac{\text{volume}}{\text{rate}} = \dfrac{400\,000}{100}$

= 4000 minutes
= 4000 ÷ 60 = 66$\frac{2}{3}$ hours
= **66 hours 40 minutes**

11 a) Volume of pencil
= vol. of cone + vol. of cylinder
+ vol. of hemisphere

Vol. of cone = $\frac{1}{3}\pi r^2 h$

= $\frac{1}{3}\pi \times 2^2 \times 5 = \frac{20\pi}{3}$ cm^3

Vol. of cylinder = area of cross-section
× length
= $\pi r^2 \times l = \pi \times 2^2 \times 28 = 112\pi$ cm^3

Volume of hemisphere = $\frac{2}{3}\pi r^3$

= $\frac{2}{3}\pi \times 2^3 = \frac{16\pi}{3}$ cm^3

Volume of pencil = $\frac{20\pi}{3} + 112\pi + \frac{16\pi}{3}$

= **124π cm^3**

b) (i) $l^2 = r^2 + h^2 = 2^2 + 5^2 = 29$,
⇒ $l = \sqrt{29}$ **cm**

(ii) Surface area of pencil
= area of curved face of cone
+ area of curved face of cylinder
+ area of curved face of hemisphere
Area of curved face of cone
= $\pi r l = \pi \times 2 \times \sqrt{29} = 33.83...$ cm^2
Area of curved face of cylinder
= $2\pi r \times l = 2\pi \times 2 \times 28 = 351.85...$ cm^2
Area of curved face of hemisphere
= $2\pi r^2 = 2\pi \times 2^2 = 25.13...$ cm^2
Surface area of pencil
= 33.83... + 351.85... + 25.13...
= **410.8 cm^2** (1 d.p.)

Page 376   Exam-Style Questions

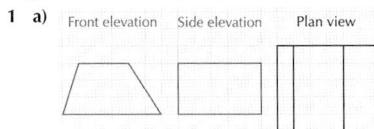

1 a)

Front elevation   Side elevation   Plan view

[2 marks available — 1 mark for 6 × 5 cm rectangle, 1 mark for correct additional lines to represent the changes in depth]

b) Volume of prism
= area of cross-section × length
Area of cross-section
= area of trapezium = $\frac{1}{2}(3 + 6) \times 3$
= 13.5 cm^2
⇒ volume = 13.5 × 5 = **67.5 cm^3**
[3 marks available — 1 mark for the correct area of the cross-section, 1 mark for using the correct formula for the volume of the prism, 1 mark for the correct final answer]

2 The rate of flow of the water filling the tank is the difference between the rate of flow of the water pouring in and the water leaking out:
7 − 2 = 5 m^3/min *[1 mark]*.
The volume of the cylindrical tank
is $\pi \times 5^2 \times 10 = 250\pi$ m^3 *[1 mark]*
so $250\pi \div 2 = 125\pi$ m^3 of water is needed to half-fill the tank *[1 mark]*.

Time taken = $\dfrac{\text{volume}}{\text{rate}} = \dfrac{125\pi}{5}$

= **25π minutes** *[1 mark]*
You could have worked out the time taken to fill the entire container = 250π ÷ 5 = 50π minutes and then halved this to get the correct answer.

3 The tube has a height of 4 × 2r = 8r *[1 mark]* so the volume of the cylinder is
$\pi r^2 h = \pi r^2 \times 8r = 8\pi r^3$ *[1 mark]*.

Volume of one sphere = $\frac{4}{3}\pi r^3$, so volume of

4 spheres = $4 \times \frac{4}{3}\pi r^3 = \frac{16}{3}\pi r^3$ *[1 mark]*
Fraction of space taken up by the balls

is $\dfrac{\frac{16}{3}\pi r^3}{8\pi r^3}$ *[1 mark]* = $\frac{16}{3} \times \frac{1}{8} = \frac{2}{3}$ *[1 mark]*

4 The ratio of the radiuses is 15 : 12 = 5 : 4.
Let h be the height of the removed cone.
By similarity the ratio of the heights
will be $(h + 30) : h = 5 : 4$.
So $5h = 4(h + 30) = 4h + 120 \Rightarrow h = 120$
Capacity of bucket
= vol. of original cone − vol. of removed cone
= $\frac{1}{3}\pi R^2 H - \frac{1}{3}\pi r^2 h$
= $\frac{1}{3}\pi \times 15^2 \times (30 + 120) - \frac{1}{3}\pi \times 12^2 \times 120$
= 17 247.34... cm^3
= 17 247.34... ÷ 1000 = **17.2 litres** (3 s.f.)
[5 marks available — 1 mark for the correct ratio of the radiuses, 1 mark for using similarity to find the height of the removed cone, 1 mark for the correct height of the removed cone or the full cone, 1 mark for subtracting volumes to find the volume of the frustum, 1 mark for the correct final answer in litres]

Section 29 — Transformations

29.1 Reflections

Page 377   Exercise 1

1 For each vertex count how many squares it is to the right of the y-axis. The image point is the same number of squares to the left of the y-axis.

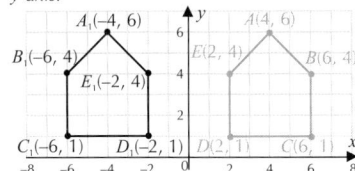

2 The point (x, y) is sent to $(-x, y)$ under a reflection in the y-axis.

a) (−4, 5)   **b)** (−7, 2)

c) (1, 3)   **d)** (3, −1)

3 a) For each vertex count how many squares it is away from the y-axis. The image point is the same number of squares on the other side of the y-axis.

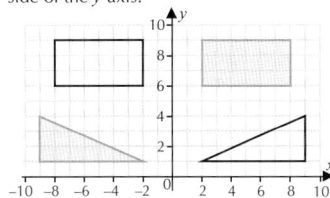

Using the same method for b):

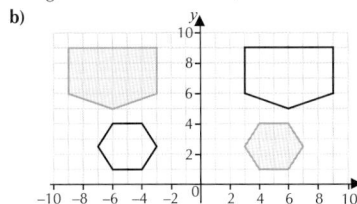

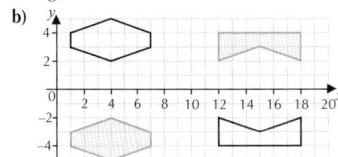

b)

4 The point (x, y) is sent to $(x, -y)$ under a reflection in the x-axis.

a) (1, −2)   **b)** (3, 0)   **c)** (−2, −4)

d) (−1, 3)   **e)** (−2, 2)

5 a) For each vertex count how many squares it is away from the x-axis. The image point is the same number of squares on the other side of the x-axis.

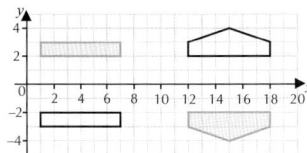

Using the same method for b):

b)

6 a)-c) Start by drawing the vertical mirror line. For each vertex, count the number of squares it is from the mirror line. Draw the new vertex the same number of squares on the other side of the mirror line. The join the new vertices up.

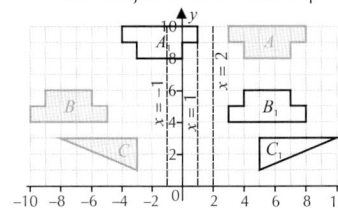

Lower case letters show the mirror lines for the shape with the corresponding upper case letter.

d) Choose a pair of corresponding vertices on shapes A and B_1, e.g. (7, 8) on A and (7, 6) on B_1. They are 2 vertical units apart and the mirror line must be the same distance from each (2 ÷ 2 = 1 unit). So the mirror line is **y = 7**.

e) Invariant points are on the mirror line $x = -7$. The only points on shape B that are on this line are **(−7, 4)** and **(−7, 6)**.

7 a)-d) Start by drawing the horizontal mirror line. Then reflect the vertices and join up the image points — see below.

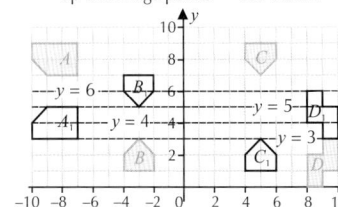

Lower case letters show the mirror lines for the shape with the corresponding upper case letter.

e) Pick a pair of corresponding vertices on shapes B and C_1, e.g. (−3, 3) on B and (5, 3) on C_1. They are 8 horizontal units apart and the mirror line must be the same distance from each (8 ÷ 2 = 4 units). So the mirror line is **x = 1**.

8 a) Pick a pair of corresponding vertices on triangles X and Y, e.g. (−6, 0) on X and (2, 0) on Y. They are 8 horizontal units apart and the mirror line must be the same distance from each (8 ÷ 2 = 4 units). So the mirror line is **x = −2**.

b) Triangle X is a reflection of triangle Z because the orientation has been flipped. Pick a pair of corresponding points, e.g. (−6, 0) and (9, 0). They are 15 horizontal units apart and the mirror line must be the same distance from each (15 ÷ 2 = 7.5 units). So the mirror line is **x = 1.5**.

c) Points on the mirror line $y = 5$ will be invariant, i.e. the whole horizontal side of X. These are points $(x, 5)$, where $-8 \le x \le -3$.

Page 379 Exercise 2

1 The point (x, y) is sent to (y, x) under a reflection in the line $y = x$.

a) $(2, 1)$ **b)** $(0, 3)$ **c)** $(4, -2)$

d) $(-3, -1)$ **e)** $(-2, -2)$

2 a) Each image point is the same perpendicular distance from the mirror line as the vertex but on the other side of it.

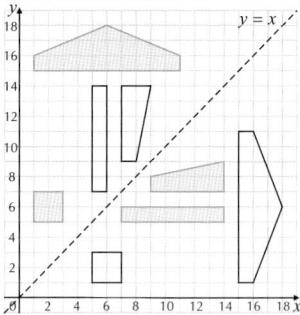

Make sure to reflect perpendicular to the mirror line. Turn your book so that the mirror line is horizontal or vertical if it helps.

Using the same method for b):

b)

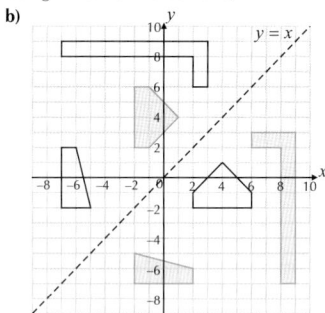

3 a) Pick a pair of corresponding vertices on A and B, e.g. $(-4, 7)$ and $(7, -4)$. Since they follow the rule '(x, y) goes to (y, x)', the reflection is in the line $y = x$.

b) Pick a pair of corresponding vertices on C and A, e.g. $(-7, 4)$ and $(-4, 7)$. Since they follow the rule '(x, y) goes to $(-y, -x)$', the reflection is in the line $y = -x$.

4 a) For each vertex find the perpendicular distance from the mirror line. The image point is the same perpendicular distance on the other side of the mirror line.

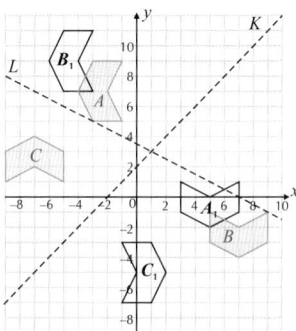

b) The points of invariance will be on the mirror line. The only points on the mirror line L and the shapes are $(-3, 5)$ on A and $(9, -1)$ on B.

29.2 Rotations

Page 380 Exercise 1

1 a) Trace the shapes on tracing paper then put your pencil on point P and turn the paper $180°$ (that's a half turn). Draw the rotated shape in its new position.

(i) **(ii)**

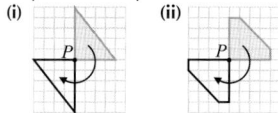

Since these rotations are by an angle of 180°, you could go clockwise or anticlockwise.

b) Trace the shapes on tracing paper then put your pencil on point P and turn the paper $90°$ (that's a quarter turn). Draw the rotated shape in its new position.

(i) **(ii)**

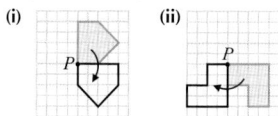

2 Draw the points on a grid and then rotate them using tracing paper around the origin $(0, 0)$.

a) (i) $(0, 1)$ **(ii)** $(-1, 0)$ **(iii)** $(0, -1)$

b) (i) $(-2, 0)$ **(ii)** $(0, -2)$ **(iii)** $(2, 0)$

c) (i) $(-1, 3)$ **(ii)** $(-3, -1)$ **(iii)** $(1, -3)$

For the rotations in Q3-6, use tracing paper to draw the shapes then put your pencil on the centre of rotation and rotate the tracing paper by the required angle in the direction given. Draw the rotated shape in its new position.

3 a)-b)

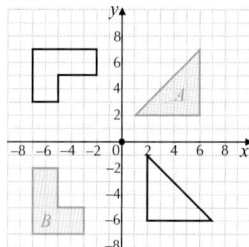

4 a), b)

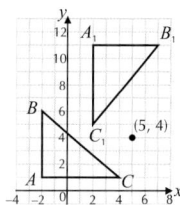

c) $A_1(2, 11)$ $B_1(7, 11)$ $C_1(2, 5)$

5 a)

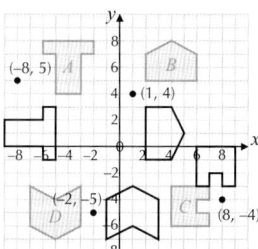

b)

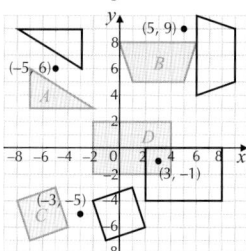

6 DEF and its image $D_1E_1F_1$ are shown below.

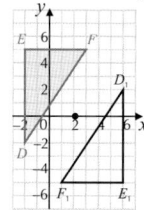

So the coordinates of the vertices are:

$D_1(6, 2)$ $E_1(6, -5)$ $F_1(1, -5)$

Page 382 Exercise 2

To describe the transformations, draw the shape being rotated on tracing paper. Then place your pencil at different centres of rotation and rotate the tracing paper until you find a point and rotation that takes it on top of the new shape.

1 a) (i) **Rotation, 90° clockwise about $(0, 0)$**
Alternatively, this is a rotation 270° anticlockwise about (O, O).

(ii) **Rotation, 90° anticlockwise about $(0, 0)$**
Alternatively, this is a rotation 270° clockwise about (O, O).

b) (i) **Rotation, 180° about $(0, 0)$**

(ii) **Rotation, 180° about $(0, 2)$**

c) (i) **Rotation, 180° about $(0, 3)$**

(ii) **Rotation, 90° anticlockwise about $(0, 2)$**

d) (i) **Rotation, 90° anticlockwise about $(-1, 7)$**

(ii) **Rotation, 90° clockwise about $(1, -2)$**

2 a) **Rotation, 180° about $(0, 6)$**

b) **Rotation, 90° anticlockwise about $(-2, 7)$**

c) **Rotation, 90° clockwise about $(1, 8)$**

3 a)

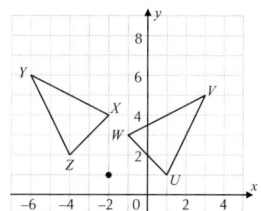

b) **Rotation, 90° anticlockwise about $(-2, 1)$**

29.3 Translations

Page 383 Exercise 1

1 a) Shape A moves 3 units to the right.
Shape B moves 2 units to the right.
Shape C moves 2 units to the right.
Shape D moves 1 unit up.
Shape E moves 2 units up.

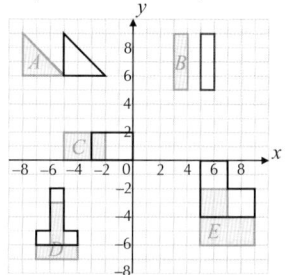

b) Shape P moves 2 units right and 1 up.
Shape Q moves 4 units left and 3 up.
Shape R moves 5 units left.
Shape S moves 4 units right and 1 down.
Shape T moves 4 units left and 3 down.

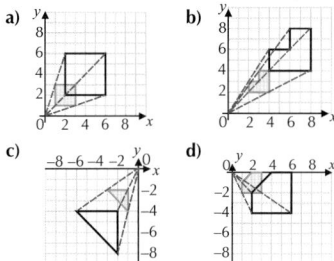

2 a), b) Translate each of A, B and C 10 units to the left and 1 unit down. Then join them up to create $A_1B_1C_1$.

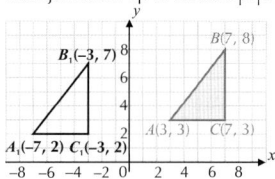

c) To get from the original point to its image, **subtract 10 from the x-coordinate**, and **subtract 1 from the y-coordinate**.
You could also write this as (x − 10, y − 1).

3 To get from D, E and F to D_1, E_1 and F_1, subtract 3 from the x-coordinates and add 2 to the y-coordinates. So the coordinates are:
$D_1(–2, 3)$ $E_1(0, 0)$ $F_1(1, 2)$

Page 384   Exercise 2

1 a) (i) A moves 4 units to the right and 3 units down so the vector is $\begin{pmatrix} 4 \\ -3 \end{pmatrix}$.

(ii) A moves 9 units to the right so the vector is $\begin{pmatrix} 9 \\ 0 \end{pmatrix}$.

Using the same method for (iii)-(vi):

(iii) $\begin{pmatrix} -5 \\ -3 \end{pmatrix}$ (iv) $\begin{pmatrix} 4 \\ -5 \end{pmatrix}$

(v) $\begin{pmatrix} -13 \\ 5 \end{pmatrix}$ (vi) $\begin{pmatrix} -9 \\ 2 \end{pmatrix}$

b) (i) P moves 5 units down so the vector is $\begin{pmatrix} 0 \\ -5 \end{pmatrix}$.

(ii) R moves 8 units to the right so the vector is $\begin{pmatrix} 8 \\ 0 \end{pmatrix}$.

Using the same method for (iii)-(vi):

(iii) $\begin{pmatrix} 13 \\ 1 \end{pmatrix}$ (iv) $\begin{pmatrix} -8 \\ 0 \end{pmatrix}$

(v) $\begin{pmatrix} -13 \\ -6 \end{pmatrix}$ (vi) $\begin{pmatrix} -8 \\ 5 \end{pmatrix}$

2 To get from the coordinates of D to G or from E to H or from F to I, the x-coordinates increase by 3 and the y-coordinates increase by 4. So the vector representing the translation is $\begin{pmatrix} 3 \\ 4 \end{pmatrix}$.

3 To get from Z to W, you add 1 to the x-coordinate and subtract 4 from the y-coordinate. So to get back to Z from W, subtract 1 from the x-coordinate and add 4 to the y-coordinate. This gives the vector $\begin{pmatrix} -1 \\ 4 \end{pmatrix}$.
To reverse any translation, change the signs of the components of the vector.

29.4 Enlargements

Page 386   Exercise 1

1 Draw lines from (0, 0) through the vertices of the shapes. The scale factor is 2, so extend the lines so they are twice as long. Mark the new vertices at the end of these lines and join up.

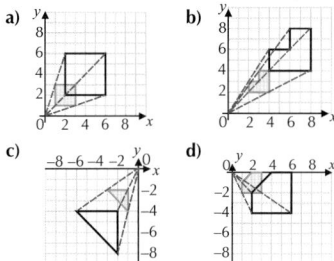

For each enlargement in Q2-6, draw lines from the centre of enlargement to each vertex of the shape. Extend these lines depending on the scale factor (e.g. for a scale factor of 3 the lines should be 3 times as long). Mark the new vertices at the end of these lines and join them up to create the new shape.

2 a)-d)

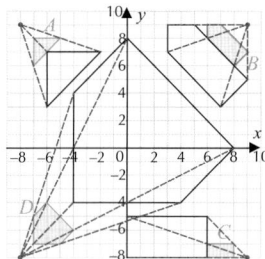

3 a), b)

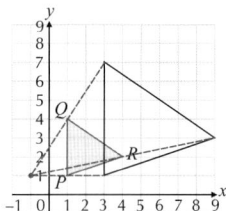

4 a)

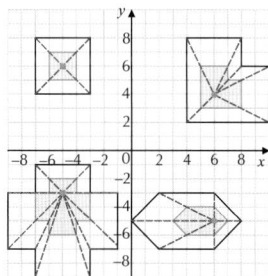

b)

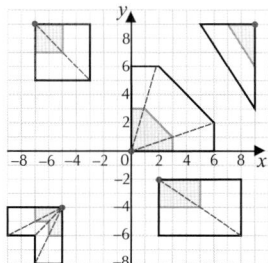

Page 387   Exercise 2

5 a), b)

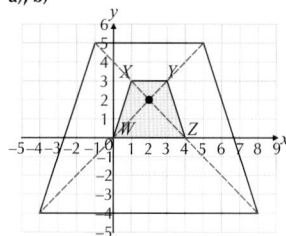

6 a), b)

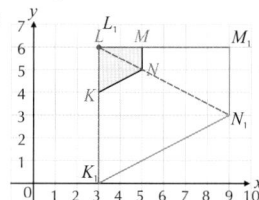

c)

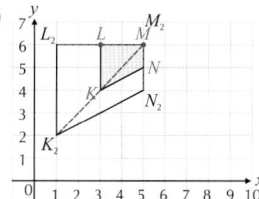

1 Draw lines from (0, 0) to the vertices of the shape. The scale factor is $\frac{1}{2}$, so mark the vertices of the new shapes halfway along these lines and then join them up to create the new shape.

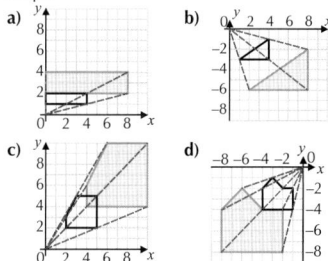

For each enlargement in Q2-3, draw lines from the centre of enlargement to each vertex of the shape. Mark the new vertices a fraction of the way along these lines (e.g. for a scale factor of $\frac{1}{4}$ the vertices should be a quarter of the way along the lines from the centre of enlargement). Join the vertices up to create the new shape.

2 a)-d)

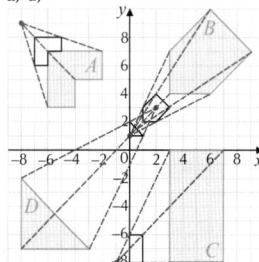

3 a)-d)

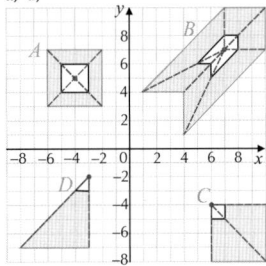

If the centre of enlargement is inside the shape you just follow the same method — your new shape will be inside the old shape.

Page 388   Exercise 3

For each enlargement in Q1-3, draw lines from each vertex of the shape to the centre of enlargement. Extend these lines out of the other side of the centre of enlargement based on the scale factor (e.g. if the scale factor is −3 they should be extended 3 times further out the other side). Mark the new vertices at the ends of the lines and join them up to create the new shape.

1 a)-c)

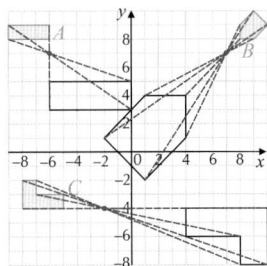

2 a)-c)

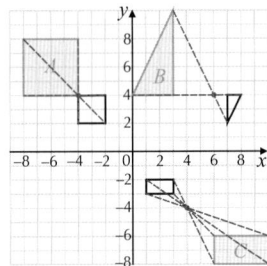

3 a)-b)

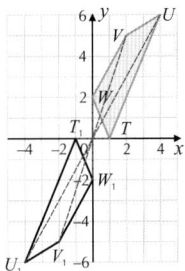

c) This is the same as a **rotation** about the **origin** by 180°.

Page 389   Exercise 4

1 a) Measure corresponding sides of shapes A and B — e.g. the vertical side of A is 3 units long and of B is 6 units long. So the scale factor is $6 \div 3 = 2$. Draw lines through corresponding vertices of shapes A and B. These intersect at the point (2, 10). So the transformation is an **enlargement, scale factor 2, centre (2, 10)**.

Using the same method for b)-c):

b) **Enlargement, scale factor 2, centre (6, 10)**

c) **Enlargement, scale factor 2, centre (1, 1)**

2 a) (i) Measure corresponding sides of shapes A and B — e.g. the bottom horizontal side of shape A is 2 units long and of B is 4 units long. So the scale factor is $4 \div 2 = 2$. Draw lines through corresponding vertices of shapes A and B. These intersect at the point (6, 5). So the transformation is an **enlargement, scale factor 2, centre (6, 5)**.

(ii) The new shape is A, so using the same side lengths as in (i), the scale factor is $\frac{2}{4} = \frac{1}{2}$. The centre of enlargement stays the same, so is (6, 5). So this is an **enlargement, scale factor $\frac{1}{2}$, centre (6, 5)**.

Using the same method for b):

b) (i) **Enlargement, scale factor 3, centre (5, 4)**

(ii) **Enlargement, scale factor $\frac{1}{3}$, centre (5, 4)**

c) (i) Measure corresponding sides of shapes A and B — e.g. the horizontal side of shape A is 6 units long and the horizontal side of B is 2 units long. The centre of enlargement is between the two shapes so the scale factor is $-(2 \div 6) = -\frac{1}{3}$. Draw lines through corresponding vertices of shapes A and B. These intersect at the point (8, 7). So the transformation is an **enlargement, scale factor $-\frac{1}{3}$, centre (8, 7)**.

(ii) The new shape is A, so using the same side lengths as in (i), the scale factor is $-(6 \div 2) = -3$. The centre stays the same, so it's (8, 7). So this is an **enlargement, scale factor −3, centre (8, 7)**.

29.5 Combinations of Transformations

Page 390   Exercise 1

1 a) (i), (ii)

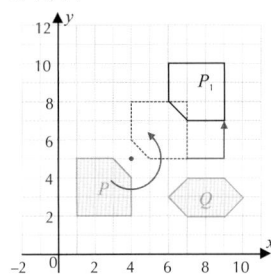

b)

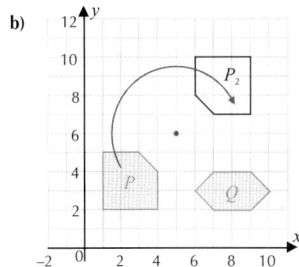

---

c) (i), (ii)

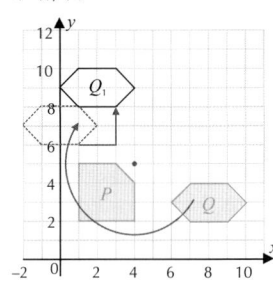

d)

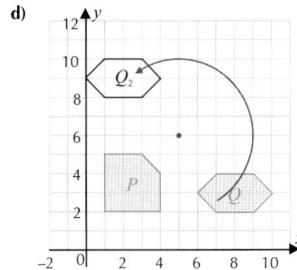

e) P_1 and P_2 are **identical** and Q_1 and Q_2 are **identical**.

2 a), b), c)

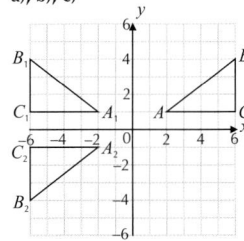

d) Use the diagram to recognise the single transformation. It looks like a rotation by a half-turn so use tracing paper to find the centre of rotation: **rotation, 180° about (0, 0)**.

3 a) (i)

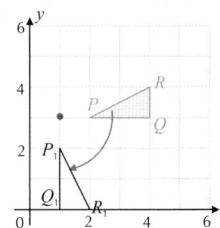

(ii) There are **no invariant points** on PQR under this transformation.

b) (i)

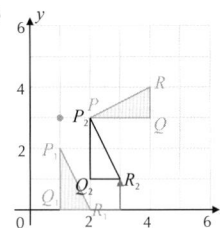

(ii) There are **no invariant points** on $P_1Q_1R_1$ under this transformation.

c) (i) It looks like a rotation by a quarter-turn clockwise so use tracing paper to find the centre of rotation: **rotation, 90° clockwise about P(2, 3)**. *Alternatively, this is a rotation 270° anticlockwise about P.*

(ii) P is the same point as P_2 so **P(2, 3)** is a point of invariance under this transformation.

4 a)

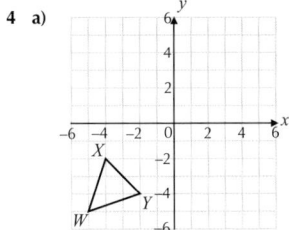

b)

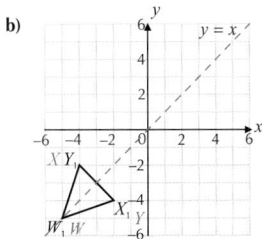

c)

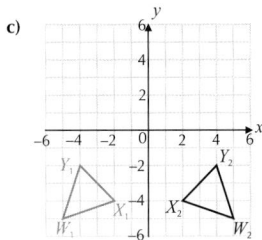

d) Use the diagram to recognise the single transformation. It looks like a rotation by a quarter-turn so use tracing paper to find the centre of rotation: **rotation, 90° anticlockwise about (0, 0)**.
The shape looks like it could be a reflection in the y-axis — but that would map e.g. Y to X_2, not to Y_2

5 a)

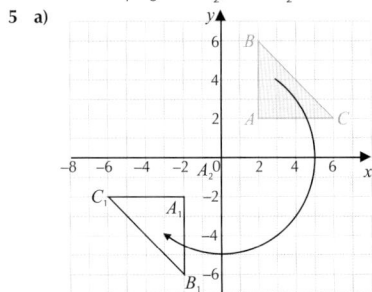

b) If B is invariant under the combination then B_1 must be mapped to B by the translation. For this to happen, B_1 must move 4 units to the right and 12 units up. So the translation is described by the vector $\binom{4}{12}$.

6 a), b), c)

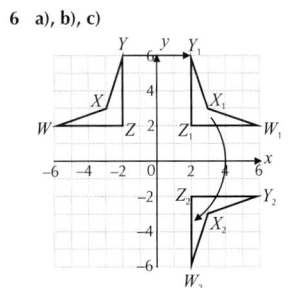

d) The shape has been flipped to the opposite quadrant so it's a **reflection in the line y = x**.
The shape looks like it could have been rotated 180° about the origin. But that would map e.g. Y to W_2 instead of Y_2

7 a)

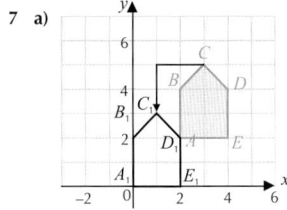

b)

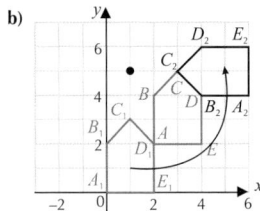

It looks like a rotation by a quarter-turn so use tracing paper to find the centre of rotation. The rotation is **90° clockwise about (1, 5)**.

Page 392   Review Exercise

1 a)

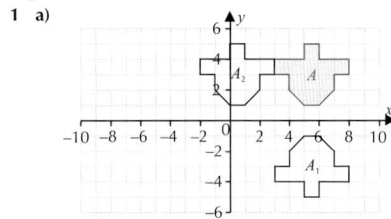

b) (i) None since A is not on the x-axis.
(ii) The points **(3, y) for $3 \leq y \leq 4$** are invariant since these are on the mirror line $x = 3$.
All the points on the line x = 3 are invariant under the reflection but you're asked to give the invariant points on the shape A — so you need to give the appropriate values of y.
c) For a point (x, y) on A, the corresponding point on U is $(-y, -x)$. E.g. $(6, 1)$ on A goes to $(-1, -6)$ on U so this is a reflection in **y = –x**.

2 a) (i), (ii)

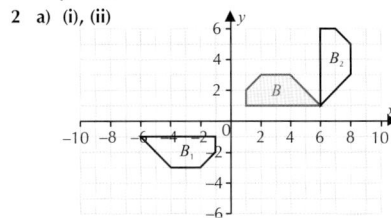

b) (i) None since B is not on the centre of rotation.
(ii) The point **(6, 1)** is invariant since this is the centre of rotation.
c) Use tracing paper and try different centres of rotation to find that this is a rotation by **90° anticlockwise about (–3, –4)**.

3 a) (i), (ii)

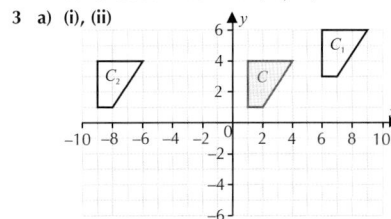

b) (i) None — no points are invariant under a translation.
(ii) None — no points are invariant under a translation.

c) Pick a pair of corresponding points on C and W and count the number of units between them, both horizontally and vertically. W is 7 units to the left and 6 units down so this is described by the vector $\binom{-7}{-6}$.

4 a) (i), (ii), (iii)

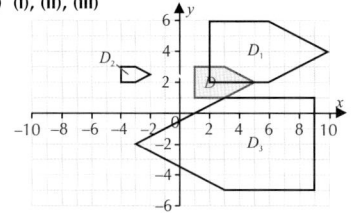

b) (i) None since D is not on the centre of enlargement.
(ii) None since D is not on the centre of enlargement.
(iii) The point **(3, 1)** is invariant since this is the centre of enlargement.
c) Measure corresponding sides to find the scale factor — e.g. the top horizontal is 2 units long on D and 4 on X so the scale factor is $4 \div 2 = 2$.
Draw lines through corresponding points and see where they intersect to find the centre of enlargement: $(10, 0)$.
So this is an enlargement by **scale factor 2 centred at (10, 0)**.

5 Use a diagram (see below). The final image looks like a rotation of XYZ by 180°. The point $(2, 2)$ is invariant so this must be the centre of the rotation. The full description is a **rotation, 180° about (2, 2)**.

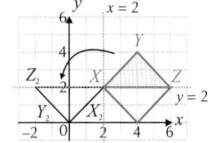

6 Use a diagram to perform each of the three transformations (see below). Labelling a vertex (e.g. B as shown) will help you to keep track of the shape's orientation. Point (x, y) maps to (y, x), e.g. $(4, -2)$ maps to $(-2, 4)$ so it's a **reflection in the line y = x**.

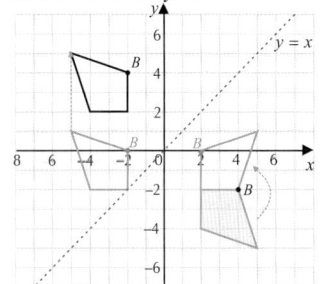

It might look like a 180° rotation about the origin but that wouldn't map point B correctly.

Page 393   Exam-Style Questions

1 a)

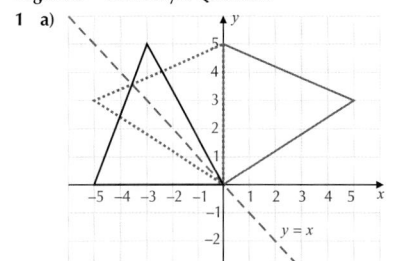

The combination of transformations is equivalent to a **rotation** *[1 mark]*, **anticlockwise by 90°** *[1 mark]*, centred on the **origin (0, 0)** *[1 mark]*.
You'd also get the mark for describing it as a rotation clockwise by 270°.

b) **(0, 0)** *[1 mark]*
The centre of rotation is always invariant.

2 Working backwards, translate point $C(3, 6)$ 10 units in the negative y-direction to get $(3, -4)$. Then reflect this in $y = k$ to return it to $C(3, 6)$. The value of k must be halfway between $y = 6$ and $y = -4$, so $k = -4 + \frac{6 - (-4)}{2} = 1$.
[2 marks available — 1 mark for a suitable method, 1 mark for the correct answer]

3 Work backwards through the transformations. First, enlarge shape R by scale factor $\frac{1}{2}$ with centre of enlargement $(-2, 3)$ to produce shape Q. Then translate shape Q by $\begin{pmatrix} -1 \\ 4 \end{pmatrix}$ to produce shape P.

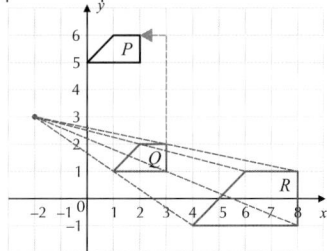

[3 marks available — 1 mark for attempting to enlarge R by scale factor $\frac{1}{2}$ centred on (-2, 3), 1 mark for Q drawn in the correct position, 1 mark for translating Q to the correct position to get P]

Section 30 — Congruence and Similarity

30.1 Congruence and Similarity

Page 395   Exercise 1

1 a) The triangles are congruent, so match the sides and angles: a = **5 cm**, b = **69°**, c = **69°**, d = **42°**, e = **7 cm**, f = **7 cm**

b) $g = 180° - 90° - 53° = 37°$
The triangles are congruent, so match the remaining sides and angles: h = **6 cm**, i = **10 cm**, j = **8 cm**, k = **37°**, l = **53°**

2 a) Two sides and the angle between on one triangle are the same as two sides and the angle between on the other triangle. Condition **SAS** holds so the triangles **are congruent.**

b) The two angles shown are the same, but the corresponding sides are not equal, so the triangles are **not congruent.**

c) Two angles and a side on one triangle are the same as two angles and the corresponding side on the other triangle. Condition **AAS** holds so the triangles **are congruent.**

d) The three sides on one triangle are the same as the three sides on the other triangle. Condition **SSS** holds so the triangles **are congruent.**

e) The hypotenuse of the triangles is different, so the triangles are **not congruent.**

f) They are both right-angled triangles $(180° - 46° - 44° = 90°)$, they have the same hypotenuse and one other side is the same. Condition **RHS** holds so the triangles **are congruent.**

Page 396   Exercise 2

1 Using Pythagoras' theorem the hypotenuse of the triangle on the right is $\sqrt{3^2 + 4^2} = 5$ cm. Both triangles have a right angle, the same hypotenuse and another side the same, which means condition **RHS** holds so the triangles **are congruent.**

2 E.g. Opposite sides and angles in a parallelogram are equal so $AB = CD$, $AD = BC$ and side AC is common to both triangles. Three sides are the same in both triangles which means condition **SSS** holds so the triangles **are congruent.**
There might be different ways to approach the questions in this exercise — you just need to show that one of the congruence conditions holds.

3 E.g. All sides and angles in a regular pentagon are equal. Angle CDE = angle ABC and $CD = DE = BC = AB$. Two sides and the angle between them are the same in both triangles which means condition **SAS** holds so the triangles **are congruent.**

4 E.g. Angle BOC = angle DOC = 90° because the diagonals of a kite cross at right angles. $BC = DC$ because a kite has two pairs of equal sides. The triangles share side CO. Both triangles have a right angle, the same hypotenuse and another side the same, which means condition **RHS** holds so the triangles **are congruent.**

5 E.g. $AC = BD$ as they are both diameters of the circle. The triangles also share a common side BC. Angle ABC = angle DCB = 90° as they are both angles in a semicircle. Condition **RHS** holds, so the triangles **are congruent.**
You could also show that angle CAB = angle BDC as angles subtended by an arc in the same segment are equal. Then use AAS to prove that the triangles are congruent.

Page 397   Exercise 3

1 a) Find the ratio of the sides in each triangle, $4:4:6 = 2:2:3$ and $6:6:9 = 2:2:3$. The sides of both triangles are in the same ratio $(2:2:3)$, so the triangles **are similar.**

b) The side lengths can't be in the same ratio as two sides of the first triangle are equal, but the sides in the second triangle are all different, so the triangles are **not similar.**

c) Find the ratio of the sides in each triangle, $7.5:9 = 15:18 = 5:6$ and $2.5:3 = 5:6$, so the corresponding sides in each triangle are in the same ratio $(5:6)$. The angle between the two sides is the same on both triangles, so the triangles **are similar.**

d) The missing angle on the left-hand triangle is $180° - 33° - 52° = 95°$ and the missing angle on the right-hand triangle is $180° - 33° - 95° = 52°$. All the angles on one triangle are the same as the angles on the other triangle, so the triangles **are similar.**

2 E.g. Using Pythagoras' theorem the missing side of triangle A is $\sqrt{13^2 - 12^2} = 5$ cm. Find the ratio of the sides in each triangle, $5:12$ and $10:24 = 5:12$, so the corresponding sides in each triangle are in the same ratio $(5:12)$. The angle between these sides is 90° in both triangles. So the triangles **are similar.**
There might be different ways to approach the questions in this exercise — you just need to show that one of the conditions for similarity holds.

3 E.g. Angle BEC = angle AED (vertically opposite angles).
Angle ADE = angle CBE (alternate angles).
Angle DAE = angle BCE (alternate angles).
The 3 angles in BEC are the same as the 3 angles in AED so the triangles **are similar.**

4 a) E.g. Angle ABC = angle DBE as it is a common angle to both triangles. Angle BDE = angle BAC and angle BED = angle BCA using corresponding angles. Since all the angles are the same, the triangles **are similar.**

b) AOC and EOD.
Angle AOC = angle EOD (vertically opposite angles).
Angle ACO = angle EDO (alternate angles).
Angle CAO = angle DEO (alternate angles).
The 3 angles in AOC are the same as the 3 angles in EOD so they **are similar.**
AC and DE are corresponding sides but $AC \neq DE$ so the triangles are **not congruent.**

Page 398   Exercise 4

1 a) Scale factor = $12 \div 4 = $ **3**

b) $YZ = QR \times 3 = 3 \times 3 = $ **9 m**

2 Scale factor from QRU to $QST = 34 \div 17 = 2$
a) $ST = RU \times 2 = 8 \times 2 = $ **16 cm**

b) $QT = QU \times 2 = 15 \times 2 = $ **30 cm**

c) $UT = QT - QU = 30 - 15 = $ **15 cm**

3 Scale factor from STW to $SUV = (2 + 4) \div 2 = 3$
a) $TW = UV \div 3 = 4.5 \div 3 = $ **1.5 m**

b) $SU = ST \times 3 = 1.8 \times 3 = $ **5.4 m**

c) $TU = SU - ST = 5.4 - 1.8 = $ **3.6 m**

4 a) Since the parallelograms are similar, angle x and 84° are allied angles, so $x = 180° - 84° = $ **96°**

b) The kites are similar.
Scale factor = $24 \div 6 = 4$
$y = 16 \div 4 = $ **4 cm**
The angle opposite the 106° angle is also 106° and angles in a quadrilateral add up to 360°, so $z = 360° - 106° - 106° - 92°$
$= $ **56°**

30.2 Areas and Volumes of Similar Shapes

Page 400   Exercise 1

1 Area of original shape = $10 \times 10 = 100$ cm^2
Area of enlarged shape = 100×4^2
$= $ **1600 cm^2**

2 Scale factor = $9 \div 3 = 3$
Area of $B = 5 \times 3^2 = $ **45 cm^2**

3 (scale factor)$^2 = 400 \div 25 = 16$, so scale factor = $\sqrt{16} = 4$
Base of $Q = 4 \times 4 = $ **16 cm**

4 Scale factor = $5 \div 2 = 2.5$
Area of $A = 125 \div 2.5^2 = $ **20 cm^2**

5 Perimeter = $12 \times 3 = $ **36 cm**
Area = $6 \times 3^2 = $ **54 cm^2**

6 (scale factor)$^2 = 540 \div 15 = 36$, so scale factor = $\sqrt{36} = 6$
Dimensions of B are $3 \times 6 = $ **18 cm** and $5 \times 6 = $ **30 cm**

Page 400   Exercise 2

1 a) Side length of $B = (8 \div 2) \times 3 = $ **12 cm**

b) Area of $B = 12^2 = $ **144 cm^2**

c) Ratio of areas = $2^2 : 3^2 = $ **4 : 9**

2 Ratio of side lengths
$= \sqrt{2.25} : \sqrt{6.25} = 1.5 : 2.5 = $ **3 : 5**

3 Ratio of areas = $4^2 : 5^2 = 16 : 25$, so area of larger shape = $(20 \div 16) \times 25$
$= $ **31.25 cm^2**

4 Ratio of areas = $1^2 : 50^2 = 1 : 2500$, so area of real bench = $(3 \div 1) \times 2500$
$= $ **7500 cm^2 (or 0.75 m^2)**

Page 401   Exercise 3

1 Scale factor = $12 \div 6 = 2$
Surface area of $A = 1440 \div 2^2 = $ **360 in^2**

2 Surface area = $400\pi \times 3^2 = $ **3600π mm^2**

3 (scale factor)^2 = 1500 ÷ 60 = 25,
so scale factor = $\sqrt{25}$ = **5**
Corresponding side of Q = 3 × 5 = **15 cm**

4 Scale factor = 3 ÷ 2 = 1.5
Surface area = 63π × 1.5^2 = **28π cm^2**
Be careful here — cylinder A is actually bigger than cylinder B, so make sure you do the calculations the correct way round.

5 (scale factor)^2 = 72 ÷ 32 = 2.25,
so scale factor = $\sqrt{2.25}$ = 1.5
Height of B = 5 × 1.5 = **7.5 cm**

6 Ratio of sides = $\sqrt{4}$: $\sqrt{9}$: $\sqrt{12.25}$
= **2 : 3 : 3.5 (or 4 : 6 : 7)**

Page 402   Exercise 4

1 (scale factor)^3 = 1920 ÷ 30 = 64,
so scale factor = $\sqrt[3]{64}$ = 4
Dimensions of B are 2 × 4 = **8 cm**,
3 × 4 = **12 cm** and 5 × 4 = **20 cm**

2 Volume = 18 × 5^3 = **2250 m^3**

3 Scale factor = 3 ÷ 1.5 = 2
Volume = 84 ÷ 2^3 = **10.5 cm^3**
Make sure you check whether you need to multiply or divide by the scale factor each time.

4 (scale factor)^3 = 3.375,
so scale factor = $\sqrt[3]{3.375}$ = 1.5
Side length = 3 × 1.5 = **4.5 cm**

5 Scale factor = 15 ÷ 6 = 2.5
Volume of B = 60π × 2.5^3 = **937.5π cm^3**

6 a) Ratio of volumes = 2^3 : 5^3 = **8 : 125**
b) Volume = (100 ÷ 8) × 125 = **1562.5 mm^3**

7 (scale factor)^3 = 16 384 ÷ 32 = 512,
so scale factor = $\sqrt[3]{512}$ = 8
Vertical height = 8 × 8 = **64 cm**

8 Uranus is approximately 4^3 = **64 times** the volume of the Earth.

9 Scale factor = 12 ÷ 1.2 = 10
Volume = 3600 ÷ 10^3 = **3.6 cm^3**

10 (scale factor)^3 = 1280 ÷ 20 = 64,
so scale factor = $\sqrt[3]{64}$ = 4
Then (scale factor)^2 = 4^2 = 16 so the surface areas will be 16 times larger, which means **James** is correct.

Page 403   Review Exercise

1 a) The missing angle on right-hand triangle is 180° − 105° − 40° = 35°, so two sides and the angle between on one triangle are the same as two sides and the angle between on the other triangle. Condition **SAS** holds so the triangles are **congruent**.
b) Find the ratio of the sides in each triangle, 3 : 3 : 6 = 1 : 1 : 2 and 1 : 1 : 3. The lengths are not in the same ratio, so the triangles are **neither** similar nor congruent.
c) The missing angle on left-hand triangle is 180° − 30° − 45° = 105° and the missing angle on the right-hand triangle is 180° − 105° − 30° = 45°. All the angles on one triangle are the same as the angles on the other triangle but corresponding sides are not the same length so the triangles are **similar**.
d) The missing angle on left-hand triangle is 180° − 38° − 112° = 30° and the missing angle on the right-hand triangle is 180° − 38° − 40° = 102°. The angles in one triangle are not all the same as the angles in the other triangle, so the triangles are **neither** similar nor congruent.

2 Scale factor = (18 + 9) ÷ 18 = 1.5
a) KN = 15 ÷ 1.5 = **10 cm**
b) JK = 36 ÷ 1.5 = 24,
so $KL = JL − JK$ = 36 − 24 = **12 cm**

3 Scale factor = 12 ÷ 6 = 2
a) Surface area of B = 96 × 2^2 = **384 cm^2**
b) Volume of B = 48 × 2^3 = **384 cm^3**

4 a) (i) Ratio of surface areas = 3^2 : 5^2 = **9 : 25**
(ii) Ratio of volumes = 3^3 : 5^3 = **27 : 125**
b) (i) Ratio of side lengths = $\sqrt{49}$: $\sqrt{81}$ = **7 : 9**
(ii) Ratio of volumes = 7^3 : 9^3 = **343 : 729**

5 (scale factor)^3 = 1620 ÷ 60 = 27,
so scale factor = $\sqrt[3]{27}$ = 3
So the dimensions of the smaller box are
9 ÷ 3 = **3 in**, 15 ÷ 3 = **5 in**, 12 ÷ 3 = **4 in**

6 a) Surface area = 22 × 1.5^2 = **49.5 cm^2**
b) Volume = 18 × 1.5^3 = **60.75 cm^3**

7 (scale factor)^3 = 10π ÷ 1.25π = 8,
so scale factor = $\sqrt[3]{8}$ = 2
The ladder needs to be 5 × 2 = **10 m** long

Page 404   Exam-Style Questions

1 a) Scale factor = 8 ÷ 3 = $\frac{8}{3}$ *[1 mark]*
BC = 10 ÷ $\frac{8}{3}$ = **3.75 m** *[1 mark]*
b) The triangles are similar, so angle ACB = angle AED = 180° − 29° − 99°
= **52°** *[1 mark]*

2 a) Area of base = 144π = πr^2
⇒ $r = \sqrt{144}$ = 12 cm *[1 mark]*
⇒ d = 2 × 12 = **24 cm** *[1 mark]*
b) Scale factor = 24 ÷ 16 = 1.5 *[1 mark]*
Height of larger cake = 6 × 1.5
= 9 cm *[1 mark]*
Volume of larger cake = 144π × 9
= **1296π cm^3** *[1 mark]*
Alternatively, you could have worked out the volume of the smaller cake and then multiplied it by the (scale factor)^3.

3 (scale factor)^2 = 162 ÷ 18 = 9,
so scale factor = $\sqrt{9}$ = 3 *[1 mark]*
Volume of larger prism = 270 × 3^3 *[1 mark]*
= **7290 cm^3** *[1 mark]*

4 $AC = BC$ (since triangle ABC is isosceles).
$CD = CE$ (since $AC = BC$ and $AD = BE$ was given).
Angle ECA = angle DCB (since this is the angle at C common to both triangles).
Condition **SAS** holds, so triangle AEC is **congruent** to triangle BDC.
[4 marks available — 1 mark for matching one pair of sides, 1 mark for matching the other pair of sides, 1 mark for matching angles between the sides, 1 mark for stating the correct congruence condition and conclusion]

5 E.g. The ratio of the surface areas is 50 : 450 = 1 : 9 (= 1 : 3^2) and the ratio of the volumes is 88 : 792 = 1 : 9 (= 1 : 3^2).
If the prisms were similar, the ratio of their volumes would be 1 : 27 (= 1 : 3^3), so they are **not similar**.
[3 marks available — 1 mark for finding the ratio of the surface areas, 1 mark for finding the ratio of the volumes, 1 mark for comparing to the ratio necessary for similarity and stating the conclusion]
Here, you could have worked out the scale factors instead of the ratios — you'll end up with the same conclusion.

6 Scale factor = 6 ÷ 3 = 2, so the volume of the larger bag is 2^3 = 8 times the volume of the smaller bag. However, the larger bag only costs 2.5 ÷ 1.1 = 2.27... times as much as the smaller bag (despite containing 8 times as much bird food), so the larger ball is **better value for money**.
[3 marks available — 1 mark for finding the scale factor, 1 mark for a statement comparing the volumes, 1 mark for the correct conclusion]

Section 31 — Collecting Data

31.1 Using Different Types of Data

Page 405   Exercise 1

1 a) Data needed — **girls' answers to some questions about school dinners.**
Method of collecting — **e.g. Nikita could ask all the girls in her class to fill in a questionnaire.**
b) **Primary data**

2 a) Data needed — **colours of cars passing Dan's house in the 30-minute interval.**
Method of collecting — **e.g. Dan could observe cars passing his house and note the colour of each car in a tally chart.**
b) **Primary data**

3 a) Data needed — **daily rainfall figures for London and Manchester last August.**
Method of collecting — **e.g. Anne could look for rainfall figures on the internet.**
b) **Secondary data**

4 a) Data needed — **the distance an identical ball can be thrown by the boys and girls in his class.**
Method of collecting — **e.g. Rohan could ask everyone in the class to throw the same ball as far as they can. He could measure the distances and record them in a table, along with whether each thrower was male or female.**
b) **Primary data**

5 a) Data needed — **one set of data consists of the temperature readings in Jim's garden taken at 10 am each day. The other set consists of the Met Office's temperatures recorded for Jim's area at the same time.**
Method of collecting — **e.g. Jim could collect the data from his garden by taking readings from a thermometer. He can get the Met Office temperatures from their website. He should record both temperatures for each day in a table.**
b) Data collected in Jim's garden is **primary data**. The Met Office data is **secondary data**.

Page 406   Exercise 2

1 a) The number of words is numerical and it can only be whole values so this is **discrete quantitative**.
b) Foods are descriptive so this is **qualitative**.
c) The number of pets is numerical and it can only take whole values so this is **discrete quantitative**.
d) Heights are numerical and they can take any value in the range $x > 0$ so this is **continuous quantitative**.
e) Nationalities are descriptive so this is **qualitative**.
f) Lengths are numerical and they can take any value in the range $x > 0$ so this is **continuous quantitative**.
g) Distances are numerical and they can take any value in the range $x > 0$ so this is **continuous quantitative**.
h) Colours are descriptive so this is **qualitative**.

2 a) (i) The data is descriptive so **qualitative**.
(ii) The data is numerical and can only take the whole number values 0, 1, 2, ..., 10 so this is **discrete quantitative**.
b) E.g. Advantages — **you can collect a lot more information / the information collected is more detailed.**
Disadvantages — **it takes more time to collect / the data is harder to analyse.**

Page 408   Exercise 1

1 a) E.g.

Siblings	Tally	Frequency
0		
1		
2		
3		
More than 3		

b) E.g.

Transport	Tally	Frequency
Walking		
Car		
Train		
Bus		
Bicycle		
Other		

c) E.g.

Fruit	Tally	Frequency
Apple		
Orange		
Banana		
Peach		
Other		

d) E.g.

Days	Tally	Frequency
28		
29		
30		
31		

2 a)

Age (years)	Tally	Frequency
0-9	JHT II	7
10-19	IIII	4
20-29	JHT II	7
30-39	JHT III	8
40-49	JHT II	7
50-59	JHT IIII	9
60-69	III	3
70-79	III	3
80-89	I	1

b) The **50-59 age group** is the most common.

3 a) E.g. **There is no row for someone who went to the cinema 0 times — the lowest possible answer is 1. The classes are overlapping.**

No. of trips	Tally	Frequency
0-10		
11-20		
21-30		
31-40		
41 or more		

b) E.g. **There are gaps between data classes. The classes don't cover all the values up to the maximum venue capacity.**

No. of people	Tally	Frequency
0-5000		
5001-10 000		
10 001-15 000		
15 001-20 000		
20 001-25 000		

c) E.g. **The classes are overlapping. There are not enough classes. There is no lower limit on t.**

Time (t mins)	Tally	Frequency
$0 < t \le 1$		
$1 < t \le 2$		
$2 < t \le 3$		
$3 < t \le 4$		
$4 < t$		

d) E.g. **There are gaps between classes. The classes don't cover all possible results. The classes don't allow for much variation in weights. There is no lower limit on w.**

Weight (w kg)	Tally	Frequency
$0 < w \le 3$		
$3 < w \le 5$		
$5 < w \le 7$		
$7 < w \le 9$		
$9 < w \le 11$		
$11 < w$		

4 a) E.g.

No. of pairs	Tally	Frequency
0-4		
5-8		
9-12		
13-16		
17 or more		

b) E.g.

Length (s cm)	Tally	Frequency
$0 < s \le 15$		
$15 < s \le 20$		
$20 < s \le 25$		
$25 < s \le 30$		
$30 < s$		

c) E.g.

Distance (d km)	Tally	Frequency
$0 \le d \le 5$		
$5 < d \le 10$		
$10 < d \le 20$		
$20 < d \le 40$		
$40 < d$		

5 Jay's question is better. E.g. two from:
Amber's question is **too vague** because people will have different ideas about how to answer (e.g. "1 hour", "a lot") / Amber's question has **no time span** / Amber's data could be **difficult to analyse** because there are no options to limit the answers / Jay's question is **clearer** because it has a time span / Jay's data will be **easier to analyse** since there are only 5 options to choose from.

Page 409   Exercise 2

1 a) There are not enough hair colour data classes. The age classes overlap. The data is for adults and so does not need a data class for 0-15 years, and the classes should start at 18.

b) E.g.

Hair Colour	Age in whole years				
	18-30	31-45	46-60	61-75	76+
Blonde					
Light brown					
Dark brown					
Ginger					
Grey					
Other					

2 a) E.g.

Music	Age Group	
	Adult	Child
Pop		
Classical		
Rock		

b) E.g.

Time spent (t hours)	School Year				
	7	8	9	10	11
$0 \le t \le 1$					
$1 < t \le 2$					
$2 < t \le 3$					
$3 < t \le 4$					

3 a) E.g.

Cats or dogs	Male or female	
	Male	Female
Cats		
Dogs		

b) E.g.

TV time (t hours)	Age Group	
	Adult	Child
$0 \le t \le 1$		
$1 < t \le 2$		
$2 < t \le 3$		
$3 < t \le 4$		
$4 < t$		

c) E.g.

Height (h cm)	No. of fruit portions eaten				
	0-2	3-4	5-6	7-10	11+
$h \le 120$					
$120 < h \le 140$					
$140 < h \le 160$					
$160 < h \le 180$					
$180 < h$					

31.3 Sampling and Bias

Page 410   Exercise 1

1 E.g. It would take far too long to time all 216 pupils, and a smaller sample would create less disruption in the school routine.

2 E.g. **Nikhil's idea is best**, as only tasting one cake would not be accurate, and tasting 50 out of the 200 would take too long and use up a quarter of their cakes.

3 Both Melissa and Karen have the same number of coin tosses in their sample (and use the same coin), so their results are **equally reliable**.

4 E.g. It would be highly impractical and time-consuming to interview everyone in the town.

5 Alfie's sample was bigger and would be expected to be more accurate — therefore, "**chocolate**" is more likely to be the most popular flavour.

Page 411   Exercise 2

1 a) People using the library on a Monday are unlikely to want it to close on a Monday, so the sample will be biased away from a Monday closure.

b) The proportions of men and women in the sample are different to the proportions of men and women that work for the company, so the sample is biased in favour of men.

c) People at work probably won't be able to answer the phone in the afternoon, so the only replies they will get will be from people who don't work, home workers, people with unusual working hours and people with the day off.

2 E.g. The manager could assign each of the female members on her database a number, generate 40 random numbers with a computer or calculator, and match the numbers to the members to create the sample.

3 George's sample will be a much higher percentage of the total number of pupils, and Stuart's sample is not random. For example, it could be a group of friends who may all support the same football team.

4 E.g. Seema's sample, being taken on a Sunday morning, may include a lower than usual percentage of churchgoers (if they are at church at the time) or a higher than usual percentage (if they are going to/from church). It would be better to conduct the survey on e.g. a Monday evening when far fewer people will be at work or at (or going to/from) a place of worship. Also, people chosen at random in the street may not even live in that street. So instead, Seema could pick 20 house numbers from her street at random and knock on the doors, asking one person from each house.

5 It would be unfair since there are more vegetarians than non-vegetarians in the group. So an equal sample like this would not be representative of the population.

6 Random sampling might give all the sampled sections of pipe in one area. But it's important to the water company that the sections they sample cover the whole stretch of the pipe.

Page 413   Exercise 3

1 $7 \div 20 = 0.35$ of the sampled rooms have damp so $0.35 \times 450 = 157.5 \approx$ **158** rooms are expected to have damp in the castle.

2 a) (i) $34 \div 50 = 0.68$ so you expect $0.68 \times 650 =$ **442** politicians to be in favour of policy X.

(ii) $1 \div 50 = 0.02$ so you expect $0.02 \times 650 =$ **13** politicians to be neutral about policy Y.

(iii) $14 \div 50 = 0.28$ so you expect $0.28 \times 650 =$ **182** politicians to be against policy Z.

b) The sample of 50 is **representative** of the population of 650 politicians.

3 $\frac{110}{N} = \frac{44}{130} \Rightarrow N = \frac{110 \times 130}{44} =$ **325 newts**

4 a) $\frac{20}{N} = \frac{3}{23} \Rightarrow N = \frac{20 \times 23}{3} = 153.33...$ $\approx$ **153 wombats**

b) E.g. The two samples are representative of the population, one week is long enough for the wombats to mix back in with the population, the population of wombats has not changed in the week between samples.

Page 414   Review Exercise

1 a) One set of data consists of the average number of chocolate bars eaten each week by each pupil. The other set consists of the times it takes these pupils to run 100 m.

b) Gemma could ask each pupil how many chocolate bars they eat on average each week. She could time how long it takes each pupil to run 100 m and record the data in a table, along with the chocolate bar data.

c) The chocolate data is **discrete quantitative** data, and the running times data is **continuous quantitative** data.

d) Both sets of data are **primary data**.

2 a) E.g. 0-4, 5-8, 9-12, 13-16, 17-20

b) E.g. $0 \leq w \leq 180$, $180 < w \leq 190$, $190 < w \leq 200$, $200 < w \leq 210$, $w > 210$

c) E.g. $0 \leq v \leq 260$, $260 < v \leq 270$, $270 < v \leq 280$, $280 < v \leq 290$, $290 < v \leq 300$

3 a) E.g.

No. of people	Tally	Frequency
1		
2		
3		
4		
5 or more		

b) E.g.

Hair rating	Age Group	
	Adult	Child
1		
2		
3		
4		
5		

c) E.g.

Pocket Money	Tally	Frequency
0 - £2.00		
£2.01 - £4.00		
£4.01 - £6.00		
£6.01 - £8.00		
£8.01 or more		

d) E.g.

TV time per week (t hours)	Sport time per week (s hours)			
	$0 < s \leq 5$	$5 < s \leq 10$	$10 < s \leq 20$	$20 < s$
$0 < t \leq 5$				
$5 < t \leq 10$				
$10 < t \leq 20$				
$20 < t$				

4 a) E.g. Asking every one of the 6000 residents their opinion will likely be costly and time-consuming. Not all the residents may be available or willing to give their opinion.

b) E.g. The views given are likely to be biased against the turbines since the respondents will agree with their friends' anti-turbine views. A sample size of 4 is not large enough for a population of 6000.

5 E.g. If the factory runs for 24 hours a day, test one freshly made component roughly every 30 minutes (or at 50 random times throughout the day).

6 The proportion in the sample that like red velvet the most is $6 \div 15 = 0.4$. In the population, you'd expect $0.4 \times 75 =$ **30** people to like red velvet the most.

7 If the population still contains the same number of dyed sheep when the second sample is taken then $\frac{20}{N} = \frac{3}{40} \Rightarrow N = \frac{20 \times 40}{3} = 266.66...$ So an estimate would be **267** sheep.

Page 415   Exam-Style Questions

1 a) (i) **Continuous** [1 mark]
The rainfall is measured in millimetres which is a clue this is continuous data.

(ii) **Secondary** [1 mark]
The car accidents haven't been reported to her personally so she needs to gather the information from some other source.

b) The data she has collected is only from her town so it is **unlikely to be representative** of the whole country — so she **cannot draw conclusions** about the whole country.
[2 marks available — 1 mark for a correct comment about the data not being representative, 1 mark for a correct conclusion that she can't apply the hypothesis to the country]

2 a) (i) **3000** [1 mark]

(ii) E.g. It's too **time-consuming/impractical** to look at all 3000 cars
[1 mark]

b) The proportion of silver cars in the sample is $12 \div 50 = 0.24$ [1 mark]. So the proportion in the whole dealership is estimated to be $0.24 \times 3000 =$ **720** [1 mark].

c) The sample would give an estimate of 0 yellow cars in the dealership but this is **insufficient** to claim there are none — if there are very few yellow cars then they may get left out of a sample [1 mark].

3 a) 30 minutes **isn't enough time** for the birds to redistribute within the population [1 mark].

b) A sample of 2 birds is **too small** to be representative of the population [1 mark].

c) Assume the same proportion of birds are tagged in the whole population $\frac{150}{5000}$ and in his second sample $\frac{6}{b}$.
$\frac{150}{5000} = \frac{6}{b}$ [1 mark]
$\Rightarrow b = \frac{6 \times 5000}{150} =$ **200** [1 mark]

Section 32 — Averages and Ranges

32.1 Averages and Ranges

Page 417   Exercise 1

1 a) (i) 6 appears three times in the data set, which is more than any other value, so mode = **6**

(ii) Largest value = 9, Smallest value = 2
Range = $9 - 2 =$ **7**

Using the same method for b)-c):

b) (i) **8**   **(ii)** $17 - 8 =$ **9**

c) (i) **8.2**   **(ii)** $8.2 - 8.1 =$ **0.1**

2 a) The data set is in numerical order and there are 11 data values, so the median is the $(11 + 1) \div 2 = $ 6th value.
Median = 6th value = **3**

b) The data set in numerical order is:
0, 1, 2, 2, 4, 5, 7, 7, 8, 9
There are 10 data values, so the position of the median is $(10 + 1) \div 2 = 5.5$.
The median is halfway between the 5th value = 4 and 6th value = 5, so median = $(4 + 5) \div 2 =$ **4.5**.

Using the same method for c):

c) **6.93**

3 a) Largest value = 95, Smallest value = 78
Range = $95 - 78 =$ **17 seconds**

b) The data set in numerical order is:
78, 78, 78, 79, 81, 84, 84, 90, 95
There are 9 data values, so the median is the $(9 + 1) \div 2 = $ 5th value.
Median = 5th value = **81 seconds**

4 Total = $34 + 67 + 86 + 58 + 51 + 52 + 71 + 65 + 58 = 542$
Mean = $542 \div 9 = 60.222... =$ **60.2** (3 s.f.)

5 a) 20 appears six times in the data set, which is more than any other value, so mode = **20**

b) (i) Largest value = 22, smallest value = 9
Range = $22 - 9 =$ **13**

(ii) The data set in numerical order is:
9, 12, 13, 15, 15, 15, 15, 16, 16, 16, 16, 17, 18, 18, 18, 18, 18, 19, 20, 20, 20, 20, 20, 20, 21, 21, 21, 22
There are 28 data values, so the position of the median is $(28 + 1) \div 2 = 14.5$. The median is halfway between the 14th value = 18 and the 15th value = 18, so median = $(18 + 18) \div 2 =$ **18**.

(iii) Total = 489
Mean = $489 \div 28 =$ **17.5** (to 3 s.f.)

6 a) If the range is 6 then the missing value must be 6 less than 8, which is the largest value in the data set, or 6 more than 5, which is the lowest value in the data set. So the missing value is $8 - 6 = $ **2** or $5 + 6 = $ **11**

b) There are 6 numbers in the data set, so the total must be $6 \times 7 = 42$
Subtract the known values from the total to find the missing value:
$42 - (6 + 5 + 8 + 8 + 5) = $ **10**

7 The data set in numerical order is:
1.29, 1.30, 1.36, 1.40, 1.40, 1.42, 1.45, 1.50, 1.60, 1.63, 1.63, 1.65, 1.65, 1.67, 1.68, 1.69, 1.69, 1.70, 1.70, 1.72, 1.72, 1.72, 1.75, 1.78, 1.80
There are 25 data values, so the median is the $(25 + 1) \div 2 = $ 13th value $= 1.65$ m
Total $= 39.9 \Rightarrow$ Mean $= 39.9 \div 25 = 1.596$ m
So, the **median** is greater.

Page 418   Exercise 2

1 a) The frequency is highest for 4 people, so mode = **4 people**

b) There are 30 data values, so the position of the median is $(30 + 1) \div 2 = 15.5$.
The median is halfway between the 15th value = 3 and 16th value = 3, so median $= (3 + 3) \div 2 = $ **3 people**

c) Total $= (1 \times 4) + (2 \times 5) + (3 \times 8) + (4 \times 10) + (5 \times 3) = 93$
Mean $= 93 \div 30 = $ **3.1 people**

d) Smallest value = 1, largest value = 5 so the range is $5 - 1 = $ **4**.

2 a) The frequency is highest for 3 goals, so mode = **3 goals**

b) Total $= (0 \times 1) + (1 \times 3) + (2 \times 4) + (3 \times 5) + (4 \times 3) + (5 \times 2) = 48$
Mean $= 48 \div 18 = 2.66...$
$= $ **2.7 goals** (1 d.p.)

c) $2.4 < 2.66...$, so the mean number of goals scored this year is **higher** than in the same week last year.

3 a) Total frequency
$= 18 + 27 + 7 + 15 + 12 + 11 = 90$
There are 90 data values, so the position of the median is $(90 + 1) \div 2 = 45.5$.
So the median is halfway between the 45th value = 17 and 46th value = 18, so median $= (17 + 18) \div 2 = $ **17.5 °C**.

b) Total $= (16 \times 18) + (17 \times 27) + (18 \times 7) + (19 \times 15) + (20 \times 12) + (21 \times 11) = 1629$
Mean $= 1629 \div 90 = $ **18.1 °C**

c) $18.5 > 18.1$, so the average temperature in the student's garden is **lower** than the UK average.

4 a) $77 + p + q + 11 + 3 = 200$
$p + q = 200 - (77 + 11 + 3) = 109$

b) Total number of televisions
$= 1.88 \times 200 = 376$
$(1 \times 77) + 2p + 3q + (4 \times 11) + (5 \times 3) = 376$
$2p + 3q = 376 - (77 + 44 + 15) = 240$

c) (1) $2p + 3q = 240$, (2) $p + q = 109$
$2 \times$ (2): $2p + 2q = 218$
(1) $- 2\times$ (2): $2p + 3q - 2p - 2q = 240 - 218$
$q = 22$
$p + 22 = 109 \Rightarrow p = 109 - 22 = 87$
So **87 people** own 2 televisions and **22 people** own 3 televisions.

Page 419   Exercise 3

1 a) Q_1 position $= (11 + 1) \div 4 = 3$
Q_3 position $= 3(11 + 1) \div 4 = 9$
$Q_1 = 9$, $Q_3 = 16$
Interquartile range $= 16 - 9 = $ **7**

b) The data set in numerical order is:
17, 20, 21, 34, 45, 56, 70, 75, 80, 84, 87, 89

Q_1 position $= (13 + 1) \div 4 = 3.5$
Q_3 position $= 3(13 + 1) \div 4 = 10.5$
$Q_1 = (21 + 21) \div 2 = 21$
$Q_3 = (80 + 84) \div 2 = 82$
Interquartile range $= 82 - 21 = $ **61**
Using the same method for c)-d):
c) 0.3   **d) 6.5**

2 Q_1 position $= (79 + 1) \div 4 = $ 20th value
Q_3 position $= 3(79 + 1) \div 4 = $ 60th value
$Q_1 = $ 20th value $= 4$
$Q_3 = $ 60th value $= 7$
Interquartile range $= 7 - 4 = $ **3**

3 a) The data set in numerical order is:
13, 20, 20, 20, 23, 24, 24, 24, 30, 34, 35, 36, 38, 40, 53, 54, 55, 56, 56, 57, 67, 67, 72, 76, 76, 78, 86, 88, 89, 89, 90, 92, 98
Q_3 position $= 3(33 + 1) \div 4 = 25.5$
Q_3 is halfway between the 25th value = 76 and 26th value = 78, so median $= (76 + 78) \div 2 = $ **77**

b) 8 students scored above 77.

c) (i) Q_1 position $= (33 + 1) \div 4 = 8.5$
Q_1 is halfway between the 8th value = 24 and 9th value = 30, so $Q_1 = (24 + 30) \div 2 = 27$
From part a), $Q_3 = 77$.
Interquartile range $= 77 - 27 = $ **50**

(ii) Class 3B scored **more consistently** than Class 3A as their interquartile range is lower.

Page 420   Exercise 4

1 a) (i) 14 appears in the data set six times, which is more than any other value, so mode = **14**

(ii) The data set in numerical order is:
4, 6, 8, 10, 10, 12, 12, 12, 12, 12, 14, 14, 14, 14, 14, 16, 16, 18, 18
There are 20 data values, so the position of the median is $(20 + 1) \div 2 = 10.5$. The median is halfway between the 10th value = 12 and 11th value = 14, so median $= (12 + 14) \div 2 = $ **13**.

(iii) Total $= 250$
Mean $= 250 \div 20 = $ **12.5**

b) The **mode is a data value** — clothes as sizes are usually whole even numbers, so 12.5 and 13 are not clothes sizes.

2 a) (i) 16 appears in the data set six times, which is more than any other value, so mode = **16**

(ii) The data set in numerical order is:
10, 12, 12, 13, 14, 14, 14, 14, 15, 16, 16, 16, 16, 16, 16, 17, 17, 17, 17, 17, 18, 19, 19, 19, 29, 60, 60, 60, 60, 60
There are 30 data values so the position of the median is $(30 + 1) \div 2 = 15.5$. The median is halfway between the 15th value = 16 and 16th value = 17, so median $= (16 + 17) \div 2 = $ **16.5**

(iii) Total $= 703$
Mean $= 703 \div 30 = 23.433...$
$= $ **23.4** (3 s.f.)

b) No — the majority of people were younger than 29 years old.

3 a) The frequency is highest for a mark of 7, so mode = **7**
There are 190 data values, so the position of the median is $(190 + 1) \div 2 = 95.5$. The median is halfway between the 95th value = 3 and 96th value = 3, so median $= (3 + 3) \div 2 = $ **3**
Total $= (1 \times 31) + (2 \times 34) + (3 \times 35) + (4 \times 34) + (5 \times 4) + (6 \times 6) + (7 \times 36) + (8 \times 7) + (9 \times 2) + (10 \times 1) = 732$
Mean $= 732 \div 190 = 3.852... = $ **3.85** (3 s.f.)

b) (i) Mode
(ii) No, the majority of marks are between 1 and 4.

Page 421   Exercise 5

1 a) The range **is** a good measure of spread in this case as the data values are evenly spread — there are no obvious outliers.

b) The range **is not** a good measure of spread in this case as 93 is a clear outlier and will have a big impact on the range.

c) The range **is not** a good measure of spread in this case as 55 and 2222 are clear outliers and will have a big impact on the range.

d) The range **is** a good measure of spread in this case as the data values are evenly spread and there are no obvious outliers.

2 a) (i) Range $= 893 - 9 = $ **884**
(ii) You'll need to put the numbers in numerical order.
Q_1 position $= (45 + 1) \div 4 = 11.5$
Q_3 position $= 3(45 + 1) \div 4 = 34.5$
$Q_1 = (38 + 43) \div 2 = 40.5$
$Q_3 = (89 + 90) \div 2 = 89.5$
Interquartile range $= $ **49**

b) The **interquartile range** — it isn't affected by the extreme data values and so reflects the spread of the majority of the data better than the range.

32.2 Averages for Grouped Data

Page 423   Exercise 1

1

Weight (w)	Freq.	Midpoint	Midpoint × freq.
$0 \leq w < 20$	1	$(0 + 20) \div 2 = 10$	$10 \times 1 = 10$
$20 \leq w < 40$	6	$(20 + 40) \div 2 = 30$	$30 \times 6 = 180$
$40 \leq w < 60$	9	$(40 + 60) \div 2 = 50$	$50 \times 9 = 450$
$60 \leq w < 80$	24	$(60 + 80) \div 2 = 70$	$70 \times 24 = 1680$
		Total	2320

Total frequency $= 1 + 6 + 9 + 24 = 40$
Mean $= 2320 \div 40 = $ **58 grams**

2 a) The group $10 \leq t < 15$ has the highest frequency, so the modal group is **$10 \leq t < 15$**.

b) There are $3 + 8 + 11 + 4 = 26$ data values, so the position of the median is $(26 + 1) \div 2 = 13.5$. The median is halfway between the 13th and 14th values. Both values are in the group $10 \leq t < 15$, so the median is in **$10 \leq t < 15$**.

c)

Time (t)	Freq.	Midpoint	Midpoint × freq.
$0 \leq t < 5$	3	$(0 + 5) \div 2 = 2.5$	$2.5 \times 3 = 7.5$
$5 \leq t < 10$	8	$(5 + 10) \div 2 = 7.5$	$7.5 \times 8 = 60$
$10 \leq t < 15$	11	$(10 + 15) \div 2 = 12.5$	$12.5 \times 11 = 137.5$
$15 \leq t < 20$	4	$(15 + 20) \div 2 = 17.5$	$17.5 \times 4 = 70$
		Total	275

Mean $= 275 \div 26 = $ **10.6 hours** (3 s.f.)

3 a) The group $1.60 \leq h < 1.70$ has the highest frequency, so the modal group is **$1.60 \leq h < 1.70$**.

b) There are 200 data values, so the position of the median is $(200 + 1) \div 2 = 100.5$. The median is halfway between the 100th and 101st values. Both values are in the group $1.60 \leq h < 1.70$, so the median is in **$1.60 \leq h < 1.70$**.

c)

Height (h) in m	Freq.	Midpoint	Midpoint × freq.
$1.50 \leq h < 1.60$	27	$(1.5 + 1.6) \div 2 = 1.55$	$1.55 \times 27 = 41.85$
$1.60 \leq h < 1.70$	92	$(1.6 + 1.7) \div 2 = 1.65$	$1.65 \times 92 = 151.8$
$1.70 \leq h < 1.80$	63	$(1.7 + 1.8) \div 2 = 1.75$	$1.75 \times 63 = 110.25$
$1.80 \leq h < 1.90$	18	$(1.8 + 1.9) \div 2 = 1.85$	$1.85 \times 18 = 33.3$
		Total	337.2

Mean $= 337.2 \div 200 = $ **1.69 m** (3 s.f.)

d) Lower bound of smallest group = 1.50 m
Upper bound of largest group = 1.90 m
Range $= 1.90 - 1.50 = $ **0.4 m**
Using the same methods for Q4-5:

4 a) $10 \leq t < 15$ **b) $10 \leq t < 15$**
c) 13.4 minutes   **d) 30 minutes**

e) $82 + 34 + 18 = 134$ took longer than 15 minutes to be delivered.
Total number of pizzas:
$40 + 64 + 89 + 82 + 34 + 18 = 327$
$(134 \div 327) \times 100 = \mathbf{41.0\%}$ (3 s.f.)

5 a) $\mathbf{7 \leq x < 10}$ **b)** $\mathbf{7 \leq x < 10}$
c) **10.74**   **d)** **30**
e) $11 > 10.74$, so on average the bird watchers saw **more species** on the next day.

Page 424   Review Exercise

1 a) (i) 8 appears in the data set eight times, which is more often than any other value, so mode = **8**

(ii) The data set in numerical order is:
1, 1, 2, 2, 2, 2, 2, 2, 2, 3, 3, 4, 4, 5, 5, 5, 5, 5, 5, 5, 6, 6, 6, 7, 7, 7, 7, 7, 7, 8, 8, 8, 8, 8, 8, 8, 9, 9, 9
There are 40 data values, so the position of the median is
$(40 + 1) \div 2 = 20.5$. The median is halfway between the 20th value = 5 and 21st value = 6, so
median = $(5 + 6) \div 2 = \mathbf{5.5}$

(iii) Total = 216
Mean = $216 \div 40 = \mathbf{5.4}$

(iv) Range = $9 - 1 = \mathbf{8}$

b) The **mean** — it takes all data values into account.
The median would also provide a good representation of the data in this case, even though it does not take into account all the data values.

c) The **mean is lower** than both the mode and median, so the product might sell better if another average was used.

2 a) 2 has the highest frequency, so mode = **2**
There are 30 data values, so the position of the median is $(30 + 1) \div 2 = 15.5$. So the median is halfway between the 15th and 16th values.
15th value = 3, 16th value = 4,
median = $(3 + 4) \div 2 = \mathbf{3.5}$
Total = $(1 \times 2) + (2 \times 9) + (3 \times 4) + (4 \times 6) + (5 \times 5) + (6 \times 0) + (7 \times 1) = 88$
Mean = $88 \div 30 = 2.933... = \mathbf{2.93}$ (3 s.f.)

b) The mean is a **decimal number**, shoe sizes come in whole numbers (or half sizes).

3 a) (i) 1600 appears twice in the data set, which is more than any other value, so the mode = **1600 kg**

(ii) The data set in numerical order is:
195, 245, 1340, 1400, 1525, 1600, 1600, 1750, 1950
There are 9 data values, so the median is the $(9 + 1) \div 2 = 5$th value.
Median = **1525**

(iii) Total = 11 605
Mean = $11\,605 \div 9 = 1289.4...$
= **1289 kg** (to nearest kg)

b) (i) Range = $1950 - 195 = \mathbf{1755\ kg}$

(ii) Q_1 position = $(9 + 1) \div 4 = 2.5$
Q_3 position = $3(9 + 1) \div 4 = 7.5$
$Q_1 = (245 + 1340) \div 2 = 792.5$
$Q_3 = (1600 + 1750) \div 2 = 1675$
IQR = $1675 - 792.5 = \mathbf{882.5\ kg}$

c) The range and interquartile range will **both have increased** as a result, but the interquartile range would be least affected.

4 For the pantomime horse:
Total = 502.6
Mean = $502.6 \div 17 = 29.564...$ seconds
The data set in numerical order is:
14.0, 15.3, 17.3, 17.8, 18.9, 19.2, 20.5, 22.7, 26.4, 30.5, 32.1, 34.2, 36.2, 40.2, 41.1, 56.0, 60.2

Q_1 position = $(17 + 1) \div 4 = 4.5$
Q1 is halfway between the 4th value = 17.8 and the 5th value = 18.9, so
$Q_1 = (17.8 + 18.9) \div 2 = 18.35$
Q_3 position = $3(17 + 1) \div 4 = 13.5$
Q_3 is halfway between the 13th value = 36.2 and the 14th value = 40.2, so
$Q_3 = (36.2 + 40.2) \div 2 = 38.2$
IQR = $38.2 - 18.35 = 19.85$ seconds
The mean time it took a kitten to run 100 m is **less than that of a pantomime horse**. However the interquartile range for the pantomime horses is smaller than that for the kittens. This means the **horses ran at a more consistent pace** than the kittens.

5 a) The group $700 \leq p < 800$ has the highest frequency, so the modal group is $\mathbf{700 \leq p < 800}$.

b) There are $2 + 14 + 25 + 9 = 50$ data values, so the position of the median is $(50 + 1) \div 2 = 25.5$. The median is halfway between the 25th and 26th values. Both values are in the group $700 \leq p < 800$, so the median is in $\mathbf{700 \leq p < 800}$.

c)

No. of potatoes (p)	Freq.	Midpoint	Midpoint × freq.
$500 \leq p < 600$	2	$(500 + 600) \div 2 = 550$	$550 \times 2 = 1100$
$600 \leq p < 700$	14	$(600 + 700) \div 2 = 650$	$650 \times 14 = 9100$
$700 \leq p < 800$	25	$(700 + 800) \div 2 = 750$	$750 \times 25 = 18\,750$
$800 \leq p < 1000$	9	$(800 + 1000) \div 2 = 900$	$900 \times 9 = 8100$
		Total	37 050

Mean = $37\,050 \div 50 = \mathbf{741}$

d) Lower bound of smallest group = 500
Upper bound of largest group = 1000
Range = $1000 - 500 = \mathbf{500}$

Page 425   Exam-Style Questions

1 The data set in numerical order is:
17, 19, 28, 29, 83, 106
There are 6 data values, so the position of the median is $(6 + 1) \div 2 = 3.5$. So the median is halfway between the 3rd and 4th values.
3rd value = 28, 4th value = 29,
median = $(28 + 29) \div 2 = \mathbf{28.5}$
[1 mark for the correct answer]

2 Total frequency = $7 + 23 + 45 + 109 + 541 + 1894 + 3561 + 2670 = 8850$
Total no. of eggs = $(24 \times 7) + (25 \times 23) + (26 \times 45) + (27 \times 109) + (28 \times 541) + (29 \times 1894) + (30 \times 3561) + (31 \times 2670) = 264\,530$
Mean = $264\,530 \div 8850 = \mathbf{29.89}$
[2 marks available — 1 mark for the total number of eggs and total frequency, 1 mark for the correct answer]

3

Weight (m kg)	Freq.	Midpoint	Midpoint × freq.
$3.0 \leq m < 4.0$	11	$(3.0 + 4.0) \div 2 = 3.5$	$3.5 \times 11 = 38.5$
$4.0 \leq m < 5.0$	9	$(4.0 + 5.0) \div 2 = 4.5$	$4.5 \times 9 = 40.5$
$5.0 \leq m < 6.0$	4	$(5.0 + 6.0) \div 2 = 5.5$	$5.5 \times 4 = 22$
$6.0 \leq m < 6.5$	1	$(6.0 + 6.5) \div 2 = 6.25$	$6.25 \times 1 = 6.25$
		Total	107.25

Mean = $107.25 \div 25 = 4.29$ kg
Marrows within the group $4.0 \leq m < 5.0$ could be below 4.29 kg, so only the 5 marrows over 5.0 kg are definitely above the mean mass.
$(5 \div 25) \times 100 = \mathbf{20\%}$
[3 marks available — 1 mark for using the midpoints to calculate the mean, 1 mark for the mean mass, 1 mark for the correct answer]

4 Range = $9.5 - 7 = 2.5$
Q_1 position = $(7 + 1) \div 4 = 2$
Q_3 position = $3(7 + 1) \div 4 = 6$
$Q_1 = 7$, $Q_3 = 8.5$
IQR = $8.5 - 7 = 1.5$
Interquartile range : range $\Rightarrow$ $1.5 : 2.5$
$\Rightarrow$ **3 : 5**
[3 marks available — 1 mark for finding the range, 1 mark for finding the interquartile range, 1 mark for the correct answer]

5 Let x = the number sold each day for the first four days
Total = $3.3 \times 10 = 33$
so $1 \times 0 + 3 \times 1 + 2 \times 3 + 4x = 33$
$4x + 9 = 33 \Rightarrow 4x = 24 \Rightarrow x = 6$
Mode = 6, because 6 has the highest frequency.
[3 marks available — 1 mark for total number of motorcycles, 1 mark for correct answer, 1 mark for explanation of why answer is the mode]

6 As the median and mode are both 7, the middle two numbers must be 7.
As the range = 10 and the mean = 9:
Let x = the lowest number.
Total = $9 \times 4 = 36$
$x + 7 + 7 + (x + 10) = 36$
$2x + 24 = 36 \Rightarrow 2x = 12 \Rightarrow x = 6$
$6 + 10 = 16$, so the numbers are **6, 7, 7 and 16**
[3 marks available — 1 mark for finding the total of the integers, 1 mark for using the median to find the values 7 and 7, 1 mark for using the range to find the values 6 and 16]

Section 33 — Displaying Data

33.1 Tables and Charts

Page 426   Exercise 1

1 a) Number of red vans = $12 - 8 - 2 = 2$
Number of black vehicles = $7 + 2 + 1 = 10$
Number of blue motorbikes = $6 - 4 - 1 = 1$
Number of white cars = $22 - 8 - 7 - 4 = 3$
Number of vans = $2 + 2 + 1 + 10 = 15$
Number of motorbikes = $2 + 1 + 1 + 2 = 6$
Number of white vehicles = $3 + 10 + 2 = 15$
Total number of vehicles = $22 + 15 + 6 = 43$
Use these values to fill in the table:

	Red	Black	Blue	White	Total
Cars	8	7	4	**3**	22
Vans	**2**	2	1	10	15
Motorbikes	2	1	**1**	2	**6**
Total	12	**10**	6	15	**43**

b) Reading from the table:
1 motorbike was blue.

c) (i) $(22 \div 43) \times 100 = \mathbf{51.2\%}$ (to 3 s.f.)
(ii) $(15 \div 43) \times 100 = \mathbf{34.9\%}$ (to 3 s.f.)
(iii) $(12 \div 43) \times 100 = \mathbf{27.9\%}$ (to 3 s.f.)

2

	Have been	Haven't been	Total
Boys	$10 - 8 = \mathbf{2}$	$12 - 4 = \mathbf{8}$	$20 \div 2 = \mathbf{10}$
Girls	$20 - 4 = \mathbf{16}$	$20 \div 5 = \mathbf{4}$	**20**
Total	**18**	$30 - 18 = \mathbf{12}$	$10 + 20 = \mathbf{30}$

3 a)

	$h < 160$	$160 \leq h < 170$	$170 \leq h < 180$
Women	4	$9 - 2 = \mathbf{7}$	11
Men	$4 - 4 = \mathbf{0}$	2	6
Total	4	9	$11 + 6 = \mathbf{17}$

	$180 \leq h < 190$	$190 \leq h$	Total
Women	$24 - 4 - 7 - 11 - 0 = \mathbf{2}$	0	24
Men	$26 - 0 - 2 - 6 - 5 = \mathbf{13}$	5	$50 - 24 = \mathbf{26}$
Total	$2 + 13 = \mathbf{15}$	$0 + 5 = \mathbf{5}$	50

b) $4 + 9 = \mathbf{13}$ **c)** $\dfrac{2}{24} = \dfrac{1}{12}$

d) E.g. More of the men's heights are in the taller height intervals, which suggests **the men are generally taller than the women.**

Page 428   Exercise 2

1 a)

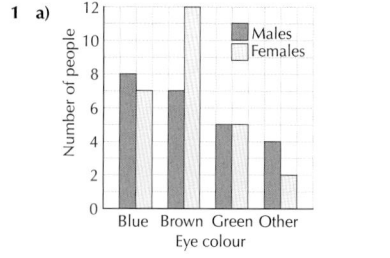

b)

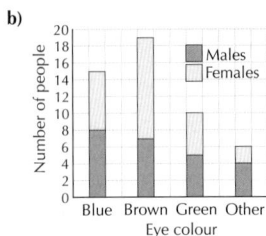

c) The **dual bar chart** is the best for comparing male and female eye colours — it is easy to see the difference in the heights of the pair of bars for each eye colour.

d) The most common eye colour is **brown**.

e) The **composite bar chart** is easier to use to find the modal eye colour, as you can just look for the tallest bar.

2 a) The orange bar (representing children's scores) for a score of 1 has a height of **2**.

b) $10 + 12 + 5 + 2 = $ **29**

c) The score that has no orange bar is **2**.

d) 10 children rated it as 3 and 5 adults rated it 3. $5 \times 2 = 10$, so the score is **3**.

e) E.g. The scores given by adults were **generally lower** than the scores given by children.

3 a) The height of the composite bar for the 26-35 age group is **12**.

b) $12 + 12 + 9 + 6 = $ **39**

c) **16-25 yrs**
   Modal = most common, so it's the tallest bar for women.

d) 8 men aged 26-35 use the gym. 4 women aged 26-35 use the gym. So $8 - 4 = $ **4** more men aged 26-35 use the gym than women aged 26-35.

e) 16-25: Difference = $6 - 6 = 0$
   26-35: Difference = 4 (from part d) above)
   36-45: Difference = $7 - 2 = 5$
   46-55: Difference = $5 - 1 = 4$
   So the age group with the greatest difference is **36-45 yrs**.

Page 429   Exercise 3

1 $360° \div 90 = 4°$, so each person is represented by $4°$
   UK: $22 \times 4° = 88°$, Europe: $31 \times 4° = 124°$,
   USA: $8 \times 4° = 32°$, Other: $11 \times 4° = 44°$,
   Nowhere: $18 \times 4° = 72°$
   Use the angles to draw and label the pie chart:

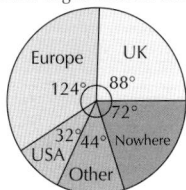

2 Total = $33 + 52 + 21 + 14 = 120$
   $360° \div 120 = 3°$, so each person is represented by $3°$
   Squash: $33 \times 3° = 99°$, Gym: $52 \times 3° = 156°$,
   Swimming: $21 \times 3° = 63°$,
   Tennis: $14 \times 3° = 42°$
   Use the angles to draw and label the pie chart:

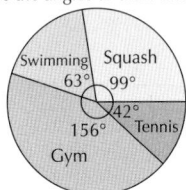

---

3 Total = $36 + 16 + 12 + 16 = 80$
   $360° \div 80 = 4.5°$, so each person is represented by $4.5°$
   Walking: $36 \times 4.5° = 162°$,
   Bus/Bike: $16 \times 4.5° = 72°$, Car: $12 \times 4.5° = 54°$

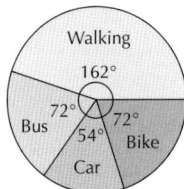

4 Total = $348 + 297 + 195 + 87 + 108 + 45 = 1080$
   $360° \div 1080 = \frac{1}{3}$, so each pupil is represented by $\frac{1}{3}°$
   Maths: $348 \times \frac{1}{3}° = 116°$, Art: $297 \times \frac{1}{3}° = 99°$,
   PE: $195 \times \frac{1}{3}° = 65°$, English: $87 \times \frac{1}{3}° = 29°$,
   Science: $108 \times \frac{1}{3}° = 36°$, Other: $45 \times \frac{1}{3}° = 15°$
   Use the angles to draw and label the pie chart:

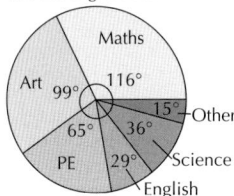

Page 430   Exercise 4

1 a) **Pepperoni**

b) The fraction of pupils who said cheese and tomato is equal to the fraction of the pie chart that cheese and tomato occupies:
   $\frac{90°}{360°} = \frac{1}{4}$

c) $360° \div 60 = 6°$ per person
   $102° \div 6 = $ **17 pupils**

2 a) $54° \div 18 = 3°$ per person
   $360° \div 3° = $ **120 people**

b) Under 11: $72° \div 3° = $ **24 people**
   11-16: $36° \div 3° = $ **12 people**
   30-49: $72° \div 3° = $ **24 people**
   50+: $126° \div 3° = $ **42 people**

c) Measure the angle with a protractor to get 60°: $\frac{60°}{360°} = \frac{1}{6}$

d) The fraction of people aged 11-16 in the first survey is $\frac{36°}{360°} = \frac{1}{10}$. This is **smaller** than the fraction of people aged 11-16 in the second survey.

e) **No**, they only show that a bigger proportion of people in the second survey were aged 11-16. You don't know whether there were more people aged 11-16 without knowing how many people were asked in each survey.

33.2 Stem and Leaf Diagrams

Page 431   Exercise 1

1 a)

4	1 8
5	1 4 9
6	5
7	4
8	0 6 9

Key: 4 | 1 means 41

b)

1	2 5 5 7
2	4 6 7
3	1 6 9
4	1 1

Key: 1 | 2 means 12

Don't forget to order the leaves.

---

c)

3	1 4
4	0 4
5	3 7 9
6	0
7	7

Key: 3 | 1 means 3.1

d)

20	3 3 5
21	1
22	1
23	2 4 6
24	0

Key: 20 | 3 means 203

2 a) Mode = most common value = **61 marks**
   There are 11 data values, so the median is the $12 \div 2 = $ 6th value. Median = **61 marks**
   Range = $77 - 52 = $ **25 marks**
   Position of $Q_1 = 12 \div 4 = 3$, $Q_1 = 55$
   Position of $Q_3 = 3 \times (12 \div 4) = 9$, $Q_3 = 69$
   Interquartile range = $69 - 55 = $ **14 marks**

b) The values between 53 and 63 are 54, 55, 57, 58, 61 and 61, so **6 pupils** scored between 53 and 63.

Page 432   Exercise 2

1 E.g.

	Set 2		Set 1
9 8 5 3 3		1	2 8
7 5 3 2 2		2	4 8 9
2		3	2 3 7 8
		4	1 8

Key: 1 | 2 for Set 1 means 12
Key: 3 | 1 for Set 2 means 13
You might have put Set 1 on the left and Set 2 on the right.

2 a) There are 15 data values, so the median is the 8th value.
   Median for 'at rest' = **72 bpm**
   Median for 'after exercise' = **77 bpm**

b) Position of $Q_1 = (15 + 1) \div 4 = 4$
   Position of $Q_3 = 3(15 + 1) \div 4 = 12$
   Q_1 for 'at rest' = 64, Q_3 for 'at rest' = 78
   so IQR for 'at rest' = $78 - 64 = $ **14 bpm**
   Q_1 for 'after exercise' = 69
   Q_3 for 'after exercise' = 87
   so IQR for 'after exercise' = $87 - 69$
   $= $ **18 bpm**

c) The higher median for the 'after exercise' data suggests that, on average, **people's heart rate was faster after they'd exercised**. The greater IQR for the 'after exercise' data shows that there was **more variety in the heart rates after exercise**, than at rest.

3 a)

	London		Dundee
9 7 4		0	1 2 3 3 4 5 6
8 6 5 3 2 2 1		1	2 5 7 9
4 1		2	3

Key: 4 | 0 for London means 4 °C
Key: 0 | 1 for Dundee means 1 °C

b) There are 12 data values for each data set, so the medians are halfway between the 6th and 7th values.
   Median for London = $\frac{12 + 13}{2} = 12.5$ °C
   Median for Dundee = $\frac{5 + 6}{2} = 5.5$ °C
   Range for London = 24 °C – 4 °C = 20 °C
   Range for Dundee = 23 °C – 1 °C = 22 °C
   The median temperature for London was higher than for Dundee, suggesting **temperatures in London were generally higher**. The range for temperatures in Dundee was higher than in London, suggesting the **temperatures in Dundee were more varied**.

33.3 Frequency Polygons

Page 433 Exercise 1

1

Floor area (a)	Midpoint	Freq.
$9 \le a < 13$	$(9 + 13) \div 2 = 11$	3
$13 \le a < 14$	$(13 + 14) \div 2 = 13.5$	7
$14 \le a < 15$	$(14 + 15) \div 2 = 14.5$	5
$15 \le a < 16$	$(15 + 16) \div 2 = 15.5$	2

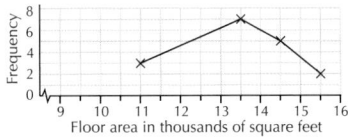

2 a) You want to find the frequency of categories above 4 hours, so look at the points at 4.5 and 5.5 hours on the horizontal axis. Both have a frequency of 3, so the number of days in February with 4 or more hours of sunshine is 3 + 3 = **6 days**

   b) The point at 1.5 hours has the highest frequency, so the modal class for February is e.g. **$1 \le h < 2$.**
   You don't know what classes were used to group the temperatures, so you can't be certain of the class — $1 < h \le 2$ would also be correct.

   c) The point at 6.5 hours has the highest frequency, so the modal class for July is e.g. **$6 \le h < 7$.**

   d) The modal classes suggest that there were generally **more daily hours of sunshine in July** than in February. The graphs show that the range of daily hours of sunshine was much **greater in July** than in February.

33.4 Histograms

Page 434 Exercise 1

1 a) The class widths are all 5.

Height (h) in cm	Frequency	Frequency Density
$0 < h \le 5$	4	$4 \div 5 = \textbf{0.8}$
$5 < h \le 10$	6	$6 \div 5 = \textbf{1.2}$
$10 < h \le 15$	3	$3 \div 5 = \textbf{0.6}$
$15 < h \le 20$	2	$2 \div 5 = \textbf{0.4}$

   b) The class widths are 10, 5, 5 and 10.

Height (h) in cm	Frequency	Frequency Density
$10 < h \le 20$	5	$5 \div 10 = \textbf{0.5}$
$20 < h \le 25$	15	$15 \div 5 = \textbf{3}$
$25 < h \le 30$	12	$12 \div 5 = \textbf{2.4}$
$30 < h \le 40$	8	$8 \div 10 = \textbf{0.8}$

2 a)

Volume (v) in ml	Frequency	Frequency Density
$0 \le v < 500$	50	$50 \div 500 = \textbf{0.1}$
$500 \le v < 1000$	75	$75 \div 500 = \textbf{0.15}$
$1000 \le v < 1500$	70	$70 \div 500 = \textbf{0.14}$
$1500 \le v < 2000$	55	$55 \div 500 = \textbf{0.11}$

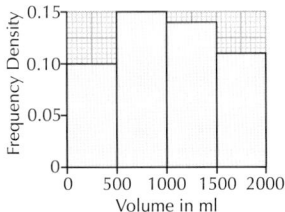

   b)

Volume (v) in ml	Frequency	Frequency Density
$0 \le v < 300$	30	$30 \div 300 = \textbf{0.1}$
$300 \le v < 600$	15	$15 \div 300 = \textbf{0.05}$
$600 \le v < 900$	24	$24 \div 300 = \textbf{0.08}$
$900 \le v < 1500$	42	$42 \div 600 = \textbf{0.07}$

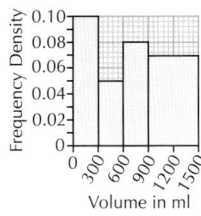

3

Mass (m) in kg	Frequency	Frequency Density
$10 \le m < 20$	5	$5 \div 10 = 0.5$
$20 \le m < 30$	7	$7 \div 10 = 0.7$
$30 \le m < 50$	10	$10 \div 20 = 0.5$

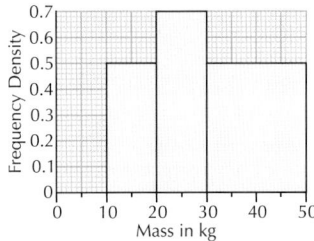

4 a), b) The bar $2 \le t < 4$ has a frequency of 16 and a class width of 2, so the frequency density is $16 \div 2 = 8$. The height of the bar is 4 divisions, so each division on the vertical axis represents $8 \div 4 = 2$. Use this information to label the vertical axis.

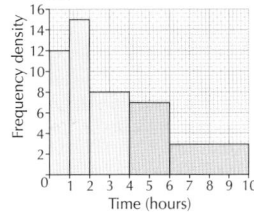

Time (t) in hours	Frequency	Frequency Density
$0 \le t < 1$	$12 \times 1 = \textbf{12}$	12
$1 \le t < 2$	$15 \times 1 = \textbf{15}$	15
$2 \le t < 4$	16	8
$4 \le t < 6$	14	$14 \div 2 = \textbf{7}$
$6 \le t < 10$	12	$12 \div 4 = \textbf{3}$

Page 436 Exercise 2

1 The inequality signs in the tables below are examples. Yours might be different (e.g. $120 < h < 130$) — but there shouldn't be any overlaps between the classes.

   a)

Height (h) in cm	Frequency density	Frequency
$120 \le h < 130$	0.5	$10 \times 0.5 = \textbf{5}$
$130 \le h < 140$	0.7	$10 \times 0.7 = \textbf{7}$
$140 \le h < 145$	1.8	$5 \times 1.8 = \textbf{9}$
$145 \le h < 150$	2.0	$5 \times 2.0 = \textbf{10}$
$150 \le h < 165$	0.6	$15 \times 0.6 = \textbf{9}$

   b)

Mass (m) in g	Frequency density	Frequency
$50 \le m < 70$	0.5	$20 \times 0.5 = \textbf{10}$
$70 \le m < 90$	0.6	$20 \times 0.6 = \textbf{12}$
$90 \le m < 100$	1.5	$10 \times 1.5 = \textbf{15}$
$100 \le m < 110$	1.8	$10 \times 1.8 = \textbf{18}$
$110 \le m < 120$	1.6	$10 \times 1.6 = \textbf{16}$

2 a) Frequency of 150-165 cm = 9 (from Q1a)) 150-155 cm is one third of 150-165 cm, so frequency of 150-155 cm = $9 \div 3 = \textbf{3}$

   b) Frequency of 150-165 cm = 9 150-160 cm is two thirds of 150-165 cm, so frequency of 150-160 cm = $2 \times (9 \div 3) = \textbf{6}$

   c) Frequency of 145-150 cm = 10 (from Q1a)) Using your answer from part b): Frequency of 145-160 cm = 10 + 6 = **16**

   d) Frequency of 120-130 cm = 5 (from Q1a)) 120-126 cm is six tenths of 120-130 cm, so frequency of 120-126 cm = $6 \times (5 \div 10) = \textbf{3}$

---

3 a) Frequency density of 0-10 cm = 8 Frequency of 0-10 cm = $10 \times 8 = 80$ Frequency density of 10-20 cm = 4.5 Frequency of 10-20 cm = $10 \times 4.5 = 45$ Frequency of 10-16 cm = $6 \times (45 \div 10) = 27$ Frequency of <16 cm = 80 + 27 = **107 cakes**

   b) Frequency density of 20-25 cm = 7 Frequency of 20-25 cm = $5 \times 7 = 35$ Frequency density of 25-40 cm = 2 Frequency of 25-40 cm = $15 \times 2 = 30$ Frequency of 25-30 cm = $30 \div 3 = 10$ Frequency of 20-30 cm = 35 + 10 = **45 cakes**

   c)

Diameter (d) in cm	Frequency Density	Frequency
$0 < d \le 10$	8	$10 \times 8 = 80$
$10 < d \le 20$	4.5	$10 \times 4.5 = 45$
$20 < d \le 25$	7	$5 \times 7 = 35$
$25 < d \le 40$	2	$15 \times 2 = 30$

Number of cakes = 80 + 45 + 35 + 30 = **190**
A cake can't have a diameter of 0 cm (if it did, there'd be no cake!) so the first inequality should be strict, i.e. $0 < d$.

   d)

Diameter (d) in cm	Freq.	Mid.	Freq. × mid.
$0 < d \le 10$	80	5	$80 \times 5 = 400$
$10 < d \le 20$	45	15	$45 \times 15 = 675$
$20 < d \le 25$	35	22.5	$35 \times 22.5 = 787.5$
$25 < d \le 40$	30	32.5	$30 \times 32.5 = 975$
Totals	190		2837.5

Mean ≈ $2837.5 \div 190 = \textbf{14.9 cm}$ (3 s.f.)

   e) E.g. To find the midpoints you assumed that the data is **evenly spread** throughout each interval. This is **unlikely** for $0 < d \le 10$ because it would be hard to bake cakes with very small diameters in practice. So there are likely to be **more cakes** towards the **upper end** of this interval, which would increase the mean.

4 a) Frequency density of 50-54 cm = 1.5 Frequency of 50-54 cm = $4 \times 1.5 = 6$ Frequency density of 54-58 cm = 2.5 Frequency of 54-58 cm = $4 \times 2.5 = 10$ Frequency of 54-56 cm = $10 \div 2 = 5$ Frequency of <56 cm = 6 + 5 = **11 penguins**

   b)

Height (h) in cm	Frequency Density	Frequency
$50 \le h < 54$	1.5	$1.5 \times 4 = 6$
$54 \le h < 58$	2.5	$2.5 \times 4 = 10$
$58 \le h < 60$	8	$8 \times 2 = 16$
$60 \le h < 62$	6	$6 \times 2 = 12$
$62 \le h < 66$	1	$1 \times 4 = 4$

Number of penguins = 6 + 10 + 16 + 12 + 4 = **48**

   c) 16 penguins have a height less than 58 cm. 32 penguins have a height less than 60 cm. There are 32 – 16 = 16 penguins with a height between 58-60 cm. 24 is halfway between 16 and 32 and the midpoint of this class is 59 cm, so $H = \textbf{59}$.

   d) There are 48 penguins, so the position of the median is $(48 + 1) \div 2 = 24.5$. The 24th penguin measures <59 cm and so the 25th penguin measures >59 cm. As you do not know the exact values, you have to assume the penguin's heights are equally spaced, so an estimate for the median is **59 cm**.

33.5 Cumulative Frequency Diagrams

Page 438 Exercise 1

1 a) The last point on the graph has a cumulative frequency of 84, so there were **84 16-year-olds**.

   b) Draw a vertical line from £20 to the curve to read off a cumulative frequency of 3.

   c) Draw a vertical line from £80 to the curve to read off a cumulative frequency of 64.

   d) Draw a vertical lines from £40 and £100 to the curve to read off cumulative frequencies of 12 and 82. 82 – 12 = **70**

e) Position of the median = 84 ÷ 2 = 42
Draw a line from a cumulative frequency of 42 to the curve to get the median **£67** (accept £66-£68).

f) Position of lower quartile = 84 ÷ 4 = 21
Draw a line from a cumulative frequency of 21 to the curve to get the lower quartile **£51** (accept £50-£52).
Position of upper quartile = 3(84 ÷ 4) = 63
Draw a line from a cumulative frequency of 63 to the curve to get the upper quartile **£79** (accept £78-£80).
Interquartile range = £79 – £51 = **£28** (accept answers based on your quartiles)

2 a)

Marks, m	Frequency	Cumulative Frequency
$0 < m \leq 10$	0	**0**
$10 < m \leq 20$	2	0 + 2 = **2**
$20 < m \leq 30$	4	2 + 4 = **6**
$30 < m \leq 40$	5	6 + 5 = **11**
$40 < m \leq 50$	19	11 + 19 = **30**
$50 < m \leq 60$	33	30 + 33 = **63**
$60 < m \leq 70$	43	63 + 43 = **106**
$70 < m \leq 80$	10	106 + 10 = **116**
$80 < m \leq 90$	3	116 + 3 = **119**
$90 < m \leq 100$	1	119 + 1 = **120**

b)

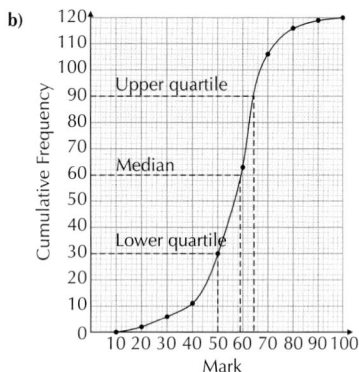

c) Position of the median = 120 ÷ 2 = 60
Draw a line from a cumulative frequency of 60 to the curve to get the median **59** (accept 58-60).

d) Position of lower quartile = 120 ÷ 4 = 30
Draw a line from a cumulative frequency of 30 to the curve to get the lower quartile **50**. (accept 49-51).
Position of upper quartile = 3(120 ÷ 4) = 90
Draw a line from a cumulative frequency of 90 to the curve to get the upper quartile **64** (accept 63-65).
Interquartile range = 64 – 50 = **14** (accept answers based on your quartiles).

e) Draw a line from a mark of 45 to the curve to get a cumulative frequency of **18** (accept 16-20).

f) Draw a line from a mark of 55 to the curve to get a cumulative frequency of 44. Total frequency = 120, so the number of students sitting the higher tier exam will be 120 – 44 = **76**.

3 a) Cumulative frequencies for each group are: 2, 2 + 3 = 5, 5 + 4 = 9, 9 + 7 = 16, 16 + 25 = 41, 41 + 51 = 92, 92 + 31 = 123, 123 + 9 = 132, 132 + 5 = 137, 137 + 3 = 140.

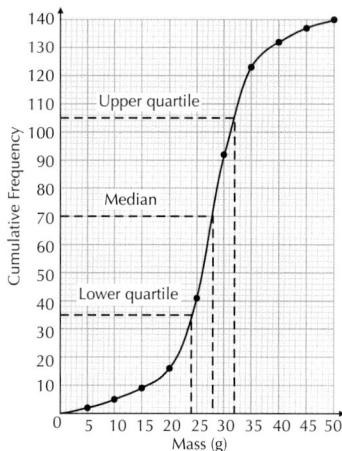

b) Position of the median = 140 ÷ 2 = 70.
Draw a line from a cumulative frequency of 70 to the curve to get the median **28 g** (accept 27-29 g).
Position of lower quartile = 140 ÷ 4 = 35
Draw a line from a cumulative frequency of 35 to the curve to get the lower quartile 24 g (accept 23-25 g).
Position of upper quartile = 3 × (140 ÷ 4) = 105
Draw a line from a cumulative frequency of 105 to the curve to get the upper quartile 32 g (accept 31-33 g).
Interquartile range = 32 g – 24 g = **8 g** (accept answers based on your quartiles)

c) Median = 28 g, so you're looking for samples that weighed between 18-38 g. Draw lines from masses of 18 g and 38 g to get cumulative frequencies of 12 (accept 11-13) and 130 (accept 129-131). So the number of samples between 18-38 g is 130 – 12 = **118** (accept answers based on your cumulative frequencies)

d) E.g. The students' estimates were **quite good** — the interquartile range is small and a lot of the students were within 10 g of the median. Alternatively, the students' estimates were **not very good** as a lot of them were higher than the true mass — the median 28 g > 25 g.

Page 439   Exercise 2

1 a) (i) The line within the box is at 48, so the median is **48 marks**.
(ii) The upper edge of the box is at 67, so the upper quartile is **67 marks**.
(iii) The lower edge of the box is at 26, so the lower quartile is **26 marks**.
(iv) Interquartile range = 67 – 26 = **41 marks**
(v) The mark for the lowest value is at 13, so the lowest value is **13 marks**.
(vi) The mark for the highest value is at 83, so the highest value is **83 marks**.

Using the same methods for b):
b) (i) 10 hours   **(ii) 18 hours**
(iii) 6 hours   **(iv) 12 hours**
(v) 2 hours   **(vi) 29 hours**

2 Busybodies   Chatterboxes
Median = 30 s   Median = 20 s
Upper quartile = 45 s   Upper quartile = 30 s
Lower quartile = 10 s   Lower quartile = 10 s
IQR = 35 s   IQR = 20 s
Lowest time = 6 s   Lowest time = 4 s
Highest time = 50 s   Highest time = 59 s
E.g. The median and upper quartile times for the busybodies are greater than those for the chatterboxes, so a **greater number of busybodies stayed quieter for longer** than the chatterboxes.

However, the maximum time a chatterbox was able to stay silent was greater than the maximum time a busybody managed to stay silent. So the statement is **not definitely true**.

33.6 Time Series

Page 441   Exercise 1

1

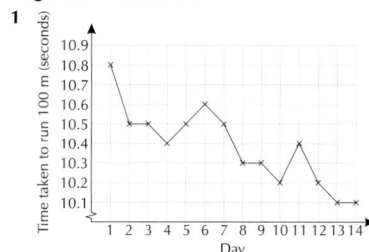

The athlete's running times showed a **downward trend** — there was a general **decrease over time**.

2 a)

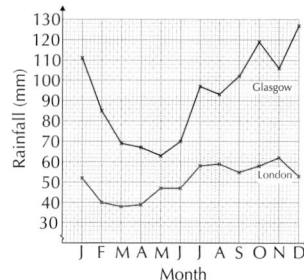

b) E.g. The average rainfall in Glasgow is **higher** than the average rainfall in London for every month of the year. Glasgow's average rainfall is **more varied** throughout the year.

c) Glasgow:
Highest value = 127 mm
Lowest value = 63 mm
Range = 127 – 63 = **64 mm**
London:
Highest value = 62 mm,
Lowest value = 38 mm
Range = 62 – 38 = **24 mm**

3

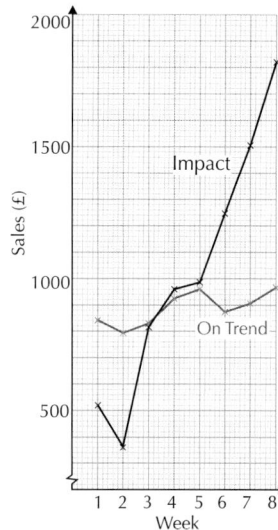

E.g. On Trend had **fairly consistent sales** over the 8-week period that generally increased slowly. Sales of Impact started off **much lower**, but **increased rapidly**, overtaking On Trend in the fourth week and selling almost twice as much by week 8.

4 a)

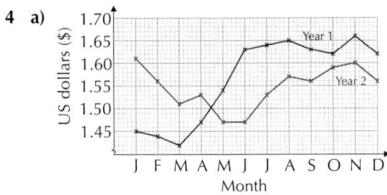

b) The highest value of £1 in dollars in Year 2 was **$1.61**.

c) The highest value of £1 in dollars in Year 1 was **$1.66 in November**.

d) June Year 1: £500 = 500 × 1.63 = $815
June Year 2: £500 = 500 × 1.47 = $735
$815 − $735 = **$80**

33.7 Scatter Graphs

Page 443   Exercise 1

1 a) It's more likely that ice cream sales depend on the temperature than the other way round, so put ice cream sales on the vertical axis. Then plot the data as points — e.g. the first point is (28, 30), the second is (25, 22) etc.

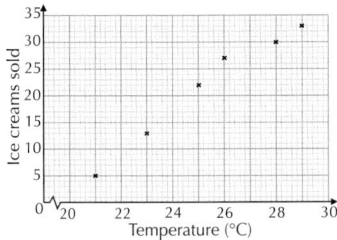

b) The points form an upward slope so there is **positive correlation** — the higher the temperature, the more ice creams were sold.

2 a) The number of baby teeth will depend on age, so put number of baby teeth on the vertical axis. Then plot the data as points — e.g. (5, 20), (6, 17), etc.

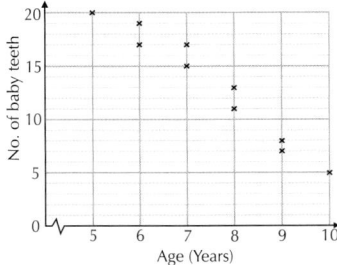

b) The points form a downward slope so there is **negative correlation** — the older the child, the fewer baby teeth they have.

Page 443   Exercise 2

1 a) The points form a downward slope fairly close to a straight line so there is **strong negative correlation**.

b) The points form an upward slope loosely spread around a straight line so there is **moderate positive correlation**.

2 a)

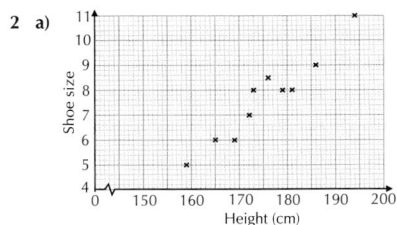

b) The graph shows moderate positive correlation (i.e. as height increases so does shoe size), so **yes** there is evidence that they are correlated.

Page 445   Exercise 3

1 a) The points form an upward slope fairly close to a straight line so there is **strong positive correlation**.

b) E.g. **The tree was not measured correctly** or **it is a different species to all the other trees**.

c) Draw a line from 13 m on the vertical axis to the line of best fit and read off the value on the horizontal axis, which is **65 cm**.

d) Draw a line from 100 cm on the horizontal axis to the line of best fit and read off the value on the vertical axis, which is **20 m**.

2 a), b)

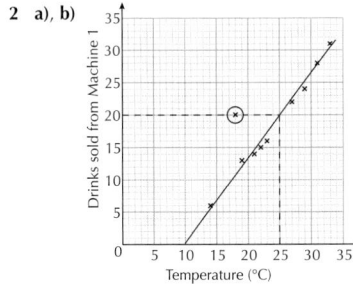

c) See the graph for Machine 1 in part a) for the outlier. E.g. **More people** may have **passed by the machine that day** because of a reason other than the temperature, this could have been a large group of tourists all wanting a drink.
There aren't any outliers for Machine 2.

d) Draw a line from 25°C to the line of best fit on both graphs and read off the value for each. The predicted number of drinks sold from Machine 1 is **20**. The predicted number of drinks sold from Machine 2 is **22**.
Your readings might be slightly different, depending on how you've drawn your line of best fit.

e) 3 °C is well below the temperature range for this data — you'd need to **extrapolate** to find your answer, which is **unreliable**. In this case, it would give you a negative number of drinks sold which is impossible.

f) **No** — just because the two things are correlated does not mean that one thing causes the other. They both could be influenced by a third variable.

3 a), c)

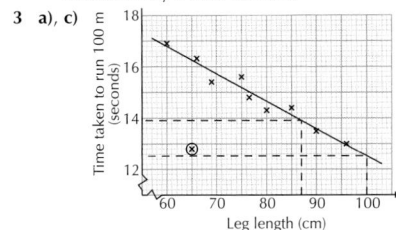

b) The points form a downward slope in a fairly straight line, so there is **strong negative correlation** between leg length and time taken to run 100 m.

d) Draw a line from 87 cm on the horizontal axis to the line of best fit and read off the value on the vertical axis, which is **13.9 s** (accept 13.6-14.2 s).

e) (i) Draw a line from 100 cm on the horizontal axis to the line of best fit and read off the value on the vertical axis, which is **12.5 s** (accept 12.2-12.8 s).

(ii) **No**, the answer is not reliable. 100 cm is outside the range of the data so you need to extrapolate to find the answer. This can lead to unreliable estimates.

33.8 Appropriate Representation of Data

Page 446   Exercise 1

1 E.g. **No** — a box plot is not a good diagram as it only shows the median, quartiles and range of temperatures throughout the day rather than the actual measurements. A **time series graph** would be a better diagram as it would show how the temperature of the fridge changed throughout the day.

2 E.g. **No** — pie charts show the proportions of discrete categories, but the grouped frequency table shows a grouped continuous variable. Also, the groups are of unequal size, which would be impossible to show using the sectors of a pie chart. A **histogram** would be a better way to show this data.

Page 447   Review Exercise

1 a) Count up the number of players with scores in each category in Round 1 — e.g. no player scored between 0-3 points in Round 1 so write 0. Repeat for Round 2.

Scores	0-3	4-7	8-11	12-15	16-20
Round 1	0	1	2	4	3
Round 2	1	2	4	2	1

b) Total frequency for Round 1 = 10
Frequency of ≥12 points = 4 + 3 = 7
So the percentage of people who scored 12 or more points = (7 ÷ 10) × 100 = **70%**

c) The data suggests that **Round 1 was easier** — e.g. 70% of people got a score of 12 or more, compared to only 30% in round 2.

2 a) Total frequency of girls:
8 + 6 + 5 + 2 + 3 + 6 = **30**

b) 9 boys chose chocolate and 6 girls chose chocolate, so 9 − 6 = **3** more boys than girls chose chocolate.

c) The highest bar for girls is vanilla, so **vanilla** is the modal flavour for girls.

d) E.g. 360° ÷ 30 girls = **12° per girl**

e) Total frequency of boys:
4 + 9 + 1 + 3 + 3 = 20
360° ÷ 20 = 18° per boy
Vanilla: 4 × 18° = 72°
Chocolate: 9 × 18° = 162°
Mint: 1 × 18° = 18°
Rocky road: 3 × 18 = 54°
Other = 3 × 18° = 54°

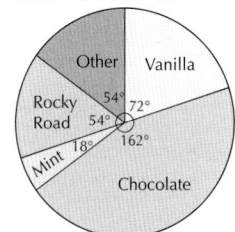

f) There are **fewer boys than girls** — so even though the number who chose rocky road is the same, the proportion of boys is larger, so the sector on the pie chart is larger.

3 a)

10 am		3 pm
	1	5 8
9 7	2	2 3 3 7
5	3	1 2 5
8	4	
9 8 4 1 1	5	
5 0	6	

Key: 7 | 2 for 10 am means 27
Key: 1 | 5 for 3 pm means 15

b) There are 11 people queuing at 10 am so the median is in the $12 \div 2 = 6$th position. That person has an age of **51**.
There are 9 people queuing at 3 pm so the median is in the $10 \div 2 = 5$th position. That person has an age of **23**.
So the median is **much lower for people queuing at 3 pm**.

4 a)

Height (h) in cm	Frequency
$150 \le h < 155$	4
$155 \le h < 160$	5
$160 \le h < 165$	9
$165 \le h < 170$	6
$170 \le h < 175$	3
$175 \le h < 180$	1

Your inequality signs might be different but there shouldn't be any overlaps between the classes.

b) Using the table from part a), the group with the highest frequency is $160 \le h < 165$, so this is the modal class.
The modal class could also be found by looking at the peak of the frequency polygon.

c) Total frequency = $4 + 5 + 9 + 6 + 3 + 1 = $ **28**

d) There are $9 + 6 + 3 + 1 = 19$ members that are 160 cm or taller, which is $(19 \div 28) \times 100 = $ **67.9%** (1 d.p.)

5 a)

Height (h) in cm	Freq. Density	Freq.
$155 \le h < 165$	0.3	$10 \times 0.3 = $ **3**
$165 \le h < 170$	0.8	$5 \times 0.8 = $ **4**
$170 \le h < 175$	2.0	$5 \times 2 = $ **10**
$175 \le h < 180$	2.4	$5 \times 2.4 = $ **12**
$180 \le h < 185$	0.6	$5 \times 0.6 = $ **3**

b) 172 is $\frac{2}{5}$ of the way through the $170 \le h < 175$ class, which has a frequency of 10. So there are $\frac{2}{5} \times 10 = 4$ players between 170 cm and 172 cm and $10 - 4 = 6$ players between 172 cm and 175 cm.
There are 12 players between 175 cm and 180 cm.
There are 3 players between 180 cm and 185 cm.
$6 + 12 + 3 = $ **21 players**

6 a) Position of median = $100 \div 2 = 50$
Draw a line from a cumulative frequency of 50 to the curve to get the median **7.2 km** (accept 7.1 km-7.3 km)

b) Position of lower quartile = $100 \div 4 = 25$
Position of upper quartile = $3(100 \div 4) = 75$
Draw lines from cumulative frequencies of 25 and 75 to the curve to get the lower quartile 4.6 km (accept 4.5 km-4.7 km) and upper quartile 9.6 km (accept 9.5 km-9.7 km).
Interquartile range = $9.6 - 4.6 = $ **5 km** (accept answers based on your quartiles)

c) Draw a line from 4 km on the horizontal axis and read off the cumulative frequency 21. So $100 - 21 = $ **79 children** get the school bus.

d) Draw a vertical line at the median and then build the box around it — it should go as far as Q_1 on the left and Q_3 on the right. Then add lines to the left and right that reach the minimum value and maximum value respectively.

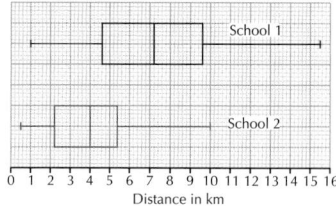
School 1 / School 2 box plots
Distance in km

e) E.g. The maximum distance a child travels is **greater for students at school 1** than school 2. The range and interquartile range are larger for school 1 than school 2, so there is a **greater spread of distances** travelled by children attending school 1 than school 2. The median distance travelled to school 1 is also greater than the median distance travelled to school 2, so on average **children travel further to get to school 1** than to school 2. The shortest distance travelled by a child to school is **shorter for school 2** than school 1.

7 a), b)

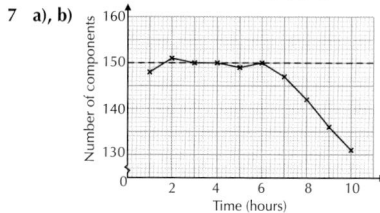
Number of components vs Time (hours)

c) E.g. **I agree** — the number of components produced has dropped significantly from the target amount over the last 4 hours shown by the data.

8 a)

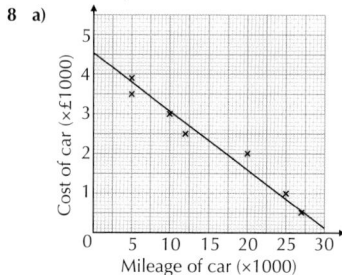
Cost of car (×£1000) vs Mileage of car (×1000)

b) Draw a line from 15 on the horizontal axis to the line of best fit and read off the value on the vertical axis, which is 2.3. So the car will cost **£2300** (accept £2100-£2500).

Page 449   Exam-Style Questions

1 a) Year 11 marks show a **strong positive correlation** with the number of revision classes. Year 10 marks show a **weak positive correlation** with the number of revision classes.
[1 mark for the correct observations]

b) **No** — to predict what a year 11 student would get if they didn't attend any revision classes requires extrapolation and this is unreliable.
[1 mark for a correct explanation of why Duane is incorrect]

2 Range = 32 bpm
Highest value = $44 + 32 = 76$ bpm
Upper quartile = 71 bpm
Lower quartile = $71 - 19 = 52$ bpm
Median = $(71 + 52) \div 2 = 61.5$ bpm

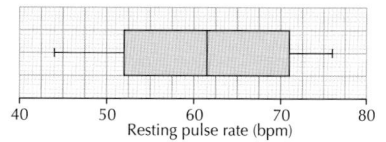
Resting pulse rate (bpm) box plot

[3 marks available — 1 mark for the highest value and lower quartile, 1 mark for the median, 1 mark for the correctly drawn box plot]

3 a) The price of gold per gram showed an **upward trend** — there was a general **increase over time**.
[1 mark for the correct observation]

b) The highest price per gram was £29.80 in November. 12 kg = $12 \times 1000 = 12\ 000$ g. So Midas could have sold his gold bar for $29.8 \times 12\ 000 = $ **£357 600**
[2 marks available — 1 mark for finding the highest price per gram, 1 mark for the correct answer]

4

Time (t) in seconds	Frequency	Frequency density
$0 \le t < 4$	18	$18 \div 4 = 4.5$
$4 \le t < 8$	12	$12 \div 4 = 3$
$8 \le t < 12$	6	$6 \div 4 = 1.5$
$12 \le t < 20$	4	$4 \div 8 = 0.5$
$20 \le t < 30$	1	$1 \div 10 = 0.1$

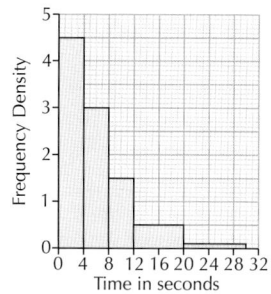
Frequency Density vs Time in seconds

[3 marks available — 1 mark for working out the frequency densities, 1 mark for both a continuous horizontal axis scale and frequency density on the vertical axis, 1 mark for correctly plotted bars]

5 a)

Score	Frequency
$0 < s \le 4$	1
$4 < s \le 8$	$4 - 1 = 3$
$8 < s \le 12$	$15 - 4 = 11$
$12 < s \le 16$	$21 - 15 = 6$
$16 < s \le 20$	$25 - 21 = 4$

[2 marks available — 2 mark for all entries correct, otherwise 1 mark for 2-3 entries correct or correct working shown for each but error in reading from graph]

b) Draw a line from 14 on the horizontal axis to the curve to get a cumulative frequency of 18. So the number of bands scoring more than 14 was $25 - 18 = 7$ bands.
$\frac{7}{25} = \frac{28}{100} = $ **28%**
[2 marks available — 1 mark for the number of bands scoring above 14, 1 mark for the correct percentage]
You could also have used the frequency table, taking all of group $16 < s \le 20$ and half of group $12 < s \le 16$, which is $4 + 3 = 7$. The percentage is then worked out in the same way.

6 Using a protractor, the sectors on the 11R pie chart measure 75° for gold, 120° for silver and 165° for bronze.
1 medal = $360° \div 24 = 15°$, so $75° \div 15°$ = 5 medals won by 11R were gold, which means 5 medals won by 11S were gold.

$\frac{120°}{360°} = \frac{1}{3}$ of the medals won by 11R
were silver so $\frac{1}{3} = \frac{6}{18} = 6$ medals won by 11S
were silver. This means that $18 - 5 - 6 = 7$
medals won by 11S were bronze.
11S won 18 medals, so
1 medal $= 360° \div 18 = 20°$, so the bronze
sector is $7 \times 20° = \mathbf{140°}$
*[4 marks available — 1 mark for the correct
number of gold medals won by 11R/11S,
1 mark for the correct proportion of silver
medals won by 11R/11S, 1 mark for the
correct number of bronze medals won by 11S,
1 mark for the correct angle]*

Section 34 — Probability

34.1 Calculating Probabilities

Page 452   Exercise 1

1 a) 6 possible outcomes,
      1 way of getting a 6, so P(6) $= \frac{1}{6}$

   b) No way of getting a 7, so P(7) $= \mathbf{0}$

   c) (4 or 5) has two ways of happening,
      so P(4 or 5) $= \frac{2}{6} = \frac{1}{3}$

   d) 4 of the numbers are factors of 6
      (1, 2, 3 and 6), so P(factor of 6) $= \frac{4}{6} = \frac{2}{3}$

2 a) 52 possible outcomes, 13 are clubs,
      so P(club) $= \frac{13}{52} = \frac{1}{4}$

   b) There are four aces, so P(ace) $= \frac{4}{52} = \frac{1}{13}$

   c) 26 of the cards are red, so P(red) $= \frac{26}{52} = \frac{1}{2}$

   d) There is only 1 two of hearts,
      so P(two of hearts) $= \frac{1}{52}$

   e) There are 13 spades, so $52 - 13 = 39$
      are not spades. So P(not spade) $= \frac{39}{52} = \frac{3}{4}$

   f) There are four 4s and four 5s, so a total of 8.
      So P(4 or 5) $= \frac{8}{52} = \frac{2}{13}$

3 a) 16 possible outcomes, 2 are green,
      so P(green) $= \frac{2}{16} = \frac{1}{8}$

   b) There are 4 blue and 2 green, so a total
      of 6. So P(blue or green) $= \frac{6}{16} = \frac{3}{8}$

   c) There is 1 purple, so $16 - 1 = 15$ not purple.
      So P(not purple) $= \frac{15}{16}$

   d) There are 3 red, 2 green
      and 1 brown, so a total of 6.
      So P(red, green or brown) $= \frac{6}{16} = \frac{3}{8}$

4 a) 8 possible outcomes, P(2) $= \frac{3}{8}$, so label
      exactly 3 sections with '2'.

   b) P(3) $= \frac{1}{2} = \frac{4}{8}$, so label **exactly 4 sections**
      with '3'.

   c) P(5) = P(6) $= \frac{1}{4} = \frac{2}{8}$,
      so label **exactly 2 sections** with '5'
      and **exactly 2 sections** with '6'.

5 After the first sock is picked, there are 39 socks
   left, and only one that matches the first sock,
   so P(match) $= \frac{1}{39}$.

6 a) There are $8 + 6 + 4 = 18$ possible outcomes.
      There are 4 pralines, so 2 white chocolate
      ones. So P(white choc praline) $= \frac{2}{18} = \frac{1}{9}$

   b) She only likes caramels or truffles, of which
      there are $8 + 6 = 14$. Half of those are
      white chocolate, so there are 7 chocolates
      she likes. So P(choc she likes) $= \frac{7}{18}$

7 a) Total $= 14 + 2 + 6 + 8 = 30$ books,
      so P(paperback fiction) $= \frac{14}{30} = \frac{7}{15}$

   b) There are $2 + 8 = 10$ non-fiction books.
      So P(non-fiction) $= \frac{10}{30} = \frac{1}{3}$

Page 453   Exercise 2

1 P(no snow) $= 1 - $ P(snow) $= 1 - \frac{5}{8} = \frac{3}{8}$

2 P(no prize) $= 100\% - 25\% = \mathbf{75\%}$
   *The question gives the probability as a percentage,
   so use 100% here instead of 1.*

3 P(finishes) $= 1 - 0.74 = \mathbf{0.26}$

4 P(win) $= 2 \times $ P(lose) and P(win) + P(lose) $= 1$
   $\Rightarrow$ 2P(lose) + P(lose) $= 1 \Rightarrow$ 3P(lose) $= 1$
   $\Rightarrow$ P(lose) $= \frac{1}{3} \Rightarrow$ P(win) $= 1 - \frac{1}{3} = \frac{2}{3}$

5 a) $0.4 + 0.1 + $ P(cuddly toy) $+ 0.2 = 1$
      $\Rightarrow 0.7 + $ P(cuddly toy) $= 1$
      $\Rightarrow$ P(cuddly toy) $= 1 - 0.7 = \mathbf{0.3}$

   b) P(not pen) $= 1 - $ P(pen) $= 1 - 0.1 = \mathbf{0.9}$
      *You could also add up the probabilities for the
      prizes that are not pens: 0.4 + 0.3 + 0.2 = 0.9.*

6 $0.4 + 0.15 + $ P(draw) $= 1$
   $\Rightarrow$ P(draw) $= 1 - 0.55 = \mathbf{0.45}$

7 P(red) $= 1 - (0.5 + 0.4) = 0.1$
   P(red) $= \dfrac{\text{number of red counters}}{\text{number of counters}}$
   $\Rightarrow 0.1 = \dfrac{4}{\text{number of counters}}$
   $\Rightarrow$ number of counters $= \frac{4}{0.1} = \mathbf{40}$

8 P(blue) $= \frac{1}{2}$P(green) and
   P(pink) + P(blue) + P(green) $= 1$
   $\Rightarrow 0.1 + \frac{1}{2}$P(green) + P(green) $= 1$
   $\Rightarrow \frac{3}{2}$P(green) $= 0.9$
   $\Rightarrow$ P(green) $= \frac{2}{3} \times 0.9 = \mathbf{0.6}$

34.2 Listing Outcomes

Page 454   Exercise 1

1

Burger	Drink
Hamburger	Cola
Hamburger	Lemonade
Hamburger	Coffee
Cheeseburger	Cola
Cheeseburger	Lemonade
Cheeseburger	Coffee
Veggie burger	Cola
Veggie burger	Lemonade
Veggie burger	Coffee

   a) 9 possible outcomes, only one (veggie
      and cola), so P(veggie and cola) $= \frac{1}{9}$

   b) There are five outcomes that feature either
      cheeseburger or coffee (or both),
      so P(cheeseburger or coffee) $= \frac{5}{9}$

2

First	Second
1	1
1	2
1	3
2	1
2	2
2	3
3	1
3	2
3	3

   a) 9 possible outcomes, only one (1, 1),
      so P(1, 1) $= \frac{1}{9}$

   b) There are four outcomes where both results
      are less than 3, so P(both less than 3) $= \frac{4}{9}$

3

First	Second	Third
H	H	H
H	H	T
H	T	H
H	T	T
T	H	H
T	H	T
T	T	H
T	T	T

   a) 8 possible outcomes, only one where
      all three are tails, so P(3T) $= \frac{1}{8}$

   b) Only one outcome where none are tails,
      so P(no tails) $= \frac{1}{8}$

   c) There are 3 outcomes with one head
      and two tails, so P(1H, 2T) $= \frac{3}{8}$

Page 455   Exercise 2

1 a)

	1	2	3	4	5	6
1	2	3	4	5	6	7
2	3	4	5	6	7	8
3	4	5	6	7	8	9
4	5	6	7	8	9	10
5	6	7	8	9	10	11
6	7	8	9	10	11	12

   b) (i) 36 possible outcomes, five 6s,
           so P(6) $= \frac{5}{36}$

      (ii) Only one outcome scores 12,
           so P(12) $= \frac{1}{36}$

      (iii) 21 outcomes score less than 8,
           so P(less than 8) $= \frac{21}{36} = \frac{7}{12}$

      (iv) 10 outcomes score more than 8,
           so P(more than 8) $= \frac{10}{36} = \frac{5}{18}$

2 a) E.g.

      B: boiled rice, L: lemon rice,
      P: pilau rice, V: vegetable rice
      *You could use a different type of diagram as long
      as it clearly shows the 16 possible outcomes.*

   b) (i) 16 possible outcomes, only one for
           (P, P), so P(pilau both times) $= \frac{1}{16}$

      (ii) There are four outcomes where
           the same rice is picked both times,
           so P(same both times) $= \frac{4}{16} = \frac{1}{4}$

      (iii) There are seven outcomes that
           have lemon rice at least once,
           so P(lemon at least once) $= \frac{7}{16}$

3

		Hayley's Score				
		1	2	3	4	5
Asha's Score	1	0	H1	H1	H1	H2
	2	A1	0	H1	H1	H2
	3	A1	A1	0	H1	H2
	4	A1	A1	A1	0	H2
	5	A2	A2	A2	A2	0

H1: Hayley gets 1 point, H2: Hayley gets
2 points, A1: Asha gets 1 point, A2: Asha
gets 2 points, 0: no one gets any points

   a) There are 25 possible outcomes, six where
      Hayley scores 1 point, so P(H1) $= \frac{6}{25}$

b) There are 4 outcomes where Hayley scores 2 points, so P(H2) = $\frac{4}{25}$

Page 456 Exercise 3

1 a) $4 \times 8 \times 3 = \textbf{96}$

b) There are 8 options for the main course, and $4 + 3 = 7$ options for the other course.
$8 \times 7 = \textbf{56}$
You could also work out the number of combinations separately — if you have a starter and a main, there are $4 \times 8 = 32$ combinations, and if you have a main and a dessert, there are $8 \times 3 = 24$ combinations, so there are $32 + 24 = 56$ combinations altogether.

2 a) There are 7 possible digits (0, 1, 2, 3, 4, 5 and 6) and five wheels, so:
$7 \times 7 \times 7 \times 7 \times 7 = \textbf{16 807}$ combinations

b) There are three possible odd digits (1, 3 and 5). $3 \times 3 \times 3 \times 3 \times 3 = \textbf{243}$

c) There are 16 807 possible outcomes, 243 where all digits are odd.
So P(all odd) = $\frac{243}{16\ 807}$

3 She is **not correct**. E.g. $6 \times 6 \times 6 \times 6 = 1296$, but if the different flavours of fruit juice are A, B, C, D, E, F then the combination A, A, A, B is the same as the combination A, A, B, A. So there will be fewer than 1296 different combinations.

34.3 Probability from Experiments

Page 457 Exercise 1

1 a) P(Red) = $49 \div 100 = \textbf{0.49}$
P(Green) = $34 \div 100 = \textbf{0.34}$
P(Yellow) = $8 \div 100 = \textbf{0.08}$
P(Blue) = $9 \div 100 = \textbf{0.09}$

b) The estimates would be more accurate if the spinner is spun a **greater number of times**.

2 a) $13 \div 50 = \textbf{0.26}$

b) $18 \div 100 = \textbf{0.18}$

c) **Jason's** estimate should be more accurate as he repeated the experiment more times.

3 Total = $452 + 124 + 237 + 98 + 89 = 1000$

a) P(silver) = $452 \div 1000 = \textbf{0.452}$

b) P(red) = $237 \div 1000 = \textbf{0.237}$

c) P(not silver, black, red or blue)
= P(other) = $89 \div 1000 = \textbf{0.089}$

4 a) P(Jack wins) = $\frac{8}{15}$

b) P(dad wins) = $1 - \frac{8}{15} = \frac{7}{15}$

5 E.g. Lilia could examine the records of her team's recent matches, count the number of wins and divide it by the total number of matches in those records.

Page 458 Exercise 2

1 a) $20 \times 0.75 = \textbf{15}$

b) $60 \times 0.75 = \textbf{45}$

c) $100 \times 0.75 = \textbf{75}$

d) $1000 \times 0.75 = \textbf{750}$

2 The spinner has 3 equally likely outcomes, so the probability of each is $\frac{1}{3}$.

a) $60 \times \frac{1}{3} = \textbf{20}$

b) $300 \times \frac{1}{3} = \textbf{100}$

c) $480 \times \frac{1}{3} = \textbf{160}$

3 a) P(5) = $\frac{1}{6}$, $120 \times \frac{1}{6} = \textbf{20}$

b) P(6) = $\frac{1}{6}$, $120 \times \frac{1}{6} = \textbf{20}$

c) P(even) = P(2, 4 or 6) = $\frac{3}{6} = \frac{1}{2}$,
$120 \times \frac{1}{2} = \textbf{60}$

d) P(higher than 1) = P(2, 3, 4, 5 or 6) = $\frac{5}{6}$,
$120 \times \frac{5}{6} = \textbf{100}$

Page 459 Exercise 3

1 a) Draw the frequency tree and fill in the ends of the branches with the data given in the table, then work backwards to the total.
Brown hair = $220 + 335 = 555$
Not brown hair = $105 + 140 = 245$
Total = $555 + 245 = 800$

E.g.   Brown hair?   Brown eyes?

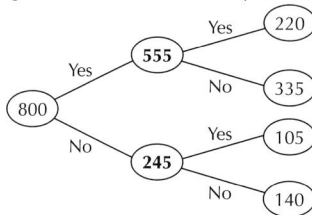

Hair colour and eye colour could be put the other way around in the diagram.

b) P(brown hair and brown eyes)
= $220 \div 800 = \textbf{0.275}$

c) P(not brown hair and not brown eyes)
= $140 \div 800 = 0.175$
Expected frequency = $2000 \times 0.175 = \textbf{350}$

2 Draw the frequency tree, then fill the data given in the question into the correct places, as below. You can then use the information given to fill out the rest of the values.

E.g.   Said their vision   Passed test?
         was fine?

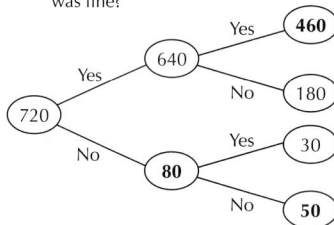

Page 460 Exercise 4

1 a) Divide the frequency of each colour by the total (100) to find the relative frequency.
Blue = $22 \div 100 = \textbf{0.22}$
Green = $21 \div 100 = \textbf{0.21}$
White = $18 \div 100 = \textbf{0.18}$
Pink = $39 \div 100 = \textbf{0.39}$

b) If the spinner was fair, you would expect the relative frequencies of all the colours to be about the same (roughly 0.25). The relative frequency of pink is around twice the size of the others. This suggests that the spinner is **biased**.

2 a) P(4) = $\frac{1}{6}$, $120 \times \frac{1}{6} = \textbf{20}$

b) 4 came up many more times than you would expect on a fair dice, which suggests the dice may be **biased**.

3 a) The relative frequency for Amy is $12 \div 20 = 0.6$. Using the same method, the completed table is:

	Amy	Steve	Hal
No. of tosses	20	60	100
No. of heads	12	33	49
Relative freq.	**0.6**	**0.55**	**0.49**

b) **Hal's** results are the most reliable as he repeated the experiment the most times.

c) The results suggest the coin is **fair**, as the relative frequency from Hal's results was very close to 0.5, the result you would expect for a fair coin.
When discussing if something is fair or biased, you can sometimes argue either way, as long as your answer is sensible and makes sense.

1 a) P(girl and geography) = $\frac{56}{280} = \frac{1}{5}$

b) P(boy and geography) = $\frac{84}{280} = \frac{3}{10}$

c) P(girl) = $\frac{77 + 56}{280} = \frac{133}{280} = \frac{19}{40}$

d) P(boy) = $\frac{63 + 84}{280} = \frac{147}{280} = \frac{21}{40}$
You could have subtracted your answer to c) from 1 as 'boy' = 'not girl'.

e) P(history) = $\frac{77 + 63}{280} = \frac{140}{280} = \frac{1}{2}$

f) P(not history) = $1 - $ P(history) = $1 - \frac{1}{2} = \frac{1}{2}$

2 a) $0.45 + 0.15 + 0.1 + $ P(compilations) = 1
$\Rightarrow$ P(compilations) = $1 - 0.7 = \textbf{0.3}$

b) P(not MV) = $1 - $ P(MV) = $1 - 0.15 = \textbf{0.85}$

c) $\frac{n(FV)}{80} = 0.1 \Rightarrow$ n(FV) = $80 \times 0.1 = \textbf{8}$

3 The 1st digit can be 1, 3, 5, 7 or 9.
The 2nd digit can be 3, 6 or 9.
The 3rd digit can be any digit.
The 4th digit can be 0, 2, 4, 6 or 8.
So total number of possible numbers is
$5 \times 3 \times 10 \times 5 = \textbf{750}$.

4 a) P(not faulty) = $1 - 0.002 = \textbf{0.998}$

b) (i) $60\ 000 \times 0.002 = \textbf{120}$

(ii) $60\ 000 \times 0.998 = \textbf{59 880}$
You could also work this out by doing $60\ 000 - 120 = 59\ 880$.

5 a) E.g.

Another type of diagram would be fine, as long as it clearly shows all 20 possible outcomes.

b) (i) 20 possible outcomes, only one for (4, 4), so P(double 4) = $\frac{1}{20}$

(ii) Two outcomes, (2, 3) and (3, 2), so P(2 and 3) = $\frac{2}{20} = \frac{1}{10}$

(iii) Four outcomes, (1, 1), (2, 2), (3, 3) and (4, 4), so P(both the same) = $\frac{4}{20} = \frac{1}{5}$

(iv) P(different) = $1 - $ P(same) = $1 - \frac{1}{5} = \frac{4}{5}$

(v) Four outcomes, (1, 5), (2, 5), (3, 5) and (4, 5), so P(at least one 5) = $\frac{4}{20} = \frac{1}{5}$

(vi) P(no fives) = $1 - $ P(at least one five)
= $1 - \frac{1}{5} = \frac{4}{5}$

c) P(3) = $\frac{1}{5}$,
so expected frequency = $100 \times \frac{1}{5} = \textbf{20}$

6 a) Divide each frequency by the total (200) to find the relative frequency of each score.

Score	1	2	3	4
Frequency	56	34	54	56
Relative frequency	0.28	0.17	0.27	0.28

b) If the dice was fair, you would expect all the relative frequencies to be about the same. The relative frequency of 2 is much lower than the others, which suggests the dice is **biased**.
You could also argue that the dice is fair (e.g. because 0.17 isn't that far from 0.25) — as long as your answer is sensible and makes sense, you would get the marks.

c) P(1 or 2) = $0.28 + 0.17 = \textbf{0.45}$

1 a) Boys = 5 + 2 = 7
Girls = 16 − 7 = 9
Girls who said yes = $\frac{2}{3} \times 9 = 6$

Girls who said no = $\frac{1}{3} \times 9 = 3$

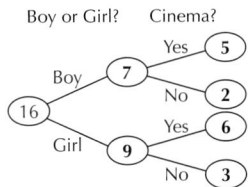

Boy or Girl?   Cinema?

[3 marks available in total — 1 mark for correct number of boys and girls, 1 mark for correct numbers of yeses and nos for the boys, 1 mark for correct numbers of yeses and nos for the girls]

b) Total number who went to the cinema
= 5 + 6 = 11 *[1 mark]*
So P(went to the cinema) = $\frac{11}{16}$ *[1 mark]*

2 a) 4 letters, each with 6 different possibilities:
Number of sequences
= 6 × 6 × 6 × 6 *[1 mark]*
= **1296 sequences** *[1 mark]*

b) 'BEAD' is one of the 1296 possibilities,
so P('BEAD') = $\frac{1}{1296}$ *[1 mark]*

c) Count the number of times each letter appears to find the frequency. To find the relative frequency divide each frequency by the total number of letters (10 × 4 = 40).

Letter	A	B	C	D	E	F
Frequency	16	2	7	5	6	4
Relative frequency	0.4	0.05	0.175	0.125	0.15	0.1

[3 marks available — 1 mark for the correct frequencies, 1 mark for the correct method to find relative frequencies, 1 mark for the correct relative frequencies]

d) E.g. The dice is **biased**, because the relative frequency of A is much higher than the relative frequency of the other letters (especially B), suggesting that each letter is not equally likely.
[2 marks available — 1 mark for linking frequency data from part c) to probabilities, 1 mark for sensible conclusion based on the data]

e) Use the relative frequencies to estimate the probability of rolling A (0.4) and B (0.05).
Expected frequency of A = 200 × 0.4 = 80
Expected frequency of B = 200 × 0.05 = 10
So you would estimate that it would land on A 80 − 10 = **70 more times** than it would land on B.
[2 marks available — 1 mark for the correct method to work out expected frequency, 1 mark for the correct answer]

Section 35 — Probability for Combined Events

35.1 The AND Rule for Independent Events

Page 464   Exercise 1

1 a) The two coin tosses do not affect each other, so the events are **independent**.
b) Taking the first chocolate will change the numbers of remaining chocolates, so the events are **not independent**.
c) Rolling the dice and drawing the card do not affect each other, so the events are **independent**.

---

2 a) P(heads ∩ 6) = $\frac{1}{2} \times \frac{1}{6} = \frac{1}{12}$
b) P(heads ∩ odd) = P(heads ∩ (1, 3 or 5))
= $\frac{1}{2} \times \frac{3}{6} = \frac{1}{4}$
c) P(tails ∩ square) = P(tails ∩ (1 or 4))
= $\frac{1}{2} \times \frac{2}{6} = \frac{1}{6}$
d) P(tails ∩ prime) = P(tails ∩ (2, 3 or 5))
= $\frac{1}{2} \times \frac{3}{6} = \frac{1}{4}$
e) P(heads ∩ multiple of 3)
= P(heads ∩ (3 or 6)) = $\frac{1}{2} \times \frac{2}{6} = \frac{1}{6}$
f) P(tails ∩ factor of 5) = P(tails ∩ (1 or 5))
= $\frac{1}{2} \times \frac{2}{6} = \frac{1}{6}$

3 a) P(right-handed) = 100% − 10% = **90%**
b) P(no glasses) = 100% − 15% = **85%**
c) P(glasses and left-handed) = 15% × 10%
= **1.5%**
Be careful here — 10% × 15% ≠ 150% (think about the decimal equivalents if you're not sure why — 0.1 × 0.15 = 0.015 = 1.5%).
d) P(no glasses and right-handed)
= 85% × 90% = **76.5%**
It's not 100% − 1.5% as there are also pupils that do wear glasses and are right-handed (13.5%) and pupils that do not wear glasses and are left-handed (8.5%).

4 a) P(red, red) = $\frac{5}{10} \times \frac{5}{10} = \frac{25}{100} = \frac{1}{4}$
b) P(not red, not red) = $\frac{5}{10} \times \frac{5}{10} = \frac{25}{100} = \frac{1}{4}$
c) P(red, blue) = $\frac{5}{10} \times \frac{3}{10} = \frac{15}{100} = \frac{3}{20}$

5 a) P(English and Maths) = 0.6 × 0.8 = **0.48**
b) P(Maths and Geography) = 0.8 × 0.3 = **0.24**
c) P(Maths, not Science) = 0.8 × (1 − 0.4)
= 0.8 × 0.6 = **0.48**

6 Len is **wrong**. When one ace has been removed, there are 3 aces remaining out of 51 cards, so the probability is $\frac{4}{52} \times \frac{3}{51} = \frac{1}{221}$.

35.2 The OR Rule

Page 465   Exercise 1

1 a) The spinner cannot land on both 6 and 3, so the events are **mutually exclusive**.
b) 2 is a factor of 6, so the events are **not mutually exclusive**.
c) There are no numbers on the spinner that are both less than 4 and greater than 3, so the events are **mutually exclusive**.

2 a) P(pink ∪ orange) = 0.5 + 0.1 = **0.6**
b) P(pink ∪ red) = 0.5 + 0.4 = **0.9**
c) P(red ∪ orange) = 0.4 + 0.1 = **0.5**

3 a) P(red or gold) = 0.14 + 0.4 = **0.54**
b) P(silver or red) = 0.26 + 0.14 = **0.4**
c) P(gold or blue) = 0.4 + 0.2 = **0.6**
d) P(silver or gold) = 0.26 + 0.4 = **0.66**

4 a) P(Eagle) = $\frac{1}{3}$
b) P(Eagle or Falcon) = $\frac{1}{3} + \frac{1}{3} = \frac{2}{3}$
c) P(Falcon or Osprey) = $\frac{1}{3} + \frac{1}{3} = \frac{2}{3}$

5 **Disagree** — having glasses and having blonde hair are not mutually exclusive (some people could have both) so you can't add the probabilities unless you know that none of the pupils both wear glasses and have blonde hair.

6 a) P(pickle or mayo) = 0.22 + 0.28 = **0.5**
b) The probabilities add up to more than 1 because some people chose more than one sandwich filling. Only mutually exclusive events sum to 1.

Page 467   Exercise 2

1 a) P(multiple of 3)
= P(3, 6, 9, 12, 15 or 18) = $\frac{6}{20}$
P(odd)
= P(1, 3, 5, 7, 9, 11, 13, 15, 17 or 19)
= $\frac{10}{20}$
P(multiple of 3 and odd) = P(3, 9 or 15)
= $\frac{3}{20}$
So P(multiple of 3 or odd)
= $\frac{6}{20} + \frac{10}{20} - \frac{3}{20} = \frac{13}{20}$

b) P(factor of 20) = P(1, 2, 4, 5, 10 or 20)
= $\frac{6}{20}$
P(multiple of 4) = P(4, 8, 12, 16 or 20)
= $\frac{5}{20}$
P(factor of 20 and multiple of 4)
= P(4 or 20) = $\frac{2}{20}$
So P(factor of 20 or multiple of 4)
= $\frac{6}{20} + \frac{5}{20} - \frac{2}{20} = \frac{9}{20}$

2 a) P(red suit) = P(heart or diamond) = $\frac{26}{52}$
P(queen) = P(Q♠, Q♥, Q♣ or Q♦) = $\frac{4}{52}$
P(red suit and queen) = P(Q♥ or Q♦) = $\frac{2}{52}$
So P(red suit or queen)
= $\frac{26}{52} + \frac{4}{52} - \frac{2}{52} = \frac{28}{52} = \frac{7}{13}$
b) P(club) = $\frac{13}{52}$
P(picture card) = P(J, Q or K) = $\frac{12}{52}$
P(club and picture card)
= P(J♣, Q♣ or K♣) = $\frac{3}{52}$
So P(club or picture card)
= $\frac{13}{52} + \frac{12}{52} - \frac{3}{52} = \frac{22}{52} = \frac{11}{26}$

3 Work out the totals:

	Boys	Girls	Total
Year 9	240	310	550
Year 10	305	287	592
Year 11	212	146	358
Total	757	743	1500

a) P(Y11 or G) = P(Y11) + P(G) − P(Y11 and G)
= $\frac{358}{1500} + \frac{743}{1500} - \frac{146}{1500} = \frac{955}{1500} = \frac{191}{300}$
b) P(B or not Y9)
= P(B) + P(not Y9) − P(B and not Y9)
= $\frac{757}{1500} + \frac{950}{1500} - \frac{517}{1500} = \frac{1190}{1500} = \frac{119}{150}$
You could also find the probabilities from the table by adding up all the required groups — e.g. n(Y11 or G) = 212 + 146 + 310 + 287 = 955.

4 P(A ∪ B) = P(A) + P(B) − P(A ∩ B)
= $\frac{n(A)}{n(\xi)} + \frac{n(B)}{n(\xi)} - \frac{n(A \cap B)}{n(\xi)}$
= $\frac{28}{100} + \frac{34}{100} - \frac{12}{100} = \frac{50}{100} = \frac{1}{2}$

5 a) P(girl) = $\frac{160}{300}$
P(≤ 70%) = $\frac{62 + 52}{300} = \frac{114}{300}$
P(girl or ≤ 70%) = $\frac{160}{300} + \frac{114}{300} - \frac{62}{300}$
= $\frac{212}{300} = \frac{53}{75}$
b) P(boy) = $\frac{140}{300}$
P(> 70%) = $\frac{98 + 88}{300} = \frac{186}{300}$
P(boy or > 70%) = $\frac{140}{300} + \frac{186}{300} - \frac{88}{300}$
= $\frac{238}{300} = \frac{119}{150}$
Like for Q3, you could also answer these parts by just adding up the right numbers from the frequency tree.
c) P((boy and ≤ 70%) or (girl and > 70%))
= P(boy and ≤ 70%) + P(girl and > 70%)
= $\frac{52}{300} + \frac{98}{300} = \frac{150}{300} = \frac{1}{2}$

6 $P(A \cup B \cup C)$ is the total white area in the diagram. C does not overlap with A or B (as they are mutually exclusive), so this is just $P(A \cup B) + P(C)$.

Using the general OR rule:
$P(A \cup B) = P(A) + P(B) - P(A \cap B)$

A and B are independent,
so $P(A \cap B) = P(A) \times P(B)$

So $P(A \cup B \cup C) = \mathbf{P(A) + P(B) - P(A)P(B) + P(C)}$

35.3 Using the AND/OR Rules

Page 469   Exercise 1

1 a) $P(H, H) = 0.5 \times 0.5 = \mathbf{0.25}$

b) $P(T, T) = 0.5 \times 0.5 = \mathbf{0.25}$

c) $P(\text{both the same}) = P((H, H) \text{ or } (T, T))$
$= P(H, H) + P(T, T) = 0.25 + 0.25 = \mathbf{0.5}$

2 a) $P(\text{white shirt and black trousers})$
$= 0.8 \times 0.55 = \mathbf{0.44}$

b) $P(\text{black shirt and black trousers})$
$= (1 - 0.8) \times 0.55 = 0.2 \times 0.55 = \mathbf{0.11}$

c) $P(\text{black shirt and grey trousers})$
$= (1 - 0.8) \times (1 - 0.55) = 0.2 \times 0.45 = \mathbf{0.09}$

d) $P(\text{black shirt or trousers, but not both})$
$= P(\text{black shirt and grey trousers})$
$\quad + P(\text{white shirt and black trousers})$
$= 0.09 + 0.44 = \mathbf{0.53}$

3 a) $P(A, A) = 0.5 \times 0.5 = \mathbf{0.25}$

b) $P(B, B) = 0.15 \times 0.15 = \mathbf{0.0225}$

c) $P(\text{not C, not C}) = (1 - 0.05) \times (1 - 0.05)$
$= 0.95 \times 0.95 = \mathbf{0.9025}$

d) $P(C, \text{not C}) = 0.05 \times 0.95 = \mathbf{0.0475}$

e) $P(\text{not C, C}) = 0.95 \times 0.05 = \mathbf{0.0475}$

f) $P(\text{exactly one C}) = P(C, \text{not C}) + P(\text{not C, C})$
$= 0.0475 + 0.0475 = \mathbf{0.095}$

4 Work out the totals:

	Car	Bus	Walk	Cycle	Total
Girls	5	6	3	1	15
Boys	2	7	2	4	15

a) $P(\text{both bus}) = P(\text{girl travels by bus})$
$\times P(\text{boy travels by bus}) = \frac{6}{15} \times \frac{7}{15}$
$= \frac{42}{225} = \mathbf{\frac{14}{75}}$

b) $P(\text{girl walks and boy travels by car})$
$= P(\text{girl walks}) \times P(\text{boy travels by car})$
$= \frac{3}{15} \times \frac{2}{15} = \frac{6}{225} = \mathbf{\frac{2}{75}}$

c) $P(\text{exactly one cycles})$
$= P(\text{girl cycles and boy doesn't})$
$\quad + P(\text{boy cycles and girl doesn't})$
$= \left(\frac{1}{15} \times \frac{(2+7+2)}{15}\right) + \left(\frac{4}{15} \times \frac{(5+6+3)}{15}\right)$
$= \left(\frac{1}{15} \times \frac{11}{15}\right) + \left(\frac{4}{15} \times \frac{14}{15}\right)$
$= \frac{11}{225} + \frac{56}{225} = \mathbf{\frac{67}{225}}$

d) $P(\text{at least one cycles})$
$= P(\text{exactly one cycles}) + P(\text{both cycle})$
$= \frac{67}{225} + \left(\frac{1}{15} \times \frac{4}{15}\right) = \frac{67}{225} + \frac{4}{225} = \mathbf{\frac{71}{225}}$
You could also do
$P(\text{at least one cycles}) = 1 - P(\text{neither cycles})$.

35.4 Tree Diagrams

Page 470   Exercise 1

1 a) $P(R) = \frac{3}{5} = 0.6$, $P(B) = \frac{2}{5} = 0.4$

1st   2nd

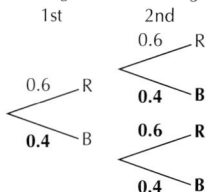

b) $P(R) = \frac{5}{10} = 0.5$, $P(B) = \frac{3}{10} = 0.3$,
$P(G) = \frac{2}{10} = 0.2$

1st   2nd

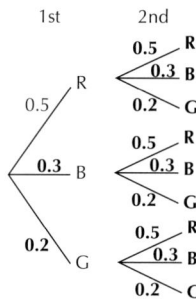

2 a) $P(H) = 0.65$, $P(T) = 1 - 0.65 = 0.35$

1st   2nd

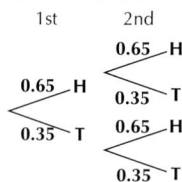

b) $P(1) = P(2) = P(3) = \frac{1}{3}$

1st   2nd

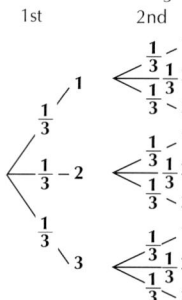

c) 1st   2nd   3rd

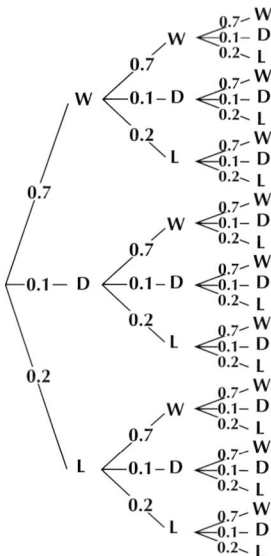

3 a) $P(\text{Freddie wins}) = 0.8$,
$P(\text{James wins}) = 1 - 0.8 = 0.2$

1st   2nd

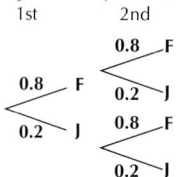

b) (i) $P(F, F) = 0.8 \times 0.8 = \mathbf{0.64}$

(ii) $P(\text{F exactly once}) = P((F, J) \text{ or } (J, F))$
$= (0.8 \times 0.2) + (0.2 \times 0.8)$
$= 0.16 + 0.16 = \mathbf{0.32}$

4 $P(\text{goes to cinema}) = 0.7$
$P(\text{doesn't go to cinema}) = 1 - 0.7 = 0.3$

1st   2nd   3rd

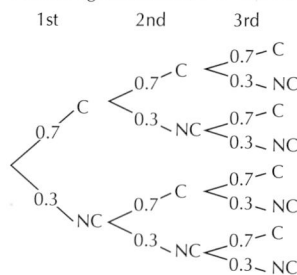

$P(C, C, C \text{ or } NC, NC, NC)$
$= P(C, C, C) + P(NC, NC, NC)$
$= (0.7 \times 0.7 \times 0.7) + (0.3 \times 0.3 \times 0.3)$
$= 0.343 + 0.027 = \mathbf{0.37}$

5 $P(< 3) = P(1 \text{ or } 2) = \frac{2}{6} = \frac{1}{3}$
$P(\geq 3) = P(3, 4, 5 \text{ or } 6) = \frac{4}{6} = \frac{2}{3}$

1st   2nd   3rd

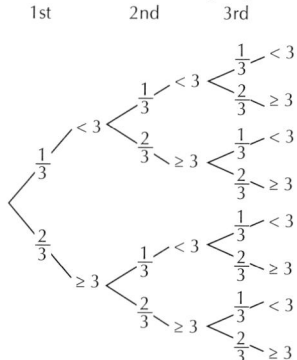

a) $P(< 3, < 3, < 3) = \frac{1}{3} \times \frac{1}{3} \times \frac{1}{3} = \mathbf{\frac{1}{27}}$

b) $P(< 3 \text{ once}, \geq 3 \text{ twice}) = P(< 3, \geq 3, \geq 3)$
$+ P(\geq 3, < 3, \geq 3) + P(\geq 3, \geq 3, < 3)$
$= \left(\frac{1}{3} \times \frac{2}{3} \times \frac{2}{3}\right) + \left(\frac{2}{3} \times \frac{1}{3} \times \frac{2}{3}\right) + \left(\frac{2}{3} \times \frac{2}{3} \times \frac{1}{3}\right)$
$= \frac{4}{27} + \frac{4}{27} + \frac{4}{27} = \frac{12}{27} = \mathbf{\frac{4}{9}}$

Page 471   Exercise 2

1 a) $P(\text{Sally chooses a comedy}) = \frac{4}{12} = \frac{1}{3}$
$P(\text{Sally doesn't}) = 1 - \frac{1}{3} = \frac{2}{3}$
$P(\text{Jesse chooses a comedy}) = \frac{8}{20} = \frac{2}{5}$
$P(\text{Jesse doesn't}) = 1 - \frac{2}{5} = \frac{3}{5}$

Sally   Jesse

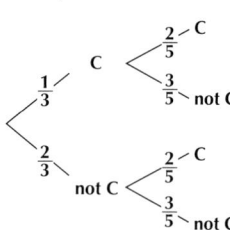

b) (i) $P(\text{not C, not C}) = \frac{2}{3} \times \frac{3}{5} = \frac{6}{15} = \mathbf{\frac{2}{5}}$

(ii) $P(\text{at least one comedy})$
$= 1 - P(\text{not C, not C}) = 1 - \frac{2}{5} = \mathbf{\frac{3}{5}}$

2 $P(R) = \frac{3}{6} = \frac{1}{2}$, $P(B) = \frac{2}{6} = \frac{1}{3}$, $P(Y) = \frac{1}{6}$

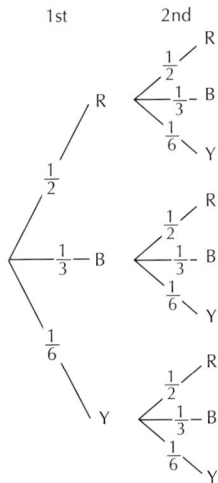

1st   2nd

(Tree diagram: 1st branches R ($\frac{1}{2}$), B ($\frac{1}{3}$), Y ($\frac{1}{6}$); each 2nd branches to R ($\frac{1}{2}$), B ($\frac{1}{3}$), Y ($\frac{1}{6}$))

a) $P(B, R) = \frac{1}{3} \times \frac{1}{2} = \frac{1}{6}$

b) P(red once and blue once)
$= P(R, B) + P(B, R)$
$= \left(\frac{1}{2} \times \frac{1}{3}\right) + \left(\frac{1}{3} \times \frac{1}{2}\right) = \frac{1}{6} + \frac{1}{6} = \frac{1}{3}$

c) P(same colour twice)
$= P(R, R) + P(B, B) + P(Y, Y)$
$= \left(\frac{1}{2} \times \frac{1}{2}\right) + \left(\frac{1}{3} \times \frac{1}{3}\right) + \left(\frac{1}{6} \times \frac{1}{6}\right)$
$= \frac{1}{4} + \frac{1}{9} + \frac{1}{36} = \frac{7}{18}$

d) P(two different colours) $= 1 - $ P(same colour)
$= 1 - \frac{7}{18} = \frac{11}{18}$

e) $P(Y', Y') = \left(1 - \frac{1}{6}\right) \times \left(1 - \frac{1}{6}\right) = \frac{5}{6} \times \frac{5}{6} = \frac{25}{36}$

f) P(yellow at least once)
$= 1 - $ P(no yellow) $= 1 - \frac{25}{36} = \frac{11}{36}$

35.5 Conditional Probability

Page 473   Exercise 1

1 There are 9 remaining balls and only one 7,
so P(7, given first was 8) $= \frac{1}{9}$

2 a) There are $16 + 14 - 1 = 29$ students left,
13 are girls, so P(girl, given first was a girl)
$= \frac{13}{29}$

b) There are still 16 boys,
so P(boy, given first was a girl) $= \frac{16}{29}$

3 a) There are 8 remaining characters
— 4 wizards, 2 elves and 2 toadstools.
  (i) P(elf, given first was an elf) $= \frac{2}{8} = \frac{1}{4}$
  (ii) P(wizard, given first was an elf)
    $= \frac{4}{8} = \frac{1}{2}$
  (iii) P(toadstool, given first was an elf)
    $= \frac{2}{8} = \frac{1}{4}$

b) There are 8 remaining characters,
and only 1 toadstool, so:
P(toadstool, given first was a toadstool) $= \frac{1}{8}$

4 There are 51 cards remaining — 12 clubs
and 13 of the other suits.

a) P(club, given first was a club) $= \frac{12}{51} = \frac{4}{17}$

b) P(not club, given first was a club)
$= 1 - \frac{4}{17} = \frac{13}{17}$

c) P(diamond, given first was a club) $= \frac{13}{51}$

d) P(black card, given first was a club)
$= $ P(club or spade, given first was a club)
$= \frac{12}{51} + \frac{13}{51} = \frac{25}{51}$

e) P(red card, given first was a club)
$= $ P(heart or diamond, given first was a club)
$= 1 - \frac{25}{51} = \frac{26}{51}$

f) There are $4 - 1 = 3$ remaining cards with
the same value as the chosen card, so:
P(same value as first card) $= \frac{3}{51} = \frac{1}{17}$

5 $P(4) = 0.1$, P(score 10, given 4) $= 0.4$
So P(4 and score 10) $= 0.1 \times 0.4 = 0.04$

6 P(stress) $= 0.6$, so P(no stress) $= 1 - 0.6 = 0.4$
P(ice cream given no stress) $= 0.2$
So P(ice cream and no stress) $= 0.4 \times 0.2$
$= 0.08$

7 a) There are $12 + 8 - 1 = 19$ students left and
7 girls, so P(girl given first was a girl) $= \frac{7}{19}$

b) There are 19 students left and 8 are girls,
so P(girl given first was a boy) $= \frac{8}{19}$

c) $P(G) = \frac{8}{20} = \frac{2}{5}$, $P(B) = \frac{12}{20} = \frac{3}{5}$
P(G given first was G) $= \frac{7}{19}$,
so P(B given first was G) $= 1 - \frac{7}{19} = \frac{12}{19}$
P(G given first was B) $= \frac{8}{19}$,
so P(B given first was B) $= 1 - \frac{8}{19} = \frac{11}{19}$

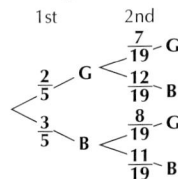

1st   2nd

(Tree diagram: G ($\frac{2}{5}$) → G ($\frac{7}{19}$), B ($\frac{12}{19}$); B ($\frac{3}{5}$) → G ($\frac{8}{19}$), B ($\frac{11}{19}$))

d) (i) $P(G, G) = \frac{2}{5} \times \frac{7}{19} = \frac{14}{95}$
  (ii) $P(B, B) = \frac{3}{5} \times \frac{11}{19} = \frac{33}{95}$

8 a) To begin with there are 10 cards, 5 of
which are even. Each time an even card
is picked, both of these numbers will go
down by 1: P(first is even) $= \frac{5}{10}$
P(second is even, given first was even) $= \frac{4}{9}$
P(third is even,
given first two were even) $= \frac{3}{8}$
So P(E, E, E) $= \frac{5}{10} \times \frac{4}{9} \times \frac{3}{8} = \frac{60}{720} = \frac{1}{12}$

b) For each situation, consider how many
odd cards and even cards are left to be
picked from.
P(1 odd, 2 even)
$= P(O, E, E) + P(E, O, E) + P(E, E, O)$
$= \left(\frac{5}{10} \times \frac{5}{9} \times \frac{4}{8}\right) + \left(\frac{5}{10} \times \frac{5}{9} \times \frac{4}{8}\right)$
$\qquad\qquad\qquad + \left(\frac{5}{10} \times \frac{4}{9} \times \frac{5}{8}\right)$
$= \frac{100}{720} + \frac{100}{720} + \frac{100}{720} = \frac{300}{720} = \frac{5}{12}$

9 a) Since he had pizza last time, P(pizza) $= 0.5$
This doesn't change each time, so:
P(pizza, pizza, pizza) $= 0.5 \times 0.5 \times 0.5$
$= 0.125$

b) Since he had pizza last time,
P(pasta) $= 1 - 0.5 = 0.5$
P(pasta, given he had pasta last time)
$= 1 - 0.9 = 0.1$
So P(pasta, pasta, pasta) $= 0.5 \times 0.1 \times 0.1$
$= 0.005$

c) There are three ways he can have pizza
twice and pasta once: (pizza, pizza, pasta),
(pizza, pasta, pizza) or (pasta, pizza, pizza).
P(pizza, pizza, pasta) $= 0.5 \times 0.5 \times 0.5$
$= 0.125$
P(pizza, pasta, pizza) $= 0.5 \times 0.5 \times 0.9$
$= 0.225$
P(pasta, pizza, pizza) $= 0.5 \times 0.9 \times 0.5$
$= 0.225$
So P(pizza twice, pasta once)
$= 0.125 + 0.225 + 0.225 = 0.575$
Drawing tree diagrams might help you with
Q8 and Q9, but they'll be quite big — you'll get
8 final outcomes, rather than the 4 you got in Q7.

Page 474   Exercise 2

1 a) $P(\text{ski}) = \frac{118 + 32}{274} = \frac{150}{274} = \frac{75}{137}$

b) $P(\text{snowboard}) = \frac{32 + 88}{274} = \frac{120}{274} = \frac{60}{137}$

c) $P(\text{ski and snowboard}) = \frac{32}{274} = \frac{16}{137}$

d) P(ski, given snowboard)
$= \frac{n(\text{ski and snowboard})}{n(\text{snowboard})}$
$= \frac{32}{32 + 88} = \frac{32}{120} = \frac{4}{15}$

e) P(can't snowboard, given can't ski)
$= \frac{n(\text{can't snowboard and can't ski})}{n(\text{can't ski})}$
$= \frac{36}{36 + 88} = \frac{36}{124} = \frac{9}{31}$

2 a) P(A given D) $= \frac{n(\text{A and D})}{n(\text{D})}$
$= \frac{9 + 11}{9 + 11 + 2 + 8} = \frac{20}{30} = \frac{2}{3}$

b) P(S given not D) $= \frac{n(\text{S and not D})}{n(\text{not D})}$
$= \frac{10 + 15}{8 + 10 + 15} = \frac{25}{33}$

c) P(not D given A) $= \frac{n(\text{not D and A})}{n(\text{A})}$
$= \frac{8 + 10}{8 + 9 + 10 + 11} = \frac{18}{38} = \frac{9}{19}$

d) P(not A given not S) $= \frac{n(\text{not A and not S})}{n(\text{not S})}$
$= \frac{2}{8 + 9 + 2} = \frac{2}{19}$

e) P((A and D) given S) $= \frac{n(\text{A and D and S})}{n(\text{S})}$
$= \frac{11}{10 + 11 + 8 + 15} = \frac{11}{44} = \frac{1}{4}$

f) P((S and A) given not D)
$= \frac{n(\text{S and A and not D})}{n(\text{not D})}$
$= \frac{10}{8 + 10 + 15} = \frac{10}{33}$

g) P((A or D) given not S)
$= \frac{n(\text{(A or D) and not S})}{n(\text{not S})}$
$= \frac{8 + 9 + 2}{8 + 9 + 2} = \frac{19}{19} = 1$
You'd expect it to be 1, as every person that
cannot sing can either act or dance.

Page 475   Review Exercise

1 a) $P(W, W) = 0.8 \times 0.8 = 0.64$

b) $P(L, L) = (1 - 0.8) \times (1 - 0.8)$
$= 0.2 \times 0.2 = 0.04$

c) $P(W, L) = 0.8 \times 0.2 = 0.16$

d) P(wins one) $= P(W, L) + P(L, W)$
$= 0.16 + 0.16 = 0.32$

2 a) $0.20 + 0.10 + 0.15 + 0.07 + 0.12 + 0.08$
$+ 0.15 + x = 1 \Rightarrow x = 1 - 0.87 = 0.13$

b) P(easy TV) is the largest
of the easy probabilities,
so **TV** has the most easy questions.

c) P(easy) $= 0.20 + 0.15 + 0.12 + 0.15 = 0.62$
P(challenging) $= 1 - 0.62 = 0.38$
  (i) P(music) $= 0.15 + 0.07 = 0.22$
  (ii) P(sport) $= 0.12 + 0.08 = 0.2$
  (iii) P(music or sport) $= 0.22 + 0.2 = 0.42$
  (iv) P(music or challenging)
    $= $ P(music) $+$ P(challenging)
         $-$ P(music and challenging)
    $= 0.22 + 0.38 - 0.07 = 0.53$
    You can also find this by doing
    P(challenging) + P(easy and music)
    = 0.38 + 0.15 = 0.53.
  (v) P(easy or sport)
    $= $ P(easy) $+$ P(sport) $-$ P(easy and sport)
    $= 0.62 + 0.2 - 0.12 = 0.7$
    You can also find this by doing
    P(easy) + P(challenging and sport)
    = 0.62 + 0.08 = 0.7.

d) (i) P(challenging music twice)
= $0.07 \times 0.07 =$ **0.0049**

(ii) P(easy literature twice)
= $0.15 \times 0.15 =$ **0.0225**

3 a) P(Y10, Y10, Y10) = $\frac{5}{8} \times \frac{4}{7} \times \frac{3}{6} = \frac{5}{28}$

b) P(Y11, Y11, Y11) = $\frac{3}{8} \times \frac{2}{7} \times \frac{1}{6} = \frac{1}{56}$

c) P(two Y10, one Y11) = P(Y10, Y10, Y11)
+ P(Y10, Y11, Y10) + P(Y11, Y10, Y10)
= $\left(\frac{5}{8} \times \frac{4}{7} \times \frac{3}{6}\right) + \left(\frac{5}{8} \times \frac{3}{7} \times \frac{4}{6}\right)$
$+ \left(\frac{3}{8} \times \frac{5}{7} \times \frac{4}{6}\right) = \frac{15}{28}$

d) P(two Y11, one Y10) = P(Y11, Y11, Y10)
+ P(Y11, Y10, Y11) + P(Y10, Y11, Y11)
= $\left(\frac{3}{8} \times \frac{2}{7} \times \frac{5}{6}\right) + \left(\frac{3}{8} \times \frac{5}{7} \times \frac{2}{6}\right)$
$+ \left(\frac{5}{8} \times \frac{3}{7} \times \frac{2}{6}\right) = \frac{15}{56}$

4 a) After the first pick, there are only 9 marbles left, and 1 fewer of the colour that was picked, so:

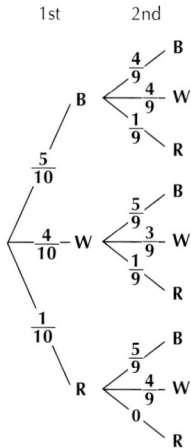

1st   2nd

$\frac{5}{10}$ → B $\left\{ \begin{array}{l} \frac{4}{9} \ \text{B} \\ \frac{4}{9} \ \text{W} \\ \frac{1}{9} \ \text{R} \end{array} \right.$

$\frac{4}{10}$ → W $\left\{ \begin{array}{l} \frac{5}{9} \ \text{B} \\ \frac{3}{9} \ \text{W} \\ \frac{1}{9} \ \text{R} \end{array} \right.$

$\frac{1}{10}$ → R $\left\{ \begin{array}{l} \frac{5}{9} \ \text{B} \\ \frac{4}{9} \ \text{W} \\ \frac{0}{9} \ \text{R} \end{array} \right.$

b) (i) P(B, B) = $\frac{5}{10} \times \frac{4}{9} = \frac{20}{90} = \frac{2}{9}$

(ii) P(not B, not B)
= P(W, not B) + P(R, not B)
= $\left(\frac{4}{10} \times \left(1 - \frac{5}{9}\right)\right) + \left(\frac{1}{10} \times \left(1 - \frac{5}{9}\right)\right)$
= $\left(\frac{4}{10} \times \frac{4}{9}\right) + \left(\frac{1}{10} \times \frac{4}{9}\right)$
= $\frac{16}{90} + \frac{4}{90} = \frac{20}{90} = \frac{2}{9}$

(iii) P(exactly one blue)
= 1 − P((B, B) or (not B, not B))
= $1 - \left(\frac{2}{9} + \frac{2}{9}\right) = 1 - \frac{4}{9} = \frac{5}{9}$

(iv) P(at least one blue)
= 1 − P(not B, not B) = $1 - \frac{2}{9} = \frac{7}{9}$

(v) P(one red, one white)
= P((R, W) or (W, R))
= $\left(\frac{1}{10} \times \frac{4}{9}\right) + \left(\frac{4}{10} \times \frac{1}{9}\right) = \frac{4}{45}$

(vi) P(different colours)
= 1 − P(same colours)
= 1 − P((B, B) or (W, W) or (R, R))
= $1 - \left(\left(\frac{5}{10} \times \frac{4}{9}\right) + \left(\frac{4}{10} \times \frac{3}{9}\right) + \left(\frac{1}{10} \times 0\right)\right)$
= $\frac{29}{45}$

5 a) P(fries given salad) = $\frac{\text{n(fries and salad)}}{\text{n(salad)}}$
= $\frac{27}{27 + 76} = \frac{27}{103}$

b) P(no salad given fries) =
$\frac{\text{n(fries and no salad)}}{\text{n(fries)}}$
= $\frac{87}{87 + 27} = \frac{87}{114} = \frac{29}{38}$

c) P(no side given no salad)
= $\frac{\text{n(no fries and no salad)}}{\text{n(no salad)}} = \frac{10}{87 + 10} = \frac{10}{97}$

---

1 a) $x = 30 - (8 + 4 + 7 + 2 + 3) =$ **6** *[1 mark]*

b) "Mutually exclusive" means the events of choosing Art and Chemistry **can't happen together** *[1 mark]*, and this is clear from the Venn diagram since the circles for A and C **do not intersect** *[1 mark]*.

c) n((A or C) and B) = 4 + 2 = 6
n(B) = 4 + 7 + 2 = 13 *[1 mark for both]*
P((A or C) given B) = $\frac{6}{13}$ *[1 mark]*

2 a) P(pass on third attempt) = P(fail, fail, pass)
= $(1 - 0.5) \times (1 - 0.6) \times 0.8$ *[1 mark]*
= $0.5 \times 0.4 \times 0.8 =$ **0.16** *[1 mark]*

b) P(both pass on third attempt)
= P(pass on third attempt)
     × P(pass on third attempt)
= $0.16 \times 0.16 =$ **0.0256** *[1 mark]*

3 a) To begin with, there are 7 tiles: 4 worth 1 point, 2 worth 4 points and 1 worth 5 point. After the first tile is picked, there will 6 tiles left, and 1 fewer of the type that was picked.

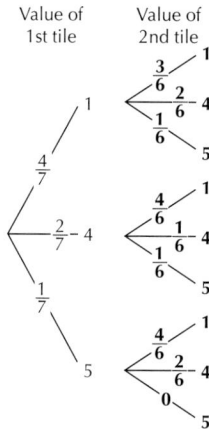

Value of   Value of
1st tile   2nd tile

$\frac{4}{7}$ → 1 $\left\{ \begin{array}{l} \frac{3}{6} \ 1 \\ \frac{2}{6} \ 4 \\ \frac{1}{6} \ 5 \end{array} \right.$

$\frac{2}{7}$ → 4 $\left\{ \begin{array}{l} \frac{4}{6} \ 1 \\ \frac{1}{6} \ 4 \\ \frac{1}{6} \ 5 \end{array} \right.$

$\frac{1}{7}$ → 5 $\left\{ \begin{array}{l} \frac{4}{6} \ 1 \\ \frac{2}{6} \ 4 \\ \frac{0}{6} \ 5 \end{array} \right.$

[3 marks available — 1 mark for each set of branches labelled with the correct probabilities]
Award marks if the probabilities on the branches have been simplified.

b) P(less than 6) = P((1, 1) or (1, 4) or (4, 1))
= $\left(\frac{4}{7} \times \frac{3}{6}\right) + \left(\frac{4}{7} \times \frac{2}{6}\right) + \left(\frac{2}{7} \times \frac{4}{6}\right) = \frac{2}{3}$
[3 marks available – 1 mark for any correct product, 1 mark for the sum of three products from the tree diagram, 1 mark for the correct final answer]

Mixed Exam-Style Questions

Page 477

There's often more than one way to tackle a problem solving question. You'll still get full marks if you show a suitable method and get the correct answer, even if your method is different to the one used here.

1 Special offer standard box contains:
$500 \times 1.25 = 625$ g
Family box costs:
225p ÷ 750 g = 0.3p per gram
Special offer standard box costs:
200p ÷ 625 g = 0.32p per gram
OR
Family box contains:
750 g ÷ 225p = 3.333... grams per penny
Special offer standard box contains:
625 g ÷ 200p = 3.125 grams per penny
The cost per gram for the family box is less than the cost per gram for the standard box OR the amount of cereal per pence (or £) for the family box is more than that for the standard box, so **the family box is better value for money**.

[4 marks available — 1 mark for finding the amount of cereal in the special offer standard box, 1 mark for each correct cost in pence (or £) per gram OR each correct amount in grams per pence (or £), where the units are the same for both, 1 mark for correct conclusion based on the calculations]

2 Number of parts in ratio 3 : 4 = 3 + 4 = 7 parts *[1 mark]*. There are 28 ÷ 7 = 4 children per part, so there are 3 × 4 = **12 boys** and 4 × 4 = **16 girls** *[1 mark for both]*.
12 ÷ 2 = **6 boys** have school dinners and **6 boys** have packed lunches.
75% = $\frac{3}{4}$, so $\frac{3 \times 16}{4} =$ **12 girls** *[1 mark]* have school dinners, and 16 − 12 = **4 girls** have packed lunches. On a frequency tree:

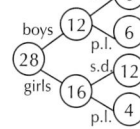

28 → { boys 12 → { s.d. 6, p.l. 6 }, girls 16 → { s.d. 12, p.l. 4 } }

[1 mark for fully correct and labelled frequency tree]
You could also have drawn the frequency tree by first splitting the class into school dinners and packed lunches, and then splitting both of these into boys and girls.

3 Pre-tax cost from America
= \$432 + \$30 = \$462
Post-tax cost = \$462 × 1.2 = \$554.40 *[1 mark]*
\$1 = £0.75, so the cost in £ is
\$554.40 × 0.75 = £415.80 *[1 mark]*.
This is greater than the price from the UK seller (£415.80 > £415) so **Jemima should buy the computer from the UK seller** *[1 mark]*.

4 Opposite sides of a parallelogram are equal.
So 5x + 2 = 20 − 10x *[1 mark]*
⇒ 15x = 18 ⇒ x = 1.2 cm *[1 mark]*.
Perimeter = (5x + 2) + (20 − 10x) + 3x + 3x
[1 mark] = x + 22 = 1.2 + 22
     = **23.2 cm as required** *[1 mark]*.

5 width : height = 3 : 2 = 90 : ?
90 ÷ 3 = 30, so height of the flag is 2 × 30 = 60 cm *[1 mark]*. The middle stripe is half of the total height: 60 ÷ 2 = 30 cm. So the diameter of the circle is $\frac{4}{5} \times 30 = 24$ cm *[1 mark]*. The radius of the circle is 24 ÷ 2 = 12 cm, and so the area = $\pi \times 12^2$ *[1 mark]* = **144π cm²** *[1 mark]*.

6 Andrea's efficiency = 81.9 ÷ 2.6
     = 31.5 km/litre *[1 mark]*
Zac's efficiency = $\frac{2}{3} \times 31.5$
     = 21 km/litre *[1 mark]*
Zac's fuel usage = $\frac{\text{distance}}{\text{efficiency}} = \frac{64.7}{21}$
= 3.0809... = **3.1 litres** (2 s.f.) *[1 mark]*

7 Per month, the FF deal will cost (in £):
£51.99 + [(5 − 4) × £4.99] +
[(7 − 5) × 60 × £0.07] = £65.38 per month.
The H2 costs for calls and texts would be:
(7 × 60 × £0.08) + (60 × £0.04) = £36.00.
£65.38 − £36.00 = £29.38
So H2 need to charge less than £29.38 for 5 GB data, i.e. £29.38 ÷ 5 = £5.876 for 1 GB of data. Rounding down, the maximum price is **£5.87**.
[5 marks available — 1 mark for a correct method to find the cost per month for FF, 1 mark for the correct cost per month for FF, 1 mark for finding the cost per month of calls and texts for H2, 1 mark for a correct method to find the price per GB for H2, 1 mark for the correct answer]
If you round up at the end then you'd get £5.88 × 5 = £29.40 for 5 GB of data, which leads to a greater price than FF are charging.

8 The sum of the interior angles of an n-sided polygon is $(n-2) \times 180°$. For a pentagon $(n = 5)$ this is $3 \times 180° = 540°$ *[1 mark]*. So there are $540° - (3 \times 90°) = 270°$ left for the other two angles *[1 mark]*. Number of parts in the ratio $4:11 = 4 + 11 = 15$ parts, so there are $270° \div 15 = 18°$ per part *[1 mark]* and so the largest angle is $11 \times 18° = \mathbf{198°}$ *[1 mark]*.

9 0.1 litres of paint covers 1 m², so 1 litre will cover 10 m², and 5 litres will cover $5 \times 10 = 50$ m² *[1 mark]*.
Surface area of the 4 vertical faces $= 2 \times (2.6 \times 2.4) + 2 \times (6 \times 2.4) = 41.28$ m² *[1 mark]*. So she can paint $50 - 41.28 = 8.72$ m² of the top face *[1 mark]*. The total area of the top face is $6 \times 2.6 = 15.6$ m² so the percentage of the top face that she can paint is $\frac{8.72}{15.6} \times 100\% = 55.897... = \mathbf{56\%}$ (to the nearest whole number) *[1 mark]*.

10 The teacher can choose from n boys, $(n - 2)$ girls for the first girl and $(n - 2) - 1 = (n - 3)$ girls for the second girl. Multiply these together to get the total number of possible combinations:
$n(n - 2)(n - 3) = n(n^2 - 3n - 2n + 6)$
$= n(n^2 - 5n + 6) = \mathbf{n^3 - 5n^2 + 6n}$ **as required**
[3 marks available — 1 mark for setting up the multiplication, 1 mark for a correct expansion of two factors, 1 mark for the correct expansion of all three factors]
You may have multiplied in a different order, e.g. $(n^2 - 2n)(n - 3) \Rightarrow n^3 - 3n^2 - 2n^2 + 6n$
$\Rightarrow n^3 - 5n^2 + 6n$

11 A sketch will help here:

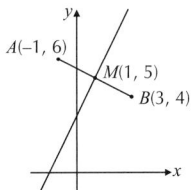

Midpoint of $AB = \left(\frac{-1 + 3}{2}, \frac{6 + 4}{2}\right)$
$= (1, 5)$ *[1 mark]*
Gradient of $AB = \frac{6 - 4}{-1 - 3} = -\frac{1}{2}$ *[1 mark]*
Gradient of line perpendicular to AB
$= -1 \div -\frac{1}{2} = 2$ *[1 mark]*
Using the point $(1, 5)$ and gradient 2 in $y = mx + c$ gives:
$5 = (2 \times 1) + c \Rightarrow c = 3$ and so the equation of the line is $\mathbf{y = 2x + 3}$ *[1 mark]*

12 8 litres $= 8 \times 1000 = 8000$ ml *[1 mark]* so area of lawn $= 8000 \div 20 = 400$ m² *[1 mark]*. If the width of the lawn is x, then the length is $4x$ and the area is $x \times 4x = 4x^2$.
So $4x^2 = 400$ *[1 mark]* $\Rightarrow x^2 = 100$
$\Rightarrow x = 10$. So the length of the lawn is $4x = 4 \times 10 = \mathbf{40\ m}$ *[1 mark]*.
You can ignore the negative square root as you're dealing with a length, which must be positive.

13 a) Events are independent if one event happening has **no effect** on the probability of the other happening *[1 mark]*.
b) The probability that Roger wins on grass is $\frac{5}{5 + 2} = \frac{5}{7}$ *[1 mark]* and the probability that he wins on clay is $\frac{4}{5 + 4} = \frac{4}{9}$ *[1 mark]*.
The probability that they win one set each is the probability that Roger wins on grass and loses on clay OR Roger loses on grass and wins on clay:
P(win one set each)
$= \left(\frac{5}{7} \times \left(1 - \frac{4}{9}\right)\right) + \left(\left(1 - \frac{5}{7}\right) \times \frac{4}{9}\right)$ *[1 mark]*
$= \left(\frac{5}{7} \times \frac{5}{9}\right) + \left(\frac{2}{7} \times \frac{4}{9}\right)$
$= \frac{25}{63} + \frac{8}{63} = \frac{33}{63} = \mathbf{\frac{11}{21}}$ *[1 mark]*

14 Split the trapezium into a rectangle and a triangle. The triangle is similar to the triangle of water since the horizontal side of the triangle and the top edge of the water are parallel, so all the angles in the triangle are the same size as the angles in the triangle of water.

Pool:

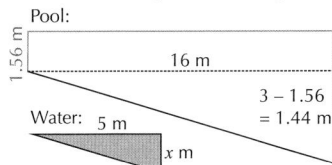

Water:

The height of the large triangle in the pool is $3 - 1.56 = 1.44$ m *[1 mark]*. The scale factor of the enlargement from water to the pool is $16 \div 5 = 3.2$ *[1 mark]*. So the height of the water is $x = 1.44 \div 3.2 = 0.45$ m *[1 mark]*. So the distance from the surface of the water to the top of the pool is $3 - 0.45 = \mathbf{2.55\ m}$ *[1 mark]*.

15 $9^{4x} = (3^2)^{4x} = 3^{8x}$ *[1 mark]*
$\Rightarrow 9^{4x} = 3^{8x} = 3^{3x + 4}$ so $8x = 3x + 4$ *[1 mark]*
$\Rightarrow 5x = 4 \Rightarrow x = \mathbf{\frac{4}{5}}$ or **0.8** *[1 mark]*

16 Original percentage of yellow sweets:
$\frac{(40 - 18)}{40} \times 100\% = \frac{22}{40} \times 100\% = 55\%$
[1 mark]
New percentage of yellow sweets:
$\frac{22 + 5}{40 + 5 - 9} \times 100\% = \frac{27}{36} \times 100\% = 75\%$
[1 mark]
So the percentage has increased by $75\% - 55\% = \mathbf{20\%}$ *[1 mark]*.

17 The sum of two consecutive triangular numbers is $u_n + u_{n+1}$
$= \frac{n(n + 1)}{2} + \frac{(n + 1)(n + 2)}{2}$ *[1 mark]*
$= \frac{n^2 + n + n^2 + 3n + 2}{2}$
$= \frac{2n^2 + 4n + 2}{2} = n^2 + 2n + 1$ *[1 mark]*
$= \mathbf{(n + 1)^2}$ **which is a square number for any n**
[1 mark]

18 a) Using Pythagoras' theorem:
$h^2 + AC^2 = AB^2$
$\Rightarrow h^2 = AB^2 - AC^2$
The lengths AC and AB may have been rounded, so $2.45 \le AB < 2.55$ and $2.25 \le AC < 2.35$. To make h^2 (and thus h) as large as possible, you need AB to be as large as possible (so 2.55) and AC to be as small as possible (so 2.25). So:
$h^2 < 2.55^2 - 2.25^2 = 1.44$
$\Rightarrow \mathbf{h < 1.2\ m}$ **as required**
[3 marks available — 1 mark for using Pythagoras' theorem to find h in terms of AB and AC, 1 mark for considering the possible values of AB and AC, 1 mark for reaching the correct conclusion]
b) The lengths are opposite and adjacent to the angle so use $\tan BDC = \frac{h}{CD}$. CD may have been rounded, so $1.25 \le CD < 1.35$. You want $\tan BDC$ to be as large as possible to make BDC as large as possible, so take the smallest possible unrounded value for CD (1.25) and the upper bound for h (1.2 from part a)).
This gives $BDC = \tan^{-1}\left(\frac{1.2}{1.25}\right)$
$= 43.830...° = \mathbf{43.8°}$ (1 d.p.).
[3 marks available — 1 mark for using the tan formula, 1 mark for considering the possible values of CD, 1 mark for reaching the correct conclusion]

19 The probability that the first team is not British $= \frac{8 - x}{8}$. The probability that the second team is not British, given that the first was not British is $\frac{7 - x}{7}$ *[1 mark for both]*. So the probability that neither team is British is:
$\frac{8 - x}{8} \times \frac{7 - x}{7} = \frac{56 - 15x + x^2}{56}$ *[1 mark]*
$\Rightarrow \frac{56 - 15x + x^2}{56} = \frac{5}{14}$ *[1 mark]* $= \frac{20}{56}$
$\Rightarrow 56 - 15x + x^2 = 20 \Rightarrow \mathbf{x^2 - 15x + 36 = 0}$ **as required** *[1 mark for correct rearrangement]*

20 To get the area of the pendant you need to subtract the area of the segment $ADCA$ from the area of the semicircle $ABCA$. The area of the segment is the area of the sector $ADCEA$ minus the area of the triangle ACE:
Area of segment $= \left(\frac{90°}{360°} \times \pi \times 5^2\right) - \left(\frac{1}{2} \times 5 \times 5\right)$
$= \frac{25\pi}{4} - \frac{25}{2}$ cm²
To find the area of the semicircle, you need to know its radius, r. Using Pythagoras' theorem:
$r = \frac{1}{2}AC = \frac{1}{2}(5^2 + 5^2) = \frac{1}{2}\sqrt{50}$
Area of semicircle $= \frac{1}{2} \times \pi \times \left(\frac{1}{2}\sqrt{50}\right)^2$
$= \frac{25\pi}{4}$ cm²
So the area of the pendant is
$\frac{25\pi}{4} - \left(\frac{25\pi}{4} - \frac{25}{2}\right) = \mathbf{\frac{25}{2}}$ **cm²**
[5 marks available — 1 mark for the correct area of the sector, 1 mark for the correct area of the triangle, 1 mark for the correct radius of the semicircle, 1 mark for the correct area of the semicircle, 1 mark for the correct final answer]
Another method would be to add the areas of the semicircle and the triangle and then subtract the area of the sector from this.

21 $k = m^{\frac{1}{4}}h^{-2} = \sqrt[4]{4096} \times \frac{1}{2^2} = 2$ *[1 mark]*
So $m^{\frac{1}{4}}h^{-2} = 2 \Rightarrow m^{\frac{1}{4}} = 2h^2$ *[1 mark]*
$\Rightarrow m = (2h^2)^4$ *[1 mark]* $\Rightarrow m = 2^4 h^8$
$= \mathbf{16h^8}$ *[1 mark]*

22 A sketch is important here:

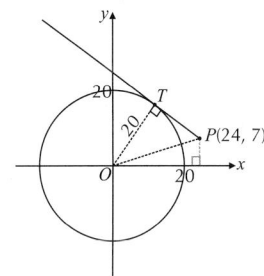

The curve $x^2 + y^2 = 400$ is a circle of radius 20 centred at the origin *[1 mark]*. Using Pythagoras' theorem, $OP^2 = 7^2 + 24^2 = 625$ *[1 mark]* (so $OP = \sqrt{625} = 25$). Angle OTP is a right angle because it's the angle between a radius and a tangent *[1 mark]*. Hence, OTP is a right-angled triangle and so $OP^2 = PT^2 + OT^2$.
$\Rightarrow 625 = PT^2 + 20^2$ *[1 mark]*
$\Rightarrow PT^2 = 625 - 20^2 = 225$
$\Rightarrow PT = \sqrt{225} = \mathbf{15}$ *[1 mark]*

Glossary

A

Acceleration The rate of change of **velocity** over time.

Adjacent Of a **right-angled triangle**, it's the side between the angle under consideration and the right angle.

Algebraic fraction A **fraction** that has an algebraic **expression** in the **numerator** and/or **denominator**.

Allied angles The pair of angles in a C- or U-shape formed when a straight line crosses two **parallel lines**. These angles always add up to 180°.

Alternate angles The pair of equal angles in a Z-shape formed when a straight line crosses two **parallel lines**.

AND rule (for dependent events) The **probability** of both A and B happening is equal to the probability of A happening multiplied by the probability of B happening given that A has happened.

AND rule (for independent events) The **probability** of both A and B happening is equal to the probability of A happening multiplied by the probability of B happening.

Angle bisector The line that cuts an angle into two equal smaller angles.

Angle of depression/elevation The angle between a horizontal line and the line of sight of an observer at the same level looking down or up, respectively.

Arc A part of the **circumference** of a circle.

Arithmetic (or linear) sequence A **sequence** where the **terms** increase or decrease by the same amount each time (the common difference).

Asymptote A straight line to show the values where the graph of a **function** is undefined.

Average A way of representing a set of data with a central or typical value of the set. The three common averages used are the **mode**, **median** and **mean**.

B

Bar chart A chart to display **discrete** or categorical data. The height of bars shows the number (or frequency) of items in different categories.

Bearing A three-figure angle measured clockwise from the **north line** to tell you the position of one object in relation to another.

Bias (in outcomes) Applies to e.g. rolling dice, where the **outcomes** are not equally likely.

Biased sample One in which some members of a **population** are more likely to be included than others.

Bisect Split a line or angle exactly in half.

BODMAS The correct order to carry out mathematical operations — it stands for Brackets, Other, Division, Multiplication, Addition, Subtraction.

Box plot A diagram that shows the distribution of data in a data set. The **lower** and **upper quartiles** form the ends of a box, the **median** is marked inside the box, and lines extend from the box to the lowest and highest values.

C

Chord A line between two points on the edge of a circle.

Circumference The distance around the outside of a circle.

Common denominator The same bottom number in two or more **fractions**.

Common factor A common **factor** of two or more numbers is a factor of both or all of those numbers.

Common multiple A common **multiple** of two or more numbers is a multiple of both or all of those numbers.

Complement of a set All **elements** of the **universal set** that aren't in the set. The complement of a set A is written A'.

Completing the square A way to write and solve a **quadratic** equation when it can't be easily **factorised**. The 'completed square form' is $(x + p)^2 + q$, where x is a variable and p and q are constants.

Composite bar chart A **bar chart** which has single bars split into different sections for each set of data displayed.

Composite function The combination of two or more **functions** — e.g. f(x) and g(x) could combine to give the composite functions gf(x) of fg(x).

Composite shape A shape made up of two or more basic shapes.

Compound decay When a quantity gets smaller over time due to successive **percentage** decreases based on the decreasing value itself.

Compound growth When a quantity gets larger over time due to successive **percentage** increases based on the increasing value itself.

Compound inequality A combination of multiple **inequalities**, e.g. $a < x < b$.

Compound interest **Compound growth** applied to money.

Compound measure A measurement made up of two or more other measurements, e.g. speed = distance ÷ time.

Conditional probability The **probability** of an **event** happening, given that another event has happened.

Cone A 3D shape which has a circular base and a curved sloping face that goes up to a point at the top.

Congruent Congruent shapes are exactly the same shape and size as each other.

Constant of proportionality Usually given the letter k, it's the constant in an **equation** where two variables are in proportion, e.g. $y = kx$ or $y = \frac{k}{x}$.

Construction An accurate drawing made using a pair of compasses, a protractor and a ruler.

Continuous data Numerical data which can take any value in a given range.

Conversion factor The number you multiply or divide by to convert a measurement from one unit to another.

Correlation How two variables are related to each other. Positive correlation means the variables increase and decrease together. Negative correlation mean that as one variable increases, the other decreases.

Corresponding angles The pair of equal angles in an F-shape formed when a straight line crosses two **parallel lines**.

Cosine The cosine of an angle x in a **right-angled triangle** is the **ratio** of the **adjacent** and **hypotenuse** sides, i.e. $\cos x = \frac{\text{adjacent}}{\text{hypotenuse}}$. The graph of $y = \cos x$ has a 'bucket' shape pattern repeated every 360°.

Cosine rule A rule connecting sides and angles in any triangle: $a^2 = b^2 + c^2 - 2bc\cos A$.

Cross-section The face exposed when cutting through a 3D shape.

Cube A **cuboid** where all six faces are squares.

Cube (power) A number multiplied by itself twice — written as the **power** of 3, x^3.

Cubic graph The graph of a cubic **function**, which has x^3 as its highest **power**.

Cuboid A 3D shape with six rectangular faces.

Cumulative frequency diagram An S-shaped curve to display the running total of frequencies for a set of data.

Cyclic quadrilateral Any **quadrilateral** that can be drawn inside a circle with all four vertices (corners) touching the **circumference**.

Cylinder A 3D shape with a constant circular **cross-section**.

D

Deceleration A negative **acceleration**.

Denominator The bottom number of a **fraction**.

Density The mass per unit **volume** of a substance.

Depreciation The loss of value over time due to **compound decay**.

Diameter The line from one side of a circle to the other through its centre. The diameter is twice the length of the **radius**.

Difference of two squares A **quadratic** expression with just two **square** terms separated by a minus sign, $a^2 - b^2$, which can be **factorised** as $(a + b)(a - b)$.

Direct proportion Two variables are in direct proportion when the **ratio** between them is always the same, i.e. $y = kx$.

Discrete data Numerical data which can only take certain values.

Disproof by counter-example Showing that a statement is false by giving an example where it doesn't work.

Distance-time graph A graph with distance travelled on the vertical axis and time taken on the horizontal axis.

Dual bar chart A **bar chart** which shows two sets of data by having two bars per category.

E

Element An item contained in a **set**, also called a member of the set. $x \in A$ means 'x is an element of set A'.

Elevation The 2D view of a 3D object looking at it either from the front or side horizontally.

Empty set A **set** that has no **elements**, denoted by $\varnothing$ or { }.

Enlargement A transformation where a shape is enlarged by a particular **scale factor**, sometimes in relation to a centre of enlargement.

Equation A way of showing that two **expressions** are equal to each other for a particular value or values of an unknown.

Equilateral triangle A triangle with 3 equal sides and equal angles (of 60°).

Equivalent fraction A **fraction** that shows the same proportion as another fraction using a different **numerator** and **denominator**.

Estimate An approximation to the answer to a calculation or the size of an amount.

Event A set of one or more **outcomes** to which a **probability** is assigned.

Exchange rate The **conversion factor** between two currencies.

Expected frequency The number of times an **event** is expected to happen — its **probability** of happening multiplied by the number of times an experiment is done.

Exponential graph The graph of an exponential **function**, $y = a^x$. They all have the same curved shape, an **asymptote** at $y = 0$ and a y-intercept at $y = 1$.

Expression An algebraic expression is a combination of **terms** separated by + and – signs.

Exterior angles The angle between a side of a **polygon** and a line that extends out from a neighbouring side.

Extrapolation Using e.g. a **line of best fit** to predict values outside the range of data you have.

F

Factor The factors of a number are the numbers that divide into it exactly.

Factorising Finding a **common factor** in the terms of an **expression** and taking it outside a pair of brackets.

Fibonacci-type sequence A **sequence** where each **term** is found by adding together the two previous terms.

Formula The mathematical relationship between different quantities.

Fraction A value written as one number divided by another.

Frequency polygon A line graph which shows the frequency of data in a grouped **frequency table**.

Frequency table A table to record the frequency of a response or **event**.

Frequency tree A diagram made up of branches to show the different possible **outcomes** of multiple **events**. The number at the end of each branch shows how many times that event or combination of events happened.

Frustum The 3D shape left once you chop off the top bit of a **cone** parallel to its circular base.

Function A rule that turns one number (the input) into another number (the output).

Function notation E.g. $f(x) = x + a$ and $f: x \rightarrow x + a$ both tell you the rule of a **function**, in this case to add a to any input value x.

G

Geometric sequence A **sequence** where the **terms** are found by multiplying by the same value each time (the common **ratio**).

Gradient The slope or steepness of a graph, which can be found by dividing the change in y by the change in x.

H

Hemisphere Half of a **sphere**.

Highest common factor (HCF) The largest number that will divide exactly into both (or all) of a given pair (or set) of numbers.

Histogram A diagram to show grouped **continuous data**. The area of each bar represents the frequency of each group.

Hypotenuse The longest side in a **right-angled triangle**, opposite the right angle.

I

Identity A way of showing that two **expressions** are always equal to each other, not just for a particular value or values. Identities use the sign '$\equiv$'.

Imperial units A set of non-metric units for measuring, e.g. inches, ounces, miles.

Improper fraction A **fraction** where the **numerator** (the top number) is bigger than the **denominator** (the bottom number).

Independent events Two (or more) **events** are independent if the **probability** of one happening has no effect on the probability of the others happening.

Index (or power) A repeated multiplication of a number or variable. a^x means 'a to the power of x' or 'x lots of a multiplied together'.

Inequality A pair of **expressions** separated by one of the symbols $<$, $>$, $\leq$, $\geq$. Like an **equation**, but with a range of solutions.

Intercept The point where a graph crosses an axis.

Interior angles The angles inside each vertex (corner) of a **polygon**.

Interpolation Using e.g. a **line of best fit** to predict values within the range of data you have.

Interquartile range The difference between the **upper** and **lower quartiles**. This gives you the range of the middle 50% of the data in a set — a measure of spread.

Intersection (of sets) The intersection of two **sets** ($A \cap B$) contains only the **elements** that are in both sets.

Invariant point A point on a shape that doesn't move under a transformation.

Inverse function A **function** $f^{-1}(x)$ that reverses the effect of the original function $f(x)$.

Inverse proportion Two variables are inversely proportional when one variable increases as the other decreases, i.e. $y = \frac{k}{x}$. The product of the two variables is constant.

Irrational number A number that cannot be written as a **fraction**, e.g. $\sqrt{2}$ or π. Irrational numbers are non-recurring decimals that go on forever.

Isometric drawing A 2D drawing of a 3D object on an isometric grid of dots or lines in a pattern of **equilateral triangles**.

Isosceles triangle A triangle with 2 equal sides and 2 equal angles.

Iteration A numerical way of solving **equations** that lets you find the approximate value of a solution by repeatedly using an iteration **formula**.

K

Kite A **quadrilateral** with two pairs of equal sides and one pair of equal angles in opposite corners.

L

Line of best fit A straight line on a **scatter graph** to show the general **trend** of the data.

Line of symmetry A mirror line on a graph or 2D shape which divides it so that each half is a **reflection** of the other.

Locus A set of points which satisfy a particular condition, e.g. points that are a fixed distance away from a point or line. The plural of locus is loci.

Lower bound The smallest actual value a rounded number can be.

Lower quartile The value 25% of the way through an ordered set of data.

Lowest common multiple (LCM) The smallest number that is a **multiple** of both (or all) of a given pair (or group) of numbers.

M

Mean The total of all the values in a set of data divided by the number of values.

Median The middle value in a set of data written in size order.

Member See **element**.

Metric units Units for measuring using a decimal-based system (so units are based on **powers** of 10), e.g. metres, kilograms.

Mixed number A number that has a whole number part and a **fraction** part.

Modal group/class The group with the highest frequency in a set of grouped data.

Mode The most common value in a set of data.

Multiple The multiples of a number are the numbers in its times table.

Mutually exclusive events Events that cannot happen at the same time. E.g. choosing a club and choosing a red card from a pack of cards are mutually exclusive events.

N

Net The 2D representation of a 3D object that can be folded up to make the object.

North line The line vertically upwards from a point, used as the start point for **bearings**.

*n***th term** A general **term** in a **sequence** in the position n, which can be used to find any term in the sequence.

Numerator The top number of a **fraction**.

O

Opposite Of a **right-angled triangle**, it's the side opposite the angle under consideration.

Order of rotational symmetry The number of positions you can **rotate** a shape into so that it still looks exactly the same.

OR rule (for mutually exclusive events) The **probability** of at least one of the **events** happening is the sum of the probabilities of each event happening. It's also called the addition rule.

OR rule (general) If **events** are not **mutually exclusive**, the **probability** that at least one event happens is equal to the sum of the probabilities of each event happening, minus the probability that both events happen.

Outcome The result of an activity in **probability**. E.g. flipping a coin and getting tails.

Outlier An extreme value in a data set.

P

Parabola The shape of a graph of a **quadratic** function.

Parallel lines Two lines that have the same **gradient**. They are always at the same distance apart and never meet.

Parallelogram A **quadrilateral** with two pairs of equal, parallel sides.

Percentage A proportion of something compared to the whole, where the whole is taken to be 100.

Perpendicular bisector The perpendicular bisector of line AB is at right angles to the line and cuts it in half.

Perpendicular lines Two lines that cross at a right angle.

Pie chart A circular chart showing the proportion of the data set in each category rather than the actual frequency.

Plan The 2D view of a 3D object looking vertically downwards on it.

Plane of symmetry A 2D shape that cuts a 3D solid into two identical halves.

Polygon A 2D shape with straight sides.

Population The whole group of people or things you want to find out about when collecting data.

Power See **index**.

Pressure A force per unit area.

Primary data Data you have collected yourself.

Prime factorisation Breaking a number down into a unique string of **prime numbers** (its prime factors) multiplied together.

Prime number A number that has no **factors** except itself and 1.

Prism A 3D shape with a constant **cross-section** in the shape of a **polygon**.

Probability How likely an **event** is to happen.

Projections **Plans** and **elevations** — 2D representations of a 3D object.

Proof A mathematical explanation to show that something is true.

Pyramid A 3D shape which has a **polygon** base and which rises to a point.

Pythagoras' theorem The rule connecting lengths of sides in **right-angled triangles**: $h^2 = a^2 + b^2$, where h is the **hypotenuse** of the triangle and a and b are the shorter sides.

Q

Quadratic An **expression**, **equation** or **function** where the highest **power** of the variable is 2. They take the form $ax^2 + bx + c$, where a, b and c are constants and x is a variable.

Quadratic formula The **formula** $x = \dfrac{-b \pm \sqrt{b^2 - 4ac}}{2a}$ which gives you all the possible solutions to the quadratic equation $ax^2 + bx + c = 0$.

Quadratic graph The graph of a quadratic **function** — a u- or n-shaped symmetrical curve.

Quadrilateral A shape that has 4 sides.

Qualitative data Data which is descriptive, so it records words instead of numbers.

Quantitative data Data which is numerical. It can be **discrete** or **continuous**.

R

Radius The line from the centre to the edge of a circle.

Range The difference between the largest and smallest value in a set of data — a measure of spread.

Ratio A way of showing proportion between quantities in the form $a : b$.

Rational number A number that can be written as a **fraction** with an integer on the top and the bottom of the fraction. **Terminating** and **recurring decimals** are rational.

Rationalising the denominator Getting rid of **surds** from the bottom of a **fraction** by multiplication.

Reciprocal The reciprocal of a number is 1 ÷ that number.

Reciprocal graph The graph of a **reciprocal** function $y = \dfrac{1}{x-a} + b$. All reciprocal graphs have the same basic shape but with different **asymptotes**.

Recurring decimal A decimal number that has a repeating pattern in its digits which goes on forever.

Recursive iteration Repeatedly putting the result from an **iteration** formula back into the formula, each time getting closer to the actual solution.

Reflection A transformation where a point, shape or graph is mirrored in a straight line.

Regular polygon A **polygon** where all its sides and angles are equal.

Relative frequency The number of times a result has occurred divided by the number of times an experiment has been done. It's used to estimate **probabilities**. Also called 'experimental probability'.

Rhombus A **parallelogram** where all sides are the same length.

Right-angled triangle A triangle with a right-angle (90°).

Root (of an equation) Another word for the solution of an equation, usually used when solving f(x) = 0.

Root The inverse operation of **squaring** and **cubing** (and raising to other **powers**). Square roots are written as $\pm\sqrt{x}$, cube roots as $\sqrt[3]{x}$.

Rotation A transformation where a shape is turned about a particular point — the centre of rotation.

S

Sample A smaller group of the **population** used to represent the population and to collect data from.

Sample space diagram A list, grid or **two-way table** to show all the possible **outcomes** of an **event** in a systematic way. Also known as a possibility diagram.

Scalar A quantity with magnitude (size) but no direction.

Scale drawing A diagram where all lengths are related to their real-life lengths by a constant **scale factor**.

Scale factor The number that tells you how many times longer the sides of an **enlarged** shape are compared to the original shape. Or how many times bigger one quantity is in relation to another proportional quantity.

Scalene triangle A triangle whose sides and angles are all different.

Scatter graph A graph of two variables plotted against each other, which can show if these variables are related or not.

Secondary data Data collected by someone else. You can get secondary data from e.g. newspapers or the internet.

Sector An area of a circle from the centre to the edge, like a "slice of pie".

Segment An area of a circle between an **arc** and a **chord**.

Sequence A list of numbers or shapes which follows a particular rule. Each number or shape is called a **term**.

Set A set is a group of items or numbers, written in a pair of curly brackets { }.

Set notation A collection of symbols used to help define the **elements** of (objects within) a **set**.

Significant figures (s.f.) The digits in a number after and including the first digit that is not a 0.

Similar Similar shapes have the same size angles (and so are the same shape) but the lengths of corresponding sides are different, related by a **scale factor**.

Simple interest A **percentage** of an initial amount of money added on at regular intervals. The amount of interest added each time doesn't change.

Simultaneous equations A pair of **equations** which are both true for particular values of the unknowns.

Sine The sine of an angle x in a **right-angled triangle** is the **ratio** of the **opposite** and **hypotenuse** sides, i.e. $\sin x = \dfrac{\text{opposite}}{\text{hypotenuse}}$. The graph of $y = \sin x$ has a 'wave' pattern that repeats every 360°.

Sine rule A rule connecting sides and angles in any triangle: $\dfrac{a}{\sin A} = \dfrac{b}{\sin B} = \dfrac{c}{\sin C}$.

Sphere A 3D shape with one curved surface, no vertices (corners) and no edges.

Square (power) A number multiplied by itself — written as the **power** of 2, x^2.

Standard form A way to write very big or small numbers as $A \times 10^n$, where $1 \le A < 10$ and n is an integer.

Stem and leaf diagram A diagram that displays data values. They consist of 'stems' (the first digit(s)) and 'leaves' (the remaining digit) of the data values arranged in numerical order.

Substituting Replacing the letters in a **formula**, **expression** or **equation** with actual values.

Surd An **expression** containing an **irrational root**.

Surface area The total area of all the faces of a 3D shape.

T

Tangent The tangent of an angle x in a **right-angled triangle** is the **ratio** of the **opposite** and **adjacent** sides, i.e. $\tan x = \dfrac{\text{opposite}}{\text{adjacent}}$. The graph of $y = \tan x$ has a pattern of a 'wiggle' and a vertical **asymptote** repeated every 180°.

Tangent (to a curve) A straight line that just touches the curve at a point and has the same **gradient** as the curve at that point.

Term An individual part of an **expression**, e.g. 3, $2x$, a^2b.

Terminating decimal A decimal number where the digits stop.

Tetrahedron A **pyramid** with a triangular base.

Time series graph A line graph that shows data collected at regular intervals over time.

Translation A transformation where a point, shape or graph is moved horizontally and/or vertically but keeps its original shape and size.

Trapezium A **quadrilateral** with one pair of parallel sides.

Tree diagram A diagram made up of branches to show the **probabilities** of different possible **outcomes** of multiple **events**.

Trend A pattern in data.

Trigonometric graph The graph of a trigonometric function: $y = \sin x$, $y = \cos x$ or $y = \tan x$.

Truncating Chopping off the digits of a decimal number after a certain number of decimal places without rounding.

Turning point The point on a curve where the **gradient** is zero, e.g. a maximum or minimum point.

Two-way table A data collection sheet which records two pieces of information about the same subject at once. It shows the frequency for two different variables.

U

Union The union of two **sets** (A ∪ B) contains all the **elements** that are in either set.

Unique factorisation theorem This states that the **prime factorisation** of every number is unique to that number.

Universal set The **set** of all things under consideration for a particular situation, denoted by ξ.

Upper bound The largest actual value a rounded number can be.

Upper quartile The value 75% of the way through an ordered set of data.

V

Vector A quantity or straight line with magnitude (size) and direction.

Velocity-time graph A graph with **velocity** on the vertical axis and time taken on the horizontal axis.

Velocity The speed of an object measured in a particular direction.

Venn diagram A diagram with two or more circles used to represent **sets**, which may overlap.

Vertically opposite angles The pair of equal angles opposite each other when two lines intersect.

Volume The amount of space a 3D shape takes up.

Y

$y = mx + c$ The **equation** for a straight line where m is the **gradient** and c is the y-intercept (the point where the line crosses the y-axis).